COURTESY OF
Clark Equipment Co.
CORPORATE ENGINEERING,
STANDARDS.

WORLD
METRIC STANDARDS
FOR ENGINEERING

Knut O. Kverneland

WORLD METRIC STANDARDS FOR ENGINEERING

INDUSTRIAL PRESS INC.
200 Madison Avenue, New York 10016

Library of Congress Cataloging in Publication Data

Kverneland, Knut O., 1937–
 World metric standards for engineering.
 Includes index.
 1. Standards, Engineering.   2. Metric system.
 I. Title.
TA368.K93   620′.002′1   77-25875
ISBN 0-8311-1113-5

WORLD METRIC STANDARDS FOR ENGINEERING

Copyright © 1978 by Industrial Press Inc., New York, N.Y. Printed in the United States of America. All rights reserved. This book, or parts thereof, may not be reproduced in any form without permission of the publishers.

Contents

	Foreword I	vi
	Foreword II	vii
	Preface	ix
1	World Standards Organizations	1-1
2	The International System of Measuring Units (SI)	2-1
3	Engineering Drawing Practice	3-1
4	Preferred Numbers	4-1
5	Surface Texture	5-1
6	The ISO System of Limits and Fits—Tolerances and Deviations	6-1
7	The ISO System of Limits and Fits—Inspection of Plain Workpieces	7-1
8	Screw Threads	8-1
9	Fasteners	9-1
10	Steel Material Data	10-1
11	Non-Ferrous Material	11-1
12	Bearings	12-1
13	Mechanical Power Transmission Systems	13-1
14	Fluid Power Systems and Components	14-1
15	Electrical Components	15-1
16	Tires, Rims, and Valves	16-1
17	Metal Cutting Tools	17-1
18	Measuring Tools and Instruments	18-1
19	Metric Conversion Tables and Factors	19-1
20	Index	20-1

Foreword I

International standards and the participation of the United States in their development are becoming increasingly important as this country moves toward metric conversion. The metric standards required by the USA to implement conversion must be compatible with international standards. If they are not, products manufactured to them may be barred from export markets that base product acceptance and certification programs on international standards. Achieving compatibility requires effective participation of USA interests in international standards development. Through such participation they can influence the content of international standards to reflect USA engineering practices, technology, and specifications and to propose new standards or projects to meet USA needs.

The world's major non-governmental standards organizations responsible for coordinating the development and approval of international standards are the International Organization for Standardization (ISO) and the International Electrotechnical Commission (IEC). USA interests are represented in their work, through the American National Standards Institute (ANSI), which manages, coordinates, finances, and administratively supports effective participation in these organizations. It is the official member of ISO. Membership in IEC is held by the USA National Committee of IEC, a fully integrated part of ANSI. Total dues to both organizations are paid by ANSI. It helps direct the work of ISO through membership on its governing bodies and actively participates in the work of some 900 technical committees and subgroups. The USNC-IEC is a member of IEC's governing Council and all of its 400 technical committees.

Now is the time for vigorous USA participation in international standardization, especially since a trend has developed in ISO and IEC technical committees to exclude inch-based modules from international measurement-sensitive standards. Relatively few voluntary international standards have been completed and fewer still are metric-dimensioned, but rapid growth is expected as the result of conversion or commitment to the metric system of all industrialized countries.

Donald L. Peyton, Managing Director
American National Standards Institute

Foreword II

The publication of Knut Kverneland's book is most timely for the USA and Canada. It should prove to be a valuable reference volume as well in other English speaking countries which are in the midst of or are completing the transition to the metric system.

National standards having a metric base have been virtually unknown in the USA. Although many standards-developing groups use dual-measurement notation, the standard sizing and rating practices are still based on the conventional inch-pound-gallon, USA customary system. USA technical committees are now coming to grips with the problem of developing metric based standards. As references in their work, they will be using the standards of ISO and IEC, as well as those of industrialized nations which are already on the metric system.

Knut Kverneland's *World Metric Standards for Engineering* will provide a very useful bridge for those engineers who are required to develop components to metric specifications in advance of availability of applicable American National Standards. The book will also be a valuable tool in guiding the many technical committees and subcommittees which will be working on the new metric American National Standards.

A native of Norway, Mr. Kverneland received his early education in that country and was graduated with a Masters of Science in Mechanical Engineering from the Technical University of Hannover, Federal Republic of Germany. He has been fully conversant with metric measurement units since childhood and is completely familiar with their use in engineering.

The author joined Massey-Ferguson in 1966 as a design engineer, and has risen through consecutive positions as Engineering Analyst, and Standards Engineer, until appointed to his current position as Supervisor for Standards. In this capacity, Mr. Kverneland is responsible for Massey-Ferguson's North American standards. He is also the chairman for the General Task Force developing similar corporate standards.

Mr. Kverneland also maintains a heavy outside professional commitment. He is a member of the Society of Automotive Engineers; Director of the Detroit Section of the Standards Engineering Society; and Chairman of the American National Standards Committee B4 on Standards for Limits and Fits.

In 1972 and 1973, Mr. Kverneland participated as a member of an ad hoc metric study committee of the SAE Off-Road Vehicle Council. He also served on the engineering standards evaluation and promotion subcommittee of the group. Because of its international manufacturing operations, Massey-Ferguson's need for world metric standards information was apparent. Mr. Kverneland was thus aided in his SAE committee work by the high degree of interest of his company's management which provided him with ready access to the computer and to standards data accumulated in its many manufacturing operations around the world. It was this work which prompted him to undertake the writing of this book.

Mr. Kverneland is to be commended for his dedication to this project, and a well-deserved vote of thanks must be given to Massey-Ferguson for the management support it provided the author in this undertaking.

Roy P. Trowbridge
Director, Engineering Standards,
General Motors Corporation
Past President, American National Standards Institute

Preface

Because of the inevitable worldwide change to the metric system, the main objective of this handbook is to assist the reader in minimizing the time and cost of conversion. With this in mind, guides have been given to help rationalize the selection of standard sizes for materials, fasteners, gages, machine components, and tooling.

The book will prove to be a valuable tool for designers and draftsmen as well as engineers, manufacturers, and others, in North America and elsewhere, who are involved with the transition to the metric system.

The selection of metric materials and components must be based on existing international and national metric standards. Therefore, in providing a foundation for this volume, the author has compared standards in the eight largest industrial countries of the free world which together produce 75 percent of the world's products.

It is the author's opinion that the use of one worldwide measurement system will save industry large sums of money, though the initial transition from the USA and English system to the metric system of measure might be costly for some companies. However, the computer is a powerful tool that can help reduce this cost of conversion; and it has been used quite actively in preparing and collating the information and data given in this work.

Nevertheless, conversion to the metric system presents an excellent and rare opportunity to reduce on going inventory costs by minimizing the number of standard sizes. An example of how this has been achieved is illustrated by the work done by the ANSI Special Committee to Study Development of an Optimum Metric Fastener System. This Committee has standardized 25 diameter/pitch fastener combinations having sufficient range to cover the more than 100 sizes in present international use. A comprehensive description of the rational selection of sizes is provided in the section on preferred numbers.

The ISO system of limits and fits can also serve in efforts at rationalization. Computer tables, specifically developed for this book, list the maximum and minimum sizes for holes and shafts for some preferred nominal sizes. They also contain the ten internationally preferred fits that tie in with commercially available material stock, gages, and tooling. These tables also make it convenient for designers to specify preferred ISO fits and nominal sizes. Rationalization of gages, tooling, and material inventory costs is, therefore, promoted by the designers using the informative tables given in this publication.

The completion of the large project of writing this book was possible only because of the extensive cooperation of top management people within the Massey-Ferguson organization. Standards engineers, working for this multinational company throughout the world, have provided substantial input to this publication in the form of national standards information and other data. The author, therefore, wishes to express his appreciation to Massey-Ferguson, his employer, for its encouragement and exceptional support in enabling him to undertake and complete this volume. Without Massey-Ferguson's worldwide resources, without access to the company's computer capabilities, without the company's generous backing in stenographic assistance, the time required for researching and preparing this manuscript would have been many times greater.

My sincerest thanks also to Mr. Roy P. Trowbridge, former president of the American National Standards Institute, who during the initial planning stages of this book, visualized the need for such a publication and gave the author encouragement and support. The same appreciation must be extended to the representatives of the American National Standards Institute, particularly Dorothy Hogan, Deputy Managing Director, Communications, and to the American National Metric Council and its President, Malcolm E. O'Hagan.

ISO and national standards are frequently changed and improved, and during the preparation of this book the publisher had to update previously submitted manuscripts on several occasions. The author appreciates all the efforts and the confidence devoted to this project by the publisher and its staff, with special thanks to Maryanne Colas and Bill Semioli.

And, most importantly, great credit goes to my wife, Inger, for the patience and devotion cheerfully given during the long working hours and under the pressures of my effort on this work.

ACKNOWLEDGMENTS

The author wishes to express his sincere appreciation to the following individuals and organizations:

Massey-Ferguson Inc., Detroit, Michigan, USA

E. J. Flewelling, Manager (immediate supervisor); E. Carr, Secretary; R. H. Morey, Manager (former supervisor); M. St. Louis, Secretary; J. W. Carson, Standards Engineer.

Massey-Ferguson Inc. International

H. Ude, K. H. Schneider, G. Koch (West Germany); P. Pezier (France); S. W. Kinzett, T. J. Rees, J. R. Stanley (United Kingdom); S. M. Lenoci, (Italy); W. L. Scott (Canada); A. B. Gerges (Australia).

Major American Contributing Organizations*

American National Standards Institute (ANSI)

D. L. Peyton, Executive Vice-President; D. Hogan, Deputy Managing Director, Communications; B. Sepaski, Distribution.

Industrial Fasteners Institute (IFI)

C. F. Roberts, Jr., President; R. B. Belford, Technical Director.

National Fluid Power Association (NFPA)

J. I. Morgan, Executive Vice President; E. D. Hoffman, International Standards Editor.

Other Contributing American Organizations*

Cemented Carbide Producers Association (CCPA); Anti-Friction Bearing Manufacturers Association (AFBMA); American Gear Manufacturers Association (AGMA); American National Metric Council (ANMC); The American Society of Mechanical Engineers (ASME); American Society for Testing and Materials (ASTM); Rubber Manufacturers Association (RMA); Society of Automotive Engineers (SAE); Socket Screw Products Bureau (SSPB); The Tire and Rim Association (TRA).

Contributing International and National Standards Organizations*

International Organization for Standardization (ISO); International Electrotechnical Commission (IEC); European Coal and Steel Community (ECSC or EURONORM); Japanese Industrial Standards Committee (JISC); German Standards Organization (DNA or DIN); French Standards Organization (AFNOR); British Standards Institute (BSI); Italian Standards Organization (UNI); Standards Council of Canada (CSA); Standards Association of Australia (SAA).

Contributing Individuals and Their Organizations

Roy P. Trowbridge—Director, Engineering Standards, General Motors Corporation; Past President, American National Standards Institute

Donald L. Peyton—Managing Director, American National Standards Institute

Section 1

Olle Sturen—Secretary General of ISO

W. A. McAdams—President, USA National Committee of the International Electrotechnical Commission, General Electric Company

R. E. Monahan—Past President, Standards Engineers Society Control Data Corporation

M. F. Hill—Author, Control Data Corporation

Section 3

PERA Training Chart, Ford of Britain

L. W. Foster—Vice-Chairman, ANSI Y14.5; Honeywell Inc.

E. Hubel—Draftsman, Massey-Ferguson Inc.

R. G. Nirva—Design Checker, Massey-Ferguson Inc.

Section 5

R. G. Lenz—Chairman, USA Technical Advisory Group (TAG) ISO/Technical Committee (TC) 57; General Motors Corp.

Section 6

F. H. Briggs—Computer Program Consultant, Massey-Ferguson Inc.

Section 8

T. C. Baumgartner—Past Chairman, ANSI B1; Standard Pressed Steel Company

Section 9

R. B. Belford—Chairman, USA TAG ISO/TC 2; Technical Director, Industrial Fasteners Institute

H. W. Ellison—USA Delegate, ISO/TC 2; Assistant Director, Engineering Standards, General Motors Corporation

G. Junker—Chairman, ISO/TC 2, UNBRAKO, Germany

R. M. Byrne—Technical Director, Socket Screw Products Bureau

J. A. Altman—Chairman, ANSI B27 Subcommittee (SC) 1; Supervisor Technical Sales, Waldes Kohinoor Inc.

Section 10

E. A. Domzal—Sr. Project Engineer, Materials, Massey-Ferguson Inc.

P. R. Wray—Director, Metallurgical Engineering, United States Steel Corporation

F. V. Kupchak—Past Chairman, USA TAG ISO/TC 62; Chairman, ANSI B32; Westinghouse Electric Corporation

Japanese Standards Association

Japan Iron and Steel Exporters' Association

Dr. Schlueter—VDEH-German Steel Trade Association

---

*Addresses of the organizations listed are shown on pages 1-7 through 1-12.

PREFACE

Section 12

P. S. Given—Director, SKF Industries Inc.

J. R. Hull—Chief Engineer, The Torrington Company

C. A. Moyer—USA Delegate, ISO/TC 4 SC 9; Assistant Chief Engineer, The Timken Company

A. O. Dehart—USA Observer ISO/TC 123; General Motors Corporation

W. G. Looft—USA Delegate, ISO/TC 4 SC 7; Manager of Engineering, REXNORD

Section 13

C. C. Cummins—Manager, The Louis Allis Company, Litton Industries

W. D. Erickson—Chief Engineer, The Gates Rubber Company

G. J. Shubat—USA Delegate, ISO/TC 100; Diamond Chain Company

Dr. F. Buchsbaum—President, Stock Drive Products

G. W. Michalec—Author and Professor

C. K. Reece—USA Delegate, ISO/TC 32; Vice-Chairman, ANSI B92; John Deere Waterloo Tractor Works

Section 14

J. I. Morgan—Chairman, USA TAG ISO/TC 131; Executive Vice President, National Fluid Power Association

Section 16

K. L. Campbell—Chairman, ISO/TC 31; The Firestone Tire & Rubber Company

Section 17

W. Janninck—Assistant Chief Engineer, Illinois/Eclipse, A Division of Illinois Tool Works, Inc.

R. W. Berry—Member, ANSI B94; Director, Cutting Tool Research, VR/Wesson Company

SPECIAL COURTESY NOTE TO CONTRIBUTING STANDARDS BODIES AND COMPANIES:

Tables and figures where a standards reference or a company abbreviation is shown in parentheses have been reprinted with the permission of the organizations that hold the copyright on the works cited.

Tables and figures without any reference have been republished courtesy of Massey-Ferguson Inc., Detroit, Michigan. Again, the author wishes to express his sincere thanks to his employer for allowing work to be done on the book during normal working hours, and for providing access to the following: national and international standards, computer usage, secretarial help, drafting and checking help, and the use of copying facilities, telephone and mailing services.

The author also wishes to express his sincere appreciation to the referenced organizations for granting permission to use their tables and figures in this publication. Special thanks go to the American National Standards Institute (ANSI), the International Organization for Standardization (ISO), and Industrial Fasteners Institute (IFI). Without their comprehensive support, the publication of the *World Metric Standards for Engineering* with its extensive standards material would not have been possible.

Finally, the author's sincere appreciation is extended to the following organizations and companies for granting permission to republish their standards or figures: the Socket Screw Products Bureau, 331 Madison Avenue, New York, N.Y. 10017; the American Society of Mechanical Engineers (ASME), 345 East 47th Street, New York, N.Y. 10017; the American Society for Testing and Materials (ASTM), 1916 Race Street, Philadelphia, Pennsylvania 19103; General Motors Corporation, Warren, Michigan; Control Data Corporation, Minneapolis, Minnesota; Ford of Britain, Brentwood, Essex, England; Regal-Beloit Corporation, South Beloit, Illinois; Chrysler Corporation, Detroit, Michigan; Truarc, Waldes Kohinoor, Inc., Long Island City, N.Y. SKF Industries, Inc., King of Prussia, Pennsylvania; the Torrington Company, Torrington, Connecticut; the Timkin Company, Canton, Ohio; the Gates Rubber Company, Denver, Colorado; Stock Drive Products, New Hyde Park, N.Y.; the Metric & Multistandard Components Corporation, Hawthorne, N.Y.; Hahn & Kolb, Stuttgart, West Germany.

Please always refer to the most recent edition of the referenced standards. In the United States, American National Standards, International Standards, and national standards of other countries may be obtained from the American National Standards Institute (ANSI), 1430 Broadway, New York, N.Y. 10018. Outside of the United States, sales of standards are transacted through the national standardizing body of the particular country.

Knut O. Kverneland

1 World Standards Organizations

ROLE OF STANDARDIZATION—PAST, PRESENT, AND FUTURE

By definition, standards are rules set up and established by authority for the measure of quantity, weight, extent, value, or quality. Monetary standards, used in determining the weight of silver and gold pieces for the exchange of goods, were among the first to be developed.

During the industrialization period, manufacturing plants developed and became more and more specialized. A need for standards to control such simple parts as fasteners evolved, thereby making them industrially interchangeable. The demand for company and trade organization standards grew apace with the advent of larger plants and the wider distribution of manufactured products.

The basis for most standards is a uniform unit of measure to check mass, length, volume, and time. Many systems were developed over the years, and the original metric system was developed in France after the French Revolution. Since 1875 all international matters concerning the metric system have been the responsibility of the Conférence Générale des Poids et Mesures (CGPM), which was constituted following the Metric Convention signed in Paris that same year.

Before the invention of the metric system a number of inch systems were used throughout the world, one of which is commonly known as the customary inch system. National and international standards were developed, however, based on *both* measuring systems. This made the worldwide interchangeability of simple standard domponents, such as fasteners, impossible.

Metric and Inch Standards

The increasing number of multinational corporations and their local suppliers have found it increasingly expensive to operate with two systems of measures and standards. In order to use available expertise in a central location, one machine might be designed in an "inch" nation only to be produced later in a "metric" country, or vice versa. This obviously generates additional costs in the conversion of drawings, substitution of standard steel sizes and fasteners, the conversion of testing and material specifications, etc.

Increase in Demand for Mandatory World Standards

The most powerful force in the development and use of international standards is the "harmonization" program for standardization now being carried out in Western Europe. More urgency is added to the interchangeability problem with the increasing use of nontariff barriers. The program had its start in the mid-1950's with a UN-sponsored report showing the differences in standards among European countries and the adverse effects of these differences on trade. When the Common Market and the European Free Trade Association were formed a few years later, both recognized the need to eliminate such differences in standards. Working together they immediately formed two European standards coordinating committees, one called CENELEC[1] for electrical standards, the other, CEN[2] for standards in all other fields.

Until recently, the European Committee for Standardization (CEN) and European Electrical Standards Coordinating Committee (CENELEC) have tried to avoid creating still another new set of European standards. Instead, they have identified the points of difference in the existing standards of the 14 member countries and have drafted "harmonization" documents, hopefully to bring about more uniformity. Where practicable, they used the International Organization for Standardization (ISO), and International Electrical Commission (IEC) standards as a base, but in many cases these were not adequate in providing the desired degree of uniformity.

Because of this, CEN and CENELEC committees are now developing position papers for changes in the ISO and IEC standards that will be of help in their unification programs. This has produced a greater sense of urgency in ISO and IEC and a faster output of international standards. Since all the CEN/CENELEC members are also members of ISO and IEC, they constitute a strong, well-prepared voting bloc. As a result, more and more of the standards coming out of ISO and IEC are being accepted by Western European countries as their national standards. It will be some time before this huge market

---

[1] CENELEC stands for the French title: Comité Européen de Coordination des Normes Electrique.
[2] CEN is Comité Européen de Coordination des Normes.

area will achieve the uniformity in standards practiced here in the United States, but the commitment there to make maximum use of international standards is beginning to work.

Regional trade agreements and related activities in other parts of the world are similarly leading to the greater use of international standards. For example, the Pan American Standards Commission, which is trying to develop uniform standards for the Latin-American Free Trade Association, has now agreed to use the ISO and IEC standards wherever possible. Also, the countries of Eastern Europe have become increasingly active in the development of international standards; apparently they are using them as the basis for trade in that region and in opening trade channels with the rest of the world as well.

Another factor in the use of international standards is the increasing number of international cooperation programs. One of the best is the North Atlantic Treaty Organization (NATO) which has been ordering a great deal of its equipment in terms of ISO and IEC standards. The same thing is happening within many of the social and economic programs of the UN and other world organizations. One result of this activity is the adoption of many ISO and IEC standards by the developing countries.

There is still another force on the horizon that may have an effect on the development and use of international standards, an effect even greater than any of the others listed. This is the "Code of Conduct for Preventing Technical Barriers to Trade," now being formulated by GATT (General Agreement on Tariffs and Trade).

Accelerating Pace in Publication of ISO Standards

The above-mentioned factors have accelerated the speed with which ISO develops world standards.[1] As an example, only 100 ISO Recommendations were published in the fifties, yet approximately 1400 international standards agreements were reached in the following decade.

Today, there are nearly 3000 ISO recommendations and standards, half of which have been published only in the last five years. A further 3000 drafts and proposals are in preparation, and the total number of ISO agreements is confidently expected to double within the next four or five years.

There are differences of opinion as to just how many international standards are needed. In the main, however, informed persons agree that in a highly industrialized society the total requirement for national and international standards is on the order of 15,000, or a maximum of 20,000. When more than that number is found in a single country, there is usually either some duplication and overlapping, which is the case in the United States, or, as is the case in socialist countries, what could be called "company standards" are listed as "national standards."

How much of the total need should be met purely by international standards? Here, opinions differ widely. Some people claim that the ideal would be to replace all national standards by international standards and that, consequently, the scope of ISO is to prepare some 15,000 to 20,000 international standards. Others suggest that in the future, as industries merge and multinational companies further develop, some of the present national standards will become company standards, but that there will always be a demand for some national standards to cater to specific local needs.

ISO, therefore, has started a study with a view to determining the total request for international standards and until this study is completed and the results presented, only guesswork can be used. If the total need is for only 5000 ISO standards, some 40 percent already has been met; if, on the other hand, the need is close to 15,000 we have so far produced only 15 percent of what society is requesting. Nevertheless, a safe prediction is that ISO has published up to the present, perhaps some 20 percent of the international standards needed, with another 20 percent in preparation.

ISO Definition of Standardization and Standard

The definitions of standardization and standard differ in the many various publications on the subject. The following are the excerpts from the *ISO Standardization Vocabulary* with comprehensive definitions of "standardization" and "standard."

Standardization—This process is to formulate and apply rules for an orderly approach to specific activity for the benefit and with the cooperation of all

---

[1] Since ISO documents are constantly being upgraded, for simplicity the author refers to all ISO publications in the text of this book as "standards," designating a particular document as "ISO . . ." followed by the appropriate identification number.

However, the actual status of a particular ISO document might be: (a) *Recommendation*, in which case it would be officially designated by "ISO/R . . ." preceding the appropriate identification number. (b) *Draft International Standard*, ("ISO DIS . . ."). (c) An officially adopted international standard, in which case the initials "ISO" followed by the identification number is the appropriate designation. The ISO references given at the end of each section describe the current status of standards. Information on the various designations is given in ISO catalogs and supplements available from: American National Standards Institute, 1430 Broadway, New York, New York 10018.

concerned and in particular for the promotion of optimum overall economy taking due account of functional conditions and safety requirements. It is based on the consolidated results of science, technique, and experience. It determines not only the basis for the present, but also for future development and therefore should keep pace with progress.

Some particular applications in standardization are:

1. Units of measurement
2. Terminology and symbolic representation
3. Products and processes (definition and selection of characteristics of products, testing and measuring methods, specification of characteristics of products for defining their quality, regulation of variety, interchangeability, etc.)
4. Safety of persons and goods.

A Standard—This is the result of a particular standardization effort, approved by a recognized authority. It may take the form of:

1. A document containing a set of conditions to be fulfilled (in French, "norme").
2. A fundamental unit or physical constant, for example: ampere, meter or metre,[1] absolute zero (Kelvin). (In French, "étalon").

Important Objectives of Standardization

The purpose of standardization is to manufacture goods for less direct and indirect incurred costs and to adapt the finished products to the demands of the marketplace.

A more detailed description of the objectives could be as follows:

Lower the production costs, when the aim is to:

1. Facilitate and systematize the skilled work of designing;
2. Ensure optimum selection of materials, components, and semifinished products;
3. Reduce stocks of materials, semifinished products and finished products;
4. Minimize the number of different products sold; and
5. Facilitate and economize the procurement of purchased goods.

Meet the demands of the marketplace, when the objective is to:

1. Conform to regulations imposed by governments and trade organizations;
2. Stay within safety regulations set forth by governments; and
3. Facilitate interchangeability requirements with existing products.

Development of Standards

The Conditions for a Standard—When there is a question of working out a standard the conditions must first be analyzed before actual technical standardization work can be carried out. Preparatory analysis must be as comprehensive as possible and must take into account both technical and economic conditions. It is not sufficient to study only the internal circumstances. It must also be understood that, with regard to standards, the company is dependent on such external factors as the suppliers' stocks of products, the production program of competitors, the customers' wishes, existing standards, governmental requirements, etc.

Waiting for the right moment to begin a particular standardization is most important. Investigation should be made as to whether an intended standard could possibly impede any technical development already under way. Lack of a standard is more often the condition, and it is important to engage in standardization at an early stage, at least to the extent of working out an experimental standard of a temporary nature.

A certain type of regularly recurrent part may, for example, be used in many products in functionally equivalent, but constructionally different, forms. In such a case the task for the standardization work will be to create some order out of chaos through variety-reduction, size standardization, etc. The work should be started as soon as the possibilities of direct cost savings in purchasing, production, inventory, etc., and indirect cost savings in engineers' time can be established.

Standardization Technique

Two basic principles for the preparation of a standard are commonly used; these are as follows:

1. *Analytical standardization*—Standard developed from scratch.
2. *Conservative standardization*—Standard based, as far as possible, on existing practice.

In practice, it appears that a standard cannot be completely prepared in one or the other of these two methods,

---

[1] Meter, liter, millimeter, etc. are accepted spellings in the United States. However, where European tables and illustrations have been reproduced in this book metre and litre have been retained.

but emerges from a compromise between the two. The quintessence of the standardization technique should be to utilize the basic material, the rules and the aids available, in such a way that a valid and practical compromise solution is reached.

The basic material could be comprised of such items as: Former company standards; vendor catalogs; national and international standards; requirements of the company's customers; and competitors' material.

Increasingly important are the national and international standards in existence on the subject; they should always play an important role in any conservative standardization work. It would be foolish to create a unique, new metric standard without first considering some existing European metric standards.

Normal Development Levels of a Standard

The most common standardization levels are: 1. Company Standard; 2. Professional Society or Trade Standard; 3. National Standard; 4. Regional Standard; and 5. World Standard.

The normal path through which a standard must pass in the developmental stages depends on the organization level and the standardization technique applied. A new world standard generated by applying the analytical principle follows the organization levels in a numerical order, while a company standard prepared after the conservative principle might be based directly on the applicable world standard.

User Acceptance of Standards

The development cycle of the standards is completed when the user applies the standards in his work. The designer should, to the degree possible, use internationally standardized parts and components. This would result in an increase of the demand for the standard sizes and a decrease in manufacturing costs for the parts. With the above principle applied to the increasing world flow of material and products, a substantial increase in worldwide productivity can be visualized.

STANDARDS ORGANIZATIONS

World Standards Organizations

ISO, the International Organization for Standardization—The history of international standardization dates back to 1926 when the International Federation of the National Standardizing Association (ISA) was formed. In 1942 the ISA officially ceased to exist, and the ISO, as we know it today, was established in London, in 1946, by 25 participating countries and was officially founded the following year. Within 25 years the number of national standards bodies involved with the ISO had grown from 25 to 70.

The objectives of ISO are to promote the development of standards in the world with a view to facilitating the international exchange of goods and services, and to developing mutual cooperation in the spheres of intellectual, scientific, technological, and economic activity.

As a means to these ends the ISO may: 1. Take action to facilitate the coordination and unification of national standards and issue the necessary recommendations to Member Bodies for this purpose; 2. Set up international standards; 3. Encourage and facilitate, as occasion demands, the development of new standards having common requirements for use in national or international spheres; 4. Arrange for the exchange of information regarding the work of its Member Bodies and of its Technical Committees; and 5. Cooperate with other international organizations interested in related matters, particularly by undertaking, at their request, studies relating to standardization projects.

ISO Member Structure—The ISO membership is composed of member countries and liaison members. A member country is represented by a Member Body, which is the national body most representative of standardization in that country.

In 1964, the Council decided to create a new category of membership, that of Correspondent Member. Such a member is normally an organization in countries under development which do not themselves have a national standards body.

The eight standards organizations shown in Fig. 1-1 have the largest Gross National Product (GNP) of the free world's ISO members, and represent more than 75 percent of the world's GNP. The national standards in these countries are referred to in many areas of this publication.

Organization—The principal officers of ISO are the President, the Vice-President, the Treasurer, and the Secretary-General.

The general organization chart for the ISO is shown in Fig. 1-2.

Central Secretariat—The Secretary-General, who is appointed by the Council, is in charge of the Central Secretariat. He represents the ISO in its relations with other international organizations and maintains overall responsibility for the application of the Constitution, the Rules of Procedure, and the Directives for the Technical Work of ISO. Under his direction the Central Secretariat coordinates the work carried out by the ISO technical

WORLD STANDARDS ORGANIZATIONS

1-5

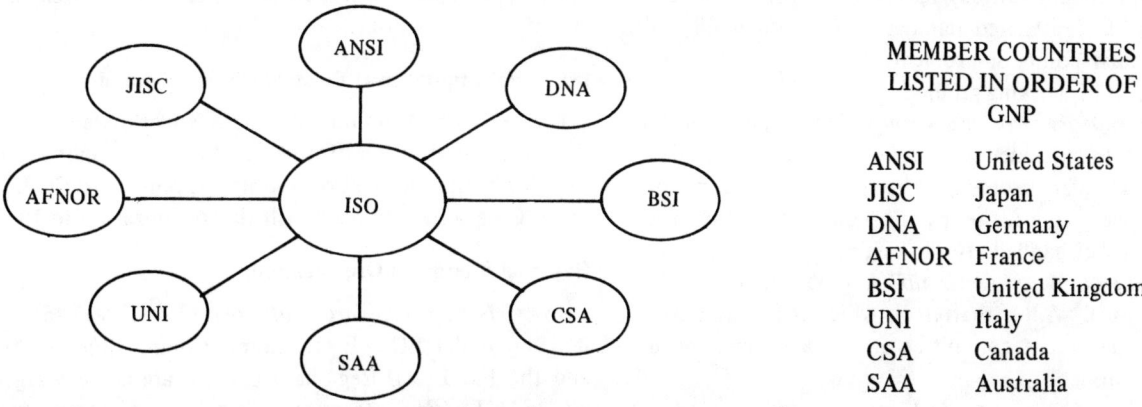

Fig. 1-1. Partial ISO membership structure.

MEMBER COUNTRIES LISTED IN ORDER OF GNP

ANSI	United States
JISC	Japan
DNA	Germany
AFNOR	France
BSI	United Kingdom
UNI	Italy
CSA	Canada
SAA	Australia

Courtesy Control Data Corporation

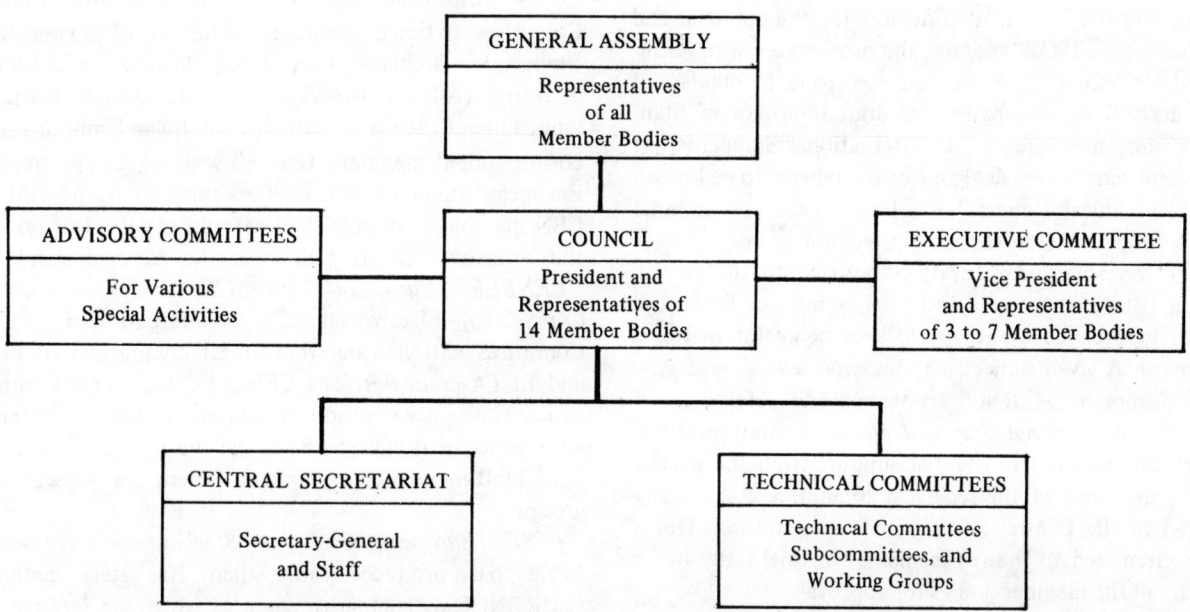

Courtesy Control Data Corporation

Fig. 1-2. International Organization for Standardization (ISO).

committes, convenes all meetings of technical committees and subcommittees, institutes the voting procedures, circulates documents to member bodies, and publishes all documents accepted by the Council as International Standards. He also keeps the Member Bodies and Council informed of the work of the various technical committees. He must inform the technical committees of work undertaken by other international organizations in related fields.

Technical Committees (ISO/TC)—The technical committees are composed of a delegation from each of the member bodies wishing to take part in their work. Each technical committee has a secretariat which is undertaken by a member body acting impartially and designated by the Council. Membership in technical committees is open to interested member bodies as participating (P) or observer (O) members.

Subcommittees (ISO/TC/SC)—This is the level at which most of the technical decisions are made and is also the level at which much of the technical liaison takes place. Subcommittees are charged with the study of one or several items within the program of work of the technical committee.

Working Groups (ISO/TC/SC/WG)—The technical committees and subcommittees may set up working groups composed of a restricted number of individuals to deal with particular points or problems that might arise. There are two kinds of working groups:

1. *A Preparatory Working Group*—This type of group may remain in existence for as long as is necessary to deal with a particular problem. The group may function between meetings of the parent committee, but it is automatically disbanded after having made its final report.

2. *An Ad-Hoc Working Group*—A group that may be formed to deal with a matter on which it is required to report to the parent committee at the same meeting in which it is formed.

Evolution of an International Standard

Prior to July 1971, subjects proposed and approved by ISO were known as "ISO Recommendations." Although the ISO charter included provision for the approval and publication as ISO Standards, the procedure was never invoked. In mid-1971 a decision was made to publish all ISO draft Recommendations as draft International Standards and, subsequently, as International Standards. At the same time a new category of document, to be known as a "Technical Report," was introduced. The material which follows reflects the current technology.

Draft Proposals—Any draft (submitted to the participating (P) members of a technical committee for study) which is intended eventually to become an International Standard. A given subject may undergo several successive draft proposals, i.e., first draft, second draft, etc.

Draft International Standard (DIS)—A draft proposal which has received substantial support from the participating members of the technical committee and is transmitted to the Central Secretariat for registration. This is then circulated to the (P) members for final letter ballot and to all the member bodies for approval.

International Standard—A draft International Standard which has been adopted by a majority of the (P) members of the technical committee and approved by 75 percent of the member bodies is submitted to the ISO Council for acceptance and publication as an International Standard.

Technical Report—A draft proposed International Standard which does not receive the required vote may be published as an ISO Technical Report. Or, the technical committee concerned may request its publication as an ISO Technical Report.

Finance

ISO is maintained by the financial contributions of its members; the amount varies according to the circumstances of the country concerned. Additional revenue, however, is gained from the sale of International Standards and other publications.

IEC, The International Electrotechnical Commission

The IEC, which produces numerous world standards in the electrical field, was formed in 1906, making it one of the oldest international organizations. Similar to the ISO, the IEC became affiliated with this organization in 1947.

Regional Standards Organizations

ASAC (Asian Standards Advisory Committee)—Set up in 1966 under ECAFE (Economic Commission for Asia and the Far East). Regional standards are not envisaged.

COPANT (Pan American Standards Commission)—Comprises national standards bodies of U.S.A. and 11 Latin American countries. Regional standards published and available from ANSI in the Spanish language.

CEN (European Committee for Standardization)—Comprises national standards bodies of EEC (Benelux, France, W. Germany, Italy, U.K., Denmark) and EFTA countries (Austria, Norway, Portugal, Sweden, Switzerland, Finland, Iceland), with Greece, Ireland, and Spain as correspondent members. Over 40 working groups prepare European Standards which, if accepted by 67 percent of CEN members, are published without variation of text in the countries accepting them as the national standard.

CENELEC (European Electrical Standards Coordinating Committee)—Electrotechnical counterpart of CEN. Comprises national electrotechnical committees of EEC and EFTA countries. The CENELEC Electronic Components Committee produces harmonization documents with which national standards can be brought into line, with built-in quality assessment. There are 12 working groups.

ECSC (European Coal and Steel Community)—The ECSC has produced more than 100 steel material (EURONORM) standards, some of which are referred to in this publication.

PASC (Pacific Area Standards Congress)—Comprises national standards bodies from U.S.A., Canada, Japan, Australia, and New Zealand.

National Standards Organizations

ANSI (American National Standards Institute)—Only a brief review of the various functions ANSI fulfills will be presented in this book. More details about the ANSI organization are available from the institute upon request.

ANSI provides the machinery for creating voluntary standards. It serves to eliminate duplication of standards activities and to weld conflicting standards into single,

WORLD STANDARDS ORGANIZATIONS

nationally accepted standards under the designation "American National Standards."

The Standards Institute, under whose auspices this work is being done, is the United States clearinghouse and coordinating body for standards activity on the national level. It is a federation of trade associations, technical societies, professional groups, and consumer organizations. Some 1000 companies are affiliated with the Institute as company members.

The American National Standards Institute is the United States member of the International Organization for Standardization (ISO), the International Electrotechnical Commission (IEC), the Pan American Standards Commision (COPANT), and the Pacific Area Standards Congress (PASC). Through these channels American interests make their position felt on the international level.

Other National Standards Bodies at the ISO—The eight national standards organizations with the largest Gross National Product in the free world are listed in Fig. 1-1 and their addresses are given on the following pages. ANSI has the distribution of their national standards in the U.S.A.

ACRONYMS AND ADDRESSES TO IMPORTANT STANDARDS ORGANIZATIONS

Acronyms	Standard Prefix	Founded	Nation	Address
ABCA	No Stds	1944	International	American British Canadian Australian Converence on Unification of Engineering Standards For information, contact: Director of Engineering Standards GM Engineering Staff General Motors Technical Center Warren, Michigan 48090
AFBMA	AFBMA	1933	U.S.A.	The Anti-Friction Bearing Manufacturers Association, Inc. 60 East 42nd Street New York, N.Y. 10017
AFNOR	NF	1926	France	Association Francaise De Normalisation French Standards Organization Tour Europe, Cedex 7, LaDefense, 92080 Paris, France
AGMA	AGMA	1916	U.S.A.	American Gear Manufacturers Association 1330 Massachusetts Avenue, N.W. Washington, D.C. 20005
AISI	Manuals	1908	U.S.A.	American Iron & Steel Institute 1000 16th Street, N.W. Washington, D.C. 20036
ANMC	No Stds	1973	U.S.A.	American National Metric Council 1625 Massachusetts Avenue, N.W. Washington, D.C. 20036

ACRONYMS AND ADDRESSES TO IMPORTANT STANDARDS ORGANIZATIONS (cont.)

Acronyms	Standard Prefix	Founded	Nation	Address
ANSI	ANSI	1918	U.S.A.	American National Standards Institute 1430 Broadway New York, New York 10018
API	API	1919	U.S.A.	American Petroleum Institute 2101 L Street, N.W. Washington, D.C. 20037
ASAC	...	1966	International	Asian Standards Advisory Committee (Check information through ECAFE— United Nations Economic Commission for Asia and the Far East) Sala Santitham Rajadamnern Ave. Bangkok 2, Thailand
ASAE	ASAE	1907	U.S.A.	American Society Agricultural Engineers 2950 Niles Road St. Joseph, Michigan 49085
ASM	Books	1913	U.S.A.	American Society for Metals Metals Park Ohio 44073
ASME	ASME	1880	U.S.A.	American Society of Mechanical Engineers 345 East 47th Street New York, N.Y. 10017
ASTM	ASTM	1898	U.S.A.	American Society for Testing and Materials 1916 Race Street Philadelphia, Pennsylvania 19103
BSI	BS	1901	Great Britain	British Standards Institute 2 Park Street London, W1A 2BS England
CCPA	...	1954	U.S.A.	Cemented Carbide Producers Association 2130 Keith Building Cleveland, Ohio 44115

WORLD STANDARDS ORGANIZATIONS

ACRONYMS AND ADDRESSES TO IMPORTANT STANDARDS ORGANIZATIONS (*cont.*)

Acronyms	Standard Prefix	Founded	Nation	Address
CDA	Books	1962	U.S.A.	Copper Development Association, Inc. 405 Lexington Avenue New York, N.Y. 10017
CEE	CEE	1946	International	International Commission on Rules for the Approval of Electrical Equipment 310 Utrechtseweg Arnhem, Netherlands
CEN	CEN	1960	International	European Committee for Standardization Tour Europe, Cedex 7, La Defense, 92080 Paris, France
CENELEC	CENELEC	1960	International	European Committee for Electrotechnical Standardization 4, Galerie Ravenstein 1000 Brussels, Belgium
CGPM	. . .	1875	International	General Conference of Weights and Measures (Paris)
CIPM	. . .	1875	International	International Converence on Weights and Measures (part of OIML) International Organization for Legal Metrology 11 Rue Turgot 75009 Paris, France
COPANT	COPANT	1947	International	Pan American Standards Commission (Refer to ANSI)
CSA	CSA	1919	Canada	Standards Council of Canada 178 Rexdale Boulevard Rexdale M9W 1R3, Ontario, Canada
DNA	DIN	1917	Germany	Deutscher Normenausschuss (German Standards Organization) 4-7 Burggrafenstrasse 1 Berlin 30 W. Germany

ACRONYMS AND ADDRESSES TO IMPORTANT STANDARDS ORGANIZATIONS (cont.)

Acronyms	Standard Prefix	Founded	Nation	Address
ECSC	EURO-NORM	1951	International	European Coal and Steel Community International Standards issued in: German, French, Italian, Dutch German contact: Beuth-Vertrieb GmbH 4-7 Burggrafenstrasse 1 Berlin 30, W. Germany
FIEI	...	1893	U.S.A.	Farm and Industrial Equipment Institute 410 N. Michigan Avenue Chicago, Illinois 60611
IEC	IEC	1906	World	International Electrotechnical Commission Central Office, 1, rue de Varembe 1211 Geneva 20, Switzerland
IFI	IFI	1931	U.S.A.	Industrial Fasteners Institute 1505 East Ohio Building Cleveland, Ohio 44114
ISO	ISO ISA	1947 1926	World	International Organization for Standardization Central Secretariat, Case Postale 56 1211 Geneva 20, Switzerland
JISC	JIS	1921	Japan	Japanese Industrial Standards Committee 3-1, Kasumigaseki 1 Chiyodaku, Tokyo, Japan
JSA	No Stds	1921	Japan	Japanese Standards Association 1-24 Akasaka 4 Chome Minato-ku, Tokyo, 107 Japan Japanese Standards Association 16, Chemin de la Voie-Greuse 1202 Geneva, Switzerland
MCTI	Books	1932	U.S.A.	Metal Cutting Tool Institute 331 Madison Avenue New York, N.Y. 10017
NBS	NBS	1901	U.S.A.	National Bureau of Standards Metric Information Office U.S. Department of Commerce Washington, D.C. 20234

WORLD STANDARDS ORGANIZATIONS

ACRONYMS AND ADDRESSES TO IMPORTANT STANDARDS ORGANIZATIONS (cont.)

Acronyms	Standard Prefix	Founded	Nation	Address
NEMA	NEMA	1926	U.S.A.	National Electrical Manufacturers Association 155 East 44th Street New York, N.Y. 10017
NFPA	NFPA	1953	U.S.A.	National Fluid Power Association 3333 North Mayfair Road Milwaukee, Wisconsin 53222
NSC	...	1913	U.S.A.	National Safety Council P.O. Box 11171 Chicago, Illinois 60611
PASC	...	1973	International	Pacific Area Standards Congress (U.S.A., Canada, Japan, Australia, New Zealand)
RMA	Books	1915	U.S.A.	Rubber Manufacturers Association 1901 Pennsylvania Avenue, N.W. Washington, D.C. 20006
SAA	AS	1922	Australia	Standards Association of Australia Standards House 80-86 Arthur Street North Sydney, N.S.W. 2060
SAE	SAE	1905	U.S.A.	Society of Automotive Engineers, Inc. 400 Commonwealth Drive Warrendale, Pennsylvania 15096
SES	No Stds	1942	U.S.A.	Standards Engineers Society 6700 Penn Avenue South Minneapolis, Minnesota 55423
SME	Books	1932	U.S.A.	Society of Manufacturing Engineers 20501 Ford Road Dearborn, Michigan 48128
TRA	Books	1903	U.S.A.	The Tire and Rim Association, Inc. 3200 W. Market Street Akron, Ohio 44313
ULI	ULI	1894	U.S.A.	Underwriters Laboratories, Inc. Corporate Headquarters 207 East Ohio Street Chicago, Illinois 60611

ACRONYMS AND ADDRESSES TO IMPORTANT STANDARDS ORGANIZATIONS (cont.)

Acronyms	Standard Prefix	Founded	Nation	Address
UNI	UNI	1928	Italy	Ente Nazionale Italiano de Unificazione Italian Standards Organization Piazza Armando Diaz 2 1 20123 Milano, Italy

RELATED PUBLICATIONS AVAILABLE FROM ANSI

ANSI Catalog — Contains listing with prices of ANSI, ISO, IEC, COPANT, CEE standards and recommendations; published annually.

ISO Catalog — Contains listing of ISO standards grouped by TC-numbers with numerical and alphabetical cross references of ISO standards; published annually with supplements.

ISO Status Report — On Draft International Standards—Reports on the progress of Draft International Standards at the various ballot levels of ISO; published quarterly.

ISO Directives — Contains the latest procedures relating to the technical work of ISO. The ISO Directives for the technical work of ISO are modified and published as required.

ISO Memento — Contains names and addresses of Member Bodies, general information on the ISO and its Technical Committees and Subcommittees; published annually.

ISO Participation — Lists participants or observer countries in ISO technical committees; published yearly.

ISO Bulletin — Calendar of ISO meetings together with revision sheets to the ISO Memento; 12 monthly issues. Subscription.

ISO/The Aims and Principles of Standardization

ISO Annual Review

2 The International System of Measuring Units (SI)

The following section describes, in detail, the SI (Système International d'Unités) metric system of measure and how it is used. The SI metric system is the internationally adopted measuring system eventually to be used to describe all physical quantities throughout the world.

The ISO (International Organization for Standardization) Technical Committee TC12 has developed the SI system in cooperation with CIPM (International Committee of Weights and Measures) and CGPM (General Conference of Weights and Measures). Representative standards[1] throughout the world for the SI system of measure are:

	ISO	1000 :	1973
U. S. A.	ANSI	Z210.1 :	1973
JAPAN	JIS	Z8203 :	1974
GERMANY	DIN	1301 :	1971
FRANCE	NF	X02-004 :	1974
UNITED KINGDOM	BS	3763 :	1970
ITALY	UNI	10003 :	1972
CANADA	CSA	Z234.2 :	1973
AUSTRALIA	AS	1000 :	1970

SI Units and Symbols

SI consists of seven base units, two supplementary units, a series of derived units consistent with the base, and supplementary units. It has a series of approved prefixes for the formation of multiples and submultiples of the various units. A number of derived units are listed in Table 2-1, including all approved units with special names. Additional derived units without special names are formed as needed from base units or other derived units, or both. (See Table 2-2.) Table 2-3 gives a list of recommended units.

metre or meter (m)— Unit of length equal to 1 650 763.73 wavelengths in a vacuum of the orange-red radiation of krypton-86 atom.

second(s)— The duration of 9 192 631 770 cycles of the radiation associated with a specified transition of the caesium-133 atom.

kilogram (kg)— Unit of mass equal to the cylinder of platinum-iridium alloy kept by the International Bureau of Weights and Measures, at Paris.

kelvin (K)— Unit for thermodynamic temperature. The kelvin scale has its zero point at absolute zero and has a fixed point at the triple point of water, defined as 273.16 kelvins, and the Celsius scale has the triple point defined as 0.01°C, which is approximately 32°F on the Fahrenheit scale.

ampere (A)— Unit for electric current which, if maintained in two straight parallel conductors of infinite length, of negligible circular cross-section, and placed one meter apart in vacuum, would produce between these conductors a force equal to 2×10^{-7} newtons per meter of length.

candela (cd)— The luminous intensity in the perpendicular direction of a surface of 1/600 000 square meter of a black body at the temperature of freezing platinum under a pressure of 101 325 newtons per square meter.

mole (mol)— The amount of a substance which contains as many elementary entities as there are atoms in 0.012 kilogram of carbon 12.

Definition of Supplementary Units

radian (rad)— The unit of measure of a plane angle with its vertex at the center of a circle that is subtended by an arc equal in length to the radius.

steradian (sr)— The unit of measure of a solid angle with its vertex at the center of a sphere and enclosing an area of the spherical surface equal to that of a square with sides equal in length to the radius.

[1] For information about the term "standard" as used in this book, please see page 1-2.

Table 2-1. Derived Units

Quantity	Unit	SI Symbol	Formula
acceleration	meter per second squared	---	m/s^2
activity (of a radioactive source)	disintegration per second	---	(disintegration)/s
angular acceleration	radian per second squared	---	rad/s^2
angular velocity	radian per second	---	rad/s
area	square meter	---	m^2
density	kilogram per cubic meter	---	kg/m^3
electric capacitance	farad	F	A·s/V
electric conductance	siemens	S	A/V
electric field strength	volt per meter	---	V/m
electric inductance	henry	H	V·s/A
electric potential difference	volt	V	W/A
electric resistance	ohm	Ω	V/A
electromotive force	volt	V	W/A
energy	joule	J	N·m
entropy	joule per kelvin	---	J/K
force	newton	N	kg·m/s^2
frequency	hertz	Hz	(cycle)/s
illuminance	lux	lx	lm/m^2
luminance	candela per square meter	---	cd/m^2
luminous flux	lumen	lm	cd·sr
magnetic field strength	ampere per meter	---	A/m
magnetic flux	weber	Wb	V·s
magnetic flux density	tesla	T	Wb/m^2
magnetomotive force	ampere	A	---
power	watt	W	J/s
pressure	pascal	Pa	N/m^2
quantity of electricity	coulomb	C	A·s
quantity of heat	joule	J	N·m
radiant intensity	watt per steradian	---	W/sr
specific heat	joule per kilogram-kelvin	---	J/kg·K
stress	pascal	Pa	N/m^2
thermal conductivity	watt per meter-kelvin	---	W/m·K
velocity	meter per second	---	m/s
viscosity, dynamic	pascal-second	---	Pa·s
viscosity, kinematic	square meter per second	---	m^2/s
voltage	volt	V	W/A
volume	cubic meter	---	m^3
wavenumber	reciprocal meter	---	(wave)/m
work	joule	J	N·m

Table 2-2. SI Prefixes

MULTIPLICATION FACTORS	EXPONENTIAL		PREFIX	SI SYMBOL
1 000 000 000 000	= 10^{12}	= E^{+12}	tera	T
1 000 000 000	= 10^9	= E^{+09}	giga	G
1 000 000	= 10^6	= E^{+06}	mega	M
1 000	= 10^3	= E^{+03}	kilo	k
100	= 10^2	= E^{+02}	hecto*	h
10	= 10^1	= E^{+01}	deka*	da
0.1	= 10^{-1}	= E^{-01}	deci*	d
0.01	= 10^{-2}	= E^{-02}	centi*	c
0.001	= 10^{-3}	= E^{-03}	milli	m
0.000 001	= 10^{-6}	= E^{-06}	micro	μ
0.000 000 001	= 10^{-9}	= E^{-09}	nano	n
0.000 000 000 001	= 10^{-12}	= E^{-12}	pico	p
0.000 000 000 000 001	= 10^{-15}	= E^{-15}	femto	f
0.000 000 000 000 000 001	= 10^{-18}	= E^{-18}	atto	a

*The preferred units are in multiples of 10 to the 3rd power (10^3 ... 10^6 ... 10^9, etc.). To promote seldom-used units such as those marked with an asterisk (*) is not recommended. Examples of preferred usage are: 10 mm instead of 1 cm; 100 grams instead of 1 hectogram.

THE INTERNATIONAL SYSTEM OF MEASURING UNITS (SI)

Table 2-3. List of Recommended Units*

Quantity	SI Unit	SI Multiples (Preferred Underlined)	Non-SI Metric Units
Plane Angle	rad (radian)	mrad (milliradian) μrad (microradian)	degree (°) $\quad 1° = \pi/180$ rad minute (') $\quad 1' = 1°/60$ second (") $\quad 1'' = 1'/60$ grade or gon$^{(g)}$, $1^g = \pi/200$ rad $400^g = 360°$
Solid Angle	sr (steradian)		
Length	m (meter)†	km (kilometer) = 10^3 m cm (centimeter) = 10^{-2} m <u>mm (millimeter)</u> = 10^{-3} m <u>μm (micrometer)</u> = 10^{-6} m nm (nanometer) = 10^{-9} m	
Area	m^2 (square meter)	<u>km^2 (square kilometer)</u> = 10^6 m^2 <u>dm^2 (square decimeter)</u> = 10^{-2} m^2 cm^2 (square centimeter) = 10^{-4} m^2 mm^2 (square millimeter) = 10^{-6} m^2	hectare (ha), 1 ha = 10^4 m^2 are (a), 1 a = 10^2 m^2
Volume	m^3 (cubic meter)	dm^3 (cubic decimeter) = 1 liter* = 10^{-3} m^3 <u>cm^3 (cubic centimeter)</u> = 1 milliliter = 10^{-6} m^3 <u>mm^3 (cubic millimeter)</u> = 10^{-9} m^3	hectoliter (hℓ), 1 hℓ = 10^{-1} m^3 liter (ℓ), 1 ℓ = 10^{-3} m^3 deciliter (dℓ), 1 dℓ = 10^{-4} m^3 centiliter (cℓ), 1 cℓ = 10^{-5} m^3 milliliter (mℓ), 1 mℓ = 1 cm^3 = 10^{-6} m^3
Time††	s (second)	ks (kilosecond) = 10^3 s ms (millisecond) = 10^{-3} s μs (microsecond) = 10^{-6} s ns (nanosecond) = 10^{-9} s	year, month, week day = 86400 s hour (h) = 3600 s minute = 60 s
Angular Velocity	rad/s (radians per second)		
Velocity	m/s (meter per second)		km/h, 1 km/h = $\frac{1}{3.6}$ m/s
Acceleration	m/s^2 (meter per square second)		
Frequency	Hz (hertz) = 1/s (cycles per second)	THz (terahertz) = 10^{12} Hz GHz (gigahertz) = 10^9 Hz <u>MHz (megahertz)</u> = 10^6 Hz <u>kHz (kilohertz)</u> = 10^3 Hz	
Rotational Frequency	s^{-1} (revolutions per second)		1 RPM = 1/60 s^{-1}
Mass§	kg (kilogram)	<u>Mg (megagram)</u> = 10^3 kg <u>g (gram)</u> = 10^{-3} kg mg (milligram) = 10^{-6} kg μg (microgram) = 10^{-9} kg	metric ton, 1 t = 10^3 kg
Linear Density	kg/m (kilogram per meter)	mg/m (milligram per meter) = 10^{-6} kg/m	1 tex = 10^{-6} kg/m The tex is used in the textile industry.
Density (mass density)	kg/m^3 (kilogram per cubic meter)	Mg/m^3 = 1000 kg/m^3 = 1 kg/dm^3 = 1 kg/ℓ = 1 g/cm^3 = 1 g/mℓ = density of water	

*For footnotes, see end of table.

Table 2-3 (*Continued*). List of Recommended Units*

Quantity	SI Unit	SI Multiples (Preferred Underlined)	Non-SI Metric Units
Momentum	kg · m/s		
Moment of Momentum, Angular Momentum	kg · m²/s		
Moment of Inertia	kg · m²		
Force	N (newton)	<u>MN (meganewton)</u> = 10^6 N <u>kN (kilonewton)</u> = 10^3 N daN (dekanewton) = 10 N** mN (millinewton) = 10^{-3} N μN (micronewton) = 10^{-6} N	See NOTE § for "Mass"
Moment of Force	N · m	MN · m (meganewton × meter) = 10^{-6} N · m <u>kN · m (kilonewton × meter)</u> = 10^3 N · m MN · m (meganewton × meter) = 10^{-3} N · m μN · m (micronewton × meter) = 10^{-6} N · m	
Pressure and Stress	Pa (pascal) = N/m²	GPa (gigapascal) = 10^9 N/m² <u>MPa or N/mm² (megapascal)</u> = 10^6 N/m² <u>kPa (kilopascal)</u> = 10^3 N/m² <u>mPa (millipascal)</u> = 10^{-3} N/m² μPa (micropascal) = 10^{-6} N/m²	10 bar = 1 MPa = MN/m² bar, 1 bar = 10^5 N/m² = 100 kPa millibar, 1 mbar = 10^2 N/m²
Viscosity (dynamic)	Pa · s	mPa · s (millipascal × second) = 10^{-3} Pa · s	centipoise, 1 cP = 1 mPa · s
Viscosity (kinematic)	m²/s	<u>mm²/s (square millimeter per second)</u> = 10^{-6} m²/s	stokes (St) centistokes, 1 cSt = 1 mm²/s
Surface Tension	N/m	<u>mN/m (millinewton per meter)</u> = 10^{-3} N/m	
Energy, Work, Heat	J (joule) = N · m	GJ (gigajoule) = 10^9 J <u>MJ (megajoule)</u> = 10^6 J <u>kJ (kilojoule)</u> = 10^3 J mJ (millijoule) = 10^{-3} J	kilowatt hour, 1 kWh = (3600 s) × (1000 W) = 3.6 × 10^6 J = 3.6 MJ

*For footnotes, see end of table.

THE INTERNATIONAL SYSTEM OF MEASURING UNITS (SI)

Table 2-3 (*Continued*). List of Recommended Units*

Quantity	SI Unit	SI Multiples (Preferred Underlined)	Non-SI Metric Units
Power	W (watt) = J/s	GW (gigawatt) = 10^9 W MW (megawatt) = 10^6 W kW (kilowatt) = 10^3 W mW (milliwatt) = 10^{-3} W μW (microwatt) = 10^{-6} W	1 W = 1 J/s = 1 N·m/s metric horsepower, 1 hp (metric) = 75 kgf·m/s = 735 W
Thermo-dynamic Temperature	K (kelvin)	The absolute temperature $T = T_0 + t$	
Celsius Temperature	°C (degree Celsius)		

NOTES:
1. The temperature difference of 1°C (degree Celsius) = 1 K (kelvin) (exactly)
2. Thermodynamic temperatures are designated with a capital T.
3. Celsius degrees are designated with a lower-case t.
4. The relationship between kelvins and degrees Celsius are as follows:

T (kelvins) = T_0 + t (degrees Celsius)
= 273.15 + t (exactly)

---

*See ISO referenced standards for metric quantities and units not listed.

**The dekanewton has some usage in Europe since 1daN = 1.02 kg (force).

†*-meter* and *-liter* are preferred spellings in the United States, although *-metre* and *-litre* are commonly used in many countries.

††TIME: The international designation is: 13:32 (instead of 1:32 P.M.); 11:15 (instead of 11:15 A.M.)

DATE: The all-numeric writing of dates varies in different parts of the world. The date April 2, 1962 is written as follows:

ISO 2014 standard	: 1962-04-02
North America	: 4-2-1962
Europe	: 2-4-1962

The ISO standard for the writing of all-numeric dates is used in other countries, and should be adopted in all international communications.

USE OF SECOND (s): The SI base unit s (second) is recommended in all applications where energy or power might be calculated (torque, flow, speed).

§In the SI system there are separate and distinct units for mass and force. The kilogram is restricted to mass. The newton is the unit of force and should be used in place of the "kilogram-force."

The newton instead of the kilogram-force should be used in combination units which include force, for example, pressure or stress (N/m² = Pa), energy (N·m = J), and power (N·m/s = W).

Considerable confusion exists in the use of the term "mass" and "weight." Mass is the property of matter to which it owes its inertia. If a body at rest on the earth's surface is released from the forces holding it at rest, it will experience the acceleration of free fall (acceleration of gravity, *g*). The force required to restrain it against free fall is commonly called weight. The acceleration of free fall varies in time and space, and weight is proportional to it. While at any point in time and space, weight can therefore vary, mass does not. Observed *g* can differ by over 0.5% between various points on the earth's surface. Therefore, the difference of local *g* from the agreed standard value, 9.80665 m/s², must be taken into account for precise measurements where *g* is involved, such as in delicate weighing.

The term "mass" should be used to indicate the quantity of matter in an object. The term "weight" is commonly used where the technically correct word is mass. Because of this widespread nontechnical use, the word *weight* should be avoided in technical reports. In converting quantities that have been presented as weight, care must be taken to determine whether force or mass is intended.

Practical Application of the SI Units

The established SI units: base, supplementary, derived, and their combinations—with appropriate multiple or submultiple prefixes—should be used as indicated in Table 2-3.

Application of Prefixes

1. The name of decimal multiples and submultiples of the unit of mass are formed by attaching prefixes to the word 'gram': 1 Mg (megagram) *not* 1 kkg (kilokilogram) for 1000 kg.
2. In addition, an SI prefix may be attached to the name of an SI base unit, supplementary unit, or to a derived unit having a special name and symbol whether above or in combination: cm (centimeter), MHz (megahertz), mA/s (milliampere per second).
3. Compound prefixes should not be used: nm (nanometer but not mµm (millimicrometer) for 10^{-9} m.
4. The prefix becomes a part of the symbol or name with no separation: 10 kilograms (kg), 3 milliseconds (ms). Note that 1 km (kilometer) is not equal to 1 K m (kelvin × meter) or 1 Mg (megagram) is not equal to 1 mg (milligram).
5. A symbol for a unit with a prefix attached is regarded as a single symbol which may be raised to a power without the use of parentheses; mm^2 means $(0.001m)^2$ and not $0.001m^2$.
6. Prefixes should be applied to the numerator of compound units, except when using kilograms (kg) in the denominator. Since kilogram (kg) is a base unit of SI, this particular multiple is not a violation and should be used in preference to the gram (g): MN/m^2 (meganewton per square meter = megapascal) and not N/mm^2 (newton per square millimeter), kg/m (kilogram per meter) and not g/mm (gram per millimeter), J/kg (joule per kilogram = specific energy) and not mJ/g (millijoule per gram).
7. Effort should be made, through selection of appropriate prefixes, to limit the number of digits to the left or right of a decimal point to four or less; 20 kN (kilonewton) and not 20000 N (newton), 7.2 µm (micrometer) and not 0.0072 mm (millimeter).

CAPITALIZATION—Unabbreviated units are not capitalized, for example, hertz, newton, kelvin. Symbols for SI units are only capitalized when the unit is derived from a proper name; N from Isaac *N*ewton. Numerical prefix symbols given previously in Table 2-2 are not capitalized; except for the symbols M (mega), G (giga), T (tera).

PLURALS—Unabbreviated SI units form their plurals in the usual manner. SI symbols are always written in singular form: 50 newtons or 50 N, and 25 grams or 25 g.

PUNCTUATION—English speaking countries use a period for the decimal point, others use a comma. Whenever a numerical value is less than one, a zero should precede the decimal point. Periods are not used after any SI unit symbol, except at the end of a sentence.

DERIVED UNITS—In symbols for derived metric units, a center dot or a blank separation is used to indicate multiplication and a slash to indicate division; N·m or N m (newton × meter), kg/m^3 (kilogram per cubic meter).

NUMBER GROUPING—A comma is used to indicate a decimal point in continental Europe. Commas to separate multidigit metric or inch values into groupings of three should be avoided; 33541 mm and not 33,541 mm.

DECIMAL POINT—The use of a period (.) or a comma (,) to indicate a decimal point for customary units and metric units is considered optional by most US standards organizations. The use of a comma on all metric values will help identify the SI dimension from the inch dimension. A center dot is used to indicate multiplication in most countries outside North America and a period for decimal place could be mistakenly read as a multiplication sign.

Example: Area of one rectangular cross-section
(U.S.) Area = width × height = 6.3 in. × 1.6 in. = 10.08 sq. in. = 10 sq. in. approx.
(Continental Europe) Area = b · h = 6,3 mm · 1,6 mm = 10,08 mm^2 = 10 mm^2 approx.

NON-SIGNIFICANT ZEROS—Where limit dimensioning is used, and where either the maximum or minimum dimension has digits following the decimal point, the other value shall have zeros added for uniformity of inscription.

Examples: 25.00   25
           not
          24.46   24.46

Non-significant zeros are generally not shown after the decimal point in the composition of a millimeter value except as noted above.

Example: 25   not   25.0

STANDARD CONDITIONS AND PHYSICAL CONSTANTS

Standard Conditions

Standard gravity acceleration g = 9.80665 m/s^2
                                     = 32.1740 ft/s^2

Absolute temperature
 (Thermodynamic temperature) K(kelvin)
    = (°C + 273.15)
    = (Celsius degrees + 273.15 exactly)

THE INTERNATIONAL SYSTEM OF MEASURING UNITS (SI)

Miscellaneous Pressure Bases

International standard atmosphere
= 0.101325 MPa (megapascal)
= 1.01325 bar
= 1.01325 10^5 N/m^2
= 1.0332 kgf/cm^2
= 14.697 lbf/in.^{2}

1 technical atmospheric pressure[1] = 1 at
= 0.98067 bar
= 1 kgf/cm^2
= 1 kp/cm^2
= 14.223 lbf/in.^{2}
= 735.6 mm Hg
= 28.96 in. Hg

Absolute pressure = atmospheric pressure + recorded pressure
ata = at + atü (gauge)
atü = atmospheric overpressure (Germany)

RELATED ISO STANDARDS[2] (ISO TC 12)

ISO 31/0-1974	General introduction to ISO 31—General principles concerning quantities, units and symbols
ISO/R 31/I-1965	Basic quantities and units of the SI
ISO/R 31/II-1958	Quantities and units of periodic and related phenomena
ISO/R 31/III-1960	Quantities and units of mechanics (Y10.3-1968)
ISO/R 31/IV-1960	Quantities and units of heat (Y10.4-1957)
ISO/R 31/V-1965	Quantities and units of electricity and magnetism
ISO 31/VI-1973	Quantities and units of light and related electromagnetic radiations
ISO/R 31/VII-1965	Quantities and units of acoustics
ISO 31/VIII-1973	Quantities and units of physical chemistry and molecular physics
ISO 31/IX-1973	Quantities and units of atomic and nuclear physics
ISO 31/X-1973	Quantities and units of nuclear reactions and ionizing radiations
ISO/R 31/XI-1961	Mathematical signs and symbols for use in the physical sciences and technology
ISO 31/XII-1975	Dimensionless parameters
ISO 31/XIII-1975	Quantities and units of solid state physics
ISO 1000-1973	SI units and recommendations for the use of their multiples and of certain other units
ISO 2014-1976	Writing of calendar dates in all numeric form
ISO 2955-1974	Information processing—Representation of SI and other units for use in systems with limited character sets

The ASME[3] have published or plan to publish the following guides to the SI system:

SI - 1 Guide for Use of SI (Metric) Units
SI - 2 SI Units in Strength of Materials
SI - 3 SI Units in Dynamics
SI - 4 SI Units in Thermodynamics
SI - 5 SI Units in Fluid Mechanics
SI - 6 SI Units in Kinematics
SI - 7 SI Units in Heat Transfer
SI - 8 SI Units in Vibration
SI - 9 SI Units in Energy Conversion
SI -10 Steam Charts
SI -11 SI Units in Air Conditioning and Refrigeration

---

[1] The technical atmospheric pressure is defined in the German standard DIN 1314 as 1 kilogram–force per square centimeter, and it approximates the barometric pressure at sea level.

[2] Since ISO documents are constantly being upgraded, for simplicity the author refers to all ISO publications in the text of this book as "standards," designating a particular document as "ISO . . ." followed by the appropriate identification number.

However, the actual status of a particular ISO document might be: (a) *Recommendation*, in which case it would be officially designated by "ISO/R . . ." preceding the appropriate identification number. (b) *Draft International Standard*, ("ISO DIS . . ."). (c) An officially adopted international standard, in which case the initials "ISO" followed by the identification number is the appropriate designation. This page describes the current status of standards. Information on the various designations is given in ISO catalogs and supplements available from: American National Standards Institute, 1430 Broadway, New York, New York 10018.

[3] See page 1-8 for ASME address.

3 Engineering Drawing Practice

GENERAL INFORMATION ON ENGINEERING DRAWINGS

Introduction

The material presented in this section is intended to serve as a general guide in finding national and international standards on the subject of Engineering Drawing. It is of great importance for multinational companies to use internationally recognized drawing practices. Where a machine is initially designed and manufactured in one country, and at a future date, must be produced in another nation, the company will avoid substantial extra expense by producing the machine from the initial drawings.

ISO Paper Sizes (ISO 216)

The familiar letter-paper size, 8½ x 11 inches, used in the United States captures a large percent of the world market. The ISO paper size A4 is principally used in Europe and in most of the emerging nations, and is estimated to represent an increasing portion of the world usage of letter-size paper. Some of the considerations made before including the ISO paper sizes in the ISO 216 standard were as follows:

> In many countries, far more sizes of paper are used than are really necessary. Many of them came into existence under conditions different from those prevailing today, while the origin of others was due to chance. Consequently they do not fulfil the present need for consistency between the sizes of paper and printed matter for various purposes, nor do they meet the many requirements for a coherent relationship between the sizes of paper, printed or not, papermaking, printing and converting machinery and equipment, and storage and filing equipment.
>
> The purpose of the ISO standard[1] is to improve the present position by providing a rationally designed *ISO System* of trimmed sizes which can bring about a reduction in the number of sizes and create more rational, clear and consistent ranges. This will simplify and cheapen ordering, production, use, dispatch and storage and will also provide a sound basis for standardization in related fields.
>
> One of the considerations leading to the present ISO standard was the fact that the standards bodies in the following countries had already adopted this system of sizes in their national standards, and others were known to be using the sizes:

Date of adoption	Countries	Standard No.
1922	Germany	DIN 476
1924	Belgium	NBN 18
1925	Netherlands	NEN 381
1926	Norway	NS 20
1927	Finland	SFS P.I. 1
1929	Switzerland	SNV 10120
1934	U.S.S.R.	GOST 9327
1938	Hungary	MOSz 16
1939	Italy	UNI 923-924
1941	Sweden	SIS 73 01 01
1943	Argentina	IRAM 3001-N.P.
1943	Brazil	--
1947	Spain	UNE 1011
1948	Austria	A 1001
1949	Romania	STAS 570-52
1951	Japan	JIS P 0138
1953	Denmark	DS 910
1953	Czechoslovakia	CSN 01 0402
1954	Israel	S.I. 117
1954	Portugal	NP-4 and NP-17
1957	India	IS:1064
1957	Poland	PN-55/P-02001
1959	United Kingdom	B.S. 3176

The Universal Postal Union has specified certain of these sizes.

Range of ISO Trimmed Paper Sizes

The range of ISO standard paper sizes is based on a rectangle of 1 square meter area, the sides of which are in the ratio $1:\sqrt{2}$. The sizes are obtained by dividing the next larger size into two equal parts, the division being parallel to the shorter side, so that the area of two successive sizes is in the ratio of 2:1. (See Figs. 3-1 and 3-2.)

The ISO-A Series of standard paper sizes are given in Table 3-1.

---

[1] Since ISO documents are constantly being upgraded, for simplicity the author refers to all ISO publications in the text of this book as "standards," designating a particular document as "ISO..." followed by the appropriate identification number. However, the actual status of a particular ISO document might be: (a) *Recommendation*, in which case it would be officially designated by "ISO/R..." preceding the appropriate identification number. (b) *Draft International Standard* "ISO DIS..." (c) An officially adopted international standard, in which case the initials "ISO..." followed by the identification number are the appropriate designation. Page 3-18 describes the current status of standards. Information on the various designations is given in ISO catalogs and supplements available from: American National Standards Institute, 1430 Broadway, New York, New York 10018.

Table 3-1. ISO Standard Trimmed Paper Sizes (A-Series)* (ISO 216)

Designation	Millimeters	Inches
A0	841 X 1189	33.11 X 46.81
A1	594 X 841	23.39 X 33.11
A2	420 X 594	16.54 X 23.39
A3	297 X 420	11.69 X 16.54
A4	210 X 297	8.27 X 11.69
A5	148 X 210	5.83 X 8.27
A6	105 X 148	4.13 X 5.83
A7	74 X 105	2.91 X 4.13
A8	52 X 74	2.05 X 2.91
A9	37 X 52	1.46 X 2.05
A10	26 X 37	1.02 X 1.46

*These rarely used trimmed paper sizes also belong to the series:
4A0: 1682 mm X 2378 mm   66.22 in. X 93.62 in.
2A0: 1189 mm X 1682 mm   46.81 in. X 66.22 in.

General (ISO 128)

This portion of the drawing practice section defines line conventions to be used on engineering drawings which are universally recognized and are acceptable for microfilm.

The most important requirements for line work are legibility and consistency. Lines should be correctly formed, clean, and dense enough to assure good reproduction. In the preparation of a drawing, it is necessary to employ various widths of lines to clarify different features and elements of a component. Three widths of lines are commonly used: (1) thin (0.35 mm nominal), (2) medium (0.5 mm nominal), and (3) thick (0.7 mm nominal).

Actual line widths should be governed by the size and style of the drawings; however, the relative width of lines should be approximately as above. Uniform line widths should be maintained throughout the drawing.

Continuous Thick Lines

The outline of the object is represented by a *visible line* and should be the most prominent part of the drawing. Therefore, it must be drawn with a dense, heavy width line. (See Example A in Table 3-2.)

Continuous Thin Lines

Continuous thin lines are used for dimension and leader lines and hatching. The ISO recommends the use of a continuous thin line to represent a fictitious (phantom) outline. These and other usages of continuous thin lines are shown in Table 3-2, Examples B and C.

Short Dashes (Medium)

Hidden Lines—These are medium-width, dashed lines used to show the hidden features of an object, as in Fig. 3-3. The dashes are approximately 3 mm long and the spaces approximately 0.7 mm, but may vary slightly according to the size of the drawing. (See Example D in Table 3-2.)

Hidden lines should always begin and end with a dash in contact with the visible or hidden line at which they start or end, except where such a dash would form a continuation of a visible line. Dashes should join at corners. Arcs should start with dashes at the tangent points, as in Fig. 3-3. Hidden lines should be omitted whenever they are not needed for clarity.

Long Dashes

Thin chain lines—These should be comprised of long dashes alternating with short dashes. The proportions should be generally as shown in E, in Table 3-2, but the

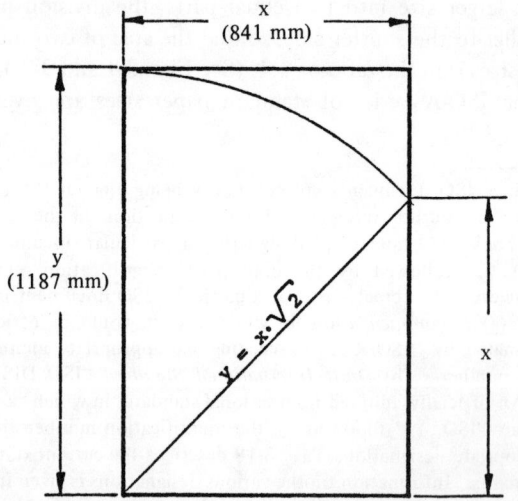

Fig. 3-1. Ratio between sides of ISO trimmed paper.

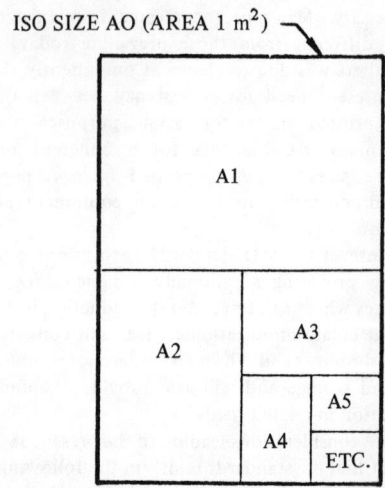

Fig. 3-2. Division of ISO trimmed paper sizes.

ENGINEERING DRAWING PRACTICE

Table 3-2. Types of Line (ISO 128)

Refer to figure 3-3	Type of line	Examples of application
A ————————	Continuous thick	Visible outlines and edges
B ————————	Continuous thin	Fictitious outlines and edges Dimension and leader lines Hatching Outlines of adjacent parts Outlines of revolved sections
C ~~~~~~~~		Limits of partial views or sections, if the line is not an axis
D ‑ ‑ ‑ ‑ ‑ ‑ ‑	Short dashes (medium)	Hidden outlines and edges
E —·—·—·—·—	Long chain thin	Centre lines Extreme positions of movable parts Parts situated in front of the cutting plane
F ━━·——·——·━━	Long chain thick at ends, thin elsewhere	Cutting planes
G ━━·━━·━━·━━	Long chain thick	Indication of surfaces which are to receive additional treatment

lengths and spacing may be increased when very lengthy lines are needed. Note special instructions in F, of Table 3-2, for cutting planes.

Thick chain lines—The lengths and spacing of the elements of thick chain lines, as in G, in Table 3-2, should be similar to those of thin chain lines.

General—All chain lines should start and finish with a long dash; when thin chain lines are used as center lines they should cross one another at solid portions of the line. Center lines should extend only a short distance beyond the feature unless required for dimensioning or other purposes. They should not extend through the spaces between views and should not terminate at another line of the drawing. Where angles are formed in chain lines, long dashes should meet at corners. Arcs should join at tangent points. Dashed lines should also meet at corners and at tangent points with dashes. See Fig. 3-3.

Lettering (ISO 3098/I)

It is important that characters should be simple, uniform, and capable of being produced by hand, stencil, machine, or other means at reasonable speed. The characters on the drawing should remain legible not only in the direct photocopy print but in the form of reduced copy or as an image on a microfilm viewing screen.

Clarity, style, size, and spacing are important, particularly for figures. Unlike letters, figures rarely fall into identifiable patterns and must be read individually. Characters should be of open form and devoid of serifs and other embellishments. All strokes should be black and of consistent density compatible with the line work. Care should be taken that sufficient space exists between characters and parts of characters to ensure that "filling in" will not take place during reproduction.

Style

No particular style for hand lettering is recommended; the aim should be to produce legible and unambiguous

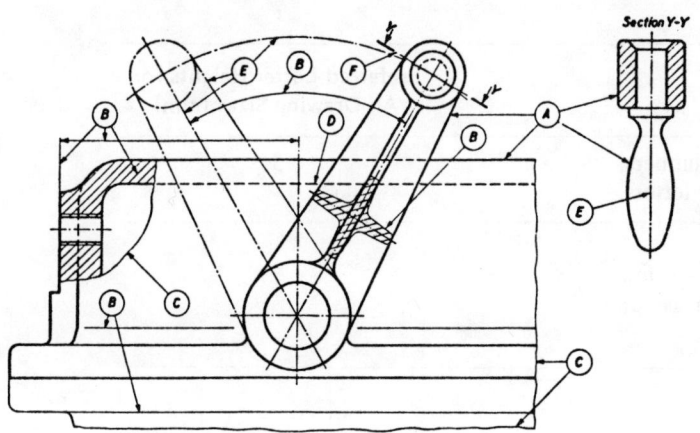

Fig. 3-3. Application of various types of lines (ISO 128).

characters. Vertical or sloping characters are suitable for general use but the presentation should be consistent on any one drawing, i.e., vertical and sloping letters should not be mixed. The examples in Fig. 3-4 are provided as a guide only.

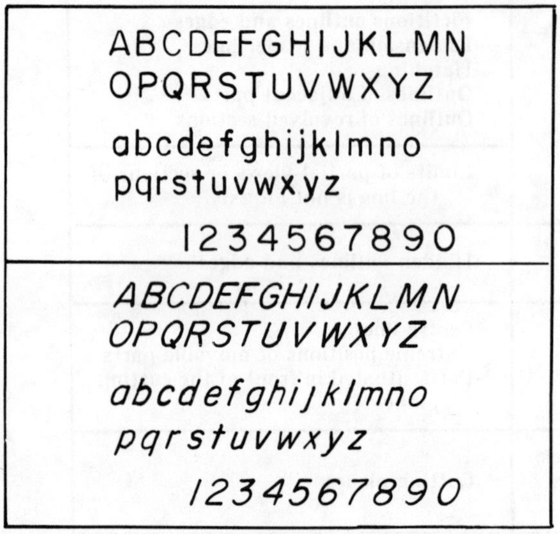

Fig. 3-4. Examples of suitable letters for drawings.

Capital letters are preferred to lower case as they are less congested and are less likely to be misread when reduced in size. It is recommended that lower case letters be restricted to instances where they form part of a standard symbol, code, or abbreviation (for example, in numerous metric units—mm, kg, MPa, etc.).

Character Height—Figure 3-5 lists the recommended minimum character heights. It is stressed that these recommendations are for minimum sizes. When lower case letters are used they should be so proportioned that the body height is approximately 0.7 times the capital letter height. The stroke thickness should be approximately 0.1 times the character height and the clear space between characters and parts of characters should be approximately 0.2 times the character height.

Space between lines of lettering should be not less than 1.8 the character height but, in the case of titles, closer spacing may be sometimes unavoidable.[1]

Orientation of Lettering

To facilitate reading, all notes should be placed so that they can be read in the same direction as the format of the drawing. Underlining of notes is not recommended. Where it is required to emphasize a note or heading, larger characters should be used. Underlining of dimensions indicates it is out of proportion or not to scale.

Recommended Scale Ratios

Scale multipliers and divisors of 2, 5, and 10 are recommended. The resultant representative fractions will be:

1000:1	50:1	1:1	1:50
500:1	20:1	1:2	1:100
200:1	10:1	1:5	1:200
100:1	5:1	1:10	1:500
	2:1	1:20	1:1000

The scale of the drawing should be indicated in the same manner, e.g., 10:1 on a drawing made at ten times full size.

Title Block

The title block is generally preprinted and contains the essential information required for the identification, administration, and interpretation of the drawing.

---

[1] While ISO 3098/I specifies 1.4 times the character height, the author recommends the larger factor of 1.8 for greater clarity.

Item	Preferred Letter Heights for All Drawing Sizes (mm)
All drawing dimensions, notes, tables, change numbers, change record entries, and component callout letters.	5
Part numbers in drawing number block and section, tabulation and flag letters for features such as surfaces, intersections or special diameters.	7

Fig. 3-5. Recommended minimum character heights (ISO 3098/I).

ENGINEERING DRAWING PRACTICE

It is recommended that the title block be at the bottom of the sheet with the drawing number in the lower right-hand corner. Adjacent to this drawing number should be the title and issue (alteration) information. For convenience the drawing number may also appear elsewhere on the drawing. (See ISO drawing example in Fig. 3-9 for the general layout of a typical title block.)

Basic Information in the Title Block

It is recommended that spaces be provided in title blocks for the following basic information:
1. Name of firm
2. Drawing number
3. Descriptive title of depicted part or assembly
4. Original scale
5. Date of the drawing
6. Signature(s)
7. Issue information
8. Copyright clause
9. Projection (third or first angle and/or symbol)
10. Unit of measurement
11. Reference to drawing practice standards.

Projection

Symbols of Projection—Two systems of projection, known respectively as FIRST ANGLE (ISO METHOD E) and THIRD ANGLE (ISO METHOD A), are approved internationally and are regarded as being of equal status in the ISO technical drawing standards.

The system of projections used on a drawing should be indicated by the appropriate symbol (Figs. 3-6, 3-7 or 3-8), and this is placed clearly in a space provided for the purpose in the title block of the drawing near the indication of the scale; otherwise the direction in which the views are taken should be clearly indicated.

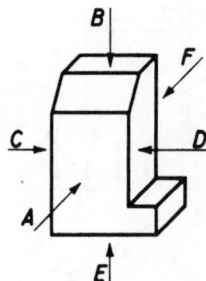

Fig. 3-6. Designation of views: View in direction A = view from the front; view in direction B = view from above; view in direction C = view from the left; view in direction D = view from the right; view in direction E = view from below; view in direction F = view from the rear. The front view (principal view) having been chosen, the other usual views make with it and between themselves angles of 90° or multiples of 90° (ISO 128).

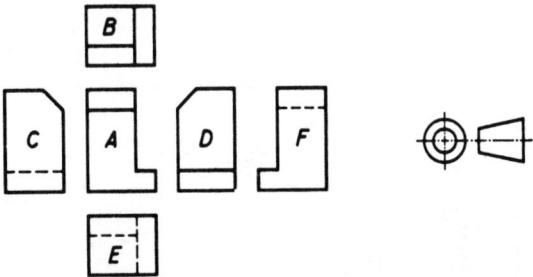

Fig. 3-7. American practice, symbol to right denotes third angle projection. With reference to the front view the other views are arranged as follows: The view from above is placed above; the view from below is placed underneath; the view from the left is placed on the left; the view from the right is placed on the right; the view from the rear may normally be placed on the left or on the right, as may be found convenient. The distinctive symbol of this method is shown to the right in this figure (ISO 128).

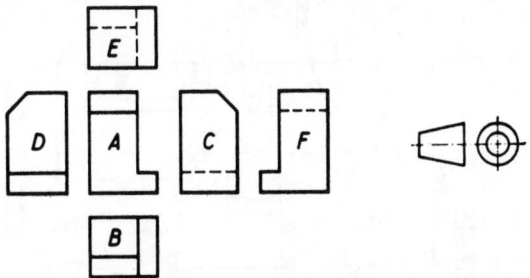

Fig. 3-8. European practice, symbol to right denotes first angle projection. With reference to the front view the other views are arranged as follows: The view from above is placed underneath; the view from below is placed above; the view from the left is placed on the right; the view from the right is placed on the left; the view from the rear may normally be placed on the left or on the right, as may be found convenient. The distinctive symbol of this method is shown to the right in this figure (ISO 128).

Systems of Measure for Engineering Drawings

Linear dimensions on drawings could be stated either in millimeters, in inches, or both. Most engineering drawings at some future date will be expressed in millimeters. At the present time, however, most of the drawings in North America are in inches. How do we convert to millimeters with the least amount of effort? An answer to this question, and also some practical examples of how the conversion is handled by some large, multinational companies, follow.

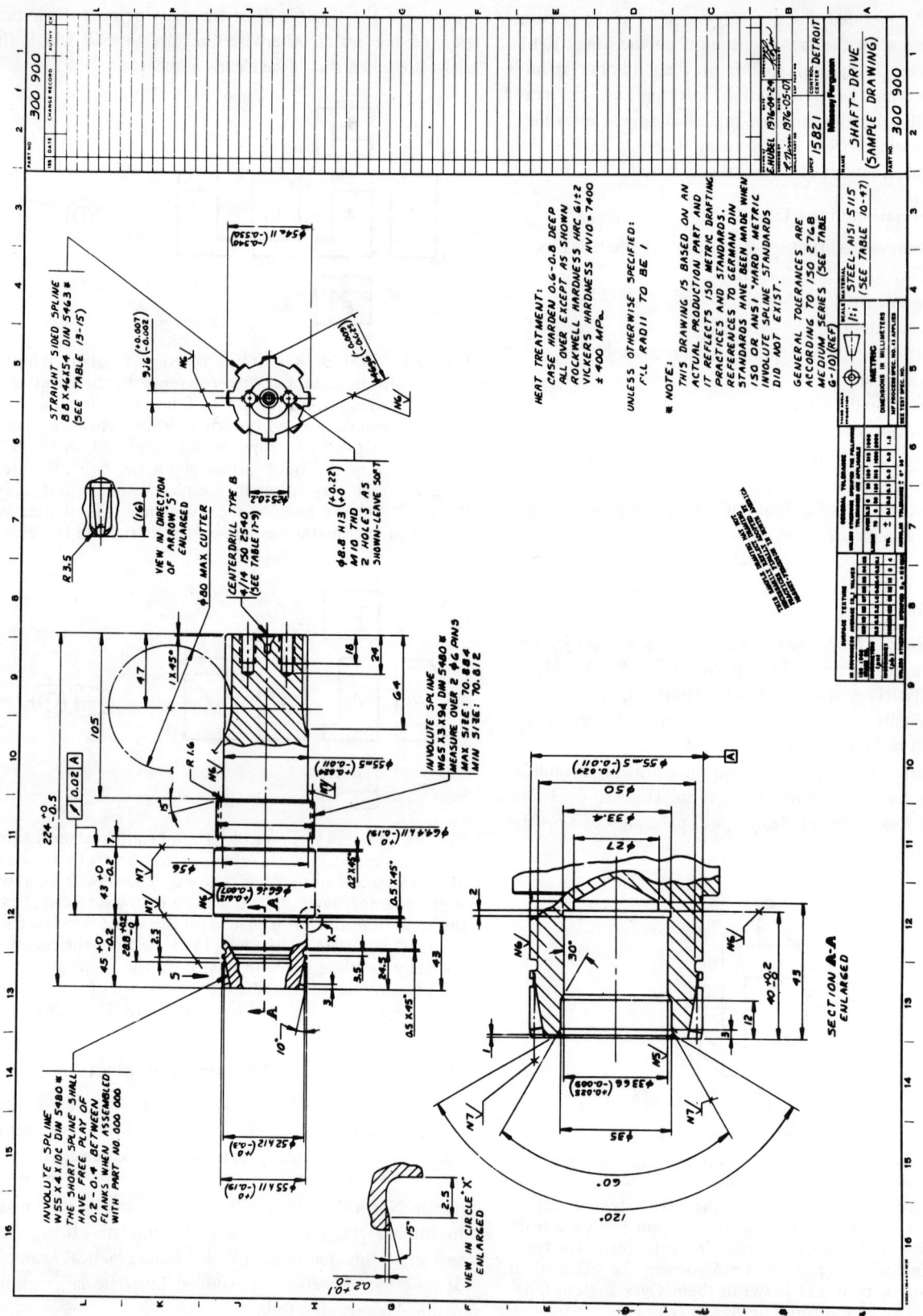

Fig. 3-9. Typical example of ISO drawing.

ENGINEERING DRAWING PRACTICE

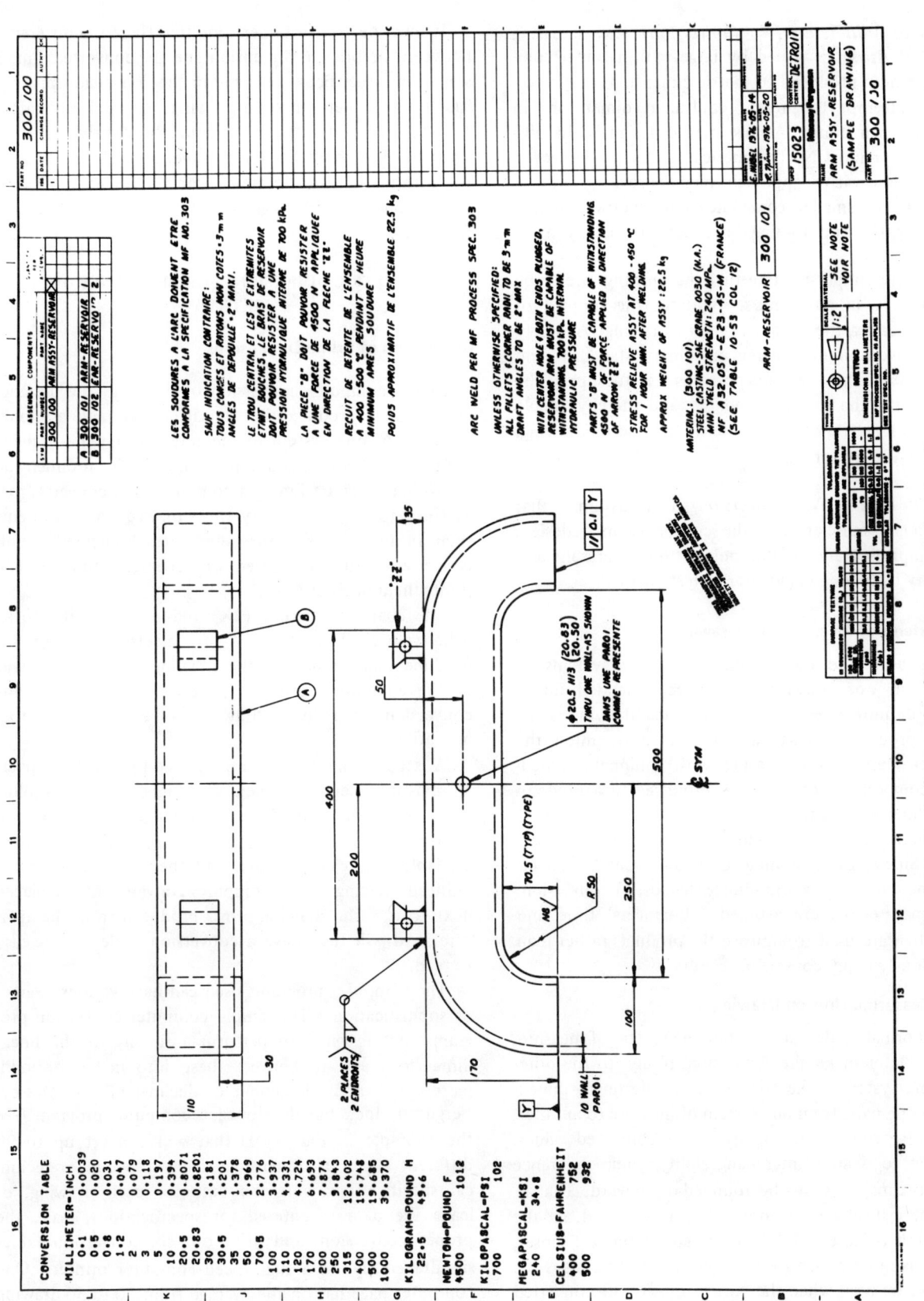

Fig. 3-10. Typical example of computer conversion drawing.

Design in Millimeters (No Conversion)—An increasing number of drawings issued by large multinational companies are now produced to metric dimensions only without any attempt to present equivalent dimensions in customary inch units. A pure metric design is the ultimate goal for most North American industry. However, a more conservative approach might prove to be advantageous for smaller companies until metric standard material, parts, and tooling have become readily available.

It is quite simple to state that the drawings are to be designed only in millimeters. Without the support of metric standards, the so-called "metric drawing" is merely a soft conversion of existing customary standards and components. The main purpose here is to supply the designer with the metric standards, and thereby enable him to think metric and to produce a true metric drawing. (See Fig. 3-9, 3-10, or 3-21.)

Design in Inches (No Conversion)—It is assumed that the reader is familiar with the customary inch design. No attempt is made in this publication to present any customary inch design guide nor inch standards.

Dual Systems of Measure on a Drawing

Using both the metric and customary systems of measure on one drawing is not recommended unless there is definite need for it. The inch drawing of an existing production part can be left as is, unless the particular part is interphasing with millimeter-dimensioned components or is to be produced in a predominantly "metric" country.

The most economical and practical conversion process is Single Dimensioning—Computer Chart Conversion. The computer can minimize the cost of conversion, and human errors are avoided. Engineers' time and talents then are used to improve the products rather than to laboriously apply conversion factors.

Prime Measuring Unit on Drawing

It is normal to design a part in one system of measure, and then later make the conversion, if any, to the other measuring system. The conversion of the linear dimensions on one part, from one system of measure to another, will involve some rounding-off of the converted values. In order to secure interchangeability some tolerance limits also may have to be rounded-off inward. That is, the numerical value of the upper limit is decreased, and the lower limit is increased. When tolerance limits are closer, the part is more expensive to produce. Existing gages can no longer be used when the part is to meet the converted dimensions.

It must be clearly understood that there is a difference between the original dimension and the converted dimension—some sacrifices are made in the conversion process. The prime dimension on a drawing, to which the part is to be made and inspected, should be clearly identified as such. The converted linear dimensions should be used for information and references only. An example of a note applied to drawings with two systems of measure follows:

NON-METRIC DIMENSIONS FOR REFERENCE ONLY METRIC

Conversion of Prime Dimensions

In the computer conversion of drawings the designer and detailer produce the layout, detail, and assembly drawings in one system of measure. When the drawing is finished, checked, and approved, each linear dimension and other units are fed to a computer that converts from SI Units to U.S. Customary Units or vice versa; it orients them in numerical sequence; eliminates duplicates; rounds converted value to degree of accuracy required; and prints them in chart form.

The chart is enlarged (2X) and is transferred to an adhesive-backed mylar via an electrostatic copying machine and applied to the drawing. The enlargement of the computer output is necessary when reading the conversion on reduced drawing copies produced from microfilm.

A second method of using a computer to convert an existing metric or inch drawing shown on a blueprint or original would be to produce a half-size print from the available copy (the Xerox Company has machines available that can do this), and then to microfilm the reduced drawing with a computer conversion table placed next to it. The drawing prints produced from the aperture card will then have a conversion table that is easy to read.

The computer programs used can have various degrees of sophistication. The simple computer conversion programs are designed to perform a sorting of the linear dimensions and to convert these into inches or millimeters with a fixed number of decimal places. Massey-Ferguson, Inc., has developed a computer program (see the example in Fig. 3-10) that will convert up to 90 different types of dimensions from drawings or technical specifications, sort the data for each drawing, or leave the data as entered for specifications, make the proper conversion, and print out six conversion tables simultaneously on a high-speed computer printer. One computer page has on the average from 20 to 30 drawing conversion tables, and the computer time for processing

ENGINEERING DRAWING PRACTICE

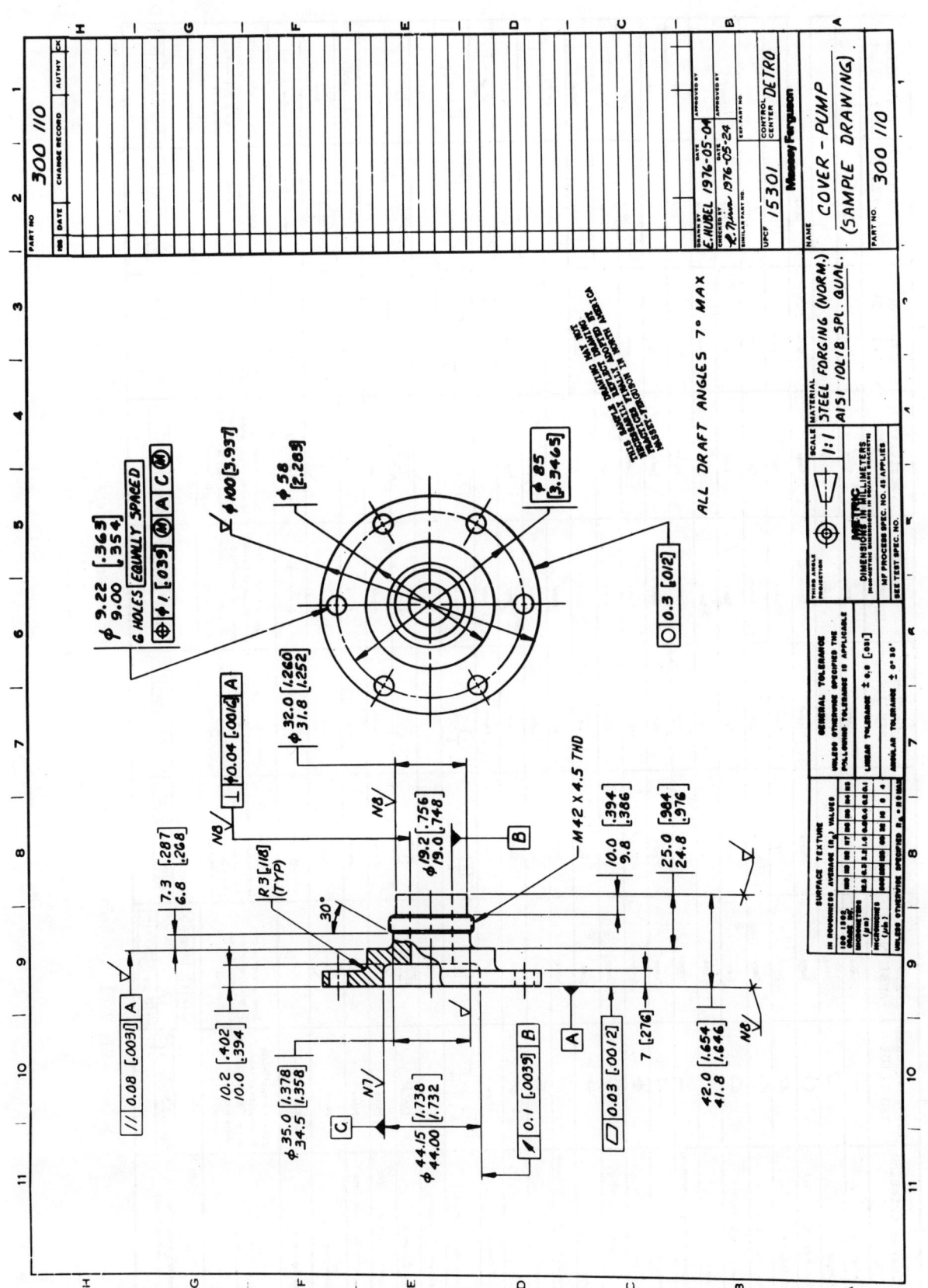

Fig. 3-11. Typical example of dual dimensioning drawing.

ENGINEERING DRAWING PRACTICE

Characteristics	International ISO R1101	American ANSI Y14.5	Japanese JIS B0021	German DIN 7184	French NF E04-121	British BS 308	Italian UNI 7226	Canadian CSA B78.2	Australian AS1100 Part 10
Straightness	—	Same	Same	Same	Same	Same	Same	Same	Same
Flatness	◻	Same	Same	Same	Same	Same	Same	Same	Same
Roundness (Circularity)	○	Same	Same	Same	Same	Same	Same	Same	Same
Cylindricity	⌭	Same	Same	Same	Same	Same	Same	Same	Same
Profile of a Line	⌒	Same	Same	Same	Same	Same	Same	Same	Same
Profile of a Surface	⌓	Same	Same	Same	Same	Same	Same	Same	Same
Parallelism	∥	Same	Same	Same	Same	Same	Same	Same	Same
Perpendicularity (Squareness)	⊥	Same	Same	Same	Same	Same	Same	Same	Same
Angularity	∠	Same	Same	Same	Same	Same	Same	Same	Same
Position	⊕	Same	Same	Same	Same	Same	Same	Same	Same
Concentricity (Coaxiality)	◎	Same	Same	Same	Same	Same	Same	Same	Same
Symmetry	⌖	Same	Same	Same	Same	Same	Same	Same	Same
Maximum Material Condition	Ⓜ	Same	Same	Same	Same	Same	Same	Same	Same
Diameter	⌀	Same	Same	Same	Same	Same	Same	Same	Same
Circular Runout	↗	Same	Same	Same	Same	Same	Same	Same	Same
Total Runout	None	↗↗	None	None	None	None	None	None	↗↗
Datum Identification	-A- or 🏴	-A-	🏴 or A	🏴 or A	🏴 or A	🏴 or A	🏴 or A	-A- or 🏴	A or 🏴
Reference Dimension	(127)	(5.000)	(127)	(127)	(127)	(127)	(127)	(5.000)	(127)
Basic Dimension	▯127▯	▯5.000▯	▯127▯	▯127▯	▯127▯	▯127▯	▯127▯	▯5.000▯	▯127▯
Regardless of Feature Size	None	Ⓢ	None	None	None	None	None	None	None
Projected Tolerance Zone	None	Ⓟ	None	None	None	None	None	Ⓟ	None
Datum Target	None	⊕	None	None	⊕	⊕	None	⊕	See Standard
Part Symmetry	⌖	None	⌖	⌖	⌖	⌖	⌖	⌖	⌖
Shape of the tolerance zone	Zone is a total width in direction of leader arrow. ⌀ specified where zone is circular or cylindrical.	Zone is a total width in direction of leader arrow. ⌀ specified where zone is circular or cylindrical.	Zone is a total width in direction of leader arrow. ⌀ specified where zone is circular or cylindrical.	Zone is a total width in direction of leader arrow. ⌀ specified where zone is circular or cylindrical.	Zone is a total width in direction of leader arrow. ⌀ specified where zone is circular or cylindrical.	Zone is a total width in direction of leader arrow. ⌀ specified where zone is circular or cylindrical.	Zone is a total width in direction of leader arrow. ⌀ specified where zone is circular or cylindrical.	Zone shape evident from characteristic being controlled.	Zone is a total width in direction of leader arrow. ⌀ specified where zone is circular or cylindrical.
Sequence within the feature control symbol	⊕ ⌀0.5 Ⓜ A B C	A B C ⌀.02 Ⓜ or ⌀.02 Ⓜ A B C	⊕ ⌀0.5 Ⓜ A B C	⊕ ⌀0.5 Ⓜ A B C	⊕ ⌀0.5 Ⓜ A B C	⊕ ⌀0.5 Ⓜ A B C	⊕ ⌀0.5 Ⓜ A B C	⊕ .02 Ⓜ A B C	⊕ ⌀0.5 Ⓜ A B C

Fig. 3-12. Comparison of international and national geometric tolerancing symbols (ANSI Y14.5).

ENGINEERING DRAWING PRACTICE

the data is kept to a minimum. The subsequent enlargement on adhesive-backed mylar is also easily processed by Xerox. Since computers are normally not equipped with lower-case letters and therefore would misprint most metric units, the names of units are spelled out whenever possible. The proper conversion factor and number of decimal places are coded with a one, two or three-digit alpha-numeric name such as: MM for conversion of millimeters into inches with three decimal places; KG for the conversion of kilograms into pound mass and rounded-off to three-number accuracy (rounding-off error less than 0.5%); N for the conversion of newtons into pound force, etc. Non-significant zeros are eliminated by the computer, to conform to metric practices.

A more complex computer programming effort could follow the SAE Standard J390 for Dual Dimensioning. The computer would test the total tolerance for a given dimension and print out the converted value to the number of decimal places shown in the above-mentioned standard.

The conversion of dimensions in one system of measure to the other requires no engineering skill and can be processed by clerical help, key-punching operators, and computer personnel. The dimensions to which the part is to be made and inspected should be shown on the drawing with a conversion for information shown on the conversion chart. If a dimension on a part is changed, the computer conversion chart might be re-run with the added data cards for the new dimensions, or might simply be omitted.

Dual Dimensioning (Designer Conversion)—The practice of dual dimensioning is not recommended; but if it is considered necessary to quote a dimension in both metric and inch units, the dimension to which the part is to be made and inspected should be shown first with a conversion for information shown after it in parentheses. (See Fig. 3-11.) Dual dimensioning is time-consuming for the engineers when producing the drawings, and confusing for the factory personnel to read. In this case, no real metric conversion is achieved since both engineers and machine operators will use customary units only.

Tolerancing

Tolerancing of Size—Section 6 of this book will describe in detail the ISO system of limits and fits, and is recommended to be used for tolerancing of size.

Tolerancing of Form and Position Using Symbols—(ISO Geometric Tolerancing)—A tolerance of size, the traditional method of linear tolerancing, when specified alone, effects a degree of control of form. However, in some circumstances, dimensions and tolerances of size, no matter how well applied, would not impose the desired control. If a different degree of control of form is required, form tolerances should be specified and such form tolerances take precedence over the form control implied by the size tolerance. (See Fig. 3-18). Geometrical tolerances should be specified for all requirements critical to functioning and interchangeability except when it is certain that the machinery and techniques which will be used can be relied upon to achieve the required standard of accuracy. How far it is necessary to specify geometrical tolerances in any particular instance can only be decided in the light of functional requirements, interchangeability, and probable manufacturing circumstances. Drawings prepared for widespread quantity production at home or abroad, or for sub-contracting in workshops of widely varying equipment and experience are particular instances where the most complete and explicit tolerancing is necessary. This demands that the information given on the drawing be so complete in dimensional and geometrical requirements that the part may be made and inspected to suit the full requirements of the designer. On the other hand, such detail may be unnecessary when adequate control is exercised by other means, for instance, where the method of production has been proved to produce parts to the required tolerances for satisfactory functioning.

A sample drawing with numerous examples of how the form and position tolerancing symbols are used is shown in Fig. 3-21. The drawing sample has been provided by the Massey Ferguson Inc. in the U.S.A. Geometrical tolerancing symbols are given in the ISO1101 standard. The ISO symbols are adopted in a number of national standards (see Fig. 3-12). The geometrical tolerancing system is replacing the written instruction which, in turn, may have required translation. (See Fig. 3-21.)

Geometric Tolerancing Training Charts

It is of the greatest importance for multinational companies to use symbols instead of written notes on drawings whenever possible. The writer recommends the use of the ISO Datum Identification symbol shown in Fig. 3-12, since the U.S. represents the only major country in the world with a special symbol for datum identification, and this deviation, in my opinion, is made without a significant reason.

Figures 3-13 through 3-20 were developed in the U.K. for the Ford Motor Company, Ltd., for training their engineers in the use of the geometric tolerancing system. The illustrations are self-explanatory. However, if the meaning of the various symbols is desired, please refer to any of the international or national standards shown in Fig. 3-12. The ANSI Y14.50-1973 provides a complete description with more than 100 pages on dimensioning and tolerancing of engineering drawings.

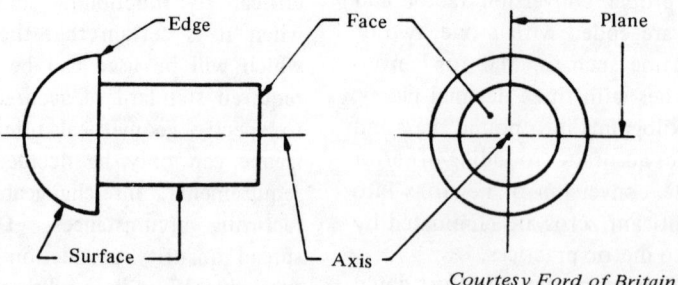

Fig. 3-13. Features of a component.

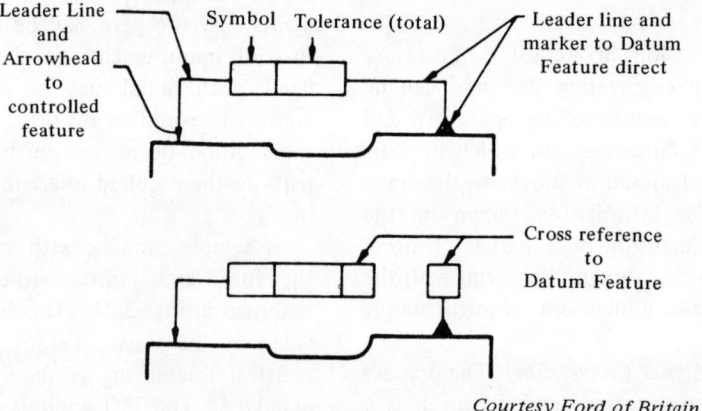

Fig. 3-14. Use of control frames.

The dimensions determining the true position or true profile (T.P.) of a feature are enclosed in boxes, thus:

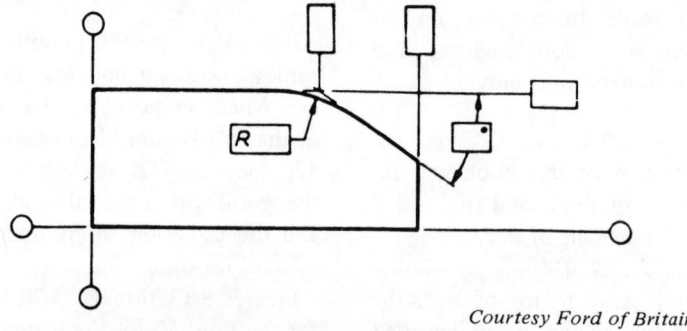

Fig. 3-15. True position, true profile (T. P.).

ENGINEERING DRAWING PRACTICE

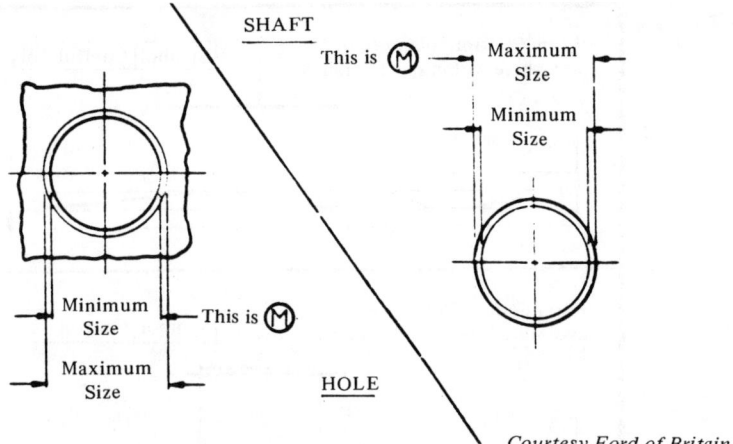

Fig. 3-16. Maximum material conditions.

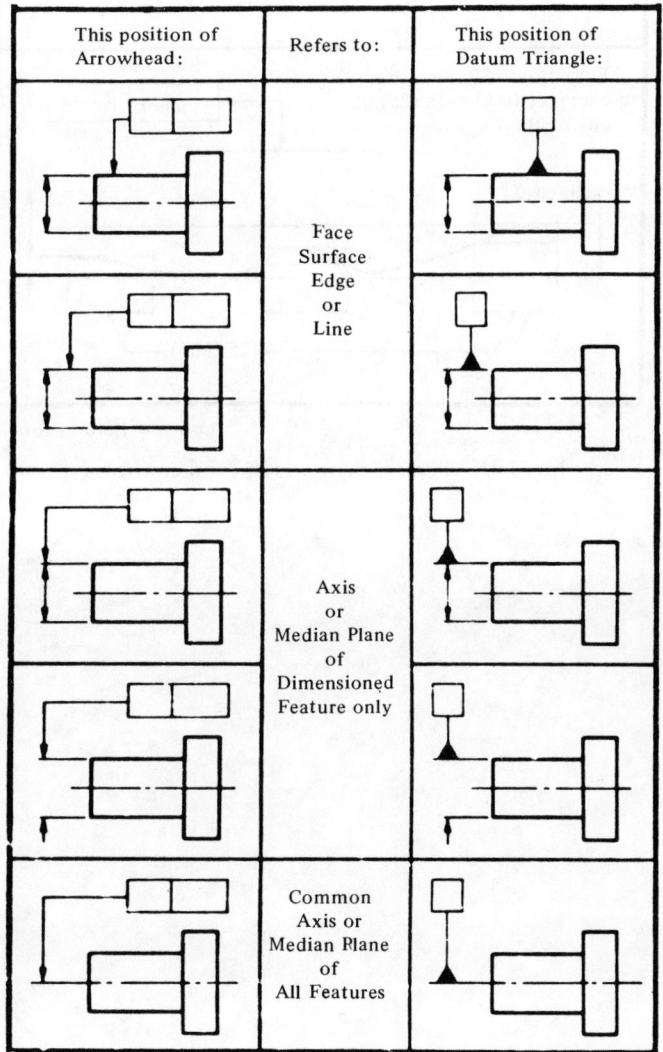

Fig. 3-17. Significance of arrowhead position.

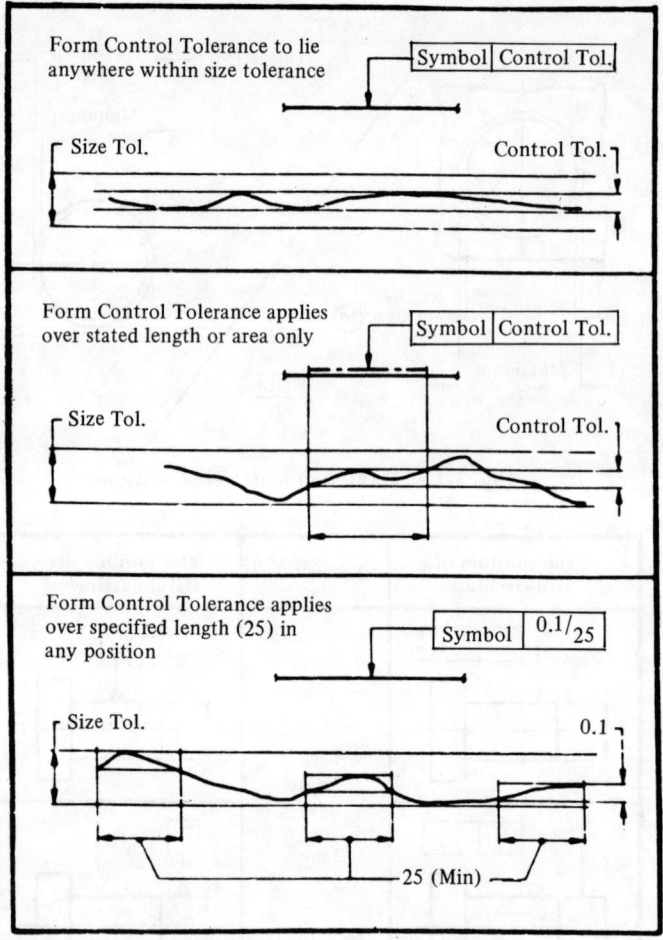

Courtesy Ford of Britain

Fig. 3-18. Typical examples of form control.

ENGINEERING DRAWING PRACTICE

3-15

TYPE OF TOLERANCE		CHARACTERISTIC TO BE TOLERANCED	SYMBOL	EXAMPLE	INTERPRETATION	EXPLANATION
INDIVIDUAL FEATURES	FORM TOLERANCES	STRAIGHTNESS	—	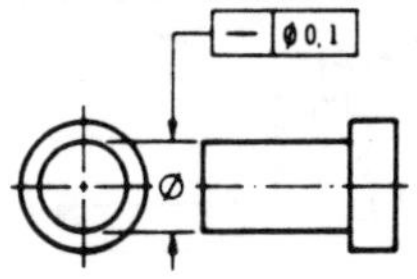 ⌀0.1	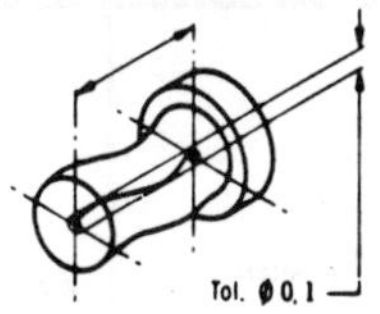 Tol. ⌀0.1	The axis of the dimensioned feature is required to be contained within a cylindrical zone of 0,1 diameter
		FLATNESS	▱	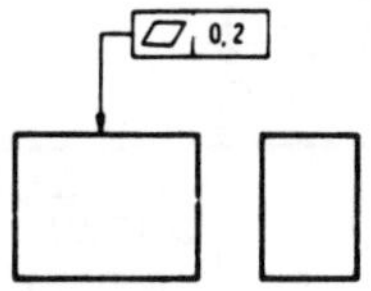 0,2	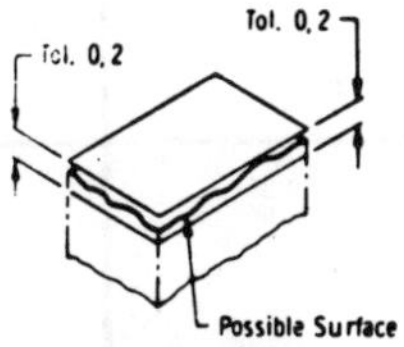 Tol. 0,2 / Possible Surface	The specified surface is required to lie between two planes 0,2 apart
		ROUNDNESS	○	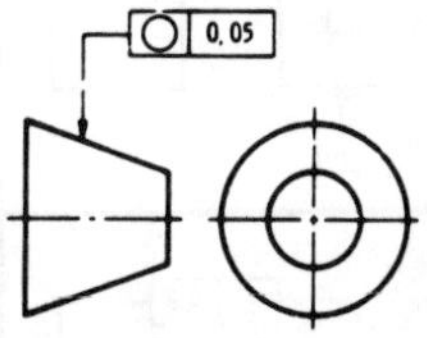 0,05	 Possible Form / Tol. 0,05	The circumference at any cross section square to the axis is required to lie between two circles concentric with each other a radial distance of 0,05 apart
		CYLINDRICITY (NON-PREFERRED)	⌭	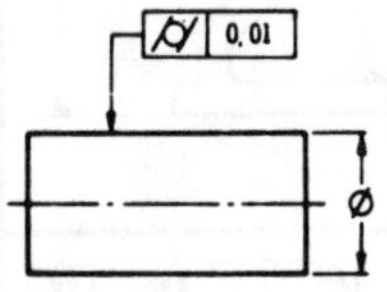 0,01	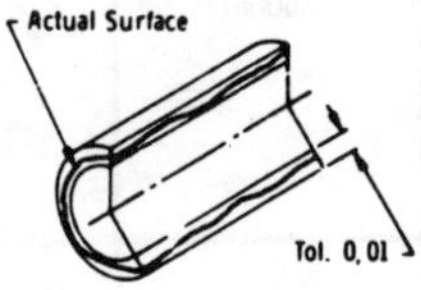 Actual Surface / Tol. 0,01	The curved surface is required to lie between two cylindrical surfaces co-axial with each other a radial distance 0,01 apart
		PROFILE OF A LINE	⌒	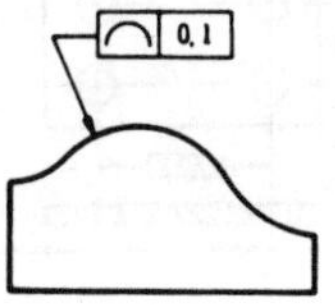 0,1	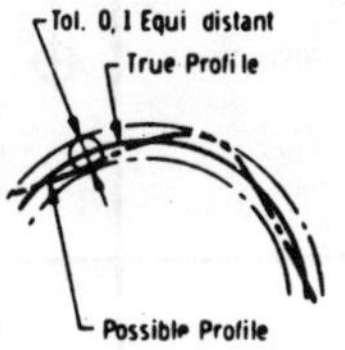 Tol. 0,1 Equi distant / True Profile / Possible Profile	The required profile to lie between two lines which envelop a series of circles of diameter 0,1 the centres of which are situated on the line representing the True Profile
		PROFILE OF A SURFACE	⌓	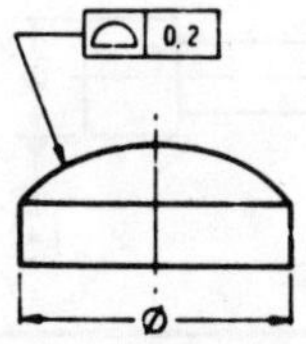 0,2	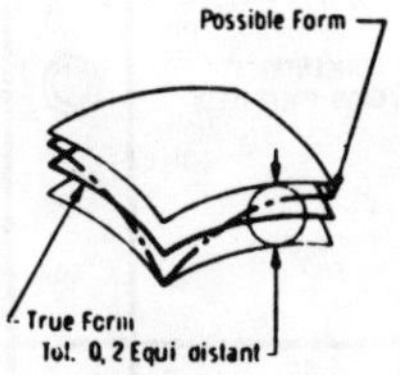 Possible Form / True Form / Tol. 0,2 Equi distant	The required surface profile to lie between two surfaces which envelop a series of spheres 0,2 diameter the centres of which are situated on the True Profile of surface
RELATED FEATURES	COMPOSITE TOLERANCE	RUN-OUT	↗	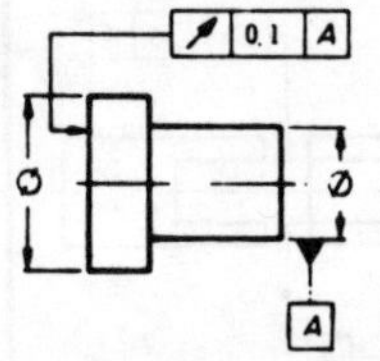 0,1 A	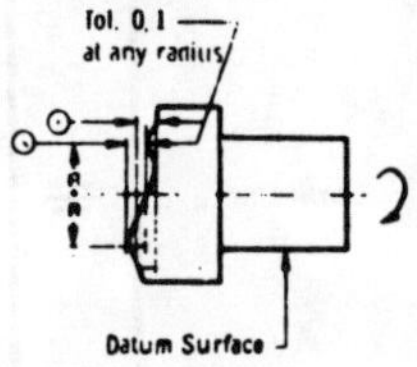 Tol. 0,1 at any radius / Datum Surface	The run-out must not exceed 0,1 measured parallel to the datum axis at any radius (two readings shown)

Courtesy Ford of Britain.

Fig. 3-19. Application of symbols (individual features).

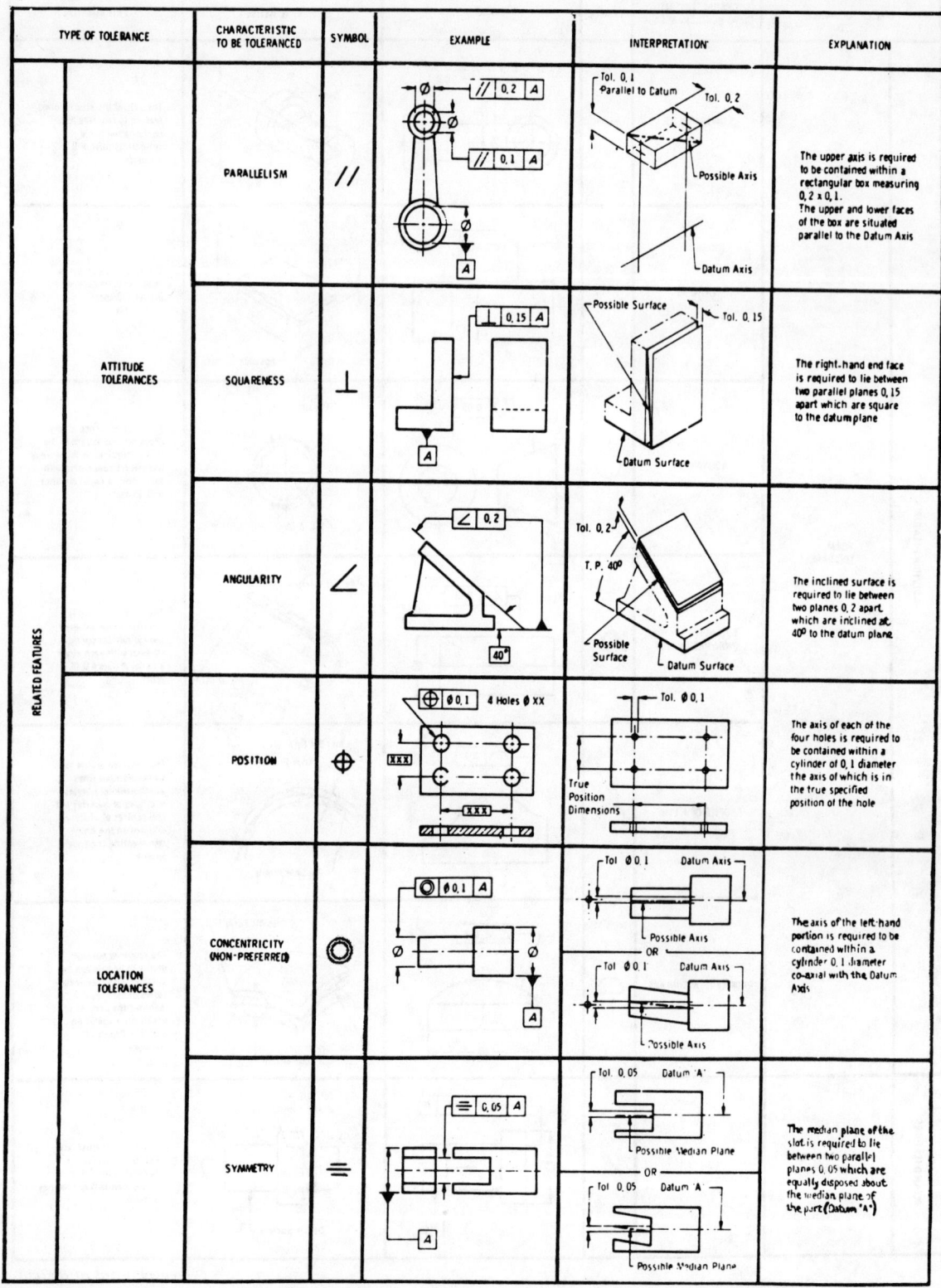

Fig. 3-20. Application of symbols (related features).

Courtesy Ford of Britain.

ENGINEERING DRAWING PRACTICE

3-17

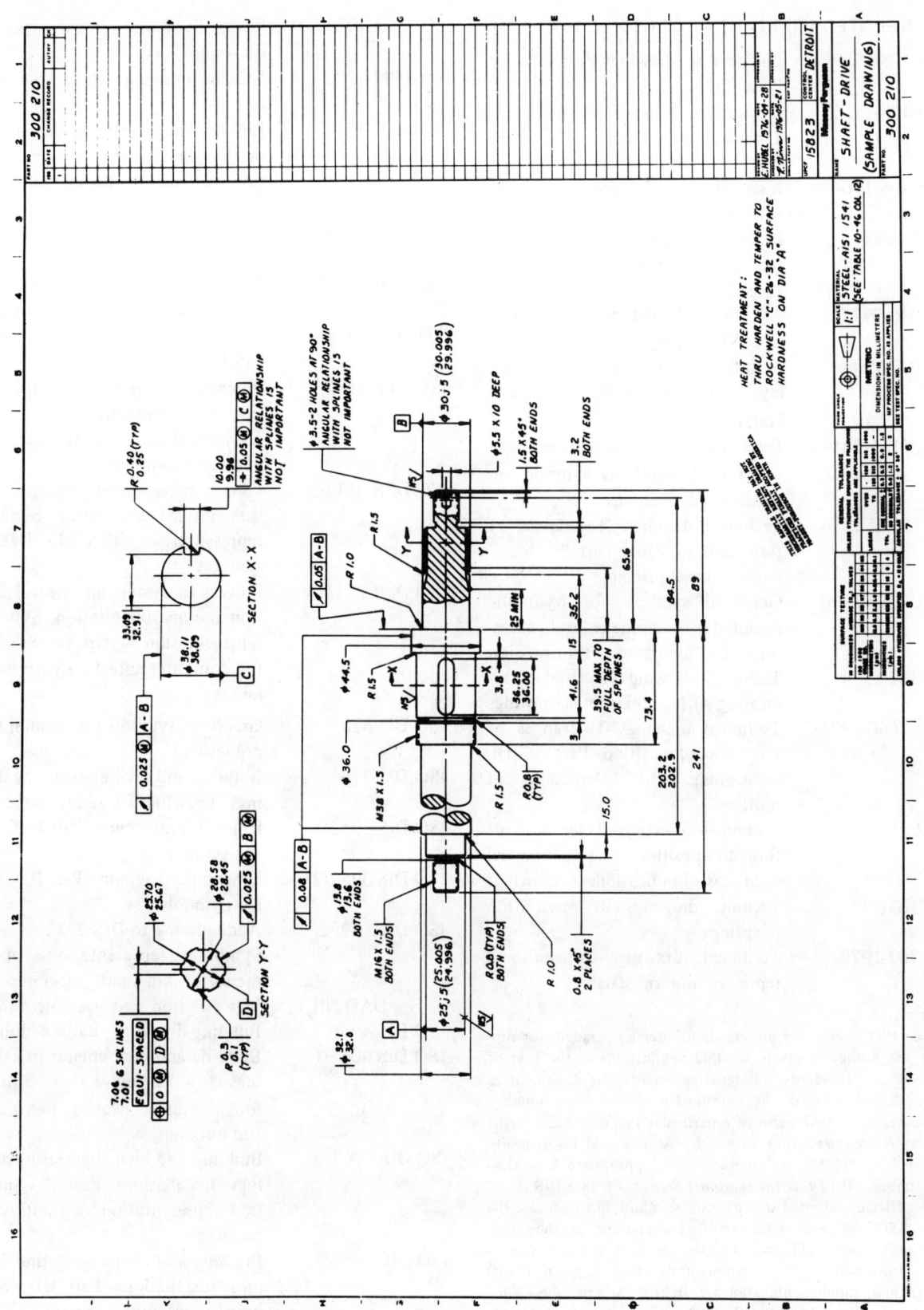

Fig. 3-21. Drawing example of geometric tolerancing.

RELATED ISO STANDARDS (ISO TC 10)[1]

ISO/R 128-1959	Engineering drawing—Principles of Presentation
ISO/R 129-1959	Engineering drawing—Dimensioning
ISO 216-1975	Trimmed sizes of writing paper and certain classes of printed matter
ISO/R 406-1964	Inscription of linear and angular tolerances
ISO 478-1974	Paper: untrimmed stock sizes for the ISO—A Series—ISO primary range
ISO 1046-1973	Architectural and building drawings—Vocabulary
ISO 1047-1973	Architectural and building drawings — Presentation of drawings — Scales
ISO/R 1101/I-1969	Tolerances of form and of position—Part I: Generalities, symbols, indications on drawings
ISO 1101/II-1974	Technical drawings—Tolerances of form and position—Part 2: Maximum material principle
ISO/R 1219-1970	Graphical symbols for hydraulic pneumatic equipment and accessories for fluid power transmission
ISO 1302-1974	Technical drawings—Method of indicating surface texture on drawings
ISO/R 1660-1971	Technical drawings—Tolerances of form and of position—Part 3: Dimensioning and tolerancing of profiles
ISO/R 1661-1971	Technical drawings—Tolerances of form and position—Part 4: Practical examples of indications on drawings
ISO 2162-1973	Technical drawings—Representation of springs
ISO 2203-1973	Technical drawings—Conventional representation of gears
ISO 2553-1974	Welds—Symbolic representation on drawings
ISO 2594-1972	Building drawings—Projection methods
ISO 2595-1973	Building drawings—Dimensioning of production drawings—Representation of manufacturing and work sizes
ISO 3040-1974	Technical drawings—Dimensions and tolerancing cones
ISO 3098/I-1974	Technical drawings—LETTERING—Part I: Currently used characters
ISO DIS 3511/I	Control functions representations, Part I
ISO DPR 3511/2	Process measurement control function and instrumentation—Symbolic representation — Part II—Extension of basic requirements
ISO DPR 3511/3	Process measurement control function and instrumentation—Symbolic representation — Part III — Detailed symbols
ISO DPR 3511/4	Process measurement control function and instrumentation—Symbolic representation — Part IV—Symbols for computer interface with instruments
ISO DIS 3753	Graphical symbols for vacuum technology
ISO DIS 3766	Building and civil engineering drawings—Drawings for reinforcements
ISO DIS 3952/I	Kinematic diagrams—Part I: Graphical symbols
ISO DIS 3952/2	Kinematic diagrams—Part I: Graphical symbols
ISO DAD 3952/ DAD NR 1	Addendum 1 to DIS 3952 Symbolic representations of elements of kinematic diagrams—Part V: Friction and gear mechanisms
ISO DIS 4066	Building drawings—Bar scheduling
ISO DIS 4067/I	Building and civil engineering drawings—Installations—Part I: Symbols for plumbing, heating, ventilation and ducting
ISO DIS 4067/2	Building and civil engineering drawings—Installations—Part II: Simplified representation of sanitary appliances
ISO DIS 4067/3	Building and civil engineering drawings—Installations—Part III: Symbols for automatic control

---

[1] Since ISO documents are constantly being upgraded, for simplicity the author refers to all ISO publications in the text of this book as "standards," designating a particular document as ISO . . ." followed by the appropriate identification number. However, the actual status of a particular ISO document might be: (a) *Recommendation*, in which case it would be officially designated by "ISO/R . . ." preceding the appropriate identification number. (b) *Draft International Standard* ("ISO DIS . . ."). (c) An officially adopted international standard, in which case the initials "ISO" followed by the identification number are the appropriate designation. This page describes the current status of standards. Information on the various designations is given in ISO catalogs and supplements available from: American National Standards Institute, 1430 Broadway, New York, New York 10018.

ISO DIS 4068	Building and civil engineering drawings—Reference lines	ISO DPR 5457	Technical drawings—Sizes of drawings
ISO DIS 4069	Building and civil engineering drawings—Representation of areas—General principles	ISO DPR 5458	Technical drawings—Positional tolerances
ISO DPR 5261	Technical drawings for structural metal work	ISO DPR 5459	Technical drawings—Datums and datum systems
ISO DIS 5455	Technical drawings—Scales of drawings	ISO DPR 5460	Technical drawings—Measuring principles
ISO DPR 5456	Technical drawings — Axonometric projection methods	ISO DPR 5461	Technical drawings—Limits of size
		ISO DPR 5462	Technical drawings — Relation between tolerances
		ISO DPR 5463	Technical drawings—Terminology

4 Preferred Numbers

Introduction

Preferred numbers are a geometrical series of numbers adopted worldwide for standardization[1] purposes. They have been in use for many years, but were used in connection with standardization for the first time during the 1870's by Charles Renard, a French army captain, who was able to reduce the number of different dimensions of rope for military balloons from 425 to 17, with the aid of the series. In today's world it has become increasingly important, from a cost standpoint, to reduce the number of different standard parts, material, and components used in products. And the preferred numbering system can provide the tool needed to achieve this goal.

The preferred numbering system is independent of the system of measure used. However, it has played an important role in those countries now on the metric system.

Design in millimeters differs from the customary inch design in that one inch has 25 integer (whole) sizes to choose from. The inch fractions have functioned to some extent as a preferred numbering system. The fraction 1/2 is preferred over 1/4; 1/4 is preferred over 1/8; 1/8 is preferred over 1/16, etc.

This feature has helped to reduce the number of standard inch fastener sizes, inch tube sizes, etc., currently in use in America. Since a guide similar to inch fractions does not exist in the metric system, *you should think preferred numbers when you THINK METRIC.*

Specific areas where the use of preferred numbers can be applied to your advantage, are as follows:

INVENTORY REDUCTION by applying preferred numbers to sizes for such items as: holes, pipes, cylinders, shafts, fasteners, steel material, drills, reamers, motors, pumps, tanks, pressure gages, wires, etc.

PRODUCT LINE SIMPLIFICATION AND PLANNING by choice of preferred numbers in planning production of model sizes to cover a given range of performance such as: lift capacity, fill capacity, rotating speeds, power ratings, etc.

Efforts to minimize cost by reducing the number of manufactured sizes help reduce inventory for the consumer of semifinished products, down the line to inventory at the hardware store.

Scope

This publication offers preferred numbers in four principal series,[2] R5; R10; R20; and R40; as well as in the additional R80 series, intended for special applications, and gives their derivation, together with definitions of the terms used.

These numbers are those internationally agreed upon by the International Organization for Standardization (ISO) in New York, June 1952, and published as ISO 3-1973. The same numbers are specified in the ANSI Z17.1-1973 standard and corresponding standards in all major industrial countries of the world. (JAPAN: Z8601-1954; GERMANY: DIN 323; FRANCE: R 962-06; U.K.: BS 2045; ITALY: UNI 2016; AUSTRALIA: SAA MP19.)

Derivation

Preferred numbers are derived from a geometric series having one of the following common ratios:

$\sqrt[5]{10}$, $\sqrt[10]{10}$, $\sqrt[20]{10}$, $\sqrt[40]{10}$, or $\sqrt[80]{10}$.

These ratios are approximately equal to 1.58, 1.26, 1.12, 1.06 and 1.03, respectively. Thus, successive terms in the respective series increase by approximately 58% for the R5 series, 26% for R10, 12% for R20, 6% for R40, and 3% for R80. The calculated values of these progressions are arbitrarily rounded off to give terms which are, in general, doubled every three terms in the R10 series, every six terms in the R20 series, and every 12 terms in the R40 series. The maximum roundings off are +1.26% and −1.01%.

It should be noted that any series can be extended indefinitely upward or downward by multiplying or dividing repeatedly by 10.

Nomenclature and Definitions

The terms used in this section and defined below, are in accordance with ISO 3 and ISO 17.

---

[1] For information about the term "standard" as used in this book, please see page 1-2.

[2] R stands for Renard.

Table 4-1. Basic Series of Preferred Numbers with Deviations (ISO 3)

1	2	3	4	5	6	7	8
\multicolumn{4}{Basic Series}	Serial Number	\multicolumn{2}{Theoretical Values}	Percentage Difference between Basic Series and Calculated Values
R5	R10	R20	R40		Calculated Values	Mantissae of Logarithms	
1.00	1.00	1.00	1.00	0	1.0000	000	0
			1.06	1	1.0593	025	+ 0.07
		1.12	1.12	2	1.1220	050	− 0.18
			1.18	3	1.1885	075	− 0.71
	1.25	1.25	1.25	4	1.2589	100	− 0.71
			1.32	5	1.3335	125	− 1.01
		1.40	1.40	6	1.4125	150	− 0.88
			1.50	7	1.4962	175	+ 0.25
1.60	1.60	1.60	1.60	8	1.5849	200	+ 0.95
			1.70	9	1.6788	225	+ 1.26
		1.80	1.80	10	1.7783	250	+ 1.22
			1.90	11	1.8836	275	+ 0.87
	2.00	2.00	2.00	12	1.9953	300	+ 0.24
			2.12	13	2.1135	325	+ 0.31
		2.24	2.24	14	2.2387	350	+ 0.06
			2.36	15	2.3714	375	− 0.48
2.50	2.50	2.50	2.50	16	2.5119	400	− 0.47
			2.65	17	2.6607	425	− 0.40
		2.80	2.80	18	2.8184	450	− 0.65
			3.00	19	2.9854	475	+ 0.49
	3.15	3.15	3.15	20	3.1623	500	− 0.39
			3.35	21	3.3497	525	+ 0.01
		3.55	3.55	22	3.5481	550	+ 0.05
			3.75	23	3.7584	575	− 0.22
4.00	4.00	4.00	4.00	24	3.9811	600	+ 0.47
			4.25	25	4.2170	625	+ 0.78
		4.50	4.50	26	4.4668	650	+ 0.74
			4.75	27	4.7315	675	+ 0.39
	5.00	5.00	5.00	28	5.0119	700	− 0.24
			5.30	29	5.3088	725	− 0.17
		5.60	5.60	30	5.6234	750	− 0.42
			6.00	31	5.9566	775	+ 0.73
6.30	6.30	6.30	6.30	32	6.3096	800	− 0.15
			6.70	33	6.6834	825	+ 0.25
		7.10	7.10	34	7.0795	850	+ 0.29
			7.50	35	7.4989	875	+ 0.01
	8.00	8.00	8.00	36	7.9433	900	+ 0.71
			8.50	37	8.4140	925	+ 1.02
		9.00	9.00	38	8.9125	950	+ 0.98
			9.50	39	9.4406	975	+ 0.63
10.00	10.00	10.00	10.00	40	10.0000	000	0

PREFERRED NUMBER	0.8	1.18	1.25	1.4	2.5	3.15	6.3	8	10
CLOSE CONSTANT	$\pi/4 =$ 0.7854	$\sqrt[4]{2} =$ 1.1892	$\sqrt[3]{2} =$ 1.2599	$\sqrt{2} =$ 1.4142	2.54	$\pi =$ 3.1416	$2\pi =$ 6.2832	$\pi/4 =$ 0.7854	$\pi^2 =$ 9.8697

PREFERRED NUMBERS

Theoretical values — The values of the terms of $(\sqrt[5]{10})^N$, $(\sqrt[10]{10})^N$, etc. These values have an infinite number of decimal places and are not suitable for practical use.

Calculated values — Values approximating to the theoretical values, expressed to five significant figures and having a relative error in comparison with the theoretical values of less than 1/20 000.

Preferred numbers — Values rounded off as explained in "Derivation" above, and given in Tables 4-1 and 4-2.

Serial numbers — An arithmetic series of consecutive numbers indicating the preferred numbers starting with 0 for the preferred number 1.00.

Designation

The series of preferred numbers are designated respectively: R5, R10, R20, R40, and R80, in which the "R" stands for Renard and the number indicates the particular root of 10 on which the series is based.

Example: R20 is based on the series having the ratio $\sqrt[20]{10}$.

Series of Preferred Numbers

Basic Series — The basic series of preferred numbers, R5, R10, R20, and R40 are given in Table 4-1; their relation to the calculated values in the corresponding geometric series is shown in Table 4-2.

Order of preferred values for usage is as follows:

    R5   Series — First Choice
    R10 Series — Second Choice
    R20 Series — Third Choice
    R40 Series — Fourth Choice
    R80 Series — Not-Preferred

R80 Series — The values for the R80 series, which is intended for use only in exceptional cases, are given in Table 4-2.

Derived Series — Additional series can be obtained by taking the terms at every second, third, fourth step, etc., of the basic series. These series are designated R5/2, R10/3, R20/4, etc., where the step number is designated after a slash separating it from the basic series designation. *Example:* R10/3 denotes a series which is at the third step in the basic R10. (See Choice of Series and Table 4-3.

Shifted Series — A series having the same gradations as the basic series, but beginning with a term not belonging to that series.

GUIDANCE IN USE OF PREFERRED NUMBERS AND SERIES OF PREFERRED NUMBERS

Choice of Series

Basic Series — It is of the greatest importance to use the internationally accepted preferred numbers in the design and planning of products as shown in Table 4-1, in the order of preference: R5, R10, R20, and R40. A number of national standards for rounding off the preferred numbers are in existence throughout the world. These are not recommended for use, and any deviations from the basic series should follow values specified in Table 4-4 which is according to ANSI Z17.1-1973 or ISO 497-1973. The use of other rounded-off number series will only defeat the purpose of preferred numbers.

Derived Series — A derived or supplementary series should only be used when none of the scales of the basic series is satisfactory and preference should be given to such of those series as, whether extended upwards or downwards, include the number 1. Further, as in the case of the basic series, derived series should be selected in the same order of preference, e.g., R5, R10, etc.

As stated under "Series of Preferred Numbers," these series are obtained by taking every third step, every fourth step, etc. of a basic series. For example, convenient series with a step ratio of 2 can be derived from every third step in the R10 series, or with a ratio of 1.4 from every third step in the R20 series.

Table 4-2. R80 Series of Preferred Numbers for Exceptional Use (ISO 3)

1.00	1.80	3.15	5.60
1.03	1.85	3.25	5.80
1.06	1.90	3.35	6.00
1.09	1.95	3.45	6.15
1.12	2.00	3.55	6.30
1.15	2.06	3.65	6.50
1.18	2.12	3.75	6.70
1.22	2.18	3.87	6.90
1.25	2.24	4.00	7.10
1.28	2.30	4.12	7.30
1.32	2.36	4.25	7.50
1.36	2.43	4.37	7.75
1.40	2.50	4.50	8.00
1.45	2.58	4.62	8.25
1.50	2.65	4.75	8.50
1.55	2.72	4.87	8.75
1.60	2.80	5.00	9.00
1.65	2.90	5.15	9.25
1.70	3.00	5.30	9.50
1.75	3.07	5.45	9.75

Table 4-3. Percentage Increments

INCREMENT† NUMERICAL	INCREMENT† EXPONENT	BASIC SERIES	DERIVED SERIES	PERCENTAGE INCREASE
10	q^{40}	(R1)*		900
4	q^{24}		R5/3 R10/6	300
2.5	q^{16}		R5/2 R10/4	150
2	q^{12}		R10/3 R20/6	100
1.6	$f_5 = q^8$	R5	R10/2 R20/4 (R40/8)*	60
1.5	q^7		(R40/7)*	50
1.4	q^6		R20/3 R40/6	40
1.32	q^5		(R40/5)*	32
1.25	$f_{10} = q^4$	R10	R20/2 R40/4	25
1.18	q^3		R40/3	18
1.12	$f_{20} = q^2$	R20	R40/2	12
1.09			(R80/3)*	9
1.06	$f_{40} = q$	R40		6
1.03		(R80)*		3

NOTES: *Series shown in brackets are nonpreferred.

†$q = \sqrt[40]{10} = f_{40} = 1.06$, where '$f$' designates the ratio for the Renard basic series given by the subscript number.

It will be noted that there is always more than one supplementary series with a given ratio, other than the normal series containing the number 1, the additional series being based on one or other of the numbers in the primary series which are omitted from the normal supplementary series. Thus, there are three supplementary series (R10/3), derived from the R10 series, as follows:

Normal series: 1 , 2 , 4 , 8 , 16...
Other series   1.25, 2.5 , 5 , 10 , 20 , 40...
                1.6 , 3.15, 6.3, 12.5, 25 , 50...

each with an increment step of approximately 100%.

It will be apparent that a supplementary series with any desired step ratio can be derived provided the ratio required corresponds to a preferred number. For example, if a ratio of 1.25 (25%) is required, this will be seen to correspond to every second term in the R20 series.

Percentage Increments—The percentage steps desired will determine the choice of series. The available percentage increments for the various series are as shown in Table 4-3.

Multiplication or Division

All preferred numbers in the R40 series can be expressed as an exponential function where the base number is q (or q^1) = $\sqrt[40]{10}$ = f_{40} and the exponent applied to q equals the serial number shown in Table 4-1. Multiplication or division of two preferred numbers will yield a third preferred number.

Numerical   *Exponential*
$1.6 \times 2.5 = 4$ $q^8 \times q^{16} = q^{(8+16)} = q^{24} = 4$
$6.3/4 = 1.6$ $q^{32}/q^{24} = q^{(32-24)} = q^8 = 1.6$

Example:
Cylindrical containers dimensioned to preferred numbers:

Size number	1	2	3	4	......
Diameter d (R10) mm	100	125	160	200	......
Height h (R10) mm	125	160	200	250	......
Volume v (R10/3) liter	1	2	4	8	......

Note:— The constant π is approximately equal to $3.15 = q^{20}$ (a preferred number in the basic R10 series, see Table 4-1.)

Single Numerical Value

In the selection of a single value, irrespective of any idea of scaling, a number of the R5, R10, R20, R40 basic series

PREFERRED NUMBERS

should be chosen, in the same order of preference as listed. The first prototype machine might be designed to the lift capacity of 4 megagrams (1 Mg = 1000 kg). The lift capacity for other models in the same production series could be, from the R5 basic series, 4 Mg, 6.3 Mg, 10 Mg, etc., or the R10 series could be used giving: 4 Mg, 5 Mg, 6.3 Mg, 10 Mg, etc.

Designation of Series—A lower or upper limit, or both, might be added as follows:

R5 (160 ....) is the series 160 250 400 630 ...
R10/3 (... 16) is the series 1 2 4 8 16
R20/3 (2 .... 8) is the series 2 2.8 4 5.6 8
R10/3 (.... 5 ....) is the series .... 1.25 2.5 5 10 20 ....

LOGARITHMIC PAPER—The preferred numbering system can be used to produce a logarithmic scale from an equal-distance marked paper, or scale. (See Fig. 4-1.) The desired measuring range and the spread of the logarithmic scale can be adjusted to fit each application.

Grading by Means of Preferred Numbers—Since preferred numbers may differ from the calculated values by +1.26% to −1.01%, it follows that sizes—graded according to preferred numbers—are not exactly proportional to each other.

More Rounded Values of Preferred Numbers

In certain special applications it may be found that the standard preferred numbers are unacceptable; for example, they may appear to imply a precision regarded as either unnecessary or impossible to achieve, such as 1/31.5 seconds instead of 1/30 second for photographic time exposures, or it may be impossible to retain all the significant figures where a whole number is necessary, e.g., 32 instead of 31.5 for the number of teeth in a gear. There may also be reasons of an economic or psychological nature which makes them unacceptable to some section of industry or the general public at the present time. In such circumstances, it is better to use more rounded numbers than nonpreferred numbers (and in some cases, this may lead to the adoption of preferred numbers in the future). Every effort should be made to use only the more rounded values given in Table 4-4.

When these more rounded values are used in a series of numbers, those which yield the most rational grading should be chosen. However, when there is a possibility that intermediate values may subsequently have to be introduced in the series, the use of these more rounded values should be avoided.

Preferred Metric Sizes

The preferred metric sizes which are shown in Table 4-5 are based on RENARD'S series of preferred numbers and existing world metric standards for engineering. Leading American companies have now adopted or are in the process of adopting the metric sizes shown in Table 4-5 in their design standards.

The first choice sizes shown in Table 4-5 approximately follow the preferred number series R10, where succeeding numbers in the series increase by 25 percent. The second choice series shown are rounded off from the R20 series of preferred numbers (12 percent increments). If other sizes are required they should be listed in a third choice column.

If you wish, you can rationalize the selection of first choice sizes by selecting every second number in the series such as 1 1.6 2.5 4 6 10 16, etc., and this number series is rounded off from the R5 series of preferred numbers (60 percent increments).

You might want to extend the preferred metric size range, which is shown from 1 to 1000, simply by multiplying or dividing the numbers shown in Table 4-5 by ten or multiples of ten.

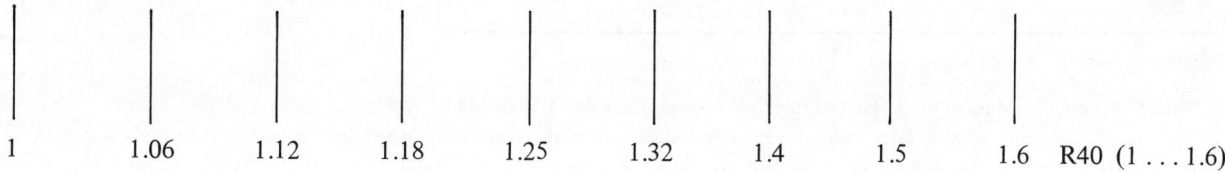

Fig. 4-1. How to make logarithmic scale paper with the help of preferred numbers.

Table 4-4. More Rounded Values of Preferred Numbers (ISO 497)

Column	1		2			3			4		5	6	7	8	9	10
Number of terms or index	5		10			20			40		Serial number	Calculated values ***	\multicolumn{4}{c}{Percentage differences between the calculated values and each value in the series}
Approximate ratio	1.6		1.25			1.12			1.06							
Series	R5	R″5	R10	R′10	R″10	R20	R′20	R″20	R40	R′40			R 5 to 40	R′ 10 to 40	R″ 20	R″ 5 and 10
	1		1			1.0			1.0		0	1.0000	0			
									1.06	1.05	1	1.0593	+ 0.07	− 0.88		
						1.12		1.1	1.12	1.1	2	1.1220	− 0.18	− 1.96	− 1.96	
						1.18		1.2 **	1.18	1.2 **	3	1.1885	− 0.71	+ 0.97		
			1.25		(1.2)	1.25		(1.2)	1.25		4	1.2589	− 0.71		− 4.68	− 4.68
									1.32	1.3	5	1.3335	− 1.01	− 2.51		
									1.4		6	1.4125	− 0.88			
									1.5		7	1.4962	+ 0.25			
	1.6	(1.5)*	1.6		(1.5)*	1.6			1.6		8	1.5849	+ 0.95			− 5.36
									1.7		9	1.6788	+ 1.26			
						1.8			1.8		10	1.7783	+ 1.22			
									1.9		11	1.8836	+ 0.87			
			2			2.0			2.0		12	1.9953	+ 0.24			
									2.12	2.1	13	2.1135	+ 0.31	− 0.64		
						2.24		2.2	2.24	2.2	14	2.2387	+ 0.06	− 1.73	− 1.73	
									2.36	2.4	15	2.3714	− 0.48	+ 1.21		
	2.5		2.5			2.5			2.5		16	2.5119	− 0.47			
									2.65	2.6	17	2.6607	− 0.40	− 2.28		
						2.8			2.8		18	2.8184	− 0.65			
									3.0		19	2.9854	+ 0.49			
			3.15	3.2	(3)	3.15	3.2	(3.0)	3.15	3.2	20	3.1623	− 0.39	+ 1.19	− 5.13	− 5.13
									3.35	3.4	21	3.3497	+ 0.01	+ 1.50		
						3.55	3.6	(3.5)	3.55	3.6	22	3.5481	+ 0.05	− 1.46	− 1.38	
									3.75	3.8	23	3.7584	− 0.22	+ 1.11		
	4		4			4.0			4.0		24	3.9811	+ 0.47			
									4.25	4.2	25	4.2170	+ 0.78	− 0.40		
						4.5			4.5		26	4.4668	+ 0.74			
									4.75	4.8	27	4.7315	+ 0.39	− 1.45		
			5			5.0			5.0		28	5.0119	− 0.24			
									5.3		29	5.3088	− 0.17			
						5.6		(5.5)	5.6		30	5.6234	− 0.42		− 2.19	
									6.0		31	5.9566	+ 0.73			
	6.3	(6)	6.3		(6)	6.3		(6.0)	6.3		32	6.3096	− 0.15		− 4.90	− 4.90
									6.7		33	6.6834	+ 0.25			
						7.1		(7.0)	7.1		34	7.0795	+ 0.29		− 1.11	
									7.5		35	7.4989	+ 0.01			
			8			8.0			8.0		36	7.9433	+ 0.71			
									8.5		37	8.4140	+ 1.02			
						9.0			9.0		38	8.9125	+ 0.98			
									9.5		39	9.4405	+ 0.63			
	10		10			10.0			10.0		40	10.0000	0			
Maximum irregularity of ratio, %	+ 1.42	− 5.37	+ 1.66	+ 1.66	− 5.61	− 1.83	− 1.97	− 4.48	+ 1.15	+ 2.94						

Preferred numbers ▎   More rounded values: 1st rounding ▎   2nd rounding ┆

*These "R Series" (values in brackets), and most particularly the value 1.5, should be avoided.

**In exceptional cases, when a series without regression is necessary in this region, for an application requiring a simple scaling of values unrelated to other data, and the preferred numbers themselves are not applicable, adopt the alternative of 1.15 for 1.18, and 1.20 for 1.25, which gives for the start of the series:

$$1 - 1.05 - 1.10 - 1.15 - 1.20 - 1.30$$

***In certain exceptional cases (for example, for the manufacture of turbine blades), when very great precision is necessary, use the calculated values (Column 6 of the table.)

PREFERRED NUMBERS

Practical Examples of Preferred Numbers Used Internationally (ISO 17)

Preferred Surface Finish Values (R_a) are as follows:

micrometers† R10/3 (0.025 ... 12.5)	0.025	0.05	0.1	0.2	0.4	0.8	1.6	3.2*	6.3	12.5
microinches R10/3 (1 .... 500)	1	2	4	8	16	32*	63	125	250	500

*The value 32 is rounded off from 31.5.

†The conversion factor from inches to millimeters is equal to 25.4. The number $25 = q^{16}$ is a preferred number in the R5 series. Conversion of a preferred inch size approximately gives a preferred millimeter dimension with a serial number 16 greater than that of the original number (see Preferred Surface Finish Values above or on page 5–4.)

Table 4-5. Preferred Metric Sizes (ANSI B4.2)

NOMINAL SIZE, mm		NOMINAL SIZE, mm		NOMINAL SIZE, mm	
FIRST	SECOND	FIRST	SECOND	FIRST	SECOND
1		10		100	
	1.1		11		110
1.2		12		120	
	1.4		14		140
1.6		16		160	
	1.8		18		180
2		20		200	
	2.2		22		220
2.5		25		250	
	2.8		28		280
3		30		300	
	3.5		35		350
4		40		400	
	4.5		45		450
5		50		500	
	5.5		55		550
6		60		600	
	7		70		700
8		80		800	
	9		90		900
				1000	

Standard Currents—Publication 59 of the International Electrotechnical Commission gives, for standard currents, the following ratings in amperes:

1	1.25	1.6	2	2.5	3.15	4	5	6.3	8
10	12.5	16	20	25	31.5	40	50	63	80
100	125	160	200	250	315	400	500	630	800
1000	1250	1600	2000	2500	3150	4000	5000	6300	8000
10000									

Copper Wire—Normal diameters of bare wires of annealed copper with circular cross-section, the French publication C31-111 has adopted the following values, expressed in millimeters:

0.0315	0.112
0.0355	0.118
0.040	0.125
0.045	0.132
0.050	0.140
0.056	0.150
0.063	0.160
0.071	0.170
0.080	0.180
0.090	0.190
0.100	0.200

Rotating Speeds of Machine-Tool Spindles—the Belgian standard NBN 123-1950 has adopted the R20 series (and the R20/2, R20/3, R20/4, R20/6 derived series) for the rotating speeds of machine-tool spindles, expressed in revolutions per minute:

1	1.12	1.25	1.4	1.6	1.8	2	2.24	2.5
10	11.2	12.5	14	16	18	20	22.4	25
100	112	125	140	160	180	200	224	250
1000								

2.8	3.15	3.55	4	4.5	5	5.6	6.3
28	31.5	35.5	40	45	50	56	63
280	315	355	400	450	500	560	630

7.1	8	9
71	80	90
710	800	900

PREFERRED NUMBERS

Pressure Containers—Tanks for water under pressure. The German standard DIN 2760 has adopted the following values for the nominal capacities, expressed in liters:

4		6.3		10		16		25
40		63		100	125	160	200	250	315
400	500	630	800	1000	1250	1600	2000	2500	3150
4000	5000	6300	8000	10000	12500	16000	20000	25000

Lifting Capacity of Cranes—the Norwegian standard NS 300 has adopted the following values for the lifting capacity of cranes, expressed in megagrams (1 Mg = 1000 kg).

1	1.25	1.6	2	2.5	3.15	4	5	6.3	8
10	12.5	16	20	25	31.5	40	50	63

RELATED ISO STANDARDS (ISO TC 19)

ISO 3-1973 Preferred numbers—Series of preferred numbers.

ISO 17-1973 Guide to the use of preferred numbers and of series of preferred numbers.

ISO 497-1973 Guide to the choice of series of preferred numbers and of series containing more rounded values of preferred numbers.

5 Surface Texture

Introduction

The internationally adopted system of measuring surface texture is to use the roughness average parameter designated R_a which is also referred to as AA (Arithmetical Average) or CLA (Center Line Average). The roughness average system is covered in national standards[1] of all the major industrial countries of the world.

Also recognized on an international level is the Ten Point Height of Irregularities system designated by the symbol R_z.

The RMS-value (Root Mean Square = Geometrical Average) is not recommended, being from 10 to 30 percent larger than the R_a value recorded for the same surface.

Some European drawings might specify the maximum height, R_{max}, of the roughness irregularities, as shown later in Table 5-5.

Surface Texture Definitions

There are two basic systems mentioned in ISO 468, designated "M" and "E," of indicating and defining the surface finish on drawings. The difference between the two systems lies in reference being made to two distinct lines for measuring the ordinates characterizing the roughness of each point of the surface profile, the criteria of roughness being defined with reference to the mean line in the "M" system (see Fig. 5-2) and with reference to the envelope line in the "E" system (see Table 5-4). The preferred system in ISO 468 is the "M" method. German drawings sometimes specify surface roughness in the "E" system.

Definitions

Real Surface—The surface limiting the body; separating it from the surrounding space.

Geometrical Surface—The surface determined by the design or by the process of manufacture, neglecting errors of form (see Fig. 5-1). NOTE: The terms "ideal geometrical surface," "design form," and "nominal surface" are used in certain national standards with the sense of "geometrical surface," as defined above.

Effective Surface—The close representation of a real surface obtained by instrumental means (see Fig. 5-1). NOTE: The term "measured surface" is used in certain national standards and has the same meaning as "effective surface," as just defined.

Real Profile—The contour that results from the intersection of the real surface by a plane conventionally defined with respect to the geometrical surface.

Geometrical Profile—The contour that results from the intersection of the geometrical surface by a plane conventionally defined with respect to this surface (see Fig. 5-1). NOTE: The terms "ideal geometrical profile," "design profile," and "nominal profile" are used in certain national standards with the sense of "geometrical profile," as defined above.

Effective Profile—The contour that results from the intersection of the effective surface by a plane conventionally defined with respect to the geometrical surface (see Fig. 5-1). NOTE: The term "measured profile" is used in certain national standards with the sense of "effective profile," as defined above. This is the desired profile.

Reference Line—A line chosen by convention to serve for the quantitative evaluation of the roughness of the effective profile.

Irregularities—The peaks and valleys of a real surface.

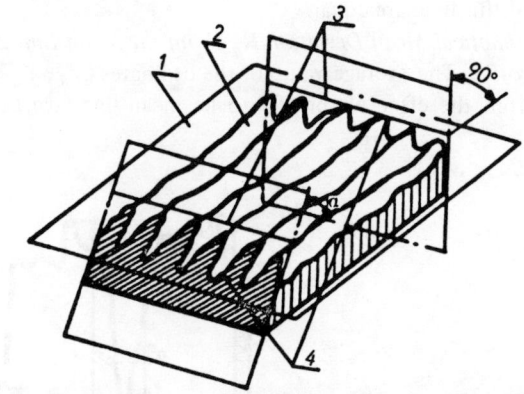

Fig. 5-1. Definitions: (1) geometrical surface, (2) effective surface, (3) geometrical profile, (4) effective profile (ISO 468).

[1] For information about the term "standard" as used in this book, please see page 1-2.

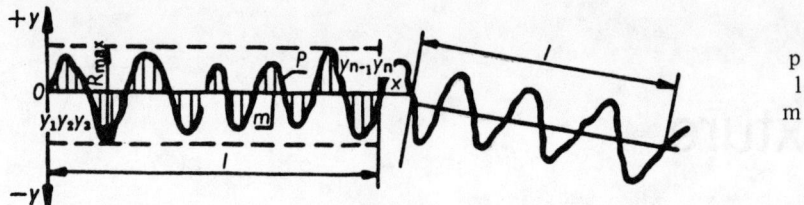

Fig. 5-2. Sampling length, "l" (ISO 468).

Surface Roughness—All those irregularities of the surface which are conventionally defined within a section of the area where deviations of form and waviness are eliminated.

Traversing Length—The length of the effective profile traversed by instrumentation for the evaluation of the surface-roughness parameters for the surface inspected. NOTE: The traversing length usually includes more than one sampling length (see Fig. 5-2).

Spacing of Irregularities—The mean distance between the more prominent irregularities of the effective profile.

Sampling Length "L"—The length of the effective profile selected for the evaluation of the surface roughness, without taking into account other types of irregularities. (See Fig. 5-2 and Table 5-8.)

Mean Line "M" of the Profile—A line having the form of the geometrical profile and dividing the effective profile so that, within the sampling length, the sum of the squares of distances $(Y_1, Y_2, ... Y_n)$ from the effective profile points and the mean line is a minimum. NOTE: The particular case of the mean line of the profile "M" is the central line, having the same form as the profile and located parallel to the general direction of the profile within the sampling length, so that the sums of the areas contained between this line and the effective profile, at both sides of this line, are equal.

Arithmetical Mean Deviation R_a From the Mean Line of the Profile—The average value of the ordinates $(Y_1, Y_2 ... Y_n)$ from the effective profile to its mean line (see Fig. 5-2). The ordinates are summed without considering their algebraic sign:

$$R_a = \frac{1}{l} \int_0^l |y| \, dx$$

approximately
$$R_a = \frac{\Sigma_1^n |Y|}{n}$$

Ten Point Height R_z of Irregularities—The average distance between the five highest peaks and the five deepest valleys within the sampling length measured from a line parallel to the mean line and not crossing the profile (see Fig. 5-3).

$$R_z = \frac{(R_1 + R_3 + ... R_9) - (R_2 + R_4 + ... R_{10})}{5}$$

Maximum Height R_{max} of Irregularities—The distance between two lines parallel to the mean line and touching the profile at the highest and lowest points, respectively, within the sampling length (see Fig. 5-2).

The following method of indicating surface texture by symbols on drawings is in accordance with the ISO 1302 standard.

Symbols Used for Indication of Surface Texture

The basic surface texture symbol, as in Fig. 5-4A, consists of two legs of unequal length inclined at approximately 60 degrees to the line representing the considered

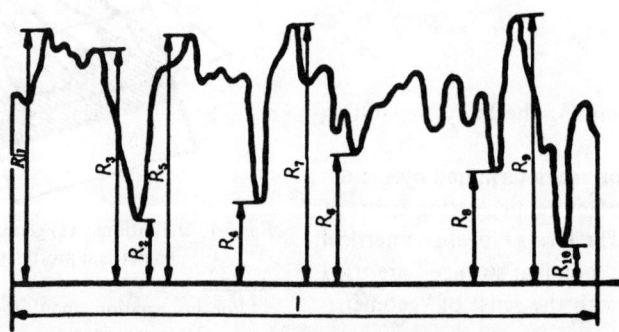

Fig. 5-3. Ten point height R_z of irregularities (ISO 468).

SURFACE TEXTURE

surface. Machining of the surface is optional. A horizontal bar, as in Fig. 5-4B, indicates that removal of material by machining is required. A circle as in Fig. 5-4C indicates that removal of material by machining is not permitted; and Fig. 5-4D indicates the position of the surface-texture specifications (listed below) in the symbol.

- a = Roughness value R_a (CLA) or in grade numbers N1 through N12 (see Table 5-1). Maximum (a_1) and minimum (a_2) surface roughness limits (if required) are shown here.
- b = Production method, treatment or coating
- c = Sampling length (see Table 5-8)
- d = Direction of lay (see Table 5-2)
- e = Machining allowance
- f = Other roughness values (in brackets, as in Fig. 5-4D).

Use of the N Series of Roughness Numbers[1]

Instead of the micrometer values, roughness numbers, N1 to N12, may be quoted on drawings. The use of the "N" series of roughness numbers is recommended to avoid possible misinterpretation on drawings that are apt to be internationally exchanged.

Table 5-1 shows the relationship between the roughness numbers and the corresponding values in the micrometer and microinch series.

Symbols for the Direction of Lay

The direction of lay is the direction of the predominant surface pattern, ordinarily determined by the production method employed. The series of symbols given in Table 5-2 specifies the common direction of lay.

Typical Examples of Symbols

Table 5-3 gives typical examples of symbols used with micrometer texture values and roughness numbers. See also Fig. 5-5. All surface-texture R_a values are in micrometers (1 μm = 0.001 mm).

German Method of Indicating Surface Texture on Drawings (System "E"—Non-Preferred)

The surface roughness indication system used in some European countries is measured in the maximum height, R_{max}, of irregularities. (German: R_t = Rautiefe). It is not the preferred method, and this information is for

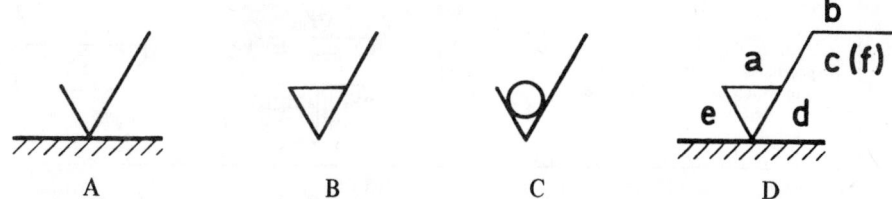

Fig. 5-4. Surface texture symbols: A. Machining of surface is optional; B. Machining of surface is required; C. Machining of surface is not permitted; and D. Position of specifications in the symbol (ISO 1302). *Note:* The symbols in "D" deviate slightly from the ANSI B46.1 standard.

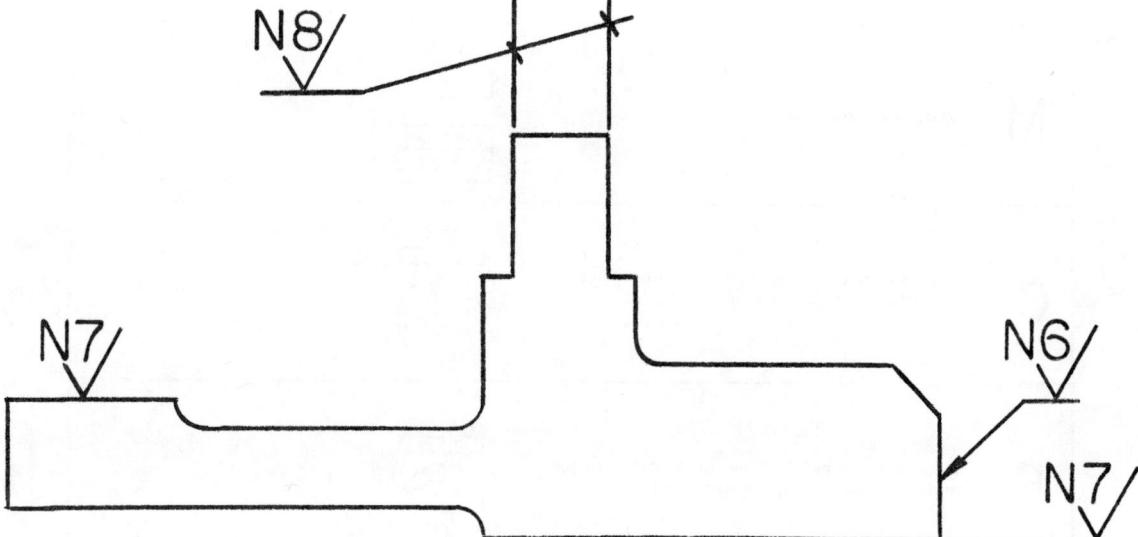

Fig. 5-5. Drawing example.

---

[1] The roughness numbers in the N Series are not recognized by the ANSI B46 standard.

reference only. The German standard DIN 3141 defines the symbols in Table 5-4.

Comparison of R_{max} to Roughness Number

A rough comparison of the height of irregularities R_{max} to the roughness number is shown in Table 5-5.

Choice of Surface Finish and Tolerances

The dimensional tolerances applied to the boundaries of a surface are a rough indication of the type of machining operations that normally would be employed. Table 5-6 is a rough comparison between dimensional tolerances and

Table 5-1. Roughness Numbers (ISO 1302)

Nominal value of R_a	micrometers:	50	25	12.5	6.3	3.2	1.6	0.8	0.4	0.2	0.1	0.05	0.025
	microinches:	2000	1000	500	250	125	63	32	16	8	4	2	1
ISO Roughness number:		N12	N11	N10	N9	N8	N7	N6	N5	N4	N3	N2	N1

Table 5-2. Direction of Lay Symbols (ISO 1302)

Symbol	Interpretation
=	Parallel to the plane of projection of the view in which the symbol is used
⊥	Perpendicular to the plane of projection of the view in which the symbol is used
X	Crossed in two slant directions relative to the plane of projection of the view in which the symbol is used
M	Multi-directional
C	Approximately circular relative to the centre of the surface to which the symbol is applied
R	Approximately radial relative to the centre of the surface to which the symbol is applied

NOTE: Should it be necessary to specify a direction of lay not clearly defined by these symbols then this must be achieved by a suitable note on the drawing.

SURFACE TEXTURE

Table 5-3. Examples of Surface-Texture Symbols

Symbol			Meaning
Removal of material by machining is			
optional	obligatory	prohibited	
3,2 ∇ N8 or ∇	3,2 ▽ N8 or ▽	3,2 ○∇ N8 or ○∇	A surface with a maximum surface roughness value R_a of 3,2 μm.
6,3 N9 / 1,6 N7 ∇ or ∇	6,3 N9 / 1,6 N7 ▽ or ▽	6,3 N9 / 1,6 N7 ○∇ or ○∇	A surface with a maximum surface roughness value R_a of 6,3 μm and a minimum of 1,6 μm.

Courtesy of British Standards BS 1134; 1972

Table 5-4. Maximum Roughness Height System

Remarks	Drawing Symbols	Remarks	DIN 3141 (March 1960) Maximum height of irregularities R_{max} in micrometers (1 μm = 0.001 mm)			
			SERIES*			
			1	2	3	4
Surface without roughness symbol produced by common manufacturing processes; rolled, forged, drawn, extruded, flame cutting, casting, etc.	//////////	Surface without special requirements.	1 = coarse series 2 = medium coarse series 3 = normal series 4 = fine series			
Surfaces produced without removal of material but with special care in the manufacturing process.	⌢//////////	Surface with some degree of clean finish.				
Surfaces produced by removal of material. The tool marks are clearly visible and can be felt.	∇ //////////	The maximum height of irregularities R_{max} must not exceed the micrometer values shown.	160	100	63	25
Surface produced by removal of material. The direction of lay can be visually identified.	∇∇ //////////		40	25	16	10
Surface produced by removal of material. The direction of lay cannot be visually identified.	∇∇∇ //////////		16	6.3	4	2.5
	∇∇∇∇ //////////		-	1	1	0.4

*The title block of the drawing should contain the following note: Finish series 3 DIN 3141 (German: Oberflächen Reihe 3 DIN 3141).

Commonly Produced Surface Roughness

Results obtained from common production processes in terms of R_a values (micrometers) are shown in Table 5-7. Relative cost of finish (production time) is shown in Fig. 5-6. The tables must be regarded as a guide to designers and engineers.

surface roughness values. It is a general reference only as the R_a values required frequently may be different.

Surface Texture Versus Production Costs

Typical relationship of surface texture and production time (cost). The chart, Fig. 5-6, shows a series of curves displaying production time in relation to R_a values for the range of common machining processes and is based on research carried out on machine tools from 1 to 10 years old. The chart is not intended to be used for making comparisons between different processes, however.

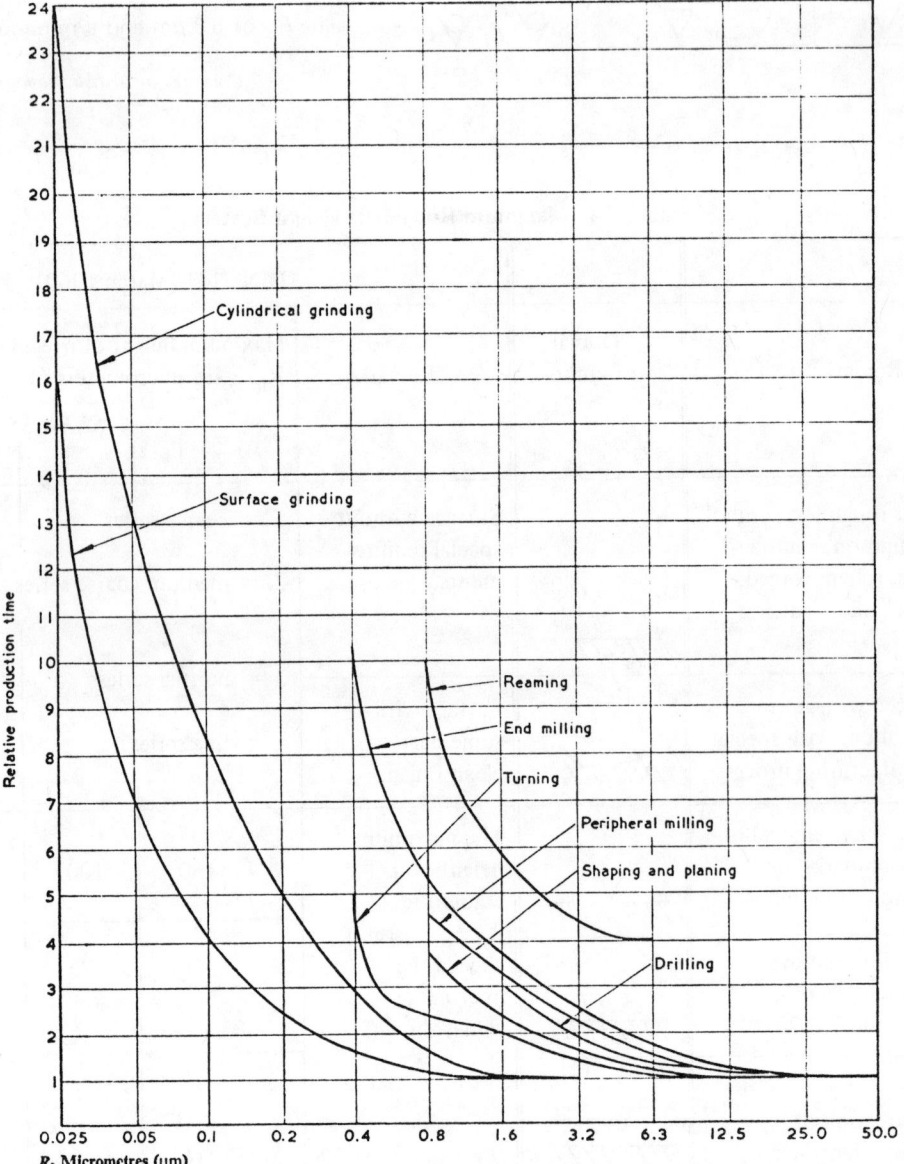

(Courtesy of British Standard BS 1134: 1972)

Fig. 5-6. Surface Texture Versus Production Time

Example: Should a given area of surface be surface ground to 3.2 μm R_a taking approximately 1 minute then to achieve 0.2 μm would take approximately 2.5 minutes.

SURFACE TEXTURE

Sampling Lengths

The specifications of the sampling length normally not required by the engineer. The range of standard sampling lengths associated with R_a values are as follows (in mm):

0.08	0.8	8.0
0.25	2.5	25.0

Table 5-8 gives typical process designations and gives some guidance by indicating the meter cut-offs (equal to the sampling lengths) found by experience to be suitable, for each process listed.

Table 5-5. Maximum Height of Irregularities (R_{max}) Compared with Roughness Numbers (DIN 4767)

R_{max} in micrometers	160	100	50	25	16	8	5	2.5	1.25	0.63	0.4	0.2
ISO Roughness number	N12	N11	N10	N9	N8	N7	N6	N5	N4	N3	N2	N1

NOTE:
The comparisons shown in Table 5-5 represent mean values. When converting a maximum height irregularity (R_{max}) to a roughness average (R_a) number or vice versa, select the value shown to the right.

    Example 1: $R_{max} = 100 \, \mu m$ Select $R_a = N10$
    Example 2: $R_a = N10$ Select $R_{max} = 25 \, \mu m$

Table 5-6. Tolerances Compared to Finishes (UNI 3963)

Tolerance	ISO grade*		IT2	IT3	IT4	IT5	IT6	IT7	IT8	IT9	IT10	IT11	IT12
	(μm)**		2.5	4	7	11	16	25	39	62	100	160	250
Finish	values R_a (μm)		0.1	0.2	0.4	0.8		1.6		3.2	6.3		12.5
	grade number		N3	N4	N5	N6		N7		N8	N9		N10

*IT stands for International Tolerance grade.
**Values shown are for nominal sizes over 18 mm up to and including 80 mm.

Table 5-7. Surface Roughness Produced by Common Production Methods and Materials

Process	Roughness values (μm R_a): 50 – 25 – 12.5 – 6.3 – 3.2 – 1.6 – 0.8 – 0.4 – 0.2 – 0.1 – 0.05 – 0.025 – 0.0125
Flame cutting	25 – 12.5 (avg); less frequent 50–25 and 12.5–6.3
Snagging	25–12.5 (avg); less frequent 50–25 and 12.5–6.3
Sawing	25–12.5 (avg); less frequent 50–25 and 12.5–3.2
Planing, shaping	12.5–3.2 (avg); less frequent 25–12.5 and 3.2–0.8
Drilling	6.3–1.6 (avg); less frequent 12.5–6.3 and 1.6–0.8
Chemical milling	6.3–1.6 (avg); less frequent 12.5–6.3 and 1.6–0.8
Electro-discharge machining	6.3–1.6 (avg); less frequent 12.5–6.3 and 1.6–0.8
Milling	6.3–0.8 (avg); less frequent 12.5–6.3 and 0.8–0.2
Broaching	3.2–0.8 (avg); less frequent 6.3–3.2
Reaming	3.2–0.8 (avg); less frequent 6.3–3.2 and 0.8–0.4
Boring, turning	6.3–0.4 (avg); less frequent 12.5–6.3 and 0.4–0.1
Barrel finishing	0.8–0.1 (avg); less frequent 1.6–0.8 and 0.1–0.05
Electrolytic grinding	0.8–0.2 (avg); less frequent 1.6–0.8 and 0.2–0.1
Roller burnishing	0.4–0.1 (avg); less frequent 0.8–0.4
Grinding	1.6–0.1 (avg); less frequent 6.3–1.6 and 0.1–0.025
Honing	0.8–0.1 (avg); less frequent 1.6–0.8 and 0.1–0.025
Polishing	0.4–0.05 (avg); less frequent 0.8–0.4 and 0.05–0.0125
Lapping	0.2–0.05 (avg); less frequent 0.4–0.2 and 0.05–0.0125
Superfinishing	0.2–0.025 (avg); less frequent 0.4–0.2 and 0.025–0.0125
Sand casting	25–12.5 (avg); less frequent 50–25 and 12.5–6.3
Hot rolling	25–12.5 (avg); less frequent 50–25 and 12.5–6.3
Forging	12.5–3.2 (avg); less frequent 25–12.5 and 3.2–1.6
Permanent mould casting	6.3–1.6 (avg); less frequent 12.5–6.3 and 1.6–0.8
Investment casting	6.3–1.6 (avg); less frequent 3.2–1.6
Extruding	6.3–0.8 (avg); less frequent 12.5–6.3 and 0.8–0.4
Cold rolling, drawing	3.2–0.8 (avg); less frequent 6.3–3.2
Die casting	1.6–0.8 (avg); less frequent 3.2–1.6 and 0.8–0.4

Courtesy of British Standard BS 1134: 1972

Key:
■ Average application
▨ Less frequent application

NOTE: The ranges shown above are typical of the processes listed. Higher or lower values may be obtained under special conditions.

SURFACE TEXTURE

Table 5-8. Sampling Lengths for Various Machining Processes*

Typical finishing process	Designation	Meter cut-off (mm) 0.25	0.8	2.5	8.0	25.0
Milling	Mill		X	X	X	
Boring	Bore		X	X	X	
Turning	Turn		X	X		
Grinding	Grind	X	X	X		
Planing	Plane			X	X	X
Reaming	Ream		X	X		
Broaching	Broach		X	X		
Diamond boring	D. bore	X	X			
Diamond turning	D. turn	X	X			
Honing	Hone	X	X			
Lapping	Lap	X	X			
Superfinishing	S. fin.	X	X			
Buffing	Buff	X	X			
Polishing	Pol.	X	X			
Shaping	Shape		X	X	X	
Electro-discharge machining	EDM	X	X			
Burnishing	Burnish		X	X		
Drawing	Drawn		X	X		
Extruding	Extrude		X	X		
Moulding	Mould		X	X		
Electro-polishing	El-pol.		X	X		

Courtesy of British Standard BS 1134: 1972

RELATED ISO AND OTHER STANDARDS (TC 57)

ISO/R 468-1966	Surface roughness
ISO 1302-1974	Method of indicating surface texture on drawings
ISO 1878-1974	Classification of instruments and devices for measurement and evaluation of the geometrical parameters of surface finish
ISO 1879-1974	Instruments for the measurement of surface roughness by the profile method—General statements—Terms and definitions
ISO 1880-1974	Instruments for the measurement of surface roughness by the profile method—Contact (stylus) instruments of progressive profile transformation—Profile recording instruments
ISO 2632/I-1974	Roughness comparison specimens—Part I: Turned, ground, bored, milled, shaped and planed
ISO DIS 2632/2	Roughness comparison specimens—Part 2
ISO 3274-1975	Instruments for the measurement of surface roughness by the profile method—Contact (stylus) instruments of consecutive profile transformation—Contact profile meters, system M

U.S.A.

ANSI B 46.1 (Draft Revision May 1975) Surface texture

United Kingdom

BS 1134: Part 1: 1972	Method for the assessment of surface texture—Method and instrumentation
BS 1134: Part 2: 1972	Method for the assessment of surface texture—General information and guidance

Germany

DIN 140: SHEET 1, 3, 4, 7	Drawing practice—Surfaces
DIN 3141: 1960	Surface symbols on drawings—Peak-to-valley height
DIN 3142: 1960	Indication of surfaces on drawings by means of roughness grades
DIN 4767: 1970	Relationship between R_a (CLA, AA) and R_t (Rmax) surface texture values

Japan

JIS B0601: 1970	Surface roughness

Australia

AS B131: 1962	Center-line-average height method (M-system) for the assessment of surface texture
AS 1100: Part 11: 1974	Indication of surface texture

Italy

UNI 3963: 1960	Surface texture definitions
UNI 4600: 1960	Surface texture drawing symbols

6 The ISO System of Limits and Fits — Tolerances and Deviations

Introduction[1]

The ISO System of Limits and Fits (referred to as the ISO system) is covered in national standards throughout the world. See the following list:

	ISO	286
U.S.A.	ANSI	B4.2
Japan	JIS	B0401
Germany	DIN	7160/61
France	NF	E 02-100-122
U.K.	BSI	4500
Italy	UNI	6388
Australia	AS	1654

Designed to provide a comprehensive range of limits and fits for engineering purposes, this system is based on a series of tolerances graded to suit all classes of work from the finest to the coarsest. The ISO system might appear quite complicated before it is understood. It is, in fact, a very simple system, and a few selected tolerances will satisfy most of the requirements. The values of the ISO system are expressed in metric units of measure only. Converted customary dimensions are published in the ISO related standards listed at the end of this section.

History of the ISO System

The present ISO system is based on the ISA System of Limits and Fits published in ISA Bulletin 25 (1940), and on comments included in the Draft Final Report of ISA Committee 3, December, 1935. The unification of the various national systems of limits and fits was one of the essential tasks discussed at the initial Conference of the ISA in New York, in April, 1926. The same year the Secretariat of ISA Committee 3, Limits and Fits, was entrusted to the German Standardizing Association, and needless to say, the system was all metric from the start.

Usage

The ISO System of Limits and Fits is now in extensive use in Europe. Some of the areas in which the system is used in the United States are as follows: ISO publications, bearings, and metric fasteners. For companies with overseas manufacturing facilities, this system must be understood in order to improve communication among the various plants. The system has been published by ANSI, and some major U.S. corporations have the ISO System of Limits and Fits included in their published standards.

Bases

Temperature—The standard reference temperature for industrial length measurement is 20 degrees Celsius (68°F).

Terminology—The recommendations made here relate to tolerances and limits of size for parts or components and to fits obtained by their assembly. In view of their particular importance, only cylindrical parts (briefly designated as "holes" and "shafts") are referred to explicitly. It should be clearly understood, however, that these recommendations apply equally well to other sections and that the general term "hole" or "shaft" can be taken as referring to the space contained by or containing two parallel faces or tangent planes of any part, such as the width of a slot, the thickness of a key, etc.

Definitions

For the purpose of the standard, the definitions given below and illustrated in Fig. 6-1 apply.

Actual size (of a part)—The size of a part as obtained by measurement.

Limits of size—The maximum and minimum sizes permitted for a feature.

Maximum limit of size—The greater of the two limits of size.

Minimum limit of size—The smaller of the two limits of size.

Basic size—The size by reference to which the limits of size are fixed. The basic size (in this section the same as *nominal size*) is the same for both members of a fit.

Deviation—The algebraic difference between a size (actual, maximum, etc.) and the corresponding basic size.

---

[1] For information about the term "standard" as used in this book, please see page 1-2.

THE ISO SYSTEM OF LIMITS AND FITS—TOLERANCES AND DEVIATIONS

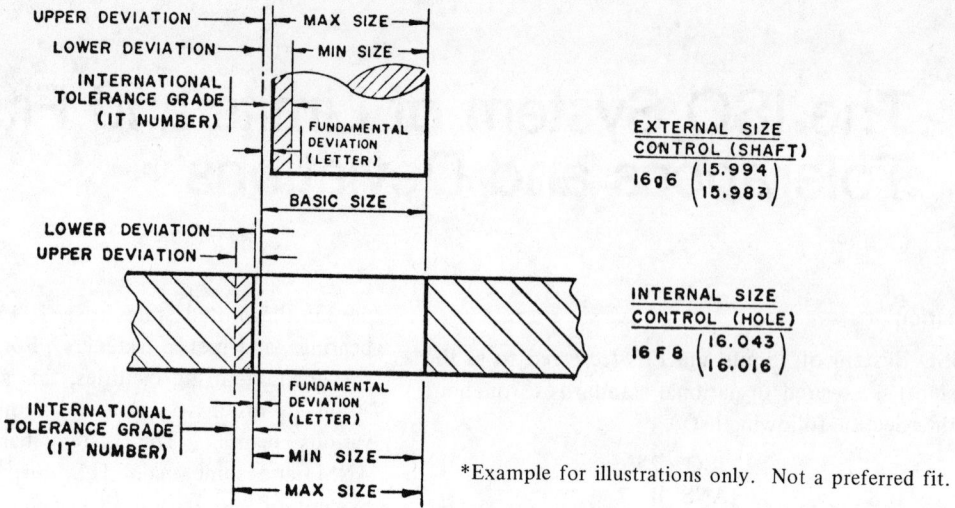

*Example for illustrations only. Not a preferred fit.

Fig. 6-1. Illustrations of definitions.

Actual deviation—The algebraic difference between the actual size and the corresponding basic size.

Upper deviation—The algebraic difference between the maximum limit of size and the corresponding basic size. This is designated "ES" for a hole and "es" for a shaft, these letters standing for the French term "écart supérieur."

Lower deviation—The algebraic difference between the minimum limit of size and the corresponding basic size. This is designated "EI" for a hole and "ei" for a shaft, the letters standing for the French term, "écart inférieur."

Zero line—The straight line in a graphical representation of limits and fits, to which the deviations are referred. The zero line is the line of zero deviation and represents the basic size. By convention, when the zero line is drawn horizontally, positive deviations are shown above and negative deviations below it.

Tolerance—The difference between the maximum limit of size and the minimum limit of size (or in other words, the algebraic difference between the upper deviation and the lower deviation). The tolerance is an absolute value without a sign.

Tolerance zone—In a graphical representation of tolerances, this is the area between the two lines representing the limits of tolerance and is defined by its magnitude (tolerance) and by its position in relation to the zero line.

Fundamental deviation—That one of the two deviations, nearest to the zero line, which is conventionally chosen to define the position of the tolerance zone in relation to the zero line.

Grade of tolerance—In a standardized system of limits and fits, a group of tolerances considered to correspond to the same level of accuracy for all basic sizes.

Fit—The relationship resulting from the difference, before assembly, between the sizes of the two parts that are to be assembled.

Clearance—The difference between the sizes of the hole and the shaft, before assembly, when this difference is positive.

Interference—The magnitude of the difference between the sizes of the hole and the shaft, before assembly, when this difference is negative.

Clearance fit—A fit which always provides a clearance. (The tolerance zone of the hole is entirely above that of the shaft.)

Interference fit—A fit which always provides an interference. (The tolerance zone of the hole is entirely below that of the shaft.)

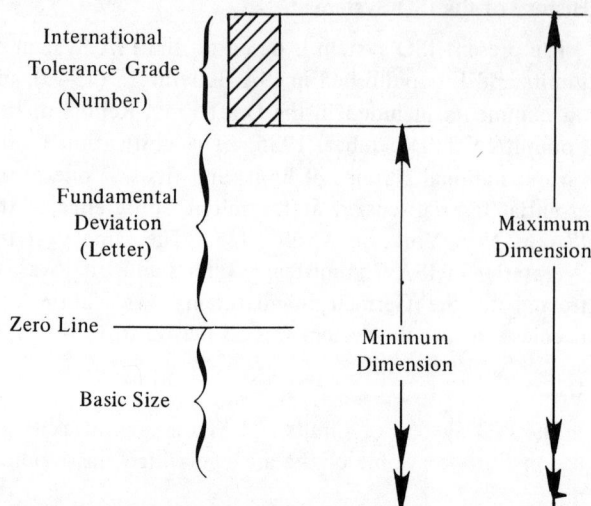

Fig. 6-2. International tolerance grade and fundamental deviation.

THE ISO SYSTEM OF LIMITS AND FITS—TOLERANCES AND DEVIATIONS

Table 6-1. International Tolerance Grades (BS 4500)

Tolerance unit 0.001 mm

Nominal sizes		Tolerance grades†																	
Over mm	Up to and including mm	IT 01	IT 0	IT 1	IT 2	IT 3	IT 4	IT 5	IT 6†	IT 7	IT 8	IT 9	IT 10	IT 11	IT 12	IT 13	IT 14*	IT 15*	IT 16*
—	3	0.3	0.5	0.8	1.2	2	3	4	6	10	14	25	40	60	100	140	250	400	600
3	6	0.4	0.6	1	1.5	2.5	4	5	8	12	18	30	48	75	120	180	300	480	750
6	10	0.4	0.6	1	1.5	2.5	4	6	9	15	22	36	58	90	150	220	360	580	900
10	18	0.5	0.8	1.2	2	3	5	8	11	18	27	43	70	110	180	270	430	700	1 100
18	30	0.6	1	1.5	2.5	4	6	9	13	21	33	52	84	130	210	330	520	840	1 300
30	50	0.6	1	1.5	2.5	4	7	11	16	25	39	62	100	160	250	390	620	1000	1 600
50	80	0.8	1.2	2	3	5	8	13	19	30	46	74	120	190	300	460	740	1200	1 900
80	120	1	1.5	2.5	4	6	10	15	22	35	54	87	140	220	350	540	870	1400	2 200
120	180	1.2	2	3.5	5	8	12	18	25	40	63	100	160	250	400	630	1000	1600	2 500
180	250	2	3	4.5	7	10	14	20	29	46	72	115	185	290	460	720	1150	1850	2 900
250	315	2.5	4	6	8	12	16	23	32	52	81	130	210	320	520	810	1300	2100	3 200
315	400	3	5	7	9	13	18	25	36	57	89	140	230	360	570	890	1400	2300	3 600
400	500	4	6	8	10	15	20	27	40	63	97	155	250	400	630	970	1550	2500	4 000
500	630	—	—	—	—	—	—	—	44	70	110	175	280	440	700	1100	1750	2800	4 400
630	800	—	—	—	—	—	—	—	50	80	125	200	320	500	800	1250	2000	3200	5 000
800	1000	—	—	—	—	—	—	—	56	90	140	230	360	560	900	1400	2300	3600	5 600
1000	1250	—	—	—	—	—	—	—	66	105	165	260	420	660	1050	1650	2600	4200	6 600
1250	1600	—	—	—	—	—	—	—	78	125	195	310	500	780	1250	1950	3100	5000	7 800
1600	2000	—	—	—	—	—	—	—	92	150	230	370	600	920	1500	2300	3700	6000	9 200
2000	2500	—	—	—	—	—	—	—	110	175	280	440	700	1100	1750	2800	4400	7000	11 000
2500	3150	—	—	—	—	—	—	—	135	210	330	540	860	1350	2100	3300	5400	8600	13 500

*Not applicable to sizes below 1 mm. Not recommended for fits in sizes above 500 mm.
†IT values for tolerance grades larger than IT 16 can be calculated by using the following formulas: IT 17 = IT 12 × 10; IT 18 = IT 13 × 10, etc.

6-3

Table 6-2. Fundamental Deviations for Shafts (BS 4500)

Dimensions in micrometers (1 μm = 0.001 mm)

Fundamental deviation		Upper deviation es										
Letter		a*	b*	c	cd	d	e	ef	f	fg	g	h	j,†
Grade		01 to 16											
Nominal sizes													
Over mm	Up to and including mm												
—	3	−270	−140	−60	−34	−20	−14	−10	−6	−4	−2	0	
3	6	−270	−140	−70	−46	−30	−20	−14	−10	−6	−4	0	
6	10	−280	−150	−80	−56	−40	−25	−18	−13	−8	−5	0	
10	14	−290	−150	−95	—	−50	−32	—	−16	—	−6	0	±IT/2
14	18	−290	−150	−95	—	−50	−32	—	−16	—	−6	0	
18	24	−300	−160	−110	—	−65	−40	—	−20	—	−7	0	
24	30	−300	−160	−110	—	−65	−40	—	−20	—	−7	0	
30	40	−310	−170	−120	—	−80	−50	—	−25	—	−9	0	
40	50	−320	−180	−130	—	−80	−50	—	−25	—	−9	0	
50	65	−340	−190	−140	—	−100	−60	—	−30	—	−10	0	
65	80	−360	−200	−150	—	−100	−60	—	−30	—	−10	0	
80	100	−380	−220	−170	—	−120	−72	—	−36	—	−12	0	
100	120	−410	−240	−180	—	−120	−72	—	−36	—	−12	0	
120	140	−460	−260	−200	—	−145	−85	—	−43	—	−14	0	
140	160	−520	−280	−210	—	−145	−85	—	−43	—	−14	0	
160	180	−580	−310	−230	—	−145	−85	—	−43	—	−14	0	

* Not applicable to sizes up to 1 mm.
† In grades 7–11, the two symmetrical deviations ±IT/2 should be rounded, if the IT value in micrometres is an odd value, by replacing it by the even value immediately beneath.

THE ISO SYSTEM OF LIMITS AND FITS—TOLERANCES AND DEVIATIONS

Table 6-2 (*Continued*). Fundamental Deviations for Shafts (BS 4500)

Dimensions in micrometers (1 μm = 0.001 mm)

Fundamental deviation		Upper deviation es										
Letter		a*	b*	c	cd	d	e	ef	f	fg	g	h	j,†
Grade		01 to 16											
Nominal sizes													
Over	Up to and including												
180	200	−660	−340	−240									
200	225	−740	−380	−260	—	−170	−100	—	−50	—	−15	0	
225	250	−820	−420	−280									
250	280	−920	−480	−300	—	−190	−110	—	−56	—	−17	0	
280	315	−1050	−540	−330									
315	355	−1200	−600	−360	—	−210	−125	—	−62	—	−18	0	
355	400	−1350	−680	−400									
400	450	−1500	−760	−440	—	−230	−135	—	−68	—	−20	0	±IT/2
450	500	−1650	−840	−480									
Grade		6 to 16											
500	630	—	—	—	—	−260	−145	—	−76	—	−22	0	
630	800	—	—	—	—	−290	−160	—	−80	—	−24	0	
800	1000	—	—	—	—	−320	−170	—	−86	—	−26	0	
1000	1250	—	—	—	—	−350	−195	—	−98	—	−28	0	
1250	1600	—	—	—	—	−390	−220	—	−110	—	−30	0	
1600	2000	—	—	—	—	−430	−240	—	−120	—	−32	0	
2000	2500	—	—	—	—	−480	−260	—	−130	—	−34	0	
2500	3150	—	—	—	—	−520	−290	—	−145	—	−38	0	

* Not applicable to sizes up to 1 mm.
† In grades 7–11, the two symmetrical deviations ±IT/2 should be rounded, if the IT value in micrometres is an odd value, by replacing it by the even value immediately beneath.

THE ISO SYSTEM OF LIMITS AND FITS—TOLERANCES AND DEVIATIONS

Table 6-2 (*Continued*). **Fundamental Deviations for Shafts (BS 4500)**

Dimensions in micrometers (1 μm = 0.001 mm)

Fundamental deviation		Lower deviation ei																	
Letter		j			k		m	n	p	r	s	t	u	v	x	y	z	za	zb	zc
Grade		5–6	7	8	4–7	≤3 >7	01 to 16													
Nominal size																				
Over	Up to and including																			
mm	mm																			
—	3	−2	−4	−6	0	0	+2	+4	+6	+10	+14	—	+18	—	+20	—	+26	+32	+40	+60
3	6	−2	−4	—	+1	0	+4	+8	+12	+15	+19	—	+23	—	+28	—	+35	+42	+50	+80
6	10	−2	−5	—	+1	0	+6	+10	+15	+19	+23	—	+28	—	+34	—	+42	+52	+67	+97
10	14	−3	−6	—	+1	0	+7	+12	+18	+23	+28	—	+33	—	+40	—	+50	+64	+90	+130
14	18	−3	−6	—	+1	0	+7	+12	+18	+23	+28	—	+33	+39	+45	—	+60	+77	+108	+150
18	24	−4	−8	—	+2	0	+8	+15	+22	+28	+35	—	+41	+47	+54	+63	+73	+98	+136	+188
24	30	−4	−8	—	+2	0	+8	+15	+22	+28	+35	+41	+48	+55	+64	+75	+88	+118	+160	+218
30	40	−5	−10	—	+2	0	+9	+17	+26	+34	+43	+48	+60	+68	+80	+94	+112	+148	+200	+274
40	50	−5	−10	—	+2	0	+9	+17	+26	+34	+43	+54	+70	+81	+97	+114	+136	+180	+242	+325
50	65	−7	−12	—	+2	0	+11	+20	+32	+41	+53	+66	+87	+102	+122	+144	+172	+226	+300	+405
65	80	−7	−12	—	+2	0	+11	+20	+32	+43	+59	+75	+102	+120	+146	+174	+210	+274	+360	+480
80	100	−9	−15	—	+3	0	+13	+23	+37	+51	+71	+91	+124	+146	+178	+214	+258	+335	+445	+585
100	120	−9	−15	—	+3	0	+13	+23	+37	+54	+79	+104	+144	+172	+210	+254	+310	+400	+525	+690
120	140	−11	−18	—	+4	0	+15	+27	+43	+63	+92	+122	+170	+202	+248	+300	+365	+470	+620	+800
140	160	−11	−18	—	+4	0	+15	+27	+43	+65	+100	+134	+190	+228	+280	+340	+415	+535	+700	+900
160	180	−11	−18	—	+4	0	+15	+27	+43	+68	+108	+146	+210	+252	+310	+380	+465	+600	+780	+1000
180	200	+13	−21	—	+4	0	+17	+31	+50	+77	+122	+166	+236	+284	+350	+425	+520	+670	+880	+1150
200	225	+13	−21	—	+4	0	+17	+31	+50	+80	+130	+180	+258	+310	+385	+470	+575	+740	+960	+1250
225	250	+13	−21	—	+4	0	+17	+31	+50	+84	+140	+196	+284	+340	+425	+520	+640	+820	+1050	+1350
250	280	−16	−26	—	+4	0	+20	+34	+56	+94	+158	+218	+315	+385	+475	+580	+710	+920	+1200	+1550
280	315	−16	−26	—	+4	0	+20	+34	+56	+98	+170	+240	+350	+425	+525	+650	+790	+1000	+1300	+1700

THE ISO SYSTEM OF LIMITS AND FITS—TOLERANCES AND DEVIATIONS

Table 6-2 (*Continued*). **Fundamental Deviations for Shafts (BS 4500)**

Dimensions in micrometers (1μm = 0.001 mm)

Fundamental deviation		Lower deviation ei																
Letter		j			k		m	n	p	r	s	t	u	v	x	y	z	za	zb	zc
Grade		5-6	7	8	4-7	≤3 >7	01 to 16													
Nominal size																				
Over	Up to and including																			
315	355	−18	−28	—	+4	0	+21	+37	+62	+108	+190	+268	+390	+475	+590	+730	+900	+1150	+1500	+1900
355	400	−18	−28	—	+4	0	+21	+37	+62	+114	+208	+294	+435	+530	+660	+820	+1000	+1300	+1650	+2100
400	450	−20	−32	—	+5	0	+23	+40	+68	+126	+232	+330	+490	+595	+740	+920	+1100	+1450	+1850	+2400
450	500	−20	−32	—	+5	0	+23	+40	+68	+132	+252	+360	+540	+660	+820	+1000	+1250	+1600	+2100	+2600
Grade						6 to 16														
500	560					0	+26	+44	+78	+150	+280	+400	+600							
560	630					0	+26	+44	+78	+155	+310	+450	+660							
630	710					0	+30	+50	+88	+175	+340	+500	+740							
710	800					0	+30	+50	+88	+185	+380	+560	+840							
800	900					0	+34	+56	+100	+210	+430	+620	+940							
900	1000					0	+34	+56	+100	+220	+470	+680	+1050							
1000	1120					0	+40	+66	+120	+250	+520	+780	+1150							
1120	1250					0	+40	+66	+120	+260	+580	+840	+1300							
1250	1400					0	+48	+78	+140	+300	+640	+960	+1450							
1400	1600					0	+48	+78	+140	+330	+720	+1050	+1600							
1600	1800					0	+58	+92	+170	+370	+820	+1200	+1850							
1800	2000					0	+58	+92	+170	+400	+920	+1350	+2000							
2000	2240					0	+68	+110	+195	+440	+1000	+1500	+2300							
2240	2500					0	+68	+110	+195	+460	+1100	+1650	+2500							
2500	2800					0	+76	+135	+240	+550	+1250	+1900	+2900							
2800	3150					0	+76	+135	+240	+580	+1400	+2100	+3200							

Table 6-3. Fundamental Deviations for Holes (BS 4500)

Dimensions in micrometers ($1\mu m = 0.001$ mm)

Fundamental deviation	Lower deviation EI										
Letter	A*	B*	C	CD	D	E	EF	F	FG	G	H	J,†
Grade	01 to 16											
Nominal sizes												
Over — Up to and including (mm)												
— 3	+270	+140	+60	+34	+20	+14	+10	+6	+4	+2	0	
3 — 6	+270	+140	+70	+46	+30	+20	+14	+10	+6	+4	0	
6 — 10	+280	+150	+80	+56	+40	+25	+18	+13	+8	+5	0	
10 — 14	+290	+150	+95	—	+50	+32	—	+16	—	+6	0	±IT/2
14 — 18	+290	+150	+95	—	+50	+32	—	+16	—	+6	0	
18 — 24	+300	+160	+110	—	+65	+40	—	+20	—	+7	0	
24 — 30	+300	+160	+110	—	+65	+40	—	+20	—	+7	0	
30 — 40	+310	+170	+120	—	+80	+50	—	+25	—	+9	0	
40 — 50	+320	+180	+130	—	+80	+50	—	+25	—	+9	0	
50 — 65	+340	+190	+140	—	+100	+60	—	+30	—	+10	0	
65 — 80	+360	+200	+150	—	+100	+60	—	+30	—	+10	0	
80 — 100	+380	+220	+170	—	+120	+72	—	+36	—	+12	0	
100 — 120	+410	+240	+180	—	+120	+72	—	+36	—	+12	0	
120 — 140	+460	+260	+200	—	+145	+85	—	+43	—	+14	0	
140 — 160	+520	+280	+210	—	+145	+85	—	+43	—	+14	0	
160 — 180	+580	+310	+230	—	+145	+85	—	+43	—	+14	0	

* Not applicable to sizes up to 1 mm.
† In grades 7–11, the two symmetrical deviations ±IT/2 should be rounded, if the IT value in micrometres is an odd value, by replacing it by the even value immediately beneath.

THE ISO SYSTEM OF LIMITS AND FITS—TOLERANCES AND DEVIATIONS

Table 6-3 (*Continued*). Fundamental Deviations for Holes (BS 4500)

Dimensions in micrometers (1 μm = 0.001 mm)

Fundamental deviation	Lower deviation EI											
Letter	A*	B*	C	CD	D	E	EF	F	FG	G	H	J,†
Grade	01 to 16											
Nominal sizes												
Over / Up to and including												
180 / 200	+660	+340	+240		+170	+100	—	+50	—	+15	0	
200 / 225	+740	+380	+260	—								
225 / 250	+820	+420	+280	—								
250 / 280	+920	+480	+300		+190	+110	—	+56	—	+17	0	
280 / 315	+1050	+540	+330	—								
315 / 355	+1200	+600	+360		+210	+125	—	+62	—	+18	0	±IT/2
355 / 400	+1350	+680	+400	—								
400 / 450	+1500	+760	+440		+230	+135	—	+68	—	+20	0	
450 / 500	+1650	+840	+480	—								
Grade	**6 to 16**											
500 / 630	—	—	—	—	+260	+145	—	+76	—	+22	0	
630 / 800	—	—	—	—	+290	+160	—	+80	—	+24	0	
800 / 1000	—	—	—	—	+320	+170	—	+86	—	+26	0	
1000 / 1250	—	—	—	—	+350	+195	—	+98	—	+28	0	
1250 / 1600	—	—	—	—	+390	+220	—	+110	—	+30	0	
1600 / 2000	—	—	—	—	+430	+240	—	+120	—	+32	0	
2000 / 2500	—	—	—	—	+480	+260	—	+130	—	+34	0	
2500 / 3150	—	—	—	—	+520	+290	—	+145	—	+38	0	

* Not applicable to sizes up to 1 mm.
† In grades 7–11, the two symmetrical deviations ±IT/2 should be rounded, if the IT value in micrometres is an odd value, by replacing it by the even value immediately beneath.

Table 6-3 (Continued). Fundamental Deviations for Holes (BS 4500)

Dimensions in micrometers (1μm = 0.001 mm)

Fundamental deviation	Upper deviation ES																					Values for Δ‡						
Letter	J			K		M			N		P to ZC	P	R	S	T	U	V	X	Y	Z	ZA	ZB	ZC	Grades:					
Grade	6	7	8	≤8	>8	≤8*	>8		≤8	>8	≤7	>8†	Above 7												3	4	5	6	7	8
Nominal sizes																														
Over — Up to and including mm																														
— 3	+2	+4	+6	0	0	−2	−2		−4	−4		−6	−10	−14	—	−18	—	−20	—	−26	−32	−40	−60		0	0	0	0	0	0
3 6	+5	+6	+10	−1+Δ	—	−4+Δ	−4		−8+Δ	0		−12	−15	−19	—	−23	—	−28	—	−35	−42	−50	−80		1	1.5	1	3	4	6
6 10	+5	+8	+12	−1+Δ	—	−6+Δ	−6		−10+Δ	0		−15	−19	−23	—	−28	—	−34	—	−42	−52	−67	−97		1	1.5	2	3	6	7
10 14 / 14 18	+6	+10	+15	−1+Δ	—	−7+Δ	−7		−12+Δ	0		−18	−23	−28	—	−33	−39	−40 / −45	—	−50 / −60	−64 / −77	−90 / −108	−130 / −150		1	2	3	3	7	9
18 24 / 24 30	+8	+12	+20	−2+Δ	—	−8+Δ	−8		−15+Δ	0		−22	−28	−35	−41	−41 / −48	−47 / −55	−54 / −64	−63 / −75	−73 / −88	−98 / −118	−136 / −160	−188 / −218		1.5	2	3	4	8	12
30 40 / 40 50	+10	+14	+24	−2+Δ	—	−9+Δ	−9		−17+Δ	0		−26	−34	−43	−48 / −54	−60 / −70	−68 / −81	−80 / −97	−94 / −114	−112 / −136	−148 / −180	−200 / −242	−274 / −325		1.5	3	4	5	9	14
50 65 / 65 80	+13	+18	+28	−2+Δ	—	−11+Δ	−11		−20+Δ	0		−32	−41 / −43	−53 / −59	−66 / −75	−87 / −102	−102 / −120	−122 / −146	−144 / −174	−172 / −210	−226 / −274	−300 / −360	−405 / −480		2	3	5	6	11	16
80 100 / 100 120	+16	+22	+34	−3+Δ	—	−13+Δ	−13		−23+Δ	0		−37	−51 / −54	−71 / −79	−91 / −104	−124 / −144	−146 / −172	−178 / −210	−214 / −254	−258 / −310	−335 / −400	−445 / −525	−585 / −690		2	4	5	7	13	19
120 140 / 140 160 / 160 180	+18	+26	+41	−3+Δ	—	−15+Δ	−15		−27+Δ	0		−43	−63 / −65 / −68	−92 / −100 / −108	−122 / −130 / −140	−170 / −190 / −210	−202 / −228 / −252	−248 / −280 / −310	−300 / −340 / −380	−365 / −415 / −465	−470 / −535 / −600	−620 / −700 / −780	−800 / −900 / −1000		3	4	6	7	15	23
180 200 / 200 225 / 225 250	+22	+30	+47	−4+Δ	—	−17+Δ	−17		−31+Δ	0		−50	−77 / −80 / −84	−122 / −130 / −140	−166 / −180 / −196	−236 / −258 / −284	−284 / −310 / −340	−350 / −385 / −425	−425 / −470 / −520	−520 / −575 / −640	−670 / −740 / −820	−880 / −960 / −1050	−1150 / −1250 / −1350		3	4	6	9	17	26
250 280 / 280 315	+25	+36	+55	−4+Δ	—	−20+Δ	−20		−34+Δ	0		−56	−94 / −98	−158 / −170	−218 / −240	−315 / −350	−385 / −425	−475 / −525	−580 / −650	−710 / −790	−920 / −1000	−1200 / −1300	−1550 / −1700		4	4	7	9	20	29

Note: "Same deviation as for grades above 7 increased by Δ"

THE ISO SYSTEM OF LIMITS AND FITS—TOLERANCES AND DEVIATIONS

Table 6-3 (*Continued*). Fundamental Deviations for Holes (BS 4500)

Dimensions in micrometers (1 μm = 0.001 mm)

Fundamental deviation	Upper deviation ES																					Values for Δ‡						
Letter	J			K			M			N			P to ZC	P	R	S	T	U	V	X	Y	Z	ZA	ZB	ZC	Grades:					
Grade	6	7	8	≤8	>8	≤8*	>8	≤8	>8†	≤7	Above 7											3	4	5	6	7	8
Nominal sizes																												
Over — Up to and including																												
315 — 355	+29	+39	+60	−4+Δ		−21+Δ		−21	−37+Δ		Same deviation as for grades above 7 increased by Δ	−62	−108	−190	−268	−390	−475	−730	−900	−1150	−1500	−1900	4	5	7	11	21	32
355 — 400													−114	−208	−294	−435	−530	−820	−1000	−1300	−1650	−2100						
400 — 450	+33	+43	+66	−5+Δ		−23+Δ		−23	−40+Δ			−68	−126	−232	−330	−490	−595	−920	−1100	−1450	−1850	−2400	5	5	7	13	23	34
450 — 500													−132	−252	−360	−540	−660	−1000	−1250	−1600	−2100	−2600						
Grade	6 to 16																											
500 — 560				0		−26		−44		0		−78	−150	−280	−400	−600												
560 — 630													−155	−310	−450	−660												
630 — 710				0		−30		−50		0		−88	−175	−340	−500	−740												
710 — 800													−185	−380	−560	−840												
800 — 900				0		−34		−56		0		−100	−210	−430	−620	−940												
900 — 1000													−220	−470	−680	−1050												
1000 — 1120				0		−40		−66		0		−120	−250	−520	−780	−1150												
1120 — 1250													−260	−580	−840	−1300												
1250 — 1400				0		−48		−78		0		−140	−300	−640	−960	−1450												
1400 — 1600													−330	−720	−1050	−1600												
1600 — 1800				0		−58		−92		0		−170	−370	−820	−1200	−1850												
1800 — 2000													−400	−920	−1350	−2000												
2000 — 2240				0		−68		−110		0		−195	−440	−1000	−1500	−2300												
2240 — 2500													−460	−1100	−1650	−2500												
2500 — 2800				0		−76		−135		0		−240	−550	−1250	−1900	−2900												
2800 — 3150													−580	−1400	−2100	−3200												

* Special case: for M6, ES = −9 from 250 to 315 (instead of −11).

† Not applicable to sizes up to 1 mm

‡ In determining K, M, N up to grade 8 and P to ZC up to grade 7, add the Δ value appropriate to the grade as indicated, e.g. for P7 from 18 to 30, Δ = 8 ∴ ES = −14.

THE ISO SYSTEM OF LIMITS AND FITS—TOLERANCES AND DEVIATIONS

Transition fit—A fit which may provide either a clearance or an interference. (The tolerance zones of the hole and the shaft overlap.)

Shaft-basis system of fits—A system of fits in which the different clearances and interferences are obtained by associating various holes with a single shaft (or, possibly, with shafts of different grades, but having the same fundamental deviation).

In the ISO System, the basic shaft is that with an upper deviation of zero.

Hole-basis system of fits—A system of fits in which the different clearances and interferences are obtained by associating various shafts with a single hole (or, possibly, with holes of different grades, but always having the same fundamental deviation).

In the ISO System, the basic hole is that with a lower deviation of zero.

Unilateral tolerance system—The shaft-basis and hole-basis systems of fits are unilateral tolerancing systems (unilateral = one-sided).

Bilateral tolerance system—A tolerancing system specifying + or − a given value is called a bilateral tolerancing system (bilateral = two-sided).

Description of the ISO System of Limits and Fits

The ISO system of limits and fits is based on a series of standard tolerances[1] and deviations covering sizes up to 3150 mm. The ISO tolerance system is shown in principle in Fig. 6-2.

The International Tolerance Grade is found in Table 6-1. It is dependent on the IT (*I*nternational *T*olerance) grade number and the nominal size.

The Fundamental Deviation indicates, with the help of one or two letters, the relative position of the Standard Tolerance measured from the Zero Line. Numerical values for the Fundamental Deviations are shown in Table 6-2,

---

[1] Standard tolerances are synonymous with International Tolerance (IT) grades. The ISO 286 and some national standards use both designations. The ANSI B4.2 refers to IT grades only.

for shafts (a, b, c, cd, d, etc.) and Table 6-3 for holes (A, B, C, CD, D, etc.).

Tolerances—The system provides a series of qualities of tolerance, called "tolerance grades," to cater to different classes of work: fine tolerance grades for fine work, coarse tolerance grades for coarse work. In each grade the tolerance values increase with size according to a formula which relates the value of a given constant to the mean diameter of a particular size range.

Eighteen tolerance grades are provided and they are designated IT01, IT0, IT1, IT2 ... IT16.

The numerical values for standard tolerances are given in Table 6-1. The practical usage ranges are shown in Fig. 6-3. The IT grade number is, to some extent, dependent on the surface texture and the machining operation chosen. (See Section 5, Table 5-6.)

Fundamental Deviations

The system provides 27 different fundamental deviations for sizes up to and including 500 mm, and 14 for larger sizes, to give different types of fit ranging from coarse clearance to heavy interference. The values associated with each of the deviations vary with size so as to maintain the same fit characteristics and, as in the case of the tolerances, each deviation value is related to the mean diameter of particular size range.

Each deviation is designated by a letter "a" representing a large negative deviation and the letter "z," a large positive deviation. The deviations apply both to shafts and holes; the same letters are used for the designation in each case, but upper case (capital) letters are used to designate hole deviations and lower case (small) letters to define shaft deviations. Shaft deviation signs are opposite those for holes.

Figure 6-4 illustrates the relative positions of the fundamental deviations for holes and shafts.

The exception to the arrangement described above is the deviation js (JS for holes) which has been provided to meet the need for symmetrical bilateral tolerances. In

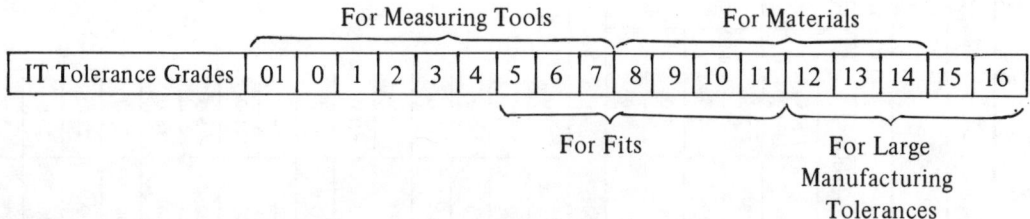

Fig. 6-3. Practical use of the tolerance grades. (ANSI B4.2)

THE ISO SYSTEM OF LIMITS AND FITS—TOLERANCES AND DEVIATIONS

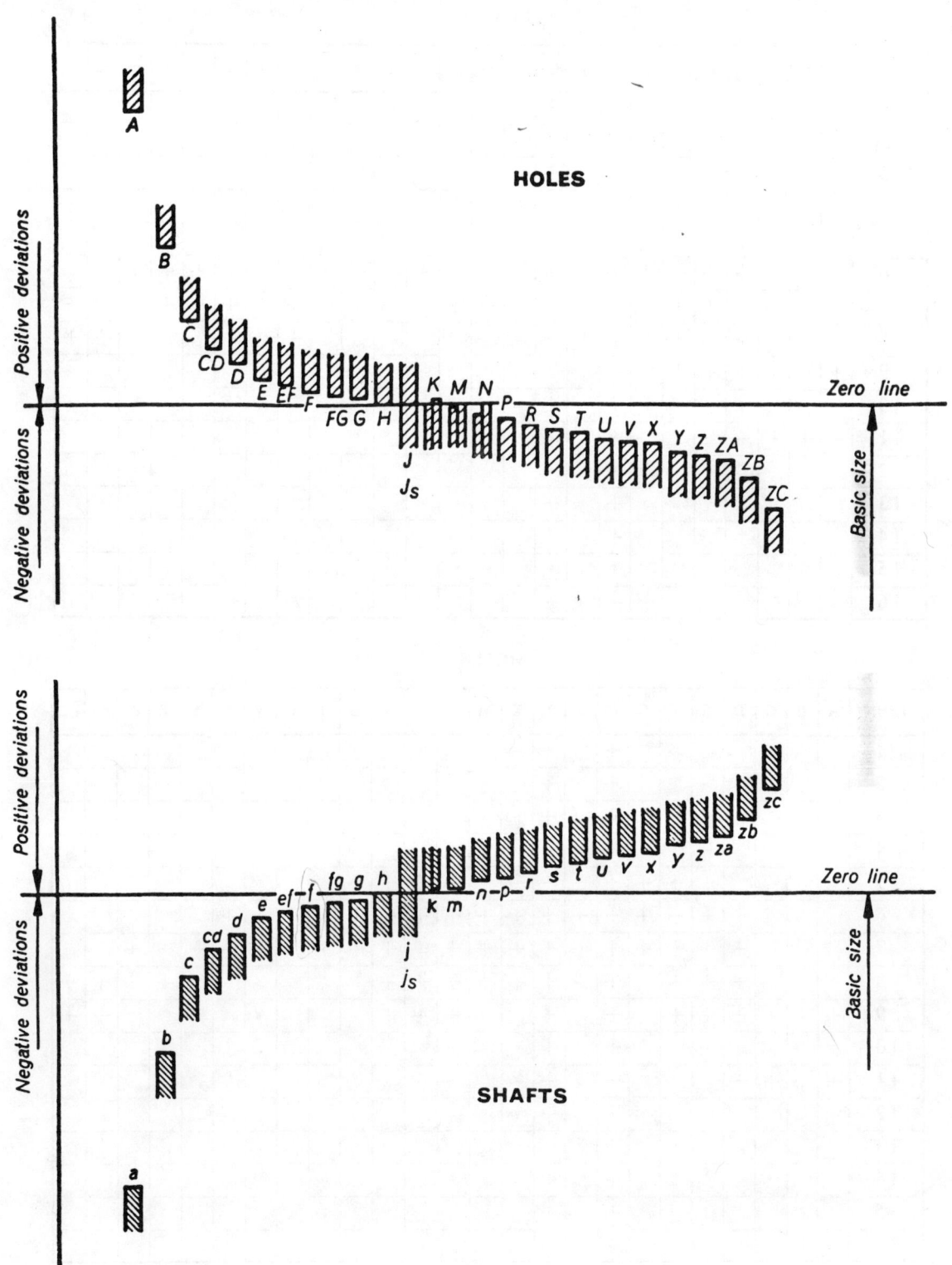

Fig. 6-4. Respective positions of the various tolerance zones for a given diameter step (ISO 286).

THE ISO SYSTEM OF LIMITS AND FITS—TOLERANCES AND DEVIATIONS

Table 6-4. General Purpose Standard Tolerance (Number) and Fundamental Deviation (Letter) Combinations* (ISO 286)

SHAFTS

Letter	a	b	c	d	e	f	g	h	j	js	k	m	n	p	r	s	t	u	v	x	y	z	za	zb	zc
1								+		+															
2								+		+															
3								+		+															
4						+	+	+		+	+	+	+	+	+	+									
5				+	+	+	+	+	+	+	+	+	+	+	+	+	+	+	+	+					
6				+	+	+	+	+	+	+	+	+	+	+	+	+	+	+	+	+	+	+	+		
7				+	+	+	+	+	+	+	+	+	+	+	+	+	+	+	+	+	+	+	+	+	+
8		+	+	+	+	+		+		+								+		+				+	+
9	+	+	+	+	+	+		+		+															
10				+				+		+															
11	+	+	+	+				+		+															
12								+		+															
13								+		+															
14								+		+															
15								+		+															
16								+		+															

HOLES

Letter	A	B	C	D	E	F	G	H	J	Js	K	M	N	P	R	S	T	U	V	X	Y	Z	ZA	ZB	ZC
1								+		+															
2								+		+															
3								+		+															
4								+		+															
5					+	+	+	+		+	+	+	+	+	+	+									
6				+	+	+	+	+	+	+	+	+	+	+	+	+	+	+	+	+					
7				+	+	+	+	+	+	+	+	+	+	+	+	+	+	+	+	+	+	+	+		
8		+	+	+	+	+		+	+	+	+	+	+	+							+	+		+	+
9	+	+	+	+	+	+		+		+			+	+										+	+
10				+	+			+		+			+												
11	+	+	+	+				+		+			+												
12								+		+															
13								+		+															
14								+		+															
15								+		+															
16								+		+															

* The symbols retained are those for shafts and holes which might be considered as the most commonly used (taking account, in particular, of existing national standards) either in the whole range of diameter steps up to 500 mm (19.69 in.) or, for some of the diameter steps, in a more limited field see tables.

THE ISO SYSTEM OF LIMITS AND FITS—TOLERANCES AND DEVIATIONS

Table 6-5. Description of Preferred Fits (ANSI B4.2)

ISO SYMBOL		DESCRIPTION
HOLE BASIS	SHAFT BASIS	
H11/c11	C11/h11	Loose running fit for wide commercial tolerances or allowances on external members.
H9/d9	D9/h9	Free running fit not for use where accuracy is essential, but good for large temperature variations, high running speeds, or heavy journal pressures.
H8/f7	F8/h7	Close running fit for running on accurate machines and for accurate location at moderate speeds and journals pressures.
H7/g6	G7/h6	Sliding fit not intended to run freely, but to move and turn freely and locate accurately.
H7/h6	H7/h6	Locational clearance fit provides snug fit for locating stationary parts, but can be freely assembled and disassembled.
H7/k6	K7/h6	Locational transition fit for accurate location, a compromise between clearance and interference.
H7/n6	N7/h6	Locational transition fit for more accurate location where greater interference is permissible.
H7/p6*	P7/h6	Locational interference fit for parts requiring rigidity and alignment with prime accuracy of location but without special bore pressure requirements.
H7/s6	S7/h6	Medium drive fit for ordinary steel parts or shrink fits on light sections, the tightest fit usable with cast iron.
H7/u6	U7/h6	Force fit suitable for parts which can be highly stressed or for shrink fits where the heavy pressing forces required are impractical.

Left margin bracket: Clearance Fits (H11/c11 through H7/h6); Transition Fits (H7/k6, H7/n6); Interference Fits (H7/p6* through H7/u6).

Right margin: More Clearance ↑ / More Interference ↓

NOTE: Transition fit for basic sizes in range from 1 through 3 mm.

6-15

THE ISO SYSTEM OF LIMITS AND FITS—TOLERANCES AND DEVIATIONS

Table 6-6. Preferred Hole Basis Clearance Fits (ANSI B4.2)

NOMINAL SIZE		LOOSE RUNNING HOLE H11	LOOSE RUNNING SHAFT c11	DIFF	FREE RUNNING HOLE H9	FREE RUNNING SHAFT d9	DIFF	CLOSE RUNNING HOLE H8	CLOSE RUNNING SHAFT f7	DIFF	SLIDING HOLE H7	SLIDING SHAFT g6	DIFF	LOCATIONAL CLEARANCE HOLE H7	LOCATIONAL CLEARANCE SHAFT h6	DIFF
F 1	MAX / MIN	1.060 / 1.000	0.940 / 0.880	0.180 / 0.060	1.025 / 1.000	0.980 / 0.955	0.070 / 0.020	1.014 / 1.000	0.994 / 0.984	0.030 / 0.006	1.010 / 1.000	0.998 / 0.992	0.018 / 0.002	1.010 / 1.000	1.000 / 0.994	0.016 / 0.000
S 1.1	MAX / MIN	1.160 / 1.100	1.040 / 0.980	0.180 / 0.060	1.125 / 1.100	1.080 / 1.055	0.070 / 0.020	1.114 / 1.100	1.094 / 1.084	0.030 / 0.006	1.110 / 1.100	1.098 / 1.092	0.018 / 0.002	1.110 / 1.100	1.100 / 1.094	0.016 / 0.000
F 1.2	MAX / MIN	1.260 / 1.200	1.140 / 1.080	0.180 / 0.060	1.225 / 1.200	1.180 / 1.155	0.070 / 0.020	1.214 / 1.200	1.194 / 1.184	0.030 / 0.006	1.210 / 1.200	1.198 / 1.192	0.018 / 0.002	1.210 / 1.200	1.200 / 1.194	0.016 / 0.000
S 1.4	MAX / MIN	1.460 / 1.400	1.340 / 1.280	0.180 / 0.060	1.425 / 1.400	1.380 / 1.355	0.070 / 0.020	1.414 / 1.400	1.394 / 1.384	0.030 / 0.006	1.410 / 1.400	1.398 / 1.392	0.018 / 0.002	1.410 / 1.400	1.400 / 1.394	0.016 / 0.000
F 1.6	MAX / MIN	1.660 / 1.600	1.540 / 1.480	0.180 / 0.060	1.625 / 1.600	1.580 / 1.555	0.070 / 0.020	1.614 / 1.600	1.594 / 1.584	0.030 / 0.006	1.610 / 1.600	1.598 / 1.592	0.018 / 0.002	1.610 / 1.600	1.600 / 1.594	0.016 / 0.000
S 1.8	MAX / MIN	1.860 / 1.800	1.740 / 1.680	0.180 / 0.060	1.825 / 1.800	1.780 / 1.755	0.070 / 0.020	1.814 / 1.800	1.794 / 1.784	0.030 / 0.006	1.810 / 1.800	1.798 / 1.792	0.018 / 0.002	1.810 / 1.800	1.800 / 1.794	0.016 / 0.000
F 2	MAX / MIN	2.060 / 2.000	1.940 / 1.880	0.180 / 0.060	2.025 / 2.000	1.980 / 1.955	0.070 / 0.020	2.014 / 2.000	1.994 / 1.984	0.030 / 0.006	2.010 / 2.000	1.998 / 1.992	0.018 / 0.002	2.010 / 2.000	2.000 / 1.994	0.016 / 0.000
S 2.2	MAX / MIN	2.260 / 2.200	2.140 / 2.080	0.180 / 0.060	2.225 / 2.200	2.180 / 2.155	0.070 / 0.020	2.214 / 2.200	2.194 / 2.184	0.030 / 0.006	2.210 / 2.200	2.198 / 2.192	0.018 / 0.002	2.210 / 2.200	2.200 / 2.194	0.016 / 0.000
F 2.5	MAX / MIN	2.560 / 2.500	2.440 / 2.380	0.180 / 0.060	2.525 / 2.500	2.480 / 2.455	0.070 / 0.020	2.514 / 2.500	2.494 / 2.484	0.030 / 0.006	2.510 / 2.500	2.498 / 2.492	0.018 / 0.002	2.510 / 2.500	2.500 / 2.494	0.016 / 0.000
S 2.8	MAX / MIN	2.860 / 2.800	2.740 / 2.680	0.180 / 0.060	2.825 / 2.800	2.780 / 2.755	0.070 / 0.020	2.814 / 2.800	2.794 / 2.784	0.030 / 0.006	2.810 / 2.800	2.798 / 2.792	0.018 / 0.002	2.810 / 2.800	2.800 / 2.794	0.016 / 0.000
F 3	MAX / MIN	3.060 / 3.000	2.940 / 2.880	0.180 / 0.060	3.025 / 3.000	2.980 / 2.955	0.070 / 0.020	3.014 / 3.000	2.994 / 2.984	0.030 / 0.006	3.010 / 3.000	2.998 / 2.992	0.018 / 0.002	3.010 / 3.000	3.000 / 2.994	0.016 / 0.000
S 3.5	MAX / MIN	3.575 / 3.500	3.430 / 3.355	0.220 / 0.070	3.530 / 3.500	3.470 / 3.440	0.090 / 0.030	3.518 / 3.500	3.490 / 3.478	0.040 / 0.010	3.512 / 3.500	3.496 / 3.488	0.024 / 0.004	3.512 / 3.500	3.500 / 3.492	0.020 / 0.000
F 4	MAX / MIN	4.075 / 4.000	3.930 / 3.855	0.220 / 0.070	4.030 / 4.000	3.970 / 3.940	0.090 / 0.030	4.018 / 4.000	3.990 / 3.978	0.040 / 0.010	4.012 / 4.000	3.996 / 3.988	0.024 / 0.004	4.012 / 4.000	4.000 / 3.992	0.020 / 0.000
S 4.5	MAX / MIN	4.575 / 4.500	4.430 / 4.355	0.220 / 0.070	4.530 / 4.500	4.470 / 4.440	0.090 / 0.030	4.518 / 4.500	4.490 / 4.478	0.040 / 0.010	4.512 / 4.500	4.496 / 4.488	0.024 / 0.004	4.512 / 4.500	4.500 / 4.492	0.020 / 0.000
F 5	MAX / MIN	5.075 / 5.000	4.930 / 4.855	0.220 / 0.070	5.030 / 5.000	4.970 / 4.940	0.090 / 0.030	5.018 / 5.000	4.990 / 4.978	0.040 / 0.010	5.012 / 5.000	4.996 / 4.988	0.024 / 0.004	5.012 / 5.000	5.000 / 4.992	0.020 / 0.000
S 5.5	MAX / MIN	5.575 / 5.500	5.430 / 5.355	0.220 / 0.070	5.530 / 5.500	5.470 / 5.440	0.090 / 0.030	5.518 / 5.500	5.490 / 5.478	0.040 / 0.010	5.512 / 5.500	5.496 / 5.488	0.024 / 0.004	5.512 / 5.500	5.500 / 5.492	0.020 / 0.000
F 6	MAX / MIN	6.075 / 6.000	5.930 / 5.855	0.220 / 0.070	6.030 / 6.000	5.970 / 5.940	0.090 / 0.030	6.018 / 6.000	5.990 / 5.978	0.040 / 0.010	6.012 / 6.000	5.996 / 5.988	0.024 / 0.004	6.012 / 6.000	6.000 / 5.992	0.020 / 0.000
T 6.5	MAX / MIN	6.590 / 6.500	6.420 / 6.330	0.260 / 0.080	6.536 / 6.500	6.460 / 6.424	0.112 / 0.030	6.522 / 6.500	6.487 / 6.472	0.050 / 0.013	6.515 / 6.500	6.495 / 6.486	0.029 / 0.005	6.515 / 6.500	6.500 / 6.491	0.024 / 0.000

NOTES: 1. Select nominal sizes to preference rating as follows: F = First Choice, S = Second Choice, T = Third Choice.
2. ANSI B4.2 lists limit dimensions for nominal sizes marked "F" (First Choice) only. A cost penalty for material stock tooling and gages is anticipated for sizes marked with "S" (Second Choice) and "T" (Third Choice).

DIMENSIONS IN MILLIMETERS

THE ISO SYSTEM OF LIMITS AND FITS—TOLERANCES AND DEVIATIONS

Table 6-6 (Continued). Preferred Hole Basis Clearance Fits (ANSI B4.2)

NOMINAL SIZE		LOOSE RUNNING HOLE H11	LOOSE RUNNING SHAFT c11	DIFF	FREE RUNNING HOLE H9	FREE RUNNING SHAFT d9	DIFF	CLOSE RUNNING HOLE H8	CLOSE RUNNING SHAFT f7	DIFF	SLIDING HOLE H7	SLIDING SHAFT g6	DIFF	LOCATIONAL CLEARANCE HOLE H7	LOCATIONAL CLEARANCE SHAFT h6	DIFF
7	MAX / MIN	7.090 / 7.000	6.920 / 6.830	0.260 / 0.080	7.036 / 7.000	6.960 / 6.924	0.112 / 0.040	7.022 / 7.000	6.987 / 6.972	0.050 / 0.013	7.015 / 7.000	6.995 / 6.986	0.029 / 0.005	7.015 / 7.000	7.000 / 6.991	0.024 / 0.000
8	MAX / MIN	8.090 / 8.000	7.920 / 7.830	0.260 / 0.080	8.036 / 8.000	7.960 / 7.924	0.112 / 0.040	8.022 / 8.000	7.987 / 7.972	0.050 / 0.013	8.015 / 8.000	7.995 / 7.986	0.029 / 0.005	8.015 / 8.000	8.000 / 7.991	0.024 / 0.000
9	MAX / MIN	9.090 / 9.000	8.920 / 8.830	0.260 / 0.080	9.036 / 9.000	8.960 / 8.924	0.112 / 0.040	9.022 / 9.000	8.987 / 8.972	0.050 / 0.013	9.015 / 9.000	8.995 / 8.986	0.029 / 0.005	9.015 / 9.000	9.000 / 8.991	0.024 / 0.000
10	MAX / MIN	10.090 / 10.000	9.920 / 9.830	0.260 / 0.080	10.036 / 10.000	9.960 / 9.924	0.112 / 0.040	10.022 / 10.000	9.987 / 9.972	0.050 / 0.013	10.015 / 10.000	9.995 / 9.986	0.029 / 0.005	10.015 / 10.000	10.000 / 9.991	0.024 / 0.000
11	MAX / MIN	11.110 / 11.000	10.905 / 10.795	0.315 / 0.095	11.043 / 11.000	10.950 / 10.907	0.136 / 0.050	11.027 / 11.000	10.984 / 10.966	0.061 / 0.016	11.018 / 11.000	10.994 / 10.983	0.035 / 0.006	11.018 / 11.000	11.000 / 10.989	0.029 / 0.000
12	MAX / MIN	12.110 / 12.000	11.905 / 11.795	0.315 / 0.095	12.043 / 12.000	11.950 / 11.907	0.136 / 0.050	12.027 / 12.000	11.984 / 11.966	0.061 / 0.016	12.018 / 12.000	11.994 / 11.983	0.035 / 0.006	12.018 / 12.000	12.000 / 11.989	0.029 / 0.000
13	MAX / MIN	13.110 / 13.000	12.905 / 12.795	0.315 / 0.095	13.043 / 13.000	12.950 / 12.907	0.136 / 0.050	13.027 / 13.000	12.984 / 12.966	0.061 / 0.016	13.018 / 13.000	12.994 / 12.983	0.035 / 0.006	13.018 / 13.000	13.000 / 12.989	0.029 / 0.000
14	MAX / MIN	14.110 / 14.000	13.905 / 13.795	0.315 / 0.095	14.043 / 14.000	13.950 / 13.907	0.136 / 0.050	14.027 / 14.000	13.984 / 13.966	0.061 / 0.016	14.018 / 14.000	13.994 / 13.983	0.035 / 0.006	14.018 / 14.000	14.000 / 13.989	0.029 / 0.000
15	MAX / MIN	15.110 / 15.000	14.905 / 14.795	0.315 / 0.095	15.043 / 15.000	14.950 / 14.907	0.136 / 0.050	15.027 / 15.000	14.984 / 14.966	0.061 / 0.016	15.018 / 15.000	14.994 / 14.983	0.035 / 0.006	15.018 / 15.000	15.000 / 14.989	0.029 / 0.000
16	MAX / MIN	16.110 / 16.000	15.905 / 15.795	0.315 / 0.095	16.043 / 16.000	15.950 / 15.907	0.136 / 0.050	16.027 / 16.000	15.984 / 15.966	0.061 / 0.016	16.018 / 16.000	15.994 / 15.983	0.035 / 0.006	16.018 / 16.000	16.000 / 15.989	0.029 / 0.000
17	MAX / MIN	17.110 / 17.000	16.905 / 16.795	0.315 / 0.095	17.043 / 17.000	16.950 / 16.907	0.136 / 0.050	17.027 / 17.000	16.984 / 16.966	0.061 / 0.016	17.018 / 17.000	16.994 / 16.983	0.035 / 0.006	17.018 / 17.000	17.000 / 16.989	0.029 / 0.000
18	MAX / MIN	18.110 / 18.000	17.905 / 17.795	0.315 / 0.095	18.043 / 18.000	17.950 / 17.907	0.136 / 0.050	18.027 / 18.000	17.984 / 17.966	0.061 / 0.016	18.018 / 18.000	17.994 / 17.983	0.035 / 0.006	18.018 / 18.000	18.000 / 17.989	0.029 / 0.000
19	MAX / MIN	19.130 / 19.000	18.890 / 18.760	0.370 / 0.110	19.052 / 19.000	18.935 / 18.883	0.169 / 0.065	19.033 / 19.000	18.980 / 18.959	0.074 / 0.020	19.021 / 19.000	18.993 / 18.980	0.041 / 0.007	19.021 / 19.000	19.000 / 18.987	0.034 / 0.000
20	MAX / MIN	20.130 / 20.000	19.890 / 19.760	0.370 / 0.110	20.052 / 20.000	19.935 / 19.883	0.169 / 0.065	20.033 / 20.000	19.980 / 19.959	0.074 / 0.020	20.021 / 20.000	19.993 / 19.980	0.041 / 0.007	20.021 / 20.000	20.000 / 19.987	0.034 / 0.000
21	MAX / MIN	21.130 / 21.000	20.890 / 20.760	0.370 / 0.110	21.052 / 21.000	20.935 / 20.883	0.169 / 0.065	21.033 / 21.000	20.980 / 20.959	0.074 / 0.020	21.021 / 21.000	20.993 / 20.980	0.041 / 0.007	21.021 / 21.000	21.000 / 20.987	0.034 / 0.000
22	MAX / MIN	22.130 / 22.000	21.890 / 21.760	0.370 / 0.110	22.052 / 22.000	21.935 / 21.883	0.169 / 0.065	22.033 / 22.000	21.980 / 21.959	0.074 / 0.020	22.021 / 22.000	21.993 / 21.980	0.041 / 0.007	22.021 / 22.000	22.000 / 21.987	0.034 / 0.000
23	MAX / MIN	23.130 / 23.000	22.890 / 22.760	0.370 / 0.110	23.052 / 23.000	22.935 / 22.883	0.169 / 0.065	23.033 / 23.000	22.980 / 22.959	0.074 / 0.020	23.021 / 23.000	22.993 / 22.980	0.041 / 0.007	23.021 / 23.000	23.000 / 22.987	0.034 / 0.000
24	MAX / MIN	24.130 / 24.000	23.890 / 23.760	0.370 / 0.110	24.052 / 24.000	23.935 / 23.883	0.169 / 0.065	24.033 / 24.000	23.980 / 23.959	0.074 / 0.020	24.021 / 24.000	23.993 / 23.980	0.041 / 0.007	24.021 / 24.000	24.000 / 23.987	0.034 / 0.000

DIMENSIONS IN MILLIMETERS

NOTES:
1. Select nominal sizes to preference rating as follows: F = First Choice, S = Second Choice, T = Third Choice.
2. ANSI B4.2 lists limit dimensions for nominal sizes marked "F" (First Choice) only. A cost penalty for material stock tooling and gages is anticipated for sizes marked with "S" (Second Choice) and "T" (Third Choice).

Table 6-6 (Continued). Preferred Hole Basis Clearance Fits (ANSI B4.2)

NOMINAL SIZE		LOOSE RUNNING HOLE H11	LOOSE RUNNING SHAFT c11	DIFF	FREE RUNNING HOLE H9	FREE RUNNING SHAFT d9	DIFF	CLOSE RUNNING HOLE H8	CLOSE RUNNING SHAFT f7	DIFF	SLIDING HOLE H7	SLIDING SHAFT g6	DIFF	LOCATIONAL CLEARANCE HOLE H7	LOCATIONAL CLEARANCE SHAFT h6	DIFF
25 F	MAX	25.130	24.890	0.370	25.052	24.935	0.169	25.033	24.980	0.074	25.021	24.993	0.041	25.021	25.000	0.034
	MIN	25.000	24.760	0.110	25.000	24.883	0.065	25.000	24.959	0.020	25.000	24.980	0.007	25.000	24.987	0.000
26 T	MAX	26.130	25.890	0.370	26.052	25.935	0.169	26.033	25.980	0.074	26.021	25.993	0.041	26.021	26.000	0.034
	MIN	26.000	25.760	0.110	26.000	25.883	0.065	26.000	25.959	0.020	26.000	25.980	0.007	26.000	25.987	0.000
28 S	MAX	28.130	27.890	0.370	28.052	27.935	0.169	28.033	27.980	0.074	28.021	27.993	0.041	28.021	28.000	0.034
	MIN	28.000	27.760	0.110	28.000	27.883	0.065	28.000	27.959	0.020	28.000	27.980	0.007	28.000	27.987	0.000
30 F	MAX	30.130	29.890	0.370	30.052	29.935	0.169	30.033	29.980	0.074	30.021	29.993	0.041	30.021	30.000	0.034
	MIN	30.000	29.760	0.110	30.000	29.883	0.065	30.000	29.959	0.020	30.000	29.980	0.007	30.000	29.987	0.000
32 T	MAX	32.160	31.880	0.440	32.062	31.920	0.204	32.039	31.975	0.089	32.025	31.991	0.050	32.025	32.000	0.041
	MIN	32.000	31.720	0.120	32.000	31.858	0.080	32.000	31.950	0.025	32.000	31.975	0.009	32.000	31.984	0.000
35 S	MAX	35.160	34.880	0.440	35.062	34.920	0.204	35.039	34.975	0.089	35.025	34.991	0.050	35.025	35.000	0.041
	MIN	35.000	34.720	0.120	35.000	34.858	0.080	35.000	34.950	0.025	35.000	34.975	0.009	35.000	34.984	0.000
38 T	MAX	38.160	37.880	0.440	38.062	37.920	0.204	38.039	37.975	0.089	38.025	37.991	0.050	38.025	38.000	0.041
	MIN	38.000	37.720	0.120	38.000	37.858	0.080	38.000	37.950	0.025	38.000	37.975	0.009	38.000	37.984	0.000
40 F	MAX	40.160	39.880	0.440	40.062	39.920	0.204	40.039	39.975	0.089	40.025	39.991	0.050	40.025	40.000	0.041
	MIN	40.000	39.720	0.120	40.000	39.858	0.080	40.000	39.950	0.025	40.000	39.975	0.009	40.000	39.984	0.000
42 T	MAX	42.160	41.870	0.450	42.062	41.920	0.204	42.039	41.975	0.089	42.025	41.991	0.050	42.025	42.000	0.041
	MIN	42.000	41.710	0.130	42.000	41.858	0.080	42.000	41.950	0.025	42.000	41.975	0.009	42.000	41.984	0.000
45 S	MAX	45.160	44.870	0.450	45.062	44.920	0.204	45.039	44.975	0.089	45.025	44.991	0.050	45.025	45.000	0.041
	MIN	45.000	44.710	0.130	45.000	44.858	0.080	45.000	44.950	0.025	45.000	44.975	0.009	45.000	44.984	0.000
48 T	MAX	48.160	47.870	0.450	48.062	47.920	0.204	48.039	47.975	0.089	48.025	47.991	0.050	48.025	48.000	0.041
	MIN	48.000	47.710	0.130	48.000	47.858	0.080	48.000	47.950	0.025	48.000	47.975	0.009	48.000	47.984	0.000
50 F	MAX	50.160	49.870	0.450	50.062	49.920	0.204	50.039	49.975	0.089	50.025	49.991	0.050	50.025	50.000	0.041
	MIN	50.000	49.710	0.130	50.000	49.858	0.080	50.000	49.950	0.025	50.000	49.975	0.009	50.000	49.984	0.000
55 S	MAX	55.190	54.860	0.520	55.074	54.900	0.248	55.046	54.970	0.106	55.030	54.990	0.059	55.030	55.000	0.049
	MIN	55.000	54.670	0.140	55.000	54.826	0.100	55.000	54.940	0.030	55.000	54.971	0.010	55.000	54.981	0.000
60 F	MAX	60.190	59.860	0.520	60.074	59.900	0.248	60.046	59.970	0.106	60.030	59.990	0.059	60.030	60.000	0.049
	MIN	60.000	59.670	0.140	60.000	59.826	0.100	60.000	59.940	0.030	60.000	59.971	0.010	60.000	59.981	0.000
65 T	MAX	65.190	64.860	0.520	65.074	64.900	0.248	65.046	64.970	0.106	65.030	64.990	0.059	65.030	65.000	0.049
	MIN	65.000	64.670	0.140	65.000	64.826	0.100	65.000	64.940	0.030	65.000	64.971	0.010	65.000	64.981	0.000
70 S	MAX	70.190	69.850	0.530	70.074	69.900	0.248	70.046	69.970	0.106	70.030	69.990	0.059	70.030	70.000	0.049
	MIN	70.000	69.660	0.150	70.000	69.826	0.100	70.000	69.940	0.030	70.000	69.971	0.010	70.000	69.981	0.000
75 T	MAX	75.190	74.850	0.530	75.074	74.900	0.248	75.046	74.970	0.106	75.030	74.990	0.059	75.030	75.000	0.049
	MIN	75.000	74.660	0.150	75.000	74.826	0.100	75.000	74.940	0.030	75.000	74.971	0.010	75.000	74.981	0.000
80 F	MAX	80.190	79.850	0.530	80.074	79.900	0.248	80.046	79.970	0.106	80.030	79.990	0.059	80.030	80.000	0.049
	MIN	80.000	79.660	0.150	80.000	79.826	0.100	80.000	79.940	0.030	80.000	79.971	0.010	80.000	79.981	0.000

DIMENSIONS IN MILLIMETERS

NOTES: 1. Select nominal sizes to preference rating as follows: F = First Choice, S = Second Choice, T = Third Choice.
2. ANSI B4.2 lists limit dimensions for nominal sizes marked "F" (First Choice) only. A cost penalty for material stock tooling and gages is anticipated for sizes marked with "S" (Second Choice) and "T" (Third Choice).

THE ISO SYSTEM OF LIMITS AND FITS—TOLERANCES AND DEVIATIONS

Table 6-6 (Continued). Preferred Hole Basis Clearance Fits (ANSI B4.2)

P F D	NOMINAL SIZE		LOOSE RUNNING HOLE H11	SHAFT c11	DIFF	FREE RUNNING HOLE H9	SHAFT d9	DIFF	CLOSE RUNNING HOLE H8	SHAFT f7	DIFF	SLIDING HOLE H7	SHAFT g6	DIFF	LOCATIONAL CLEARANCE HOLE H7	SHAFT h6	DIFF
S	90	MAX MIN	90.220 90.000	89.830 89.610	0.610 0.170	90.087 90.000	89.880 89.793	0.294 0.120	90.054 90.000	89.964 89.929	0.125 0.036	90.035 90.000	89.988 89.966	0.069 0.012	90.035 90.000	90.000 89.978	0.057 0.000
F	100	MAX MIN	100.220 100.000	99.830 99.610	0.610 0.170	100.087 100.000	99.880 99.793	0.294 0.120	100.054 100.000	99.964 99.929	0.125 0.036	100.035 100.000	99.988 99.966	0.069 0.012	100.035 100.000	100.000 99.978	0.057 0.000
S	110	MAX MIN	110.220 110.000	109.820 109.600	0.620 0.180	110.087 110.000	109.880 109.793	0.294 0.120	110.054 110.000	109.964 109.929	0.125 0.036	110.035 110.000	109.988 109.966	0.069 0.012	110.035 110.000	110.000 109.978	0.057 0.000
F	120	MAX MIN	120.220 120.000	119.820 119.600	0.620 0.180	120.087 120.000	119.880 119.793	0.294 0.120	120.054 120.000	119.964 119.929	0.125 0.036	120.035 120.000	119.988 119.966	0.069 0.012	120.035 120.000	120.000 119.978	0.057 0.000
T	130	MAX MIN	130.250 130.000	129.800 129.550	0.700 0.200	130.100 130.000	129.855 129.755	0.345 0.145	130.063 130.000	129.957 129.917	0.146 0.043	130.040 130.000	129.986 129.961	0.079 0.014	130.040 130.000	130.000 129.975	0.065 0.000
S	140	MAX MIN	140.250 140.000	139.800 139.550	0.700 0.200	140.100 140.000	139.855 139.755	0.345 0.145	140.063 140.000	139.957 139.917	0.146 0.043	140.040 140.000	139.986 139.961	0.079 0.014	140.040 140.000	140.000 139.975	0.065 0.000
T	150	MAX MIN	150.250 150.000	149.790 149.540	0.710 0.210	150.100 150.000	149.855 149.755	0.345 0.145	150.063 150.000	149.957 149.917	0.146 0.043	150.040 150.000	149.986 149.961	0.079 0.014	150.040 150.000	150.000 149.975	0.065 0.000
F	160	MAX MIN	160.250 160.000	159.790 159.540	0.710 0.210	160.100 160.000	159.855 159.755	0.345 0.145	160.063 160.000	159.957 159.917	0.146 0.043	160.040 160.000	159.986 159.961	0.079 0.014	160.040 160.000	160.000 159.975	0.065 0.000
T	170	MAX MIN	170.250 170.000	169.770 169.520	0.730 0.230	170.100 170.000	169.855 169.755	0.345 0.145	170.063 170.000	169.957 169.917	0.146 0.043	170.040 170.000	169.986 169.961	0.079 0.014	170.040 170.000	170.000 169.975	0.065 0.000
S	180	MAX MIN	180.250 180.000	179.770 179.520	0.730 0.230	180.100 180.000	179.855 179.755	0.345 0.145	180.063 180.000	179.957 179.917	0.146 0.043	180.040 180.000	179.986 179.961	0.079 0.014	180.040 180.000	180.000 179.975	0.065 0.000
T	190	MAX MIN	190.290 190.000	189.760 189.470	0.820 0.240	190.115 190.000	189.830 189.715	0.400 0.170	190.072 190.000	189.950 189.904	0.168 0.050	190.046 190.000	189.985 189.956	0.090 0.015	190.046 190.000	190.000 189.971	0.075 0.000
F	200	MAX MIN	200.290 200.000	199.760 199.470	0.820 0.240	200.115 200.000	199.830 199.715	0.400 0.170	200.072 200.000	199.950 199.904	0.168 0.050	200.046 200.000	199.985 199.956	0.090 0.015	200.046 200.000	200.000 199.971	0.075 0.000
S	220	MAX MIN	220.290 220.000	219.740 219.450	0.840 0.260	220.115 220.000	219.830 219.715	0.400 0.170	220.072 220.000	219.950 219.904	0.168 0.050	220.046 220.000	219.985 219.956	0.090 0.015	220.046 220.000	220.000 219.971	0.075 0.000
F	250	MAX MIN	250.290 250.000	249.720 249.430	0.860 0.280	250.115 250.000	249.830 249.715	0.400 0.170	250.072 250.000	249.950 249.904	0.168 0.050	250.046 250.000	249.985 249.956	0.090 0.015	250.046 250.000	250.000 249.971	0.075 0.000
S	280	MAX MIN	280.320 280.000	279.700 279.380	0.940 0.300	280.130 280.000	279.810 279.680	0.450 0.190	280.081 280.000	279.944 279.892	0.189 0.056	280.052 280.000	279.983 279.951	0.101 0.017	280.052 280.000	280.000 279.968	0.084 0.000
F	300	MAX MIN	300.320 300.000	299.670 299.350	0.970 0.330	300.130 300.000	299.810 299.680	0.450 0.190	300.081 300.000	299.944 299.892	0.189 0.056	300.052 300.000	299.983 299.951	0.101 0.017	300.052 300.000	300.000 299.968	0.084 0.000
S	350	MAX MIN	350.360 350.000	349.640 349.280	1.080 0.360	350.140 350.000	349.790 349.650	0.490 0.210	350.089 350.000	349.938 349.881	0.208 0.062	350.057 350.000	349.982 349.946	0.111 0.018	350.057 350.000	350.000 349.964	0.093 0.000
F	400	MAX MIN	400.360 400.000	399.600 399.240	1.120 0.400	400.140 400.000	399.790 399.650	0.490 0.210	400.089 400.000	399.938 399.881	0.208 0.062	400.057 400.000	399.982 399.946	0.111 0.018	400.057 400.000	400.000 399.964	0.093 0.000

DIMENSIONS IN MILLIMETERS

NOTES: 1. Select nominal sizes to preference rating as follows: F = First Choice, S̃ = Second Choice, T = Third Choice.
2. ANSI B4.2 lists limit dimensions for nominal sizes marked "F" (First Choice) only. A cost penalty for material stock tooling and gages is anticipated for sizes marked with "S" (Second Choice) and "T" (Third Choice).

THE ISO SYSTEM OF LIMITS AND FITS—TOLERANCES AND DEVIATIONS

Table 6-7. Preferred Hole Basis Transition and Interference Fits (ANSI B4.2)

P F D	NOMINAL SIZE		LOCATIONAL TRANSN HOLE H7	LOCATIONAL TRANSN SHAFT k6	DIFF	LOCATIONAL TRANSN HOLE H7	LOCATIONAL TRANSN SHAFT n6	DIFF	LOCATIONAL INTERF HOLE H7	LOCATIONAL INTERF SHAFT p6	DIFF	MEDIUM DRIVE HOLE H7	MEDIUM DRIVE SHAFT s6	DIFF	FORCE HOLE H7	FORCE SHAFT u6	DIFF
F	1	MAX MIN	1.010 1.000	1.006 1.000	0.010 -0.006	1.010 1.000	1.010 1.004	0.006 -0.010	1.010 1.000	1.012 1.006	0.004 -0.012	1.010 1.000	1.020 1.014	-0.004 -0.020	1.010 1.000	1.024 1.018	-0.008 -0.024
S	1.1	MAX MIN	1.110 1.100	1.106 1.100	0.010 -0.006	1.110 1.100	1.110 1.104	0.006 -0.010	1.110 1.100	1.112 1.106	0.004 -0.012	1.110 1.100	1.120 1.114	-0.004 -0.020	1.110 1.100	1.124 1.118	-0.008 -0.024
F	1.2	MAX MIN	1.210 1.200	1.206 1.200	0.010 -0.006	1.210 1.200	1.210 1.204	0.006 -0.010	1.210 1.200	1.212 1.206	0.004 -0.012	1.210 1.200	1.220 1.214	-0.004 -0.020	1.210 1.200	1.224 1.218	-0.008 -0.024
S	1.4	MAX MIN	1.410 1.400	1.406 1.400	0.010 -0.006	1.410 1.400	1.410 1.404	0.006 -0.010	1.410 1.400	1.412 1.406	0.004 -0.012	1.410 1.400	1.420 1.414	-0.004 -0.020	1.410 1.400	1.424 1.418	-0.008 -0.024
F	1.6	MAX MIN	1.610 1.600	1.606 1.600	0.010 -0.006	1.610 1.600	1.610 1.604	0.006 -0.010	1.610 1.600	1.612 1.606	0.004 -0.012	1.610 1.600	1.620 1.614	-0.004 -0.020	1.610 1.600	1.624 1.618	-0.008 -0.024
S	1.8	MAX MIN	1.810 1.800	1.806 1.800	0.010 -0.006	1.810 1.800	1.810 1.804	0.006 -0.010	1.810 1.800	1.812 1.806	0.004 -0.012	1.810 1.800	1.820 1.814	-0.004 -0.020	1.810 1.800	1.824 1.818	-0.008 -0.024
F	2	MAX MIN	2.010 2.000	2.006 2.000	0.010 -0.006	2.010 2.000	2.010 2.004	0.006 -0.010	2.010 2.000	2.012 2.006	0.004 -0.012	2.010 2.000	2.020 2.014	-0.004 -0.020	2.010 2.000	2.024 2.018	-0.008 -0.024
S	2.2	MAX MIN	2.210 2.200	2.206 2.200	0.010 -0.006	2.210 2.200	2.210 2.204	0.006 -0.010	2.210 2.200	2.212 2.206	0.004 -0.012	2.210 2.200	2.220 2.214	-0.004 -0.020	2.210 2.200	2.224 2.218	-0.008 -0.024
F	2.5	MAX MIN	2.510 2.500	2.506 2.500	0.010 -0.006	2.510 2.500	2.510 2.504	0.006 -0.010	2.510 2.500	2.512 2.506	0.004 -0.012	2.510 2.500	2.520 2.514	-0.004 -0.020	2.510 2.500	2.524 2.518	-0.008 -0.024
S	2.8	MAX MIN	2.810 2.800	2.806 2.800	0.010 -0.006	2.810 2.800	2.810 2.804	0.006 -0.010	2.810 2.800	2.812 2.806	0.004 -0.012	2.810 2.800	2.820 2.814	-0.004 -0.020	2.810 2.800	2.824 2.818	-0.008 -0.024
F	3	MAX MIN	3.010 3.000	3.006 3.000	0.010 -0.006	3.010 3.000	3.010 3.004	0.006 -0.010	3.010 3.000	3.012 3.006	0.004 -0.012	3.010 3.000	3.020 3.014	-0.004 -0.020	3.010 3.000	3.024 3.018	-0.008 -0.024
S	3.5	MAX MIN	3.512 3.500	3.509 3.501	0.011 -0.009	3.512 3.500	3.516 3.508	0.004 -0.016	3.512 3.500	3.520 3.512	0.000 -0.020	3.512 3.500	3.527 3.519	-0.007 -0.027	3.512 3.500	3.531 3.523	-0.011 -0.031
F	4	MAX MIN	4.012 4.000	4.009 4.001	0.011 -0.009	4.012 4.000	4.016 4.008	0.004 -0.016	4.012 4.000	4.020 4.012	0.000 -0.020	4.012 4.000	4.027 4.019	-0.007 -0.027	4.012 4.000	4.031 4.023	-0.011 -0.031
S	4.5	MAX MIN	4.512 4.500	4.509 4.501	0.011 -0.009	4.512 4.500	4.516 4.508	0.004 -0.016	4.512 4.500	4.520 4.512	0.000 -0.020	4.512 4.500	4.527 4.519	-0.007 -0.027	4.512 4.500	4.531 4.523	-0.011 -0.031
F	5	MAX MIN	5.012 5.000	5.009 5.001	0.011 -0.009	5.012 5.000	5.016 5.008	0.004 -0.016	5.012 5.000	5.020 5.012	0.000 -0.020	5.012 5.000	5.027 5.019	-0.007 -0.027	5.012 5.000	5.031 5.023	-0.011 -0.031
S	5.5	MAX MIN	5.512 5.500	5.509 5.501	0.011 -0.009	5.512 5.500	5.516 5.508	0.004 -0.016	5.512 5.500	5.520 5.512	0.000 -0.020	5.512 5.500	5.527 5.519	-0.007 -0.027	5.512 5.500	5.531 5.523	-0.011 -0.031
F	6	MAX MIN	6.012 6.000	6.009 6.001	0.011 -0.009	6.012 6.000	6.016 6.008	0.004 -0.016	6.012 6.000	6.020 6.012	0.000 -0.020	6.012 6.000	6.027 6.019	-0.007 -0.027	6.012 6.000	6.031 6.023	-0.011 -0.031
T	6.5	MAX MIN	6.515 6.500	6.510 6.501	0.014 -0.010	6.515 6.500	6.519 6.510	0.005 -0.019	6.515 6.500	6.524 6.515	0.000 -0.024	6.515 6.500	6.532 6.523	-0.008 -0.032	6.515 6.500	6.537 6.528	-0.013 -0.037

DIMENSIONS IN MILLIMETERS

NOTES: 1. Select nominal sizes to preference rating as follows: F = First Choice, S = Second Choice, T = Third Choice.
2. ANSI B4.2 lists limit dimensions for nominal sizes marked "F" (First Choice) only. A cost penalty for material stock tooling and gages is anticipated for sizes marked with "S" (Second Choice) and "T" (Third Choice).

THE ISO SYSTEM OF LIMITS AND FITS—TOLERANCES AND DEVIATIONS   6-21

Table 6-7 (Continued). Preferred Hole Basis Transition and Interferance Fits (ANSI B4.2)

P F D	NOMINAL SIZE		LOCATIONAL TRANSN HOLE H7	LOCATIONAL TRANSN SHAFT k6	LOCATIONAL TRANSN DIFF	LOCATIONAL TRANSN HOLE H7	LOCATIONAL TRANSN SHAFT n6	LOCATIONAL TRANSN DIFF	LOCATIONAL INTERF HOLE H7	LOCATIONAL INTERF SHAFT p6	LOCATIONAL INTERF DIFF	MEDIUM DRIVE HOLE H7	MEDIUM DRIVE SHAFT s6	MEDIUM DRIVE DIFF	FORCE HOLE H7	FORCE SHAFT u6	FORCE DIFF
S	7	MAX MIN	7.015 7.000	7.010 7.001	0.014 -0.010	7.015 7.000	7.019 7.010	0.005 -0.019	7.015 7.000	7.024 7.015	0.000 -0.024	7.015 7.000	7.032 7.023	-0.008 -0.032	7.015 7.000	7.037 7.028	-0.013 -0.037
F	8	MAX MIN	8.015 8.000	8.010 8.001	0.014 -0.010	8.015 8.000	8.019 8.010	0.005 -0.019	8.015 8.000	8.024 8.015	0.000 -0.024	8.015 8.000	8.032 8.023	-0.008 -0.032	8.015 8.000	8.037 8.028	-0.013 -0.037
S	9	MAX MIN	9.015 9.000	9.010 9.001	0.014 -0.010	9.015 9.000	9.019 9.010	0.005 -0.019	9.015 9.000	9.024 9.015	0.000 -0.024	9.015 9.000	9.032 9.023	-0.008 -0.032	9.015 9.000	9.037 9.028	-0.013 -0.037
F	10	MAX MIN	10.015 10.000	10.010 10.001	0.014 -0.010	10.015 10.000	10.019 10.010	0.005 -0.019	10.015 10.000	10.024 10.015	0.000 -0.024	10.015 10.000	10.032 10.023	-0.008 -0.032	10.015 10.000	10.037 10.028	-0.013 -0.037
S	11	MAX MIN	11.018 11.000	11.012 11.001	0.017 -0.012	11.018 11.000	11.023 11.012	0.006 -0.023	11.018 11.000	11.029 11.018	0.000 -0.029	11.018 11.000	11.039 11.028	-0.010 -0.039	11.018 11.000	11.044 11.033	-0.015 -0.044
F	12	MAX MIN	12.018 12.000	12.012 12.001	0.017 -0.012	12.018 12.000	12.023 12.012	0.006 -0.023	12.018 12.000	12.029 12.018	0.000 -0.029	12.018 12.000	12.039 12.028	-0.010 -0.039	12.018 12.000	12.044 12.033	-0.015 -0.044
T	13	MAX MIN	13.018 13.000	13.012 13.001	0.017 -0.012	13.018 13.000	13.023 13.012	0.006 -0.023	13.018 13.000	13.029 13.018	0.000 -0.029	13.018 13.000	13.039 13.028	-0.010 -0.039	13.018 13.000	13.044 13.033	-0.015 -0.044
S	14	MAX MIN	14.018 14.000	14.012 14.001	0.017 -0.012	14.018 14.000	14.023 14.012	0.006 -0.023	14.018 14.000	14.029 14.018	0.000 -0.029	14.018 14.000	14.039 14.028	-0.010 -0.039	14.018 14.000	14.044 14.033	-0.015 -0.044
T	15	MAX MIN	15.018 15.000	15.012 15.001	0.017 -0.012	15.018 15.000	15.023 15.012	0.006 -0.023	15.018 15.000	15.029 15.018	0.000 -0.029	15.018 15.000	15.039 15.028	-0.010 -0.039	15.018 15.000	15.044 15.033	-0.015 -0.044
F	16	MAX MIN	16.018 16.000	16.012 16.001	0.017 -0.012	16.018 16.000	16.023 16.012	0.006 -0.023	16.018 16.000	16.029 16.018	0.000 -0.029	16.018 16.000	16.039 16.028	-0.010 -0.039	16.018 16.000	16.044 16.033	-0.015 -0.044
T	17	MAX MIN	17.018 17.000	17.012 17.001	0.017 -0.012	17.018 17.000	17.023 17.012	0.006 -0.023	17.018 17.000	17.029 17.018	0.000 -0.029	17.018 17.000	17.039 17.028	-0.010 -0.039	17.018 17.000	17.044 17.033	-0.015 -0.044
S	18	MAX MIN	18.018 18.000	18.012 18.001	0.017 -0.012	18.018 18.000	18.023 18.012	0.006 -0.023	18.018 18.000	18.029 18.018	0.000 -0.029	18.018 18.000	18.039 18.028	-0.010 -0.039	18.018 18.000	18.044 18.033	-0.015 -0.044
T	19	MAX MIN	19.021 19.000	19.015 19.002	0.019 -0.015	19.021 19.000	19.028 19.015	0.006 -0.028	19.021 19.000	19.035 19.022	-0.001 -0.035	19.021 19.000	19.048 19.035	-0.014 -0.048	19.021 19.000	19.054 19.041	-0.020 -0.054
F	20	MAX MIN	20.021 20.000	20.015 20.002	0.019 -0.015	20.021 20.000	20.028 20.015	0.006 -0.028	20.021 20.000	20.035 20.022	-0.001 -0.035	20.021 20.000	20.048 20.035	-0.014 -0.048	20.021 20.000	20.054 20.041	-0.020 -0.054
T	21	MAX MIN	21.021 21.000	21.015 21.002	0.019 -0.015	21.021 21.000	21.028 21.015	0.006 -0.028	21.021 21.000	21.035 21.022	-0.001 -0.035	21.021 21.000	21.048 21.035	-0.014 -0.048	21.021 21.000	21.054 21.041	-0.020 -0.054
S	22	MAX MIN	22.021 22.000	22.015 22.002	0.019 -0.015	22.021 22.000	22.028 22.015	0.006 -0.028	22.021 22.000	22.035 22.022	-0.001 -0.035	22.021 22.000	22.048 22.035	-0.014 -0.048	22.021 22.000	22.054 22.041	-0.020 -0.054
T	23	MAX MIN	23.021 23.000	23.015 23.002	0.019 -0.015	23.021 23.000	23.028 23.015	0.006 -0.028	23.021 23.000	23.035 23.022	-0.001 -0.035	23.021 23.000	23.048 23.035	-0.014 -0.048	23.021 23.000	23.054 23.041	-0.020 -0.054
T	24	MAX MIN	24.021 24.000	24.015 24.002	0.019 -0.015	24.021 24.000	24.028 24.015	0.006 -0.028	24.021 24.000	24.035 24.022	-0.001 -0.035	24.021 24.000	24.048 24.035	-0.014 -0.048	24.021 24.000	24.054 24.041	-0.020 -0.054

DIMENSIONS IN MILLIMETERS

NOTES: 1. Select nominal sizes to preference rating as follows: F = First Choice, S = Second Choice, T = Third Choice.
2. ANSI B4.2 lists limit dimensions for nominal sizes marked "F" (First Choice) only. A cost penalty for material stock tooling and gages is anticipated for sizes marked with "S" (Second Choice) and "T" (Third Choice).

6-22 THE ISO SYSTEM OF LIMITS AND FITS—TOLERANCES AND DEVIATIONS

Table 6-7 (Continued). Preferred Hole Basis Transition and Interference Fits (ANSI B4.2)

P F D	NOMINAL SIZE		LOCATIONAL TRANSN HOLE H7	LOCATIONAL TRANSN SHAFT k6	DIFF	LOCATIONAL TRANSN HOLE H7	LOCATIONAL TRANSN SHAFT n6	DIFF	LOCATIONAL INTERF HOLE H7	LOCATIONAL INTERF SHAFT p6	DIFF	MEDIUM DRIVE HOLE H7	MEDIUM DRIVE SHAFT s6	DIFF	FORCE HOLE H7	FORCE SHAFT u6	DIFF*
F	25	MAX MIN	25.021 25.000	25.015 25.002	0.019 -0.015	25.021 25.000	25.028 25.015	0.006 -0.028	25.021 25.000	25.035 25.022	-0.001 -0.035	25.021 25.000	25.048 25.035	-0.014 -0.048	25.021 25.000	25.061 25.048	-0.027 -0.061
T	26	MAX MIN	26.021 26.000	26.015 26.002	0.019 -0.015	26.021 26.000	26.028 26.015	0.006 -0.028	26.021 26.000	26.035 26.022	-0.001 -0.035	26.021 26.000	26.048 26.035	-0.014 -0.048	26.021 26.000	26.061 26.048	-0.027 -0.061
S	28	MAX MIN	28.021 28.000	28.015 28.002	0.019 -0.015	28.021 28.000	28.028 28.015	0.006 -0.028	28.021 28.000	28.035 28.022	-0.001 -0.035	28.021 28.000	28.048 28.035	-0.014 -0.048	28.021 28.000	28.061 28.048	-0.027 -0.061
F	30	MAX MIN	30.021 30.000	30.015 30.002	0.019 -0.015	30.021 30.000	30.028 30.015	0.006 -0.028	30.021 30.000	30.035 30.022	-0.001 -0.035	30.021 30.000	30.048 30.035	-0.014 -0.048	30.021 30.000	30.061 30.048	-0.027 -0.061
T	32	MAX MIN	32.025 32.000	32.018 32.002	0.023 -0.018	32.025 32.000	32.033 32.017	0.008 -0.033	32.025 32.000	32.042 32.026	-0.001 -0.042	32.025 32.000	32.059 32.043	-0.018 -0.059	32.025 32.000	32.076 32.060	-0.035 -0.076
S	35	MAX MIN	35.025 35.000	35.018 35.002	0.023 -0.018	35.025 35.000	35.033 35.017	0.008 -0.033	35.025 35.000	35.042 35.026	-0.001 -0.042	35.025 35.000	35.059 35.043	-0.018 -0.059	35.025 35.000	35.076 35.060	-0.035 -0.076
T	38	MAX MIN	38.025 38.000	38.018 38.002	0.023 -0.018	38.025 38.000	38.033 38.017	0.008 -0.033	38.025 38.000	38.042 38.026	-0.001 -0.042	38.025 38.000	38.059 38.043	-0.018 -0.059	38.025 38.000	38.076 38.060	-0.035 -0.076
F	40	MAX MIN	40.025 40.000	40.018 40.002	0.023 -0.018	40.025 40.000	40.033 40.017	0.008 -0.033	40.025 40.000	40.042 40.026	-0.001 -0.042	40.025 40.000	40.059 40.043	-0.018 -0.059	40.025 40.000	40.076 40.060	-0.035 -0.076
T	42	MAX MIN	42.025 42.000	42.018 42.002	0.023 -0.018	42.025 42.000	42.033 42.017	0.008 -0.033	42.025 42.000	42.042 42.026	-0.001 -0.042	42.025 42.000	42.059 42.043	-0.018 -0.059	42.025 42.000	42.086 42.070	-0.045 -0.086
S	45	MAX MIN	45.025 45.000	45.018 45.002	0.023 -0.018	45.025 45.000	45.039 45.017	0.008 -0.033	45.025 45.000	45.042 45.026	-0.001 -0.042	45.025 45.000	45.059 45.043	-0.018 -0.059	45.025 45.000	45.086 45.070	-0.045 -0.086
T	48	MAX MIN	48.025 48.000	48.018 48.002	0.023 -0.018	48.025 48.000	48.033 48.017	0.008 -0.033	48.025 48.000	48.042 48.026	-0.001 -0.042	48.025 48.000	48.059 48.043	-0.018 -0.059	48.025 48.000	48.086 48.070	-0.045 -0.086
F	50	MAX MIN	50.025 50.000	50.018 50.002	0.023 -0.018	50.025 50.000	50.033 50.017	0.008 -0.033	50.025 50.000	50.042 50.026	-0.001 -0.042	50.025 50.000	50.059 50.043	-0.018 -0.059	50.025 50.000	50.086 50.070	-0.045 -0.086
S	55	MAX MIN	55.030 55.000	55.021 55.002	0.028 -0.021	55.030 55.000	55.039 55.020	0.010 -0.039	55.030 55.000	55.051 55.032	-0.002 -0.051	55.030 55.000	55.072 55.053	-0.023 -0.072	55.030 55.000	55.106 55.087	-0.057 -0.106
F	60	MAX MIN	60.030 60.000	60.021 60.002	0.028 -0.021	60.030 60.000	60.039 60.020	0.010 -0.039	60.030 60.000	60.051 60.032	-0.002 -0.051	60.030 60.000	60.072 60.053	-0.023 -0.072	60.030 60.000	60.106 60.087	-0.057 -0.106
T	65	MAX MIN	65.030 65.000	65.021 65.002	0.028 -0.021	65.030 65.000	65.039 65.020	0.010 -0.039	65.030 65.000	65.051 65.032	-0.002 -0.051	65.030 65.000	65.072 65.053	-0.023 -0.072	65.030 65.000	65.106 65.087	-0.057 -0.106
S	70	MAX MIN	70.030 70.000	70.021 70.002	0.028 -0.021	70.030 70.000	70.039 70.020	0.010 -0.039	70.030 70.000	70.051 70.032	-0.002 -0.051	70.030 70.000	70.078 70.059	-0.029 -0.078	70.030 70.000	70.121 70.102	-0.072 -0.121
T	75	MAX MIN	75.030 75.000	75.021 75.002	0.028 -0.021	75.030 75.000	75.039 75.020	0.010 -0.039	75.030 75.000	75.051 75.032	-0.002 -0.051	75.030 75.000	75.078 75.059	-0.029 -0.078	75.030 75.000	75.121 75.102	-0.072 -0.121
F	80	MAX MIN	80.030 80.000	80.021 80.002	0.028 -0.021	80.030 80.000	80.039 80.020	0.010 -0.039	80.030 80.000	80.051 80.032	-0.002 -0.051	80.030 80.000	80.078 80.059	-0.029 -0.078	80.030 80.000	80.121 80.102	-0.072 -0.121

DIMENSIONS IN MILLIMETERS

NOTES: 1. Select nominal sizes to preference rating as follows: F = First Choice, S = Second Choice, T = Third Choice.
2. ANSI B4.2 lists limit dimensions for nominal sizes marked "F" (First Choice) only. A cost penalty for material stock tooling and gages is anticipated for sizes marked with "S" (Second Choice) and "T" (Third Choice).

THE ISO SYSTEM OF LIMITS AND FITS—TOLERANCES AND DEVIATIONS

Table 6-7 (Continued). Preferred Hole Basis Transition and Interference Fits (ANSI B4.2)

P F D	NOMINAL SIZE		LOCATIONAL TRANSN HOLE H7	LOCATIONAL TRANSN SHAFT k6	DIFF	LOCATIONAL TRANSN HOLE H7	LOCATIONAL TRANSN SHAFT n6	DIFF	LOCATIONAL INTERF HOLE H7	LOCATIONAL INTERF SHAFT p6	DIFF	MEDIUM DRIVE HOLE H7	MEDIUM DRIVE SHAFT s6	DIFF	FORCE HOLE H7	FORCE SHAFT u6	DIFF
S	90	MAX MIN	90.035 90.000	90.025 90.003	0.032 -0.025	90.035 90.000	90.045 90.023	0.012 -0.045	90.035 90.000	90.059 90.037	-0.002 -0.059	90.035 90.000	90.093 90.071	-0.036 -0.093	90.035 90.000	90.146 90.124	-0.089 -0.146
F	100	MAX MIN	100.035 100.000	100.025 100.003	0.032 -0.025	100.035 100.000	100.045 100.023	0.012 -0.045	100.035 100.000	100.059 100.037	-0.002 -0.059	100.035 100.000	100.093 100.071	-0.036 -0.093	100.035 100.000	100.146 100.124	-0.089 -0.146
S	110	MAX MIN	110.035 110.000	110.025 110.003	0.032 -0.025	110.035 110.000	110.045 110.023	0.012 -0.045	110.035 110.000	110.059 110.037	-0.002 -0.059	110.035 110.000	110.101 110.079	-0.044 -0.101	110.035 110.000	110.166 110.144	-0.109 -0.166
F	120	MAX MIN	120.035 120.000	120.025 120.003	0.032 -0.025	120.035 120.000	120.045 120.023	0.012 -0.045	120.035 120.000	120.059 120.037	-0.002 -0.059	120.035 120.000	120.101 120.079	-0.044 -0.101	120.035 120.000	120.166 120.144	-0.109 -0.166
T	130	MAX MIN	130.040 130.000	130.028 130.003	0.037 -0.028	130.040 130.000	130.052 130.027	0.013 -0.052	130.040 130.000	130.068 130.043	-0.003 -0.068	130.040 130.000	130.117 130.092	-0.052 -0.117	130.040 130.000	130.195 130.170	-0.130 -0.195
S	140	MAX MIN	140.040 140.000	140.028 140.003	0.037 -0.028	140.040 140.000	140.052 140.027	0.013 -0.052	140.040 140.000	140.068 140.043	-0.003 -0.068	140.040 140.000	140.117 140.092	-0.052 -0.117	140.040 140.000	140.195 140.170	-0.130 -0.195
T	150	MAX MIN	150.040 150.000	150.028 150.003	0.037 -0.028	150.040 150.000	150.052 150.027	0.013 -0.052	150.040 150.000	150.068 150.043	-0.003 -0.068	150.040 150.000	150.125 150.100	-0.060 -0.125	150.040 150.000	150.215 150.190	-0.150 -0.215
F	160	MAX MIN	160.040 160.000	160.028 160.003	0.037 -0.028	160.040 160.000	160.052 160.027	0.013 -0.052	160.040 160.000	160.068 160.043	-0.003 -0.068	160.040 160.000	160.125 160.100	-0.060 -0.125	160.040 160.000	160.215 160.190	-0.150 -0.215
T	170	MAX MIN	170.040 170.000	170.028 170.003	0.037 -0.028	170.040 170.000	170.052 170.027	0.013 -0.052	170.040 170.000	170.068 170.043	-0.003 -0.068	170.040 170.000	170.133 170.108	-0.068 -0.133	170.040 170.000	170.235 170.210	-0.170 -0.235
S	180	MAX MIN	180.040 180.000	180.028 180.003	0.037 -0.028	180.040 180.000	180.052 180.027	0.013 -0.052	180.040 180.000	180.068 180.043	-0.003 -0.068	180.040 180.000	180.133 180.108	-0.068 -0.133	180.040 180.000	180.235 180.210	-0.170 -0.235
T	190	MAX MIN	190.046 190.000	190.033 190.004	0.042 -0.033	190.046 190.000	190.060 190.031	0.015 -0.060	190.046 190.000	190.079 190.050	-0.004 -0.079	190.046 190.000	190.151 190.122	-0.076 -0.151	190.046 190.000	190.265 190.236	-0.190 -0.265
F	200	MAX MIN	200.046 200.000	200.033 200.004	0.042 -0.033	200.046 200.000	200.060 200.031	0.015 -0.060	200.046 200.000	200.079 200.050	-0.004 -0.079	200.046 200.000	200.151 200.122	-0.076 -0.151	200.046 200.000	200.265 200.236	-0.190 -0.265
S	220	MAX MIN	220.046 220.000	220.033 220.004	0.042 -0.033	220.046 220.000	220.060 220.031	0.015 -0.060	220.046 220.000	220.079 220.050	-0.004 -0.079	220.046 220.000	220.159 220.130	-0.084 -0.159	220.046 220.000	220.287 220.258	-0.212 -0.287
F	250	MAX MIN	250.046 250.000	250.033 250.004	0.042 -0.033	250.046 250.000	250.060 250.031	0.015 -0.060	250.046 250.000	250.079 250.050	-0.004 -0.079	250.046 250.000	250.169 250.140	-0.094 -0.169	250.046 250.000	250.313 250.284	-0.238 -0.313
S	280	MAX MIN	280.052 280.000	280.036 280.004	0.048 -0.036	280.052 280.000	280.066 280.034	0.018 -0.066	280.052 280.000	280.088 280.056	-0.004 -0.088	280.052 280.000	280.190 280.158	-0.106 -0.190	280.052 280.000	280.347 280.315	-0.263 -0.347
F	300	MAX MIN	300.052 300.000	300.036 300.004	0.048 -0.036	300.052 300.000	300.066 300.034	0.018 -0.066	300.052 300.000	300.088 300.056	-0.004 -0.088	300.052 300.000	300.202 300.170	-0.118 -0.202	300.052 300.000	300.382 300.350	-0.298 -0.382
S	350	MAX MIN	350.057 350.000	350.040 350.004	0.053 -0.040	350.057 350.000	350.073 350.037	0.020 -0.073	350.057 350.000	350.098 350.062	-0.005 -0.098	350.057 350.000	350.226 350.190	-0.133 -0.226	350.057 350.000	350.426 350.390	-0.333 -0.426
F	400	MAX MIN	400.057 400.000	400.040 400.004	0.053 -0.040	400.057 400.000	400.073 400.037	0.020 -0.073	400.057 400.000	400.098 400.062	-0.005 -0.098	400.057 400.000	400.244 400.208	-0.151 -0.244	400.057 400.000	400.471 400.435	-0.378 -0.471

DIMENSIONS IN MILLIMETERS

NOTES: 1. Select nominal sizes to preference rating as follows: F = First Choice, S = Second Choice, T = Third Choice.
2. ANSI B4.2 lists limit dimensions for nominal sizes marked "F" (First Choice) only. A cost penalty for material stock tooling and gages is anticipated for sizes marked with "S" (Second Choice) and "T" (Third Choice).

THE ISO SYSTEM OF LIMITS AND FITS—TOLERANCES AND DEVIATIONS

Table 6-8. Preferred Shaft Basis Clearance Fits (ANSI B4.2)

PFD	NOMINAL SIZE		LOOSE RUNNING HOLE C11	LOOSE RUNNING SHAFT h11	LOOSE RUNNING DIFF	FREE RUNNING HOLE D9	FREE RUNNING SHAFT h9	FREE RUNNING DIFF	CLOSE RUNNING HOLE F8	CLOSE RUNNING SHAFT h7	CLOSE RUNNING DIFF	SLIDING HOLE G7	SLIDING SHAFT h6	SLIDING DIFF	LOCATIONAL CLEARANCE HOLE H7	LOCATIONAL CLEARANCE SHAFT h6	LOCATIONAL CLEARANCE DIFF
F	1	MAX	1.120	1.000	0.180	1.045	1.000	0.070	1.020	1.000	0.030	1.012	1.000	0.018	1.010	1.000	0.016
		MIN	1.060	0.940	0.060	1.020	0.975	0.020	1.006	0.990	0.006	1.002	0.994	0.002	1.000	0.994	0.000
S	1.1	MAX	1.220	1.100	0.180	1.145	1.100	0.070	1.120	1.100	0.030	1.112	1.100	0.018	1.110	1.100	0.016
		MIN	1.160	1.040	0.060	1.120	1.075	0.020	1.106	1.090	0.006	1.102	1.094	0.002	1.100	1.094	0.000
F	1.2	MAX	1.320	1.200	0.180	1.245	1.200	0.070	1.220	1.200	0.030	1.212	1.200	0.018	1.210	1.200	0.016
		MIN	1.260	1.140	0.060	1.220	1.175	0.020	1.206	1.190	0.006	1.202	1.194	0.002	1.200	1.194	0.000
S	1.4	MAX	1.520	1.400	0.180	1.445	1.400	0.070	1.420	1.400	0.030	1.412	1.400	0.018	1.410	1.400	0.016
		MIN	1.460	1.340	0.060	1.420	1.375	0.020	1.406	1.390	0.006	1.402	1.394	0.002	1.400	1.394	0.000
F	1.6	MAX	1.720	1.600	0.180	1.645	1.600	0.070	1.620	1.600	0.030	1.612	1.600	0.018	1.610	1.600	0.016
		MIN	1.660	1.540	0.060	1.620	1.575	0.020	1.606	1.590	0.006	1.602	1.594	0.002	1.600	1.594	0.000
S	1.8	MAX	1.920	1.800	0.180	1.845	1.800	0.070	1.820	1.800	0.030	1.812	1.800	0.018	1.810	1.800	0.016
		MIN	1.860	1.740	0.060	1.820	1.775	0.020	1.806	1.790	0.006	1.802	1.794	0.002	1.800	1.794	0.000
F	2	MAX	2.120	2.000	0.180	2.045	2.000	0.070	2.020	2.000	0.030	2.012	2.000	0.018	2.010	2.000	0.016
		MIN	2.060	1.940	0.060	2.020	1.975	0.020	2.006	1.990	0.006	2.002	1.994	0.002	2.000	1.994	0.000
S	2.2	MAX	2.320	2.200	0.180	2.245	2.200	0.070	2.220	2.200	0.030	2.212	2.200	0.018	2.210	2.200	0.016
		MIN	2.260	2.140	0.060	2.220	2.175	0.020	2.206	2.190	0.006	2.202	2.194	0.002	2.200	2.194	0.000
F	2.5	MAX	2.620	2.500	0.180	2.545	2.500	0.070	2.520	2.500	0.030	2.512	2.500	0.018	2.510	2.500	0.016
		MIN	2.560	2.440	0.060	2.520	2.475	0.020	2.506	2.490	0.006	2.502	2.494	0.002	2.500	2.494	0.000
S	2.8	MAX	2.920	2.800	0.180	2.845	2.800	0.070	2.820	2.800	0.030	2.812	2.800	0.018	2.810	2.800	0.016
		MIN	2.860	2.740	0.060	2.820	2.775	0.020	2.806	2.790	0.006	2.802	2.794	0.002	2.800	2.794	0.000
F	3	MAX	3.120	3.000	0.180	3.045	3.000	0.070	3.020	3.000	0.030	3.012	3.000	0.018	3.010	3.000	0.016
		MIN	3.060	2.940	0.060	3.020	2.975	0.020	3.006	2.990	0.006	3.002	2.994	0.002	3.000	2.994	0.000
S	3.5	MAX	3.645	3.500	0.220	3.560	3.500	0.090	3.528	3.500	0.040	3.516	3.500	0.024	3.512	3.500	0.020
		MIN	3.570	3.425	0.070	3.530	3.470	0.030	3.510	3.488	0.010	3.504	3.492	0.004	3.500	3.492	0.000
F	4	MAX	4.145	4.000	0.220	4.060	4.000	0.090	4.028	4.000	0.040	4.016	4.000	0.024	4.012	4.000	0.020
		MIN	4.070	3.925	0.070	4.030	3.970	0.030	4.010	3.988	0.010	4.004	3.992	0.004	4.000	3.992	0.000
S	4.5	MAX	4.645	4.500	0.220	4.560	4.500	0.090	4.528	4.500	0.040	4.516	4.500	0.024	4.512	4.500	0.020
		MIN	4.570	4.425	0.070	4.530	4.470	0.030	4.510	4.488	0.010	4.504	4.492	0.004	4.500	4.492	0.000
F	5	MAX	5.145	5.000	0.220	5.060	5.000	0.090	5.028	5.000	0.040	5.016	5.000	0.024	5.012	5.000	0.020
		MIN	5.070	4.925	0.070	5.030	4.970	0.030	5.010	4.988	0.010	5.004	4.992	0.004	5.000	4.992	0.000
S	5.5	MAX	5.645	5.500	0.220	5.560	5.500	0.090	5.528	5.500	0.040	5.516	5.500	0.024	5.512	5.500	0.020
		MIN	5.570	5.425	0.070	5.530	5.470	0.030	5.510	5.488	0.010	5.504	5.492	0.004	5.500	5.492	0.000
F	6	MAX	6.145	6.000	0.220	6.060	6.000	0.090	6.028	6.000	0.040	6.016	6.000	0.024	6.012	6.000	0.020
		MIN	6.070	5.925	0.070	6.030	5.970	0.030	6.010	5.988	0.010	6.004	5.992	0.004	6.000	5.992	0.000
T	6.5	MAX	6.670	6.500	0.260	6.576	6.500	0.112	6.535	6.500	0.050	6.520	6.500	0.029	6.515	6.500	0.024
		MIN	6.580	6.410	0.080	6.540	6.464	0.040	6.513	6.485	0.013	6.505	6.491	0.005	6.500	6.491	0.000

DIMENSIONS IN MILLIMETERS

NOTES: 1. Select nominal sizes to preference rating as follows: F = First Choice, S = Second Choice, T = Third Choice.
2. ANSI B4.2 lists limit dimensions for nominal sizes marked "F" (First Choice) only. A cost penalty for material stock tooling and gages is anticipated for sizes marked with "S" (Second Choice) and "T" (Third Choice).

THE ISO SYSTEM OF LIMITS AND FITS—TOLERANCES AND DEVIATIONS 6-25

Tables 6-8 *(Continued).* Preferred Shaft Basis Clearance Fits (ANSI B4.2)

P F D	NOMINAL SIZE		LOOSE RUNNING HOLE C11	LOOSE RUNNING SHAFT h11	DIFF	FREE RUNNING HOLE D9	FREE RUNNING SHAFT h9	DIFF	CLOSE RUNNING HOLE F8	CLOSE RUNNING SHAFT h7	DIFF	SLIDING HOLE G7	SLIDING SHAFT h6	DIFF	LOCATIONAL CLEARANCE HOLE H7	LOCATIONAL CLEARANCE SHAFT h6	DIFF
S	7	MAX MIN	7.170 7.080	7.000 6.910	0.260 0.080	7.076 7.040	7.000 6.964	0.112 0.040	7.035 7.013	7.000 6.985	0.050 0.013	7.020 7.005	7.000 6.991	0.029 0.005	7.015 7.000	7.000 6.991	0.024 0.000
F	8	MAX MIN	8.170 8.080	8.000 7.910	0.260 0.080	8.076 8.040	8.000 7.964	0.112 0.040	8.035 8.013	8.000 7.985	0.050 0.013	8.020 8.005	8.000 7.991	0.029 0.005	8.015 8.000	8.000 7.991	0.024 0.000
S	9	MAX MIN	9.170 9.080	9.000 8.910	0.260 0.080	9.076 9.040	9.000 8.964	0.112 0.040	9.035 9.013	9.000 8.985	0.050 0.013	9.020 9.005	9.000 8.991	0.029 0.005	9.015 9.000	9.000 8.991	0.024 0.000
F	10	MAX MIN	10.170 10.080	10.000 9.910	0.260 0.080	10.076 10.040	10.000 9.964	0.112 0.040	10.035 10.013	10.000 9.985	0.050 0.013	10.020 10.005	10.000 9.991	0.029 0.005	10.015 10.000	10.000 9.991	0.024 0.000
S	11	MAX MIN	11.205 11.095	11.000 10.890	0.315 0.095	11.093 11.050	11.000 10.957	0.136 0.050	11.043 11.016	11.000 10.982	0.061 0.016	11.024 11.006	11.000 10.989	0.035 0.006	11.018 11.000	11.000 10.989	0.029 0.000
F	12	MAX MIN	12.205 12.095	12.000 11.890	0.315 0.095	12.093 12.050	12.000 11.957	0.136 0.050	12.043 12.016	12.000 11.982	0.061 0.016	12.024 12.006	12.000 11.989	0.035 0.006	12.018 12.000	12.000 11.989	0.029 0.000
T	13	MAX MIN	13.205 13.095	13.000 12.890	0.315 0.095	13.093 13.050	13.000 12.957	0.136 0.050	13.043 13.016	13.000 12.982	0.061 0.016	13.024 13.006	13.000 12.989	0.035 0.006	13.018 13.000	13.000 12.989	0.029 0.000
S	14	MAX MIN	14.205 14.095	14.000 13.890	0.315 0.095	14.093 14.050	14.000 13.957	0.136 0.050	14.043 14.016	14.000 13.982	0.061 0.016	14.024 14.006	14.000 13.989	0.035 0.006	14.018 14.000	14.000 13.989	0.029 0.000
T	15	MAX MIN	15.205 15.095	15.000 14.890	0.315 0.095	15.093 15.050	15.000 14.957	0.136 0.050	15.043 15.016	15.000 14.982	0.061 0.016	15.024 15.006	15.000 14.989	0.035 0.006	15.018 15.000	15.000 14.989	0.029 0.000
F	16	MAX MIN	16.205 16.095	16.000 15.890	0.315 0.095	16.093 16.050	16.000 15.957	0.136 0.050	16.043 16.016	16.000 15.982	0.061 0.016	16.024 16.006	16.000 15.989	0.035 0.006	16.018 16.000	16.000 15.989	0.029 0.000
T	17	MAX MIN	17.205 17.095	17.000 16.890	0.315 0.095	17.093 17.050	17.000 16.957	0.136 0.050	17.043 17.016	17.000 16.982	0.061 0.016	17.024 17.006	17.000 16.989	0.035 0.006	17.018 17.000	17.000 16.989	0.029 0.000
S	18	MAX MIN	18.205 18.095	18.000 17.890	0.315 0.095	18.093 18.050	18.000 17.957	0.136 0.050	18.043 18.016	18.000 17.982	0.061 0.016	18.024 18.006	18.000 17.989	0.035 0.006	18.018 18.000	18.000 17.989	0.029 0.000
T	19	MAX MIN	19.240 19.110	19.000 18.870	0.370 0.110	19.117 19.065	19.000 18.948	0.169 0.065	19.053 19.020	19.000 18.979	0.074 0.020	19.028 19.007	19.000 18.987	0.041 0.007	19.021 19.000	19.000 18.987	0.034 0.000
F	20	MAX MIN	20.240 20.110	20.000 19.870	0.370 0.110	20.117 20.065	20.000 19.948	0.169 0.065	20.053 20.020	20.000 19.979	0.074 0.020	20.028 20.007	20.000 19.987	0.041 0.007	20.021 20.000	20.000 19.987	0.034 0.000
T	21	MAX MIN	21.240 21.110	21.000 20.870	0.370 0.110	21.117 21.065	21.000 20.948	0.169 0.065	21.053 21.020	21.000 20.979	0.074 0.020	21.028 21.007	21.000 20.987	0.041 0.007	21.021 21.000	21.000 20.987	0.034 0.000
S	22	MAX MIN	22.240 22.110	22.000 21.870	0.370 0.110	22.117 22.065	22.000 21.948	0.169 0.065	22.053 22.020	22.000 21.979	0.074 0.020	22.028 22.007	22.000 21.987	0.041 0.007	22.021 22.000	22.000 21.987	0.034 0.000
T	23	MAX MIN	23.240 23.110	23.000 22.870	0.370 0.110	23.117 23.065	23.000 22.948	0.169 0.065	23.053 23.020	23.000 22.979	0.074 0.020	23.028 23.007	23.000 22.987	0.041 0.007	23.021 23.000	23.000 22.987	0.034 0.000
T	24	MAX MIN	24.240 24.110	24.000 23.870	0.370 0.110	24.117 24.065	24.000 23.948	0.169 0.065	24.053 24.020	24.000 23.979	0.074 0.020	24.028 24.007	24.000 23.987	0.041 0.007	24.021 24.000	24.000 23.987	0.034 0.000

DIMENSIONS IN MILLIMETERS

NOTES: 1. Select nominal sizes to preference rating as follows: F = First Choice, S = Second Choice, T = Third Choice.
2. ANSI B4.2 lists limit dimensions for nominal sizes marked "F" (First Choice) only. A cost penalty for material stock tooling and gages is anticipated for sizes marked with "S" (Second Choice) and "T" (Third Choice).

6-26 THE ISO SYSTEM OF LIMITS AND FITS—TOLERANCES AND DEVIATIONS

Tables 6-8 (Continued). Preferred Shaft Basis Clearance Fits (ANSI B4.2)

	NOMINAL SIZE		LOOSE RUNNING HOLE C11	SHAFT h11	DIFF	FREE RUNNING HOLE D9	SHAFT h9	DIFF	CLOSE RUNNING HOLE F8	SHAFT h7	DIFF	SLIDING HOLE G7	SHAFT h6	DIFF	LOCATIONAL CLEARANCE HOLE H7	SHAFT h6	DIFF
F	25	MAX MIN	25.240 25.110	25.000 24.870	0.370 0.110	25.117 25.065	25.000 24.948	0.169 0.065	25.053 25.020	25.000 24.979	0.074 0.020	25.028 25.007	25.000 24.987	0.041 0.007	25.021 25.000	25.000 24.987	0.034 0.000
T	26	MAX MIN	26.240 26.110	26.000 25.870	0.370 0.110	26.117 26.065	26.000 25.948	0.169 0.065	26.053 26.020	26.000 25.979	0.074 0.020	26.028 26.007	26.000 25.987	0.041 0.007	26.021 26.000	26.000 25.987	0.034 0.000
S	28	MAX MIN	28.240 28.110	28.000 27.870	0.370 0.110	28.117 28.065	28.000 27.948	0.169 0.065	28.053 28.020	28.000 27.979	0.074 0.020	28.028 28.007	28.000 27.987	0.041 0.007	28.021 28.000	28.000 27.987	0.034 0.000
F	30	MAX MIN	30.240 30.110	30.000 29.870	0.370 0.110	30.117 30.065	30.000 29.948	0.169 0.065	30.053 30.020	30.000 29.979	0.074 0.020	30.028 30.007	30.000 29.987	0.041 0.007	30.021 30.000	30.000 29.987	0.034 0.000
T	32	MAX MIN	32.280 32.120	32.000 31.840	0.440 0.120	32.142 32.080	32.000 31.938	0.204 0.080	32.064 32.025	32.000 31.975	0.089 0.025	32.034 32.009	32.000 31.984	0.050 0.009	32.025 32.000	32.000 31.984	0.041 0.000
S	35	MAX MIN	35.280 35.120	35.000 34.840	0.440 0.120	35.142 35.080	35.000 34.938	0.204 0.080	35.064 35.025	35.000 34.975	0.089 0.025	35.034 35.009	35.000 34.984	0.050 0.009	35.025 35.000	35.000 34.984	0.041 0.000
T	38	MAX MIN	38.280 38.120	38.000 37.840	0.440 0.120	38.142 38.080	38.000 37.938	0.204 0.080	38.064 38.025	38.000 37.975	0.089 0.025	38.034 38.009	38.000 37.984	0.050 0.009	38.025 38.000	38.000 37.984	0.041 0.000
F	40	MAX MIN	40.280 40.120	40.000 39.840	0.440 0.120	40.142 40.080	40.000 39.938	0.204 0.080	40.064 40.025	40.000 39.975	0.089 0.025	40.034 40.009	40.000 39.984	0.050 0.009	40.025 40.000	40.000 39.984	0.041 0.000
T	42	MAX MIN	42.290 42.130	42.000 41.840	0.450 0.130	42.142 42.080	42.000 41.938	0.204 0.080	42.064 42.025	42.000 41.975	0.089 0.025	42.034 42.009	42.000 41.984	0.050 0.009	42.025 42.000	42.000 41.984	0.041 0.000
S	45	MAX MIN	45.290 45.130	45.000 44.840	0.450 0.130	45.142 45.080	45.000 44.938	0.204 0.080	45.064 45.025	45.000 44.975	0.089 0.025	45.034 45.009	45.000 44.984	0.050 0.009	45.025 45.000	45.000 44.984	0.041 0.000
T	48	MAX MIN	48.290 48.130	48.000 47.840	0.450 0.130	48.142 48.080	48.000 47.938	0.204 0.080	48.064 48.025	48.000 47.975	0.089 0.025	48.034 48.009	48.000 47.984	0.050 0.009	48.025 48.000	48.000 47.984	0.041 0.000
F	50	MAX MIN	50.290 50.130	50.000 49.840	0.450 0.130	50.142 50.080	50.000 49.938	0.204 0.080	50.064 50.025	50.000 49.975	0.089 0.025	50.034 50.009	50.000 49.984	0.050 0.009	50.025 50.000	50.000 49.984	0.041 0.000
S	55	MAX MIN	55.330 55.140	55.000 54.810	0.520 0.140	55.174 55.100	55.000 54.926	0.248 0.100	55.076 55.030	55.000 54.970	0.106 0.030	55.040 55.010	55.000 54.981	0.059 0.010	55.030 55.000	55.000 54.981	0.049 0.000
F	60	MAX MIN	60.330 60.140	60.000 59.810	0.520 0.140	60.174 60.100	60.000 59.926	0.248 0.100	60.076 60.030	60.000 59.970	0.106 0.030	60.040 60.010	60.000 59.981	0.059 0.010	60.030 60.000	60.000 59.981	0.049 0.000
T	65	MAX MIN	65.330 65.140	65.000 64.810	0.520 0.140	65.174 65.100	65.000 64.926	0.248 0.100	65.076 65.030	65.000 64.970	0.106 0.030	65.040 65.010	65.000 64.981	0.059 0.010	65.030 65.000	65.000 64.981	0.049 0.000
S	70	MAX MIN	70.340 70.150	70.000 69.810	0.530 0.150	70.174 70.100	70.000 69.926	0.248 0.100	70.076 70.030	70.000 69.970	0.106 0.030	70.040 70.010	70.000 69.981	0.059 0.010	70.030 70.000	70.000 69.981	0.049 0.000
T	75	MAX MIN	75.340 75.150	75.000 74.810	0.530 0.150	75.174 75.100	75.000 74.926	0.248 0.100	75.076 75.030	75.000 74.970	0.106 0.030	75.040 75.010	75.000 74.981	0.059 0.010	75.030 75.000	75.000 74.981	0.049 0.000
F	80	MAX MIN	80.340 80.150	80.000 79.810	0.530 0.150	80.174 80.100	80.000 79.926	0.248 0.100	80.076 80.030	80.000 79.970	0.106 0.030	80.040 80.010	80.000 79.981	0.059 0.010	80.030 80.000	80.000 79.981	0.049 0.000

DIMENSIONS IN MILLIMETERS

NOTES: 1. Select nominal sizes to preference rating as follows: F = First Choice, S = Second Choice, T = Third Choice.
2. ANSI B4.2 lists limit dimensions for nominal sizes marked "F" (First Choice) only. A cost penalty for material stock tooling and gages is anticipated for sizes marked with "S" (Second Choice) and "T" (Third Choice).

THE ISO SYSTEM OF LIMITS AND FITS—TOLERANCES AND DEVIATIONS

Tables 6-8 (Continued). Preferred Shaft Basis Clearance Fits (ANSI B4.2)

NOMINAL SIZE		LOOSE RUNNING HOLE C11	LOOSE RUNNING SHAFT h11	DIFF	FREE RUNNING HOLE D9	FREE RUNNING SHAFT h9	DIFF	CLOSE RUNNING HOLE F8	CLOSE RUNNING SHAFT h7	DIFF	SLIDING HOLE G7	SLIDING SHAFT h6	DIFF	LOCATIONAL CLEARANCE HOLE H7	LOCATIONAL CLEARANCE SHAFT h6	DIFF
90	MAX	90.390	90.000	0.610	90.207	90.000	0.294	90.090	90.000	0.125	90.047	90.000	0.069	90.035	90.000	0.057
S	MIN	90.170	89.780	0.170	90.120	89.913	0.120	90.036	89.965	0.036	90.012	89.978	0.012	90.000	89.978	0.000
100	MAX	100.390	100.000	0.610	100.207	100.000	0.294	100.090	100.000	0.125	100.047	100.000	0.069	100.035	100.000	0.057
F	MIN	100.170	99.780	0.170	100.120	99.913	0.120	100.036	99.965	0.036	100.012	99.978	0.012	100.000	99.978	0.000
110	MAX	110.400	110.000	0.620	110.207	110.000	0.294	110.090	110.000	0.125	110.047	110.000	0.069	110.035	110.000	0.057
S	MIN	110.180	109.780	0.180	110.120	109.913	0.120	110.036	109.965	0.036	110.012	109.978	0.012	110.000	109.978	0.000
120	MAX	120.400	120.000	0.620	120.207	120.000	0.294	120.090	120.000	0.125	120.047	120.000	0.069	120.035	120.000	0.057
F	MIN	120.180	119.780	0.180	120.120	119.913	0.120	120.036	119.965	0.036	120.012	119.978	0.012	120.000	119.978	0.000
130	MAX	130.450	130.000	0.700	130.245	130.000	0.345	130.106	130.000	0.146	130.054	130.000	0.079	130.040	130.000	0.065
T	MIN	130.200	129.750	0.200	130.145	129.900	0.145	130.043	129.960	0.043	130.014	129.975	0.014	130.000	129.975	0.000
140	MAX	140.450	140.000	0.700	140.245	140.000	0.345	140.106	140.000	0.146	140.054	140.000	0.079	140.040	140.000	0.065
S	MIN	140.200	139.750	0.200	140.145	139.900	0.145	140.043	139.960	0.043	140.014	139.975	0.014	140.000	139.975	0.000
150	MAX	150.460	150.000	0.710	150.245	150.000	0.345	150.106	150.000	0.146	150.054	150.000	0.079	150.040	150.000	0.065
T	MIN	150.210	149.750	0.210	150.145	149.900	0.145	150.043	149.960	0.043	150.014	149.975	0.014	150.000	149.975	0.000
160	MAX	160.460	160.000	0.710	160.245	160.000	0.345	160.106	160.000	0.146	160.054	160.000	0.079	160.040	160.000	0.065
F	MIN	160.210	159.750	0.210	160.145	159.900	0.145	160.043	159.960	0.043	160.014	159.975	0.014	160.000	159.975	0.000
170	MAX	170.480	170.000	0.730	170.245	170.000	0.345	170.106	170.000	0.146	170.054	170.000	0.079	170.040	170.000	0.065
T	MIN	170.230	169.750	0.230	170.145	169.900	0.145	170.043	169.960	0.043	170.014	169.975	0.014	170.000	169.975	0.000
180	MAX	180.480	180.000	0.730	180.245	180.000	0.345	180.106	180.000	0.146	180.054	180.000	0.079	180.040	180.000	0.065
S	MIN	180.230	179.750	0.230	180.145	179.900	0.145	180.043	179.960	0.043	180.014	179.975	0.014	180.000	179.975	0.000
190	MAX	190.530	190.000	0.820	190.285	190.000	0.400	190.122	190.000	0.168	190.061	190.000	0.090	190.046	190.000	0.075
T	MIN	190.240	189.710	0.240	190.170	189.885	0.170	190.050	189.954	0.050	190.015	189.971	0.015	190.000	189.971	0.000
200	MAX	200.530	200.000	0.820	200.285	200.000	0.400	200.122	200.000	0.168	200.061	200.000	0.090	200.046	200.000	0.075
F	MIN	200.240	199.710	0.240	200.170	199.885	0.170	200.050	199.954	0.050	200.015	199.971	0.015	200.000	199.971	0.000
220	MAX	220.550	220.000	0.840	220.285	220.000	0.400	220.122	220.000	0.168	220.061	220.000	0.090	220.046	220.000	0.075
S	MIN	220.260	219.710	0.260	220.170	219.885	0.170	220.050	219.954	0.050	220.015	219.971	0.015	220.000	219.971	0.000
250	MAX	250.570	250.000	0.860	250.285	250.000	0.400	250.122	250.000	0.168	250.061	250.000	0.090	250.046	250.000	0.075
F	MIN	250.280	249.710	0.280	250.170	249.885	0.170	250.050	249.954	0.050	250.015	249.971	0.015	250.000	249.971	0.000
280	MAX	280.620	280.000	0.940	280.320	280.000	0.450	280.137	280.000	0.189	280.069	280.000	0.101	280.052	280.000	0.084
S	MIN	280.300	279.680	0.300	280.190	279.870	0.190	280.056	279.948	0.056	280.017	279.968	0.017	280.000	279.968	0.000
300	MAX	300.650	300.000	0.970	300.320	300.000	0.450	300.137	300.000	0.189	300.069	300.000	0.101	300.052	300.000	0.084
F	MIN	300.330	299.680	0.330	300.190	299.870	0.190	300.056	299.948	0.056	300.017	299.968	0.017	300.000	299.968	0.000
350	MAX	350.720	350.000	1.080	350.350	350.000	0.490	350.151	350.000	0.208	350.075	350.000	0.111	350.057	350.000	0.093
S	MIN	350.360	349.640	0.360	350.210	349.860	0.210	350.062	349.943	0.062	350.018	349.964	0.018	350.000	349.964	0.000
400	MAX	400.760	400.000	1.120	400.350	400.000	0.490	400.151	400.000	0.208	400.075	400.000	0.111	400.057	400.000	0.093
F	MIN	400.400	399.640	0.400	400.210	399.860	0.210	400.062	399.943	0.062	400.018	399.964	0.018	400.000	399.964	0.000

DIMENSIONS IN MILLIMETERS

NOTES:
1. Select nominal sizes to preference rating as follows: F = First Choice, S = Second Choice, T = Third Choice.
2. ANSI B4.2 lists limit dimensions for nominal sizes marked "F" (First Choice) only. A cost penalty for material stock tooling and gages is anticipated for sizes marked with "S" (Second Choice) and "T" (Third Choice).

6-28 THE ISO SYSTEM OF LIMITS AND FITS—TOLERANCES AND DEVIATIONS

Table 6-9. Preferred Shaft Basis Transition and Interference Fits (ANSI B4.2)

P F D	NOMINAL SIZE		LOCATIONAL TRANSN HOLE K7	SHAFT h6	DIFF	LOCATIONAL TRANSN HOLE N7	SHAFT h6	DIFF	LOCATIONAL INTERF HOLE P7	SHAFT h6	DIFF	MEDIUM DRIVE HOLE S7	SHAFT h6	DIFF	HOLE U7	FORCE SHAFT h6	DIFF
F	1	MAX MIN	1.000 0.990	1.000 0.994	0.006 -0.010	0.996 0.986	1.000 0.994	0.002 -0.014	0.994 0.984	1.000 0.994	0.000 -0.016	0.986 0.976	1.000 0.994	-0.008 -0.024	0.982 0.972	1.000 0.994	-0.012 -0.028
S	1.1	MAX MIN	1.100 1.090	1.100 1.094	0.006 -0.010	1.096 1.086	1.100 1.094	0.002 -0.014	1.094 1.084	1.100 1.094	0.000 -0.016	1.086 1.076	1.100 1.094	-0.008 -0.024	1.082 1.072	1.100 1.094	-0.012 -0.028
F	1.2	MAX MIN	1.200 1.190	1.200 1.194	0.006 -0.010	1.196 1.186	1.200 1.194	0.002 -0.014	1.194 1.184	1.200 1.194	0.000 -0.016	1.186 1.176	1.200 1.194	-0.008 -0.024	1.182 1.172	1.200 1.194	-0.012 -0.028
S	1.4	MAX MIN	1.400 1.390	1.400 1.394	0.006 -0.010	1.396 1.386	1.400 1.394	0.002 -0.014	1.394 1.384	1.400 1.394	0.000 -0.016	1.386 1.376	1.400 1.394	-0.008 -0.024	1.382 1.372	1.400 1.394	-0.012 -0.028
F	1.6	MAX MIN	1.600 1.590	1.600 1.594	0.006 -0.010	1.596 1.586	1.600 1.594	0.002 -0.014	1.594 1.584	1.600 1.594	0.000 -0.016	1.586 1.576	1.600 1.594	-0.008 -0.024	1.582 1.572	1.600 1.594	-0.012 -0.028
S	1.8	MAX MIN	1.800 1.790	1.800 1.794	0.006 -0.010	1.796 1.786	1.800 1.794	0.002 -0.014	1.794 1.784	1.800 1.794	0.000 -0.016	1.786 1.776	1.800 1.794	-0.008 -0.024	1.782 1.772	1.800 1.794	-0.012 -0.028
F	2	MAX MIN	2.000 1.990	2.000 1.994	0.006 -0.010	1.996 1.986	2.000 1.994	0.002 -0.014	1.994 1.984	2.000 1.994	0.000 -0.016	1.986 1.976	2.000 1.994	-0.008 -0.024	1.982 1.972	2.000 1.994	-0.012 -0.028
S	2.2	MAX MIN	2.200 2.190	2.200 2.194	0.006 -0.010	2.196 2.186	2.200 2.194	0.002 -0.014	2.194 2.184	2.200 2.194	0.000 -0.016	2.186 2.176	2.200 2.194	-0.008 -0.024	2.182 2.172	2.200 2.194	-0.012 -0.028
F	2.5	MAX MIN	2.500 2.490	2.500 2.494	0.006 -0.010	2.496 2.486	2.500 2.494	0.002 -0.014	2.494 2.484	2.500 2.494	0.000 -0.016	2.486 2.476	2.500 2.494	-0.008 -0.024	2.482 2.472	2.500 2.494	-0.012 -0.028
S	2.8	MAX MIN	2.800 2.790	2.800 2.794	0.006 -0.010	2.796 2.786	2.800 2.794	0.002 -0.014	2.794 2.784	2.800 2.794	0.000 -0.016	2.786 2.776	2.800 2.794	-0.008 -0.024	2.782 2.772	2.800 2.794	-0.012 -0.028
F	3	MAX MIN	3.000 2.990	3.000 2.994	0.006 -0.010	2.996 2.986	3.000 2.994	0.002 -0.014	2.994 2.984	3.000 2.994	0.000 -0.016	2.986 2.976	3.000 2.994	-0.008 -0.024	2.982 2.972	3.000 2.994	-0.012 -0.028
S	3.5	MAX MIN	3.503 3.491	3.500 3.492	0.011 -0.009	3.496 3.484	3.500 3.492	0.004 -0.016	3.492 3.480	3.500 3.492	0.000 -0.020	3.485 3.473	3.500 3.492	-0.007 -0.027	3.481 3.469	3.500 3.492	-0.011 -0.031
F	4	MAX MIN	4.003 3.991	4.000 3.992	0.011 -0.009	3.996 3.984	4.000 3.992	0.004 -0.016	3.992 3.980	4.000 3.992	0.000 -0.020	3.985 3.973	4.000 3.992	-0.007 -0.027	3.981 3.969	4.000 3.992	-0.011 -0.031
S	4.5	MAX MIN	4.503 4.491	4.500 4.492	0.011 -0.009	4.496 4.484	4.500 4.492	0.004 -0.016	4.492 4.480	4.500 4.492	0.000 -0.020	4.485 4.473	4.500 4.492	-0.007 -0.027	4.481 4.469	4.500 4.492	-0.011 -0.031
F	5	MAX MIN	5.003 4.991	5.000 4.992	0.011 -0.009	4.996 4.984	5.000 4.992	0.004 -0.016	4.992 4.980	5.000 4.992	0.000 -0.020	4.985 4.973	5.000 4.992	-0.007 -0.027	4.981 4.969	5.000 4.992	-0.011 -0.031
S	5.5	MAX MIN	5.503 5.491	5.500 5.492	0.011 -0.009	5.496 5.484	5.500 5.492	0.004 -0.016	5.492 5.480	5.500 5.492	0.000 -0.020	5.485 5.473	5.500 5.492	-0.007 -0.027	5.481 5.469	5.500 5.492	-0.011 -0.031
F	6	MAX MIN	6.003 5.991	6.000 5.992	0.011 -0.009	5.996 5.984	6.000 5.992	0.004 -0.016	5.992 5.980	6.000 5.992	0.000 -0.020	5.985 5.973	6.000 5.992	-0.007 -0.027	5.981 5.969	6.000 5.992	-0.011 -0.031
T	6.5	MAX MIN	6.505 6.490	6.500 6.491	0.014 -0.010	6.496 6.481	6.500 6.491	0.005 -0.019	6.491 6.476	6.500 6.491	0.000 -0.024	6.483 6.468	6.500 6.491	-0.008 -0.032	6.478 6.463	6.500 6.491	-0.013 -0.037

DIMENSIONS IN MILLIMETERS

NOTES: 1. Select nominal sizes to preference rating as follows: F = First Choice, S = Second Choice, T = Third Choice.
2. ANSI B4.2 lists limit dimensions for nominal sizes marked "F" (First Choice) only. A cost penalty for material stock tooling and gages is anticipated for sizes marked with "S" (Second Choice) and "T" (Third Choice).

THE ISO SYSTEM OF LIMITS AND FITS—TOLERANCES AND DEVIATIONS

Table 6-9 *(Continued)*. **Preferred Shaft Basis Transition and Interferance Fits (ANSI B4.2)**

P F D	NOMINAL SIZE		LOCATIONAL TRANSN HOLE K7	LOCATIONAL TRANSN SHAFT h6	LOCATIONAL TRANSN DIFF	LOCATIONAL TRANSN HOLE N7	LOCATIONAL TRANSN SHAFT h6	LOCATIONAL TRANSN DIFF	LOCATIONAL INTERF HOLE P7	LOCATIONAL INTERF SHAFT h6	LOCATIONAL INTERF DIFF	MEDIUM DRIVE HOLE S7	MEDIUM DRIVE SHAFT h6	MEDIUM DRIVE DIFF	FORCE HOLE U7	FORCE SHAFT h6	FORCE DIFF
S	7	MAX MIN	7.005 6.990	7.000 6.991	0.014 -0.010	6.996 6.981	7.000 6.991	0.005 -0.019	6.991 6.976	7.000 6.991	0.000 -0.024	6.983 6.968	7.000 6.991	-0.008 -0.032	6.978 6.963	7.000 6.991	-0.013 -0.037
F	8	MAX MIN	8.005 7.990	8.000 7.991	0.014 -0.010	7.996 7.981	8.000 7.991	0.005 -0.019	7.991 7.976	8.000 7.991	0.000 -0.024	7.983 7.968	8.000 7.991	-0.008 -0.032	7.978 7.963	8.000 7.991	-0.013 -0.037
S	9	MAX MIN	9.005 8.990	9.000 8.991	0.014 -0.010	8.996 8.981	9.000 8.991	0.005 -0.019	8.991 8.976	9.000 8.991	0.000 -0.024	8.983 8.968	9.000 8.991	-0.008 -0.032	8.978 8.963	9.000 8.991	-0.013 -0.037
F	10	MAX MIN	10.005 9.990	10.000 9.991	0.014 -0.010	9.996 9.981	10.000 9.991	0.005 -0.019	9.991 9.976	10.000 9.991	0.000 -0.024	9.983 9.968	10.000 9.991	-0.008 -0.032	9.978 9.963	10.000 9.991	-0.013 -0.037
S	11	MAX MIN	11.006 10.988	11.000 10.989	0.017 -0.012	10.995 10.977	11.000 10.989	0.006 -0.023	10.989 10.971	11.000 10.989	0.000 -0.029	10.979 10.961	11.000 10.989	-0.010 -0.039	10.974 10.956	11.000 10.989	-0.015 -0.044
F	12	MAX MIN	12.006 11.988	12.000 11.989	0.017 -0.012	11.995 11.977	12.000 11.989	0.006 -0.023	11.989 11.971	12.000 11.989	0.000 -0.029	11.979 11.961	12.000 11.989	-0.010 -0.039	11.974 11.956	12.000 11.989	-0.015 -0.044
T	13	MAX MIN	13.006 12.988	13.000 12.989	0.017 -0.012	12.995 12.977	13.000 12.989	0.006 -0.023	12.989 12.971	13.000 12.989	0.000 -0.029	12.979 12.961	13.000 12.989	-0.010 -0.039	12.974 12.956	13.000 12.989	-0.015 -0.044
S	14	MAX MIN	14.006 13.988	14.000 13.989	0.017 -0.012	13.995 13.977	14.000 13.989	0.006 -0.023	13.989 13.971	14.000 13.989	0.000 -0.029	13.979 13.961	14.000 13.989	-0.010 -0.039	13.974 13.956	14.000 13.989	-0.015 -0.044
T	15	MAX MIN	15.006 14.988	15.000 14.989	0.017 -0.012	14.995 14.977	15.000 14.989	0.006 -0.023	14.989 14.971	15.000 14.989	0.000 -0.029	14.979 14.961	15.000 14.989	-0.010 -0.039	14.974 14.956	15.000 14.989	-0.015 -0.044
F	16	MAX MIN	16.006 15.988	16.000 15.989	0.017 -0.012	15.995 15.977	16.000 15.989	0.006 -0.023	15.989 15.971	16.000 15.989	0.000 -0.029	15.979 15.961	16.000 15.989	-0.010 -0.039	15.974 15.956	16.000 15.989	-0.015 -0.044
T	17	MAX MIN	17.006 16.988	17.000 16.989	0.017 -0.012	16.995 16.977	17.000 16.989	0.006 -0.023	16.989 16.971	17.000 16.989	0.000 -0.029	16.979 16.961	17.000 16.989	-0.010 -0.039	16.974 16.956	17.000 16.989	-0.015 -0.044
S	18	MAX MIN	18.006 17.988	18.000 17.989	0.017 -0.012	17.995 17.977	18.000 17.989	0.006 -0.023	17.989 17.971	18.000 17.989	0.000 -0.029	17.979 17.961	18.000 17.989	-0.010 -0.039	17.974 17.956	18.000 17.989	-0.015 -0.044
T	19	MAX MIN	19.006 18.985	19.000 18.987	0.019 -0.015	18.993 18.972	19.000 18.987	0.006 -0.028	18.986 18.965	19.000 18.987	-0.001 -0.035	18.973 18.952	19.000 18.987	-0.014 -0.048	18.967 18.946	19.000 18.987	-0.020 -0.054
F	20	MAX MIN	20.006 19.985	20.000 19.987	0.019 -0.015	19.993 19.972	20.000 19.987	0.006 -0.028	19.986 19.965	20.000 19.987	-0.001 -0.035	19.973 19.952	20.000 19.987	-0.014 -0.048	19.967 19.946	20.000 19.987	-0.020 -0.054
T	21	MAX MIN	21.006 20.985	21.000 20.987	0.019 -0.015	20.993 20.972	21.000 20.987	0.006 -0.028	20.986 20.965	21.000 20.987	-0.001 -0.035	20.973 20.952	21.000 20.987	-0.014 -0.048	20.967 20.946	21.000 20.987	-0.020 -0.054
S	22	MAX MIN	22.006 21.985	22.000 21.987	0.019 -0.015	21.993 21.972	22.000 21.987	0.006 -0.028	21.986 21.965	22.000 21.987	-0.001 -0.035	21.973 21.952	22.000 21.987	-0.014 -0.048	21.967 21.946	22.000 21.987	-0.020 -0.054
T	23	MAX MIN	23.006 22.985	23.000 22.987	0.019 -0.015	22.993 22.972	23.000 22.987	0.006 -0.028	22.986 22.965	23.000 22.987	-0.001 -0.035	22.973 22.952	23.000 22.987	-0.014 -0.048	22.967 22.946	23.000 22.987	-0.020 -0.054
T	24	MAX MIN	24.006 23.985	24.000 23.987	0.019 -0.015	23.993 23.972	24.000 23.987	0.006 -0.028	23.986 23.965	24.000 23.987	-0.001 -0.035	23.973 23.952	24.000 23.987	-0.014 -0.048	23.967 23.946	24.000 23.987	-0.020 -0.054

DIMENSIONS IN MILLIMETERS

NOTES: 1. Select nominal sizes to preference rating as follows: F = First Choice, S = Second Choice, T = Third Choice.
2. ANSI B4.2 lists limit dimensions for nominal sizes marked "F" (First Choice) only. A cost penalty for material stock tooling and gages is anticipated for sizes marked with "S" (Second Choice) and "T" (Third Choice).

6-30 THE ISO SYSTEM OF LIMITS AND FITS—TOLERANCES AND DEVIATIONS

Table 6-9 (Continued). Preferred Shaft Basis Transition and Interference Fits (ANSI B4.2)

NOMINAL SIZE		LOCATIONAL TRANSN HOLE K7	SHAFT h6	DIFF *	LOCATIONAL TRANSN HOLE N7	SHAFT h6	DIFF *	LOCATIONAL INTERF HOLE P7	SHAFT h6	DIFF *	MEDIUM DRIVE HOLE S7	SHAFT h6	DIFF *	FORCE HOLE U7	SHAFT h6	DIFF **
F 25	MAX MIN	25.006 24.985	25.000 24.987	0.019 -0.015	24.993 24.972	25.000 24.987	0.006 -0.028	24.986 24.965	25.000 24.987	-0.001 -0.035	24.973 24.952	25.000 24.987	-0.014 -0.048	24.960 24.939	25.000 24.987	-0.027 -0.061
T 26	MAX MIN	26.006 25.985	26.000 25.987	0.019 -0.015	25.993 25.972	26.000 25.987	0.006 -0.028	25.986 25.965	26.000 25.987	-0.001 -0.035	25.973 25.952	26.000 25.987	-0.014 -0.048	25.960 25.939	26.000 25.987	-0.027 -0.061
S 28	MAX MIN	28.006 27.985	28.000 27.987	0.019 -0.015	27.993 27.972	28.000 27.987	0.006 -0.028	27.986 27.965	28.000 27.987	-0.001 -0.035	27.973 27.952	28.000 27.987	-0.014 -0.048	27.960 27.939	28.000 27.987	-0.027 -0.061
F 30	MAX MIN	30.006 29.985	30.000 29.987	0.019 -0.015	29.993 29.972	30.000 29.987	0.006 -0.028	29.986 29.965	30.000 29.987	-0.001 -0.035	29.973 29.952	30.000 29.987	-0.014 -0.048	29.960 29.939	30.000 29.987	-0.027 -0.061
T 32	MAX MIN	32.007 31.982	32.000 31.984	0.023 -0.018	31.992 31.967	32.000 31.984	0.008 -0.033	31.983 31.958	32.000 31.984	-0.001 -0.042	31.966 31.941	32.000 31.984	-0.018 -0.059	31.949 31.924	32.000 31.984	-0.035 -0.076
S 35	MAX MIN	35.007 34.982	35.000 34.984	0.023 -0.018	34.992 34.967	35.000 34.984	0.008 -0.033	34.983 34.958	35.000 34.984	-0.001 -0.042	34.966 34.941	35.000 34.984	-0.018 -0.059	34.949 34.924	35.000 34.984	-0.035 -0.076
T 38	MAX MIN	38.007 37.982	38.000 37.984	0.023 -0.018	37.992 37.967	38.000 37.984	0.008 -0.033	37.983 37.958	38.000 37.984	-0.001 -0.042	37.966 37.941	38.000 37.984	-0.018 -0.059	37.949 37.924	38.000 37.984	-0.035 -0.076
F 40	MAX MIN	40.007 39.982	40.000 39.984	0.023 -0.018	39.992 39.967	40.000 39.984	0.008 -0.033	39.983 39.958	40.000 39.984	-0.001 -0.042	39.966 39.941	40.000 39.984	-0.018 -0.059	39.949 39.924	40.000 39.984	-0.035 -0.076
T 42	MAX MIN	42.007 41.982	42.000 41.984	0.023 -0.018	41.992 41.967	42.000 41.984	0.008 -0.033	41.983 41.958	42.000 41.984	-0.001 -0.042	41.966 41.941	42.000 41.984	-0.018 -0.059	41.939 41.914	42.000 41.984	-0.045 -0.086
S 45	MAX MIN	45.007 44.982	45.000 44.984	0.023 -0.018	44.992 44.967	45.000 44.984	0.008 -0.033	44.983 44.958	45.000 44.984	-0.001 -0.042	44.966 44.941	45.000 44.984	-0.018 -0.059	44.939 44.914	45.000 44.984	-0.045 -0.086
T 48	MAX MIN	48.007 47.982	48.000 47.984	0.023 -0.018	47.992 47.967	48.000 47.984	0.008 -0.033	47.983 47.958	48.000 47.984	-0.001 -0.042	47.966 47.941	48.000 47.984	-0.018 -0.059	47.939 47.914	48.000 47.984	-0.045 -0.086
F 50	MAX MIN	50.007 49.982	50.000 49.984	0.023 -0.018	49.992 49.967	50.000 49.984	0.008 -0.033	49.983 49.958	50.000 49.984	-0.001 -0.042	49.966 49.941	50.000 49.984	-0.018 -0.059	49.939 49.914	50.000 49.984	-0.045 -0.086
S 55	MAX MIN	55.009 54.979	55.000 54.981	0.028 -0.021	54.991 54.961	55.000 54.981	0.010 -0.039	54.979 54.949	55.000 54.981	-0.002 -0.051	54.958 54.928	55.000 54.981	-0.023 -0.072	54.924 54.894	55.000 54.981	-0.057 -0.106
F 60	MAX MIN	60.009 59.979	60.000 59.981	0.028 -0.021	59.991 59.961	60.000 59.981	0.010 -0.039	59.979 59.949	60.000 59.981	-0.002 -0.051	59.958 59.928	60.000 59.981	-0.023 -0.072	59.924 59.894	60.000 59.981	-0.057 -0.106
T 65	MAX MIN	65.009 64.979	65.000 64.981	0.028 -0.021	64.991 64.961	65.000 64.981	0.010 -0.039	64.979 64.949	65.000 64.981	-0.002 -0.051	64.958 64.928	65.000 64.981	-0.023 -0.072	64.924 64.894	65.000 64.981	-0.057 -0.106
S 70	MAX MIN	70.009 69.979	70.000 69.981	0.028 -0.021	69.991 69.961	70.000 69.981	0.010 -0.039	69.979 69.949	70.000 69.981	-0.002 -0.051	69.952 69.922	70.000 69.981	-0.029 -0.078	69.909 69.879	70.000 69.981	-0.072 -0.121
T 75	MAX MIN	75.009 74.979	75.000 74.981	0.028 -0.021	74.991 74.961	75.000 74.981	0.010 -0.039	74.979 74.949	75.000 74.981	-0.002 -0.051	74.952 74.922	75.000 74.981	-0.029 -0.078	74.909 74.879	75.000 74.981	-0.072 -0.121
F 80	MAX MIN	80.009 79.979	80.000 79.981	0.028 -0.021	79.991 79.961	80.000 79.981	0.010 -0.039	79.979 79.949	80.000 79.981	-0.002 -0.051	79.952 79.922	80.000 79.981	-0.029 -0.078	79.909 79.879	80.000 79.981	-0.072 -0.121

DIMENSIONS IN MILLIMETERS

NOTES:
1. Select nominal sizes to preference rating as follows: F = First Choice, S = Second Choice, T = Third Choice.
2. ANSI B4.2 lists limit dimensions for nominal sizes marked "F" (First Choice) only. A cost penalty for material stock tooling and gages is anticipated for sizes marked with "S" (Second Choice) and "T" (Third Choice).

THE ISO SYSTEM OF LIMITS AND FITS—TOLERANCES AND DEVIATIONS 6-31

Table 6-9 (Continued). Preferred Shaft Basis Transition and Interference Fits (ANSI B4.2)

P F D	NOMINAL SIZE		LOCATIONAL TRANSN HOLE K7	SHAFT h6	DIFF	LOCATIONAL TRANSN HOLE N7	SHAFT h6	DIFF	LOCATIONAL INTERF HOLE P7	SHAFT h6	DIFF	MEDIUM DRIVE HOLE S7	SHAFT h6	DIFF	FORCE HOLE U7	SHAFT h6	DIFF
S	90	MAX MIN	90.010 89.975	90.000 89.978	0.032 -0.025	89.990 89.955	90.000 89.978	0.012 -0.045	89.976 89.941	90.000 89.978	-0.002 -0.059	89.942 89.907	90.000 89.978	-0.036 -0.093	89.889 89.854	90.000 89.978	-0.089 -0.146
F	100	MAX MIN	100.010 99.975	100.000 99.978	0.032 -0.025	99.990 99.955	100.000 99.978	0.012 -0.045	99.976 99.941	100.000 99.978	-0.002 -0.059	99.942 99.907	100.000 99.978	-0.036 -0.093	99.889 99.854	100.000 99.978	-0.089 -0.146
S	110	MAX MIN	110.010 109.975	110.000 109.978	0.032 -0.025	109.990 109.955	110.000 109.978	0.012 -0.045	109.976 109.941	110.000 109.978	-0.002 -0.059	109.934 109.899	110.000 109.978	-0.044 -0.101	109.869 109.834	110.000 109.978	-0.109 -0.166
F	120	MAX MIN	120.010 119.975	120.000 119.978	0.032 -0.025	119.990 119.955	120.000 119.978	0.012 -0.045	119.976 119.941	120.000 119.978	-0.002 -0.059	119.934 119.899	120.000 119.978	-0.044 -0.101	119.869 119.834	120.000 119.978	-0.109 -0.166
T	130	MAX MIN	130.012 129.972	130.000 129.975	0.037 -0.028	129.988 129.948	130.000 129.975	0.013 -0.052	129.972 129.932	130.000 129.975	-0.003 -0.068	129.923 129.883	130.000 129.975	-0.052 -0.117	129.845 129.805	130.000 129.975	-0.130 -0.195
S	140	MAX MIN	140.012 139.972	140.000 139.975	0.037 -0.028	139.988 139.948	140.000 139.975	0.013 -0.052	139.972 139.932	140.000 139.975	-0.003 -0.068	139.923 139.883	140.000 139.975	-0.052 -0.117	139.845 139.805	140.000 139.975	-0.130 -0.195
T	150	MAX MIN	150.012 149.972	150.000 149.975	0.037 -0.028	149.988 149.948	150.000 149.975	0.013 -0.052	149.972 149.932	150.000 149.975	-0.003 -0.068	149.915 149.875	150.000 149.975	-0.060 -0.125	149.825 149.785	150.000 149.975	-0.150 -0.215
F	160	MAX MIN	160.012 159.972	160.000 159.975	0.037 -0.028	159.988 159.948	160.000 159.975	0.013 -0.052	159.972 159.932	160.000 159.975	-0.003 -0.068	159.915 159.875	160.000 159.975	-0.060 -0.125	159.825 159.785	160.000 159.975	-0.150 -0.215
T	170	MAX MIN	170.012 169.972	170.000 169.975	0.037 -0.028	169.988 169.948	170.000 169.975	0.013 -0.052	169.972 169.932	170.000 169.975	-0.003 -0.068	169.907 169.867	170.000 169.975	-0.068 -0.133	169.805 169.765	170.000 169.975	-0.170 -0.235
S	180	MAX MIN	180.012 179.972	180.000 179.975	0.037 -0.028	179.988 179.948	180.000 179.975	0.013 -0.052	179.972 179.932	180.000 179.975	-0.003 -0.068	179.907 179.867	180.000 179.975	-0.068 -0.133	179.805 179.765	180.000 179.975	-0.170 -0.235
T	190	MAX MIN	190.013 189.967	190.000 189.971	0.042 -0.033	189.986 189.940	190.000 189.971	0.015 -0.060	189.967 189.921	190.000 189.971	-0.004 -0.079	189.895 189.849	190.000 189.971	-0.076 -0.151	189.781 189.735	190.000 189.971	-0.190 -0.265
F	200	MAX MIN	200.013 199.967	200.000 199.971	0.042 -0.033	199.986 199.940	200.000 199.971	0.015 -0.060	199.967 199.921	200.000 199.971	-0.004 -0.079	199.895 199.849	200.000 199.971	-0.076 -0.151	199.781 199.735	200.000 199.971	-0.190 -0.265
S	220	MAX MIN	220.013 219.967	220.000 219.971	0.042 -0.033	219.986 219.940	220.000 219.971	0.015 -0.060	219.967 219.921	220.000 219.971	-0.004 -0.079	219.887 219.841	220.000 219.971	-0.084 -0.159	219.759 219.713	220.000 219.971	-0.212 -0.287
F	250	MAX MIN	250.013 249.967	250.000 249.971	0.042 -0.033	249.986 249.940	250.000 249.971	0.015 -0.060	249.967 249.921	250.000 249.971	-0.004 -0.079	249.877 249.831	250.000 249.971	-0.094 -0.169	249.733 249.687	250.000 249.971	-0.238 -0.313
S	280	MAX MIN	280.016 279.964	280.000 279.968	0.048 -0.036	279.986 279.934	280.000 279.968	0.018 -0.066	279.964 279.912	280.000 279.968	-0.004 -0.088	279.862 279.810	280.000 279.968	-0.106 -0.190	279.705 279.653	280.000 279.968	-0.263 -0.347
F	300	MAX MIN	300.016 299.964	300.000 299.968	0.048 -0.036	299.986 299.934	300.000 299.968	0.018 -0.066	299.964 299.912	300.000 299.968	-0.004 -0.088	299.850 299.798	300.000 299.968	-0.118 -0.202	299.670 299.618	300.000 299.968	-0.298 -0.382
S	350	MAX MIN	350.017 349.960	350.000 349.964	0.053 -0.040	349.984 349.927	350.000 349.964	0.020 -0.073	349.959 349.902	350.000 349.964	-0.005 -0.098	349.831 349.774	350.000 349.964	-0.133 -0.226	349.631 349.574	350.000 349.964	-0.333 -0.426
F	400	MAX MIN	400.017 399.960	400.000 399.964	0.053 -0.040	399.984 399.927	400.000 399.964	0.020 -0.073	399.959 399.902	400.000 399.964	-0.005 -0.098	399.813 399.756	400.000 399.964	-0.151 -0.244	399.586 399.529	400.000 399.964	-0.378 -0.471

DIMENSIONS IN MILLIMETERS

NOTES:
1. Select nominal sizes to preference rating as follows: F = First Choice, S = Second Choice, T = Third Choice. Nominal sizes marked "F" (First Choice) only. A cost penalty for material stock tooling and gages is anticipated for sizes marked with "S" (Second Choice) and "T" (Third Choice).
2. ANSI B4.2 lists limit dimensions for nominal sizes marked "F" (First Choice) only.

this case there is no fundamental deviation and the tolerance zone (of whatever magnitude) is disposed equally around the zero line.

The values for the fundamental deviations are given in Tables 6-2 and 6-3.

Designation

The complete designation of the limits of tolerance for a shaft or hole requires the use of the appropriate letter to indicate the fundamental deviation, followed by a suffix number denoting the tolerance grade. For this latter purpose the numerical part of the tolerance grade designation is used.

For example, a hole tolerance with deviation "H" and tolerance grade IT 7 is designated as "H7."

Similarly, a shaft tolerance with deviation "p" and tolerance grade IT 6 is designated "p6."

The limits of size for a feature are defined by the basic size of the feature, say, 45 mm, followed by the appropriate tolerance designation, for example, 45 H7 or 45 p6.

A fit is indicated by combining the basic size common to both features with the designations appropriate to each. The designation of the hole limits should always be quoted first.

Example: 45 H7–p6 or 45 H7/p6.

Practical Application of the ISO System of Limits and Fits

The information presented in the previous section will suffice to calculate all specified fits in the ISO system.

The ISO 286 has the upper and lower deviations for the tolerances marked with a + in Table 6-4 and in millimeters and inches tabulated for dimensions up to 500 mm. The table is a good indication of the various standard tolerance (number) and fundamental deviation (letter) combinations used throughout the world.

Tolerance systems now in use can, in most cases, be converted to the ISO system. Bilateral and unilateral (hole basis or shaft basis) tolerancing is covered in the ISO standard, and practical examples below will show how this is done. In Table 6-4 fundamental deviations marked "js" and "Js" represent the bilateral tolerance system and h (shaft basis) and H (hole basis) unilateral tolerancing.

The ISO system provides a great many hole and shaft tolerances, so as to cater to a wide range of conditions. However, experience shows that the majority of fit conditions required for normal engineering products can be provided by a rather limited selection of tolerances. It is important to keep the number of preferred fit combinations to a minimum in order to achieve the greatest savings in tooling, gages, fixtures, etc.

Calculation of Limits*

The upper and lower deviation for a nominal size can readily be derived from the tables of standard tolerances (Table 6-1) and fundamental deviations (Tables 6-2 and 6-3).

Examples—Shafts (dimension in mm)

Fundamental deviations a through h

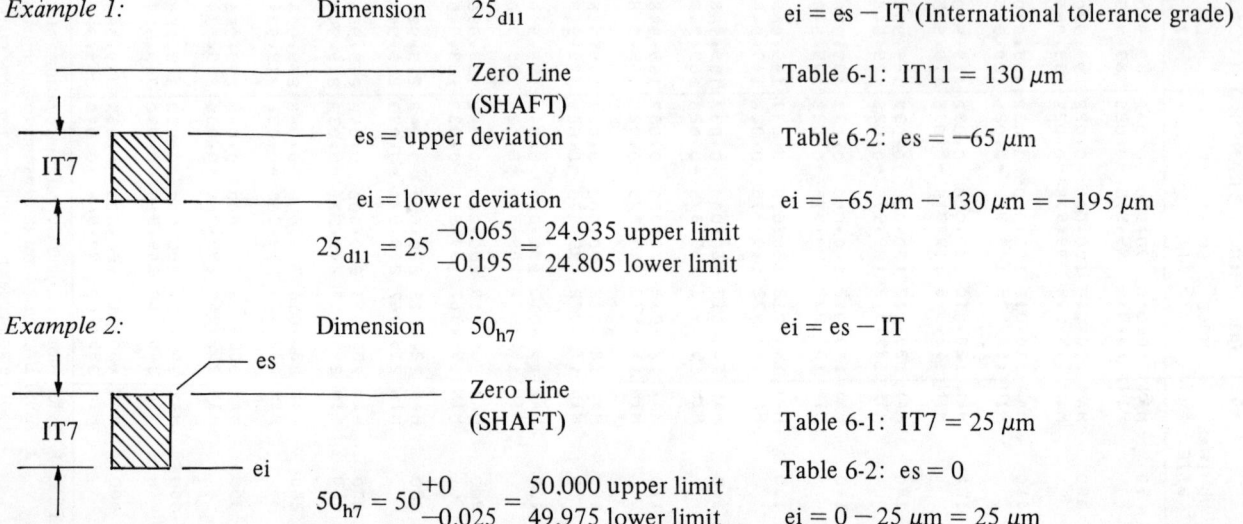

Example 1: Dimension 25_{d11}

$ei = es - IT$ (International tolerance grade)

Table 6-1: $IT11 = 130 \ \mu m$

Table 6-2: $es = -65 \ \mu m$

$ei = -65 \ \mu m - 130 \ \mu m = -195 \ \mu m$

$25_{d11} = 25 \begin{array}{c} -0.065 \\ -0.195 \end{array} = \begin{array}{l} 24.935 \ \text{upper limit} \\ 24.805 \ \text{lower limit} \end{array}$

Example 2: Dimension 50_{h7}

$ei = es - IT$

Table 6-1: $IT7 = 25 \ \mu m$

Table 6-2: $es = 0$

$ei = 0 - 25 \ \mu m = 25 \ \mu m$

$50_{h7} = 50 \begin{array}{c} +0 \\ -0.025 \end{array} = \begin{array}{l} 50.000 \ \text{upper limit} \\ 49.975 \ \text{lower limit} \end{array}$

*NOTE: Numerical values for deviations from basic sizes up to 500 mm for most tolerance zones used are shown in Tables 6-10 through 6-33.

THE ISO SYSTEM OF LIMITS AND FITS—TOLERANCES AND DEVIATIONS

Fundamental deviations js

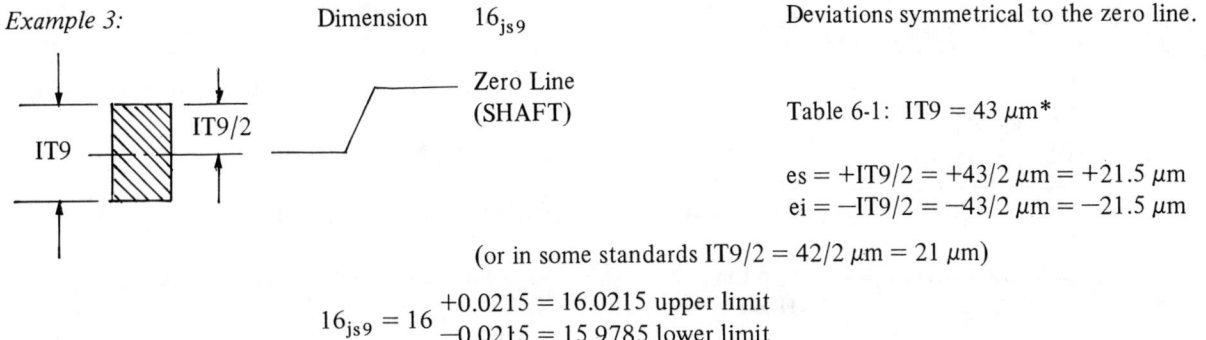

Example 3:

Dimension 16_{js9}

Deviations symmetrical to the zero line.

Table 6-1: IT9 = 43 μm*

es = +IT9/2 = +43/2 μm = +21.5 μm
ei = −IT9/2 = −43/2 μm = −21.5 μm

(or in some standards IT9/2 = 42/2 μm = 21 μm)

$$16_{js9} = 16 \begin{matrix} +0.0215 \\ -0.0215 \end{matrix} = \begin{matrix} 16.0215 \text{ upper limit} \\ 15.9785 \text{ lower limit} \end{matrix}$$

Fundamental deviations j through zc

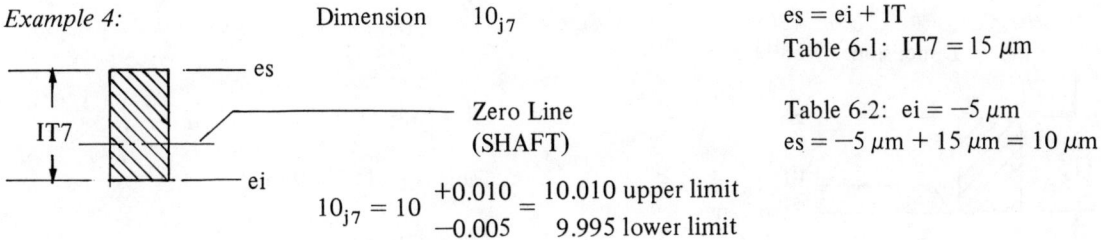

Example 4:

Dimension 10_{j7}

es = ei + IT
Table 6-1: IT7 = 15 μm

Table 6-2: ei = −5 μm
es = −5 μm + 15 μm = 10 μm

$$10_{j7} = 10 \begin{matrix} +0.010 \\ -0.005 \end{matrix} = \begin{matrix} 10.010 \text{ upper limit} \\ 9.995 \text{ lower limit} \end{matrix}$$

NOTE: For js in the particular grades 7 to 11, the two symmetrical deviations $\pm \frac{IT}{2}$ may possibly be rounded in national standards, if the IT value, in micrometers, is an odd value, by replacing it by the even value immediately below.

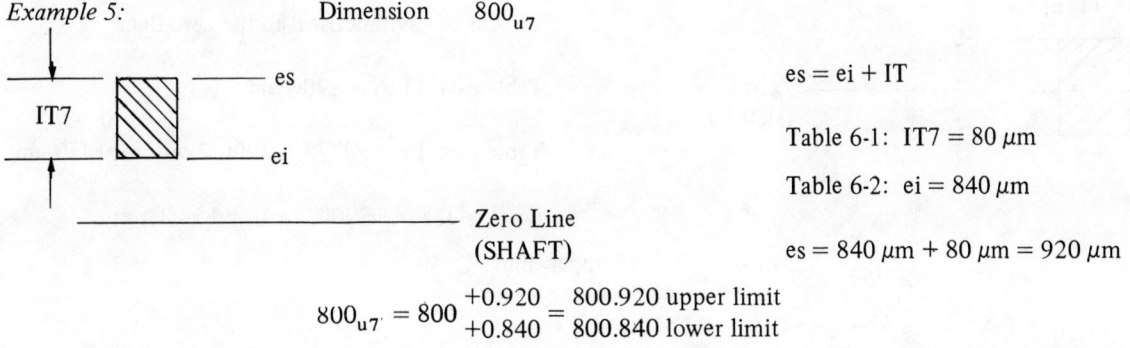

Example 5:

Dimension 800_{u7}

es = ei + IT

Table 6-1: IT7 = 80 μm

Table 6-2: ei = 840 μm

es = 840 μm + 80 μm = 920 μm

$$800_{u7} = 800 \begin{matrix} +0.920 \\ +0.840 \end{matrix} = \begin{matrix} 800.920 \text{ upper limit} \\ 800.840 \text{ lower limit} \end{matrix}$$

THE ISO SYSTEM OF LIMITS AND FITS—TOLERANCES AND DEVIATIONS

Example-Holes (dimensions in mm)

Fundamental deviations A through H

Example 6:   Dimension 63^{C9}

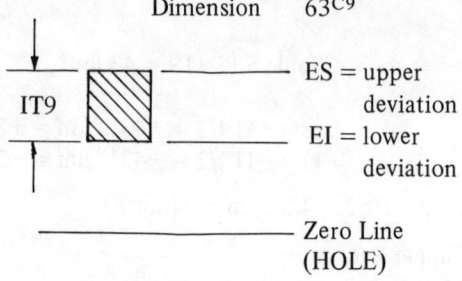

ES = upper deviation
EI = lower deviation

Zero Line (HOLE)

ES = EI + IT (International tolerance grade)

Table 6-1: IT9 = 74 μm

Table 6-3: EI = 140 μm

ES = 140 μm + 74 μm = 214 μm

$63^{C9} = 63 \begin{matrix} +0.214 \\ +0.140 \end{matrix} = \begin{matrix} 63.214 \text{ upper limit} \\ 63.140 \text{ lower limit} \end{matrix}$

Example 7:   Dimension 28^{H5}

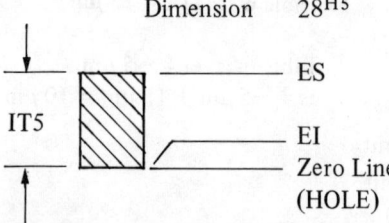

ES
EI
Zero Line (HOLE)

ES = EI + IT

Table 6-1: IT5 = 9 μm

Table 6-3: EI = 0

ES = 0 μm + 9 μm = 9 μm

$28^{H5} = 28 \begin{matrix} +0.009 \\ -0 \end{matrix} = \begin{matrix} 28.009 \text{ upper limit} \\ 28.000 \text{ lower limit} \end{matrix}$

Fundamental deviation JS

Example 8:   Dimension 250^{JS16}

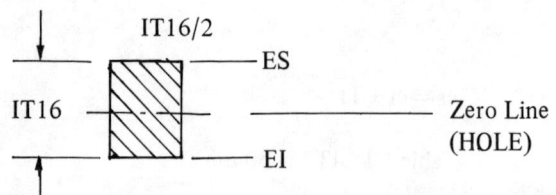

ES
Zero Line (HOLE)
EI

Dimensions symmetrical to the zero line.

Table 6-1: IT16 = 2900 μm

Table 6-3: EI = $-$IT/2 = 2900/2 μm = $-$1450 μm

ES = +IT/2 = +2900/2 μm = +1450 μm

$250^{JS16} = 250 \begin{matrix} +1.45 \\ -1.45 \end{matrix} = \begin{matrix} 251.45 \text{ upper limit} \\ 248.55 \text{ lower limit} \end{matrix}$

THE ISO SYSTEM OF LIMITS AND FITS—TOLERANCES AND DEVIATIONS

Fundamental deviations J through ZC

Example 9: Dimension 40^{J7}

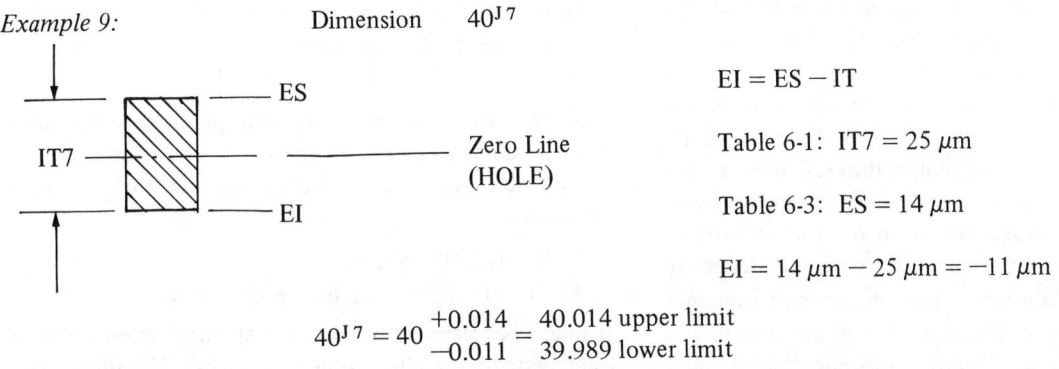

$EI = ES - IT$

Table 6-1: $IT7 = 25\ \mu m$

Table 6-3: $ES = 14\ \mu m$

$EI = 14\ \mu m - 25\ \mu m = -11\ \mu m$

$$40^{J7} = 40 \begin{array}{l} +0.014 \\ -0.011 \end{array} = \begin{array}{l} 40.014 \text{ upper limit} \\ 39.989 \text{ lower limit} \end{array}$$

Example 10: Dimension 125^{Z7}

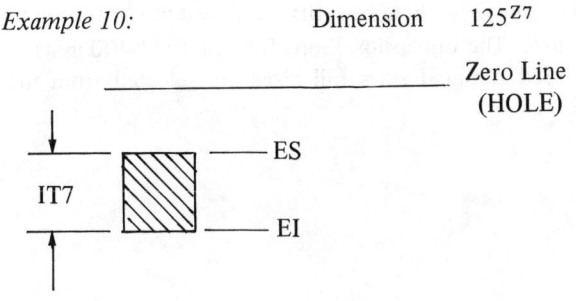

$EI = ES - IT$

Table 6-1: $IT7 = 40\ \mu m$

Table 6-3: $ES = -365\ \mu m + \Delta = -365\ \mu m + 15\ \mu m$

$EI = -350\ \mu m - 40\ \mu m = -390\ \mu m$

$$125^{Z7} = 125 \begin{array}{l} -0.35 \\ -0.39 \end{array} = \begin{array}{l} 124.65 \text{ upper limit} \\ 124.61 \text{ lower limit} \end{array}$$

Bilateral Tolerance System

The ISO system of limits and fits has a full range of bilateral (two-sided) tolerances designated js and JS. The two symmetrical deviations are $\pm \frac{IT}{2}$ (one half of the standard tolerance, see Table 6-1). The odd standard tolerances for IT grades 7 through 11 are rounded in some national standards to the even value immediately below. (See previous Example 3.)

Unilateral Tolerance System

The ISO system of limits and fits, with all its possible combinations, include two unilateral (one-sided) tolerancing methods that are in common use. One tolerancing practice is based on a nominal hole (H, hole basis) the other on a nominal shaft (h, shaft basis). The ISO system has been in use in Germany for nearly 50 years, and standards for selected fits have influenced the European market for metric standard material sizes, measuring tools, couplings, collars, bearings, etc.

Hole-basis or Shaft-basis Fits

The nominal H, hole-basis fit and h, shaft-basis fit tolerancing systems are both used, depending on each specific application. The hole-basis system is used with stepped shaft designs. Standard gages for checking the hole-basis fits cost less than those required for checking shaft-basis fits. In designs where a uniform-diameter shaft is used it is advantageous to employ the shaft-basis system. For example, in the case of driving shafts a single shaft may have to accommodate a variety of accessories—such as couplings, bearings, collars, etc. Steel products toleranced to the shaft-basis system are supplied in a number of steel grades and finishes throughout Europe. Both types of fits might be used on the same design.

Selected Fits

There are three types of fits in the national standards:
1. Clearance fits with the fundamental deviation ranging from a through h for hole-basis fits and from A through H

for shaft-basis fits; 2. Transition fits from js through n for hole-basis fits and JS through N for shaft-basis fits; and 3. Interference fits from p through zc for hole-basis fits and P through ZC for shaft-basis fits. The following recommended fundamental deviations can be used as a rough guide when choosing a hole- or shaft-basis fit: g or G require well-lined-up bearings; F E D or f e d is recommended for normal operating temperatures; C B A or c b a is used for fast rotating shafts with high bearing temperatures; and H J or h j is recommended for a shaft without continuous operation or applications where shafts make less than one complete turn. The selections of tolerance zones shown framed in Fig. 6-5 and 6-6 are specified in the ISO 1829 standard. Whenever possible, the tolerance zones should be chosen from the selected ISO symbols shown in Fig. 6-5 and 6-6. Those enclosed in the circles are first choices.

The ten preferred fits described in Table 6-5 are specified in the national standards as follows:

U.S.A. ANSI B4.2
Japan JIS B0401 (all except 2)
Germany DIN 7157 (most fits)
U.K. BS 4500 (most fits)
Australia AS 1654 (most fits)

Each of the ten hole base fits corresponds to a shaft base fit with equal clearances for the same nominal size.

The ten hole base selected fits use the following tolerances:

Hole: H7, H8, H9, H11
Shaft: c11, d9, f7, g6, h6, k6, n6, p6, s6, u6

It might be desirable for a user to standardize on three, or even two, of the above hole tolerances. The shaft tolerances might also be reduced to fit the requirements of certain types of products.

The above ten hole base fits are shown in illustration in Fig. 6-7. The dimension limits for a range (1-400 mm) of preferred nominal sizes (all sizes are selected from the

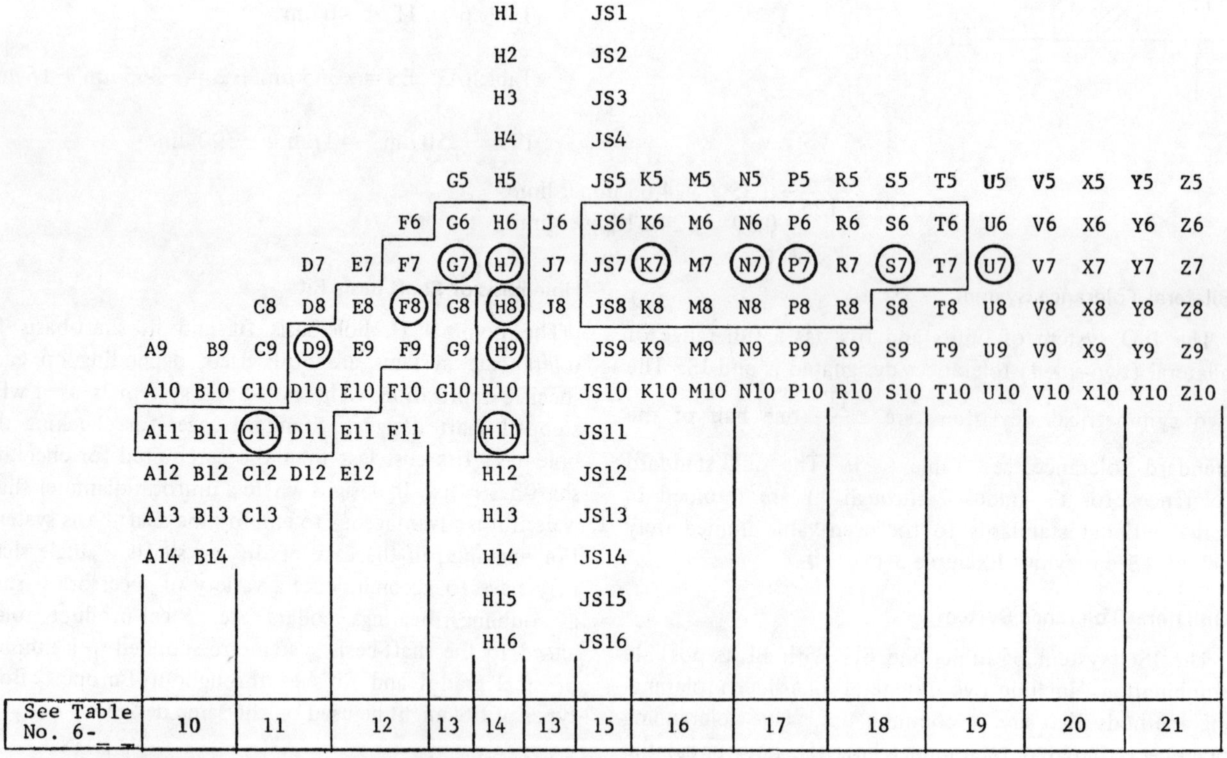

Legend: First choice tolerance zones encircled (ANSI B4.2 preferred)
Second choice tolerance zones framed (ISO 1829 selected)
Third choice tolerance zones open

Fig. 6-5. Tolerance zones for internal dimensions (holes) (ANSI B4.2).

THE ISO SYSTEM OF LIMITS AND FITS—TOLERANCES AND DEVIATIONS

ANSI B32.4 standard for preferred diameters of round metal products) are computed and shown for both holes and shafts with clearances or interferences (−) in Tables 6-6 and 6-7.

The ten shaft base selected fits use the following tolerances:

Hole: C11, D9, F8, G7, H7, K7, N7, P7, S7, U7
Shaft: h6, h7, h9, h11

Note that many steel products shown in Section 10 are produced worldwide to the shaft tolerances shown above. An illustration of the ten shaft base fits is shown in Fig. 6-8, and the dimension limits for a range (1-400 mm) of preferred nominal sizes (all sizes are selected from the ANSI B32.4 standard for preferred diameters of round metal products) are shown for holes, shafts, and clearances or interferences (−) in Tables 6-8 and 6-9.

Deviations from basic sizes for all tolerance zones in Fig. 6-5 and 6-6 are tabulated in size range from 1 to 500 mm in Tables 6-10 through 6-33. The specific table number where the deviations are tabulated for a given tolerance zone is referenced on the lower line in both figures.

The preference rating is more selective here compared with the ANSI standard to help rationalize inventory of material stock, cutting tools, and gages. The Renard series of preferred numbers provided the basis for the author's selection of preference ratings such as: F = first choice (R10 series), S = second choice (R20 series), T = third choice (R40 series). See Table 4-5.

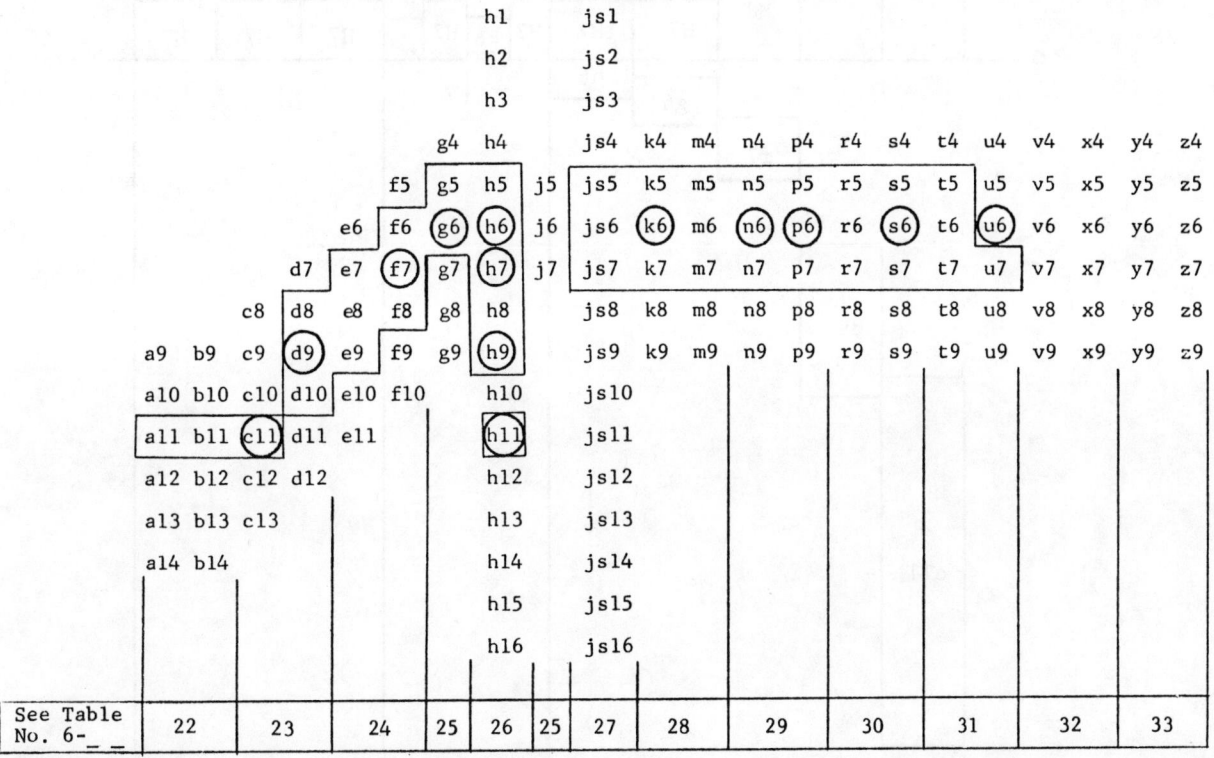

Legend: First choice tolerance zones encircled (ANSI B4.2 preferred)
Second choice tolerance zones framed (ISO 1829 selected)
Third choice tolerance zones open

Fig. 6-6. Tolerance zones for external dimensions (shafts) (ANSI B4.2).

6-38   THE ISO SYSTEM OF LIMITS AND FITS—TOLERANCES AND DEVIATIONS

Conversion of Fits

(1) General Rule

$$EI = -es \quad \text{for A to H}$$
$$ES = -ei \quad \text{for J to ZC}$$

This rule is applicable to all deviations, except
(a) those to which the special rule given below applies,
(b) holes N for grades 9 to 16, above 3 mm (or 0.12 in.), for which the fundamental deviation ES = 0.

Conversions: clearance fit   C10/f10 = F10/c10
                hole-basis fit   H11/c11 = C11/h11
                interference fit   P9/t9   = T9/p9

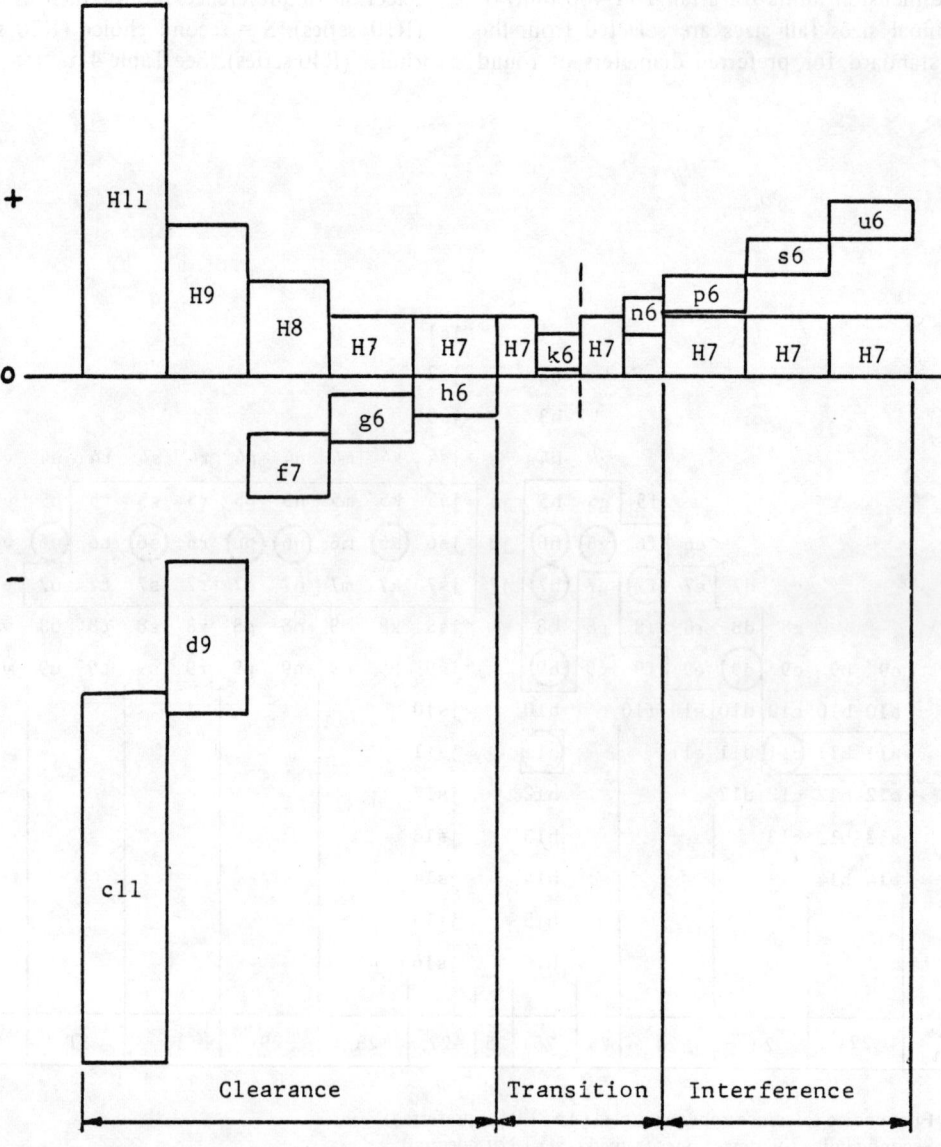

Fig. 6-7. Preferred hole base fits (to scale for 25 mm diameter) (ANSI B4.2).

THE ISO SYSTEM OF LIMITS AND FITS—TOLERANCES AND DEVIATIONS 6-39

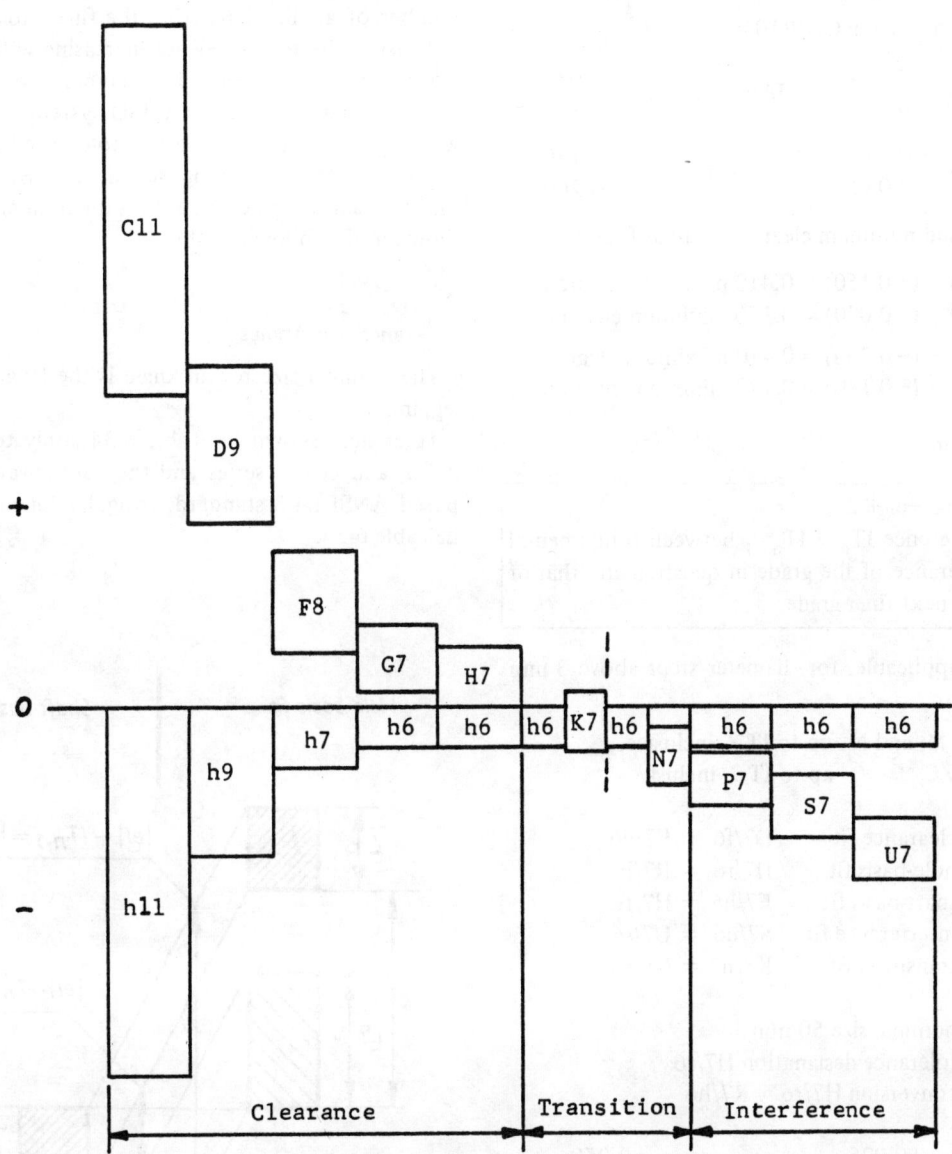

Fig. 6-8. Preferred shaft base fits (to scale for 25 mm diameter) (ANSI B4.2).

Example: Nominal size 60 mm

Tolerance designation C10/f10

Conversion C10/f10 = F10/c10

HOLE C10: $60 \begin{array}{c} +0.260 \\ +0.140 \end{array}$ SHAFT f10: $60 \begin{array}{c} -0.030 \\ -0.150 \end{array}$

F10: $60 \begin{array}{c} +0.150 \\ +0.030 \end{array}$ c10: $60 \begin{array}{c} -0.140 \\ -0.260 \end{array}$

The maximum and minimum clearances are as follows:

C10/f10   0.260 − (−0.150) = 0.410 maximum clearance
          0.140 − (−0.030) = 0.170 minimum clearance

F10/c10   0.150 − (−0.260) = 0.410 maximum clearance
          0.030 − (−0.140) = 0.170 minimum clearance

(2) Special rule

$$ES = -ei + \Delta$$
where Δ = difference $IT_n - IT_{n-1}$ between fundamental tolerance of the grade in question and that of the next finer grade

This rule is applicable, for diameter steps above 3 mm (or 0.12 in.), to:

J, K, M, and N   up to IT 8, inclusive,
P to ZC   up to IT 7, inclusive.

Conversions: clearance fit   D7/f6 = F7/d6
               hole-basis fit   H7/r6 = R7/h6
               shaft-basis fit   F7/h6 = H7/f6
               interference fit   S7/u6 = U7/s6
               transition fit   K8/n7 = N8/k7

Example: nominal size 50 mm
             tolerance designation H7/r6
             conversion H7/r6 = R7/h6

Hole H7: $50 \begin{array}{c} +0.025 \\ 0 \end{array}$ Shaft r6: $50 \begin{array}{c} +0.050 \\ +0.034 \end{array}$

R7: $50 \begin{array}{c} -0.025 \\ -0.050 \end{array}$ h6: $50 \begin{array}{c} 0 \\ -0.016 \end{array}$

The maximum and minimum interferences are as follows:

H7/r6   0 − (+0.050) = −0.050 maximum interference
         0.025 − (+0.034) = −0.009 minimum interference

R7/h6   −0.050 (+0) = −0.050 maximum interference
        −0.025 − (−0.016) = −0.009 minimum interference

Non-Toleranced Dimensions

The ISO system of limits and fits can be used on a great number of applications from the finest tolerances to the coarsest. The tolerances are increasing with the nominal size in the ISO system. It is common practice in those countries that have used the ISO system to apply a somewhat related system to the non-toleranced dimensions on a drawing. It is now proposed as an ANSI B4.3 standard. The tolerances are dependent on the nominal sizes and are shown in the following table.

Tolerances on Angles

The nominal size for an angle is the length of the short leg, in mm.

Tolerances shown in Table 6-34 apply to the fine, medium, and coarse series and they are covered in the proposed ANSI B4.3 standard. Angular tolerances are given in Table 6-35.

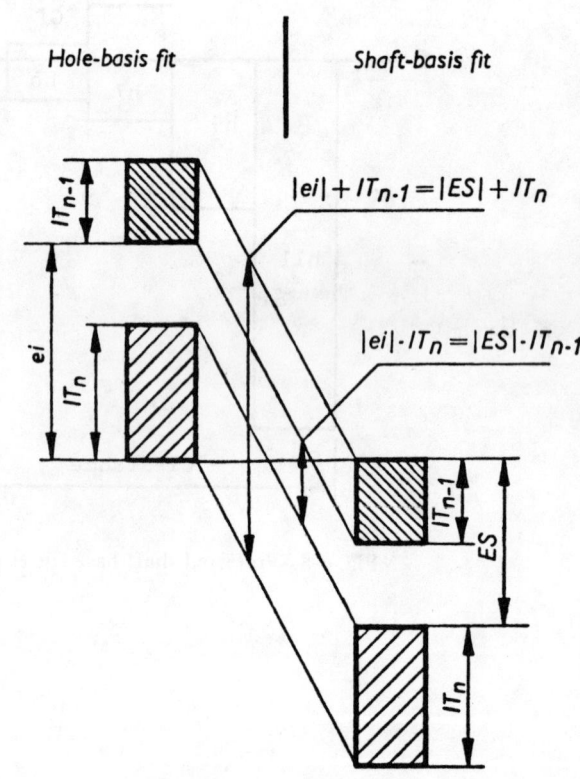

Fig. 6-9 Conversion of fits (ISO 286).

THE ISO SYSTEM OF LIMITS AND FITS—TOLERANCES AND DEVIATIONS

Table 6-10. Tolerance Zones—Internal Dimensions (Holes) (A14...A9, B14...B9) Dimensions in mm (ANSI B4.2)

SIZE	A14	A13	A12	A11	A10	A9	B14	B13	B12	B11	B10	B9
OVER 0 TO 3	+0.520 +0.270	+0.410 +0.270	+0.370 +0.270	+0.330 +0.270	+0.310 +0.270	+0.295 +0.270	+0.390 +0.140	+0.280 +0.140	+0.240 +0.140	+0.200 +0.140	+0.180 +0.140	+0.165 +0.140
OVER 3 TO 6	+0.570 +0.270	+0.450 +0.270	+0.390 +0.270	+0.345 +0.270	+0.318 +0.270	+0.300 +0.270	+0.440 +0.140	+0.320 +0.140	+0.260 +0.140	+0.215 +0.140	+0.188 +0.140	+0.170 +0.140
OVER 6 TO 10	+0.640 +0.280	+0.500 +0.280	+0.430 +0.280	+0.370 +0.280	+0.338 +0.280	+0.316 +0.280	+0.510 +0.150	+0.370 +0.150	+0.300 +0.150	+0.240 +0.150	+0.208 +0.150	+0.186 +0.150
OVER 10 TO 14	+0.720 +0.290	+0.560 +0.290	+0.470 +0.290	+0.400 +0.290	+0.360 +0.290	+0.338 +0.290	+0.580 +0.150	+0.420 +0.150	+0.330 +0.150	+0.260 +0.150	+0.220 +0.150	+0.193 +0.150
OVER 14 TO 18	+0.720 +0.290	+0.560 +0.290	+0.470 +0.290	+0.400 +0.290	+0.360 +0.290	+0.333 +0.290	+0.580 +0.150	+0.420 +0.150	+0.330 +0.150	+0.260 +0.150	+0.220 +0.150	+0.193 +0.150
OVER 18 TO 24	+0.820 +0.300	+0.630 +0.300	+0.510 +0.300	+0.430 +0.300	+0.384 +0.300	+0.352 +0.300	+0.680 +0.160	+0.490 +0.160	+0.370 +0.160	+0.290 +0.160	+0.244 +0.160	+0.212 +0.160
OVER 24 TO 30	+0.820 +0.300	+0.630 +0.300	+0.510 +0.300	+0.430 +0.300	+0.384 +0.300	+0.352 +0.300	+0.680 +0.160	+0.490 +0.160	+0.370 +0.160	+0.290 +0.160	+0.244 +0.160	+0.212 +0.160
OVER 30 TO 40	+0.930 +0.310	+0.700 +0.310	+0.560 +0.310	+0.470 +0.310	+0.410 +0.310	+0.372 +0.310	+0.790 +0.170	+0.560 +0.170	+0.420 +0.170	+0.330 +0.170	+0.270 +0.170	+0.232 +0.170
OVER 40 TO 50	+0.940 +0.320	+0.710 +0.320	+0.570 +0.320	+0.480 +0.320	+0.420 +0.320	+0.382 +0.320	+0.800 +0.180	+0.570 +0.180	+0.430 +0.180	+0.340 +0.180	+0.280 +0.180	+0.242 +0.180
OVER 50 TO 65	+1.080 +0.340	+0.800 +0.340	+0.640 +0.340	+0.530 +0.340	+0.460 +0.340	+0.414 +0.340	+0.930 +0.190	+0.650 +0.190	+0.490 +0.190	+0.380 +0.190	+0.310 +0.190	+0.264 +0.190
OVER 65 TO 80	+1.100 +0.360	+0.820 +0.360	+0.660 +0.360	+0.550 +0.360	+0.480 +0.360	+0.434 +0.360	+0.940 +0.200	+0.660 +0.200	+0.500 +0.200	+0.390 +0.200	+0.320 +0.200	+0.274 +0.200
OVER 80 TO 100	+1.250 +0.380	+0.920 +0.380	+0.730 +0.380	+0.600 +0.380	+0.520 +0.380	+0.467 +0.380	+1.090 +0.220	+0.760 +0.220	+0.570 +0.220	+0.440 +0.220	+0.360 +0.220	+0.307 +0.220
OVER 100 TO 120	+1.280 +0.410	+0.950 +0.410	+0.760 +0.410	+0.630 +0.410	+0.550 +0.410	+0.497 +0.410	+1.110 +0.240	+0.780 +0.240	+0.590 +0.240	+0.460 +0.240	+0.380 +0.240	+0.327 +0.240
OVER 120 TO 140	+1.460 +0.460	+1.090 +0.460	+0.860 +0.460	+0.710 +0.460	+0.620 +0.460	+0.560 +0.460	+1.260 +0.260	+0.890 +0.260	+0.660 +0.260	+0.510 +0.260	+0.420 +0.260	+0.360 +0.260
OVER 140 TO 160	+1.520 +0.520	+1.150 +0.520	+0.920 +0.520	+0.770 +0.520	+0.680 +0.520	+0.620 +0.520	+1.280 +0.280	+0.910 +0.280	+0.680 +0.280	+0.530 +0.280	+0.440 +0.280	+0.380 +0.280
OVER 160 TO 180	+1.580 +0.580	+1.210 +0.580	+0.980 +0.580	+0.830 +0.580	+0.740 +0.580	+0.680 +0.580	+1.310 +0.310	+0.940 +0.310	+0.710 +0.310	+0.560 +0.310	+0.470 +0.310	+0.410 +0.310
OVER 180 TO 200	+1.810 +0.660	+1.380 +0.660	+1.120 +0.660	+0.950 +0.660	+0.845 +0.660	+0.775 +0.660	+1.490 +0.340	+1.060 +0.340	+0.800 +0.340	+0.630 +0.340	+0.525 +0.340	+0.455 +0.340
OVER 200 TO 225	+1.890 +0.740	+1.460 +0.740	+1.200 +0.740	+1.030 +0.740	+0.925 +0.740	+0.855 +0.740	+1.530 +0.380	+1.100 +0.380	+0.840 +0.380	+0.670 +0.380	+0.565 +0.380	+0.495 +0.380
OVER 225 TO 250	+1.970 +0.820	+1.540 +0.820	+1.280 +0.820	+1.110 +0.820	+1.005 +0.820	+0.935 +0.820	+1.570 +0.420	+1.140 +0.420	+0.880 +0.420	+0.710 +0.420	+0.605 +0.420	+0.535 +0.420
OVER 250 TO 280	+2.220 +0.920	+1.730 +0.920	+1.440 +0.920	+1.240 +0.920	+1.130 +0.920	+1.050 +0.920	+1.780 +0.480	+1.290 +0.480	+1.000 +0.480	+0.800 +0.480	+0.690 +0.480	+0.610 +0.480
OVER 280 TO 315	+2.350 +1.050	+1.860 +1.050	+1.570 +1.050	+1.370 +1.050	+1.260 +1.050	+1.180 +1.050	+1.840 +0.540	+1.350 +0.540	+1.060 +0.540	+0.860 +0.540	+0.750 +0.540	+0.670 +0.540
OVER 315 TO 355	+2.600 +1.200	+2.090 +1.200	+1.770 +1.200	+1.560 +1.200	+1.430 +1.200	+1.340 +1.200	+2.000 +0.600	+1.490 +0.600	+1.170 +0.600	+0.960 +0.600	+0.830 +0.600	+0.740 +0.600
OVER 355 TO 400	+2.750 +1.350	+2.240 +1.350	+1.920 +1.350	+1.710 +1.350	+1.580 +1.350	+1.490 +1.350	+2.080 +0.680	+1.570 +0.680	+1.250 +0.680	+1.040 +0.680	+0.910 +0.680	+0.820 +0.680
OVER 400 TO 450	+3.050 +1.500	+2.470 +1.500	+2.130 +1.500	+1.900 +1.500	+1.750 +1.500	+1.655 +1.500	+2.310 +0.760	+1.730 +0.760	+1.390 +0.760	+1.160 +0.760	+1.010 +0.760	+0.915 +0.760
OVER 450 TO 500	+3.200 +1.650	+2.620 +1.650	+2.280 +1.650	+2.050 +1.650	+1.900 +1.650	+1.805 +1.650	+2.390 +0.840	+1.810 +0.840	+1.470 +0.840	+1.240 +0.840	+1.090 +0.840	+0.995 +0.840

Table 6-11. Tolerance Zones—Internal Dimensions (Holes) (C13 . . . C8, D12 . . . D7) Dimensions in mm (ANSI B4.2)

SIZE	C13	C12	C11	C10	C9	C8	D12	D11	D10	D9	D8	D7
OVER 0 TO 3	+0.200 / +0.060	+0.160 / +0.060	+0.120 / +0.060	+0.100 / +0.060	+0.085 / +0.060	+0.074 / +0.060	+0.120 / +0.020	+0.080 / +0.020	+0.060 / +0.020	+0.045 / +0.020	+0.034 / +0.020	+0.030 / +0.020
OVER 3 TO 6	+0.250 / +0.070	+0.190 / +0.070	+0.145 / +0.070	+0.118 / +0.070	+0.100 / +0.070	+0.088 / +0.070	+0.150 / +0.030	+0.105 / +0.030	+0.078 / +0.030	+0.060 / +0.030	+0.048 / +0.030	+0.042 / +0.030
OVER 6 TO 10	+0.300 / +0.080	+0.230 / +0.080	+0.170 / +0.080	+0.138 / +0.080	+0.116 / +0.080	+0.102 / +0.080	+0.190 / +0.040	+0.130 / +0.040	+0.098 / +0.040	+0.076 / +0.040	+0.062 / +0.040	+0.055 / +0.040
OVER 10 TO 14	+0.365 / +0.095	+0.275 / +0.095	+0.205 / +0.095	+0.165 / +0.095	+0.138 / +0.095	+0.122 / +0.095	+0.230 / +0.050	+0.160 / +0.050	+0.120 / +0.050	+0.093 / +0.050	+0.077 / +0.050	+0.068 / +0.050
OVER 14 TO 18	+0.365 / +0.095	+0.275 / +0.095	+0.205 / +0.095	+0.165 / +0.095	+0.138 / +0.095	+0.122 / +0.095	+0.230 / +0.050	+0.160 / +0.050	+0.120 / +0.050	+0.093 / +0.050	+0.077 / +0.050	+0.068 / +0.050
OVER 18 TO 24	+0.440 / +0.110	+0.320 / +0.110	+0.240 / +0.110	+0.194 / +0.110	+0.162 / +0.110	+0.143 / +0.110	+0.275 / +0.065	+0.195 / +0.065	+0.149 / +0.065	+0.117 / +0.065	+0.098 / +0.065	+0.086 / +0.065
OVER 24 TO 30	+0.440 / +0.110	+0.320 / +0.110	+0.240 / +0.110	+0.194 / +0.110	+0.162 / +0.110	+0.143 / +0.110	+0.275 / +0.065	+0.195 / +0.065	+0.149 / +0.065	+0.117 / +0.065	+0.098 / +0.065	+0.086 / +0.065
OVER 30 TO 40	+0.510 / +0.120	+0.370 / +0.120	+0.280 / +0.120	+0.220 / +0.120	+0.182 / +0.120	+0.159 / +0.120	+0.330 / +0.080	+0.240 / +0.080	+0.180 / +0.080	+0.142 / +0.080	+0.119 / +0.080	+0.105 / +0.080
OVER 40 TO 50	+0.520 / +0.130	+0.380 / +0.130	+0.290 / +0.130	+0.230 / +0.130	+0.192 / +0.130	+0.169 / +0.130	+0.330 / +0.080	+0.240 / +0.080	+0.180 / +0.080	+0.142 / +0.080	+0.119 / +0.080	+0.105 / +0.080
OVER 50 TO 65	+0.600 / +0.140	+0.440 / +0.140	+0.330 / +0.140	+0.260 / +0.140	+0.214 / +0.140	+0.186 / +0.140	+0.400 / +0.100	+0.290 / +0.100	+0.220 / +0.100	+0.174 / +0.100	+0.146 / +0.100	+0.130 / +0.100
OVER 65 TO 80	+0.610 / +0.150	+0.450 / +0.150	+0.340 / +0.150	+0.270 / +0.150	+0.224 / +0.150	+0.196 / +0.150	+0.400 / +0.100	+0.290 / +0.100	+0.220 / +0.100	+0.174 / +0.100	+0.146 / +0.100	+0.130 / +0.100
OVER 80 TO 100	+0.710 / +0.170	+0.520 / +0.170	+0.390 / +0.170	+0.310 / +0.170	+0.257 / +0.170	+0.224 / +0.170	+0.470 / +0.120	+0.340 / +0.120	+0.260 / +0.120	+0.207 / +0.120	+0.174 / +0.120	+0.155 / +0.120
OVER 100 TO 120	+0.720 / +0.180	+0.530 / +0.180	+0.400 / +0.180	+0.320 / +0.180	+0.267 / +0.180	+0.234 / +0.180	+0.470 / +0.120	+0.340 / +0.120	+0.260 / +0.120	+0.207 / +0.120	+0.174 / +0.120	+0.155 / +0.120
OVER 120 TO 140	+0.830 / +0.200	+0.600 / +0.200	+0.450 / +0.200	+0.360 / +0.200	+0.300 / +0.200	+0.263 / +0.200	+0.545 / +0.145	+0.395 / +0.145	+0.305 / +0.145	+0.245 / +0.145	+0.208 / +0.145	+0.185 / +0.145
OVER 140 TO 160	+0.840 / +0.210	+0.610 / +0.210	+0.460 / +0.210	+0.370 / +0.210	+0.310 / +0.210	+0.273 / +0.210	+0.545 / +0.145	+0.395 / +0.145	+0.305 / +0.145	+0.245 / +0.145	+0.208 / +0.145	+0.185 / +0.145
OVER 160 TO 180	+0.860 / +0.230	+0.630 / +0.230	+0.480 / +0.230	+0.390 / +0.230	+0.330 / +0.230	+0.293 / +0.230	+0.545 / +0.145	+0.395 / +0.145	+0.305 / +0.145	+0.245 / +0.145	+0.208 / +0.145	+0.185 / +0.145
OVER 180 TO 200	+0.960 / +0.240	+0.700 / +0.240	+0.530 / +0.240	+0.425 / +0.240	+0.355 / +0.240	+0.312 / +0.240	+0.630 / +0.170	+0.460 / +0.170	+0.355 / +0.170	+0.285 / +0.170	+0.242 / +0.170	+0.216 / +0.170
OVER 200 TO 225	+0.980 / +0.260	+0.720 / +0.260	+0.550 / +0.260	+0.445 / +0.260	+0.375 / +0.260	+0.332 / +0.260	+0.630 / +0.170	+0.460 / +0.170	+0.355 / +0.170	+0.285 / +0.170	+0.242 / +0.170	+0.216 / +0.170
OVER 225 TO 250	+1.000 / +0.280	+0.740 / +0.280	+0.570 / +0.280	+0.465 / +0.280	+0.395 / +0.280	+0.352 / +0.280	+0.630 / +0.170	+0.460 / +0.170	+0.355 / +0.170	+0.285 / +0.170	+0.242 / +0.170	+0.216 / +0.170
OVER 250 TO 280	+1.110 / +0.300	+0.820 / +0.300	+0.620 / +0.300	+0.510 / +0.300	+0.430 / +0.300	+0.381 / +0.300	+0.710 / +0.190	+0.510 / +0.190	+0.400 / +0.190	+0.320 / +0.190	+0.271 / +0.190	+0.242 / +0.190
OVER 280 TO 315	+1.140 / +0.330	+0.850 / +0.330	+0.650 / +0.330	+0.540 / +0.330	+0.460 / +0.330	+0.411 / +0.330	+0.710 / +0.190	+0.510 / +0.190	+0.400 / +0.190	+0.320 / +0.190	+0.271 / +0.190	+0.242 / +0.190
OVER 315 TO 355	+1.250 / +0.360	+0.930 / +0.360	+0.720 / +0.360	+0.590 / +0.360	+0.500 / +0.360	+0.449 / +0.360	+0.780 / +0.210	+0.570 / +0.210	+0.440 / +0.210	+0.350 / +0.210	+0.299 / +0.210	+0.267 / +0.210
OVER 355 TO 400	+1.290 / +0.400	+0.970 / +0.400	+0.760 / +0.400	+0.630 / +0.400	+0.540 / +0.400	+0.489 / +0.400	+0.780 / +0.210	+0.570 / +0.210	+0.440 / +0.210	+0.350 / +0.210	+0.299 / +0.210	+0.267 / +0.210
OVER 400 TO 450	+1.410 / +0.440	+1.070 / +0.440	+0.840 / +0.440	+0.690 / +0.440	+0.595 / +0.440	+0.537 / +0.440	+0.860 / +0.230	+0.630 / +0.230	+0.480 / +0.230	+0.385 / +0.230	+0.327 / +0.230	+0.293 / +0.230
OVER 450 TO 500	+1.450 / +0.480	+1.110 / +0.480	+0.880 / +0.480	+0.730 / +0.480	+0.635 / +0.480	+0.577 / +0.480	+0.860 / +0.230	+0.630 / +0.230	+0.480 / +0.230	+0.385 / +0.230	+0.327 / +0.230	+0.293 / +0.230

THE ISO SYSTEM OF LIMITS AND FITS—TOLERANCES AND DEVIATIONS

Table 6-12. Tolerance Zones—Internal Dimensions (Holes) (E12 ... E7, F11 ... F6) Dimensions in mm (ANSI B4.2)

SIZE	E12	E11	E10	E9	E8	E7	F11	F10	F9	F8	F7	F6
OVER 0 TO 3	+0.114 +0.014	+0.074 +0.014	+0.054 +0.014	+0.039 +0.014	+0.028 +0.014	+0.024 +0.014	+0.066 +0.006	+0.046 +0.006	+0.031 +0.006	+0.020 +0.006	+0.016 +0.006	+0.012 +0.006
OVER 3 TO 6	+0.140 +0.020	+0.095 +0.020	+0.068 +0.020	+0.050 +0.020	+0.038 +0.020	+0.032 +0.020	+0.085 +0.010	+0.058 +0.010	+0.040 +0.010	+0.028 +0.010	+0.022 +0.010	+0.018 +0.010
OVER 6 TO 10	+0.175 +0.025	+0.115 +0.025	+0.083 +0.025	+0.061 +0.025	+0.047 +0.025	+0.040 +0.025	+0.103 +0.013	+0.071 +0.013	+0.049 +0.013	+0.035 +0.013	+0.028 +0.013	+0.022 +0.013
OVER 10 TO 14	+0.212 +0.032	+0.142 +0.032	+0.102 +0.032	+0.075 +0.032	+0.059 +0.032	+0.050 +0.032	+0.126 +0.016	+0.086 +0.016	+0.059 +0.016	+0.043 +0.016	+0.034 +0.016	+0.027 +0.016
OVER 14 TO 18	+0.212 +0.032	+0.142 +0.032	+0.102 +0.032	+0.075 +0.032	+0.059 +0.032	+0.050 +0.032	+0.126 +0.016	+0.086 +0.016	+0.059 +0.016	+0.043 +0.016	+0.034 +0.016	+0.027 +0.016
OVER 18 TO 24	+0.250 +0.040	+0.170 +0.040	+0.124 +0.040	+0.092 +0.040	+0.073 +0.040	+0.061 +0.040	+0.150 +0.020	+0.104 +0.020	+0.072 +0.020	+0.053 +0.020	+0.041 +0.020	+0.033 +0.020
OVER 24 TO 30	+0.250 +0.040	+0.170 +0.040	+0.124 +0.040	+0.092 +0.040	+0.073 +0.040	+0.061 +0.040	+0.150 +0.020	+0.104 +0.020	+0.072 +0.020	+0.053 +0.020	+0.041 +0.020	+0.033 +0.020
OVER 30 TO 40	+0.300 +0.050	+0.210 +0.050	+0.150 +0.050	+0.112 +0.050	+0.089 +0.050	+0.075 +0.050	+0.185 +0.025	+0.125 +0.025	+0.087 +0.025	+0.064 +0.025	+0.050 +0.025	+0.041 +0.025
OVER 40 TO 50	+0.300 +0.050	+0.210 +0.050	+0.150 +0.050	+0.112 +0.050	+0.089 +0.050	+0.075 +0.050	+0.185 +0.025	+0.125 +0.025	+0.087 +0.025	+0.064 +0.025	+0.050 +0.025	+0.041 +0.025
OVER 50 TO 65	+0.360 +0.060	+0.250 +0.060	+0.180 +0.060	+0.134 +0.060	+0.106 +0.060	+0.090 +0.060	+0.220 +0.030	+0.150 +0.030	+0.104 +0.030	+0.076 +0.030	+0.060 +0.030	+0.049 +0.030
OVER 65 TO 80	+0.360 +0.060	+0.250 +0.060	+0.180 +0.060	+0.134 +0.060	+0.106 +0.060	+0.090 +0.060	+0.220 +0.030	+0.150 +0.030	+0.104 +0.030	+0.076 +0.030	+0.060 +0.030	+0.049 +0.030
OVER 80 TO 100	+0.422 +0.072	+0.292 +0.072	+0.212 +0.072	+0.159 +0.072	+0.126 +0.072	+0.107 +0.072	+0.256 +0.036	+0.176 +0.036	+0.123 +0.036	+0.090 +0.036	+0.071 +0.036	+0.058 +0.036
OVER 100 TO 120	+0.422 +0.072	+0.292 +0.072	+0.212 +0.072	+0.159 +0.072	+0.126 +0.072	+0.107 +0.072	+0.256 +0.036	+0.176 +0.036	+0.123 +0.036	+0.090 +0.036	+0.071 +0.036	+0.058 +0.036
OVER 120 TO 140	+0.485 +0.085	+0.335 +0.085	+0.245 +0.085	+0.185 +0.085	+0.148 +0.085	+0.125 +0.085	+0.293 +0.043	+0.203 +0.043	+0.143 +0.043	+0.106 +0.043	+0.083 +0.043	+0.068 +0.043
OVER 140 TO 160	+0.485 +0.085	+0.335 +0.085	+0.245 +0.085	+0.185 +0.085	+0.148 +0.085	+0.125 +0.085	+0.293 +0.043	+0.203 +0.043	+0.143 +0.043	+0.106 +0.043	+0.083 +0.043	+0.068 +0.043
OVER 160 TO 180	+0.485 +0.085	+0.335 +0.085	+0.245 +0.085	+0.185 +0.085	+0.148 +0.085	+0.125 +0.085	+0.293 +0.043	+0.203 +0.043	+0.143 +0.043	+0.106 +0.043	+0.083 +0.043	+0.068 +0.043
OVER 180 TO 200	+0.560 +0.100	+0.390 +0.100	+0.285 +0.100	+0.215 +0.100	+0.172 +0.100	+0.146 +0.100	+0.340 +0.050	+0.235 +0.050	+0.165 +0.050	+0.122 +0.050	+0.096 +0.050	+0.079 +0.050
OVER 200 TO 225	+0.560 +0.100	+0.390 +0.100	+0.285 +0.100	+0.215 +0.100	+0.172 +0.100	+0.146 +0.100	+0.340 +0.050	+0.235 +0.050	+0.165 +0.050	+0.122 +0.050	+0.096 +0.050	+0.079 +0.050
OVER 225 TO 250	+0.560 +0.100	+0.390 +0.100	+0.285 +0.100	+0.215 +0.100	+0.172 +0.100	+0.146 +0.100	+0.340 +0.050	+0.235 +0.050	+0.165 +0.050	+0.122 +0.050	+0.096 +0.050	+0.079 +0.050
OVER 250 TO 280	+0.630 +0.110	+0.430 +0.110	+0.320 +0.110	+0.240 +0.110	+0.191 +0.110	+0.162 +0.110	+0.376 +0.056	+0.266 +0.056	+0.186 +0.056	+0.137 +0.056	+0.108 +0.056	+0.088 +0.056
OVER 280 TO 315	+0.630 +0.110	+0.430 +0.110	+0.320 +0.110	+0.240 +0.110	+0.191 +0.110	+0.162 +0.110	+0.376 +0.056	+0.266 +0.056	+0.186 +0.056	+0.137 +0.056	+0.108 +0.056	+0.088 +0.056
OVER 315 TO 355	+0.695 +0.125	+0.485 +0.125	+0.355 +0.125	+0.265 +0.125	+0.214 +0.125	+0.182 +0.125	+0.422 +0.062	+0.292 +0.062	+0.202 +0.062	+0.151 +0.062	+0.119 +0.062	+0.098 +0.062
OVER 355 TO 400	+0.695 +0.125	+0.485 +0.125	+0.355 +0.125	+0.265 +0.125	+0.214 +0.125	+0.182 +0.125	+0.422 +0.062	+0.292 +0.062	+0.202 +0.062	+0.151 +0.062	+0.119 +0.062	+0.098 +0.062
OVER 400 TO 450	+0.765 +0.135	+0.535 +0.135	+0.385 +0.135	+0.290 +0.135	+0.232 +0.135	+0.198 +0.135	+0.468 +0.068	+0.318 +0.068	+0.223 +0.068	+0.165 +0.068	+0.131 +0.068	+0.108 +0.068
OVER 450 TO 500	+0.765 +0.135	+0.535 +0.135	+0.385 +0.135	+0.290 +0.135	+0.232 +0.135	+0.198 +0.135	+0.468 +0.068	+0.318 +0.068	+0.223 +0.068	+0.165 +0.068	+0.131 +0.068	+0.108 +0.068

THE ISO SYSTEM OF LIMITS AND FITS—TOLERANCES AND DEVIATIONS

Table 6-13. Tolerance Zones—Internal Dimensions (Holes) (G10 ... G5 , J8 ... J6) Dimensions in mm (ANSI B4.2)

SIZE	G10	G9	G8	G7	G6	G5	J8	J7	J6
OVER 0 TO 3	+0.042 +0.002	+0.027 +0.002	+0.016 +0.002	+0.012 +0.002	+0.008 +0.002	+0.006 +0.002	+0.006 -0.008	+0.004 -0.006	+0.002 -0.004
OVER 3 TO 6	+0.052 +0.004	+0.034 +0.004	+0.022 +0.004	+0.016 +0.004	+0.012 +0.004	+0.009 +0.004	+0.010 -0.008	+0.006 -0.006	+0.005 -0.003
OVER 6 TO 10	+0.063 +0.005	+0.041 +0.005	+0.027 +0.005	+0.020 +0.005	+0.014 +0.005	+0.011 +0.005	+0.012 -0.010	+0.008 -0.007	+0.005 -0.004
OVER 10 TO 14	+0.076 +0.006	+0.049 +0.006	+0.033 +0.006	+0.024 +0.006	+0.017 +0.006	+0.014 +0.006	+0.015 -0.012	+0.010 -0.008	+0.006 -0.005
OVER 14 TO 18	+0.076 +0.006	+0.049 +0.006	+0.033 +0.006	+0.024 +0.006	+0.017 +0.006	+0.014 +0.006	+0.015 -0.012	+0.010 -0.008	+0.006 -0.005
OVER 18 TO 24	+0.091 +0.007	+0.059 +0.007	+0.040 +0.007	+0.028 +0.007	+0.020 +0.007	+0.016 +0.007	+0.020 -0.013	+0.012 -0.009	+0.008 -0.005
OVER 24 TO 30	+0.091 +0.007	+0.059 +0.007	+0.040 +0.007	+0.028 +0.007	+0.020 +0.007	+0.016 +0.007	+0.020 -0.013	+0.012 -0.009	+0.008 -0.005
OVER 30 TO 40	+0.109 +0.009	+0.071 +0.009	+0.048 +0.009	+0.034 +0.009	+0.025 +0.009	+0.020 +0.009	+0.024 -0.015	+0.014 -0.011	+0.010 -0.006
OVER 40 TO 50	+0.109 +0.009	+0.071 +0.009	+0.048 +0.009	+0.034 +0.009	+0.025 +0.009	+0.020 +0.009	+0.024 -0.015	+0.014 -0.011	+0.010 -0.006
OVER 50 TO 65	+0.130 +0.010	+0.084 +0.010	+0.056 +0.010	+0.040 +0.010	+0.029 +0.010	+0.023 +0.010	+0.028 -0.018	+0.018 -0.012	+0.013 -0.006
OVER 65 TO 80	+0.130 +0.010	+0.084 +0.010	+0.056 +0.010	+0.040 +0.010	+0.029 +0.010	+0.023 +0.010	+0.028 -0.018	+0.018 -0.012	+0.013 -0.006
OVER 80 TO 100	+0.152 +0.012	+0.099 +0.012	+0.066 +0.012	+0.047 +0.012	+0.034 +0.012	+0.027 +0.012	+0.034 -0.020	+0.022 -0.013	+0.016 -0.006
OVER 100 TO 120	+0.152 +0.012	+0.099 +0.012	+0.066 +0.012	+0.047 +0.012	+0.034 +0.012	+0.027 +0.012	+0.034 -0.020	+0.022 -0.013	+0.016 -0.006
OVER 120 TO 140	+0.174 +0.014	+0.114 +0.014	+0.077 +0.014	+0.054 +0.014	+0.039 +0.014	+0.032 +0.014	+0.041 -0.022	+0.026 -0.014	+0.018 -0.007
OVER 140 TO 160	+0.174 +0.014	+0.114 +0.014	+0.077 +0.014	+0.054 +0.014	+0.039 +0.014	+0.032 +0.014	+0.041 -0.022	+0.026 -0.014	+0.018 -0.007
OVER 160 TO 180	+0.174 +0.014	+0.114 +0.014	+0.077 +0.014	+0.054 +0.014	+0.039 +0.014	+0.032 +0.014	+0.041 -0.022	+0.026 -0.014	+0.018 -0.007
OVER 180 TO 200	+0.200 +0.015	+0.130 +0.015	+0.087 +0.015	+0.061 +0.015	+0.044 +0.015	+0.035 +0.015	+0.047 -0.025	+0.030 -0.016	+0.022 -0.007
OVER 200 TO 225	+0.200 +0.015	+0.130 +0.015	+0.087 +0.015	+0.061 +0.015	+0.044 +0.015	+0.035 +0.015	+0.047 -0.025	+0.030 -0.016	+0.022 -0.007
OVER 225 TO 250	+0.200 +0.015	+0.130 +0.015	+0.087 +0.015	+0.061 +0.015	+0.044 +0.015	+0.035 +0.015	+0.047 -0.025	+0.030 -0.016	+0.022 -0.007
OVER 250 TO 280	+0.227 +0.017	+0.147 +0.017	+0.098 +0.017	+0.069 +0.017	+0.049 +0.017	+0.040 +0.017	+0.055 -0.026	+0.036 -0.016	+0.025 -0.007
OVER 280 TO 315	+0.227 +0.017	+0.147 +0.017	+0.098 +0.017	+0.069 +0.017	+0.049 +0.017	+0.040 +0.017	+0.055 -0.026	+0.036 -0.016	+0.025 -0.007
OVER 315 TO 355	+0.248 +0.018	+0.158 +0.018	+0.107 +0.018	+0.075 +0.018	+0.054 +0.018	+0.043 +0.018	+0.060 -0.029	+0.039 -0.018	+0.029 -0.007
OVER 355 TO 400	+0.248 +0.018	+0.158 +0.018	+0.107 +0.018	+0.075 +0.018	+0.054 +0.018	+0.043 +0.018	+0.060 -0.029	+0.039 -0.018	+0.029 -0.007
OVER 400 TO 450	+0.270 +0.020	+0.175 +0.020	+0.117 +0.020	+0.083 +0.020	+0.060 +0.020	+0.047 +0.020	+0.066 -0.031	+0.043 -0.020	+0.033 -0.007
OVER 450 TO 500	+0.270 +0.020	+0.175 +0.020	+0.117 +0.020	+0.083 +0.020	+0.060 +0.020	+0.047 +0.020	+0.066 -0.031	+0.043 -0.020	+0.033 -0.007

THE ISO SYSTEM OF LIMITS AND FITS—TOLERANCES AND DEVIATIONS

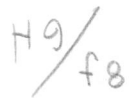

Table 6-14. Tolerance Zones—Internal Dimensions (Holes) (H16 ... H1) Dimensions in mm (ANSI B4.2)

SIZE	H16	H15	H14	H13	H12	H11	H10	H9	H8	H7	H6	H5	H4	H3	H2	H1
OVER 0 TO 3	+0.600 / 0.000	+0.400 / 0.000	+0.250 / 0.000	+0.140 / 0.000	+0.100 / 0.000	+0.060 / 0.000	+0.040 / 0.000	+0.025 / 0.000	+0.014 / 0.000	+0.010 / 0.000	+0.006 / 0.000	+0.004 / 0.000	+0.003 / 0.000	+0.002 / 0.000	+0.0012 / 0.0000	+0.0008 / 0.0000
OVER 3 TO 6	+0.750 / 0.000	+0.480 / 0.000	+0.300 / 0.000	+0.180 / 0.000	+0.120 / 0.000	+0.075 / 0.000	+0.048 / 0.000	+0.030 / 0.000	+0.018 / 0.000	+0.012 / 0.000	+0.008 / 0.000	+0.005 / 0.000	+0.004 / 0.000	+0.0025 / 0.0000	+0.0015 / 0.0000	+0.001 / 0.000
OVER 6 TO 10	+0.900 / 0.000	+0.580 / 0.000	+0.360 / 0.000	+0.220 / 0.000	+0.150 / 0.000	+0.090 / 0.000	+0.058 / 0.000	+0.036 / 0.000	+0.022 / 0.000	+0.015 / 0.000	+0.009 / 0.000	+0.006 / 0.000	+0.004 / 0.000	+0.0025 / 0.0000	+0.0015 / 0.0000	+0.001 / 0.000
OVER 10 TO 14	+1.100 / 0.000	+0.700 / 0.000	+0.430 / 0.000	+0.270 / 0.000	+0.180 / 0.000	+0.110 / 0.000	+0.070 / 0.000	+0.043 / 0.000	+0.027 / 0.000	+0.018 / 0.000	+0.011 / 0.000	+0.008 / 0.000	+0.005 / 0.000	+0.003 / 0.000	+0.002 / 0.000	+0.0012 / 0.0000
OVER 14 TO 18	+1.100 / 0.000	+0.700 / 0.000	+0.430 / 0.000	+0.270 / 0.000	+0.180 / 0.000	+0.110 / 0.000	+0.070 / 0.000	+0.043 / 0.000	+0.027 / 0.000	+0.018 / 0.000	+0.011 / 0.000	+0.008 / 0.000	+0.005 / 0.000	+0.003 / 0.000	+0.002 / 0.000	+0.0012 / 0.0000
OVER 18 TO 24	+1.300 / 0.000	+0.840 / 0.000	+0.520 / 0.000	+0.330 / 0.000	+0.210 / 0.000	+0.130 / 0.000	+0.084 / 0.000	+0.052 / 0.000	+0.033 / 0.000	+0.021 / 0.000	+0.013 / 0.000	+0.009 / 0.000	+0.006 / 0.000	+0.004 / 0.000	+0.0025 / 0.0000	+0.0015 / 0.0000
OVER 24 TO 30	+1.300 / 0.000	+0.840 / 0.000	+0.520 / 0.000	+0.330 / 0.000	+0.210 / 0.000	+0.130 / 0.000	+0.084 / 0.000	+0.052 / 0.000	+0.033 / 0.000	+0.021 / 0.000	+0.013 / 0.000	+0.009 / 0.000	+0.006 / 0.000	+0.004 / 0.000	+0.0025 / 0.0000	+0.0015 / 0.0000
OVER 30 TO 40	+1.600 / 0.000	+1.000 / 0.000	+0.620 / 0.000	+0.390 / 0.000	+0.250 / 0.000	+0.160 / 0.000	+0.100 / 0.000	+0.062 / 0.000	+0.039 / 0.000	+0.025 / 0.000	+0.016 / 0.000	+0.011 / 0.000	+0.007 / 0.000	+0.004 / 0.000	+0.0025 / 0.0000	+0.0015 / 0.0000
OVER 40 TO 50	+1.600 / 0.000	+1.000 / 0.000	+0.620 / 0.000	+0.390 / 0.000	+0.250 / 0.000	+0.160 / 0.000	+0.100 / 0.000	+0.062 / 0.000	+0.039 / 0.000	+0.025 / 0.000	+0.016 / 0.000	+0.011 / 0.000	+0.007 / 0.000	+0.004 / 0.000	+0.0025 / 0.0000	+0.0015 / 0.0000
OVER 50 TO 65	+1.900 / 0.000	+1.200 / 0.000	+0.740 / 0.000	+0.460 / 0.000	+0.300 / 0.000	+0.190 / 0.000	+0.120 / 0.000	+0.074 / 0.000	+0.046 / 0.000	+0.030 / 0.000	+0.019 / 0.000	+0.013 / 0.000	+0.008 / 0.000	+0.005 / 0.000	+0.003 / 0.000	+0.002 / 0.000
OVER 65 TO 80	+1.900 / 0.000	+1.200 / 0.000	+0.740 / 0.000	+0.460 / 0.000	+0.300 / 0.000	+0.190 / 0.000	+0.120 / 0.000	+0.074 / 0.000	+0.046 / 0.000	+0.030 / 0.000	+0.019 / 0.000	+0.013 / 0.000	+0.008 / 0.000	+0.005 / 0.000	+0.003 / 0.000	+0.002 / 0.000
OVER 80 TO 100	+2.200 / 0.000	+1.400 / 0.000	+0.870 / 0.000	+0.540 / 0.000	+0.350 / 0.000	+0.220 / 0.000	+0.140 / 0.000	+0.087 / 0.000	+0.054 / 0.000	+0.035 / 0.000	+0.022 / 0.000	+0.015 / 0.000	+0.010 / 0.000	+0.006 / 0.000	+0.004 / 0.000	+0.0025 / 0.0000
OVER 100 TO 120	+2.200 / 0.000	+1.400 / 0.000	+0.870 / 0.000	+0.540 / 0.000	+0.350 / 0.000	+0.220 / 0.000	+0.140 / 0.000	+0.087 / 0.000	+0.054 / 0.000	+0.035 / 0.000	+0.022 / 0.000	+0.015 / 0.000	+0.010 / 0.000	+0.006 / 0.000	+0.004 / 0.000	+0.0025 / 0.0000
OVER 120 TO 140	+2.500 / 0.000	+1.600 / 0.000	+1.000 / 0.000	+0.630 / 0.000	+0.400 / 0.000	+0.250 / 0.000	+0.160 / 0.000	+0.100 / 0.000	+0.063 / 0.000	+0.040 / 0.000	+0.025 / 0.000	+0.018 / 0.000	+0.012 / 0.000	+0.008 / 0.000	+0.005 / 0.000	+0.0035 / 0.0000
OVER 140 TO 160	+2.500 / 0.000	+1.600 / 0.000	+1.000 / 0.000	+0.630 / 0.000	+0.400 / 0.000	+0.250 / 0.000	+0.160 / 0.000	+0.100 / 0.000	+0.063 / 0.000	+0.040 / 0.000	+0.025 / 0.000	+0.018 / 0.000	+0.012 / 0.000	+0.008 / 0.000	+0.005 / 0.000	+0.0035 / 0.0000
OVER 160 TO 180	+2.500 / 0.000	+1.600 / 0.000	+1.000 / 0.000	+0.630 / 0.000	+0.400 / 0.000	+0.250 / 0.000	+0.160 / 0.000	+0.100 / 0.000	+0.063 / 0.000	+0.040 / 0.000	+0.025 / 0.000	+0.018 / 0.000	+0.012 / 0.000	+0.008 / 0.000	+0.005 / 0.000	+0.0035 / 0.0000
OVER 180 TO 200	+2.900 / 0.000	+1.850 / 0.000	+1.150 / 0.000	+0.720 / 0.000	+0.460 / 0.000	+0.290 / 0.000	+0.185 / 0.000	+0.115 / 0.000	+0.072 / 0.000	+0.046 / 0.000	+0.029 / 0.000	+0.020 / 0.000	+0.014 / 0.000	+0.010 / 0.000	+0.007 / 0.000	+0.0045 / 0.0000
OVER 200 TO 225	+2.900 / 0.000	+1.850 / 0.000	+1.150 / 0.000	+0.720 / 0.000	+0.460 / 0.000	+0.290 / 0.000	+0.185 / 0.000	+0.115 / 0.000	+0.072 / 0.000	+0.046 / 0.000	+0.029 / 0.000	+0.020 / 0.000	+0.014 / 0.000	+0.010 / 0.000	+0.007 / 0.000	+0.0045 / 0.0000
OVER 225 TO 250	+2.900 / 0.000	+1.850 / 0.000	+1.150 / 0.000	+0.720 / 0.000	+0.460 / 0.000	+0.290 / 0.000	+0.185 / 0.000	+0.115 / 0.000	+0.072 / 0.000	+0.046 / 0.000	+0.029 / 0.000	+0.020 / 0.000	+0.014 / 0.000	+0.010 / 0.000	+0.007 / 0.000	+0.0045 / 0.0000
OVER 250 TO 280	+3.200 / 0.000	+2.100 / 0.000	+1.300 / 0.000	+0.810 / 0.000	+0.520 / 0.000	+0.320 / 0.000	+0.210 / 0.000	+0.130 / 0.000	+0.081 / 0.000	+0.052 / 0.000	+0.032 / 0.000	+0.023 / 0.000	+0.016 / 0.000	+0.012 / 0.000	+0.008 / 0.000	+0.006 / 0.000
OVER 280 TO 315	+3.200 / 0.000	+2.100 / 0.000	+1.300 / 0.000	+0.810 / 0.000	+0.520 / 0.000	+0.320 / 0.000	+0.210 / 0.000	+0.130 / 0.000	+0.081 / 0.000	+0.052 / 0.000	+0.032 / 0.000	+0.023 / 0.000	+0.016 / 0.000	+0.012 / 0.000	+0.008 / 0.000	+0.006 / 0.000
OVER 315 TO 355	+3.600 / 0.000	+2.300 / 0.000	+1.400 / 0.000	+0.890 / 0.000	+0.570 / 0.000	+0.360 / 0.000	+0.230 / 0.000	+0.140 / 0.000	+0.089 / 0.000	+0.057 / 0.000	+0.036 / 0.000	+0.025 / 0.000	+0.018 / 0.000	+0.013 / 0.000	+0.009 / 0.000	+0.007 / 0.000
OVER 355 TO 400	+3.600 / 0.000	+2.300 / 0.000	+1.400 / 0.000	+0.890 / 0.000	+0.570 / 0.000	+0.360 / 0.000	+0.230 / 0.000	+0.140 / 0.000	+0.089 / 0.000	+0.057 / 0.000	+0.036 / 0.000	+0.025 / 0.000	+0.018 / 0.000	+0.013 / 0.000	+0.009 / 0.000	+0.007 / 0.000
OVER 400 TO 450	+4.000 / 0.000	+2.500 / 0.000	+1.550 / 0.000	+0.970 / 0.000	+0.630 / 0.000	+0.400 / 0.000	+0.250 / 0.000	+0.155 / 0.000	+0.097 / 0.000	+0.063 / 0.000	+0.040 / 0.000	+0.027 / 0.000	+0.020 / 0.000	+0.015 / 0.000	+0.010 / 0.000	+0.008 / 0.000
OVER 450 TO 500	+4.000 / 0.000	+2.500 / 0.000	+1.550 / 0.000	+0.970 / 0.000	+0.630 / 0.000	+0.400 / 0.000	+0.250 / 0.000	+0.155 / 0.000	+0.097 / 0.000	+0.063 / 0.000	+0.040 / 0.000	+0.027 / 0.000	+0.020 / 0.000	+0.015 / 0.000	+0.010 / 0.000	+0.008 / 0.000

THE ISO SYSTEM OF LIMITS AND FITS—TOLERANCES AND DEVIATIONS

SIZE		JS16	JS15	JS14	JS13	JS12	JS11	JS10	JS9	JS8	JS7	JS6	JS5	JS4	JS3	JS2	JS
OVER	0	+0.300	+0.200	+0.125	+0.070	+0.050	+0.030	+0.020	+0.012	+0.007	+0.005	+0.003	+0.002	+0.0015	+0.001	+0.0006	+0.0
TO	3	-0.300	-0.200	-0.125	-0.070	-0.050	-0.030	-0.020	-0.012	-0.007	-0.005	-0.003	-0.002	-0.0015	-0.001	-0.0006	-0.0
OVER	3	+0.375	+0.240	+0.150	+0.090	+0.060	+0.037	+0.024	+0.015	+0.009	+0.006	+0.004	+0.0025	+0.002	+0.00125	+0.00075	+0.0
TO	6	-0.375	-0.240	-0.150	-0.090	-0.060	-0.037	-0.024	-0.015	-0.009	-0.006	-0.004	-0.0025	-0.002	-0.00125	-0.00075	-0.0
OVER	6	+0.450	+0.290	+0.180	+0.110	+0.075	+0.045	+0.029	+0.018	+0.011	+0.007	+0.0045	+0.003	+0.002	+0.00125	+0.00075	+0.0
TO	10	-0.450	-0.290	-0.180	-0.110	-0.075	-0.045	-0.029	-0.018	-0.011	-0.007	-0.0045	-0.003	-0.002	-0.00125	-0.00075	-0.0
OVER	10	+0.550	+0.350	+0.215	+0.135	+0.090	+0.055	+0.035	+0.021	+0.013	+0.009	+0.0055	+0.004	+0.0025	+0.0015	+0.001	+0.0
TO	14	-0.550	-0.350	-0.215	-0.135	-0.090	-0.055	-0.035	-0.021	-0.013	-0.009	-0.0055	-0.004	-0.0025	-0.0015	-0.001	-0.0
OVER	14	+0.550	+0.350	+0.215	+0.135	+0.090	+0.055	+0.035	+0.021	+0.013	+0.009	+0.0055	+0.004	+0.0025	+0.0015	+0.001	+0.0
TO	18	-0.550	-0.350	-0.215	-0.135	-0.090	-0.055	-0.035	-0.021	-0.013	-0.009	-0.0055	-0.004	-0.0025	-0.0015	-0.001	-0.0
OVER	18	+0.650	+0.420	+0.260	+0.165	+0.105	+0.065	+0.042	+0.026	+0.016	+0.010	+0.0065	+0.0045	+0.003	+0.002	+0.00125	+0.0
TO	24	-0.650	-0.420	-0.260	-0.165	-0.105	-0.065	-0.042	-0.026	-0.016	-0.010	-0.0065	-0.0045	-0.003	-0.002	-0.00125	-0.0
OVER	24	+0.650	+0.420	+0.260	+0.165	+0.105	+0.065	+0.042	+0.026	+0.016	+0.010	+0.0065	+0.0045	+0.003	+0.002	+0.00125	+0.0
TO	30	-0.650	-0.420	-0.260	-0.165	-0.105	-0.065	-0.042	-0.026	-0.016	-0.010	-0.0065	-0.0045	-0.003	-0.002	-0.00125	-0.0
OVER	30	+0.800	+0.500	+0.310	+0.195	+0.125	+0.080	+0.050	+0.031	+0.019	+0.012	+0.008	+0.0055	+0.0035	+0.002	+0.00125	+0.0
TO	40	-0.800	-0.500	-0.310	-0.195	-0.125	-0.080	-0.050	-0.031	-0.019	-0.012	-0.008	-0.0055	-0.0035	-0.002	-0.00125	-0.0
OVER	40	+0.800	+0.500	+0.310	+0.195	+0.125	+0.080	+0.050	+0.031	+0.019	+0.012	+0.008	+0.0055	+0.0035	+0.002	+0.00125	+0.0
TO	50	-0.800	-0.500	-0.310	-0.195	-0.125	-0.080	-0.050	-0.031	-0.019	-0.012	-0.008	-0.0055	-0.0035	-0.002	-0.00125	-0.0
OVER	50	+0.950	+0.600	+0.370	+0.230	+0.150	+0.095	+0.060	+0.037	+0.023	+0.015	+0.0095	+0.0065	+0.004	+0.0025	+0.0015	+0.0
TO	65	-0.950	-0.600	-0.370	-0.230	-0.150	-0.095	-0.060	-0.037	-0.023	-0.015	-0.0095	-0.0065	-0.004	-0.0025	-0.0015	-0.0
OVER	65	+0.950	+0.600	+0.370	+0.230	+0.150	+0.095	+0.060	+0.037	+0.023	+0.015	+0.0095	+0.0065	+0.004	+0.0025	+0.0015	+0.0
TO	80	-0.950	-0.600	-0.370	-0.230	-0.150	-0.095	-0.060	-0.037	-0.023	-0.015	-0.0095	-0.0065	-0.004	-0.0025	-0.0015	-0.0
OVER	80	+1.100	+0.700	+0.435	+0.270	+0.175	+0.110	+0.070	+0.043	+0.027	+0.017	+0.011	+0.0075	+0.005	+0.003	+0.002	+0.
TO	100	-1.100	-0.700	-0.435	-0.270	-0.175	-0.110	-0.070	-0.043	-0.027	-0.017	-0.011	-0.0075	-0.005	-0.003	-0.002	-0.
OVER	100	+1.100	+0.700	+0.435	+0.270	+0.175	+0.110	+0.070	+0.043	+0.027	+0.017	+0.011	+0.0075	+0.005	+0.003	+0.002	+0.
TO	120	-1.100	-0.700	-0.435	-0.270	-0.175	-0.110	-0.070	-0.043	-0.027	-0.017	-0.011	-0.0075	-0.005	-0.003	-0.002	-0.
OVER	120	+1.250	+0.800	+0.500	+0.315	+0.200	+0.125	+0.080	+0.050	+0.031	+0.020	+0.0125	+0.009	+0.006	+0.004	+0.0025	+0.
TO	140	-1.250	-0.800	-0.500	-0.315	-0.200	-0.125	-0.080	-0.050	-0.031	-0.020	-0.0125	-0.009	-0.006	-0.004	-0.0025	-0.
OVER	140	+1.250	+0.800	+0.500	+0.315	+0.200	+0.125	+0.080	+0.050	+0.031	+0.020	+0.0125	+0.009	+0.006	+0.004	+0.0025	+0.
TO	160	-1.250	-0.800	-0.500	-0.315	-0.200	-0.125	-0.080	-0.050	-0.031	-0.020	-0.0125	-0.009	-0.006	-0.004	-0.0025	-0.
OVER	160	+1.250	+0.800	+0.500	+0.315	+0.200	+0.125	+0.080	+0.050	+0.031	+0.020	+0.0125	+0.009	+0.006	+0.004	+0.0025	+0.
TO	180	-1.250	-0.800	-0.500	-0.315	-0.200	-0.125	-0.080	-0.050	-0.031	-0.020	-0.0125	-0.009	-0.006	-0.004	-0.0025	-0.
OVER	180	+1.450	+0.925	+0.575	+0.360	+0.230	+0.145	+0.092	+0.057	+0.036	+0.023	+0.0145	+0.010	+0.007	+0.005	+0.0035	+0.
TO	200	-1.450	-0.925	-0.575	-0.360	-0.230	-0.145	-0.092	-0.057	-0.036	-0.023	-0.0145	-0.010	-0.007	-0.005	-0.0035	-0.
OVER	200	+1.450	+0.925	+0.575	+0.360	+0.230	+0.145	+0.092	+0.057	+0.036	+0.023	+0.0145	+0.010	+0.007	+0.005	+0.0035	+0.
TO	225	-1.450	-0.925	-0.575	-0.360	-0.230	-0.145	-0.092	-0.057	-0.036	-0.023	-0.0145	-0.010	-0.007	-0.005	-0.0035	-0.
OVER	225	+1.450	+0.925	+0.575	+0.360	+0.230	+0.145	+0.092	+0.057	+0.036	+0.023	+0.0145	+0.010	+0.007	+0.005	+0.0035	+0.
TO	250	-1.450	-0.925	-0.575	-0.360	-0.230	-0.145	-0.092	-0.057	-0.036	-0.023	-0.0145	-0.010	-0.007	-0.005	-0.0035	-0.
OVER	250	+1.600	+1.050	+0.650	+0.405	+0.260	+0.160	+0.105	+0.065	+0.040	+0.026	+0.016	+0.0115	+0.008	+0.006	+0.004	+0.
TO	280	-1.600	-1.050	-0.650	-0.405	-0.260	-0.160	-0.105	-0.065	-0.040	-0.026	-0.016	-0.0115	-0.008	-0.006	-0.004	-0.
OVER	280	+1.600	+1.050	+0.650	+0.405	+0.260	+0.160	+0.105	+0.065	+0.040	+0.026	+0.016	+0.0115	+0.008	+0.006	+0.004	+0.
TO	315	-1.600	-1.050	-0.650	-0.405	-0.260	-0.160	-0.105	-0.065	-0.040	-0.026	-0.016	-0.0115	-0.008	-0.006	-0.004	-0.
OVER	315	+1.800	+1.150	+0.700	+0.445	+0.285	+0.180	+0.115	+0.070	+0.044	+0.028	+0.018	+0.0125	+0.009	+0.0065	+0.0045	+0.
TO	355	-1.800	-1.150	-0.700	-0.445	-0.285	-0.180	-0.115	-0.070	-0.044	-0.028	-0.018	-0.0125	-0.009	-0.0065	-0.0045	-0.
OVER	355	+1.800	+1.150	+0.700	+0.445	+0.285	+0.180	+0.115	+0.070	+0.044	+0.028	+0.018	+0.0125	+0.009	+0.0065	+0.0045	+0.
TO	400	-1.800	-1.150	-0.700	-0.445	-0.285	-0.180	-0.115	-0.070	-0.044	-0.028	-0.018	-0.0125	-0.009	-0.0065	-0.0045	-0.
OVER	400	+2.000	+1.250	+0.775	+0.485	+0.315	+0.200	+0.125	+0.077	+0.048	+0.031	+0.020	+0.0135	+0.010	+0.0075	+0.005	+0.
TO	450	-2.000	-1.250	-0.775	-0.485	-0.315	-0.200	-0.125	-0.077	-0.048	-0.031	-0.020	-0.0135	-0.010	-0.0075	-0.005	-0.
OVER	450	+2.000	+1.250	+0.775	+0.485	+0.315	+0.200	+0.125	+0.077	+0.048	+0.031	+0.020	+0.0135	+0.010	+0.0075	+0.005	+0.
TO	500	-2.000	-1.250	-0.775	-0.485	-0.315	-0.200	-0.125	-0.077	-0.048	-0.031	-0.020	-0.0135	-0.010	-0.0075	-0.005	-0.

THE ISO SYSTEM OF LIMITS AND FITS—TOLERANCES AND DEVIATIONS 6-47

Table 6-16. Tolerance Zones—Internal Dimensions (Holes) (K10 ... K5, M10 ... M5) Dimensions in mm (ANSI B4.2)

SIZE	K10	K9	K8	K7	K6	K5	M10	M9	M8	M7	M6	M5
OVER 0 TO 3	0.000 -0.040	0.000 -0.025	0.000 -0.014	0.000 -0.010	0.000 -0.006	0.000 -0.004	-0.002 -0.042	-0.002 -0.027	-0.002 -0.016	-0.002 -0.012	-0.002 -0.008	-0.002 -0.006
OVER 3 TO 6			+0.005 -0.013	+0.003 -0.009	+0.002 -0.006	0.000 -0.005	-0.004 -0.052	-0.004 -0.034	+0.002 -0.016	0.000 -0.012	-0.001 -0.009	-0.003 -0.008
OVER 6 TO 10			+0.006 -0.016	+0.005 -0.010	+0.002 -0.007	+0.001 -0.005	-0.006 -0.064	-0.006 -0.042	+0.001 -0.021	0.000 -0.015	-0.003 -0.012	-0.004 -0.010
OVER 10 TO 14			+0.008 -0.019	+0.006 -0.012	+0.002 -0.009	+0.002 -0.006	-0.007 -0.077	-0.007 -0.050	+0.002 -0.025	0.000 -0.018	-0.004 -0.015	-0.004 -0.012
OVER 14 TO 18			+0.008 -0.019	+0.006 -0.012	+0.002 -0.009	+0.002 -0.006	-0.007 -0.077	-0.007 -0.050	+0.002 -0.025	0.000 -0.018	-0.004 -0.015	-0.004 -0.012
OVER 18 TO 24			+0.010 -0.023	+0.006 -0.015	+0.002 -0.011	+0.001 -0.008	-0.008 -0.092	-0.008 -0.060	+0.004 -0.029	0.000 -0.021	-0.004 -0.017	-0.005 -0.014
OVER 24 TO 30			+0.010 -0.023	+0.006 -0.015	+0.002 -0.011	+0.001 -0.008	-0.008 -0.092	-0.008 -0.060	+0.004 -0.029	0.000 -0.021	-0.004 -0.017	-0.005 -0.014
OVER 30 TO 40			+0.012 -0.027	+0.007 -0.018	+0.003 -0.013	+0.002 -0.009	-0.009 -0.109	-0.009 -0.071	+0.005 -0.034	0.000 -0.025	-0.004 -0.020	-0.005 -0.016
OVER 40 TO 50		NUMERICAL VALUES FOR TOLERANCE ZONES IN THIS AREA NOT DEFINED	+0.012 -0.027	+0.007 -0.018	+0.003 -0.013	+0.002 -0.009	-0.009 -0.109	-0.009 -0.071	+0.005 -0.034	0.000 -0.025	-0.004 -0.020	-0.005 -0.016
OVER 50 TO 65			+0.014 -0.032	+0.009 -0.021	+0.004 -0.015	+0.003 -0.010	-0.011 -0.131	-0.011 -0.085	+0.005 -0.041	0.000 -0.030	-0.005 -0.024	-0.006 -0.019
OVER 65 TO 80			+0.014 -0.032	+0.009 -0.021	+0.004 -0.015	+0.003 -0.010	-0.011 -0.131	-0.011 -0.085	+0.005 -0.041	0.000 -0.030	-0.005 -0.024	-0.006 -0.019
OVER 80 TO 100			+0.016 -0.038	+0.010 -0.025	+0.004 -0.018	+0.002 -0.013	-0.013 -0.153	-0.013 -0.100	+0.006 -0.048	0.000 -0.035	-0.006 -0.028	-0.008 -0.023
OVER 100 TO 120			+0.016 -0.038	+0.010 -0.025	+0.004 -0.018	+0.002 -0.013	-0.013 -0.153	-0.013 -0.100	+0.006 -0.048	0.000 -0.035	-0.006 -0.028	-0.008 -0.023
OVER 120 TO 140			+0.020 -0.043	+0.012 -0.028	+0.004 -0.021	+0.003 -0.015	-0.015 -0.175	-0.015 -0.115	+0.008 -0.055	0.000 -0.040	-0.008 -0.033	-0.009 -0.027
OVER 140 TO 160			+0.020 -0.043	+0.012 -0.028	+0.004 -0.021	+0.003 -0.015	-0.015 -0.175	-0.015 -0.115	+0.008 -0.055	0.000 -0.040	-0.008 -0.033	-0.009 -0.027
OVER 160 TO 180			+0.020 -0.043	+0.012 -0.028	+0.004 -0.021	+0.003 -0.015	-0.015 -0.175	-0.015 -0.115	+0.008 -0.055	0.000 -0.040	-0.008 -0.033	-0.009 -0.027
OVER 180 TO 200			+0.022 -0.050	+0.013 -0.033	+0.005 -0.024	+0.002 -0.018	-0.017 -0.202	-0.017 -0.132	+0.009 -0.063	0.000 -0.046	-0.008 -0.037	-0.011 -0.031
OVER 200 TO 225			+0.022 -0.050	+0.013 -0.033	+0.005 -0.024	+0.002 -0.018	-0.017 -0.202	-0.017 -0.132	+0.009 -0.063	0.000 -0.046	-0.008 -0.037	-0.011 -0.031
OVER 225 TO 250			+0.022 -0.050	+0.013 -0.033	+0.005 -0.024	+0.002 -0.018	-0.017 -0.202	-0.017 -0.132	+0.009 -0.063	0.000 -0.046	-0.008 -0.037	-0.011 -0.031
OVER 250 TO 280			+0.025 -0.056	+0.016 -0.036	+0.005 -0.027	+0.003 -0.020	-0.020 -0.230	-0.020 -0.150	+0.009 -0.072	0.000 -0.052	-0.009 -0.041	-0.013 -0.036
OVER 280 TO 315			+0.025 -0.056	+0.016 -0.036	+0.005 -0.027	+0.003 -0.020	-0.020 -0.230	-0.020 -0.150	+0.009 -0.072	0.000 -0.052	-0.009 -0.041	-0.013 -0.036
OVER 315 TO 355			+0.028 -0.061	+0.017 -0.040	+0.007 -0.029	+0.003 -0.022	-0.021 -0.251	-0.021 -0.161	+0.011 -0.078	0.000 -0.057	-0.010 -0.046	-0.014 -0.039
OVER 355 TO 400			+0.028 -0.061	+0.017 -0.040	+0.007 -0.029	+0.003 -0.022	-0.021 -0.251	-0.021 -0.161	+0.011 -0.078	0.000 -0.057	-0.010 -0.046	-0.014 -0.039
OVER 400 TO 450			+0.029 -0.068	+0.018 -0.045	+0.008 -0.032	+0.002 -0.025	-0.023 -0.273	-0.023 -0.178	+0.011 -0.086	0.000 -0.063	-0.010 -0.050	-0.016 -0.043
OVER 450 TO 500			+0.029 -0.068	+0.018 -0.045	+0.008 -0.032	+0.002 -0.025	-0.023 -0.273	-0.023 -0.178	+0.011 -0.086	0.000 -0.063	-0.010 -0.050	-0.016 -0.043

Table 6-17. Tolerance Zones—Internal Dimensions (Holes) (N10 ... N5, P10 ... P5) Dimensions in mm (ANSI B4.2)

SIZE	N10	N9	N8	N7	N6	N5	P10	P9	P8	P7	P6	P5
OVER 0 TO 3	-0.004 -0.044	-0.004 -0.029	-0.004 -0.018	-0.004 -0.014	-0.004 -0.010	-0.004 -0.008	-0.006 -0.046	-0.006 -0.031	-0.006 -0.020	-0.006 -0.016	-0.006 -0.012	-0.006 -0.010
OVER 3 TO 6	0.000 -0.048	0.000 -0.030	-0.002 -0.020	-0.004 -0.016	-0.005 -0.013	-0.007 -0.012	-0.012 -0.060	-0.012 -0.042	-0.012 -0.030	-0.008 -0.020	-0.009 -0.017	-0.011 -0.016
OVER 6 TO 10	0.000 -0.058	0.000 -0.036	-0.003 -0.025	-0.004 -0.019	-0.007 -0.016	-0.008 -0.014	-0.015 -0.073	-0.015 -0.051	-0.015 -0.037	-0.009 -0.024	-0.012 -0.021	-0.013 -0.019
OVER 10 TO 14	0.000 -0.070	0.000 -0.043	-0.003 -0.030	-0.005 -0.023	-0.009 -0.020	-0.009 -0.017	-0.018 -0.088	-0.018 -0.061	-0.018 -0.045	-0.011 -0.029	-0.015 -0.026	-0.015 -0.023
OVER 14 TO 18	0.000 -0.070	0.000 -0.043	-0.003 -0.030	-0.005 -0.023	-0.009 -0.020	-0.009 -0.017	-0.018 -0.088	-0.018 -0.061	-0.018 -0.045	-0.011 -0.029	-0.015 -0.026	-0.015 -0.023
OVER 18 TO 24	0.000 -0.084	0.000 -0.052	-0.003 -0.036	-0.007 -0.028	-0.011 -0.024	-0.012 -0.021	-0.022 -0.106	-0.022 -0.074	-0.022 -0.055	-0.014 -0.035	-0.018 -0.031	-0.019 -0.028
OVER 24 TO 30	0.000 -0.084	0.000 -0.052	-0.003 -0.036	-0.007 -0.028	-0.011 -0.024	-0.012 -0.021	-0.022 -0.106	-0.022 -0.074	-0.022 -0.055	-0.014 -0.035	-0.018 -0.031	-0.019 -0.028
OVER 30 TO 40	0.000 -0.100	0.000 -0.062	-0.003 -0.042	-0.008 -0.033	-0.012 -0.028	-0.013 -0.024	-0.026 -0.126	-0.026 -0.088	-0.026 -0.065	-0.017 -0.042	-0.021 -0.037	-0.022 -0.033
OVER 40 TO 50	0.000 -0.100	0.000 -0.062	-0.003 -0.042	-0.008 -0.033	-0.012 -0.028	-0.013 -0.024	-0.026 -0.126	-0.026 -0.088	-0.026 -0.065	-0.017 -0.042	-0.021 -0.037	-0.022 -0.033
OVER 50 TO 65	0.000 -0.120	0.000 -0.074	-0.004 -0.050	-0.009 -0.039	-0.014 -0.033	-0.015 -0.028	-0.032 -0.152	-0.032 -0.106	-0.032 -0.078	-0.021 -0.051	-0.026 -0.045	-0.027 -0.040
OVER 65 TO 80	0.000 -0.120	0.000 -0.074	-0.004 -0.050	-0.009 -0.039	-0.014 -0.033	-0.015 -0.028	-0.032 -0.152	-0.032 -0.106	-0.032 -0.078	-0.021 -0.051	-0.026 -0.045	-0.027 -0.040
OVER 80 TO 100	0.000 -0.140	0.000 -0.087	-0.004 -0.058	-0.010 -0.045	-0.016 -0.038	-0.018 -0.033	-0.037 -0.177	-0.037 -0.124	-0.037 -0.091	-0.024 -0.059	-0.030 -0.052	-0.032 -0.047
OVER 100 TO 120	0.000 -0.140	0.000 -0.087	-0.004 -0.058	-0.010 -0.045	-0.016 -0.038	-0.018 -0.033	-0.037 -0.177	-0.037 -0.124	-0.037 -0.091	-0.024 -0.059	-0.030 -0.052	-0.032 -0.047
OVER 120 TO 140	0.000 -0.160	0.000 -0.100	-0.004 -0.067	-0.012 -0.052	-0.020 -0.045	-0.021 -0.039	-0.043 -0.203	-0.043 -0.143	-0.043 -0.106	-0.028 -0.068	-0.036 -0.061	-0.037 -0.055
OVER 140 TO 160	0.000 -0.160	0.000 -0.100	-0.004 -0.067	-0.012 -0.052	-0.020 -0.045	-0.021 -0.039	-0.043 -0.203	-0.043 -0.143	-0.043 -0.106	-0.028 -0.068	-0.036 -0.061	-0.037 -0.055
OVER 160 TO 180	0.000 -0.160	0.000 -0.100	-0.004 -0.067	-0.012 -0.052	-0.020 -0.045	-0.021 -0.039	-0.043 -0.203	-0.043 -0.143	-0.043 -0.106	-0.028 -0.068	-0.036 -0.061	-0.037 -0.055
OVER 180 TO 200	0.000 -0.185	0.000 -0.115	-0.005 -0.077	-0.014 -0.060	-0.022 -0.051	-0.025 -0.045	-0.050 -0.235	-0.050 -0.165	-0.050 -0.122	-0.033 -0.079	-0.041 -0.070	-0.044 -0.064
OVER 200 TO 225	0.000 -0.185	0.000 -0.115	-0.005 -0.077	-0.014 -0.060	-0.022 -0.051	-0.025 -0.045	-0.050 -0.235	-0.050 -0.165	-0.050 -0.122	-0.033 -0.079	-0.041 -0.070	-0.044 -0.064
OVER 225 TO 250	0.000 -0.185	0.000 -0.115	-0.005 -0.077	-0.014 -0.060	-0.022 -0.051	-0.025 -0.045	-0.050 -0.235	-0.050 -0.165	-0.050 -0.122	-0.033 -0.079	-0.041 -0.070	-0.044 -0.064
OVER 250 TO 280	0.000 -0.210	0.000 -0.130	-0.005 -0.086	-0.014 -0.066	-0.025 -0.057	-0.027 -0.050	-0.056 -0.266	-0.056 -0.186	-0.056 -0.137	-0.036 -0.088	-0.047 -0.079	-0.049 -0.072
OVER 280 TO 315	0.000 -0.210	0.000 -0.130	-0.005 -0.086	-0.014 -0.066	-0.025 -0.057	-0.027 -0.050	-0.056 -0.266	-0.056 -0.186	-0.056 -0.137	-0.036 -0.088	-0.047 -0.079	-0.049 -0.072
OVER 315 TO 355	0.000 -0.230	0.000 -0.140	-0.005 -0.094	-0.016 -0.073	-0.026 -0.062	-0.030 -0.055	-0.062 -0.292	-0.062 -0.202	-0.062 -0.151	-0.041 -0.098	-0.051 -0.087	-0.055 -0.080
OVER 355 TO 400	0.000 -0.230	0.000 -0.140	-0.005 -0.094	-0.016 -0.073	-0.026 -0.062	-0.030 -0.055	-0.062 -0.292	-0.062 -0.202	-0.062 -0.151	-0.041 -0.098	-0.051 -0.087	-0.055 -0.080
OVER 400 TO 450	0.000 -0.250	0.000 -0.155	-0.006 -0.103	-0.017 -0.080	-0.027 -0.067	-0.033 -0.060	-0.068 -0.318	-0.068 -0.223	-0.068 -0.165	-0.045 -0.108	-0.055 -0.095	-0.061 -0.088
OVER 450 TO 500	0.000 -0.250	0.000 -0.155	-0.006 -0.103	-0.017 -0.080	-0.027 -0.067	-0.033 -0.060	-0.068 -0.318	-0.068 -0.223	-0.068 -0.165	-0.045 -0.108	-0.055 -0.095	-0.061 -0.088

THE ISO SYSTEM OF LIMITS AND FITS—TOLERANCES AND DEVIATIONS

Table 6-18. Tolerance Zones—Internal Dimensions (Holes) (R10 . . . R5, S10 . . . S5) Dimensions in mm (ANSI B4.2)

SIZE	R10	R9	R8	R7	R6	R5	S10	S9	S8	S7	S6	S5
OVER 0 TO 3	-0.010 -0.050	-0.010 -0.035	-0.010 -0.024	-0.010 -0.020	-0.010 -0.016	-0.010 -0.014	-0.014 -0.054	-0.014 -0.039	-0.014 -0.028	-0.014 -0.024	-0.014 -0.020	-0.014 -0.018
OVER 3 TO 6	-0.015 -0.063	-0.015 -0.045	-0.015 -0.033	-0.011 -0.023	-0.012 -0.020	-0.014 -0.019	-0.019 -0.067	-0.019 -0.049	-0.019 -0.037	-0.015 -0.027	-0.016 -0.024	-0.018 -0.023
OVER 6 TO 10	-0.019 -0.077	-0.019 -0.055	-0.019 -0.041	-0.013 -0.028	-0.016 -0.025	-0.017 -0.023	-0.023 -0.081	-0.023 -0.059	-0.023 -0.045	-0.017 -0.032	-0.020 -0.029	-0.021 -0.027
OVER 10 TO 14	-0.023 -0.093	-0.023 -0.066	-0.023 -0.050	-0.016 -0.034	-0.020 -0.031	-0.020 -0.028	-0.028 -0.098	-0.028 -0.071	-0.028 -0.055	-0.021 -0.039	-0.025 -0.036	-0.025 -0.033
OVER 14 TO 18	-0.023 -0.093	-0.023 -0.066	-0.023 -0.050	-0.016 -0.034	-0.020 -0.031	-0.020 -0.028	-0.028 -0.098	-0.028 -0.071	-0.028 -0.055	-0.021 -0.039	-0.025 -0.036	-0.025 -0.033
OVER 18 TO 24	-0.028 -0.112	-0.028 -0.080	-0.028 -0.061	-0.020 -0.041	-0.024 -0.037	-0.025 -0.034	-0.035 -0.119	-0.035 -0.087	-0.035 -0.068	-0.027 -0.048	-0.031 -0.044	-0.032 -0.041
OVER 24 TO 30	-0.028 -0.112	-0.028 -0.080	-0.028 -0.061	-0.020 -0.041	-0.024 -0.037	-0.025 -0.034	-0.035 -0.119	-0.035 -0.087	-0.035 -0.068	-0.027 -0.048	-0.031 -0.044	-0.032 -0.041
OVER 30 TO 40	-0.034 -0.134	-0.034 -0.096	-0.034 -0.073	-0.025 -0.050	-0.029 -0.045	-0.030 -0.041	-0.043 -0.143	-0.043 -0.105	-0.043 -0.082	-0.034 -0.059	-0.038 -0.054	-0.039 -0.050
OVER 40 TO 50	-0.034 -0.134	-0.034 -0.096	-0.034 -0.073	-0.025 -0.050	-0.029 -0.045	-0.030 -0.041	-0.043 -0.143	-0.043 -0.105	-0.043 -0.082	-0.034 -0.059	-0.038 -0.054	-0.039 -0.050
OVER 50 TO 65	-0.041 -0.161	-0.041 -0.115	-0.041 -0.087	-0.030 -0.060	-0.035 -0.054	-0.036 -0.049	-0.053 -0.173	-0.053 -0.127	-0.053 -0.099	-0.042 -0.072	-0.047 -0.066	-0.048 -0.061
OVER 65 TO 80	-0.043 -0.163	-0.043 -0.117	-0.043 -0.089	-0.032 -0.062	-0.037 -0.056	-0.038 -0.051	-0.059 -0.179	-0.059 -0.133	-0.059 -0.105	-0.048 -0.078	-0.053 -0.072	-0.054 -0.067
OVER 80 TO 100	-0.051 -0.191	-0.051 -0.138	-0.051 -0.105	-0.038 -0.073	-0.044 -0.066	-0.046 -0.061	-0.071 -0.211	-0.071 -0.158	-0.071 -0.125	-0.058 -0.093	-0.064 -0.086	-0.066 -0.081
OVER 100 TO 120	-0.054 -0.194	-0.054 -0.141	-0.054 -0.108	-0.041 -0.076	-0.047 -0.069	-0.049 -0.064	-0.079 -0.219	-0.079 -0.166	-0.079 -0.133	-0.066 -0.101	-0.072 -0.094	-0.074 -0.089
OVER 120 TO 140	-0.063 -0.223	-0.063 -0.163	-0.063 -0.126	-0.048 -0.088	-0.056 -0.081	-0.057 -0.075	-0.092 -0.252	-0.092 -0.192	-0.092 -0.155	-0.077 -0.117	-0.085 -0.110	-0.086 -0.104
OVER 140 TO 160	-0.065 -0.225	-0.065 -0.165	-0.065 -0.128	-0.050 -0.090	-0.058 -0.083	-0.059 -0.077	-0.100 -0.260	-0.100 -0.200	-0.100 -0.163	-0.085 -0.125	-0.093 -0.118	-0.094 -0.112
OVER 160 TO 180	-0.068 -0.228	-0.068 -0.168	-0.068 -0.131	-0.053 -0.093	-0.061 -0.086	-0.062 -0.080	-0.108 -0.268	-0.108 -0.208	-0.108 -0.171	-0.093 -0.133	-0.101 -0.126	-0.102 -0.120
OVER 180 TO 200	-0.077 -0.262	-0.077 -0.192	-0.077 -0.149	-0.060 -0.106	-0.068 -0.097	-0.071 -0.091	-0.122 -0.307	-0.122 -0.237	-0.122 -0.194	-0.105 -0.151	-0.113 -0.142	-0.116 -0.136
OVER 200 TO 225	-0.080 -0.265	-0.080 -0.195	-0.080 -0.152	-0.063 -0.109	-0.071 -0.100	-0.074 -0.094	-0.130 -0.315	-0.130 -0.245	-0.130 -0.202	-0.113 -0.159	-0.121 -0.150	-0.124 -0.144
OVER 225 TO 250	-0.084 -0.269	-0.084 -0.199	-0.084 -0.156	-0.067 -0.113	-0.075 -0.104	-0.078 -0.098	-0.140 -0.325	-0.140 -0.255	-0.140 -0.212	-0.123 -0.169	-0.131 -0.160	-0.134 -0.154
OVER 250 TO 280	-0.094 -0.304	-0.094 -0.224	-0.094 -0.175	-0.074 -0.126	-0.085 -0.117	-0.087 -0.110	-0.158 -0.368	-0.158 -0.288	-0.158 -0.239	-0.138 -0.190	-0.149 -0.181	-0.151 -0.174
OVER 280 TO 315	-0.098 -0.308	-0.098 -0.228	-0.098 -0.179	-0.078 -0.130	-0.089 -0.121	-0.091 -0.114	-0.170 -0.380	-0.170 -0.300	-0.170 -0.251	-0.150 -0.202	-0.161 -0.193	-0.163 -0.186
OVER 315 TO 355	-0.108 -0.338	-0.108 -0.248	-0.108 -0.197	-0.087 -0.144	-0.097 -0.133	-0.101 -0.126	-0.190 -0.420	-0.190 -0.330	-0.190 -0.279	-0.169 -0.226	-0.179 -0.215	-0.183 -0.208
OVER 355 TO 400	-0.114 -0.344	-0.114 -0.254	-0.114 -0.203	-0.093 -0.150	-0.103 -0.139	-0.107 -0.132	-0.208 -0.438	-0.208 -0.348	-0.208 -0.297	-0.187 -0.244	-0.197 -0.233	-0.201 -0.226
OVER 400 TO 450	-0.126 -0.376	-0.126 -0.281	-0.126 -0.223	-0.103 -0.166	-0.113 -0.153	-0.119 -0.146	-0.232 -0.482	-0.232 -0.387	-0.232 -0.329	-0.209 -0.272	-0.219 -0.259	-0.225 -0.252
OVER 450 TO 500	-0.132 -0.382	-0.132 -0.287	-0.132 -0.229	-0.109 -0.172	-0.119 -0.159	-0.125 -0.152	-0.252 -0.502	-0.252 -0.407	-0.252 -0.349	-0.229 -0.292	-0.239 -0.279	-0.245 -0.272

Table 6-19. Tolerance Zones—Internal Dimensions (Holes) (T10 ... T5, U10 ... U5) Dimensions in mm (ANSI B4.2)

SIZE	T10	T9	T8	T7	T6	T5	U10	U9	U8	U7	U6	U5
OVER 0 TO 3							-0.018 -0.058	-0.018 -0.043	-0.018 -0.032	-0.018 -0.028	-0.018 -0.024	-0.018 -0.022
OVER 3 TO 6							-0.023 -0.071	-0.023 -0.053	-0.023 -0.041	-0.019 -0.031	-0.020 -0.028	-0.022 -0.027
OVER 6 TO 10		NUMERICAL VALUES FOR					-0.028 -0.086	-0.028 -0.064	-0.028 -0.050	-0.022 -0.037	-0.025 -0.034	-0.026 -0.032
OVER 10 TO 14		TOLERANCE ZONES IN					-0.033 -0.103	-0.033 -0.076	-0.033 -0.060	-0.026 -0.044	-0.030 -0.041	-0.030 -0.038
OVER 14 TO 18		THIS AREA NOT DEFINED.					-0.033 -0.103	-0.033 -0.076	-0.033 -0.060	-0.026 -0.044	-0.030 -0.041	-0.030 -0.038
OVER 18 TO 24							-0.041 -0.125	-0.041 -0.093	-0.041 -0.074	-0.033 -0.054	-0.037 -0.050	-0.038 -0.047
OVER 24 TO 30	-0.041 -0.125	-0.041 -0.093	-0.041 -0.074	-0.033 -0.054	-0.037 -0.050	-0.038 -0.047	-0.048 -0.132	-0.048 -0.100	-0.048 -0.081	-0.040 -0.061	-0.044 -0.057	-0.045 -0.054
OVER 30 TO 40	-0.048 -0.148	-0.048 -0.110	-0.048 -0.087	-0.039 -0.064	-0.043 -0.059	-0.044 -0.055	-0.060 -0.160	-0.060 -0.122	-0.060 -0.099	-0.051 -0.076	-0.055 -0.071	-0.056 -0.067
OVER 40 TO 50	-0.054 -0.154	-0.054 -0.116	-0.054 -0.093	-0.045 -0.070	-0.049 -0.065	-0.050 -0.061	-0.070 -0.170	-0.070 -0.132	-0.070 -0.109	-0.061 -0.086	-0.065 -0.081	-0.066 -0.077
OVER 50 TO 65	-0.066 -0.186	-0.066 -0.140	-0.066 -0.112	-0.055 -0.085	-0.060 -0.079	-0.061 -0.074	-0.087 -0.207	-0.087 -0.161	-0.087 -0.133	-0.076 -0.106	-0.081 -0.100	-0.082 -0.095
OVER 65 TO 80	-0.075 -0.195	-0.075 -0.149	-0.075 -0.121	-0.064 -0.094	-0.069 -0.088	-0.070 -0.088	-0.102 -0.222	-0.102 -0.176	-0.102 -0.148	-0.091 -0.121	-0.096 -0.115	-0.097 -0.110
OVER 80 TO 100	-0.091 -0.231	-0.091 -0.178	-0.091 -0.145	-0.078 -0.113	-0.084 -0.106	-0.086 -0.101	-0.124 -0.264	-0.124 -0.211	-0.124 -0.178	-0.111 -0.146	-0.117 -0.139	-0.119 -0.134
OVER 100 TO 120	-0.104 -0.244	-0.104 -0.191	-0.104 -0.158	-0.091 -0.126	-0.097 -0.119	-0.099 -0.114	-0.144 -0.284	-0.144 -0.231	-0.144 -0.198	-0.131 -0.166	-0.137 -0.159	-0.139 -0.154
OVER 120 TO 140	-0.122 -0.282	-0.122 -0.222	-0.122 -0.185	-0.107 -0.147	-0.115 -0.140	-0.116 -0.134	-0.170 -0.330	-0.170 -0.270	-0.170 -0.233	-0.155 -0.195	-0.163 -0.188	-0.164 -0.182
OVER 140 TO 160	-0.134 -0.294	-0.134 -0.234	-0.134 -0.197	-0.119 -0.159	-0.127 -0.152	-0.128 -0.146	-0.190 -0.350	-0.190 -0.290	-0.190 -0.253	-0.175 -0.215	-0.183 -0.208	-0.184 -0.202
OVER 160 TO 180	-0.146 -0.306	-0.146 -0.246	-0.146 -0.209	-0.131 -0.171	-0.139 -0.164	-0.140 -0.158	-0.210 -0.370	-0.210 -0.310	-0.210 -0.273	-0.195 -0.235	-0.203 -0.228	-0.204 -0.222
OVER 180 TO 200	-0.166 -0.351	-0.166 -0.281	-0.166 -0.238	-0.149 -0.195	-0.157 -0.186	-0.160 -0.180	-0.236 -0.421	-0.236 -0.351	-0.236 -0.308	-0.219 -0.265	-0.227 -0.256	-0.230 -0.250
OVER 200 TO 225	-0.180 -0.365	-0.180 -0.295	-0.180 -0.252	-0.163 -0.209	-0.171 -0.200	-0.174 -0.194	-0.258 -0.443	-0.258 -0.373	-0.258 -0.330	-0.241 -0.287	-0.249 -0.278	-0.252 -0.272
OVER 225 TO 250	-0.196 -0.381	-0.196 -0.311	-0.196 -0.268	-0.179 -0.225	-0.187 -0.216	-0.190 -0.210	-0.284 -0.469	-0.284 -0.399	-0.284 -0.356	-0.267 -0.313	-0.275 -0.304	-0.278 -0.298
OVER 250 TO 280	-0.218 -0.428	-0.218 -0.348	-0.218 -0.299	-0.198 -0.250	-0.209 -0.241	-0.211 -0.234	-0.315 -0.525	-0.315 -0.445	-0.315 -0.396	-0.295 -0.347	-0.306 -0.338	-0.308 -0.331
OVER 280 TO 315	-0.240 -0.450	-0.240 -0.370	-0.240 -0.321	-0.220 -0.272	-0.231 -0.263	-0.233 -0.256	-0.350 -0.560	-0.350 -0.480	-0.350 -0.431	-0.330 -0.382	-0.341 -0.373	-0.343 -0.366
OVER 315 TO 355	-0.268 -0.498	-0.268 -0.408	-0.268 -0.357	-0.247 -0.304	-0.257 -0.293	-0.261 -0.286	-0.390 -0.620	-0.390 -0.530	-0.390 -0.479	-0.369 -0.426	-0.379 -0.415	-0.383 -0.408
OVER 355 TO 400	-0.294 -0.524	-0.294 -0.434	-0.294 -0.383	-0.273 -0.330	-0.283 -0.319	-0.287 -0.312	-0.435 -0.665	-0.435 -0.575	-0.435 -0.524	-0.414 -0.471	-0.424 -0.460	-0.428 -0.453
OVER 400 TO 450	-0.330 -0.580	-0.330 -0.485	-0.330 -0.427	-0.307 -0.370	-0.317 -0.357	-0.323 -0.350	-0.490 -0.740	-0.490 -0.645	-0.490 -0.587	-0.467 -0.530	-0.477 -0.517	-0.483 -0.510
OVER 450 TO 500	-0.360 -0.610	-0.360 -0.515	-0.360 -0.457	-0.337 -0.400	-0.347 -0.387	-0.353 -0.380	-0.540 -0.790	-0.540 -0.695	-0.540 -0.637	-0.517 -0.580	-0.527 -0.567	-0.533 -0.560

THE ISO SYSTEM OF LIMITS AND FITS—TOLERANCES AND DEVIATIONS

Table 6-20. Tolerance Zones—Internal Dimensions (Holes) (V10 . . . V5, X10 . . . X5) Dimensions in mm (ANSI B4.2)

SIZE	V10	V9	V8	V7	V6	V5	X10	X9	X8	X7	X6	X5
OVER 0 TO 3			NUMERICAL VALUES FOR TOLERANCE ZONES IN THIS AREA NOT DEFINED.				-0.020 -0.060	-0.020 -0.045	-0.020 -0.034	-0.020 -0.030	-0.020 -0.026	-0.020 -0.024
OVER 3 TO 6							-0.028 -0.076	-0.028 -0.058	-0.028 -0.046	-0.024 -0.036	-0.025 -0.033	-0.027 -0.032
OVER 6 TO 10							-0.034 -0.092	-0.034 -0.070	-0.034 -0.056	-0.028 -0.043	-0.031 -0.040	-0.032 -0.038
OVER 10 TO 14							-0.040 -0.110	-0.040 -0.083	-0.040 -0.067	-0.033 -0.051	-0.037 -0.048	-0.037 -0.045
OVER 14 TO 18	-0.039 -0.109	-0.039 -0.082	-0.039 -0.066	-0.032 -0.050	-0.036 -0.047	-0.036 -0.044	-0.045 -0.115	-0.045 -0.088	-0.045 -0.072	-0.038 -0.056	-0.042 -0.053	-0.042 -0.050
OVER 18 TO 24	-0.047 -0.131	-0.047 -0.099	-0.047 -0.080	-0.039 -0.060	-0.043 -0.056	-0.044 -0.053	-0.054 -0.138	-0.054 -0.106	-0.054 -0.087	-0.046 -0.067	-0.050 -0.063	-0.051 -0.060
OVER 24 TO 30	-0.055 -0.139	-0.055 -0.107	-0.055 -0.088	-0.047 -0.068	-0.051 -0.064	-0.052 -0.061	-0.064 -0.148	-0.064 -0.116	-0.064 -0.097	-0.056 -0.077	-0.060 -0.073	-0.061 -0.070
OVER 30 TO 40	-0.068 -0.168	-0.068 -0.130	-0.068 -0.107	-0.059 -0.084	-0.063 -0.079	-0.064 -0.075	-0.080 -0.180	-0.080 -0.142	-0.080 -0.119	-0.071 -0.096	-0.075 -0.091	-0.076 -0.087
OVER 40 TO 50	-0.081 -0.181	-0.081 -0.143	-0.081 -0.120	-0.072 -0.097	-0.076 -0.092	-0.077 -0.088	-0.097 -0.197	-0.097 -0.159	-0.097 -0.136	-0.088 -0.113	-0.092 -0.108	-0.093 -0.104
OVER 50 TO 65	-0.102 -0.222	-0.102 -0.176	-0.102 -0.148	-0.091 -0.121	-0.096 -0.115	-0.097 -0.110	-0.122 -0.242	-0.122 -0.196	-0.122 -0.168	-0.111 -0.141	-0.116 -0.135	-0.117 -0.130
OVER 65 TO 80	-0.120 -0.240	-0.120 -0.194	-0.120 -0.166	-0.109 -0.139	-0.114 -0.133	-0.115 -0.128	-0.146 -0.266	-0.146 -0.220	-0.146 -0.192	-0.135 -0.165	-0.140 -0.159	-0.141 -0.154
OVER 80 TO 100	-0.146 -0.286	-0.146 -0.233	-0.146 -0.200	-0.133 -0.168	-0.139 -0.161	-0.141 -0.156	-0.178 -0.318	-0.178 -0.265	-0.178 -0.232	-0.165 -0.200	-0.171 -0.193	-0.173 -0.188
OVER 100 TO 120	-0.172 -0.312	-0.172 -0.259	-0.172 -0.226	-0.159 -0.194	-0.165 -0.187	-0.167 -0.182	-0.210 -0.350	-0.210 -0.297	-0.210 -0.264	-0.197 -0.232	-0.203 -0.225	-0.205 -0.220
OVER 120 TO 140	-0.202 -0.362	-0.202 -0.302	-0.202 -0.265	-0.187 -0.227	-0.195 -0.220	-0.196 -0.214	-0.248 -0.408	-0.248 -0.348	-0.248 -0.311	-0.233 -0.273	-0.241 -0.266	-0.242 -0.260
OVER 140 TO 160	-0.228 -0.388	-0.228 -0.328	-0.228 -0.291	-0.213 -0.253	-0.221 -0.246	-0.222 -0.240	-0.280 -0.440	-0.280 -0.380	-0.280 -0.343	-0.265 -0.305	-0.273 -0.298	-0.274 -0.292
OVER 160 TO 180	-0.252 -0.412	-0.252 -0.352	-0.252 -0.315	-0.237 -0.277	-0.245 -0.270	-0.246 -0.264	-0.310 -0.470	-0.310 -0.410	-0.310 -0.373	-0.295 -0.335	-0.303 -0.328	-0.304 -0.322
OVER 180 TO 200	-0.284 -0.469	-0.284 -0.399	-0.284 -0.356	-0.267 -0.313	-0.275 -0.304	-0.278 -0.298	-0.350 -0.535	-0.350 -0.465	-0.350 -0.422	-0.333 -0.379	-0.341 -0.370	-0.344 -0.364
OVER 200 TO 225	-0.310 -0.495	-0.310 -0.425	-0.310 -0.382	-0.293 -0.339	-0.301 -0.330	-0.304 -0.324	-0.385 -0.570	-0.385 -0.500	-0.385 -0.457	-0.368 -0.414	-0.376 -0.405	-0.379 -0.399
OVER 225 TO 250	-0.340 -0.525	-0.340 -0.455	-0.340 -0.412	-0.323 -0.369	-0.331 -0.360	-0.334 -0.354	-0.425 -0.610	-0.425 -0.540	-0.425 -0.497	-0.408 -0.454	-0.416 -0.445	-0.419 -0.439
OVER 250 TO 280	-0.385 -0.595	-0.385 -0.515	-0.385 -0.466	-0.365 -0.417	-0.376 -0.408	-0.378 -0.401	-0.475 -0.685	-0.475 -0.605	-0.475 -0.556	-0.455 -0.507	-0.466 -0.498	-0.468 -0.491
OVER 280 TO 315	-0.425 -0.635	-0.425 -0.555	-0.425 -0.506	-0.405 -0.457	-0.416 -0.448	-0.418 -0.441	-0.525 -0.735	-0.525 -0.655	-0.525 -0.606	-0.505 -0.557	-0.516 -0.548	-0.518 -0.541
OVER 315 TO 355	-0.475 -0.705	-0.475 -0.615	-0.475 -0.564	-0.454 -0.511	-0.464 -0.500	-0.468 -0.493	-0.590 -0.820	-0.590 -0.730	-0.590 -0.679	-0.569 -0.626	-0.579 -0.615	-0.583 -0.608
OVER 355 TO 400	-0.530 -0.760	-0.530 -0.670	-0.530 -0.619	-0.509 -0.566	-0.519 -0.555	-0.523 -0.548	-0.660 -0.890	-0.660 -0.800	-0.660 -0.749	-0.639 -0.696	-0.649 -0.685	-0.653 -0.678
OVER 400 TO 450	-0.595 -0.845	-0.595 -0.750	-0.595 -0.692	-0.572 -0.635	-0.582 -0.622	-0.588 -0.615	-0.740 -0.990	-0.740 -0.895	-0.740 -0.837	-0.717 -0.780	-0.727 -0.767	-0.733 -0.760
OVER 450 TO 500	-0.660 -0.910	-0.660 -0.815	-0.660 -0.757	-0.637 -0.700	-0.647 -0.687	-0.653 -0.680	-0.820 -1.070	-0.820 -0.975	-0.820 -0.917	-0.797 -0.860	-0.807 -0.847	-0.813 -0.840

THE ISO SYSTEM OF LIMITS AND FITS—TOLERANCES AND DEVIATIONS

Table 6-21. Tolerance Zones—Internal Dimensions (Holes) (Y10 ... Y5, Z10 ... Z5) Dimensions in mm (ANSI B4.2)

SIZE	Y10	Y9	Y8	Y7	Y6	Y5	Z10	Z9	Z8	Z7	Z6	Z5
OVER 0 TO 3							-0.026 / -0.066	-0.026 / -0.051	-0.026 / -0.040	-0.026 / -0.036	-0.026 / -0.032	-0.026 / -0.030
OVER 3 TO 6							-0.035 / -0.083	-0.035 / -0.065	-0.035 / -0.053	-0.031 / -0.043	-0.032 / -0.040	-0.034 / -0.039
OVER 6 TO 10	NUMERICAL VALUES FOR TOLERANCE ZONES IN THIS AREA NOT DEFINED.						-0.042 / -0.100	-0.042 / -0.078	-0.042 / -0.064	-0.036 / -0.051	-0.039 / -0.048	-0.040 / -0.046
OVER 10 TO 14							-0.050 / -0.120	-0.050 / -0.093	-0.050 / -0.077	-0.043 / -0.061	-0.047 / -0.058	-0.047 / -0.055
OVER 14 TO 18							-0.060 / -0.130	-0.060 / -0.103	-0.060 / -0.087	-0.053 / -0.071	-0.057 / -0.068	-0.057 / -0.065
OVER 18 TO 24	-0.063 / -0.147	-0.063 / -0.115	-0.063 / -0.096	-0.055 / -0.076	-0.059 / -0.072	-0.060 / -0.069	-0.073 / -0.157	-0.073 / -0.125	-0.073 / -0.106	-0.065 / -0.086	-0.069 / -0.082	-0.070 / -0.079
OVER 24 TO 30	-0.075 / -0.159	-0.075 / -0.127	-0.075 / -0.108	-0.067 / -0.088	-0.071 / -0.084	-0.072 / -0.081	-0.088 / -0.172	-0.088 / -0.140	-0.088 / -0.121	-0.080 / -0.101	-0.084 / -0.097	-0.085 / -0.094
OVER 30 TO 40	-0.094 / -0.194	-0.094 / -0.156	-0.094 / -0.133	-0.085 / -0.110	-0.089 / -0.105	-0.090 / -0.101	-0.112 / -0.212	-0.112 / -0.174	-0.112 / -0.151	-0.103 / -0.128	-0.107 / -0.123	-0.108 / -0.119
OVER 40 TO 50	-0.114 / -0.214	-0.114 / -0.176	-0.114 / -0.153	-0.105 / -0.130	-0.109 / -0.125	-0.110 / -0.121	-0.136 / -0.236	-0.136 / -0.198	-0.136 / -0.175	-0.127 / -0.152	-0.131 / -0.147	-0.132 / -0.143
OVER 50 TO 65	-0.144 / -0.264	-0.144 / -0.218	-0.144 / -0.190	-0.133 / -0.163	-0.138 / -0.157	-0.139 / -0.152	-0.172 / -0.292	-0.172 / -0.246	-0.172 / -0.218	-0.161 / -0.191	-0.166 / -0.185	-0.167 / -0.180
OVER 65 TO 80	-0.174 / -0.294	-0.174 / -0.248	-0.174 / -0.220	-0.163 / -0.193	-0.168 / -0.187	-0.169 / -0.182	-0.210 / -0.330	-0.210 / -0.284	-0.210 / -0.256	-0.159 / -0.229	-0.204 / -0.223	-0.205 / -0.218
OVER 80 TO 100	-0.214 / -0.354	-0.214 / -0.301	-0.214 / -0.268	-0.201 / -0.236	-0.207 / -0.229	-0.209 / -0.224	-0.258 / -0.398	-0.258 / -0.345	-0.258 / -0.312	-0.245 / -0.280	-0.251 / -0.273	-0.253 / -0.268
OVER 100 TO 120	-0.254 / -0.394	-0.254 / -0.341	-0.254 / -0.308	-0.241 / -0.276	-0.247 / -0.269	-0.249 / -0.264	-0.310 / -0.450	-0.310 / -0.397	-0.310 / -0.364	-0.297 / -0.332	-0.303 / -0.325	-0.305 / -0.320
OVER 120 TO 140	-0.300 / -0.460	-0.300 / -0.400	-0.300 / -0.363	-0.285 / -0.325	-0.293 / -0.318	-0.294 / -0.312	-0.365 / -0.525	-0.365 / -0.465	-0.365 / -0.428	-0.350 / -0.390	-0.358 / -0.383	-0.359 / -0.377
OVER 140 TO 160	-0.340 / -0.500	-0.340 / -0.440	-0.340 / -0.403	-0.325 / -0.365	-0.333 / -0.358	-0.334 / -0.352	-0.415 / -0.575	-0.415 / -0.515	-0.415 / -0.478	-0.400 / -0.440	-0.408 / -0.433	-0.409 / -0.427
OVER 160 TO 180	-0.380 / -0.540	-0.380 / -0.480	-0.380 / -0.443	-0.365 / -0.405	-0.373 / -0.398	-0.374 / -0.392	-0.465 / -0.625	-0.465 / -0.565	-0.465 / -0.528	-0.450 / -0.490	-0.458 / -0.483	-0.459 / -0.477
OVER 180 TO 200	-0.425 / -0.610	-0.425 / -0.540	-0.425 / -0.497	-0.408 / -0.454	-0.416 / -0.445	-0.419 / -0.439	-0.520 / -0.705	-0.520 / -0.635	-0.520 / -0.592	-0.503 / -0.549	-0.511 / -0.540	-0.514 / -0.534
OVER 200 TO 225	-0.470 / -0.655	-0.470 / -0.585	-0.470 / -0.542	-0.453 / -0.499	-0.461 / -0.490	-0.464 / -0.484	-0.575 / -0.760	-0.575 / -0.690	-0.575 / -0.647	-0.558 / -0.604	-0.566 / -0.595	-0.569 / -0.589
OVER 225 TO 250	-0.520 / -0.705	-0.520 / -0.635	-0.520 / -0.592	-0.503 / -0.549	-0.511 / -0.540	-0.514 / -0.534	-0.640 / -0.825	-0.640 / -0.755	-0.640 / -0.712	-0.623 / -0.669	-0.631 / -0.660	-0.634 / -0.654
OVER 250 TO 280	-0.580 / -0.790	-0.580 / -0.710	-0.580 / -0.661	-0.560 / -0.612	-0.571 / -0.603	-0.573 / -0.596	-0.710 / -0.920	-0.710 / -0.840	-0.710 / -0.791	-0.690 / -0.742	-0.701 / -0.733	-0.703 / -0.726
OVER 280 TO 315	-0.650 / -0.860	-0.650 / -0.780	-0.650 / -0.731	-0.630 / -0.682	-0.641 / -0.673	-0.643 / -0.666	-0.790 / -1.000	-0.790 / -0.920	-0.790 / -0.871	-0.770 / -0.822	-0.781 / -0.813	-0.783 / -0.806
OVER 315 TO 355	-0.730 / -0.960	-0.730 / -0.870	-0.730 / -0.819	-0.709 / -0.766	-0.719 / -0.755	-0.723 / -0.748	-0.900 / -1.130	-0.900 / -1.040	-0.900 / -0.989	-0.879 / -0.936	-0.889 / -0.925	-0.893 / -0.918
OVER 355 TO 400	-0.820 / -1.050	-0.820 / -0.960	-0.820 / -0.909	-0.799 / -0.856	-0.809 / -0.845	-0.813 / -0.838	-1.000 / -1.230	-1.000 / -1.140	-1.000 / -1.089	-0.979 / -1.036	-0.989 / -1.025	-0.993 / -1.018
OVER 400 TO 450	-0.920 / -1.170	-0.920 / -1.075	-0.920 / -1.017	-0.897 / -0.960	-0.907 / -0.947	-0.913 / -0.940	-1.100 / -1.350	-1.100 / -1.255	-1.100 / -1.197	-1.077 / -1.140	-1.087 / -1.127	-1.093 / -1.120
OVER 450 TO 500	-1.000 / -1.250	-1.000 / -1.155	-1.000 / -1.097	-0.977 / -1.040	-0.987 / -1.027	-0.993 / -1.020	-1.250 / -1.500	-1.250 / -1.405	-1.250 / -1.347	-1.227 / -1.250	-1.237 / -1.277	-1.243 / -1.270

THE ISO SYSTEM OF LIMITS AND FITS—TOLERANCES AND DEVIATIONS

Table 6-22. Tolerance Zones—External Dimensions (Shafts) (a14 ... a9, b14 ... b9) Dimensions in mm (ANSI B4.2)

SIZE	a14	a13	a12	a11	a10	a9	b14	b13	b12	b11	b10	b9
OVER 0 TO 3	-0.270 -0.520	-0.270 -0.410	-0.270 -0.370	-0.270 -0.330	-0.270 -0.310	-0.270 -0.295	-0.140 -0.390	-0.140 -0.280	-0.140 -0.240	-0.140 -0.200	-0.140 -0.180	-0.140 -0.165
OVER 3 TO 6	-0.270 -0.570	-0.270 -0.450	-0.270 -0.390	-0.270 -0.345	-0.270 -0.318	-0.270 -0.300	-0.140 -0.440	-0.140 -0.320	-0.140 -0.260	-0.140 -0.215	-0.140 -0.188	-0.140 -0.170
OVER 6 TO 10	-0.280 -0.640	-0.280 -0.500	-0.280 -0.430	-0.280 -0.370	-0.280 -0.338	-0.280 -0.316	-0.150 -0.510	-0.150 -0.370	-0.150 -0.300	-0.150 -0.240	-0.150 -0.208	-0.150 -0.186
OVER 10 TO 14	-0.290 -0.720	-0.290 -0.560	-0.290 -0.470	-0.290 -0.400	-0.290 -0.360	-0.290 -0.333	-0.150 -0.580	-0.150 -0.420	-0.150 -0.330	-0.150 -0.260	-0.150 -0.220	-0.150 -0.193
OVER 14 TO 18	-0.290 -0.720	-0.290 -0.560	-0.290 -0.470	-0.290 -0.400	-0.290 -0.360	-0.290 -0.333	-0.150 -0.580	-0.150 -0.420	-0.150 -0.330	-0.150 -0.260	-0.150 -0.220	-0.150 -0.193
OVER 18 TO 24	-0.300 -0.820	-0.300 -0.630	-0.300 -0.510	-0.300 -0.430	-0.300 -0.384	-0.300 -0.352	-0.160 -0.680	-0.160 -0.490	-0.160 -0.370	-0.160 -0.290	-0.160 -0.244	-0.160 -0.212
OVER 24 TO 30	-0.300 -0.820	-0.300 -0.630	-0.300 -0.510	-0.300 -0.430	-0.300 -0.384	-0.300 -0.352	-0.160 -0.680	-0.160 -0.490	-0.160 -0.370	-0.160 -0.290	-0.160 -0.244	-0.160 -0.212
OVER 30 TO 40	-0.310 -0.930	-0.310 -0.700	-0.310 -0.560	-0.310 -0.470	-0.310 -0.410	-0.310 -0.372	-0.170 -0.790	-0.170 -0.560	-0.170 -0.420	-0.170 -0.330	-0.170 -0.270	-0.170 -0.232
OVER 40 TO 50	-0.320 -0.940	-0.320 -0.710	-0.320 -0.570	-0.320 -0.480	-0.320 -0.420	-0.320 -0.382	-0.180 -0.800	-0.180 -0.570	-0.180 -0.430	-0.180 -0.340	-0.180 -0.280	-0.180 -0.242
OVER 50 TO 65	-0.340 -1.080	-0.340 -0.800	-0.340 -0.640	-0.340 -0.530	-0.340 -0.460	-0.340 -0.414	-0.190 -0.930	-0.190 -0.650	-0.190 -0.490	-0.190 -0.380	-0.190 -0.310	-0.190 -0.264
OVER 65 TO 80	-0.360 -1.100	-0.360 -0.820	-0.360 -0.660	-0.360 -0.550	-0.360 -0.480	-0.360 -0.434	-0.200 -0.940	-0.200 -0.660	-0.200 -0.500	-0.200 -0.390	-0.200 -0.320	-0.200 -0.274
OVER 80 TO 100	-0.380 -1.250	-0.380 -0.920	-0.380 -0.730	-0.380 -0.600	-0.380 -0.520	-0.380 -0.467	-0.220 -1.090	-0.220 -0.760	-0.220 -0.570	-0.220 -0.440	-0.220 -0.360	-0.220 -0.307
OVER 100 TO 120	-0.410 -1.280	-0.410 -0.950	-0.410 -0.760	-0.410 -0.630	-0.410 -0.550	-0.410 -0.497	-0.240 -1.110	-0.240 -0.780	-0.240 -0.590	-0.240 -0.460	-0.240 -0.380	-0.240 -0.327
OVER 120 TO 140	-0.460 -1.460	-0.460 -1.090	-0.460 -0.860	-0.460 -0.710	-0.460 -0.620	-0.460 -0.560	-0.260 -1.260	-0.260 -0.890	-0.260 -0.660	-0.260 -0.510	-0.260 -0.420	-0.260 -0.360
OVER 140 TO 160	-0.520 -1.520	-0.520 -1.150	-0.520 -0.920	-0.520 -0.770	-0.520 -0.680	-0.520 -0.620	-0.280 -1.280	-0.280 -0.910	-0.280 -0.680	-0.280 -0.530	-0.280 -0.440	-0.280 -0.380
OVER 160 TO 180	-0.580 -1.580	-0.580 -1.210	-0.580 -0.980	-0.580 -0.830	-0.580 -0.740	-0.580 -0.680	-0.310 -1.310	-0.310 -0.940	-0.310 -0.710	-0.310 -0.560	-0.310 -0.470	-0.310 -0.410
OVER 180 TO 200	-0.660 -1.810	-0.660 -1.380	-0.660 -1.120	-0.660 -0.950	-0.660 -0.845	-0.660 -0.775	-0.340 -1.490	-0.340 -1.060	-0.340 -0.800	-0.340 -0.630	-0.340 -0.525	-0.340 -0.455
OVER 200 TO 225	-0.740 -1.890	-0.740 -1.460	-0.740 -1.200	-0.740 -1.030	-0.740 -0.925	-0.740 -0.855	-0.380 -1.530	-0.380 -1.100	-0.380 -0.840	-0.380 -0.670	-0.380 -0.565	-0.380 -0.495
OVER 225 TO 250	-0.820 -1.970	-0.820 -1.540	-0.820 -1.280	-0.820 -1.110	-0.820 -1.005	-0.820 -0.935	-0.420 -1.570	-0.420 -1.140	-0.420 -0.880	-0.420 -0.710	-0.420 -0.605	-0.420 -0.535
OVER 250 TO 280	-0.920 -2.220	-0.920 -1.730	-0.920 -1.440	-0.920 -1.240	-0.920 -1.130	-0.920 -1.050	-0.480 -1.780	-0.480 -1.290	-0.480 -1.000	-0.480 -0.800	-0.480 -0.690	-0.480 -0.610
OVER 280 TO 315	-1.050 -2.350	-1.050 -1.860	-1.050 -1.570	-1.050 -1.370	-1.050 -1.260	-1.050 -1.180	-0.540 -1.840	-0.540 -1.350	-0.540 -1.060	-0.540 -0.860	-0.540 -0.750	-0.540 -0.670
OVER 315 TO 355	-1.200 -2.600	-1.200 -2.090	-1.200 -1.770	-1.200 -1.560	-1.200 -1.430	-1.200 -1.340	-0.600 -2.000	-0.600 -1.490	-0.600 -1.170	-0.600 -0.960	-0.600 -0.830	-0.600 -0.740
OVER 355 TO 400	-1.350 -2.750	-1.350 -2.240	-1.350 -1.920	-1.350 -1.710	-1.350 -1.580	-1.350 -1.490	-0.680 -2.080	-0.680 -1.570	-0.680 -1.250	-0.680 -1.040	-0.680 -0.910	-0.680 -0.820
OVER 400 TO 450	-1.500 -3.050	-1.500 -2.470	-1.500 -2.130	-1.500 -1.900	-1.500 -1.750	-1.500 -1.655	-0.760 -2.310	-0.760 -1.730	-0.760 -1.390	-0.760 -1.160	-0.760 -1.010	-0.760 -0.915
OVER 450 TO 500	-1.650 -3.200	-1.650 -2.620	-1.650 -2.280	-1.650 -2.050	-1.650 -1.900	-1.650 -1.805	-0.840 -2.390	-0.840 -1.810	-0.840 -1.470	-0.840 -1.240	-0.840 -1.090	-0.840 -0.995

Table 6-23. Tolerance Zones—External Dimensions (Shafts) (c13 ... c8, d12 ... d7) Dimensions in mm (ANSI B4.2)

SIZE	c13	c12	c11	c10	c9	c8	d12	d11	d10	d9	d8	d7
OVER 0 TO 3	-0.060 -0.200	-0.060 -0.160	-0.060 -0.120	-0.060 -0.100	-0.060 -0.085	-0.060 -0.074	-0.020 -0.120	-0.020 -0.080	-0.020 -0.060	-0.020 -0.045	-0.020 -0.034	-0.020 -0.030
OVER 3 TO 6	-0.070 -0.250	-0.070 -0.190	-0.070 -0.145	-0.070 -0.118	-0.070 -0.100	-0.070 -0.088	-0.030 -0.150	-0.030 -0.105	-0.030 -0.078	-0.030 -0.060	-0.030 -0.048	-0.030 -0.042
OVER 6 TO 10	-0.080 -0.300	-0.080 -0.230	-0.080 -0.170	-0.080 -0.138	-0.080 -0.116	-0.080 -0.102	-0.040 -0.190	-0.040 -0.130	-0.040 -0.098	-0.040 -0.076	-0.040 -0.062	-0.040 -0.055
OVER 10 TO 14	-0.095 -0.365	-0.095 -0.275	-0.095 -0.205	-0.095 -0.165	-0.095 -0.138	-0.095 -0.122	-0.050 -0.230	-0.050 -0.160	-0.050 -0.120	-0.050 -0.093	-0.050 -0.077	-0.050 -0.068
OVER 14 TO 18	-0.095 -0.365	-0.095 -0.275	-0.095 -0.205	-0.095 -0.165	-0.095 -0.138	-0.095 -0.122	-0.050 -0.230	-0.050 -0.160	-0.050 -0.120	-0.050 -0.093	-0.050 -0.077	-0.050 -0.068
OVER 18 TO 24	-0.110 -0.440	-0.110 -0.320	-0.110 -0.240	-0.110 -0.194	-0.110 -0.162	-0.110 -0.143	-0.065 -0.275	-0.065 -0.195	-0.065 -0.149	-0.065 -0.117	-0.065 -0.098	-0.065 -0.086
OVER 24 TO 30	-0.110 -0.440	-0.110 -0.320	-0.110 -0.240	-0.110 -0.194	-0.110 -0.162	-0.110 -0.143	-0.065 -0.275	-0.065 -0.195	-0.065 -0.149	-0.065 -0.117	-0.065 -0.098	-0.065 -0.086
OVER 30 TO 40	-0.120 -0.510	-0.120 -0.370	-0.120 -0.280	-0.120 -0.220	-0.120 -0.182	-0.120 -0.159	-0.080 -0.330	-0.080 -0.240	-0.080 -0.180	-0.080 -0.142	-0.080 -0.119	-0.080 -0.105
OVER 40 TO 50	-0.130 -0.520	-0.130 -0.380	-0.130 -0.290	-0.130 -0.230	-0.130 -0.192	-0.130 -0.169	-0.080 -0.330	-0.080 -0.240	-0.080 -0.180	-0.080 -0.142	-0.080 -0.119	-0.080 -0.105
OVER 50 TO 65	-0.140 -0.600	-0.140 -0.440	-0.140 -0.330	-0.140 -0.260	-0.140 -0.214	-0.140 -0.186	-0.100 -0.400	-0.100 -0.290	-0.100 -0.220	-0.100 -0.174	-0.100 -0.146	-0.100 -0.130
OVER 65 TO 80	-0.150 -0.610	-0.150 -0.450	-0.150 -0.340	-0.150 -0.270	-0.150 -0.224	-0.150 -0.196	-0.100 -0.400	-0.100 -0.290	-0.100 -0.220	-0.100 -0.174	-0.100 -0.146	-0.100 -0.130
OVER 80 TO 100	-0.170 -0.710	-0.170 -0.520	-0.170 -0.390	-0.170 -0.310	-0.170 -0.257	-0.170 -0.224	-0.120 -0.470	-0.120 -0.340	-0.120 -0.260	-0.120 -0.207	-0.120 -0.174	-0.120 -0.155
OVER 100 TO 120	-0.180 -0.720	-0.180 -0.530	-0.180 -0.400	-0.180 -0.320	-0.180 -0.267	-0.180 -0.234	-0.120 -0.470	-0.120 -0.340	-0.120 -0.260	-0.120 -0.207	-0.120 -0.174	-0.120 -0.155
OVER 120 TO 140	-0.200 -0.830	-0.200 -0.600	-0.200 -0.450	-0.200 -0.360	-0.200 -0.300	-0.200 -0.263	-0.145 -0.545	-0.145 -0.395	-0.145 -0.305	-0.145 -0.245	-0.145 -0.208	-0.145 -0.185
OVER 140 TO 160	-0.210 -0.840	-0.210 -0.610	-0.210 -0.460	-0.210 -0.370	-0.210 -0.310	-0.210 -0.273	-0.145 -0.545	-0.145 -0.395	-0.145 -0.305	-0.145 -0.245	-0.145 -0.208	-0.145 -0.185
OVER 160 TO 180	-0.230 -0.860	-0.230 -0.630	-0.230 -0.480	-0.230 -0.390	-0.230 -0.330	-0.230 -0.293	-0.145 -0.545	-0.145 -0.395	-0.145 -0.305	-0.145 -0.245	-0.145 -0.208	-0.145 -0.185
OVER 180 TO 200	-0.240 -0.960	-0.240 -0.700	-0.240 -0.530	-0.240 -0.425	-0.240 -0.355	-0.240 -0.312	-0.170 -0.630	-0.170 -0.460	-0.170 -0.355	-0.170 -0.285	-0.170 -0.242	-0.170 -0.216
OVER 200 TO 225	-0.260 -0.980	-0.260 -0.720	-0.260 -0.550	-0.260 -0.445	-0.260 -0.375	-0.260 -0.332	-0.170 -0.630	-0.170 -0.460	-0.170 -0.355	-0.170 -0.285	-0.170 -0.242	-0.170 -0.216
OVER 225 TO 250	-0.280 -1.000	-0.280 -0.740	-0.280 -0.570	-0.280 -0.465	-0.280 -0.395	-0.280 -0.352	-0.170 -0.630	-0.170 -0.460	-0.170 -0.355	-0.170 -0.285	-0.170 -0.242	-0.170 -0.216
OVER 250 TO 280	-0.300 -1.110	-0.300 -0.820	-0.300 -0.620	-0.300 -0.510	-0.300 -0.430	-0.300 -0.381	-0.190 -0.710	-0.190 -0.510	-0.190 -0.400	-0.190 -0.320	-0.190 -0.271	-0.190 -0.242
OVER 280 TO 315	-0.330 -1.140	-0.330 -0.850	-0.330 -0.650	-0.330 -0.540	-0.330 -0.460	-0.330 -0.411	-0.190 -0.710	-0.190 -0.510	-0.190 -0.400	-0.190 -0.320	-0.190 -0.271	-0.190 -0.242
OVER 315 TO 355	-0.360 -1.250	-0.360 -0.930	-0.360 -0.720	-0.360 -0.590	-0.360 -0.500	-0.360 -0.449	-0.210 -0.780	-0.210 -0.570	-0.210 -0.440	-0.210 -0.350	-0.210 -0.299	-0.210 -0.267
OVER 355 TO 400	-0.400 -1.290	-0.400 -0.970	-0.400 -0.760	-0.400 -0.630	-0.400 -0.540	-0.400 -0.489	-0.210 -0.780	-0.210 -0.570	-0.210 -0.440	-0.210 -0.350	-0.210 -0.299	-0.210 -0.267
OVER 400 TO 450	-0.440 -1.410	-0.440 -1.070	-0.440 -0.840	-0.440 -0.690	-0.440 -0.595	-0.440 -0.537	-0.230 -0.860	-0.230 -0.630	-0.230 -0.480	-0.230 -0.385	-0.230 -0.327	-0.230 -0.293
OVER 450 TO 500	-0.480 -1.450	-0.480 -1.110	-0.480 -0.880	-0.480 -0.730	-0.480 -0.635	-0.480 -0.577	-0.230 -0.860	-0.230 -0.630	-0.230 -0.480	-0.230 -0.385	-0.230 -0.327	-0.230 -0.293

THE ISO SYSTEM OF LIMITS AND FITS—TOLERANCES AND DEVIATIONS

Table 6-24. Tolerance Zones—External Dimensions (Shafts) (e11 ... e6, f10 ... f5) Dimensions in mm (ANSI B4.2)

SIZE	e11	e10	e9	e8	e7	e6	f10	f9	f8	f7	f6	f5
OVER 0 TO 3	-0.014 -0.074	-0.014 -0.054	-0.014 -0.039	-0.014 -0.028	-0.014 -0.024	-0.014 -0.020	-0.006 -0.046	-0.006 -0.031	-0.006 -0.020	-0.006 -0.016	-0.006 -0.012	-0.006 -0.010
OVER 3 TO 6	-0.020 -0.095	-0.020 -0.068	-0.020 -0.050	-0.020 -0.038	-0.020 -0.032	-0.020 -0.028	-0.010 -0.058	-0.010 -0.040	-0.010 -0.028	-0.010 -0.022	-0.010 -0.018	-0.010 -0.015
OVER 6 TO 10	-0.025 -0.115	-0.025 -0.083	-0.025 -0.061	-0.025 -0.047	-0.025 -0.040	-0.025 -0.034	-0.013 -0.071	-0.013 -0.049	-0.013 -0.035	-0.013 -0.028	-0.013 -0.022	-0.013 -0.019
OVER 10 TO 14	-0.032 -0.142	-0.032 -0.102	-0.032 -0.075	-0.032 -0.059	-0.032 -0.050	-0.032 -0.043	-0.016 -0.086	-0.016 -0.059	-0.016 -0.043	-0.016 -0.034	-0.016 -0.027	-0.016 -0.024
OVER 14 TO 18	-0.032 -0.142	-0.032 -0.102	-0.032 -0.075	-0.032 -0.059	-0.032 -0.050	-0.032 -0.043	-0.016 -0.086	-0.016 -0.059	-0.016 -0.043	-0.016 -0.034	-0.016 -0.027	-0.016 -0.024
OVER 18 TO 24	-0.040 -0.170	-0.040 -0.124	-0.040 -0.092	-0.040 -0.073	-0.040 -0.061	-0.040 -0.053	-0.020 -0.104	-0.020 -0.072	-0.020 -0.053	-0.020 -0.041	-0.020 -0.033	-0.020 -0.029
OVER 24 TO 30	-0.040 -0.170	-0.040 -0.124	-0.040 -0.092	-0.040 -0.073	-0.040 -0.061	-0.040 -0.053	-0.020 -0.104	-0.020 -0.072	-0.020 -0.053	-0.020 -0.041	-0.020 -0.033	-0.020 -0.029
OVER 30 TO 40	-0.050 -0.210	-0.050 -0.150	-0.050 -0.112	-0.050 -0.089	-0.050 -0.075	-0.050 -0.066	-0.025 -0.125	-0.025 -0.087	-0.025 -0.064	-0.025 -0.050	-0.025 -0.041	-0.025 -0.036
OVER 40 TO 50	-0.050 -0.210	-0.050 -0.150	-0.050 -0.112	-0.050 -0.089	-0.050 -0.075	-0.050 -0.066	-0.025 -0.125	-0.025 -0.087	-0.025 -0.064	-0.025 -0.050	-0.025 -0.041	-0.025 -0.036
OVER 50 TO 65	-0.060 -0.250	-0.060 -0.180	-0.060 -0.134	-0.060 -0.106	-0.060 -0.090	-0.060 -0.079	-0.030 -0.150	-0.030 -0.104	-0.030 -0.076	-0.030 -0.060	-0.030 -0.049	-0.030 -0.043
OVER 65 TO 80	-0.060 -0.250	-0.060 -0.180	-0.060 -0.134	-0.060 -0.106	-0.060 -0.090	-0.060 -0.079	-0.030 -0.150	-0.030 -0.104	-0.030 -0.076	-0.030 -0.060	-0.030 -0.049	-0.030 -0.043
OVER 80 TO 100	-0.072 -0.292	-0.072 -0.212	-0.072 -0.159	-0.072 -0.126	-0.072 -0.107	-0.072 -0.094	-0.036 -0.176	-0.036 -0.123	-0.036 -0.090	-0.036 -0.071	-0.036 -0.058	-0.036 -0.051
OVER 100 TO 120	-0.072 -0.292	-0.072 -0.212	-0.072 -0.159	-0.072 -0.126	-0.072 -0.107	-0.072 -0.094	-0.036 -0.176	-0.036 -0.123	-0.036 -0.090	-0.036 -0.071	-0.036 -0.058	-0.036 -0.051
OVER 120 TO 140	-0.085 -0.335	-0.085 -0.245	-0.085 -0.185	-0.085 -0.148	-0.085 -0.125	-0.085 -0.110	-0.043 -0.203	-0.043 -0.143	-0.043 -0.106	-0.043 -0.083	-0.043 -0.068	-0.043 -0.061
OVER 140 TO 160	-0.085 -0.335	-0.085 -0.245	-0.085 -0.185	-0.085 -0.148	-0.085 -0.125	-0.085 -0.110	-0.043 -0.203	-0.043 -0.143	-0.043 -0.106	-0.043 -0.083	-0.043 -0.068	-0.043 -0.061
OVER 160 TO 180	-0.085 -0.335	-0.085 -0.245	-0.085 -0.185	-0.085 -0.148	-0.085 -0.125	-0.085 -0.110	-0.043 -0.203	-0.043 -0.143	-0.043 -0.106	-0.043 -0.083	-0.043 -0.068	-0.043 -0.061
OVER 180 TO 200	-0.100 -0.390	-0.100 -0.285	-0.100 -0.215	-0.100 -0.172	-0.100 -0.146	-0.100 -0.129	-0.050 -0.235	-0.050 -0.165	-0.050 -0.122	-0.050 -0.096	-0.050 -0.079	-0.050 -0.070
OVER 200 TO 225	-0.100 -0.390	-0.100 -0.285	-0.100 -0.215	-0.100 -0.172	-0.100 -0.146	-0.100 -0.129	-0.050 -0.235	-0.050 -0.165	-0.050 -0.122	-0.050 -0.096	-0.050 -0.079	-0.050 -0.070
OVER 225 TO 250	-0.100 -0.390	-0.100 -0.285	-0.100 -0.215	-0.100 -0.172	-0.100 -0.146	-0.100 -0.129	-0.050 -0.235	-0.050 -0.165	-0.050 -0.122	-0.050 -0.096	-0.050 -0.079	-0.050 -0.070
OVER 250 TO 280	-0.110 -0.430	-0.110 -0.320	-0.110 -0.240	-0.110 -0.191	-0.110 -0.162	-0.110 -0.142	-0.056 -0.266	-0.056 -0.186	-0.056 -0.137	-0.056 -0.108	-0.056 -0.088	-0.056 -0.079
OVER 280 TO 315	-0.110 -0.430	-0.110 -0.320	-0.110 -0.240	-0.110 -0.191	-0.110 -0.162	-0.110 -0.142	-0.056 -0.266	-0.056 -0.186	-0.056 -0.137	-0.056 -0.108	-0.056 -0.088	-0.056 -0.079
OVER 315 TO 355	-0.125 -0.485	-0.125 -0.355	-0.125 -0.265	-0.125 -0.214	-0.125 -0.182	-0.125 -0.161	-0.062 -0.292	-0.062 -0.202	-0.062 -0.151	-0.062 -0.119	-0.062 -0.098	-0.062 -0.087
OVER 355 TO 400	-0.125 -0.485	-0.125 -0.355	-0.125 -0.265	-0.125 -0.214	-0.125 -0.182	-0.125 -0.161	-0.062 -0.292	-0.062 -0.202	-0.062 -0.151	-0.062 -0.119	-0.062 -0.098	-0.062 -0.087
OVER 400 TO 450	-0.135 -0.535	-0.135 -0.385	-0.135 -0.290	-0.135 -0.232	-0.135 -0.198	-0.135 -0.175	-0.068 -0.318	-0.068 -0.223	-0.068 -0.165	-0.068 -0.131	-0.068 -0.108	-0.068 -0.095
OVER 450 TO 500	-0.135 -0.535	-0.135 -0.385	-0.135 -0.290	-0.135 -0.232	-0.135 -0.198	-0.135 -0.175	-0.068 -0.318	-0.068 -0.223	-0.068 -0.165	-0.068 -0.131	-0.068 -0.108	-0.068 -0.095

Table 6-25. Tolerance Zones—External Dimensions (Shafts) (g9 ... g4, j7 ... j5) Dimensions in mm (ANSI B4.2)

SIZE	g9	g8	g7	g6	g5	g4	j7	j6	j5
OVER 0 TO 3	-0.002 / -0.027	-0.002 / -0.016	-0.002 / -0.012	-0.002 / -0.008	-0.002 / -0.006	-0.002 / -0.005	+0.006 / -0.004	+0.004 / -0.002	+0.002 / -0.002
OVER 3 TO 6	-0.004 / -0.034	-0.004 / -0.022	-0.004 / -0.016	-0.004 / -0.012	-0.004 / -0.009	-0.004 / -0.008	+0.008 / -0.004	+0.006 / -0.002	+0.003 / -0.002
OVER 6 TO 10	-0.005 / -0.041	-0.005 / -0.027	-0.005 / -0.020	-0.005 / -0.014	-0.005 / -0.011	-0.005 / -0.009	+0.010 / -0.005	+0.007 / -0.002	+0.004 / -0.002
OVER 10 TO 14	-0.006 / -0.049	-0.006 / -0.033	-0.006 / -0.024	-0.006 / -0.017	-0.006 / -0.014	-0.006 / -0.011	+0.012 / -0.006	+0.008 / -0.003	+0.005 / -0.003
OVER 14 TO 18	-0.006 / -0.049	-0.006 / -0.033	-0.006 / -0.024	-0.006 / -0.017	-0.006 / -0.014	-0.006 / -0.011	+0.012 / -0.006	+0.008 / -0.003	+0.005 / -0.003
OVER 18 TO 24	-0.007 / -0.059	-0.007 / -0.040	-0.007 / -0.028	-0.007 / -0.020	-0.007 / -0.016	-0.007 / -0.013	+0.013 / -0.008	+0.009 / -0.004	+0.005 / -0.004
OVER 24 TO 30	-0.007 / -0.059	-0.007 / -0.040	-0.007 / -0.028	-0.007 / -0.020	-0.007 / -0.016	-0.007 / -0.013	+0.013 / -0.008	+0.009 / -0.004	+0.005 / -0.004
OVER 30 TO 40	-0.009 / -0.071	-0.009 / -0.048	-0.009 / -0.034	-0.009 / -0.025	-0.009 / -0.020	-0.009 / -0.016	+0.015 / -0.010	+0.011 / -0.005	+0.006 / -0.005
OVER 40 TO 50	-0.009 / -0.071	-0.009 / -0.048	-0.009 / -0.034	-0.009 / -0.025	-0.009 / -0.020	-0.009 / -0.016	+0.015 / -0.010	+0.011 / -0.005	+0.006 / -0.005
OVER 50 TO 65	-0.010 / -0.084	-0.010 / -0.056	-0.010 / -0.040	-0.010 / -0.029	-0.010 / -0.023	-0.010 / -0.018	+0.018 / -0.012	+0.012 / -0.007	+0.006 / -0.007
OVER 65 TO 80	-0.010 / -0.084	-0.010 / -0.056	-0.010 / -0.040	-0.010 / -0.029	-0.010 / -0.023	-0.010 / -0.018	+0.018 / -0.012	+0.012 / -0.007	+0.006 / -0.007
OVER 80 TO 100	-0.012 / -0.099	-0.012 / -0.066	-0.012 / -0.047	-0.012 / -0.034	-0.012 / -0.027	-0.012 / -0.022	+0.020 / -0.015	+0.013 / -0.009	+0.006 / -0.009
OVER 100 TO 120	-0.012 / -0.099	-0.012 / -0.066	-0.012 / -0.047	-0.012 / -0.034	-0.012 / -0.027	-0.012 / -0.022	+0.020 / -0.015	+0.013 / -0.009	+0.006 / -0.009
OVER 120 TO 140	-0.014 / -0.114	-0.014 / -0.077	-0.014 / -0.054	-0.014 / -0.039	-0.014 / -0.032	-0.014 / -0.026	+0.022 / -0.018	+0.014 / -0.011	+0.007 / -0.011
OVER 140 TO 160	-0.014 / -0.114	-0.014 / -0.077	-0.014 / -0.054	-0.014 / -0.039	-0.014 / -0.032	-0.014 / -0.026	+0.022 / -0.018	+0.014 / -0.011	+0.007 / -0.011
OVER 160 TO 180	-0.014 / -0.114	-0.014 / -0.077	-0.014 / -0.054	-0.014 / -0.039	-0.014 / -0.032	-0.014 / -0.026	+0.022 / -0.018	+0.014 / -0.011	+0.007 / -0.011
OVER 180 TO 200	-0.015 / -0.130	-0.015 / -0.087	-0.015 / -0.061	-0.015 / -0.044	-0.015 / -0.035	-0.015 / -0.029	+0.025 / -0.021	+0.016 / -0.013	+0.007 / -0.013
OVER 200 TO 225	-0.015 / -0.130	-0.015 / -0.087	-0.015 / -0.061	-0.015 / -0.044	-0.015 / -0.035	-0.015 / -0.029	+0.025 / -0.021	+0.016 / -0.013	+0.007 / -0.013
OVER 225 TO 250	-0.015 / -0.130	-0.015 / -0.087	-0.015 / -0.061	-0.015 / -0.044	-0.015 / -0.035	-0.015 / -0.029	+0.025 / -0.021	+0.016 / -0.013	+0.007 / -0.013
OVER 250 TO 280	-0.017 / -0.147	-0.017 / -0.098	-0.017 / -0.069	-0.017 / -0.049	-0.017 / -0.040	-0.017 / -0.033	+0.026 / -0.026	+0.016 / -0.016	+0.007 / -0.016
OVER 280 TO 315	-0.017 / -0.147	-0.017 / -0.098	-0.017 / -0.069	-0.017 / -0.049	-0.017 / -0.040	-0.017 / -0.033	+0.026 / -0.026	+0.016 / -0.016	+0.007 / -0.016
OVER 315 TO 355	-0.018 / -0.158	-0.018 / -0.107	-0.018 / -0.075	-0.018 / -0.054	-0.018 / -0.043	-0.018 / -0.036	+0.029 / -0.028	+0.018 / -0.018	+0.007 / -0.018
OVER 355 TO 400	-0.018 / -0.158	-0.018 / -0.107	-0.018 / -0.075	-0.018 / -0.054	-0.018 / -0.043	-0.018 / -0.036	+0.029 / -0.028	+0.018 / -0.018	+0.007 / -0.018
OVER 400 TO 450	-0.020 / -0.175	-0.020 / -0.117	-0.020 / -0.083	-0.020 / -0.060	-0.020 / -0.047	-0.020 / -0.040	+0.031 / -0.032	+0.020 / -0.020	+0.007 / -0.020
OVER 450 TO 500	-0.020 / -0.175	-0.020 / -0.117	-0.020 / -0.083	-0.020 / -0.060	-0.020 / -0.047	-0.020 / -0.040	+0.031 / -0.032	+0.020 / -0.020	+0.007 / -0.020

THE ISO SYSTEM OF LIMITS AND FITS—TOLERANCES AND DEVIATIONS

Table 6-26. Tolerance Zones—External Dimensions (Shafts) (h16 ... h1) Dimensions in mm (ANSI B4.2)

SIZE	h16	h15	h14	h13	h12	h11	h10	h9	h8	h7	h6	h5	h4	h3	h2	h1
OVER 0 TO 3	0.000 -0.600	0.000 -0.400	0.000 -0.250	0.000 -0.140	0.000 -0.100	0.000 -0.060	0.000 -0.040	0.000 -0.025	0.000 -0.014	0.000 -0.010	0.000 -0.006	0.000 -0.004	0.000 -0.003	0.000 -0.002	0.0000 -0.0012	0.0000 -0.0008
OVER 3 TO 6	0.000 -0.750	0.000 -0.480	0.000 -0.300	0.000 -0.180	0.000 -0.120	0.000 -0.075	0.000 -0.048	0.000 -0.030	0.000 -0.018	0.000 -0.012	0.000 -0.008	0.000 -0.005	0.000 -0.004	0.0000 -0.0025	0.0000 -0.0015	0.000 -0.001
OVER 6 TO 10	0.000 -0.900	0.000 -0.580	0.000 -0.360	0.000 -0.220	0.000 -0.150	0.000 -0.090	0.000 -0.058	0.000 -0.036	0.000 -0.022	0.000 -0.015	0.000 -0.009	0.000 -0.006	0.000 -0.004	0.0000 -0.0025	0.0000 -0.0015	0.000 -0.001
OVER 10 TO 14	0.000 -1.100	0.000 -0.700	0.000 -0.430	0.000 -0.270	0.000 -0.180	0.000 -0.110	0.000 -0.070	0.000 -0.043	0.000 -0.027	0.000 -0.018	0.000 -0.011	0.000 -0.008	0.000 -0.005	0.000 -0.003	0.000 -0.002	0.0000 -0.0012
OVER 14 TO 18	0.000 -1.100	0.000 -0.700	0.000 -0.430	0.000 -0.270	0.000 -0.180	0.000 -0.110	0.000 -0.070	0.000 -0.043	0.000 -0.027	0.000 -0.018	0.000 -0.011	0.000 -0.008	0.000 -0.005	0.000 -0.003	0.000 -0.002	0.0000 -0.0012
OVER 18 TO 24	0.000 -1.300	0.000 -0.840	0.000 -0.520	0.000 -0.330	0.000 -0.210	0.000 -0.130	0.000 -0.084	0.000 -0.052	0.000 -0.033	0.000 -0.021	0.000 -0.013	0.000 -0.009	0.000 -0.006	0.000 -0.004	0.0000 -0.0025	0.0000 -0.0015
OVER 24 TO 30	0.000 -1.300	0.000 -0.840	0.000 -0.520	0.000 -0.330	0.000 -0.210	0.000 -0.130	0.000 -0.084	0.000 -0.052	0.000 -0.033	0.000 -0.021	0.000 -0.013	0.000 -0.009	0.000 -0.006	0.000 -0.004	0.0000 -0.0025	0.0000 -0.0015
OVER 30 TO 40	0.000 -1.600	0.000 -1.000	0.000 -0.620	0.000 -0.390	0.000 -0.250	0.000 -0.160	0.000 -0.100	0.000 -0.062	0.000 -0.039	0.000 -0.025	0.000 -0.016	0.000 -0.011	0.000 -0.007	0.000 -0.004	0.0000 -0.0025	0.0000 -0.0015
OVER 40 TO 50	0.000 -1.600	0.000 -1.000	0.000 -0.620	0.000 -0.390	0.000 -0.250	0.000 -0.160	0.000 -0.100	0.000 -0.062	0.000 -0.039	0.000 -0.025	0.000 -0.016	0.000 -0.011	0.000 -0.007	0.000 -0.004	0.0000 -0.0025	0.0000 -0.0015
OVER 50 TO 65	0.000 -1.900	0.000 -1.200	0.000 -0.740	0.000 -0.460	0.000 -0.300	0.000 -0.190	0.000 -0.120	0.000 -0.074	0.000 -0.046	0.000 -0.030	0.000 -0.019	0.000 -0.013	0.000 -0.008	0.000 -0.005	0.000 -0.003	0.000 -0.002
OVER 65 TO 80	0.000 -1.900	0.000 -1.200	0.000 -0.740	0.000 -0.460	0.000 -0.300	0.000 -0.190	0.000 -0.120	0.000 -0.074	0.000 -0.046	0.000 -0.030	0.000 -0.019	0.000 -0.013	0.000 -0.008	0.000 -0.005	0.000 -0.003	0.000 -0.002
OVER 80 TO 100	0.000 -2.200	0.000 -1.400	0.000 -0.870	0.000 -0.540	0.000 -0.350	0.000 -0.220	0.000 -0.140	0.000 -0.087	0.000 -0.054	0.000 -0.035	0.000 -0.022	0.000 -0.015	0.000 -0.010	0.000 -0.006	0.000 -0.004	0.0000 -0.0025
OVER 100 TO 120	0.000 -2.200	0.000 -1.400	0.000 -0.870	0.000 -0.540	0.000 -0.350	0.000 -0.220	0.000 -0.140	0.000 -0.087	0.000 -0.054	0.000 -0.035	0.000 -0.022	0.000 -0.015	0.000 -0.010	0.000 -0.006	0.000 -0.004	0.0000 -0.0025
OVER 120 TO 140	0.000 -2.500	0.000 -1.600	0.000 -1.000	0.000 -0.630	0.000 -0.400	0.000 -0.250	0.000 -0.160	0.000 -0.100	0.000 -0.063	0.000 -0.040	0.000 -0.025	0.000 -0.018	0.000 -0.012	0.000 -0.008	0.000 -0.005	0.000 -0.0035
OVER 140 TO 160	0.000 -2.500	0.000 -1.600	0.000 -1.000	0.000 -0.630	0.000 -0.400	0.000 -0.250	0.000 -0.160	0.000 -0.100	0.000 -0.063	0.000 -0.040	0.000 -0.025	0.000 -0.018	0.000 -0.012	0.000 -0.008	0.000 -0.005	0.000 -0.0035
OVER 160 TO 180	0.000 -2.500	0.000 -1.600	0.000 -1.000	0.000 -0.630	0.000 -0.400	0.000 -0.250	0.000 -0.160	0.000 -0.100	0.000 -0.063	0.000 -0.040	0.000 -0.025	0.000 -0.018	0.000 -0.012	0.000 -0.008	0.000 -0.005	0.0000 -0.0035
OVER 180 TO 200	0.000 -2.900	0.000 -1.850	0.000 -1.150	0.000 -0.720	0.000 -0.460	0.000 -0.290	0.000 -0.185	0.000 -0.115	0.000 -0.072	0.000 -0.046	0.000 -0.029	0.000 -0.020	0.000 -0.014	0.000 -0.010	0.000 -0.007	0.0000 -0.0045
OVER 200 TO 225	0.000 -2.900	0.000 -1.850	0.000 -1.150	0.000 -0.720	0.000 -0.460	0.000 -0.290	0.000 -0.185	0.000 -0.115	0.000 -0.072	0.000 -0.046	0.000 -0.029	0.000 -0.020	0.000 -0.014	0.000 -0.010	0.000 -0.007	0.0000 -0.0045
OVER 225 TO 250	0.000 -2.900	0.000 -1.850	0.000 -1.150	0.000 -0.720	0.000 -0.460	0.000 -0.290	0.000 -0.185	0.000 -0.115	0.000 -0.072	0.000 -0.046	0.000 -0.029	0.000 -0.020	0.000 -0.014	0.000 -0.010	0.000 -0.007	0.0000 -0.0045
OVER 250 TO 280	0.000 -3.200	0.000 -2.100	0.000 -1.300	0.000 -0.810	0.000 -0.520	0.000 -0.320	0.000 -0.210	0.000 -0.130	0.000 -0.081	0.000 -0.052	0.000 -0.032	0.000 -0.023	0.000 -0.016	0.000 -0.012	0.000 -0.008	0.000 -0.006
OVER 280 TO 315	0.000 -3.200	0.000 -2.100	0.000 -1.300	0.000 -0.810	0.000 -0.520	0.000 -0.320	0.000 -0.210	0.000 -0.130	0.000 -0.081	0.000 -0.052	0.000 -0.032	0.000 -0.023	0.000 -0.016	0.000 -0.012	0.000 -0.008	0.000 -0.006
OVER 315 TO 355	0.000 -3.600	0.000 -2.300	0.000 -1.400	0.000 -0.890	0.000 -0.570	0.000 -0.360	0.000 -0.230	0.000 -0.140	0.000 -0.089	0.000 -0.057	0.000 -0.036	0.000 -0.025	0.000 -0.018	0.000 -0.013	0.000 -0.009	0.000 -0.007
OVER 355 TO 400	0.000 -3.600	0.000 -2.300	0.000 -1.400	0.000 -0.890	0.000 -0.570	0.000 -0.360	0.000 -0.230	0.000 -0.140	0.000 -0.089	0.000 -0.057	0.000 -0.036	0.000 -0.025	0.000 -0.018	0.000 -0.013	0.000 -0.009	0.000 -0.007
OVER 400 TO 450	0.000 -4.000	0.000 -2.500	0.000 -1.550	0.000 -0.970	0.000 -0.630	0.000 -0.400	0.000 -0.250	0.000 -0.155	0.000 -0.097	0.000 -0.063	0.000 -0.040	0.000 -0.027	0.000 -0.020	0.000 -0.015	0.000 -0.010	0.000 -0.008
OVER 450 TO 500	0.000 -4.000	0.000 -2.500	0.000 -1.550	0.000 -0.970	0.000 -0.630	0.000 -0.400	0.000 -0.250	0.000 -0.155	0.000 -0.097	0.000 -0.063	0.000 -0.040	0.000 -0.027	0.000 -0.020	0.000 -0.015	0.000 -0.010	0.000 -0.008

Table 6-27. Tolerance Zones—External Dimensions (Shafts) (js16 ... js1) Dimensions in mm (ANSI B4.2)

SIZE	js16	js15	js14	js13	js12	js11	js10	js9	js8	js7	js6	js5	js4	js3	js2	js1
0 3	+0.300 -0.300	+0.200 -0.200	+0.125 -0.125	+0.070 -0.070	+0.050 -0.050	+0.030 -0.030	+0.020 -0.020	+0.012 -0.012	+0.007 -0.007	+0.005 -0.005	+0.003 -0.003	+0.002 -0.002	+0.0015 -0.0015	+0.001 -0.001	+0.0006 -0.0006	+0.0004 -0.0004
3 6	+0.375 -0.375	+0.240 -0.240	+0.150 -0.150	+0.090 -0.090	+0.060 -0.060	+0.037 -0.037	+0.024 -0.024	+0.015 -0.015	+0.009 -0.009	+0.006 -0.006	+0.004 -0.004	+0.0025 -0.0025	+0.002 -0.002	+0.00125 -0.00125	+0.00075 -0.00075	+0.0005 -0.0005
6 10	+0.450 -0.450	+0.290 -0.290	+0.180 -0.180	+0.110 -0.110	+0.075 -0.075	+0.045 -0.045	+0.029 -0.029	+0.018 -0.018	+0.011 -0.011	+0.007 -0.007	+0.0045 -0.0045	+0.003 -0.003	+0.002 -0.002	+0.00125 -0.00125	+0.00075 -0.00075	+0.0005 -0.0005
10 14	+0.550 -0.550	+0.350 -0.350	+0.215 -0.215	+0.135 -0.135	+0.090 -0.090	+0.055 -0.055	+0.035 -0.035	+0.021 -0.021	+0.013 -0.013	+0.009 -0.009	+0.0055 -0.0055	+0.004 -0.004	+0.0025 -0.0025	+0.0015 -0.0015	+0.001 -0.001	+0.0006 -0.0006
14 18	+0.550 -0.550	+0.350 -0.350	+0.215 -0.215	+0.135 -0.135	+0.090 -0.090	+0.055 -0.055	+0.035 -0.035	+0.021 -0.021	+0.013 -0.013	+0.009 -0.009	+0.0055 -0.0055	+0.004 -0.004	+0.0025 -0.0025	+0.0015 -0.0015	+0.001 -0.001	+0.0006 -0.0006
18 24	+0.650 -0.650	+0.420 -0.420	+0.260 -0.260	+0.165 -0.165	+0.105 -0.105	+0.065 -0.065	+0.042 -0.042	+0.026 -0.026	+0.016 -0.016	+0.010 -0.010	+0.0065 -0.0065	+0.0045 -0.0045	+0.003 -0.003	+0.002 -0.002	+0.00125 -0.00125	+0.00075 -0.00075
24 30	+0.650 -0.650	+0.420 -0.420	+0.260 -0.260	+0.165 -0.165	+0.105 -0.105	+0.065 -0.065	+0.042 -0.042	+0.026 -0.026	+0.016 -0.016	+0.010 -0.010	+0.0065 -0.0065	+0.0045 -0.0045	+0.003 -0.003	+0.002 -0.002	+0.00125 -0.00125	+0.00075 -0.00075
30 40	+0.800 -0.800	+0.500 -0.500	+0.310 -0.310	+0.195 -0.195	+0.125 -0.125	+0.080 -0.080	+0.050 -0.050	+0.031 -0.031	+0.019 -0.019	+0.012 -0.012	+0.008 -0.008	+0.0055 -0.0055	+0.0035 -0.0035	+0.002 -0.002	+0.00125 -0.00125	+0.00075 -0.00075
40 50	+0.800 -0.800	+0.500 -0.500	+0.310 -0.310	+0.195 -0.195	+0.125 -0.125	+0.080 -0.080	+0.050 -0.050	+0.031 -0.031	+0.019 -0.019	+0.012 -0.012	+0.008 -0.008	+0.0055 -0.0055	+0.0035 -0.0035	+0.002 -0.002	+0.00125 -0.00125	+0.00075 -0.00075
50 65	+0.950 -0.950	+0.600 -0.600	+0.370 -0.370	+0.230 -0.230	+0.150 -0.150	+0.095 -0.095	+0.060 -0.060	+0.037 -0.037	+0.023 -0.023	+0.015 -0.015	+0.0095 -0.0095	+0.0065 -0.0065	+0.004 -0.004	+0.0025 -0.0025	+0.0015 -0.0015	+0.001 -0.001
65 80	+0.950 -0.950	+0.600 -0.600	+0.370 -0.370	+0.230 -0.230	+0.150 -0.150	+0.095 -0.095	+0.060 -0.060	+0.037 -0.037	+0.023 -0.023	+0.015 -0.015	+0.0095 -0.0095	+0.0065 -0.0065	+0.004 -0.004	+0.0025 -0.0025	+0.0015 -0.0015	+0.001 -0.001
80 100	+1.100 -1.100	+0.700 -0.700	+0.435 -0.435	+0.270 -0.270	+0.175 -0.175	+0.110 -0.110	+0.070 -0.070	+0.043 -0.043	+0.027 -0.027	+0.017 -0.017	+0.011 -0.011	+0.0075 -0.0075	+0.005 -0.005	+0.003 -0.003	+0.002 -0.002	+0.00125 -0.00125
100 120	+1.100 -1.100	+0.700 -0.700	+0.435 -0.435	+0.270 -0.270	+0.175 -0.175	+0.110 -0.110	+0.070 -0.070	+0.043 -0.043	+0.027 -0.027	+0.017 -0.017	+0.011 -0.011	+0.0075 -0.0075	+0.005 -0.005	+0.003 -0.003	+0.002 -0.002	+0.00125 -0.00125
120 140	+1.250 -1.250	+0.800 -0.800	+0.500 -0.500	+0.315 -0.315	+0.200 -0.200	+0.125 -0.125	+0.080 -0.080	+0.050 -0.050	+0.031 -0.031	+0.020 -0.020	+0.0125 -0.0125	+0.009 -0.009	+0.006 -0.006	+0.004 -0.004	+0.0025 -0.0025	+0.00175 -0.00175
140 160	+1.250 -1.250	+0.800 -0.800	+0.500 -0.500	+0.315 -0.315	+0.200 -0.200	+0.125 -0.125	+0.080 -0.080	+0.050 -0.050	+0.031 -0.031	+0.020 -0.020	+0.0125 -0.0125	+0.009 -0.009	+0.006 -0.006	+0.004 -0.004	+0.0025 -0.0025	+0.00175 -0.00175
160 180	+1.250 -1.250	+0.800 -0.800	+0.500 -0.500	+0.315 -0.315	+0.200 -0.200	+0.125 -0.125	+0.080 -0.080	+0.050 -0.050	+0.031 -0.031	+0.020 -0.020	+0.0125 -0.0125	+0.009 -0.009	+0.006 -0.006	+0.004 -0.004	+0.0025 -0.0025	+0.00175 -0.00175
180 200	+1.450 -1.450	+0.925 -0.925	+0.575 -0.575	+0.360 -0.360	+0.230 -0.230	+0.145 -0.145	+0.092 -0.092	+0.057 -0.057	+0.036 -0.036	+0.023 -0.023	+0.0145 -0.0145	+0.010 -0.010	+0.007 -0.007	+0.005 -0.005	+0.0035 -0.0035	+0.00225 -0.00225
200 225	+1.450 -1.450	+0.925 -0.925	+0.575 -0.575	+0.360 -0.360	+0.230 -0.230	+0.145 -0.145	+0.092 -0.092	+0.057 -0.057	+0.036 -0.036	+0.023 -0.023	+0.0145 -0.0145	+0.010 -0.010	+0.007 -0.007	+0.005 -0.005	+0.0035 -0.0035	+0.00225 -0.00225
225 250	+1.450 -1.450	+0.925 -0.925	+0.575 -0.575	+0.360 -0.360	+0.230 -0.230	+0.145 -0.145	+0.092 -0.092	+0.057 -0.057	+0.036 -0.036	+0.023 -0.023	+0.0145 -0.0145	+0.010 -0.010	+0.007 -0.007	+0.005 -0.005	+0.0035 -0.0035	+0.00225 -0.00225
250 280	+1.600 -1.600	+1.050 -1.050	+0.650 -0.650	+0.405 -0.405	+0.260 -0.260	+0.160 -0.160	+0.105 -0.105	+0.065 -0.065	+0.040 -0.040	+0.026 -0.026	+0.016 -0.016	+0.0115 -0.0115	+0.008 -0.008	+0.006 -0.006	+0.004 -0.004	+0.003 -0.003
280 315	+1.600 -1.600	+1.050 -1.050	+0.650 -0.650	+0.405 -0.405	+0.260 -0.260	+0.160 -0.160	+0.105 -0.105	+0.065 -0.065	+0.040 -0.040	+0.026 -0.026	+0.016 -0.016	+0.0115 -0.0115	+0.008 -0.008	+0.006 -0.006	+0.004 -0.004	+0.003 -0.003
315 355	+1.800 -1.800	+1.150 -1.150	+0.700 -0.700	+0.445 -0.445	+0.285 -0.285	+0.180 -0.180	+0.115 -0.115	+0.070 -0.070	+0.044 -0.044	+0.028 -0.028	+0.018 -0.018	+0.0125 -0.0125	+0.009 -0.009	+0.0065 -0.0065	+0.0045 -0.0045	+0.0035 -0.0035
355 400	+1.800 -1.800	+1.150 -1.150	+0.700 -0.700	+0.445 -0.445	+0.285 -0.285	+0.180 -0.180	+0.115 -0.115	+0.070 -0.070	+0.044 -0.044	+0.028 -0.028	+0.018 -0.018	+0.0125 -0.0125	+0.009 -0.009	+0.0065 -0.0065	+0.0045 -0.0045	+0.0035 -0.0035
400 450	+2.000 -2.000	+1.250 -1.250	+0.775 -0.775	+0.485 -0.485	+0.315 -0.315	+0.200 -0.200	+0.125 -0.125	+0.077 -0.077	+0.048 -0.048	+0.031 -0.031	+0.020 -0.020	+0.0135 -0.0135	+0.010 -0.010	+0.0075 -0.0075	+0.005 -0.005	+0.004 -0.004
450 500	+2.000 -2.000	+1.250 -1.250	+0.775 -0.775	+0.485 -0.485	+0.315 -0.315	+0.200 -0.200	+0.125 -0.125	+0.077 -0.077	+0.048 -0.048	+0.031 -0.031	+0.020 -0.020	+0.0135 -0.0135	+0.010 -0.010	+0.0075 -0.0075	+0.005 -0.005	+0.004 -0.004

THE ISO SYSTEM OF LIMITS AND FITS—TOLERANCES AND DEVIATIONS

Table 6-28. Tolerance Zones—External Dimensions (Shafts) (k9 ... k4, m9 ... m4) Dimensions in mm (ANSI B4.2)

SIZE	k9	k8	k7	k6	k5	k4	m9	m8	m7	m6	m5	m4
OVER 0 TO 3	+0.025 0.000	+0.014 0.000	+0.010 0.000	+0.006 0.000	+0.004 0.000	+0.003 0.000	+0.027 +0.002	+0.016 +0.002	+0.012 +0.002	+0.008 +0.002	+0.006 +0.002	+0.005 +0.002
OVER 3 TO 6	+0.030 0.000	+0.018 0.000	+0.013 +0.001	+0.009 +0.001	+0.006 +0.001	+0.005 +0.001	+0.034 +0.004	+0.022 +0.004	+0.016 +0.004	+0.012 +0.004	+0.009 +0.004	+0.008 +0.004
OVER 6 TO 10	+0.036 0.000	+0.022 0.000	+0.016 +0.001	+0.010 +0.001	+0.007 +0.001	+0.005 +0.001	+0.042 +0.006	+0.028 +0.006	+0.021 +0.006	+0.015 +0.006	+0.012 +0.006	+0.010 +0.006
OVER 10 TO 14	+0.043 0.000	+0.027 0.000	+0.019 +0.001	+0.012 +0.001	+0.009 +0.001	+0.006 +0.001	+0.050 +0.007	+0.034 +0.007	+0.025 +0.007	+0.018 +0.007	+0.015 +0.007	+0.012 +0.007
OVER 14 TO 18	+0.043 0.000	+0.027 0.000	+0.019 +0.001	+0.012 +0.001	+0.009 +0.001	+0.006 +0.001	+0.050 +0.007	+0.034 +0.007	+0.025 +0.007	+0.018 +0.007	+0.015 +0.007	+0.012 +0.007
OVER 18 TO 24	+0.052 0.000	+0.033 0.000	+0.023 +0.002	+0.015 +0.002	+0.011 +0.002	+0.008 +0.002	+0.060 +0.008	+0.041 +0.008	+0.029 +0.008	+0.021 +0.008	+0.017 +0.008	+0.014 +0.008
OVER 24 TO 30	+0.052 0.000	+0.033 0.000	+0.023 +0.002	+0.015 +0.002	+0.011 +0.002	+0.008 +0.002	+0.060 +0.008	+0.041 +0.008	+0.029 +0.008	+0.021 +0.008	+0.017 +0.008	+0.014 +0.008
OVER 30 TO 40	+0.062 0.000	+0.039 0.000	+0.027 +0.002	+0.018 +0.002	+0.013 +0.002	+0.009 +0.002	+0.071 +0.009	+0.048 +0.009	+0.034 +0.009	+0.025 +0.009	+0.020 +0.009	+0.016 +0.009
OVER 40 TO 50	+0.062 0.000	+0.039 0.000	+0.027 +0.002	+0.018 +0.002	+0.013 +0.002	+0.009 +0.002	+0.071 +0.009	+0.048 +0.009	+0.034 +0.009	+0.025 +0.009	+0.020 +0.009	+0.016 +0.009
OVER 50 TO 65	+0.074 0.000	+0.046 0.000	+0.032 +0.002	+0.021 +0.002	+0.015 +0.002	+0.010 +0.002	+0.085 +0.011	+0.057 +0.011	+0.041 +0.011	+0.030 +0.011	+0.024 +0.011	+0.019 +0.011
OVER 65 TO 80	+0.074 0.000	+0.046 0.000	+0.032 +0.002	+0.021 +0.002	+0.015 +0.002	+0.010 +0.002	+0.085 +0.011	+0.057 +0.011	+0.041 +0.011	+0.030 +0.011	+0.024 +0.011	+0.019 +0.011
OVER 80 TO 100	+0.087 0.000	+0.054 0.000	+0.038 +0.003	+0.025 +0.003	+0.018 +0.003	+0.013 +0.003	+0.100 +0.013	+0.067 +0.013	+0.048 +0.013	+0.035 +0.013	+0.028 +0.013	+0.023 +0.013
OVER 100 TO 120	+0.087 0.000	+0.054 0.000	+0.038 +0.003	+0.025 +0.003	+0.018 +0.003	+0.013 +0.003	+0.100 +0.013	+0.067 +0.013	+0.048 +0.013	+0.035 +0.013	+0.028 +0.013	+0.023 +0.013
OVER 120 TO 140	+0.100 0.000	+0.063 0.000	+0.043 +0.003	+0.028 +0.003	+0.021 +0.003	+0.015 +0.003	+0.115 +0.015	+0.078 +0.015	+0.055 +0.015	+0.040 +0.015	+0.033 +0.015	+0.027 +0.015
OVER 140 TO 160	+0.100 0.000	+0.063 0.000	+0.043 +0.003	+0.028 +0.003	+0.021 +0.003	+0.015 +0.003	+0.115 +0.015	+0.078 +0.015	+0.055 +0.015	+0.040 +0.015	+0.033 +0.015	+0.027 +0.015
OVER 160 TO 180	+0.100 0.000	+0.063 0.000	+0.043 +0.003	+0.028 +0.003	+0.021 +0.003	+0.015 +0.003	+0.115 +0.015	+0.078 +0.015	+0.055 +0.015	+0.040 +0.015	+0.033 +0.015	+0.027 +0.015
OVER 180 TO 200	+0.115 0.000	+0.072 0.000	+0.050 +0.004	+0.033 +0.004	+0.024 +0.004	+0.018 +0.004	+0.132 +0.017	+0.089 +0.017	+0.063 +0.017	+0.046 +0.017	+0.037 +0.017	+0.031 +0.017
OVER 200 TO 225	+0.115 0.000	+0.072 0.000	+0.050 +0.004	+0.033 +0.004	+0.024 +0.004	+0.018 +0.004	+0.132 +0.017	+0.089 +0.017	+0.063 +0.017	+0.046 +0.017	+0.037 +0.017	+0.031 +0.017
OVER 225 TO 250	+0.115 0.000	+0.072 0.000	+0.050 +0.004	+0.033 +0.004	+0.024 +0.004	+0.018 +0.004	+0.132 +0.017	+0.089 +0.017	+0.063 +0.017	+0.046 +0.017	+0.037 +0.017	+0.031 +0.017
OVER 250 TO 280	+0.130 0.000	+0.081 0.000	+0.056 +0.004	+0.036 +0.004	+0.027 +0.004	+0.020 +0.004	+0.150 +0.020	+0.101 +0.020	+0.072 +0.020	+0.052 +0.020	+0.043 +0.020	+0.036 +0.020
OVER 280 TO 315	+0.130 0.000	+0.081 0.000	+0.056 +0.004	+0.036 +0.004	+0.027 +0.004	+0.020 +0.004	+0.150 +0.020	+0.101 +0.020	+0.072 +0.020	+0.052 +0.020	+0.043 +0.020	+0.036 +0.020
OVER 315 TO 355	+0.140 0.000	+0.089 0.000	+0.061 +0.004	+0.040 +0.004	+0.029 +0.004	+0.022 +0.004	+0.161 +0.021	+0.110 +0.021	+0.078 +0.021	+0.057 +0.021	+0.046 +0.021	+0.039 +0.021
OVER 355 TO 400	+0.140 0.000	+0.089 0.000	+0.061 +0.004	+0.040 +0.004	+0.029 +0.004	+0.022 +0.004	+0.161 +0.021	+0.110 +0.021	+0.078 +0.021	+0.057 +0.021	+0.046 +0.021	+0.039 +0.021
OVER 400 TO 450	+0.155 0.000	+0.097 0.000	+0.068 +0.005	+0.045 +0.005	+0.032 +0.005	+0.025 +0.005	+0.178 +0.023	+0.120 +0.023	+0.086 +0.023	+0.063 +0.023	+0.050 +0.023	+0.043 +0.023
OVER 450 TO 500	+0.155 0.000	+0.097 0.000	+0.068 +0.005	+0.045 +0.005	+0.032 +0.005	+0.025 +0.005	+0.178 +0.023	+0.120 +0.023	+0.086 +0.023	+0.063 +0.023	+0.050 +0.023	+0.043 +0.023

Table 6-29. Tolerance Zones—External Dimensions (Shafts) (n9 ... n4, p9 ... p4) Dimensions in mm (ANSI B4.2)

SIZE	n9	n8	n7	n6	n5	n4	p9	p8	p7	p6	p5	p4
OVER 0 TO 3	+0.029 +0.004	+0.018 +0.004	+0.014 +0.004	+0.010 +0.004	+0.008 +0.004	+0.007 +0.004	+0.031 +0.006	+0.020 +0.006	+0.016 +0.006	+0.012 +0.006	+0.010 +0.006	+0.009 +0.006
OVER 3 TO 6	+0.038 +0.008	+0.026 +0.008	+0.020 +0.008	+0.016 +0.008	+0.013 +0.008	+0.012 +0.008	+0.042 +0.012	+0.030 +0.012	+0.024 +0.012	+0.020 +0.012	+0.017 +0.012	+0.016 +0.012
OVER 6 TO 10	+0.046 +0.010	+0.032 +0.010	+0.025 +0.010	+0.019 +0.010	+0.016 +0.010	+0.014 +0.010	+0.051 +0.015	+0.037 +0.015	+0.030 +0.015	+0.024 +0.015	+0.021 +0.015	+0.019 +0.015
OVER 10 TO 14	+0.055 +0.012	+0.039 +0.012	+0.030 +0.012	+0.023 +0.012	+0.020 +0.012	+0.017 +0.012	+0.061 +0.018	+0.045 +0.018	+0.036 +0.018	+0.029 +0.018	+0.026 +0.018	+0.023 +0.018
OVER 14 TO 18	+0.055 +0.012	+0.039 +0.012	+0.030 +0.012	+0.023 +0.012	+0.020 +0.012	+0.017 +0.012	+0.061 +0.018	+0.045 +0.018	+0.036 +0.018	+0.029 +0.018	+0.026 +0.018	+0.023 +0.018
OVER 18 TO 24	+0.067 +0.015	+0.048 +0.015	+0.036 +0.015	+0.028 +0.015	+0.024 +0.015	+0.021 +0.015	+0.074 +0.022	+0.055 +0.022	+0.043 +0.022	+0.035 +0.022	+0.031 +0.022	+0.028 +0.022
OVER 24 TO 30	+0.067 +0.015	+0.048 +0.015	+0.036 +0.015	+0.028 +0.015	+0.024 +0.015	+0.021 +0.015	+0.074 +0.022	+0.055 +0.022	+0.043 +0.022	+0.035 +0.022	+0.031 +0.022	+0.028 +0.022
OVER 30 TO 40	+0.079 +0.017	+0.056 +0.017	+0.042 +0.017	+0.033 +0.017	+0.028 +0.017	+0.024 +0.017	+0.088 +0.026	+0.065 +0.026	+0.051 +0.026	+0.042 +0.026	+0.037 +0.026	+0.033 +0.026
OVER 40 TO 50	+0.079 +0.017	+0.056 +0.017	+0.042 +0.017	+0.033 +0.017	+0.028 +0.017	+0.024 +0.017	+0.088 +0.026	+0.065 +0.026	+0.051 +0.026	+0.042 +0.026	+0.037 +0.026	+0.033 +0.026
OVER 50 TO 65	+0.094 +0.020	+0.066 +0.020	+0.050 +0.020	+0.039 +0.020	+0.033 +0.020	+0.028 +0.020	+0.106 +0.032	+0.078 +0.032	+0.062 +0.032	+0.051 +0.032	+0.045 +0.032	+0.040 +0.032
OVER 65 TO 80	+0.094 +0.020	+0.066 +0.020	+0.050 +0.020	+0.039 +0.020	+0.033 +0.020	+0.028 +0.020	+0.106 +0.032	+0.078 +0.032	+0.062 +0.032	+0.051 +0.032	+0.045 +0.032	+0.040 +0.032
OVER 80 TO 100	+0.110 +0.023	+0.077 +0.023	+0.058 +0.023	+0.045 +0.023	+0.038 +0.023	+0.033 +0.023	+0.124 +0.037	+0.091 +0.037	+0.072 +0.037	+0.059 +0.037	+0.052 +0.037	+0.047 +0.037
OVER 100 TO 120	+0.110 +0.023	+0.077 +0.023	+0.058 +0.023	+0.045 +0.023	+0.038 +0.023	+0.033 +0.023	+0.124 +0.037	+0.091 +0.037	+0.072 +0.037	+0.059 +0.037	+0.052 +0.037	+0.047 +0.037
OVER 120 TO 140	+0.127 +0.027	+0.090 +0.027	+0.067 +0.027	+0.052 +0.027	+0.045 +0.027	+0.039 +0.027	+0.143 +0.043	+0.106 +0.043	+0.083 +0.043	+0.068 +0.043	+0.061 +0.043	+0.055 +0.043
OVER 140 TO 160	+0.127 +0.027	+0.090 +0.027	+0.067 +0.027	+0.052 +0.027	+0.045 +0.027	+0.039 +0.027	+0.143 +0.043	+0.106 +0.043	+0.083 +0.043	+0.068 +0.043	+0.061 +0.043	+0.055 +0.043
OVER 160 TO 180	+0.127 +0.027	+0.090 +0.027	+0.067 +0.027	+0.052 +0.027	+0.045 +0.027	+0.039 +0.027	+0.143 +0.043	+0.106 +0.043	+0.083 +0.043	+0.068 +0.043	+0.061 +0.043	+0.055 +0.043
OVER 180 TO 200	+0.146 +0.031	+0.103 +0.031	+0.077 +0.031	+0.060 +0.031	+0.051 +0.031	+0.045 +0.031	+0.165 +0.050	+0.122 +0.050	+0.096 +0.050	+0.079 +0.050	+0.070 +0.050	+0.064 +0.050
OVER 200 TO 225	+0.146 +0.031	+0.103 +0.031	+0.077 +0.031	+0.060 +0.031	+0.051 +0.031	+0.045 +0.031	+0.165 +0.050	+0.122 +0.050	+0.096 +0.050	+0.079 +0.050	+0.070 +0.050	+0.064 +0.050
OVER 225 TO 250	+0.146 +0.031	+0.103 +0.031	+0.077 +0.031	+0.060 +0.031	+0.051 +0.031	+0.045 +0.031	+0.165 +0.050	+0.122 +0.050	+0.096 +0.050	+0.079 +0.050	+0.070 +0.050	+0.064 +0.050
OVER 250 TO 280	+0.164 +0.034	+0.115 +0.034	+0.086 +0.034	+0.066 +0.034	+0.057 +0.034	+0.050 +0.034	+0.186 +0.056	+0.137 +0.056	+0.108 +0.056	+0.088 +0.056	+0.079 +0.056	+0.072 +0.056
OVER 280 TO 315	+0.164 +0.034	+0.115 +0.034	+0.086 +0.034	+0.066 +0.034	+0.057 +0.034	+0.050 +0.034	+0.186 +0.056	+0.137 +0.056	+0.108 +0.056	+0.088 +0.056	+0.079 +0.056	+0.072 +0.056
OVER 315 TO 355	+0.177 +0.037	+0.126 +0.037	+0.094 +0.037	+0.073 +0.037	+0.062 +0.037	+0.055 +0.037	+0.202 +0.062	+0.151 +0.062	+0.119 +0.062	+0.098 +0.062	+0.087 +0.062	+0.080 +0.062
OVER 355 TO 400	+0.177 +0.037	+0.126 +0.037	+0.094 +0.037	+0.073 +0.037	+0.062 +0.037	+0.055 +0.037	+0.202 +0.062	+0.151 +0.062	+0.119 +0.062	+0.098 +0.062	+0.087 +0.062	+0.080 +0.062
OVER 400 TO 450	+0.195 +0.040	+0.137 +0.040	+0.103 +0.040	+0.080 +0.040	+0.067 +0.040	+0.060 +0.040	+0.223 +0.068	+0.165 +0.068	+0.131 +0.068	+0.108 +0.068	+0.095 +0.068	+0.088 +0.068
OVER 450 TO 500	+0.195 +0.040	+0.137 +0.040	+0.103 +0.040	+0.080 +0.040	+0.067 +0.040	+0.060 +0.040	+0.223 +0.068	+0.165 +0.068	+0.131 +0.068	+0.108 +0.068	+0.095 +0.068	+0.088 +0.068

THE ISO SYSTEM OF LIMITS AND FITS—TOLERANCES AND DEVIATIONS

Table 6-30. Tolerance Zones—External Dimensions (Shafts) (r9 ... r4, s9 ... s4) Dimensions in mm (ANSI B4.2)

SIZE	r9	r8	r7	r6	r5	r4	s9	s8	s7	s6	s5	s4
OVER 0 TO 3	+0.035 +0.010	+0.024 +0.010	+0.020 +0.010	+0.016 +0.010	+0.014 +0.010	+0.013 +0.010	+0.039 +0.014	+0.028 +0.014	+0.024 +0.014	+0.020 +0.014	+0.018 +0.014	+0.017 +0.014
OVER 3 TO 6	+0.045 +0.015	+0.033 +0.015	+0.027 +0.015	+0.023 +0.015	+0.020 +0.015	+0.019 +0.015	+0.049 +0.019	+0.037 +0.019	+0.031 +0.019	+0.027 +0.019	+0.024 +0.019	+0.023 +0.019
OVER 6 TO 10	+0.055 +0.019	+0.041 +0.019	+0.034 +0.019	+0.028 +0.019	+0.025 +0.019	+0.023 +0.019	+0.059 +0.023	+0.045 +0.023	+0.038 +0.023	+0.032 +0.023	+0.029 +0.023	+0.027 +0.023
OVER 10 TO 14	+0.066 +0.023	+0.050 +0.023	+0.041 +0.023	+0.034 +0.023	+0.031 +0.023	+0.028 +0.023	+0.071 +0.028	+0.055 +0.028	+0.046 +0.028	+0.039 +0.028	+0.036 +0.028	+0.033 +0.028
OVER 14 TO 18	+0.066 +0.023	+0.050 +0.023	+0.041 +0.023	+0.034 +0.023	+0.031 +0.023	+0.028 +0.023	+0.071 +0.028	+0.055 +0.028	+0.046 +0.028	+0.039 +0.028	+0.036 +0.028	+0.033 +0.028
OVER 18 TO 24	+0.080 +0.028	+0.061 +0.028	+0.049 +0.028	+0.041 +0.028	+0.037 +0.028	+0.034 +0.028	+0.087 +0.035	+0.068 +0.035	+0.056 +0.035	+0.048 +0.035	+0.044 +0.035	+0.041 +0.035
OVER 24 TO 30	+0.080 +0.028	+0.061 +0.028	+0.049 +0.028	+0.041 +0.028	+0.037 +0.028	+0.034 +0.028	+0.087 +0.035	+0.068 +0.035	+0.056 +0.035	+0.048 +0.035	+0.044 +0.035	+0.041 +0.035
OVER 30 TO 40	+0.096 +0.034	+0.073 +0.034	+0.059 +0.034	+0.050 +0.034	+0.045 +0.034	+0.041 +0.034	+0.105 +0.043	+0.082 +0.043	+0.068 +0.043	+0.059 +0.043	+0.054 +0.043	+0.050 +0.043
OVER 40 TO 50	+0.096 +0.034	+0.073 +0.034	+0.059 +0.034	+0.050 +0.034	+0.045 +0.034	+0.041 +0.034	+0.105 +0.043	+0.082 +0.043	+0.068 +0.043	+0.059 +0.043	+0.054 +0.043	+0.050 +0.043
OVER 50 TO 65	+0.115 +0.041	+0.087 +0.041	+0.071 +0.041	+0.060 +0.041	+0.054 +0.041	+0.049 +0.041	+0.127 +0.053	+0.099 +0.053	+0.083 +0.053	+0.072 +0.053	+0.066 +0.053	+0.061 +0.053
OVER 65 TO 80	+0.117 +0.043	+0.089 +0.043	+0.073 +0.043	+0.062 +0.043	+0.056 +0.043	+0.051 +0.043	+0.133 +0.059	+0.105 +0.059	+0.089 +0.059	+0.078 +0.059	+0.072 +0.059	+0.067 +0.059
OVER 80 TO 100	+0.138 +0.051	+0.105 +0.051	+0.086 +0.051	+0.073 +0.051	+0.066 +0.051	+0.061 +0.051	+0.158 +0.071	+0.125 +0.071	+0.106 +0.071	+0.093 +0.071	+0.086 +0.071	+0.081 +0.071
OVER 100 TO 120	+0.141 +0.054	+0.108 +0.054	+0.089 +0.054	+0.076 +0.054	+0.069 +0.054	+0.064 +0.054	+0.166 +0.079	+0.133 +0.079	+0.114 +0.079	+0.101 +0.079	+0.094 +0.079	+0.089 +0.079
OVER 120 TO 140	+0.163 +0.063	+0.126 +0.063	+0.103 +0.063	+0.088 +0.063	+0.081 +0.063	+0.075 +0.063	+0.192 +0.092	+0.155 +0.092	+0.132 +0.092	+0.117 +0.092	+0.110 +0.092	+0.104 +0.092
OVER 140 TO 160	+0.165 +0.065	+0.128 +0.065	+0.105 +0.065	+0.090 +0.065	+0.083 +0.065	+0.077 +0.065	+0.200 +0.100	+0.163 +0.100	+0.140 +0.100	+0.125 +0.100	+0.118 +0.100	+0.112 +0.100
OVER 160 TO 180	+0.168 +0.068	+0.131 +0.068	+0.108 +0.068	+0.093 +0.068	+0.086 +0.068	+0.080 +0.068	+0.208 +0.108	+0.171 +0.108	+0.148 +0.108	+0.133 +0.108	+0.126 +0.108	+0.120 +0.108
OVER 180 TO 200	+0.192 +0.077	+0.149 +0.077	+0.123 +0.077	+0.106 +0.077	+0.097 +0.077	+0.091 +0.077	+0.237 +0.122	+0.194 +0.122	+0.168 +0.122	+0.151 +0.122	+0.142 +0.122	+0.136 +0.122
OVER 200 TO 225	+0.195 +0.080	+0.152 +0.080	+0.126 +0.080	+0.109 +0.080	+0.100 +0.080	+0.094 +0.080	+0.245 +0.130	+0.202 +0.130	+0.176 +0.130	+0.159 +0.130	+0.150 +0.130	+0.144 +0.130
OVER 225 TO 250	+0.199 +0.084	+0.156 +0.084	+0.130 +0.084	+0.113 +0.084	+0.104 +0.084	+0.098 +0.084	+0.255 +0.140	+0.212 +0.140	+0.186 +0.140	+0.169 +0.140	+0.160 +0.140	+0.154 +0.140
OVER 250 TO 280	+0.224 +0.094	+0.175 +0.094	+0.146 +0.094	+0.126 +0.094	+0.117 +0.094	+0.110 +0.094	+0.288 +0.158	+0.239 +0.158	+0.210 +0.158	+0.190 +0.158	+0.181 +0.158	+0.174 +0.158
OVER 280 TO 315	+0.228 +0.098	+0.179 +0.098	+0.150 +0.098	+0.130 +0.098	+0.121 +0.098	+0.114 +0.098	+0.300 +0.170	+0.251 +0.170	+0.222 +0.170	+0.202 +0.170	+0.193 +0.170	+0.186 +0.170
OVER 315 TO 355	+0.248 +0.108	+0.197 +0.108	+0.165 +0.108	+0.144 +0.108	+0.133 +0.108	+0.126 +0.108	+0.330 +0.190	+0.279 +0.190	+0.247 +0.190	+0.226 +0.190	+0.215 +0.190	+0.208 +0.190
OVER 355 TO 400	+0.254 +0.114	+0.203 +0.114	+0.171 +0.114	+0.150 +0.114	+0.139 +0.114	+0.132 +0.114	+0.348 +0.208	+0.297 +0.208	+0.265 +0.208	+0.244 +0.208	+0.233 +0.208	+0.226 +0.208
OVER 400 TO 450	+0.281 +0.126	+0.223 +0.126	+0.189 +0.126	+0.166 +0.126	+0.153 +0.126	+0.146 +0.126	+0.387 +0.232	+0.329 +0.232	+0.295 +0.232	+0.272 +0.232	+0.259 +0.232	+0.252 +0.232
OVER 450 TO 500	+0.287 +0.132	+0.229 +0.132	+0.195 +0.132	+0.172 +0.132	+0.159 +0.132	+0.152 +0.132	+0.407 +0.252	+0.349 +0.252	+0.315 +0.252	+0.292 +0.252	+0.279 +0.252	+0.272 +0.252

Table 6-31. Tolerance Zones—External Dimensions (Shafts) (t9 ... t4, u9 ... u4) Dimensions in mm (ANSI B4.2)

SIZE	t9	t8	t7	t6	t5	t4	u9	u8	u7	u6	u5	u4
OVER 0 TO 3							+0.043 +0.018	+0.032 +0.018	+0.028 +0.018	+0.024 +0.018	+0.022 +0.018	+0.021 +0.018
OVER 3 TO 6							+0.053 +0.023	+0.041 +0.023	+0.035 +0.023	+0.031 +0.023	+0.028 +0.023	+0.027 +0.023
OVER 6 TO 10		NUMERICAL VALUES FOR					+0.064 +0.028	+0.050 +0.028	+0.043 +0.028	+0.037 +0.028	+0.034 +0.028	+0.032 +0.028
OVER 10 TO 14		TOLERANCE ZONES IN					+0.076 +0.033	+0.060 +0.033	+0.051 +0.033	+0.044 +0.033	+0.041 +0.033	+0.038 +0.033
OVER 14 TO 18		THIS AREA NOT DEFINED.					+0.076 +0.033	+0.060 +0.033	+0.051 +0.033	+0.044 +0.033	+0.041 +0.033	+0.038 +0.033
OVER 18 TO 24							+0.093 +0.041	+0.074 +0.041	+0.062 +0.041	+0.054 +0.041	+0.050 +0.041	+0.047 +0.041
OVER 24 TO 30	+0.093 +0.041	+0.074 +0.041	+0.062 +0.041	+0.054 +0.041	+0.050 +0.041	+0.047 +0.041	+0.100 +0.048	+0.081 +0.048	+0.069 +0.048	+0.061 +0.048	+0.057 +0.048	+0.054 +0.048
OVER 30 TO 40	+0.110 +0.048	+0.087 +0.048	+0.073 +0.048	+0.064 +0.048	+0.059 +0.048	+0.055 +0.048	+0.122 +0.060	+0.099 +0.060	+0.085 +0.060	+0.076 +0.060	+0.071 +0.060	+0.067 +0.060
OVER 40 TO 50	+0.116 +0.054	+0.093 +0.054	+0.079 +0.054	+0.070 +0.054	+0.065 +0.054	+0.061 +0.054	+0.132 +0.070	+0.109 +0.070	+0.095 +0.070	+0.086 +0.070	+0.081 +0.070	+0.077 +0.070
OVER 50 TO 65	+0.140 +0.066	+0.112 +0.066	+0.096 +0.066	+0.085 +0.066	+0.079 +0.066	+0.074 +0.066	+0.161 +0.087	+0.133 +0.087	+0.117 +0.087	+0.106 +0.087	+0.100 +0.087	+0.095 +0.087
OVER 65 TO 80	+0.149 +0.075	+0.121 +0.075	+0.105 +0.075	+0.094 +0.075	+0.088 +0.075	+0.083 +0.075	+0.176 +0.102	+0.148 +0.102	+0.132 +0.102	+0.121 +0.102	+0.115 +0.102	+0.110 +0.102
OVER 80 TO 100	+0.178 +0.091	+0.145 +0.091	+0.126 +0.091	+0.113 +0.091	+0.106 +0.091	+0.101 +0.091	+0.211 +0.124	+0.178 +0.124	+0.159 +0.124	+0.146 +0.124	+0.139 +0.124	+0.134 +0.124
OVER 100 TO 120	+0.191 +0.104	+0.158 +0.104	+0.139 +0.104	+0.126 +0.104	+0.119 +0.104	+0.114 +0.104	+0.231 +0.144	+0.198 +0.144	+0.179 +0.144	+0.166 +0.144	+0.159 +0.144	+0.154 +0.144
OVER 120 TO 140	+0.222 +0.122	+0.185 +0.122	+0.162 +0.122	+0.147 +0.122	+0.140 +0.122	+0.134 +0.122	+0.270 +0.170	+0.233 +0.170	+0.210 +0.170	+0.195 +0.170	+0.188 +0.170	+0.182 +0.170
OVER 140 TO 160	+0.234 +0.134	+0.197 +0.134	+0.174 +0.134	+0.159 +0.134	+0.152 +0.134	+0.146 +0.134	+0.290 +0.190	+0.253 +0.190	+0.230 +0.190	+0.215 +0.190	+0.208 +0.190	+0.202 +0.190
OVER 160 TO 180	+0.246 +0.146	+0.209 +0.146	+0.186 +0.146	+0.171 +0.146	+0.164 +0.146	+0.158 +0.146	+0.310 +0.210	+0.273 +0.210	+0.250 +0.210	+0.235 +0.210	+0.228 +0.210	+0.222 +0.210
OVER 180 TO 200	+0.281 +0.166	+0.238 +0.166	+0.212 +0.166	+0.195 +0.166	+0.186 +0.166	+0.180 +0.166	+0.351 +0.236	+0.308 +0.236	+0.282 +0.236	+0.265 +0.236	+0.256 +0.236	+0.250 +0.236
OVER 200 TO 225	+0.295 +0.180	+0.252 +0.180	+0.226 +0.180	+0.209 +0.180	+0.200 +0.180	+0.194 +0.180	+0.373 +0.258	+0.330 +0.258	+0.304 +0.258	+0.287 +0.258	+0.278 +0.258	+0.272 +0.258
OVER 225 TO 250	+0.311 +0.196	+0.268 +0.196	+0.242 +0.196	+0.225 +0.196	+0.216 +0.196	+0.210 +0.196	+0.399 +0.284	+0.356 +0.284	+0.330 +0.284	+0.313 +0.284	+0.304 +0.284	+0.298 +0.284
OVER 250 TO 280	+0.348 +0.218	+0.299 +0.218	+0.270 +0.218	+0.250 +0.218	+0.241 +0.218	+0.234 +0.218	+0.445 +0.315	+0.396 +0.315	+0.367 +0.315	+0.347 +0.315	+0.338 +0.315	+0.331 +0.315
OVER 280 TO 315	+0.370 +0.240	+0.321 +0.240	+0.292 +0.240	+0.272 +0.240	+0.263 +0.240	+0.256 +0.240	+0.480 +0.350	+0.431 +0.350	+0.402 +0.350	+0.382 +0.350	+0.373 +0.350	+0.366 +0.350
OVER 315 TO 355	+0.408 +0.268	+0.357 +0.268	+0.325 +0.268	+0.304 +0.268	+0.293 +0.268	+0.286 +0.268	+0.530 +0.390	+0.479 +0.390	+0.447 +0.390	+0.426 +0.390	+0.415 +0.390	+0.408 +0.390
OVER 355 TO 400	+0.434 +0.294	+0.383 +0.294	+0.351 +0.294	+0.330 +0.294	+0.319 +0.294	+0.312 +0.294	+0.575 +0.435	+0.524 +0.435	+0.492 +0.435	+0.471 +0.435	+0.460 +0.435	+0.453 +0.435
OVER 400 TO 450	+0.485 +0.330	+0.427 +0.330	+0.393 +0.330	+0.370 +0.330	+0.357 +0.330	+0.350 +0.330	+0.645 +0.490	+0.587 +0.490	+0.553 +0.490	+0.530 +0.490	+0.517 +0.490	+0.510 +0.490
OVER 450 TO 500	+0.515 +0.360	+0.457 +0.360	+0.423 +0.360	+0.400 +0.360	+0.387 +0.360	+0.380 +0.360	+0.695 +0.540	+0.637 +0.540	+0.603 +0.540	+0.580 +0.540	+0.567 +0.540	+0.560 +0.540

THE ISO SYSTEM OF LIMITS AND FITS—TOLERANCES AND DEVIATIONS

Table 6-32. Tolerance Zones—External Dimensions (Shafts) (v9 . . . v4, x9 . . . x4) Dimensions in mm (ANSI B4.2)

SIZE	v9	v8	v7	v6	v5	v4	x9	x8	x7	x6	x5	x4
OVER 0 TO 3							+0.045 +0.020	+0.034 +0.020	+0.030 +0.020	+0.026 +0.020	+0.024 +0.020	+0.023 +0.020
OVER 3 TO 6		NUMERICAL VALUES FOR					+0.058 +0.028	+0.046 +0.028	+0.040 +0.028	+0.036 +0.028	+0.033 +0.028	+0.032 +0.028
OVER 6 TO 10		TOLERANCE ZONES IN THIS AREA NOT DEFINED.					+0.070 +0.034	+0.056 +0.034	+0.049 +0.034	+0.043 +0.034	+0.040 +0.034	+0.038 +0.034
OVER 10 TO 14							+0.083 +0.040	+0.067 +0.040	+0.058 +0.040	+0.051 +0.040	+0.048 +0.040	+0.045 +0.040
OVER 14 TO 18	+0.082 +0.039	+0.066 +0.039	+0.057 +0.039	+0.050 +0.039	+0.047 +0.039	+0.044 +0.039	+0.088 +0.045	+0.072 +0.045	+0.063 +0.045	+0.056 +0.045	+0.053 +0.045	+0.050 +0.045
OVER 18 TO 24	+0.099 +0.047	+0.080 +0.047	+0.068 +0.047	+0.060 +0.047	+0.056 +0.047	+0.053 +0.047	+0.106 +0.054	+0.087 +0.054	+0.075 +0.054	+0.067 +0.054	+0.063 +0.054	+0.060 +0.054
OVER 24 TO 30	+0.107 +0.055	+0.088 +0.055	+0.076 +0.055	+0.068 +0.055	+0.064 +0.055	+0.061 +0.055	+0.116 +0.064	+0.097 +0.064	+0.085 +0.064	+0.077 +0.064	+0.073 +0.064	+0.070 +0.064
OVER 30 TO 40	+0.130 +0.068	+0.107 +0.068	+0.093 +0.068	+0.084 +0.068	+0.079 +0.068	+0.075 +0.068	+0.142 +0.080	+0.119 +0.080	+0.105 +0.080	+0.096 +0.080	+0.091 +0.080	+0.087 +0.080
OVER 40 TO 50	+0.143 +0.081	+0.120 +0.081	+0.106 +0.081	+0.097 +0.081	+0.092 +0.081	+0.088 +0.081	+0.159 +0.097	+0.136 +0.097	+0.122 +0.097	+0.113 +0.097	+0.108 +0.097	+0.104 +0.097
OVER 50 TO 65	+0.176 +0.102	+0.148 +0.102	+0.132 +0.102	+0.121 +0.102	+0.115 +0.102	+0.110 +0.102	+0.196 +0.122	+0.168 +0.122	+0.152 +0.122	+0.141 +0.122	+0.135 +0.122	+0.130 +0.122
OVER 65 TO 80	+0.194 +0.120	+0.166 +0.120	+0.150 +0.120	+0.139 +0.120	+0.133 +0.120	+0.128 +0.120	+0.220 +0.146	+0.192 +0.146	+0.176 +0.146	+0.165 +0.146	+0.159 +0.146	+0.154 +0.146
OVER 80 TO 100	+0.233 +0.146	+0.200 +0.146	+0.181 +0.146	+0.168 +0.146	+0.161 +0.146	+0.156 +0.146	+0.265 +0.178	+0.232 +0.178	+0.213 +0.178	+0.200 +0.178	+0.193 +0.178	+0.188 +0.178
OVER 100 TO 120	+0.259 +0.172	+0.226 +0.172	+0.207 +0.172	+0.194 +0.172	+0.187 +0.172	+0.182 +0.172	+0.297 +0.210	+0.264 +0.210	+0.245 +0.210	+0.232 +0.210	+0.225 +0.210	+0.220 +0.210
OVER 120 TO 140	+0.302 +0.202	+0.265 +0.202	+0.242 +0.202	+0.227 +0.202	+0.220 +0.202	+0.214 +0.202	+0.348 +0.248	+0.311 +0.248	+0.288 +0.248	+0.273 +0.248	+0.266 +0.248	+0.260 +0.248
OVER 140 TO 160	+0.328 +0.228	+0.291 +0.228	+0.268 +0.228	+0.253 +0.228	+0.246 +0.228	+0.240 +0.228	+0.380 +0.280	+0.343 +0.280	+0.320 +0.280	+0.305 +0.280	+0.298 +0.280	+0.292 +0.280
OVER 160 TO 180	+0.352 +0.252	+0.315 +0.252	+0.292 +0.252	+0.277 +0.252	+0.270 +0.252	+0.264 +0.252	+0.410 +0.310	+0.373 +0.310	+0.350 +0.310	+0.335 +0.310	+0.328 +0.310	+0.322 +0.310
OVER 180 TO 200	+0.399 +0.284	+0.356 +0.284	+0.330 +0.284	+0.313 +0.284	+0.304 +0.284	+0.298 +0.284	+0.465 +0.350	+0.422 +0.350	+0.396 +0.350	+0.379 +0.350	+0.370 +0.350	+0.364 +0.350
OVER 200 TO 225	+0.425 +0.310	+0.382 +0.310	+0.356 +0.310	+0.339 +0.310	+0.330 +0.310	+0.324 +0.310	+0.500 +0.385	+0.457 +0.385	+0.431 +0.385	+0.414 +0.385	+0.405 +0.385	+0.399 +0.385
OVER 225 TO 250	+0.455 +0.340	+0.412 +0.340	+0.386 +0.340	+0.369 +0.340	+0.360 +0.340	+0.354 +0.340	+0.540 +0.425	+0.497 +0.425	+0.471 +0.425	+0.454 +0.425	+0.445 +0.425	+0.439 +0.425
OVER 250 TO 280	+0.515 +0.385	+0.466 +0.385	+0.437 +0.385	+0.417 +0.385	+0.408 +0.385	+0.401 +0.385	+0.605 +0.475	+0.556 +0.475	+0.527 +0.475	+0.507 +0.475	+0.498 +0.475	+0.491 +0.475
OVER 280 TO 315	+0.555 +0.425	+0.506 +0.425	+0.477 +0.425	+0.457 +0.425	+0.448 +0.425	+0.441 +0.425	+0.655 +0.525	+0.606 +0.525	+0.577 +0.525	+0.557 +0.525	+0.548 +0.525	+0.541 +0.525
OVER 315 TO 355	+0.615 +0.475	+0.564 +0.475	+0.532 +0.475	+0.511 +0.475	+0.500 +0.475	+0.493 +0.475	+0.730 +0.590	+0.679 +0.590	+0.647 +0.590	+0.626 +0.590	+0.615 +0.590	+0.608 +0.590
OVER 355 TO 400	+0.670 +0.530	+0.619 +0.530	+0.587 +0.530	+0.566 +0.530	+0.555 +0.530	+0.548 +0.530	+0.800 +0.660	+0.749 +0.660	+0.717 +0.660	+0.696 +0.660	+0.685 +0.660	+0.678 +0.660
OVER 400 TO 450	+0.750 +0.595	+0.692 +0.595	+0.658 +0.595	+0.635 +0.595	+0.622 +0.595	+0.615 +0.595	+0.895 +0.740	+0.837 +0.740	+0.803 +0.740	+0.780 +0.740	+0.767 +0.740	+0.760 +0.740
OVER 450 TO 500	+0.815 +0.660	+0.757 +0.660	+0.723 +0.660	+0.700 +0.660	+0.687 +0.660	+0.680 +0.660	+0.975 +0.820	+0.917 +0.820	+0.883 +0.820	+0.860 +0.820	+0.847 +0.820	+0.840 +0.820

THE ISO SYSTEM OF LIMITS AND FITS—TOLERANCES AND DEVIATIONS

Table 6-33. Tolerance Zones—External Dimensions (Shafts) (y9 ... y4, z9 ... z4) Dimensions in mm (ANSI B4.2)

SIZE	y9	y8	y7	y6	y5	y4	z9	z8	z7	z6	z5	z4
OVER 0 TO 3							+0.051 +0.026	+0.040 +0.026	+0.036 +0.026	+0.032 +0.026	+0.030 +0.026	+0.029 +0.026
OVER 3 TO 6							+0.065 +0.035	+0.053 +0.035	+0.047 +0.035	+0.043 +0.035	+0.040 +0.035	+0.039 +0.035
OVER 6 TO 10	NUMERICAL VALUES FOR TOLERANCE ZONES IN THIS AREA NOT DEFINED.						+0.078 +0.042	+0.064 +0.042	+0.057 +0.042	+0.051 +0.042	+0.048 +0.042	+0.046 +0.042
OVER 10 TO 14							+0.093 +0.050	+0.077 +0.050	+0.068 +0.050	+0.061 +0.050	+0.058 +0.050	+0.055 +0.050
OVER 14 TO 18							+0.103 +0.060	+0.087 +0.060	+0.078 +0.060	+0.071 +0.060	+0.068 +0.060	+0.065 +0.060
OVER 18 TO 24	+0.115 +0.063	+0.096 +0.063	+0.084 +0.063	+0.076 +0.063	+0.072 +0.063	+0.069 +0.063	+0.125 +0.073	+0.106 +0.073	+0.094 +0.073	+0.086 +0.073	+0.082 +0.073	+0.079 +0.073
OVER 24 TO 30	+0.127 +0.075	+0.108 +0.075	+0.096 +0.075	+0.088 +0.075	+0.084 +0.075	+0.081 +0.075	+0.140 +0.088	+0.121 +0.088	+0.109 +0.088	+0.101 +0.088	+0.097 +0.088	+0.094 +0.088
OVER 30 TO 40	+0.156 +0.094	+0.133 +0.094	+0.119 +0.094	+0.110 +0.094	+0.105 +0.094	+0.101 +0.094	+0.174 +0.112	+0.151 +0.112	+0.137 +0.112	+0.128 +0.112	+0.123 +0.112	+0.119 +0.112
OVER 40 TO 50	+0.176 +0.114	+0.153 +0.114	+0.139 +0.114	+0.130 +0.114	+0.125 +0.114	+0.121 +0.114	+0.198 +0.136	+0.175 +0.136	+0.161 +0.136	+0.152 +0.136	+0.147 +0.136	+0.143 +0.136
OVER 50 TO 65	+0.218 +0.144	+0.190 +0.144	+0.174 +0.144	+0.163 +0.144	+0.157 +0.144	+0.152 +0.144	+0.246 +0.172	+0.218 +0.172	+0.202 +0.172	+0.191 +0.172	+0.185 +0.172	+0.180 +0.172
OVER 65 TO 80	+0.248 +0.174	+0.220 +0.174	+0.204 +0.174	+0.193 +0.174	+0.187 +0.174	+0.182 +0.174	+0.284 +0.210	+0.256 +0.210	+0.240 +0.210	+0.229 +0.210	+0.223 +0.210	+0.218 +0.210
OVER 80 TO 100	+0.301 +0.214	+0.268 +0.214	+0.249 +0.214	+0.236 +0.214	+0.229 +0.214	+0.224 +0.214	+0.345 +0.258	+0.312 +0.258	+0.293 +0.258	+0.280 +0.258	+0.273 +0.258	+0.268 +0.258
OVER 100 TO 120	+0.341 +0.254	+0.308 +0.254	+0.289 +0.254	+0.276 +0.254	+0.269 +0.254	+0.264 +0.254	+0.397 +0.310	+0.364 +0.310	+0.345 +0.310	+0.332 +0.310	+0.325 +0.310	+0.320 +0.310
OVER 120 TO 140	+0.400 +0.300	+0.363 +0.300	+0.340 +0.300	+0.325 +0.300	+0.318 +0.300	+0.312 +0.300	+0.465 +0.365	+0.428 +0.365	+0.405 +0.365	+0.390 +0.365	+0.383 +0.365	+0.377 +0.365
OVER 140 TO 160	+0.440 +0.340	+0.403 +0.340	+0.380 +0.340	+0.365 +0.340	+0.358 +0.340	+0.352 +0.340	+0.515 +0.415	+0.478 +0.415	+0.455 +0.415	+0.440 +0.415	+0.433 +0.415	+0.427 +0.415
OVER 160 TO 180	+0.480 +0.380	+0.443 +0.380	+0.420 +0.380	+0.405 +0.380	+0.398 +0.380	+0.392 +0.380	+0.565 +0.465	+0.528 +0.465	+0.505 +0.465	+0.490 +0.465	+0.483 +0.465	+0.477 +0.465
OVER 180 TO 200	+0.540 +0.425	+0.497 +0.425	+0.471 +0.425	+0.454 +0.425	+0.445 +0.425	+0.439 +0.425	+0.635 +0.520	+0.592 +0.520	+0.566 +0.520	+0.549 +0.520	+0.540 +0.520	+0.534 +0.520
OVER 200 TO 225	+0.585 +0.470	+0.542 +0.470	+0.516 +0.470	+0.499 +0.470	+0.490 +0.470	+0.484 +0.470	+0.690 +0.575	+0.647 +0.575	+0.621 +0.575	+0.604 +0.575	+0.595 +0.575	+0.589 +0.575
OVER 225 TO 250	+0.635 +0.520	+0.592 +0.520	+0.566 +0.520	+0.549 +0.520	+0.540 +0.520	+0.534 +0.520	+0.755 +0.640	+0.712 +0.640	+0.686 +0.640	+0.669 +0.640	+0.660 +0.640	+0.654 +0.640
OVER 250 TO 280	+0.710 +0.580	+0.661 +0.580	+0.632 +0.580	+0.612 +0.580	+0.603 +0.580	+0.596 +0.580	+0.840 +0.710	+0.791 +0.710	+0.762 +0.710	+0.742 +0.710	+0.733 +0.710	+0.726 +0.710
OVER 280 TO 315	+0.780 +0.650	+0.731 +0.650	+0.702 +0.650	+0.682 +0.650	+0.673 +0.650	+0.666 +0.650	+0.920 +0.790	+0.871 +0.790	+0.842 +0.790	+0.822 +0.790	+0.813 +0.790	+0.806 +0.790
OVER 315 TO 355	+0.870 +0.730	+0.819 +0.730	+0.787 +0.730	+0.766 +0.730	+0.755 +0.730	+0.748 +0.730	+1.040 +0.900	+0.989 +0.900	+0.957 +0.900	+0.936 +0.900	+0.925 +0.900	+0.918 +0.900
OVER 355 TO 400	+0.960 +0.820	+0.909 +0.820	+0.877 +0.820	+0.856 +0.820	+0.845 +0.820	+0.838 +0.820	+1.140 +1.000	+1.089 +1.000	+1.057 +1.000	+1.036 +1.000	+1.025 +1.000	+1.018 +1.000
OVER 400 TO 450	+1.075 +0.920	+1.017 +0.920	+0.983 +0.920	+0.960 +0.920	+0.947 +0.920	+0.940 +0.920	+1.255 +1.100	+1.197 +1.100	+1.163 +1.100	+1.140 +1.100	+1.127 +1.100	+1.120 +1.100
OVER 450 TO 500	+1.155 +1.000	+1.097 +1.000	+1.063 +1.000	+1.040 +1.000	+1.027 +1.000	+1.020 +1.000	+1.405 +1.250	+1.347 +1.250	+1.313 +1.250	+1.290 +1.250	+1.277 +1.250	+1.270 +1.250

THE ISO SYSTEM OF LIMITS AND FITS—TOLERANCES AND DEVIATIONS

Table 6-34. Permissible Machinery for Linear Dimensions Without Tolerance Indication (ISO 2768)

Variations in millimetres

Nominal dimensions mm		0,5 to 3	over 3 to 6	over 6 to 30	over 30 to 120	over 120 to 315	over 315 to 1 000	over 1 000 to 2 000
Permissible variations	Fine series	± 0,05	± 0,05	± 0,1	± 0,15	± 0,2	± 0,3	± 0,5
	Medium series	± 0,1	± 0,1	± 0,2	± 0,3	± 0,5	± 0,8	± 1,2
	Coarse series		± 0,2	± 0,5	± 0,8	± 1,2	± 2	± 3

Table 6-35. Angular Tolerances Other than Specified (ISO 2768)

Length of the shorter side mm		up to 10	over 10 to 50	over 50 to 120	over 120 to 400
Permissible variations	in degrees and minutes	± 1°	± 30'	± 20'	± 10'
	in millimetres per 100 mm	± 1,8	± 0,9	± 0,6	± 0,3

Related ISO Standards (TC3) are shown at the end of Section 18.

7 The ISO System of Limits and Fits — Inspection of Plain Workpieces

1. SCOPE*

This part of the ISO System of Limits and Fits relates to the inspection of plain workpieces. It specifies the interpretation to be given to the limits of dimensions to be inspected, and gives the essential details concerning limit gauges and indicating measuring instruments necessary for the inspection of tolerances of the ISO system.

2. GENERAL RULES OF INSPECTION

2.1 Reference temperature and measuring force

ISO 1-1975 Standard, *Standard reference temperature for industrial length measurements* fixes this temperature at

$$20\,°C$$

This is the temperature at which dimensions specified for workpieces and their inspection instruments are defined and at which the inspection should normally be carried out.

In addition all measuring operations provided for in this standard are understood as referred to a zero measuring force.

If the measurement is carried out with a measuring force different from zero its result should be corrected accordingly. This correction however is not required for comparative measurements carried out with the same comparing means and the same comparing force between similar elements of identical material and identical surface roughness.

2.2 Interpretation of size limits (Taylor principle)

In order to guarantee, so far as is practicable, that the functional requirements of the ISO system of limits and fits are attained, the limits of size should be interpreted in the following way within the prescribed length.

For *holes*, the diameter of the largest perfect imaginary cylinder which can be inscribed within the hole so that it just contacts the highest points of the surface should not be a diameter smaller than the GO limit of size. In addition the maximum diameter at any position in the hole must not exceed the NOT GO limit of size.

---

*NOTE: The material in Section 7 has been adopted from the ISO recommendation 1938-1971 with permission of the American National Standards Institute and the International Organization for Standardization. The proposed ANSI B4.4 standard on inspection of plain workpieces might not agree with the material presented here, which reflects European gaging practices.

For *shafts*, the diameter of the smallest perfect imaginary cylinder which can be circumscribed about the shaft so that it just contacts the highest points of the surface should not be a diameter larger than the GO limit of size. In addition the minimum diameter at any position on the shaft must not be less than the NOT GO limit of size.

The above interpretation means that if the size of the hole or shaft is everywhere at its GO limit then the hole or shaft should be perfectly round and straight.

Unless otherwise specified, and subject to the above requirements, departures from true roundness and straightness may reach the full value of the diametral tolerance specified. Typical extreme errors of form permitted by this interpretation are illustrated in Figures 1 and 2. Such extreme errors are unlikely to arise in practice.

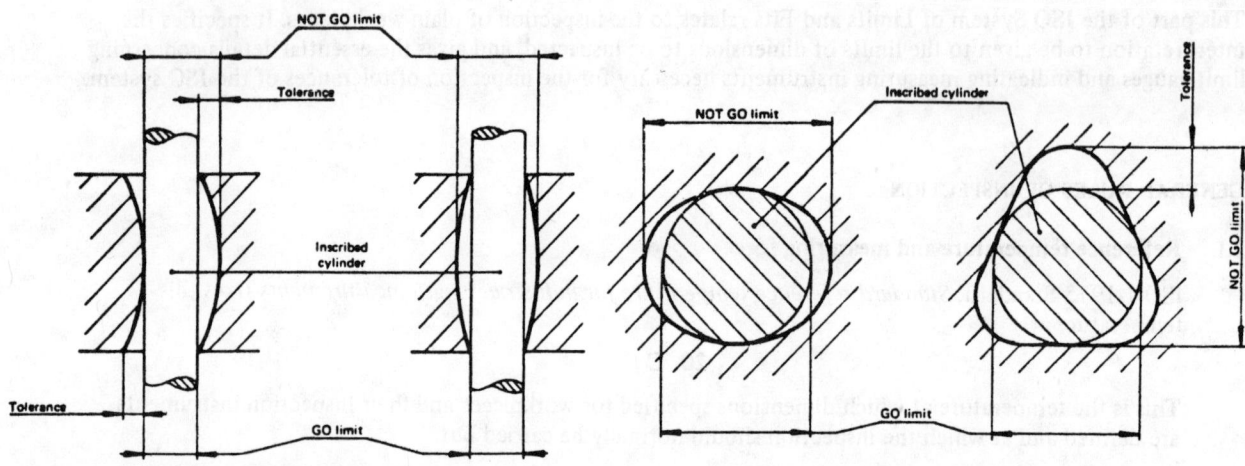

FIG. 1 — Extreme errors of form of hole allowed by the recommended interpretation of the limits of size

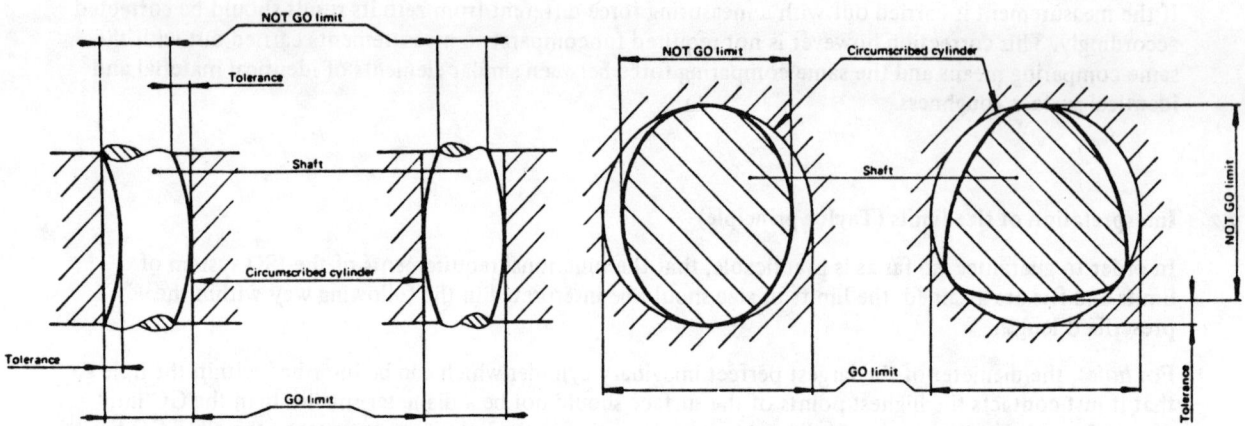

FIG. 2 — Extreme errors of form of shaft allowed by the recommended interpretation of the limits of size

The above interpretation of the size limits results from the "Taylor principle", named after the late W. TAYLOR who first laid it down in 1905. It is based on the use of a correct system of limit gauges to inspect shafts and holes. According to this principle a hole should completely assemble with a GO cylindrical plug gauge made to the specified GO limit of the hole, having a length at least equal to the length of engagement of the hole and shaft. In addition the hole is measured or gauged to check that its maximum diameter is not larger than the NOT GO limit. The shaft should assemble completely with a ring gauge made to the specified GO limit of the shaft and of a length at least equal to that of the length of engagement of the shaft and hole. Finally the shaft is measured or gauged to check that its minimum diameter is not smaller than the NOT GO limit.

In special cases the maximum errors of form permitted by the above interpretation may be too large to allow satisfactory functioning of the assembled parts; in such cases separate tolerances should be given for the form, e.g. separate tolerances on circularity or straightness.

2.3 Exceeding the limits

The above mentioned size limits are those specified in *ISO system of limits and fits – General, tolerances and deviations*. However, in order to take account of the existing manufacturing techniques of gauges, tolerances of manufacture and wear of limit gauges are such that the limits of specified dimensions for grades 6 to 8 may be exceeded in some cases (y or y_1 margin, see clause 3.9.2.1). In this case if workpieces are inspected during the manufacture by means of indicating measuring instruments instead of limit gauges, the manufacturer may also take into account the same y or y_1 margin as for gauges, in order to establish a uniform acceptance principle.

If inspection should exceptionally be carried out without any margin (y or $y_1 = 0$) in grades 6 to 8 which normally require it, this should be explicitly specified by writing the letter N following the grade number.

2.4 Choice of inspection method

Workpieces may be inspected either by means of fixed limit gauges or by means of indicating measuring instruments.

Both methods have advantages and drawbacks of their own which it is important to know before selecting one of them.

A system of limit gauges designed in strict conformity with the Taylor principle has the advantage of checking the geometry as well as the sizes of workpieces. However, for practical reasons departures from this principle may be made as stated in clause 3.3 and so the inspection is possibly not so satisfactory as might be expected theoretically.

Furthermore gauges themselves have errors of form and size, and their necessary manufacturing and wear tolerances further reduce the amount of tolerance that remains available on the workpiece.

Measuring instruments give the workpiece size in the measuring position only and do not check the geometry, which requires separate measurements the result of which, in theory, should be correlated to that of the dimensional measurement. This tedious procedure is not necessary provided sufficient reliance can be placed on the manufacturing accuracy to ensure that, in practice, form errors can be ignored.

On the other hand, contrary to what occurs with limit gauges, the use of such instruments has the advantage, when workpieces have very small tolerances, of not reducing the amount of tolerance that remains available on the workpieces. Finally the use of such instruments allows sampling inspection which gives warning when the sizes approach one of the limits during a continuous manufacturing process.

In order to avoid disputes it is recommended that the type of inspection to be used for acceptance be specified on the order.

Unless strictly specified to the contrary a workpiece should be considered as good when the manufacturer can prove that it was recognized as such by the inspection method he chose in conformity with this standard.

3. LIMIT GAUGES

3.1. Gauge types

3.1.1 *Limit gauges* are used to inspect the workpieces. For gauging internal diameters they may be of the following types:

- full form cylindrical plug gauge;
- full form spherical plug or disk gauge;
- segmental cylindrical bar gauge;
- segmental spherical plug gauge;
- segmental cylindrical bar gauge with reduced measuring faces;
- rod gauge with spherical ends.

For gauging external diameters the following gauges may be used:

- full form cylindrical ring gauge;
- gap gauge.

3.1.2 *Reference gauges* or *block gauges* may be used to inspect or adjust limit gauges:

(a) *reference gauges* are either reference disks intended for setting gap gauges, or cylindrical ring or plug gauges used for calibrating gauges or indicating measuring instruments;

(b) *block gauges* are standards of length having parallel plane end surfaces which are used for calibrating gauges or indicating measuring instruments.

3.2 Application of the Taylor principle

Except for allowable deviations (see clause 3.3) strict application of the Taylor principle leads to using

for checking the GO limit of the workpiece:

a plug gauge or a ring gauge having exactly the GO limit diameter and a length equal to the workpiece length (or the engagement length of the fit to be made);

for checking the NOT GO limit:

a gauge contacting the workpiece surface only at two diametrically opposite points and having exactly the NOT GO limit diameter.

The GO gauge should perfectly assemble with the workpiece to be inspected and the NOT GO gauge should not be able to pass over or in the workpiece in any consecutive position in the various diametrical directions on the workpiece length.

3.3 Allowable deviation from the Taylor principle

As the application of the Taylor principle is not always strictly compulsory or comes up against difficulties in the convenient use of gauges, certain deviations may be allowed (see clause 3.4).

At the GO limit a full form gauge is not always necessary or used, as for instance in the following cases:

The length of a GO cylindrical plug or ring gauge may be less than the length of engagement of the mating workpieces if it is known that with the manufacturing process used the error of straightness of the hole or shaft is so small that it does not affect the character of fit of the assembled workpieces. This deviation from the ideal facilitates the use of standard gauge blanks.

For gauging a large hole a GO cylindrical plug gauge may be too heavy for convenient use, and it is permissible to use a segmental cylindrical bar or spherical gauge if it is known that with the manufacturing process used the error of roundness or straightness of the hole is so small that it does not affect the character of fit of the assembled workpieces.

A GO cylindrical ring gauge is often inconvenient for gauging shafts and may be replaced by a gap gauge if it is known that with the manufacturing process used the errors of roundness (especially lobing) and straightness of the shaft are so small that they do not affect the character of fit of the assembled workpieces. The straightness of long shafts which have a small diameter should be checked separately.

At the NOT GO limit a two-point checking device is not always necessary or used, as for instance in the following cases:

Point contacts are subject to rapid wear, and in most cases may be replaced where appropriate by small plane, cylindrical or spherical surfaces.

For gauging very small holes a two-point checking device is difficult to design and manufacture. NOT GO plug gauges of full cylindrical form have to be used but the user must be aware that there is a possibility of accepting workpieces having diameters outside the NOT GO limit.

Non-rigid workpieces may be deformed to an oval by a two-point mechanical contact device operating under a finite contact force. If it is not possible to reduce the contact force to almost zero, then it is necessary to use NOT GO ring or plug gauges of full cylindrical form.

Such thin-walled workpieces may be out of round (due to internal stresses or heat treatment). In these cases the NOT GO limit has the meaning that the circumference of the cylinder corresponding to that limit must not be transgressed. Therefore NOT GO gauges of full cylindrical form have to be applied with a force that just suffices to convert the elastic deformation into circularity but does not expand or compress the wall of the workpiece.

Lastly, the sizes of gauges cannot be made exactly to the appropriate workpiece limit: they have to be made to specified tolerances.

3.4 Field of utilization of the various types of limit gauges

Taking account of the above remarks, recommended types of gauges for various ranges of workpiece nominal dimensions are given in Figures 5 and 6, the meaning of symbols used being given in Figures 3 and 4.

3.5 General design features of limit gauges

No recommendation is given for the details of the designs; these are left to the initiative of the gauge makers or the national standards organizations.

The various types of gauges are illustrated in Figures 3 and 4. The recommended types of gauges for the different ranges of nominal size of the workpieces are shown in Figures 5 and 6. The key to the symbols used in Figures 5 and 6 is given in Figures 3 and 4.

A *full form cylindrical plug gauge* (Fig. 3 A) has a gauging surface in the form of an external cylinder. The method of attaching the gauge to the handle should not affect the size and form of the gauge by producing an undesirable stress.

A small circumferential groove near the leading end of the gauge and a slight reduction in diameter of the remaining short cylindrical surface at the end are recommended to serve as a pilot to facilitate the insertion of the gauge into the workpiece hole.

A *full form spherical plug* or *disk gauge* (Fig. 3 B) has a gauging surface in the form of a sphere from which two equal segments are cut off by planes normal to the axis of the handle.

A *segmental cylindrical bar gauge* (Fig. 3 C) has a gauging surface in the form of an external cylinder from which two axial segments are either relieved (Fig. 3 C (i)) or removed (Fig. 3 C (ii)). This gauge may have reduced measuring faces (Fig. 3 E).

THE ISO SYSTEM OF LIMITS AND FITS—INSPECTION OF PLAIN WORKPIECES

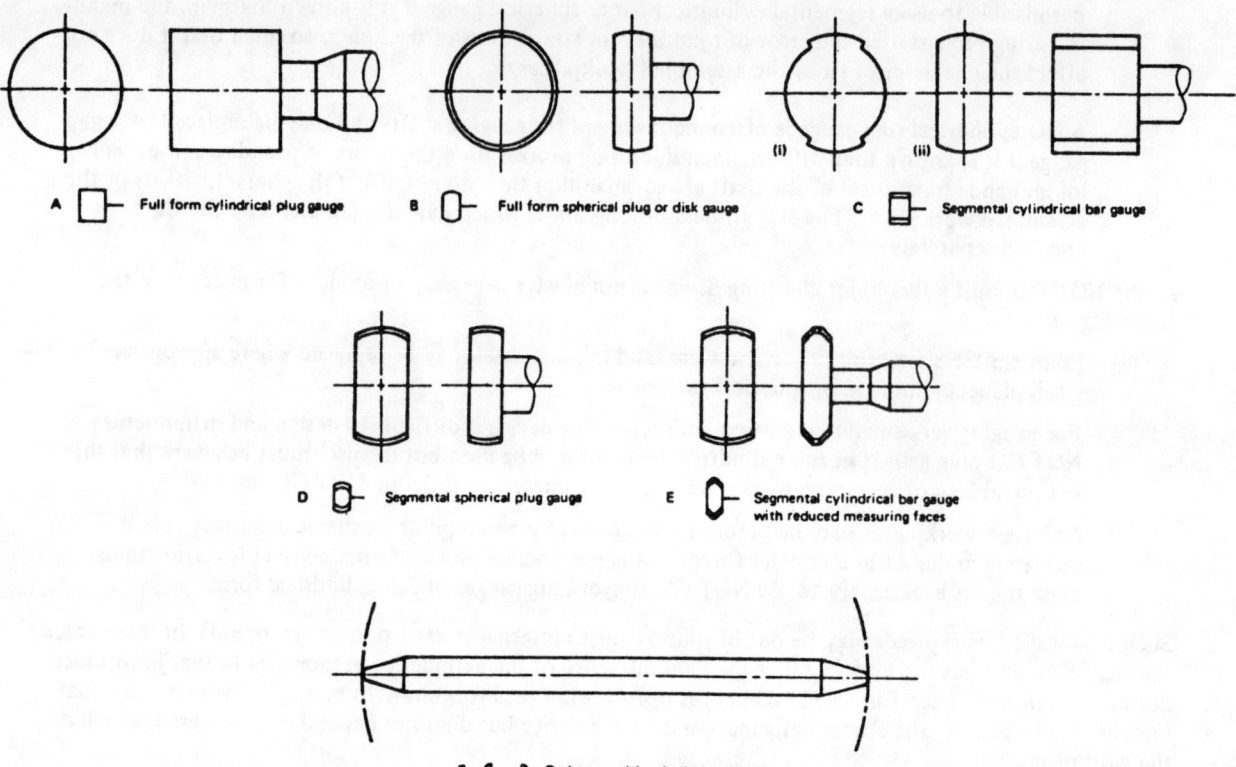

FIG. 3 – Recommended types of gauges for holes (and their corresponding symbols for Fig. 5)

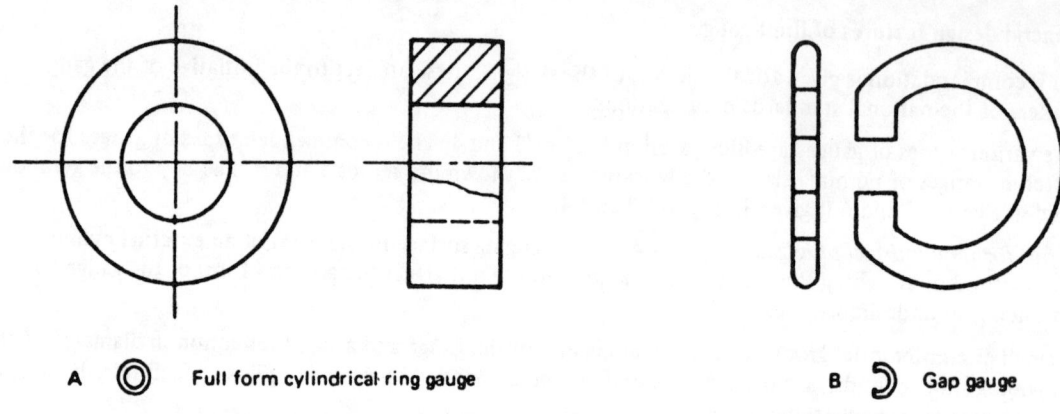

FIG. 4 – Recommended types of gauges for shafts (and their corresponding symbols for Fig. 6)

THE ISO SYSTEM OF LIMITS AND FITS—INSPECTION OF PLAIN WORKPIECES

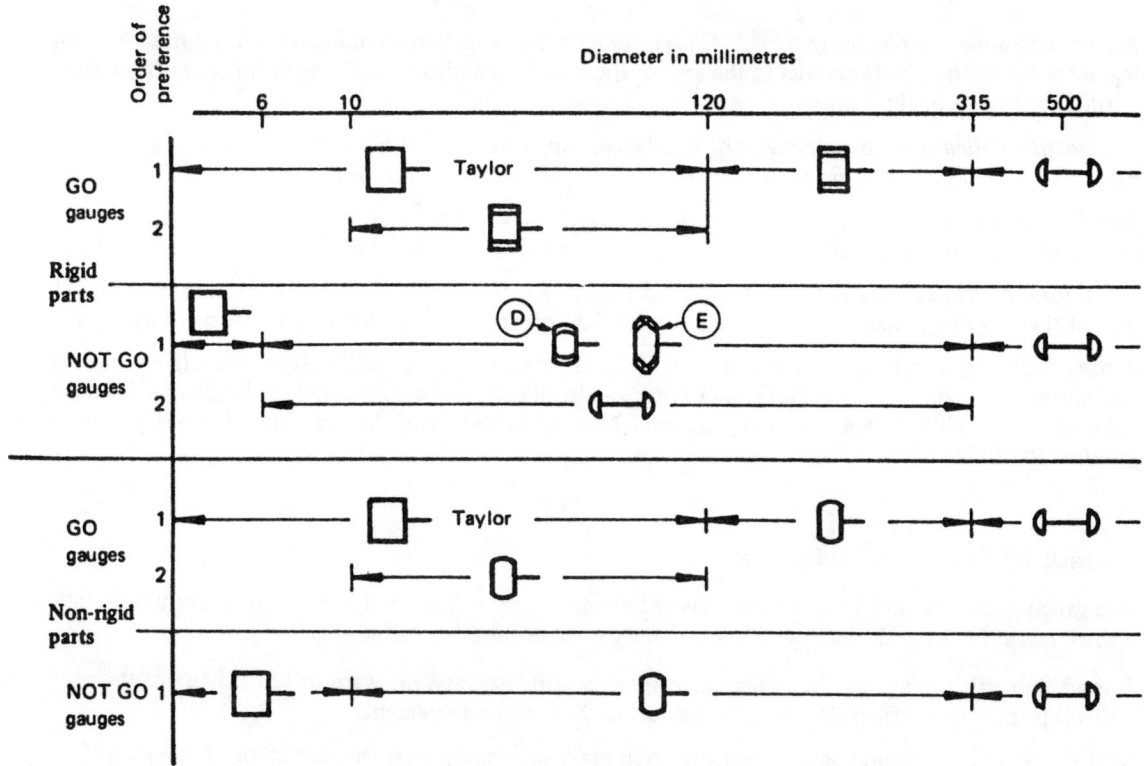

FIG. 5 – Types of gauges used to check holes, in order of preference

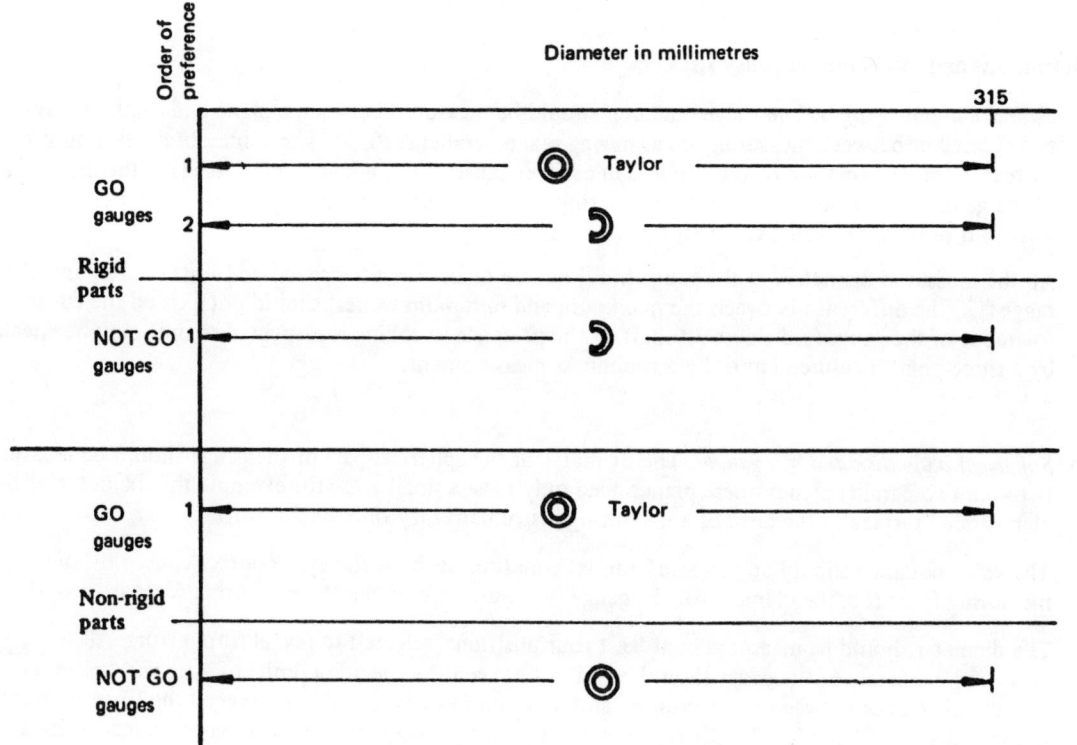

FIG. 6 – Types of gauges used to check shafts, in order of preference

A *segmental spherical plug gauge* (Fig. 3 D) is similar to the gauge shown in Figure 3 B but has two equal segments cut off by planes parallel to the axis of the handle in addition to the segments cut off by planes normal to the axis of the handle.

A *segmental cylindrical bar gauge with reduced measuring faces* (Fig. 3 E) is similar to the gauge shown in Figure 3 C but has reduced measuring faces in a plane parallel to the axis of the handle.

A *rod gauge with spherical ends* (Fig. 3 F) has spherical end surfaces the radius of which should not be greater than half the length of the gauge. The rod may be fixed or adjustable.

A *full form cylindrical ring gauge* (Fig. 4 A) has a gauging surface in the form of an internal cylinder. The wall of the ring gauge shall be thick enough to avoid deformation under normal conditions of use.

A *gap gauge* (Fig. 4 B) has for its working size flat and parallel gauging surfaces (or, alternatively one flat and one spherical or cylindrical surface, or two cylindrical surfaces being parallel to the axis of the shaft being checked). The GO and NOT GO gaps may lie on the same side of the gap gauge. The gap gauge may be fixed or adjustable.

3.6 Materials and further details of gauges

The gauging surfaces shall be of a wear-resistant material such as hardened steel, hard chromium plating of a thickness at least equal to the wear zone of the gauge, or tungsten carbide.

It is advisable that the gauges be insulated, as far as possible, against the warmth of the hand of the user, when this is likely to affect significantly the accuracy of the measurement.

NOT GO gap gauges should bear an identification mark such as a groove or a red color or an easily visible reduction in the length of the gauging surface. This identification is not necessary if the NOT GO side is self-evident as with the progressive type GO and NOT GO gap gauge.

3.7 Definitions and inspection of gauge sizes

3.7.1 *Cylindrical plug gauges*.

The gauge diameter should be measured between a plane and a spherically-ended anvil or between measuring anvils having plane parallel surfaces. The value obtained should be corrected for deformation of the surfaces in contact caused by the measuring force (i.e. the diameter of the gauge is the diameter when the measuring force is zero). The diameter should be measured in at least four positions selected to reveal form errors.

All the measured diameters of the gauge should be on or between the specified limits of size, and the range (i.e. the difference between the maximum and minimum values) should not exceed the form tolerance of the gauge (see clause 3.9.2.3). If the presence of lobing is suspected then it may be checked by a three-point measurement or by a roundness measurement.

3.7.2 *Spherical plug, disk and rod gauges*.

The diameter of the spherical part of the gauge should be measured between two parallel planes; these planes need only have a small area (for example the diameter of the plane-ended surface of the anvil of a measuring instrument may only be 5 mm).

The value obtained should be corrected for deformation of the surfaces in contact caused by the measuring force (i.e. the diameter of the gauge is the diameter when the measuring force is zero).

The diameter should be measured in at least four positions, selected to reveal form errors. All the measured diameters of the gauge should be on or between the specified limits of size, and the range (i.e. the difference between the maximum and minimum values) should not exceed the form tolerance of the gauge (see clause 3.9.2.3). If the presence of lobing is suspected then it may be checked by a three-point measurement or by a roundness measurement.

3.7.3 *Cylindrical ring gauges.* The diameter should be measured by means of two spherically-ended anvils positioned in a plane normal to the axis of the ring gauge. When moving the measuring instrument in this plane the greatest distance apart of the two anvils determines the diameter. The value obtained should be corrected for deformation of the surfaces in contact caused by the measuring force (i.e. the diameter of the gauge is the diameter when the measuring force is zero).

The diameter should be measured in at least four positions, selected to reveal form errors. All the measured diameters of the gauge should be on or between the specified limits of size, and the range (i.e. the difference between the maximum and minimum values) should not exceed the form tolerance of the gauge (see clause 3.9.2.3). If the presence of lobing is suspected then it may be checked by a three-point measurement or by a roundness measurement.

3.7.4 *Gap gauges.* The *actual size* of a gap gauge is defined as the perpendicular distance between the gauging surfaces, when no force is exerted on the gauge.

The *working size* of a gap gauge is defined as the diameter of a reference disk over which the gap gauge just passes in a vertical direction under the working load marked on it, or, if this is not indicated, under its own weight. Beforehand, the disk should be greased with a thin film of petroleum jelly and then carefully wiped but not rubbed. The gauging surfaces of the gap gauge should be cleaned. The gap gauge should slide over the disk after having been brought carefully to rest in contact with the disk and then released : inertia forces are thus avoided.

For heavier gap gauges it is recommended that the working load should be less than the weight of the gauge, so that the working size may be determined more accurately. The positions of the places where the forces counter-balancing part of the weight of the gauge are to be applied (see Fig. 7) should be marked on gauges of nominal sizes above 100 mm.

The working size of a gap gauge is not defined with a zero measuring force, as for the other definitions, because the size of a reference disk is defined with a zero measuring force and the gap gauge may be regarded as a comparator which is intended to transfer (on the particular limit) the size of the reference disk to the workpiece.

In practice a reference disk may be used directly to accept a gap gauge in the case where the disk and the gap gauge are supplied together and the gap gauge has been adjusted to the disk. In other cases the following two alternative procedures are recommended.

(*a*) Determine the successive loads under which the gap gauge will pass over two reference disks of different diameter under the conditions specified in the definition of the working size. The difference in these two loads is taken as a basis for calculating the working size of the gap gauge at its working load.

(*b*) Take a reference disk with a diameter smaller* than the smallest permissible size of the gap gauge. Place gauge blocks successively on the gauging surfaces of the gap gauge, if possible equally distributed so that in one case the sum of the diameter of the reference disk and the gauge blocks is equal to the lowest permissible working size and in the other case to the highest permissible working size of the gap gauge.**

In the first case the gap gauge should pass over the reference disk, and in the second case it should not pass over the reference disk, under the conditions specified in the definition of the working size.

Doubtful cases are decided according to method (*b*).*

---
*For gauges up to 100 mm it is advisable to make the diameter of the disk 5 mm smaller than the nominal size, and for gap gauges over 100 mm smaller than the nominal size.

Where the new gap gauge is successively applied on two combinations of block gauges corresponding to the limits for this gap gauge, the gap gauge should be capable of being raised carrying one of these combinations but without carrying the other.

**Reference disks made to the lowest and highest permissible sizes may also be used.

When the gap gauge is used in a horizontal position, with the axis of the workpiece vertical, its working size is defined as the largest size of a reference disk or gauge block combination over which it can just be moved by hand without excessive force.

The difference between the working size and the actual size of a gap gauge is equal to the amount by which the gauge is deformed by the force applied when determining the working size. The design of the gap gauge should be as rigid as possible in relation to the weight of the gauge so as to keep this difference in size to a minimum.

3.8 Method of use of gauges

The following recommendations relate to the general use of the gauges in the workshop as well as borderline cases.

3.8.1 *Gauges for holes*

3.8.1.1 GO GAUGE. A GO gauge should assemble completely with the hole when applied by hand without using excessive force, and the total length of the hole should be checked. When gauging non-rigid workpieces, such as thin-walled parts, the application of too great a force will enlarge the diameter of the hole. A GO segmental gauge should be applied to the hole in at least two or three axial planes uniformly distributed around the circumference.

3.8.1.2 NOT GO GAUGE. A cylindrical NOT GO plug gauge should not enter the hole when applied by hand without using excessive force. The hole should be checked from both ends if possible.

A NOT GO gauge with spherical measuring surfaces should be introduced into the hole by tilting it. When it is erected in the hole, contacting the hole on a diameter, it should not be possible to pass it through the hole by hand without using excessive force. This test should be performed at not less than four positions around and along the cylindrical surface of the hole.

3.8.2 *Gauges for shafts*

3.8.2.1 GO GAUGE. The GO gap gauge should pass over a shaft, the axis of which is horizontal, under its own weight or the force marked on the gauge, under the conditions of clause 3.7.4.

The GO gap gauge should pass over a shaft, the axis of which is vertical, when applied by hand without using excessive force. It is recommended that the corresponding reference disk should be used to assess the measuring force.

The above test should be carried out at not less than four positions around and along the shaft.

A cylindrical GO ring gauge should pass over the complete length of the shaft, when applied by hand without using excessive force.

3.8.2.2 NOT GO GAUGE. The NOT GO gap gauge should not pass over a shaft, the axis of which is horizontal, under its own weight or the force marked on the gauge, under the conditions of clause 3.7.4.

The NOT GO gap gauge should not pass over a shaft, the axis of which is vertical, when applied by hand without using excessive force.

The above test should be applied at not less than four positions around and along the shaft.

THE ISO SYSTEM OF LIMITS AND FITS—INSPECTION OF PLAIN WORKPIECES

3.9 Manufacturing tolerances and permissible wear of gauges

3.9.1 *Symbols.* The following symbols are used in this standard:

D	=	nominal diameter of workpiece, in millimetres
H	=	tolerance on cylindrical plug or cylindrical bar gauges
H_s	=	tolerance on spherical gauges
H_1	=	tolerance on gauges for shafts
H_p	=	tolerance on reference disks for gap gauges
y	=	margin, outside the GO workpiece limit, of the wear limit of gauges for holes
y_1	=	margin, outside the GO workpiece limit, of the wear limit of gauges for shafts
z	=	distance between center of tolerance zone of new GO gauges for holes and GO workpiece limit
z_1	=	distance between center of tolerance zone of new GO gauges for shafts and GO workpiece limit
α	=	safety zone provided for compensating measuring uncertainties of gauges for holes of nominal diameter over 180 mm
α_1	=	safety zone provided for compensating measuring uncertainties of gauges for shafts of nominal diameter over 180 mm
y' and y'_1	=	difference in absolute value between y and α, and y_1 and α_1.

3.9.2 *Limit gauges*

3.9.2.1 POSITIONS OF TOLERANCE ZONES AND WEAR LIMITS IN RELATION TO WORKPIECE LIMITS (shown diagrammatically in Figure 9).

3.9.2.1.1 NOT GO LIMIT OF WORKPIECES.
The tolerance zone of new NOT GO gauges for nominal sizes up to and including 180 mm is symmetrical to the NOT GO limit. For sizes above 180 mm the tolerance zone is symmetrical to a line lying inside the workpiece tolerance zone at a distance α or α_1 from the NOT GO limit.

3.9.2.1.2 GO LIMIT OF WORKPIECES.
A reasonable life for GO gauges is obtained in two ways:

(a) by moving the tolerance zone of a new GO gauge inside the workpiece tolerance by an amount z or z_1;

(b) by allowing a tolerance of wear of the GO gauge outside the GO limit of the workpiece by an amount y or y_1 when this value is not zero.

In the range of sizes of nominal diameter above 180 mm the values of y and y_1 are reduced by the amount of the safety zones α and α_1 respectively so that in these cases the actual wear of the GO gauges is limited to y' and y'_1 respectively outside the GO limit of the workpiece (or to α and α_1 within this limit if y and y_1 are equal to zero).

The values of the y or y_1 margin have been taken as small as possible in order to reduce to a minimum the risk that workpieces with sizes outside the prescribed GO limit be accepted. This margin is therefore provided only in the case of smaller tolerances on workpieces, grades 6 to 8 up to 500 mm, the deletion of this margin being contemplated as a possibility in the future (in connection with the development of low cost low wear gauges).

However, inspection without the y or y_1 margin is now permitted provided that it is clearly specified. In this case, in order to avoid any mistake, the workpiece tolerance should be conventionally designated by adding the letter N to its designation (for example 10H6N, 15G7N, 20f8N).

In the range of sizes of nominal diameter above 180 mm, the workpiece tolerance has been reduced, at the GO (with y or y_1 margin) and NOT GO limits, by the amount of the safety zones α and α_1. Manufacturers and users should not forget that due to errors of measurement the sizes of workpieces may fall outside the limits of the gauges by the amount of the safety zones α and α_1 and that the extreme workpiece limits, given by y and y_1, may be reached.

3.9.2.2 TOLERANCES ON SIZE OF WORKING GAUGES. The tolerances on size of working gauges are based on the fundamental tolerances of grades 1 to 7, as given in Table 1.

Values of α, y, z, etc. for the gauges are given in Tables 2 and 2A.

3.9.2.3 TOLERANCES ON FORM OF WORKING GAUGES. The tolerances on form of working gauges are based on the fundamental tolerances of grades 1 to 5, as given in Table 1.

3.9.2.4 ADJUSTABLE GAUGES OF SIZES ABOVE 180 mm. New GO gauges may be adjusted at any desired value within the limits of α and z or of α_1 and z_1 according to the permissible wear value.

Any trespassing of the GO workpiece limit may be easily avoided by adjusting the new GO gauge within the limit of z or z_1 and by adjusting it anew at the same limit as soon as wear brings its size to the size of the GO workpiece limit with a shift of α or α_1 inwards.

It is advisable to adjust NOT GO gauges to the size of the NOT GO workpiece limit with a shift of α or α_1 inwards.

3.9.3 Reference disks for gap gauges

3.9.3.1 POSITIONS OF TOLERANCE ZONES WITH RESPECT TO THE WORKPIECE LIMITS (shown diagrammatically in Figure 9).

3.9.3.1.1 NOT GO LIMIT OF WORKPIECE. For nominal sizes up to and including 180 mm the tolerance zone of the reference disk is symmetrical to the NOT GO limit. For sizes above 180 mm the tolerance zone is symmetrical to a line lying inside the workpiece tolerance zone at a distance α_1 from the NOT GO limit.

3.9.3.1.2 GO LIMIT OF WORKPIECE. The tolerance zone of the reference disk for a new gap gauge is symmetrical to the z_1 value.

The tolerance zone of the reference disk for checking wear is located in the following manner:

(a) Workpieces of nominal sizes up to and including 180 mm : For tolerance grades 6 to 8 the tolerance zone of the reference disk is symmetrical to the y_1 value. For tolerance grades 9 to 16, y_1 is zero and hence the tolerance zone of the reference disk is symmetrical to the GO limit of the workpiece.

(b) Workpieces of nominal sizes above 180 mm : For tolerance grades 6 to 8 the tolerance zone of the reference disk is symmetrical to the y'_1 value ($y'_1 = y_1 - \alpha_1$). For tolerance grades 9 to 16, y_1 is zero and hence the tolerance zone of the reference disk is symmetrical to the α_1 value inside the GO limit of the workpiece.

3.9.3.2 TOLERANCES ON SIZE OF REFERENCE DISKS. The tolerances on size of the reference disks are based on the fundamental tolerances of grades 1 to 3, as given in Table 1.

Values of α_1, y_1, y'_1, z_1, etc. for the reference disks are given in Tables 2 and 2 A.

3.9.3.3 TOLERANCES ON FORM OF REFERENCE DISKS. The tolerances on form of the reference disks are based on the fundamental tolerances of grades 1 to 2, as given in Table 1.

3.9.3.4 RELATION BETWEEN TOLERANCES ON GAP GAUGES AND THEIR REFERENCE DISKS. The relation between the tolerance H_1 of the gap gauge and the tolerance H_p of its reference disk is as follows:

H_1 determines the limits of the value of the working size (see clause 3.7.4) of a gap gauge. The difference between the limits of size given by H_1 for the gap gauge and by H_p for the reference disk represents a safety zone on both sides of H_p to compensate for errors of measurement, in the same way as α and α_1 compensate for errors of measurement for workpieces of diameters over 180 mm. H_1 and H_p are therefore symmetrical (see Fig. 8). Therefore if, according to the definition of working size, gap gauges lie outside the zone H_p but within the zone H_1 they are still to be regarded as correct.

3.9.4 *Reference ring and plug gauges for setting measuring instruments.* The gauges should be made to tolerances on size and form equal to those for reference disks. The tolerance on size is disposed bilaterally with respect to the appropriate test limit of the workpiece. The size of each gauge should be measured across a diameter halfway through the gauge; the axial plane in which this diameter occurs and the measured size of the gauge should be marked on the end face of the gauge.

3.9.5 *Surface finish of gauges.* It is recommended that the arithmetical mean deviation R_a for the surface roughness of a gauge should not exceed 10 % of the corresponding tolerance on the size of the gauge, with preferably a maximum value of 0.2 μm.

3.10 Settlement of disputes

Unless the contrary is clearly specified, inspection by limit gauges should be recognized as authoritative for acceptance, and it is agreed that a workpiece is satisfactory if it is recognized as good by a gauge conforming with the requirements of this standard.

To avoid any dispute requiring checking of the conformity of the gauges of the manufacturer, the following procedure is recommended in the use of gauges of the manufacturer and the purchaser.

3.10.1 *Inspection by the manufacturer.* Generally the inspection department that checks the workpieces made in the workshop can use the same types of gauges as those used in the workshop. In order to avoid differences between the results obtained by the workshop and inspection department it is recommended that the workshop uses new or only slightly worn GO gauges while the inspection department uses GO gauges having sizes nearer the permissible wear limit.

3.10.2 *Inspection by the purchaser.* There are three possible procedures for inspection on behalf of the purchaser by an inspector who does not belong to the manufacturing plant concerned:

(a) the inspector may gauge the workpieces with the manufacturer's own gauges, provided that he first checks the accuracy of these gauges;

(b) the inspector may use his own gauges, made in accordance with this standard, for inspecting workpieces. It is recommended that the GO gauges should have sizes near the wear limit in order to avoid differences between the results obtained by the manufacturer and inspector;

(c) the inspector may use his own inspection gauges for checking the workpieces. The disposition of the tolerance zones for these gauges should be such as to ensure that the inspector does not reject workpieces the sizes of which are within the specified limits.

3.11 Marking and designation of gauges

The recommendations relating to the marking and designation of the gauges are limited to the most essential information and it is assumed that there is room enough on the gauges.

The information that is recommended is:

(a) the nominal size of the workpiece and the ISO symbols for deviation and tolerance (or, if these are not available, the value of the tolerance limit to be checked);

(b) a method of distinguishing the GO and NOT GO sides:

red coloring is recommended for the NOT GO side, and it is also possible to distinguish between the GO and NOT GO sides by using different shapes of measuring elements;

(c) the working load, if necessary, for a gap gauge (see clause 3.7.4).

Room should be left for the following information: manufacturer's name or trademark, purchaser's mark and special remarks such as the serial number, workshop where the gauge is used, etc.

In the case of plug gauges of the renewable end type, the marking should appear both on the handle and on the renewable end.

It is unnecessary to indicate the reference temperature of 20 °C.

4. INDICATING MEASURING INSTRUMENTS

4.1 Definitions relating to measurements

4.1.1 *True size*. The size of a dimension which would be obtained by a measurement without any errors.

NOTE. – The nearest value of the true size at a given position on the workpiece is the average of a great number of measurements carried out with the greatest possible accuracy.

4.1.2 *Error of measurement*. The algebraical difference between the measured size and the true size.

NOTE. – Errors of measurement may be caused especially by the measuring equipment, the method of measurement, the operator or the environmental conditions.

Errors of measurement may be separated into systematic errors and random errors, defined as follows:

(a) the systematic error remains the same during a single series of measurements and may theoretically be eliminated by a corresponding correction of the measurement result;

(b) on the contrary, random error variations are indefinite and may not be eliminated. The resulting measurement uncertainty for a single series of measurements is represented by the standard deviation, as defined below, of the dispersion due to these errors.

4.1.3 *Systematic error*. The algebraical difference between the average of measured values and the true size, in a series of measurements of the size of a dimension, made at one position on the workpiece under the same experimental conditions.

4.1.4 *Random error*. The algebraical difference between the result of one individual measurement and the average of measured values, in a series of measurements made at one position on the workpiece under the same experimental conditions.

4.1.5 *Average*. The arithmetic mean value ($\bar{x}$) of a certain number (n) of values ($x_1, x_2, \ldots x_n$).

4.1.6 *Standard deviation*. A value representing the measuring uncertainty due to random error dispersion and estimated as the square root of the quotient of the sum of the squared differences of the individual results x_i and of their average $\bar{x}$ by the number of measurements minus one :

$$s = \sqrt{\frac{\sum_{i=1}^{n}(x_i - \bar{x})^2}{n-1}}$$

4.2 Types of measuring instruments

Measuring instruments may be divided into two main classes according to the way in which the measurement is carried out, either by comparison with a standard of length built into the instrument itself (for example a micrometer screw as in a hand micrometer, or a linear scale as in a vernier caliper), or by means of a comparator, the size of the workpiece being determined by comparing its size with that of a reference gauge of closely the same size.

4.3 Measuring uncertainty of the instrument

Any measuring instrument has its inherent error, independent of the part to be measured and of exterior conditions of measurement.

This inherent error may be subdivided into a systematic error likely to be compensated by a correction, and a random error giving rise to dispersion.

The random error is an irregular error the value of which cannot be estimated from one single reading but may be estimated as most probable from a number of successive readings. The frequency distribution of successive readings is close to a normal distribution, for which standardized tables are available allowing the determination of the percentage of readings within certain limits on both sides of the average reading (see Fig. 10).

However, it is not practical in the workshop to make the correction for suppressing the systematic error at each position (even if a curve of graduation errors has been plotted initially for each position of the measuring anvil) nor to carry out a sufficient number of readings at this position to deduce an average reading and thus suppress the random error.

It is therefore preferable to assume that each reading is correct within some assumedly constant error margin in the whole range of measurement of the instrument.

It is considered that for a very great number of measurements the systematic errors due to any basic elements of the instrument have an equal probability of being positive or negative and therefore these are considered as random errors.

Their being considered simultaneously with true random errors results in enlarging the distribution curve which thus represents the inherent errors of the instrument.

The corresponding error margin, i.e. the instrument measuring uncertainty, may be expressed in terms of the standard deviations of this curve, as being equal to $\pm 2s$; for a normal distribution, 95.45 % of readings will not depart from the mean size (true value) by more than twice the standard deviation s.

To obtain a representative value of the measuring uncertainty of a single design of instrument, it is necessary to test a complete lot of instruments, generally in a laboratory : the values thus obtained represent the inherent uncertainty of the instrument.

It is recommended that manufacturers should specify the uncertainty at ± 2s in the directions or specifications for use of newly supplied instruments, and state whether the specified measuring uncertainty is that of the indicating head or of the complete equipment including the indicating head.

4.4 Total measuring uncertainty

In order to be used in the workshop, the above-defined value of the inherent uncertainty of the instrument should be multiplied by a coefficient w taking into account the other systematic and random errors caused by the poorer environmental conditions in the workshop, and the inferior care, experience and ability of the operator.

Hence, for the standard deviation : $s_m = ws$, the value of s_m being the overall measuring uncertainty.

As a rule, $w = 2$ may be chosen (a value greater than 2 would correspond to especially poor conditions, while a lesser value would on the contrary correspond to exceptionally fine conditions, as for example in a room having controlled constant temperature).

4.5 Inspection limits

Due to uncertainty of measurement, if workpiece dimensions are very close to the prescribed limits, the user of the measuring instrument may run the risk of accepting workpieces outside the limits or of rejecting workpieces within these limits.

To reduce this risk to a minimum, the inspection limits should be placed within the specified limits by a conventional amount s_m, equal to the value of the standard deviation of the measuring process (see Fig. 11).

In consequence, in order not to reduce the manufacturing tolerance zone of the workpiece, the measuring instrument should be chosen so that the corresponding value of s_m does not exceed a given maximum s_M for each diameter and tolerance of a part.

The standardized values of s_M are given for guidance in Tables 3 and 3 A for each grade and diameter range.

4.6 Influence of form errors

If the usual type of measuring instrument having two diametrically opposed anvils is used it should be borne in mind that this inspection method satisfies the Taylor principle only for the minimum material limit (NOT GO side of the workpiece).

For this limit it is enough to carry out the inspection at a sufficient number of positions on the workpiece.

On the contrary, at the maximum material limit (GO side of the workpiece), two-point measurements even at several successive positions do not ensure, if the workpiece has form errors (especially lobing), that no point of the surface infringes the perfect inscribed or circumscribed cylinder specified in the Taylor principle.

THE ISO SYSTEM OF LIMITS AND FITS—INSPECTION OF PLAIN WORKPIECES

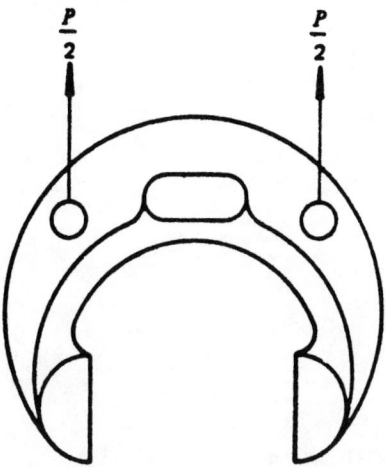

FIG. 7 – Location of points where forces counterbalancing parts of the weight of the gauge should be applied

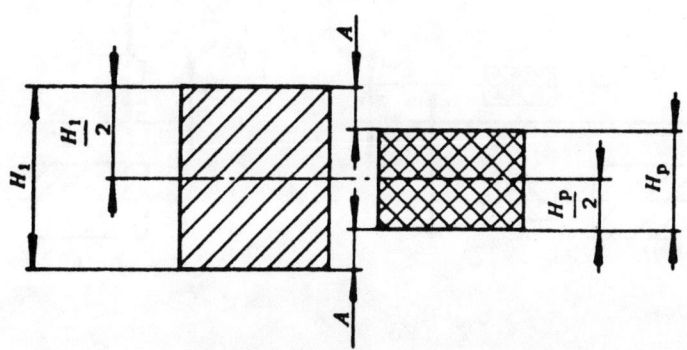

H_1 = Manufacturing tolerance of the gap-gauge
H_p = Manufacturing tolerance of the reference disk
A = Safety zone

FIG. 8 – Relation between manufacturing tolerances of gap gauges and of reference disks

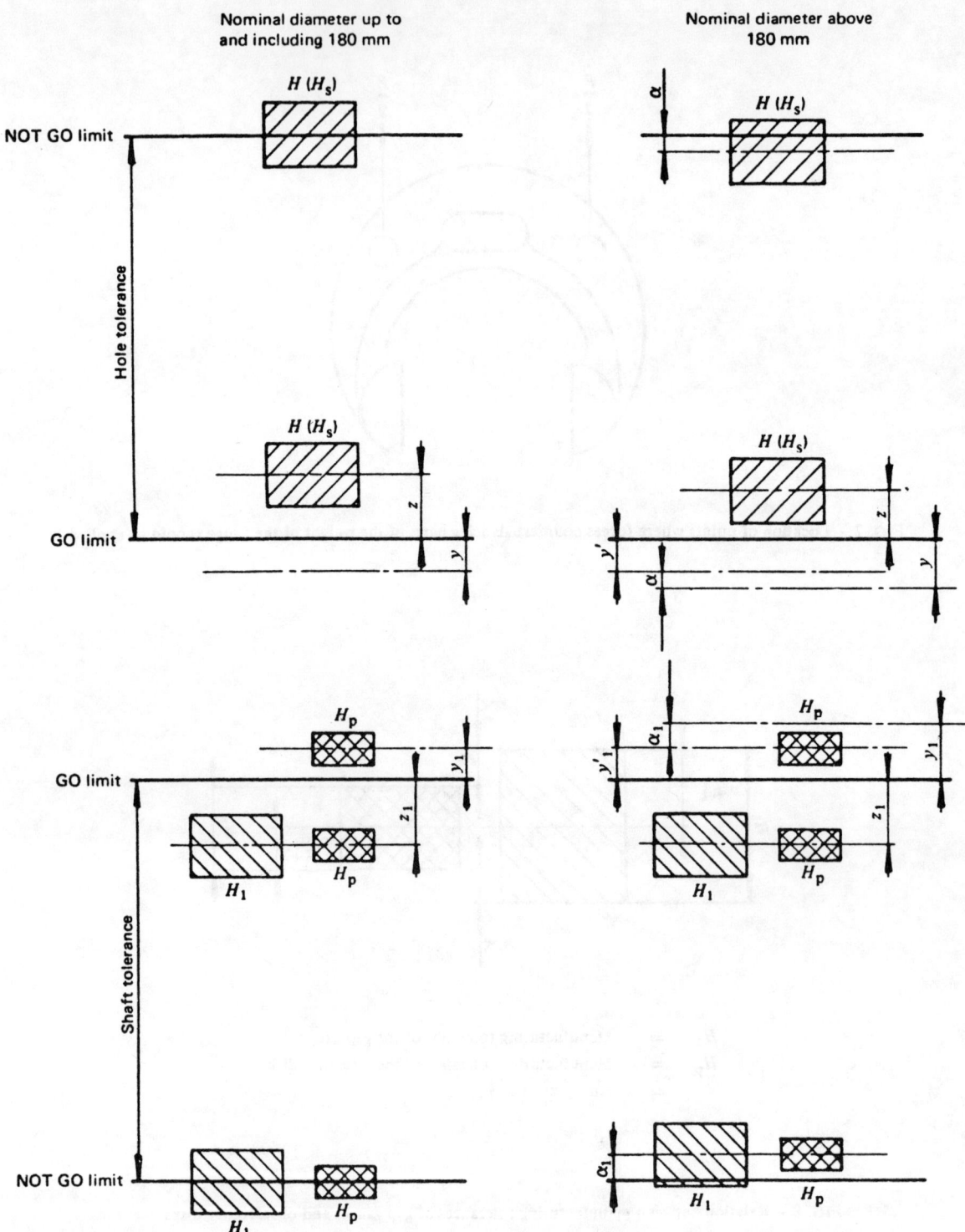

FIG. 9 – Tolerance zones of limit gauges and reference disks

THE ISO SYSTEM OF LIMITS AND FITS—INSPECTION OF PLAIN WORKPIECES

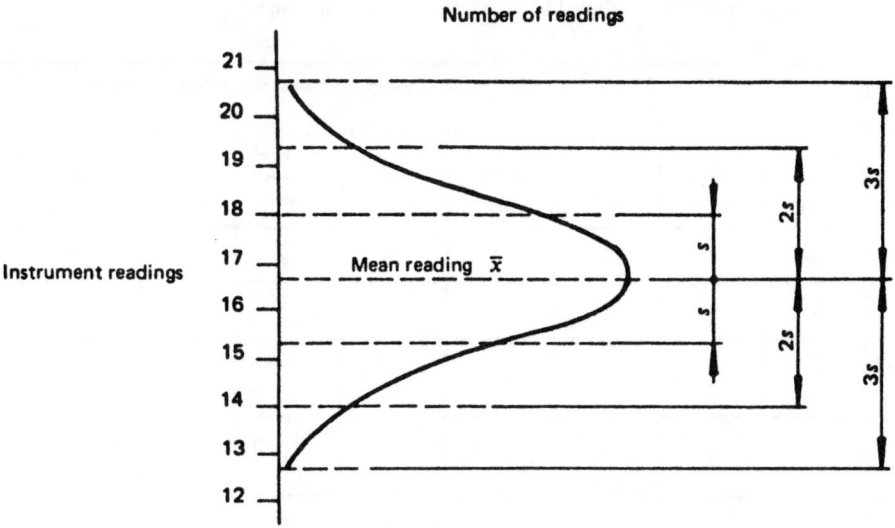

$\bar{x}$ = mean reading $\qquad s$ = standard deviation

68.27 % of readings fall between limits $\bar{x} \pm s$
95.45 % of readings fall between limits $\bar{x} \pm 2s$
99.73 % of readings fall between limits $\bar{x} \pm 3s$

FIG. 10 – Normal distribution curve of instrument readings

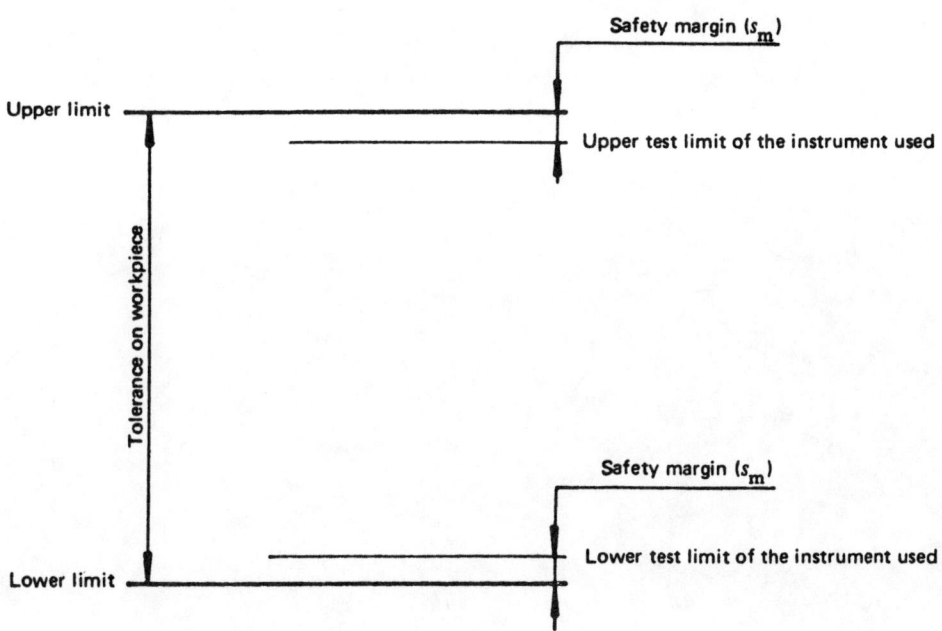

FIG. 11 – Relation between "test limits" and the "limits of size"

THE ISO SYSTEM OF LIMITS AND FITS—INSPECTION OF PLAIN WORKPIECES

Table 7-1. Manufacturing Tolerances for Gages

Type of gauge	\multicolumn{10}{c}{Tolerance grade of gauge for workpiece tolerance grade}
	\multicolumn{2}{c}{6}	\multicolumn{2}{c}{7}	\multicolumn{2}{c}{8 to 10}	\multicolumn{2}{c}{11 and 12}	\multicolumn{2}{c}{13 to 16}
	Size (IT)	Form (IT)	Size (IT)	Form (IT)	Size (IT)	Form (IT)	Size (IT)	Form (IT)	Size (IT)	Form (IT)
Cylindrical plug gauge	2	1	3	2	3	2	5	4	7	5
Cylindrical bar gauge	2	1	3	2	3	2	5	4	7	5
Spherical plug or disk gauge	2	1	2	1	2	1	4	3	6	5
Spherically-ended rod gauge	2	1	2	1	2	1	4	3	6	5
Cylindrical ring gauge	3	2	3	2	4	3	5	4	7	5
Gap gauge	3	2	3	2	4	3	5	4	7	5
Reference disk for gap gauge	1	1	1	1	2	1	2	1	3	2
Reference cylindrical setting plug gauge	1	1	1	1	2	1	2	1	3	2
Reference cylindrical setting ring gauge	1	1	1	1	2	1	2	1	3	2

THE ISO SYSTEM OF LIMITS AND FITS—INSPECTION OF PLAIN WORKPIECES

Table 7-2. Location of the Gage Tolerances and the Limit of Maximum Permissible Gage Wear in Relation to the Nominal Limit of the Workpiece, for Grades 6 to 16 (for $D \leqslant 500$ mm)

Values in micrometres

D mm	IT 0	IT 1	IT 2	IT 3	IT 4	IT 5	IT 6	6* z / z_1	6* y / y_1	6* y' / y'_1	6* α / $α_1$	6* z_1	6* y_1	6* y'_1	IT 7	7* z / z_1	7* y / y_1	7* y' / y'_1	7* α / $α_1$	IT 8	8* z / z_1	8* y / y_1	8* y' / y'_1	8* α / $α_1$
≤ 3	0,5	0,8	1,2	2	3	4	6	1	1	—	—	1,5	1,5	—	10	1,5	1,5	—	—	14	2	3	—	—
> 3 to 6	0,6	1	1,5	2,5	4	5	8	1,5	1	—	—	2	1,5	—	12	2	1,5	—	—	18	3	3	—	—
> 6 to 10	0,6	1	1,5	2,5	4	6	9	1,5	1	—	—	2	1,5	—	15	2	1,5	—	—	22	3	3	—	—
> 10 to 18	0,8	1,2	2	3	5	8	11	2	1,5	—	—	2,5	2	—	18	2,5	2	—	—	27	4	4	—	—
> 18 to 30	1	1,5	2,5	4	6	9	13	2	1,5	—	—	3	3	—	21	3	3	—	—	33	5	4	—	—
> 30 to 50	1	1,5	2,5	4	7	11	16	2,5	2	—	—	3,5	3	—	25	3,5	3	—	—	39	6	5	—	—
> 50 to 80	1,2	2	3	5	8	13	19	2,5	2	—	—	4	3	—	30	4	3	—	—	46	7	5	—	—
> 80 to 120	1,5	2,5	4	6	10	15	22	3	3	—	—	5	4	—	35	5	4	—	—	54	8	6	—	—
> 120 to 180	2	3,5	5	8	12	18	25	4	3	—	—	6	4	—	40	6	4	—	—	63	9	6	—	—
> 180 to 250	3	4,5	7	10	14	20	29	5	4	2	2	7	5	3	46	7	6	3	3	72	12	7	3	4
> 250 to 315	4	6	8	12	16	23	32	6	5	2	3	8	6	3	52	8	7	3	4	81	14	9	3	6
> 315 to 400	5	7	9	13	18	25	36	7	6	2	4	10	6	2	57	10	8	2	6	89	16	9	2	7
> 400 to 500	6	8	10	15	20	27	40	8	7	2	5	11	7	2	63	11	9	2	7	97	18	11	2	9

* For grades 6N, 7N and 8N the values of y and y', y_1 and y'_1 are zero.

Values in micrometres

D mm	IT 9	9 z/z_1	9 y/y_1	9 y'/y'_1	9 α/$α_1$	IT 10	10 z/z_1	10 y/y_1	10 y'/y'_1	10 α/$α_1$	IT 11	11 z/z_1	11 y/y_1	11 y'/y'_1	11 α/$α_1$	IT 12	12 z/z_1	12 y/y_1	12 y'/y'_1	12 α/$α_1$
≤ 3	25	5	0	—	—	40	5	0	—	—	60	10	0	—	—	100	10	0	—	—
> 3 to 6	30	6	0	—	—	48	6	0	—	—	75	12	0	—	—	120	12	0	—	—
> 6 to 10	36	7	0	—	—	58	7	0	—	—	90	14	0	—	—	150	14	0	—	—
> 10 to 18	43	8	0	—	—	70	8	0	—	—	110	16	0	—	—	180	16	0	—	—
> 18 to 30	52	9	0	—	—	84	9	0	—	—	130	19	0	—	—	210	19	0	—	—
> 30 to 50	62	11	0	—	—	100	11	0	—	—	160	22	0	—	—	250	22	0	—	—
> 50 to 80	74	13	0	—	—	120	13	0	—	—	190	25	0	—	—	300	25	0	—	—
> 80 to 120	87	15	0	—	—	140	15	0	—	—	220	28	0	—	—	350	28	0	—	—
> 120 to 180	100	18	0	—	—	160	18	0	—	—	250	32	0	—	—	400	32	0	—	—
> 180 to 250	115	21	0	4	4	185	24	0	7	7	290	40	0	10	10	460	45	0	15	15
> 250 to 315	130	24	0	6	6	210	27	0	9	9	320	45	0	15	15	520	50	0	20	20
> 315 to 400	140	28	0	7	7	230	32	0	11	11	360	50	0	15	15	570	65	0	30	30
> 400 to 500	155	32	0	9	9	250	37	0	14	14	400	55	0	20	20	630	70	0	35	35

Values in micrometres

D mm	IT 13	13 z/z_1	13 y/y_1	13 y'/y'_1	13 α/$α_1$	IT 14	14* z/z_1	14* y/y_1	14* y'/y'_1	14* α/$α_1$	IT 15	15* z/z_1	15* y/y_1	15* y'/y'_1	15* α/$α_1$	IT 16	16* z/z_1	16* y/y_1	16* y'/y'_1	16* α/$α_1$
≤ 3	140	20	0	—	—	250	20	0	—	—	400	40	0	—	—	600	40	0	—	—
> 3 to 6	180	24	0	—	—	300	24	0	—	—	480	48	0	—	—	750	48	0	—	—
> 6 to 10	220	28	0	—	—	360	28	0	—	—	580	56	0	—	—	900	56	0	—	—
> 10 to 18	270	32	0	—	—	430	32	0	—	—	700	64	0	—	—	1100	64	0	—	—
> 18 to 30	330	36	0	—	—	520	36	0	—	—	840	72	0	—	—	1300	72	0	—	—
> 30 to 50	390	42	0	—	—	620	42	0	—	—	1000	80	0	—	—	1600	80	0	—	—
> 50 to 80	460	48	0	—	—	740	48	0	—	—	1200	90	0	—	—	1900	90	0	—	—
> 80 to 120	540	54	0	—	—	870	54	0	—	—	1400	100	0	—	—	2200	100	0	—	—
> 120 to 180	630	60	0	—	—	1000	60	0	—	—	1600	110	0	—	—	2500	110	0	—	—
> 180 to 250	720	80	0	25	25	1150	100	0	45	45	1850	170	0	70	70	2900	210	0	110	110
> 250 to 315	810	90	0	35	35	1300	110	0	55	55	2100	190	0	90	90	3200	240	0	140	140
> 315 to 400	890	100	0	45	45	1400	125	0	70	70	2300	210	0	110	110	3600	280	0	180	180
> 400 to 500	970	110	0	55	55	1550	145	0	90	90	2500	240	0	140	140	4000	320	0	220	220

* Up to 1 mm grades 14 to 16 are not provided.

Table 2A. Location of the Gauge Tolerances and the Limit of Maximum Permissible Gauge Wear in Relation to the Nominal Limit of the Workpiece, for Grades 6 to 16 (for $D > 500$ mm)

Values in micrometres

D mm	IT 01	IT 0	IT 1	IT 2	IT 3	IT 4	IT 5	IT 6	z	z_1	α α_1	IT 7	z_1	α α_1	IT 8	z_1	α α_1
> 500 to 630	4.5	6	9	11	16	22	30	44	9	12	5.5	70	12	8	110	14	11
> 630 to 800	5	7	10	13	18	25	35	50	10	13	6	80	13	9	125	15	12
> 800 to 1000	5.5	8	11	15	21	29	40	56	11	14	7	90	14	10	140	16	14
> 1000 to 1250	6.5	9	13	18	24	34	46	66	12	15	8.5	105	15	12	165	17	17
> 1250 to 1600	8	11	15	21	29	40	54	78	13	16	10	125	16	14	195	18	20
> 1600 to 2000	9	13	18	25	35	48	65	92	14	17	12	150	17	17	230	19	24
> 2000 to 2500	11	15	22	30	41	57	77	110	16	18	14	175	18	20	280	20	28
> 2500 to 3150	13	18	26	36	50	69	93	135	17	19	16	210	19	23	330	22	32

The values of y and y_1 are zero even for grades 6, 7 and 8.

NOTE. – Provisional table, for experimental purposes, regarding in particular the α values and their incidence to z values which must allow in all cases a wear margin for the gauge (condition not entirely assured in this table, unless some exceeding is still implicitly admissible).

Values in micrometres

D mm	IT 9	z z_1	α α_1	IT 10	z z_1	α α_1	IT 11	z z_1	α α_1	IT 12	z z_1	α α_1	IT 13	z z_1	α α_1	IT 14	z z_1	α α_1	IT 15	z z_1	α α_1	IT 16	z z_1	α α_1
> 500 to 630	175	20	16	280	20	22	440	43	32	700	43	45	1100	78	70	1750	78	110	2800	145	180	4 400	145	280
> 630 to 800	200	22	18	320	22	25	500	45	36	800	45	50	1250	82	80	2000	82	120	3200	150	200	5 000	150	300
> 800 to 1000	230	23	20	360	23	28	560	47	40	900	47	56	1400	85	90	2300	85	140	3600	155	220	5 600	155	360
> 1000 to 1250	260	24	24	420	24	34	660	49	48	1050	49	67	1650	86	105	2600	86	170	4200	160	260	6 600	160	420
> 1250 to 1600	310	25	28	500	25	40	780	50	56	1250	50	80	1950	88	120	3100	88	200	5000	165	300	7 800	165	500
> 1600 to 2000	370	27	34	600	27	48	920	52	67	1500	52	95	2300	90	150	3700	90	230	6000	170	370	9 200	170	600
> 2000 to 2500	440	28	40	700	28	56	1100	54	80	1750	54	110	2800	92	180	4400	92	280	7000	175	450	11 000	175	700
> 2500 to 3150	540	30	47	860	30	65	1350	55	95	2100	55	130	3300	95	210	5400	95	330	8600	180	530	13 500	180	850

The values of y and y_1 are zero.

THE ISO SYSTEM OF LIMITS AND FITS—INSPECTION OF PLAIN WORKPIECES

Table 7-3. Values of s_M (for $D \leqslant 500$ mm)

Values in micrometres

D mm	\multicolumn{14}{c}{Values of s_M for workpiece tolerance grade}
	3	4	5	6	7	8	9	10	11	12	13	14	15	16
⩽ 3 to	0.4	0.4	0.6	0.8	1.1	1.5	2.1	3	4.2	6	9.5	15	24	38
> 3 to 6	0.4	0.6	0.7	1	1.3	1.9	2.6	3.8	5.3	7.5	12	19	30	48
> 6 to 10	0.5	0.7	0.8	1.1	1.6	2.2	3.2	4.5	6.3	9	14	22	36	56
> 10 to 18	0.7	0.8	1.1	1.4	2	2.8	4	5.6	8	11	18	28	45	70
> 18 to 30	0.8	1	1.2	1.7	2.4	3.4	4.8	6.7	10	13	21	33	53	85
> 30 to 50	1	1.2	1.5	2	2.8	4	5.6	8	11	16	25	40	63	100
> 50 to 80	1.1	1.4	1.8	2.4	3.4	4.8	6.7	9.5	13	19	30	48	75	120
> 80 to 120	1.3	1.7	2.1	2.8	4	5.6	8	11	16	22	36	56	90	140
> 120 to 180	1.5	1.9	2.4	3.2	4.5	6.3	9	12	18	25	40	63	100	160
> 180 to 250	1.7	2.1	2.6	3.6	5	7.1	10	14	20	28	45	70	110	180
> 250 to 315	1.9	2.4	3	4	5.6	8	11	16	22	32	50	80	125	200
> 315 to 400	2.1	2.6	3.4	4.5	6.3	9	12	18	25	36	56	90	140	220
> 400 to 500	2.4	3	3.8	5	7.1	10	14	20	28	40	63	100	160	250

Table 7-3A. Values of s_M (for $D > 500$ mm)

Values in micrometres

D mm	\multicolumn{11}{c}{Values of s_M for workpiece tolerance grade}
	6	7	8	9	10	11	12	13	14	15	16
> 500 to 630	5.5	8	11	16	22	32	45	70	110	180	280
> 630 to 800	6	9	12	18	25	36	50	80	120	200	300
> 800 to 1000	7	10	14	20	28	40	56	90	140	220	360
> 1000 to 1250	8.5	12	17	24	34	48	67	105	170	260	420
> 1250 to 1600	10	14	20	28	40	56	80	120	200	300	500
> 1600 to 2000	12	17	24	34	48	67	95	150	230	370	600
> 2000 to 2500	14	20	28	40	56	80	110	180	280	450	700
> 2500 to 3150	16	23	32	47	65	95	130	210	330	530	850

Related ISO (TC 3) Standards are shown at the end of Section 18.

8 Screw Threads

Development of ISO General Metric Screw Threads

The ISO diameter pitch and basic profile of screw threads were agreed upon in 1955; the basic profile is defined in the ISO 68 standard.[1] This worldwide approved standard was a compromise of existing profiles and generally permitted interchangeability between existing product and new product with a minimum of interference. Subsequently, the ISO technical committee TC1 standardized diameter-pitch combinations for both inch and metric series threads, and designed tolerancing and gaging systems.

The nominal diameters in the ISO 261 standard range from 1 mm through 300 mm (see Table 8-1) and are grouped into three preference ratings as follows; first choice, second choice, and third choice. The ISO standard coarse and fine thread pitches are specified for each nominal size and listed in Table 8-1. The selected sizes for screws, bolts, and nuts are specified in the world (ISO 262), regional (CEN 11/U1), and national standards as shown in Table 8-1 and in the section on related standards. The ISO metric screw threads are described in detail in the ANSI B1 40-page report available from ANSI, 1430 Broadway, New York, New York, 10018.

Development of ANSI-OMFS Fastener Screw Threads

The most important area in which screw threads are used is in the threaded fasteners. The 580 manufacturing plants in North America alone produce 200 billion fasteners yearly with an annual worth of $2,000,000,000. This industry saw the opportunity to simplify and improve threaded fasteners based on advancements in the technology and on experience gained with threaded fasteners over the last 20 years. The economically beneficial simplifications of the fasteners were all done, without any exceptions, within the diameter-thread pitch series outlined in the ISO standards.

The Industrial Fastener Institute (IFI) is an association of leading North American manufacturers of bolts, nuts, screws, rivets, and all types of special industrial fasteners. IFI member companies combine their technical knowledge to advance the technology and application engineering of fasteners through planned programs of research and education. The institute and its member companies work closely with leading national and international technical organizations in developing fastener standards and other technical practices. The IFI Fastener Standards are in wide use by North American industries.

In May 1970, the IFI Governing Board met to establish a policy position relative to metric. A carefully selected Task Group of Fastener Design and Simplification was appointed, and their first report was issued in January, 1971.

The American National Standards Institute (ANSI) appointed a Special Committee to Study Development of an Optimum Metric Fastener System (OMFS). This committee was strengthened with the addition of a number of leading fastener people from the automotive industries and other companies and organizations that used fasteners. After some years of intensive research, and with many meetings here and abroad, the OMFS committee developed a fastener system with many advantages over the existing ISO system. The new OMFS fastener system is now called the "modified ISO fastener system."

On April 17-18, 1975, the Ad Hoc Panel for ISO/TC1/TC2 met in Munich, Germany. The panel completely evaluated the merits of the U.S.A. proposal, compared them against the current ISO metric screw thread system, and developed a series of statements recommending appropriate action to ISO/TC1 which would resolve differences between screw thread systems for commercial and industrial fasteners, and lead to a single international standard. On May 1, 1975, the American Standards Committee B1 reviewed the statements of the Munich meeting, endorsed them unanimously, and instructed that ANSI B1.18 be re-drafted to be in conformance.

Existing Versus Modified ISO Screw Threads

As can be seen by Table 8-1 the existing ISO metric screw threads and fasteners are available throughout the

---

[1] For information about the term "standard" as used in this book, please see page 1-2.

8-2   SCREW THREADS

Table 8-1. World Standards for General Purpose Metric Screw Threads (ISO 261) with Selected Sizes for Screws, Bolts and Nuts (ISO 262 and CEN 11/U1).

I S O	NOM. SIZE $d=D$	PITCH P	C O N	PITCH DIA. $d_2=D_2$	MINOR DIAMETER d_3 D_1	THREAD HEIGHT h_3 H_1	RADIUS MIN. R	TENSILE STR. AREA AS(MM**2)	U.S. AUSTRAL ANSI SAA B1.18 AS1275	JAPAN JIS B0205-7	FRANCE AFNOR E03-013	U.K. BSI BS3643	GERMANY ITALY DNA UNI DIN13 4534-6
F*	1	0.25	COARSE	0.838	0.693 0.729	0.153 0.135	0.031	0.460E+00		F* F	F* S	F* F	F* F
		0.2		0.870	0.755 0.783	0.123 0.108	0.025	0.518E+00					
S*	1.1	0.25	COARSE	0.938	0.793 0.829	0.153 0.135	0.031	0.588E+00		S* S	S* T	S* S	S* S
		0.2		0.970	0.855 0.883	0.123 0.108	0.025	0.654E+00					
F*	1.2	0.25	COARSE	1.038	0.893 0.929	0.153 0.135	0.031	0.732E+00		F* F	F* S	F* F	F* F
		0.2		1.070	0.955 0.983	0.123 0.108	0.025	0.805E+00					
F*	1.4	0.3	COARSE	1.205	1.032 1.075	0.184 0.162	0.038	0.983E+00		S* F	S* T	S* F	S* F
		0.2		1.270	1.155 1.183	0.123 0.108	0.025	0.115E+01					
F*	1.6	0.35	COARSE	1.373	1.171 1.221	0.215 0.189	0.044	0.127E+01	F*	F* F	F* S	F* F	F* F
		0.2		1.470	1.355 1.383	0.123 0.108	0.025	0.157E+01					
S*	1.8	0.35	COARSE	1.573	1.371 1.421	0.215 0.189	0.044	0.170E+01		S* F	S* T	S* F	S* F
		0.2		1.670	1.555 1.583	0.123 0.108	0.025	0.204E+01					
F*	2	0.4	COARSE	1.740	1.509 1.567	0.245 0.217	0.050	0.207E+01	F*	F* F	F* S	F* F	F* F
		0.25		1.838	1.693 1.729	0.153 0.135	0.031	0.245E+01					
S*	2.2	0.45	COARSE	1.908	1.648 1.713	0.276 0.244	0.056	0.248E+01		S* F	S* T	S* F	S* F
		0.25		2.038	1.893 1.929	0.153 0.135	0.031	0.303E+01					
F*	2.5	0.45	COARSE	2.208	1.948 2.013	0.307 0.244	0.056	0.339E+01	F*	F* F	F* S	F* F	F* F
		0.35		2.273	2.071 2.121	0.215 0.189	0.044	0.370E+01					
F*	3	0.5	COARSE	2.675	2.387 2.459	0.307 0.271	0.063	0.503E+01	F*	F* F	F* S	F* F	F* F
		0.35		2.773	2.571 2.621	0.215 0.189	0.044	0.561E+01					
S*	3.5	0.6	COARSE	3.110	2.764 2.850	0.368 0.325	0.075	0.678E+01	S*	S* F	S* T	S* F	S* F
		0.35		3.273	3.071 3.121	0.215 0.189	0.044	0.790E+01					
F*	4	0.7	COARSE	3.545	3.141 3.242	0.429 0.379	0.087	0.678E+01	F*	F* F	F* S	F* F	F* F
		0.5		3.675	3.387 3.459	0.307 0.271	0.063	0.979E+01					
S*	4.5	0.75	COARSE	4.013	3.580 3.688	0.460 0.406	0.094	0.113E+02		S* F	S* T	S* F	S* F
		0.5		4.175	3.887 3.959	0.307 0.271	0.063	0.128E+02					
F*	5	0.8	COARSE	4.480	4.018 4.134	0.491 0.433	0.100	0.142E+02	F*	F* F	F* S	F* F	F* F
		0.5		4.675	4.387 4.459	0.307 0.271	0.063	0.161E+02					
T	5.5	0.5		5.175	4.887 4.959	0.307	0.063	0.199E+02		T	T	T	T
F*	6	1	COARSE	5.350	4.773 4.917	0.613 0.541	0.125	0.201E+02	F*	F* F	F* S	F* F	F* F
		0.75		5.513	5.080 5.188	0.460 0.466	0.094	0.220E+02					

NOTES:
1. ALL DIMENSIONS ARE IN MILLIMETERS
2. SEE FIGURE 8-2 AND PAGE 8-12 FOR KEY TO SYMBOLS
3. ALL VALUES FOR THREADS IN THEIR MAX. METAL CONDITION
4. THE NOMINAL SIZE IS NATIONAL STANDARD AS INDICATED
   F = FIRST CHOICE, S = SECOND CHOICE, T = THIRD CHOICE
   * = SELECTED SIZE FOR SCREWS, BOLTS AND NUTS

SCREW THREADS

Table 8-1. *(Continued)* World Standards for General Purpose Metric Screw Threads (ISO 261) with Selected Sizes for Screws, Bolts and Nuts (ISO 262 and CEN 11/U1).

ISO	NOM. SIZE d=D	PITCH P	C E N	PITCH DIA. d₂=D₂	MINOR DIAMETER d₃	D₁	THREAD HEIGHT h₃ H₁	RADIUS MIN. R	TENSILE STR. AREA AS(MM**2)	U.S. ANSI B1.18	AUSTRAL SAA AS1275	JAPAN JIS B0205-7	FRANCE AFNOR E03-013	U.K. BSI BS3643	GERMANY DNA DIN13	ITALY UNI 4534-6
T*	7	1 0.75		6.350 6.513	5.773 6.080	5.917 6.188	0.613 0.541 0.460 0.406	0.125 0.094	0.289E+02 0.311E+02			T* T	T T	T* T	T* T	T* T
F*	8	1.25 1 0.75	COARSE FINE	7.188 7.350 7.513	6.466 6.647 7.080	6.647 6.917 7.188	0.767 0.677 0.613 0.541 0.460 0.406	0.156 0.125 0.094	0.366E+02 0.392E+02 0.418E+02	F*	F*	F* F* F	F* F* S	F* F* F	F* F*	F* F* F
T	9	1.25 1 0.75		8.188 8.350 8.513	7.466 7.773 8.080	7.647 7.917 8.188	0.767 0.677 0.613 0.541 0.460 0.406	0.156 0.125 0.094	0.481E+02 0.510E+02 0.541E+02			T T T		T T T	T	T T T
F*	10	1.5 1.25 1 0.75	COARSE FINE EX FIN	9.026 9.188 9.350 9.513	8.160 8.466 8.773 9.080	8.376 8.647 8.917 9.188	0.920 0.812 0.767 0.677 0.613 0.541 0.460 0.406	0.188 0.156 0.125 0.094	0.580E+02 0.612E+02 0.645E+02 0.679E+02	F*	F*	F* F* F* F	F* F* S F*	F* F* F F	F* F*	F* F* F F
T	11	1.5 1 0.75		10.026 10.350 10.513	9.160 9.773 10.080	9.376 9.917 10.188	0.920 0.812 0.613 0.541 0.460 0.406	0.188 0.125 0.094	0.723E+02 0.795E+02 0.833E+02			T T T		T T T	T	T T T
F*	12	1.75 1.5 1.25 1	COARSE FINE EX FIN	10.863 11.026 11.188 11.350	9.853 10.160 10.446 10.773	10.106 10.376 10.647 10.917	1.074 0.947 0.920 0.812 0.767 0.677 0.613 0.541	0.219 0.188 0.156 0.125	0.843E+02 0.881E+02 0.921E+02 0.961E+02	F*	F*	F* F F F	F* S F* F*	F* F F F	F*	F* S* F* F
S*	14	2 1.5 1.25 1	COARSE FINE EX FIN	12.701 13.026 13.188 13.350	11.546 12.160 12.466 12.773	11.835 12.376 12.647 12.917	1.227 1.083 0.920 0.812 0.767 0.677 0.613 0.541	0.250 0.188 0.156 0.125	0.115E+03 0.125E+03 0.129E+03 0.134E+03	F*	S*	S* S* S S	S* S* T S*	S* S* S S	S* S*	S* S* S S
T	15	1.5 1		14.026 14.350	13.160 13.773	13.376 13.917	0.920 0.812 0.613 0.541	0.188 0.125	0.145E+03 0.155E+03			T T	T* T	T T	T*	T T
F*	16	2 1.5 1	COARSE FINE EX FIN	14.701 15.026 15.350	13.546 14.160 14.773	13.835 14.376 14.917	1.227 1.083 0.920 0.812 0.613 0.541	0.250 0.188 0.125	0.157E+03 0.167E+03 0.178E+03	F*	F*	F* F* F	F* F* F*	F* F* F	F* F*	F* S* F
T	17	1.5 1		16.026 16.350	15.160 15.773	15.376 15.917	0.920 0.812 0.613 0.541	0.188 0.125	0.191E+03 0.203E+03			T T	T T*	T T	T	T T
S*	18	2.5 2 1.5 1	COARSE FINE EX FIN	16.376 16.701 17.026 17.350	14.933 15.546 16.160 16.773	15.294 15.835 16.376 16.917	1.534 1.353 1.227 1.083 0.920 0.812 0.613 0.541	0.313 0.250 0.188 0.125	0.192E+03 0.204E+03 0.216E+03 0.229E+03		S*	S* S* S* S	S* T S* S*	S* S S S	S* S*	S* S* S S
F*	20	2.5 2 1.5 1	COARSE FINE EX FIN	18.376 18.701 19.026 19.350	16.933 17.546 18.160 18.773	17.294 17.835 18.376 18.917	1.534 1.353 1.227 1.083 0.920 0.812 0.613 0.541	0.313 0.250 0.188 0.125	0.245E+03 0.258E+03 0.272E+03 0.285E+03	F*	F*	F* F F* F	F* F* F* F*	F* F F F	F* F*	F* F* F* F

NOTES:
1. ALL DIMENSIONS ARE IN MILLIMETERS
2. SEE FIGURE 8-2 AND PAGE 8-12 FOR KEY TO SYMBOLS
3. ALL VALUES FOR THREADS IN THEIR MAX. METAL CONDITION
4. THE NOMINAL SIZE IS NATIONAL STANDARD AS INDICATED
   F = FIRST CHOICE, S = SECOND CHOICE, T = THIRD CHOICE
   * = SELECTED SIZE FOR SCREWS, BOLTS AND NUTS

SCREW THREADS

Table 8-1. (*Continued*) World Standards for General Purpose Metric Screw Threads (ISO 261) with Selected Sizes for Screws, Bolts and Nuts (ISO 262 and CEN 11/U1).

I S O	NOM. SIZE d =D	PITCH P	C E N	PITCH DIA. $d_2=D_2$	MINOR DIAMETER d_3 D_1	THREAD HEIGHT h_3 H_1	RADIUS MIN. R	TENSILE STR. AREA AS(MM**2)	U.S. ANSI B1.18	AUSTRAL SAA AS1275	JAPAN JIS B0205-7	FRANCE AFNOR E03-013	U.K. BSI BS3643	GERMANY DNA DIN13	ITALY UNI 4534-6
S*	22	2.5	COARSE	20.376	18.933 19.294	1.534 1.353	0.313	0.303E+03		S*	S	S*	S*	S*	S*
		2	FINE	20.701	19.546 19.835	1.227 1.227	0.250	0.318E+03			S	T	S	S*	S
*		1.5		21.026	20.160 20.376	0.920 0.812	0.188	0.333E+03			S*	S*	S*	S*	S*
		1	EX FIN	21.350	20.773 20.917	0.613 0.541	0.125	0.348E+03			S	S*	S	S*	S
F*	24	3	COARSE	22.051	20.319 20.752	1.840 1.624	0.375	0.353E+03	F*	F*	F*	F*	F*	F*	F*
*		2	FINE	22.701	21.546 21.835	1.227 1.227	0.250	0.384E+03			F*	F*	F*	F*	F
		1.5	EX FIN	23.026	22.160 22.376	0.920 0.812	0.188	0.401E+03			F	S	F	F*	F
		1		23.350	22.773 22.917	0.613 0.541	0.125	0.418E+03			F	F	F	F*	F
T	25	2		23.701	22.546 22.835	1.227 1.083	0.250	0.420E+03			T	T	T		T
		1.5		24.026	23.160 23.376	0.920 0.812	0.188	0.437E+03			T	T*	T	T*	T
		1	EX FIN	24.350	23.773 23.917	0.613 0.541	0.125	0.455E+03			T	T	T		T
T	26	1.5		25.026	24.160 24.376	0.920 0.812	0.188	0.475E+03			T	T	T		T
S*	27	3	COARSE	25.051	23.319 23.752	1.840 1.624	0.375	0.459E+03		S*	S*	S*	S*	S*	S*
*		2	FINE	25.701	24.546 24.835	1.227 1.227	0.250	0.496E+03			S	T	S*	S*	S
		1.5	EX FIN	26.026	25.160 25.376	0.920 0.812	0.188	0.514E+03			S	T	S	S*	S
		1		26.350	25.773 25.917	0.613 0.541	0.125	0.533E+03			S	T	S	S*	S
T	28	2		26.701	25.546 25.835	1.227 1.227	0.250	0.536E+03			T	T	T	T*	T
		1.5		27.026	26.160 26.376	0.920 0.812	0.188	0.555E+03			T	T	T	T*	T
		1		27.350	26.773 26.917	0.613 0.541	0.125	0.575E+03			T	T	T	T*	T
F*	30	3.5	COARSE	27.727	25.706 26.211	2.147 1.894	0.438	0.561E+03	F*	F*	F*	F*	F*	F*	F*
*		3		28.051	26.319 26.752	1.840 1.624	0.375	0.580E+03			S	F*	F*	F*	F
		2	FINE	28.701	27.546 27.835	1.227 1.227	0.250	0.621E+03			F	F*	F	F*	F
		1.5	EX FIN	29.026	28.160 28.376	0.920 0.812	0.188	0.642E+03			F	F	F	F*	F
		1		29.350	28.773 28.917	0.613 0.541	0.125	0.663E+03			F	T	F	F*	F
T	32	2		30.701	29.546 29.835	1.227 1.227	0.250	0.713E+03			T	T	T		T
		1.5		31.026	30.160 30.376	0.920 0.812	0.188	0.735E+03			T	T	T		T
S*	33	3.5	COARSE	30.727	28.706 29.211	2.147 1.894	0.438	0.694E+03		S*	S*	S*	S*	S*	S*
*		3		31.051	29.319 29.752	1.840 1.624	0.375	0.716E+03			S	T	S*	S*	S
		2	FINE	31.701	30.546 30.835	1.227 1.227	0.250	0.761E+03			S	S	S	S*	S
		1.5	EX FIN	32.026	31.160 31.376	0.920 0.812	0.188	0.784E+03			S	T	S	S*	S
T	35	1.5	EX FIN	34.026	33.160 33.376	0.920 0.812	0.188	0.886E+03			T	T*	T	T*	T
F*	36	4	COARSE	33.402	31.092 31.670	2.454 2.165	0.500	0.817E+03	F*	F*	F*	F*	F*	F*	F*
*		3	FINE	34.051	32.319 32.752	1.840 1.624	0.375	0.865E+03			F*	F*	F*	F*	F
		2		34.701	33.546 33.835	1.227 1.227	0.250	0.915E+03			F	S	F	F*	F
		1.5	EX FIN	35.026	34.160 34.376	0.920 0.812	0.188	0.940E+03			F	F	F	F*	F
T	38	1.5		37.026	36.160 36.376	0.920 0.812	0.188	0.105E+04				T	T		

NOTES:
1. ALL DIMENSIONS ARE IN MILLIMETERS
2. SEE FIGURE 8-2 AND PAGE 8-12 FOR KEY TO SYMBOLS
3. ALL VALUES FOR THREADS IN THEIR MAX. METAL CONDITION
4. THE NOMINAL SIZE IS NATIONAL STANDARD AS INDICATED
   F = FIRST CHOICE, S = SECOND CHOICE, T = THIRD CHOICE
   * = SELECTED SIZE FOR SCREWS, BOLTS AND NUTS

SCREW THREADS

Table 8-1. *(Continued)* World Standards for General Purpose Metric Screw Threads (ISO 261) with Selected Sizes for Screws, Bolts and Nuts (ISO 262 and CEN 11/U1).

I S O	NOM. SIZE d=D	PITCH P	C O A R S E	PITCH DIA. $d_2=D_2$	MINOR DIAMETER d_3 / D_1	THREAD HEIGHT h_3 / H_1	RADIUS MIN. R	TENSILE STR. AREA AS(MM**2)	U.S. ANSI B1.18	AUSTRAL SAA AS1275	JAPAN JIS B0205-7	FRANCE AFNOR E03-013	U.K. BSI BS3643	GERMANY DNA DIN13	ITALY UNI 4534-6
S*	39	4	COARSE	36.402	34.092 / 34.670	2.454 / 2.165	0.500	0.976E+03		S*	S*	S*	S*	S*	S*
*		3	FINE	37.051	35.319 / 35.752	1.840 / 1.624	0.375	0.103E+04			S*	S*	S*	S*	S*
		2		37.701	36.546 / 36.835	1.227 / 1.083	0.250	0.108E+04			S	S	S	S	S
		1.5	EX FIN	38.026	37.160 / 37.376	0.920 / 0.812	0.188	0.111E+04				S*			
T	40	3		38.051	36.319 / 36.752	1.840 / 1.624	0.375	0.109E+04			T	T	T	T	T
		2		38.701	37.546 / 37.835	1.227 / 1.083	0.250	0.114E+04			T	T	T	T	T
		1.5	EX FIN	39.026	38.160 / 38.376	0.920 / 0.812	0.188	0.117E+04			T	T*	T	T*	T
F	42	4.5	COARSE	39.077	36.479 / 37.129	2.760 / 2.436	0.563	0.112E+04	F*	F*	F	F*	F	F*	F
		4		39.402	37.092 / 37.670	2.454 / 2.165	0.500	0.115E+04			F	S	F	S	S
		3	FINE	40.051	38.319 / 38.752	1.840 / 1.624	0.375	0.121E+04			F	F	F	S*	S
		2		40.701	39.546 / 39.835	1.227 / 1.083	0.250	0.126E+04			F	S	F	S*	S
		1.5	EX FIN	41.026	40.160 / 40.376	0.920 / 0.812	0.188	0.129E+04			F	F*	F	F*	F
S	45	4.5	COARSE	42.077	39.479 / 40.129	2.760 / 2.436	0.563	0.131E+04			S	S*	S	S*	S
		4		42.402	40.092 / 40.670	2.454 / 2.165	0.500	0.134E+04			S	T	S	S*	S
		3	FINE	43.051	41.319 / 41.752	1.840 / 1.624	0.375	0.140E+04			S	S*	S	S*	S
		2		43.701	42.546 / 42.835	1.227 / 1.083	0.250	0.146E+04			S	T	S	S*	S
		1.5	EX FIN	44.026	43.160 / 43.376	0.920 / 0.812	0.188	0.149E+04			S	S*	S	S*	S
F	48	5	COARSE	44.752	41.866 / 42.587	3.067 / 2.706	0.625	0.147E+04	F*	F*	F	F*	F	F*	F
		4		45.402	43.092 / 43.670	2.454 / 2.165	0.500	0.154E+04			F	S	F	S	S
		3	FINE	46.051	44.319 / 44.752	1.840 / 1.624	0.375	0.160E+04			F	F	F	F*	F
		2		46.701	45.546 / 45.835	1.227 / 1.083	0.250	0.167E+04			F	S	F	F*	F
		1.5	EX FIN	47.026	46.160 / 46.376	0.920 / 0.812	0.188	0.171E+04			F	F*	F	F*	F
T	50	3		48.051	46.319 / 46.752	1.840 / 1.624	0.375	0.175E+04			T	T	T	T	T
		2		48.701	47.546 / 47.835	1.227 / 1.083	0.250	0.182E+04			T	T	T	T	T
		1.5	EX FIN	49.026	48.160 / 48.376	0.920 / 0.812	0.188	0.185E+04			T	T*	T	T*	T
S	52	5	COARSE	48.752	45.866 / 46.587	3.067 / 2.706	0.625	0.176E+04			S	S*	S	S*	S
		4		49.402	47.092 / 47.670	2.454 / 2.165	0.500	0.183E+04			S	T	S	S*	S
		3	FINE	50.051	48.319 / 48.752	1.840 / 1.624	0.375	0.190E+04			S	S*	S	S*	S
		2	EX FIN	50.701	49.546 / 49.835	1.227 / 1.083	0.250	0.197E+04			S	S*	S	S*	S
		1.5		51.026	50.160 / 50.376	0.920 / 0.812	0.188	0.201E+04			S	T	S		S
T	55	4		52.402	50.092 / 50.670	2.454 / 2.165	0.500	0.206E+04			T	T	T	T	T
		3		53.051	51.319 / 51.752	1.840 / 1.624	0.375	0.214E+04			T	T	T	T	T
		2		53.701	52.546 / 52.835	1.227 / 1.083	0.250	0.222E+04			T	T	T	T	T
		1.5	EX FIN	54.026	53.160 / 53.376	0.920 / 0.812	0.188	0.226E+04			T	T*	T	T*	T
F	56	5.5	COARSE	52.428	49.252 / 50.046	3.374 / 2.977	0.688	0.203E+04	F*	F*	F	F*	F	F*	F
		4	FINE	53.402	51.092 / 51.670	2.454 / 2.165	0.500	0.214E+04			F	F	F	F*	F
		3		54.051	52.319 / 52.752	1.840 / 1.624	0.375	0.222E+04			F	S	F	F*	F
		2	EX FIN	54.701	53.546 / 53.835	1.227 / 1.083	0.250	0.230E+04			F	F	F		F
		1.5		55.026	54.160 / 54.376	0.920 / 0.812	0.188	0.234E+04			F	S	F	F*	F

NOTES:
1. ALL DIMENSIONS ARE IN MILLIMETERS
2. SEE FIGURE 8-2 AND PAGE 8-12 FOR KEY TO SYMBOLS
3. ALL VALUES FOR THREADS IN THEIR MAX. METAL CONDITION
4. THE NOMINAL SIZE IS NATIONAL STANDARD AS INDICATED
   F = FIRST CHOICE, S = SECOND CHOICE, T = THIRD CHOICE
   * = SELECTED SIZE FOR SCREWS, BOLTS AND NUTS

8-6   SCREW THREADS

Table 8-1. *(Continued)* World Standards for General Purpose Metric Screw Threads (ISO 261) with Selected Sizes for Screws, Bolts and Nuts (ISO 262 and CEN 11/U1).

ISO	NOM. SIZE $d=D$	PITCH P	C E N	PITCH DIA. $d_2=D_2$	MINOR DIAMETER d_3 D_1	THREAD HEIGHT h_3 H_1	RADIUS MIN. R	TENSILE STR. AREA AS(MM**2)	U.S. ANSI B1.18	AUSTRAL SAA AS1275	JAPAN JIS B0205-7	FRANCE AFNOR E03-013	U.K. BSI BS3643	GERMANY DNA DIN13	ITALY UNI 4534-6
T	58	4		55.402	53.670	2.454	0.500	0.231E+04			T	T	T	T	T
		3		56.051	54.319	1.840	0.375	0.239E+04			T	T	T	T	T
		2		56.701	55.546	1.227	0.250	0.247E+04			T	T	T	T	T
		1.5		57.026	56.160	0.920	0.188	0.252E+04			T	T	T	T	T
S	60	5.5	COARSE	56.428	53.252	3.374	0.688	0.236E+04			S	S*	S	S*	S
		4	FINE	57.402	55.092	2.454	0.500	0.248E+04			S	S*	S	S*	S
		3		58.051	56.319	1.840	0.375	0.257E+04			S	T	S	S	S
		2	EX FIN	58.701	57.546	1.227	0.250	0.265E+04			S	S*	S	S*	S
		1.5		59.026	58.160	0.920	0.188	0.270E+04			S	T	S	S	S
T	62	4		59.402	57.670	2.454	0.500	0.266E+04			T	T	T	T	T
		3		60.051	58.319	1.840	0.375	0.275E+04			T	T	T	T	T
		2		60.701	59.546	1.227	0.250	0.284E+04			T	T	T	T	T
		1.5		61.026	60.160	0.920	0.188	0.288E+04			T	T	T	T	T
F	64	6	COARSE	60.103	56.639	3.681	0.750	0.268E+04	F*	F*	F	F*	F	F*	F
		4	FINE	61.402	59.092	2.454	0.500	0.285E+04			F	F*	F	F*	F
		3		62.051	60.319	1.840	0.375	0.294E+04			F	T	F	F	F
		2	EX FIN	62.701	61.546	1.227	0.250	0.303E+04			F	F*	F	F*	F
		1.5		63.026	62.160	0.920	0.188	0.308E+04			F	S	F	F	F
T	65	4		62.402	60.670	2.454	0.500	0.295E+04			T	T	T	T	T
		3		63.051	61.319	1.840	0.375	0.304E+04			T	T	T	T	T
		2		63.701	62.546	1.227	0.250	0.313E+04			T	T*	T	T*	T
		1.5		64.026	63.160	0.920	0.188	0.318E+04			T	T	T	T	T
S	68	6	COARSE	64.103	60.639	3.681	0.750	0.306E+04			S	S*	S	S*	S
		4	FINE	65.402	63.092	2.454	0.500	0.324E+04			S	S*	S	S*	S
		3		66.051	64.319	1.840	0.375	0.334E+04			S	T	S	S	S
		2	EX FIN	66.701	65.546	1.227	0.250	0.344E+04			S	S*	S	S*	S
		1.5		67.026	66.160	0.920	0.188	0.348E+04			S	T	S	S	S
T	70	6		66.103	62.639	3.681	0.750	0.325E+04			T	T	T	T	T
		4		67.402	65.092	2.454	0.500	0.345E+04			T	T	T	T	T
		3		68.051	66.319	1.840	0.375	0.355E+04			T	T	T	T	T
		2	EX FIN	68.701	67.546	1.227	0.250	0.364E+04			T	T*	T	T*	T
		1.5		69.026	68.160	0.920	0.188	0.370E+04			T	T	T	T	T
F	72	6	FINE 1	68.103	64.639	3.681	0.750	0.346E+04			F	F*	F	F*	F
		4	FINE 2	69.402	67.092	2.454	0.500	0.366E+04			F	F*	F	F*	F
		3		70.051	68.319	1.840	0.375	0.376E+04			F	S	F	F	F
		2	EX FIN	70.701	69.546	1.227	0.250	0.386E+04			F	F*	F	F*	F
		1.5		71.026	70.160	0.920	0.188	0.391E+04			F	S	F	F	F
T	75	4		72.402	70.670	2.454	0.500	0.399E+04			T	T	T	T	T
		3		73.051	71.319	1.840	0.375	0.409E+04			T	T	T	T	T
		2		73.701	72.546	1.227	0.250	0.420E+04			T	T*	T	T*	T
		1.5		74.026	73.160	0.920	0.188	0.425E+04			T	T	T	T	T

NOTES:
1. ALL DIMENSIONS ARE IN MILLIMETERS
2. SEE FIGURE 8-2 AND PAGE 8-12 FOR KEY TO SYMBOLS
3. ALL VALUES FOR THREADS IN THEIR MAX. METAL CONDITION
4. THE NOMINAL SIZE IS NATIONAL STANDARD AS INDICATED
    F = FIRST CHOICE, S = SECOND CHOICE, T = THIRD CHOICE
    * = SELECTED SIZE FOR SCREWS, BOLTS AND NUTS

SCREW THREADS

Table 8-1. *(Continued)* World Standards for General Purpose Metric Screw Threads (ISO 261) with Selected Sizes for Screws, Bolts and Nuts (ISO 262 and CEN 11/U1).

I S O	NOM. SIZE $d=D$	PITCH P	C E N	PITCH DIA. $d_2=D_2$	MINOR DIAMETER d_3	D_1	THREAD HEIGHT h_3	H_1	RADIUS MIN. R	TENSILE STR. AREA AS(MM**2)	U.S. ANSI B1.18	AUSTRAL SAA AS1275	JAPAN JIS B0205-7	FRANCE AFNOR E03-013	U.K. BSI BS3643	GERMANY DNA DIN13	ITALY UNI 4534-6
S	76	6	FINE 1	72.103	68.639	69.505	3.681	3.248	0.750	0.389E+04			S	S*	S	S*	S
		4	FINE 2	73.402	71.092	71.670	2.454	2.165	0.500	0.410E+04			S	S*	S	S*	S
		3		74.051	72.319	72.752	1.840	1.624	0.375	0.421E+04			S	S	S	S	S
		2	EX FIN	74.701	73.546	73.835	1.227	1.083	0.250	0.432E+04			S	S*	S	S*	S
		1.5		75.026	74.160	74.376	0.920	0.812	0.188	0.437E+04							
T	78	2		76.701	75.546	75.835	1.227	1.083	0.250	0.455E+04			T		T		
F	80	6	FINE 1	76.103	72.639	73.505	3.681	3.248	0.750	0.434E+04	F*		F	F*	F	F*	F
		4	FINE 2	77.402	75.092	75.670	2.454	2.165	0.500	0.457E+04			F	F*	F	F*	F
		3		78.051	76.319	76.752	1.840	1.624	0.375	0.468E+04				S	F		
		2	EX FIN	78.701	77.546	77.835	1.227	1.083	0.250	0.479E+04			F	F*	F	F*	F
		1.5		79.026	78.160	78.376	0.920	0.812	0.188	0.485E+04			F	S			
T	82	2		80.701	79.546	79.835	1.227	1.083	0.250	0.504E+04			T		T		
S	85	6	FINE 1	81.103	77.639	78.505	3.681	3.248	0.750	0.495E+04			S	S*	S	S*	S
		4	FINE 2	82.402	80.092	80.670	2.454	2.165	0.500	0.518E+04			S	S*	S	S*	S
		3		83.051	81.319	81.752	1.840	1.624	0.375	0.530E+04			S	T	S		
		2	EX FIN	83.701	82.546	82.835	1.227	1.083	0.250	0.543E+04			S	S*	S	S*	S
F	90	6	FINE 1	86.103	82.639	83.505	3.681	3.248	0.750	0.559E+04	F*		F	F*	F	F*	F
		4	FINE 2	87.402	85.092	85.670	2.454	2.165	0.500	0.584E+04			F	F*	F	F*	F
		3		88.051	86.319	86.752	1.840	1.624	0.375	0.597E+04			F	S	F		
		2	EX FIN	88.701	87.546	87.835	1.227	1.083	0.250	0.610E+04			F	F*	F	F*	F
S	95	6	FINE 1	91.103	87.639	88.505	3.681	3.248	0.750	0.627E+04			S	S*	S	S*	S
		4	FINE 2	92.402	90.092	90.670	2.454	2.165	0.500	0.653E+04			S	S*	S	S*	S
		3		93.051	91.319	91.752	1.840	1.624	0.375	0.667E+04			S	T	S		
		2	EX FIN	93.701	92.546	92.835	1.227	1.083	0.250	0.681E+04			S	S*	S	S*	S
F	100	6	FINE 1	96.103	92.639	93.505	3.681	3.248	0.750	0.699E+04	F*		F	F*	F	F*	F
		4	FINE 2	97.402	95.092	95.670	2.454	2.165	0.500	0.728E+04			F	F*	F	F*	F
		3		98.051	96.319	96.752	1.840	1.624	0.375	0.742E+04			F	S	F		
		2	EX FIN	98.701	97.546	97.835	1.227	1.083	0.250	0.755E+04			F	F*	F	F*	F
S	105	6	FINE 1	101.103	97.639	98.505	3.681	3.248	0.750	0.776E+04			S	S*	S	S*	S
		4	FINE 2	102.402	100.092	100.670	2.454	2.165	0.500	0.805E+04			S	S*	S	S*	S
		3		103.051	101.319	101.752	1.840	1.624	0.375	0.820E+04			S	T	S		
		2	EX FIN	103.701	102.546	102.835	1.227	1.083	0.250	0.835E+04			S	S*	S	S*	S
F	110	6	FINE 1	106.103	102.639	103.505	3.681	3.248	0.750	0.856E+04			F	F*	F	F*	F
		4	FINE 2	107.402	105.092	105.670	2.454	2.165	0.500	0.887E+04			F	F*	F	F*	F
		3		108.051	106.319	106.752	1.840	1.624	0.375	0.902E+04			F	S	F		
		2	EX FIN	108.701	107.546	107.835	1.227	1.083	0.250	0.918E+04			F	F*	F	F*	F
S	115	6	FINE 1	111.103	107.639	108.505	3.681	3.248	0.750	0.939E+04			S	S*	S	S*	S
		4	FINE 2	112.402	110.092	110.670	2.454	2.165	0.500	0.972E+04			S	S*	S	S*	S
		3		113.051	111.319	111.752	1.840	1.624	0.375	0.988E+04			S	T	S		
		2	EX FIN	113.701	112.546	112.835	1.227	1.083	0.250	0.101E+05			S	S*	S	S*	S

NOTES:
1. ALL DIMENSIONS ARE IN MILLIMETERS
2. SEE FIGURE 8-2 AND PAGE 8-12 FOR KEY TO SYMBOLS
3. ALL VALUES FOR THREADS IN THEIR MAX. METAL CONDITION
4. THE NOMINAL SIZE IS NATIONAL STANDARD AS INDICATED
   F = FIRST CHOICE, S = SECOND CHOICE, T = THIRD CHOICE
   * = SELECTED SIZE FOR SCREWS, BOLTS AND NUTS

8-7

8-8 SCREW THREADS

Table 8-1. (Continued) World Standards for General Purpose Metric Screw Threads (ISO 261) with Selected Sizes for Screws, Bolts and Nuts (ISO 262 and CEN 11/U1).

ISO SIZE d=D	NOM. SIZE	PITCH P	C E N	PITCH DIA. $d_2=D_2$	MINOR DIAMETER d_3 D_1	THREAD HEIGHT h_3 H_1	RADIUS MIN. R	TENSILE STR. AREA AS(MM**2)	U.S. ANSI B1.18	AUSTRAL SAA AS1275	JAPAN JIS B0205-7	FRANCE AFNOR E03-013	U.K. BSI BS3643	GERMANY DNA DIN13	ITALY UNI 4534-6
S	120	6	FINE 1	116.103	112.639 113.505	3.681	0.750	0.103E+05			S	S*	S	S*	S
		4	FINE 2	117.402	115.092 115.670	2.454	0.500	0.106E+05			S	S*	S	S*	S
		3		118.051	116.319 116.752	1.840	0.375	0.108E+05			S	T	S	S*	S
		2	EX FIN	118.701	117.546 117.835	1.227	0.250	0.110E+05			S	S*	S	S*	S
F	125	6	FINE 1	121.103	117.639 118.505	3.681	0.750	0.112E+05			F	F*	F	F*	F
		4	FINE 2	122.402	120.092 120.670	2.454	0.500	0.115E+05			F	F*	F	F*	F
		3		123.051	121.319 121.752	1.840	0.375	0.117E+05			F	F*	F	F*	F
		2	EX FIN	123.701	122.546 122.835	1.227	0.250	0.119E+05			F	F*	F	F*	F
S	130	6	FINE 1	126.103	122.639 123.505	3.681	0.750	0.121E+05			S	S*	S	S*	S
		4	FINE 2	127.402	125.092 125.670	2.454	0.500	0.125E+05			S	S*	S	S*	S
		3		128.051	126.319 126.752	1.840	0.375	0.127E+05			S	S	S	S*	S
		2	EX FIN	128.701	127.546 127.835	1.227	0.250	0.129E+05			S	S*	S	S*	S
T	135	6	FINE 1	131.103	127.639 128.505	3.681	0.750	0.131E+05			T	T*	T	T*	T
		4	FINE 2	132.402	130.092 130.670	2.454	0.500	0.135E+05			T	T	T		T
		3		133.051	131.319 131.752	1.840	0.375	0.137E+05			T	T	T		T
		2	EX FIN	133.701	132.546 132.835	1.227	0.250	0.139E+05			T	T*	T	T*	T
F	140	6	FINE 1	136.103	132.639 133.505	3.681	0.750	0.142E+05			F	F*	F	F*	F
		4	FINE 2	137.402	135.092 135.670	2.454	0.500	0.146E+05			F	F*	F	F*	F
		3		138.051	136.319 136.752	1.840	0.375	0.148E+05			F	F	F	F*	F
		2	EX FIN	138.701	137.546 137.835	1.227	0.250	0.150E+05			F	F*	F	F*	F
T	145	6	FINE 1	141.103	137.639 138.505	3.681	0.750	0.153E+05			T	T	T		T
		4	FINE 2	142.402	140.092 140.670	2.454	0.500	0.157E+05			T	T	T		T
		3		143.051	141.319 141.752	1.840	0.375	0.159E+05			T	T	T		T
		2	EX FIN	143.701	142.546 142.835	1.227	0.250	0.161E+05			T	T*	T	T*	T
S	150	6	FINE 1	146.103	142.639 143.505	3.681	0.750	0.164E+05			S	S*	S	S*	S
		4	FINE 2	147.402	145.092 145.670	2.454	0.500	0.168E+05			S	S*	S	S*	S
		3		148.051	146.319 146.752	1.840	0.375	0.170E+05			S	T	S	S*	S
		2	EX FIN	148.701	147.546 147.835	1.227	0.250	0.172E+05			S	S*	S	S*	S
T	155	6	FINE 1	151.103	147.639 148.505	3.681	0.750	0.175E+05			T	T	T		T
		4	FINE 2	152.402	150.092 150.670	2.454	0.500	0.180E+05			T	T	T		T
		3	EX FIN	153.051	151.319 151.752	1.840	0.375	0.182E+05			T	T*	T	T*	T
F	160	6	FINE 2	156.103	152.639 153.505	3.681	0.750	0.187E+05			F	F*	F	F*	F
		4		157.402	155.092 155.670	2.454	0.500	0.192E+05			F	S	F	F*	F
		3	EX FIN	158.051	156.319 156.752	1.840	0.375	0.194E+05			F	F*	F	F*	F
T	165	6	FINE 2	161.103	157.639 158.505	3.681	0.750	0.199E+05			T	T	T		T
		4		162.402	160.092 160.670	2.454	0.500	0.204E+05			T	T	T		T
		3	EX FIN	163.051	161.319 161.752	1.840	0.375	0.207E+05			T	T*	T	T*	T

NOTES:
1. ALL DIMENSIONS ARE IN MILLIMETERS
2. SEE FIGURE 8-2 AND PAGE 8-12 FOR KEY TO SYMBOLS
3. ALL VALUES FOR THREADS IN THEIR MAX. METAL CONDITION
4. THE NOMINAL SIZE IS NATIONAL STANDARD AS INDICATED
   F = FIRST CHOICE, S = SECOND CHOICE, T = THIRD CHOICE
   * = SELECTED SIZE FOR SCREWS, BOLTS AND NUTS

SCREW THREADS

Table 8-1. *(Continued)* World Standards for General Purpose Metric Screw Threads (ISO 261) with Selected Sizes for Screws, Bolts and Nuts (ISO 262 and CEN 11/U1).

NOM. SIZE $d=D$	PITCH P	C E N	PITCH DIA. $d_2=D_2$	MINOR DIAMETER d_3 D_1	THREAD HEIGHT h_3 H_1	RADIUS MIN. R	TENSILE STR. AREA AS(MM**2)	U.S. ANSI B1.18	AUSTRAL SAA AS1275	JAPAN JIS B0205-7	FRANCE AFNOR E03-013	U.K. BSI BS3643	GERMANY DNA DIN13	ITALY UNI 4534-6
S 170	6	FINE 2	166.103 167.402 168.051	162.639 165.092 166.319	163.505 165.670 166.752	3.681 2.454 1.840	3.248 2.165 1.624	0.750 0.500 0.375	0.212E+05 0.217E+05 0.220E+05			S S S	S* T S*	S S S	S* S*	S S
T 175	6 4 EX FIN 3	FINE 2	171.103 172.402 173.051	167.639 170.092 171.319	168.505 170.670 171.752	3.681 2.454 1.840	3.248 2.165 1.624	0.750 0.500 0.375	0.225E+05 0.230E+05 0.233E+05			T T T	T T T	T T T		T T T
F 180	6 4 EX FIN 3	FINE 2	176.103 177.402 178.051	172.639 175.092 176.319	173.505 175.670 176.752	3.681 2.454 1.840	3.248 2.165 1.624	0.750 0.500 0.375	0.239E+05 0.244E+05 0.247E+05			F F F	F* S F*	F F F	F* F*	F F F
T 185	6 4 EX FIN 3	FINE 2	181.103 182.402 183.051	177.639 180.092 181.319	178.505 180.670 181.752	3.681 2.454 1.840	3.248 2.165 1.624	0.750 0.500 0.375	0.253E+05 0.258E+05 0.261E+05			T T T	T T T	T T T		T T T
S 190	6 4 EX FIN 3	FINE 2	186.103 187.402 188.051	182.639 185.092 186.319	183.505 185.670 186.752	3.681 2.454 1.840	3.248 2.165 1.624	0.750 0.500 0.375	0.267E+05 0.272E+05 0.275E+05			S S S	S* T S*	S S S	S* S*	S S
T 195	6 4 EX FIN 3	FINE 2	191.103 192.402 193.051	187.639 190.092 191.319	188.505 190.670 191.752	3.681 2.454 1.840	3.248 2.165 1.624	0.750 0.500 0.375	0.282E+05 0.287E+05 0.290E+05			T T T	T T T	T T T		T T T
F 200	6 4 EX FIN 3	FINE 2	196.103 197.402 198.051	192.639 195.092 196.319	193.505 195.670 196.752	3.681 2.454 1.840	3.248 2.165 1.624	0.750 0.500 0.375	0.297E+05 0.302E+05 0.305E+05			F F F	F* S F*	F F F	F* F*	F F F
T 205	6 4 EX FIN 3	FINE 2	201.103 202.402 203.051	197.639 200.092 201.319	198.505 200.670 201.752	3.681 2.454 1.840	3.248 2.165 1.624	0.750 0.500 0.375	0.312E+05 0.318E+05 0.321E+05			T T T	T T T	T T T		T T T
S 210	6 4 EX FIN 3	FINE 2	206.103 207.402 208.051	202.639 205.092 206.319	203.505 205.670 206.752	3.681 2.454 1.840	3.248 2.165 1.624	0.750 0.500 0.375	0.328E+05 0.334E+05 0.337E+05			S S S	S* S* T	S S S	S* S*	S S
T 215	6 4 EX FIN 3	FINE 2	211.103 212.402 213.051	207.639 210.092 211.319	208.505 210.670 211.752	3.681 2.454 1.840	3.248 2.165 1.624	0.750 0.500 0.375	0.344E+05 0.350E+05 0.354E+05			T T T	T T T	T T T		T T T
F 220	6 4 EX FIN 3	FINE 2	216.103 217.402 218.051	212.639 215.092 216.319	213.505 215.670 216.752	3.681 2.454 1.840	3.248 2.165 1.624	0.750 0.500 0.375	0.361E+05 0.367E+05 0.370E+05			F F F	F* F* S	F F F	F* F*	F F F
T 225	6 4 EX FIN 3	FINE 2	221.103 222.402 223.051	217.639 220.092 221.319	218.505 220.670 221.752	3.681 2.454 1.840	3.248 2.165 1.624	0.750 0.500 0.375	0.378E+05 0.384E+05 0.388E+05			T T T	T T T	T T T		T T T

NOTES:
1. ALL DIMENSIONS ARE IN MILLIMETERS
2. SEE FIGURE 8-2 AND PAGE 8-12 FOR KEY TO SYMBOLS
3. ALL VALUES FOR THREADS IN THEIR MAX. METAL CONDITION
4. THE NOMINAL SIZE IS NATIONAL STANDARD AS INDICATED
   F = FIRST CHOICE, S = SECOND CHOICE, T = THIRD CHOICE
   * = SELECTED SIZE FOR SCREWS, BOLTS AND NUTS

8-10 SCREW THREADS

Table 8-1. *(Continued)* World Standards for General Purpose Metric Screw Threads (ISO 261) with Selected Sizes for Screws, Bolts and Nuts (ISO 262 and CEN 11/U1).

ISO	NOM. SIZE d=D	PITCH P	C E N	PITCH DIA. d₂=D₂	MINOR DIAMETER d₃ D₁	THREAD HEIGHT h₃ H₁	RADIUS MIN. R	TENSILE STR. AREA AS(MM**2)	U.S. ANSI B1.18	AUSTRAL SAA AS1275	JAPAN JIS B0205-7	FRANCE AFNOR E03-013	U.K. BSI BS3643	GERMANY DNA DIN13	ITALY UNI 4534-6
T	230	6	FINE 2	226.103 227.402	222.639 225.092	223.505 225.670	3.681 2.454	3.248 2.165	0.750 0.500	0.395E+05 0.402E+05			T T	T T	T* T*	T T
		4	EX FIN	228.051	226.319	226.752	1.840	1.624	0.375	0.405E+05			T	T		T
		3														
T	235	6	FINE 2	231.103 232.402	227.639 230.092	228.505 230.670	3.681 2.454	3.248 2.165	0.750 0.500	0.413E+05 0.420E+05			T T	T T	T* T*	T T
		4	EX FIN	233.051	231.319	231.752	1.840	1.624	0.375	0.423E+05			T	T		T
		3														
S	240	6	FINE 2	236.103 237.402	232.639 235.092	233.505 235.670	3.681 2.454	3.248 2.165	0.750 0.500	0.431E+05 0.438E+05		S S	S S	S* S*	S S
		4	EX FIN	238.051	236.319	236.752	1.840	1.624	0.375	0.442E+05			S	T		S
		3														
T	245	6	FINE 2	241.103 242.402	237.639 240.092	238.505 240.670	3.681 2.454	3.248 2.165	0.750 0.500	0.450E+05 0.457E+05			T T	T T	T T
		4	EX FIN	243.051	241.319	241.752	1.840	1.624	0.375	0.461E+05			T	T		T
		3														
F	250	6	FINE 2	246.103 247.402	242.639 245.092	243.505 245.670	3.681 2.454	3.248 2.165	0.750 0.500	0.469E+05 0.476E+05		F F	F* F*	F F	F* F*	F F
		4	EX FIN	248.051	246.319	246.752	1.840	1.624	0.375	0.480E+05			F	S		F
		3														
T	255	6	FINE 2	251.103 252.402	247.639 250.092	248.505 250.670	3.681 2.454	3.248 2.165	0.750 0.500	0.488E+05 0.496E+05			T T	T T	T T
		4	EX FIN													
S	260	6	FINE 2	256.103 257.402	252.639 255.092	253.505 255.670	3.681 2.454	3.248 2.165	0.750 0.500	0.508E+05 0.516E+05		S S	S* S*	S S	S* S*	S S
		4	EX FIN													
T	265	6	FINE 2	261.103 262.402	257.639 260.092	258.505 260.670	3.681 2.454	3.248 2.165	0.750 0.500	0.526E+05 0.536E+05			T T	T T	T T
		4	EX FIN													
T	270	6	FINE 2	266.103 267.402	262.639 265.092	263.505 265.670	3.681 2.454	3.248 2.165	0.750 0.500	0.545E+05 0.557E+05			T* T*	T T	T* T*	T T
		4	EX FIN													
T	275	6	FINE 2	271.103 272.402	267.639 270.092	268.505 270.670	3.681 2.454	3.248 2.165	0.750 0.500	0.570E+05 0.578E+05			T T	T T	T T
		4	EX FIN													
F	280	6	FINE 2	276.103 277.402	272.639 275.092	273.505 275.670	3.681 2.454	3.248 2.165	0.750 0.500	0.591E+05 0.599E+05		F F	F* F*	F F	F* F*	F F
		4	EX FIN													
T	285	6	FINE 2	281.103 282.402	277.639 280.092	278.505 280.670	3.681 2.454	3.248 2.165	0.750 0.500	0.613E+05 0.621E+05			T T	T T	T T
		4	EX FIN													
T	290	6	FINE 2	286.103 287.402	282.639 285.092	283.505 285.670	3.681 2.454	3.248 2.165	0.750 0.500	0.635E+05 0.644E+05			T* T*	T T	T* T*	T T
		4	EX FIN													
T	295	6	FINE 2	291.103 292.402	287.639 290.092	288.505 290.670	3.681 2.454	3.248 2.165	0.750 0.500	0.658E+05 0.666E+05			T T	T T	T T
		4	EX FIN													
S	300	6	FINE 2	296.103 297.402	292.639 295.092	293.505 295.670	3.681 2.454	3.248 2.165	0.750 0.500	0.681E+05 0.689E+05		S S	S* S*	S S	S* S*	S S
		4	EX FIN													

NOTES:
1. ALL DIMENSIONS ARE IN MILLIMETERS
2. SEE FIGURE 8-2 AND PAGE 8-12 FOR KEY TO SYMBOLS
3. ALL VALUES FOR THREADS IN THEIR MAX. METAL CONDITION
4. THE NOMINAL SIZE IS NATIONAL STANDARD AS INDICATED
   F = FIRST CHOICE, S = SECOND CHOICE, T = THIRD CHOICE
   * = SELECTED SIZE FOR SCREWS, BOLTS AND NUTS

SCREW THREADS

world. However, some of the advantages with the modified ISO system (OMFS) are as follows:

1. Interchangeability with the existing ISO threaded fasteners. The 6.3* mm size is added as an ISO third choice size.
2. Standardization on one, the ISO coarse thread pitch. The ISO coarse thread pitch is slightly finer than the customary UNC series (see Table 8-8). The number of standard fasteners are reduced to 25 in the range from 1.6 to 100 mm, and it is less than 25 percent of the existing standard fasteners in the metric and customary inch system combined. The economic benefits derived from the reduction of standard fasteners to be installed and inventoried by manufacturing companies throughout the world could very well run into the millions of dollars in savings for some companies.
3. The 6.3* mm thread size replaces the 6 and the 7 mm sizes and it is ideally located strengthwise, between the 5 and 8 mm sizes. (See Preferred Numbers, Section 4.)
4. Simplified thread profile definitions. The thread pitch diameter is no longer used, and instead the basic major diameter is its base line.
5. The modified ISO has an improved gaging system for screw threads by which an improvement of fastener uniformity and quality is expected.
6. Upgrading the strength properties for the most commonly used fastener grades without increasing the manufacturing cost.
7. Increasing the proof load and the fatigue life for the fasteners by increasing the thread root radius.

The two fastener systems will exist side-by-side for some years to come. A rapid adaptation of the new ISO fastener system is anticipated in North America, and since it also has some manufacturer and consumer advantages, the writer predicts that the modified ISO system will become an ISO standard and take precedence over the existing ISO fastener system, worldwide.

It is therefore recommended that the modified ISO fasteners be specified for North American production and that temporarily optional usage of the existing ISO fasteners be allowed in other countries for availability reasons only.

It might be necessary to specify other ISO standard metric screw threads than those recognized by the ANSI B1.18 standard as preferred (see Table 8-1). However, the user should take this as an opportunity to rationalize the selection of sizes and thereby reduce cost of inventory and tooling. It is anticipated that fasteners and tools to the preferred ANSI diameter-pitch combinations will cost less than other ISO sizes.

ISO General Metric Screw Threads

The limited number of thread diameters and pitches (25) in the modified ISO or Optimum Metric Fastener System (OMFS) should be used for all fasteners. However, the OMFS is not sufficiently comprehensive to satisfy all general engineering requirements for screw threads. As was shown in Table 8-1, the ISO describes more than 300 diameter-pitch combinations in the nominal size range from 1 mm through 300 mm. The above ISO screw threads are standardized in the national standards of major industrial countries, as shown in Table 8-1, and the standards listed have effected the availability of toolings and fasteners throughout the world.

The existing ISO metric screw thread system and the modified system introduced to ISO by ANSI are dimensionally identical except for the minimum thread root radius. The existing ISO 965/1-1973 standard specifies a minimum allowed thread root radius which equals 0.1 × pitch, and the modified thread system has the mandatory minimum thread root radius equal to 0.125 × pitch as shown in Table 8-1. The maximum thread root radius possible without interference ranges from 0.15 to 0.16 times pitch depending on thread clearance.

Other non-dimensional differences between the existing and the modified ISO metric thread systems are as follows:

Definition. Threads are no longer defined around a fictitious pitch diameter in the new system. The basic major diameter forms the reference surface around which all other screw thread surfaces are defined.

Designation. ISO metric fine screw threads are described by the letter "M" (see page 8-15), followed by the nominal diameter and pitch—e.g., M 10 × 1.25. However, ISO metric coarse threads have the "M" designation followed only by the nominal diameter: M 10. But, the coarse thread is specified in the ISO 261 and other foreign national standards.

Nevertheless, while the modified ISO system does not recognize any fine threads, its coarse thread pitch is always given with the letter and nominal diameter designations: M 10 × 1.5.

In this section, related ISO and national standards are referred to where they have been used, and all the ISO standards developed by the Technical Committee TC 1-Screw Threads are listed at the end of this section.

---
*The metric fastener size M6.3 × 1 has been replaced by the M6 × 1 in the ANSI B1.18 proposed standards on screw threads for threaded fasteners.

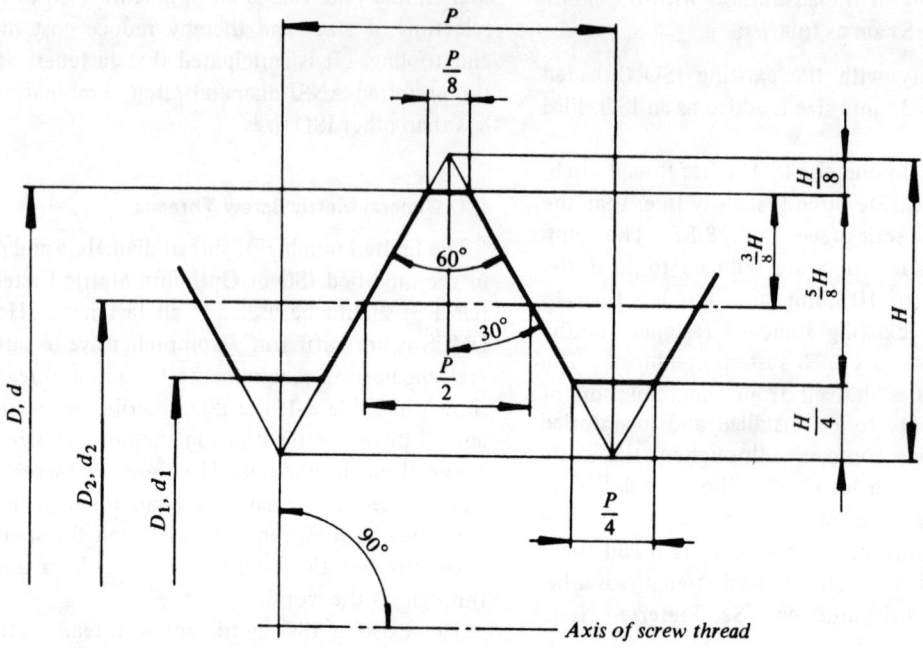

Fig. 8-1. ISO basic thread profile (IFO 68).

ISO Basic Thread Profile

- D = major diameter of internal thread
- d = major diameter of external thread
- D_2 = pitch diameter of internal thread
- d_2 = pitch diameter of external thread
- D_1 = minor diameter of internal thread
- d_1 = minor diameter of external thread
- P = pitch
- H = height of fundamental triangle

The basic thread profile is specified in the ISO 68 standard and shown in Fig. 8-1. The tolerances and deviations for external and internal screw threads are applied to the basic sizes, for which the numerical values are listed in Table 8-1 for all ISO general metric screw thread diameter-pitch combinations.

World Metric Screw Thread Standards

Table 8-1 is developed with the objective of specifying the ISO screw thread basic dimensions, and at the same time relating the ISO standards to the national standards in the major industrial countries of the world. This is an important piece of information to use when designing for possible export or foreign production.

The international standards used to develop Table 8-1 are as follows: ISO general purpose screw threads—basic profile and metric diameters and pitches as given in ISO 261, and the selected sizes for screws, bolts, and nuts in ISO 262 for nominal diameters less than 40 mm.

The thread pitch name, e.g., Coarse, Fine, etc., indicates the nominal size is a selected size within the European Standards Coordinating Committee (CEN) CEN 11/U1 standard, and the national screw thread standards numbers are shown for each country listed.

The design profiles for ISO metric internal and external threads are shown in Fig. 8-2, and the numerical dimensions for the various ISO screw threads in Table 8-1. These represent the profiles of the threads in their maximum metal conditions. It will be noted that the root of each thread is deepened so as to clear the basic flat crest of the other thread. The contact between the threads is thus confined to their sloping flanks.

Key to Symbols and Formulas Used to Compute Values in Table 8-1

Nominal Diameter	d	= D (from Standard)
Thread Pitch	P	(from Standard)
Height of Fundamental Triangle	H	= $0.86603P$
Pitch Diameter	$d2$	= $D2$ = $d - 0.75H$ = $d - 0.64953P$
Minor Diameter (Bolt)	$d3$	= $d - 2h3$ = $d - 1.22687P$
Minor Diameter (Nut)	$D1$	= $d - 2H1$ = $d - 1.08253P$
Thread Depth (Bolt)	$h3$	= $(d-d3)/2$
Thread Depth (Nut)	$H1$	= $(D-D1)/2$
Thread Root Radius	R	= $0.125P$
Tensile Stress Area	AS	= $(\pi/4)((d2 + d3)/2)^2$
		= $0.7854(D - 0.9382P)^2$

SCREW THREADS

Classes of Thread Fit (ISO)

The system of classes of fit described here is used throughout the world for controlling the dimensions for ISO general metric screw threads and fasteners, and it uses a system similar to the ISO system of limits and fits (see Section 6) applied to the basic sizes for screw threads listed in Table 8-1.

The ISO general purpose metric screw threads tolerances are described in the ISO 965 Part I through III standards, which form the basis of the national standards in the major industrial countries of the world. See Table 8-1 for the national screw threads standards number from the country of your choice.

A brief introduction to the system of limits and fits as it applies to fasteners is as follows:

A tolerance zone must be specified both in magnitude and position in relation to the basic size. The nature of a fit in dependent on both the magnitudes of the tolerances and the positions of the tolerance zone for the two members. The position of a tolerance zone is defined by the distance between the basic size and the nearest end of the tolerance zone. This distance is known as the "fundamental deviation." In the ISO metric screw thread system fundamental deviations are designated by letters—capitals for internal threads and small letters for external threads. The magnitudes of tolerance zones are designated by tolerance grades (figures). A combination of a tolerance grade (figure) and a fundamental deviation (letter) forms a tolerance class designation, e.g., "6g."

Three classes of fits similar to the unified thread classes 1A/1B; 2A/2B; 3A/3B are in general use by countries on the metric system, and they are as follows:

For external threads (bolts): 8g, 6g, 4h (Unified Class; 1A, 2A, 3A)

For internal threads (nuts): 7H, 6H, 5H (Unified Class; 1B, 2B, 3B)

The medium fit (6H/6g) is approximately equivalent to the customary unified class 2 (2A/2B) fit, and it is used in most screw thread applications. Figure 8-3 shows the relationship between classes of fits and the tolerance zones.

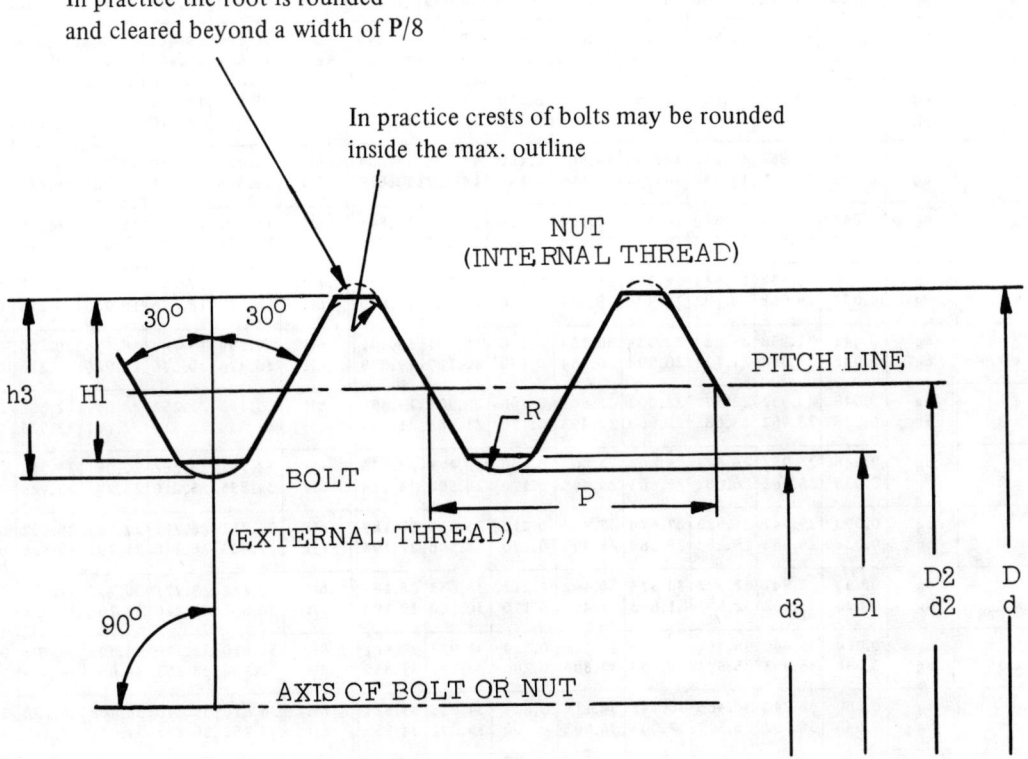

Fig. 8-2. Design forms of internal and external ISO metric screw threads (maximum metal conditions).

SCREW THREADS

Table 8-2. Limiting Dimensions of Standard Series Threads for Commercial Screws, Bolts and Nuts (mm)

Nominal Size Diam.	Pitch P	Basic Thread Designation	Tol Class	Allowance	Major Diameter Max	Major Diameter Min	Pitch Diameter Max	Pitch Diameter Min	Pitch Diameter Tol	Minor Diameter Max[a]	Minor Diameter Min[b]	Tol Class	Minor Diameter Min	Minor Diameter Max	Pitch Diameter Min	Pitch Diameter Max	Pitch Diameter Tol	Major Dia Min
1.6	0.35	M1.6	6g	0.019	1.581	1.496	1.354	1.291	0.063	1.151	1.063	6H	1.221	1.321	1.373	1.458	0.085	1.600
1.8	0.35	M1.8	6g	0.019	1.781	1.696	1.554	1.491	0.063	1.351	1.263	6H	1.421	1.521	1.573	1.658	0.085	1.800
2	0.4	M2	6g	0.019	1.981	1.886	1.721	1.654	0.067	1.490	1.394	6H	1.567	1.679	1.740	1.830	0.090	2.000
2.2	0.45	M2.2	6g	0.020	2.180	2.080	1.888	1.817	0.071	1.628	1.525	6H	1.713	1.838	1.908	2.003	0.095	2.200
2.5	0.45	M2.5	6g	0.020	2.480	2.380	2.188	2.117	0.071	1.928	1.825	6H	2.013	2.138	2.208	2.303	0.095	2.500
3	0.5	M3	6g	0.020	2.980	2.874	2.655	2.580	0.075	2.367	2.256	6H	2.459	2.599	2.675	2.775	0.100	3.000
3.5	0.6	M3.5	6g	0.021	3.479	3.354	3.089	3.004	0.085	2.742	2.614	6H	2.850	3.010	3.110	3.222	0.112	3.500
4	0.7	M4	6g	0.022	3.978	3.838	3.523	3.433	0.090	3.119	2.979	6H	3.242	3.422	3.545	3.663	0.118	4.000
4.5	0.75	M4.5	6g	0.022	4.478	4.338	3.991	3.901	0.090	3.558	3.414	6H	3.688	3.878	4.013	4.131	0.118	4.500
5	0.8	M5	6g	0.024	4.976	4.826	4.456	4.361	0.095	3.994	3.841	6H	4.134	4.334	4.480	4.605	0.125	5.000
6	1	M6	6g	0.026	5.974	5.794	5.324	5.212	0.112	4.747	4.563	6H	4.917	5.153	5.350	5.500	0.150	6.000
7	1	M7	6g	0.026	6.974	6.794	6.324	6.212	0.112	5.747	5.563	6H	5.917	6.153	6.350	6.500	0.150	7.000
8	1.25	M8	6g	0.028	7.972	7.760	7.160	7.042	0.118	6.439	6.231	6H	6.647	6.912	7.188	7.348	0.160	8.000
	1	M8 x 1	6g	0.026	7.974	7.794	7.324	7.212	0.112	6.747	6.563	6H	6.917	7.153	7.350	7.500	0.150	8.000
10	1.5	M10	6g	0.032	9.968	9.732	8.994	8.862	0.132	8.127	7.879	6H	8.376	8.676	9.026	9.206	0.180	10.000
	1.25	M10 x 1.25	6g	0.028	9.972	9.760	9.160	9.042	0.118	8.439	8.231	6H	8.647	8.912	9.188	9.348	0.160	10.000
12	1.75	M12	6g	0.034	11.966	11.701	10.829	10.679	0.150	9.819	9.543	6H	10.106	10.441	10.863	11.063	0.200	12.000
	1.25	M12 x 1.25	6g	0.028	11.972	11.760	11.160	11.028	0.118	10.439	10.217	6H	10.647	10.912	11.188	11.368	0.180	12.000
14	2	M14	6g	0.038	13.962	13.682	12.663	12.503	0.160	11.508	11.204	6H	11.835	12.210	12.701	12.913	0.212	14.000
	1.5	M14 x 1.5	6g	0.032	13.968	13.732	12.994	12.854	0.140	12.127	11.879	6H	12.376	12.676	13.026	13.216	0.190	14.000
16	2	M16	6g	0.038	15.962	15.682	14.663	14.503	0.160	13.508	13.204	6H	13.835	14.210	14.701	14.913	0.212	16.000
	1.5	M16 x 1.5	6g	0.032	15.968	15.732	14.994	14.854	0.140	14.127	13.879	6H	14.376	14.676	15.026	15.216	0.190	16.000
18	2.5	M18	6g	0.042	17.958	17.623	16.334	16.164	0.170	14.891	14.541	6H	15.294	15.744	16.376	16.600	0.224	18.000
	1.5	M18 x 1.5	6g	0.032	17.968	17.732	16.994	16.854	0.140	16.127	15.879	6H	16.376	16.676	17.026	17.216	0.190	18.000
20	2.5	M20	6g	0.042	19.958	19.623	18.334	18.164	0.170	16.891	16.541	6H	17.294	17.744	18.376	18.600	0.224	20.000
	1.5	M20 x 1.5	6g	0.032	19.968	19.732	18.994	18.854	0.140	18.127	17.879	6H	18.376	18.676	19.026	19.216	0.190	20.000
22	2.5	M22	6g	0.042	21.958	21.623	20.334	20.164	0.170	18.891	18.541	6H	19.294	19.744	20.376	20.600	0.224	22.000
	1.5	M22 x 1.5	6g	0.032	21.968	21.732	20.994	20.854	0.140	20.127	19.879	6H	20.376	20.676	21.026	21.216	0.190	22.000
24	3	M24	6g	0.048	23.952	23.577	22.003	21.803	0.200	20.271	19.855	6H	20.752	21.252	22.051	22.316	0.265	24.000
	2	M24 x 2	6g	0.038	23.962	23.682	22.663	22.493	0.170	21.508	21.194	6H	21.835	22.210	22.701	22.925	0.224	24.000
27	3	M27	6g	0.048	26.952	26.577	25.003	24.803	0.200	23.271	22.855	6H	23.752	24.252	25.051	25.316	0.265	27.000
	2	M27 x 2	6g	0.038	26.962	26.682	25.663	25.493	0.170	24.508	24.194	6H	24.835	25.210	25.701	25.925	0.224	27.000
30	3.5	M30	6g	0.053	29.947	29.522	27.674	27.462	0.212	25.653	25.189	6H	26.211	26.771	27.727	28.007	0.280	30.000
	2	M30 x 2	6g	0.038	29.962	29.682	28.663	28.493	0.170	27.508	27.194	6H	27.835	28.210	28.701	28.925	0.224	30.000
33	3.5	M33	6g	0.053	32.947	32.522	30.674	30.462	0.212	28.653	28.189	6H	29.211	29.771	30.727	31.007	0.280	33.000
	2	M33 x 2	6g	0.038	32.962	32.682	31.663	31.493	0.170	30.508	30.194	6H	30.835	31.210	31.701	31.925	0.224	33.000
36	4	M36	6g	0.060	35.940	35.465	33.342	33.118	0.224	31.033	30.521	6H	31.670	32.270	33.402	33.702	0.300	36.000
	3	M36 x 3	6g	0.048	35.952	35.577	34.003	33.803	0.200	32.271	31.855	6H	32.752	33.252	34.051	34.316	0.265	36.000
39	4	M39	6g	0.060	38.940	38.465	36.342	36.118	0.224	34.033	33.521	6H	34.670	35.270	36.402	36.702	0.300	39.000
	3	M39 x 3	6g	0.048	38.952	38.577	37.003	36.803	0.200	35.271	34.855	6H	35.752	36.252	37.051	37.316	0.265	39.000

(Reproduced courtesy of ASME)

NOTES: Metric coarse screw threads are designated with nominal size and pitch in the ANSI standards; M10 x 1.5.
    a. Design Form, see Figure 8-1 and 8-2.
    b. Required for high strength applications where rounded root is specified.

SCREW THREADS

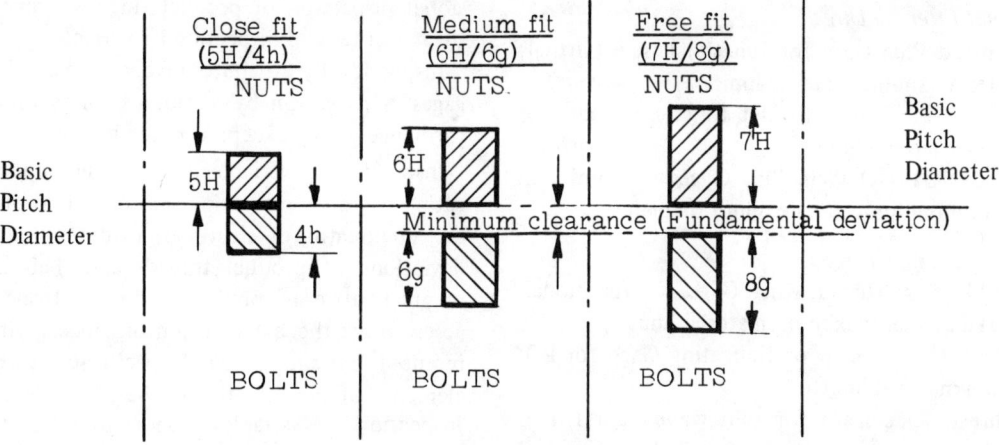

Fig. 8-3. Diagram showing relationship between tolerance zones and classes of fits. *

*The above thread fits are not recognized in American standards, which specify 6H/6g for normal applications and 6H/5g6g for close fits.

Table 8-2 shows the upper and lower limits for the medium fit metric fasteners in the range from 1 through 39 mm nominal sizes.

Thread Designations (ISO)[1]

Coarse threads are designated in accordance with the following examples:

For an internal thread (nut) M 5 -6H
For an external thread (bolt) M 8 -6g

- Thread system symbol for ISO metric
- Nominal size (mm)
- Thread tolerance class symbol

A fit between a pair of threaded parts is indicated by the internal thread (nut) tolerance class designation followed by the external thread (bolt) class designation, the two separated by a stroke, e.g.,

M8 − 6H/6g
M5 − 6H/6g

Fine threads are designated in accordance with the following examples:

For an internal thread (nut) M 5 × 0.5 − 6H
For an external thread (bolt) M 8 × 1.0 − 6g

- Thread system symbol for ISO metric
- Nominal size
- Pitch
- Thread tolerance class symbol

[1] ANSI thread designations may show the thread pitch.

A fit between a pair of threaded parts is indicated by the internal thread (nut) tolerance class designation followed by the external thread (bolt) class designation, the two separated by a stroke; e.g.,

M8 × 1.0 − 6H/6g
M16 × 1.5 − 6H/6g

In the example for the bolt M8-6g the tolerance 6g applies to both the pitch and the crest diameters. If the tolerances are different for the two diameters, they are designated as follows:

M8 - 5g6g
- class designation for pitch diameter tolerance
- class designation for crest diameter tolerance

Gages for ISO Metric Screw Threads

Fasteners and screw threads produced to the existing ISO and corresponding national standards will be in existence for some time to come. It is therefore necessary to describe gages and methods used to control product thread acceptability.

Companies that purchase fasteners or products with metric screw threads to the former ISO recommendations should refer to the comprehensive American standard ANSI B1.16–American Gaging Practice for metric screw threads and the ISO 1502-ISO general purpose metric screw threads.

The ANSI B1.16 provides for the following type of gages:

Gaging

For Product Internal Thread:

1. GO Thread Plug Gage for functional size (virtual) diameter maximum-material-limit.
2. HI Thread Plug Gage for HI functional diameter minimum-material-limit.
3. GO and NOT GO Plain Plug Gages for minimum and maximum limits of the minor diameter.

For Product External Thread:

1. Adjustable GO Thread Ring Gage for functional (virtual) diameter maximum-material-limit.
2. LO Limit Thread Snap or Indicating Gage for LO minimum-material-limit.
3. LO Thread Ring Gage for thin walled parts that could be deformed by snap or indicating gage pressures.
4. HI and LO Thread Setting Plug Gages for (1), (2) and (3) above.
5. Plain gages for minimum and maximum limits of the major diameter.

For measurement of pitch diameter for thread plug gages, refer to ANSI B1.2.

Reference Documents

1. B1 Report on ISO Metric Screw Threads[1]
2. Unified Screw Threads ANSI B1.1
3. Gages and Gaging for Unified Screw Threads ANSI B1.2
4. Nomenclature, Definitions, and Letter Symbols for Screw Threads ANSI B1.7[2]
5. Gage Blanks ANSI B47.1 (Commercial Standard CS8-61).

Reference sheets identifying the 60° inch screw threads covered by ANSI and comparing conformance requirements of UN, UNR, UNK, UNJ and ISO Metric Screw Threads are given in Figures 5 and 6 at the end of this standard.

BASIC PRINCIPLES (ANSI B1.16)

Object of Conformance Gaging

The object of conformance gaging of product threads is to determine the extent they conform dimensionally to prescribed limits of size, and to segregate or reject product threads that are outside of prescribed limitations.

There are two general methods of approach to dimensional inspection of product threads, namely, inspection by attributes and inspection by variables.

Inspection by attributes involves the application of limit gages. Inspection by attributes forms the basis of conformance gaging except as noted below.

Inspection by variables involves the application of indicating gages or measuring instruments (optical, mechanical, pneumatic, or electrical) to determine the extent of deviations of product threads and their individual elements relative to prescribed limits. Inspection by variables forms the basis of conformance gaging where it is required by supplemental specifications that individual elements of product threads be controlled. Dimensional inspection by variables is most useful in the control of manufacturing tools and processes and to collect manufacturing data for the analysis of product thread deviations.

Screw Thread Conformance

Dimensional acceptance of product threads shall be in accordance with the limits of size as determined by the conformance gages recommended herein. It is important that the method of conformance gaging be understood by both the producer and user. See page 8-17 for gaging and verification of product threads.

Accuracy in Gaging—Thread plug gages are controlled by *direct* measuring methods. Thread ring, thread snap limit gages, and indicating thread gages are controlled by reference to the appropriate setting gages.

Limitations of Gaging—Product threads accepted by a gage of one type may be verified by other types. It is possible, however, that parts which are near either rejection limit may be accepted by one type and rejected by another. Also it is possible for two individual limit gages of the same type to be at the opposite extremes of the gage tolerances permitted, and borderline product threads accepted by one gage could be rejected by another (see page 8-17).

Large product external and internal threads above 158.8 mm (6.25 inches) nominal size may present additional problems for technical and economic reasons. In these instances verification may be based on use of gages or measurement of thread elements. Various types of gages or measuring devices (refer to ANSI B1.2) in addition to those defined in this document are available and acceptable when properly correlated to this document. It is essential to achieve agreement between producer and consumer with respect to method and equipment used.

Surveillance of Gages—Periodic rechecking and surveillance of gages are necessary precautions to assure satisfactory product thread conformance.

---

[1] This Report is not an ANSI Standard but can be purchased from the American Society of Mechanical Engineers, 345 East 47th Street, New York, N.Y. 10017.

[2] ANSI B1.7 includes several revised definitions. Those pertinent to this document are included as Appendix A in the ANSI B1.2 Standard.

SCREW THREADS

Determining Size of Gages

Determining Pitch Diameter. The three-wire method of determining pitch diameter of thread plug gages is standard for gages to this specification. Refer to ANSI B1.2 Appendix B.

Sizes of ring thread gages are determined by their fit on their respective setting plugs so measured. Other thread gages for product external threads are controlled by reference to appropriate setting plugs so measured.

Standard Temperature (ISO 1)

20°C (68°F) is the standard temperature used internationally for linear measurements. Nominal dimensions of gages and product as specified, and actual dimensions as measured shall be within specified limits at this temperature.

As product threads are frequently checked at temperatures which are not controlled it is desirable that the coefficient of thermal expansion of gages be the same as that of the product on which they are used. Inasmuch as the majority of threaded product consists of iron or steel, and screw-thread gages are ordinarily made of hardened steel, this condition is usually fulfilled without special attention. When the materials of the product thread and the gage are dissimilar, the differing thermal coefficients can cause serious complications and must be taken into account.

Measuring Force for Wire Measurements of 60 Degree Threads

In measuring the pitch diameter of screw thread gages by means of wires the following measuring forces shall be used:

Pitch mm	Measuring Force	
1.25 or larger	11.1 N	(2½ lbs force)
Below 1.25 and including 0.6	4.45 N	(1 lb force)
Below 0.6 and including 0.35	2.22 N	(8 oz force)

The thread wires should be calibrated by the procedure specified in ANSI B1.2.

GAGING AND VERIFICATION OF PRODUCT THREADS (ANSI B1.16)

Types of Gages

Gages are classified as to type and use in this section, together with specific details of gaging practice applicable to each type.

GO thread gages check the maximum-material-size, to assure interchangeable assembly. HI and LO thread gages check the minimum-material-size.

The thread form of GO thread gages corresponds to maximum product thread depth of engagement to assure clearance at the major diameter of the product internal thread or the minor diameter of the product external thread.

GO and NOT GO plain cylindrical plug or ring gages, snap or indicating gages, check the limits of size of the minor diameter of product internal threads and the major diameter of product external threads respectively.

At the product thread maximum-material-limit the gages used for final conformance gaging are within the limits of size of the product thread. At the product thread minimum-material-limit the usual practice for gages used for final conformance gaging is to have the gage tolerance within the extreme limits of size of the product thread. However, to assure that usable product thread at the extreme limit of size (minimum-material-limit) is not rejected, in borderline cases, the consumer may elect to use HI/LO gages having pitch diameter tolerances outside the product thread limits.

Use of Gages: Threaded and Plain Gages for Verification of Product Internal Threads

Unless otherwise specified all thread gages which directly check the product thread shall be X tolerance for all classes.

GO Thread Plug Gages—GO thread plug gages must enter the full threaded length of the product freely. The GO thread plug gage is a cumulative check of all thread elements except the minor diameter.

HI Thread Plug Gages—HI thread plug gages when applied to the product internal thread may engage only the end threads (which may not be representative of the complete thread). Entering threads on product are incomplete and permit gage to start. Starting threads on HI plugs are subject to greater wear than the remaining threads. Such wear in combination with the incomplete product threads permits further entry of the gage. Surveillance facilities ordinarily available in the field are often inadequate for fully determining such gage wear. Also, it is not practical to control nor limit the torque applied by operators, nor that utilized by a specific operator at various times and under varying conditions. For these reasons the following standard practice has been adopted with respect to permissible entry. Threads are acceptable when the HI thread plug gage is applied to the product internal thread if: (a) it does not enter, or if (b) all complete product threads can be entered, provided that a *definite* drag from contact with the product material results on or before the second turn of entry. The gage should not be forced after the drag is definite. Special requirements such as exceptionally thin or ductile material, small number of threads, may necessitate modification of this practice.

GO and NOT GO Plain Plug Gages for Minor Diameter of Product Internal Thread—GO plain plug gages must completely enter the product internal thread to assure that the minor diameter does not exceed the maximum-material-limit. NOT GO plain plug gages must not enter the product internal thread to provide adequate assurance that the minor diameter does not exceed the minimum-material-limit.

Thread Setting Plug Gages

GO and LO Truncated Setting Plugs—W tolerance truncated setting plugs are recommended for setting adjustable thread ring gages to and including 158.8 mm (6.25 inches) nominal size and may be used for setting thread snap gages and indicating thread gages. Above 158.8 mm (6.25 inches) nominal size, the difference in feel between the full form and truncated sections in setting thread ring gages is insignificant, and the basic crest setting plug may be used.

When setting adjustable thread ring gages to size, the truncated portion of the setting plug controls the functional size, and the full form portion assures that adequate clearance is provided at the major diameter of the ring gage. The full form portion in conjunction with the truncated portion checks—to some degree—the half-angle accuracy of the gage. The same procedure may be applied to detect uneven angle wear of ring gages in use.

GO and LO Basic-crest (Full Form) Setting Plugs—W tolerance basic crest setting plugs are frequently used for setting thread snap limit gages and indicating thread gages. They may also be used for setting large adjustable thread ring gages, especially those above 158.8 mm (6.25 inches) nominal size. When they are so used it may be desirable to take a cast of the ring thread form to check the half-angle and profile.

GO and NOT GO Plain Plug Acceptance Check Gages for Checking Minor Diameter of Thread Ring Gages—The GO plain plug gage is made to the minimum minor diameter specified for the thread ring gage (GO or LO), while the NOT GO gage is made to maximum minor diameter specified for the thread ring gage (GO or LO). After the adjustable thread ring gages have been set to the applicable thread setting plugs, the GO and NOT GO plain plug acceptance check gages are applied to check the minor diameter of the ring gage to assure that it is within the specified limits. An alternate method for checking minor diameter of thread ring gages is by the use of measuring equipment.

Threaded and Plain Ring, Snap and Indicating Thread Gages for Verification of Product External Thread

Adjustable GO Thread Ring Gages—Adjustable GO thread ring gages must be set to the applicable W tolerance setting plugs. The product thread must freely enter the GO thread ring gage for the entire length of the threaded portion. The GO thread ring gage is a cumulative check of all thread elements except the major diameter.

LO Thread Snap Limit Gages or Indicating Thread Gages—LO thread snap limit gages (or indicating thread gages) must be set to the applicable W tolerance setting plugs. The gage is then applied to the product thread at various points around the circumference, and over the entire length of complete product threads. In applying the thread snap limit gage, threads are dimensionally acceptable when the gaging elements do not pass over the product thread or just pass over the product thread with perceptible drag from contact with the product material and the gage. Indicating thread gages provide a numerical value for the product thread size. Product external threads are dimensionally acceptable when the value derived in applying the gage (as described above) is not less than the specified minimum-material-limit.

LO Thread Ring Gages—LO thread ring gages must be set to the applicable W tolerance setting plugs. LO thread ring gages when applied to the product external thread may engage only the end threads (which may not be representative of the complete product thread).

Starting threads on LO rings are subject to greater wear than the remaining threads. Such wear in combination with the incomplete threads at the end of the product thread permit further entry in the gage. Surveillance facilities ordinarily available in the field are often inadequate for fully determining such gage wear. Also, it is not practical to control nor limit the torque applied by operators, nor that utilized by a specific operator at various times and under varying conditions. For these reasons the following standard practice has been adopted with respect to permissible entry. Threads are acceptable when the LO thread ring gage is applied to the product external thread if (a) it is not entered, or if (b) all complete product threads can enter, provided that a *definite* drag from contact with the product material results on or before the second turn of entry. The gage should not be forced after the drag is definite. Special requirements such as exceptionally thin or ductile material, small number of threads, etc., may necessitate modification of this practice.

Check of Effect of Lead and Flank Angle Deviations on Product Thread—When this check is specified there are two general methods available for the inspection procedures involved, as follows:

SCREW THREADS

Direct Measurement of Deviations—The lead and flank angle of the product thread may be measured by means of available measuring equipment such as projection comparators, measuring microscopes, graduated cone points, lead measuring machines; helix variation measuring machines, thread flank charting equipment. Diameter equivalents of such deviations are calculated by applying well-known formulas.[1]

Differential Gaging Utilizing Indicating Thread Gages with appropriate gaging elements as outlined in ANSI B1.2 may be used.

GO and NOT GO Plain Rings and Adjustable Snap Limit and Indicating Gages for Checking Major Diameter of Product External Thread—The GO gage must completely receive or pass over the major diameter of the product external thread to assure that the major diameter does not exceed the maximum-material-limit. The NOT GO gage must not pass over the major diameter of the product external thread to assure that the major diameter is not less than the minimum-material-limit.

Limitations

Product threads accepted by a gage of one type may be verified by other types. It is possible, however, that parts which are near either rejection limit may be accepted by one type and rejected by another. Also it is possible for two individual limit gages of the same type to be at the opposite extremes of the gage tolerances permitted, and borderline product threads accepted by one gage could be rejected by another. In such instances, the applicable gages outlined in this document (limit plug and ring, LO thread snap or indicating thread gages) that approximate as closely as practicable the extreme maximum-material-product-limit and minimum-material-product-limit shall be used to determine whether or not the product threads under inspection are within the specified limits of size.

Large product external and internal threads above 158.8 mm (6.25 inches) nominal size may present additional problems for technical and economic reasons. In these instances verification may be based on use of gages or measurement of thread elements. Various types of gages or measuring devices (refer to ANSI B1.2) in addition to those defined in this standard are available and acceptable when properly correlated to this standard. Producer and user should agree on the method and equipment used.

---

[1] Each 0.0001 variation in lead amounts to 0.00017 (1.732 × 0.0001) increase in effective pitch diameter, (differential reading) for 60° screw threads.

The tangent of half angle variation times 1.5p equals the increase in effective pitch diameter, based on a height of thread engagement of 0.625H (Differential reading-diameter equivalent).

Surveillance—Gages are subject to wear and/or damage from normal usage. Periodic rechecking and surveillance are necessary precautions to assure product thread conformance.

Marking of Gages

Each gage shall be plainly and permanently marked with the minimum marking essential for positive identification. In the case of plug gages of the renewable-end type, in addition to marking the handle, the marking shall also appear on the face of the gaging member where practicable.

Unless otherwise specified by the purchaser, the following particulars shall be included in the gage marking:

The designation of the corresponding product thread in accordance with the recommendations in ANSI B1 Report "ISO Metric Screw Threads." *NOTE:* In the case of left-hand screw gages, the symbol "L.H." follows the designation.

Examples of Gage Marking (ANSI B1.16)

The GO thread plug should be identified by metric nominal size, pitch (except for the standard coarse thread series where the pitch designation is omitted), tolerance class and GO pitch diameter in inches.
Example:
M6 – 6H GO PD 0.2106
M8 × 1 – 6H GO PD 0.2897

The HI thread plug gage should be marked with the nominal size, pitch and tolerance class in metric and HI and pitch diameter in inches.
Example:
M6 – 6H HI PD 0.2165
M8 × 1 – 6H HI PD 0.2953

The GO plain plug gage member for metric threads should be marked with: Nominal size, pitch and tolerance class, in millimeters and GO and minor diameter in inches.
Example:
M8 × 1 – 6H GO 0.2724

The NOT GO plain plug gage should be marked with: Nominal size, pitch and tolerance class in metric, NOT GO and minor diameter in inches.
Example:
M8 × 1 – 6H NOT GO 0.2817

The GO Thread Ring Gage should be identified by nominal size, pitch and tolerance class in metric and GO and pitch diameter in inches.
Example:
M8 × 1 – 6g GO PD 0.2883

Where practicable *gages and gaging elements* should be marked with the minimum marking essential for identification. When space available for marking is inadequate and the gages and gaging elements are packaged separately, the containers should be suitably marked and/or the gaging elements suitable tagged.

The LO Thread Gage should be identified by nominal size, pitch and tolerance class in metric and LO pitch diameter (inches).

Example:

M8 × 1 – 6g LO PD 0.2839

Where practicable *comparators and gaging elements* should be marked with the minimum marking essential for identification. When space available for marking is inadequate and the comparators and gaging elements are packaged separately, the containers should be suitably marked and/or the gaging elements suitably tagged.

Fixed limit gages for major diameter of product external threads are to be identified by GO and the major diameter (in inches) as follows: GO 0.3155.

The GO thread setting plug gage should be identified by set plug, nominal size, pitch and tolerance series in metric, and GO pitch diameter in inches.

Example:

M8 × 1 – 6g GO PD 0.2883

The LO thread setting plug gage should be identified by set plug, nominal size, pitch, tolerance class in metric, and LO pitch diameter (in inches).

Example:

M8 × 1 – 6g LO PD 0.2839

The GO and NOT GO plain plug acceptance check gages for the GO thread ring gage should be identified as GO and NOT GO Acceptance Checks for GO Thread Ring Minor Dia XXXX – XXXX (in inches).

The GO and NOT GO plain plug acceptance check gages for the LO thread ring gage should be identified as GO and NOT GO Acceptance Checks for LO Thread Ring Minor Dia XXXX – XXXX (in inches).

MODIFIED ISO FASTENER SCREW THREADS

The material presented in this chapter is based on the IFI-500 Standard covering Screw Threads for Metric Series Mechanical Fasteners.

The permission to use the material in the IFI-500 Standard was granted to the author by Industrial Fastener Institute (IFI) of Cleveland, Ohio. This organization has provided industry throughout the world with valuable metric fastener standards information, and has published a book on metric fasteners issued in 1976.

The modified ISO screw thread system described in this section introduces several improvements over the existing ISO thread system. It is fully interchangeable with the former ISO system and has received approval among leading industrial countries within ISO. It is anticipated that the new system will be included in national and international standards during the coming years. Initially the modified ISO thread system was developed by an ANSI-Optimum Metric Fastener System (OMFS) committee, and the most valuable features have been retained in the modified ISO system. The 25 metric coarse diameter-pitch combinations which range in size from 1.6 through 100 millimeters are the only standard fastener sizes recognized by the ANSI B1.18 committee. This alone should produce savings for American industry in the millions of dollars.

The new referee gaging system described in the IFI-500 standard is innovative and more discriminating than current referee gaging practices in identifying non-conforming products. It is suggested that during the next two to three year period while American industry develops the capacity for the new referee gaging practice as presented in this section, the use of other standard methods for gaging screw threads, such as the practice outlined in ANSI B1.16, might be found more convenient.

Mechanical fasteners in sizes 1.6 to 100 mm, inclusive, are described in this chapter. It establishes the basic thread profile, the diameter-pitch series, the maximum and minimum boundary profiles for gaging, and acceptance criteria.

Modified ISO Basic Thread Profile

The basic profile of the thread is identical to ISO 68. The basic major diameter is the reference datum for all dimensioning. Figure 8-4 shows the simple outline of the modified ISO basic thread profile.

Series of Threads

The ANSI B1.18 committee recognizes only one series of diameter-pitch combinations, as given in Table 8-3.

Of the 25 diameter-pitch combinations in the series, all are standard in the existing ISO metric coarse thread series as detailed in ISO 261 and ISO 262.

Classes of Thread Fit (OMFS)

The IFI-500 describes two classes of thread fit; one class, designated 6H/6g, is for general-purpose applications; and the other class, designated as 6H/5g6g, is used where closer thread fits are required. The thread fit for the general-purpose applications approximates Unified Class 2A/2B. The closer thread-fit class approximates the tolerances of Unified Class 3A/3B with the addition of an allowance. The Optimum Metric Fastener System (OMFS) committee recommended the use of one class of fit only, and the writer

SCREW THREADS

8-21

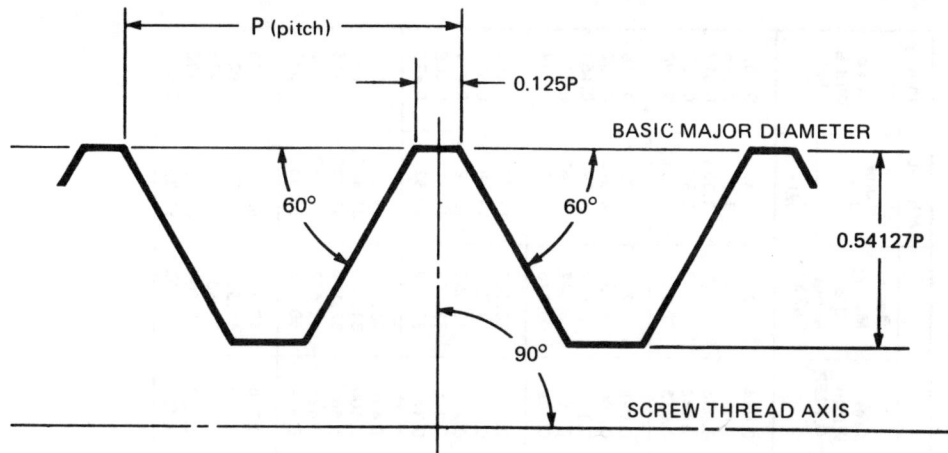

Fig. 8-4. Modified ISO basic thread profile (IFI-500).

is convinced that one thread fit suffices in nearly all applications. This is of advantage to the designer, since it provides a good opportunity to use only one class of fit in design, and thereby reduces inventory of fasteners and tooling.

The boundary profiles for gaging are based on tolerance grades and tolerance position selected from ISO 965, Part I.

External Threads

General Purpose—For diameters of 5 mm, and larger, the tolerance position (fundamental deviation) is "g" (small allowance). For diameters of 1.6 mm through 4 mm there is a constant tolerance position (fundamental deviation) of 0.024 mm. The flank diametral displacement of the boundary profiles for gaging is ISO tolerance grade 6 for pitch-diameters. The major diameter displacement is ISO tolerance grade 6 for major diameters. See Table 8-3.

For close tolerance the same tolerance position as for the general-purpose thread applies. The flank diametral displacement of the boundary profiles for gaging is equal to the ISO tolerance grade 5 for pitch-diameters. The major diameter displacement equals the ISO tolerance grade 6 for major diameters. See Table 8-4.

Internal Threads

For internal threads, tolerance position, "H" (no allowance) applies. The minor diameter displacement of the boundary profiles for gaging is ISO tolerance grade 6 for minor diameters, and the flank diametral displacement is ISO tolerance grade 6 for pitch diameters. (Table 8-5.)

Basic Designation—These metric screw threads are designated by the letter "M" followed by the nominal size (basic major diameter) in millimeters and the pitch in millimeters separated by the symbol "X." Example: M8 X 1.25.

General Purpose—General purpose screw threads shall be designated by the basic designation.

Example: M5 X 0.8—A thread with a nominal size (basic major diameter) of 5 mm and a pitch of 0.8 mm and with a "g" tolerance position and a 6 tolerance grade for the external thread and/or an "H" tolerance position with a 6 tolerance grade for the internal thread. (These threads may be substituted for those which are designated M5-6g or M5-6H in the existing ISO system.)

Close Tolerance. Close tolerance screw threads shall be designated by the basic designation with the addition of the letter "C" after the pitch.

Example: M10 X 1.5C—A thread with a nominal size (basic major diameter) of 10 mm and a pitch of 1.5 mm, and with a "g" tolerance position, a 5 flank diametral tolerance grade, and a 6 major diameter tolerance grade. (These threads may be substituted for those which are designated M10-5g6g.)

When it is not obvious whether the basic designation applies to an external or an internal thread, the letter A shall be used to identify an external thread, and the letter B shall be used to identify an internal thread.

Example: M14 X 2A—Externally threaded part; M20 X 2.5B—internally threaded part.

Table 8-3. Boundary Profiles for Gaging General Purpose External Threads (IFI-500)

Basic Major Dia and Thread Pitch (P)	Allowance	Flank Diametral Displacement	Min Root Radius = 0.125P	Non-Plated Threads Major Dia Max	Non-Plated Threads Y GO Gage Crest Width	Plated Threads Major Dia Max	Plated Threads Y GO Gage Crest Width	Crest Width = 0.125P	Basic Major Dia Minus 1.08253 P	Minimum Boundary Major Dia Min	Minimum Boundary Crest Width W
M1.6x0.35	0.024	0.063	0.044	1.576	0.101	1.600	0.088	0.044	1.221	1.491	0.057
M2x0.4	0.024	0.067	0.050	1.976	0.114	2.000	0.100	0.050	1.567	1.881	0.066
M2.5x0.45	0.024	0.071	0.056	2.476	0.126	2.500	0.113	0.056	2.013	2.376	0.073
M3x0.5	0.024	0.075	0.062	2.976	0.139	3.000	0.125	0.062	2.459	2.870	0.080
M3.5x0.6	0.024	0.085	0.075	3.476	0.164	3.500	0.150	0.075	2.850	3.351	0.098
M4x0.7	0.024	0.090	0.088	3.976	0.189	4.000	0.175	0.088	3.242	3.836	0.116
M5x0.8	0.024	0.095	0.100	4.976	0.214	5.000	0.200	0.100	4.134	4.826	0.132
M6x1	0.026	0.112	0.125	5.974	0.265	6.000	0.250	0.125	4.917	5.794	0.164
M8x1.25	0.028	0.118	0.156	7.972	0.329	8.000	0.313	0.156	6.647	7.760	0.211
M10x1.5	0.032	0.132	0.188	9.968	0.393	10.000	0.375	0.188	8.376	9.732	0.248
M12x1.75	0.034	0.150	0.219	11.966	0.457	12.000	0.438	0.219	10.106	11.701	0.285
M14x2	0.038	0.160	0.250	13.962	0.522	14.000	0.500	0.250	11.835	13.682	0.319
M16x2	0.038	0.160	0.250	15.962	0.522	16.000	0.500	0.250	13.835	15.682	0.319
M20x2.5	0.042	0.170	0.312	19.958	0.649	20.000	0.625	0.312	17.294	19.623	0.408
M24x3	0.048	0.200	0.375	23.952	0.778	24.000	0.750	0.375	20.752	23.577	0.476
M30x3.5	0.053	0.212	0.438	29.947	0.906	30.000	0.875	0.438	26.211	29.522	0.560
M36x4	0.060	0.224	0.500	35.940	1.035	36.000	1.000	0.500	31.670	35.465	0.645
M42x4.5	0.063	0.236	0.562	41.937	1.161	42.000	1.125	0.562	37.129	41.437	0.715
M48x5	0.071	0.250	0.625	47.929	1.291	48.000	1.250	0.625	42.587	47.399	0.787
M56x5.5	0.075	0.265	0.688	55.925	1.418	56.000	1.375	0.688	50.046	55.365	0.858
M64x6	0.080	0.280	0.750	63.920	1.546	64.000	1.500	0.750	57.505	63.320	0.935
M72x6	0.080	0.280	0.750	71.920	1.546	72.000	1.500	0.750	65.505	71.320	0.935
M80x6	0.080	0.280	0.750	79.920	1.546	80.000	1.500	0.750	73.505	79.320	0.935
M90x6	0.080	0.280	0.750	89.920	1.546	90.000	1.500	0.750	83.505	89.320	0.935
M100x6	0.080	0.300	0.750	99.920	1.546	100.000	1.500	0.750	93.505	99.320	0.923

NOTE: All dimensions are in millimeters.

SCREW THREADS

Table 8-4. Boundary Profiles for Gaging Close Tolerance External Threads (IFI-500)

Basic Major Dia and Thread Pitch (P)	Allowance	Flank Diametral Displacement	Min Root Radius = 0.125P	Maximum Boundary — Non-Plated Threads Major Dia Max	Maximum Boundary — Non-Plated Threads Y GO Gage Crest Width	Maximum Boundary — Plated Threads Major Dia Max	Maximum Boundary — Plated Threads Y GO Gage Crest Width	Crest Width = 0.125P	Basic Major Dia Minus 1.08253P	Minimum Boundary Major Dia Min	Minimum Boundary Crest Width W
M1.6x0.35	0.024	0.050	0.044	1.576	0.101	1.600	0.088	0.044	1.221	1.491	0.064
M2x0.4	0.024	0.053	0.050	1.976	0.114	2.000	0.100	0.050	1.567	1.881	0.074
M2.5x0.45	0.024	0.056	0.056	2.476	0.126	2.500	0.113	0.056	2.013	2.376	0.082
M3x0.5	0.024	0.060	0.062	2.976	0.139	3.000	0.125	0.062	2.459	2.870	0.089
M3.5x0.6	0.024	0.067	0.075	3.476	0.164	3.500	0.150	0.075	2.850	3.351	0.108
M4x0.7	0.024	0.071	0.088	3.976	0.189	4.000	0.175	0.088	3.242	3.836	0.127
M5x0.8	0.024	0.075	0.100	4.976	0.214	5.000	0.200	0.100	4.134	4.826	0.143
M6x1	0.026	0.090	0.125	5.974	0.265	6.000	0.250	0.125	4.917	5.794	0.177
M8x1.25	0.028	0.095	0.156	7.972	0.329	8.000	0.313	0.156	6.647	7.760	0.224
M10x1.5	0.032	0.106	0.188	9.968	0.393	10.000	0.375	0.188	8.376	9.732	0.263
M12x1.75	0.034	0.118	0.219	11.966	0.457	12.000	0.438	0.219	10.106	11.701	0.304
M14x2	0.038	0.125	0.250	13.962	0.522	14.000	0.500	0.250	11.835	13.682	0.339
M16x2	0.038	0.125	0.250	15.962	0.522	16.000	0.500	0.250	13.835	15.682	0.339
M20x2.5	0.042	0.132	0.312	19.958	0.649	20.000	0.625	0.312	17.294	19.623	0.430
M24x3	0.048	0.160	0.375	23.952	0.778	24.000	0.750	0.375	20.752	23.577	0.499
M30x3.5	0.053	0.170	0.438	29.947	0.906	30.000	0.875	0.438	26.211	29.522	0.585
M36x4	0.060	0.180	0.500	35.940	1.035	36.000	1.000	0.500	31.670	35.465	0.670
M42x4.5	0.063	0.190	0.562	41.937	1.161	42.000	1.125	0.562	37.129	41.437	0.741
M48x5	0.071	0.200	0.625	47.929	1.291	48.000	1.250	0.625	42.587	47.399	0.816
M56x5.5	0.075	0.212	0.688	55.925	1.418	56.000	1.375	0.688	50.046	55.365	0.888
M64x6	0.080	0.224	0.750	63.920	1.546	64.000	1.500	0.750	57.505	63.320	0.967
M72x6	0.080	0.224	0.750	71.920	1.546	72.000	1.500	0.750	65.505	71.320	0.967
M80x6	0.080	0.224	0.750	79.920	1.546	80.000	1.500	0.750	73.505	79.320	0.967
M90x6	0.080	0.224	0.750	89.920	1.546	90.000	1.500	0.750	83.505	89.320	0.967
M100x6	0.080	0.236	0.750	99.920	1.546	100.000	1.500	0.750	93.505	99.320	0.960

NOTE: All dimensions are in millimeters.

Table 8-5. Boundary Profiles for Gaging General Purpose Internal Threads. (IFI-500)

Basic Major Dia and Thread Pitch (P)	Flank Diametral Displacement	Maximum Boundary Minor Dia Min	Maximum Boundary Major Dia Min	Root Width = 0.125P	Minimum Boundary Minor Dia Max	Minimum Boundary Major Dia Max	Crest Width U
M1.6x0.35	0.085	1.221	1.600	0.044	1.321	1.736	0.096
M2x0.4	0.090	1.567	2.000	0.050	1.679	2.148	0.113
M2.5x0.45	0.095	2.013	2.500	0.056	2.138	2.660	0.130
M3x0.5	0.100	2.459	3.000	0.062	2.599	3.172	0.148
M3.5x0.6	0.112	2.850	3.500	0.075	3.010	3.699	0.178
M4x0.7	0.118	3.242	4.000	0.088	3.422	4.219	0.211
M5x0.8	0.125	4.134	5.000	0.100	4.334	5.240	0.243
M6x1	0.150	4.917	6.000	0.125	5.153	6.294	0.300
M8x1.25	0.160	6.647	8.000	0.156	6.912	8.340	0.373
M10x1.5	0.180	8.376	10.000	0.188	8.676	10.397	0.444
M12x1.75	0.200	10.106	12.000	0.219	10.441	12.453	0.515
M14x2	0.212	11.835	14.000	0.250	12.210	14.501	0.594
M16x2	0.212	13.835	16.000	0.250	14.210	16.501	0.594
M20x2.5	0.224	17.294	20.000	0.312	17.744	20.585	0.755
M24x3	0.265	20.752	24.000	0.375	21.252	24.698	0.886
M30x3.5	0.280	26.211	30.000	0.438	26.771	30.785	1.037
M36x4	0.300	31.670	36.000	0.500	32.270	36.877	1.173
M42x4.5	0.315	37.129	42.000	0.562	37.799	42.965	1.330
M48x5	0.335	42.587	48.000	0.625	43.297	49.057	1.466
M56x5.5	0.355	50.046	56.000	0.688	50.796	57.149	1.603
M64x6	0.375	57.505	64.000	0.750	58.305	65.241	1.745
M72x6	0.375	65.505	72.000	0.750	66.305	73.241	1.745
M80x6	0.375	73.505	80.000	0.750	74.305	81.241	1.745
M90x6	0.375	83.505	90.000	0.750	84.305	91.241	1.745
M100x6	0.400	93.505	100.000	0.750	94.305	101.266	1.731

NOTE: All dimensions are in millimeters.

SCREW THREADS

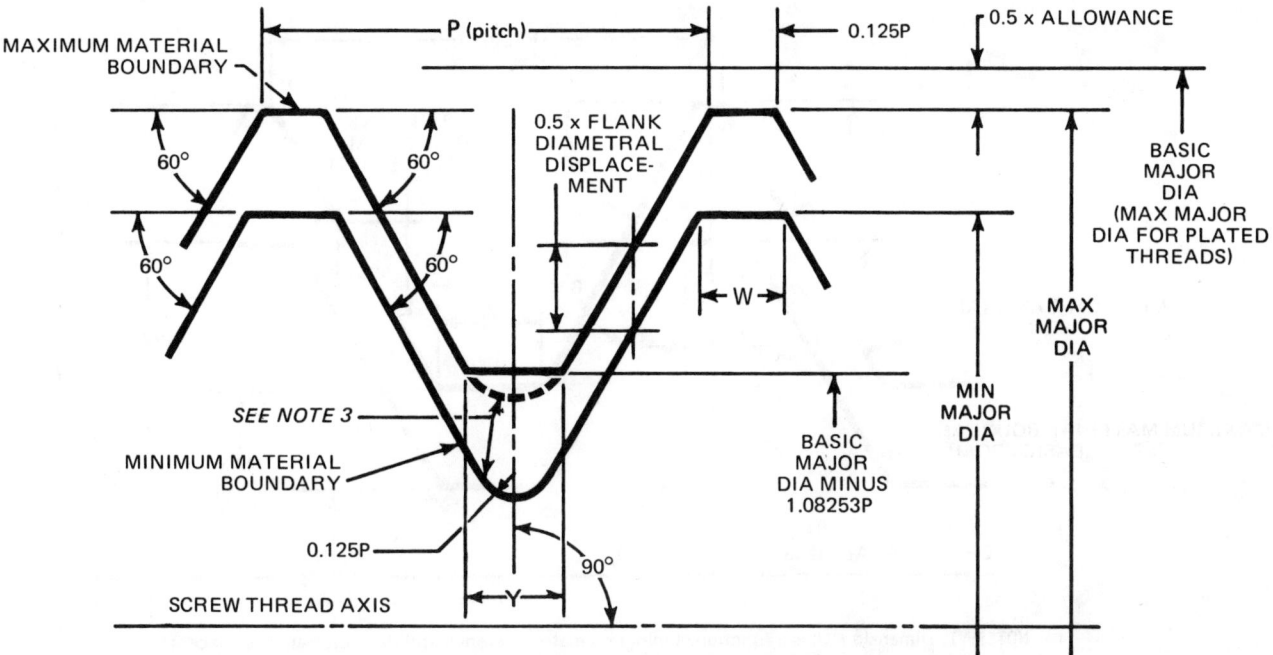

NOTES:
1. Dimension 'W' is a function of major diameter tolerance and flank diametral displacement.
2. Dimension 'Y' is the crest width at the minor diameter of a GO thread ring gage with flank set for tolerance position 'g', and with minor diameter at basic.
3. The thread root of property class 8.8 and higher strength externally threaded fasteners shall have a non-reversing curvature, no portion of which shall have a radius less than 0.125P, and blend tangentially into the flanks and any flat portion if present. The maximum root radius is limited by the boundary profiles. The thread root of lower strength externally threaded fasteners shall preferably have a non-reversing curvature, no portion of which shall have a radius less than 0.125P, however, a flat root is optional if permitted by the purchaser.
4. See page 8-29 for formulas.

Fig. 8-5. Boundary profiles for gaging external threads (IFI-500).

Designations for Coated Threads

Unless the basic designation is qualified to indicate otherwise, the allowance on the external thread may be used to accommodate the coating or plating thickness on coated or plated threads; i.e., the thread after coating or plating is subject to acceptance using a basic size GO thread ring gage. When the allowance must be retained on coated or plated external threads the basic designation must be followed by "after plating."

Examples: M24 × 3 AFTER PLATING and M10 × 1.5C AFTER PLATING—Threads having an allowance after plating or coating.

Boundary Profiles for Gaging

The following principle establishes boundary profiles for gaging. The acceptability of product threads is based entirely on gages conforming to these boundary profiles.

Definitions—Boundary profiles for gaging establish the boundaries of the gaging system which determines product acceptance. The maximum boundary profile for gaging establishes the maximum boundary for the gaging system (GO gages). The minimum boundary profile for gaging establishes the minimum boundary for the gaging system (NOT GO gages).

Construction—The boundary profiles are derived from the basic thread profile.

For gaging external threads the boundary profile is constructed as illustrated in Fig. 8-5, using the values given in Tables 8-3 and 8-4, as applicable. Unless otherwise specified, the root of the external thread shall have a non-reversing curvature, no portion of which shall have a radius smaller than 0.125 P. The maximum root radius is limited by the boundary profiles.

The boundary profiles for gaging internal threads shall be constructed as illustrated in Fig. 8-6, using the values given in Table 8-5:

The derivations of values appearing in Tables 8-3 through 8-5 are given later under "Formulas for the Derivation of Dimensions." (See page 8-29.)

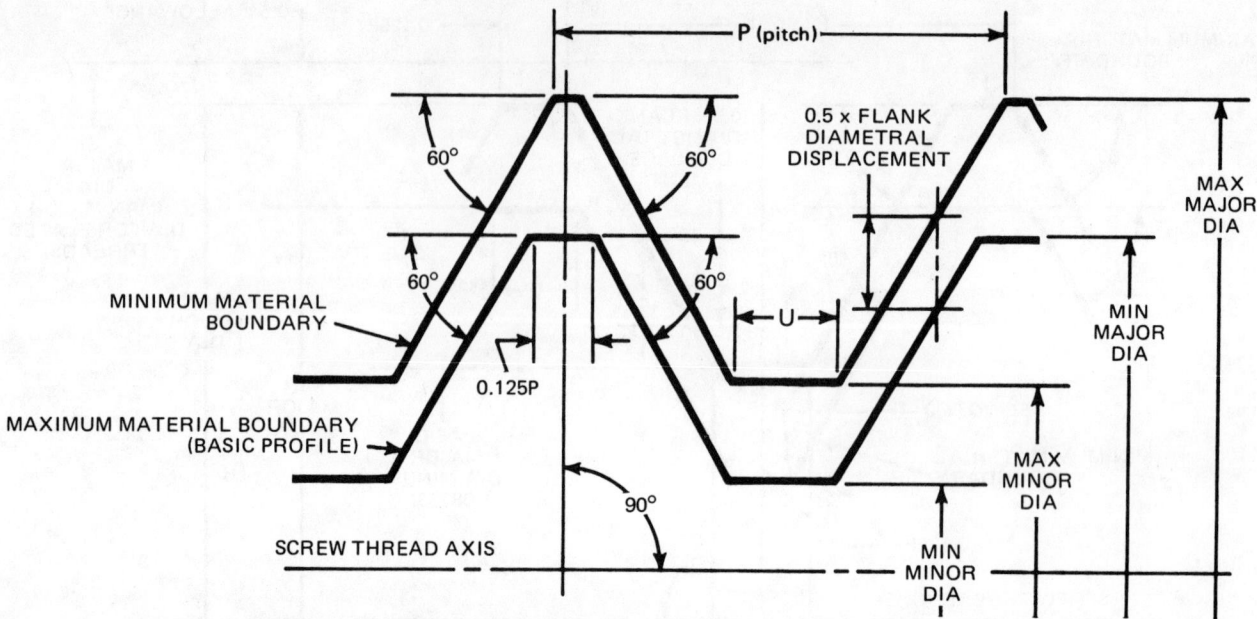

NOTES: 1. Dimension 'U' is a function of minor diameter tolerance and flank diametral displacement.
2. See page 8-29 for formulas.

Fig. 8-6. Boundary profiles for gaging internal threads (IFI-500).

Product Thread Acceptability

General Inspection—Any system conforming to the boundary profiles for gaging may be used to inspect product threads.

Referee Inspection—For referee purposes, acceptability is based on conformance to the criteria specified in the following paragraphs on external and internal threads. These criteria apply only to complete (full form) threads. Criteria for incomplete threads are specified in fastener product standards.

External Threads—Criteria

Acceptability at maximum material of the thread flanks of the product shall be based on gaging with adjustable GO thread ring gages. The GO thread ring gage shall assemble freely over the entire threaded length of the product.

For products of sizes 1.6 through 4 mm acceptability at minimum material of the thread flanks of the product shall be based on gaging with Type B (Fig. 8-9) NOT GO thread snap gage. The gage shall be applied to the product thread at various points around the circumference and over the entire length of complete threads. The gage shall not pass over the product thread in any position, except for the permitted length of incomplete threads.

For products of sizes 5 mm and larger, acceptability at minimum material of the thread flanks of the product shall be based on gaging with Type B (Fig. 8-9) and Type C (Fig. 8-10) NOT GO thread snap gages. The two gages shall be applied to the product thread at various points around the circumference and over the entire length of complete threads. Neither of the two NOT GO gages shall pass over the product thread in any position, except for the permitted length of incomplete threads.

Acceptability at the maximum major diameter of the product thread shall be based on gaging with a solid cylindrical GO ring gage. The GO ring gage shall pass over the entire threaded length of the product.

Acceptability at the minimum major diameter of the product thread shall be based on measuring with a micrometer or other direct measuring device. When measured, the major diameter shall not be less than the major diameter of the minimum boundary profile (Tables 8-3 and 8-4), except for the permitted length of incomplete thread.

Acceptability of the contour of the thread at the minor diameter of fasteners of property class 8.8 and higher strengths shall be based on inspecting by optical projection of 100X for thread pitches 0.35 to 0.6 mm inclusive, 50X for thread pitches 0.7 to 1.5 mm inclusive, and 20X for thread pitches 1.75 mm and coarser. Unless otherwise specified, the contour shall have a non-reversing curvature, no portion of which shall have a radius smaller than 0.125P. Contour of the thread at the minor diameter of fasteners of lower strengths than property class 8.8 need not conform to the root radius requirements.

SCREW THREADS

Table 8-6. Nominal Gage Dimensions (IFI-500)

Basic Major Dia and Thread Pitch (P)	S	X	V General Purpose	V Close Tolerance	P-W General Purpose	P-W Close Tolerance
M5x0.8	0.172	0.100	0.255	0.243	0.668	0.657
M6x1	0.212	0.100	0.315	0.302	0.836	0.823
M8x1.25	0.249	0.125	0.381	0.367	1.039	1.026
M10x1.5	0.291	0.150	0.451	0.436	1.252	1.237
M12x1.75	0.334	0.175	0.524	0.506	1.465	1.446
M14x2	0.372	0.200	0.592	0.572	1.681	1.661
M16x2	0.372	0.200	0.592	0.572	1.681	1.661
M20x2.5	0.442	0.250	0.723	0.701	2.092	2.070
M24x3	0.528	0.300	0.865	0.842	2.524	2.501
M30x3.5	0.599	0.350	0.997	0.973	2.940	2.915
M36x4	0.673	0.400	1.129	1.104	3.355	3.330
M42x4.5	0.744	0.450	1.261	1.235	3.785	3.759
M48x5	0.818	0.500	1.394	1.365	4.213	4.184
M56x5.5	0.892	0.550	1.528	1.497	4.642	4.612
M64x6	0.967	0.600	1.662	1.629	5.065	5.033
M72x6	0.967	0.600	1.662	1.629	5.065	5.033
M80x6	0.967	0.600	1.662	1.629	5.065	5.033
M90x6	0.967	0.600	1.662	1.629	5.065	5.033
M100x6	0.981	0.600	1.673	1.636	5.077	5.040

NOTES: All dimensions are in millimeters.
Gage tolerances are detailed in a separate document.

Internal Thread Criteria

Acceptability at maximum material of the thread flanks of the product shall be based on gaging with GO thread plug gages. The entire thread length of the product shall freely assemble with the GO thread plug gage. For products of sizes 1.6 through 4 mm, acceptability at minimum material of the thread flanks of the product shall be based on gaging with a Type C (Fig. 8-12) NOT GO thread plug gage. The gage shall not enter the product thread, except for the permitted length of incomplete threads.

For products of sizes 5 mm and larger, acceptability at minimum material of the thread flanks of the product shall be based on gaging with Type B (Fig. 8-11) and Type C (Fig. 8-12) NOT GO thread plug gages. Neither of the two gages shall enter the product thread, except for the permitted length of incomplete threads. Acceptability at minor diameter of the product thread shall be based on gaging with cylindrical GO and NOT GO plug gages. The GO plug gage shall enter the product for the entire threaded length and the NOT GO plug gage shall not enter the threads of the product except for the permitted length of incomplete thread.

Thread Lead and Flank Angles

Thread lead and flank angles and deviations of form of external and internal threads do not require inspection. When product threads are acceptable to both maximum and minimum boundary gages, as specified in external and internal thread criteria, the lead and flank angles and derivations of form of product thread are acceptable.

Gage Design

The design of gages is specified only to the extent of those portions which must conform to the boundary profiles for gaging. This includes the following gages:

For External Threads

a. GO thread ring gage, as shown in Fig. 8-7.
b. Basic GO thread ring gage.
c. NOT GO thread cone and vee snap gage, Type B, as shown in Fig. 8-9.
d. NOT GO thread cone and vee snap gage, Type C, as shown in Fig. 8-10.
e. GO plain ring gages.
f. Micrometer or equivalent direct measurement device.
g. Optical comparator.

For Internal Threads

a. GO thread plug gage, as shown in Fig. 8-8.
b. NOT GO thread plug gage, Type B, as shown in Fig. 8-11.
c. NOT GO thread plug gage, Type C, as shown in Fig. 8-12.
d. GO and NOT GO plain plug gages.

Gage Thread Profile Dimensions

Dimensions necessary for establishing the gage thread profiles are given in Table 8-3 through 8-6. Derivations of values in Tables 8-3 through 8-6 are outlined later in "Formulas for the Derivation of Dimensions" (see page 8-29).

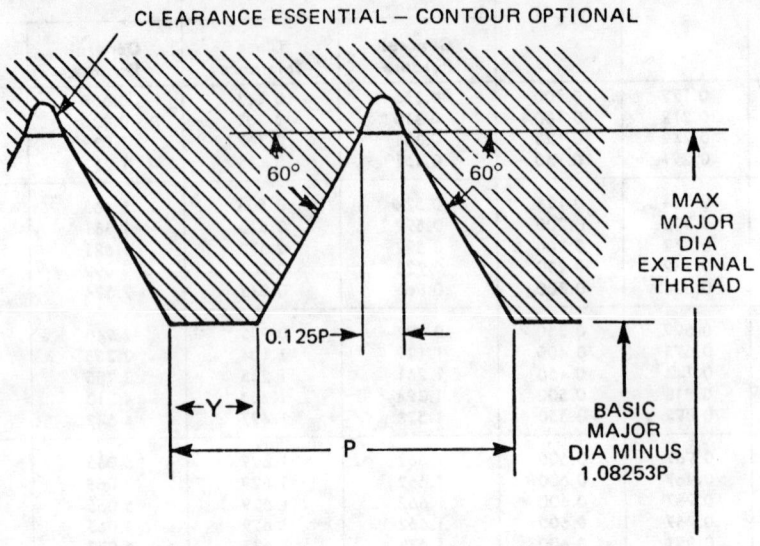

Fig. 8-7. Thread profile for go thread ring gage (IFI-500).

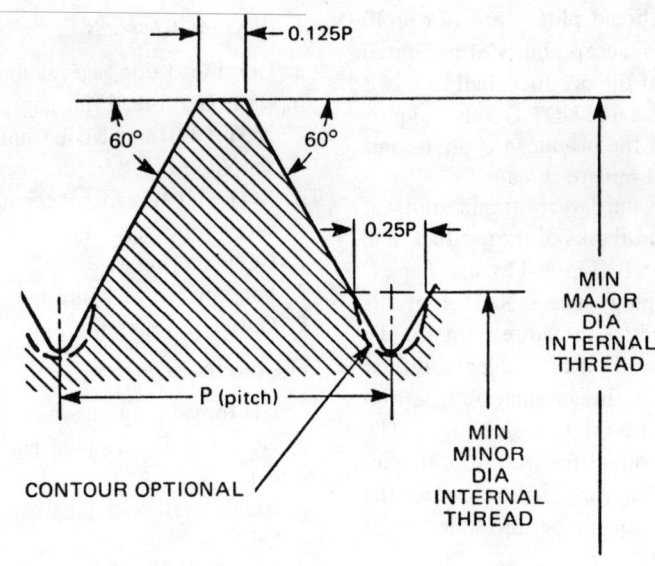

Fig. 8-8. Thread profile for go thread plug gage (IFI-500).

Designs of standard gage blanks for thread ring and plug gages and plain cylindrical ring and plug gages are outlined in ANSI B47.1.

Tolerances—Gage tolerances shall be positioned within the specified maximum and minimum boundary profiles for gaging.

SCREW THREADS

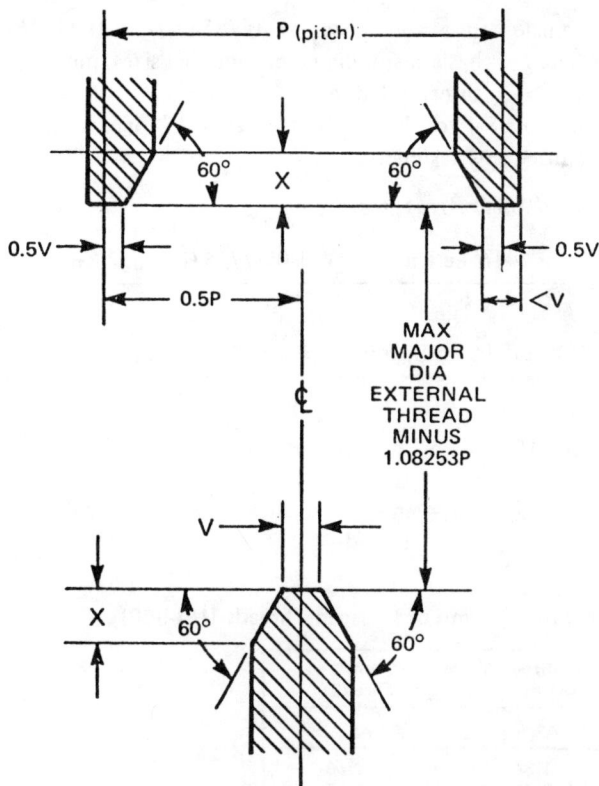

NOTE: Dimension 'V' is a function of minimum boundary profile.

Fig. 8-9. Thread profile for type B NOT GO cone and Vee snap gage (IFI-500).

FORMULAS FOR DERIVATION OF DIMENSIONS

The values given in Tables 8-3, 8-4, 8-5, and 8-6 are derived as follows:

1. Basic major diameter and thread pitch series = selected from ISO 261 and ISO 262.
2. Allowance (external thread) = in accordance with ISO 965/I, Table 1, fundamental deviation 'g' for sizes 5 mm, and larger. For sizes 1.6 through 4 mm allowance is 0.024 mm.
3. Major diameter, max (external thread) = basic major diameter minus allowance for non-plated threads, and basic major diameter for plated threads.
4. Major diameter, min (external thread) = max major diameter minus tolerance grade 6 for major diameters, as given in ISO 965/I, Table 4.
5. Minor diameter, min (internal thread) = basic major diameter minus 1.08253P.
6. Minor diameter, max (internal thread) = min minor diameter plus tolerance grade 6 for minor diameters, as given in ISO 965/I, Table 3.
7. Major diameter, min (internal thread) = basic major diameter.
8. Major diameter, max (internal thread) = basic major diameter plus 0.14434P plus tolerance grade 6 for pitch diameters as given in ISO 965/I, Table 5.
9. Flank diametral displacement (external thread) = tolerance grade 6 or 5, as applicable, for pitch diameters, as given in ISO 965/I, Table 6.
10. Flank diametral displacement (internal thread) = tolerance grade 6 for pitch diameters, as given in ISO 965/I, Table 5.
11. Root radius, min (external thread) = 0.125P.
12. W, crest width of min material boundary profile, external thread = 0.125P + 0.57735 (major diameter tolerance, external thread minus flank diametral displacement, external thread). (Figs. 8-5, 8-10.)

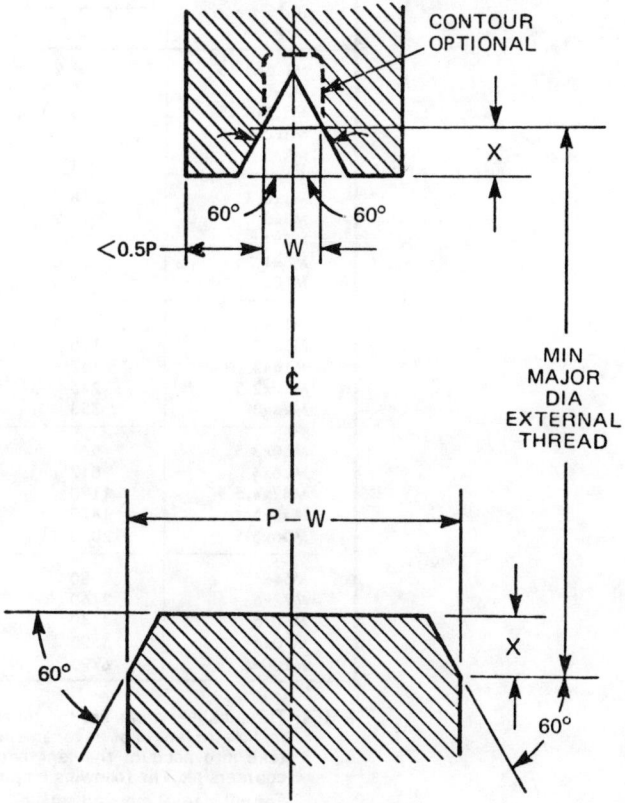

NOTE: Dimension 'W' is a function of minimum boundary profile.

Fig. 8-10. Thread profile for type C NOT GO cone and Vee snap gage (IFI-500).

13. *U*, crest width of min boundary profile, internal thread = 0.25P − 0.57735 (minor diameter tolerance internal thread minus flank diametral displacement, internal thread) (Figs. 8-6, 8-7, 8-12.)
14. *V*, gage width for Type B NOT GO cone and vee snap gage, external thread = 0.25P plus 0.57735 (flank diametral displacement, external thread). (Fig. 8-9.)
15. *S*, gage width for Type B NOT GO thread plug gage, internal thread = 0.125P + 0.57735 (flank diametral displacement, external thread). (Fig. 8-11).
16. *X*, gage flank height = 0.100 mm for sizes M5 and M6, and 0.10*P* for sizes 8 mm and larger. (Figs. 8-9, 8-10, 8-11, and 8-12.)
17. *Y*, crest width of GO thread ring gage, external thread = 0.250*P* + 0.57735 (allowance, external thread) for non-plated threads, and 0.250*P* for plated threads.

Thread Strength Data

Tensile stress area = $A_s = 0.7854 (D - 0.9382P)^2$
where D = basic major diameter (nominal size), mm
P = thread pitch, mm

See Table 8-1 for the tensile stress area AS for existing and modified ISO fasteners.

Thread Shear Areas (see Table 8-7):

$$AS_s = \frac{3.1416 \, Le \, Kn \, max \, [W + 0.57735 \, (Ds \, min - Kn \, max)]}{P}$$

$$AS_n = \frac{3.1416 \, Le \, Ds \, min \, [U + 0.57735 \, (Ds \, min - Kn \, max)]}{P}$$

where AS_s = min thread shear area for external threads
AS_n = min thread shear area for internal threads

Table 8-7. Thread Tensile Stress Shear Areas (mm²) per mm Length of Engaged Threads (IFI-500)

Basic Major Dia and Thread Pitch	Tensile Stress Area mm² A_s	Thread Shear Area, mm² per mm of Engaged Threads AS_s	AS_n
M1.6x0.35	1.27	1.84	2.60
M2x0.4	2.07	2.41	3.37
M2.5x0.45	3.39	3.14	4.44
M3x0.5	5.03	3.86	5.49
M3.5x0.6	6.78	4.65	6.58
M4x0.7	8.78	5.45	7.75
M5x0.8	14.2	7.08	9.99
M6x1	20.1	8.86	12.2
M8x1.25	36.6	12.2	16.8
M10x1.5	58.0	15.6	21.5
M12x1.75	84.3	19.0	26.1
M14x2	115	22.4	31.0
M16x2	157	26.1	35.6
M20x2.5	245	33.3	45.4
M24x3	353	40.5	55.0
M30x3.5	561	51.6	69.6
M36x4	817	63.1	84.1
M42x4.5	1120	74.3	99.2
M48x5	1470	85.8	114
M56x5.5	2030	101	134
M64x6	2680	117	154
M72x6	3460	133	173
M80x6	4340	149	193
M90x6	5590	169	217
M100x6	6990	189	241

NOTES: Thread shear areas are for general purpose threads and were computed per millimetre length of engagement. For nuts the length of engagement should take into account the length of incomplete thread and the influence of countersink. The following empirically determined formulas apply:

For nut sizes 8 mm and smaller —
Le = Nut height − 0.42 (D + 0.5 − Kn min)

For nut sizes 10 mm and larger —
Le = Nut height − 0.42 (1.06D − Kn min)

SCREW THREADS

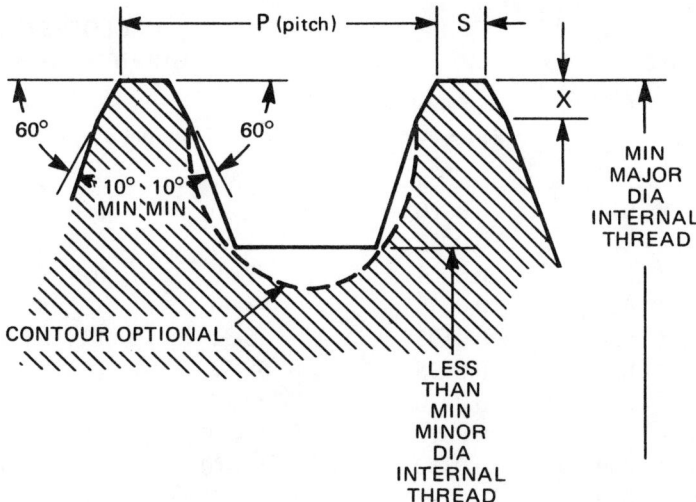

NOTE: Dimension 'S' is a function of minimum boundary profile.

Fig. 8-11. Thread profile for type B Not GO thread plug gage (IFI-500).

D = basic major diameter (Table 8-3)
Le = length of thread engagement
$K_n$max = max minor diameter of internal thread (Table 8-5)
$D_s$min = min major diameter of external thread (Tables 8-3 and 8-4)
P = thread pitch (Table 8-3)
W = crest width of minimum boundary profile for gaging, external threads (Tables 8-3 and 8-4)
U = crest width of minimum boundary profile for gaging, internal thread (Table 8-5)

Conversion Data

A comparison of diameters, pitches, and tensile stress areas for metric fasteners with customary inch sizes is shown in Table 8-8. The metric fastener sizes have the diameters and pitches converted to inches, and the closest customary inch thread sizes have their threads per inch converted to a pitch in decimal inches.

Trapezoidal Screw Threads

The ISO Technical Committee (TC1) has developed four International Standards ISO 2901, 2902, 2903, and 2904 on metric module trapezoidal screw threads. The material in these international and national standards from other major industrial countries serves as a basis for the material in this sub-section.

Trapezoidal screw threads are used for lead spindles in machine tool design, and the ISO general plan of diameter-pitch combinations with references to other national standards should prove to be useful when customary inch module machines are converted or new metric module lead spindles are produced. The above standards drafts could be subject to some future changes in the tolerancing area; however, they are useful in their present form.

Basic Trapezoidal Thread Profile—The basic trapezoidal thread profile is specified in the ISO 2901 standard and shown in Fig. 8-13. The tolerances and deviations for external and internal trapezoidal screw threads are applied to the basic sizes, for which the numerical values are listed in Table 8-9.

World Metric Trapezoidal Screw Threads Standards—Table 8-9 lists the general dimensions for the ISO metric trapezoidal screw threads, and at the same time, relates the ISO international standards to the national standards in some major industrial countries.

The ISO 2902 specifies the nominal diameters and pitches for metric module trapezoidal screw threads, and the table is based on the material presented in the above standard. The dimensions shown in the table are valid for the national standard in the U.K., and are slightly different for the nut crest clearances and therefore also for the minor nut diameters in the German and Italian standards. The basic thread profile is identical, however, and trapezoidal screw threads produced to one standard should interchange with those produced to the other national standards listed.

Table 8-8. Comparison of Metric Screw Threads with Customary Inch Sizes

METRIC SCREW THREAD SIZE			CLOSEST UNIFIED INCH SIZE	
Nominal Size-Pitch mm	Nominal Size-Pitch in.	Tensile Stress Area in^2	Nominal Size-Pitch in.	Tensile Stress Area in^2
M 1.6 × 0.35	.063 × .014	0.0020	#0-80	.060 × .013	0.0018
M 2 × 0.4	.079 × .016	0.0032	#1-64	.073 × .016	0.0026
M 2.5 × 0.45	.098 × .018	0.0053	#3-48	.099 × .021	0.0049
M 3 × 0.5	.118 × .020	0.0078	#5-40	.125 × .025	0.0081
M 3.5 × 0.6	.138 × .024	0.0105	#6-32	.138 × .031	0.0093
M 4 × 0.7	.157 × .028	0.0136	#8-32	.164 × .031	0.0143
M 5 × 0.8	.197 × .031	0.0220	#10-24	.190 × .042	0.0178
M 6 × 1	.236 × .039	0.0312	1/4-20	.250 × .050	0.0324
M 6.3 × 1	.248 × .039	0.0350	1/4-20	.250 × .050	0.0324
M 8 × 1.25	.315 × .049	0.0567	5/16-18	.313 × .056	0.0532
M 10 × 1.5	.394 × .059	0.0899	3/8-16	.375 × .063	0.0786
M 12 × 1.75	.472 × .069	0.131	1/2-13	.500 × .077	0.1438
M 14 × 2	.551 × .079	0.178	9/16-12	.562 × .083	0.184
M 16 × 2	.630 × .079	0.243	5/8-11	.625 × .091	0.229
M 20 × 2.5	.787 × .098	0.380	3/4-10	.750 × .100	0.338
M 24 × 3	.945 × .118	0.547	1-8	1.000 × .125	0.612
M 30 × 3.5	1.181 × .138	0.870	1-3/16	1.188	
M 36 × 4	1.417 × .157	1.27	1-7/16	1.438	
M 42 × 4.5	1.654 × .177	1.74	1-5/8	1.625	
M 48 × 5	1.890 × .197	2.28	1-7/8	1.875	
M 56 × 5.5	2.205 × .217	3.15	2-1/4	2.250	
M 64 × 6	2.520 × .236	4.15	2-1/2	2.500	
M 72 × 6	2.835 × .236	5.36	2-7/8	2.875	
M 80 × 6	3.150 × .236	6.73	3-1/8	3.125	
M 90 × 6	3.543 × .236	8.66	3-1/2	3.500	
M 100 × 6	3.937 × .236	10.83	4	4.000	

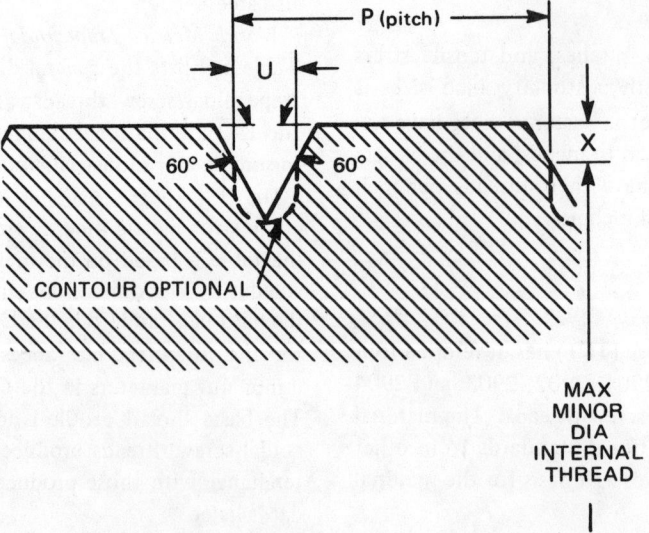

Fig. 8-12. Thread profile for type C Not GO thread plug gage (IFI-500).

SCREW THREADS
8-33

Table 8-9. World Standards for Metric Trapezoidal Screw Threads (ISO 2901, 2902, 2903, and 2904)

NOM. SIZE d	PITCH P		PITCH DIA d2=D2	MINOR DIAMETER d3	MINOR DIAMETER D1	MAJOR DIA D4	RADIUS AC	RADIUS R1MAX	RADIUS R2MAX	THREAD HEIGHT H4=h3	STR AREA AT MINER ϕ (MM**2)	U.S. ANSI	AUSTRAL SAA	JAPAN JISC	FRANCE AFNOR E03-616	U.K. BSI DRAFT	GERMANY DNA DIN103	ITALY UNI 124-126
F 8	1.5	NORMAL	7.25	6.2	6.5	8.3	0.15	0.075	0.15	0.90	0.302E+02				F	F	F	F
S 9	2	NORMAL	8.00	6.5	7.0	9.5	0.25	0.125	0.25	1.25	0.333E+02				S	S	S	
	1.5	FINE	8.25	7.2	7.5	9.3	0.15	0.075	0.15	0.90	0.407E+02					S	S	
F 10	2	NORMAL	9.00	7.5	8.0	10.5	0.25	0.125	0.25	1.25	0.442E+02				F	F	F	F
	1.5	FINE	9.25	8.2	8.5	10.3	0.15	0.075	0.15	0.90	0.528E+02				S	F	F	F
S 11	3	COARSE	9.50	7.5	8.0	11.5	0.25	0.125	0.25	1.75	0.442E+02				1.5T	S	S	
	2	NORMAL	10.00	8.5	9.0	11.5	0.25	0.125	0.25	1.25	0.567E+02				S	S	S	
F 12	3	NORMAL	10.50	8.5	9.0	12.5	0.25	0.125	0.25	1.75	0.567E+02				1.5S	F	F	F
	2	FINE	11.00	9.5	10.0	12.5	0.25	0.125	0.25	1.25	0.709E+02				F	F	F	F
S 14	3	NORMAL	12.50	10.5	11.0	14.5	0.25	0.125	0.25	1.75	0.866E+02				1.5T	S	S	
	2	FINE	13.00	11.5	12.0	14.5	0.25	0.125	0.25	1.25	0.104E+03				S	S.	S	F
F 16	4	NORMAL	14.00	11.5	12.0	16.5	0.25	0.125	0.25	2.25	0.104E+03				3F	F	F	F
	2	FINE	15.00	13.5	14.0	16.5	0.25	0.125	0.25	1.25	0.143E+03				S	F	F	F
S 18	4	NORMAL	16.00	13.5	14.0	18.5	0.25	0.125	0.25	2.25	0.143E+03				3S	S	S	F
	2	FINE	17.00	15.5	16.0	18.5	0.25	0.125	0.25	1.25	0.189E+03				T	S	S	F
F 20	4	NORMAL	18.00	15.5	16.0	20.5	0.25	0.125	0.25	2.25	0.189E+03				S,3F	F	F	F
	2	FINE	19.00	17.5	18.0	20.5	0.25	0.125	0.25	1.25	0.241E+03				S	F	F	F
S 22	8	COARSE	18.00	13.0	14.0	23.0	0.50	0.250	0.50	4.50	0.133E+03				4T	S	S	S
	5	NORMAL	19.50	16.5	17.0	22.5	0.25	0.125	0.25	2.75	0.214E+03				2T	S	S	S
	3	FINE	20.50	18.5	19.0	22.5	0.25	0.125	0.25	1.75	0.269E+03				S	S	S	S
F 24	8	COARSE	20.00	15.0	16.0	25.0	0.50	0.250	0.50	4.50	0.177E+03				4F	F	F	
	5	NORMAL	21.50	18.5	19.0	24.5	0.25	0.125	0.25	2.75	0.269E+03				S	F	F	
	3	FINE	22.50	20.5	21.0	24.5	0.25	0.125	0.25	1.75	0.330E+03				S	F	F	
S 26	8	COARSE	22.00	17.0	18.0	27.0	0.50	0.250	0.50	4.50	0.227E+03					S	S	
	5	NORMAL	23.50	20.5	21.0	26.5	0.25	0.125	0.25	2.75	0.330E+03					S	S	
	3	FINE	24.50	22.5	23.0	26.5	0.25	0.125	0.25	1.75	0.398E+03					S	S	
F 28	8	COARSE	24.00	19.0	20.0	29.0	0.50	0.250	0.50	4.50	0.284E+03				4S	S	S	S
	5	NORMAL	25.50	22.5	23.0	28.5	0.25	0.125	0.25	2.75	0.398E+03				T	F	F	F
	3	FINE	26.50	24.5	25.0	28.5	0.25	0.125	0.25	1.75	0.471E+03				T	F	F	F
S 30	10	COARSE	25.00	19.0	20.0	31.0	0.50	0.250	0.50	5.50	0.284E+03				5F	S	S	S
	6	NORMAL	27.00	23.0	24.0	31.0	0.50	0.250	0.50	3.50	0.415E+03					F	F	F
	3	FINE	28.50	26.5	27.0	30.5	0.25	0.125	0.25	1.75	0.552E+03					S	S	S
F 32	10	COARSE	27.00	21.0	22.0	33.0	0.50	0.250	0.50	5.50	0.346E+03				5F	F	F	F
	6	NORMAL	29.00	25.0	26.0	33.0	0.50	0.250	0.50	3.50	0.491E+03					F	F	F
	3	FINE	30.50	28.5	29.0	32.5	0.25	0.125	0.25	1.75	0.638E+03				4S	S	S	S

NOTES:
1. ALL DIMENSIONS ARE IN MILLIMETERS
2. SEE FIGURE 8-14 AND PAGE 8-39 FOR KEY TO SYMBOLS
3. THE NOMINAL SIZE IS NATIONAL STANDARD AS INDICATED
    F = FIRST CHOICE, S = SECOND CHOICE, T = THIRD CHOICE

8-34 SCREW THREADS

Table 8-9. *(Continued)* World Standards for Metric Trapezoidal Screw Threads (ISO 2901, 2902, 2903, and 2904)

I NOM. S SIZE PITCH O d P	PITCH DIA $d_2=D_2$	MINOR DIAMETER d_3	MAJOR DIA D_4	AC	RADIUS R1MAX R2MAX	THREAD HEIGHT AT MINOR $H_4=h_3$	STR AREA ϕ(MM**2)	U.S. ANSI	AUSTRAL SAA	JAPAN JISC	FRANCE AFNOR E03-616	U.K. BSI DRAFT	GERMANY DNA DIN103	ITALY UNI 124-126
S 34 10 COARSE 6 NORMAL 3 FINE	29.00 31.00 32.50	23.0 27.0 30.5	35.0 35.0 34.5	0.50 0.50 0.25	0.250 0.50 0.250 0.50 0.125 0.25	5.50 3.50 1.75	0.415E+03 0.573E+03 0.731E+03					S S S	S S S	S S
F 36 10 COARSE 6 NORMAL 3 FINE	31.00 33.00 34.50	25.0 29.0 32.5	37.0 37.0 36.5	0.50 0.50 0.25	0.250 0.50 0.250 0.50 0.125 0.25	5.50 3.50 1.75	0.491E+03 0.661E+03 0.830E+03				5S T 4T	F F F	F F F	F F
S 38 10 COARSE 7 NORMAL 3 FINE	33.00 34.50 36.50	27.0 30.0 34.5	39.0 39.0 38.5	0.50 0.50 0.25	0.250 0.50 0.250 0.50 0.125 0.25	5.50 4.00 1.75	0.573E+03 0.707E+03 0.935E+03					S S S	S S S	S S
F 40 10 COARSE 7 NORMAL 3 FINE	35.00 36.50 38.50	29.0 32.5 36.5	41.0 41.0 40.5	0.50 0.50 0.25	0.250 0.50 0.250 0.50 0.125 0.25	5.50 4.00 1.75	0.661E+03 0.804E+03 0.105E+04				8S 6F 4S	F F F	F F F	F F
S 42 10 COARSE 7 NORMAL 3 FINE	37.00 38.50 40.50	31.0 34.0 38.5	43.0 43.0 42.5	0.50 0.50 0.25	0.250 0.50 0.250 0.50 0.125 0.25	5.50 4.00 1.75	0.755E+03 0.908E+03 0.116E+04					S S S	S S S	S S
F 44 12 COARSE 7 NORMAL 3 FINE	38.00 40.50 42.50	31.0 36.0 40.5	45.0 45.0 44.5	0.50 0.50 0.25	0.250 0.50 0.250 0.50 0.125 0.25	6.50 4.00 1.75	0.755E+03 0.102E+04 0.129E+04					F F F	F F F	F F
S 46 12 COARSE 8 NORMAL 3 FINE	40.00 42.00 44.50	33.0 37.0 42.5	47.0 47.0 46.5	0.50 0.50 0.25	0.250 0.50 0.250 0.50 0.125 0.25	6.50 4.50 1.75	0.855E+03 0.108E+04 0.142E+04				45X8T 45X6S 45X4T	S S S	S S S	S S S
F 48 12 COARSE 8 NORMAL 3 FINE	42.00 44.00 46.50	35.0 39.0 44.5	49.0 49.0 48.5	0.50 0.50 0.25	0.250 0.50 0.250 0.50 0.125 0.25	6.50 4.50 1.75	0.962E+03 0.119E+04 0.156E+04					F F F	F F F	F F
S 50 12 COARSE 8 NORMAL 3 FINE	44.00 46.00 48.50	37.0 41.0 46.5	51.0 51.0 50.5	0.50 0.50 0.25	0.250 0.50 0.250 0.50 0.125 0.25	6.50 4.50 1.75	0.108E+04 0.132E+04 0.170E+04				10S F 5S	S S S	S S S	S S
F 52 12 COARSE 8 NORMAL 3 FINE	46.00 48.00 50.50	39.0 43.0 48.5	53.0 53.0 52.5	0.50 0.50 0.25	0.250 0.50 0.250 0.50 0.125 0.25	6.50 4.50 1.75	0.119E+04 0.145E+04 0.185E+04					F F F	F F F	F F
S 55 14 COARSE 9 NORMAL 3 FINE	48.00 50.50 53.50	41.0 45.0 51.5	57.0 56.0 55.5	1.00 0.50 0.25	0.500 1.00 0.250 0.50 0.125 0.25	8.00 5.00 1.75	0.119E+04 0.159E+04 0.208E+04				56X10T 56X8S 56X5T	S S S	S S S	S S
F 60 14 COARSE 9 NORMAL 3 FINE	53.00 55.50 58.50	44.0 50.0 56.5	62.0 61.0 60.5	1.00 0.50 0.25	0.500 1.00 0.250 0.50 0.125 0.25	8.00 5.00 1.75	0.152E+04 0.196E+04 0.251E+04					F F F	F F F	F F
S 65 16 COARSE 10 NORMAL 4 FINE	57.00 60.00 63.00	47.0 54.0 60.5	67.0 66.0 65.5	1.00 0.50 0.50	0.500 1.00 0.250 0.50 0.125 0.25	9.00 5.50 2.25	0.173E+04 0.229E+04 0.287E+04				63X12S 63X8F 63X5S	S S S	S S S	S S

NOTES:
1. ALL DIMENSIONS ARE IN MILLIMETERS
2. SEE FIGURE 8-14 AND PAGE 8-39 FOR KEY TO SYMBOLS
3. THE NOMINAL SIZE IS NATIONAL STANDARD AS INDICATED
F = FIRST CHOICE, S = SECOND CHOICE, T = THIRD CHOICE

SCREW THREADS

Table 8-9. (Continued) World Standards for Metric Trapezoidal Screw Threads (ISO 2901, 2902, 2903, and 2904)

ISO SIZE d	PITCH P		PITCH DIA d2=D2	MINOR DIAMETER d3 D1	MAJOR DIA D4	RADIUS AC R1MAX R2MAX	THREAD HEIGHT AT MINOR H4=h3	STR AREA ⌀(MM**2)	FRANCE AFNOR E03-616	U.K. BSI DRAFT	GERMANY DNA DIN103	ITALY UNI 124-126
F	70	16	62.00	52.0 54.0	72.0 71.0 70.5	1.00 0.500 1.00 0.50 0.250 0.50 0.25 0.125 0.25	9.00 5.50 2.25	0.212E+04 0.273E+04 0.337E+04	12T 8S 5T	F F F	F F F	F F F
		10	65.00	59.0	71.0							
		4	68.00	65.5	70.5							
S	75	16	67.00	57.0 59.0	77.0 76.0 75.5	1.00 0.500 1.00 0.50 0.250 0.50 0.25 0.125 0.25	9.00 5.50 2.25	0.255E+04 0.322E+04 0.390E+04		S S S	S S S	S S S
		10	70.00	64.0 65.0	76.0							
		4	73.00	70.5 71.0	75.5							
F	80	16	72.00	62.0 64.0	82.0 81.0 80.5	1.00 0.500 1.00 0.50 0.250 0.50 0.25 0.125 0.25	9.00 5.50 2.25	0.302E+04 0.374E+04 0.448E+04	S F 5S	F F F	F F F	F F F
		10	75.00	69.0 70.0	81.0							
		4	78.00	75.5 76.0	80.5							
S	85	18	76.00	65.0 67.0	87.0 86.0 85.5	1.00 0.500 1.00 0.50 0.250 0.50 0.25 0.125 0.25	10.00 6.50 2.25	0.332E+04 0.407E+04 0.509E+04		S S S	S S S	S S S
		12	79.00	72.0 73.0	86.0							
		4	83.00	80.5 81.0	85.5							
F	90	18	81.00	70.0 72.0	92.0 91.0 90.5	1.00 0.500 1.00 0.50 0.250 0.50 0.25 0.125 0.25	10.00 6.50 2.25	0.385E+04 0.466E+04 0.574E+04	16T 10S 5T	F F F	F F F	F F F
		12	84.00	77.0 78.0	91.0							
		4	88.00	85.5 86.0	90.5							
S	95	18	86.00	75.0 77.0	97.0 96.0 95.5	1.00 0.500 1.00 0.50 0.250 0.50 0.25 0.125 0.25	10.00 6.50 2.25	0.442E+04 0.528E+04 0.643E+04		S S S	S S S	S S S
		12	89.00	82.0 83.0	96.0							
		4	93.00	90.5 91.0	95.5							
F	100	20	90.00	78.0 80.0	102.0 101.0 100.5	1.00 0.500 1.00 0.50 0.250 0.50 0.25 0.125 0.25	11.00 6.50 2.25	0.478E+04 0.594E+04 0.716E+04	S F 6S	F F F	F F F	F F F
		12	94.00	87.0 88.0	101.0							
		4	98.00	95.5 96.0	100.5							
T	105	20	95.00	83.0 85.0	107.0 106.0 105.5	1.00 0.500 1.00 0.50 0.250 0.50 0.25 0.125 0.25	11.00 6.50 2.25	0.541E+04 0.665E+04 0.793E+04		T T T	T T T	T T T
		12	99.00	92.0 93.0	106.0							
		4	103.00	100.5 101.0	105.5							
S	110	20	100.00	88.0 90.0	112.0 111.0 110.5	1.00 0.500 1.00 0.50 0.250 0.50 0.25 0.125 0.25	11.00 6.50 2.25	0.608E+04 0.739E+04 0.874E+04	T S 6T	S S S	S S S	S S S
		12	104.00	97.0 98.0	111.0							
		4	108.00	105.5 106.0	110.5							
T	115	22	104.00	91.0 93.0	117.0 117.0 116.0	1.00 0.500 1.00 0.50 0.250 0.50 0.50 0.250 0.50	12.00 8.00 3.50	0.650E+04 0.770E+04 0.916E+04		T T T	T T T	T T T
		14	108.00	99.0 101.0	117.0							
		6	112.00	108.0 109.0	116.0							
F	120	22	109.00	98.0 103.0	122.0 122.0 121.0	1.00 0.500 1.00 0.50 0.250 0.50 0.50 0.250 0.50	12.00 8.00 3.50	0.724E+04 0.849E+04 0.100E+05		F F F	F F F	F F F
		14	113.00	104.0 106.0	122.0							
		6	117.00	113.0 114.0	121.0							
T	125	22	114.00	101.0 103.0	127.0 127.0 126.0	1.00 0.500 1.00 1.00 0.500 1.00 0.50 0.250 0.50	12.00 8.00 3.50	0.801E+04 0.931E+04 0.109E+05	20S 12F S	T T T	T T T	T T T
		14	118.00	109.0 111.0	127.0							
		6	122.00	118.0 119.0	126.0							
S	130	22	119.00	106.0 108.0	132.0 132.0 131.0	1.00 0.500 1.00 1.00 0.500 1.00 0.50 0.250 0.50	12.00 8.00 3.50	0.882E+04 0.102E+05 0.119E+05		S S S	S S S	S S S
		14	123.00	114.0 116.0	132.0							
		6	127.00	123.0 124.0	131.0							

NOTES:
1. ALL DIMENSIONS ARE IN MILLIMETERS
2. SEE FIGURE 8-14 AND PAGE 8-39 FOR KEY TO SYMBOLS
3. THE NOMINAL SIZE IS NATIONAL STANDARD AS INDICATED
   F = FIRST CHOICE, S = SECOND CHOICE, T = THIRD CHOICE

SCREW THREADS

Table 8-9 (*Continued*). World Standards for Metric Trapezoidal Screw Threads (ISO 2901, 2902, 2903, and 2904)

I NOM. S SIZE PITCH O d P	PITCH DIA d2=D2	MINOR DIAMETER d3 D1	MAJOR DIA D4	RADIUS AC R1MAX R2MAX	THREAD HEIGHT AT MINOR H4=h3 Ø	STR AREA MINER Ø(MM**2)	U.S. ANSI	AUSTRAL SAA	JAPAN JISC	FRANCE AFNOR E03-616	U.K. BSI DRAFT	GERMANY DNA DIN103	ITALY UNI 124-126
T 135 24 COARSE	123.00	109.0 111.0	137.0	1.00 0.500 1.00	13.00	0.933E+04					T	T	
14 NORMAL	128.00	119.0 121.0	137.0	1.00 0.500 1.00	8.00	0.111E+05					T	T	
6 FINE	132.00	128.0 129.0	136.0	0.50 0.250 0.50	3.50	0.129E+05					T	T	
F 140 24 COARSE	128.00	114.0 116.0	142.0	1.00 0.500 1.00	13.00	0.102E+05				20T	F	F	F
14 NORMAL	133.00	124.0 126.0	142.0	1.00 0.500 1.00	8.00	0.121E+05				12S	F	F	F
6 FINE	137.00	133.0 134.0	141.0	0.50 0.250 0.50	3.50	0.139E+05				T	F	F	F
T 145 24 COARSE	133.00	119.0 121.0	147.0	1.00 0.500 1.00	13.00	0.111E+05					T	T	
14 NORMAL	138.00	129.0 131.0	147.0	1.00 0.500 1.00	8.00	0.131E+05					T	T	
6 FINE	142.00	138.0 139.0	146.0	0.50 0.250 0.50	3.50	0.150E+05					T	T	
S 150 24 COARSE	138.00	124.0 126.0	152.0	1.00 0.500 1.00	13.00	0.120E+05					S	S	S
16 NORMAL	142.00	132.0 134.0	152.0	1.00 0.500 1.00	9.00	0.137E+05					S	S	S
6 FINE	147.00	143.0 144.0	151.0	0.50 0.250 0.50	3.50	0.161E+05					S	S	S
T 155 24 COARSE	143.00	129.0 131.0	157.0	1.00 0.500 1.00	13.00	0.131E+05					T	T	
16 NORMAL	147.00	137.0 139.0	157.0	1.00 0.500 1.00	9.00	0.147E+05					T	T	
6 FINE	152.00	148.0 149.0	156.0	0.50 0.250 0.50	3.50	0.172E+05					T	T	
F 160 28 COARSE	146.00	130.0 132.0	162.0	1.00 0.500 1.00	15.00	0.133E+05				24S	F	F	F
16 NORMAL	152.00	142.0 144.0	162.0	1.00 0.500 1.00	9.00	0.158E+05				F	F	F	F
6 FINE	157.00	153.0 154.0	161.0	0.50 0.250 0.50	3.50	0.184E+05				8S	F	F	F
T 165 28 COARSE	151.00	135.0 137.0	167.0	1.00 0.500 1.00	15.00	0.143E+05					T	T	
16 NORMAL	157.00	147.0 149.0	167.0	1.00 0.500 1.00	9.00	0.170E+05					T	T	
6 FINE	162.00	158.0 159.0	166.0	0.50 0.250 0.50	3.50	0.196E+05					T	T	
S 170 28 COARSE	156.00	140.0 142.0	172.0	1.00 0.500 1.00	15.00	0.154E+05					S	S	S
16 NORMAL	162.00	152.0 154.0	172.0	1.00 0.500 1.00	9.00	0.181E+05					S	S	S
6 FINE	167.00	163.0 164.0	171.0	0.50 0.250 0.50	3.50	0.209E+05					S	S	S
T 175 28 COARSE	161.00	145.0 147.0	177.0	1.00 0.500 1.00	15.00	0.165E+05					T	T	
16 NORMAL	167.00	157.0 159.0	177.0	1.00 0.500 1.00	9.00	0.194E+05					T	T	
8 FINE	171.00	166.0 167.0	176.0	0.50 0.250 0.50	4.50	0.216E+05					T	T	
F 180 28 COARSE	166.00	150.0 152.0	182.0	1.00 0.500 1.00	15.00	0.177E+05				20T	F	F	F
18 NORMAL	171.00	160.0 162.0	182.0	1.00 0.500 1.00	10.00	0.201E+05				16S	F	F	F
8 FINE	176.00	171.0 172.0	181.0	0.50 0.250 0.50	4.50	0.230E+05				T	F	F	F
T 185 32 COARSE	169.00	151.0 153.0	187.0	1.00 0.500 1.00	17.00	0.179E+05					T	T	
18 NORMAL	176.00	165.0 167.0	187.0	1.00 0.500 1.00	10.00	0.214E+05					T	T	
8 FINE	181.00	176.0 177.0	186.0	0.50 0.250 0.50	4.50	0.243E+05					T	T	
S 190 32 COARSE	174.00	156.0 158.0	192.0	1.00 0.500 1.00	17.00	0.191E+05					S	S	S
18 NORMAL	181.00	170.0 172.0	192.0	1.00 0.500 1.00	10.00	0.227E+05					S	S	S
8 FINE	186.00	181.0 182.0	191.0	0.50 0.250 0.50	4.50	0.257E+05					S	S	S
T 195 32 COARSE	179.00	161.0 163.0	197.0	1.00 0.500 1.00	17.00	0.204E+05					T	T	
18 NORMAL	186.00	175.0 177.0	197.0	1.00 0.500 1.00	10.00	0.241E+05					T	T	
8 FINE	191.00	186.0 187.0	196.0	0.50 0.250 0.50	4.50	0.272E+05					T	T	

NOTES:
1. ALL DIMENSIONS ARE IN MILLIMETERS
2. SEE FIGURE 8-14 AND PAGE 8-39 FOR KEY TO SYMBOLS
3. THE NOMINAL SIZE IS NATIONAL STANDARD AS INDICATED
   F = FIRST CHOICE, S = SECOND CHOICE, T = THIRD CHOICE

SCREW THREADS 8-37

Table 8-9 (*Continued*). World Standards for Metric Trapezoidal Screw Threads (ISO 2901, 2902, 2903, and 2904)

I S NOM. O SIZE d	PITCH p		PITCH DIA $d_2=D_2$	MINOR DIAMETER d_3 D_1	MAJOR DIA D_4	AC	RADIUS R1MAX R2MAX	THREAD HEIGHT AT MINOR $H_4=h_3$	STR AREA $\emptyset$(MM**2)	U.S. ANSI	AUSTRAL SAA	JAPAN JISC	FRANCE AFNOR E03-616	U.K. BSI DRAFT	GERMANY DNA DIN103	ITALY UNI 124-126
F	200	32 COARSE	184.00	166.0 168.0	202.0	1.00	0.500 1.00	17.00	0.216E+05				S	F	F	F
		18 NORMAL	191.00	180.0 182.0	202.0	1.00	0.500 1.00	10.00	0.254E+05				20F	F	F	F
		8 FINE	196.00	190.0 192.0	201.0	0.50	0.250 0.50	4.50	0.287E+05				10S	F	F	F
S	210	36 COARSE	192.00	172.0 174.0	212.0	1.00	0.500 1.00	19.00	0.232E+05					S	S	
		20 NORMAL	200.00	188.0 190.0	212.0	1.00	0.500 1.00	11.00	0.278E+05				S	S	S	
		8 FINE	206.00	199.0 201.0	211.0	0.50	0.250 0.50	4.50	0.317E+05				S	S	S	
F	220	36 COARSE	202.00	182.0 184.0	222.0	1.00	0.500 1.00	19.00	0.260E+05				32T	F	F	F
		20 NORMAL	210.00	198.0 200.0	222.0	1.00	0.500 1.00	11.00	0.308E+05				S	F	F	F
		8 FINE	216.00	209.0 211.0	221.0	0.50	0.250 0.50	4.50	0.350E+05				10T	F	F	F
S	230	36 COARSE	212.00	192.0 194.0	232.0	1.00	0.500 1.00	19.00	0.290E+05					S	S	
		20 NORMAL	220.00	208.0 210.0	232.0	1.00	0.500 1.00	11.00	0.340E+05					S	S	
		8 FINE	226.00	219.0 221.0	231.0	0.50	0.250 0.50	4.50	0.384E+05					S	S	
F	240	36 COARSE	222.00	202.0 204.0	242.0	1.00	0.500 1.00	19.00	0.320E+05					F	F	
		22 NORMAL	229.00	216.0 218.0	242.0	1.00	0.500 1.00	12.00	0.366E+05					F	F	
		8 FINE	236.00	229.0 231.0	241.0	0.50	0.250 0.50	4.50	0.419E+05					F	F	
S	250	40 COARSE	230.00	208.0 210.0	252.0	1.00	0.500 1.00	21.00	0.340E+05				S	S	S	F
		22 NORMAL	239.00	226.0 228.0	252.0	1.00	0.500 1.00	12.00	0.401E+05				24F	S	S	F
		12 FINE	244.00	235.0 237.0	251.0	0.50	0.250 0.50	6.50	0.441E+05				S	S	S	
F	260	40 COARSE	240.00	218.0 220.0	262.0	1.00	0.500 1.00	21.00	0.373E+05					F	F	
		22 NORMAL	249.00	236.0 238.0	262.0	1.00	0.500 1.00	12.00	0.437E+05					F	F	
		12 FINE	254.00	245.0 247.0	261.0	0.50	0.250 0.50	6.50	0.479E+05					F	F	
S	270	40 COARSE	250.00	228.0 230.0	272.0	1.00	0.500 1.00	21.00	0.408E+05				T	S	S	
		24 NORMAL	258.00	244.0 246.0	272.0	1.00	0.500 1.00	13.00	0.468E+05				S	S	S	
		12 FINE	264.00	255.0 257.0	271.0	0.50	0.250 0.50	6.50	0.519E+05				T	S	S	
F	280	40 COARSE	260.00	238.0 240.0	282.0	1.00	0.500 1.00	21.00	0.445E+05					F	F	F
		24 NORMAL	268.00	254.0 256.0	282.0	1.00	0.500 1.00	13.00	0.507E+05					F	F	F
		12 FINE	274.00	265.0 267.0	281.0	0.50	0.250 0.50	6.50	0.560E+05					F	F	F
S	290	44 COARSE	268.00	244.0 246.0	292.0	1.00	0.500 1.00	23.00	0.468E+05					S	S	
		24 NORMAL	278.00	264.0 266.0	292.0	1.00	0.500 1.00	13.00	0.547E+05					S	S	
		12 FINE	284.00	275.0 277.0	291.0	0.50	0.250 0.50	6.50	0.603E+05					S	S	
F	300	44 COARSE	278.00	254.0 256.0	302.0	1.00	0.500 1.00	23.00	0.507E+05					F	F	F
		24 NORMAL	288.00	274.0 276.0	302.0	1.00	0.500 1.00	13.00	0.590E+05					F	F	F
		12 FINE	294.00	285.0 287.0	301.0	0.50	0.250 0.50	6.50	0.647E+05					F	F	F

NOTES:
1. ALL DIMENSIONS ARE IN MILLIMETERS
2. SEE FIGURE 8-14 AND PAGE 8-39 FOR KEY TO SYMBOLS
3. THE NOMINAL SIZE IS NATIONAL STANDARD AS INDICATED
   F = FIRST CHOICE, S = SECOND CHOICE, T = THIRD CHOICE

Table 8-10. Length of Trapezoidal Thread Engagement* (ISO 2903)

Basic major diameter d		Pitch P	Groups of lengths of thread engagement		
			N		L
over	up to and incl.		over	up to and incl.	over
5.6	11.2	1.5	5	15	15
		2	6	19	19
		3	10	28	28
11.2	22.4	2	8	24	24
		3	11	32	32
		4	15	43	43
		5	18	53	53
		8	30	85	85
22.4	45	3	12	36	36
		5	21	63	63
		6	25	75	75
		7	30	85	85
		8	34	100	100
		10	42	125	125
		12	50	150	150
45	90	3	15	45	45
		4	19	56	56
		8	38	118	118
		9	43	132	132
		10	50	140	140
		12	60	170	170
		14	67	200	200
		16	75	236	236
		18	85	265	265
90	180	4	24	71	71
		6	36	106	106
		8	45	132	132
		12	67	200	200
		14	75	236	236
		16	90	265	265
		18	100	300	300
		20	112	335	335
		22	118	355	355
		24	132	400	400
		28	150	450	450
180	355	8	50	150	150
		12	75	224	224
		18	112	335	335
		20	125	375	375
		22	140	425	425
		24	150	450	450
		32	200	600	600
		36	224	670	670
		40	250	750	750
		44	280	850	850

*Dimensions in millimeters.

SCREW THREADS

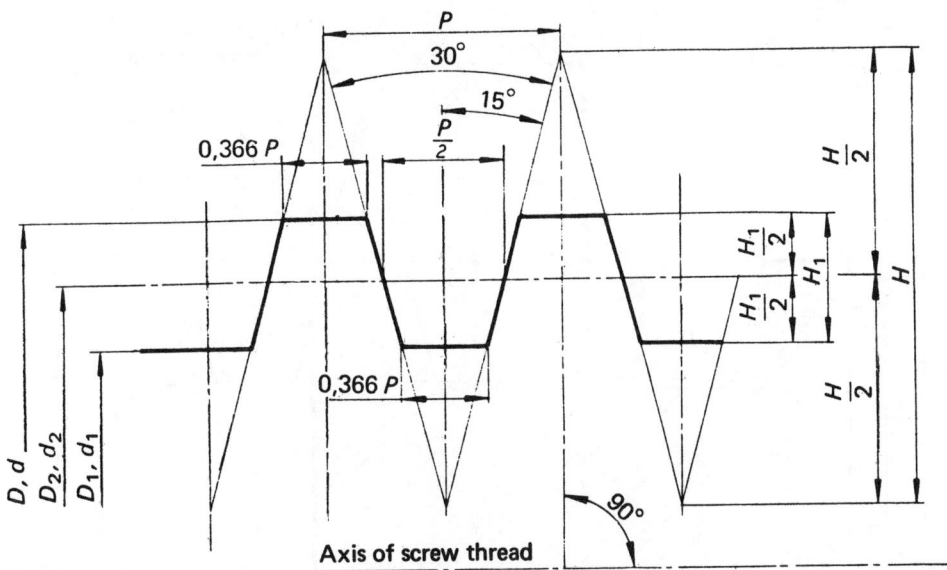

Fig. 8-13. Basic trapezoidal thread profile. D = major diameter of internal thread; d = major diameter of external thread; D_2 = pitch diameter of internal thread; d_2 = pitch diameter of external thread; D_1 = minor diameter of internal thread; d_1 = minor diameter of external thread; P = pitch; H = height of fundamental triangle; and H_1 = height of basic profile (ISO 2901).

The design profiles for the ISO metric internal and external trapezoidal screw threads are shown in Fig. 8-14, and the most important general dimensions are computed and shown in Table 8-9. All dimensions represent the profiles of the threads in the maximum metal conditions.

Key to Symbols and Formulas used to compute values in Table 8-9

Nominal Diameter	$d = D$ (from Standard)
Thread Pitch	P (from Standard)
Pitch Diameter	$d_2 = D_2 = d - 0.5P$
Minor Diameter (Bolt)	$d_3 = d - 2h_3$
Minor Diameter (Nut)	$D_1 = d - 2H_1 = d - P$
Major Diameter (Nut)	$D_4 = d + 2a_c$
Crest Clearance	a_c (from Standard)
Radius	$R_{1\,max} = 0.5\,a_c$
Radius	$R_{2\,max} = a_c$
Thread Height (Bolt)	$h_3 = 0.5P + a_c$
Thread Height (Nut)	$H_4 = 0.5P + a_c$
Area of Minor Diameter Section (Bolt)	$A - d_3 = (d_3)^2 \cdot \pi/4$

Classes of Trapezoidal Thread Fit—The tolerance system for trapezoidal screw threads is similar to the one used on ISO general metric fasteners, and it is described in the ISO 2903 standard, which has been based on the ISO 965 Part 1 standard completed with tolerance positions c and e. The recommended tolerance classes are, however, not the same for the two tolerance systems.

The position of a tolerance zone is defined by the distance between the basic size and the nearest end of the tolerance zone. This distance is known as the fundamental deviation, and it is designated with capital letters for nuts and small letters for bolts. The magnitudes of tolerance zones are designated by tolerance grades (figures). A combination of a tolerance grade (figure) and a fundamental deviation (letter) forms a tolerance class designation; e.g., 7H. (See Fig. 8-15.)

Symbols used, other than those shown above, are as follows:

N	= designation for thread engagement group Normal (see Table 8-10)
L	= designation for thread engagement group Long (see Table 8-10)
T	= tolerance
$TD_1\,TD_2$	= tolerances for D_1, D_2, d, d_3, d_2 (see Key to symbols above)
$Td\,Td_3\,Td_2$	= (for D_4 no tolerances are specified)
ei EI	= lower deviations (EI for the nut threads is equal to zero)
es ES	= upper deviations

Recommended Tolerances

The following general rules can be formulated for the choice of tolerance quality:

1. Medium: For general use.
2. Coarse: For cases where manufacturing difficulties can arise.

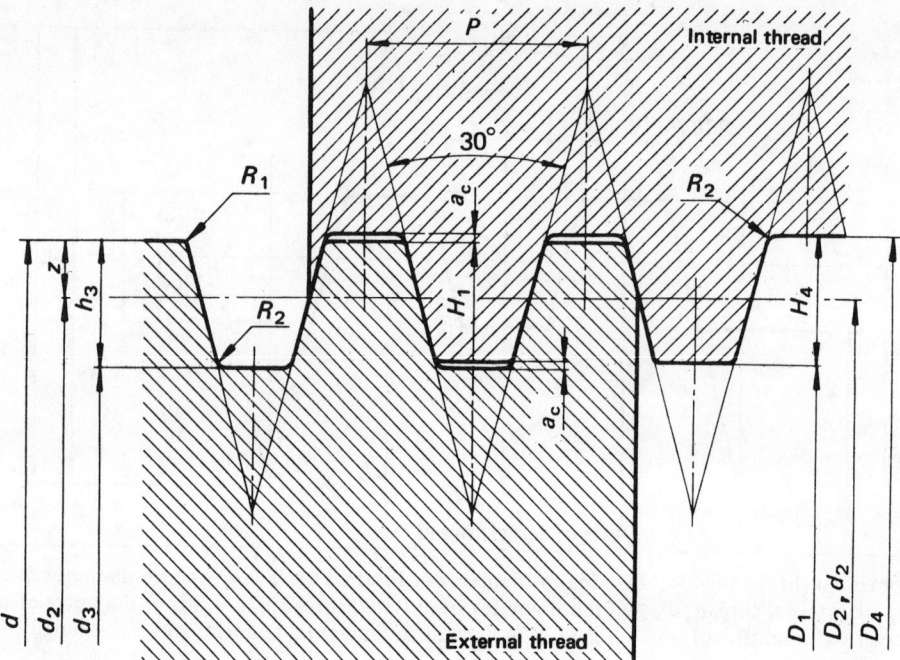

Fig. 8-14. Design forms of internal and external trapezoidal threads (maximum metal conditions), (ISO 2901).

If the actual length of thread engagement is unknown, group N is recommended.

Recommended Tolerances for Nut Threads (ISO 2903):

Tolerance quality	Tolerances for the pitch diameter	
	N	L
Medium	7H	8H
Coarse	8H	9H

Recommended Tolerances for Bolt Threads (ISO 2903):

Tolerance quality	Tolerances for the pitch diameter	
	N	L
Medium	7e	8e
Coarse	8c	9c

Tolerance grades (ISO 2903):

The following tolerance grades are established in the ISO standard.

Dimension	Tolerance grade
Minor diameter of nut threads D_1:	4
Major diameter of bolt threads d:	4
Pitch diameter of nut threads D_2:	7 8 9
Pitch diameter of bolt threads d_2:	7 8 9
Minor diameter of bolt threads d_3:	7 8 9

The classification of the trapezoidal thread engagement as long (L), and normal (N), is dependent on the nominal diameter and the pitch. It is necessary to determine this differentiation before selecting the tolerances for bolt and nut threads. The length of thread engagement is classified into the groups N or L, as shown in Table 8-10.

Calculation of Limits—In order to calculate the upper and lower limits for a trapezoidal screw thread size, proceed as follows:

1. Determine nominal thread size and tolerance designation.
2. Find the basic dimension in Table 8-9.
3. Use the guide shown in Fig. 8-16 and find the fundamental deviation (see Table 8-11) and tolerance grade in referenced Tables 8-12 through 8-16.
4. Calculate upper and lower limits.

SCREW THREADS

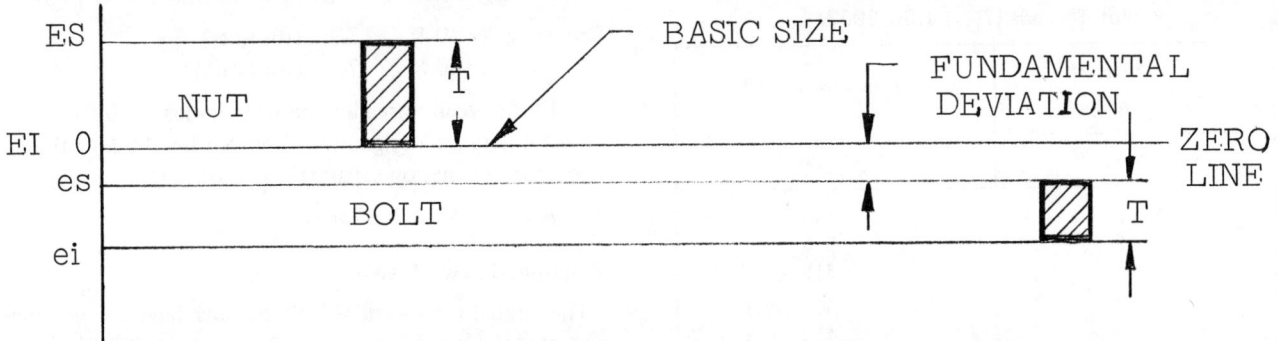

Fig. 8-15. Tolerance positions for trapezoidal screw threads with respect to zero line (basic size).

Dimension name (symbol)	Fundamental deviation (letter)	Tolerance grade (number)
Minor diameter of nut threads (D_1)	zero	T_{D_1} see Table 8-12
Pitch diameter of nut threads (D_2)	zero	T_{D_2} see Table 8-15
Major diameter of bolt threads (d)	zero	T_d see Table 8-13
Minor diameter of bolt threads (d_3)	use d_2 values - see Table 8-11	T_{d_3} see Table 8-14
Pitch diameter of bolt threads (d_2)	see Table 8-11	T_{d_2} see Table 8-16

Fig. 8-16. Guide to Calculating Thread Limits

Example Nut: Tr 40 × 7 − 7H

The minor diameter of nut threads (D_1):

Basic size (see Table 8-9): $D_1 = 33$
Fundamental deviation (see Fig. 8-16): $EI = 0$
Tolerance grade (see Table 8-12): $T_{D_1} = 0.560$

thus: $33 \begin{array}{l} +0.560 \\ +0 \end{array} = \begin{array}{l} 33.560 \text{ upper limit} \\ 33.000 \text{ lower limit} \end{array}$

Example Bolt: Tr 16 × 4 − 7e

The minor diameter of bolt threads (d_3):

Basic size (see Table 8-9): $d_3 = 11.5$
Fundamental deviation (see Table 8-11): $es = -0.095$
Tolerance grade (see Table 8-14): $T_{d_3} = 0.426$

thus: $11.5 \begin{array}{l} -0.095 \\ -0.521 \end{array} = \begin{array}{l} 11.405 \text{ upper limit} \\ 10.979 \text{ lower limit} \end{array}$

Table 8-11. Fundamental Deviations for the Pitch Diameter of Nut Threads and Bolt Threads (ISO 2903)

	Fundamental deviation		
	Nut thread	Bolt thread	
Pitch	D_2	d_2	
P	H EI	c es	e es
mm	μm	μm	μm*
1.5	0	−140	− 67
2	0	−150	− 71
3	0	−170	− 85
4	0	−190	− 95
5	0	−212	−106
6	0	−236	−118
7	0	−250	−125
8	0	−265	−132
9	0	−280	−140
10	0	−300	−150
12	0	−335	−170
14	0	−355	−180
16	0	−375	−190
18	0	−400	−200
20	0	−425	−212
22	0	−450	−224
24	0	−475	−236
28	0	−500	−250
32	0	−530	−265
36	0	−560	−280
40	0	−600	−300
44	0	−630	−315

*1μm = 0.001 mm

Table 8-12. Minor Diameter Tolerance of Trapezoidal Nut Threads (T_{D_1}) (ISO 2903)

Pitch P	Tolerance grade 4
mm	μm**
1.5	190
2	236
3	315
4	375
5	450
6	500
7	560
8	630
9	670
10	710
12	800
14	900
16	1000
18	1120
20	1180
22	1250
24	1320
28	1500
32	1600
36	1800
40	1900
44	2000

*The fundamental deviation for the minor diameter of nut threads is zero.
**1 μm = 0.001 mm.

Trapezoidal Thread Designation—The one-start metric trapezoidal screw thread conforming to the ISO standard is designated by the letters *Tr*, followed by the values of the nominal diameter and of the pitch expressed in millimeters and separated by the sign ×.

Example: Tr 40 × 7

The multiple-start metric trapezoidal screw threads conforming to the ISO standard are designated by the letters *Tr* followed by the values of the nominal diameter and of the lead for the multiple-start threads, separated by the sign ×, and, in brackets, the letter *P* and the value of the pitch, all expressed in millimeters.

Example: Tr 40 × 14 (P7)

(Number of starts = $\frac{\text{Lead}}{\text{Pitch}} = \frac{14}{7} = 2$ defines a screw thread of 40 diameters with 2 starts)

For left-hand metric trapezoidal screw threads, the letters *LH* should be added to the thread designation.

Example: Tr 40 × 14 (P7) LH

The designations for thread tolerances are as follows:

Example: Tr 40 × 7 — 7H (for nuts)
           Tr 40 × 7 — 7e (for bolts)

A fit between threaded parts is indicated by the nut thread tolerance designation followed by the bolt thread tolerance designation separated by a stroke.

Example: Tr 40 × 7 — 7H/7e

Miniature Screw Threads

The miniature metric screw threads have 14 nominal diameter sizes ranging from 0.3 through 1.4 millimeters, and they are specified in the ISO 1501 standard. The basic thread profile is identical to the ISO general screw thread form (ISO 68) except for the thread height, which equals 0.48 × pitch.

The ISO diameter pitch combinations are standardized in the ANSI-B1-10 standard, but the basic thread height is slightly different and it equals 0.52P. (ISO 68 thread height is equal to 0.54P). It is expected that the diameter-

Table 8-13. Major Diameter Tolerance of Trapezoidal Bolt Thread (T_d)* (ISO 2903)

Pitch P	Tolerance grade 4
mm	μm**
1.5	150
2	180
3	236
4	300
5	335
6	375
7	425
8	450
9	500
10	530
12	600
14	670
16	710
18	800
20	850
22	900
24	950
28	1060
32	1120
36	1250
40	1320
44	1400

*The fundamental deviation for the major diameter of the bolt thread is zero.
**1 μm = 0.001 mm.

SCREW THREADS

Table 8-14. Minor Diameter Tolerance of Trapezoidal Bolt Thread (T_{d_3}) (ISO 2903)

Basic major diameter d		Pitch P	Tolerance position c of the pitch diameter tolerance			Tolerance position e of the pitch diameter tolerance		
over	up to		Tolerance grade			Tolerance grade		
			7	8	9	7	8	9
mm	mm	mm	μm	μm	μm	μm	μm	μm*
5.6	11.2	1.5	352	405	471	279	332	398
		2	388	445	525	309	366	446
		3	435	501	589	350	416	504
11.2	22.4	2	400	462	544	321	383	465
		3	450	520	614	365	435	529
		4	521	609	690	426	514	595
		5	562	656	775	456	550	669
		8	709	828	965	576	695	832
22.4	45	3	482	564	670	397	479	585
		5	587	681	806	481	575	700
		6	655	767	899	537	649	781
		7	694	813	950	569	688	825
		8	734	859	1015	601	726	882
		10	800	925	1087	650	775	937
		12	866	998	1223	691	823	1048
45	90	3	501	589	701	416	504	616
		4	565	659	784	470	564	689
		8	765	890	1052	632	757	919
		9	811	943	1118	671	803	978
		10	831	963	1138	681	813	988
		12	929	1085	1273	754	910	1098
		14	970	1142	1355	805	967	1180
		16	1038	1213	1438	853	1028	1253
		18	1100	1288	1525	900	1088	1320
90	180	4	584	690	815	489	595	720
		6	705	830	986	587	712	868
		8	796	928	1103	663	795	970
		12	960	1122	1335	785	947	1160
		14	1018	1193	1418	843	1018	1243
		16	1075	1263	1500	890	1078	1315
		18	1150	1338	1588	950	1138	1388
		20	1175	1363	1613	962	1150	1400
		22	1232	1450	1700	1011	1224	1474
		24	1313	1538	1800	1074	1299	1561
		28	1388	1625	1900	1138	1375	1650
180	355	8	828	965	1153	695	832	1020
		12	998	1173	1398	823	998	1223
		18	1187	1400	1650	987	1200	1450
		20	1263	1488	1750	1050	1275	1537
		22	1288	1513	1775	1062	1287	1549
		24	1363	1600	1875	1124	1361	1636
		32	1530	1780	2092	1265	1515	1827
		36	1623	1885	2210	1343	1605	1930
		40	1663	1925	2250	1363	1625	1950
		44	1755	2030	2380	1440	1715	2065

*1μm = 0.001 mm

Table 8-15. Pitch Diameter Tolerance of Trapezoidal Nut Thread (T_{D_2}) (ISO 2903)

Basic major diameter d		Pitch P	Tolerance grade		
over	up to and incl.		7	8	9
mm	mm	mm	µm	µm	µm*
5.6	11.2	1.5 2 3	224 250 280	280 315 355	355 400 450
11.2	22.4	2 3 4	265 300 355	335 375 450	425 475 560
		5 8	375 475	475 600	600 750
22.4	45	3 5 6	335 400 450	425 500 560	530 630 710
		7 8 10 12	475 500 530 560	600 630 670 710	750 800 850 900
45	90	3 4 8	355 400 530	450 500 670	550 630 850
		9 10 12	560 560 630	710 710 800	900 900 1000
		14 16 18	670 710 750	850 900 950	1060 1120 1180
90	180	4 6 8 12	425 500 560 670	530 630 710 850	670 800 900 1060
		14 16 18	710 750 800	900 950 1000	1120 1180 1250
		20 22 24 28	800 850 900 950	1000 1060 1120 1180	1250 1320 1400 1500
180	355	8 12 18	600 710 850	750 900 1060	950 1120 1320
		20 22 24	900 900 950	1120 1120 1180	1400 1400 1500
		32 36 40 44	1060 1120 1120 1250	1320 1400 1400 1500	1700 1800 1800 1900

*1 µm = 0.001 mm

SCREW THREADS

Table 8-16. Pitch Diameter Tolerance of Trapezoidal Bolt Thread (T_{D_2}) (ISO 2903)

Basic major diameter d over mm	up to and incl. mm	Pitch P mm	Tolerance grade 6 μm	7 μm	8 μm	9 μm*
5.6	11.2	1.5	132	170	212	265
		2	150	190	236	300
		3	170	212	265	335
11.2	22.4	2	160	200	250	315
		3	180	224	280	355
		4	212	265	335	400
		5	224	280	355	450
		8	280	355	450	560
22.4	45	3	200	250	315	400
		5	236	300	375	475
		6	265	335	425	530
		7	280	355	450	560
		8	300	375	475	600
		10	315	400	500	630
		12	335	425	530	710
45	90	3	212	265	335	425
		4	236	300	375	475
		8	315	400	500	630
		9	335	425	530	670
		10	335	425	530	670
		12	375	475	600	750
		14	400	500	630	800
		16	425	530	670	850
		18	450	560	710	900
90	180	4	250	315	400	500
		6	300	375	475	600
		8	335	425	530	670
		12	400	500	630	800
		14	425	530	670	850
		16	450	560	710	900
		18	475	600	750	950
		20	475	600	750	950
		22	500	630	800	1000
		24	530	670	850	1060
		28	560	710	900	1120
180	355	8	355	450	560	710
		12	425	530	670	850
		18	500	630	800	1000
		20	530	670	850	1060
		22	530	670	850	1060
		24	560	710	900	1120
		32	630	800	1000	1250
		36	670	850	1060	1320
		40	670	850	1060	1320
		44	710	900	1120	1400

*1 μm = 0.001 mm

pitch combinations outlined in ISO 1501 will remain as is. However, one standard thread height is anticipated. This should not result in any difficulty in interchangeability of threaded fasteners with the same diameter-pitch combination in the miniature sizes.

Basic Miniature Thread Profile—The basic miniature thread profile is specified in the ISO 1501 standard and is shown in Fig. 8-17. The tolerances and deviations for external and internal miniature screw threads are applied to the basic sizes, for which the numerical values are listed in Table 8-17.

World Metric Miniature Screw Threads Standard

Table 8-17 tabulates the basic general dimensions for the ISO metric miniature screw threads, and at the same time, relates the ISO standards to the national standards in some major industrial countries.

Designation of Miniature Screw Threads—A complete designation of a miniature screw thread is shown in the following example:

ISO miniature screw thread S-0.6 mm diameter designation (Nut/Bolt combination)

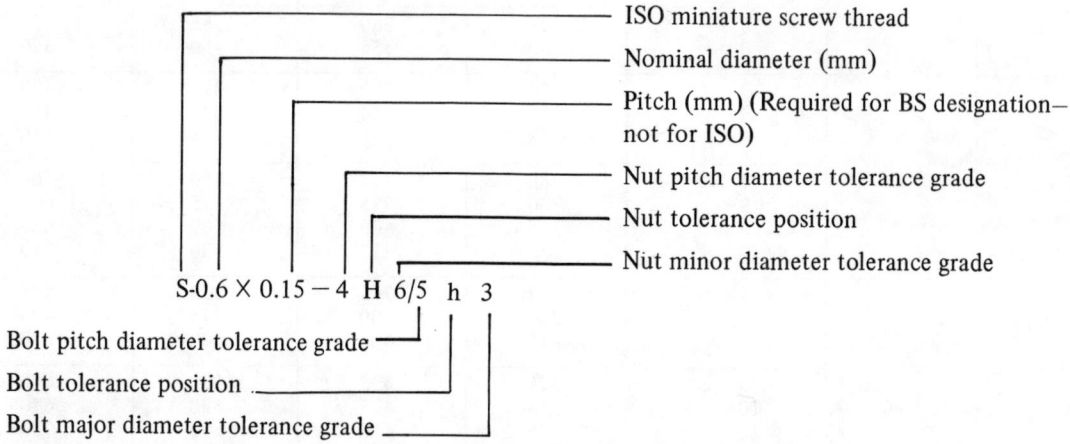

Refer to the ISO 1501, or other national standards listed for thread-fit details.

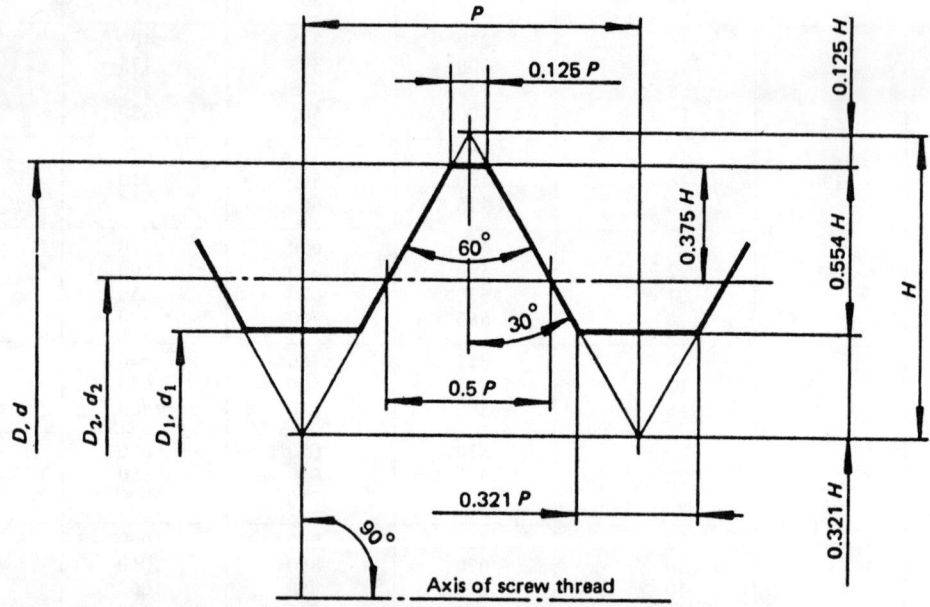

Fig. 8-17. Basic miniature thread profile. D = major diameter of internal thread; d = major diameter of external thread; D_2 = pitch diameter of internal thread; d_2 = pitch diameter of external thread; D_1 = minor diameter of internal thread; d_1 = minor diameter of external thread; P = pitch; and H = height of fundamental triangle (ISO 1501).

SCREW THREADS

Table 8-17. World Standards for Metric Miniature Screw Threads (ISO 1501)

ISO	NOM. SIZE d=D	PITCH P	PITCH DIA. d2=D2	MINOR DIAMETER d3 D1	THREAD HEIGHT h3 H1	RADIUS MAX. R	TENSILE STR. AREA AS(MM**2)	U.S. ANSI B1.10	AUSTRAL SAA	JAPAN JIS	FRANCE AFNOR E03-502	U.K. BSI BS4827	GERMANY DNA DIN14	ITALY UNI
F	0.3	0.08	0.248	0.223	0.038 0.038	0.012	0.436E-01	F			F	F	F	
S	0.35	0.09	0.292	0.264	0.043 0.043	0.013	0.605E-01	S			S	S	S	
F	0.4	0.1	0.335	0.304	0.048 0.048	0.014	0.802E-01	F			F	F	F	
S	0.45	0.1	0.385	0.354	0.048 0.048	0.014	0.107E+00	S			S	S		
F	0.5	0.125	0.419	0.380	0.060 0.060	0.018	0.125E+00	F			F	F	F	
S	0.55	0.125	0.469	0.430	0.060 0.060	0.018	0.159E+00	S			S	S		
F	0.6	0.15	0.503	0.456	0.072 0.072	0.022	0.180E+00	F			F	F	F	
S	0.7	0.175	0.586	0.532	0.084 0.084	0.025	0.246E+00	S			S	S	F	
F	0.8	0.2	0.670	0.608	0.096 0.096	0.029	0.321E+00	F			F	F	F	
S	0.9	0.225	0.754	0.684	0.108 0.108	0.032	0.406E+00	S			S	S	F	
F	1	0.25	0.838	0.760	0.120 0.120	0.036	0.501E+00	F			F	F		
S	1.1	0.25	0.938	0.860	0.120 0.120	0.036	0.635E+00	S			S	S		
F	1.2	0.25	1.038	0.960	0.120 0.120	0.036	0.784E+00	F			F	F		
S	1.4	0.3	1.205	1.112	0.144 0.144	0.043	0.105E+01	S			S	S		

NOTES:
1. ALL DIMENSIONS ARE IN MILLIMETERS
2. SEE FIGURE 8-17 AND PAGE 8-46 FOR KEY TO SYMBOLS
3. THE NOMINAL SIZE IS NATIONAL STANDARD AS INDICATED
   F = FIRST CHOICE, S = SECOND CHOICE, T = THIRD CHOICE

International Pipe Threads

The material presented here is based on a report from the ISO Technical Committee (TC5) Metal Pipes and Fittings chairman, and a study conducted by the Society of Automotive Engineers—Off Road Vehicle Council (SAE-ORVC) metric standards study group.

Inch standard pipes have been supplied to the world market for more than 150 years. Most of the pipe thread and fitting standards are therefore based on the inch system.

Non-Sealing Pipe Threads

ISO 228, "Pipe Threads Where Pressure-Tight Joints Are Not Made on the Threads," is based on the British Whitworth standard. The thread angle is 55° and additional sealing compound or gasket is required to stop leakage through threads. The ISO thread is only suitable for transmitting axial pipe loads. The American standards (ANSI B2.1 "Straight Pipe Threads in Pipe Couplings," and B2.2 "Straight Pipe Threads for Mechanical Joints") differ from the ISO 228, in thread pitch and thread angle, and the threads are not interchangeable.

Dryseal Pipe Threads

There are three major national standards covering the above type screw thread, and they are as follows: American (ANSI B2.2), British (BS 21), and German (DIN 158).

The above standards have a 1:16 taper, and the thread angle is 55° for the British and 60° for the other standards. The thread pitches for the American and the British dryseal pipe threads differ slightly, and the threads are not interchangeable. ISO 7 "Pipe Threads for Gas List Tubes and Screwed Fittings Where Pressure-Tight Joints Are Made on the Threads," is based on the BS 21 standard.

British (BS 21)—The Whitworth standard is the basis for the ISO 7 dryseal standard. The British standard is adopted in many national standards in Europe, and it is in general use in the Common Market countries as well as in many British Commonwealth countries. The ISO Technical Committee (TC5) is in the process of revising the ISO 7 standard and will include a complete gaging specification for this screw thread in the revised issue.

German (DIN 158)—This standard is most frequently used on tapered male threaded parts screwed into ISO 261 general screw thread tapped holes. Both threaded parts can be tapered for improved mechanical support. It is used for lubricating fittings, plugs, and connector ends screwed into machine parts. The following ISO 261 thread sizes are standardized: M6, M8 ×1, M10 × 1, M12 × 1, M12 × 1.5, M14 × 1.5, M16 × 1.5, M18 × 1.5, M20 × 1.5, M22 × 1.5, M24 × 1.5, M26 × 1.5, M30 × 1.5, M36 × 1.5, M38 × 1.5, M42 × 1.5, M45 × 1.5, M48 × 1.5, M52 × 1.5, M27 × 2, M30 × 2, M33 × 2, M36 × 2, M39 × 2, M42 × 2, M45 × 2, M48 × 2, M52 × 2, M56 × 2, and M60 × 2.

The DIN 158 thread has a 1:16 taper, and the male screw thread is available in two standard lengths. Metric valves and lubricating fittings produced to this standard are available in Europe.

American (ANSI B2.2)

American National Standard ANSI B2.2 covers four types of Dryseal Threads. They are designated:

NPTF—Dryseal ANSI Standard Pipe Thread
PTF-SAE SHORT—Dryseal SAE Short Taper Pipe Thread
NPSF—Dryseal ANSI Standard Fuel Internal Straight Pipe Thread
NPSI—Dryseal ANSI Standard Intermediate Internal Straight Pipe Thread

The full designation gives in sequence the nominal size, number of threads per inch, form (Dryseal), and symbol, as: 1/8–27 DRYSEAL NPTF.

Type 1—Dryseal ANSI Standard Taper Pipe Thread, NPTF: This series of threads applies to both external and internal threads of full length and is suitable for pipe joints in practically every type of service. These threads are generally conceded to be superior for strength and seal. Use of the internal tapered thread in hard or brittle materials having thin sections will minimize trouble from fracture.

Type 2—Dryseal SAE Short Taper Pipe Thread, PTF-SAE SHORT: External threads of this series conform in all respects to the NPTF threads except that the full thread length has been shortened by eliminating one thread at the small end for increased clearance and economy of material.

Internal threads of this series conform in all respects to NPTF threads except that the full thread length has been shortened by one thread at the large end.

Type 3—Dryseal ANSI Standard Fuel Internal Straight Pipe Threads, NPSF: Threads of this series are straight instead of tapered. They are generally used in soft or ductile materials which will adjust at assembly to the taper of external threads but may also be used in hard or brittle materials where the section is heavy.

Type 4—Dryseal ANSI Standard Intermediate Internal Straight Pipe Threads, NPSI: Threads of this series are straight instead of tapered. They are generally used in hard or brittle materials where the section is heavy and where there is little expansion at assembly with the external taper threads.

The SAE-ORVC metric standards study group conducted a survey in 1973 among its industry groups' licensed com-

SCREW THREADS

panies located throughout the world. All countries included in the survey indicated they use NPTF (Dryseal) threads per ANSI B 2.2. Components and tooling for the ANSI dryseal pipe thread are readily available throughout the world according to the above survey. This is partially due to the strong, worldwide influence from industry groups such as the American Petroleum Institute and the Society of Automotive Engineers.

Future ISO standards to cover these important screw thread standards will be needed.

RELATED STANDARDS SCREW THREADS (TC 1)

ISO 68-1973	ISO general-purpose screw threads—Basic profile—2nd Ed.
ISO 261-1973	ISO general-purpose metric screw threads—General plan—2nd Ed.
ISO 262-1973	ISO general-purpose metric screw threads—Selected sizes for screws, bolts, and nuts—2nd Ed.
ISO/R 724-1968	ISO general-purpose metric screw threads—Basic dimensions
ISO 965/Part 1-1973	ISO general-purpose metric screw threads—Tolerances—Principles and basic data
ISO 965/Part 2-1973	ISO general-purpose metric screw threads—Tolerances—Limits of sizes for commercial bolt and nut threads—Medium quality
ISO 965/Part 3-1973	ISO general-purpose metric screw threads—Tolerances—Deviations for constructional threads
ISO/R 1501-1970	ISO miniature screw threads
ISO 1502-1970	ISO general-purpose metric screw threads—Gaging
ISO 2901-1977	ISO metric trapezoidal screw threads—Basic profile and maximum profiles
ISO 2902-1977	ISO metric trapezoidal screw threads—General plan
ISO 2903-1977	ISO metric trapezoidal screw threads—Tolerances
ISO 2904-1977	ISO metric trapezoidal screw threads—Nominal sizes
ISO/DIS 5408	Screw threads—Terms and definitions
ISO/DIS 5410	ISO high fatigue purpose metric screw threads
ISO/DPR 5411	Stub metric trapezoidal screw threads
ISO/DPR 5412	Buttress threads

METAL PIPES AND FITTINGS (TC 5)

ISO/R 7-1954	Pipe threads for gas list tubes and screwed fillings where pressure-tight joints are made on the threads
ISO/R 228-1961	Pipe threads where pressure-tight joints are not made on the threads

OTHER NATIONAL SCREW THREAD STANDARDS

U. S. A.

ANSI B1.10-1958	Unified miniature screw threads (ISO/R 1501)
ANSI B1.16-1972	Metric screw threads, American gaging practice
ANSI B1 Report	—ISO metric screw threads
ANSI B2.1-1968	Pipe threads (except Dryseal)
ANSI B2.2-1968	Dryseal pipe threads

JAPAN

JIS B 0205-1973	Metric coarse screw threads
JIS B 0207-1973	Metric fine screw threads
JIS B 0123-1970	Designation of screw threads
JIS B 0202-1966	Parallel pipe threads
JIS B B203-1966	Taper pipe threads

NOTE: All national standards are available in language listed from ANSI or the National Standards Association for each country (see Section 1).

GERMANY

METRIC SCREW THREADS

DIN 13-1973	Sheet 1: ISO metric threads; coarse threads from 1 to 68 mm thread diameters, basic sizes
DIN 13-1970	Sheet 2: Fine threads with pitches 0.2 - 0.25 - 0.35 mm and thread diameters from 1 to 50 mm, basic sizes
DIN 13-1970	Sheet 3: Fine threads with pitch 0.5 mm and thread diameters from 3.5 to 90 mm, basic sizes
DIN 13-1970	Sheet 6: Fine thread with 1.5 mm pitch in thread diameters from 12 to 300 mm, basic sizes
DIN 13-1970	Sheet 7: ISO metric threads; fine thread with 2 mm pitch in thread diameters from 17 to 300 mm, basic sizes
DIN 13-1970	Sheet 8: Fine thread with 3 mm pitch in thread diameters from 28 to 300 mm, basic sizes

SCREW THREADS

DIN 13-1970	Sheet 9: Fine thread with 4 mm pitch in thread diameters from 40 to 300 mm, basic sizes
DIN 13-1970	Sheet 10: Fine thread with 6 mm pitch in thread diameters from 70 to 500 mm, basic sizes
DIN 13-1969	Sheet 12: Coarse and fine pitches from 1 to 300 mm diameter, selection of diameters and pitches
DIN 13-1972	Sheet 13: Review of threads for bolts and nuts from 1 to 52 mm thread diameter, and limiting sizes (5)
DIN 13-1972	Sheet 14: Bases of the tolerance system for threads of 1 mm diameter and larger (6)
DIN 13-1972	Sheet 15: Basic allowances and tolerances for threads of 1 mm diameter and larger (7)
DIN 13-1972	Sheet 19: Basic profile and production profiles
DIN 13-1972	Sheet 20: Limiting sizes for coarse threads from 1 to 68 mm nominal diameters with commonly used tolerance zones
DIN 13-1944	Suppl. Sheet 14: Metric screw threads; screw threads for interference fit fastenings without sealing action (for inserted ends of studs)
DIN 13-1944	Suppl. Sheet 15: Metric screw threads; screw threads for interference fit fastenings with sealing action (for inserted ends of studs)
DIN 14-1971	Sheet 1: ISO metric screw thread; threads under 1 mm diameter, basic profile

WHITWORTH SCREW THREADS

DIN 259-1966	Sheet 1: Whitworth pipe thread; parallel internal and parallel external thread, nominal dimensions
DIN 259-1966	Sheet 2: Parallel internal and parallel external thread, tolerances
DIN 259-1966	Sheet 3: Parallel internal and parallel external thread, limits
DIN 2999-1960	Whitworth pipe threads for threaded tubes and fittings; parallel internal thread and tapered external thread
DIN 3858-1970	Whitworth pipe threads; parallel internal thread and tapered external thread for pipe unions

ACME SCREW THREADS, BUTTRESS THREADS AND KNUCKLE THREADS

DIN 103-1970	Sheet 1: ISO metric trapezoidal screw thread; profiles
DIN 103-1970	Sheet 2: General plan
DIN 103-1970	Sheet 3: Allowances and tolerances for general purpose trapezoidal screw threads (6)
DIN 103-1971	Sheet 4: Basic sizes
DIN 405-1928	Knuckle thread
DIN 513-1927	Buttress thread; single-start
DIN 514-1927	Sheet 2: Fine pitch, single-start
DIN 515-1927	Coarse pitch, single-start
DIN 7273-1970	Sheet 1: Knuckle threads for steel sheet pieces up to 0.5 mm thickness and appropriated couplings; dimensions, tolerances

U. K.

BS 21-1973	Pipe threads for tubes and fittings where pressure-tight joints are made on the threads
BS 919 Part 3-1968	Gages for ISO metric screw threads
BS 2779-1973	Pipe threads where pressure-tight joints are made on the threads
BS 3643	ISO metric screw threads
Part 1-1963	Thread data and standard thread series
Part 2-1966	Limits and tolerances for coarse pitch series threads
Part 3-1967	Limits and tolerances for fine pitch threads (constant pitch series)
BS 4827-1972	ISO Miniature screw threads

AUSTRALIA

AS 1014-1971	Gaging of metric screw threads
AS 1098-1972	Roller-type screw calliper gages
AS 1275-1972	Metric screw threads for fasteners

9 Fasteners

Introduction

The national standards[1] for fasteners are shown in a World Metric Fastener Standards Index, which follows; and the comparison of standards on metric screw threads around the world can be found in Table 8-1. A comprehensive description of both the Modified ISO Fastener System (OMFS) and the existing ISO fasteners will be given later in this chapter. The fasteners from the two systems can be interchanged, although the inspection of parts must be as specified in the appropriate national standard since the minimum thread root radius is larger in the new ISO system.

The OMFS fasteners are generally stronger and more economical to produce than the existing ISO hardware, and therefore are recommended for use wherever they are available. The following index provides an illustration of each type of fastener with its name in English (E), German (G), French (F), and Italian (I). The applicable national standard number is shown with references as to where to find specific information.

GENERAL SPECIFICATIONS FOR FASTENERS

How to Order Metric Hardware

An Order Check List—The proper designation of modified or existing ISO metric bolts, screws, and nuts should include the following information:

1. General product description such as: bolts, screws, nuts, slotted nuts, rivets, etc., and material if other than steel. Refer to the world index for the fastener name in the required language.
2. The letter *M* is used for a product with modified or existing ISO screw threads (Table 8-3), and the thread pitch is specified for modified ISO fasteners. The designations are as follows:

   ISO Modified: M10 × 1.5 10 mm nominal diameter modified ISO metric screw thread with standard 1.5 mm coarse thread pitch.

   ISO Existing: M10 10 mm nominal diameter ISO metric screw thread with standard 1.5 mm coarse thread pitch.

   M10 × 1.25 10 mm nominal diameter ISO metric screw thread with standard 1.25 mm fine thread pitch.

3. Thread fit designation. Refer to Section 8, page 8-20 (modified ISO) and page 8-13 (existing ISO) for thread fit details.

   Modified ISO:
   M12 × 1.75   Designates standard thread fit 6H/6g approximately equal to SAE class 2 fit.

   M12 × 1.75C   Designates close thread fit—6H/5g6g approximately equal to SAE class 3 fit.

   Existing ISO:
   M12-6H/6g   Designates medium thread fit—6H nut thread tolerance/6g pitch diameter tolerance.

   M12-5H/4h   Designates close thread fit—5H nut thread tolerance/4h pitch diameter tolerance.

   M12-7H/8g   Designates loose thread fit—7H nut thread tolerance/8g pitch diameter tolerance.

For modified ISO fasteners no thread fit call-out is made for standard fit hardware. For close thread fit fasteners a "C" must be added adjacent to the nominal size and thread pitch designations. The medium thread fit is customary in most countries, and need not be specified when ordering fasteners to a specific standard.

---

[1] For information about the term "standard" as used in this book, please see page 1-2.

World Metric Fastener Standards Index

STANDARD	FIGURE	NAME	REFERENCE
NO. 1 ISO 4014,4015 4016,272 U.S.A. ANSI B18.2.3 JAPAN JIS B1180 GERMANY DIN 913,960 FRANCE NF E27-311 U.K. BS 3692 ITALY UNI 5737,5738 AUSTRALIA AS 1110		(E) HEX HEAD CAP SCREW (G) SECHSKANTSCHRAUBE MIT SCHAFT (F) VIS A TETE HEXAGONALE (I) VITE A TESTA ESAGONALE	FOR DESIGN DETAILS SEE TABLE 9-37 (CAP SCREW) TABLE 9-42 (BOLT) NOTE: THE ISO DESIGNATES PRODUCTS PARTIALLY THREADED AS BOLTS.
NO. 2 ISO 4017,4018 272 U.S.A. ANSI B18.2.3 JAPAN JIS GERMANY DIN 933,961 FRANCE NF E27-310 U.K. BS 4190 (BOLTS) ITALY UNI 5739,5740 AUSTRALIA AS 1111(BOLTS)		(E) HEX HEAD CAP SCREW THREADED TO HEAD (G) SECHSKANTSCHRAUBE MIT GEWINDE BIS KOPF (F) VIS A TETE HEXAGONALE FILETEE JUSQU'A PROXIMITE DE LA TETE (I) VITE A TESTA ESAGONALE	FOR DESIGN DETAILS SEE TABLE 9-37 NOTE: THE ISO DESIGNATES PRODUCTS THREADED TO HEAD AS SCREWS
NO. 3 ISO 272 U.S.A. ANSI B18.2.3 JAPAN JIS B1186 GERMANY DIN 6914 FRANCE NF E27-711 U.K. BS 4395 ITALY UNI 5712 AUSTRALIA AS 1252		(E) HEX HEAD BOLT WITH LARGE HEAD (HIGH STRENGTH STRUCTURAL) (G) SECHSKANTSCHRAUBE MIT GROSSER SCHLUSSEL-WEITE (F) VIS A TETE HEXAGONALE LARGE, A COLERETTE (I) VITE A TESTA ESAGONALE LARGA	FOR DESIGN DETAILS SEE IFI-526 AVAILABLE FROM INDUSTRIAL FASTENER INSTITUTE, 1505 EAST OHIO BUILDING CLEVELAND, OHIO U.S.A. 44114
NO. 4 ISO U.S.A. ANSI B18.5.1 JAPAN JIS B1171 GERMANY DIN 603 FRANCE WF E27-350 U.K. BS 4933 ITALY UNI 5731,5732 AUSTRALIA AS 1390		(E) ROUND HEAD SQUARE NECK BOLT (CARRIAGE BOLT) (G) FLACHRUNDSCHRAUBE MIT VIERKANTANSATZ (F) VIS A TETE BOMBEE A COLLET CARRE (I) VITE A TESTA TONDA LARGA CON QUARDRA SOTTOTESTA	FOR DESIGN DETAILS SEE TABLE 9-44
NO. 5 ISO 5713 U.S.A. ANSI B18.5.1 JAPAN JIS B1179 GERMANY DIN 608 FRANCE NF E27-354 U.K. BS 4933 ITALY UNI 5735,6104 AUSTRALIA AS		(E) ROUND HEAD COUNTERSUNK SQUARE NECK BOLT (PLOW BOLT) (G) SENKSCHRAUBE MIT VIERKANTANSATZ (F) VIS A TETE FRAISEE COLLET CARRE (I) VITE A TESTA SVASTA CON QUADRA SOTTOTESTA	FOR DESIGN DETAILS SEE TABLE 9-46

FASTENERS

World Metric Fastener Standards Index

STANDARD	FIGURE	NAME	REFERENCE
NO. 6 ISO 272 U.S.A. ANSI JAPAN JIS B1182 GERMANY DIN 479 FRANCE NF E27-311 U.K. BS ITALY UNI 5726,5728 AUSTRALIA AS		(E) SQUARE HEAD BOLT (G) VIERKANTSCHRAUBE (F) VIS A TETE CARREE (I) VITE A TESTA QUADRA	FOR DESIGN DETAILS SEE NATIONAL STANDARDS AVAILABLE FROM ANSI 1430 BROADWAY NEW YORK, N.Y. U.S.A. 10018
NO. 7 ISO U.S.A. ANSI B18.2.5 JAPAN JIS GERMANY DIN FRANCE NF U.K. BS ITALY UNI AUSTRALIA AS		(E) 12 POINT SPLINE FLANGE SCREW (G) 12 PUNKT SCHRAUBE MIT BUND (F) VIS A EMBASE TETE CRANTEE (I) VITE A TESTA CON BORDINO	FOR DESIGN DETAILS SEE TABLE 9-47
NO. 8 ISO 861,4762 U.S.A. ANSI B18.3.1 JAPAN JIS B1176 GERMANY DIN 912 FRANCE NF E27-161 U.K. BS 4168 ITALY UNI 5931,5932 AUSTRALIA AS 1420		(E) HEX SOCKET HEAD (NORMAL) CAP SCREW (G) ZYLINDERSCHRAUBE MIT INNENSECHSKANT (F) VIS A TETE CYLINDRIQUE A SIX PANS CREUX (I) VITE A TESTA CILINDRICA CON ESAGONO INCASSATO	FOR DESIGN DETAILS SEE TABLE 9-52
NO. 9 ISO U.S.A. ANSI B18.3.1 JAPAN JIS GERMANY DIN 7984,6912 FRANCE NF U.K. BS ITALY UNI AUSTRALIA AS		(E) HEX SOCKET HEAD (LOW) CAP SCREW (G) ZYLINDERSCHRAUBE MIT INNENSECHSKANT UND NIEDRIGEM KOPF (F) VIS A TETE CYLINDRIQUE A SIX PANS CREUX (TETE REDUITE) (I) VITE A TESTA CILINDRICA CON ESAGONO INCASSATO (TESTA RIDOTTA)	FOR DESIGN DETAILS SEE NATIONAL STANDARDS AVAILABLE FROM ANSI 1430 BROADWAY NEW YORK, N.Y. U.S.A. 10018
NO. 10 ISO U.S.A. ANSI B18.3.1 JAPAN JIS GERMANY DIN 7991 FRANCE NF U.K. BS 4168 ITALY UNI 5933,5934 AUSTRALIA AS		(E) HEX SOCKET COUNTERSUNK (FLAT) HEAD SCREW (G) SENKSCHRAUBE MIT INNENSECHSKANT (F) VIS A TETE FRAISEE A SIX PANS CREUX (I) VITE A TESTA SVASATA CON ESAGONO INCASSATO	FOR DESIGN DETAILS SEE SS-205 AVAILABLE FROM SOCKET SCREW PRODUCTS BUREAU 331 MADISON AVE. NEW YORK, N.Y. U.S.A. 10017

World Metric Fastener Standards Index

STANDARD	FIGURE	NAME	REFERENCE
NO. 11 ISO 2009 U.S.A. ANSI B18.6.7 JAPAN JIS B1101 / B1111 GERMANY DIN 963,965 FRANCE NF E27-113 U.K. BS 4183 ITALY UNI 6109 AUSTRALIA AS 1427	SLOTTED CROSS RECESSED	(E) COUNTERSUNK FLAT HEAD MACHINE SCREW WITH SLOT OR CROSS RECESS (G) SENKSCHRAUBEN MIT SCHLITZ ODER KREUZ-SCHLITZ (F) VIS A TETE FRAISEE AVEC FENDUE OU EMPREINTE CRUCIFORME (I) VITE A TESTA SVASATA CON INTAGLIO O INTAGLIO A CROCE	FOR DESIGN DETAILS SEE TABLE 9-54A (SLOTTED) TABLE 9-54B (CROSS RECESSED)
NO. 12 ISO 2010 U.S.A. ANSI B18.6.7 JAPAN JIS B1101 / B1111 GERMANY DIN 964,966 FRANCE NF E27-113 U.K. BS 4183 ITALY UNI 6110 AUSTRALIA AS 1427	SLOTTED CROSS RECESSED	(E) COUNTERSUNK OVAL HEAD MACHINE SCREW WITH SLOT OR CROSS RECESS (G) LINSENSENKSCHRAUBE MIT SCHLITZ ODER KREUZSCHLITZ (F) VIS A TETE FRAISEE BOMBEE AVEL FENDUE OU EMPREINTE CRUCIFORME (I) VITE A TESTA SVASATA CON CALOTTA INTAGLIO O INTAGLIO A CROCE	FOR DESIGN DETAILS SEE TABLE 9-55A (SLOTTED) TABLE 9-55B (CROSS RECESSED)
NO. 13 ISO 1580 U.S.A. ANSI B18.6.7 JAPAN JIS B1101 / B1111 GERMANY DIN 85,7986 FRANCE NF E27-116 U.K. BS 4183 ITALY UNI 6107 AUSTRALIA AS 1427	SLOTTED CROSS RECESSED	(E) PAN HEAD MACHINE SCREW WITH SLOT OR CROSS RECESS (G) FLACHKOPFSCHRAUBE MIT SCHLITZ ODER KREUZSCHLITZ (F) VIS A TETE CYLINDRI-QUE A DEPOUILLE AVEC FENDUE OU EMP CRUCIFORME (I) VITE A TESTA CILINDRICA CON CALOTTA INTAGLIO O INTAGLIO A CROCE	FOR DESIGN DETAILS SEE TABLE 9-56A (SLOTTED) TABLE 9-56B (CROSS RECESSED)
NO. 14 ISO 272 U.S.A. ANSI B18.6.7 JAPAN JIS GERMANY DIN FRANCE NF U.K. BS ITALY UNI AUSTRALIA AS 1427	PLAIN HEX HEX WASHER	(E) HEX HEAD MACHINE SCREW WITH PLAIN OR WASHER HEAD (G) SECHSKANTSCHRAUBE OHNE BUND ODER MIT BUND (F) VIS A TETE HEXAGONALE SANS EMBASE OU AVEC EMBASE (I) VITE A TESTA ESAGONALE SENZA BORDINA O CON BORDINA	FOR DESIGN DETAILS SEE TABLE 9-57A (PLAIN HEX) TABLE 9-57B (HEX WASHER)
NO. 15 ISO U.S.A. ANSI B18.17.1 JAPAN JIS B1184 GERMANY DIN 316 FRANCE NF U.K. BS ITALY UNI 5449 AUSTRALIA AS		(E) WING SCREW (G) FLUGELSCHRAUBE (F) VIS A OREILLES (I) VITE AD ALETTE	FOR DESIGN DETAILS SEE NATIONAL STANDARDS AVAILABLE FROM ANSI 1430 BROADWAY NEW YORK, N.Y. U.S.A. 10018

FASTENERS

World Metric Fastener Standards Index

STANDARD	FIGURE	NAME	REFERENCE
NO. 16 ISO 272 U.S.A. ANSI B18.13.1 JAPAN JIS GERMANY DIN 6900 FRANCE NF U.K. BS ITALY UNI AUSTRALIA AS		(E) HEX HEAD SCREW AND WASHER ASSEMBLY (SEMS) (G) KOMBI-SCHRAUBEN (F) VIS A TETE HEXAGO-NALE AVEC RONDELLE (I) VITE A TESTA ESAGONALE CON ROSETTA	FOR DESIGN DETAILS SEE IFI-531 AVAILABLE FROM INDUSTRIAL FASTENER INSTITUTE 1505 EAST OHIO BLDG CLEVELAND, OHIO U.S.A. 44114
NO. 17 ISO U.S.A. ANSI JAPAN JIS B1173 GERMANY DIN 938,939 FRANCE NF E27-241 U.K. BS 4439 ITALY UNI 5909-5919 AUSTRALIA AS		(E) STUD (G) STIFTSCHRAUBE (F) GOUJON (I) VITE PRIGIONIERA	FOR DESIGN DETAILS SEE TABLE 9-58
NO. 18 ISO 272 U.S.A. ANSI JAPAN JIS B1118 GERMANY DIN 478,479 480 FRANCE NF E27-110 U.K. BS ITALY UNI 6050-6053 AUSTRALIA AS	FOR POINT TYPES AND DETAILS SEE BELOW	(E) SQUARE HEAD SET SCREW (G) VIERKANTSCHRAUBE (F) VIS A TETE CARREE (I) VITE A TESTA QUADRA	FOR DESIGN DETAILS SEE NATIONAL STANDARDS AVAILABLE FROM ANSI 1430 BROADWAY NEW YORK, N.Y. U.S.A. 10018
NO. 19 ISO 2342 U.S.A. ANSI JAPAN JIS B1117 GERMANY DIN 551,553 417,438 FRANCE NF E27-110 U.K. BS 4219 ITALY UNI 6113-6119 AUSTRALIA AS	FOR POINT TYPES AND DETAILS SEE BELOW	(E) SLOTTED HEADLESS SET SCREW (G) SCHAFTSCHRAUBE MIT SCHLITZ (F) VIS SANS TETE FENDUE (I) VITE SENZA TESTA CON INTAGLIO	FOR DESIGN DETAILS SEE NATIONAL STANDARDS AVAILABLE FROM ANSI 1430 BROADWAY NEW YORK, N.Y. U.S.A. 10018
NO. 20 ISO 2343 4026-4029 U.S.A. ANSI B18.3.1 JAPAN JIS B1177 GERMANY DIN 913,914 915,916 FRANCE NF E27-162 U.K. BS 4168 ITALY UNI 5923-5930 AUSTRALIA AS 1421	ISO POINT TYPES FP TP DP CP FLAT CONE DOG CUP	(E) HEX SOCKET HEADLESS SET SCREW (G) GEWINDESTIFT MIT INNENSECHSKANT (F) VIS SANS TETE A SIX PANS CREUX (I) VITE SENZA TESTA CON ESAGONO INCASSATO	FOR DESIGN DETAILS SEE TABLE 9-59 (FLAT POINT) TABLE 9-60 (CONE POINT) TABLE 9-61 (DOG POINT) TABLE 9-62 (CUP POINT)

World Metric Fastener Standards Index

STANDARD	FIGURE	NAME	REFERENCE

NO. 21
U.S.A.   ANSI B18.6.5, 8
              B18.6.6
JAPAN   JIS
GERMANY DIN 7970, 7513
              7516
FRANCE NF E27-131-3
U.K.   BS 4174
ITALY   UNI
AUSTRALIA AS B194
	ANSI TYPE
	AB B
	BF BT
	F T
	SF TT
		(E) TAPPING SCREW TYPE DESIGNATION, THREAD AND POINT DETAILS
		(G) BLECH UND SCHNEID-SCHRAUBE BENENNUNG, GEWINDE UND ANSATZ EINZELHEITEN
		(F) VIS A TOLE-DESIGNATION, FILETAGE ET POINTU DETAILS
		(I) VITI AUTOFILETTANTI TIPO, FILETTATURA E PUNTE PARTICOLARI
			FOR DESIGN DETAILS SEE
			TABLE 9-63 (DESIGNATION)
			TABLE 9-73 (LARGE PITCH)
			TABLE 9-74 (ISO METRIC PITCH)
			TABLE 9-75 (THREAD ROLLING)
			TABLE 9-76 (SELF DRILLING)

NO. 22 U.S.A. ANSI B18.6.5 JAPAN JIS B1115, B1122 GERMANY DIN 7972, 7982 7513 FRANCE NF E27-132 U.K. BS 4174 ITALY UNI 6952, 6955 AUSTRALIA AS B194	SLOTTED CROSS RECESSED	(E) COUNTERSUNK FLAT HEAD TAPPING SCREW WITH SLOT OR CROSS RECESS (G) SENK-BLECHSCHRAUBE MIT SCHLITZ ODER KREUZSCHLITZ (F) VIS A TOLE A TETE FRAISEE AVEC FENDUE OU EMPREINTE CRUCIFORME (I) VITI AUTOFILETTANTI A TESTA SVASATA CON INTAGLIO O INTAGLIO A CROCE	FOR DESIGN DETAILS SEE TABLE 9-66 (SLOTTED) TABLE 9-67 (CROSS RECESSED)
NO. 23 U.S.A. ANSI B18.6.5 JAPAN JIS B1115 B1122 GERMANY DIN 7973, 7983 7513 FRANCE NF E27-132 U.K. BS 4174 ITALY UNI 6953, 6956 AUSTRALIA AS B194	SLOTTED CROSS RECESSED	(E) COUNTERSUNK OVAL HEAD TAPPING SCREW WITH SLOT OR CROSS RECESS (G) LINSENSENK-BLECH-SCHRAUBE MIT SCHLITZ ODER KREUZSCHLITZ (F) VIS A TOLE A TETE FRAISEE BOMBEE AVEC FENDUE OU EMP CRUCIFORME (I) VITI AUTOFILETTANTI A TESTA SVASATA CON CALOTTA INTAGLIO O A CROCE	FOR DESIGN DETAILS SEE TABLE 9-68 (SLOTTED) TABLE 9-69 (CROSS RECESSED)
NO. 24 U.S.A. ANSI B18.6.5 JAPAN JIS B1115 B1122 GERMANY DIN 7971, 7981 7513 FRANCE NF E27-131 U.K. BS 4174 ITALY UNI 6951, 6954 AUSTRALIA AS B194	SLOTTED CROSS RECESSED	(E) PAN HEAD TAPPING SCREW WITH SLOT OR CROSS RECESS (G) FLACHKOPF-BLECH-SCHRAUBE MIT SCHLITZ ODER KREUZSCHLITZ (F) VIS A TOLE A TETE CYLINDRIQUE AVEC FENDUE OU EMPREINTE CRUCIFORME (I) VITI AUTOFILETTANTI A TESTA CILINDRICA CON CALOTTA INTAGLIO O A CROCE	FOR DESIGN DETAILS SEE TABLE 9-70 (SLOTTED) TABLE 9-71 (CROSS RECESSED)
NO. 25 U.S.A. ANSI B18.6.5 JAPAN JIS B1123 GERMANY DIN 7976, 7513 FRANCE NF E27-133 U.K. BS 4174 ITALY UNI 6949 AUSTRALIA AS B194	PLAIN HEX HEX WASHER	(E) HEX HEAD TAPPING SCREW WITH PLAIN OR WASHER HEAD (G) SECHSKANT-BLECH-SCHRAUBE OHNE BUND ODER MIT BUND (F) VIS A TOLE A TETE HEXAGONALE SANS EMBASE OU AVEC EMBASE (I) VITI AUTOFILETTANTI A TESTA ESAGONALE SENZA BORDINA O CON BORDINA	FOR DESIGN DETAILS SEE TABLE 9-72A (PLAIN HEX) TABLE 9-72B (HEX WASHER)

FASTENERS

World Metric Fastener Standards Index

STANDARD	FIGURE	NAME	REFERENCE
NO. 26 ISO U.S.A. ANSI B18.6.4 JAPAN JIS GERMANY DIN 7514 FRANCE NF U.K. BS 4174 ITALY UNI 5241 AUSTRALIA AS		(E) ROUND HEAD DRIVE SCREW (G) HALBRUND-NAGEL-SCHRAUBE (F) FAUSSE-VIS A TETE RONDE (I) CHIODO FILETTATO A TESTA TONDA	FOR DESIGN DETAILS SEE NATIONAL STANDARDS AVAILABLE FROM ANSI 1430 BROADWAY NEW YORK, N.Y. U.S.A. 10018
NO. 27 ISO U.S.A. ANSI JAPAN JIS GERMANY DIN FRANCE NF U.K. BS ITALY UNI AUSTRALIA AS		(E) PROJECTION WELD STUD - ANSI TYPE U3 (G) SCHWEISS-BOLZEN (F) VIS SOUDABLE (I) VITE SALDATURA	FOR DESIGN DETAILS CONTACT INDUSTRIAL FASTENER INSTITUTE, 1505 EAST OHIO BUILDING CLEVELAND, OHIO U.S.A. 44114
NO. 28 ISO U.S.A. ANSI JAPAN JIS GERMANY DIN FRANCE NF U.K. BS ITALY UNI AUSTRALIA AS		(E) PROJECTION WELD STUD - ANSI TYPE T3 (G) SCHWEISS-BOLZEN (F) VIS SOUDABLE (I) VITE SALDATURA	FOR DESIGN DETAILS CONTACT INDUSTRIAL FASTENER INSTITUTE, 1505 EAST OHIO BUILDING CLEVELAND, OHIO U.S.A. 44114
NO. 29 ISO U.S.A. ANSI JAPAN JIS B1135 B1112 GERMANY DIN 97,7997 95,7995 FRANCE NF E27-142,3 U.K. BS ITALY UNI 702 AUSTRALIA AS 1476	FLAT SLOTTED OVAL SLOTTED CROSS RECESSED	(E) COUNTERSUNK FLAT AND OVAL HEAD WOOD SCREW WITH SLOT OR CR RECESS (G) SENK-UND LINSENSENK HOLZSCHRAUBE MIT SCHLITZ ODER KREUZSCHLITZ (F) VIS A BOIS A TETE FRAISEE OU BOMBEE AVEC FENDUE OU CRUCIFORME (I) VITE PER LEGNO A TESTA SVASATA O CON CALOTTA CON INTAGLIO O INTAGLIO O A CROCE	FOR DESIGN DETAILS SEE NATIONAL STANDARDS AVAILABLE FROM ANSI 1430 BROADWAY NEW YORK, N.Y. U.S.A. 10018
NO. 30 ISO U.S.A. ANSI JAPAN JIS B1135 B1112 GERMANY DIN 96,7996 FRANCE NF E27-141 U.K. BS ITALY UNI 701 AUSTRALIA AS 1476	SLOTTED CROSS RECESSED	(E) ROUND HEAD WOOD SCREWS WITH SLOT OR CROSS RECESS (G) HALBRUND-HOLZSCHR-AUBE MIT SCHLITZ ODER KREUZSCHLITZ (F) VIS A BOIS A TETE RONDE AVEL FENDUE OU EMPREINTE CRUCIFORME (I) VITE PER LEGNO A TESTA TONDA CON INTAGLIO O INTAGLIO A CROCE	FOR DESIGN DETAILS SEE NATIONAL STANDARDS AVAILABLE FROM ANSI 1430 BROADWAY NEW YORK, N.Y. U.S.A. 10018

World Metric Fastener Standards Index

STANDARD	FIGURE	NAME	REFERENCE
NO. 31 ISO U.S.A. ANSI JAPAN JIS GERMANY DIN 571,570 FRANCE NF E27-140,4 U.K. BS ITALY NUI 704,705 AUSTRALIA AS 1393	PLAIN HEX SQUARE	(E) HEX AND SQUARE HEAD WOOD SCREW (G) SECHS-UND VIER-KANT HOLZSCHRAUBE (F) VIS A BOIS A TETE HEXAGONAL ET CARREE (I) VITE PER LEGNO A TESTA ESAGONALE O QUADRA	FOR DESIGN DETAILS SEE NATIONAL STANDARDS AVAILABLE FROM ANSI 1430 BROADWAY NEW YORK, N.Y. U.S.A. 10018
NO. 32 ISO 4032,4033 4034 U.S.A. ANSI B18.2.4 JAPAN JIS B1181 GERMANY DIN 934,555 FRANCE NF E27-411 U.K. BS 3692 ITALY UNI 5587,5588 AUSTRALIA AS 1112		(E) HEXAGON NUT (G) SECHSKANTMUTTER (F) ECROU HEXAGONALE (I) DADO ESAGONALE	FOR DESIGN DETAILS SEE TABLE 9-79
NO. 33 ISO 4035,4036 U.S.A. ANSI B18.2.4 JAPAN JIS GERMANY DIN 439,936 FRANCE NF E27-411 U.K. BS 3692 ITALY UNI 5589,5590 AUSTRALIA AS 1112		(E) HEXAGON JAM (LOW) NUT (G) FLACHE SECHSKANT-MUTTER (F) ECROU HEXAGONALE BAS (I) DADO ESAGONALE BASSO	FOR DESIGN DETAILS SEE NATIONAL STANDARDS AVAILABLE FROM ANSI 1430 BROADWAY NEW YORK, N.Y. U.S.A. 10018
NO. 34 ISO U.S.A. ANSI JAPAN JIS B1186 GERMANY DIN 6915 FRANCE NF E27-711 U.K. BS 4395 ITALY UNI 5713 AUSTRALIA AS 1252		(E) LARGE HEX NUTS (HIGH STRENGTH STRUCTURAL) (G) SECHSKANTMUTTER MIT GROSSER SCHLUSSEL-WEITE (F) ECROU HEXAGONALE LARGE (I) DADO ESAGONALE LARGO	FOR DESIGN DETAILS SEE IFI-526 AVAILABLE FROM INDUSTRIAL FASTENER INSTITUTE, 1505 EAST OHIO BUILDING CLEVELAND, OHIO U.S.A. 44114
NO. 35 ISO U.S.A. ANSI B18.2.4 JAPAN JIS B1183 GERMANY DIN 6331 FRANCE NF E27-452 U.K. BS ITALY UNI AUSTRALIA AS		(E) HEXAGON NUT WITH FLANGE (G) SECHSKANTMUTTER MIT BUND (F) ECROU HEXAGONALE A EMBASE (I) DADO ESAGONALE ALTO CON BORDINO	FOR DESIGN DETAILS SEE IFI-507 AVAILABLE FROM INDUSTRIAL FASTENER INSTITUTE, 1505 EAST OHIO BUILDING CLEVELAND, OHIO U.S.A. 44114

FASTENERS

World Metric Fastener Standards Index

STANDARD	FIGURE	NAME	REFERENCE
NO. 36 ISO U.S.A. ANSI JAPAN JIS B1163 GERMANY DIN 557,562 FRANCE NF E27-411 U.K. BS ITALY UNI 5596,5597 AUSTRALIA AS		(E) SQUARE NUT (G) VIERKANTMUTTER (F) ECROU CARRE (I) DADO QUADRO	FOR DESIGN DETAILS SEE NATIONAL STANDARDS AVAILABLE FROM ANSI 1430 BROADWAY NEW YORK, N.Y. U.S.A. 10018
NO. 37 ISO 2358,2320 U.S.A. ANSI B18.16.1 JAPAN JIS GERMANY DIN 980-V FRANCE NF U.K. BS 4929 ITALY UNI AUSTRALIA AS 1285		(E) PREVAILING TORQUE STEEL HEX LOCKNUT (G) SELBSTSICHERENDE (STAHL) SECHSKANTMUTTER (F) ECROUS DE BLOCAGE (DEFORME) (I) DADO ESAGONALE DI BLOCCAGGIO	FOR DESIGN DETAILS SEE TABLE 9-81
NO. 38 ISO 2358,2320 U.S.A. ANSI JAPAN JIS GERMANY DIN 980-N FRANCE NF U.K. BS 4929 ITALY UNI AUSTRALIA AS		(E) HEX LOCKNUT WITH NYLON INSERT (G) SELBSTSICHERENDE SECHSKANTMUTTER (SICHER-UNGSTEIL AUS NICHTMETAL) (F) ECROU DE SECURITE (NYLON) (I) DADO ESAGONALE DI BLOCCAGGIO CON INSERTO DI NYLON	FOR DESIGN DETAILS SEE TABLE 9-81
NO. 39 ISO 288 U.S.A. ANSI B18.2.4 JAPAN JIS B1170 GERMANY DIN 935,979 FRANCE NF E27-414 U.K. BS 3692 ITALY UNI 5593,5594 AUSTRALIA AS 1112		(E) HEX SLOTTED AND CASTLE NUT (G) KRONENMUTTER (F) ECROU A CRENEAUX (I) DADO ESAGONALE AD INTAGLI	FOR DESIGN DETAILS SEE TABLE 9-83
NO. 40 ISO U.S.A. ANSI B18.17.1 JAPAN JIS B1185 GERMANY DIN 315 FRANCE NF E27-454 U.K. BS 856 ITALY UNI 5448 AUSTRALIA AS		(E) WING NUT (G) FLUGELMUTTER (F) ECROU A OREILLES (I) DADO AD ALETTE	FOR DESIGN DETAILS SEE NATIONAL STANDARDS AVAILABLE FROM ANSI 1430 BROADWAY NEW YORK, N.Y. U.S.A. 10018

World Metric Fastener Standards Index

STANDARD	FIGURE	NAME	REFERENCE
NO. 41 ISO U.S.A. ANSI JAPAN JIS GERMANY DIN 929 FRANCE NF U.K. BS ITALY UNI AUSTRALIA AS		(E) PROJECTION HEX WELD NUT (G) SECHSKANT-SCHWEISSMUTTER (F) ECROU SOUDABLE (I) DADO SALDATURA ESAGONALE	FOR DESIGN DETAILS SEE TABLE 9-84
NO. 42 ISO 887 U.S.A. ANSI B18.22.2 JAPAN JIS B1256 GERMANY DIN 125,126 FRANCE NF E27-611 U.K. BS 4320 ITALY UNI 6592,6593 AUSTRALIA AS 1237		(E) FLAT WASHER (G) SCHEIBE (F) ROUNDELLE PLATE (I) ROSETTA PIANA	FOR DESIGN DETAILS SEE TABLE 9-85
NO. 43 ISO U.S.A. ANSI B18.21.2 JAPAN JIS B1251 GERMANY DIN 127,7980 FRANCE NF E27-622 U.K. BS 4464,856 ITALY UNI 1751,1752 AUSTRALIA AS		(E) SPRING LOCK WASHER (G) FEDERRING (F) RONDELLE A RESSORT (GROWER) (I) ROSETTA ELASTICA	FOR DESIGN DETAILS SEE TABLE 9-86 (NORMAL) TABLE 9-87 (HEAVY)
NO. 44 ISO U.S.A. ANSI B18.23.2 JAPAN JIS B1252 GERMANY DIN 6796,6908 FRANCE NF U.K. BS ITALY UNI AUSTRALIA AS		(E) CONICAL SPRING LOCK WASHER (G) SPANNSCHEIBE (F) RONDELLE RESSORT (RONDELLE BELLEVILLE) (I) ROSETTA ELASTICA DI FORMA CONICA	FOR DESIGN DETAILS SEE NATIONAL STANDARDS AVAILABLE FROM ANSI 1430 BROADWAY NEW YORK, N.Y. U.S.A. 10018
NO. 45 ISO U.S.A. ANSI B18.23.2 JAPAN JIS GERMANY DIN 137 FRANCE NF U.K. BS 4463 ITALY UNI AUSTRALIA AS		(E) CURVED SPRING WASHER (G) GEWELLTE FEDER-SCHEIBE (F) RONDELLE ELASTIQUE ONDULEE (I) ROSETTA ELASTICA ONDULATA	FOR DESIGN DETAILS SEE NATIONAL STANDARDS AVAILABLE FROM ANSI 1430 BROADWAY NEW YORK, N.Y. U.S.A. 10018

FASTENERS

World Metric Fastener Standards Index

STANDARD	FIGURE	NAME	REFERENCE
NO. 46 ISO U.S.A. ANSI B18.21.2 JAPAN JIS B1255 GERMANY DIN 6797,6798 FRANCE NF E27-618 U.K. BS ITALY UNI 3703,3704 3705,3706 AUSTRALIA AS		(E) LOCK WASHER WITH INTERNAL OR EXTERNAL TEETH (G) FEDERNDE ZAHNSCHEIBE INNENGEZAHNT ODER AUSSENGEZAHNT (F) ROUNDELLE ELASTIQUE A DENTURE INTERIEURE OU EXTERIEURE (I) ROSETTA ELASTICA PIANA CON DENTURA INTERNA O ESTERNA	FOR DESIGN DETAILS SEE IFI-532 AVAILABLE FROM INDUSTRIAL FASTENER INSTITUTE, 1505 EAST OHIO BUILDING CLEVELAND, OHIO U.S.A 44114
NO. 47 ISO U.S.A. ANSI JAPAN JIS GERMANY DIN 93,463 FRANCE NF E27-614 U.K. BS ITALY UNI 6599,6560 AUSTRALIA AS		(E) WASHER WITH ONE OR TWO LOCKING TABS (G) SICHERUNGSBLECH MIT LAPPEN (EIN ODER ZWEI LAPPEN) (F) FREIN D'ECROU A AILERON (UN OU DEUX AILERONS) (I) ROSETTA DI SICUREZZA CON UNA O DUE LINGUETTE	FOR DESIGN DETAILS SEE NATIONAL STANDARDS AVAILABLE FROM ANSI 1430 BROADWAY NEW YORK, N.Y. U.S.A. 10018
NO. 48 ISO U.S.A. ANSI JAPAN JIS GERMANY DIN 988 FRANCE NF U.K. BS ITALY UNI AUSTRALIA AS		(E) PRECISION FLAT SHIM WASHER (G) PASS-SCHEIBE (F) CALE DE REGLAGE (I) ROSETTA PIANA DI PRICISIONE	FOR DESIGN DETAILS SEE NATIONAL STANDARDS AVAILABLE FROM ANSI 1430 BROADWAY NEW YORK, N.Y. U.S.A. 10018
NO. 49 ISO U.S.A. ANSI B18.23.2 JAPAN JIS GERMANY DIN 434,435 6917,6918 FRANCE NF E27-681 U.K. BS 4395 ITALY UNI 5716,5715 AUSTRALIA AS 1252		(E) SQUARE TAPER WASHER FOR U AND I STRUCTURAL SECTION (G) VIERKANTSCHEIBE FUR U- AND I- TRAGER (F) PLAQUETTE OBLIQUE (POUR PROFILES U ET I) (I) PIASTRINA DE APPOGGIO SU ALI DI TRAVE	FOR DESIGN DETAILS SEE IFI-526 AVAILABLE FROM INDUSTRIAL FASTENER INSTITUTE, 1505 EAST OHIO BUILDING CLEVELAND, OHIO U.S.A. 44114
NO. 50 ISO U.S.A. ANSI JAPAN JIS GERMANY DIN 436 FRANCE NF E27-682 U.K. BS ITALY UNI 6596 AUSTRALIA AS		(E) SQUARE WASHER FOR WOOD CONNECTION (G) VIERKANTSCHEIBE FUR HOLZ VERBINDUNGEN (F) PLAQUETTE CARREE (I) ROSETTA QUADRA	FOR DESIGN DETAILS SEE NATIONAL STANDARDS AVAILABLE FROM ANSI 1430 BROADWAY NEW YORK, N.Y. U.S.A. 10018

World Metric Fastener Standards Index

STANDARD	FIGURE	NAME	REFERENCE
NO. 51 ISO 1051 U.S.A. ANSI JAPAN JIS B1213 GERMANY DIN 661,675 FRANCE NF E27-154 U.K. BS 4620 ITALY UNI 139,2513 752,753 AUSTRALIA AS		(E) FLAT COUNTERSUNK HEAD RIVET (G) SENKNIETE (F) RIVET TETE FRAISEE (I) RIBATTINI A TESTA SVASATA PIANA	FOR DESIGN DETAILS SEE TABLE 9-88 AND NATIONAL STANDARDS AVAILABLE FROM ANSI 1430 BROADWAY NEW YORK, N.Y. U.S.A. 10018
NO. 52 ISO 1051 U.S.A. ANSI JAPAN JIS B1213 GERMANY DIN 302,662 FRANCE NF E27-154 U.K. BS 4620 ITALY UNI 140,2514 754,755 AUSTRALIA AS		(E) OVAL COUNTERSUNK HEAD RIVET (G) LINSEN SENKNIETE (F) RIVET TETE FRAISEE BOMBEE (I) RIBATTINI A TESTA SVASATA CON CALOTTA	FOR DESIGN DETAILS SEE TABLE 9-88 AND NATIONAL STANDARDS AVAILABLE FROM ANSI 1430 BROADWAY NEW YORK, N.Y. U.S.A. 10018
NO. 53 ISO 1051 U.S.A. ANSI JAPAN JIS B1213 GERMANY DIN 124,123 FRANCE NF E27-153 U.K. BS 4620 ITALY UNI 134,136 748,749 AUSTRALIA AS		(E) BUTTON HEAD RIVET (G) HALBRUND NIETE (F) RIVET TETE RONDE (I) RIBATTINI A TESTA TONDA	FOR DESIGN DETAILS SEE TABLE 9-88 AND NATIONAL STANDARDS AVAILABLE FROM ANSI 1430 BROADWAY NEW YORK, N.Y. U.S.A. 10018
NO. 54 ISO 1051 U.S.A. ANSI JAPAN JIS GERMANY DIN 7338 FRANCE NF E27-151 U.K. BS 4620 ITALY UNI 756 AUSTRALIA AS		(E) FLAT HEAD RIVET (G) NIETE FUR BREMS- UND KUPPLUNGSBELAG (F) RIVET A TETE PLATE (I) RIBATTINI A TESTA CILINDRICA	FOR DESIGN DETAILS SEE TABLE 9-88 AND NATIONAL STANDARDS AVAILABLE FROM ANSI 1430 BROADWAY NEW YORK, N.Y. U.S.A. 10018
NO. 55 ISO U.S.A. ANSI B18.7.1 JAPAN JIS GERMANY DIN FRANCE NF U.K. BS ITALY UNI AUSTRALIA AS	PROTRUDING COUNTERSUNK	(E) PROTRUDING OR COUNTERSUNK HEAD BREAK MANDREL BLIND RIVET (G) BLIND ROHRNIETE MIT HALBRUND ODER SENKKOPF (F) RIVET AVEUGLES (POP) AVEC TETE RONDE OU FRAISEE (I) RIBATTINI (CIECHI) CON TESTA TONDA O SVASTA	FOR DESIGN DETAILS SEE TABLE 9-90 (PROTRUDING) TABLE 9-92 (COUNTERSUNK)

FASTENERS

World Metric Fastener Standards Index

STANDARD	FIGURE	NAME	REFERENCE
NO. 56　ISO U.S.A.　ANSI B18.7.1 JAPAN　JIS GERMANY　DIN FRANCE　NF U.K.　BS ITALY　UNI AUSTRALIA AS	PROTRUDING COUNTERSUNK	(E) PROTRUDING OR COUNTERSUNK HEAD BREAK MANDREL CLOSED END BLIND RIVET (G) GESCHLOSSEN BLIND NIETE MIT HALBRUND ODER SENKKOPF (F) RIVET AVEUGLE (POP FERME) AVEC TETE RONDE OU FRAISEE (I) RIBATTINI (CIECHI) CON TESTA TONDA O SVASATA	FOR DESIGN DETAILS SEE TABLE 9-97 (PROTRUDING) TABLE 9-99 (COUNTERSUNK)
NO. 57　ISO 1234 U.S.A.　ANSI JAPAN　JIS B1351 GERMANY　DIN 94 FRANCE　NF E27-487 U.K.　BS 1574 ITALY　UNI 1336 AUSTRALIA AS 1236		(E) SPLIT COTTER PIN (G) SPLINTE (F) GOUPILLE CYLINDRIQUE FENDUE (I) COPIGLIA	FOR DESIGN DETAILS SEE TABLE 9-103
NO. 58　ISO U.S.A.　ANSI IFI-512-S JAPAN　JIS GERMANY　DIN 7346,1481 FRANCE　NF U.K.　BS ITALY　UNI 6873,6874 AUSTRALIA AS		(E) ROLL PIN (G) SPANNHULSE (F) GOUPILLE ELASTIQUE (I) SPINA ELASTICA	FOR DESIGN DETAILS SEE TABLE 9-104A
NO. 59　ISO U.S.A.　ANSI IFI-512-C JAPAN　JIS GERMANY　DIN 7343 FRANCE　NF U.K.　BS ITALY　UNI 6875,6876 AUSTRALIA AS		(E) COILED SPRING PIN (G) SPIRAL-SPANNSTIFT (F) GOUPILLE SPIRALE (I) SPINA ELASTICA A SPIRALE	FOR DESIGN DETAILS SEE TABLE 9-104B
NO. 60　ISO 2338 U.S.A.　ANSI JAPAN　JIS B1354 GERMANY　DIN 7,6325 FRANCE　NF U.K.　BS 1804 ITALY　UNI 1707 AUSTRALIA AS		(E) PARALLEL STEEL DOWEL PIN (G) ZYLINDERSTIFT (F) GOUPILLE CYLINDRIQUE (I) SPINA CILINDRICA	FOR DESIGN DETAILS SEE TABLE 9-108

World Metric Fastener Standards Index

STANDARD	FIGURE	NAME	REFERENCE
NO. 61 ISO 2339 U.S.A. ANSI JAPAN JIS B1352 GERMANY DIN 1 FRANCE NF E27-481 U.K. BS ITALY UNI 129 AUSTRALIA AS		(E) TAPER PIN (G) KEGELSTIFT (F) GOUPILLE CONIQUE (I) SPINA CONICA	FOR DESIGN DETAILS SEE TABLE 9-109
NO. 62 ISO U.S.A. ANSI JAPAN JIS GERMANY DIN 1472,1473 FRANCE NF U.K. BS ITALY UNI AUSTRALIA AS		(E) GROOVE PIN (G) PASSKERBSTIFT (F) GOUPILLE CANNELEE (I) SPINA CON INTAGLIO	FOR DESIGN DETAILS SEE NATIONAL STANDARDS AVAILABLE FROM ANSI 1430 BROADWAY NEW YORK, N.Y. U.S.A. 10018
NO. 63 ISO 2340 U.S.A. ANSI JAPAN JIS GERMANY DIN 1433 FRANCE NF R126.09 U.K. BS ITALY UNI 1707,1709 AUSTRALIA AS		(E) CYLINDRICAL CLEVIS PIN WITHOUT HEAD (G) BOLZEN OHNE KOPF (F) AXE LISSE (I) PERNO SENZA TESTA	FOR DESIGN DETAILS SEE TABLE 9-110
NO. 64 ISO 2341 U.S.A. ANSI JAPAN JIS GERMANY DIN 1434-1436 FRANCE NF R126.09 U.K. BS ITALY UNI 1710-1715 AUSTRALIA AS		(E) CYLINDRICAL CLEVIS PIN WITH HEAD (G) BOLZEN MIT KOPF (F) AXE EPAULE (I) PERNO CON TESTA	FOR DESIGN DETAILS SEE TABLE 9-111
NO. 65 ISO U.S.A. ANSI B27.7 JAPAN JIS B2804 GERMANY DIN 472 FRANCE NF E22-165 U.K. BS 3673 ITALY UNI 3654 AUSTRALIA AS		(E) BASIC RETAINING RING FOR HOLE (G) SICHERUNGSRING FUR BOHRUNG (F) SEGMENT D'ARRET POUR ALESAGE (I) ANELLO ELASTICO DI SICUREZZA PER FORO	FOR DESIGN DETAILS SEE TABLE 9-112

FASTENERS 9-15

World Metric Fastener Standards Index

STANDARD	FIGURE	NAME	REFERENCE
NO. 66 ISO U.S.A. ANSI B27.7 JAPAN JIS B2804 GERMANY DIN 471 FRANCE NF E22-164 U.K. BS 3673 ITALY UNI 3653 AUSTRALIA AS		(E) BASIC RETAINING RING FOR SHAFT (G) SICHERUNGSRING FUR WELLE (F) SEGMENT D'ARRET POUR ARBRE (I) ANELLO ELASTICO DI SICUREZZA PER ALBERO	FOR DESIGN DETAILS SEE TABLE 9-113
NO. 67 ISO U.S.A. ANSI B27.7 JAPAN JIS B2805 GERMANY DIN 6799 FRANCE NF PRL 23203 U.K. BS 3673 ITALY UNI AUSTRALIA AS		(E) RETAINING RING FOR SHAFT TYPE E-RING (G) SICHERUNGSSCHEIBE FUR WELLE (F) SEGMENT D'ARRET POUR ARBRE (I) ANELLO DI SICUREZZA PER PERNO TIPO E	FOR DESIGN DETAILS SEE TABLE 9-114
NO. 68 ISO U.S.A. ANSI JAPAN JIS GERMANY DIN 9045,7993 FRANCE NF U.K. BS ITALY UNI 3656 AUSTRALIA AS		(E) RETAINING RING ROUND SECTION (G) RUNDDRAHTSPRENG-RING (F) JONC D'ARRET SECTION CYLINDRIQUE (I) ANELLO ELASTICO DI ARRESTO	FOR DESIGN DETAILS SEE NATIONAL STANDARDS AVAILABLE FROM ANSI 1430 BROADWAY NEW YORK, N.Y. U.S.A. 10018
NO. 69 ISO U.S.A. ANSI JAPAN JIS GERMANY DIN 470 FRANCE NF E29-584 U.K. BS ITALY UNI AUSTRALIA AS		(E) EXPANSION PLUG (G) VERSCHLUSS-SCHEIBE (F) BOUCHON EXPANSIBLE (I) TAPPO SPANSIBILE	FOR DESIGN DETAILS SEE NATIONAL STANDARDS AVAILABLE FROM ANSI 1430 BROADWAY NEW YORK, N.Y. U.S.A. 10018
NO. 70 ISO U.S.A. ANSI JAPAN. JIS GERMANY DIN 442,443 FRANCE NF R939-10 U.K. BS ITALY UNI AUSTRALIA AS		(E) CUPPED PLUG (G) VERSCHLUSS-DECKEL (F) BOUCHON CUVETTE (I) TAPPO SVASATO	FOR DESIGN DETAILS SEE NATIONAL STANDARDS AVAILABLE FROM ANSI 1430 BROADWAY NEW YORK, N.Y. U.S.A. 10018

World Metric Fastener Standards Index

STANDARD	FIGURE	NAME	REFERENCE
NO. 71 ISO 3290 U.S.A. ANSI B3.12 JAPAN JIS B1501 GERMANY DIN 5401 FRANCE NF E22-381 U.K. BS ITALY UNI AUSTRALIA AS		(E) STEEL BALL (G) STAHL KUGEL (F) BILLE ACIER (I) PALLA D'ACCIAIO	FOR DESIGN DETAILS SEE NATIONAL STANDARDS AVAILABLE FROM ANSI 1430 BROADWAY NEW YORK, N.Y. U.S.A. 10018
NO. 72 ISO U.S.A. ANSI JAPAN JIS GERMANY DIN 906 FRANCE NF E29-583 U.K. BS ITALY UNI AUSTRALIA AS		(E) HEX SOCKET PIPE PLUG (G) VERSCHLUSS-SCHRAUBE MIT KEGLIGEM GEWINDE (F) BOUCHON FILETE CONIQUE A SIX PANS CREUX (I) TAPPO CONICO CON ESAGONO INCASSATO	FOR DESIGN DETAILS SEE NATIONAL STANDARDS AVAILABLE FROM ANSI 1430 BROADWAY NEW YORK, N.Y. U.S.A. 10018
NO. 73 ISO U.S.A. ANSI JAPAN JIS GERMANY DIN 909 FRANCE NF E29-583 U.K. BS ITALY UNI 5210 AUSTRALIA AS		(E) HEX HEAD PIPE PLUG (G) VERSCHLUSS-SCHRAUBE MIT AUSSENSECHSKANT UND KEGLIGEM GEWINDE (F) BOUCHON FILETE CONIQUE A TETE HEXAGONALE (I) TAPPO CONICO A TESTA ESAGONALE	FOR DESIGN DETAILS SEE NATIONAL STANDARDS AVAILABLE FROM ANSI 1430 BROADWAY NEW YORK, N.Y. U.S.A. 10018
NO. 74 ISO U.S.A. ANSI JAPAN JIS GERMANY DIN 910,7604 FRANCE NF U.K. BS ITALY UNI AUSTRALIA AS		(E) HEX HEAD GASKET PLUG (G) VERSCHLUSS-SCHRAUBE MIT BUND UND AUSSENSECHSKANT (F) BOUCHON FILETE A EPAULEMENT A TETE HEXAGONALE (I) TAPPO A TESTA ESAGONALE CON BORDINO	FOR DESIGN DETAILS SEE NATIONAL STANDARDS AVAILABLE FROM ANSI 1430 BROADWAY NEW YORK, N.Y. U.S.A. 10018
NO. 75 ISO U.S.A. ANSI JAPAN JIS B1575 GERMANY DIN 71412 FRANCE NF R16-521 U.K. BS ITALY UNI AISTRALIA AS		(E) LUBRICATION FITTING (G) KEGEL SCHMIERNIPPEL (F) GRAISSEUR (I) INGRASSATORE	FOR DESIGN DETAILS SEE NATIONAL STANDARDS AVAILABLE FROM ANSI 1430 BROADWAY NEW YORK, N.Y. U.S.A. 10018

FASTENERS

4. The length designation is shown in millimeters. Refer to Table 9-1 for the preferred fastener lengths.
5. The standard thread length conforms to ISO recommendations worldwide, and no special call-out for thread length is required (Table 9-3).
6. National standards reference. The threaded fasteners details are defined in industry or national standards, and a reference to the desired standards must be made when ordering fasteners.
7. The strength grade which applies to steel products only, must be specified. See page 9-18 for details on strength properties.
8. The surface protection (if required) should be in accordance with company practice or other standards.

Examples for Ordering Modified ISO (OMFS) Fasteners:
1. Bolts 10 mm diameter, with standard coarse threads 50 mm long, produced from grade 9.8 steel and having a medium thread fit, are designated as follows: Bolt M10 × 1.5 × 50 to IFI 506–9.8.
2. Nuts 12 mm diameter with standard coarse threads, produced from steel of grade 5 strength, and having a close thread fit should be designated as follows: Nut M12 × 1.75C to IFI 507–5.

Examples for Ordering Existing ISO Fasteners:
1. Bolts 8 mm diameter, with coarse threads 25 mm long, produced from grade 8.8 steel and having a medium thread fit, are designated: Bolt M8 × 25 to BS3692–8-8.

Table 9-1. Nominal Fastener Lengths (ISO 888)

Nominal Length		Nominal Length	
Millimeters	Inches (REF)	Millimeters	Inches (REF)
2	.079	60	2.36
2.5	.098	65*	2.56
3	.12	70	2.76
4	.16	75*	2.95
5	.20	80	3.15
6	.24	85*	3.35
7S	.28	90	3.54
8	.31	95S	3.74
9S	.35	100	3.94
10	.39	105S	4.13
11S	.43	110	4.33
12	.47	115S	4.53
14	.55	120	4.72
16	.63	125S	4.92
18S	.71	130	5.12
20	.79	140	5.51
22S	.87	150	5.91
25	.98	160	6.30
28S	1.10	170*	6.69
30	1.18	180	7.1
32S	1.26	190*	7.5
35	1.38	200	7.9
38S	1.50	220	8.7
40	1.57	240	9.4
45	1.77	260	10.2
50	1.97	280	11.0
55*	2.17	300	11.8

NOTES:
S = ISO 888 second choice fastener lengths and not standard lengths in the American standards.
* = American second choice fastener lengths.

2. Brass screws 6 mm diameter, with slotted pan head and coarse threads 20 mm long would be designated: Brass Screw M6 × 20 to DIN 84.

Fastener Length Specifications

Nominal Lengths for Bolts, Screws, and Studs—The preferred metric fastener lengths are shown in Table 9-1. The customary inch values shown are rounded off to two decimal places, and should be used for reference only. The recommended minimum bolt lengths are $2d$ ($2 \times$ nominal diameter). A further reduction of the preferred fastener lengths should be considered by the user, in order to minimize the number of different standard parts used in the product.

Standard Length Tolerances—The standard tolerances used on bolt and screw lengths are not standardized in the ISO, and the tolerances differ in the referenced national standards. The most frequently used tolerance outside North America for bolt lengths is the ISO tolerance designation ±1/2 IT15 or js15. For nut heights the ISO tolerance h14 is used and for self-tapping screws ±1/2 IT16. The OMFS defines the tolerance on the nominal length as +0 −the millimeter tolerance shown in Table 9-2.

Table 9-2. American Length Tolerances (IFI-506)

	Nom Size of Screw		
Nom Length	5 through 8	10 through 16	20
to 10 mm, incl	0.6	0.6	—
over 10 to 18 mm, incl	0.7	0.7	—
over 18 to 30 mm, incl	0.8	0.8	1.3
over 30 to 50 mm, incl	1.0	1.6	2.5
over 50 to 80 mm, incl	1.2	1.9	3.0
over 80 to 120 mm, incl	1.4	2.2	3.5
over 120 to 180 mm, incl	2.5	2.5	4.0
over 180 to 240 mm, incl	4.6	4.6	4.6

This manner in defining the tolerance for fastener lengths is similar to the tolerance system used on the customary inch fasteners.

Threaded Lengths—The standard minimum thread lengths for general purpose bolts are based on the ISO 888 recommended formula given in Table 9-3.

Strength Properties for Threaded Fasteners

General—In 1968, the International Organization for Standardization released ISO Recommendation 898 Part I, "Mechanical Properties of Fasteners—Bolts, Screws and Studs." Parts II, III, and IV, which dealt with strength

Table 9-3. ISO Standard Thread Lengths (ISO 888)

Nominal Length of Bolt L	Length of Thread b
Up to and including 125 mm	$2d + 6$ mm
Over 125 mm up to and including 200 mm	$2d + 12$ mm
Over 200 mm	$2d + 25$ mm

d = nominal diameter of the bolt or screw.

grades for nuts and marking of fasteners, were released during the subsequent years. A formal request to update the above standards was given to the subcommittee ISO/TC2/SC1 in 1970, and the present ISO/DIS 898 Part I was refined as a result of four meetings and unanimously approved in Toronto (1973). A proposal for a new strength grade 9.8 is included in the draft. The material in this section is based on the latest tentatively approved proposals, ISO standards, and the ISO 898 Part I through IV standards and drafts for bolts, screws, studs, and nuts. Certain additional requirements have been included to reflect practices for manufacturing fasteners in North America. To identify these additional requirements which are not, as of this writing, recognized as ISO standards, an asterisk (*) has been placed in the left margin.

Bolts, Screws, and Studs—Mechanical and material requirements for externally threaded steel fasteners of the metric series are shown in IFI-501 and in ISO 898 Part I standards. A brief description of the new fastener grading system is as follows:

Designation System—Property classes are designated by numbers, whereas increasing numbers generally represent increasing tensile strengths. The designation symbol consists of two parts: (1) the first numeral of a two digit symbol or the first two numerals of a three digit symbol will approximate 1/100 of the minimum tensile strengths in MPa; (2) the last numeral approximates 1/10 of the ratio expressed as a percentage between minimum yield stress and minimum tensile stress.

Property Classes[1]—The IFI-501 standard recognizes seven property classes out of the ten ISO grades as follows; 4.6, 4.8, 5.8, 8.8, 9.8, 10.9, and 12.9. Omitted are ISO

---

[1] CONVERSION GUIDANCE. For guidance purposes only, to assist designers in selecting a property class: Class 4.6 is approximately equivalent to SAE Grade 1 and ASTM A307, Grade A. Class 5.8 is approximately equivalent to SAE Grade 2. Class 8.8 is approximately equivalent to SAE Grade 5, and ASTM A449. Class 9.8 has properties approximately 9 percent stronger than SAE Grade 5, and ASTM A449. Class 10.9 is approximately equivalent to SAE Grade 8 and ASTM A354 Grade BD.

FASTENERS

classes 3.6, 5.6, and 6.8. Machine screws are normally available only in classes 4.8 and 9.8; other bolts, screws, and studs are available in all classes within the specified product size limitations given in Table 9-4. At the option of the manufacturer, class 5.8 may be supplied when either classes 4.6 or 4.8 are ordered, and class 4.8 may be supplied when class 4.6 is ordered.

Materials and Processes

Steel Characteristics—Bolts, screws, and studs are made of steel conforming to the description and chemical composition requirements specified in Table 9-5 for the applicable property class.

Heading Practice—Methods other than upsetting and/or extrusion are permitted only by special agreement between purchaser and producer. Class 4.6 may be hot- or cold-headed at the option of the manufacturer. Classes 4.8, 5.8, 8.8, 9.8, 10.9, and 12.9 bolts and screws in sizes up to 20 mm inclusive and of lengths up to 10 times the nominal product size or 150 mm, whichever is shorter, are cold-headed at the option of the manufacturer.

Threading Practice—Classes 4.8, 5.8, 8.8, 9.8, 10.9, and 12.9 bolts and screws in sizes up to 20 mm inclusive and of lengths up to 150 mm inclusive, are roll-threaded, except by special agreement. Threads of all sizes of classes 4.8, 5.8, 8.8, 9.8, 10.9, and 12.9 bolts and screws in sizes over 20 mm and/or lengths longer than 150 mm may be rolled, cut or ground, at the option of the manufacturer. Threads of all classes and sizes of studs may be rolled, cut or ground at the option of the manufacturer.

Heat Treatment Practice—Class 4.6 bolts and screws and classes 4.6, 4.8, and 5.8 studs need not be heat-treated. Classes 4.8 and 5.8 bolts and screws shall be stress-relieved if necessary to assure the soundness of the head to shank junction. When specified by the purchaser, class 5.8 bolts and screws will be stress-relieved at a minimum stress relief temperature of 470°C. Where higher temperatures are necessary to relieve stress in severely upset heads, mechanical requirements will be agreed upon by producer and purchaser.

Classes 8.8 and 9.8 bolts, screws, and studs are heat-treated, oil- or water-quenched at the option of the manufacturer, and tempered at a minimum tempering temperature of 425°C for class 8.8 and 410°C for class 9.8. Medium carbon alloy steel class 10.9 bolts, screws, and studs are heat-treated, oil-quenched and tempered at a minimum tempering temperature of 425°C. Low carbon martensite steel class 10.9 bolts, screws, and studs are heat-treated, quenched in oil or water, and tempered at a minimum temperature of 340°C. Class 12.9 bolts, screws, and studs are heat-treated, oil-quenched, and tempered at a minimum tempering temperature of 345°C.

Requirements

Mechanical—Bolts, screws, and studs are tested in accordance with the mechanical testing requirements for the applicable type, property class, size, and length of product as specified in Table 9-6; and they must meet the mechanical requirements specified in Tables 9-4 and 9-7.

Methods of Testing

Product Hardness—For routine inspection, hardness of bolts, screws, and studs may be determined on head, end, or shank after removal of any plating or other coating. For referee purposes, the hardness of bolts, screws, and studs is determined at mid-radius of a transverse section through the threaded portion of the product taken at a distance of one diameter from the end of the product.

Table 9-4. Mechanical Requirements for Bolts, Screws, and Studs (IFI-501)

Property Class	Nominal Dia	Full Size Bolts, Screws and Studs Proof Load MPa (4)	Full Size Bolts, Screws and Studs Tensile Strength Min MPa (4)	Machined Test Specimens Yield Strength Min (1) MPa	Machined Test Specimens Tensile Strength Min MPa	Elongation Min %	Reduction of Area Min %	Surface Hardness Rockwell 30N Max	Product Hardness Rockwell Min	Product Hardness Rockwell Max	Product Hardness Vickers Min	Product Hardness Vickers Max
4.6	M5 thru M36	225	400	240 (2)	400	22	35	—	B67	B100	120	255
4.8	M1.6 thru M16	310	420	340	420	14	35	—	B71	B100	130	255
5.8	M5 thru M24 (3)	380	520	420	520	10	35	—	B82	B100	160	255
8.8	M16 thru M36	600	830	660	830	12	35	54	C24	C34	260	335
9.8	M1.6 thru M16	650	900	720	900	10	35	56	C27	C36	280	360
10.9	M5 thru M36	830	1040	940	1040	9	35	59	C33	C39	330	385
12.9	M1.6 thru M36	970	1220	1100	1220	8	35	63	C39	C44	385	435

NOTES:
1. Yield strength is stress at which a permanent set of 0.2% of gage length occurs.
2. Yield point shall apply instead of yield strength at 0.2% offset for class 4.6 products.
3. Class 5.8 requirements apply to bolts and screws with lengths 150 mm and shorter, and to studs of all lengths.
4. Proof load and tensile strength values for full size products of each property class are given in Table 9-7.

The reported hardness is the average of four hardness readings located at 90 deg to one another. The preparation of test specimens and the performance of hardness tests should conform with the requirements of SAE J417.

Surface Hardness—Tests to determine surface hardness conditions are conducted on the ends, hexagon flats, or unthreaded shanks which have been prepared by lightly grinding or polishing to insure accurate reproducible readings in accordance with SAE J417. Proper correction factors are made on curved surfaces, per ASTM E18.

Proof Load. The proof load test consists of stressing the bolt, screw, or stud with a specified load which the product must withstand without permanent set. The proof load and tensile strength values are shown in Table 9-7, and a detailed description of test methods can be found in the IFI-523 standard.

Basis for Calculating Proof and Ultimate Bolt Load— The proof and ultimate bolt loads for nominal diameter-pitch combinations not shown in Table 9-7 can be calculated by following these simple steps:

1. Read the tensile stress area (AS) for the applicable diameter pitch combination in Table 8-1 (Section 8).
2. Find the appropriate proof stress or ultimate stress value for the bolt strength grade to be determined in Table 9-4.
3. Multiply the tensile stress area (AS) with the proof or ultimate stress, and the product is the load in newtons (N).

Example: Calculate the proof load for a grade 8.8 bolt with 7 mm nominal diameter and 0.75 mm pitch.

$AS = 31.1$ mm^2 (Table 8-1)
$Sp = 600$ MPa (MPa = N/mm^2) (Table 9-4)
Proof Load = $AS \times Sp = 31.1 \times 600$ N = 18660 N
$= 18.7$ kN

Marking

Bolts and Screws—All bolts and screws except slotted and cross recessed head screws, with nominal diameters of 5 mm and larger, are marked to identify the property class. The symbols used are those given in Table 9-8. In addition, bolts and screws are marked with the manufacturer's identification symbol. Markings are located on the top of the head of bolts and screws, and may be either raised or depressed at the option of the manufacturer. Alternatively, for hex head products, the markings may be indented on the side of the head.

Studs—All studs with nominal diameters of 5 mm and larger are marked to identify the property class. The symbols used are those given in Table 9-8. Markings are located at the extreme end of the stud, and may be raised or depressed. Interference fit threads are marked at the nut end. Studs smaller than 12 mm nominal diameter may be marked using the property class symbols given in Table 9-8.

Table 9-5. Chemical Composition Requirements for Bolts (IFI-501)

Property Class	Material and Treatment	C Min	C Max	P Max	S Max
4.6	Low or medium carbon steel	—	0.55	.048	.058
4.8	Low or medium carbon steel, partially or fully annealed as required	—	0.55	.048	.058
5.8	Low or medium carbon steel, cold worked	0.13	0.55	.048	.058 (2)
8.8	Medium carbon steel, product is quenched and tempered (6) (7)	0.28	0.55	.048	.058 (3)
9.8	Medium carbon steel, product is quenched and tempered (6)	0.28	0.55	.048	.058 (3)
10.9	Medium carbon alloy steel, product is quenched and tempered (4), (6)	0.28	0.55	.040	.045
12.9	Alloy steel, product is quenched and tempered (5)	0.31	0.65	.045	.045

Element Percent (1)

NOTES:
1. All values are for product analysis (per cent by weight).
2. For studs only, sulfur content may be 0.33 per cent max.
3. For studs only, sulfur content may be 0.13 per cent max.
4. Medium carbon alloy steel shall be fine grain, with hardenability that will produce a minimum hardness of Rockwell C47 at the center of the threaded section one diameter from the end of the bolt, screw or stud after oil quenching (see SAE J407). Carbon steel may be used by agreement between producer and consumer, for sizes thru M20. SAE 1541 steel, oil quenched and tempered may be used at option of the producer for products of sizes M12 and smaller.
5. One or more of the alloying elements chromium, nickel, molybdenum or vanadium shall be present in the steel in sufficient quantity to assure that the specified strength properties are met after oil quenching and tempering.
6. Unless otherwise specified by the customer, the manufacturer may use for sizes thru M24 a low carbon martensite steel with 0.15 to 0.40 per cent carbon, 0.74 per cent min manganese, 0.048 per cent max phosphorus, 0.058 per cent max sulfur, and 0.0005 per cent min boron. The steel for class 10.9 products, shall have a hardenability that will produce a minimum hardness of Rockwell C38 at the center of a transverse section one diameter from the threaded end of the bolt, screw or stud after quenching. Products made using this material shall be specially identified.
7. At manufacturer's option, medium carbon alloy steel may be used for sizes over M24.

FASTENERS

Table 9-6. Mechanical Testing Requirements for Bolts, Screws, and Studs (IFI-501)

Product	Property Class	Specified Min Tensile Strength of Product (See Table 5) kN	Length of Product (2)	Product Hardness Max	Product Hardness Min	Surface Hardness (4) Max	Proof Load	Wedge Tensile Strength (5)	Axial Tensile Strength	Yield Strength	Tensile Strength	Elongation	Reduction of Area	Decarburization in Threaded Section (4)
Short bolts and screws	all	all	less than 3D	●	●	●	—	—	—	—	—	—	—	○
Special head bolts and screws (3)	all	all	all	●	●	●	—	—	—	—	—	—	—	○
Hex bolts and screws	all	450 and less	3D to 8D or 200 mm, whichever is greater	●	—	●	○	●	—	—	—	—	—	○
			over 8D or 200 mm, whichever is greater thru and incl 300 mm	●	—	●	○	●	—	—	—	—	—	○
			over 300 mm	●	—	●	○	A	—	B	B	B	B	○
		over 450	3D and longer	●	—	●	○	A	—	B	B	B	B	○
All other bolts and screws	all	450 and less	3D to 8D or 200 mm, whichever is greater	●	—	●	○	—	●	—	—	—	—	○
			over 8D or 200 mm, whichever is greater	●	—	●	○	—	A	B	B	B	B	○
Short studs	all	all	less than 3D	●	●	●	—	—	—	—	—	—	—	○
All other studs	all	450 and less	3D to 8D or 200 mm, whichever is greater	●	—	●	○	●	—	—	—	—	—	○
			over 8D or 200 mm, whichever is greater	●	—	●	○	A	—	B	B	B	B	○
		over 450	3D and longer	●	—	●	○	A	—	B	B	B	B	○
Tests to be conducted in accordance with paragraph				3.1		3.2	3.3	3.6	3.5 IFI-523 (page B-33)		3.7			5.3

NOTES:
1. (●) denotes a mandatory test. For each product all mandatory tests (●) shall be performed. In addition, either all tests denoted (A), which apply to full size products, or all tests denoted (B), which apply to machine test specimens, shall be performed. (●) denotes tests to be performed when specifically required in the original inquiry and purchase order. In case arbitration is necessary, a test shall be performed. (—) indicates tests which are not required.
2. (D) equals nominal diameter of product.
3. Special head bolts and screws are those with special configuration or with drilled heads which are weaker than the threaded section.
4. Surface hardness and decarburization requirements apply only to property classes 8.8, 9.8, 10.9 and 12.9.

Table 9-7. Proof Load and Tensile Strength Values for Bolts (IFI-501)

Nominal Thread Dia and Thread Pitch	Stress Area (2) mm²	Class 4.6 Proof Load kN	Class 4.6 Tensile Strength Min kN	Class 4.8 Proof Load kN	Class 4.8 Tensile Strength Min kN	Class 5.8 Proof Load kN	Class 5.8 Tensile Strength Min kN	Class 8.8 Proof Load kN	Class 8.8 Tensile Strength Min kN	Class 9.8 Proof Load kN	Class 9.8 Tensile Strength Min kN	Class 10.9 Proof Load kN	Class 10.9 Tensile Strength Min kN	Class 12.9 Proof Load kN	Class 12.9 Tensile Strength Min kN
M1.6x0.35	1.27			0.39	0.53					0.83	1.14			1.23	1.55
M2x0.4	2.07			0.64	0.87					1.35	1.86			2.01	2.53
M2.5x0.45	3.39			1.05	1.42					2.20	3.05			3.29	4.14
M3x0.5	5.03			1.56	2.11					3.27	4.53			4.88	6.14
M3.5x0.6	6.78			2.10	2.85					4.41	6.10			6.58	8.27
M4x0.7	8.78	3.20	5.68	2.72	3.69					5.71	7.90			8.52	10.7
M5x0.8	14.2			4.40	5.96	5.40	7.38			9.23	12.8	11.8	14.8	13.8	17.3
M6x1	20.1	4.52	8.04	6.23	8.44	7.64	10.5			13.1	18.1	16.7	20.9	19.5	24.5
M8x1.25	36.6	8.24	14.6	11.3	15.4	13.9	19.0			23.8	32.9	30.4	38.1	35.5	44.7
M10x1.5	58.0	13.1	23.2	18.0	24.4	22.0	30.2			37.7	52.2	48.1	60.3	56.3	70.8
M12x1.75	84.3	19.0	33.7	26.1	35.4	32.0	43.8			54.8	75.9	70.0	87.7	81.8	103
M14x2	115	25.9	46.0	35.7	48.3	43.7	59.8			74.8	104	95.4	120	112	140
M16x2	157	35.3	62.8	48.7	65.9	59.7	81.6	94.2	130	102	141	130	163	152	192
M20x2.5	245	55.1	98.0			93.1	127	147	203			203	255	238	299
M24x3	353	79.4	141			134	184	212	293			293	367	342	431
M30x3.5	561	126	224					337	466			466	583	544	684
M36x4	817	184	327					490	678			678	850	792	997

NOTES:
1. Proof loads and tensile strengths are computed by multiplying the stresses given in Table 9-4 by the stress area of the thread shown above (or in Table 8-1).
2. Stress area = $0.7854(D - 0.9382P)^2$ where D is nominal thread diameter in mm and P is thread pitch in mm. For stress areas other than those shown, see Table 8-1.

FASTENERS

Table 9-8. Property Class Identification Symbols for Bolts, Screws, and Studs (IFI-501)

Property Class	Identification Symbol	
	Bolts, Screws and Studs	Studs Smaller Than M12
4.6	4.6	—
4.8	4.8	—
5.8	5.8	—
8.8 (1)	8.8	○
9.8 (1)	9.8	+
10.9 (1)	10.9	□
12.9	12.9	△

NOTE: Products made of low carbon martensite steel should be additionally identified by underlining the numerals.

SCREWS AND WASHER ASSEMBLIES (SEMS)

Mechanical and material requirements for screw and washer assemblies (SEMS) are specified in IFI-501: Part II standard in nominal sizes from 1.6 mm to 12 mm. The strength requirements for SEMS are found in IFI-501: Part I except as noted below.

Strength Grades—Two strength grades are standard: 4.8 and 9.8 and only the 9.8 grade requires heat treatment. The mechanical requirements for SEMS are shown in Table 9-9, and the chemical composition in Table 9-10.

Tapping Screws—Material and performance requirements for thread forming, thread cutting, thread rolling, and self-drilling tapping screws are specified in the IFI-502, 503, and 504 standards. These IFI publications specify requirements for "hard" metric fasteners.

Material—Screws shall be made from cold heading quality, killed steel wire, and should conform to the composition limits shown below:

(IFI-502)

Tapping Screw Size (Dia) mm	Analysis (1)	Chemical Composition, Per Cent By Wgt	
		Carbon	Manganese
2 thru 3	Ladle	0.13–0.25	0.60–1.65
	Check	0.11–0.27	0.57–1.71
3.5 thru 12	Ladle	0.15–0.25	0.70–1.65
	Check	0.13–0.27	0.64–1.71

NOTE: Ladle analyses are shown for informational purposes. Check analyses are mandatory and refer to individual determination on the uncarburized or core portion of screws.

Heat Treatment—Shall be in carbonitriding or gas carburizing system. Screws shall be quenched in a liquid medium and then tempered by reheating to 340°C min. (CSD and BSD screws are reheated to 330°C). Cyaniding systems may be approved by a purchaser when the producer shows that a continuous flow (no batch) quenching process is employed which consistently produces uniform case and core. When cyaniding systems are approved, the minimum tempering temperature shall be 230°C.

Total Case Depth—Shall conform to the following, as measured at thread flank midpoint between crest and root:

(IFI-502)

Size	Thickness, mm
2 thru 3.5	0.05–0.18
4 thru 5.5	0.10–0.23
6 thru 12	0.13–0.28

Surface Hardness after Tempering—Shall be equivalent to Rockwell C45 minimum. For routine quality control purposes (where case depth and geometry of screws permit), measurements may be made on end, shank, or head using Rockwell 15 N. As an alternate, or where this method is not applicable, a micro-hardness instrument with a Knoop or diamond pyramid indenter and a 5 N load may be used. In such cases, measurements shall be made on the thread profile of a properly prepared longitudinal metallographic specimen.

Core Hardness after Tempering[1]—Shall be Rockwell C28-38, as determined at mid-radius of a transverse section through the screw, taken at a distance sufficiently behind the screw point to be through the full minor diameter.

Microstructure—Shall show no band of free ferrite between case and core, as determined by metallographic examination.

Torsional Strength Requirements—The torsional strength requirements for thread-forming, thread-cutting and self-drilling tapping screws are shown in Table 9-11, and for thread rolling (high performance) screws in Table 9-12.

Table 9-13 provides the mechanical and performance requirements for customary inch threaded rolling screws. Since the data in Table 9-12 have not yet been completely developed, Table 9-13 is given for reference.

NUTS[2]

Mechanical and material requirements for internally threaded steel fasteners of the metric series are specified in the IFI-508, the ISO 898 Part II, and the ISO 898 Part IV standards. The ISO 898 Part II standards specify seven strength grades for nuts with effective heights of thread

---

[1] Hardness shall not exceed maximum shown and preferably should be no higher than Rockwell C36 to insure against failure in assembly and service.

[2] The total nut height is approximately equal to 0.8 times the nominal diameter in traditional metric standards. The above ratio is increased in the IFI-508 standard to 0.9 for Style 1 nuts and to 1 for Style 2 nuts in order to ensure detectable stripping of the bolt rather than the nut.

Table 9-9. Mechanical Requirements for SEMS (IFI-501)

Property Class	Nominal Dia	Full Size Bolts, Screws and Studs		Surface Hardness Rockwell 30N Max (4)	Product Hardness			
		Proof Load MPa	Tensile Strength Min MPa		Rockwell Min	Rockwell Max	Vickers Min	Vickers Max
4.8	M1.6 thru M12	310	420	—	B71	B100	130	255
9.8	M1.6 thru M12	650	900	60	C25	C40	270	390

Table 9-10. Chemical Composition Requirements for SEMS (IFI-501)

Property Class	Material and Treatment	Element % (1)			
		C Min	C Max	P Max	S Max
4.8	Low carbon steel, partially or fully annealed as required	—	0.55	0.048	0.058
9.8	Low or medium carbon steel, product is quenched and tempered	0.15	0.40	0.048	0.058

NOTES:
1. All values are for product analysis (percentage by weight).
2. Unless otherwise specified by the customer, the manufacturer may use for class 9.8 SEMS a low carbon martensite steel with 0.15 to 0.40% carbon, 0.74% min manganese, 0.048% max phosphorus, 0.058% max sulfur, and 0.0005% min boron. Products made using this material shall be specially identified as specified in the notes to Table 9-8.

Table 9-11. Torsional Strength Requirements for Tapping and Self-Drilling Screws (IFI-502, 504)

NOMINAL SCREW SIZE, mm		TORSIONAL STRENGTH MIN, N·m				HYDROGEN EMBRIT TEST TORQUE, N·m	
THREAD TYPES AB, B, BF, BT, BSD	THREAD TYPES D, F, T, CSD	TYPES AB, B, BF, BT	TYPES D, F, T	TYPE BSD	TYPE CSD	CAD PL BSD*	ZINC PL BSD*
2.2	2	0.4	0.6				
-	2.5	-	1.2				
2.9	3	1.5	2.2	1.6	2.2	1.2	1.3
3.5	3.5	2.7	3.4	2.7	3.5	2	2.3
4.2	4	4.4	5	4.7	5.2	4.1	4.6
4.8	5	6.3	10	6.9	10.5	5.5	6.2
5.5	-	9.9	-	10.4	-	8.1	9.6
6.3	6.3	16	21	16.9	21	12.9	14.9
8	8	33	43				
9.5	10	67	86				
-	12	-	150				

FASTENERS

Table 9-12. Mechanical and Performance Requirements for Thread Rolling Screws (IFI-503)

Nom Screw Size and Thread Pitch mm	Tensile Strength Min kN	Torsional Strength Min N·m	Drive Torque For PC & CP Screws Max N·m	Drive Torque For ZP Screws Max N·m	Clamp Load kN	Clamp Load Torque For PC & CP Screws Max N·m	Clamp Load Torque For ZP Screws Max N·m	Proof Torque For PC & CP Screws N·m	Proof Torque For ZP Screws N·m	Hydrogen Embrittlement Torque For CP Screws N·m	Hydrogen Embrittlement Torque For ZP Screws N·m
2x0.4	1.86	0.6									
2.5x0.45	3.04	1.2									
3x0.5	4.51	2.2									
3.5x0.6	6.08	3.4									
4x0.7	7.87	5.0				VALUES ARE UNDER DEVELOPMENT					
5x0.8	12.7	10									
6x1	18.1	18									
8x1.25	32.8	43									
10x1.5	52.0	86									
12x1.75	75.6	150									

NOTES:
1. Legend: CP = cadmium-plated, ZP = zinc-plated, PC = phosphate-coated.
2. Marking: Self-trapping screws do not require any strength grade marking.
3. Thread rolling screws to "hard" metric sizes are covered in the IFI-503 standard from M2 × 0.4 to M12 × 1.75, and the torque and clamp load values are under development by IFI for the metric series: screws.

Table 9-13. Mechanical and Performance Requirements for Thread Rolling Screws

Nom Screw Size and Thread Pitch mm	Tensile Strength Min kN	Torsional Strength Min N·m	Drive Torque For PC & CP Screws Max N·m	Drive Torque For ZP Screws Max N·m	Clamp Load kN	Clamp Load Torque For PC & CP Screws Max N·m	Clamp Load Torque For ZP Screws Max N·m	Proof Torque For PC & GP Screws N·m	Proof Torque For ZP Screws N·m	Hydrogen Embrittlement Torque For CP Screws N·m	Hydrogen Embrittlement Torque For ZP Screws N·m
2.2 × 0.45	2.22	0.7	0.51	0.68	—	—	—	0.79	0.90	0.51	0.56
2.8 × 0.64	3.60	1.6	1.02	1.47	—	—	—	1.92	2.15	1.19	1.36
3.5 × 0.79	5.56	2.7	1.58	2.26	2.05	2.15	2.82	3.16	3.73	2.03	2.26
4.2 × 0.79	8.45	5.4	2.82	3.62	3.11	4.18	5.42	5.65	6.44	4.07	4.63
4.8 × 1.06	10.5	7.3	3.95	5.88	4.00	6.21	7.68	7.68	8.70	5.54	6.21
5.5 × 1.06	14.5	13	7.30	9.60	5.34	11.0	14.0	14.0	16.0	9.70	11.0
6.3 × 1.27	19.1	17	10.2	13.6	7.12	13.6	16.3	18.3	21.0	12.9	14.9
7.9 × 1.41	31.6	37	20.3	27.1	11.6	28.5	35.2	38.6	42.0	28.5	31.2
9.5 × 1.59	46.7	68	27.1	33.9	17.8	54.2	69.1	71.8	78.0	51.5	57.6
11.1 × 1.81	64.1	95	40.7	54.2	24.0	84.1	102	100	108	71.2	81.3
12.7 × 1.95	85.0	122	61.0	75.6	32.0	113	129	132	142	92.2	105

NOTES:
1. Legend: CP = cadmium-plated, ZP = zinc-plated, PC = phosphate-coated.
2. Marking: Self-tapping screws do not require any strength grade marking.

greater than or equal to 0.6 times the nominal diameter, and the ISO 898 Part IV standards specify requirements of nuts with effective heights of thread from 0.4d to < 0.6d. The IFI-508 covers the mechanical and material requirements for three property classes of metric series steel nuts in sizes from 1.6 to 36 mm, inclusive, and suitable for general engineering applications. The material presented in this chapter is based on IFI-508 standard with the permission from the Industrial Fastener Institute, 1505 East Ohio Building; Cleveland, Ohio 44114.

Designation

Property Classes—The three property classes are designated by the numbers 5, 9, and 10, and are normally available in the sizes specified in Table 9-15. The following ISO classes have been omitted; 4, 6, 8, 12, 14, and for low nuts 04 and 06. When agreed upon by supplier and customer, nuts of a higher strength property class may be substituted when nuts of a lower strength property class are ordered.

Application Guidance—Class 5 nuts are suitable for use with classes 5.8, 4.8 and 4.6 bolts, screws, and studs; class 9 nuts are suitable for use with classes 9.8, 8.8, 5.8 and 4.8 bolts, screws, and studs; and class 10 nuts are suitable for use with classes 10.9, 9.8, and 8.8 bolts, screws, and studs.

Materials and Processes

Steel Characteristics—Nuts shall be made of steel conforming to the chemical composition limits specified in Table 9-14 for the applicable property class.

Heat Treatment—Classes 5, and 9 nuts need not be heat treated. Class 10 nuts are heat treated. Case-hardening of any class of nut is not permitted.

Finish—Nuts may be furnished with a natural as processed finish or with a protective coating as specified by the customer.

Requirements

Dimensional—Classes 5 and 10 nuts must conform to the dimensions of Style 1 hex nuts, and class 9 nuts should conform to the dimensions of Style 2 hex nuts as shown in Table 9-79 and specified in the IFI-507 standard.

Mechanical

Proof Load—Nuts shall withstand the proof load stress specified for the applicable property class in Table 9-15, and the proof load values as shown in Table 9-16. (For details concerning test methods refer to IFI-523 or other standards shown on pages 9-8 and 9-9.

Class 9 nuts, in size 1.6 through 4 mm, are not subject to proof load testing unless specified by the purchaser in the original inquiry or purchase order.

Hardness—Nuts shall have a hardness conforming to the limits specified for the applicable property class in Table 9-15.

Table 9-14. Chemical Composition Requirements for Nuts (IFI-508)

PROPERTY CLASS	ELEMENT, %*			
	C Max	Mn Min	P Max	S Max
5 †	0.55	—	0.12‡	0.15
9	0.55	0.30	0.05§	0.15§
10	0.55	0.30	0.04	0.05‖

NOTES:
*All values are for ladle analysis (percent by weight) and are subject to standard variations for check analysis as given in SAE J409.
†Free cutting steel may be used only if agreed by manufacturer and customer. In such cases, permissible chemical compositions may have a maximum content of 0.34% sulfur, 0.12% phosphorus, and 0.35% lead.
‡Resulphurized and rephosphorized material is not subject to rejection based on check analysis for sulfur.
§If agreed by manufacturer and customer, sulfur content may be 0.35% max, and phosphorus content may be 0.12% max provided that manganese content is 0.70% min.
‖If agreed to by manufacturer and customer, sulfur content may be 0.33% max provided that manganese content is 1.35% min.

Table 9-15. Mechanical Requirements for Hex Nuts (IFI-508)

Property Class	Nominal Nut Size, mm	Proof Load Stress MPa	Nut Hardness	
			Vickers	Rockwell
5	M5 thru M36	570	300 max	C 30 max
9	M1.6 thru M4 M5 thru M16 M20 thru M36	900 990 910	300 max	C 30 max
10	M5 thru M36	1040	275/360	C 26/36

FASTENERS

Table 9-16. Proof Load Values of Hex Nuts (IFI-508)

Nom Nut Size and Thread Pitch	Stress Area mm²	Proof Load kN Class 5	Proof Load kN Class 9	Proof Load kN Class 10
M1.6x0.35	1.27	—	1.14	—
M2x0.4	2.07	—	1.86	—
M2.5x0.45	3.39	—	3.05	—
M3x0.5	5.03	—	4.53	—
M3.5x0.6	6.78	—	6.10	—
M4x0.7	8.78	—	7.90	—
M5x0.8	14.2	8.09	14.1	14.8
M6x1	20.1	11.4	19.9	20.9
M8x1.25	36.6	20.9	36.2	38.1
M10x1.5	58.0	33.1	57.4	60.3
M12x1.75	84.3	48.1	83.5	87.7
M14x2	115	65.6	114	120
M16x2	157	89.5	155	163
M20x2.5	245	140	223	255
M24x3	353	201	321	367
M30x3.5	561	320	511	583
M36x4	817	466	743	850

NOTE:
Proof load in kilonewtons is computed by multiplying the proof load stress (MPa) for the nut property class as given in Table 9-15, and the stress area (mm²) for the applicable nut size and dividing by 1000. Stress areas for other metric fastener diameter-pitch combinations are shown in Table 8-1.

Marking

Class 9 nuts, in sizes 1.6 through 4 mm, need not be marked. All other nuts shall be marked for property class and for manufacturer identification. Markings shall preferably be located on the top surface, but when practicable may be located on a wrench face. Markings on the top surface shall be of a size such that not more than 10 percent of the top surface area may be used for property class and manufacturer's markings. In the case of double chamfered nuts, one face is considered a top surface. Markings may be raised or depressed at manufacturer's option providing that raised marking shall not project beyond the actual width of the nut, or above its top surface.

Classes 5, 9, and 10 nuts are marked with the corresponding numeral 5, 9, or 10 and the manufacturer's identification.

Low Nuts

The ISO 898 IV standard covers nuts with effective heights of thread from 0.4 to 0.6 times the nominal thread diameter. The two property classes 04 and 06 are specified in this standard, and the minimum proof load stresses are shown in Table 9-17 opposite.

Nuts with property class 04 are not required to be marked for grade or trade identification.

Nuts with property class 06 should be marked with the symbol of the property class and the trade mark of the manufacturer in accordance with ISO 898/III.

Nuts Without Specified Load Requirements

The classification of nuts according to hardness is covered in DIN 267. Four hardness classes for nuts with their minimum Vicker HV5 hardness are shown in Table 9-21. Tables 9-21 through 9-23 have been reproduced with permission from DIN (German Standards Institute). Refer to the most recent edition of the standard.

Table 9-17. Designation of Property Classes in Relation to Proof Load Stresses for Low Nuts (ISO 898 IV)

Property class	04	06
Proof load stress S_p (MPa)	400	600

The chemical composition of low nuts is shown for the two property classes in Table 9-18.

Table 9-18. Chemical Composition Requirements for Low Nuts (ISO 898 IV)

Property Class	Chemical Composition Limits (Check Analysis)			
	Carbon max. %	Manganese min. %	Phosphorus max. %	Sulphur max. %
04*	0.50	–	0.110	0.150
06†	0.58	0.30	0.060	0.150

NOTES:

*Free-cutting steel may be used only by special agreement between customer and supplier. In such cases the following maximum sulphur, phosphorus, and lead contents respectively are permissible: sulphur 0.34% phosphorus 0.12% lead 0.35%.

†Alloying elements may be added if necessary to develop the mechanical properties of the nuts.

Mechanical requirements for low nuts are given in Table 9-19. Metric series nuts with specified proof loads in excess of 350 kN may be exempted from proof load testing. Such nuts must meet minimum hardness requirements as determined between customer and supplier.

Table 9-19. Mechanical Requirements for Low Nuts (ISO 898 IV)

Mechanical Requirements		Property Class	
		04	06
Proof load stress	MPa (N/mm²)	400	600
S_p	1000 lbf/in²	60	90
Brinell hardness HB max		302	353
Rockwell hardness HRC max		30	36

NOTE:

Proof load stresses are in MPa which equals N/mm². The listed stresses in 1000 lbf/in² are rounded values and not exact conversions.

Proof load values for metric coarse threaded low nuts are shown in Table 9-20. Other thread types, failure loads and test methods for low nuts are specified in the ISO 898 IV standard.

Table 9-20. Proof Load Values for Low Nuts (ISO 898 IV)

Nominal Thread Diameter	Pitch of the Thread	Nominal Stress Area of Test Mandrel A_s	Property Class of Nut	
			04	06
			Proof Load ($A_s \times S_p$)	
mm	mm	mm²	N	
1.6	0.35	1.27	510	760
2	0.4	2.07	830	1 200
2.5	0.45	3.39	1 350	2 000
3	0.5	5.03	2 000	3 000
3.5	0.6	6.78	2 700	4 050
4	0.7	8.78	3 500	5 250
5	0.8	14.2	5 700	8 500
6	1	20.1	8 000	12 000
7	1	28.9	11 500	17 300
8	1.25	36.6	14 500	22 000
10	1.5	58.0	23 000	35 000
12	1.75	84.3	33 500	50 500
14	2	115	46 000	69 000
16	2	157	63 000	94 000
18	2.5	192	77 000	115 000
20	2.5	245	98 000	147 000
22	2.5	303	121 000	182 000
24	3	353	141 000	212 000
27	3	459	184 000	276 000
30	3.5	561	224 000	336 000
33	3.5	694	278 000	416 000
36	4	817	327 000	490 000
39	4	976	390 000	585 000

NOTE: The proof load is calculated by multiplying the proof load given in Table 9-19 by the nominal stress area of corresponding male thread as shown in Table 8-1.

FASTENERS

Table 9-21. Designation of Property of Nuts Without Specified Proof Loads (DIN 267)

Hardness Class	11H	14H	17H	22H
Vickers Hardness HV5 min	110	140	170	220

A chemical analysis of material for nuts with rolled threads is shown in Table 9-22A and for nuts made out of free cutting steels in Table 9-22 B.

Table 9-22A. Chemical Composition Requirements for Nuts Without Specified Proof Loads (Rolled Threads) (DIN 267)

Hardness Class	Chemical Composition by Weight % (Sample Test)			
	C max.	Mn min.	P max.	S max.
11H	0.50	—	0.110	0.150
14H	0.50	—	0.110	0.150
17H	0.58	0.30	0.060	0.150
22H	0.58	0.30	0.048	0.058

Table 9-22B. Chemical Composition Requirements for Nuts Without Specified Proof Loads (Free Cutting Materials) (DIN 267)

Hardness Class	Chemical Composition by Weight % (Sample Test)			
	C max.	P max.	Pb max.	S max.
11H, 14H, 17H	0.50	0.12	0.35	0.34

Vicker and Brinell hardness limits for nuts in the four hardness classes are shown in Table 9-23.

INSTALLATION OF THREADED FASTENERS

Clearance Holes for Metric Bolts and Screws

The clearance holes for bolts and screws are based on the ISO 273: Parts I and II standards and the ANSI-OMFS-15 recommendation. Clearance hole diameters for bolts, screws, and studs are given in Table 9-25, and for tapping screws in Table 9-24.

Normal Clearance—Normal clearance hole sizes are preferred for general purpose applications, and are recommended unless special design considerations dictate the need for either a close or loose clearance.

Close Clearance—Close clearance hole sizes should be specified only where conditions such as critical alignment of assembled parts, wall thickness, or other limitations necessitate the use of a minimal hole.

Loose Clearance—Loose clearance hole sizes should be specified only for applications where maximum adjustment capability between components being assembled is necessary.

Table 9-23. Required Surface Hardness for Nuts Without Specified Proof Load (DIN 267)

Mechanical Property		Hardness Class			
		11H	14H	17H	22H
Vicker Hardness HV 5	min.	110	140	170	220
	max.	185	215	245	300
Brinell Hardness HB 30	min.	110	140	170	220
	max.	185	215	245	300

Table 9-24. Clearance Holes for Tapping Screws (IFI-527)

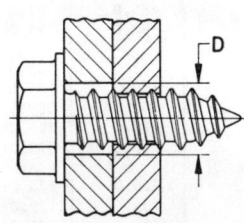

Nom Screw Size	D — Clearance Hole Diameter, Basic		
	Close Clearance	Normal Clearance (Preferred)	Loose Clearance
2.2	2.4	2.6	2.8
2.9	3.1	3.3	3.6
3.5	3.8	4.0	4.4
4.2	4.5	4.8	5.2
4.8	5.2	5.4	5.8
5.5	5.8	6.2	6.5
6.3	6.7	7.1	7.5
8.0	8.5	9.0	9.5
9.5	10.0	10.5	11.5

All dimensions are in millimetres.

Table 9-25. Clearance Holes for Metric Bolts and Screws (ISO 273: Part I and II)

NOM Fastener Size	Close Clearance Min.	Close Clearance Max.	Normal Clearance Min.	Normal Clearance Max.	Loose Clearance Min.	Loose Clearance Max.
1.6	1.7(1.75)	1.8	1.8(1.9)	1.94	2.0(2.1)	2.25
2.0	2.2	2.3	2.4	2.54	2.6	2.85
2.5	2.7	2.8	2.9	3.04	3.1	3.3
3.0	3.2	3.32	3.4	3.58	3.6	3.9
4.0	4.3(4.2)	4.42	4.5	4.68	4.8	5.1
5.0	5.3	5.42	5.5(5.6)	5.68	5.8(6.0)	6.1
6.0	6.4	6.55	6.6	6.82	7.0	7.36
6.3	(6.7)		(7.1)		(7.5)	
7.0	7.4	7.55	7.6	7.82	8.0	8.36
8.0	8.4(8.5)	8.55	9.0	9.22	10.0(9.5)	10.36
10.0	10.5	10.68	11.0	11.27	12.0	12.43
12.0	13.0(12.5)	13.18	14.0(13.0)	14.27	15.0(14.0)	15.43
14.0	15.0(14.5)	15.18	16.0(15.0)	16.27	17.0(16.0)	17.43
16.0	17.0(16.5)	17.18	18.0(17.5)	18.27	19.0(18.5)	19.52
18.0	19.0	19.21	20.0	20.33	21.0	21.52
20.0	21.0	21.21	22.0	22.33	24.0(23.0)	24.52
22.0	23.0	23.21	24.0	24.33	26.0	26.52
24.0	25.0	25.21	26.0	26.33	28.0(27.0)	28.52
27.0	28.0	28.21	30.0	30.33	32.0	32.62
30.0	31.0	31.25	33.0	33.39	35.0	35.62
33.0	34.0	34.25	36.0	36.39	38.0	38.62
36.0	37.0	37.25	39.0	39.39	42.0(41.0)	42.62
39.0	40.0	40.25	42.0	42.39	45.0	45.62
42.0	43.0	43.25	45.0	45.39	48.0	48.62
45.0	46.0	46.25	48.0	48.39	52.0	52.74
48.0	50.0	50.25	52.0	52.46	56.0	56.74
52.0	54.0	54.3	56.0	56.46	62.0	62.74
56.0	58.0	58.3	62.0	62.46	66.0	66.74
60.0	62.0	62.3	66.0	66.46	70.0	70.74
64.0	66.0	66.3	70.0	70.46	74.0	74.74
68.0	70.0	70.3	74.0	74.46	78.0	78.74
72.0	74.0	74.3	78.0	78.46	82.0	82.87
76.0	78.0	78.3	82.0	82.54	86.0	86.87
80.0	82.0	80.35	86.0	86.54	91.0	91.87
85.0	87.0	87.35	91.0	91.54	96.0	96.87
90.0	93.0	93.35	96.0	96.54	101.0	101.87
95.0	98.0	98.35	101.0	101.54	107.0	107.87
100.0	104.0	104.35	107.0	107.54	112.0	112.87
105.0	109.0	109.35	112.0	112.54	117.0	117.87
110.0	114.0	114.35	117.0	117.54	122.0	123.0
115.0	119.0	119.35	122.0	122.63	127.0	128.0
120.0	124.0	124.4	127.0	127.63	132.0	133.0
125.0	129.0	129.4	132.0	132.63	137.0	138.0
130.0	134.0	134.4	137.0	137.63	144.0	145.0
140.0	144.0	144.4	147.0	147.63	155.0	156.0
150.0	155.0	155.4	158.0	158.63	165.0	166.0

NOTES:
1. ANSI-OMFS-15 recommended basic clearance hole diameters are shown in parentheses ( ).
2. The maximum hole limits shown are based on ISO 273 recommended tolerances H12-close clearance, H13-normal clearance and H14-coarse clearance.
3. All dimensions are in millimeters.

Recommended Metric Tap Drill Sizes

The metric tap drill sizes are specified in the ISO 2306 standard and in the ANSI-OMFS-15 publication and shown for a 75% (normal) and 60% thread. The recommended metric drill sizes and the closest customary inch sizes are also tabulated (Table 9-26 is courtesy of the Regal-Beloit Corporation).

Holes for Tapping Screws

The recommended hole sizes for tapping screws to be fastened into steel plates and other materials are based on the British standard BS 4174, and the basic hole sizes are shown in Tables 9-27 through 9-35.

The tables do not cover hole sizes for the "hard metric" thread cutting screw types D, F. and T nor for the thread

FASTENERS

Table 9-26. Recommended Metric Tap Drill Sizes with 60% and 75% (Normal) Thread.
(Courtesy of the Regal-Beloit Corporation)

METRIC TAP SIZE & PITCH	BASED ON APPROX. 60% THREAD Recommended metric drill size/mm	Theoretical drill size mm	Closest American drill size	mm equiv.	BASED ON APPROX. 75% THREAD Recommended metric drill size/mm	Theoretical drill size mm	Closest American drill size	mm equiv.
M1.5 × 0.35	1.20	1.2272	56	1.1913	1.15	1.1590	57	1.0922
M1.6 × 0.35	1.30	1.3272	55	1.3208	1.25	1.2590	⁵⁄₁₀₄	1.1913
M1.8 × 0.35	1.50	1.5272	53	1.5113	1.45	1.4590	54	1.3970
M2 × 0.45	1.65	1.6492	52	1.6129	1.55	1.5612	53	1.5113
M2.2 × 0.45	1.85	1.8492	49	1.8542	1.75	1.7615	51	1.7018
M2.3 × 0.4	1.95	1.9882	⁵⁄₆₄	1.9837	1.90	1.9102	49	1.8542
M2.5 × 0.45	2.15	2.1492	45	2.0828	2.05	2.0615	46	2.0574
M2.6 × 0.45	2.25	2.2492	44	2.1844	2.15	2.1615	45	2.0828
M3 × 0.5	2.50	2.5324	39	2.5273	2.40	2.4155	³⁄₃₂	2.3825
M3 × 0.6	2.60	2.6147	38	2.5184	2.50	2.5184	40	2.4892
M3.5 × 0.6	3.00	3.0324	32	2.9464	2.90	2.9155	33	2.8702
M4 × 0.75	3.40	3.4154	30	3.2639	3.25	3.2693	30	3.2639
M4 × 0.7	3.40	3.4544	29	3.4544	3.30	3.3180	30	3.2639
M4.5 × 0.75	3.90	3.9154	23	3.9116	3.75	3.7693	26	3.7338
M5 × 1	4.20	4.2206	19	4.2164	4.00	4.0257	22	3.9878
M5 × 0.9	4.25	4.2988	19	4.2164	4.10	4.1235	20	4.0894
M5 × 0.8	4.30	4.3764	¹¹⁄₆₄	4.3662	4.20	4.2204	19	4.2164
M5.5 × 0.9	4.75	4.7988	³⁄₁₆	4.7625	4.60	4.6235	14	4.6228
M6 × 1	5.20	5.2206	5	5.2197	5.00	5.0257	9	4.9784
M6 × 0.75	5.40	5.4154	3	5.4102	5.25	5.2693	5	5.2197
M7 × 1	6.20	6.2206	C	6.1468	6.00	6.0257	¹⁵⁄₆₄	5.9537
M7 × 0.75	6.40	6.4154	D	6.2484	6.25	6.2693	D	6.2484
M8 × 1.25	7.00	7.0258	I	6.9088	6.75 (6-80)	6.7823	H	6.7564
M8 × 1	7.20	7.2206	⁹⁄₃₂	7.1450	7.00	7.0257	I	6.9088
M9 × 1.25	8.00	8.0258	⁵⁄₁₆	7.9375	7.75	7.7823	N	7.6708
M9 × 1	8.20	8.2206	P	8.2042	8.00	8.0257	⁵⁄₁₆	7.9375
M10 × 1.5	8.50	8.8308	¹¹⁄₃₂	8.7325	8.50	8.5385	Q	8.4328
M10 × 1.25	9.00	9.0258	S	8.8392	8.75	8.7823	¹¹⁄₃₂	8.7325
M10 × 1	9.20	9.2206	²³⁄₆₄	9.1286	9.00	9.0257	S	8.8392
M11 × 1.5	9.80	9.8308	W	9.8044	9.50	9.5385	³⁄₈	9.5250
M12 × 1.75	10.50	10.6360	Z	10.4902	10.25 (10.30)	10.2950	Y	10.2616
M12 × 1.5	10.80	10.8308	²⁷⁄₆₄	10.7162	10.50	10.5385	Z	10.4902
M12 × 1.25	11.00	11.0258	²⁷⁄₆₄	10.7162	10.50	10.7823	²⁷⁄₆₄	10.7162
M14 × 2	12.40	12.4413	³¹⁄₆₄	12.3037	12.00	12.0516	²⁷⁄₆₄	11.9075
M14 × 1.5	12.50	12.8308	½	12.7000	12.50	12.5385	³¹⁄₆₄	12.3037
M14 × 1.25	13.00	13.0258	½	12.7000	12.75	12.7823	½	12.7000
M15 × 1.5	13.80	13.8308	¹⁷⁄₃₂	13.4950	13.50	13.5385	¹⁷⁄₃₂	13.4950
M16 × 2	14.25	14.4413	³⁶⁄₆₄	14.2875	14.00	14.0516	³⁶⁄₆₄	13.8912
M16 × 1.5	14.75	14.8308	³⁷⁄₆₄	14.6837	14.50	14.5385	³⁶⁄₆₄	14.2875
M17 × 1.5	15.75	15.8308	³⁹⁄₆₄	15.4787	15.50	15.5385	³⁹⁄₆₄	15.4787
M18 × 2.5	16.00	16.0513	⁵⁄₈	15.8750	15.50	15.5643	³⁹⁄₆₄	15.4787
M18 × 2	16.25	16.4413	⁴¹⁄₆₄	16.2712	16.00	16.0516	⁵⁄₈	15.8750
M18 × 1.5	16.75	16.8308	²¹⁄₃₂	16.6700	16.50	16.5385	⁴¹⁄₆₄	16.2712

METRIC TAP SIZE & PITCH	BASED ON APPROX. 60% THREAD Theoretical drill size mm	Recommended metric drill size/mm	Closest American drill size	mm equiv.	BASED ON APPROX. 75% THREAD Theoretical drill size mm	Recommended metric drill size/mm	Closest American drill size	mm equiv.
M19 × 2.5	17.0513	17.00	²¹⁄₃₂	16.6700	16.5643	16.50	⁴¹⁄₆₄	16.2712
M20 × 2.5	18.0513	18.00	⁴⁵⁄₆₄	17.8587	17.5643	17.50	¹¹⁄₁₆	17.4625
M20 × 2	18.4413	18.25	²³⁄₃₂	18.2557	18.0516	18.00	⁴⁵⁄₆₄	17.8587
M20 × 1.5	18.8308	18.75	⁴⁷⁄₆₄	18.6538	18.5385	18.50	²³⁄₃₂	18.2557
M22 × 2.5	20.0513	20.00	²⁵⁄₃₂	19.8425	19.5643	19.50	⁴⁹⁄₆₄	19.4462
M22 × 2	20.4413	20.25	⁵¹⁄₆₄	20.2413	20.0516	20.00	²⁵⁄₃₂	19.8425
M22 × 1.5	20.8308	20.75	¹³⁄₁₆	20.6358	20.5385	20.50	⁵¹⁄₆₄	20.2413
M24 × 3	21.6619	21.50	²⁷⁄₃₂	21.4325	21.0773	21.00	³³⁄₆₄	21.0337
M24 × 2	22.4413	22.25	⁷⁄₈	22.2250	22.0516	22.00	⁵⁵⁄₆₄	21.8288
M24 × 1.5	22.8308	22.75	⁵⁷⁄₆₄	22.6212	22.5385	22.50	⁷⁄₈	22.2250
M25 × 2	23.4413	23.25	⁵⁹⁄₆₄	23.4163	23.0516	23.00	²⁹⁄₃₂	23.0175
M25 × 1.5	23.8308	23.75	¹⁵⁄₁₆	23.8125	23.5385	23.50	⁵⁹⁄₆₄	23.4163
M26 × 3	23.6619	23.50	⁵⁹⁄₆₄	23.4163	23.0773	23.00	²⁹⁄₃₂	23.0175
M27 × 3	24.6619	24.50	³¹⁄₃₂	24.6075	24.0773	24.00	¹⁵⁄₁₆	23.8125
M27 × 2	25.4413	25.25	1	25.4000	25.0516	25.00	⁶³⁄₆₄	25.0038
M28 × 3	25.6519	25.50	1	25.4000	25.0773	25.00	⁶³⁄₆₄	25.0038
M28 × 2	26.4413	26.25	1¹⁄₃₂	26.1925	26.0516	26.00	1¹⁄₆₄	25.7962
M30 × 3.5	27.2720	27.25	1³⁄₃₂	26.9875	26.5900	26.50	1¹⁄₃₂	26.1925
M30 × 3	27.6619	27.50	1³⁄₃₂	27.3837	27.0773	27.00	1¹⁄₁₆	26.9875
M30 × 2	28.4413	28.25	1¹⁄₈	28.1788	28.0516	28.00	1¹⁄₃₂	27.7825
M32 × 3.5	29.2720	29.25	1⁹⁄₆₄	28.9712	28.5900	28.50	1⅛	28.5750
M32 × 2	30.4413	30.25	1³⁄₁₆	30.1625	30.0516	30.00	1¹¹⁄₆₄	29.7663
M33 × 3.5	30.2720	30.25	1³⁄₁₆	30.1625	29.5900	29.50	1³⁄₃₂	29.3675
M33 × 3	30.6619	30.50	1¹³⁄₆₄	30.5587	30.0773	30.00	1¹¹⁄₆₄	29.7663
M33 × 2	31.4413	31.25	1¹⁄₄	31.3538	31.0516	31.00	1¹⁵⁄₆₄	30.9575
M34 × 3.5	31.2720	31.25	1¹⁄₄	30.9575	30.5900	30.50	1³⁄₁₆	30.5587
M36 × 4	32.8822	32.75	1⁹⁄₃₂	32.5425	32.1028	32.00	1¹⁷⁄₆₄	31.7500
M36 × 3	33.6619	33.50	1⁵⁄₁₆	33.3375	33.0773	33.00	1²⁹⁄₆₄	32.9413
M36 × 2	34.4413	34.25	1²⁵⁄₆₄	34.1325	34.0516	34.00	1³¹⁄₆₄	33.7337
M38 × 4	34.8822	34.75	1²³⁄₆₄	34.5288	34.1028	34.00	1³¹⁄₆₄	33.7337
M39 × 4	35.8822	35.75	1¹³⁄₃₂	35.7175	35.1028	35.00	1¾	34.9250
M39 × 3	36.6619	36.50	1⁷⁄₁₆	36.5125	36.0773	36.00	1¹⁷⁄₆₄	35.7175
M39 × 2	37.4413	37.25	1¹⁵⁄₃₂	37.3075	37.0516	37.00	1²⁹⁄₆₄	36.9087

APPROXIMATE METRIC TAP DRILL FORMULA

$$\frac{\text{Nominal O.D. Minus (Pitch)}}{} = 77\% \text{ of Thread}$$

Nominal O.D. Minus (.65 × Pitch) = 50% of Thread	Nominal O.D. Minus (.91 × Pitch) = 70% of Thread
Nominal O.D. Minus (.71 × Pitch) = 55% of Thread	Nominal O.D. Minus (.97 × Pitch) = 75% of Thread
Nominal O.D. Minus (.78 × Pitch) = 60% of Thread	Nominal O.D. Minus (1.04 × Pitch) = 80% of Thread
Nominal O.D. Minus (.84 × Pitch) = 65% of Thread	Nominal O.D. Minus (1.10 × Pitch) = 85% of Thread

NOTES:
1. New ISO size. M6.3 × 1 5.5206 5.50 7/32 5.5550 5.3257 5.30 4 5.3086.
2. ANSI-OMFS recommended values shown in brackets ( ).

Table 9-27. Types AB and B Thread Forming Screws: Recommended Hole Sizes in Sheet Metal Steel, Brass, Aluminum Alloy, Stainless Steel, and Monel Metal (BS 4174)*

Screw size (No.)	Metal thickness mm	Metal thickness in	Pierced or extruded hole diameter in	Drilled or clean punched holes hole diameter mm	Drilled or clean punched holes hole diameter in
2	0.45	0.018	—	1.60	0.063
	0.91	0.036	—	1.85	0.073
	1.62	0.064	—	1.95	0.077
4	0.45	0.018	—	2.05	0.081
	0.91	0.036	0.098	2.30	0.091
	1.62	0.064	—	2.40	0.095
	2.03	0.080	—	2.60	0.102
6	0.45	0.018	—	2.35	0.092
	0.91	0.036	0.111	2.80	0.110
	1.62	0.064	—	2.95	0.116
	2.03	0.080	—	3.10	0.122
	2.64	0.104	—	3.20	0.126
8	0.71	0.028	—	2.90	0.114
	0.91	0.036	0.136	3.10	0.122
	1.22	0.048	—	3.20	0.126
	1.62	0.064	—	3.40	0.134
	2.64	0.104	—	3.70	0.146
	3.18	0.125	—	3.80	0.150
10	0.71	0.028	—	3.40	0.134
	1.22	0.048	—	3.60	0.142
	1.62	0.064	—	3.80	0.150
	2.64	0.104	—	4.10	0.161
	3.18	0.125	—	4.30	0.169
	4.75	0.187	—	4.50	0.177
12	0.71	0.028	—	4.10	0.161
	1.22	0.048	—	4.30	0.169
	1.62	0.064	—	4.50	0.177
	2.64	0.104	—	4.80	0.189
	3.18	0.125	—	4.90	0.193
	4.75	0.187	—	5.10	0.201
14(1/4)	1.22	0.048	—	4.80	0.189
	1.62	0.064	—	5.20	0.205
	2.03	0.080	—	5.40	0.213
	3.18	0.125	—	5.70	0.224
	4.75	0.187	—	5.90	0.232
	6.35	0.250	—	6.00	0.236
16(5/16)	4.75	0.187	—	7.60	0.299
	6.35	0.250	—	7.60	0.299
	7.93	0.312	—	7.60	0.299

*Material from British Standards is reproduced by permission of BSI, 2 Park Street, London W1A 2BS, UK

FASTENERS

Table 9-28. Type B Thread-Forming Screws: Recommended Hole Sizes (Drilled or Cored Hole) in Non-Ferrous Castings—Aluminum Magnesium, Zinc, Brass, Bronze, etc. (BS 4174)*

Screw size (No.)	Minimum penetration depth mm	in	hole dia. mm	in	Normal maximum penetration depth mm	in	hole dia. mm	in
2	3.0	1/8	1.80	0.071	6.5	1/4	2.00	0.079
4	4.0	5/32	2.45	0.096	8.0	5/16	2.65	0.104
6	5.0	3/16	3.30	0.130	9.5	3/8	3.30	0.130
8	5.5	7/32	3.90	0.153	11.0	7/16	3.90	0.153
10	6.5	1/4	4.50	0.177	12.5	1/2	4.50	0.177
12	7.0	9/32	5.10	0.201	14.5	9/16	5.10	0.201
14	8.0	5/16	6.00	0.236	16.0	5/8	6.00	0.236
16	11.0	7/16	7.50	0.295	22.0	7/8	7.50	0.295

*Material from British Standards is reproduced by permission of BSI, 2 Park Street, London W1A 2BS, UK

NOTES:
1. Cored holes, a taper of 0.25 mm (0.010 in.) in depth is permitted. A cored hole diameter should equal the nominal hole size at one-half the screw penetration depth.
2. Porous castings may require use of a smaller hole and/or increased depth of engagement.

rolling screw types SF, SW, TT, and TR-3. Refer to the IFI-502 and IFI-503 standards for recommended metric hole and drill sizes for the above screw types.

Recommended Torque Values for Metric Fasteners

The torque values for metric threaded fasteners have been based on existing inch practices, and the nominal torque values for five strength grades are shown in Table 9-36.

HEXAGON HEAD CAP SCREWS AND BOLTS (IFI-506)

World and national standards for hexagon head screws and bolts are shown in the fastener index, page 9-2. The material presented in this chapter has been based entirely on the IFI-506 standard with the permission from the Industrial Fasteners Institute, 1505 East Ohio Building, Cleveland, Ohio 44114. In this IFI standard, three widths across flats, 17, 19, and 22 mm, were changed to 15, 18, and 21 mm to offer the best use of material.

The only recognized thread pitch in the IFI standards is the ISO metric coarse threaded. This publication is in full agreement with this contribution to our rationalization efforts, and there is a clear trend to standardize on the ISO metric coarse thread worldwide.

The dimensions for hexagon head cap screws in sizes ranging from 5 to 100 mm diameters are shown in Table 9-37 and for hexagon head bolts in the same size range in Table 9-42.

General Notes on Hexagon Head Cap Screws

Dimensions—All dimensions in the tables and notes are in millimeters unless otherwise stated.

Top of Head—This should be full form and chamfered or rounded with the diameter of chamfer circle or start of rounding being equal to the maximum width across flats, within a tolerance of minus 15 percent.

Wrenching Height, J—Wrenching height is a distance measured from the bearing surface up the side of the head at the corners. The width across corners shall be within specified limits for the full wrenching height.

Head Height—The head height is the distance, as measured parallel to the axis of the screw, from the top of the head to the plane of the bearing circle diameter.

Concentricity of Head—The axis of the head shall be located at true position with respect to the axis of the screw (determined over a distance under the head equal to one screw diameter) within a tolerance zone of diameter shown in Table 9-41.

Bearing Surface—The bearing surface should be flat and washer faced. Diameter of bearing surface should not exceed the maximum specified width across flats nor be less than the specified minimum washer face diameter, M.

Table 9-29. Type B Thread-Forming Screws: Recommended Hole Sizes (BS 4174)*

Screw size (No.)	Metal thickness mm	Metal thickness in	Cadmium plated lubricated screws hole diameter mm	Cadmium plated lubricated screws hole diameter in	Self colour or non-lubricated screws hole diameter mm	Self colour or non-lubricated screws hole diameter in
6	0.91	0.036	2.60	0.102	2.60	0.102
	1.62	0.064	2.80	0.110	2.80	0.110
	2.03	0.080	2.90	0.114	2.90	0.114
	2.64	0.104	3.10	0.122	3.10	0.122
8	1.62	0.064	3.30	0.130	3.30	0.130
	2.03	0.080	3.60	0.142	3.60	0.142
	2.64	0.104	3.60	0.142	3.70	0.146
	3.18	0.125	3.70	0.146	3.80	0.150
10	1.62	0.064	3.80	0.150	3.80	0.150
	2.64	0.104	4.00	0.158	4.00	0.158
	3.18	0.125	4.10	0.161	4.10	0.161
	4.75	0.187	4.40	0.173	4.50	0.177
14	3.18	0.125	5.60	0.220	5.60	0.220
	4.75	0.187	5.90	0.232	5.90	0.232
	6.35	0.250	5.90	0.232	5.90	0.232
	7.92	0.312	5.90	0.232	5.90	0.232
16	3.18	0.125	7.20	0.283	7.30	0.287
	4.75	0.187	7.40	0.291	7.50	0.295
	6.35	0.250	7.40	0.291	7.60	0.299
	7.92	0.312	7.50	0.295	7.60	0.299
	9.52	0.375	7.50	0.295	7.70	0.303
	12.70	0.500	7.60	0.299	7.70	0.303

*Material from British Standards is reproduced by permission of BSI, 2 Park Street, London W1A 2BS, UK

NOTES
1. In very soft materials a smaller hole is required.
2. In very hard materials a larger hole is required.

Measurement of FIR shall be made as close to the periphery of the bearing surface as possible while the screw is held in a collet or other gripping device at a distance of one screw diameter from the underside of the head.

Fillet—The fillet at junction of head and shank should be a smooth concave curve within an envelope of R minimum, and a smooth multi-radius curve tangent to the underside of head at a point no greater than one-half of E_a maximum from the axis of the screw and tangent to the shank of the screw at a distance no greater than L_a maximum from the underside of head.

Body Diameter—The diameter of the body on screws which are not threaded full length should be within the limits specified (E). For screws threaded full length, the diameter of the unthreaded shank under the head should not exceed the specified max body diameter (E) nor be less than the min body diameter given in Table 9-41.

Screws with lengths longer than the lower dashed line of Table 9-38 may have bearing surface die seams and body fins or die seams to the limits specified in Table 9-45.

Length—The length of the screw should be measured parallel to the axis of the screw from the under head bear-

FASTENERS

Table 9-30. Types D and T Thread-Cutting Screws: Recommended Hole Sizes in Ferrous and Non-Ferrous Sheets and Cast Iron (BS 4174)*

SCREW SIZE (NO.)(OR IN.) AND THREADS PER INCH	METAL THICKNESS mm	METAL THICKNESS inches	SHEET METAL mm	SHEET METAL inches	STAINLESS STEEL AND CAST IRON mm	STAINLESS STEEL AND CAST IRON inches	ALUMINUM AND BRASS SHEETS mm	ALUMINUM AND BRASS SHEETS inches
4 - 40	1.25	0.048	2.25	0.089	2.4	0.094	2.25	0.089
	1.6	0.063	2.25	0.089	2.4	0.094	2.25	0.089
	2.36	0.094	2.35	0.092	2.45	0.096	2.25	0.089
6 - 32	1.25	0.048	2.8	0.110	2.95	0.116	2.8	0.110
	1.6	0.063	2.8	0.110	2.95	0.116	2.8	0.110
	2.36	0.094	2.95	0.116	3	0.118	2.85	0.112
	3.15	0.125	3	0.118	3.1	0.122	2.95	0.116
8 - 32	1.6	0.063	3.5	0.138	3.6	0.142	3.4	0.134
	2.36	0.094	3.6	0.142	3.6	0.142	3.5	0.138
	3.15	0.125	3.7	0.146	3.7	0.146	3.6	0.142
	4.75	0.187	3.8	0.150	3.9	0.153	3.7	0.146
10 - 24	1.6	0.063	3.9	0.153	4.2	0.165	3.9	0.153
	2.36	0.094	4	0.158	4.2	0.165	4	0.158
	3.15	0.125	4.2	0.165	4.3	0.169	4	0.158
	4.75	0.187	4.4	0.173	4.5	0.177	4.2	0.165
	6.3	0.250	4.5	0.177	4.6	0.181	4.4	0.173
1/4 - 20	3.15	0.125	5.6	0.220	5.8	0.228	5.4	0.213
	4.75	0.187	5.8	0.228	5.9	0.232	5.6	0.221
	6.3	0.250	5.9	0.232	6	0.236	5.8	0.228
	8	0.312	5.9	0.232	6	0.236	5.9	0.232
5/16 - 18	3.15	0.125	7	0.276	7.4	0.291	6.9	0.272
	4.75	0.187	7.4	0.291	7.5	0.295	7.1	0.280
	6.3	0.250	7.5	0.295	7.5	0.295	7.4	0.291
	8	0.312	7.5	0.295	7.5	0.295	7.5	0.295
	9.5	0.375	7.5	0.295	7.6	0.299	7.5	0.295

*Material from British Standards is reproduced by permission of BSI, 2 Park Street, London W1A 2BS, UK

NOTES:
1. For stainless steel and cast-iron applications, hole diameters should generally be slightly increased.
2. Because conditions vary, it may be necessary to change the hole size to suit the particular application.
3. A minimum engagement of two full threads above the cutting flutes is recommended.

ing surface to the extreme end of the shank. The maximum length for all screw sizes shall be equal to nominal length and minimum length shall be nominal length minus the tolerance given in Table 9-39.

Points—At manufacturer's option, the end of the screw should be chamfered from a diameter equal to or slightly less than the thread root diameter to produce a length of chamfer or incomplete thread within the limits for Z given in Table 9-40, or should have a rounded point of radius X as given in Table 9-40. The end of the screw should be reasonably square with the axis of the screw and where pointed blanks are used, the slight rim or cup resulting from roll threading shall be permissible.

Table 9-31. Types D and T Thread-Cutting Screws: Recommended Hole Sizes in Cast Aluminum, Zinc, and Aluminum Die Castings (BS 4174)*

Screw size (No.) (or in) and threads per inch	Metal thickness mm	Metal thickness in	Hole diameter mm	Hole diameter in	Screw size (No.) (or in) and threads per inch	Metal thickness mm	Metal thickness in	Hole diameter mm	Hole diameter in
4—40	3.15	0.125	2.50	0.098	4—48	3.15	0.125	2.50	0.098
	4.75	0.187	2.50	0.098		4.75	0.187	2.55	0.100
6—32	3.15	0.125	3.00	0.118	6—40	3.15	0.125	3.20	0.126
	4.75	0.187	3.10	0.122		4.75	0.187	3.20	0.126
	6.30	0.250	3.20	0.126		6.30	0.250	3.20	0.126
8—32	3.15	0.125	3.70	0.146	8—36	3.15	0.125	3.70	0.146
	4.75	0.187	3.70	0.146		4.75	0.187	3.80	0.150
	6.30	0.250	3.80	0.150		6.30	0.250	3.80	0.150
10—24	3.15	0.125	4.20	0.165	10—32	3.15	0.125	4.40	0.173
	4.75	0.187	4.20	0.165		4.75	0.187	4.40	0.173
	6.30	0.250	4.30	0.169		6.30	0.250	4.40	0.173
	8.00	0.312	4.40	0.173		8.00	0.312	4.50	0.177
1/4—20	3.15	0.125	5.60	0.221	1/4—28	3.15	0.125	5.80	0.228
	4.75	0.187	5.60	0.221		4.75	0.187	5.80	0.228
	6.30	0.250	5.80	0.228		6.30	0.250	5.90	0.232
	8.00	0.312	5.80	0.228		8.00	0.312	5.90	0.232
5/16—18	3.15	0.125	7.10	0.280	5/16—24	3.15	0.125	7.30	0.288
	4.75	0.187	7.10	0.280		4.75	0.187	7.40	0.291
	6.30	0.250	7.20	0.283		6.30	0.250	7.40	0.291
	8.00	0.312	7.40	0.291		8.00	0.312	7.50	0.295

*Material from British Standards is reproduced by permission of BSI, 2 Part Street, London W1A 2BS, UK

NOTE:
Because conditions vary, it may be necessary to change the hole size to suit the particular application.

Straightness—Shanks of screws should be straight within the following limits: for screws with nominal lengths to and including 300 mm, the maximum camber should be 0.006 mm per mm of screw length, and for screws with nominal lengths over 300 mm to and including 600 mm the maximum camber should be 0.008 mm per mm of length.

Thread Length—The length of thread on screws should be controlled by the maximum grip gaging length (L_G) and the minimum body length (L_B) as set forth in the following: Grip gaging length, (L_G) max, is the distance, measured parallel to the axis of the screw, from the under head bearing surface to the face of a non-counter-bored or non-countersunk standard GO thread ring gage assembled by hand as far as the thread will permit. Values for L_G max are given in Table 9-38. For diameter-length combinations not listed in Table 9-38 the maximum grip gaging length, as calculated and rounded to one decimal place, is equal to the minimum screw length (nominal length minus tolerance in Table 9-39 minus the minimum thread length as given in Table 9-37 (L_G max = L min − L_T min). L_G max shall be used as a criteria for inspection.

FASTENERS

Table 9-32. Type BT Thread-Cutting Screws: Recommended Hole Sizes in Plastics (BS 4174)*

Screw type and size		Phenolics		Cellulose acetate and nitrate, acrylic and styrene resins		Depth of penetration			
BT	T	Hole diameter		Hole diameter		min.		max.	
		mm	in	mm	in	mm	in	mm	in
2—32		2.00	0.079	1.95	0.077	2.5	3/32	6.5	1/4
4—24		2.65	0.104	2.55	0.100	3.0	1/8	8.0	5/16
	4—40	2.50	0.098	2.40	0.095	6.5	1/4	11.0	7/16
6—20		3.10	0.122	3.10	0.122	5.0	3/16	9.5	3/8
	6—32	3.00	0.118	2.95	0.116	6.5	1/4	11.0	7/16
8—18		3.70	0.146	3.60	0.142	6.5	1/4	12.5	1/2
	8—32	3.60	0.142	3.60	0.142	8.0	5/16	12.5	1/2
10—16		4.30	0.169	4.20	0.165	8.0	5/16	16.0	5/8
	10—24	4.10	0.161	4.10	0.161	9.5	3/8	12.5	1/2
12—14		4.90	0.193	4.80	0.189	9.5	3/8	16.0	5/8
14—14		5.80	0.228	5.70	0.224	9.5	3/8	19.0	3/4
	1/4—20	5.70	0.224	5.60	0.221	9.5	3/8	16.0	5/8

*Material from British Standards is reproduced by permission of BSI, 2 Park Street, London W1A 2BS, UK

Body length, L_B min is the distance, measured parallel to the axis of the screw, from the under head bearing surface to the last scratch of thread or the top of the extrusion angle. Values of L_B min are given in Table 9-38. For diameter-length combinations not listed in Table 9-38 the minimum body length, as calculated and rounded to one decimal place, is equal to the maximum grip gaging length (as computed) minus the maximum transition thread length as given in Table 9-37 (L_B min = L_G max − Y max). L_B min shall be used as a criterion for inspection. Screws of nominal lengths which have a calculated L_B min value equal to or less than the length of 2.5 times the thread pitch shall be threaded full length. For screws which are threaded full length, the distance from the plane formed by the bearing circle diameter to the face of a non counterbored or non countersunk standard GO thread ring gage assembled by hand as far as the thread will permit shall not exceed a length equal to 2.5 times the thread pitch.

Thread length, L_T min, as given in Table 9-37 is a reference dimension intended for calculation purposes only, and is the distance measured parallel to the axis of the screw, from the extreme end of the screw to the last complete (full form) thread.

Transition thread length, Y max, as given in Table 9-37 is a reference dimension intended for calculation purposes only. It includes the length of incomplete threads and tolerances on grip gaging length and body length. The transition from full thread to incomplete thread shall be smooth and uniform. The major diameter of the incomplete threads shall not exceed the actual major diameter of the complete (full form) threads.

Thread Concentricity—The axis of the thread shall be concentric with the axis of the screw shank within the limits given in Table 9-41 when measured at a distance of one screw diameter from the thread runout.

Thread Series—Threads shall conform to dimensions given in Section 8—Modified ISO Fastener Screw Threads (page 8-20) and shall be class 6g unless otherwise specified by the customer. The class 6g tolerance shall apply to plain finish (unplated or uncoated) screws, and to plated or coated screws before plating or coating. For screws with additive finish, the class 6g diameters may be exceeded by the amount of the allowance; i.e., the basic diameters shall apply to screws after plating or coating.

Material and Mechanical Properties—Carbon steel screws shall conform to the requirements for the applicable property class as described in the section on strength properties for threaded fasteners (page 9-18). Class 4.6 screws are available in sizes 5 through 36 mm; class 5.8 in sizes 5 through 24 mm and with lengths no longer than 150 mm; class 9.8 in sizes 5 through 16 mm; class 8.8 in sizes 20 through 36 mm; and class 10.9 in sizes 6 through 36 mm. Carbon steel screws of sizes 42 through 100 mm, and screws of other materials such as stainless steel, brass, bronze, and aluminum alloys shall have properties as agreed upon by the manufacturer and the purchaser. (For guidance, refer to IFI-516 and IFI-518).

Table 9-33. Type BT Thread Cutting Screws: Recommended Hole Sizes in Cast Aluminum, Zinc, and Aluminum, Die Castings* (BS 4174)

Screw size (No.)	Metal thickness mm	Metal thickness in	Hole diameter mm	Hole diameter in
2	1.62	0.064	1.85	0.073
	2.36	0.093	1.90	0.075
	3.18	0.125	1.95	0.077
4	2.36	0.093	2.50	0.098
	3.18	0.125	2.55	0.100
	4.75	0.187	2.55	0.100
	6.35	0.250	2.55	0.100
6	3.18	0.125	3.00	0.118
	4.75	0.187	3.10	0.122
	6.35	0.250	3.15	0.126
	7.92	0.312	3.15	0.126
8	3.18	0.125	3.80	0.150
	4.75	0.187	3.80	0.150
	6.35	0.250	3.80	0.150
	7.92	0.312	3.90	0.153
10	3.18	0.125	4.20	0.165
	4.75	0.187	4.20	0.165
	6.35	0.250	4.30	0.169
	7.92	0.312	4.30	0.169
12	4.75	0.187	4.90	0.193
	6.35	0.250	5.00	0.197
	7.92	0.312	5.00	0.197
	9.52	0.375	5.00	0.197
14	4.75	0.187	5.60	0.221
	6.35	0.250	5.80	0.228
	7.92	0.312	5.80	0.228
	9.52	0.375	5.80	0.228
16	4.75	0.187	7.20	0.283
	6.35	0.250	7.20	0.283
	7.92	0.312	7.30	0.287
	9.52	0.375	7.30	0.287

*Material from British Standards is reproduced by permission of BSI, 2 Park Street, London W1A 2BS, UK

Identification Symbols—Screws shall be marked with the property class symbol as shown in Table 9-8 and with the manufacturer's identification symbol. Markings shall be located on the top of the head and may be raised or indented at option of the manufacturer.

Designation—Hex cap screws shall be designated by the following data in the sequence shown: Nominal size and thread pitch, nominal length, property class or material, product name, and protective coating, if required.

Examples: M8 × 1.25 × 40, 9.8 hex cap screw, zinc plated. M20 × 2.5 × 80, 651 silicon bronze hex cap screw.

General Notes on Hexagon Bolts

Dimensions—All dimensions in the tables and notes are in millimeters unless otherwise stated.

Availability—Hex bolts in sizes 5 through 24 mm are standard only in lengths longer than 10 D or 150 mm, whichever is shorter, where D is nominal bolt size. When shorter lengths of these sizes are ordered, hex cap screws are normally supplied. Hex bolts in sizes 30 mm and larger are standard in all lengths. However, at the manufacturer's option, hex cap screws may be substituted for any diameter-length combination.

FASTENERS

Table 9-34. Type BT Thread-Cutting Screws: Recommended Hole Sizes in Sheet Steel (BS 4174)*

Screw size (No.)	Metal thickness mm	Metal thickness in	Hole diameter mm	Hole diameter in
2	0.79	0.031	1.75	0.069
	1.22	0.048	1.80	0.071
	1.60	0.063	1.85	0.073
4	0.79	0.031	2.35	0.093
	1.22	0.048	2.40	0.095
	1.60	0.063	2.45	0.097
6	1.22	0.048	2.85	0.112
	1.60	0.063	2.90	0.114
	2.39	0.094	3.00	0.118
8	1.22	0.048	3.30	0.130
	1.60	0.063	3.40	0.134
	2.39	0.094	3.60	0.142
	3.18	0.125	3.70	0.146
10	1.60	0.063	3.80	0.150
	2.39	0.094	3.90	0.154
	3.18	0.125	4.10	0.161
	4.75	0.187	4.20	0.165
12	1.60	0.063	4.40	0.173
	2.39	0.094	4.60	0.181
	3.18	0.125	4.70	0.185
	4.75	0.187	4.80	0.189
14	2.39	0.094	5.30	0.209
	3.18	0.125	5.40	0.213
	4.75	0.187	5.60	0.221
	6.35	0.250	5.70	0.224
16	3.18	0.125	6.70	0.264
	4.75	0.187	6.90	0.272
	6.35	0.250	7.10	0.280
	7.92	0.312	7.30	0.287

*Material from British Standards is reproduced by permission of BSI, 2 Park Street, London W1A 2BS, UK

Table 9-35. Type U—Hammer Drive Screws: Recommended Hole Sizes* (BS 4174)

For use in screw size (No.)	Thin sheet metal, non-ferrous castings, plastics, etc. hole diameter mm	in	Cast iron thick sheet metal hole diameter mm	in
00	1.30	0.051	1.40	0.055
0	1.65	0.065	1.75	0.069
2	2.20	0.087	2.30	0.091
4	2.55	0.100	2.70	0.106
6	3.10	0.122	3.30	0.130
7	3.40	0.134	3.60	0.142
8	3.70	0.146	3.90	0.154
10	4.10	0.161	4.30	0.169
12	4.80	0.189	5.00	0.197
14	5.50	0.217	5.80	0.228

Material from British Standards is reproduced by permission of BSI, 2 Park Street, London W1A 2BS, UK

NOTES:
1. The material shall be thick enough to provide adequate thread engagement, and normally should not be less than the screw diameter.
2. For applications in plastics consideration must be given to the fragility of the section and the brittleness of the plastics.

Table 9-36. Recommended Torque Values for Metric Fasteners

NOMINAL SIZE	STRESS AREA (MM**2)	CLASS 4.6 400 MEGAPASCALS MIN TENSILE STR (N·M) (LB·FT)	CLASS 8.8 830 MEGAPASCALS MIN TENSILE STR (N·M) (LB·FT)	CLASS 9.8 900 MEGAPASCALS MIN TENSILE STR (N·M) (LB·FT)	CLASS 10.9 1040 MEGAPASCALS MIN TENSILE STR (N·M) (LB·FT)	CLASS 12.9 1220 MEGAPASCALS MIN TENSILE STR (N·M) (LB·FT)
M 2.0	2.1	0.14 0.10	0.37 0.27	0.40 0.30	0.52 0.38	0.61 0.45
M 2.5	3.4	0.28 0.21	0.76 0.56	0.82 0.61	1.06 0.78	1.24 0.92
M 3.0	5.0	0.51 0.37	1.35 1.00	1.47 1.08	1.88 1.39	2.21 1.63
M 3.5	6.8	0.80 0.59	2.13 1.57	2.31 1.70	2.96 2.18	3.47 2.56
M 4.0	8.8	1.18 0.87	3.15 2.32	3.41 2.52	4.38 3.23	5.14 3.79
M 5.0	14.2	2.39 1.76	6.36 4.69	6.90 5.09	8.86 6.54	10.39 7.67
M 6.0	20.1	4.05 2.99	10.81 7.97	11.72 8.65	15.05 11.10	17.66 13.02
M 6.3	22.6	4.78 3.53	12.76 9.41	13.84 10.21	17.77 13.11	20.84 15.37
M 8.0	36.6	9.84 7.26	26.25 19.36	28.46 20.99	36.54 26.95	42.87 31.62
M10.0	58.0	19.49 14.37	51.99 38.35	56.38 41.58	72.38 53.39	84.91 62.63
M12.0	84.3	33.99 25.07	90.68 66.88	98.33 72.52	126.25 93.12	148.10 109.23
M14.0	115.0	54.10 39.90	144.32 106.45	156.49 115.42	200.93 148.20	235.70 173.85
M16.0	157.0	84.40 62.25	225.18 166.08	244.17 180.09	313.50 231.22	367.76 271.24
M20.0	245.0	164.64 121.43	439.24 323.96	476.28 351.29	611.52 451.03	717.36 529.10
M24.0	353.0	284.66 209.95	759.43 560.13	823.48 607.37	1057.31 779.83	1240.30 914.80
M30.0	561.0	565.49 417.08	1508.64 1112.72	1635.88 1206.56	2100.38 1549.16	2463.91 1817.29
M36.0	817.0	988.24 728.89	2636.49 1944.58	2858.85 2108.58	3670.62 2707.31	4305.92 3175.88
M42.0	1120.0	1580.54 1165.75	4216.66 3110.05	4572.29 3372.35	5870.59 4329.93	6886.66 5079.34

NOTES:
1. 1 lb (force) × foot = 1.355818 N · m (newtons × meter).
2. The minimum recommended torque values shown are valid for zinc coated fasteners assembled in rigid joints to 75% of proof loads. For maximum or 100% proof torques, multiply table values by 1.33. Reduce torque values for fasteners with less friction or non-rigid (gasket) joints.

FASTENERS

Table 9-37. Dimensions of Hex Cap Screws (IFI-506)

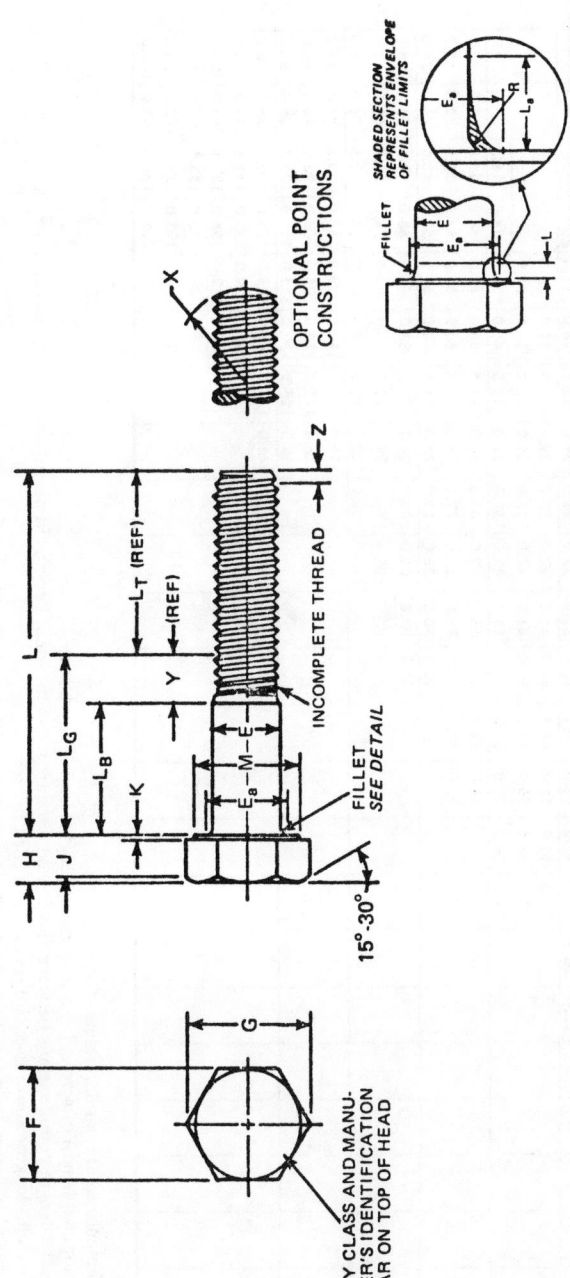

Nom Screw Size & Thread Pitch	E Body Diameter Max	E Body Diameter Min	F Width Across Flats Max	F Width Across Flats Min	G Width Across Corners Max	G Width Across Corners Min	H Head Height Max	H Head Height Min	J Wrenching Height Min	K Washer Face Thickness Max	K Washer Face Thickness Min	M Washer Face Dia Min	Runout of Bearing Surface FIR Max	E_a Fillet Transition Dia Max	L_a Fillet Transition Length Max	R Radius of Fillet Min	L_T Screw Lengths ≤125	L_T Screw Lengths >125 and ≤200	L_T Screw Lengths >200	Y (Ref) Transition Thread Length Max
M5×0.8	5.00	4.82	8.00	7.78	9.24	8.87	3.65	3.35	2.4	0.5	0.2	7.0	0.22	5.7	1.2	0.2	16	22	35	4.0
M6×1	6.00	5.82	10.00	9.76	11.55	11.13	4.47	4.13	3.0	0.5	0.2	8.9	0.25	7.0	1.8	0.3	18	24	37	5.0
M8×1.25	8.00	7.78	13.00	12.73	15.01	14.51	5.50	5.10	3.7	0.6	0.3	11.6	0.28	9.2	2.0	0.4	22	28	41	6.2
*M10×1.5	10.00	9.78	15.00	14.70	17.32	16.76	6.63	6.17	4.5	0.6	0.3	13.6	0.31	11.2	2.0	0.4	26	32	45	7.5
M12×1.75	12.00	11.73	18.00	17.67	20.78	20.14	7.76	7.24	5.2	0.6	0.3	16.6	0.35	13.2	3.0	0.4	30	36	49	8.8
M14×2	14.00	13.73	21.00	20.64	24.25	23.53	9.09	8.51	6.2	0.6	0.3	19.4	0.39	15.2	3.0	0.4	34	40	53	10.0
M16×2	16.00	15.73	24.00	23.61	27.71	26.92	10.32	9.68	7.0	0.8	0.4	22.4	0.43	17.7	3.0	0.6	38	44	57	10.0
M20×2.5	20.00	19.67	30.00	29.35	34.64	33.46	12.88	12.12	8.8	0.8	0.4	27.6	0.53	22.4	4.0	0.8	46	52	65	12.5
M24×3	24.00	23.67	36.00	35.25	41.57	40.19	15.44	14.56	10.5	0.8	0.4	32.9	0.63	26.4	4.0	0.8	54	60	73	15.0
M30×3.5	30.00	29.61	46.00	44.50	53.12	50.73	19.48	17.92	13.1	0.8	0.4	42.5	0.78	33.4	6.0	1.0	66	72	85	17.5
M36×4	36.00	35.61	55.00	53.20	63.51	60.65	23.38	21.62	15.8	0.8	0.4	50.8	0.93	39.4	6.0	1.0	78	84	97	20.0
M42×4.5	42.00	41.61	65.00	62.90	75.06	71.71	26.97	25.03	18.2	1.0	0.5	58.5	1.09	45.6	6.3	1.2	90	96	109	22.5
M48×5	48.00	47.61	75.00	72.60	86.60	82.76	31.07	28.93	21.0	1.0	0.5	67.5	1.25	52.6	8.0	1.5	102	108	121	25.0
M56×5.5	56.00	55.54	85.00	82.20	98.15	93.71	36.20	33.80	24.5	1.0	0.5	76.5	1.47	62.0	10.5	2.0	—	124	137	27.5
M64×6	64.00	63.54	95.00	91.80	109.70	104.65	41.32	38.68	28.0	1.0	0.5	85.5	1.69	70.0	10.5	2.0	—	140	153	30.0
M72×6	72.00	71.54	105.00	101.40	121.24	115.60	46.45	43.55	31.5	1.2	0.6	94.5	1.91	78.0	10.5	2.0	—	156	169	30.0
M80×6	80.00	79.54	115.00	111.00	132.79	126.54	51.58	48.42	35.0	1.2	0.6	103.5	2.13	86.0	10.5	2.0	—	172	185	30.0
M90×6	90.00	89.46	130.00	125.50	150.11	143.07	57.74	54.26	39.2	1.2	0.6	117.0	2.41	96.0	10.5	2.0	—	192	205	30.0
M100×6	100.00	99.46	145.00	140.00	167.43	159.60	63.90	60.10	43.4	1.2	0.6	130.5	2.69	107.0	12.2	2.5	—	212	225	30.0

*M10 screws with a 15 mm width across flats are currently being produced and used in substantial quantities in the U.S. as well as in many other countries. The ISO standards specify a 16 mm width across flats for the M10 size.

Table 9-38. Maximum Grip Gaging Lengths and Minimum Body Lengths for Hex Cap Screws (IFI-506)

Nom Screw Size	M5x0.8 L_G Max	M5x0.8 L_G Min	M6x1 L_G Max	M6x1 L_G Min	M8x1.25 L_G Max	M8x1.25 L_G Min	M10x1.5 L_G Max	M10x1.5 L_G Min	M12x1.75 L_G Max	M12x1.75 L_G Min	M14x2 L_G Max	M14x2 L_G Min	M16x2 L_G Max	M16x2 L_G Min	M20x2.5 L_G Max	M20x2.5 L_G Min	M24x3 L_G Max	M24x3 L_G Min	M30x3.5 L_G Max	M30x3.5 L_G Min	M36x4 L_G Max	M36x4 L_G Min
L Nom																						
8																						
10																						
12																						
14																						
16																						
20	9.0	5.0																				
25	9.0	5.0	12.0	7.0																		
30	19.0	15.0	17.0	12.0																		
35	19.0	15.0	17.0	12.0	13.0	6.8																
40	19.0	15.0	17.0	12.0	13.0	6.8	14.0	6.5														
45	29.0	25.0	27.0	22.0	23.0	16.8	19.0	11.5														
50	29.0	25.0	27.0	22.0	23.0	16.8	19.0	11.5	15.0	6.2												
(55)			37.0	32.0	33.0	26.8	29.0	21.5	25.0	16.2	16.0	6.0										
60			37.0	32.0	33.0	26.8	29.0	21.5	25.0	16.2	21.0	11.0	17.0	7.0								
(65)					43.0	36.8	39.0	31.5	35.0	26.2	21.0	11.0	17.0	7.0								
70					43.0	36.8	39.0	31.5	35.0	26.2	31.0	21.0	27.0	17.0	19.0	6.5						
(75)					53.0	46.8	49.0	41.5	45.0	36.2	31.0	21.0	27.0	17.0	19.0	6.5						
80					53.0	46.8	49.0	41.5	45.0	36.2	41.0	31.0	37.0	27.0	29.0	16.5	26.0	11.0				
(85)							59.0	51.5	55.0	46.2	41.0	31.0	37.0	27.0	29.0	16.5	31.0	16.0				
90							59.0	51.5	55.0	46.2	51.0	41.0	47.0	37.0	39.0	26.5	36.0	21.0				
100							74.0	66.5	70.0	61.2	51.0	41.0	47.0	37.0	39.0	26.5	46.0	31.0	34.0	16.5		
110									80.0	71.2	66.0	56.0	62.0	52.0	54.0	41.5	56.0	41.0	44.0	26.5	32.0	12.0
120									90.0	81.2	76.0	66.0	72.0	62.0	64.0	51.5	66.0	51.0	54.0	36.5	42.0	22.0
130											86.0	76.0	82.0	72.0	74.0	61.5	70.0	55.0	58.0	40.5	46.0	26.0
140											90.0	80.0	86.0	76.0	78.0	65.5	80.0	65.0	68.0	50.5	56.0	36.0
150											100.0	90.0	96.0	86.0	88.0	75.5	90.0	75.0	78.0	60.5	66.0	46.0
160													106.0	96.0	98.0	85.5	100.0	85.0	88.0	70.5	76.0	56.0
(170)													116.0	106.0	108.0	95.5	110.0	95.0	98.0	80.5	86.0	66.0
180															118.0	105.5	120.0	105.0	108.0	90.5	96.0	76.0
(190)															128.0	115.5	130.0	115.0	118.0	100.5	106.0	86.0
200															138.0	125.5	140.0	125.0	128.0	110.5	116.0	96.0
220															148.0	135.5	147.0	132.0	135.0	117.5	123.0	103.0
240																	167.0	152.0	155.0	137.5	143.0	123.0
260																			175.0	157.5	163.0	143.0
280																			195.0	177.5	183.0	163.0
300																			215.0	197.5	203.0	183.0

NOTES:
1. All dimensions are in millimeters.
2. L is nominal length of screw; L_G is grip gaging length; L_B is body length.
3. Diameter-length combinations between the dashed lines are recommended. Lengths in parentheses are not recommended.
4. Screws with lengths above the solid line are threaded full length.
5. For screws of larger sizes and/or with lengths longer than the lower dashed line, L_G and L_B values shall be computed from formulas given in the note on *thread length* of the General Notes.

FASTENERS

Surface Condition—Bolts need not be finished on any surface except threads.

Top of Head—Top of head shall be full form and chamfered or rounded with the diameter of chamfer circle or start of rounding being equal to the maximum width across flats, within a tolerance of minus 15 percent.

Head Taper—Maximum width across flats shall not be exceeded. No transverse section through the head between 25 and 75 percent of actual head height as measured from the bearing surface shall be less than the minimum width across flats.

Bearing Surface—A die seam across the bearing surface is permissible. Bearing surface shall be perpendicular to the axis of the body within a tolerance of 3 deg for 24 mm size and smaller, and 2 deg for sizes larger than 24 mm. Angularity measurement shall be taken at a location to avoid interference from a die seam.

Concentricity of Head—The axis of the head shall be concentric with the axis of the body (determined by one diameter length of body under head) within a tolerance equal to 3 percent (6 percent FIR) of maximum width across flats.

Body Diameter—Bolts shall be furnished with full diameter body within the limits given in Table 9-42 or shall be threaded to the head unless the purchaser specifies bolts with "reduced body diameter."

Table 9-39. Length Tolerances (IFI-506)

Nom Length	\multicolumn{5}{c}{Nom Screw Size}
	M5 thru M8	M10 thru M16	M20 and M24	M30 and M36	M42 thru M100
to 10 mm	0.6	0.6	—	—	—
over 10 to 18 mm	0.7	0.7	—	—	—
over 18 to 30 mm	0.8	0.8	1.3	—	—
over 30 to 50 mm	1.0	1.6	2.5	3.0	—
over 50 to 80 mm	1.2	1.9	3.0	3.5	4.5
over 80 to 120 mm	1.4	2.2	3.5	4.0	6.0
over 120 to 180 mm	2.5	2.5	4.0	4.5	6.0
over 180 to 240 mm	4.6	4.6	4.6	6.0	6.0
over 240 mm	6.0	6.0	6.0	6.0	6.0

NOTE: All tolerances are minus from nominal screw length.

Table 9-40. Point Lengths and Radii (IFI-506)

Nominal Screw Size	X Point Radius Approx	Z Point Length Max	Z Point Length Min
M5x0.8	7.0	1.20	0.40
M6x1	8.4	1.50	0.50
M8x1.25	11.2	1.88	0.62
M10x1.5	14.0	2.25	0.75
M12x1.75	16.8	2.62	0.88
M14x2	19.6	3.00	1.00
M16x2	22.4	3.00	1.00
M20x2.5	28.0	3.75	1.25
M24x3	33.6	4.50	1.50
M30x3.5	42.0	5.25	1.75
M36x4	50.4	6.00	2.00
M42x4.5	58.8	6.75	2.25
M48x5	67.2	7.50	2.50
M56x5.5	78.4	8.25	2.75
M64x6	89.6	9.00	3.00
M72x6	100	9.00	3.00
M80x6	112	9.00	3.00
M90x6	126	9.00	3.00
M100x6	140	9.00	3.00

X equals 1.4 times nom screw diameter.
Z max equals 1.5 times thread pitch.
Z min equals 0.5 times thread pitch.

Table 9-41. Geometric Tolerancing Details for Screws (IFI-506)

Nominal Screw Size	Head True Position Tolerance Zone Diameter	Minimum Body Dia For Prod Threaded to Head	Total Runout of Thread to Shank
M5x0.8	0.35	4.38	0.48
M6x1	0.44	5.24	0.58
M8x1.25	0.56	7.07	0.58
M10x1.5	0.70	8.89	0.58
M12x1.75	0.84	10.71	0.70
M14x2	0.98	12.54	0.70
M16x2	1.12	14.54	0.70
M20x2.5	1.40	18.20	0.84
M24x3	1.68	21.84	0.84
M30x3.5	2.10	27.50	0.84
M36x4	2.52	33.16	1.00
M42x4.5	2.94	38.82	1.00
M48x5	3.36	44.48	1.00
M56x5.5	3.92	52.14	1.20
M64x6	4.48	59.80	1.20
M72x6	5.04	67.80	1.20
M80x6	5.60	75.80	1.40
M90x6	6.30	85.80	1.40
M100x6	7.00	95.79	1.40

Table 9-42. Hexagon Bolts (IFI-506)

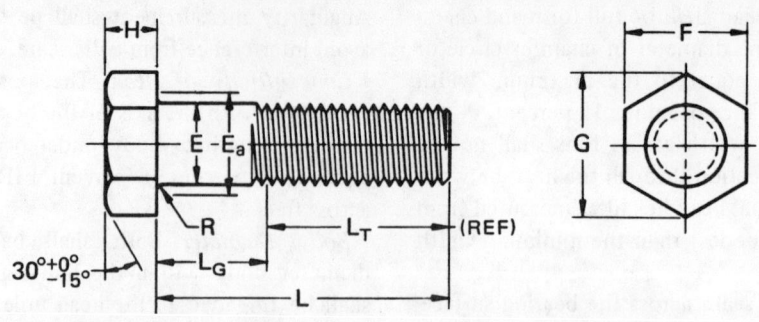

Nominal Bolt Size & Thread Pitch	E Body Diameter Max	E Body Diameter Min	F Width Across Flats Max	F Width Across Flats Min	G Width Across Corners Max	G Width Across Corners Min	H Head Height Max	H Head Height Min	Eₐ Fillet Transition Dia Max	R Radius of Fillet Min	Lₜ (Ref) Thread Length (Basic) Bolt Lengths ≤125	Lₜ (Ref) Thread Length (Basic) Bolt Lengths >125 and ≤200	Lₜ (Ref) Thread Length (Basic) Bolt Lengths >200
M5x0.8	5.48	4.52	8.00	7.75	9.24	8.84	3.88	3.35	5.8	0.2	16	22	35
M6x1	6.48	5.52	10.00	9.69	11.55	11.05	4.70	4.13	7.0	0.3	18	24	37
M8x1.25	8.58	7.42	13.00	12.60	15.01	14.36	5.73	5.10	9.2	0.4	22	28	41
*M10x1.5	10.58	9.42	15.00	14.50	17.32	16.53	6.86	6.17	11.2	0.4	26	32	45
M12x1.75	12.70	11.30	18.00	17.40	20.78	19.84	7.99	7.24	13.2	0.4	30	36	49
M14x2	14.70	13.30	21.00	20.30	24.25	23.14	9.32	8.51	15.2	0.6	—	40	53
M16x2	16.70	15.30	24.00	23.20	27.71	26.45	10.56	9.68	17.8	0.6	—	44	57
M20x2.5	20.84	19.16	30.00	29.00	34.64	33.06	13.12	12.12	22.4	0.8	—	52	65
M24x3	24.84	23.16	36.00	34.80	41.57	39.67	15.68	14.56	26.4	0.8	—	60	73
M30x3.5	30.84	29.16	46.00	44.50	53.12	50.73	19.48	17.92	33.6	1.2	66	72	85
M36x4	37.00	35.00	55.00	53.20	63.51	60.65	23.38	21.72	39.6	1.2	78	84	97
M42x4.5	43.00	41.00	65.00	62.90	75.06	71.71	26.97	25.03	45.6	1.2	90	96	109
M48x5	49.00	47.00	75.00	72.60	86.60	82.76	31.07	28.93	52.6	1.5	102	108	121
M56x5.5	57.20	54.80	85.00	82.20	98.15	93.71	36.20	33.80	62.0	2.0	—	124	137
M64x6	65.52	62.80	95.00	91.80	109.70	104.65	41.32	38.68	70.0	2.0	—	140	153
M72x6	73.84	70.80	105.00	101.40	121.24	115.60	46.45	43.55	78.0	2.0	—	156	169
M80x6	82.16	78.80	115.00	111.00	132.79	126.54	51.58	48.42	86.0	2.0	—	172	185
M90x6	92.48	88.60	130.00	125.50	150.11	143.07	57.74	54.26	96.0	2.0	—	192	205
M100x6	102.80	98.60	145.00	140.00	167.43	159.60	63.90	60.10	107.0	2.5	—	212	225

*M10 bolts with a 15 mm width across flats are currently being produced and used in substantial quantities in the U.S. as well as in many other countries. The ISO standards specify a 16 mm width across flats for the M10 size.

Table 9-43. Length Tolerances for Hexagon Bolts (IFI-506)

Nom Length	Nom Bolt Size M5 thru M8	M10 thru M16	M20 and M24	M30 and M36	M42 thru M100
to 50 mm	—	—	—	±3.0	—
over 50 to 80 mm	+0.8 / −1.2	—	—	±3.5	±4.5
over 80 to 120 mm	+0.8 / −1.4	+1.4 / −2.2	—	±4.0	±6.0
over 120 to 180 mm	+1.6 / −2.5	+1.6 / −2.5	+2.5 / −4.0	±4.5	±6.0
over 180 to 240 mm	+3.0 / −4.6	+3.0 / −4.6	+3.0 / −4.6	±6.0	±6.0
over 240 mm	+4.0 / −6.0	+4.0 / −6.0	+4.0 / −6.0	±6.0	±6.0

FASTENERS

There may be a reasonable swell, fin, or die seam on the body adjacent to the underside of the head not to exceed the basic bolt diameter by the following:

0.5 mm for size 5 mm
0.65 mm for size 6 mm
0.75 mm for sizes 8 through 14 mm
1.25 mm for size 16 mm
1.5 mm for sizes 20 through 30 mm
2.3 mm for sizes 36 through 48 mm
3 mm for sizes 56 through 72 mm
4.8 mm for sizes 80 through 100 mm

Bolts may be obtained with "reduced diameter body" if so specified. Where "reduced diameter body" is specified the minimum body diameter shall equal the minimum body diameter for products threaded to the head as given in Table 9-41. A shoulder of full body diameter under the head, of approximately 1/2 diameter in length, may be supplied at the option of the manufacturer, with "reduced body limits" between shoulder and thread.

Fillet—The fillet at junction of head and shank shall be a smooth concave curve within an envelope of R minimum and a radius tangent to the underside of head at a point equal to one-half of E_a maximum from the axis of the bolt.

Point—Bolts need not be pointed.

Straightness—Shanks of bolts shall be straight within the following limits: for bolts with nominal lengths to and including 300 mm, the maximum camber shall be 0.006 mm per mm of bolt length, and for bolts with nominal lengths over 300 mm to and including 600 mm the maximum camber shall be 0.008 mm per mm of length.

Length Tolerances—Bolt length tolerances are given in Table 9-43.

Thread Length—The length of thread on bolts shall be controlled by the grip gaging length L_G max as set forth in the following:

Grip Gaging Length, L_G max, is the distance, measured parallel to the axis of bolt, from the underside of the head to the face of a noncounterbored or non-countersunk standard GO thread ring gage assembled by hand as far as the thread will permit. The maximum grip gaging length, as calculated and rounded to one decimal place, for any bolt length shall be equal to the nominal bolt length minus the basic thread length (L_G max = L nom − L_T). It represents the minimum design grip length of the bolt and shall be used as the criterion for inspection and for determining thread availability when selecting bolt lengths even though usable threads may extend beyond this point.

All bolts of nominal lengths equal to or shorter than the basic thread length, L_T, plus a length of 2-1/2 thread pitches for sizes up to and including 24 mm, and L_T plus 3-1/2 thread pitches for sizes larger than 24 mm shall be threaded for full length. The distance from the bearing surface of the head to the first complete (full form) thread, as measured with a GO thread ring gage assembled by hand as far as the thread will permit, shall not exceed the length of 2-1/2 thread pitches for sizes up to and including 24 mm, and 3-1/2 thread pitches for sizes larger than 24 mm.

Basic Thread Length, L_T, is a reference dimension, intended for calculation purposes only, which represents the distance from the extreme end of the bolt to the last complete (full form) thread.

Incomplete Thread Diameter. The major diameter of incomplete thread shall not exceed the actual major diameter of the full form thread.

Thread Series—Threads shall conform to dimensions given on page 8-20 and shall be class 6g. The class 6g tolerance shall apply to plain finish (unplated or uncoated) bolts, and to plated or coated bolts before plating or coating. For bolts with additive finish, the class 6g diameters may be exceeded by the amount of the allowance, i.e., the basic diameters shall apply to bolts after plating or coating.

Marking—Bolts shall be marked with the property class symbol as shown in Table 9-8 and the manufacturer's identification symbol. Markings shall be located on the top of the head and may be raised or indented at the option of the manufacturer.

Material and Mechanical Properties—Carbon steel bolts shall conform to the requirements for the applicable property class as described on page 9-18. Class 4.6 bolts are available in sizes 5 through 36 mm; class 9.8 in size 5 through 16 mm; class 8.8 in size 20 through 36 mm; and class 10.9 in sizes 6 through 36 mm. Carbon steel bolts of sizes 42 through 100 mm, and bolts of other materials such as stainless steel, brass, bronze, and aluminum alloys shall have properties as agreed upon by the manufacturer and the purchaser. (See IFI-516 and IFI-518.)

Designation—Hex bolts shall be designated by the following data in the sequence shown: Nominal size and thread pitch, nominal length, property class or material, product name, and protective coating, if required.

Examples: M16 × 2 × 100, 9.8 hex bolt, zinc plated.
M36 × 4 × 240, 304 stainless steel hex bolt.

ROUND HEAD SQUARE NECK BOLTS

The ANSI special committee for studying the development of an optimum metric fastener system with members of the off-road vehicle industry in America developed this new type carriage bolt. The depth of square in this proposal has been kept to a bare minimum; and this fastener should, therefore, replace both the currently used types with short and normal depth of square. In addition to the simplification introduced by one thread pitch, we believe that this standardization will help save large sums of money for industry and consumers in the future.

The Industrial Fasteners Institute, 1505 East Ohio Building, Cleveland, Ohio 44114 has given permission to use the material presented here, which was based on the IFI-515 standard.

Carriage bolts in nominal diameters from 5 to 20 mm are shown in Table 9-44.

General Notes

Dimensions—All dimensions in the tables and notes are in millimeters unless otherwise stated.

Surface Condition—Bolts need not be finished on any surface except threads.

Height of Head—The height of head shall be measured, parallel to the axis of the bolt, from the top of the head to the bearing surface. The spherical top surface of the head may be underfilled within a circular area equal to one nominal bolt diameter concentric with the bolt axis provided the height of head is maintained within the specified limits.

Head Diameter—Because the heads of bolts normally are not machined or trimmed, the outer periphery between the bearing circle diameter (N) and head diameter (F) may be somewhat irregular and the edge may be rounded or flat.

Body Diameter—Bolts shall be furnished with full diameter body within the limits given in Table 9-44 or shall be threaded full length, unless the purchaser specifies bolts with "reduced body diameter."

There may be a reasonable swell, fin, or die seam on the body adjacent to the square neck not to exceed the basic (nominal size) bolt diameter by the following:

Table 9-44. Dimensions of Round Head Square Neck Bolts (IFI-515)

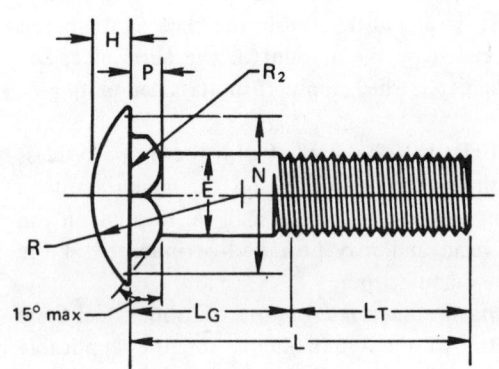 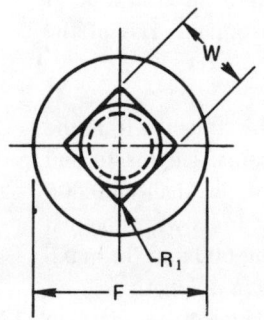 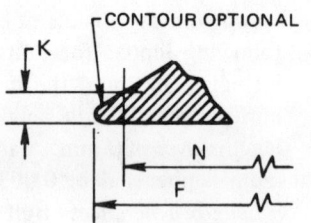

HEAD PERIPHERY DETAIL

Nom Bolt Size & Thread Pitch	E Body Diameter Max	E Body Diameter Min	F Head Dia Max	N Bearing Circle Dia Min	H Head Height Max	H Head Height Min	K Head Edge Thickness Ref	R Head Radius Approx	W Width of Square Max	W Width of Square Min	P Depth of Square Max	P Depth of Square Min	R₁ Corner Radius Max	R₂ Fillet Radius Max	R₂ Fillet Radius Min	L_T Basic Thread Length ≤125	L_T Basic Thread Length >125
M5x0.8	5.48	4.52	11.3	9.3	3.0	2.5	0.9	7.2	5.48	4.87	3.0	2.0	0.8	0.8	0.4	16	22
M6x1	6.48	5.52	13.8	11.8	3.6	3.1	1.1	8.9	6.48	5.87	3.0	2.0	0.8	0.8	0.4	18	24
M8x1.25	8.58	7.42	17.0	15.0	4.5	4.0	1.4	11.2	8.58	7.86	3.0	2.0	0.8	0.8	0.4	22	28
M10x1.5	10.58	9.42	20.6	18.6	5.5	5.0	1.8	13.7	10.58	9.83	4.0	3.0	1.2	0.8	0.4	26	32
M12x1.75	12.70	11.30	24.7	22.2	6.5	6.0	2.2	16.5	12.70	11.80	4.0	3.0	1.2	0.8	0.4	30	36
M14x2	14.70	13.30	28.3	25.8	7.5	7.0	2.6	19.1	14.70	13.80	4.0	3.0	1.2	0.8	0.4	34	40
M16x2	20.84	15.30	33.1	30.1	8.8	8.0	3.0	22.6	20.84	15.77	5.0	4.0	1.6	1.6	0.8	38	44
M20x2.5	24.84	19.16	39.8	36.8	10.8	10.0	3.9	27.3	24.84	19.77	5.0	4.0	1.6	1.6	0.8	46	52

FASTENERS

0.5 mm for size 5 mm
0.65 mm for size 6 mm
0.75 mm for sizes 8 through 14 mm
1.25 mm for 16 mm size
1.5 mm for 20 mm size

Bolts may be obtained with "reduced diameter body" if so specified. Where "reduced diameter body" is specified, the minimum body diameter shall be:

(IFI-515)

NOM BOLT SIZE	MIN BODY DIA
M 5 × 0.8	4.38
M 6 × 1	5.24
M 8 × 1.25	7.07
M 10 × 1.5	8.89
M 12 × 1.75	10.71
M 14 × 2	12.54
M 16 × 2	14.54
M 20 × 2.5	18.20

Fillet—All bolts shall have a fillet at the junction of the head and the square neck within the specified radius limits (R_2) in Table 9-44.

Square Neck—The depth of square shall be measured, parallel to the axis of the bolt at the midpoint of the flats of the square, from the bottom of the square to the under head bearing surface.

The corners of the square neck may be rounded within the limits specified (R_1) in Table 9-44. The 15° maximum corner chamfer in combination with the minimum specified depth of square neck (P) establishes the minimum corner height. This chamfer may be increased when the depth of square (P) is greater than its specified minimum provided that the minimum corner height is maintained.

Point—Bolts need not be pointed.

Straightness—Shanks of bolts shall be straight within the following limits: for bolts with nominal lengths to and including 300 mm, the maximum camber shall be 0.006 mm per mm of bolt length, and for bolts with nominal lengths over 300 mm to and including 600 mm the maximum camber shall be 0.008 mm per mm of length.

Thread Length—The length of thread on bolts shall be controlled by the grip gaging length L_G max as set forth in the following:

Grip Gaging Length, L_G max, is the distance, measured parallel to the axis of bolt, from the underside of the head to the face of a non-counterbored or non-countersunk standard GO thread ring gage assembled by hand as far as the thread will permit. The maximum grip gaging length, as calculated and rounded to one decimal place, for any bolt length shall be equal to the nominal bolt length minus the basic thread length (L_G max = L nom − L_T). It represents the minimum design grip length of the bolt and shall be used as the criterion for inspection and for determining thread availability when selecting bolt lengths even though usable threads may extend beyond this point.

All bolts of nominal lengths equal to or shorter than the basic thread length, L_T, plus the maximum depth of square (P), plus a length of 2.5 thread pitches shall be threaded for full length. Additionally, at the manufacturer's option, bolts of lengths 75 mm and shorter may be threaded for full length. The distance from the bearing surface of the head to the first complete (full form) thread, as measured with a GO thread ring gage assembled by hand as far as the thread will permit, shall not exceed the length of the maximum depth of square (P), plus 2.5 pitches. Bolts threaded full length may have minimum body diameter limits equal to those specified for "reduced diameter body" bolts, above.

Basic Thread Length, L_T, is a reference dimension, intended for calculation purposes only, which represents the distance from the extreme end of the bolt to the last complete (full form) thread.

Incomplete Thread Diameter. The major diameter of incomplete thread shall not exceed the actual major diameter of the full form thread.

Length—The length of the bolt shall be measured parallel to the axis of bolt, from the extreme end of bolt to the under head bearing surface. Bolt length tolerances are given in Table 9-45.

Table 9-45. Carriage Bolt Length Tolerances (IFI-515)

Nom Length	M5 thru M8	M10 thru M16	M20
to 10 mm	+0.4 / −0.6	+0.4 / −0.6	—
over 10 to 18 mm	+0.5 / −0.7	+0.5 / −0.7	—
over 18 to 30 mm	+0.6 / −0.8	+0.6 / −0.8	+0.8 / −1.3
over 30 to 50 mm	+0.7 / −1.0	+1.0 / −1.6	+1.6 / −2.5
over 50 to 80 mm	+0.8 / −1.3	+1.2 / −1.9	+2.0 / −3.0
over 80 to 120 mm	+0.8 / −1.4	+1.4 / −2.2	+2.3 / −3.5
over 120 to 180 mm	+1.6 / −2.5	+1.6 / −2.5	+2.5 / −4.0
over 180 to 240 mm	+3.0 / −4.6	+3.0 / −4.6	+3.0 / −4.6
over 240 mm	+4.0 / −6.0	+4.0 / −6.0	+4.0 / −6.0

Table 9-46. Plow Bolts (ISO 5713)

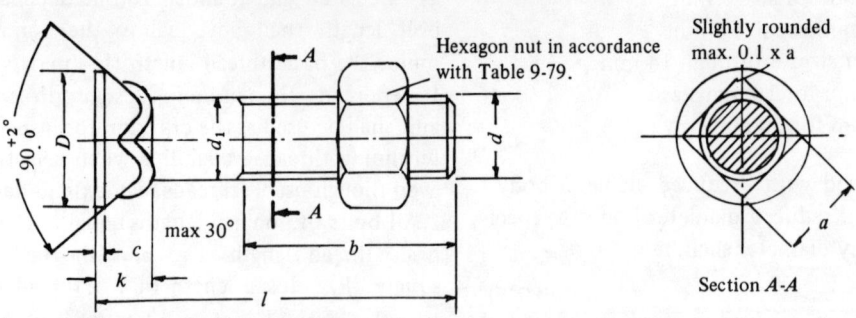

d	M8	M10	M12	M16	M20
Thread Pitch	1.25	1.5	1.75	2	2.5
d_1	\multicolumn{4}{c}{Blank rolling diameter}	20 h15
a	8 h14	10 h14	12 h14	16 h14	20 h15
b Minimum	\multicolumn{5}{c}{From square section 2x pitch max}
b Nominal	34	40	46	58	70
c	1	1.2	1.2	1.4	2.5
D h15	14	18	21	30	36
k h15	5.5	7	8	10.5	13.5
l js17	\multicolumn{5}{c}{Standard-lengths are indicated by x}
20	x				
25	x	x			
30	x	x	x		
35	x	x	x		
40		x	x	x	
45		x	x	x	x
50		x	x	x	x
60			x	x	x
70				x	x
80				x	x
90				x	x
100				x	x

NOTES:
1. All dimensions are in millimeters.
2. Screw threads are according to ISO 965. See Section 8 on modified ISO fastener screw threads (page 8-20) and thread tolerance 6g.
3. Mechanical properties to ISO 898 grade 9.8, see page 9-18.
4. Bolt marking according to ISO 898, see page 9-20.
5. Designation: Example for the designation of a plow bolt, thread size d = M12, nominal length l = 30 mm, and property class 9, 8, (Plow bolt ISO 5713 M12 × 30 − 9,8).

FASTENERS

Table 9-47. Dimensions of 12 Spline Flange Screws (IFI-511)

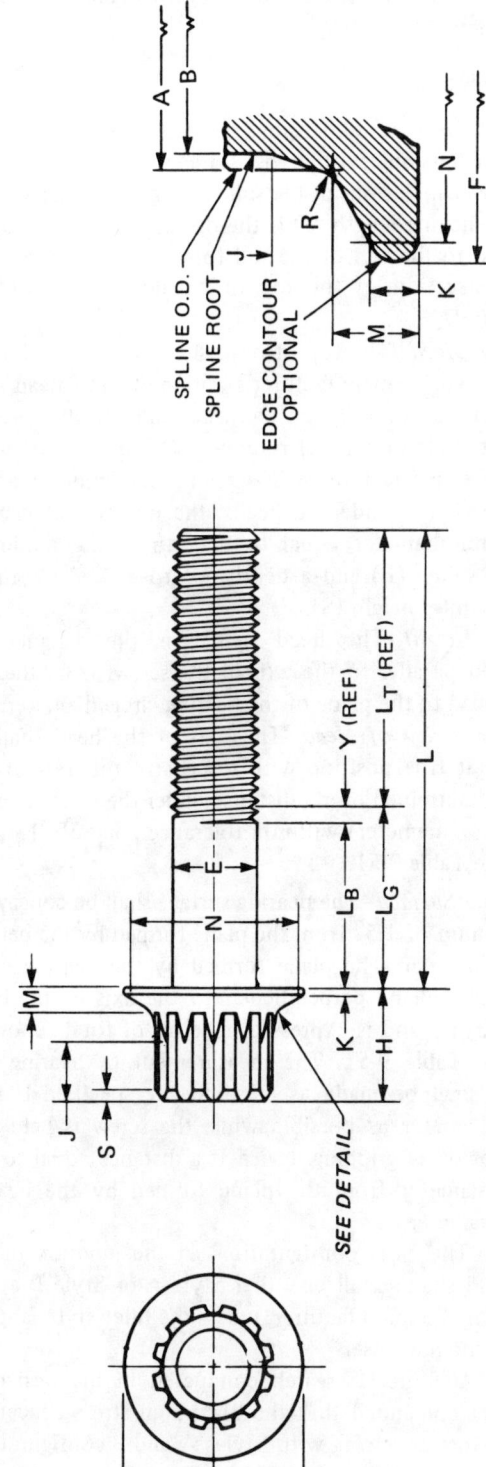

Nom Screw Size & Thread Pitch	Spline Size	E Body Dia Max	E Body Dia Min	F Flange Dia Max	N Bearing Circle Dia Min	K Flange Edge Thickness Min	M Flange Height Min	J Wrenching Height Min	H Head Height Max	S Chamfer Height Max	R Spline Junction Radius Min	L_T (Ref) Thread Length, Basic — For Screw Lengths ≤125mm	L_T (Ref) — For Screw Lengths >125mm and ≤200mm	L_T (Ref) — For Screw Lengths >200mm	Y (Ref) Transition Thread Length Max
M5×0.8	5	5.00	4.82	9.4	8.4	1.0	1.7	1.8	5	0.6	0.4	16	22	35	4.0
M6×1	6	6.00	5.82	11.8	10.7	1.2	2.2	2.3	6	0.8	0.5	18	24	37	5.0
M8×1.25	8	8.00	7.78	15.0	13.7	1.5	2.7	3.0	8	1.0	0.6	22	28	41	6.2
M10×1.5	10	10.00	9.78	18.6	17.1	2.0	3.4	3.8	10	1.2	0.7	26	32	45	7.5
M12×1.75	12	12.00	11.73	22.8	21.1	2.3	4.1	4.5	12	1.5	0.8	30	36	49	8.8
M14×2	14	14.00	13.73	26.4	24.5	2.7	4.8	5.4	14	1.8	0.9	34	40	53	10.0
M16×2	16	16.00	15.73	30.3	28.1	3.2	5.7	5.8	16	2.1	1.0	38	44	57	10.0
M20×2.5	20	20.00	19.67	37.4	34.9	4.1	7.2	7.2	20	2.5	1.2	46	52	65	12.5

Thread Series—Threads shall conform to dimensions given in Section 8 (Modified ISO Fastener Screw Threads, page 8-20) and shall be class 6g. The class 6g tolerance shall apply to plain finish (unplated or uncoated) bolts, and to plated or coated bolts before plating or coating. For bolts with additive finish, the class 6g diameters may be exceeded by the amount of the allowance; i.e., the basic diameters shall apply to bolts after plating or coating.

Material and Mechanical Properties—Carbon steel bolts shall conform to the requirements for the applicable property class as covered on page 9-18. Class 4.6 bolts are available in sizes 5 to 36 mm; class 9.8 in sizes 5 to 16 mm and class 8.8 in size 20 mm only. Bolts of other materials shall have properties as agreed upon by the manufacturer and the purchaser.

Identification Symbols—Bolts shall be marked with the property class symbol as shown in Table 9-8 and the manufacturer's identification symbol. Markings shall be located on the top of the head and may be raised or indented at the option of the manufacturer.

Designation—Round head square neck bolts shall be designated by the following data in the sequence shown: Nominal size and thread pitch, nominal length, property class or material, product name, and protective coating, if required.

Examples: M10 × 1.5 × 80, 9.8 round head square neck bolts, zinc plated. M20 × 2.5 × 120, 304 stainless steel round head square neck bolts.

FLAT COUNTERSUNK SQUARE NECK BOLTS (PLOW BOLTS)

The most important national standards for plow bolts are listed in the world fastener index, page 9-2. The ISO/TC23/subcommittee 5 published a revised draft proposal which is shown in Table 9-46. Five nominal coarse thread diameters from M8 to M20 in lengths ranging from 20 to 100 mm were included in the ISO draft international standard.

12-POINT SPLINE FLANGE HEAD SCREWS (IFI-511)

The 12-point spline flange head screw is not covered in national standards outside the United States. The unique head design was developed by the ANSI special committee to study the development of an optimum metric fastener system and the recommendations were published in the OMFS-9 publication. The above material is now covered in the IFI-511 standard. Permission to use the material presented here was granted by the Industrial Fasteners Institute, 1505 East Ohio Building, Cleveland, Ohio 44114. Flange head screws with 12-point spline drives are shown in Table 9-47 for nominal diameters from 5 to 20 mm.

General Notes

Dimensions—All dimensions in the tables and notes are in millimeters unless otherwise stated.

Spline—Dimensions of the spline are given in Table 9-48. The wrenching length (J) is the distance from the top of the head to the end of the full form spline; i.e. generally, the intersection of the top of extrude angle with the spline root.

Top of Head—The top of head shall be either full form or indented, at manufacturer's option. Top of head shall be chamfered or rounded and the minimum diameter of chamfer circle or start of rounding shall be the minimum O.D. of spline (A) minus 2 times the maximum chamfer height (S). For indented heads, the indent shall have a maximum diameter equal to 0.75 times the minimum root of spline (B) and a depth not to exceed the maximum chamfer height (S).

Head Height—The head height is the distance, as measured parallel to the axis of the screw, from the top of the head to the plane of the bearing circle diameter.

Concentricity of Head—The axis of the head shall be located at true position with respect to the axis of the screw (determined over a distance under the head equal to one screw diameter) within a tolerance zone of diameter shown in Table 9-51.

Bearing Surface—The bearing surface shall be concave to a maximum of 1.5° from the plane formed by the bearing circle diameter. The plane formed by the bearing circle diameter shall be perpendicular to the axis of the body within 1 deg and is expressed in terms of total runout as given in Table 9-51. The measurement of bearing face runout shall be made as close to the specified bearing circle diameter as possible while the screw is held in a collet or other gripping device at a distance equal to one screw diameter from the plane formed by the bearing circle diameter.

Fillet—The fillet configurations at the junction of the head and shank shall be either Style A or Style B at the option of the manufacturer, unless the fillet style is specified by the purchaser.

(NOTE: Analytical research conducted in the design of the fillet concluded that the maximum stress concentration factors occurring with Styles A and B configurations were essentially the same.) The fillet shall be a smooth and continuous curve faring smoothly into the underhead bearing surface and the shank within the limits specified.

FASTENERS

Table 9-48. Wrenching Configurations for 12 Spline Flange Screws (IFI-511)

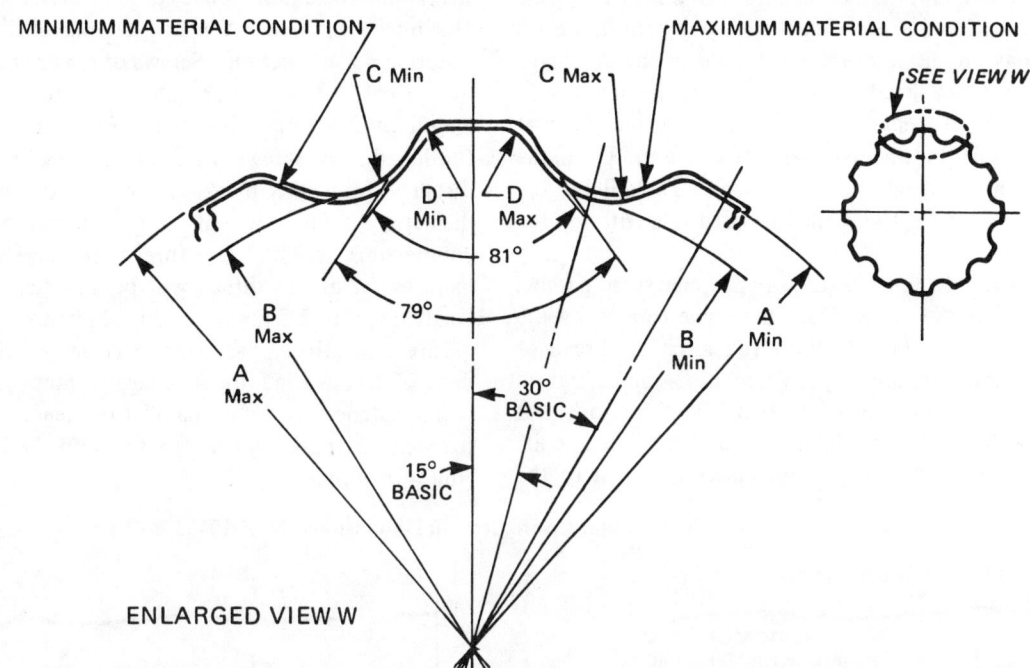

Spline Size	Max Material Condition				Min Material Condition			
	A Dia Max	B Dia Max	C Rad Min	D Rad Min	A Dia Min	B Dia Min	C Rad Max	D Rad Max
5	5.86	5.22	0.55	0.13	5.71	5.07	0.70	0.26
6	7.08	6.27	0.62	0.13	6.98	6.07	0.77	0.26
8	9.37	8.34	0.75	0.23	9.17	8.14	0.90	0.39
10	11.70	10.42	0.95	0.23	11.50	10.22	1.10	0.39
12	14.04	12.50	1.10	0.36	13.84	12.30	1.25	0.52
14	16.29	14.59	1.30	0.48	16.06	14.36	1.45	0.64
16	18.71	16.66	1.40	0.48	18.48	16.43	1.55	0.64
20	23.40	20.83	1.75	0.74	23.17	20.60	1.90	0.90

NOTE: Spline dimensions are based on SAE AS1159A.

Body Diameter—The diameter of the body on screws which are not threaded full length shall be within the limits specified (E). For screws threaded full length, the diameter of the unthreaded shank under the head shall not exceed the specified max body diameter (E) nor be less than the min body diameter given in Table 9-51.

Length—The length of the screw shall be measured parallel to the axis of the screw from the plane formed by the under head bearing circle diameter to the extreme end of the shank. The maximum length for all screw sizes shall be equal to nominal length, minimum length shall be nominal length minus the tolerance given in Table 9-39.

Points—At the manufacturer's option, the end of the screw shall be chamfered from a diameter equal to or slightly less than the thread root diameter to produce a length of chamfer or incomplete thread within the limits for Z given in Table 9-40, or shall have a rounded point of radius X as given in Table 9-40. The end of the screw shall be reasonably square with the axis of the screw and where pointed blanks are used, the slight rim or cup resulting from roll threading shall be permissible.

Straightness—Shanks of screws shall be straight within a maximum camber of 0.006 mm per millimeter of screw length.

Thread Length—The length of thread on screws shall be controlled by the maximum grip gaging length (L_G) and the minimum body length (L_B) as set forth in the following:

Grip gaging length (L_G) max, is the distance, measured parallel to the axis of the screw, from the plane formed by the bearing circle diameter to the face of a non-counterbored or non-countersunk standard GO thread ring gage assembled by hand as far as the thread will permit. Values for L_G max are given in Table 9-50. For diameter-length combinations not listed in Table 9-50 the maximum grip gaging length, as calculated and rounded to one decimal place, is equal to the nominal screw length, L, minus the basic thread length, L_T, as given in Table 9-47 (L_G max = L − L_T). L_G max shall be used as a criterion for inspection.

Body length, L_B min, is the distance, measured parallel to the axis of the screw, from the plane formed by the bearing circle diameter to the last scratch of thread or the top of the extrusion angle. Values of L_B min are given in Table 9-50. For diameter-length combinations not listed in Table 9-50 the minimum body length, as calculated and rounded to one decimal place, is equal to the maximum grip gaging length (as computed) minus the maximum transition thread length as given in Table 9-47 (L_B min = L_G max - Y max). L_B min shall be used as a criterion for inspection. Screws of nominal lengths which have a calculated L_B min value equal to or less than the length of 2.5 times the thread pitch shall be threaded full length. For screws which are threaded full length, the distance from the plane formed by the bearing circle diameter to the face of a non-counterbored or non-countersunk standard GO thread ring gage assembled by hand as far as the thread will permit shall not exceed a length equal to 2.5 times the thread pitch.

Thread length, L_T, as given in Table 9-47, is a reference dimension intended for calculation purposes only, and is the distance, measured parallel to the axis of the screw, from the extreme end of the screw to the last complete (full form) thread.

Table 9-49. Fillet Configuration Dimensions (IFI-511)

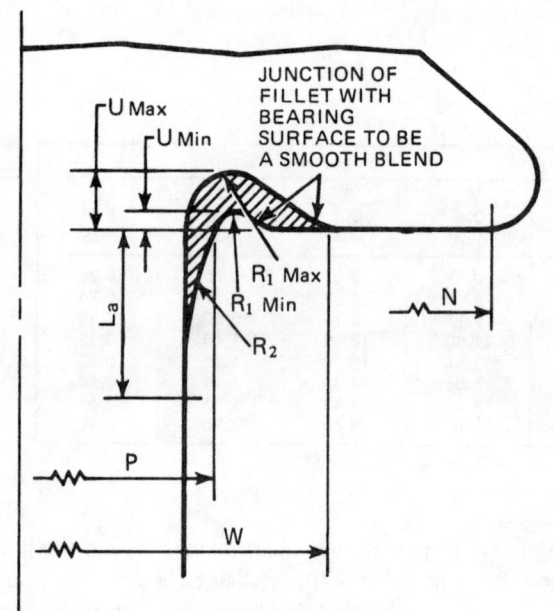

FILLET – STYLE A

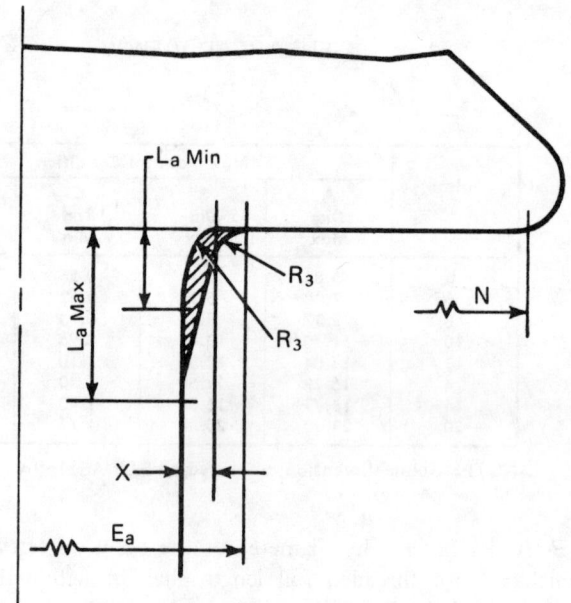

FILLET – STYLE B

Nom Screw Size	Style A							Style B				
	W	P	U		L_a	R_1		R_2	E_a	X	L_a		R_3
	Max	Max	Max	Min	Max	Max	Min	Ref	Max	Min	Max	Min	Min
M5x0.8	6.1	5.5	0.15	0.05	1.4	0.25	0.10	4.05	5.6	0.20	1.4	0.7	0.15
M6x1	7.5	6.6	0.19	0.08	1.8	0.28	0.13	5.55	6.8	0.30	1.8	0.9	0.20
M8x1.25	10.1	8.8	0.24	0.11	2.1	0.36	0.16	5.71	9.0	0.35	2.1	1.0	0.25
M10x1.5	12.5	10.8	0.31	0.13	2.1	0.45	0.20	5.71	11.0	0.35	2.1	1.0	0.25
M12x1.75	15.7	12.8	0.37	0.16	2.1	0.54	0.24	5.71	13.0	0.35	2.1	1.0	0.25
M14x2	18.1	14.8	0.43	0.19	2.1	0.63	0.28	5.71	15.0	0.35	2.1	1.0	0.25
M16x2	20.5	17.2	0.51	0.23	3.2	0.72	0.32	8.83	17.5	0.55	3.2	1.6	0.37
M20x2.5	26.1	21.6	0.65	0.29	4.2	0.90	0.40	11.42	22.0	0.80	4.2	2.1	0.50

FASTENERS

Table 9-50. Maximum Grip Gaging Lengths and Minimum Body Lengths for 12 Spline Flanged Screws (IFI-511)

Nom Screw Size →	M5 x 0.8		M6 x 1		M8 x 1.25		M10 x 1.5		M12 x 1.75		M14 x 2		M16 x 2		M20 x 2.5	
L Nom	L_G Max	L_B Min	L_G Max	L_B Min	L_G Max	L_B Min	L_G Max	L_B Min	L_G Max	L_B Min	L_G Max	L_B Min	L_G Max	L_B Min	L_G Max	L_B Min
8			—	—												
10			—	—												
12			—	—												
14					—	—										
16					—	—										
20													—	—		
25	9.0	5.0											—	—		
30	9.0	5.0	12.0	7.0									—	—		
35	19.0	15.0	17.0	12.0	13.0	6.8							—	—		
40	19.0	15.0	17.0	12.0	13.0	6.8	14.0	6.5								
45	29.0	25.0	27.0	22.0	23.0	16.8	19.0	11.5	15.0	6.2						
50	29.0	25.0	27.0	22.0	23.0	16.8	19.0	11.5	15.0	6.2	16.0	6.0				
(55)			37.0	32.0	33.0	26.8	29.0	21.5	25.0	16.2	21.0	11.0	17.0	7.0		
60			37.0	32.0	33.0	26.8	29.0	21.5	25.0	16.2	21.0	11.0	17.0	7.0		
(65)					43.0	36.8	39.0	31.5	35.0	26.2	31.0	21.0	27.0	17.0	19.0	6.5
70					43.0	36.8	39.0	31.5	35.0	26.2	31.0	21.0	27.0	17.0	19.0	6.5
(75)					53.0	46.8	49.0	41.5	45.0	36.2	41.0	31.0	37.0	27.0	29.0	16.5
80					53.0	46.8	49.0	41.5	45.0	36.2	41.0	31.0	37.0	27.0	29.0	16.5
(85)							59.0	51.5	55.0	46.2	51.0	41.0	47.0	37.0	39.0	26.5
90							59.0	51.5	55.0	46.2	51.0	41.0	47.0	37.0	39.0	26.5
100							74.0	66.5	70.0	61.2	66.0	56.0	62.0	52.0	54.0	41.5
110									80.0	71.2	76.0	66.0	72.0	62.0	64.0	51.5
120									90.0	81.2	86.0	76.0	82.0	72.0	74.0	61.5
130											90.0	80.0	86.0	76.0	78.0	65.5
140											100.0	90.0	96.0	86.0	88.0	75.5
150													106.0	96.0	98.0	85.5
160													116.0	106.0	108.0	95.5
(170)															118.0	105.5
180															128.0	115.5
(190)															138.0	125.5
200															148.0	135.5
220																
240																

NOTES:
1. All dimensions are in millimeters.
2. L is nominal length of screw; L_G is grip gaging length; L_B is body length.
3. Diameter-length combinations between the dashed lines are recommended. Lengths in parentheses are not recommended.
4. Screws with lengths above the solid line are threaded full length.
5. For screws with lengths longer than the lower dashed line, L_G and L_B values shall be computed from formulas given on page 9-51.

Transition thread length, Y max, as given in Table 9-47 is a reference dimension intended for calculation purposes only. It includes the length of incomplete threads and tolerances on grip gaging length and body length. The transition from full thread to incomplete thread shall be smooth and uniform. The major diameter of the incomplete threads shall not exceed the actual major diameter of the complete (full form) threads.

Thread Concentricity — The axis of the thread shall be concentric with the axis of the screw shank within the limits given in Table 9-51 when measured at a distance of one screw diameter from the thread runout.

Thread Series — Unless permitted otherwise by the customer, threads shall be rolled. Threads shall conform to dimensions given in Section 8 (Modified ISO Fastener Screw Threads, page 8-20). The class 6g tolerance shall apply to plain finish (unplated or uncoated) screws, and to plated or coated screws before plating or coating. For screws with additive finish, the class 6g diameters may be exceeded by the amount of the allowance; i.e., the basic diameters shall apply to screws after plating or coating.

Material and Mechanical Properties — Carbon steel screws, in sizes 5 through 16 mm inclusive, shall conform to the requirements for property class 9.8 as covered on page

Table 9-51. Geometric Tolerancing Details and Body Diameters
For Screws (IFI-511)

Nom Screw Size	Head True Position Tolerance Zone Diameter	Total Runout of Bearing Surface	Minimum Body Dia for Screws Threaded to Head	Total Runout of Thread to Shank
M5x0.8	0.35	0.15	4.38	0.48
M6x1	0.44	0.19	5.24	0.58
M8x1.25	0.56	0.24	7.07	0.58
M10x1.5	0.70	0.30	8.89	0.58
M12x1.75	0.84	0.37	10.71	0.70
M14x2	0.98	0.43	12.54	0.70
M16x2	1.12	0.49	14.54	0.70
M20x2.5	1.40	0.61	18.20	0.84

9-18; steel screws in size 20 mm shall conform to the requirements of property class 8.8. Screws of other materials such as stainless steel, brass, bronze and aluminum alloys shall have properties as agreed upon by the manufacturer and the purchaser. (For guidance refer to IFI-516 and IFI-518.)

Identification Symbols—Screws shall be marked with the property class symbol as shown in Table 9-8 and the manufacturer's identification symbol. Markings shall be located on the top of the head and may be raised or indented at the option of the manufacturer.

Designation—12 spline flange screws shall be designated by the following data in the sequence shown: Nominal size and thread pitch, nominal length, property class or material, product name, and protective coating, if required.

Examples: M10 × 1.5 × 50, 9.8 12 spline flange screw, zinc plated. M5 × 0.8 × 35, 410 stainless steel 12 spline flange screws.

SOCKET HEAD CAP SCREWS (SS-200)

Hexagon socket head cap screws covered in national standards conform to dimensions specified in the ISO 861 standard throughout the world. Tables and figures in this chapter are published in the SS-200 standard available from the Socket Screw Products Bureau, 331 Madison Avenue, New York, New York 10017.

Screws made to the SS-200 standard are interchangeable dimensionally with screws conforming to ISO 861. They have longer thread lengths than do ISO 861 screws and their shank length tolerance is negative only whereas ISO screws have a bilateral tolerance. All other dimensions are within the limits of ISO 861.

A proposed revision of the ISO 861 is near completion, and some of the material presented in the tables and notes reflect the latest ISO draft proposal.

Socket head cap screws in nominal sizes from 1.6 to 48 mm are shown in Tables 9-52 and 9-53.

MACHINE SCREWS

General

A number of different types head designs with various drive types are covered in national standards in the major industrial countries of the world. Some of the preferred head and drive types are shown in the world fastener index, and ISO standards exist for the types shown in this section. The IFI-513 standard for machine screws, on which the tables presented here have been based, courtesy Industrial Fasteners Institute, is almost in complete agreement with the ISO standards and draft proposals on the subject. The IFI-513 has been authorized for use as a first draft of ANSI B18.6.7 standard.

Flat Countersunk Head Machine Screws (IFI-513)

Machine screws of this type are specified for nominal sizes from 2 to 12 mm in the IFI-513 standard and screws with slotted head style are shown in Table 9-54A. Cross recessed head styles with Phillips (ANSI Type I and ISO Type Ph) and Pozidriv (ANSI Type IA and ISO Type Pz) drive are shown in Table 9-54B.

Oval Countersunk Head Machine Screws (IFI-513)

Oval countersunk head machine screws are specified in the IFI-513 standard ranging from 2 to 12 mm nominal sizes, and screws with slotted heads are shown in Table 9-55A and the cross recessed heads in Table 9-55B.

Pan Head Machine Screws (IFI-513)

Fasteners of this type are standard in nominal sizes from 2 to 12 mm in the IFI-513, and screws with slotted and recessed head styles are shown in Table 9-56A. Recess details for Phillips (ANSI Type I and ISO Type Ph) and

FASTENERS

Table 9-52. Socket Head Cap Screws (SS-200)

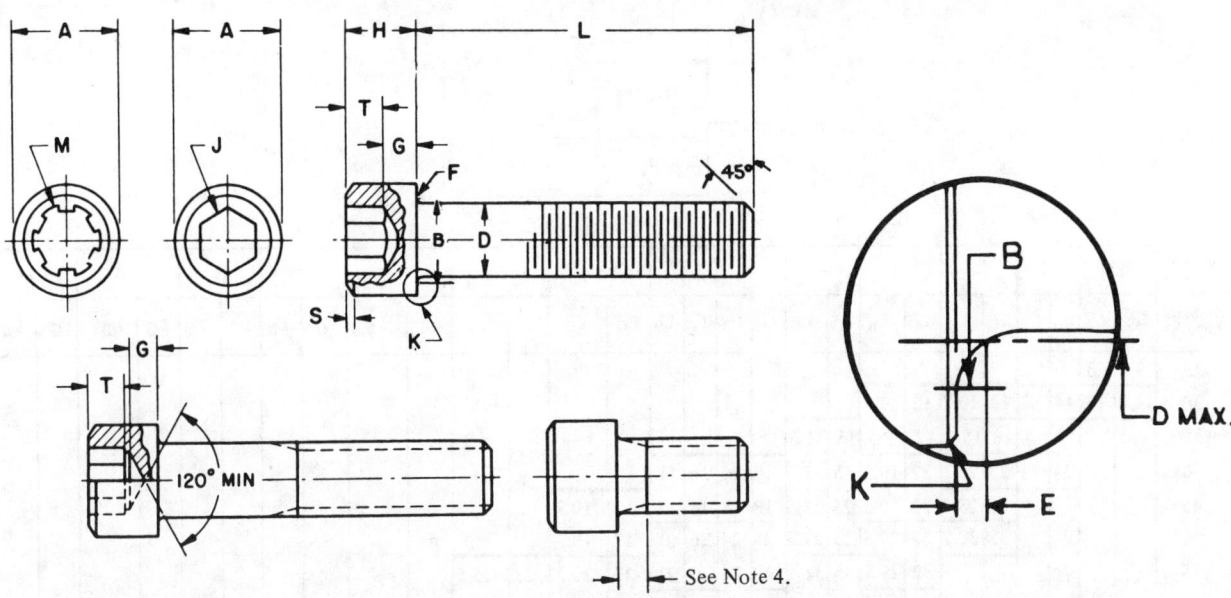

Nominal Size	D Body Diameter Max.	D Body Diameter Min.	A Head Diameter Max.	A Head Diameter Min.	H Head Height Max.	H Head Height Min.	S Chamfer or Radius Max.	J Hexagon Socket Size Nom.	M Spline Socket Size Nom.	T Key Engagement Min.	G Wall Thickness Min.	E Under Head Fillet Radius Max.	F Under Head Fillet Radius Min.	B Under Head Fillet Radius Max.	B Under Head Fillet Radius Min.	K Chamfer or Radius Max.
1.6	1.6	1.46	3	2.87	1.6	1.52	.16	1.5	.072 in.	0.8	.54	.34	0.1	2	1.8	0.08
2	2	1.86	3.8	3.65	2	1.91	.2	1.5	.072 in.	1	.68	.51	0.1	2.6	2.2	0.08
2.5	2.5	2.36	4.5	4.33	2.5	2.40	.25	2	.096 in.	1.25	.85	.51	0.1	3.1	2.7	0.08
3	3	2.86	5.5	5.32	3	2.89	.3	2.5	.111 in.	1.5	1.02	.51	0.1	3.6	3.2	0.13
4	4	3.82	7	6.80	4	3.88	.4	3	.133 in.	2	1.52	.60	0.2	4.7	4.4	0.13
5	5	4.82	8.5	8.27	5	4.86	.5	4	.183 in.	2.5	1.90	.60	0.2	5.7	5.4	0.13
6	6	5.82	10	9.74	6	5.85	.6	5	.216 in.	3	2.28	.68	0.25	6.8	6.5	0.2
8	8	7.78	13	12.70	8	7.83	.8	6	.291 in.	4	3.20	1.02	0.4	9.2	8.8	0.2
10	10	9.78	16	15.67	10	9.81	1	8		5	4.00	1.02	0.4	11.2	10.8	0.2
12	12	11.73	18	17.63	12	11.79	1.2	10		6	4.80	1.87	0.6	14.2	13.2	0.25
(14)	14	13.73	21	20.60	14	13.77	1.4	12		7	6.02	1.87	0.6	16.2	15.2	0.25
16	16	15.73	24	23.58	16	15.76	1.6	14		8	6.88	1.87	0.6	18.2	17.2	0.25
20	20	19.67	30	29.53	20	19.73	2	17		10	8.60	2.04	0.8	22.4	21.6	0.4
24	24	23.67	36	35.48	24	23.70	2.4	19		12	10.32	2.04	0.8	26.4	25.6	0.4
30	30	29.67	45	44.42	30	29.67	3	22		15	12.90	2.89	1	33.4	32.0	0.4
36	36	35.61	54	53.37	36	35.64	3.6	27		18	15.48	2.89	1	39.4	38.0	0.4
42	42	41.61	63	62.31	42	41.61	4.2	32		21	18.06	3.06	1.2	45.6	44.4	0.4
48	48	47.61	72	71.27	48	47.58	4.8	36		24	20.64	3.91	1.6	52.6	51.2	0.4

*Notes to table are given on page 9-56.

Table 9-53. Nominal Body and Grip Lengths for Hexagon Socket Screws (SS-200)

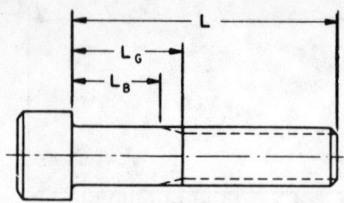

Nominal Diameter	1.6		2		2.5		3		4		5		6		8		10		12		14		16		20		24	
Nominal Lengths	L_G	L_B	L_G	L_B	L_G	L_B	L_G	L_B	L_G	L_B	L_G	L_B	L_G	L_B	L_G	L_B	L_G	L_B	L_G	L_B	L_G	L_B	L_G	L_B	L_G	L_B	L_G	L_B
25	9.8	8	9	7																								
30	14.8	13	14	12	13	10.7	12	9.5																				
35	...	...	19	17	18	15.7	12	9.5	15	11.5																		
40	...	...	24	22	23	20.7	22	19.5	15	11.5	18	14	16	11														
45	...	...	...	...	28	25.7	22	19.5	25	21.5	18	14	16	11	17	10.7												
50	...	...	...	...	33	30.7	32	29.5	25	21.5	28	24	26	21	17	10.7												
55	...	...	...	...	...	...	32	29.5	35	31.5	28	24	26	21	27	20.7	23	15.5										
60	...	...	...	...	...	...	42	39.5	35	31.5	38	34	36	31	27	20.7	23	15.5										
65	...	...	...	...	...	...	42	39.5	45	41.5	38	34	36	31	37	30.7	33	25.5	29	20.2								
70	...	...	...	...	...	...	...	...	45	41.5	48	44	46	41	37	30.7	33	25.5	29	20.2	30	20						
80	...	...	...	...	...	...	...	...	60	56.5	48	44	46	41	52	45.7	48	40.5	29	20.2	30	20	36	26				
90	...	...	...	...	...	...	...	...	...	...	68	64	66	61	52	45.7	48	40.5	54	45.2	50	40	36	26				
100	...	...	...	...	...	...	...	...	...	...	68	64	66	61	72	65.7	68	60.5	54	45.2	50	40	36	26	48	35.5		
110	...	...	...	...	...	...	...	...	...	...	86	81	72	65.7	68	60.5	74	65.2	70	60	66	56	48	35.5	50	35		
120	...	...	...	...	...	...	...	...	...	...	86	81	92	85.7	88	80.5	74	65.2	70	60	66	56	48	35.5	50	35		
130	...	...	...	...	...	...	...	...	...	...	...	...	92	85.7	88	80.5	94	85.2	90	80	66	56	78	65.5	50	35		
140	...	...	...	...	...	...	...	...	...	...	...	...	112	105.7	108	100.5	94	85.2	90	80	96	86	78	65.5	50	35		
150	...	...	...	...	...	...	...	...	...	...	...	...	112	105.7	108	100.5	114	105.2	110	100	96	86	78	65.5	90	75		
160	...	...	...	...	...	...	...	...	...	...	...	...	132	125.7	128	120.5	114	105.2	110	100	96	86	108	95.5	90	75		
180	...	...	...	...	...	...	...	...	...	...	...	...	...	...	148	140.5	144	135.2	140	130	136	126	108	95.5	90	75		
200	...	...	...	...	...	...	...	...	...	...	...	...	...	...	168	160.5	164	155.2	160	150	136	126	148	135.5	140	125		
220	...	...	...	...	...	...	...	...	...	...	...	...	...	...	...	...	184	175.2	180	170	176	166	148	135.5	140	125		
240	...	...	...	...	...	...	...	...	...	...	...	...	...	...	...	...	204	195.2	200	190	176	166	188	175.5	140	125		
260	...	...	...	...	...	...	...	...	...	...	...	...	...	...	...	...	...	...	220	210	216	206	188	175.5	200	185		
300	...	...	...	...	...	...	...	...	...	...	...	...	...	...	...	...	...	...	...	...	256	246	248	235.5	240	225		

NOTES:
1. All dimensions are in millimeters.
2. Screw thread details are shown in Table 8-1, and the thread tolerance is shown in Table 8-6 for general purpose tolerance (6g) and in Table 8-7 for close tolerance (5g6g).
3. The ISO 898 9.8 strength grade is preferred for screws with general purpose tolerance (6g), and 12.9 is used for alloy steel screws with close tolerance (5g6g) in the ISO draft proposal.
4. Nominal screw length L, grip length LG, and body length L_B are shown in Table 9-53. Screws above solid line in Table 9-53 are threaded to within two pitches (threads) of the head for diameters 1.6 to 16 mm inclusive, and must extend as close to the head as is practicable for sizes larger than 16 mm (data from SS-200 standard).
5. Product Grade A: Hexagon socket head cap screws are to product Grade A in the ISO 861 proposed revision. Product Grade A to be specified in future ISO standard.
6. Surface finish: Black oxide (thermal or chemical) unless otherwise specified.
7. Acceptability: For acceptance procedure, see ISO 3269 or SS-200 standards.
8. Designation: Example for the designation of a hexagon socket head cap screw with thread size d_1 = M5, nominal length l = 20 mm and property class 12.9:
    To ISO : Hexagon socket head cap screw ISO 861 M5 × 20 – 12.9
    To SS-200: AM5 × 20 SHCS

FASTENERS

Table 9-54A. Slotted Flat Countersunk Head Machine Screws (IFI-513)

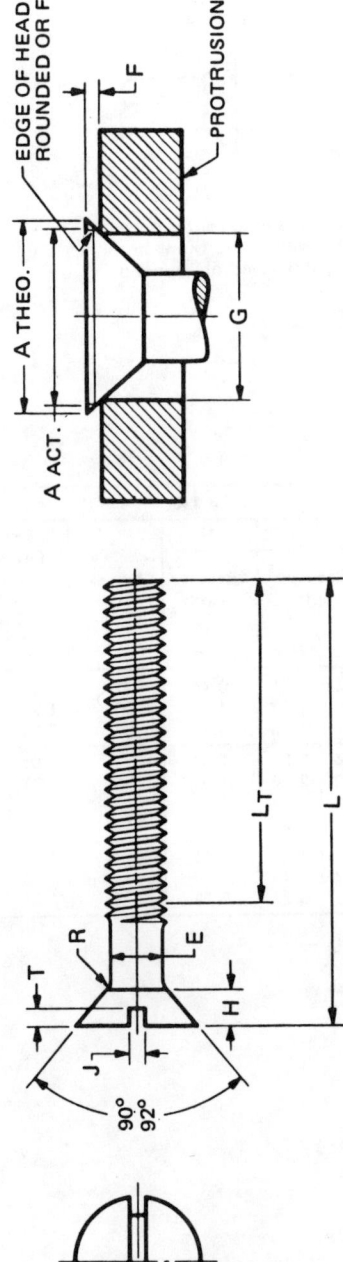

Nom Screw Size and Thread Pitch	E Body Dia Max	E Body Dia Min	A Head Diameter Theoretical Sharp Max	A Head Diameter Theoretical Sharp Min	A Head Diameter Actual Min	H Head Height Max Ref	R Fillet Radius Max	R Fillet Radius Min	J Slot Width Max	J Slot Width Min	T Slot Depth Max	T Slot Depth Min	F Protrusion of Head Above Gage Dia Max	F Protrusion of Head Above Gage Dia Min	G Gage Dia
M2x0.4	2.00	1.65	4.40	3.90	3.60	1.20	0.8	0.2	0.7	0.5	0.6	0.4	0.79	0.52	2.82
M2.5x0.45	2.50	2.12	5.50	4.90	4.60	1.50	1.0	0.3	0.8	0.6	0.7	0.5	0.88	0.56	3.74
M3x0.5	3.00	2.58	6.60	5.80	5.50	1.80	1.2	0.3	1.0	0.8	0.9	0.6	0.98	0.55	4.65
M3.5x0.6	3.50	3.00	7.70	6.80	6.44	2.10	1.4	0.4	1.2	1.0	1.0	0.7	1.07	0.59	5.57
M4x0.7	4.00	3.43	8.65	7.80	7.44	2.32	1.6	0.4	1.4	1.2	1.1	0.8	1.09	0.63	6.48
M5x0.8	5.00	4.36	10.70	9.80	9.44	2.85	2.0	0.5	1.5	1.2	1.4	1.0	1.20	0.71	8.31
M6x1	6.00	5.21	13.50	12.30	11.87	3.60	2.5	0.6	1.9	1.6	1.8	1.3	1.41	0.77	10.69
M8x1.25	8.00	7.04	16.80	15.60	15.17	4.40	3.2	0.8	2.3	2.0	2.1	1.6	1.50	0.86	13.80
M10x1.5	10.00	8.86	20.70	19.50	18.98	5.35	4.0	1.0	2.8	2.5	2.6	2.0	2.58	1.91	15.54
M12x1.75	12.00	10.68	24.70	23.50	22.88	6.35	4.8	1.2	2.8	2.5	3.1	2.5	2.79	2.11	19.12

NOTES:
1. See general notes on machine screws.
2. No tolerance for gage diameter is given. If the gage diameter of the gage used differs from the tabulated value, the protrusion will be affected accordingly, and the proper protrusion values must be recalculated using the formulas given in Appendix A of IFI-502.

Table 9-54B. Cross Recess Dimensions of Flat Countersunk Head Machine Screws (IFI-513)

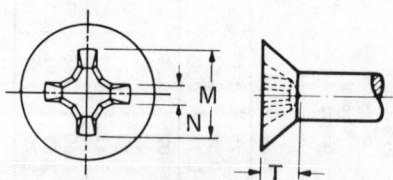

This type of recess has a large center opening, tapered wings, and blunt bottom, with all edges relieved or rounded.

TYPE 1

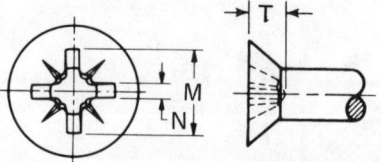

This type of recess has a large center opening, wide straight wings, and blunt bottom, with all edges relieved or rounded.

TYPE 1A

| Nom Screw Size | Type 1 ||||||||| Type 1A |||||||||
| | M (Recess Dia) || T (Recess Depth) || N (Recess Width) | Driver Size | Recess Penetration Gaging Depth || M (Recess Dia) || T (Recess Depth) || N (Recess Width) | Driver Size | Recess Penetration Gaging Depth ||
	Max	Min	Max	Min	Min		Max	Min	Max	Min	Max	Min	Min		Max	Min
M 2	1.96	1.63	1.30	0.89	0.38	0	1.12	0.71	2.39	2.06	1.75	1.35	0.46	0	1.57	1.17
M 2.5	2.97	2.64	1.98	1.57	0.46	1	1.80	1.40	2.97	2.64	1.98	1.57	0.74	1	1.73	1.32
M 3	3.25	2.92	2.26	1.85	0.46	1	2.08	1.68	3.25	2.92	2.26	1.85	0.76	1	2.01	1.60
M 3.5	4.17	3.84	2.44	1.85	0.71	2	2.16	1.57	4.17	3.84	2.46	2.01	1.04	2	2.06	1.60
M 4	4.42	4.09	2.69	2.11	0.74	2	2.41	1.83	4.42	4.09	2.72	2.26	1.04	2	2.31	1.85
M 5	4.62	4.29	2.90	2.31	0.76	2	2.62	2.03	4.62	4.29	2.90	2.44	1.04	2	2.51	2.05
M 6	6.35	6.02	3.45	2.87	0.81	3	3.02	2.44	6.30	5.97	3.48	3.02	1.42	3	2.92	2.46
M 8	8.69	8.36	4.90	4.34	1.42	4	4.39	3.84	8.69	8.36	4.98	4.52	2.16	4	4.32	3.86
M10	9.60	9.27	5.84	5.28	1.57	4	5.33	4.78	9.60	9.27	5.92	5.46	2.18	4	5.23	4.77
M12	10.39	10.06	6.63	6.07	1.73	4	6.12	5.56	10.39	10.06	6.73	6.27	2.18	4	6.05	5.59

NOTES:
1. Head dimensions not shown are the same as those of slotted heads given in Table 9-54A.
2. For penetration gaging, see Appendix C of IFI-502.
3. For wobble gaging, see Appendix D of IFI-502.
4. See general notes on machine screws.

FASTENERS

Pozidriv (ANSI Type IA and ISO Type Pz) are shown in Table 9-56B.

Hexagon Head Machine Screws (IFI-513)

Hexagon head machine screws in nominal sizes from 2 to 12 mm are specified in the IFI-513 standard. Screws with the trimmed hexagon head style are shown in Table 9-57A, and screws with hexagon washer heads are shown in Table 9-57B.

General Notes on Machine Screws

Dimensions—All dimensions in this standard are given in millimeters, unless stated otherwise.

Head Types—The head types covered by this standard include those commonly recognized as being applicable to machine screws and are enumerated and described in the following:

Flat Countersunk Head—The flat countersunk head shall have a flat top surface and a conical bearing surface with a head angle of approximately 90 degrees.

Oval Countersunk Head—The oval countersunk head shall have a rounded top surface and a conical bearing surface with a head angle of approximately 90 degrees.

Pan Head—The slotted pan head shall have a flat top surface rounded into cylindrical sides and a flat bearing surface. The recessed pan head shall have a rounded top surface blending into cylindrical sides and a flat bearing surface.

Hex Head—The hex head shall have a flat or indented top surface, six flat sides and a flat bearing surface.

Hex Washer Head—The hex washer head shall have an indented top surface and six flat sides formed integrally with a flat washer which projects beyond the sides and provides a flat bearing surface.

Table 9-55A. Slotted Oval Countersunk Head Machine Screws (IFI-513)

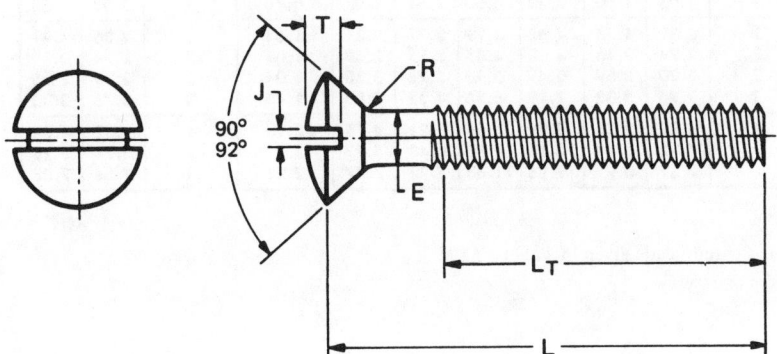
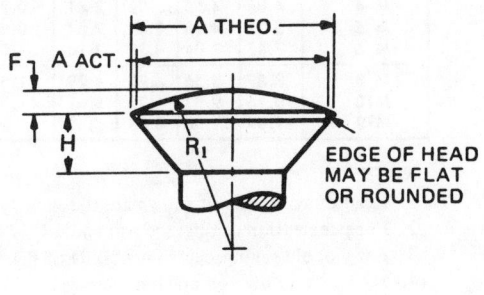

Nom Screw Size and Thread Pitch	E Body Dia Max	E Body Dia Min	A Head Diameter Theoretical Sharp Max	A Head Diameter Theoretical Sharp Min	A Head Diameter Actual Min	H Head Side Height Max Ref	F Raised Head Height Max	R₁ Head Radius Approx	R Fillet Radius Max	R Fillet Radius Min	J Slot Width Max	J Slot Width Min	T Slot Depth Max	T Slot Depth Min
M2x0.4	2.00	1.65	4.40	3.90	3.60	1.20	0.50	3.8	0.8	0.2	0.7	0.5	1.0	0.8
M2.5x0.45	2.50	2.12	5.50	4.90	4.60	1.50	0.60	5.0	1.0	0.3	0.8	0.6	1.2	1.0
M3x0.5	3.00	2.58	6.60	5.80	5.50	1.80	0.75	5.7	1.2	0.3	1.0	0.8	1.5	1.2
M3.5x0.6	3.50	3.00	7.70	6.80	6.44	2.10	0.90	6.5	1.4	0.4	1.2	1.0	1.7	1.4
M4x0.7	4.00	3.43	8.65	7.80	7.44	2.32	1.00	7.8	1.6	0.4	1.4	1.2	1.9	1.6
M5x0.8	5.00	4.36	10.70	9.80	9.44	2.85	1.25	9.9	2.0	0.5	1.5	1.2	2.3	2.0
M6x1	6.00	5.21	13.50	12.30	11.87	3.60	1.60	12.2	2.4	0.6	1.9	1.6	3.0	2.6
M8x1.25	8.00	7.04	16.80	15.60	15.17	4.40	2.00	15.8	3.2	0.8	2.3	2.0	3.7	3.2
M10x1.5	10.00	8.86	20.70	19.50	18.98	5.35	2.50	19.8	4.0	1.0	2.8	2.5	4.5	4.0
M12x1.75	12.00	10.68	24.70	23.50	22.88	6.35	3.00	23.8	4.8	1.2	2.8	2.5	5.3	4.8

NOTE: See general notes on machine screws.

Table 9-55B. Cross Recess Dimensions of Oval Countersunk Head Machine Screws (IFI-513)

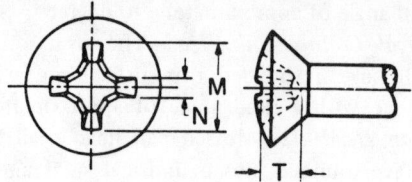

This type of recess has a large center opening, tapered wings, and blunt bottom, with all edges relieved or rounded.

TYPE 1

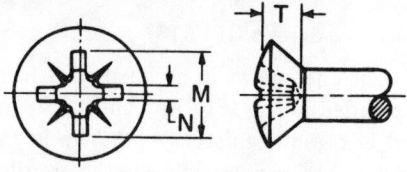

This type of recess has a large center opening, wide straight wings, and blunt bottom, with all edges relieved or rounded.

TYPE 1A

Nom Screw Size	Type 1 M Recess Dia Max	Min	T Recess Depth Max	Min	N Recess Width Min	Driver Size	Recess Penetration Gaging Depth Max	Min	Type 1A M Recess Dia Max	Min	T Recess Depth Max	Min	N Recess Width Min	Driver Size	Recess Penetration Gaging Depth Max	Min
M 2	1.96	1.63	1.22	0.76	0.38	0	1.04	0.58	2.41	2.08	1.68	1.27	0.47	0	1.52	1.12
M 2.5	2.82	2.49	1.73	1.32	0.48	1	1.55	1.14	3.00	2.67	2.01	1.60	0.74	1	1.75	1.35
M 3	3.18	2.84	2.11	1.68	0.48	1	1.93	1.50	3.33	3.00	2.34	1.93	0.76	1	2.08	1.68
M 3.5	4.52	4.19	2.67	2.03	0.76	2	2.39	1.75	4.52	4.19	2.77	2.31	1.04	2	2.36	1.91
M 4	4.88	4.55	3.02	2.41	0.79	2	2.74	2.13	4.88	4.55	3.15	2.69	1.04	2	2.74	2.29
M 5	5.31	4.98	3.48	2.87	0.84	2	3.20	2.59	5.49	5.16	3.76	3.30	1.04	2	3.35	2.90
M 6	7.37	7.04	4.39	3.76	1.02	3	3.96	3.33	7.19	6.86	4.32	3.86	1.45	3	3.76	3.30
M 8	8.89	8.56	4.90	4.29	1.50	4	4.39	3.78	8.99	8.66	5.21	4.75	2.13	4	4.52	4.06
M10	10.16	9.83	6.22	5.61	1.67	4	5.72	5.11	10.13	9.80	6.43	5.97	2.18	4	5.74	5.28
M12	11.89	11.56	8.03	7.42	1.96	4	7.52	6.91	11.96	11.63	8.23	7.77	2.21	4	7.54	7.08

NOTES:
1. Head dimensions not shown are the same as those of slotted heads given in Table 9-55A.
2. For penetration gaging, see Appendix C of IFI-502.
3. For wobble gaging, see Appendix D of IFI-502.
4. See general notes on machine screws.

Options—Options, where specified, shall be at the discretion of the manufacturer unless otherwise agreed upon by the manufacturer and the purchaser.

Terminology—For definitions of terms relating to fasteners or component features used here, refer to the American National Standard, Glossary of Terms for Mechanical Fasteners, ANSI B18.12.

Height of Head—The height of head indicated in the dimensional tables represents a metal to metal measurement. In other words, on heads having rounded top surfaces, the truncation of the rounded surface due to recess or slot is not considered part of the head height.

On countersunk type heads, the height of head is a reference dimension measured parallel to the axis of the screw from the largest diameter of the bearing surface of the head to the point of intersection of the bearing surface of the head and basic major diameter of the screw. This point of intersection may not necessarily be the same as the actual junction of head and shank.

Bearing Surface—The bearing surface of perpendicular bearing surface type screw heads shall be at right angles to the axis of the screw shanks within 2 degrees.

Depth of Recess—The depth of recess in recessed head screws shall be measured parallel to the axis of the screw from the intersection of the maximum diameter of the recess with the head surface to the bottom of the recess.

Recess penetration gaging depth values are included in the dimensional tables and the method of gaging and specifications for gages are covered in Appendix C of IFI-502.

Recess wobble gaging procedures and operating limits are given in Appendix D of IFI-502.

Depth of Slot—The depth of slot in slotted head screws shall be measured parallel to the axis of the screw from

FASTENERS

the top of the head to the intersection of the bottom of the slot with the head surface or bearing surface.

Positional Tolerances—The positional relationship between the heads and driving provisions of screws and the shanks of screws (formerly defined as Eccentricity) shall be as follows:

Position of Head—The axis of the head shall be located at true position relative to the axis of the screw shank within a tolerance zone having a diameter equivalent to 6 per cent of the maximum head diameter, or the maximum width across flats of hex and hex washer heads, regardless of feature size.

Position of Recess—The recess in cross recessed head screws shall be located at true position relative to the axis of the screw shank within a tolerance zone having a diameter equivalent to 12 percent of the basic screw diameter or 0.75 mm, whichever is greater, regardless of feature size.

Position of Slot—The slot in slotted head screws shall be located at true position relative to the axis of the screw shank within a tolerance zone having a diameter equivalent to 12 percent of the basic screw diameter or 0.50 mm, whichever is greater, regardless of feature size.

Underhead Fillets—All screws shall have a fillet radius at the junction of the head to shank within the limits as specified in the dimensional tables. For flat and oval countersunk head screws, the maximum fillet radius equals 0.4 D and minimum radius equals 0.1 D, rounded to one decimal place. For pan, hex and hex washer head screws, the maximum fillet radius equals 0.15 D and minimum radius equals 0.05 D, rounded to one decimal place. D equals nominal screw size.

Table 9-56A. Slotted and Recessed Pan Head Machine Screws (IFI-513)

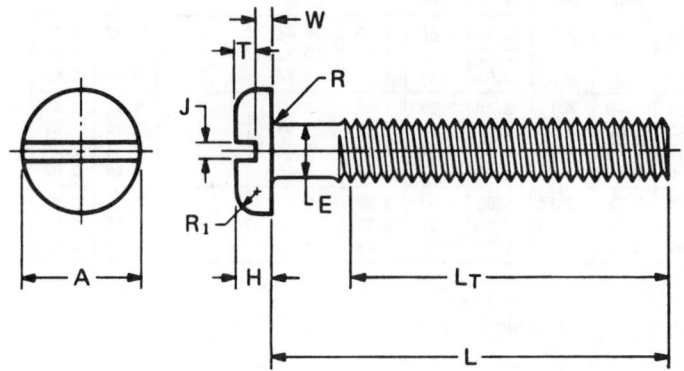
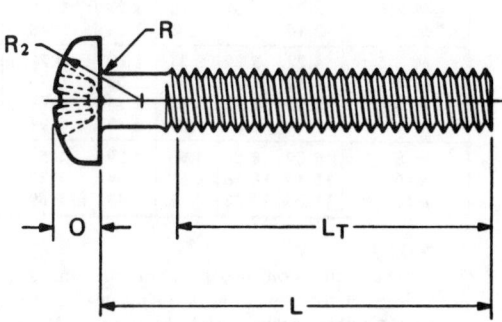

Nom Screw Size and Thread Pitch	E Body Diameter Max	E Body Diameter Min	A Head Diameter Max	A Head Diameter Min	H Head Height Slotted Head Max	H Head Height Slotted Head Min	O Head Height Recessed Head Max	O Head Height Recessed Head Min	R₁ Head Radius (Slttd) Max	R₂ Head Radius (Rcssd) Ref	R Fillet Radius Max	R Fillet Radius Min	J Slot Width Max	J Slot Width Min	T Slot Depth Min	W Un-slotted Thickness Min
M2x0.4	2.00	1.65	3.90	3.60	1.35	1.15	1.60	1.40	0.8	4	0.3	0.1	0.7	0.5	0.55	0.44
M2.5x0.45	2.50	2.12	4.90	4.60	1.65	1.45	1.95	1.75	1.0	5	0.4	0.1	0.8	0.6	0.73	0.55
M3x0.5	3.00	2.58	5.80	5.50	1.90	1.65	2.30	2.05	1.2	6	0.5	0.2	1.0	0.8	0.80	0.66
M3.5x0.6	3.50	3.00	6.80	6.44	2.25	2.00	2.50	2.25	1.4	7	0.5	0.2	1.2	1.0	0.95	0.77
M4x0.7	4.00	3.43	7.80	7.44	2.55	2.30	2.80	2.55	1.6	8	0.6	0.2	1.4	1.2	1.15	0.88
M5x0.8	5.00	4.36	9.80	9.44	3.10	2.85	3.50	3.25	2.0	10	0.8	0.3	1.5	1.2	1.35	1.10
M6x1	6.00	5.21	12.00	11.57	3.90	3.50	4.30	4.00	2.5	13	1.0	0.3	1.9	1.6	1.70	1.36
M8x1.25	8.00	7.04	15.60	15.17	5.00	4.60	5.60	5.20	3.2	16	1.2	0.4	2.3	2.0	2.20	1.76
M10x1.5	10.00	8.86	19.50	18.98	6.20	5.70	7.00	6.50	4.0	20	1.5	0.5	2.8	2.5	2.70	2.20
M12x1.75	12.00	10.68	23.40	22.88	7.50	6.90	8.30	7.80	4.8	24	1.8	0.6	2.8	2.5	3.20	2.70

NOTE: See general notes on machine screws.

Table 9-56B. Cross Recess Dimensions of Pan Head Machine Screws (IFI-513)

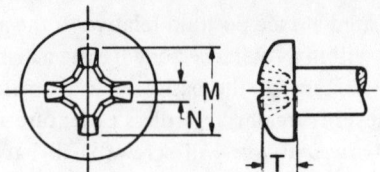
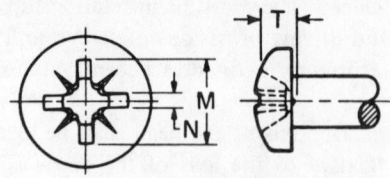

This type of recess has a large center opening, tapered wings, and blunt bottom, with all edges relieved or rounded.

This type of recess has a large center opening, wide straight wings, and blunt bottom, with all edges relieved or rounded.

TYPE 1 — TYPE 1A

Nom Screw Size	Type 1							Type 1A						
	M (Recess Dia)		T (Recess Depth)		N (Recess Width)	Driver Size	Recess Penetration Gaging Depth		M (Recess Dia)		T (Recess Depth)		N (Recess Width)	Driver Size	Recess Penetration Gaging Depth	
	Max	Min	Max	Min	Min		Max	Min	Max	Min	Max	Min	Min		Max	Min
M 2	1.88	1.55	1.14	0.64	0.36	0	1.02	0.56	2.31	1.98	1.60	1.19	0.46	0	1.42	1.02
M 2.5	2.84	2.51	1.73	1.27	0.48	1	1.55	1.09	3.10	2.77	2.03	1.63	0.74	1	1.78	1.37
M 3	3.10	2.77	1.98	1.52	0.48	1	1.80	1.35	3.35	3.02	2.31	1.91	0.74	1	2.06	1.65
M 3.5	4.22	3.89	2.31	1.68	0.71	2	2.03	1.40	4.11	3.78	2.34	1.88	1.02	2	1.93	1.47
M 4	4.62	4.29	2.74	2.08	0.76	2	2.46	1.80	4.50	4.17	2.74	2.29	1.04	2	2.34	1.88
M 5	5.05	4.72	3.15	2.54	0.79	2	2.87	2.26	4.90	4.57	3.15	2.69	1.04	2	2.74	2.29
M 6	7.14	6.81	4.09	3.43	0.91	3	3.66	3.00	6.93	6.60	4.04	3.58	1.45	3	3.48	3.02
M 8	8.89	8.56	4.90	4.29	1.50	4	4.39	3.78	8.66	8.33	4.85	4.39	2.18	4	4.17	3.71
M10	10.49	10.16	6.58	5.94	1.73	4	6.07	5.44	10.13	9.80	6.38	5.92	2.18	4	5.69	5.23
M12	11.05	10.72	7.11	6.48	1.80	4	6.60	5.97	10.67	10.34	6.93	6.48	2.18	4	6.25	5.79

NOTES:
1. Head dimensions not shown are the same as those of slotted heads given in Table 9-56A.
2. For penetration gaging, see Appendix C of IFI-502.
3. For wobble gaging, see Appendix D of IFI-502.
4. See general notes on machine screws.

Length Measurement—The length of screw shall be measured parallel to the axis of the screw from the extreme point to largest diameter of the bearing surface of the head. Recommended lengths of machine screws are given in Table 9-57C.

Tolerance on Length—The tolerance on length of screws shall conform to the following:

Nom Screw Size	M2 thru M5	M6 thru M12
Nom Screw Length	Tolerance on Length mm	
Up to 12 mm incl.	−0.5	−0.8
over 12 through 25 mm	−0.8	−0.8
over 25 through 50 mm	−1.5	−1.5
over 50 mm	−2.2	−2.2

Threads—Threads on machine screws are general purpose threads in accordance with dimensions given in Section 8. Modified ISO Fastener Screw Threads (page 8-20) class 6g thread fit unless otherwise specified by the customer.

Length of Thread—The length of thread on machine screws shall be as specified in Table 9-57D for the applicable screw size and length. For screws to which the maximum unthreaded lengths apply, the distance to the first full form thread shall be measured parallel to the axis of screw from the bearing surface of the head to the face of a non-chamfered or non-counterbored standard GO thread ring gage assembled by hand as far as the thread will permit.

Points—Unless otherwise specified, machine screws shall have plain sheared ends. When specified, header points shall be obtainable as shown in Table 9-57E. Other points or pointing of longer lengths to header point dimensions may require machining.

FASTENERS   9-63

Table 9-57A. Hexagon Head Machine Screws (IFI-513)

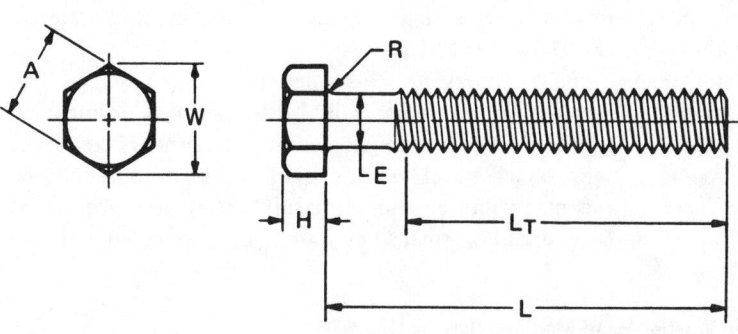

TRIMMED HEAD OR FULLY UPSET HEAD

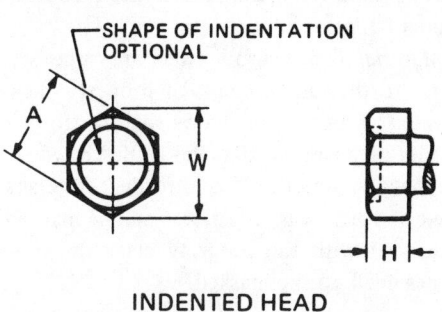

INDENTED HEAD

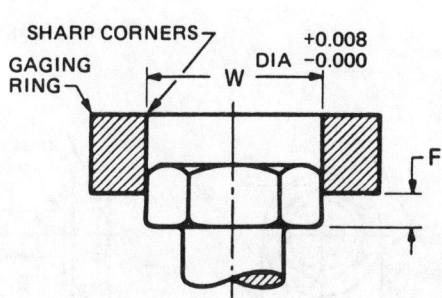

Nom Screw Size and Thread Pitch	E Body Diameter Max	E Body Diameter Min	A Width Across Flats Max	A Width Across Flats Min	W Width Across Corners Min	H Head Height Max	H Head Height Min	R Fillet Radius Max	R Fillet Radius Min	F Protrusion Beyond Gaging Ring Min
M2x0.4	2.00	1.65	3.20	3.02	3.36	1.27	1.02	0.3	0.1	0.61
M2.5x0.45	2.50	2.12	4.00	3.82	4.25	1.40	1.12	0.4	0.1	0.67
M3x0.5	3.00	2.58	5.00	4.82	5.36	1.52	1.24	0.5	0.2	0.74
M3.5x0.6	3.50	3.00	5.50	5.32	5.92	2.36	2.03	0.5	0.2	1.22
M4x0.7	4.00	3.43	7.00	6.78	7.55	2.79	2.44	0.6	0.2	1.46
M5x0.8	5.00	4.36	8.00	7.78	8.66	3.05	2.67	0.8	0.3	1.60
M6x1	6.00	5.21	10.00	9.78	10.89	4.83	4.37	1.0	0.3	2.62
M8x1.25	8.00	7.04	13.00	12.73	14.17	5.84	5.28	1.2	0.4	3.17
M10x1.5	10.00	8.86	15.00	14.73	16.41	7.49	6.86	1.5	0.5	4.12
M12x1.75	12.00	10.68	18.00	17.73	19.75	9.50	8.66	1.8	0.6	5.20

NOTES:
1. A slight rounding of all edges of the hex surfaces of indented hex heads is permissible provided the diameter of the bearing circle is not less than 90 percent of the minimum width across flats dimension.
2. Dimensions across flats and across corners of the head shall be measured at the point of maximum metal. Taper of sides of hex (angle between one side and the axis) shall not exceed 2 degrees or 0.10 mm whichever is greater, the specified width across flats being the large dimension.
3. The rounding due to lack of fill on all six corners of the head shall be reasonably uniform, and the width across corners of the head shall be such that when a sharp ring having an inside diameter equal to the specified minimum width across corners is placed on the top and the bottom of the head, the head shall protrude by an amount equal to, or greater than, the F value tabulated. For across corners gaging see Appendix B of IFI-502.
4. Heads may be idented, trimmed or fully upset at the option of the manufacturer.
5. See general notes on machine screws.
6. See note, page 9-41.

Diameter of Body—The diameter of the body of machine screws shall be within the limits specified in the dimensional tables.

Material—Low carbon steel machine screws shall conform to the requirements of property class 4.8, and heat-treated carbon steel screws shall conform to the requirements of property class 9.8 as described in the section on strength properties for threaded fasteners (page 9-18). Hex and hex washer screws shall be marked on the top of the head with the property class numerals. Other head types need not be marked.

Machine screws may also be made from higher strength steels, corrosion resistant steel, brass, monel, aluminum alloys or other materials, as agreed upon between the manufacturer and the purchaser. (For guidance refer to IFI-516 and IFI-518.)

Finish—Unless otherwise specified, machine screws shall be supplied with a naturally bright unplated or uncoated finish. When corrosion preventative treatment is required, screws shall be plated or coated as agreed upon between the manufacturer and purchaser. However, where heat treated carbon steel screws are plated or coated and sub-

Table 9-57B. Hexagon Washer Head Machine Screws (IFI-513)

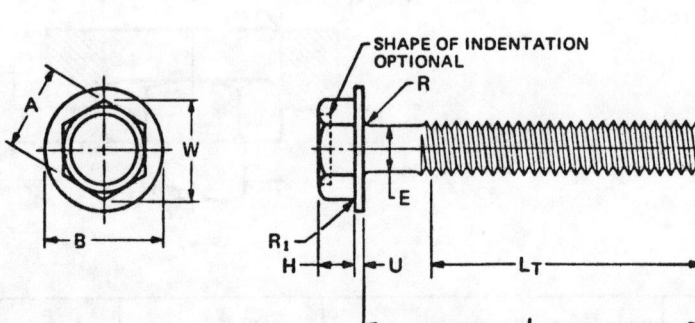
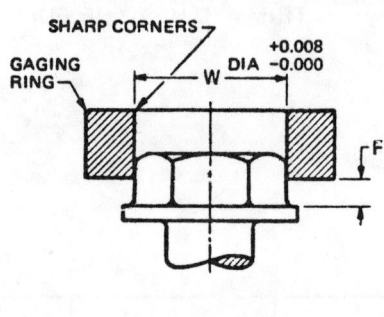

Nom Screw Size and Thread Pitch	E Body Diameter Max	E Body Diameter Min	A Width Across Flats Max	A Width Across Flats Min	W Width Across Corners Min	H Head Height Max	H Head Height Min	B Washer Diameter Max	B Washer Diameter Min	U Washer Thickness Max	U Washer Thickness Min	R Fillet Radius Max	R Fillet Radius Min	R₁ Head Fillet Radius Max	F Protrusion Beyond Gaging Dia Min
M2x0.4	2.00	1.65	3.20	3.02	3.36	1.27	1.02	4.2	3.9	0.4	0.2	0.3	0.1	0.4	0.61
M2.5x0.45	2.50	2.12	4.00	3.82	4.25	1.40	1.12	5.3	5.0	0.5	0.3	0.4	0.1	0.5	0.67
M3x0.5	3.00	2.58	5.00	4.82	5.36	1.52	1.24	6.2	5.8	0.5	0.3	0.5	0.2	0.5	0.74
M3.5x0.6	3.50	3.00	5.50	5.32	5.92	2.36	2.03	7.5	6.8	0.6	0.4	0.5	0.2	0.6	1.22
M4x0.7	4.00	3.43	7.00	6.78	7.55	2.79	2.44	9.2	8.5	0.7	0.5	0.6	0.2	0.7	1.46
M5x0.8	5.00	4.36	8.00	7.78	8.66	3.05	2.67	10.5	9.8	0.8	0.5	0.8	0.3	0.8	1.60
M6x1	6.00	5.21	10.00	9.78	10.89	4.83	4.37	13.2	12.2	1.3	0.8	1.0	0.3	1.0	2.62
M8x1.25	8.00	7.04	13.00	12.73	14.17	5.84	5.28	17.2	15.9	1.4	0.9	1.2	0.4	1.2	3.17
M10x1.5	10.00	8.86	15.00	14.73	16.41	7.49	6.86	19.8	18.3	1.6	0.9	1.5	0.5	1.5	4.12
M12x1.75	12.00	10.68	18.00	17.73	19.75	9.50	8.66	23.8	22.0	1.8	1.1	1.8	0.6	1.8	5.20

NOTES:
1. A slight rounding of all edges and corners of the hex surfaces shall be permissible.
2. Dimensions across flats and across corners of the head shall be measured at the point of maximum metal. Taper of sides of hex (angle between one side and the axis) shall not exceed 2 degrees or 0.10 mm whichever is greater, the specified width across flats being the large dimension.
3. The rounding due to lack of fill on all six corners of the head shall be reasonably uniform, and the width across corners of the head shall be such that when a sharp ring having an inside diameter equal to the specified minimum width across corners is placed on the top and the bottom of the head, the head shall protrude by an amount equal to, or greater than, the F value tabulated. For across corners gaging see Appendix B of IFI-502.
4. See general notes on machine screws.
5. See note, page 9-41.

FASTENERS

Table 9-57C. Recommended Machine Screw Lengths (IFI-513)

Nom Screw Length	\multicolumn{10}{c	}{Nominal Screw Size}
	M2	M2.5	M3	M3.5	M4	M5	M6	M8	M10	M12
2.5	PH									
3	A	PH								
4	A	A	PH							
5	A	A	A	PH	PH					
6	A	A	A	A	A	PH				
8	A	A	A	A	A	A	A			
10	A	A	A	A	A	A	A	A		
13	A	A	A	A	A	A	A	A	A	
16	A	A	A	A	A	A	A	A	A	A
20	A	A	A	A	A	A	A	A	A	A
25		A	A	A	A	A	A	A	A	A
30			A	A	A	A	A	A	A	A
35				A	A	A	A	A	A	A
40					A	A	A	A	A	A
45						A	A	A	A	A
50						A	A	A	A	A
55							A	A	A	A
60							A	A	A	A
65								A	A	A
70								A	A	A
80								A	A	A
90									A	A

NOTE:
Lengths included between the heavy lines are recommended for the applicable screw size. 'PH' means length is recommended only for pan, hex and hex washer head screws; 'A' means length is recommended for all head styles.

Table 9-57D. Length of Thread for Machine Screws (IFI-513)

Nominal Screw Size	\multicolumn{3}{c	}{L_T Full Form Thread Length}	\multicolumn{4}{c	}{Y Unthreaded Length Under Head}
	For Nominal Screw Lengths >Than	Min(1)	For Nominal Screw Lengths ≤Than(2)	Max(3)	For Nominal Screw Lengths >Than	≤Than	Max(4)
M 2	30	25	6	0.40	6	30	0.8
M 2.5	30	25	8	0.45	8	30	0.9
M 3	30	25	9	0.50	9	30	1.0
M 3.5	50	38	10	0.60	10	50	1.2
M 4	50	38	12	0.70	12	50	1.4
M 5	50	38	15	0.80	15	50	1.6
M 6	50	38	18	1.00	18	50	2.0
M 8	50	38	24	1.25	24	50	2.5
M10	50	38	30	1.50	30	50	3.0
M12	50	38	36	1.75	36	50	3.5

NOTES:
1. These lengths shall apply unless otherwise specified by purchaser.
2. Tabulated values are equal to 3 times basic screw diameter, rounded to nearest mm.
3. Tabulated values are equal to 1 thread pitch.
4. Tabulated values are equal to 2 thread pitches.

Table 9-57E. Header Points for Machine Screws Before Threading (IFI-513)

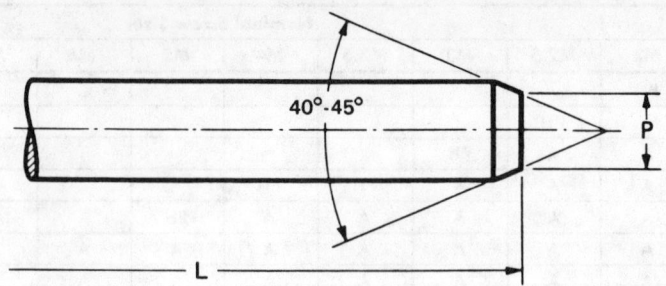

Nominal Screw Size	P Point Diameter Max	P Point Diameter Min	L Nominal Screw Length Max
M2x0.4	1.33	1.21	13
M2.5x0.45	1.73	1.57	13
M3x0.5	2.12	1.93	16
M3.5x0.6	2.46	2.24	20
M4x0.7	2.80	2.55	25
M5x0.8	3.60	3.28	30
M6x1	4.25	3.85	40
M8x1.25	5.82	5.30	40
M10x1.5	7.36	6.71	40
M12x1.75	8.90	8.11	45

NOTES:
1. Edges of point may be rounded and end of point need not be flat nor perpendicular to axis of shank.
2. Header points apply to these nominal lengths or shorter. The pointing of longer lengths may require machining to the dimensions specified.

ject to hydrogen embrittlement, they shall be suitably treated subsequent to the plating or coating operation to obviate such embrittlement.

Designation—Machine screws shall be designated by the following data in the sequence shown: Nominal size and thread pitch; nominal length; product name, including head type and driving provision; header point, if desired; material (and property class, if steel); and protective finish, if required. See examples below:

M8 × 1.25 × 30 slotted pan head machine screw, class 4.8 steel, zinc plated.

M3.5 × 0.6 × 20 Type 1A cross recessed oval countersunk head machine screw, header pointed, brass.

SET SCREWS AND DOUBLE END STUDS

General

Set screws with square, hexagon, hexagon socket and slotted heads are generally matched up with points such as flat, cone, short and long dog, and cup points in national standards listed in the world fastener index, page 9-5.

Double End Studs (IFI-528)

Double end studs of types 1, 2, 3 and 4 in nominal sizes from 6 to 36 mm are covered in the IFI-528 standard and shown in Table 9-58.

Hexagon Socket Set Screws (ISO 2343)

Set screws in sizes ranging from 1.4 to 24 mm are covered in the ISO standards, with the details of point design as shown in Table 9-59 (ISO 4026) for flat points; Table 9-60 (ISO 4027) for cone points; Table 9-61 (ISO 4028) for short and long dog points; Table 9-62 (ISO 4029) for cup points. Set screws with hexagon socket heads will be specified in the ISO standards. However, these points are generally available with other types of heads as shown in the national standards listed in the index.

The ISO 898 V standard for set screws and similar parts without specified proof load values gives three property classes as follows: 14H (140), 22H (220), and 45H (450). The values shown in parentheses are the minimum Vickers hardness values for the screws.

FASTENERS

Table 9-58. Double End Studs (IFI-528)

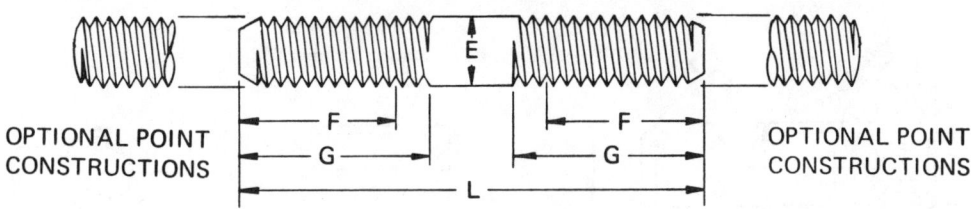

Nom Stud Size & Thread Pitch	Body Diameter E								Nut Ends						F,G
	Type 1		Type 2		Type 3		Type 4		F	G	F	G	F	G	
									For Studs Up To And Including 125 mm Length		For Studs Over 125 to 200 mm Length		For Studs Over 200 mm Length		For Short Studs
									Thread Length		Thread Length		Thread Length		
									Full Thread	Total Thread	Full Thread	Total Thread	Full Thread	Total Thread	Thread Length
	Max	Min	Max	Min	Max	Min	Max	Min	Min	Max	Min	Max	Min	Max	
M6x1	—	—	6.00	5.21	6.00	5.79	—	—	18	22	24	28	37	41	—
M8x1.25	—	—	8.00	7.04	8.00	7.76	—	—	22	27	28	33	41	46	—
M10x1.5	—	—	10.00	8.86	10.00	9.73	—	—	26	32	32	38	45	51	—
M12x1.75	—	—	12.00	10.68	12.00	11.70	—	—	30	37	36	43	49	56	—
M14x2	—	—	14.00	12.50	14.00	13.68	—	—	34	42	40	48	53	61	—
M16x2	—	—	16.00	14.50	16.00	15.68	—	—	38	46	44	52	57	65	—
M20x2.5	—	—	20.00	18.17	20.00	19.62	—	—	46	56	52	62	65	75	—
M24x3	—	—	24.00	21.80	24.00	23.58	—	—	—	—	60	72	73	85	—
M30x3.5	—	—	30.00	27.46	30.00	29.52	—	—	—	—	72	86	85	99	—
M36x4	—	—	36.00	33.12	36.00	35.46	—	—	—	—	—	—	97	113	—

Notes to Table 9-58

All dimensions are in millimeters.

Types;
Type 1—Unfinished
Type 2—Finished, Full or Undersize Body
Type 3—Finished, Full Body
Type 4—Finished, Close Body

Body Diameter Tolerances—Type 1 studs shall have an unfinished body with no specified body diameter tolerances. Type 2 studs shall have a maximum body diameter equal to basic major diameter of the thread, and a minimum body diameter equal to the rolled thread blank size. Type 3 studs shall have a maximum body diameter equal to basic major diameter of the thread, and a minimum body diameter equal to the specified minimum major diameter of the thread. Type 4 studs shall have body diameter tolerances as specified by the purchaser (milled or ground body).

Short Studs—Continuous thread studs shall be supplied for all studs too short to accommodate the standard thread length on each end. Continuous thread studs shall be supplied when the specified stud length is shorter than 4D + 8P + 12 mm for stud lengths 125 mm and shorter, 4D + 8P + 24 mm for stud lengths over 125 mm to and incl. 200 mm, and 4D + 8P + 50 mm for stud lengths over 200 mm. D equals basic stud diameter, and P is thread pitch.

Length—The length of stud, measured parallel to the axis of stud, shall be the distance from extreme end to extreme end. The tolerance on length shall be as tabulated below:

Nom Stud Size	Tolerance on Length, mm	
	For Lengths 150 mm and Shorter	For Lengths over 150 mm
M6 thru M12	±0.8	±1.5
M14 thru M24	±1.5	±2.3
M30 and M36	±2.3	±3.0

Threads—Threads are general purpose threads in accordance with dimensions given in Section 8. Modified ISO Fastener Screw Threads (page 8-20) class 6g thread fit unless otherwise specified by the customer.

Full Thread Length, F—The full thread length is the distance, measured parallel to the axis of stud, from the

Table 9-59. Hexagon Socket Set Screws with Flat Point (ISO 4026)

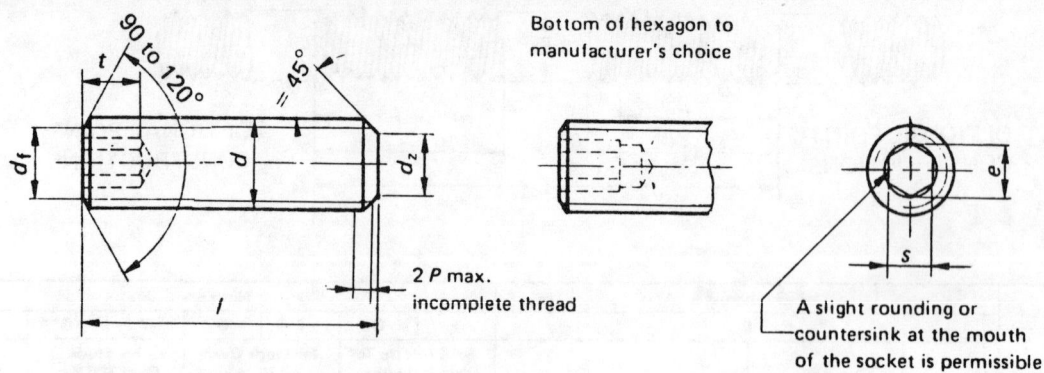

Thread size d		M 1,6	M 2	M 2,5	M 3	M 4	M 5	M 6	M 8	M 10	M 12	M 16	M 20	M 24
P		0,35	0,4	0,45	0,5	0,7	0,8	1,0	1,25	1,5	1,75	2,0	2,5	3,0
d_z	max.	0,8	1,0	1,5	2,0	2,5	3,5	4,0	5,5	7,0	8,5	12,0	15,0	18,0
	min.	0,55	0,75	1,25	1,75	2,25	3,2	3,7	5,2	6,64	8,14	11,57	14,57	17,57
d_f	≈	Minor thread diameter												
e	min.	0,803	1,003	1,427	1,73	2,30	2,87	3,44	4,58	5,72	6,86	9,15	11,43	13,72
	nominal	0,7	0,9	1,3	1,5	2,0	2,5	3,0	4,0	5,0	6,0	8,0	10,0	12,0
s	min.	0,711	0,889	1,270	1,520	2,020	2,520	3,020	4,020	5,050	6,020	8,025	10,025	12,032
	max.	0,724	0,902	1,295	1,545	2,045	2,560	3,080	4,095	5,095	6,095	8,115	10,115	12,142
t	min.	0,7	0,8	1,2	1,2	1,5	2,0	2,0	3,0	4,0	4,8	6,4	8,0	10,0
		1,5	1,7	2,0	2,0	2,5	3,0	3,5	5,0	6,0	8,0	10,0	12,0	15,0

FASTENERS

Table 9-59 (*Continued*). Hexagon Socket Set Screws with Flat Point (ISO 4026)

Thread size *d*			M1,6	M2	M2,5	M3	M4	M5	M6	M8	M10	M12	M16	M20	M24
nominal	min.	max.													
2	1,80	2,20													
2,5	2,30	2,70													
3	2,80	3,20													
4	3,76	4,24													
5	4,76	5,24			Range										
6	5,76	6,24													
8	7,71	8,29													
10	9,71	10,29					of								
12	11,65	12,35													
16	15,65	16,35													
20	19,58	20,42							popular						
25	24,58	25,42													
30	29,58	30,42													
35	34,5	35,5											lengths		
40	39,5	40,5													
45	44,5	45,5													
50	49,5	50,5													
55	54,4	55,6													
60	59,4	60,6													

NOTES:
1. All dimensions are in millimeters.
2. Minimum depth of key engagement for screws with nominal lengths above the stepped line, marked thus "-----."
3. Minimum depth of key engagement for screws with nominal lengths below the stepped line, marked thus "-----."
4. P = pitch of the thread.
5. Angle: The 120° angle is mandatory for short length screws above the dotted stepped line.
   The 45° angle applies only to the portion of the thread below the root diameter of the thread. In Table 9-60 the cone angle applies only to the portion of the thread below the root diameter of the thread and should be 120° for lengths above the dotted stepped line. For all other lengths it should be 90°.
6. Screw thread details are shown in Table 8-1. The standard thread tolerance is 5g6g, as shown in the ISO 965 standard or Table 8-4.
7. Mechanical properties Class 45H covered in the ISO 898 V standard. Class 45H screws have the minimum Vickers hardness of 450 HV or the Rockwell hardness of 45 HRC.
8. Product Group A: General tolerances for screws in product Grade A to be covered in a future ISO standard.
9. Surface finish: Black oxide (thermal or chemical).
10. Acceptability: For acceptance procedure see ISO 3269 standard.
11. Designation: Example for the designation of a hexagon socket set screw with flat point, thread size d = M8, nominal length 1 = 25 mm, and property class 45H: hexagon socket set screw, flat point–ISO 4026 M8 × 25 –45H.

Table 9-60. Hexagon Socket Set Screws with Cone Point (ISO 4027)*

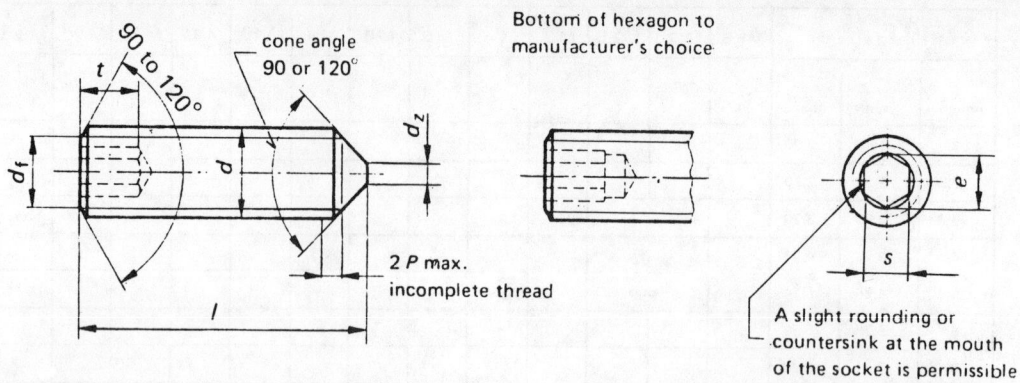

Thread size d		M 1,6	M 2	M 2,5	M 3	M 4	M 5	M 6	M 8	M 10	M 12	M 16	M 20	M 24
P		0,35	0,4	0,45	0,5	0,7	0,8	1,0	1,25	1,5	1,75	2,0	2,5	3,0
d_z	max.	0	0	0	0	0	0	5	2,0	2,5	3,0	4,0	5,0	6,0
d_f		Minor thread diameter												
e	min.	0,803	1,003	427	1,73	2,30	2,87	3,44	4,58	5,72	6,86	9,15	11,43	13,72
s	nominal	0,7	0,9	1,3	1,5	2,0	2,5	3,0	4,0	5,0	6,0	8,0	10,0	12,0
	min.	0,711	0,889	1,270	1,520	2,020	2,520	3,020	4,020	5,020	6,020	8,025	10,025	12,032
	max.	0,724	0,902	1,295	1,545	2,045	2,560	3,080	4,095	5,095	6,095	8,115	10,115	12,142
t	min.	0,7	0,8	1,2	1,2	1,5	2,0	2,0	3,0	4,0	4,8	6,4	8,0	10,0
		1,5	1,7	2,0	2,0	2,5	3,0	3,5	5,0	6,0	8,0	10,0	12,0	15,0

l nominal	min.	max.												
2	1,80	2,20												
2,5	2,30	2,70												
3	2,80	3,20												
4	3,76	4,24												
5	4,76	5,24	Range											
6	5,76	6,24												
8	7,71	8,29												
10	9,71	10,29	of											
12	11,65	12,35												
16	15,65	16,35												
20	19,58	20,42						popular						
25	24,58	25,42												
30	29,58	30,42												
35	34,5	35,5									lengths			
40	39,5	40,5												
45	44,5	45,5												
50	49,5	50,5												
55	54,4	55,6												
60	59,4	60,6												

*See notes, page 9-69.

FASTENERS

Table 9-61. Hexagon Socket Set Screws with Dog Point (ISO 4028)*

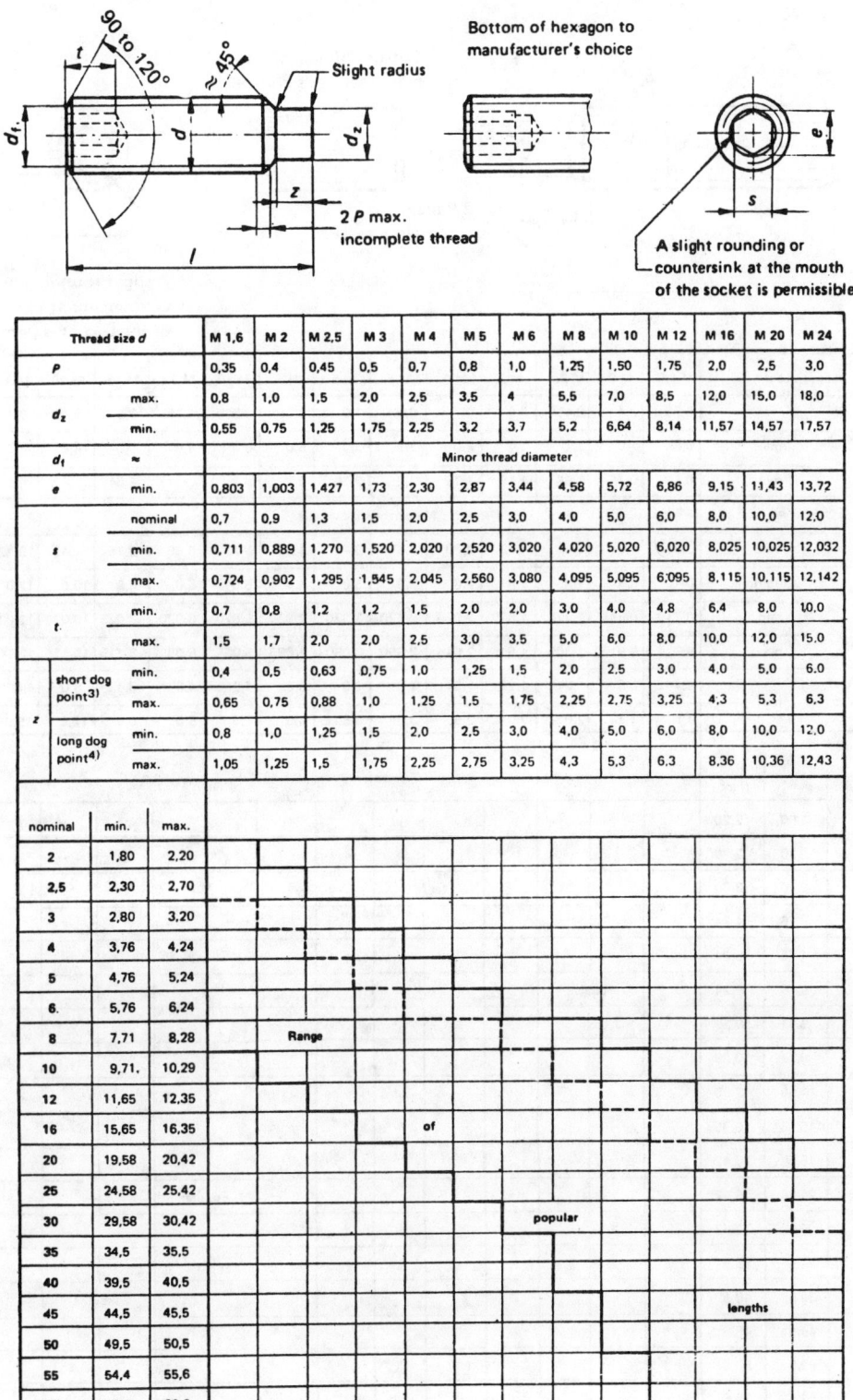

Thread size d		M 1,6	M 2	M 2,5	M 3	M 4	M 5	M 6	M 8	M 10	M 12	M 16	M 20	M 24
P		0,35	0,4	0,45	0,5	0,7	0,8	1,0	1,25	1,50	1,75	2,0	2,5	3,0
d_z	max.	0,8	1,0	1,5	2,0	2,5	3,5	4	5,5	7,0	8,5	12,0	15,0	18,0
	min.	0,55	0,75	1,25	1,75	2,25	3,2	3,7	5,2	6,64	8,14	11,57	14,57	17,57
d_f	≈	Minor thread diameter													
e	min.	0,803	1,003	1,427	1,73	2,30	2,87	3,44	4,58	5,72	6,86	9,15	11,43	13,72
s	nominal	0,7	0,9	1,3	1,5	2,0	2,5	3,0	4,0	5,0	6,0	8,0	10,0	12,0
	min.	0,711	0,889	1,270	1,520	2,020	2,520	3,020	4,020	5,020	6,020	8,025	10,025	12,032
	max.	0,724	0,902	1,295	1,545	2,045	2,560	3,080	4,095	5,095	6,095	8,115	10,115	12,142
t	min.	0,7	0,8	1,2	1,2	1,5	2,0	2,0	3,0	4,0	4,8	6,4	8,0	10,0
	max.	1,5	1,7	2,0	2,0	2,5	3,0	3,5	5,0	6,0	8,0	10,0	12,0	15,0
z short dog point[3]	min.	0,4	0,5	0,63	0,75	1,0	1,25	1,5	2,0	2,5	3,0	4,0	5,0	6,0
	max.	0,65	0,75	0,88	1,0	1,25	1,5	1,75	2,25	2,75	3,25	4,3	5,3	6,3
z long dog point[4]	min.	0,8	1,0	1,25	1,5	2,0	2,5	3,0	4,0	5,0	6,0	8,0	10,0	12,0
	max.	1,05	1,25	1,5	1,75	2,25	2,75	3,25	4,3	5,3	6,3	8,36	10,36	12,43

nominal	min.	max.
2	1,80	2,20
2,5	2,30	2,70
3	2,80	3,20
4	3,76	4,24
5	4,76	5,24
6	5,76	6,24
8	7,71	8,28
10	9,71	10,29
12	11,65	12,35
16	15,65	16,35
20	19,58	20,42
25	24,58	25,42
30	29,58	30,42
35	34,5	35,5
40	39,5	40,5
45	44,5	45,5
50	49,5	50,5
55	54,4	55,6
60	59,4	60,6

Range of popular lengths

Table 9-62. Hexagon Socket Set Screws with Cup Point (ISO 4029)*

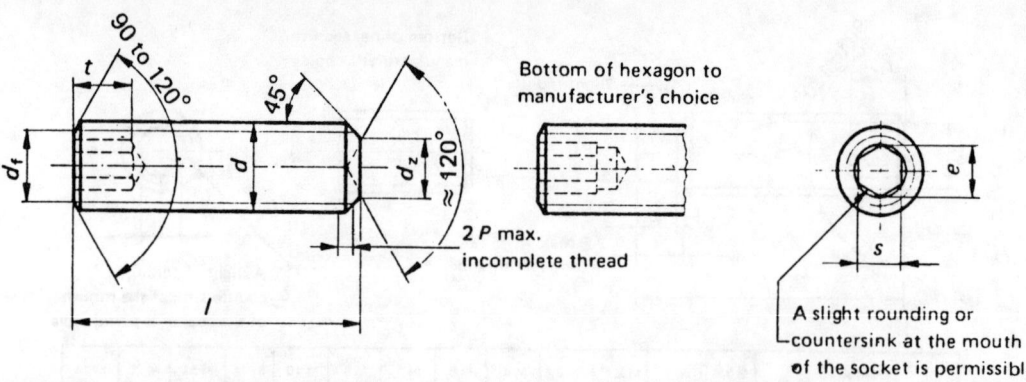

Thread size d			M 1,6	M 2	M 2,5	M 3	M 4	M 5	M 6	M 8	M 10	M 12	M 16	M 20	M 24
p			0,35	0,4	0,45	0,5	0,7	0,8	1,0	1,25	1,5	1,75	2,0	2,5	3
d_z	max.		0,8	1,0	1,2	1,4	2,0	2,5	3,0	5,0	6,0	8,0	10,0	14,0	16,0
	min.		0,55	0,75	0,95	1,15	1,75	2,25	2,75	4,7	5,7	7,64	9,64	13,57	15,57
d_f	$\approx$		Minor thread diameter												
e	min.		0,81	1,01	1,45	1,73	2,30	2,87	3,44	4,58	5,72	6,86	9,15	11,43	13,72
s	nominal		0,7	0,9	1,3	1,5	2,0	2,5	3,0	4,0	5,0	6,0	8,0	10,0	12,0
	min.		0,710	0,888	1,275	1,520	2,020	2,520	3,020	4,020	5,020	6,020	8,025	10,025	12,032
	max.		0,724	0,912	1,300	1,545	2,045	2,560	3,080	4,095	5,095	6,095	8,115	10,115	12,142
t	min.		0,7	0,8	1,2	1,2	1,5	2,0	2,0	3,0	4,0	4,8	6,4	8,0	10,0
			1,5	1,7	2,0	2,0	2,5	3,0	3,5	5,0	6,0	8,0	10,0	12,0	15,0
nominal	min.	max.													
2	1,80	2,20													
2,5	2,30	2,70													
3	2,80	3,20													
4	3,76	4,24													
5	4,76	5,24													
6	5,76	6,24		Range											
8	7,71	8,29													
10	9,71	10,29													
12	11,65	12,35					of								
16	15,65	16,35													
20	19,58	20,42													
25	24,58	25,42										popular			
30	29,58	30,42													
35	34,5	35,5													
40	39,5	40,5											lengths		
45	44,5	45,5													
50	49,5	50,5													
55	54,4	55,6													
60	59,4	60,6													

*See notes, page 9-69.

FASTENERS

Table 9-63. Type Designations of Tapping Screws (IFI-502)

THREAD FORMING		
	AB	
	B	

THREAD CUTTING		
	BF	
	BT	
	F ▲	
	D ▲	

	T ▲	

THREAD ROLLING (REPRESENTATIVE)	
	SF ▲
	SW ▲
	TT ▲
	TR-3 ▲

SELF-DRILLING (REPRESENATIVE)	
	BSD
	CSD ▲

▲ "Hard" metric standard. Fastener type interchangeable with ISO metric screw threads.

9-73

extreme end of the stud to the opposite face of a GO thread ring gage, having the chamfer and/or counterbore removed, which has been assembled by hand as far as the thread will permit.

Total Thread Length, G—The total thread length is the distance, measured parallel to the axis of stud, from the extreme end of the stud to the last scratch on cut threads or to the top of the extrusion angle on rolled threads.

Point—Both ends of the stud shall be pointed. At manufacturer's option, points may be rounded (oval) or flat and chamfered. When rounded, the stud shall have an oval point with a radius equal to approximately one times the basic stud diameter. When flat and chamfered, the end shall be chamfered from a diameter approximately 0.4 mm below the minor diameter of the thread to produce a length of chamfer or incomplete thread equivalent to 1 to 1.5 times the thread pitch.

Material and Mechanical Properties—Carbon steel studs shall conform to the requirements of the applicable property class as covered in the section strength properties for threaded fasteners (page 9-18). Class 4.6 studs are available in sizes M6 through M36; class 9.8 in sizes M6 through M16; class 8.8 in sizes M20 through M36; and class 10.9 in sizes M6 through M36. Studs of other materials such as stainless steel, brass, bronze, and aluminum alloys shall have properties as agreed upon by the manufacturer and purchaser. (For guidance refer to IFI-516 and IFI-518.)

Designation—To avoid possible misunderstanding when specifying double end studs, it is recommended that they be designated in the following sequence: Product type and name; nominal size and thread pitch; stud length; material, including grade identification; and finish (plating or coating) if required.

Example: Type 2 double end stud, M10 × 1.5 × 100, steel class 9.8, zinc plated.

TAPPING SCREWS

There are four basic types of tapping screws: thread-forming, thread-cutting, thread-rolling, and self drilling. Also, there is one type of drive screw—type U. Thread-forming screws (IFI-502), when installed and driven in preformed holes, form a mating internal thread through the displacement of material adjacent to the hole.

The tapping screws described here are based on the IFI-502, IFI-504, and ANSI B18.6.4 standards. The dimensions shown are for "hard metric" or "soft converted" inch fasteners as specified in Tables 9-63 and 9-64. The material presented here is published with permission from the Industrial Fastener Institute, 1505 East Ohio Building, Cleveland, Ohio 44114.

Thread-cutting screws (IFI-502) have cutting edges and chip cavities at their points. When installed and driven in a preformed hole, these screws cut a mating internal thread through removal of material adjacent to the hole.

Thread-rolling screws (IFI-502) have performance capabilities exceeding those of other types of self-tapping screws. When installed and driven in preformed holes, thread rolling screws form a mating internal thread through displacement of material; and because of a special cross sectional design through the threaded section, driving torques are reduced.

Self-drilling screws (IFI-504) have a special point design which permits the screw to drill its own hole through the material to be joined. As the screw is driven in, it forms or cuts a mating internal thread.

Driving screws (ANSI 18.6.4) are driven with a hammer or press into a preformed hole in ferrous and nonferrous materials. The drive screws form their own mating threads. Table 9-63 shows the type designations and illustrates the thread and point design of each tapping screw type. Table 9-64 presents the basic diameters and thread pitches for the various types of threads, and Table 9-65 outlines standard tapping screw lengths specified in the IFI-502 standard.

Table 9-64. Diameters and Thread Pitches for Steel Tapping Screws (IFI-502)

Nominal Size	Thread Types AB, B, BF, BT, BSD Dia × Pitch	Thread Types D, F, T, SW, SF, TT, CSD Dia × Pitch
2	2.2 × 0.79	2 × 0.4
4	2.8 × 1.06	2.5 × 0.45
6	3.5 × 1.27	3 × 0.5
8	4.2 × 1.41	3.5 × 0.6
10	4.8 × 1.59	4 × 0.7
12	5.5 × 1.81	5 × 0.8
14	6.3 × 1.81	6 × 1
16	7.9 × 2.12	8 × 1.25
18	9.5 × 2.12	10 × 1.5
		12 × 1.75

NOTES:

1. All dimensions are in millimeters.
2. Nominal sizes are designated as numbers which are adopted in ISO 1478.

Head Types (IFI-502)

The head types covered by IFI-502 (Industrial Fasteners Institute) include those commonly recognized as being

FASTENERS

Table 9-65. Recommended Tapping Screw Lengths (IFI-502)

Nom Screw Length	Nom Screw Size — Thread Types AB, B, BF, BT										
	2.2	—	2.9	3.5	4.2	4.8	5.5	6.3	8	9.5	—
	Nom Screw Size — Thread Types D, F, T										
	2	2.5	3	3.5	4	5	—	6	8	10	12
3											
4	PH	PH									
5	PH	PH									
6	A	A	PH								
8	A	A	A	PH	PH						
10	A	A	A	A	A	PH					
13	A	A	A	A	A	A	PH	PH			
16		A	A	A	A	A	A	A	PH		
20				A	A	A	A	A	A	PH	PH
25				A	A	A	A	A	A	A	A
30						A	A	A	A	A	A
35						A	A	A	A	A	A
40							A	A	A	A	A
45									A	A	A
50									A	A	A
55										A	A
60										A	A

NOTES:
1. Lengths included between the heavy lines are recommended. "PH" means the length is recommended only for pan, hex and hex washer head screws. "A" means the length is recommended for all head styles. When shorter length screws are required, refer to Tables 9-73 and 9-74 for guidance on minimum practical lengths.
2. All dimensions are in millimeters.

applicable to tapping screws and are described as follows:

Flat Countersunk Head—The flat countersunk head has a flat top surface and a conical bearing surface with a head angle of approximately 90 degrees.

Oval Trim Countersunk Head—The oval countersunk head shall have a rounded top surface and a conical bearing surface with a head angle of approximately 90 degrees.

Pan Head—The slotted pan head has a flat top surface rounded into cylindrical sides and a flat bearing surface. The recessed pan head has a rounded top surface blending into cylindrical sides and a flat bearing surface.

Hex Head—The hex head has a flat or indented top surface, six flat sides, and a flat bearing surface.

Hex Washer Head—The hex washer head has an indented top surface and six flat sides formed integrally with a flat washer which projects beyond the sides and provides a flat bearing surface.

Drive Types—The head styles shown in Table 9-72 transmitting the highest torques are the hexagon and the hexagon washer heads. Cross recess screws of Type Pz(Pozidriv) provide the second best screw drive condition followed by cross-recess Type Ph (Phillips)[1] and slotted heads. Tests conducted by a leading U.S. fastener producer concluded the following:

1. End-load pressure for Type Pz (Pozidriv)[1] is less than 50 percent of that required with Type Ph (Phillips) drive screws.
2. Maximum torque values achieved before cam-out with Type Pz (Pozidriv) screws are approximately three times those reached with Type Ph (Phillips) cross-recessed screws.

Length. Nominal lengths are as follows:

MEASUREMENT. The length of the screw is measured parallel to the axis of the screw from the largest diameter of the bearing surface of the head to the extreme point. Recommended lengths of tapping screws are given in Table 9-65.

---
[1] Pz is new ISO designation for Pozidriv, Ph is new ISO designation for Phillips drive. The ANSI designation for Phillips cross recess is Type I and for Pozidriv is Type IA.

Table 9-66. Slotted Flat Countersunk Head Tapping Screws (IFI-502)

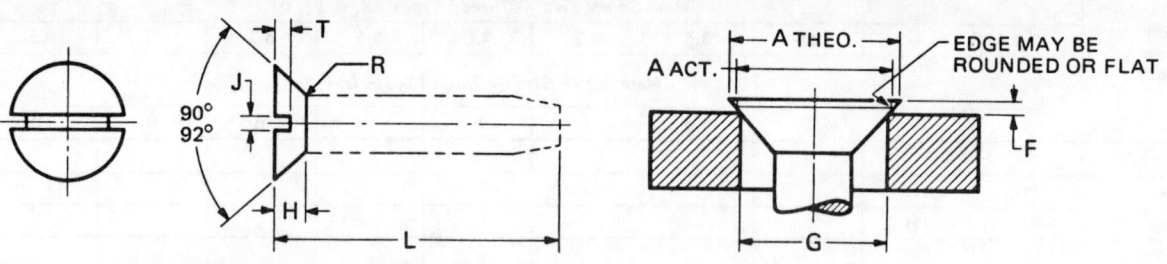

Nom Screw Size		A Head Diameter			H Head Height		R Fillet Radius		J Slot Width		T Slot Depth		F Protrusion Above Gaging Dia		G Gaging Dia
Thread Types AB, B, BF, BT	Thread Types D, F, T	Theoretical Sharp		Actual	Types AB, B BF, BT	Types D, F T									
		Max	Min	Min	Max	Ref	Max	Min	Max	Min	Max	Min	Max	Min	
2.2	2	4.40	3.90	3.60	1.11	1.20	0.8	0.2	0.7	0.5	0.6	0.4	0.79	0.52	2.82
—	2.5	5.50	4.90	4.60	—	1.50	1.0	0.3	0.8	0.6	0.7	0.5	0.88	0.56	3.74
2.9	3	6.60	5.80	5.50	1.88	1.80	1.2	0.3	1.0	0.8	0.9	0.6	0.98	0.55	4.65
3.5	3.5	7.70	6.80	6.44	2.10	2.10	1.4	0.4	1.2	1.0	1.0	0.7	1.07	0.59	5.57
4.2	4	8.65	7.80	7.44	2.24	2.32	1.6	0.4	1.4	1.2	1.1	0.8	1.09	0.63	6.48
4.8	5	10.70	9.80	9.44	2.94	2.85	2.0	0.5	1.5	1.2	1.4	1.0	1.20	0.71	8.31
5.5	—	11.80	10.90	10.27	3.16	—	2.2	0.6	1.9	1.6	1.6	1.1	1.29	0.80	9.23
6.3	6	13.50	12.30	11.87	3.58	3.60	2.5	0.6	1.9	1.6	1.8	1.3	1.41	0.77	10.69
8	8	16.80	15.60	15.17	4.43	4.40	3.2	0.8	2.3	2.0	2.1	1.6	1.50	0.86	13.80
9.5	10	20.70	19.50	18.98	5.59	5.35	4.0	1.0	2.8	2.5	2.6	2.0	2.58	1.91	15.54
—	12	24.70	23.50	22.88	—	6.35	4.8	1.2	2.8	2.5	3.1	2.5	2.79	2.11	19.12

NOTES:
1. All dimensions are in millimeters.
2. See Table 9-73 and 9-74 for thread point dimensions and minimum practical screw lengths.
3. See Table 9-65 for recommended screw lengths.

TOLERANCE ON LENGTH. The tolerance on length of tapping screws and metallic drive screws must conform to the following:

TAPPING SCREWS, TYPE AB. The tolerance on length is as tabulated below:

(IFI-502)

NOMINAL SCREW LENGTH	TOLERANCE ON LENGTH
Up to 25 mm Incl.	±0.8 mm
Over 25 mm	∓1.3 mm

TAPPING SCREWS, TYPES B, BF, BT, D, F, AND T. The tolerance on length is as tabulated below:

(IFI-502)

NOMINAL SCREW LENGTH	TOLERANCE ON LENGTH
Up to 20 mm Incl.	−0.8 mm
Over 20 mm to 40 mm	−1.35 mm
Over 40 mm	−1.5 mm

Finish. Screws may be furnished plain or with a protective coating (electrodeposited, mechanical plating, or chemical conversion coating) as specified by the user. At the option of the manufacturer screws may be provided with an additional supplementary lubricant as necessary to meet the performance requirements.

THREAD ROLLING. Screws are cadmium or zinc electroplated with a coating thickness of 0.005 to 0.010 mm or have a zinc phosphate and oil coating, as specified by the purchaser. At the option of the manufacturer, screws may be provided with an additional supplementary lubricant as necessary to meet the performance requirements. Electroplated screws must be baked for a minimum of one hour within the temperature range 190°–230°C as soon as practicable after plating to avoid hydrogen embrittlement.

Screw Threads

Thread Forming Tapping Screws. Thread-forming tapping screws are generally used for application in materials where large internal stresses are permissible. These greater stresses are often desired to increase resistance to loosening. The thread forming screw threads are of the following types:

FASTENERS

TYPE AB. Type AB tapping screws have spread threads, with the same pitches as Type B, and a gimlet point. They are primarily intended for use in thin metal, resin-impregnated plywood, and asbestos compositions.

TYPE B. Type B tapping screws have spaced threads and a blunt point with incomplete entering threads. They are intended for use in material such as thin metal, nonferrous castings, plastics, resin impregnated plywood and asbestos compositions.

Thread-Cutting and Rolling Tapping Screws. Thread-cutting tapping screws are generally for application in materials where disruptive internal stresses are undesirable or where excessive driving torques are encountered with thread-forming screws. Thread-rolling (high performance) screws are for applications requiring high clamp loads (similar to those of ISO grade 10.9 fasteners). They are of the following types:

TYPES BF AND BT. Types BF and BT tapping screws should have spaced threads with a blunt point and tapered entering threads as on Type B, with one or more cutting edges and chip cavities. These screws are intended for use in plastics, asbestos, and other similar compositions.

TYPES D, E, T, SF, SW, TR-3, TT, and CSD. Screws of these types have threads of machine screw diameter-pitch combination approximating a 60° basic thread form (not necessarily either ISO or OMFS thread profile) with a blunt point and tapered entering threads having one or more cutting edges and chip cavities (except for Types SF, SW, and TT). Tapping screws are not subject to thread gaging but must meet dimensions specified in the following Tables. These screws are intended for use in materials such as aluminum, zinc, and lead die castings, steel sheets, and shapes, cast iron, brass, plastics, weld nuts, etc.

Thread Lengths. Tapping screws must have thread lengths conforming to the following:

TYPES AB, B, BF, and BT. For screws of nominal lengths equal to or shorter than those shown in Column L of Table 9-77A, the full form threads shall extend close to the head such that the specified thread minor diameter limits are maintained to within Y distance from the under-

Table 9-67. Cross-Recess Dimensions of Flat Countersunk Head Tapping Screws (IFI-502)

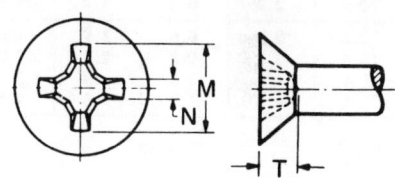

TYPE I

This type of recess has a large center opening, tapered wings, and blunt bottom, with all edges relieved or rounded.

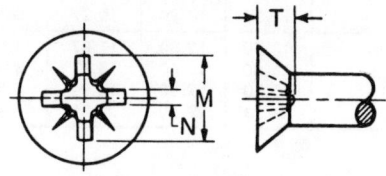

TYPE IA

This type of recess has a large center opening, wide straight wings, and blunt bottom, with all edges relieved or rounded.

Nom Screw Size		Type I							Type IA							
Thread Types AB, B, BF, BT	Thread Types D, F, T	M Recess Dia		T Recess Depth		N Recess Width	Driver Size	Recess Penetration Gaging Depth		M Recess Dia		T Recess Depth		N Recess Width	Driver Size	Recess Penetration Gaging Depth	
		Max	Min	Max	Min	Min		Max	Min	Max	Min	Max	Min	Min		Max	Min
2.2	2	1.96	1.63	1.30	0.89	0.38	0	1.12	0.71	2.39	2.06	1.75	1.35	0.46	0	1.57	1.17
—	2.5	2.97	2.64	1.98	1.57	0.46	1	1.80	1.40	2.97	2.64	1.98	1.57	0.74	1	1.73	1.32
2.9	3	3.25	2.92	2.26	1.85	0.46	1	2.08	1.68	3.25	2.92	2.26	1.85	0.76	1	2.01	1.60
3.5	3.5	4.17	3.84	2.44	1.85	0.71	2	2.16	1.57	4.17	3.84	2.46	2.01	1.04	2	2.06	1.60
4.2	4	4.42	4.09	2.69	2.11	0.74	2	2.41	1.83	4.42	4.09	2.72	2.26	1.04	2	2.31	1.85
4.8	5	4.62	4.29	2.90	2.31	0.76	2	2.62	2.03	4.62	4.29	2.90	2.44	1.04	2	2.51	2.05
5.5	—	5.92	5.59	3.07	2.49	0.76	3	2.64	2.06	5.92	5.59	3.07	2.62	1.40	3	2.54	2.08
6.3	6	6.35	6.02	3:45	2.87	0.81	3	3.02	2.44	6.30	5.97	3.48	3.02	1.42	3	2.92	2.46
8	8	8.69	8.36	4.90	4.34	1.42	4	4.39	3.84	8.69	8.36	4.98	4.52	2.16	4	4.32	3.86
9.5	10	9.60	9.27	5.84	5.28	1.57	4	5.33	4.78	9.60	9.27	5.92	5.46	2.18	4	5.23	4.77
—	12	10.39	10.06	6.63	6.07	1.73	4	6.12	5.56	10.39	10.06	6.73	6.27	2.18	4	6.05	5.59

NOTES:
1. All dimensions are in millimeters.
2. The ISO designations for Phillips cross recess is Type Ph and for Pozidriv Type Pz.

Table 9-68. Slotted Oval Countersunk Head Tapping Screws (IFI-502)

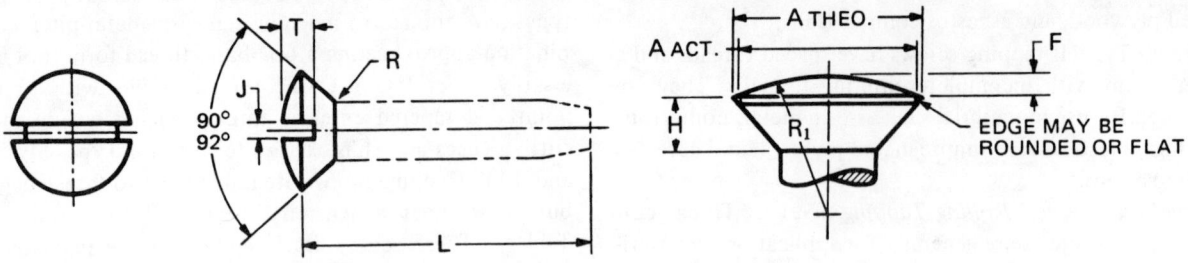

Nom Screw Size		A Head Diameter			H Head Side Height		F Raised Head Height	R₁ Head Radius	R Fillet Radius		J Slot Width		T Slot Depth	
Thread Types AB, B, BF, BT	Thread Types D, F, T	Theoretical Sharp		Actual	Types AB, B BF, BT	Types D, F T								
		Max	Min	Min	Max	Ref	Max	Approx	Max	Min	Max	Min	Max	Min
2.2	2	4.40	3.90	3.60	1.11	1.20	0.50	3.8	0.8	0.2	0.7	0.5	1.0	0.8
—	2.5	5.50	4.90	4.60	—	1.50	0.60	5.0	1.0	0.3	0.8	0.6	1.2	1.0
2.9	3	6.60	5.80	5.50	1.88	1.80	0.75	5.7	1.2	0.3	1.0	0.8	1.5	1.2
3.5	3.5	7.70	6.80	6.44	2.10	2.10	0.90	6.5	1.4	0.4	1.2	1.0	1.7	1.4
4.2	4	8.65	7.80	7.44	2.24	2.32	1.00	7.8	1.6	0.4	1.4	1.2	1.9	1.6
4.8	5	10.70	9.80	9.44	2.94	2.85	1.25	9.9	2.0	0.5	1.5	1.2	2.3	2.0
5.5	—	11.80	10.90	10.27	3.16	—	1.40	10.5	2.2	0.6	1.9	1.6	2.5	2.2
6.3	6	13.50	12.30	11.87	3.58	3.60	1.60	12.2	2.5	0.6	1.9	1.6	3.0	2.6
8	8	16.80	15.60	15.17	4.43	4.40	2.00	15.8	3.2	0.8	2.3	2.0	3.7	3.2
9.5	10	20.70	19.50	18.98	5.59	5.35	2.50	19.8	4.0	1.0	2.8	2.5	4.5	4.0
—	12	24.70	23.50	22.88	—	6.35	3.00	23.8	4.8	1.2	2.8	2.5	5.3	4.8

NOTES:
1. All dimensions are in millimeters.
2. See Tables 9-73 and 9-74 for thread and point dimensions and minimum practical screw lengths.
3. See Table 9-65 for recommended screw lengths.

side of the head, or closer if practicable. See figure in Table 9-77A. Screws of longer nominal lengths than those tabulated, unless otherwise specified by the purchaser, shall have a minimum length of full form thread as shown in Column L$_T$ of Table 9-77A.

TYPES D, F, T, SF, SW, TR-3, and TT. For screws of nominal lengths within the ranges listed under Column Y of Table 9-77B, the full form threads shall extend close to the head such that the specified thread major diameter limits are maintained to within the respective Y distance from the underside of the head, or closer if practicable. See figure in Table 9-77B. Screws of longer nominal lengths, unless otherwise specified by the purchaser, shall have a minimum length of full form thread as specified in Column L$_T$.

TYPE BSD SCREWS. For screws of nominal lengths equal to or shorter than 40 mm, the full form threads shall extend close to the head so that the specified minor diameter limits are maintained to within one pitch (thread), or closer if practicable, of the underside of the head. See the figure in Table 9-77A. For screws of nominal lengths longer than 40 mm the length of full form thread is as specified by the purchaser.

TYPE CSD SCREWS. For screws of nominal lengths equal to or shorter than 40 mm, the full form threads shall extend close to the head so that the specified major diameter limits are maintained to within two pitches (threads), or closer if practicable of the underside of the head. See the figure in Table 9-77B. For screws of nominal lengths longer than 40 mm the length of full form thread is as specified by the purchaser.

Strength Grades—For tapping screws, thread cutting screws, and self-drilling and drive screws do not follow the ISO system. As a guide, tapping and self drilling screws can be torqued to produce an ISO grade 9.8 clamping load, and the thread-rolling (high performance) screws a 10.9 clamping load.

ASSEMBLY CONSIDERATIONS. The finish (plating or coating) on tapping screws and the material composition and

FASTENERS

hardness of the mating components are factors which affect assembly torques in individual applications. It should be noted that, because of various finishes providing different degrees of lubricity, some adjustment of installation torques may be necessary to suit individual applications. Also, where exceptionally heavy finishes are involved or screws are to be assembled into materials of higher hardness, some deviation in hole sizes may be required to provide optimum assembly. The necessity and extent of such deviations can best be determined by experiment in the particular assembly environment.

TYPE BSD AND CSD SCREWS

Screw Selection Chart—Table 9-78 represents a screw selection chart recommending panel thicknesses which can be fastened with various screw types and sizes.
BITS AND SOCKETS. Magnetic bits and sockets are not recommended for sealing applications or when material thickness is near the maximum drilling limit of the particular screw (see Table 9-78) because of possible chip collection around the area being sealed and in the socket.
DRIVING TECHNIQUE. End pressures of 110 to 290 N are required for efficient drilling of self-drilling tapping screws. Excessive pressure, especially during the initial penetration of the drill point, increases the amount of material being removed by the flutes and could result in drilling torques in excess of the strength of the drill point.

Designation—Tapping screws shall be designated by the following data in the sequence shown: Nominal size; thread pitch; nominal length; thread and point type; product name, including head type and driving provision; material; and protective finish, if required. See examples below:

5.5 × 1.81 × 30 Type AB, slotted pan head tapping screw, steel.

3.5 × 1.27 × 20 Type B, Type 1A cross recessed oval countersunk head tapping screw, corrosion resistant steel.

8 × 1.25 × 40 Type F, hexagon washer head tapping screw, steel.

Table 9-69. Cross-Recess Dimensions of Oval Countersunk Head Tapping Screws (IFI-502)

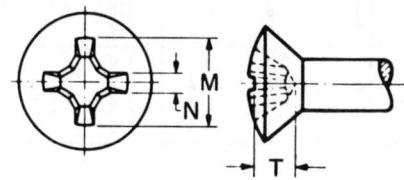

TYPE I

This type of recess has a large center opening, tapered wings, and blunt bottom, with all edges relieved or rounded.

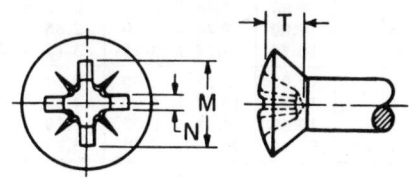

TYPE 1A

This type of recess has a large center opening, wide straight wings, and blunt bottom, with all edges relieved or rounded.

Nom Screw Size		Type I							Type 1A						
Thread Types AB, B, BF, BT	Thread Types D, F, T	M Recess Dia		T Recess Depth		N Recess Width	Driver Size	Recess Penetration Gaging Depth		M Recess Dia		T Recess Depth		N Recess Width	Driver Size	Recess Penetration Gaging Depth	
		Max	Min	Max	Min	Min		Max	Min	Max	Min	Max	Min	Min		Max	Min
2.2	2	1.96	1.63	1.22	0.76	0.38	0	1.04	0.58	2.41	2.08	1.68	1.27	0.47	0	1.52	1.12
—	2.5	2.82	2.49	1.73	1.32	0.48	1	1.55	1.14	3.00	2.67	2.01	1.60	0.74	1	1.75	1.35
2.9	3	3.18	2.84	2.11	1.68	0.48	1	1.93	1.50	3.33	3.00	2.34	1.93	0.76	1	2.08	1.68
3.5	3.5	4.52	4.19	2.67	2.03	0.76	2	2.39	1.75	4.52	4.19	2.77	2.31	1.04	2	2.36	1.91
4.2	4	4.88	4.55	3.02	2.41	0.79	2	2.74	2.13	4.88	4.55	3.15	2.69	1.04	2	2.74	2.29
4.8	5	5.31	4.98	3.48	2.87	0.84	2	3.20	2.59	5.49	5.16	3.76	3.30	1.04	2	3.35	2.90
5.5	—	6.86	6.53	3.86	3.25	0.97	3	3.43	2.82	6.86	6.53	3.99	3.53	1.42	3	3.43	2.97
6.3	6	7.37	7.04	4.39	3.76	1.02	3	3.96	3.33	7.19	6.86	4.32	3.86	1.45	3	3.76	3.30
8	8	8.89	8.56	4.90	4.29	1.50	4	4.39	3.78	8.99	8.66	5.21	4.75	2.13	4	4.52	4.06
9.5	10	10.16	9.83	6.22	5.61	1.67	4	5.72	5.11	10.13	9.80	6.43	5.97	2.18	4	5.74	5.28
—	12	11.89	11.56	8.03	7.42	1.96	4	7.52	6.91	11.96	11.63	8.23	7.77	2.21	4	7.54	7.08

NOTES:
1. All dimensions are in millimeters.
2. The ISO designation to Pozidriv cross recess is Type Pz and for Phillips, Type Ph.

Table 9-70. Slotted and Recessed Pan Head Tapping Screws (IFI-502)

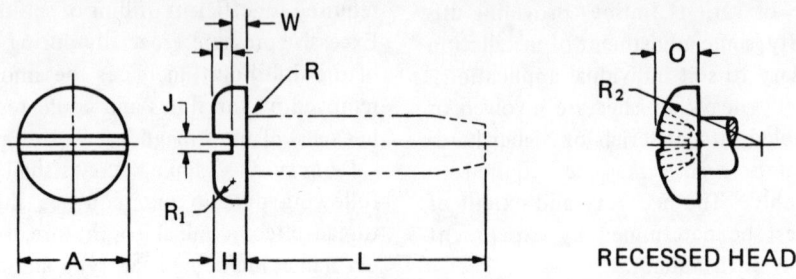

Nom Screw Size		A		H		O		R₁	R₂	R		J		T	W
Thread Types AB, B, BF, BT	Thread Types D, F, T	Head Diameter		Head Height Slotted Head		Head Height Recessed Head		Head Radius (Slotted)	Head Radius (Recessed)	Fillet Radius		Slot Width		Slot Depth	Unslotted Thickness
		Max	Min	Max	Min	Max	Min	Max	Ref	Max	Min	Max	Min	Min	Min
2.2	2	3.90	3.60	1.35	1.15	1.60	1.40	0.8	4	0.3	0.1	0.7	0.5	0.55	0.44
—	2.5	4.90	4.60	1.65	1.45	1.95	1.75	1.0	5	0.4	0.1	0.8	0.6	0.73	0.55
2.9	3	5.80	5.50	1.90	1.65	2.30	2.05	1.2	6	0.5	0.2	1.0	0.8	0.80	0.66
3.5	3.5	6.80	6.44	2.25	2.00	2.50	2.25	1.4	7	0.5	0.2	1.2	1.0	0.95	0.77
4.2	4	7.80	7.44	2.55	2.30	2.80	2.55	1.6	8	0.6	0.2	1.4	1.2	1.15	0.88
4.8	5	9.80	9.44	3.10	2.85	3.50	3.25	2.0	10	0.8	0.3	1.5	1.2	1.35	1.10
5.5	—	10.70	10.27	3.45	3.15	3.85	3.55	2.2	11	0.8	0.3	1.9	1.6	1.55	1.25
6.3	6	12.00	11.57	3.90	3.50	4.30	4.00	2.5	13	1.0	0.3	1.9	1.6	1.70	1.36
8	8	15.60	15.17	5.00	4.60	5.60	5.20	3.2	16	1.2	0.4	2.3	2.0	2.20	1.76
9.5	10	19.50	18.98	6.20	5.70	7.00	6.50	4.0	20	1.5	0.5	2.8	2.5	2.70	2.20
—	12	23.40	22.88	7.50	6.90	8.30	7.80	4.8	24	1.8	0.6	2.8	2.5	3.20	2.70

NOTES:
1. All dimensions are in millimeters.
2. The ISO designation for Pozidriv cross recess is Type Pz and for Phillips, Type Ph.
3. See Table 9-65 for recommended screw lengths, and Table 9-73 and 9-74 for thread and point dimensions.

HEXAGON NUTS (IFI-507)

National and international standards for nuts are shown in the world fastener index, page 9-8. The material presented here was based on the IFI-507 standard, with permission from Industrial Fasteners Institute, 1505 East Ohio Building, Cleveland, Ohio 44114.

Three widths across flats (17 to IFI 15 mm; 19 to IFI 18 mm; 22 to IFI 21 mm) were changed from past practices for practical utilization of the material.

The design of Style 1 and 2 nuts is based on providing sufficient nut strength to reduce the possibility of thread stripping rather than bolt or screw fracture as the failure mode in an over-tightened or over-stressed assembly. Nut proof loads (the axial load the nut can support without evidence of failure) were established equal to or higher than the minimum specified tensile strength of the highest strength property class of bolt or screw with which the nut would normally be assembled.

There are three standard property classes for hex nuts. Class 5 nuts are non-heat treated, have Style 1 dimensions, and are designed to use with bolts and screws of ISO 898 II property class 5.8 and lower strengths; class 9 nuts are non-heat treated, have Style 2 dimensions, and are designed for use with bolts and screws of property classes 9.8, 8.8, and lower strengths; class 10 nuts are heat treated, have Style 1 dimensions, and are designed for use with bolts and screws of property class 10.9 and lower strengths. Dimensions for nuts in nominal thread sizes from M1.6 to M36 are shown in Table 9-79.

HEXAGON STEEL LOCKNUTS (IFI-514)

The ISO 2320 standard covers mechanical and performance properties for steel locknuts, and the ISO 2358 standard specifies dimensions for steel locknuts. Both these standards were based on the IFI-100 inch standard.

FASTENERS

The IFI-514 standard, which has been used here, is for the most part in agreement with the ISO standards.

Property Classes—The three property classes of nuts are designated by the numbers 5, 9, and 10, and are normally available in the sizes specified in Table 9-80. The property classes of bolts, screws, and studs suitable for use with each property class of nut are also shown in the same table.

Basic Dimensions—Dimensions for nuts are given in Table 9-79. Classes 5 and 10 nuts shall conform to the dimensions of Style 1 nuts, and class 9 shall conform to the dimensions of Style 2 nuts. The portion of the nut containing the prevailing-torque feature may have a special contour within the maximum permitted width across flats and thickness. The minimum width across flats shall not apply at depressed portion of nut at prevailing-torque feature.

Proof Load—Nuts shall withstand the proof loads specified in Table 9-82 for the applicable class.

Hardness—Nuts shall have a hardness conforming to the limits specified for the applicable class in Table 9-80.

Material—Nuts shall be made of carbon or alloy steel of a grade adequate for the nut to meet the mechanical and performance requirements of this recommendation. The prevailing-torque element of insert type nuts may be of a material other than steel.

Heat Treatment—Class 5 nuts need not be heat treated. Class 9 nuts may be heat treated as necessary, and class 10 nuts shall be heat treated to meet the mechanical and performance requirements of this recommendation. Case hardening is not allowed for any property class.

Finish—Nuts may be furnished plain (bare metal) or with a protective coating (electrodeposited plating or chemical conversion coating) as specified by the user.

Table 9-71. Cross-Recess Dimensions of Pan Head Tapping Screws (IFI-502)

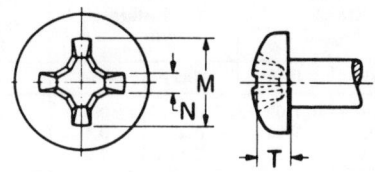

TYPE I

This type of recess has a large center opening, tapered wings, and blunt bottom, with all edges relieved or rounded.

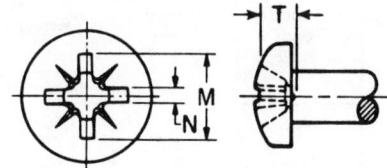

TYPE IA

This type of recess has a large center opening, wide straight wings, and blunt bottom, with all edges relieved or rounded.

Nom Screw Size		Type I							Type 1A							
Thread Types AB, B, BF, BT	Thread Types D, F, T	M Recess Dia		T Recess Depth		N Recess Width	Driver Size	Recess Penetration Gaging Depth		M Recess Dia		T Recess Depth		N Recess Width	Driver Size	Recess Penetration Gaging Depth	
		Max	Min	Max	Min	Min		Max	Min	Max	Min	Max	Min	Min		Max	Min
2.2	2	1.88	1.55	1.14	0.64	0.36	0	1.02	0.56	2.31	1.98	1.60	1.19	0.46	0	1.42	1.02
—	2.5	2.84	2.51	1.73	1.27	0.48	1	1.55	1.09	3.10	2.77	2.03	1.63	0.74	1	1.78	1.37
2.9	3	3.10	2.77	1.98	1.52	0.48	1	1.80	1.35	3.35	3.02	2.31	1.91	0.74	1	2.06	1.65
3.5	3.5	4.22	3.89	2.31	1.68	0.71	2	2.03	1.40	4.11	3.78	2.34	1.88	1.02	2	1.93	1.47
4.2	4	4.62	4.29	2.74	2.08	0.76	2	2.46	1.80	4.50	4.17	2.74	2.29	1.04	2	2.34	1.88
4.8	5	5.05	4.72	3.15	2.54	0.79	2	2.87	2.26	4.90	4.57	3.15	2.69	1.04	2	2.74	2.29
5.5	—	6.58	6.25	3.58	2.92	0.86	3	3.15	2.49	6.45	6.12	3.53	3.07	1.42	3	2.97	2.51
6.3	6	7.14	6.81	4.09	3.43	0.91	3	3.66	3.00	6.93	6.60	4.04	3.58	1.45	3	3.48	3.02
8	8	8.89	8.56	4.90	4.29	1.50	4	4.39	3.78	8.66	8.33	4.85	4.39	2.18	4	4.17	3.71
9.5	10	10.49	10.16	6.58	5.94	1.73	4	6.07	5.44	10.13	9.80	6.38	5.92	2.18	4	5.69	5.23
—	12	11.05	10.72	7.11	6.48	1.80	4	6.60	5.97	10.67	10.34	6.93	6.48	2.18	4	6.25	5.79

NOTES:
1. All dimensions are in millimeters.
2. Head dimensions not shown are the same as those of slotted heads given in Table 9-70.
3. The ISO designation for Phillips cross recess is Type Ph and for Pozidiv Type Pz.

Table 9-72A. Hex Head Tapping Screws (IFI-502)

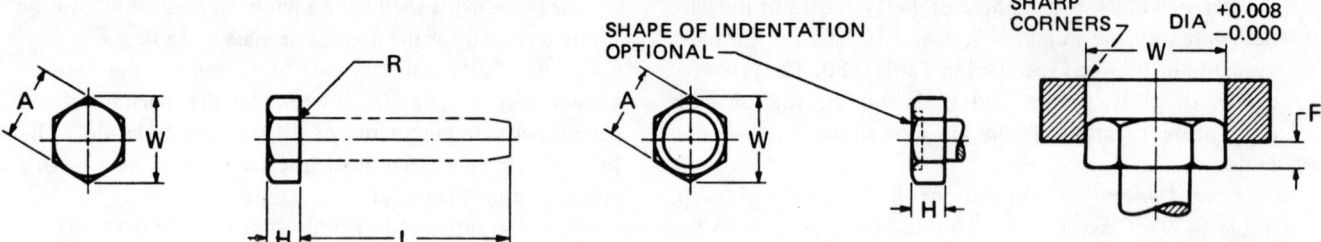

TRIMMED HEAD OR FULLY UPSET HEAD   INDENTED HEAD

Nom Screw Size		A		W	H		R		F
Thread Types AB, B, BF, BT	Thread Types D, F, T	Width Across Flats		Width Across Corners	Head Height		Fillet Radius		Protrusion Beyond Gaging Ring
		Max	Min	Min	Max	Min	Max	Min	Min
2.2	2	3.20	3.02	3.36	1.27	1.02	0.3	0.1	0.61
—	2.5	4.00	3.82	4.25	1.40	1.12	0.4	0.1	0.67
2.9	3	5.00	4.82	5.36	1.52	1.24	0.5	0.2	0.74
3.5	3.5	5.50	5.32	5.92	2.36	2.03	0.5	0.2	1.22
4.2	4	7.00	6.78	7.55	2.79	2.44	0.6	0.2	1.46
4.8	5	8.00	7.78	8.66	3.05	2.67	0.8	0.3	1.60
5.5	—	8.00	7.78	8.66	3.94	3.53	0.8	0.3	2.12
6.3	6	10.00	9.78	10.89	4.83	4.37	1.0	0.3	2.62
8	8	13.00	12.73	14.17	5.84	5.28	1.2	0.4	3.17
9.5	10	15.00	14.73	16.41	7.49	6.86	1.5	0.5	4.12
—	12	18.00	17.73	19.75	9.50	8.66	1.8	0.6	5.20

NOTES:
1. All dimensions are in millimeters.
2. See Table 9-73 and 9-74 for thread and point dimensions and minimum practical screw lengths.
3. A slight rounding of all edges and corners of the hex surfaces is permissible.
4. Dimensions across flats and across corners of the head are measured at the point of maximum metal. Taper of sides of hex (angle between one side and the axis) must not exceed 2 degrees or 0.10 mm, whichever is greater. The specified width across flats is the large dimension.
5. The rounding due to lack of fill on all six corners of the head shall be reasonably uniform, and the width across corners of the head shall be such that when a sharp ring having an inside diameter equal to the specified minimum width across corners is placed on the top and bottom of the head, the head shall protrude by an amount equal to, or greater than, the F value tabulated. For across corners gaging see Appendix B of IFI-502.
6. See note, page 9-41.

FASTENERS

Table 9-72B. Hex Washer Head Tapping Screws (IFI-502)

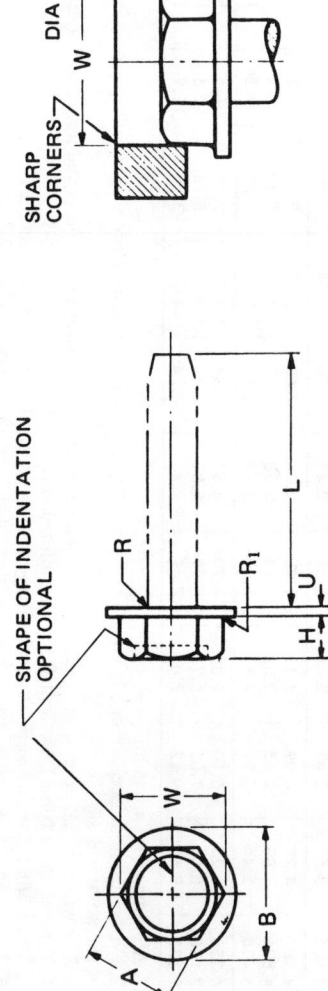

Nom Screw Size		A Width Across Flats		W Width Across Corners		H Head Height		B Washer Diameter		U Washer Thickness		R Fillet Radius		R₁ Fillet Radius		F Protrusion Beyond Gaging Ring
Thread Types AB, B, BF, BT	Thread Types D, F, T	Max.	Min	Max	Min	Max	Min	Max	Min	Max	Min	Max	Min	Max	Min	Min
2.2	2	3.20	3.02	3.36		1.27	1.02	4.2	3.9	0.4	0.2	0.3	0.1	0.4		0.61
—	2.5	4.00	3.82	4.25		1.40	1.12	5.3	5.0	0.5	0.3	0.4	0.1	0.5		0.67
2.9	3	5.00	4.82	5.36		1.52	1.24	6.2	5.8	0.5	0.3	0.5	0.2	0.5		0.74
3.5	3.5	5.50	5.32	5.92		2.36	2.03	7.5	6.8	0.6	0.4	0.5	0.2	0.6		1.22
4.2	4	7.00	6.78	7.55		2.79	2.44	9.2	8.5	0.7	0.5	0.6	0.2	0.7		1.46
4.8	5	8.00	7.78	8.66		3.05	2.67	10.5	9.8	0.8	0.5	0.8	0.3	0.8		1.60
5.5	—	8.00	7.78	8.66		3.94	3.53	10.5	9.8	0.8	0.5	0.8	0.3	0.8		2.12
6.3	6	10.00	9.78	10.89		4.83	4.37	13.2	12.2	1.3	0.8	1.0	0.3	1.0		2.62
8	8	13.00	12.73	14.17		5.84	5.28	17.2	15.9	1.4	0.9	1.2	0.4	1.2		3.17
9.5	10	15.00	14.73	16.41		7.49	6.86	19.8	18.3	1.6	0.9	1.5	0.5	1.5		4.12
—	12	18.00	17.73	19.75		9.50	8.66	23.8	22.0	1.8	1.1	1.8	0.6	1.8		5.20

NOTES:
1. All dimensions are in millimeters.
2. See Table 9-73 and 9-74 for thread and point dimensions and minimum practical screw lengths.
3. A slight rounding of all edges and corners of the hex surfaces is permissible.
4. Dimensions across flats and across corners of the head are measured at the point of maximum metal. Taper of sides of hex (angle between one side and the axis) must not exceed 2 degrees or 0.10 mm, whichever is greater. The specified width across flats is the large dimension.
5. The rounding due to lack of fill on all six corners of the head shall be reasonably uniform, and the width across corners of the head shall be such that when a sharp ring having an inside diameter equal to the specified minimum width across corners is placed on the top and bottom of the head, the head shall protrude by an amount equal to, or greater than, the F value tabulated. For across corners gaging see Appendix B of IFI-502.
6. See note, page 9-41.

Table 9-73. Threads and Points for Types AB and B Thread-Forming and Types BF and BT Thread-Cutting Tapping Screws (IFI-502)

Nom Screw Size and Thread Pitch	Basic Screw Dia	D Major Diameter Max	D Major Diameter Min	d Minor Diameter Max	d Minor Diameter Min	V Point Diameter Max	V Point Diameter Min	S Point Taper Length For Types B, BF, BT Max	S Point Taper Length For Types B, BF, BT Min	Z Point Length For Type AB Ref	L Min Type AB Pan, Hex and Hex Washer Heads	L Min Type AB Flat and Oval CTSK Heads	L Min Types B, BF, BT Pan, Hex and Hex Washer Heads	L Min Types B, BF, BT Flat and Oval CTSK Heads
	Ref													
2.2x0.79	2.184	2.24	2.13	1.63	1.52	1.47	1.37	1.57	1.19	3.0	4	6	4	5
2.9x1.06	2.845	2.90	2.79	2.18	2.08	2.01	1.88	2.11	1.60	4.0	6	8	5	7
3.5x1.27	3.505	3.53	3.43	2.64	2.51	2.41	2.26	2.54	1.90	4.9	7	9	6	8
4.2x1.41	4.166	4.22	4.09	3.10	2.95	2.84	2.69	2.82	2.11	5.7	8	10	7	9
4.8x1.59	4.826	4.80	4.65	3.58	3.43	3.30	3.12	3.18	2.39	6.5	9	12	8	11
5.5x1.81	5.486	5.46	5.31	4.17	3.99	3.86	3.68	3.63	2.72	7.5	11	14	9	12
6.3x1.81	6.350	6.25	6.10	4.88	4.70	4.55	4.34	3.63	2.72	8.5	12	16	10	13
8.0x2.12	7.938	8.00	7.82	6.20	5.99	5.84	5.64	4.24	3.18	10.6	16	20	12	16
9.5x2.12	9.525	9.65	9.42	7.85	7.59	7.44	7.24	4.24	3.18	12.9	19	25	14	19

NOTES:
1. All dimensions are in millimeters.
2. No extrusion of excess metal beyond apex of the Type AB point resulting from thread rolling is permissible; a slight rounding truncation of the point is desirable.
3. The width of flat at crest of thread shall not exceed 0.10 mm for sizes up to and including 4.2 mm and 0.15 mm for larger sizes.
4. Tapered threads on Type B shall have unfinished crests.
5. The tabulated values of point diameter apply to screw blanks before thread rolling.
6. Tabulated max values equal 2 times the pitch of the thread rounded to two places.
7. Points of types BF and BT screws shall be tapered and fluted or slotted as illustrated. Details of taper and flute design are optional with the manufacturer, provided the screws meet the performance requirements. Flutes or slots shall extend through the first full form thread, except for Type BF screws on which the length of flutes may be one thread pitch short of the first full form thread.

FASTENERS

Table 9-74. Threads and Points for Thread Cutting Tapping Screws Types D, F, and T (IFI-502)

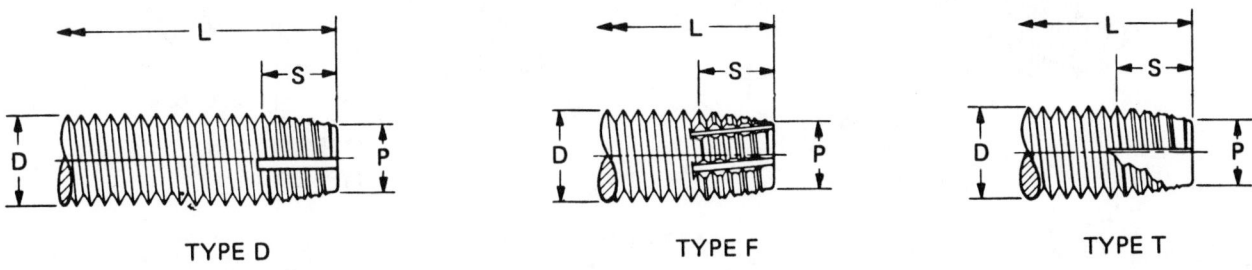

TYPE D　　　　TYPE F　　　　TYPE T

Nom Screw Size and Thread Pitch	D Major Diameter Max	D Major Diameter Min	P Point Diameter Max	P Point Diameter Min	S Point Taper Length Max	S Point Taper Length Min	Body Diameter Min	L Minimum Practical Screw Lengths Pan, Hex and Hex Washer Heads	L Minimum Practical Screw Lengths Flat and Oval CTSK Heads
2x0.4	2.00	1.88	1.61	1.54	1.4	1.0	1.65	3	5
2.5x0.45	2.50	2.37	2.07	2.00	1.6	1.1	2.12	4	6
3x0.5	3.00	2.87	2.52	2.44	1.8	1.3	2.58	5	7
3.5x0.6	3.50	3.35	2.92	2.83	2.1	1.5	3.00	6	8
4x0.7	4.00	3.83	3.33	3.24	2.5	1.8	3.43	6	9
5x0.8	5.00	4.82	4.22	4.10	2.8	2.0	4.36	8	11
6x1	6.00	5.82	5.03	4.92	3.5	2.5	5.51	10	13
8x1.25	8.00	7.76	6.79	6.67	4.4	3.1	7.04	12	16
10x1.5	10.00	9.73	8.55	8.42	5.2	3.8	8.86	15	20
12x1.75	12.00	11.70	10.31	10.16	6.1	4.4	10.68	18	24

NOTES:
1. All dimensions are in millimeters.
2. Points of screw are tapered and fluted or slotted as illustrated. Details of taper and flute design are optional with the manufacturer, provided screws meet the performance requirements. Flutes or slots shall extend through the first full form thread, except for Type F screws on which the length of flutes may be one thread pitch short of the first full form thread.
3. The tabulated values apply to screw blanks before roll threading.
4. Tabulated max values equal basic minor diameter of thread plus 20 percent of the double thread height.
5. Tabulated max values equal 3-1/2 and min values equal 2-1/2 times the pitch of the thread rounded to two places.
6. Threads of tapping screws are not normally subjected to thread gaging requirements.
7. Lengths shown are theoretical minimums. Refer to Table 9-65 for recommended diameter-length combinations.

Table 9-75. Thread and Point Dimensions of Thread Rolling Screws (IFI-503)

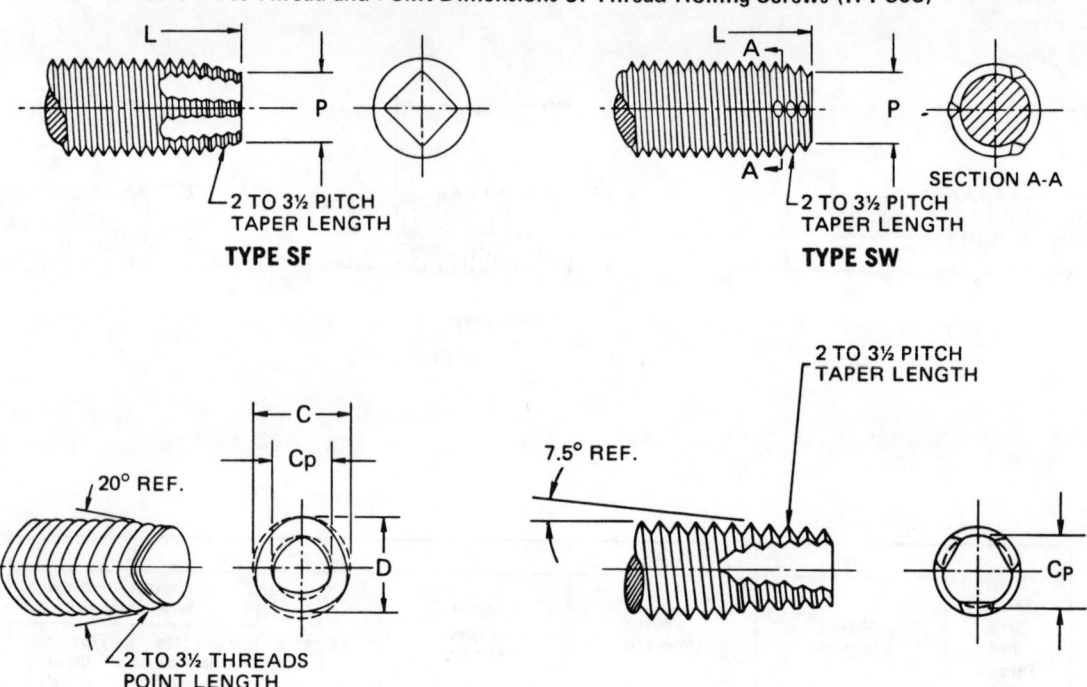

Nom Screw Size and Thread Pitch	P Major Dia	Point Dia	C Dia of Circumscribing Circle	Cp Circumscribing Circle (Point)	Point Length		L Min Practical Nom Screw Length	
	Max	Max	Max	Max	Max	Min	Pan, Hex, Hex Washer Heads	Flat and Oval Ctsk. Heads
2x0.4	2.00	1.6	—	—	1.4	0.8	4	5
2.5x0.45	2.50	2.1	2.57	2.13	1.6	0.9	4	6
3x0.5	3.00	2.5	3.07	2.58	1.8	1.0	5	8
3.5x0.6	3.50	2.9	3.58	2.99	2.1	1.2	6	8
4x0.7	4.00	3.4	4.08	3.40	2.4	1.4	8	10
5x0.8	5.00	4.4	5.09	4.31	2.8	1.6	8	10
6x1	6.00	5.5	6.10	5.12	3.5	2.0	10	13
8x1.25	8.00	7.1	8.13	6.92	4.4	2.5	10	16
10x1.5	10.00	9.0	10.15	8.69	5.2	3.0	13	16
12x1.75	12.00	10.5	12.18	10.48	6.1	3.5	16	20

NOTES:
1. All dimensions are in millimeters.
2. These dimensions are appllicable to types of screws where the periphery of the thread approximates a circle.
3. These dimensions are applicable to types of screws where some portions of the periphery of the thread are farther from the screw axis than others (lobular, triroundular, etc).
4. These values are equal to 3.5 times the pitch distance rounded off to 1 decimal place.
5. These values are equal to 2 times the pitch distance rounded off to 1 decimal place.
6. SF, SW, TR-3, and TT are representative types of thread-rolling screws.
7. Lengths shown are theoretical minimums. Refer to Table 9-65 for recommended diameter-length combinations.

FASTENERS

Table 9-76. Dimensions of Threads and Points for Types BSD and CSD Self-Drilling Tapping Screws (IFI-504)

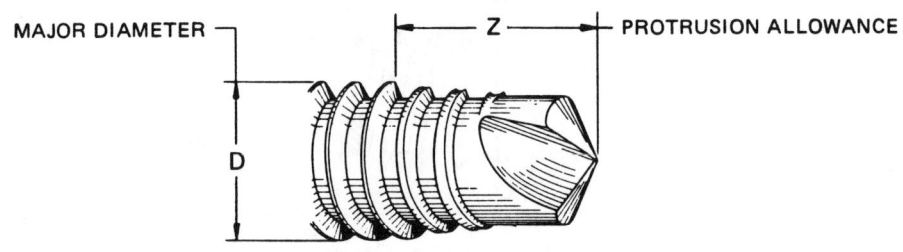

TYPE BSD

Nom Screw Size mm	Thread Pitch mm	D Major Diameter mm Max	D Major Diameter mm Min	d Minor Diameter mm Max	d Minor Diameter mm Min	Z Protrusion Allowance (Ref) mm Style 2 Point	Z Protrusion Allowance (Ref) mm Style 3 Point	L Style 2 Points Formed Pan & Hex Washer Heads	L Style 2 Points Formed Flat & Oval CTSK Heads	L Style 2 Points Milled Pan & Hex Washer Heads	L Style 2 Points Milled Flat & Oval CTSK Heads	L Style 3 Points Formed Pan & Hex Washer Heads	L Style 3 Points Formed Flat & Oval CTSK Heads	L Style 3 Points Milled Pan & Hex Washer Heads	L Style 3 Points Milled Flat & Oval CTSK Heads
2.9	1.06	2.90	2.79	2.18	2.08	4.1	—	8	9.5	9.5	11	—	—	—	—
3.5	1.27	3.53	3.43	2.64	2.51	4.8	5.6	8	9.5	9.5	11	9.5	11	11	12.5
4.2	1.41	4.22	4.09	3.10	2.95	5.4	6.4	9.5	11	11	12.5	11	12.5	12.5	14
4.8	1.59	4.80	4.65	3.58	3.43	6.0	7.6	11	12.5	12	15	12.5	14	14	16.5
5.5	1.81	5.46	5.31	4.17	3.99	7.0	9.0	12.5	16	13.5	16.5	12.5	16	16.5	20
6.3	1.81	6.25	6.10	4.88	4.70	8.1	10.0	12.5	16	13.5	17.5	12.5	16	17.5	21.5

TYPE CSD

Nom Screw Size mm	Thread Pitch mm	D Major Diameter mm Max	D Major Diameter mm Min	Z Protrusion Allowance (Ref) mm Style 2 Point	Z Protrusion Allowance (Ref) mm Style 3 Point	L Style 2 Points Formed Pan & Hex Washer Heads	L Style 2 Points Formed Flat & Oval CTSK Heads	L Style 2 Points Milled Pan & Hex Washer Heads	L Style 2 Points Milled Flat & Oval CTSK Heads	L Style 3 Points Formed Pan & Hex Washer Heads	L Style 3 Points Formed Flat & Oval CTSK Heads	L Style 3 Points Milled Pan & Hex Washer Heads	L Style 3 Points Milled Flat & Oval CTSK Heads
3	0.5	3.00	2.87	4.8	—	9.5	11.0	11.0	12.7	—	—	—	—
3.5	0.6	3.50	3.35	5.4	6.3	9.5	11.0	11.0	12.7	11.0	12.7	12.7	14.0
4	0.7	4.00	3.84	5.9	7.3	11.0	12.7	12.7	14.0	13.5	15.0	15.0	16.7
5	0.8	5.00	4.83	6.7	8.2	12.7	14.0	13.5	16.7	14.0	16.0	16.0	18.3
6	1.0	6.00	5.81	9.9	11.6	16.0	19.0	16.7	19.8	15.0	18.3	19.8	23.0

NOTES:
1. All dimensions are in millimeters.
2. Drill portion of points may be milled and/or cold formed, and details of point taper and flute design are optional with the manufacturer provided the screws meet the performance requirements specified in IFI-504 standard.
3. Protrusion allowance Z is the distance measured parallel to the axis of the screw, from the extreme end of the point to the first full form thread beyond the point. This encompasses the length of drill point and the tapered incomplete thread. It is intended for use in calculating the maximum effective design grip length Y on the screw in accordance with the following: $Y = L \text{ min} - Z$

Table 9-77A. Thread Distance from Head (Types AB, B, BF and BT) (IFI-502)

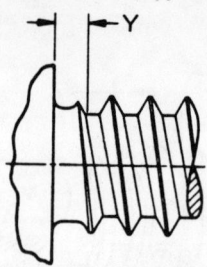

Nominal Screw Size	L Nominal Screw Length	L_T Full Form Thread Length Min[1]	Y Unthreaded Length Under Head Max[2]
2.2	16	13	0.79
2.9	20	17	1.06
3.5	25	21	1.27
4.2	30	25	1.41
4.8	35	29	1.59
5.5	40	33	1.81
6.3	45	38	1.81
8	45	38	2.12
9.5	50	38	2.12

NOTES:
1. Tabulated values through 6.3 mm size are equal to 6 times the basic screw diameter rounded to nearest millimeter.
2. Tabulated values are equal to 1 times thread pitch.

Table 9-77B. Thread Distance from Head (Types D, F, T, SF, SW, TR-3, and TT) (IFI-502)

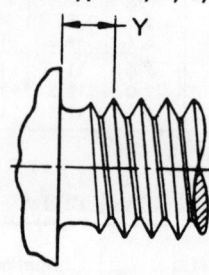

Nominal Screw Size	L_T Full Form Thread Length For Nominal Screw Lengths >Than	Min(1)	For Nominal Screw Lengths ≤Than	Max(2)	Y Unthreaded Length Under Head For Nominal Screw Lengths >Than	≤Than	Max(3)
2	16	12	6	0.40	6	16	0.8
2.5	20	15	8	0.45	8	20	0.9
3	25	18	9	0.50	9	25	1.0
3.5	30	21	10	0.60	10	30	1.2
4	35	24	12	0.70	12	35	1.4
5	40	30	15	0.80	15	40	1.6
6	45	38	19	1.00	19	45	2.0
8	45	38	24	1.25	24	45	2.5
10	45	38	30	1.50	30	45	3.0
12	50	38	36	1.75	36	50	3.5

NOTES: 1. Tabulated values through 6 mm size are equal to 6 times basic screw diameter rounded to nearest millimeter.
2. Tabulated values are equal to 1 times thread pitch. 3. Tabulated values are equal to 2 times thread pitch.

FASTENERS

All nuts shall be provided with a supplementary lubricant if necessary to meet the stated performance requirements without galling and shall be clean and dry to the touch.

The performance of nuts which are furnished with a protective coating shall not deteriorate when the nuts are stored indoors for a period of six months.

Prevailing-Torque Type Nut—A nut which is frictionally resistant to rotation due to a self-contained prevailing-torque feature, and not because of a compressive load developed against the bearing surface of the nut.

Prevailing-Torque Developed by a Nut—The torque necessary to rotate the nut on its mating externally threaded component, with the torque being measured while the nut is in motion, and with no axial load in the mating component.

Prevailing-Torque—The prevailing-torque developed by nuts during their first installation, or any subsequent installation or removal, shall not exceed the maximum first installation torque specified for the applicable class in Table 9-82. In addition, the maximum and minimum prevailing-torque developed by nuts during their first and fifth removals shall not be less than the respective "highest" and "lowest" readings removal torques specified in Table 9-82. NOTE: The purpose of this requirement is to verify that the nut's resistance to removal is at least equal to or greater than a specified prevailing-torque ("highest") in at least one location during a full 360° of rotation, and also, that at no location during that same full rotation is the resistance to removal less than a specified prevailing-torque ("lowest").

SLOTTED HEXAGON NUTS (IFI-507)

Slotted hexagon nuts for use with standard cotter pins, as shown in Table 9-103, are specified in the IFI-507 standard, and nominal thread sizes from 5 to 36 mm are shown in Table 9-83.

Chemical composition, mechanical properties and identification markings of Style 1 steel nuts shall conform with either property class 5 or property class 10 as applicable and as specified on page (9-23), except that proof load values of hex slotted nuts shall be 80 percent of those specified for hex nuts. Chemical composition, mechanical properties and identification markings of Style 2 steel nuts shall conform with property class 9 as specified on page (9-23), except that proof load values of hex slotted nuts shall be 80 percent of those specified for hex nuts. Nuts of other materials shall be of chemical composition and have mechanical properties as agreed upon by manufacturer and puchaser.

Table 9-78. Self-Drilling Tapping Screw Selection Chart (IFI-504)

Screw Type	Point Style	Nom Screw Size mm	P (a) Recommended Panel Thickness, mm
BSD and CSD	2	2.9 and 3 3.5 4 and 4.2 4.8 and 5 5.5 6 and 6.3	2.0 2.3 2.5 2.8 3.6 4.4
	3	3.5 4 and 4.2 4.8 and 5 5.5 to 6.3	2.3-2.8 2.5-3.6 2.8-4.4 2.8-5.3

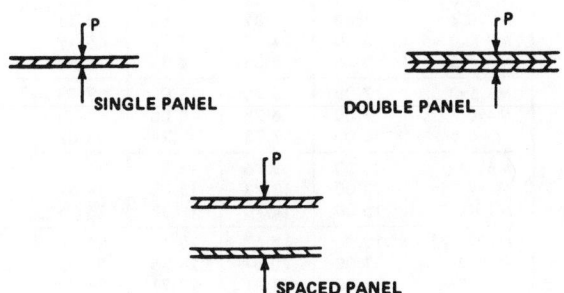

Fig. 9-1. Self-drilling screws in single panel, double panel and spaced panel. If the panel to be drilled is comprised of two or more layers, see above right. The gap between the layers (which might consist of a sealing strip, air space caused by warpage, etc., or just the separation caused by the pressure exerted by the driver) must be considered in determining the point style for the particular fastener. Using a self-drilling tapping screw as covered in this publication in a multi-layer application with an excessive gap could result in point breakage since the tapping in one layer begins completion of the drilling of the other layers and since the advancement of the screw in the tapping operation is much faster than in the drilling operation.

PROJECTION WELD NUTS (CHRYSLER CORP.)

Hexagon projection weld nuts are not covered in any national or international standards yet. The type weld nuts shown here are hard metric module projection weld nuts developed in cooperation with automotive welding experts. Dimensions for nuts with nominal thread sizes from M5 × 0.8 to M16 × 2 are shown in Table 9-84. Tables and figures for projection weld nuts are reproduced from Chrysler standards with permission from the Chrysler Corporation.

Table 9-79. Hexagon Nuts (IFI-507)

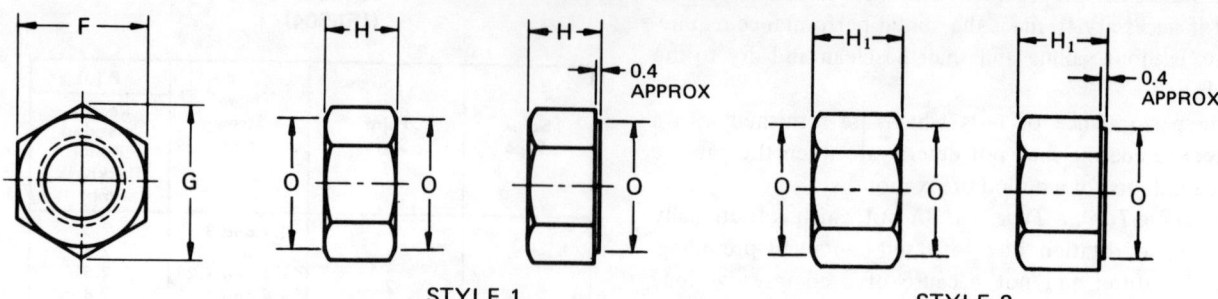

Nominal Nut Size and Thread Pitch	F Width Across Flats Max	F Width Across Flats Min	G Width Across Corners Max	G Width Across Corners Min	O Bearing Face Dia Min	H Nut Thickness Style 1 Max	H Nut Thickness Style 1 Min	H₁ Nut Thickness Style 2 Max	H₁ Nut Thickness Style 2 Min	Runout of Bearing Surface FIR Max
M1.6x0.35	3.20	3.02	3.70	3.44	2.5	—	—	1.3	1.1	—
M2x0.4	4.00	3.82	4.62	4.35	3.1	—	—	1.6	1.3	—
M2.5x0.45	5.00	4.82	5.77	5.49	4.1	—	—	2.0	1.7	—
M3x0.5	5.50	5.32	6.35	6.06	4.6	—	—	2.4	2.1	—
M3.5x0.6	7.00	6.78	8.08	7.73	6.0	—	—	2.8	2.5	—
M4x0.7	7.00	6.78	8.08	7.73	6.0	—	—	3.2	2.9	—
M5x0.8	8.00	7.78	9.24	8.87	7.0	4.5	4.2	5.3	5.0	0.30
M6x1	10.00	9.76	11.55	11.13	8.9	5.6	5.3	6.5	6.2	0.33
M8x1.25	13.00	12.73	15.01	14.51	11.6	6.6	6.2	7.8	7.4	0.36
M10x1.5	15.00	14.70	17.32	16.76	13.6	9.0	8.5	10.7	10.2	0.39
M12x1.75	18.00	17.67	20.78	20.14	16.6	10.7	10.2	12.8	12.3	0.42
M14x2	21.00	20.64	24.25	23.53	19.4	12.5	11.9	14.9	14.3	0.45
M16x2	24.00	23.61	27.71	26.92	22.4	14.5	13.9	17.4	16.8	0.48
M20x2.5	30.00	29.00	34.64	33.06	27.6	18.4	17.4	21.2	20.2	0.56
M24x3	36.00	34.80	41.57	39.67	32.9	22.0	20.9	25.4	24.3	0.64
M30x3.5	46.00	44.50	53.12	50.73	42.5	26.7	25.4	31.0	29.7	0.76
M36x4	55.00	53.20	63.51	60.65	50.8	32.0	30.5	37.6	36.1	0.89

NOTES:
1. All dimensions are in millimeters.
2. Width Across Flats. Maximum width across flats of nuts shall not be exceeded; except that for milled-from-bar non-ferrous nuts, the tabulated maximum width across flats dimensions may be exceeded to conform with the commercial tolerances of drawn or rolled stock material. For nuts of all materials, no transverse section through the nut between 25 and 75 percent of the actual nut thickness as measured from the bearing surface shall be less than the minimum width across flats.
3. Corner Fill. A rounding or lack of fill at junction of hex corners with chamfer shall be permissible provided the width across corners is within specified limits at and beyond a distance equal to 17.5 percent of the nominal nut size from the chamfered face(s).
4. Tops and Bearing Surfaces. Nuts in sizes 16 mm nominal size and smaller shall be double chamfered. Nuts of larger sizes, at the manufacturer's option, shall be either double chamfered or have a washer faced bearing surface and chamfered top. The diameter of chamfer circle on double chamfered nuts and the diameter of washer face shall be within the maximum width across flats and minimum bearing circle diameter. The tops of washer faced nuts shall be flat and the diameter of chamfer circle shall be equal to the maximum width across flats within a tolerance of minus 15 percent. The length of the chamfer at hex corners shall be from 5 to 15 percent of the nominal nut size. The surface of the chamfer may be slightly convex or rounded. Bearing surfaces shall be flat and perpendicular to the axis of the threaded hole within the FIR limit specified.
5. Concentricity of Tapped Hole. Axis of tapped hole shall be concentric with the axis of the nut body within a tolerance of 3 percent (6 percent FIR) of the maximum width across flats.
6. Countersink. The tapped hole shall be countersunk on the bearing face(s). The maximum countersink diameter shall be the thread basic major diameter (nominal nut size) plus 0.75 mm for 8 mm nominal size nuts and smaller, and 1.08 times the thread basic major diameter for 10 mm nominal size nuts and larger. No part of threaded portion shall project beyond the bearing surface.
7. Threads. Threads shall be Modified ISO internal threads, class 6H, as given in Section 8.
8. Property Class. Chemical composition, mechanical properties and identification markings of Style 1 steel nuts shall conform with either property class 5 or property class 10 as applicable and as specified in Table 9-14. Chemical composition, mechanical properties and identification markings of Style 2 steel nuts shall conform with property class 9. Nuts of other materials shall be of chemical composition and have mechanical properties as agreed upon by manufacturer and purchaser.
9. Designation: Example for the designation of a hexagon style 2 nut with thread size M8 × 1.25 (ISO designation–M8) and property class 9, finish zinc plated: M8 × 1.25, 9 Hexagon nut style 2 zinc plated.
10. See note, page 9-41.

FASTENERS

Table 9-80. Mechanical Requirements for Nuts

Property Class	Nominal Nut Size	Proof Load Stress MPa	Nut Hardness Diamond Pyramid	Nut Hardness Rockwell	Suggested Property Class of Mating Bolt, Screw or Stud
5	M5 thru M36	570	300 max	C 30 max	4.6, 4.8, 5.8
9	M5 thru M16	990	300 max	C 30 max	5.8, 9.8
	M20 thru M36	910	300 max	C 30 max	5.8, 8.8
10	M6 thru M36	1040	275/360	C 26/36	10.9

Table 9-81. Prevailing-Torque Type Steel Hex Nuts (IFI-514)

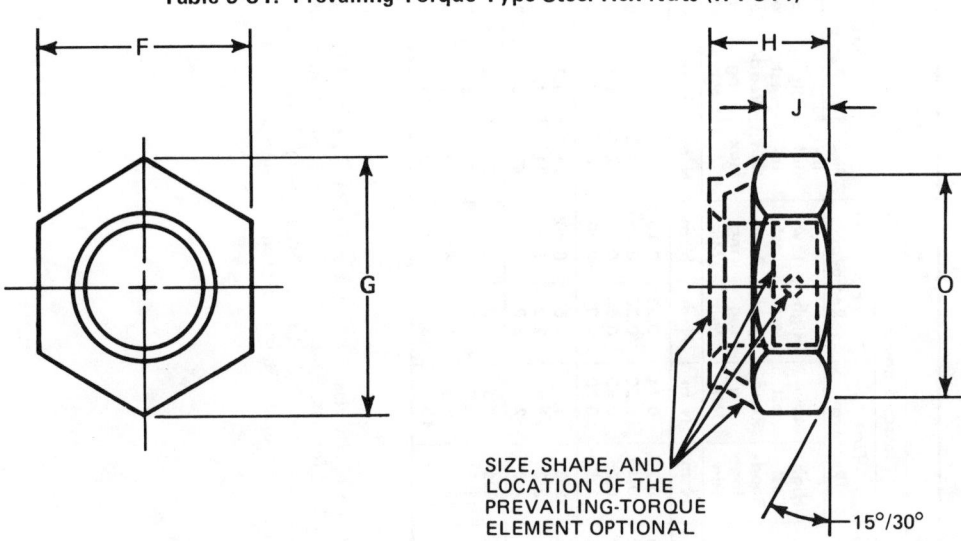

SIZE, SHAPE, AND LOCATION OF THE PREVAILING-TORQUE ELEMENT OPTIONAL

15°/30°

Nom Nut Size	F Width Across Flats Max	F Width Across Flats Min	G Width Across Corners Max	G Width Across Corners Min	H Style 1 Insert Type Nut Max	H Style 1 Insert Type Nut Min	H Style 1 All Metal Type Nut Max	H Style 1 All Metal Type Nut Min	H Style 2 Insert Type Nut Max	H Style 2 Insert Type Nut Min	H Style 2 All Metal Type Nut Max	H Style 2 All Metal Type Nut Min	J Wrenching Height Style 1 Min	J Wrenching Height Style 2 Min	O Dia of Bearing Circle Min	Runout of Bearing Circle Max
M5x0.8	8.00	7.78	9.24	8.87	6.1	5.8	6.1	4.2	7.6	6.6	7.6	5.0	2.3	2.9	7.4	0.19
M6x1	10.00	9.76	11.55	11.13	7.6	7.3	7.6	5.3	8.8	8.2	8.8	6.2	3.0	3.7	9.3	0.22
M8x1.25	13.00	12.73	15.01	14.51	9.1	8.7	9.1	6.2	10.3	9.9	10.3	7.4	3.7	4.5	12.1	0.28
M10x1.5	15.00	14.70	17.32	16.76	12.0	11.5	12.0	8.5	14.0	13.2	14.0	10.2	5.5	6.7	14.0	0.31
M12x1.75	18.00	17.67	20.78	20.14	14.2	13.7	14.2	10.2	16.8	15.8	16.8	12.3	6.7	8.2	16.8	0.33
M14x2	21.00	20.64	24.25	23.53	16.5	15.9	16.5	11.9	18.9	18.3	18.9	14.3	7.8	9.6	19.7	0.36
M16x2	24.00	23.61	27.71	26.92	18.5	17.9	18.5	13.9	21.4	20.8	21.4	16.8	9.5	11.7	22.5	0.40
M20x2.5	30.00	29.00	34.64	33.06	23.4	22.4	23.4	17.4	26.5	25.2	26.5	20.2	11.1	12.6	27.6	0.46
M24x3	36.00	34.80	41.57	39.67	28.0	26.9	28.0	20.9	31.4	30.3	31.4	24.3	13.3	15.1	33.1	0.51
M30x3.5	46.00	44.50	53.12	50.73	33.7	32.4	33.7	25.4	38.0	36.7	38.0	29.7	16.4	18.5	42.3	0.67
M36x4	55.00	53.20	63.51	60.65	40.0	38.5	40.0	30.5	45.6	44.1	45.6	36.1	20.1	22.8	50.6	0.80

NOTES:
1. All dimensions are in millimeters.
2. Except as noted, dimensions apply to all property classes of nuts.
3. Screw thread details are shown in Table 8-1, and the standard thread tolerance class is 6H as shown in Table 8-5.
4. Tapped hole shall be countersunk on the bearing face(s). The maximum countersink diameter shall be the thread basic major diameter (nominal nut size) plus 0.75 mm for 8 mm nominal size nuts and smaller, and 1.08 times the thread basic major diameter for 10 mm nominal size nuts and larger. No part of threaded portion shall project beyond the bearing surface.
5. Axis of the tapped hole shall be concentric with the axis of the nut body within a tolerance of 1.5 percent (3 percent FIR) of the maximum width across flats.
6. Designation: Example for the designation of a prevailing tongue type steel hexagon lock nuts with thread size M10 × 1.5 and property class 9, Style 1; Steel hexagon lock unit IFI-514 M10 × 1.5, 9 hex lock nuts, Style 1.
7. See note, page 9-41.

Table 9-82. Proof Loads, Clamp Loads, and Prevailing-Torques for Nuts (IFI-514)

Nom Nut Size	Proof Load Class 5 kN	Proof Load Class 9 kN	Proof Load Class 10 kN	Clamp Load (1) Class 5 kN	Clamp Load (1) Class 9 kN	Clamp Load (1) Class 10 kN	Prevailing Torque Classes 5 and 9 First Installation Max N·m	Classes 5 and 9 First Removal Highest Reading Min N·m	Classes 5 and 9 First Removal Lowest Reading Min N·m	Classes 5 and 9 Fifth Removal Highest Reading Min N·m	Classes 5 and 9 Fifth Removal Lowest Reading Min N·m	Class 10 First Installation Max N·m	Class 10 First Removal Highest Reading Min N·m	Class 10 First Removal Lowest Reading Min N·m	Class 10 Fifth Removal Highest Reading Min N·m	Class 10 Fifth Removal Lowest Reading Min N·m
M5x0.8	8.09	14.1	—	4.05	6.92	—	1.6	0.29	0.14	0.23	0.10	—	—	—	—	—
M6x1	11.4	19.9	20.9	5.73	9.83	12.5	3.0	0.45	0.20	0.30	0.15	4.0	0.55	0.25	0.40	0.20
M8x1.25	20.9	36.2	38.1	10.4	17.8	22.8	6.0	0.85	0.40	0.60	0.30	8.0	1.15	0.60	0.80	0.40
M10x1.5	33.1	57.4	60.3	16.5	28.3	36.1	10.5	1.5	0.70	1.0	0.50	14	2.0	1.0	1.4	0.70
M12x1.75	48.1	83.5	87.7	24.0	41.1	52.5	15.5	2.3	1.0	1.6	0.80	21	3.1	1.5	2.1	1.0
M14x2	65.6	114	120	32.8	56.1	71.6	24	3.3	1.5	2.3	1.0	31	4.4	2.0	3.0	1.5
M16x2	89.5	155	163	44.8	76.5	97.5	32	4.5	2.0	3.0	1.5	42	6.0	3.0	4.2	2.0
M20x2.5	140	223	255	69.8	110	152	54	7.5	3.5	5.3	2.5	72	10.5	5.0	7.0	3.5
M24x3	201	321	367	101	159	220	80	11.5	5.5	8.0	4.0	106	15	7.5	10.5	5.0
M30x3.5	320	511	583	94.5	253	350	108	16	8.0	12	6.0	140	19	9.5	14.0	7.0
M36x4	466	743	850	138	368	508	136	21	10	16	8.0	180	24	12	17.5	8.5

NOTE:
The clamp load for class 5 nuts is equal to 75 percent of the proof load of property class 5.8 bolts for sizes M5 through M24 and 75 percent of the proof load of class 4.6 bolts for sizes M30 and M36. The clamp load for class 9 nuts is equal to 75 percent of the proof load of class 9.8 bolts for sizes M5 through M16 and 75 percent of the proof load of class 8.8 bolts for sizes M20 through M36. The clamp load for class 10 nuts is equal to 75 percent of the proof load of class 10.9 bolts. Proof loads of bolts are given in Table 9-7.

FASTENERS

WASHERS

A great number of flat, locking and special purpose washers of various shapes are covered in national standards throughout the world, some of which are shown in the world fastener index on page 9-10.

Flat Washers (ISO 887)

There is no ISO standard which describes flat washers in general use in countries already on the metric system. An American standard for flat metric washers is in preparation, and its use is recommended as soon as it becomes available. The ISO 887 standard specifies the basic dimensions for several series of flat washers and their corresponding metric bolts and nuts for nominal sizes from 1.6 to 150 mm. The size range from 1.6 to 39 mm is shown in Table 9-85.

Helical Spring Lock Washers

The trapezoidal cross section of wires used for helical springs in the United States differs from the standard rectangular and square sections used in Europe. There is no ISO standard for this type of fastener. The helical spring lock washers shown in Table 9-86 are for regular usage. Those in Table 9-87 are for heavy duty applications and nominal fastener sizes from 2.5 to 36 mm. Tables and figures for helical spring lock washers are reproduced from GM standards with the permission of General Motors Corporation.

Table 9-83. Slotted and Castle Nuts (IFI-507)

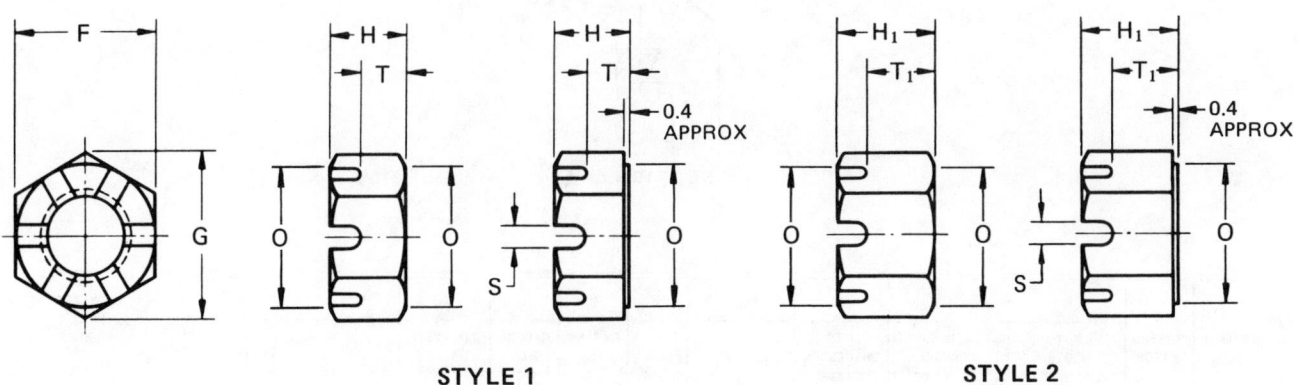

STYLE 1   STYLE 2

Nominal Nut Size and Thread Pitch	F Width Across Flats Max	F Width Across Flats Min	G Width Across Corners Max	G Width Across Corners Min	O Bearing Face Dia Min	H Nut Thickness Style 1 Max	H Nut Thickness Style 1 Min	H₁ Nut Thickness Style 2 Max	H₁ Nut Thickness Style 2 Min	T Unslotted Thickness Style 1 Max	T Unslotted Thickness Style 1 Min	T₁ Unslotted Thickness Style 2 Max	T₁ Unslotted Thickness Style 2 Min	S Width of Slot Max	S Width of Slot Min	Runout of Bearing Surface FIR Max
M5×0.8	8.00	7.78	9.24	8.87	7.0	4.5	4.2	5.3	5.0	3.2	2.7	3.7	3.2	2.2	1.4	0.30
M6×1	10.00	9.76	11.55	11.13	8.9	5.6	5.3	6.5	6.2	3.9	3.4	4.5	4.0	2.8	2.0	0.33
M8×1.25	13.00	12.73	15.01	14.51	11.6	6.6	6.2	7.8	7.4	4.5	4.0	5.3	4.8	3.3	2.5	0.36
M10×1.5	15.00	14.70	17.32	16.76	13.6	9.0	8.5	10.7	10.2	6.0	5.5	7.1	6.6	3.6	2.8	0.39
M12×1.75	18.00	17.67	20.78	20.14	16.6	10.7	10.2	12.8	12.3	7.1	6.6	8.5	8.0	4.3	3.5	0.42
M14×2	21.00	20.64	24.25	23.53	19.4	12.5	11.9	14.9	14.3	8.2	7.7	9.8	9.3	4.3	3.5	0.45
M16×2	24.00	23.61	27.71	26.92	22.4	14.5	13.9	17.4	16.8	9.5	9.0	11.4	10.9	6.0	4.5	0.48
M20×2.5	30.00	29.00	34.64	33.06	27.6	18.4	17.4	21.2	20.2	12.1	11.3	13.9	13.1	6.0	4.5	0.56
M24×3	36.00	34.80	41.57	39.67	32.9	22.0	20.9	25.4	24.3	14.4	13.6	16.6	15.8	7.0	5.5	0.64
M30×3.5	46.00	44.50	53.12	50.73	42.5	26.7	25.4	31.0	29.7	17.3	16.5	20.1	19.3	9.3	7.0	0.76
M36×4	55.00	53.20	63.51	60.65	50.8	32.0	30.5	37.6	36.1	20.8	19.8	24.5	23.5	9.3	7.0	0.89

1. All dimensions are in millimeters.
2. General purpose tolerance is 6H and gaging dimensions are shown in Table 8-5.
3. For screw thread details, see Table 8-1.
4. Strength grades for nuts from 5 through 10, see Table 9-15.
5. Designation: Example for the designation of a slotted hexagon nut with thread size M10 × 1.5 and property class 9 and style 2: M10 × 1.5, 9 hexagon nut, zinc plated, style 2.
6. See note, page 9-41.

Table 9-84. Projection Weld Nut (Chrysler Corp.)

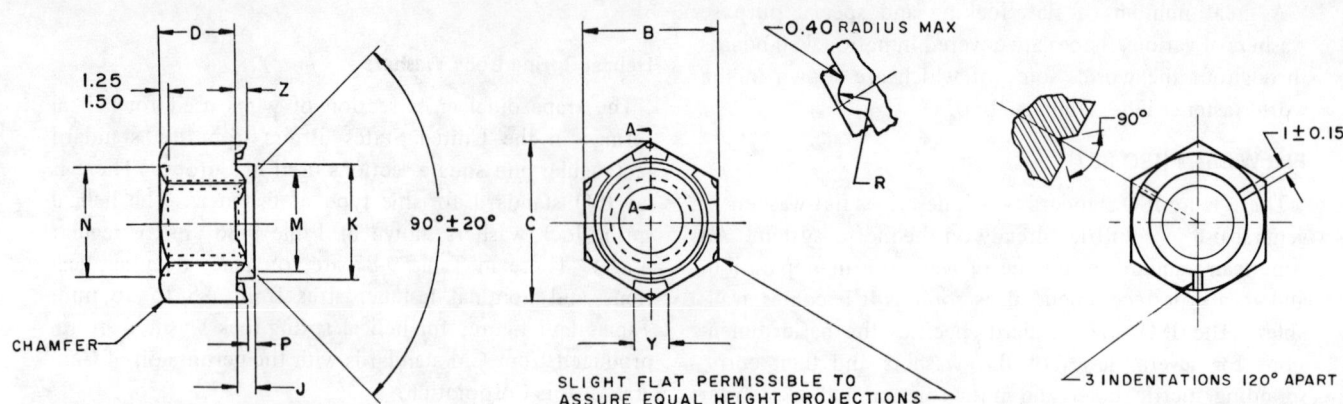

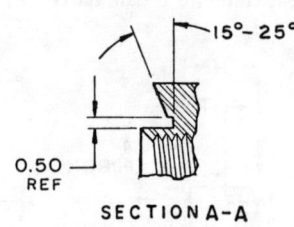

NOMINAL SIZE	THREAD PITCH	B NUT WIDTH Min	B NUT WIDTH Max	C ACROSS CORNERS Min	C ACROSS CORNERS Max Ref	D NUT THICKNESS Min	D NUT THICKNESS Max	J PILOT HEIGHT* Min	J PILOT HEIGHT* Max*	K PILOT DIA +0.0 -0.3	M C/SINK DIA Min	N C/BORE DIA +0.00 -0.25	P LOCATING HEIGHT Min	R RADIUS Ref	Y PROJ WIDTH ±0.25	Z PROJ HEIGHT ±.075	PANEL HOLE +0.0 -0.3	CHRYSLER PART NUMBERS STEEL / COPPER COAT
M5	0.80	9.76	10.0	11.03	11.55	6.0	6.5	0.55	0.80	7.0	6.0	6.6	0.25	2.8	2.75	0.35	7.5	6100057
M6	1.00	12.73	13.0	14.38	15.01	7.0	7.5	0.55	0.80	9.0	7.0	8.4	0.25	2.8	2.75	0.35	9.5	6100058
								1.75	2.0	9.0	7.0	8.4	0.65	2.8	3.5	1.00	9.5	6100059
M8	1.25	14.70	15.0	16.61	17.32	8.0	8.5	0.55	0.80	11.0	9.0	10.5	0.25	2.8	2.75	0.35	11.5	6100060
								1.75	2.0	11.0	9.0	10.5	0.65	2.8	4.0	1.00	11.5	6100061
M10	1.50	17.67	18.0	19.97	20.78	8.5	9.0	0.55	0.80	13.5	11.0	12.8	0.25	2.8	3.0	0.35	14.0	6100062
								1.75	2.0	13.5	11.0	12.8	0.65	2.8	4.25	1.00	14.0	6100063
M12	1.75	20.64	21.0	23.32	24.25	10.5	11.0	1.0	1.25	16.0	13.5	15.3	0.50	3.25	3.0	0.65	16.5	6100064
								1.75	2.0	16.0	13.5	15.3	0.65	3.25	4.75	1.00	16.5	6100065
M14	2.00	23.61	24.0	26.68	27.71	12.5	13.0	1.0	1.25	19.5	16.0	17.3	0.50	4.0	4.5	0.65	20.0	6100066
								1.75	2.0	19.5	16.0	17.3	0.65	4.0	5.0	1.00	20.0	6100067
M16	2.00	25.48	26.0	28.79	30.02	14.5	15.0	1.75	2.0	20.5	18.0	19.5	0.65	4.5	5.25	1.00	21.0	6100068

NOTES:
1. All dimensions are in millimeters.
2. Screw thread details are shown in Table 8-1, and gage dimensions for the preferred tolerance class 6H are shown in Table 8-5.
3. Material: Strength grade 9 with mechanical requirements as specified in Tables 9-15 and 9-16.
4. Surface finish: Plain or copper coated.
5. Designation—Example for the designation of a projection weld nut with thread size M10 × 1.5 and property class 9; projection weld nut (long pilot) M10 × 1.5, steel class 9, plain finish (Ref. Chrysler part no. 6100063).
6. 0.80 pilot for sheet metal thickness 0.8 min through 1.9 min.; 1.25 pilot for sheet metal thickness 1.2 min through 1.9 min.; 2.0 pilot for sheet metal thickness 2.0 min through 4.8 min.

FASTENERS

Table 9-85. Washers for Hexagon Bolts and Nuts (ISO 887)

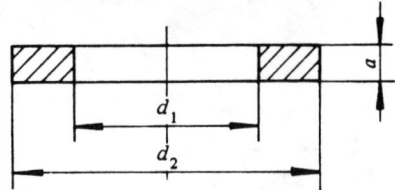

Clearance hole d_1		Diameter of the washer d_2 for widths across flats			Thickness of the washer a	For hexagon bolts and nuts	widths across flats		
fine series	medium series	normal	small	large		thread diameter	normal	small	large
1.7	—	4	—	—	0.3	1.6	3.2	—	—
2.2	—	5	—	—	0.3	2	4	—	—
2.7	—	6.5	—	—	0.5	2.5	5	—	—
3.2	—	7	—	—	0.5	3	5.5	—	—
4.3	—	9	—	—	0.8	4	7	—	—
5.3	5.5	10	—	—	1	5	8	—	—
6.4	6.6	12.5	—	—	1.6	6	10	—	—
7.4	7.6	14	—	—	1.6	7	11	—	—
8.4	9	17	15.5	21	1.6	8	13	12	17
10.5	11	21	18	24	2	10	17	14	19
13	14	24	21	28	2.5	12	19	17	22
15	16	28	24	30	2.5	14	22	19	24
17	18	30	28	34	3	16	24	22	27
19	20	34	30	37	3	18	27	24	30
21	22	37	34	39	3	20	30	27	32
23	24	39	37	44	3	22	32	30	36
25	26	44	39	50	4	24	36	32	41
28	30	50	44	56	4	27	41	36	46
31	33	56	50	60	4	30	46	41	50
34	36	60	56	66	5	33	50	46	55
37	39	66	60	72	5	36	55	50	60
40	42	72	66	78	6	39	60	55	65

NOTES:
1. Dimensions are in millimeters.
2. These clearance holes conform to ISO 273, Part 1: Clearance holes for metric bolts 1.6 up to and including 39 mm thread diameter.
3. The hexagon bolts and nuts conform to ISO 272, Hexagon bolts and nuts - Widths across flats, heights of heads, thicknesses of nuts.
4. See ISO 887 standard dor washers for nominal bolt sizes from M42 to M150.
5. Material: commercial quality - AISI 1010 through 1025 hot or cold rolled, see Table 10-42B, col. 1 and Table 10-42A, cols. 1 and 9. Hardened quality - AISI 1050 through 1065 and hardened to HRC 38-45, see Table 10-44, cols. 10, 11, 12, and 13.

Table 9-86. Regular Helical Spring Lock Washers (GM)

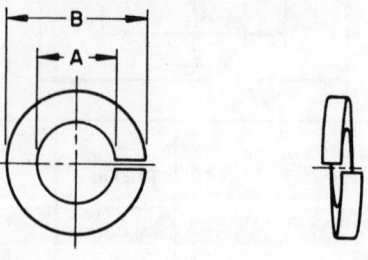

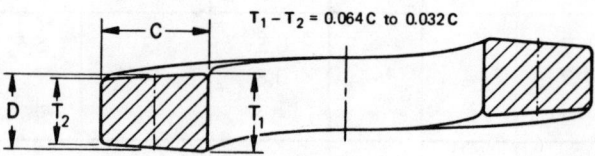

ENLARGED SECTION

Nominal Washer Size	A Inside Diameter Max	A Inside Diameter Min	B Outside Diameter Max	C Section Width Min	D Mean Section Thickness Max	D Mean Section Thickness Min	$T_1 - T_2$ Increase in Thickness, OD to ID Max	$T_1 - T_2$ Increase in Thickness, OD to ID Min	GM PART NUMBERS STEEL Plain	GM PART NUMBERS STEEL Zinc	GM PART NUMBERS STEEL Phos Coated
2.5	2.70	2.55	4.94	1.02	0.74	0.64	0.066	0.033	11500185	11500200	11500215
3	3.21	3.06	5.83	1.19	0.94	0.79	0.076	0.038	11500186	11500201	11500216
3.5	3.76	3.58	6.35	1.19	0.94	0.79	0.076	0.038	11500187	11500202	11500217
4	4.26	4.08	7.28	1.40	1.17	1.02	0.091	0.046	11500188	11500203	11500218
5	5.26	5.08	8.66	1.57	1.35	1.19	0.102	0.051	11500189	11500204	11500219
6	6.29	6.10	12.08	2.77	1.82	1.57	0.178	0.089	11503961	11503962	11503963
8	8.36	8.13	14.96	3.18	2.24	1.98	0.203	0.102	11500191	11500206	11500221
10	10.38	10.13	17.83	3.58	2.64	2.39	0.229	0.114	11500192	11500207	11500222
12	12.45	12.15	21.47	4.34	3.43	3.18	0.279	0.140	11500193	11500208	11500223
14	14.50	14.20	24.39	4.78	3.84	3.58	0.305	0.152	11500194	11500209	11500224
16	16.63	16.25	27.53	5.16	4.22	3.96	0.330	0.165	11500195	11500210	11500225
20	20.66	20.28	33.26	5.94	5.03	4.78	0.381	0.191	11500196	11500211	11500226
24	24.81	24.30	39.79	7.14	6.45	5.94	0.457	0.228	11500197	11500212	11500227
30	31.25	30.51	49.36	8.74	8.05	7.54	0.559	0.279	11500198	11500213	11500228
36	37.50	36.61	58.76	10.31	9.62	9.12	0.660	0.330	11500199	11500214	11500229

NOTES:
1. All dimensions are in millimeters.
2. Material: AISI 1060. See Table 10-42C, col. 14, or Table 10-44, col. 12.
3. Designation: Where specifying metric helical spring lock washers, the designation shall include the nominal size and the washer series. The nominal washer sizes listed in the dimensional tables indicate the washers are intended for use with metric threaded fasteners of comparable sizes.

FASTENERS

Table 9-87. Heavy Helical Spring Lock Washers (GM)

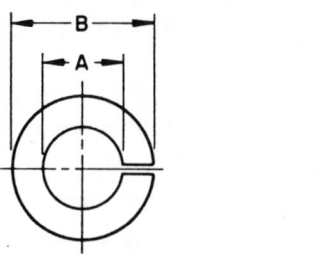

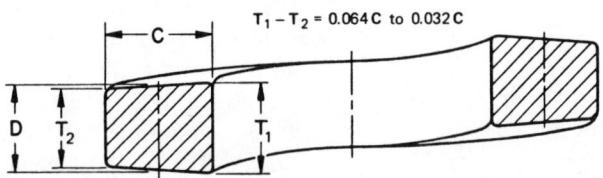

ENLARGED SECTION

Nominal Washer Size	A Inside Diameter Max	A Inside Diameter Min	B Outside Diameter Max	C Section Width Min	D Mean Section Thickness Max	D Mean Section Thickness Min	T_1-T_2 Increase in Thickness, OD to ID Max	T_1-T_2 Increase in Thickness, OD to ID Min	GM PART NUMBERS STEEL Plain	GM PART NUMBERS STEEL Zinc	GM PART NUMBERS STEEL Phos Coated
2.5	2.70	2.55	5.29	1.19	0.94	0.79	0.076	0.038	11500150	11500155	11500170
3	3.21	3.06	6.24	1.40	1.17	1.02	0.091	0.046	11500151	11500156	11500171
3.5	3.76	3.58	6.76	1.40	1.17	1.02	0.091	0.046	11500152	11500157	11500172
4	4.26	4.08	7.64	1.57	1.34	1.19	0.102	0.051	11500153	11500158	11500173
5	5.26	5.08	9.07	1.78	1.57	1.42	0.112	0.056	11500154	11500159	11500174
6	6.29	6.10	12.12	2.79	2.21	1.96	0.178	0.089	11503964	11503965	11503966
8	8.36	8.13	15.22	3.30	2.72	2.46	0.213	0.107	11500045	11500161	11500176
10	10.38	10.13	18.03	3.68	3.18	2.92	0.234	0.117	11500046	11500162	11500177
12	12.45	12.15	21.73	4.47	4.09	3.84	0.284	0.142	11500047	11500163	11500178
14	14.50	14.20	24.64	4.90	4.57	4.32	0.315	0.157	11500048	11500164	11500179
16	16.63	16.25	27.88	5.33	5.05	4.80	0.340	0.170	11500049	11500165	11500180
20	20.66	20.28	33.77	6.20	5.99	5.74	0.396	0.198	11500050	11500166	11500181
24	24.81	24.30	40.66	7.57	7.72	7.21	0.483	0.241	11500051	11500167	11500182
30	31.25	30.51	50.81	9.47	9.75	9.25	0.605	0.302	11500052	11500168	11500183
36	37.50	36.61	60.60	11.23	11.68	11.18	0.716	0.358	11500053	11500169	11500184

NOTES:

1. All dimensions are in millimeters.
2. Material: AISI 1060. See Table 10-42C, col. 14 or Table 10-44, col. 12.
3. Designation: Where specifying metric helical spring lock washers, the designation shall include the nominal size and the washer series. The nominal washer sizes listed in the dimensional tables indicate the washers are intended for use with metric threaded fasteners of comparable sizes.

RIVETS

Two basic types of rivets are in use: the tubular type most frequently used in the United States and the solid type mostly used in other countries. Very little ISO standardization work has been done in this area, and the main publication is the ISO 1051 standard which specifies the nominal rivet shank diameters from 1 to 36 mm as shown in Table 9-88. Existing national standards numbers for solid metric rivets of various types can be found in the world fastener index on page 9-12. Publications are available from ANSI, 1430 Broadway, New York, N.Y. 10018, U.S.A.

Table 9-88. Rivet Shank Diameters (ISO 1051)

Nominal diameters	
Main series	Secondary series
1	
1.2	
	1.4
1.6	
2	
2.5	
3	
	3.5
4	
5	
6	
	7
8	
10	
12	
	14
16	
	18
20	
	22
24	
	27
30	
	33
36	

NOTE:
1. Dimensions are in millimeters.

Blind Rivets

Tubular break mandrel blind rivets are specified in the IFI-505 standard for regular types and in the IFI-509 standard for closed end types. The above standards have a complete description of test methods and inspection of blind rivets. The standards are available from Industrial Fasteners Institute, 1505 East Ohio Building, Cleveland, Ohio 44114, U.S.A.

Portions of the IFI-505 and IFI-509 standards are reprinted here with permission from IFI.

Break Mandrel Blind Rivets

Break mandrel blind rivets are pull mandrel type blind rivets where during the setting operation the mandrel is pulled into or against the rivet body and breaks at or near the junction of the mandrel shank and its upset end.

The two basic styles of break mandrel blind rivets are designed as protruding head and flush head. Protruding head rivets are available in two styles: regular head and large head. Flush head rivets are available in two styles: 100 deg countersunk head and 120 deg countersunk head.

Grades—The material combinations of break mandrel blind rivets are designated as grades, with each material combination representing a different combination of rivet body material and mandrel material as given in Table 9-89.

Rivet Dimensions—Protruding and flush head break mandrel blind rivets shall conform to the dimensions given in Tables 9-90 and 9-92, respectively.

Application Data—Recommendations on the selection and application of protruding and flush head break mandrel blind rivets are given in Tables 9-91 and 9-93, respectively.

Shear Strength—Rivets, except Grades 20 and 21, shall have ultimate shear strengths not less than the minimum ultimate shear strengths specified for the applicable size and grade given in Table 9-94.

Tensile Strength—Rivets, except Grades 20 and 21, shall have ultimate tensile strengths not less than the minimum ultimate tensile strengths specified for the applicable size and grade given in Table 9-94.

Mandrel Break Load—While the rivet is being set, the axially applied load necessary to break the mandrel shall be within the limits specified for the applicable rivet size and grade in Table 9-95.

Break Mandrel Closed End Blind Rivets

Break mandrel closed end blind rivets are pull mandrel type blind rivets where during the setting operation the mandrel is pulled within the rivet body and breaks at or near the junction of the mandrel shank and its upset end.

Closed End—The end of the rivet, as manufactured, is solid and remains closed on the blind side after setting.

FASTENERS

The two basic styles of break mandrel closed end blind rivets are designated as protruding head and flush head. Flush head rivets are available only in the 120° countersunk head style.

Grades—The material combinations of break mandrel closed end blind rivets are designated as grades, with each material combination representing a different combination of rivet body material and mandrel material as given in Table 9-96.

Rivet Dimensions—Protruding and flush head break mandrel closed end blind rivets shall conform to the dimensions given in Tables 9-97 and 9-98, respectively.

Application Data—Recommendations on the selection and application of protruding and flush head break mandrel closed end blind rivets are given in Tables 9-98 and 9-100, respectively.

Shear Strength—Rivets, except those described below, shall have ultimate shear strengths not less than the minimum ultimate shear strengths specified for the applicable size and grade given in Table 9-101.

Tensile Strength—Rivets, except those described below, shall have ultimate tensile strengths not less than the minimum ultimate tensile strengths specified for the applicable size and grade given in Table 9-96.

Protruding head rivets with specified maximum grip lengths shorter than 1.0 times the nominal rivet diameter, and flush head rivets with specified maximum grip lengths shorter than 1.5 times the nominal rivet diameter shall not be subject to either shear or tensile testing.

PINS

National and international standards for pins of various types are shown in the world metric fastener standards index. Design details for the most popular type pins shown in this book are either covered in an ISO standard or a proposal.

Split Cotter Pins (ISO 1234)

Metric standards for cotter pins are in basic agreement with the ISO 1234 standard worldwide, and pins in nominal sizes from 0.6 to 10 mm are shown in Table 9-103.

Roll Pins (IFI-512-S)

Slotted spring pins are covered in the IFI-512-S standard, which has been submitted as an ISO proposal. It is a hard converted metric standard which reflects U.S. practices for this type of fastener. Nominal pin sizes from 1.5 to 12 mm are shown in Table 9-104A.

Coiled Spring Pins (IFI-512-C)

Coiled spring pins are covered in the IFI-512-C standard, which has been submitted as an ISO proposal. It is a standard based on hard metric modules which also reflects U. S. practices for this type of fastener. Nominal pin sizes from 0.8 to 20 mm are shown in Table 9-104B.

Table 9-89. Grades of Break Mandrel Blind Rivets (IFI-505)

Grade Designation	Rivet Body Material	Mandrel Material
10	Aluminum Alloy 5050	Aluminum Alloy 7178 or 2024
11	Aluminum Alloy 5052	Aluminum Alloy 7178 or 2024
17	Aluminum Alloy 5154	Carbon Steel
18	Aluminum Alloy 5052	Carbon Steel
19	Aluminum Alloy 5056	Carbon Steel
20	Copper Alloy No. 110	Carbon Steel
21	Copper Alloy No. 102	Carbon Steel
30	Low Carbon Steel	Carbon Steel
40	Nickel-Copper Alloy (Monel)	Carbon Steel
50	Stainless Steel (300 Series)	Carbon Steel
51	Stainless Steel (300 Series)	Stainless Steel (300 Series, A286 or equivalent)
52	Stainless Steel (300 Series	Stainless Steel (400 Series)

Table 9-90. Dimensions of Regular and Large Protruding Head Style Break Mandrel Blind Rivets (IFI-505)

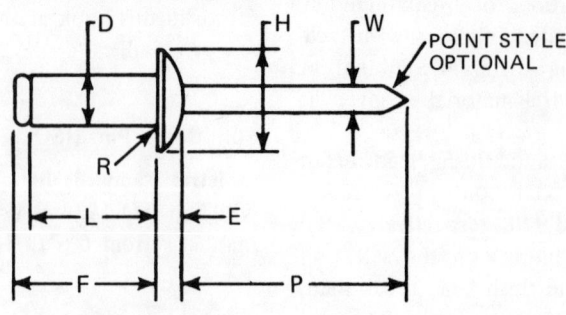

Rivet Series No.	Nom Rivet Size	D Body Dia Max	D Body Dia Min	Style 1 — Regular Head H Head Dia Max	Style 1 — Regular Head H Head Dia Min	Style 1 — Regular Head E Head Height Max	Style 2 — Large Head H Head Dia Max	Style 2 — Large Head H Head Dia Min	Style 2 — Large Head E Head Height Max	R Radius of Fillet Max	W Mandrel Dia Nom	P Mandrel Protrusion Min	F Blind Side Protrusion Max
3	2.4	2.44	2.29	5.03	4.52	0.81	7.44	6.83	1.02	0.4	1.45	25	L + 2.5
4	3.2	3.25	3.10	6.65	6.05	1.02	9.91	9.14	1.65	0.5	1.93	25	L + 3.0
5	4.0	4.04	3.89	8.33	7.52	1.27	12.40	11.38	1.90	0.5	2.41	27	L + 3.5
6	4.8	4.85	4.65	10.01	9.04	1.52	16.51	15.24	2.34	0.7	2.90	27	L + 4.0
8	6.3	6.48	6.25	13.33	12.07	2.03	19.81	18.29	2.72	0.8	3.84	31	L + 4.5

Notes:
1. All dimensions are in millimeters.
2. For application data see Table 9-91.
3. The junction of head and shank shall have a fillet with a max radius as shown. For Grades 40, 50, 51, and 52 rivets, the max fillet radius for No. 6 rivets shall be 0.9 mm, and for No. 8 rivets shall be 1.5 mm.
4. When computing the blind side protrusion (F), the max length of rivet (L), as given in Table 9-91 for the applicable grip shall be used. Minimum blind side clearance may be calculated by subtracting the actual grip (G), (i.e. total thickness of the material to be joined), from the specified blind side protrusion (F). (Example: To join two plates, each 2.5 mm thick, with a No. 5 rivet, a No. 54 rivet would be used. Minimum blind side clearance necessary to permit proper rivet setting would be L + 3.5 − G, which is 10.8 + 3.5 − 5.0, and equals 9.3 mm).

FASTENERS

Fig. 9-91. Application Data for Protruding Head Style Break Mandrel Blind Rivets (IFI-505)

Rivet Series No.	Nom Rivet Size	Recommended Metric Drill Size	Recommended Hole Size Max	Recommended Hole Size Min	Rivet No.	Grip Range	Rivet Length L Max
3	2.4	2.5	2.54	2.46	32 34 36	0.5 to 3.2 3.3 to 6.4 6.5 to 9.5	6.4 9.5 12.7
4	3.2	3.3	3.38	3.28	41 42 43 44 45 46 48 410	0.5 to 1.6 1.7 to 3.2 3.3 to 4.8 4.9 to 6.4 6.5 to 7.9 8.0 to 9.5 9.6 to 12.7 12.8 to 15.9	5.4 7.0 8.6 10.2 11.7 13.4 16.5 19.7
5	4.0	4.1	4.16	4.06	52 53 54 56 58 510	0.5 to 3.2 3.3 to 4.8 4.9 to 6.4 6.5 to 9.5 9.6 to 12.7 12.8 to 15.9	7.6 9.2 10.8 14.0 17.2 20.3
6	4.8	4.9	4.98	4.88	62 63 64 66 68 610 612 614 616 618	0.5 to 3.2 3.3 to 4.8 4.9 to 6.4 6.5 to 9.5 9.6 to 12.7 12.8 to 15.9 16.0 to 19.1 19.2 to 22.2 22.3 to 25.4 25.5 to 28.6	8.3 9.8 11.5 14.6 17.8 21.0 24.2 27.3 30.5 33.7
8	6.3	6.5	6.63	6.53	82 84 86 88 810 812 814 816 818 820	0.5 to 3.2 3.3 to 6.4 6.5 to 9.5 9.6 to 12.7 12.8 to 15.9 16.0 to 19.1 19.2 to 22.2 22.3 to 25.4 25.5 to 28.6 28.7 to 31.8	9.5 12.7 15.9 19.1 22.2 25.4 28.6 31.8 34.9 38.1
See Notes		2					

NOTES:
1. All dimensions are in millimeters.
2. Recommended drill sizes are those which normally produce holes within the specified hole size limits.

Table 9-92. Dimensions of 100 Deg and 120 Deg Flush Head Style Break Mandrel Blind Rivets (IFI-505)

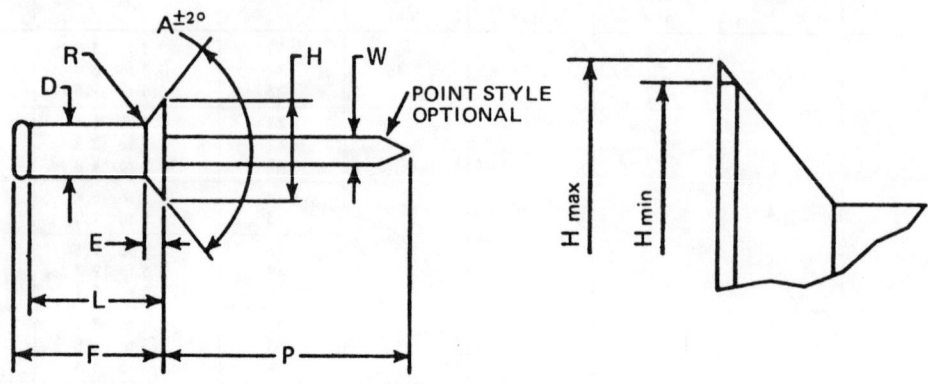

Rivet Series No.	Nom Rivet Size	D Body Dia Max	D Body Dia Min	Style 3 — 100 Deg Head A Head Angle Deg Nom	Style 3 — 100 Deg Head H Head Dia Max	Style 3 — 100 Deg Head H Head Dia Min	Style 3 — 100 Deg Head E Head Height Ref	Style 4 — 120 Deg Head A Head Angle Deg Nom	Style 4 — 120 Deg Head H Head Dia Max	Style 4 — 120 Deg Head H Head Dia Min	Style 4 — 120 Deg Head E Head Height Ref	R Radius of Fillet Max	W Mandrel Dia Nom	P Mandrel Protrusion Min	F Blind Side Protrusion Max
3	2.4	2.44	2.29	100	4.75	4.09	0.99	120	4.75	4.09	0.69	0.5	1.45	25	L + 2.5
4	3.2	3.25	3.10	100	5.92	5.26	1.14	120	5.92	5.26	0.79	0.7	1.93	25	L + 3.0
5	4.0	4.04	3.89	100	7.47	6.81	1.47	120	7.47	6.81	1.02	0.8	2.41	27	L + 3.5
6	4.8	4.85	4.65	100	9.17	8.51	1.85	120	9.17	8.51	1.27	0.9	2.90	27	L + 4.0
See Notes					3		4		3		4				5

Notes:
1. All dimensions are in millimeters.
2. For application data see Table 9-93.
3. Max head diameter is calculated on nominal rivet diameter and nominal head angle extended to sharp corner. Min head diameter is absolute.
4. Head height is given for reference purposes only. Variations in this dimension are controlled by the diameters (H) and (D) and the included angle of the head.
5. When computing the blind side protrusion (F), the max length of rivet (L) as given in Table 9-93 for the applicable grip shall be used. Minimum blind side clearance may be calculated by subtracting the actual grip (G), (i.e. total thickness of the material to be joined), from the specified blind side protrusion (F). (Example: To join two plates, each 4.7 mm thick, with a No. 6 rivet, a No. 66 rivet would be used. Minimum blind side clearance necessary to permit proper rivet setting would be L + 4.0 − G, which is 14.6 + 4.0 − 9.4 which equals 9.2 mm).

FASTENERS

Parallel Pins, Unhardened (ISO 2338)

Dowel pins of Types A, B, and C are specified in the ISO 2338 standard, and nominal diameters from 0.6 to 25 mm are shown in Table 9-108. National standards for dowel pins are in general agreement with the ISO standard worldwide.

Taper Pins (ISO 2339)

Table 9-109 shows ISO 2339 standard 1:50 taper pins with small diameters from 0.6 to 25 mm. Similar taper pins are covered in national standards throughout the world, as shown in the index.

Clevis Pins (ISO-2340)

Plain clevis pins without heads of Type A (without holes) and Type B (with holes) are specified in ISO 2340 standard, and nominal sizes from 3 to 50 mm are shown in Table 9-110.

Clevis Pins with Head (ISO 2341)

The ISO 2341 standard defines clevis pins with head of Type A (without hole) and Type B (with hole). Pins with nominal diameters from 3 to 50 mm are shown in Table 9-111.

Table 9-93. Application Data for Flush Head Style Break Mandrel Blind Rivets (IFI-505)

Rivet Series No.	Nom Rivet Size	Recommended Metric Drill Size	Recommended Hole Size Max	Recommended Hole Size Min	Rivet No.	Grip Range	Rivet Length L Max
3	2.4	2.5	2.54	2.46	32	2.0 to 3.2	6.4
					34	3.3 to 6.4	9.5
4	3.2	3.3	3.38	3.28	42	2.3 to 3.2	7.0
					43	3.3 to 4.8	8.6
					44	4.9 to 6.4	10.2
					45	6.5 to 7.9	11.7
					46	8.0 to 9.5	13.4
					48	9.6 to 12.7	16.5
5	4.0	4.1	4.16	4.06	53	3.0 to 4.8	9.2
					54	4.9 to 6.4	10.8
					56	6.5 to 9.5	14.0
					58	9.6 to 12.7	17.2
6	4.8	4.9	4.98	4.88	63	3.8 to 4.8	9.9
					64	4.9 to 6.4	11.5
					66	6.5 to 9.5	14.6
					68	9.6 to 12.7	17.8
					610	12.8 to 15.9	21.0
See Notes		2					

Notes:
1. All dimensions are in millimeters.
2. Recommended drill sizes are those which normally produce holes within the specified hole size limits.

Table 9-94. Ultimate Shear and Tensile Strengths of Break Mandrel Blind Rivets (IFI-505)

Nom Rivet Size mm	Ultimate Shear Strength newtons, min					Ultimate Tensile Strength newtons, min				
	Grades 10, 11, 18	Grade 17, 19	Grade 30	Grade 40	Grades 50, 51, 52	Grades 10, 11, 18	Grade 17, 19	Grade 30	Grade 40	Grades 50, 51, 52
2.4	310	400	580	890	1020	360	530	760	1110	1250
3.2	530	760	1160	1560	1870	670	980	1380	2000	2360
4.0	850	1160	1650	2450	2890	1020	1560	2090	3110	3650
4.8	1160	1690	2400	3560	4230	1420	2220	3020	4450	5340
6.3	2050	3110	4450	6230	7560	2490	4090	5520	8230	9340

Note:
Grades 20 and 21 rivets are not subject to shear and tensile testing.

General Notes for Slotted Spring Pins

All dimensions are in millimeters unless noted.

Diameter. The maximum diameter shall be inspected using a GO ring gage. The minimum diameter shall be determined by averaging the diameters D_1, D_2 and D_3. These diameters shall be measured at the mid length of pins with lengths 25 mm or shorter; and at a distance of approximately 6 mm from each end of pins with lengths longer than 25 mm.

Slot. Slot design shall be in accordance with the practice of the manufacturer, except that the slot width shall be narrow enough to prevent nesting or interlocking of pins and wide enough to permit satisfactory installation in the hole of the minimum diameters as recommended in Table 9-104A.

Straightness—Straightness of pins shall be within the limits necessary to permit pins to fall through a ring gage by their own weight. Ring gage dimensions shall be as shown in Table 9-106A.

Lengths—Length of pins is the distance measured parallel to the axis of the pin between the extreme ends. Length tolerances are given in Table 9-105A.

Material—Carbon steel pins shall be made of analyses AISI 1070-1095 (see Table 10-44, cols. 14, 15, 16). Corrosion resistant steel pins shall be made of AISI 420 (see Table 10-49, col. 9). Pins shall be heat treated as necessary to meet the mechanical and performance requirements.

Hardness—Carbon steel pins shall have a hardness of Rockwell C46-53. Corrosion resistant steel pins shall have a hardness of Rockwell C43-52. Hardness may be determined using any acceptable method with Tukon or Rockwell testing equipment provided that deflection of the pin under load is prevented. The proper Rockwell scale should be selected when testing pins of different wall thickness. Equivalencies to the Rockwell C scale shall be in accordance with SAE J417.

Finish—Pins shall be furnished plain unless otherwise specified. When pins are electroplated or phosphate-

Table 9-95. Mandrel Break Loads of Break Mandrel Blind Rivets (IFI-505)

Nom Rivet Size mm	Grade	10, 11	17, 18, 19	20, 21	30	40	50	51, 52
	Rivet Body Material	Aluminum	Aluminum	Copper	Steel	Monel	Stainless Steel	Stainless Steel
	Mandrel Material	Aluminum	Steel	Steel	Steel	Steel	Steel	Stainless Steel
2.4	Min	620	780	780	1160	1330	1330	1330
	Max	1070	1220	1220	1600	2000	2220	2220
3.2	Min	1110	1780	1780	2670	2890	2890	2890
	Max	1780	2670	2670	3560	3780	4230	4230
4.0	Min	1890	2670	2670	3340	4230	5120	5120
	Max	2670	3780	3780	4450	5340	6450	6450
4.8	Min	2780	3340	3340	5120	6450	6230	6230
	Max	3670	4670	4670	6450	7780	8450	8450
6.3	Min	4890	6450	6450	8670	11100	13300	13300
	Max	6230	8230	8230	10500	12900	16000	16000

NOTES:
1. All loads in newtons.
2. Mandrel break load is defined as the load in newtons necessary to break the mandrel when setting break mandrel types of pull mandrel blind rivets.

Table 9-96. Grades of Break Mandrel Closed End Blind Rivets (IFI-509)

Grade Designation	Rivet Body Material	Mandrel Material
15	Aluminum Alloy 1100	Aluminum Alloy 7178 or 2024
19	Aluminum Alloy 5056	Carbon Steel
20	Copper Alloy No. 110	Carbon Steel

FASTENERS

Table 9-97. Dimension of Protruding Head Style Break Mandrel Closed End Blind Rivets (IFI-509)

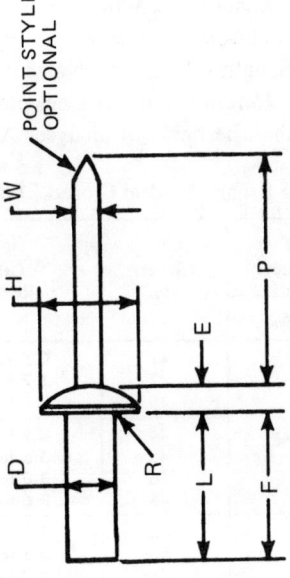

Rivet Series No.	Nom Rivet Size	D Body Dia Max	D Body Dia Min	H Style 1 — Regular Head Head Dia Max	H Style 1 — Regular Head Head Dia Min	E Head Height Max	R Radius of Fillet Max	W Mandrel Dia Max	P Mandrel Protrusion Min	F Blind Side Protrusion Max
4	3.2	3.25	3.10	6.40	5.69	1.27	0.7	1.85	25	Equal to "L" Rivet Length
5	4.0	4.04	3.89	8.33	7.52	1.65	0.7	2.31	27	
6	4.8	4.85	4.65	10.01	9.04	2.03	0.7	2.77	27	
8	6.3	6.48	6.25	13.33	12.07	2.54	0.7	3.71	27	
See Notes							3			4

Notes:
1. All dimensions are in millimeters.
2. For application data see Table 9-98.
3. The junction of head and shank shall have a fillet with a max. radius as shown.
4. The blind side protrusion (F) equals the max length of rivet (L) as given in Table 9-98 for application grip. Minimum blind side clearance may be caculated by subtracting the actual grip (G) (i.e. the total thickness of the material to be joined) from the blind side protrusion (F). (Example: To join two plates each 2.5 mm thick with a No. 5 rivet, a No. 54 rivet would be used. Minimum blind side clearance necessary to permit proper rivet setting would be L−G, which is 12.7 − 5.0 and equals 7.7 mm).

coated, appropriate plating or coating processes should be employed to avoid hydrogen embrittlement. If necessary, pins shall be suitably treated as soon as practicable after plating or coating to remove detrimental hydrogen embrittlement.

Performance—Pins shall be capable of withstanding the double shear loads specified in Table 9-107A. The shear test shall be performed in a fixture in which the pin support members and the member for applying the load shall have holes with diameters conforming to the recommended minimum hole sizes specified in Table 9-104A, and shall have a hardness of Rockwell C58 minimum. The clearance between the supporting members and the loading member shall not exceed 0.15 mm, and the shear plane shall be at least one pin diameter away from each end. Pins shall be located in the fixture so that the slot is approximately at right angles to the line of applied load. Pins too short to be tested in double shear shall be tested by shearing two pins simultaneously in single shear. Speed of testing shall not exceed 13 mm per minute.

Pins shall be tested to failure. The maximum load applied to the pin coincident with or prior to pin failure shall be recorded as the double shear strength of the pin. Pins, tested for shear strength, shall show a ductile shear without longitudinal cracks.

General Notes for Coiled Spring Pins

All dimensions are in millimeters unless noted.

Diameter—Pin diameter shall be inspected using GO and NO GO ring gages.

Straightness—Straightness of pins shall be within the limits necessary to permit pins to fall through a ring gage by their own weight. Ring gage dimensions shall be as specified in Table 9-106B.

Lengths—Length of pins is the distance measured parallel to the axis of the pin between the extreme ends. Length tolerances are given in Table 9-105B.

Material—Carbon steel pins of sizes 12 mm and smaller shall be made of analyses AISI 1070-1095 (see Table 10-

Table 9-98. Application Data for Protruding Head Style Break Mandrel Closed End Blind Rivets (IFI-509)

Rivet Series No.	Nom Rivet Size	Recommended Metric Drill Size	Recommended Hole Size Max	Recommended Hole Size Min	Rivet No.	Grip Range	Rivet Length L Max
4	3.2	3.3	3.38	3.28	41	0.5 to 1.6	7.5
					42	1.7 to 3.2	9.1
					43	3.3 to 4.8	10.7
					44	4.9 to 6.4	12.3
					45	6.5 to 7.9	13.9
					46	8.0 to 9.5	15.5
					48	9.6 to 12.7	18.7
5	4.0	4.1	4.16	4.06	52	0.5 to 3.2	9.5
					53	3.3 to 4.8	11.1
					54	4.9 to 6.4	12.7
					55	6.5 to 7.9	14.3
					56	8.0 to 9.5	15.9
					58	9.6 to 12.7	19.1
6	4.8	4.9	4.98	4.88	62	0.5 to 3.2	10.3
					63	3.3 to 4.8	11.9
					64	4.9 to 6.4	13.5
					66	6.5 to 9.5	16.7
					68	9.6 to 12.7	19.8
					610	12.8 to 15.9	23.0
					612	16.0 to 19.1	26.1
8	6.3	6.5	6.63	6.53	82	0.5 to 3.2	11.3
					84	3.3 to 6.4	14.5
					86	6.5 to 9.5	17.7
					88	9.6 to 12.7	20.8
					810	12.8 to 15.9	24.0
					812	16.0 to 19.1	27.2
					814	19.2 to 22.2	30.4
					816	22.3 to 25.4	33.5
See Notes		2					

1. All dimensions are in millimeters.
2. Recommended drill sizes are those which normally produce holes within the specified hole size limits.

FASTENERS

44, cols. 14, 15, 16). Pins of sizes larger than 12 mm shall be made of alloy steel analysis AISI 6150 (see Table 10-51, col. 13). Corrosion resistant pins shall be made of AISI 420 or AISI 302 (see Table 10-49, col. 9 or Table 10-50, col. 5), as specified by the purchaser. Pins shall be heat treated as necessary to meet the mechanical and performance requirements. Pins of other materials shall be as agreed upon by purchaser and manufacturer.

Hardness—AISI 1070-1095 pins shall have a hardness of Rockwell C 46-53. AISI 6150 pins shall have a hardness of Rockwell C 43-51. AISI 420 pins shall have a hardness of Rockwell C 46-55. AISI 302 is a work hardening material and hardness limits are not specified. Hardness may be determined using any acceptable method with Tukon or Rockwell testing equipment, provided deflection of the pin under load is prevented. Longitudinal or axial cutting and testing individual coils, or testing the end of a mounted pin is recommended to eliminate the effects of flexibility. The proper Rockwell Scale should be selected when testing pins of different wall thickness. Equivalencies to the Rockwell C Scale shall be in accordance with SAE J417.

Finish—Pins shall be furnished plain unless otherwise specified. When pins are electroplated or phosphate coated, appropriate plating or coating processes should be employed to avoid hydrogen embrittlement. If necessary,

Table 9-99. Dimensions of 120 Degree Flush Head Style Break Mandrel Closed End Blind Rivets (IFI-509)

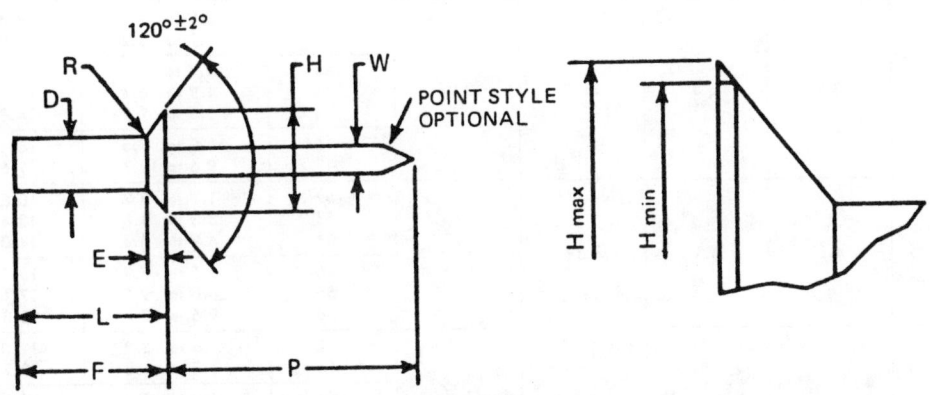

Rivet Series No.	Nom Rivet Size	D Body Dia Max	D Body Dia Min	H Style 4 — 120 Deg Head Head Dia Max	H Style 4 — 120 Deg Head Head Dia Min	E Head Height Ref	R Radius of Fillet Max	W Mandrel Dia Max	P Mandrel Protrusion Min	F Blind Side Protrusion Max
4	3.2	3.25	3.10	6.22	5.61	1.07	0.7	1.85	25	Equal to "L" Rivet Length
5	4.0	4.04	3.89	8.33	7.52	1.30	0.7	2.31	27	
6	4.8	4.85	4.65	10.01	9.04	1.52	0.7	2.77	27	
8	6.3	6.48	6.25	13.33	12.07	2.03	0.7	3.71	27	
See Notes				3		4				5

NOTES:
1. All dimensions are in millimeters.
2. For application data see Table 9-100.
3. Max head diameter is calculated on nominal rivet diameter and nominal head angle extended to sharp corner. Min. head diameter is absolute.
4. Head height is given for reference purposes only. Variations in this dimension are controlled by the diameters (H) and (D) and the included angle of the head.
5. The blind side protrusion (F) equals the max length of rivet (L) as given in Table 9-100 for the applicable grip. Minimum blind side clearance may be calculated by subtracting the actual grip (G) (i.e. the total thickness of the material to be joined) from the blind side protrusion (F). (Example: To join two plates each 2.5 mm thick with a No. 5 rivet, a No. 54 rivet would be used. Minimum blind side clearance necessary to permit proper rivet setting would be L–G, which is 14.0 – 5.0 and equals 9.0 mm).

pins shall be suitably treated as soon as practicable after plating or coating to remove detrimental hydrogen embrittlement.

Performance—Pins shall be capable of withstanding the double shear loads specified in Table 9-107B.

The shear test shall be performed in a fixture in which the pin support members and the member for applying the load shall have holes with diameters conforming to the nominal pin size as specified in Table 9-104B and shall have a hardness of Rockwell C 58 minimum. The clearance between the supporting members and the loading member shall not exceed 0.15 mm, and the shear plane shall be at least one pin diameter away from each end, and at least two diameters apart. Pins too short to be tested in double shear shall be tested by shearing two pins simultaneously in single shear. Speed of testing shall not exceed 13 mm per minute.

Pins shall be tested to failure. The maximum load applied to the pin coincident with or prior to pin failure shall be recorded as the double shear strength of the pin.

Table 9-100. Application Data for Flush Head Style Break Mandrel Closed End Blind Rivets (IFI-509)

Rivet Series No.	Nom Rivet Size	Recommended Metric Drill Size	Recommended Hole Size Max	Recommended Hole Size Min	Rivet No.	Grip Range	Rivet Length L Max
4	3.2	3.3	3.38	3.28	41	0.9 to 1.6	8.4
					42	1.7 to 3.2	10.0
					43	3.3 to 4.8	11.6
					44	4.9 to 6.4	13.2
					45	6.5 to 7.9	14.8
					46	8.0 to 9.5	16.4
					48	9.6 to 12.7	19.6
5	4.0	4.1	4.16	4.06	52	1.7 to 3.2	10.8
					53	3.3 to 4.8	12.4
					54	4.9 to 6.4	14.0
					55	6.5 to 7.9	15.5
					56	8.0 to 9.5	17.1
					58	9.6 to 12.7	20.3
6	4.8	4.9	4.98	4.88	62	1.6 to 3.2	12.0
					63	3.3 to 4.8	13.7
					64	4.9 to 6.4	15.3
					66	6.5 to 7.9	18.7
					68	8.0 to 12.7	21.6
					610	12.8 to 15.9	26.1
					612	16.0 to 19.1	28.0
See Notes		2					

Notes:
1. All dimensions are in millimeters.
2. Recommended drill sizes are those which normally produce holes within the specified hole size limits.

Table 9-101. Ultimate Shear and Tensil Strength of Break Mandrel Closed End Blind Rivets (IFI-509)

NOM RIVET SIZE mm	ULTIMATE SHEAR STRENGTH Min newtons			ULTIMATE TENSILE STRENGTH Min newtons		
	GRADE 15	GRADE 19	GRADE 20	GRADE 15	GRADE 19	GRADE 20
3.2	440	1070	980	490	1240	1330
4.0	580	1560	–	710	2130	–
4.8	930	2220	–	1110	3070	–
6.3	–	4000	–	–	4890	–

FASTENERS

Table 9-102. Mandrel Break Loads of Break Mandrel Closed End Blind Rivets (IFI-509)

NOM RIVET SIZE mm	GRADE	15	19	20
	RIVET BODY MATERIAL	ALUMINUM	ALUMINUM	COPPER
	MANDREL MATERIAL	ALUMINUM	STEEL	STEEL
3.2	Min	1110	2450	2450
	Max	1780	3340	3340
4.0	Min	1890	3110	–
	Max	2670	4890	–
4.8	Min	2670	4000	–
	Max	3560	6340	–
6.3	Min	–	8450	–
	Max	–	10230	–

NOTES:
1. All loads in newtons.
2. Mandrel break load is defined as the load in newtons necessary to break the mandrel when setting break mandrel closed end blind rivets.

Pins, tested for shear strength, shall show a ductile shear without longitudinal cracks.

Coiled type pins are designed to withstand shock and rapidly changing oscillating or intermittent dynamic loads. Due to the many factors which are involved in dynamic loading, theoretical data cannot readily be related to actual application performance. Tests which simulate the conditions of the actual application are recommended to determine dynamic loading performance.

Workmanship—Pins shall be free from burrs, loose scale, seams, notches, sharp edges and corners and any other defects affecting their serviceability.

RETAINING RINGS

There is no ISO standard for retaining rings yet. National standards covering various types of retaining rings are shown in the world fastener index. The first German standards of this type of fastener were published in 1941, and other national standards have followed the basic principles of the DIN standards.

The following is a brief analysis of some of the advantages and disadvantages of the European and American designs for retaining rings, based on technical data from countries throughout the world.

In comparison with their American counterparts, shown in this book, European-type retaining rings covered by the DIN standards are thicker, less flexible, held to closer thickness tolerances, are seated in shallower grooves and are more difficult to assemble. The American-type retaining rings generally are thinner and utilize flexibility and a hysteresis principle of design, which permits the use of deeper grooves for higher thrust load capacity.

In both instances, the rings are made of hardened material while the shaft or housing in which the rings are installed generally is of softer material, making the groove wall the limiting factor for allowable thrust loads in the assembly. The greater the area of contact between groove wall and ring, the higher the load capacity.

For optimum ring performance, groove specifications prescribed by individual retaining ring manufacturers should be adhered to. Grooves that are cut too shallow could cause a ring to deflect to a point of failure. Because of stress concentrations, grooves cut too deep could reduce the strength of the shaft or housing in areas subjected to dynamic torsional and bending moments.

Basic Retaining Rings for Holes (Truarc 10-100 Series)

Axially assembled basic internal retaining rings are shown for nominal hole diameters from 8 to 78 mm in Table 9-112.

Basic Retaining Rings for Shafts (Truarc 11-100 Series)

Axially assembled basic external retaining rings for nominal shafts from 4 to 100 mm are shown in Table 9-113.

E-Rings for Shafts (Truarc 11-420 Series)

Radially assembled E-rings used for shafts provide the ease of "clip on" assembly. Rings for nominal shaft diameters from 1 to 25 mm are shown in Table 9-114.

Table 9-103 Split Cotter Pins (ISO 1234)

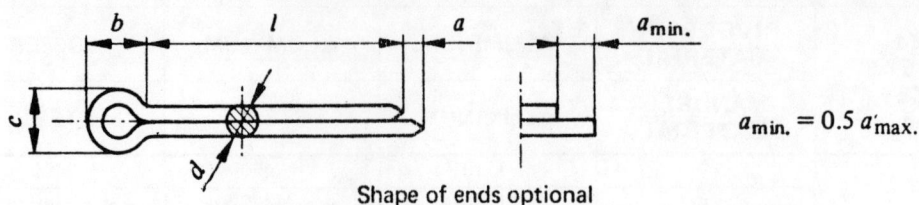

Shape of ends optional

$a_{min.} = 0.5\, a'_{max.}$

Nominal size (6)		0.6	0.8	1	1.2	1.6	2	2.5	3.2	4	5	6.3	8	10
d	max.	0.5	0.7	0.9	1	1.4	1.8	2.3	2.9	3.7	4.6	5.9	7.5	9.5
	min.	0.4	0.6	0.8	0.9	1.3	1.7	2.1	2.7	3.5	4.4	5.7	7.3	9.3
a	max.	1.6	1.6	1.6	2.5	2.5	2.5	2.5	3.2	4	4	4	4	6.3
b	≈	2	2.4	3	3	3.2	4	4	6.4	8	10	12.6	16	20
c	max.	1	1.4	1.8	2	2.8	3.6	4.6	5.8	7.4	9.2	11.8	15	19
	min.	0.9	1.2	1.6	1.7	2.4	3.2	4	5.1	6.5	8	10.3	13.1	16.6
Bolts	over	—	2.5	3.5	4.5	5.5	7	9	11	14	20	27	39	56
	to	2.5	3.5	4.5	5.5	7	9	11	14	20	27	39	56	80
Clevis pins	over	—	2	3	4	5	6	8	9	12	17	23	29	44
	to	2	3	4	5	6	8	9	12	17	23	29	44	69
(3) l	from	4	5	6	8	8	10	12	14	18	22	32	40	45
	to	12	16	20	25	32	40	50	63	80	100	125	160	200

Notes:
1. All dimensions are in millimeters.
2. For cotter pins with d = 13, 16, 20, see standard (ISO 1234).
3. Preferred length range l show minimum and maximum standard lengths inclusive.
   The standard lengths are as follows: l = 4, 5, 6, 8, 10, 12, 14, 16, 18, 20, 22, 25, 28, 32, 36, 40, 45, 50, 56, 63, 71, 80, 90, 100, 112, 125, 140, 160, 180, 200, 224, 250, 280
4. Tolerance on length is js17 (see Table 6-1). Other dimensions are held to ISO 2768 medium (see Table 6-34) in several national standards.
5. Material: Steel (St) (See Table 10-44, col. 2, Table 10-42-C, cols. 3 & 4), Brass, Copper (Cu), Aluminum (Al).
6. Nominal size = diameter of the split pin hole.
7. Typical designation: Cotter Pin 5 × 50 ISO 1234 AISI 1010.

FASTERNERS

Table 9-104A. Slotted Spring Pin Dimensions (IFI-512-S)*

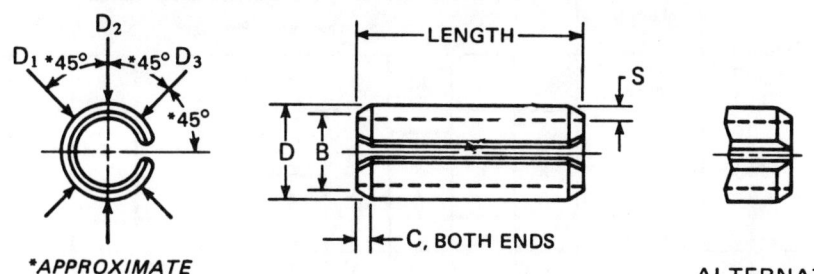

*APPROXIMATE

ALTERNATE END DESIGN

Nominal Pin Size	D Diameter Max	D Diameter Min	B Chamfer Dia Max	C Chamfer Length Max	C Chamfer Length Min	S Stock Thickness Nom	Recommended Hole Size Max	Recommended Hole Size Min
1.5	1.68	1.60	1.4	0.7	0.15	0.3	1.57	1.50
2	2.20	2.12	1.9	0.8	0.2	0.4	2.07	2.00
2.5	2.72	2.63	2.4	0.9	0.2	0.5	2.58	2.50
3	3.25	3.15	2.9	1.0	0.2	0.6	3.10	3.00
4	4.28	4.15	3.9	1.2	0.3	0.8	4.10	4.00
5	5.33	5.17	4.8	1.4	0.3	1.0	5.12	5.00
6	6.36	6.20	5.8	1.6	0.4	1.2	6.13	6.00
8	8.40	8.22	7.8	2.0	0.4	1.6	8.15	8.00
10	10.43	10.25	9.7	2.4	0.5	2.0	10.15	10.00
12	12.48	12.28	11.7	2.8	0.6	2.5	12.18	12.00

*See notes, page 9-104.

Table 9-104B. Coiled Spring Pin Dimensions (IFI-512-C)*

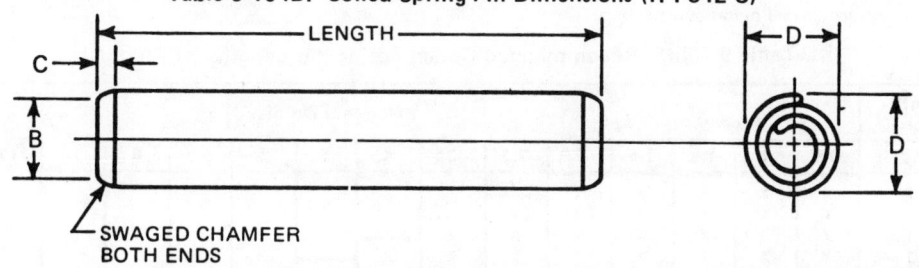

Nominal Pin Size	D Standard Duty Max	D Standard Duty Min	D Heavy Duty Max	D Heavy Duty Min	D Light Duty Max	D Light Duty Min	B Chamfer Dia Max	C Chamfer Length Ref	Recommended Hole Size Max	Recommended Hole Size Min
0.8	0.89	0.84	—	—	—	—	0.74	0.4	0.83	0.80
1	1.12	1.04	—	—	—	—	0.94	0.4	1.03	1.00
1.2	1.32	1.24	—	—	—	—	1.14	0.4	1.23	1.20
1.5	1.73	1.61	1.68	1.58	1.75	1.61	1.44	0.5	1.57	1.50
2	2.24	2.11	2.19	2.08	2.26	2.11	1.90	0.6	2.07	1.97
2.5	2.77	2.63	2.72	2.60	2.79	2.62	2.40	0.6	2.58	2.47
3	3.32	3.15	3.25	3.12	3.35	3.15	2.90	0.8	3.10	2.97
4	4.35	4.15	4.27	4.12	4.38	4.15	3.90	1.0	4.10	3.97
5	5.45	5.20	5.36	5.16	5.50	5.20	4.85	1.2	5.12	4.95
6	6.48	6.23	6.39	6.19	6.53	6.23	5.85	1.3	6.13	5.95
6.3	6.83	6.55	6.75	6.50	6.88	6.55	6.15	1.4	6.45	6.25
8	8.60	8.27	8.52	8.22	8.65	8.27	7.80	1.7	8.17	7.93
10	10.73	10.33	10.66	10.28	—	—	9.75	2.0	10.20	9.93
12	12.82	12.37	12.72	12.32	—	—	11.70	2.4	12.22	11.90
14	14.89	14.41	14.82	14.36	—	—	13.65	2.9	14.25	13.85
16	16.95	16.43	16.86	16.38	—	—	15.65	3.2	16.25	15.85
20	21.00	20.45	20.90	20.40	—	—	19.65	3.8	20.25	19.85

*See notes, page 9-106.

Table 9-105A. Recommended Slotted Spring Pin Lengths (IFI-512-S)

Nominal Length	Length Tol Plus and Minus	\multicolumn{10}{c}{Nominal Pin Size}
		1.5	2	2.5	3	4	5	6	8	10	12
4	0.40 for all lengths thru 25 mm	X	X								
5		X	X	X							
6		X	X	X	X						
8		X	X	X	X	X					
10		X	X	X	X	X	X				
12		X	X	X	X	X	X	X			
14		X	X	X	X	X	X	X			
16		X	X	X	X	X	X	X	X		
18		X	X	X	X	X	X	X	X		
20		X	X	X	X	X	X	X	X	X	
25		X	X	X	X	X	X	X	X	X	X
30	0.50 for all lengths over 25 mm thru 50 mm		X	X	X	X	X	X	X	X	X
35			X	X	X	X	X	X	X	X	X
40			X	X	X	X	X	X	X	X	X
45				X	X	X	X	X	X	X	X
50					X	X	X	X	X	X	X
55	0.60 for all lengths over 50 mm thru 75 mm					X	X	X	X	X	X
60						X	X	X	X	X	X
65								X	X	X	X
70								X	X	X	X
75								X	X	X	X
80	0.75 for all lengths over 75 mm								X	X	X
90									X	X	X
100									X	X	X

NOTE:
The lengths shown between the heavy lines for each pin size are considered standard and are generally available. Spring pins of lengths shorter, longer, or intermediate to those listed may require special manufacture.

Table 9-105B. Recommended Coiled Spring Pin Lengths (IFI-512-C)

Nominal Length	Length Tol Plus	\multicolumn{16}{c}{Nominal Pin Size}
		0.8	1	1.2	1.5	2	2.5	3	4	5	6	6.3	8	10	12	14	16	20
4	0.5 for all lengths thru 10 mm	X	X	X	X	X												
5		X	X	X	X	X	X											
6		X	X	X	X	X	X	X										
8		X	X	X	X	X	X	X	X									
10		X	X	X	X	X	X	X	X	X								
12	1.0 for all lengths over 10 mm thru 50 mm	X	X	X	X	X	X	X	X	X	X	X						
14		X	X	X	X	X	X	X	X	X	X	X						
16		X	X	X	X	X	X	X	X	X	X	X	X					
18					X	X	X	X	X	X	X	X	X					
20					X	X	X	X	X	X	X	X	X	X				
22				X	X	X	X	X	X	X	X	X	X	X				
24				X	X	X	X	X	X	X	X	X	X	X				
26				X	X	X	X	X	X	X	X	X	X	X	X			
28					X	X	X	X	X	X	X	X	X	X	X			
30						X	X	X	X	X	X	X	X	X	X			
35						X	X	X	X	X	X	X	X	X	X			
40						X	X	X	X	X	X	X	X	X	X	X		
45							X	X	X	X	X	X	X	X	X	X		
50							X	X	X	X	X	X	X	X	X	X	X
55	1.5 for all lengths over 50 mm							X	X	X	X	X	X	X	X	X	X	X
60								X	X	X	X	X	X	X	X	X	X	X
65										X	X	X	X	X	X	X	X	X
70										X	X	X	X	X	X	X	X	X
75										X	X	X	X	X	X	X	X	X
80													X	X	X	X	X	X
90													X	X	X	X	X	X
100|												X	X	X	X	X	X
120|													X	X	X	X	X
140|																X	X
160|																X	X

FASTENERS

Table 9-106A. Straightness Limits (IFI-512-S)

Pin Length mm	Gage Length mm Max	Gage Length mm Min	Gage Hole Dia, mm Min	Gage Hole Dia, mm Max
to and incl 25	25.15	24.85	Specified Pin Dia, Max plus 0.20	Specified Pin Dia, Max plus 0.21
over 25 to 50	50.15	49.85	0.40	0.42
over 50	75.15	74.85	0.60	0.63

Table 9-106B. Straightness Limits (IFI-512-C)

Pin Length mm	Gage Length mm Max	Gage Length mm Min	Gage Hole Dia, mm Min	Gage Hole Dia, mm Max
to and incl 24	25.15	24.85	Specified Pin Dia, Max plus 0.19	Specified Pin Dia, Max plus 0.20
over 24 to 50	50.15	49.85	0.32	0.34
over 50	75.15	74.85	0.45	0.48

Table 9-107A. Double Shear Strengths of Slotted Spring Pins (IFI-512-S)

Nominal Pin Size mm	Double Shear Strength Min kN
1.5	1.7
2	3.0
2.5	4.4
3	7.1
4	13.2
5	20.3
6	29.2
8	52.8
10	84.0
12	106

Table 9-107B. Double Shear Strengths of Coiled Spring Pins (IFI-512-C)

Nominal Pin Size mm	Standard Duty (1) 1070-1095 6150 420	Standard Duty 302	Heavy Duty 1070-1095 6150 420	Heavy Duty 302	Light Duty 1070-1095 6150 420	Light Duty 302
0.8	0.33	0.26	—	—	—	—
1	0.53	0.44	—	—	—	—
1.2	0.76	0.62	—	—	—	—
1.5	1.2	1.0	1.7	1.4	—	0.5
2	2.1	1.8	3.0	2.5	—	1.0
2.5	3.1	2.5	4.4	3.5	1.7	1.3
3	5.0	4.0	7.1	5.7	2.8	2.2
4	8.9	7.1	13.2	10.7	4.9	3.9
5	14.2	11.4	20.3	16.2	7.8	6.2
6	20.0	16.0	29.2	23.4	11.2	8.8
6.3	22.9	18.3	34.3	27.4	12.6	10.0
8	35.0	28.0	52.8	42.2	19.3	15.4
10	58.0	46.4	84.0	67.2	—	—
12	73.9	59.1	106	84.5	—	—
14	109	87.2	156	125	—	—
16	138	111	205	165	—	—
20	227	182	325	260	—	—

NOTE: 1. Pins of sizes 0.8 thru 1.2 mm are available in AISI 420 and 302 only.

Table 9-108. Parallel Pins, Unhardened (ISO 2338)

Type A
Tolerance on *d* : m6

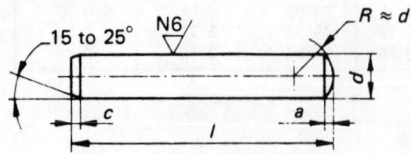

Type B
Tolerance on *d* : h8

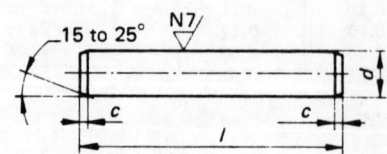

Type C
Tolerance on *d* : h11

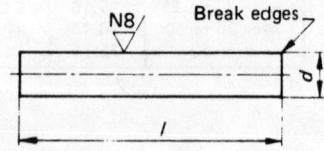

d	0.6	0.8	1	1.2	1.5	2	2.5	3	4	5	6	8	10	12	16	20	25
a ≈	0.08	0.1	0.12	0.16	0.2	0.25	0.3	0.4	0.5	0.63	0.8	1	1.2	1.6	2	2.5	3
c ≈	0.12	0.16	0.2	0.25	0.3	0.35	0.4	0.5	0.63	0.8	1.2	1.6	2	2.5	3	3.5	4
ℓ from	2	2	4	4	4	6	6	8	8	10	12	14	20	25	30	40	50
ℓ to	6	8	10	12	16	25	25	30	45	50	60	80	100	150	180	200	200

Notes:
1. All dimensions are in millimeters.
2. For parallel pins with nominal diameter d = 30, 40, 50, see ISO 2338 standard.
3. Preferred length range 1 shows minimum and maximum standard lengths inclusive. The standard lengths are as follows: 1 = 1, 2, 3, 4, 5, 6, 8, 10, 12, 14, 16, 20, 25, 30, 35, 40, 45, 50, 55, 60, 65, 70, 75, 80, 90, 100, 110, 120, 130, 140, 150, 160, 170, 180, 190, 200, and then in increments of 20 mm. For special circumstances, intermediate lengths selected from Table 9-1 are acceptable.
4. Tolerance on length is ISO js15 (see Table 6-27) and on nominal diameters m6, h6, h11 (see Tables 6-26 and 28).
5. Material. See Table 10-46 cols. 1 and 6; Table 10-43, col. 6; Table 10-42C, col. 4.
6. The ISO surface finish designations N6, N7, and N8 are specified in Table 5-1.
7. Steel bars to tolerance h11 for Type C pins are shown in Table 10-7.
8. Typical designation for a parallel pin with d = 4 to ISO tolerance m6, 1 = 20, U.S. Steel quality AISI 1211. Parallel Pin 4 m6 × 20 ISO 2338 Type A AISI 1211. Dowel pins to other national standards and tolerance classes have a similar designation.

FASTENERS 9-115

Table 9-109. Taper Pins (ISO 2339)

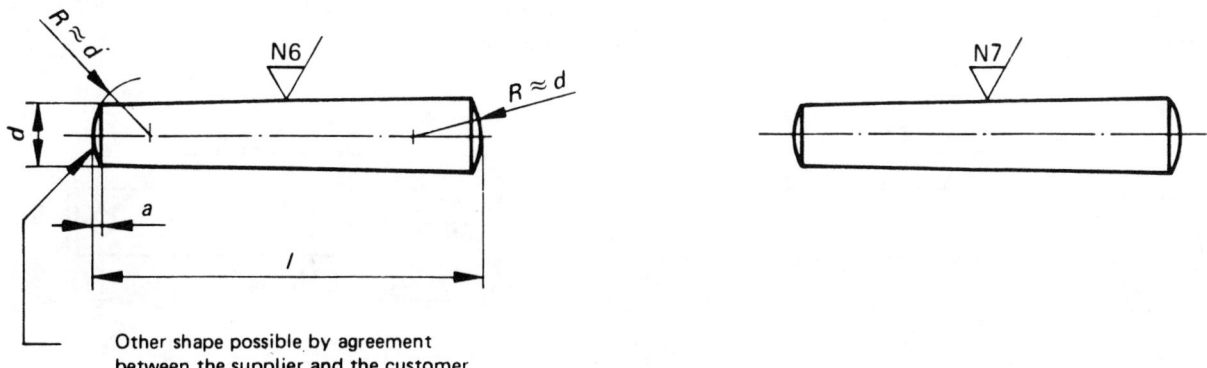

Other shape possible by agreement between the supplier and the customer

d h10	0.6	0.8	1	1.2	1.5	2	2.5	3	4	5	6	8	10	12	16	20	25
$a \approx$	0.08	0.1	0.12	0.12	0.2	0.25	0.3	0.4	0.5	0.63	0.8	1	1.2	1.6	2	2.5	3
l from	4	5	6	6	8	10	10	12	14	20	25	25	30	35	40	45	50
to	8	12	16	20	25	35	35	45	55	60	90	130	160	180	200	200	200

Notes:
1. All dimensions are in millimeters.
2. For taper pins with nominal diameters: d = 30, 40 and 50, see ISO 2339 standard.
3. Preferred length range l shows minimum and maximum standard lengths inclusive. The standard lengths are as follows: l = 2, 3, 4, 5, 6, 8, 10, 12, 14, 16, 20, 25, 30, 35, 40, 45, 50, 55, 60, 65, 70, 75, 80, 90, 100, 110, 120, 130, 140, 150, 160, 170, 180, 190, 200, and then in increments of 20 mm. For special circumstances, intermediate lengths selected from Table 9-1 are acceptable.
4. The tolerance on lengths is ISO js15 (see Table 6-27). Angle tolerances AT8 and AT10 applicable for taper pins with nominal lengths larger than 6 mm (see ISO 1947).
5. Material. See Table 10-46, col. 1 and 6; Table 10-43, col. 6; Table 10-42C, col. 4.
6. The ISO surface finish designations N6 and N7 are specified in Table 5-1.
7. Order example for a ground taper pin d − 4 mm with ISO tolerance AT8, 40 mm long, Type A, steel type AISI 1108, taper pin Type A 4 × 40 ISO 2339 AISI 1108.

Table 9-110. Clevis Pins (ISO 2340)

Type A
Without split pin holes

Type B
With split pin holes

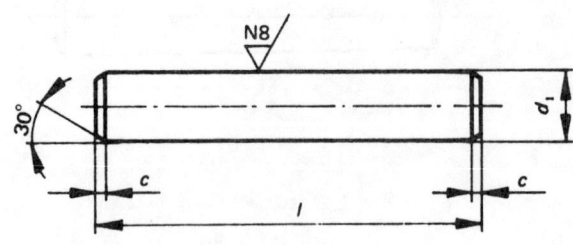
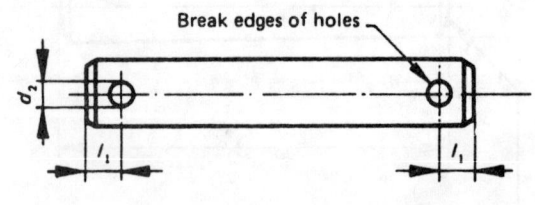

d_1		3	4	5	6	8	10	12	14	16	18	20	22	24	27	30	33	36	40	45	50
d_2	H13	0.8	1	1.2	1.6	2	3.2	3.2	4	4	5	5	5	6.3	6.3	8	8	8	8	10	10
c	max.	1	1	2	2	2	2	3	3	3	3	4	4	4	4	4	4	4	4	4	4
l_1	min.	1.6	2.2	2.9	3.2	3.5	4.5	5.5	6	6	7	8	8	9	9	10	10	10	10	12	12
ℓ	from	6	8	10	12	16	20	25	30	35	40	40	45	50	55	60	65	70	80	90	100
	to	30	40	50	60	80	100	120	140	160	180	200	200	200	200	200	200	200	200	200	200

Notes:
1. All dimensions are in millimeters.
2. For clevis pins with nominal diameters d_1 = 55, 60, 70, 80, 90, 100 (see ISO 2340 standard).
3. Preferred length range l shows minimum and maximum standard lengths inclusive. The first choice standard lengths, except as noted, (S = Second Choice) are as follows: 1 = 6, 8, 10, 12, 14, 16, 18S, 20, 22S, 25, 28S, 30, 35, 40, 45, 50, 55, 60, 65, 70, 75, 80, 85, 90, 95S, 100, 105S, 110, 115S, 120, 125S, 130, 140, 150, 160, 170, 180, 190, 200. For lengths between 200 and 300 mm, use increments of 10 mm; above 300 mm use steps of 20 mm. The ISO recommended tolerance on lengths is ISO js15. (see Table 6-27).
4. The ISO recommended tolerances for the nominal diameters are as follows: a11, c11, f8, h11 (see Section 6).
5. Material: See Table 10-42C, col. 4; Table 10-43, col. 6; Table 10-44, col. 7; Table 10-46, col. 1.
6. The ISO surface texture designation N8 is specified in Table 5-1.
7. Holes are for split cotter pins shown in Table 9-103.
8. Steel bars to tolerance h11 for Type B pins are shown in Table 10-7.
9. Typical designation for a clevis pin with two holes and d_1 = 16 mm to ISO tolerance h11, 1 = 100 mm, l_1 = 6 mm, and ISO steel quality Fe 50-1 is as follows: Clevis Pin 16 h11 × 100 × 6 ISO 2340 Type B Fe 50-1.

FASTENERS

Table 9-111. Clevis Pins with Head (ISO 2341)

TYPE A
Without split pin hole

TYPE B
With split pin hole

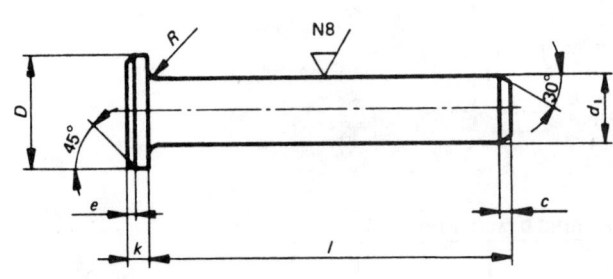
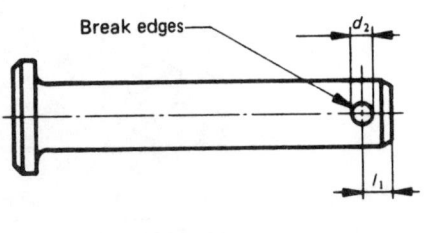

d_1 (4)		3	4	5	6	8	10	12	14	16	18	20	22	24	27	30	33	36	40	45	50
D (5)		5	6	8	10	14	18	20	22	25	28	30	33	36	40	44	47	50	55	60	66
d_2 (8)	H13	0.8	1	1.2	1.6	2	3.2	3.2	4	4	5	5	5	6.3	6.3	8	8	8	8	10	10
c	max.	1	1	2	2	2	2	3	3	3	3	4	4	4	4	4	4	4	4	4	4
e	approx.	0.5	0.5	1	1	1	1	1.6	1.6	1.6	1.6	2	2	2	2	2	2	2	2	2	2
k		1	1	1.6	2	3	4	4	4	4.5	5	5	5.5	6	6	8	8	8	8	9	9
l_1	min.	1.6	2.2	2.9	3.2	3.5	4.5	5.5	6	6	7	8	8	9	9	10	10	10	10	12	12
R		0.6	0.6	0.6	0.6	0.6	0.6	0.6	0.6	0.6	1	1	1	1	1	1	1	1	1	1	1
l	from	6	8	10	12	16	20	25	30	35	40	40	45	50	55	60	65	70	80	90	100
	to	30	40	50	60	80	100	120	140	160	180	200	200	200	200	200	200	200	200	200	200

Notes:
1. All dimensions are in millimeters.
2. For clevis pins with head and nominal diameters d = 55, 60, 70, 80, 90, 100 (see ISO 2341 standard).
3. Preferred length ranges l show minimum and maximum standard lengths inclusive. The first choice standard lengths, except as noted (S = Second Choice), are as follows: l = 6, 8, 10, 12, 14, 16 18S, 20, 22S, 25, 28S, 30, 35, 40, 45, 50, 55, 60, 65, 70, 75, 80, 85, 90, 95S, 100, 105S, 110, 115S, 120, 125S, 130, 140, 150, 160, 170, 180, 190, 200 and then in increments of 10 mm up to 300 mm, and 20 mm steps above this length. The ISO recommended tolerance on lengths is ISO js 15 (see Table 6-27).
4. The ISO recommended tolerances for the nominal diameters are as follows: a11, c11, f8, h11 (see Section 6).
5. The head diameter D for pins used without bushings may be one size smaller than specified in each case.
6. Material: See Table 10-42C, col. 4; Table 10-43, cols. 6 and 7; Table 10-44, col. 7; Table 10-46, col. 1.
7. The ISO surface finish designation N8 is specified in Table 5-1.
8. Holes are for split cotter pins shown in Table 9-103.
9. Typical designation for a clevis pin with head of Type B with hole is as follows: d_1 = 12 mm to ISO tolerance h11, l = 80 mm, l_1 = 5.5 mm, and German steel quality St50: Clevis Pin 12 h11 × 80 × 5.5 ISO 2341 Type B St50.

Table 9-112. Basic Retaining Ring for Holes (Truarc)

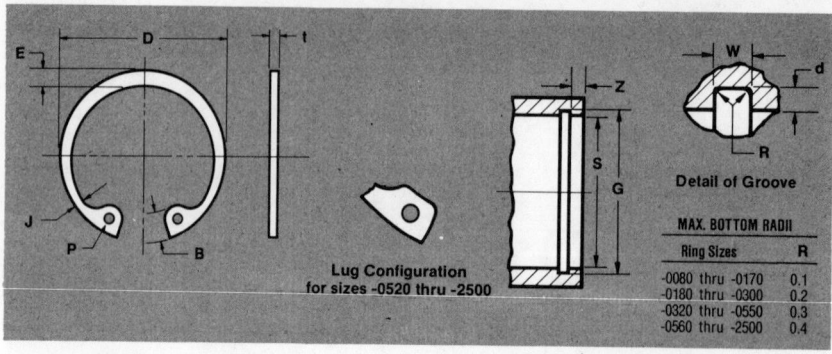

BORE DIA.		TRUARC RING DIMENSIONS (mm)								
mm S	EQUIV. INCH S	TRUARC RING SERIES and SIZE NO. 10-100	FREE DIAMETER D	tol	STANDARD THICKNESS All materials and finishes t	tol	HOLE DIA. P min	LUG B nom	LARGE SECTION E nom	SMALL SECTION J nom	Approx. mass per 1000 Pcs. kg
8	.315	10-100-0080	8.80	+0.25 −0.13	0.4	±0.06	0.8	1.7	0.85	0.45	0.05
9	.354	10-100-0090	10.00	+0.25 −0.13	0.6	±0.06	1.0	2.1	1.25	0.65	0.11
10	.393	10-100-0100	11.10	+0.25 −0.13	0.6	±0.06	1.0	2.1	1.30	0.70	0.14
11	.433	10-100-0110	12.20	+0.25 −0.13	0.6	±0.06	1.0	2.5	1.30	0.70	0.17
12	.472	10-100-0120	13.30	+0.25 −0.13	0.6	±0.06	1.0	2.5	1.35	0.75	0.19
13	.512	10-100-0130	14.25	+0.25 −0.13	0.9	±0.06	1.2	2.9	1.35	0.90	0.35
14	.551	10-100-0140	15.45	+0.25 −0.13	0.9	±0.06	1.2	3.3	1.60	0.90	0.39
15	.591	10-100-0150	16.60	+0.25 −0.13	0.9	±0.06	1.5	3.3	1.65	0.95	0.42
16	.630	10-100-0160	17.70	+0.25 −0.13	0.9	±0.06	1.5	3.4	1.70	0.95	0.47
17	.669	10-100-0170	18.90	+0.25 −0.13	0.9	±0.06	1.5	3.4	1.70	0.95	0.52
18	.708	10-100-0180	20.05	+0.25 −0.13	0.9	±0.06	1.5	3.6	1.8	1.0	0.58
19	.748	10-100-0190	21.10	+0.25 −0.13	0.9	±0.06	1.5	3.6	1.8	1.0	0.59
20	.787	10-100-0200	22.25	+0.25 −0.13	0.9	±0.06	1.5	4.0	2.0	1.1	0.70
21	.826	10-100-0210	23.30	+0.40 −0.25	0.9	±0.06	1.5	4.0	2.1	1.2	0.82
22	.866	10-100-0220	24.40	+0.40 −0.25	1.1	±0.06	1.5	4.0	2.1	1.2	0.90
23	.905	10-100-0230	25.45	+0.40 −0.25	1.1	±0.06	1.5	4.0	2.2	1.2	1.00
24	.945	10-100-0240	26.55	+0.40 −0.25	1.1	±0.06	1.5	4.0	2.3	1.3	1.09
25	.984	10-100-0250	27.75	+0.40 −0.25	1.1	±0.06	1.5	4.0	2.6	1.3	1.26
26	1.023	10-100-0260	28.85	+0.40 −0.25	1.1	±0.06	1.5	4.0	2.7	1.4	1.3
27	1.063	10-100-0270	29.95	+0.65 −0.50	1.3	±0.06	1.9	4.6	2.8	1.4	1.7
28	1.102	10-100-0280	31.10	+0.65 −0.50	1.3	±0.06	1.9	4.6	2.9	1.5	1.8
30	1.181	10-100-0300	33.40	+0.65 −0.50	1.3	±0.06	1.9	4.6	3.0	1.5	2.0
32	1.260	10-100-0320	35.35	+0.65 −0.50	1.3	±0.06	1.9	4.6	3.1	1.6	2.2
34	1.339	10-100-0340	37.75	+0.65 −0.50	1.3	±0.06	1.9	4.6	3.2	1.6	2.3
35	1.378	10-100-0350	38.75	+0.65 −0.50	1.3	±0.06	1.9	4.6	3.3	1.6	2.3
36	1.417	10-100-0360	40.00	+0.65 −0.50	1.3	±0.06	1.9	4.6	3.4	1.7	2.6
37	1.457	10-100-0370	41.05	+0.65 −0.50	1.3	±0.06	1.9	4.6	3.4	1.7	2.9
38	1.496	10-100-0380	42.15	+0.65 −0.50	1.3	±0.06	1.9	4.6	3.4	1.7	3.0
40	1.575	10-100-0400	44.25	+0.90 −0.65	1.6	±0.08	1.9	5.1	4.0	2.0	4.0
42	1.654	10-100-0420	46.60	+0.90 −0.65	1.6	±0.08	1.9	5.8	4.2	2.1	4.7
45	1.772	10-100-0450	49.95	+0.90 −0.65	1.6	±0.08	1.9	6.0	4.3	2.1	5.1
46	1.811	10-100-0460	51.05	+0.90 −0.65	1.6	±0.08	2.3	6.0	4.3	2.1	5.2
47	1.850	10-100-0470	52.15	+0.90 −0.65	1.6	±0.08	2.3	6.0	4.3	2.2	5.8
48	1.890	10-100-0480	53.30	+0.90 −0.65	1.6	±0.08	2.3	6.0	4.5	2.3	6.1
50	1.969	10-100-0500	55.35	+0.90 −0.65	1.6	±0.08	2.3	6.0	4.6	2.3	6.2
52	2.047	10-100-0520	57.90	+1.00 −0.75	2.0	±0.08	2.3	6.4	4.7	2.3	8.1
55	2.165	10-100-0550	61.10	+1.00 −0.75	2.0	±0.08	2.3	6.7	5.1	2.5	8.9
57	2.244	10-100-0570	63.25	+1.00 −0.75	2.0	±0.08	2.3	6.9	5.2	2.5	9.9
58	2.283	10-100-0580	64.4	+1.00 −0.75	2.0	±0.08	2.3	6.9	5.3	2.6	10.1
60	2.362	10-100-0600	66.8	+1.00 −0.75	2.0	±0.08	2.3	6.9	5.3	2.6	10.5
62	2.441	10-100-0620	68.6	+1.00 −0.75	2.0	±0.08	2.7	7.1	5.3	2.6	11.5
63	2.480	10-100-0630	69.9	+1.00 −0.75	2.0	±0.08	2.7	7.1	5.4	2.7	11.6
65	2.559	10-100-0650	72.2	+1.00 −0.75	2.4	±0.08	2.7	7.4	5.6	2.8	15.4
68	2.677	10-100-0680	75.7	+1.00 −0.75	2.4	±0.08	2.7	7.6	5.8	2.9	15.9
70	2.756	10-100-0700	77.5	+1.00 −0.75	2.4	±0.08	2.7	7.6	5.8	2.9	16.1
72	2.835	10-100-0720	79.6	+1.00 −0.75	2.4	±0.08	2.7	7.6	5.8	2.9	16.3
75	2.953	10-100-0750	83.3	+1.00 −0.75	2.4	±0.08	2.7	7.9	6.2	3.1	19.3
78	3.071	10-100-0780	86.8	+1.40 −1.40	2.8	±0.08	3.1	7.9	6.5	3.2	24.0

FASTENERS

Table 9-112 (Continued). Basic Retaining Ring for Holes (Truarc)

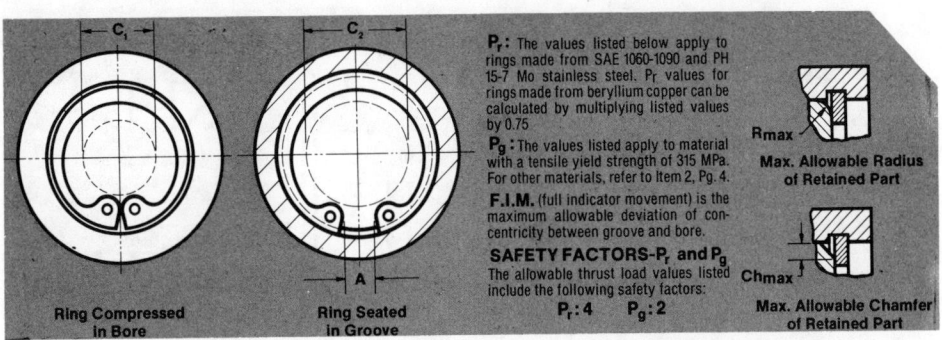

P_r: The values listed below apply to rings made from SAE 1060-1090 and PH 15-7 Mo stainless steel. P_r values for rings made from beryllium copper can be calculated by multiplying listed values by 0.75

P_g: The values listed apply to material with a tensile yield strength of 315 MPa. For other materials, refer to Item 2, Pg. 4.

F.I.M. (full indicator movement) is the maximum allowable deviation of concentricity between groove and bore.

SAFETY FACTORS—P_r and P_g
The allowable thrust load values listed include the following safety factors:
P_r: 4 P_g: 2

TRUARC RING SERIES and SIZE NO.	GROOVE DIAMETER G	tol	F.I.M.	GROOVE WIDTH W	tol	GROOVE DEPTH d ref	EDGE MARGIN Z min	CLEARANCE DIAMETER Ring compressed in bore C_1	Ring seated in groove C_2	GAP WIDTH For checking ring when seated in groove A min	ALLOW. THRUST LOADS Sharp corner abutment SAE 1060-1090 and stainless steel rings used in hardened bores Rc 50 min. P_r(kN)	All standard rings used in low carbon steel bores P_g(kN)	Maximum allowable corner radii and chamfers of retained parts R_{max}	Ch_{max}	Allowable assembly load with R_{max} or Ch_{max} P'_r(kN)
10-100-0080	8.40	+0.06	0.03	0.5	+0.10	0.20	0.6	4.4	4.8	1.40	2.4	1.0	0.4	0.3	1.2
10-100-0090	9.45	+0.06	0.03	0.7	+0.15	0.23	0.7	4.6	5.0	1.50	4.4	1.2	0.5	0.35	3.3
10-100-0100	10.50	+0.06	0.03	0.7	+0.15	0.25	0.8	5.5	6.0	1.85	4.9	1.5	0.5	0.35	3.5
10-100-0110	11.60	+0.10	0.05	0.7	+0.15	0.30	0.9	5.7	6.3	1.95	5.4	2.0	0.6	0.4	3.0
10-100-0120	12.65	+0.10	0.05	0.7	+0.15	0.33	1.0	6.7	7.3	2.25	5.8	2.4	0.6	0.4	3.2
10-100-0130	13.70	+0.10	0.05	1.0	+0.15	0.35	1.1	6.8	7.5	2.35	8.9	2.6	0.6	0.4	7.1
10-100-0140	14.80	+0.10	0.05	1.0	+0.15	0.40	1.2	6.9	7.7	2.65	9.7	3.2	0.6	0.4	6.7
10-100-0150	15.85	+0.10	0.05	1.0	+0.15	0.43	1.3	7.9	8.7	2.80	10.4	3.7	0.6	0.4	6.9
10-100-0160	16.90	+0.10	0.1	1.0	+0.15	0.45	1.4	8.8	9.7	2.80	11.0	4.2	0.6	0.4	7.1
10-100-0170	18.00	+0.10	0.1	1.0	+0.15	0.50	1.5	9.8	10.8	3.35	11.7	4.9	0.6	0.4	7.1
10-100-0180	19.05	+0.10	0.1	1.0	+0.15	0.53	1.6	10.3	11.3	3.40	12.3	5.5	0.6	0.4	7.5
10-100-0190	20.10	+0.15	0.1	1.0	+0.15	0.55	1.7	11.4	12.5	3.40	13.1	6.0	0.6	0.4	7.5
10-100-0200	21.15	+0.15	0.1	1.0	+0.15	0.57	1.7	11.6	12.7	3.8	13.7	6.6	1.0	0.8	4.2
10-100-0210	22.20	+0.15	0.1	1.0	+0.15	0.60	1.8	12.6	13.8	4.2	14.5	7.3	1.0	0.8	4.4
10-100-0220	23.30	+0.15	0.1	1.2	+0.15	0.65	1.9	13.5	14.8	4.3	22.5	8.3	1.0	0.8	6.6
10-100-0230	24.35	+0.15	0.1	1.2	+0.15	0.67	2.0	14.5	15.9	4.9	23.5	8.9	1.0	0.8	6.9
10-100-0240	25.4	+0.15	0.1	1.2	+0.15	0.70	2.1	15.5	16.9	5.2	24.8	9.7	1.0	0.8	7.2
10-100-0250	26.6	+0.15	0.1	1.2	+0.15	0.80	2.4	16.5	18.1	6.0	25.7	11.6	1.0	0.8	8.2
10-100-0260	27.7	+0.15	0.15	1.2	+0.15	0.85	2.6	17.5	19.2	5.7	26.8	12.7	1.0	0.8	8.5
10-100-0270	28.8	+0.15	0.15	1.4	+0.15	0.90	2.7	17.4	19.2	5.9	33	14.0	1.0	0.8	12.2
10-100-0280	29.8	+0.15	0.15	1.4	+0.15	0.90	2.7	18.2	20.0	6.0	34	14.6	1.0	0.8	12.7
10-100-0300	31.9	+0.20	0.15	1.4	+0.15	0.95	2.9	20.0	21.9	6.0	37	16.5	1.0	0.8	13.1
10-100-0320	33.9	+0.20	0.15	1.4	+0.15	0.95	2.9	22.0	23.9	7.3	39	17.6	1.0	0.8	13.6
10-100-0340	36.1	+0.20	0.15	1.4	+0.15	1.05	3.2	24.0	26.1	7.6	42	20.6	1.0	0.8	14.0
10-100-0350	37.2	+0.20	0.15	1.4	+0.15	1.10	3.3	25.0	27.2	8.0	43	22.3	1.0	0.8	14.4
10-100-0360	38.3	+0.20	0.15	1.4	+0.15	1.15	3.5	26.0	28.3	8.3	44	23.9	1.0	0.8	14.9
10-100-0370	39.3	+0.20	0.15	1.4	+0.15	1.15	3.5	27.0	29.3	8.4	45	24.6	1.0	0.8	14.9
10-100-0380	40.4	+0.20	0.15	1.4	+0.15	1.20	3.6	28.0	30.4	8.6	46	26.4	1.6	1.3	9.1
10-100-0400	42.4	+0.20	0.15	1.75	+0.20	1.20	3.6	29.2	31.6	9.7	62	27.7	1.6	1.3	16.3
10-100-0420	44.5	+0.20	0.15	1.75	+0.20	1.25	3.7	29.7	32.2	9.0	65	30.2	1.6	1.3	17.1
10-100-0450	47.6	+0.20	0.15	1.75	+0.20	1.30	3.9	32.3	34.9	9.6	69	33.8	1.6	1.3	17.5
10-100-0460	48.7	+0.20	0.2	1.75	+0.20	1.35	4.0	33.3	36.0	9.7	71	36	1.6	1.3	17.5
10-100-0470	49.8	+0.20	0.2	1.75	+0.20	1.40	4.2	34.3	37.1	10.0	72	38	1.6	1.3	17.5
10-100-0480	50.9	+0.20	0.2	1.75	+0.20	1.45	4.3	35.0	37.9	10.5	74	40	1.6	1.3	18.3
10-100-0500	53.1	+0.20	0.2	1.75	+0.20	1.55	4.6	36.9	40.0	12.1	77	45	2.0	1.6	15.2
10-100-0520	55.3	+0.30	0.2	2.15	+0.20	1.65	5.0	38.6	41.9	11.7	99	50	2.0	1.6	24
10-100-0550	58.4	+0.30	0.2	2.15	+0.20	1.70	5.1	40.8	44.2	11.9	105	54	2.0	1.6	26
10-100-0570	60.5	+0.30	0.2	2.15	+0.20	1.75	5.3	42.2	45.7	12.5	109	58	2.0	1.6	27
10-100-0580	61.6	+0.30	0.2	2.15	+0.20	1.80	5.4	43.2	46.8	13.0	111	60	2.0	1.6	27
10-100-0600	63.8	+0.30	0.2	2.15	+0.20	1.90	5.7	45.5	49.3	12.7	115	66	2.0	1.6	27
10-100-0620	65.8	+0.30	0.2	2.15	+0.20	1.90	5.7	47.0	50.8	14.0	119	68	2.0	1.6	27
10-100-0630	66.9	+0.30	0.2	2.15	+0.20	1.95	5.9	47.8	51.7	14.2	120	71	2.0	1.6	28
10-100-0650	69.0	+0.30	0.2	2.55	+0.20	2.00	6.0	49.4	53.4	14.2	149	75	2.0	1.6	42
10-100-0680	72.2	+0.30	0.2	2.55	+0.20	2.10	6.3	52.0	56.2	14.4	156	82	2.0	1.6	43
10-100-0700	74.4	+0.30	0.2	2.55	+0.20	2.20	6.6	53.8	58.2	16.1	161	88	2.0	1.6	43
10-100-0720	76.5	+0.30	0.2	2.55	+0.20	2.25	6.7	55.9	60.4	17.4	166	93	2.0	1.6	43
10-100-0750	79.7	+0.30	0.2	2.55	+0.20	2.35	7.1	58.2	62.9	16.8	172	101	2.5	2.0	37
10-100-0780	82.8	+0.30	0.2	2.95	+0.20	2.40	7.2	61.2	66.0	17.6	209	108	2.5	2.0	53

Table 9-113. Basic Retaining Ring for Shafts (Truarc)

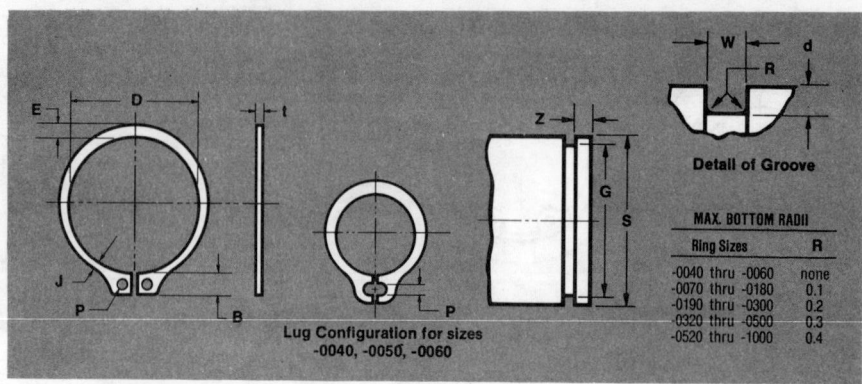

★ Available *only* in beryllium copper

SHAFT DIA. mm (S)	EQUIV. INCH (S)	TRUARC RING SERIES and SIZE NO. 11-100	FREE DIAMETER D	tol	STANDARD THICKNESS t	tol	HOLE DIA. P min	LUG B nom	LARGE SECTION E nom	SMALL SECTION J nom	Approx. mass per 1000 Pcs. kg
4	.157	11-100-0040	3.60	+0.05 −0.10	0.25	±0.05	0.6	1.35	0.65	0.40	0.017
5	.197	11-100-0050	4.55	+0.05 −0.10	0.4	±0.06	0.6	1.40	0.65	0.40	0.029
6	.236	11-100-0060	5.45	+0.05 −0.10	0.4	±0.06	0.6	1.40	0.75	0.50	0.040
7	.275	11-100-0070	6.35	+0.05 −0.15	0.6	±0.06	1.0	2.05	0.90	0.60	0.10
8	.315	11-100-0080	7.15	+0.05 −0.15	0.6	±0.06	1.0	2.20	1.00	0.65	0.12
9	.354	11-100-0090	8.15	+0.05 −0.15	0.6	±0.06	1.0	2.20	1.15	0.75	0.15
10	.393	11-100-0100	9.00	+0.05 −0.15	0.6	±0.06	1.0	2.20	1.30	0.80	0.19
11	.433	11-100-0110	10.00	+0.05 −0.15	0.6	±0.06	1.0	2.20	1.40	0.85	0.23
12	.472	11-100-0120	10.85	+0.05 −0.15	0.6	±0.06	1.0	2.20	1.50	0.90	0.24
13	.512	11-100-0130	11.90	+0.15 −0.25	0.9	±0.06	1.2	2.80	1.60	0.95	0.44
14	.551	11-100-0140	12.90	+0.15 −0.25	0.9	±0.06	1.2	2.80	1.70	1.00	0.49
15	.591	11-100-0150	13.80	+0.15 −0.25	0.9	±0.06	1.2	2.80	1.80	1.05	0.54
16	.630	11-100-0160	14.70	+0.15 −0.25	0.9	±0.06	1.2	2.80	2.05	1.15	0.59
17	.669	11-100-0170	15.75	+0.15 −0.25	0.9	±0.06	1.2	2.80	2.10	1.15	0.64
18	.708	11-100-0180	16.65	+0.15 −0.25	1.1	±0.06	1.3	3.45	2.25	1.25	0.92
19	.748	11-100-0190	17.60	+0.15 −0.25	1.1	±0.06	1.3	3.45	2.35	1.30	0.95
20	.787	11-100-0200	18.35	+0.15 −0.25	1.1	±0.06	1.3	3.45	2.40	1.35	1.00
21	.826	11-100-0210	19.40	+0.15 −0.25	1.1	±0.06	1.3	3.45	2.50	1.40	1.1
22	.866	11-100-0220	20.30	+0.15 −0.25	1.1	±0.06	1.3	3.45	2.70	1.50	1.3
23	.906	11-100-0230	21.25	+0.15 −0.25	1.1	±0.06	1.3	3.45	2.80	1.60	1.4
24	.945	11-100-0240	22.20	+0.15 −0.25	1.1	±0.06	1.9	4.2	2.9	1.6	1.5
25	.984	11-100-0250	23.10	+0.15 −0.25	1.1	±0.06	1.9	4.2	2.9	1.7	1.6
26	1.023	11-100-0260	24.05	+0.15 −0.25	1.1	±0.06	1.9	4.2	3.0	1.7	1.8
27	1.063	11-100-0270	24.95	+0.25 −0.40	1.3	±0.06	1.9	4.6	3.1	1.8	2.2
28	1.102	11-100-0280	25.80	+0.25 −0.40	1.3	±0.06	1.9	4.6	3.2	1.8	2.3
30	1.181	11-100-0300	27.90	+0.25 −0.40	1.3	±0.06	1.9	4.6	3.3	1.8	2.5
32	1.260	11-100-0320	29.60	+0.25 −0.40	1.3	±0.06	1.9	4.6	3.6	1.9	2.8
34	1.339	11-100-0340	31.40	+0.25 −0.40	1.3	±0.06	1.9	4.6	3.8	2.0	3.1
35	1.378	11-100-0350	32.30	+0.25 −0.40	1.3	±0.06	1.9	4.6	3.9	2.1	3.3
36	1.417	11-100-0360	33.25	+0.25 −0.40	1.3	±0.06	1.9	5.4	4.1	2.2	3.6
38	1.496	11-100-0380	35.20	+0.25 −0.40	1.3	±0.06	3.1	5.4	4.3	2.3	4.0
40	1.575	11-100-0400	36.75	+0.35 −0.50	1.6	±0.08	3.1	6.0	4.4	2.3	5.6
42	1.654	11-100-0420	38.80	+0.35 −0.50	1.6	±0.08	3.1	6.0	4.6	2.4	6.3
43	1.693	11-100-0430	39.65	+0.35 −0.50	1.6	±0.08	3.1	6.0	4.7	2.5	6.7
45	1.772	11-100-0450	41.60	+0.35 −0.50	1.6	±0.08	3.1	6.0	4.8	2.6	7.0
46	1.811	11-100-0460	42.55	+0.35 −0.50	1.6	±0.08	3.1	6.0	4.9	2.6	7.3
48	1.890	11-100-0480	44.40	+0.35 −0.50	1.6	±0.08	3.1	6.2	5.0	2.6	7.7
50	1.969	11-100-0500	46.20	+0.35 −0.50	1.6	±0.08	3.1	6.2	5.1	2.7	8.2
52	2.047	11-100-0520	48.40	+0.35 −0.65	2.0	±0.08	3.1	6.8	5.3	2.8	11.3
54	2.126	11-100-0540	49.9	+0.35 −0.65	2.0	±0.08	3.1	6.8	5.4	2.9	11.8
55	2.165	11-100-0550	50.6	+0.35 −0.65	2.0	±0.08	3.1	6.8	5.4	2.9	11.9
57	2.244	11-100-0570	52.9	+0.35 −0.65	2.0	±0.08	3.1	6.8	5.6	3.0	12.5
58	2.283	11-100-0580	53.6	+0.35 −0.65	2.0	±0.08	3.1	6.8	5.6	3.0	12.6
60	2.362	11-100-0600	55.8	+0.35 −0.65	2.0	±0.08	3.1	6.8	5.7	3.0	13.2
62	2.441	11-100-0620	57.3	+0.35 −0.65	2.0	±0.08	3.1	6.8	5.8	3.0	13.4
65	2.559	11-100-0650	60.4	+0.50 −0.75	2.0	±0.08	3.1	6.8	6.0	3.1	15.4
68	2.677	11-100-0680	63.1	+0.50 −0.75	2.0	±0.08	3.1	6.8	6.2	3.3	16.3
70	2.756	11-100-0700	64.6	+0.50 −0.75	2.4	±0.08	3.1	7.8	6.3	3.3	19.3
72	2.835	11-100-0720	66.6	+0.50 −0.75	2.4	±0.08	3.1	7.8	6.4	3.3	20.6
75	2.953	11-100-0750	69.0	+0.50 −0.75	2.4	±0.08	3.1	7.8	6.6	3.4	22.6
78	3.071	11-100-0780	72.0	+0.50 −0.75	2.4	±0.08	3.1	7.8	6.6	3.4	21.5
80	3.150	11-100-0800	74.2	+0.50 −0.75	2.4	±0.08	3.1	7.8	7.0	3.6	26.8
82	3.228	11-100-0820	76.4	+0.50 −0.75	2.4	±0.08	3.1	7.8	7.1	3.7	28.1
85	3.346	11-100-0850	78.6	+0.50 −0.75	2.4	±0.08	3.1	7.8	7.3	3.8	29.0
88	3.464	11-100-0880	81.4	+0.50 −0.75	2.8	±0.08	3.1	8.4	7.5	3.9	32.2
90	3.543	11-100-0900	83.2	+0.50 −0.75	2.8	±0.08	3.1	8.4	7.5	3.9	33.1
95	3.740	11-100-0950	88.1	+0.50 −0.75	2.8	±0.08	3.1	8.4	7.9	4.1	37.6
100	3.937	11-100-1000	92.5	+0.50 −0.75	2.8	±0.08	3.1	8.7	8.0	4.1	43.1

FASTENERS

Table 9-113 (*Continued*). Basic Retaining Ring for Shafts (Truarc)

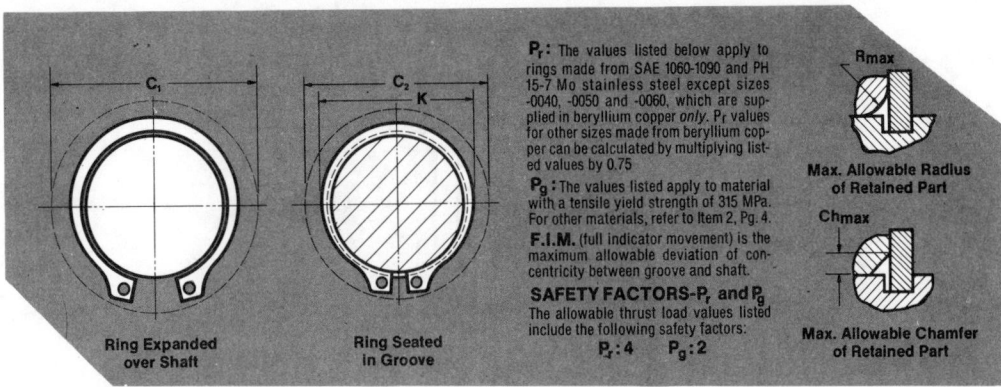

TRUARC RING SERIES and SIZE NO. 11-100	GROOVE DIMENSIONS (mm)							APPLICATION DATA							
	GROOVE DIAMETER			GROOVE WIDTH		GROOVE DEPTH	EDGE MARGIN	CLEARANCE DIAMETER		GAGING DIA.	ALLOW. THRUST LOADS Sharp corner abutment		Maximum allowable corner radii and chamfers of retained parts		Allowable assembly load with R_{max} or Ch_{max}	Calculated allowable assembly rpm (steel rings)
								Ring expanded over shaft	Ring seated in groove	For checking ring when seated in groove	SAE 1060-1090 and stainless steel rings used on hardened shafts Rc 50 min.	All standard rings used on low carbon steel shafts				
	G	tol	F.I.M.	W	tol	d ref	Z min	C_1	C_2	K max	P_r(kN)	P_g(kN)	R max	Ch max	P'_r(kN)	rpm
11-100-0040	3.80	−0.08	0.03	0.32	+0.05	0.10	0.3	7.0	6.8	4.90	0.6	0.2	0.35	0.25	0.18	70 000
11-100-0050	4.75	−0.08	0.03	0.5	+0.10	0.13	0.4	8.2	7.9	5.85	1.1	0.3	0.35	0.25	0.48	70 000
11-100-0060	5.70	−0.08	0.03	0.5	+0.10	0.15	0.5	9.1	8.8	6.95	1.4	0.4	0.35	0.25	0.55	70 000
11-100-0070	6.60	−0.10	0.05	0.7	+0.15	0.20	0.6	12.3	11.8	8.05	2.6	0.7	0.45	0.3	1.6	60 000
11-100-0080	7.50	−0.10	0.05	0.7	+0.15	0.25	0.8	13.6	13.0	9.15	3.1	1.0	0.5	0.35	1.6	55 000
11-100-0090	8.45	−0.10	0.05	0.7	+0.15	0.28	0.8	14.5	13.8	10.35	3.5	1.2	0.6	0.4	1.7	48 000
11-100-0100	9.40	−0.10	0.05	0.7	+0.15	0.30	0.9	15.5	14.7	11.50	3.9	1.5	0.7	0.45	1.6	42 000
11-100-0110	10.35	−0.12	0.05	0.7	+0.15	0.33	1.0	16.4	15.6	12.60	4.3	1.8	0.75	0.45	1.7	38 000
11-100-0120	11.35	−0.12	0.05	0.7	+0.15	0.33	1.0	17.4	16.6	13.80	4.7	2.0	0.8	0.5	1.7	34 000
11-100-0130	12.30	−0.12	0.10	1.0	+0.15	0.35	1.0	19.7	18.8	15.05	7.5	2.2	0.8	0.5	4.0	31 000
11-100-0140	13.25	−0.12	0.10	1.0	+0.15	0.38	1.2	20.7	19.7	15.60	8.1	2.6	0.9	0.55	3.8	28 000
11-100-0150	14.15	−0.12	0.10	1.0	+0.15	0.43	1.3	21.7	20.6	17.20	8.7	3.2	1.0	0.6	3.8	27 000
11-100-0160	15.10	−0.15	0.10	1.0	+0.15	0.45	1.4	22.7	21.6	18.35	9.3	3.5	1.0	0.6	4.3	25 000
11-100-0170	16.10	−0.15	0.10	1.0	+0.15	0.45	1.4	23.7	22.6	19.35	9.9	4.0	1.0	0.6	4.4	24 000
11-100-0180	17.00	−0.15	0.10	1.2	+0.15	0.50	1.5	26.2	25.0	20.60	16.0	4.4	1.2	0.7	6.1	23 000
11-100-0190	17.95	−0.15	0.10	1.2	+0.15	0.53	1.6	27.2	25.9	21.70	16.9	4.9	1.2	0.7	6.3	21 500
11-100-0200	18.85	−0.15	0.10	1.2	+0.15	0.58	1.7	28.2	26.8	22.65	17.8	5.7	1.4	0.8	5.6	20 000
11-100-0210	19.80	−0.15	0.10	1.2	+0.15	0.60	1.8	29.2	27.7	23.80	18.6	6.2	1.4	0.8	5.9	19 000
11-100-0220	20.70	−0.15	0.10	1.2	+0.15	0.65	1.9	30.3	28.7	24.90	19.6	7.0	1.4	0.8	6.4	18 500
11-100-0230	21.65	−0.15	0.10	1.2	+0.15	0.67	2.0	31.3	29.6	26.00	20.5	7.6	1.5	0.9	5.9	18 000
11-100-0240	22.60	−0.15	0.10	1.2	+0.15	0.70	2.1	34.1	32.4	27.15	21.4	8.2	1.5	0.9	6.0	17 500
11-100-0250	23.50	−0.15	0.10	1.2	+0.15	0.75	2.3	35.1	33.3	28.10	22.3	9.2	1.5	0.9	6.0	17 000
11-100-0260	24.50	−0.15	0.10	1.2	+0.15	0.75	2.3	36.0	34.2	29.25	23.2	9.6	1.6	1.0	5.6	16 500
11-100-0270	25.45	−0.20	0.10	1.4	+0.15	0.78	2.3	37.8	35.9	30.35	28.4	10.3	1.6	1.0	8.1	16 300
11-100-0280	26.40	−0.20	0.10	1.4	+0.15	0.80	2.4	38.8	36.9	31.45	28.4	11.0	1.6	1.0	8.4	15 800
11-100-0300	28.35	−0.20	0.15	1.4	+0.15	0.83	2.5	40.8	38.8	33.6	31.6	12.3	1.6	1.0	8.7	15 000
11-100-0320	30.20	−0.20	0.15	1.4	+0.15	0.90	2.7	42.8	40.7	35.9	33.6	14.1	1.8	1.1	8.6	14 800
11-100-0340	32.00	−0.20	0.15	1.4	+0.15	1.00	3.0	44.9	42.5	37.9	36	16.7	1.8	1.1	9.1	14 000
11-100-0350	32.90	−0.20	0.15	1.4	+0.15	1.05	3.1	45.9	43.4	39.0	37	18.1	2.0	1.2	8.5	13 500
11-100-0360	33.85	−0.20	0.15	1.4	+0.15	1.08	3.2	48.6	46.1	40.2	38	18.9	2.0	1.2	9.0	13 300
11-100-0380	35.8	−0.30	0.15	1.4	+0.15	1.10	3.3	50.6	48.0	42.5	40	20.5	2.0	1.2	9.4	12 700
11-100-0400	37.7	−0.30	0.15	1.75	+0.20	1.15	3.4	54.0	51.3	44.5	52	22.6	2.0	1.2	14.5	12 000
11-100-0420	39.6	−0.30	0.15	1.75	+0.20	1.20	3.6	56.0	53.2	46.9	54	24.8	2.0	1.2	15.2	11 000
11-100-0430	40.5	−0.30	0.15	1.75	+0.20	1.25	3.8	57.0	54.0	47.9	55	26.4	2.0	1.2	15.6	10 800
11-100-0450	42.4	−0.30	0.15	1.75	+0.20	1.30	3.9	59.0	55.9	50.0	58	28.8	2.0	1.2	15.9	10 000
11-100-0460	43.3	−0.30	0.15	1.75	+0.20	1.35	4.0	60.0	56.8	50.9	59	30.4	2.0	1.2	16.2	9 500
11-100-0480	45.2	−0.30	0.15	1.75	+0.20	1.40	4.2	62.4	59.1	53.0	62	33	2.0	1.2	16.6	8 800
11-100-0500	47.2	−0.30	0.15	1.75	+0.20	1.40	4.2	64.4	61.1	55.2	64	35	2.0	1.2	16.9	8 000
11-100-0520	49.1	−0.30	0.15	2.15	+0.20	1.45	4.3	67.6	64.1	57.4	84	37	2.5	1.5	22	7 700
11-100-0540	51.0	−0.30	0.15	2.15	+0.20	1.50	4.5	69.6	66.1	59.5	87	40	2.5	1.5	22	7 500
11-100-0550	51.8	−0.30	0.15	2.15	+0.20	1.60	4.8	70.6	66.9	60.4	89	44	2.5	1.5	22	7 400
11-100-0570	53.8	−0.40	0.20	2.15	+0.20	1.60	4.8	72.6	68.9	62.7	91	45	2.5	1.5	23	7 200
11-100-0580	54.7	−0.40	0.20	2.15	+0.20	1.65	4.9	73.6	69.8	63.6	93	46	2.5	1.5	23	7 100
11-100-0600	56.7	−0.40	0.20	2.15	+0.20	1.65	4.9	75.6	71.8	65.8	97	49	2.5	1.5	24	7 000
11-100-0620	58.6	−0.40	0.20	2.15	+0.20	1.70	5.1	77.6	73.6	67.9	100	52	2.5	1.5	24	6 900
11-100-0650	61.6	−0.40	0.20	2.15	+0.20	1.70	5.1	80.6	76.6	71.2	105	54	2.5	1.5	25	6 700
11-100-0680	64.5	−0.40	0.20	2.15	+0.20	1.75	5.3	83.6	79.5	74.5	110	58	2.5	1.5	26	6 500
11-100-0700	66.4	−0.40	0.20	2.55	+0.20	1.80	5.4	88.1	83.9	76.4	136	62	2.5	1.5	38	6 400
11-100-0720	68.3	−0.40	0.20	2.55	+0.20	1.85	5.5	90.1	85.8	78.5	140	65	2.5	1.5	38	6 200
11-100-0750	71.2	−0.40	0.20	2.55	+0.20	1.90	5.7	93.1	88.7	81.7	147	69	2.5	1.5	39	5 900
11-100-0780	74.0	−0.40	0.20	2.55	+0.20	2.00	6.0	95.4	92.1	84.6	151	76	2.5	1.5	30	5 600
11-100-0800	75.9	−0.40	0.20	2.55	+0.20	2.05	6.1	97.9	93.1	87.0	155	80	3.5	2.1	30	5 400
11-100-0820	77.8	−0.40	0.20	2.55	+0.20	2.10	6.3	100.0	95.1	89.0	159	84	3.5	2.1	30	5 200
11-100-0850	80.6	−0.40	0.20	2.55	+0.20	2.20	6.6	103.0	97.9	92.1	165	91	3.5	2.1	31	5 000
11-100-0880	83.5	−0.40	0.20	2.95	+0.20	2.25	6.7	107.0	100.8	95.1	199	97	3.5	2.1	43	4 800
11-100-0900	85.4	−0.40	0.20	2.95	+0.20	2.30	6.9	109.0	103.6	97.1	204	101	3.5	2.1	43	4 500
11-100-0950	90.2	−0.40	0.20	2.95	+0.20	2.40	7.2	114.0	108.6	102.7	215	112	4.0	2.4	40	4 350
11-100-1000	95.0	−0.40	0.20	2.95	+0.20	2.50	7.5	119.5	113.7	108.0	227	123	4.0	2.4	41	4 150

Table 9-114. Type E Ring, Retaining Ring for Shafts (Truarc)

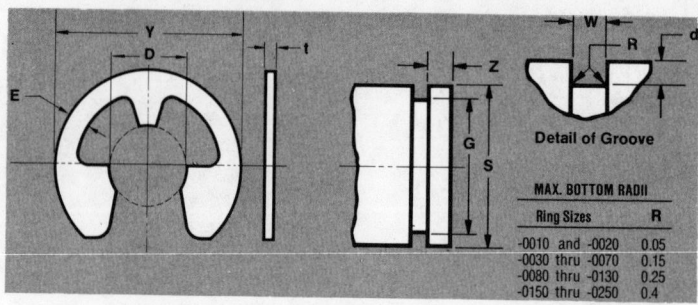

★ Available *only* in beryllium copper

SHAFT DIA. mm S	EQUIV INCH S	TRUARC RING SERIES and SIZE NO. 11-420	FREE DIAMETER D	tol	STANDARD THICKNESS All materials and finishes t	tol	OUTER DIA. Y nom	BRIDGE E nom	Approx. mass per 1000 Pcs. kg
1	.039	11-420-0010	0.64	+0.03 −0.08	0.25	±0.05	2.0	0.30	0.004
2	.079	11-420-0020	1.30	+0.03 −0.08	0.25	±0.05	4.0	0.55	0.014
3	.118	11-420-0030	2.10	+0.03 −0.08	0.4	±0.06	5.6	0.65	0.036
4	.157	11-420-0040	2.90	+0.03 −0.08	0.6	±0.06	7.2	0.85	0.095
5	.197	11-420-0050	3.70	+0.03 −0.08	0.6	±0.06	8.5	0.90	0.13
6	.236	11-420-0060	4.70	+0.03 −0.08	0.6	±0.06	11.1	1.15	0.21
7	.275	11-420-0070	5.25	+0.03 −0.08	0.6	±0.06	13.4	1.4	0.34
8	.315	11-420-0080	6.15	+0.05 −0.10	0.6	±0.06	14.6	1.4	0.35
9	.354	11-420-0090	6.80	+0.05 −0.10	0.9	±0.06	15.8	1.5	0.58
10	.393	11-420-0100	7.60	+0.05 −0.10	0.9	±0.06	16.8	1.5	0.68
11	.433	11-420-0110	8.55	+0.05 −0.10	0.9	±0.06	17.4	1.6	0.68
12	.472	11-420-0120	9.20	+0.05 −0.10	1.1	±0.06	18.6	1.8	1.00
13	.512	11-420-0130	9.95	+0.05 −0.10	1.1	±0.06	20.3	2.0	1.13
15	.591	11-420-0150	11.40	+0.10 −0.15	1.1	±0.06	22.8	2.1	1.40
16	.630	11-420-0160	12.15	+0.10 −0.15	1.1	±0.06	23.8	2.3	1.45
18	.709	11-420-0180	13.90	+0.10 −0.15	1.3	±0.06	27.2	2.5	2.3
20	.787	11-420-0200	15.60	+0.10 −0.15	1.3	±0.06	30.0	2.8	2.8
22	.866	11-420-0220	17.00	+0.10 −0.15	1.3	±0.06	33.0	3.0	3.4
25	.984	11-420-0250	19.50	+0.10 −0.15	1.3	±0.06	37.1	3.3	4.2

FASTENERS

Table 9-114 (Continued). Type E Ring, Retaining Ring for Shafts (Truarc)

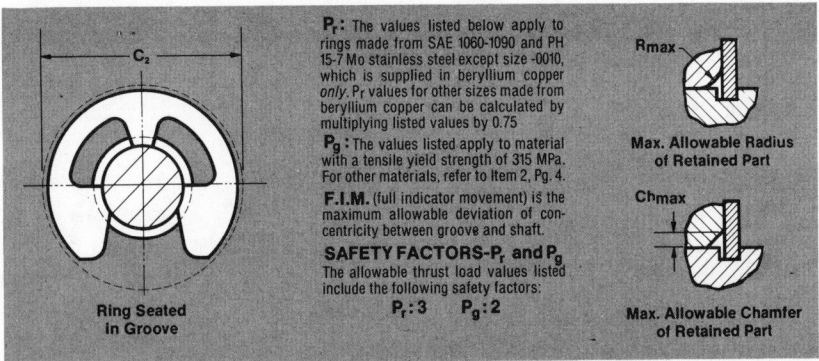

TRUARC RING SERIES and SIZE NO.	GROOVE DIAMETER G	tol	F.I.M.	GROOVE WIDTH W	tol	GROOVE DEPTH d ref	EDGE MARGIN Z min	CLEARANCE DIAMETER Ring seated in groove C_2	ALLOW. THRUST LOADS SAE 1060-1090 and stainless steel rings used on hardened shafts Rc 50 min. P_r(kN)	ALLOW. THRUST LOADS All standard rings used on low carbon steel shafts P_g(kN)	Maximum allowable corner radii and chamfers of retained parts R_{max}	Maximum allowable corner radii and chamfers of retained parts Ch_{max}	Allowable assembly load with R_{max} or Ch_{max} P'_r(kN)	Calculated allowable assembly rpm (steel rings) rpm
11-420-0010	0.72	−0.05	0.04	0.32	+0.05	0.14	0.3	2.2	0.06	0.02	0.4	0.25	0.06	40 000
11-420-0020	1.45	−0.05	0.04	0.32	+0.05	0.28	0.6	4.3	0.13	0.09	0.8	0.5	0.13	40 000
11-420-0030	2.30	−0.08	0.04	0.5	+0.10	0.35	0.7	6.0	0.3	0.17	1.1	0.7	0.3	34 000
11-420-0040	3.10	−0.08	0.05	0.7	+0.15	0.45	0.9	7.6	0.7	0.3	1.6	1.2	0.7	31 000
11-420-0050	3.90	−0.08	0.05	0.7	+0.15	0.55	1.1	8.9	0.9	0.4	1.6	1.2	0.9	27 000
11-420-0060	4.85	−0.08	0.05	0.7	+0.15	0.58	1.2	11.5	1.1	0.6	1.6	1.2	1.1	25 000
11-420-0070	5.55	−0.10	0.08	0.7	+0.15	0.73	1.5	14.0	1.2	0.8	1.6	1.2	1.2	23 000
11-420-0080	6.40	−0.10	0.08	0.7	+0.15	0.80	1.6	15.1	1.4	1.0	1.7	1.3	1.4	21 500
11-420-0090	7.20	−0.10	0.08	1.0	+0.15	0.90	1.8	16.5	3.0	1.3	1.7	1.3	3.0	19 500
11-420-0100	8.00	−0.10	0.08	1.0	+0.15	1.00	2.0	17.5	3.4	1.6	1.7	1.3	3.4	18 000
11-420-0110	8.90	−0.10	0.10	1.0	+0.15	1.05	2.1	18.0	3.7	1.9	1.7	1.3	3.7	16 500
11-420-0120	9.60	−0.15	0.10	1.2	+0.15	1.20	2.4	19.3	4.9	2.3	1.9	1.4	4.9	15 000
11-420-0130	10.30	−0.15	0.10	1.2	+0.15	1.35	2.7	21.0	5.4	2.9	2.0	1.5	5.4	13 000
11-420-0150	11.80	−0.15	0.10	1.2	+0.15	1.60	3.2	23.5	6.2	4.0	2.0	1.5	6.2	11 500
11-420-0160	12.50	−0.15	0.10	1.2	+0.15	1.75	3.5	24.5	6.6	4.5	2.0	1.5	6.6	10 000
11-420-0180	14.30	−0.15	0.10	1.4	+0.15	1.85	3.7	27.9	8.7	5.4	2.1	1.6	8.7	9 000
11-420-0200	16.00	−0.20	0.10	1.4	+0.15	2.00	4.0	30.7	9.8	6.5	2.2	1.7	9.8	8 000
11-420-0220	17.40	−0.20	0.10	1.4	+0.15	2.30	4.6	33.7	10.8	8.1	2.2	1.7	10.8	7 000
11-420-0250	20.00	−0.20	0.10	1.4	+0.15	2.50	5.0	37.9	12.2	10.1	2.4	1.9	12.2	5 000

RELATED ISO STANDARDS

Fasteners (TC 2)

ISO 225-1976	Bolts, screws, and studs. Dimensioning.
ISO/R 272-1968	Hexagon bolts and nuts. Widths across flats, heights of heads, thicknesses of nuts. Metric series, 2nd Edition.
ISO/R 273/I-1962	Clearance holes for metric bolts 1.6 up to and including 39 mm thread diameter.
ISO/R 273/II-1968	Clearance holes for metric bolts 42 up to and including 150 mm thread diameter.
ISO/R 288/I-1963	Slotted and castle nuts with metric thread.
ISO/R 288/II-1969	Slotted and castle nuts with metric thread, 42 up to and including 100 mm thread diameter.
ISO/R 733-1968	Hexagon bolts and nuts. Metric series. Tolerances on widths across flats. Widths across corners.
ISO/R 861-1968	Hexagon socket head cap screws. Metric series.
ISO 885-1976	Bolts and screws. Radii under the head for general purpose bolts and screws. Metric series.
ISO/R 887-1968	Washers for hexagon bolts and nuts. Metric series.
ISO 888-1976	Normal lengths for bolts, screws, and studs. Thread lengths for general purpose bolts.
ISO/R 898/I-1968	Mechanical properties of fasteners. Part 1: bolts, screws, and studs.
ISO/R 898/II-1969	Mechanical properties of fasteners. Nuts with specified proof load values.
ISO/R 898/III-1969	Mechanical properties of fasteners. Marking of bolts, screws, studs, and nuts.
ISO 898/IV-1972	Mechanical properties of fasteners. Nuts with specified proof load values, effective heights of threads 0.4D to 0.6D.
ISO/DIS 898/V	Mechanical properties of fasteners. Set screws and similar parts without specified proof load values.
ISO/R 1051-1969	Rivet shank diameters (diameter range from 1 to 36 mm)
ISO 1207-1976	Slotted cheese (fillister) head screws. Metric series.
ISO 1234-1976	Split pins. Metric series.
ISO/R 1478-1970	Tapping screw thread. Dimensions in millimeters and inches.
ISO/R 1479-1970	Hexagon-head tapping screws. Metric series.
ISO/R 1480-1970	Hexagon-head tapping screws. Inch series.
ISO/R 1481-1970	Slotted pan head tapping screws. Dimensions in millimeters and inches.
ISO/R 1482-1970	Slotted countersunk (flat) head tapping screws. Dimensions in millimeters and inches.
ISO/R 1483-1970	Slotted raised countersunk (oval) head tapping screws. Dimensions in millimeters and inches.
ISO/R 1580-1970	Slotted pan head screws. Metric series.
ISO 2009-1974	Slotted countersunk (flat) head screws. Metric series.
ISO 2010-1974	Slotted raised countersunk (oval) head screws. Metric series.
ISO/DIS 2318	Measurement of carbon variation in threads of hardened and tempered bolts, screws, and studs.
ISO 2320-1972	Prevailing torque-type steel hexagon locknuts. Mechanical and performance properties.
ISO 2338-1972	Parallel pins, unhardened. Metric series.
ISO 2339-1974	Taper pins, unhardened. Metric series.
ISO 2340-1972	Clevis pins. Metric series.
ISO 2341-1972	Clevis pins with heads. Metric series.
ISO 2342-1972	Slotted headless screws. Metric series.
ISO 2343-1972	Hexagon socket set screws. Metric series.
ISO 2358-1972	Prevailing torque types steel hexagon locknuts. Dimensions. Metric series.
ISO 2359-1972	Prevailing torque-type steel hexagon locknuts. Dimensions. Inch series.
ISO 2702-1974	Heat-treated steel tapping screws. Mechanical properties.
ISO 2770-1974	Tapping screws according to ISO 1478 to 1483. Minimum and maximum lengths.
ISO/DIS 3269	Acceptance inspection of fasteners.
ISO/DIS 3506	Mechanical properties of stainless steel fasteners.
ISO 3508-1976	Thread run-outs for fasteners.
ISO 3800/1-1977	Axial load fatigue testing for threaded fasteners.

FASTENERS

ISO/DIS 4014	Hexagon head bolts, product grades A and B.
ISO/DIS 4015	Hexagon head bolts, product grade B.
ISO/DIS 4016	Hexagon head bolts, product grade C.
ISO/DIS 4017	Hexagon head screws, product grades A and B.
ISO/DIS 4018	Hexagon head screws, product grade C.
ISO 4026-1977	Hexagon socket set screws with flat point.
ISO 4027-1977	Hexagon socket set screws with cone point.
ISO 4028-1977	Hexagon socket set screws with dog point.
ISO 4029-1977	Hexagon socket set screws with cup point.
ISO/DIS 4032	Hexagon nuts, product grade A and B.
ISO/DIS 4033	Hexagon nuts, product grade A and B and property class 9.
ISO/DIS 4034	Hexagon nuts, product grade C.
ISO/DIS 4035	Hexagon thin nuts, product grade A and B, chamfered.
ISO/DIS 4036	Hexagon thin nuts, product grade B, unchamfered.
ISO/DIS 4042	Electroplated coatings on threaded components.
ISO/DIS 4759/1	Tolerances for fasteners—Part I: Bolts, screws and nuts with thread diameters.
ISO 4759-1977	Tolerances for fasteners—Part III: washers for metric bolts, screws and nuts with thread diameters from 1 up to and including 150 mm—Product grades A and B.
ISO 4762-1977	Hexagon socket head cap screws.

Small Tools (TC 29)

ISO 1085-1974	Combinations of double-ended wrench gaps.
ISO 1173-1975	Assembly tools for bolts and screws. Hexagon drives ends for power tools.
ISO 1174-1975	Assembly tools for bolts and screws. Driving squares for power socket wrenches and hand socket wrenches.
ISO 1703-1975	Assembly tools for screws and nuts. Nomenclature.
ISO 1711-1975	Hand-operated wrenches and sockets. Technical specifications.
ISO 2236-1972	Assembly tools for screws and nuts. Forged and tubular socket wrenches. Metric series. Maximum outside dimensions.
ISO 2306-1972	Drills for use prior to tapping screw threads.
ISO 2351-1972	Screwdriver bits, for slotted head screws, with male hexagon drive.
ISO 2352-1972	Spiral ratchet screwdriver ends.
ISO 2380-1972	Screwdriver blades for slotted head screws.
ISO 2568-1973	Hand and machine-operated circular screwing dies and hand-operated die stocks.
ISO 2725-1973	Assembly tools for screws and nuts. Power and hand-operated square drive sockets. Metric series.
ISO 2936-1973	Assembly tools for screws and nuts. Hexagon socket screw keys. Metric series.
ISO 3109-1974	Assembly tools for screws and nuts. Hexagon insert bits for hexagon socket head screws.
ISO 3315-1975	Assembly tools for screws and nuts—Driving parts for hand-operated square drive socket wrenches—Torque testing.
ISO 3316-1975	Assembly tools for screws and nuts—Attachments for hand-operated square drive socket wrenches—Torque testing.
ISO 3317-1974	Assembly tools for screws and nuts. Square drive extension hexagon insert, for power socket wrenches.
ISO 3318-1974	Assembly tools for screws and nuts. Open-end double-head engineers' wrenches, double-head box wrenches and combination wrenches. Maximum outside dimensions of heads.
ISO/DIS 4428	Spline drive ends for power socket wrenches.
ISO/DIS 4429	Single head engineers wrenches.

ANSI STANDARDS

The following ANSI Standards are under development. When published, copies may be purchased from: American Society of Mechanical Engineers, United Engineering Center, 345 East 47th Street, New York, New York 10017.

B1.18	Interim Standard for Modified ISO Metric Screw Threads for Mechanical Fasteners

B1.19	Interim Standard for Gages for Modified ISO Metric Screw Threads for Mechanical Fasteners
B1.21	Interim Standard for "J" Form Metric Screw Threads for Aerospace Fasteners
B1.22	Interim Standard for Gages for "J" Form Metric Screw Threads for Aerospace Fasteners
B1.13	Interim Standard for ISO Metric Screw Threads (B1 Report)
B18.2.3	Hex Cap Screws and Bolts
B18.2.4	Hex Nuts
B18.2.5	12 Spline Flange Screws
B18.2.6	12 Spline Flange Nuts
B18.3.1	Socket Cap and Set Screws
B18.5.1	Round Head and Plow Bolts
B18.6.5	Thread Forming and Thread Cutting Tapping Screws
B18.6.6	Self-Drilling Screws
B18.6.7	Machine Screws
B18.6.8	Thread Rolling Screws
B18.7.1	Tubular and Split Rivets
B18.13.1	Sems
B18.15.1	Eye Bolts
B18.16.1	Prevailing-Torque Type Steel Hex Locknuts
B18.17.1	Wing Nuts, Thumb Screws and Wing Screws
B18.20.4	Chemical Type Lock Screws
B18.20.5	Free Spinning Type Lock Screws
B18.20.6	Prevailing-Torque Type Lock Screws
B18.21.2	Lock Washers
B18.22.2	Plain Washers
B18.23.2	Beveled Washers

METRIC FASTENER STANDARD HANDBOOKS

United States: *Metric Fastener Standards* (306 pages) in English. Available from: Industrial Fasteners Institute, 1505 East Ohio Building, Cleveland, Ohio 44114, U.S.A.

Germany: *DIN handbook 10—Mechanische Verbindungselemente* (470 pages) in German. Available from: Beuth-Vertrieb GMBH, Berlin, Burggrafenstrasse, West Germany

France: *Recueil de Normes de Boulonnerie–Visserie* (320 pages) in French. Available from: AFNOR, Tour Europe, Cedex 7, 92080 Paris - La Defense, France

Great Britain: *BS Handbook 18* (170 pages) in English. Available from: British Standards Institution, 2 Park Street, London W1A 2BS, England

10 Steel Material Data[1]

Introduction

Major steel-producing countries such as Japan, Germany, France, and Italy have produced steel material dimensioned to metric modules for a number of years. The national standards[2] for steel sizes reflect the available commercial sizes in the above countries. The European countries just listed have developed coordinated EURO-NORM steel standards issued by the European Coal and Steel Community. Since 1955, approximately 100 EURO-NORM standards have been published in German, French, Italian, and Dutch. These standards are available from BEUTH-VERTRIEB GMBH, Burggrafen-strasse 4-7, 1 Berlin 30, West Germany.

Recently the American National Standards Institute issued American National Standards B 32.3 and B 32.4 1974 for preferred metric sizes for flat, round, square, and hexagon metal products. The preferred metric sizes were the result of the informational input of representatives from industry and professional societies as well as the experience of other countries where the metric system has already been or is about to be used.

Description of Tables—The steel material dimension tables list standard metric sizes produced in major countries already using the metric system. The ISO and ANSI standards are also shown, and an asterisk is placed next to the preference rating in the ANSI column for the U.S. commercial sizes. The asterisk added in the U.S. column does not reflect the ANSI preference rating; it represents metric sizes published in the price list by U.S. Steel Corporation. Linear dimensions, section areas, and mass per length or area are shown both in SI units and in customary inch units. The left column in the tables marked ISO, ANS, or DIN indicates the preference rating in the standard from which the metric sizes are chosen. On the right side of the tables national standards for the applicable products are listed with their specified preference rating.

The preference ratings—F for first choice, S for second choice, and T for third choice—are used in the referenced standards to promote usage of fewer sizes. Little rationalization of sizes is achieved in a company if one freely selects from all first choice sizes listed in the steel tables. One should, at all times, keep the material presented in the section for preferred numbers in mind, and try to keep the variety of material stock sizes, tooling, and gages to a bare minimum. You can save your company large sums of money if you can do the job with fewer steel materials, fasteners, gages, and cutting tool sizes.

The tolerances in the tables are given in millimeters, and they are either ISO, EURONORM, or standard tolerances selected from the ISO system of limits and fits. Steel materials produced to the latter tolerances are standard in all major countries outside North America.

The ISO referenced standard numbers in the tables in Section 10 might be recommendations (ISO/R–), draft international standards (ISO DIS–) or standards (ISO–). The proper status of the ISO document can be found in the latest ISO Catalog or in the ISO Draft International Standards and their supplements, the ANSI Catalog, or at the end of Section 10, "Related ISO standards." The above mentioned publications are available from ANSI, 1430 Broadway, New York, New York 10018.

All conversions and calculations are processed by a computer with the data shown rounded off to the nearest number in each case. The computer exponential E-format was selected to cover a wide range of steel sizes and at the same time to present an accuracy to three significant digits for all numbers shown. The maximum error range is from 0.5 to 0.05 percent.

---

[1] The World Ferrous Materials Index is given on pages 10-150 through 10-173 of this section.

[2] For information about the term "standard" as used in this book, please see page 1-2.

Some typical examples of the use of E-format are as follows:

Computer Listing		Decimal Value
$0.427E-01 = 0.427 \times 10^{-1} =$		0.0427
$0.628E+00 = 0.628 \times 10^{0} =$		0.628
$0.243E+01 = 0.243 \times 10^{1} =$		2.43
$0.962E+03 = 0.962 \times 10^{3} =$		962

The standard density for steel used in ISO and national standards outside Canada and the United States is 1 cubic meter steel, which has a 7850 kilogram mass. The standard density factor for steel in the United States is 0.2833 lbs/in^3, and it differs from the ISO standard by approximately 0.1 percent. The mass per unit shown in the steel material tables is calculated using the ISO density factor. The conversion factor 2.767990E+04 was used to compute the pound per cubic inch equivalent. One cubic meter steel has 7842 kilogram mass using the U.S. density factor and 7850 kilogram mass with the ISO factor.

World Sheet Metal Standards

Thicknesses—The following discussion is intended to assist the designer in his choice of metric sheet metal sizes and qualities anywhere in the world. Standard thicknesses for hot-rolled steel sheets are shown in Table 10-1A and for cold-reduced sheets in Table 10-1B.

The column marked ANSI also indicates commercial sizes in addition to the standard thicknesses. The ANSI B32.3 standard specifies a large number of first and second-choice sizes. Most of the above thicknesses are available worldwide. However, this text recommends that sizes be selected on the preferred numbering system so that there is a preference rating similar to that shown in the British column.

Tolerances—The tolerances for sheet metal published in this section are specified in ISO 3573 for hot-rolled sheets and in ISO 3574 for cold-reduced carbon steel sheets.

The EURONORM 51 standard, which is extensively used in continental Europe, specifies tolerances similar to the ISO standard for hot-rolled sheets.

Material—A worldwide comparison of the designations used for hot-rolled and cold-rolled steels is shown in Tables 10-42A and B. The ISO, EURONORM, and national standards for each product are also given for easy reference to further details.

Qualities and Coatings—Customary inch qualities and finishes are specified for metric steel sheets (see Tables 10-42A and 10-42B) and the coating, if required, is specified in grams per square meter, 1 oz/ft^2 = 305 g/m^2.

Order Example—For hot-rolled steel sheets the desired thicknesses and national standards numbers are selected from Table 10-1A, material qualities from Table 10-42A, and tolerances from Table 10-1C. A typical example of designation of steel sheet 2 mm thick × 1000 mm wide × 2000 mm long to Japanese standard JIS G3193 and the commercial quality SPHC in JIS G3131 is as follows: Hot-rolled steel sheet, 2 × 1000 × 2000 JIS G3193–JIS G3131 SPHC–ISO tolerance ± 0.18.

Cold-rolled steel sheets have the worldwide standard thicknesses defined in Table 10-1B. Material qualities are selected from Table 10-42B, and tolerances from Tables 10-1D or 10-1E. A typical designation example of 1 mm thick steel sheet which is 1000 mm wide and 3000 mm long to the ANSI B 32.3 standard and quality ASTM A620 DQ is as follows: Cold-rolled steel sheet 1 × 1000 × 3000 ANSI B32.3 – ASTM A620 DQ – ISO tolerance ± 0.1.

World Steel Plate Standards

Thicknesses—The standard steel plate thicknesses in most of the major steel-producing countries are shown in Table 10-2A. Although some differences exist in the various material standards, most first-choice sizes in the ANSI B32.3 standard should be available worldwide. You will note the few thicknesses considered first choice in the German DIN 1543 standard. This text recommends that the American consumer of steel plate avoid the use of some first choice sizes such as 3.5, 4.5, 7, 35, 140 mm and all the second choice sizes in the ANSI standard. Refer to sizes recommended used in Table 4-5.

Tolerances—The EURONORM 29 standard for hot-rolled steel plates specifies tolerances for thicknesses from 3 to 100 mm, which are shown in Table 10-2B.

Material—Steel plates are generally specified with structural or carbon steel types shown in Tables 10-43 and 10-44. Other steel types are, of course, used for special applications such as high temperature, extended wear resistance, and high pressure.

Qualities and Coatings—Customary inch qualities and finishes are generally available throughout the world.

Order Example—Steel plate thicknesses can be selected from Table 10-2A, material qualities from Tables 10-43 and 10-44, and tolerances from Table 10-2B. A typical example of steel plate designation of a 20 mm thick by 2000 mm wide and 3000 mm long steel plate to the German standard DIN 1543 and the steel quality R St 34-2 is as follows: Steel Plate 20 × 2000 × 3000, DIN 1543-R St 34-2–EURONORM 29 Tolerance − 0.5 + 3.5%.

STEEL MATERIAL DATA

World Steel Wire Standards

Nominal Diameters—The ANSI B32.4 standard diameters for cold drawn steel wires are shown in Table 10-3 and for cold drawn spring steel wire in Table 10-4. Commercial wire sizes are marked with an asterisk in the U.S. column. The national standards for wire diameters in the United Kingdom and in Australia follow the ISO 388 standard, and the standards in Germany specify most of the ISO first and second choice diameters as the preferred sizes. The suggestion here is that the ISO 388 standard should be followed in order to reduce the standard sizes for wires to a minimum. This ISO 388 standard for wire diameters was based on the Renard series of preferred numbers, and groups the wire diameter sizes into three preference ratings, as follows:

First choice: In increments of 25 percent for diameters and in steps of 60 percent for section area or strength.
Second choice: In increments of 12 percent for diameters and in steps of 25 percent for section area or strength.
Third choice: In increments of 6 percent for diameters and in steps of 12 percent for section area or strength. This means the next larger first choice wire size has 25 percent bigger diameter and 60 percent larger section area, which is proportional to strength.

Tolerances—The tolerances for steel wire shown in Table 10-3 are taken from the German Standard DIN 177, which is also valid for bright, normalized, copper, zinc, and tin-coated drawn wire.

The tolerances in Table 10-4 are for spring steel wire, and they are specified in DIN 2076 standard for tolerance Classes A, B, and C. The tolerance Class C is shown in the table, and the other two classes have total tolerances that are approximately twice as large for the same wire sizes.

Material—Property designations and standard data for carbon steels for wires are given in Table 10-42C. Qualities range from low-carbon commercial to high-carbon music wire. A series of alloy spring steel types are shown with worldwide equivalents in Table 10-51. It should be noted that American steel chemistry tolerances are a lot broader than commonly used in other major industrial countries for high carbon steel wires.

Qualities and Coating—Wire products are specified in the United States as to qualities and coatings shown in tables from steel suppliers. The wire qualities shown in these tables are generally available worldwide.

Order Example—The wire diameters and national standards numbers are selected from Table 10-3 for regular qualities and from Table 10-4 for spring steel qualities. Material qualities come from Tables 10-42C or 10-51, and tolerances from Tables 10-3, 10-4 or from American steel suppliers.

A typical order example for 500 kilograms of 1.6 mm diameter steel wire to the Japanese standard is as follows: 500 kg Bright Steel Wire 1.6 JIS G3521 — SWB.

World Steel Bar Standards

Nominal Sizes

Table 10-5A. Hot-Rolled Round Steel Bars (*ISO 1035/I*) —The selection of hot rolled rounds to metric standard sizes (5-300 mm in diameter) is made simply from Table 10-5A. As can be seen, most of the ISO and ANSI standard diameters are available worldwide. It is advised to have the preferred numbering system in mind when sizes are chosen. A 16 mm diameter rod should be preferred over both 14 mm and 18 mm sizes, which is reflected in the British preference rating. However, they are all first choice sizes within the ANSI and ISO standards.

Table 10-6. Hot-Rolled Round Steel Bars for Bolts and Rivets (EURONORM 65)—The EURONORM 65 standard specifies hot-rolled steel bars suitable for manufacturing metric fasteners with a size range from 7.8 to 51.5 mm diameters. Some differences exist between national standards in countries already on the metric system. However, the EURONORM standard is in basic agreement with the German Standard DIN 59130.

Table 10-7. Bright Finish Round Steel Bars (DIN 668 & ISO Tolerance h11)—Cold-drawn steel bars are shown in Table 10-7, and a number of metric sizes (1-320 mm diameter) are available throughout the world. Again, the preferred numbering system should be used when selecting sizes in order to minimize inventory and tooling costs.

Table 10-8. Bright Finish Round Steel Bars (DIN 669 and ISO Tolerance h9)—Cold-finished metric steel bars to the ISO tolerance class h9 are shown in Table 10-8. It should be noted that the German, British, and Australian national standards follow the preferred numbering system to a large extent in specifying standard nominal diameters (3-200 mm). It is, of course, of greatest economic interest to keep the preferred bar stock sizes to a bare minimum.

Table 10-9. Ground or Polished Finish Round Steel Bars[1] *(DIN 59360 and ISO Tolerance h7)*—National standards for ground or polished finish round steel bars with the precision ISO tolerance class h7 are shown in Table 10-9 in diameters ranging from 2 to 80 mm.

Table 10-10. Round Spring Steel Bars (JIS G4801) — Round bars made from spring steel qualities are standardized throughout the world, as shown in Table 10-10, in nominal diameters from 6 to 80 mm.

---
[1] DIN 59361 covers steel bars to ISO tolerance h6.

Table 10-11. Hot Rolled Hexagon Steel Bars (EURO-NORM 61)—Table 10-11 compares standards for hot-rolled hexagon steel bars in the size range from 3.2 to 145 mm widths across flats. A number of differences exist among the various national standards. The U.S. widths across flats entered reflect the ANSI B32.4-1976 draft proposal, and sizes intended produced in the U.S. are marked with an asterisk.

Table 10-12. Bright Finish Hexagon Steel Bars (DIN 176 and ISO Tolerance h11 and h12)—Cold-drawn bright finish hexagon steel bar standard sizes from 3.2 to 105 mm widths across flats are compared in Table 10-12. The U.S. entries reflect the new hexagon sizes included in the ANSI B32.4-1976 proposal.

Table 10-13. Hot Rolled Square Steel Bars (ISO 1035/II)—Hot-rolled square steel bars in sizes from 3 to 320 mm widths across flats are shown in Table 10-13. As can be seen from this table, the ANSI B32.4 standard sizes are the same as those in the ISO 1035/II standard, with minor exceptions. The nominal sizes marked with an asterisk in the U.S. column are commercial sizes in North America, and the same sizes should be available throughout the world.

Table 10-14. Bright Finish Square Steel Bars (DIN 178 and ISO Tolerances h11 and h12)—Cold-finished square steel bars are standardized throughout the world, as shown in Table 10-14, in widths from 3 to 120 mm. The sizes in the ANSI B32.4 standard are, for the most part, selected based on the preferred numbering system, and should be commercially available in the countries listed.

Table 10-15. Hot-Rolled Rectangular Steel Bars (ISO 1035/III)—Hot-rolled rectangular steel bars with nominal sizes from 10 × 3 mm to 400 × 40 mm are specified in ISO 1035/III. The U.S. entries have been based on preferred widths and thicknesses for flat metal products as described in the ANSI B32.3 standard. A bar with either the width or the thickness shown as a second choice size is marked with an S, and when both dimensions are in the second choice classification, the nominal size is shown with a T for third choice. The table entries have been based on the ISO 1035/III standard, and most of the sizes shown for the United States can be anticipated as being available worldwide. The ANSI B32.3 standard specifies a number of intermediate thicknesses for flat metal products, and some of the ANSI sizes not covered in the ISO recommendation were omitted.

Table 10-16. Bright-Finish Rectangular Steel Bars with Sharp Corners (DIN 174)—Cold-formed steel bars with rectangular cross-sections and sharp corners in nominal sizes from 5 × 2 mm to 100 × 50 mm are specified in the national standards, as shown in Table 10-16. A systematic trend to use the preferred numbering system in the German and British standards is evident, and the second and third choice sizes in the above standards might not be commercially available in Europe. The ANSI B32.3 standard has no widths specified between 60 and 80 mm, which represents a rather large jump in size.

Table 10-17. Bright Steel for Parallel Keys—Square Cross-sections (DIN 6880)—Cold-drawn steel bars with square cross-sections intended for parallel keys given in ISO 773 are shown in Table 10-17 in nominal sizes from 2 × 2 mm to 22 × 22 mm. For the practical use of parallel keys, see Tables 13-17 and 13-18.

Table 10-18. Bright Steel for Parallel Keys—Rectangular Cross-sections (DIN 6880)—Cold-drawn steel bars with rectangular cross-sections intended for parallel keys given in ISO 773 are covered in the German standard DIN 6880 in sizes from 8 × 7 mm to 100 × 50 mm and are shown in Table 10-18. The standard steel material agrees with that shown in Table 13-17 and Table 13-18. The above keys and keyways are standard throughout the world, as shown in the referenced tables.

Table 10-19. Bright Steel for Parallel Keys—Flat Cross-sections (DIN 6880)—Cold-reduced steel bars with rectangular cross-sections used for thin parallel keys given in ISO 2491 are specified in the DIN 6880 standard and shown in Table 10-19 in sizes from 5 × 3 mm to 40 × 14 mm. Details of their usage and references to other national standards are shown in Tables 13-21 and 13-22.

Tables 10-20 and 10-21. Hot-Rolled Steel Bars for Woodruff Keys—Half Rounds and Half Ovals (EURO-NORM 66)—Hot-rolled steel bars for Woodruff keys are specified in the EURONORM 66 standard. Steel in sizes shown in Tables 10-20 and 10-21 is available in Europe.

Tolerances

Hot-Rolled Steel Bars—The tolerances for hot-rolled bar products shown in this section are based on EURONORM and ISO Standards. The EURONORM steel standards have been in use in Europe for some time, and the allowances shown in the EURONORM columns are not always in agreement with those in the ISO standard ISO 1035/IV shown in Table 10-5B.

Cold-Drawn Steel Bars—Dimensions for cold-drawn steel bars are generally held within an ISO limits and fits tolerance throughout the world. Round steel bars to ISO tolerances[1] h7, h9, and h11 are in use in major industrial countries already on the metric system, and these tolerances are also used in the selected fits within the ISO system. The tolerances for cold-reduced steel bars might be shown in one national column in the worldwide steel

---
[1] DIN 59361 covers steel bars to ISO tolerance h6.

STEEL MATERIAL DATA

Table 10-1A. Hot Rolled Steel Sheet and Strip Thicknesses (ANSI B32.3)

```
TABLE 10-1A. HOT ROLLED STEEL SHEET
AND STRIP THICKNESSES (ANSI B32.3)
BASIS: 1 IN = 25.4 MM
1 CUBIC METER STEEL = 7850 KG MASS
```

ORDER EXAMPLE:
SIZE,HOT ROLLED STEEL SHEET 2.5 DIN 1016*ST QU

THE NOMINAL SIZE IS NATIONAL STANDARD AS INDICATED
F=FIRST CHOICE,S=SECOND CHOICE,T=THIRD CHOICE,NUMBER=OTHER SIZE
* = COMMERCIAL SIZE

A S	NOMINAL SIZE = D MM IN	MASS PER UNIT KG/M**2 LB/FT**2	U.S.A. ANSI B32.3	AUSTRAL AS 1594	JAPAN JIS G3193	ISO 3573	FRANCE NF A46-501	U.K. BS 1449	GERMANY DIN 1016	ITALY UNI
F	1	0.0394	0.785E+01 0.161E+01	F	F		D+-0.17	F	F	F	
S	1.1	0.0433	0.863E+01 0.177E+01	S			D+-0.17				
F	1.2	0.0472	0.942E+01 0.193E+01	F	F	1.25F	D+-0.17				
S	1.4	0.0551	0.110E+02 0.225E+01	S	S	F	D+-0.17				
F	1.5	0.0591	0.118E+02 0.241E+01			F	D+-0.17	F			
F	1.6	0.0630	0.126E+02 0.257E+01	F*		F	D+-0.17	F	F		
F	1.75	0.0689	0.137E+02 0.281E+01			F	D+-0.17				
S	1.8	0.0709	0.141E+02 0.289E+01	S*	S	F	D+-0.18				
F	2	0.0787	0.157E+02 0.322E+01	F*	F	F	D+-0.18	F	F		
S	2.2	0.0866	0.173E+02 0.354E+01	S*	S	2.3F	D+-0.2	2.25F			
F	2.5	0.0984	0.196E+02 0.402E+01	F*	F	F	D+-0.2	F	F		
S	2.8	0.1102	0.220E+02 0.450E+01	S*	S	F ,2.65	D+-0.21				
F	3.2	0.1181	0.235E+02 0.482E+01	F*	F	2.9S	D+-0.21	F			
S	3.2	0.1260	0.251E+02 0.514E+01	S*		F	D+-0.23	F			
F	3.5	0.1378	0.275E+02 0.563E+01	F*	F	F	D+-0.23	F			
S	3.8	0.1496	0.298E+02 0.611E+01	S*	S	3.6F	D+-0.23				
F	4.2	0.1575	0.314E+02 0.643E+01	F*	F	F	D+-0.23	F			
S	4.2	0.1654	0.330E+02 0.675E+01	S*	S	F	D+-0.27	F			
F	4.5	0.1772	0.353E+02 0.724E+01	F*	F	F	D+-0.27	F			
S	4.8	0.1890	0.377E+02 0.772E+01	S*	S	F	D+-0.27	4.75F			
F	5	0.1969	0.392E+02 0.804E+01	F*	F	F	D+-0.27	F			
S	5.5	0.2165	0.432E+02 0.884E+01	S*	S	5.6F	D+-0.27	F			
F	6	0.2362	0.471E+02 0.965E+01	F*	F	F	D+-0.27	F			
F	7	0.2756	0.549E+02 0.113E+02	F*	F	F ,6.3F	D+-0.3	F			
F	8	0.3150	0.628E+02 0.129E+02	F*	F	F	D+-0.3	F			
S	9	0.3543	0.706E+02 0.145E+02	S*	S	F	D+-0.33	F			
F	10	0.3937	0.785E+02 0.161E+02	F*	F	F	D+-0.33	F			
S	11	0.4331	0.863E+02 0.177E+02	S	F	F	D+-0.36				
F	12	0.4724	0.942E+02 0.193E+02	F		F	D+-0.36	F		12.5F	
```
```

standards comparison tables. Then one could assume that similar tolerances are used in other major countries.

Material

Steel bars are generally specified as to physical and chemical properties, as shown in Table 10-43 in Europe, or to various carbon steel grades in Table 10-44 in the United States. Steel types with special physical properties, as compared in Tables 10-45 through 48, are also frequently used throughout the world for hot and cold formed steel bars.

Qualities and Coatings

The use of customary inch qualities and coatings is recommended, and protective coatings, if required, should be specified in the SI units of g/m^2 (1 oz/ft^2 = 305 g/m^2).

Order Example

Steel bars in metric standard sizes are specified with the name of the product, national standards number, tolerance, steel designation and special requirements as necessary. The nominal size and national standards number for each bar product are shown in Tables 10-5A through 10-21. Tolerances for hot-rolled steel are given in Tables 10-5B. Cold-formed steel tolerances can be found in each product dimension table, and steel designations in Tables 10-43 through 10-48. Two typical order examples are as follows:

Length, Hot-Rolled Round Steel Bar
10 JIS G3191-SS50-
ISO Tolerance ±0.4

Length, Bright-Finish Round Steel Bar
25 DIN 668-St37-
ISO Tolerance h11

Tolerances are generally specified within the referred national standard. Thus, while the examples shown above have redundant information, they should help avoid misunderstandings in the tolerancing of products ordered.

World Steel Tube Standards

Nominal Sizes—Table 10-22A. Seamless (SL) and Welded (WE) Steel Tubes for General Use (ISO 336, 134, 221, 2937)—Worldwide seamless and welded steel tubes standards for general use are shown in Table 10-22A in sizes from 10.2 to 141.3 mm diameters. National standards for both seamless (SL) and welded (WE) tubes are shown in the same table. The seamless standards numbers (TOP) and preference ratings (LEFT) are listed first.

Table 10-22A also shows the L = light, N = normal, and H = heavy series. In this table the left F applies to the seamless tube preference rating and the right F to the welded tube preference rating. ISO preferred wall thicknesses other than listed for each tube diameter are shown as follows: C = 0.6, 0.8, 1, 1.2, etc.

Table 10-23. Hollow Steel Bars for Machining (ISO 2938)—The hollow steel bars for machining are sometimes used in countries already on the metric system, and Mannesmann, producer of tubular products, supplies this type of tubing throughout the world. The diameters and wall thicknesses have been published in the ISO 2938 standard, where the outside diameters have been chosen from the preferred number series.

Table 10-24. Seamless (SL) and Welded (WE) Precision Steel Tubes (ISO 560, 3304, 3305, 3306)—The national and international standards for precision steel tubes have been developed according to preferred metric modules, and two types in sizes from 4 to 120 mm nominal diameters are shown in Table 10-24. These steel products have been made available worldwide by the above producer in Germany.*

Table 10-22B. Tolerance Classes for Tubes (ISO 5252)

For Outside Tube Diameters		
Tolerance Designation	Tolerance Nominal + or -	Minimum + or -
D_0	2%	1 mm
D_1	1.5%	0.75 mm
D_2	1%	0.5 mm
D_3	0.75%	0.3 mm
D_4	0.5%	0.1 mm

For Wall Thicknesses		
Tolerance Designation	Tolerance Nominal + or -	Minimum + or -
T_0	20%	1
T_1	15%	0.6 mm
T_2	12.5%	0.4 mm
T_3	10%	0.2 mm
T_4	7.5%	0.15 mm
T_5	5%	0.1 mm

*NOTE: The text is continued on page 10-127.

STEEL MATERIAL DATA

Table 10-1B. Cold Rolled Steel Sheet and Strip Thicknesses (ANSI B32.3)

TABLE 10-1B. COLD ROLLED STEEL SHEET
AND STRIP THICKNESSES (ANSI B32.3)
BASIS: 1 IN = 25.4 MM
1 CUBIC METER STEEL = 7850 KG MASS

ORDER EXAMPLE:
SIZE,COLD ROLLED STEEL SHEET 0.8 BS 1449*ST QU

THE NOMINAL SIZE IS NATIONAL STANDARD AS INDICATED
F=FIRST CHOICE,S=SECOND CHOICE,T=THIRD CHOICE,NUMBER=OTHER SIZE
* = COMMERCIAL SIZE

A N S	NOMINAL SIZE = D		MASS PER UNIT		U.S.A. ANSI B32.3	AUSTRAL AS 1595	JAPAN JIS G3141	ISA 3574	FRANCE NF A46-402	U.K. BS 1449	GERMANY DIN 1540	ITALY UNI
	MM	IN	KG/M**2	LB/FT**2								
F	0.1	0.0039	0.785E+00	0.161E+00		F	F ,0.11S	D+-0.07	D+-0.07		STANDARD THICK-	
F	0.12	0.0047	0.942E+00	0.193E+00		F	S	D+-0.07	D+-0.07		NESSES	
S	0.14	0.0055	0.110E+01	0.225E+00		S	F	D+-0.07	D+-0.07		FROM	
F	0.16	0.0063	0.126E+01	0.257E+00		F	F	D+-0.07	D+-0.07			
S	0.18	0.0071	0.141E+01	0.289E+00		S	S	D+-0.07	D+-0.07		0.15 TO	
F	0.2	0.0079	0.157E+01	0.322E+00		F	F	D+-0.07	D+-0.07		0.5 ARE	
S	0.22	0.0087	0.173E+01	0.354E+00		S	S	D+-0.07	D+-0.07		IN INCR	
F	0.25	0.0098	0.196E+01	0.402E+00		F	F	D+-0.07	D+-0.07		OF 0.01	
S	0.28	0.0110	0.220E+01	0.450E+00		S	S	D+-0.07				
F	0.3	0.0118	0.235E+01	0.482E+00		F	F	D+-0.07 S				
S	0.35	0.0138	0.275E+01	0.563E+00		S*	S	D+-0.07 S			DIN 1541	
S	0.35	0.0138	0.275E+01	0.563E+00		S*					F	
F	0.4	0.0157	0.314E+01	0.643E+00		F*	F	D+-0.07 S			F	
S	0.45	0.0177	0.353E+01	0.724E+00		S*	F	D+-0.08 S			F	
F	0.5	0.0197	0.392E+01	0.804E+00		F*	F	D+-0.08 S			F	
S	0.55	0.0217	0.432E+01	0.884E+00		S*	S	D+-0.08			F	
F	0.6	0.0236	0.471E+01	0.965E+00		F*	F	D+-0.08 S			F	
S	0.65	0.0256	0.510E+01	0.105E+01		S*	F	D+-0.09 S			F	
F	0.7	0.0276	0.549E+01	0.113E+01		S*	F	D+-0.09 S			F	
F	0.8	0.0315	0.628E+01	0.129E+01		F*	F	D+-0.09 F			F	
S	0.9	0.0354	0.706E+01	0.145E+01		F*	S	D+-0.1 S			F	
F	1	0.0394	0.785E+01	0.161E+01		F*	F	D+-0.1 F			F	
S	1.1	0.0433	0.863E+01	0.177E+01		S*	S	D+-0.12 F			F	
F	1.2	0.0472	0.942E+01	0.193E+01		F*	F	D+-0.12 F			F	
S	1.4	0.0551	0.110E+02	0.225E+01		S*	S	D+-0.14 F			F	
	1.5	0.0591	0.118E+02	0.241E+01			F	D+-0.14 F			F	
F	1.6	0.0630	0.126E+02	0.257E+01		F*	F	D+-0.14 F			F	
S	1.8	0.0709	0.141E+02	0.289E+01		S*	S	D+-0.16 F			F	
F	2	0.0787	0.157E+02	0.322E+01		F*	F 2.3F	D+-0.16 F			F	
S	2.2	0.0866	0.173E+02	0.354E+01		S*	S	D+-0.18 F			F	
F	2.5	0.0984	0.196E+02	0.402E+01		F*	F	D+-0.18 F			F	
S	2.8	0.1102	0.220E+02	0.450E+01		S*	S F,2.6S	D+-0.2 F			F	
F	3	0.1181	0.235E+02	0.482E+01		F*	F 2.9S	D+-0.2	2.99F		F	
S	3.2	0.1260	0.251E+02	0.514E+01		S*	F	D+-0.23			F	
F	3.5	0.1378	0.275E+02	0.563E+01		F	S	D+-0.23			F	
S	3.8	0.1496	0.298E+02	0.611E+01				D+-0.23				
F	4	0.1575	0.314E+02	0.643E+01		F	F	D+-0.23			F	

Table 10-1C. Thickness Tolerances for Coils[1] and Cut Lengths (Including Descaled Sheet) (ISO 3573)

Hot-Rolled

Values in millimetres

Specified widths	Thickness tolerances[2], over and under, for specified thicknesses							
	up to and including 1,60	over 1,60 up to and including 2,00	over 2,00 up to and including 2,50	over 2,50 up to and including 3,00	over 3,00 up to and including 4,00	over 4,00 up to and including 6,00	over 6,00 up to and including 8,00	over 8,00 up to and including 10,00	over 10,00 to 12,50 inclusive
600 up to and including 1 200	0,17	0,18	0,20	0,21	0,23	0,27	0,30	0,33	0,36
Over 1 200 up to and including 1 500	0,19	0,20	0,22	0,23	0,25	0,28	0,31	0,36	0,38
Over 1 500 up to and including 1 800	0,21	0,22	0,24	0,25	0,26	0,29	0,33	0,38	0,41
Over 1 800	—	—	0,26	0,27	0,28	0,30	0,38	0,43	0,46

1. The values specified do not apply to the uncropped ends for a length ℓ of a mill edge coil. Length ℓ would be calculated using the formula:

$$\text{Length } \ell \text{ in meters} = \frac{90}{\text{Thickness in millimeters}}$$

   provided that the result was not greater than 30 m.
2. Thickness is measured at any point on the sheet not less than 40 mm from a side edge.

STEEL MATERIAL DATA

10-9

Cold-Reduced

Table 10-1D. Standard Thickness Tolerances for Coils[1] and Cut Lengths (ISO 3574)

Values in millimetres

Specified widths	Thickness tolerances[2], over and under, for specified thicknesses								
	up to and including 0,4	over 0,4 up to and including 0,6	over 0,6 up to and including 0,8	over 0,8 up to and including 1,0	over 1,0 up to and including 1,2	over 1,2 up to and including 1,6	over 1,6 up to and including 2,0	over 2,0 up to and including 2,5	over 2,5 up to and including 3,0	over 3,0 up to and including 4
600 up to and including 1 200	0,07	0,08	0,09	0,10	0,12	0,14	0,16	0,18	0,20	0,23
Over 1 200 up to and including 1 500	0,08	0,09	0,10	0,11	0,13	0,15	0,17	0,20	0,23	0,25
Over 1 500 up to and including 1 800	—	0,10	0,11	0,13	0,14	0,17	0,19	0,22	0,23	0,27
Over 1 800	—	0,12	0,13	0,14	0,16	0,19	0,21	0,24	0,26	0,29

1. The thickness tolerances for sheets in coil form are the same as for sheets supplied in cut lengths, but in cases where welds are present, the tolerances shall be double those given over a length of 15 m in the vicinity of the weld.
2. Thickness is measured at any point on the sheet not less than 40 mm from a side edge.

Cold-Reduced

Table 10-1E. Special Thickness Tolerances for Coils[1] and Cut Lengths (ISO 3574)

Values in millimetres

Specified widths	Thickness tolerances[2], over and under, for specified thicknesses								
	up to and including 0,4	over 0,4 up to and including 0,6	over 0,6 up to and including 0,8	over 0,8 up to and including 1,0	over 1,0 up to and including 1,2	over 1,2 up to and including 1,6	over 1,6 up to and including 2,0	over 2,0 up to and including 2,5	over 2,5 up to and including 3,0	over 3,0 up to and including 4
600 up to and including 1 200	0,040	0,045	0,055	0,065	0,075	0,090	0,110	0,125	0,140	0,165
Over 1 200 up to and including 1 500	0,045	0,055	0,065	0,075	0,085	0,110	0,125	0,140	0,155	0,180
Over 1 500 up to and including 1 800	—	—	0,075	0,085	0,100	0,120	0,140	0,155	0,170	0,190
Over 1 800	—	—	0,080	0,095	0,105	0,135	0,150	0,165	0,185	0,200

1. The thickness tolerances for sheets in coil form are the same as for sheets supplied in cut lengths, but in cases where welds are present, the tolerances shall be double those given over a length of 15 m in the vicinity of the weld.
2. Thickness is measured at any point on the sheet not less than 40 mm from a side edge.

Table 10-2A. Steel Plate Thicknesses (ANSI B32.3)

```
TABLE 10-2A. STEEL PLATE                          ORDER EXAMPLE:
THICKNESSES (ANSI B32.3)                            SIZE,STEEL PLATE 5 JIS G3193 + STEEL QUALITY
BASIS: 1 IN = 25.4 MM
1 CUBIC METER STEEL = 7850 KG MASS

                                                  THE NOMINAL SIZE IS NATIONAL STANDARD AS INDICATED
                                                  F=FIRST CHOICE,S=SECOND CHOICE,T=THIRD CHOICE,NUMBER=OTHER SIZE
                                                  * = COMMERCIAL SIZE
                                                  U.S.A. AUSTRAL JAPAN  EURO-  FRANCE  U.K.  GERMANY ITALY
A                                                  ANSI    AS    JIS    NORM    NF      BS     DIN    UNI
N  NOMINAL SIZE = D   MASS PER UNIT                B32.3  1184  G3193   29    A45-005  1449   1543
S   MM     IN         KG/M**2    LB/FT**2
   3      0.118       0.235E+02  0.482E+01         F      F     F       D-0.4+%   F       F       F      F
F  3.2    0.126       0.251E+02  0.514E+01         S            F       D-0.4+%
F  3.5    0.138       0.275E+02  0.563E+01         F      S            D-0.4+%
F  4      0.157       0.314E+02  0.643E+01         F,3.8S       F,3.6F  D-0.4+%

F  4.5    0.177       0.353E+02  0.724E+01         F,4.2S       F       D-0.4+%                4.76F
S  4.8    0.189       0.377E+02  0.772E+01         S            F       D-0.4+%
F  5      0.197       0.392E+02  0.804E+01         F*           F       D-0.4+%                        THICK-
S  5.5    0.217       0.432E+02  0.884E+01         S*           F       D-0.4+%                        NESS

F  6      0.236       0.471E+02  0.965E+01         F*           F,5.6F  D-0.4+%                         NOT
F  7      0.276       0.549E+02  0.113E+02         F*           F,6.3F  D-0.4+% F,6.5S                 SPECI-
F  8      0.315       0.628E+02  0.129E+02         F*           F       D-0.4+%                        FIED
S  9      0.354       0.706E+02  0.145E+02         S*           F       D-0.4+%                        FOR
                                                                                                       PLATE
F  10     0.394       0.785E+02  0.161E+02         F*           F       D-0.4+%  F       F       F      F
S  11     0.433       0.863E+02  0.177E+02         S*           F       D-0.4+%  F       T       F      F
F  12     0.472       0.942E+02  0.193E+02         F*           F       D-0.4+%
S  14     0.551       0.110E+03  0.225E+02         S*           F,12.7F D-0.4+% F,13T           12.5F   F

   15     0.591       0.118E+03  0.241E+02         F*           F,13F   D-0.4+%  F       F       F      F     TOLE-
F  16     0.630       0.126E+03  0.257E+02         F*           F,17S   D-0.4+%                               RANCE
S  18     0.709       0.141E+03  0.289E+02         F*           F,19F   D-0.4+%                               SEE
F  20     0.787       0.157E+03  0.322E+02         F*           F       D-0.5+%                               UNI
                                                                                                              6669
S  22     0.866       0.173E+03  0.354E+02         S*           S,25.4F D-0.5+%  T
F  25     0.984       0.196E+03  0.402E+02         F*           F       D-0.6+%  F
F  28     1.102       0.220E+03  0.450E+02         F*           F       D-0.6+%  F
F  30     1.181       0.235E+03  0.482E+02         F*           F       D-0.7+%  F

S  32     1.260       0.251E+03  0.514E+02         S*           F       D-0.7+%  S
F  35     1.378       0.275E+03  0.563E+02         F*           F       D-0.7+%
F  36     1.417       0.283E+03  0.579E+02                      S       D-0.7+%
S  38     1.496       0.298E+03  0.611E+02         S*           F       D-0.7+%  S

F  40     1.575       0.314E+03  0.643E+02         F*           F       D-0.9+%  F       F       F      F
F  45     1.772       0.353E+03  0.724E+02         F*           F       D-0.9+%  F
F  50     1.969       0.392E+03  0.804E+02         F*           F       D-1.0+%  F       S       F      F
S  55     2.165       0.432E+03  0.884E+02         S*           F       D-1.0+%  F

F  60     2.362       0.471E+03  0.965E+02         F*           F       D-1.0+%  F
S  70     2.756       0.549E+03  0.113E+03         S*           F       D-1.0+%  F                     F,65F  F
F  80     3.150       0.628E+03  0.129E+03         F*           F       D-1.0+%  F                     F,75F  F
S  90     3.543       0.706E+03  0.145E+03         S*           F       D-1.0+%  F

F  100    3.937       0.785E+03  0.161E+03         F*           F               D-1.0+%  F       F             F
S  110    4.331       0.863E+03  0.177E+03         F*           F                        F
F  120    4.724       0.942E+03  0.193E+03         F*           F                        F
S  130    5.118       0.102E+04  0.209E+03         S*           S                        F

F  140    5.512       0.110E+04  0.225E+03         F*           F                        F
S  150    5.906       0.118E+04  0.241E+03         S*           S
F  160    6.299       0.126E+04  0.257E+03         F*           F                        F
```

10-10   STEEL MATERIAL DATA

STEEL MATERIAL DATA

Table 10-2B. Thickness Tolerances for Steel Plates (EURONORM 29)*

Nominal Thickness Ranges	Under (−)	% Over Theoretical Weight for Width Ranges								
		< 1500	≧ 1500 < 1750	≧ 1750 < 2000	≧ 2000 < 2250	≧ 2250 < 2500	≧ 2500 < 2750	≧ 2750 < 3000	≧ 3000 < 3500	≧ 3500 ≦ 4000
≧ 3 < 4	− 0,4	5	6	7	8	9	—	—	—	—
≧ 4 < 4,76	− 0,4	5	6	7	8	9	9	—	—	—
≧ 4,76 < 6	− 0,4	4,5	5	6	7	7,5	9	—	—	—
≧ 6 < 8	− 0,4	4,5	5	6	7	7,5	9	10	—	—
≧ 8 < 10	− 0,4	4,5	5	5,5	6,5	7,5	8,5	9,5	10,5	—
≧ 10 < 12,5	− 0,4	3,5	4	4,5	5	6	7	7,5	8,5	—
≧ 12,5 < 15	− 0,4	3,5	4	4	5	5,5	6	7	7,5	—
≧ 15 < 20	− 0,4	3	3	3,5	4	4,5	5,5	6	6,5	7
≧ 20 < 25	− 0,5	3	3	3	3,5	3,5	4	5	5,5	6
≧ 25 < 30	− 0,6	3	3	3	3,5	3,5	4	5	5,5	6
≧ 30 < 40	− 0,7	3	3	3	3	3	3,5	4	4,5	5
≧ 40 < 50	− 0,9	3	3	3	3	3	3,5	4	4,5	5
≧ 50 < 60	− 1,0	3	3	3	3	3	3,5	4	4,5	5
≧ 60 < 80	− 1,0	3	3	3	3	3	3,5	4	4,5	5
≧ 80 ≦ 100	− 1,0	3	3	3	3	3	3,5	4	4,5	5

*Dimensions in Millimeters.

10-12

STEEL MATERIAL DATA

TABLE 10-3
COLD DRAWN ROUND STEEL WIRE (ISO 388)
BASIS; 1 IN = 25.4 MM
1 CUBIC METER STEEL = 7850 KG MASS

ORDER EXAMPLE;
LENGTH,STEEL WIRE 0.1 BS4391 + STEEL QUALITY
FOR STEEL MATL DESIGNATIONS SEE PAGE

PAGE NO. 1

THE NOMINAL SIZE IS NATIONAL STANDARD AS INDICATED
F=FIRST CHOICE,S=SECOND CHOICE,T=THIRD CHOICE,NUMBER=OTHER SIZE
* = COMMERCIAL SIZE

	NOMINAL SIZE = D		SECTION AREA		MASS PER UNIT		U.S.A. ANSI B32.4	AUSTRAL SAA AS1153	JAPAN JISC G3521	EURO- NORM	FRANCE AFNOR A47.411	U.K. BSI BS4391	GERMANY DNA DIN177	ITALY UNI UNI467
	MM	IN	MM**2	IN**2	KG/M	LB/FT								
F	0.02	.00079	0.314E-03	0.487E-06	0.247E-05	0.166E-05								
T	0.021	.00083	0.346E-03	0.537E-06	0.272E-05	0.183E-05								
S	0.022	.00087	0.380E-03	0.589E-06	0.298E-05	0.201E-05						T		
T	0.024	.00094	0.452E-03	0.701E-06	0.355E-05	0.239E-05	S					T		
F	0.025	.00098	0.491E-03	0.761E-06	0.385E-05	0.259E-05						T		
T	0.026	.00102	0.531E-03	0.823E-06	0.417E-05	0.280E-05	F					S		
S	0.028	.00110	0.616E-03	0.954E-06	0.483E-05	0.325E-05						S		
T	0.03	.00118	0.707E-03	0.110E-05	0.555E-05	0.373E-05	F					T		
F	0.032	.00126	0.804E-03	0.125E-05	0.631E-05	0.424E-05						F		
T	0.034	.00134	0.908E-03	0.141E-05	0.713E-05	0.479E-05						T		
S	0.035	.00138	0.962E-03	0.149E-05	0.755E-05	0.508E-05						S		
S	0.036	.00142	0.102E-02	0.158E-05	0.799E-05	0.537E-05	S							
T	0.038	.00150	0.113E-02	0.176E-05	0.890E-05	0.598E-05						F		
F	0.04	.00157	0.126E-02	0.195E-05	0.986E-05	0.663E-05						T		
S	0.042	.00165	0.139E-02	0.215E-05	0.109E-04	0.731E-05	F					F		
T	0.045	.00177	0.159E-02	0.247E-05	0.125E-04	0.839E-05	S					S		
T	0.048	.00189	0.181E-02	0.280E-05	0.142E-04	0.955E-05						T		
F	0.05	.00197	0.196E-02	0.304E-05	0.154E-04	0.104E-04						F		
S	0.053	.00209	0.221E-02	0.342E-05	0.173E-04	0.116E-04	F					T		
S	0.055	.00217	0.238E-02	0.368E-05	0.187E-04	0.125E-04	S							
T	0.056	.00220	0.246E-02	0.382E-05	0.193E-04	0.130E-04						S		
F	0.06	.00236	0.283E-02	0.438E-05	0.222E-04	0.149E-04						T		
S	0.063	.00248	0.312E-02	0.483E-05	0.245E-04	0.164E-04						F		
T	0.065	.00256	0.332E-02	0.514E-05	0.260E-04	0.175E-04						T		
S	0.067	.00264	0.353E-02	0.546E-05	0.277E-04	0.186E-04						F		
F	0.07	.00276	0.385E-02	0.597E-05	0.302E-04	0.203E-04						T		
S	0.071	.00280	0.396E-02	0.614E-05	0.311E-04	0.209E-04						S		
T	0.075	.00295	0.442E-02	0.685E-05	0.347E-04	0.233E-04						T		
F	0.08	.00315	0.503E-02	0.779E-05	0.395E-04	0.265E-04	F					F		
T	0.085	.00335	0.567E-02	0.880E-05	0.445E-04	0.299E-04						T		
S	0.09	.00354	0.636E-02	0.986E-05	0.499E-04	0.336E-04						S		
T	0.095	.00374	0.709E-02	0.110E-04	0.556E-04	0.374E-04						T		
F	0.1	.00394	0.785E-02	0.122E-04	0.617E-04	0.414E-04	F					F	F +-0.01	
T	0.106	.00417	0.882E-02	0.137E-04	0.693E-04	0.466E-04						T	+-0.01	
S	0.11	.00433	0.950E-02	0.147E-04	0.746E-04	0.501E-04	S					T	F +-0.01	
T	0.112	.00441	0.985E-02	0.153E-04	0.773E-04	0.520E-04						S	+-0.01	
F	0.118	.00465	0.109E-01	0.170E-04	0.858E-04	0.577E-04						F	+-0.01	
T	0.12	.00472	0.113E-01	0.175E-04	0.888E-04	0.597E-04	F					F	F +-0.01	
F	0.125	.00492	0.123E-01	0.190E-04	0.963E-04	0.647E-04						F	F +-0.01	
T	0.132	.00520	0.137E-01	0.212E-04	0.107E-03	0.722E-04						T	+-0.01	
S	0.14	.00551	0.154E-01	0.239E-04	0.121E-03	0.812E-04	S					S	F +-0.01	

STEEL MATERIAL DATA

Table 10-3 (Continued). Cold Drawn Round Steel Wire (ISO 388)

```
TABLE 10-3.
COLD DRAWN ROUND STEEL WIRE (ISO 388)                                    ORDER EXAMPLE:
BASIS; 1 IN = 25.4 MM                                                    LENGTH,STEEL WIRE 0.1 BS 4391 + STEEL QUALITY
1 CUBIC METER STEEL = 7850 KG MASS                                       PAGE NO. 2

                                                THE NOMINAL SIZE IS NATIONAL STANDARD AS INDICATED
                                                F=FIRST CHOICE,S=SECOND CHOICE,T=THIRD CHOICE,NUMBER=OTHER SIZE
                                                * = COMMERCIAL SIZE
                                                U.S.A.   AUSTRAL JAPAN   EURO-   FRANCE    U.K.     GERMANY  ITALY
         NOMINAL SIZE = D  SECTION AREA  MASS PER UNIT  ANSI     AS     JIS     NORM    NF       BS       DIN     UNI
          MM      IN       MM**2   IN**2  KG/M   LB/FT  B32.4   1153   G3521           A47-411  4391     177     467

     T   0.15   0.0059   0.177E-01 0.274E-04  0.135E-03 0.932E-04                                       F +-0.01
     F   0.16   0.0063   0.201E-01 0.312E-04  0.158E-03 0.106E-03        F                       T      F +-0.01     F
     T   0.17   0.0067   0.227E-01 0.352E-04  0.178E-03 0.120E-03                                 T      F +-0.01
     S   0.18   0.0071   0.254E-01 0.394E-04  0.200E-03 0.134E-03        S                       S      F +-0.01

     T   0.19   0.0075   0.284E-01 0.439E-04  0.223E-03 0.150E-03                                 T      F +-0.01
     F   0.2    0.0079   0.314E-01 0.487E-04  0.247E-03 0.166E-03        F                       F      F +-0.01     F
     T   0.212  0.0083   0.353E-01 0.547E-04  0.277E-03 0.186E-03                                 T      F +-0.015
     S   0.22   0.0087   0.380E-01 0.589E-04  0.298E-03 0.201E-03        S                              F +-0.015

     S   0.224  0.0088   0.394E-01 0.611E-04  0.309E-03 0.208E-03        S                       S      F +-0.015
     T   0.236  0.0093   0.437E-01 0.678E-04  0.343E-03 0.231E-03                                 T      F +-0.015
     F   0.25   0.0098   0.491E-01 0.761E-04  0.385E-03 0.259E-03        F      0.23F             F      F +-0.015  0.24F
     T   0.265  0.0104   0.552E-01 0.855E-04  0.433E-03 0.291E-03                                 T      F +-0.015  0.26F

     S   0.28   0.0110   0.616E-01 0.954E-04  0.483E-03 0.325E-03        S      0.26F             S      F +-0.015
     T   0.3    0.0118   0.707E-01 0.110E-03  0.555E-03 0.373E-03                                 T      F +-0.015   F
     F   0.315  0.0124   0.779E-01 0.121E-03  0.612E-03 0.411E-03        F      0.29F             F      F +-0.015  0.31F
     T   0.335  0.0132   0.881E-01 0.137E-03  0.692E-03 0.465E-03                                 T      0.32F-0.015

         0.35   0.0138   0.962E-01 0.149E-03  0.755E-03 0.508E-03               0.32F                    F +-0.015  0.34F
     S   0.355  0.0140   0.990E-01 0.153E-03  0.777E-03 0.522E-03        S                       S      F +-0.015
     T   0.375  0.0148   0.110E+00 0.171E-03  0.867E-03 0.583E-03                                 T      0.36F-0.015  0.37F
     F   0.4    0.0157   0.126E+00 0.195E-03  0.986E-03 0.663E-03        F       F                F      F +-0.02     F

     T   0.425  0.0167   0.142E+00 0.220E-03  0.111E-02 0.748E-03                                 T      F +-0.02
     S   0.45   0.0177   0.159E+00 0.247E-03  0.125E-02 0.839E-03        S                       T      F +-0.02     F
     T   0.475  0.0187   0.177E+00 0.275E-03  0.139E-02 0.935E-03                                 T      F +-0.02
     F   0.5    0.0197   0.196E+00 0.304E-03  0.154E-02 0.104E-02        F       F                F      F +-0.02     F

     T   0.53   0.0209   0.221E+00 0.342E-03  0.173E-02 0.116E-02                                 T      F +-0.02
     S   0.55   0.0217   0.238E+00 0.368E-03  0.187E-02 0.125E-02        S                              F +-0.02     F
     T   0.56   0.0220   0.246E+00 0.382E-03  0.193E-02 0.130E-02                                 T      F +-0.02
     S   0.6    0.0236   0.283E+00 0.438E-03  0.222E-02 0.149E-02        S                       S      F +-0.03     F

     F   0.63   0.0248   0.312E+00 0.483E-03  0.245E-02 0.164E-02        F       F                F      F +-0.03
     F   0.65   0.0256   0.332E+00 0.514E-03  0.260E-02 0.175E-02        F                              F +-0.03     F
     T   0.67   0.0264   0.353E+00 0.546E-03  0.277E-02 0.184E-02                                 T      F +-0.03
         0.7    0.0276   0.385E+00 0.597E-03  0.302E-02 0.203E-02                                        F +-0.03     F

     S   0.71   0.0280   0.396E+00 0.614E-03  0.311E-02 0.209E-02        S                       S      F +-0.03
     T   0.75   0.0295   0.442E+00 0.685E-03  0.347E-02 0.233E-02                                 T      F +-0.03     F
     F   0.8    0.0315   0.503E+00 0.779E-03  0.395E-02 0.265E-02        F       F                F      F +-0.03
     T   0.85   0.0335   0.567E+00 0.880E-03  0.445E-02 0.299E-02                                 T      F +-0.03     F

     S   0.9    0.0354   0.636E+00 0.986E-03  0.499E-02 0.336E-02        S                       S        F +-0.03
     T   0.95   0.0374   0.709E+00 0.110E-02  0.556E-02 0.374E-02                                 T      F +-0.04     F
     F   1      0.0394   0.785E+00 0.122E-02  0.617E-02 0.414E-02        F       F                F      F +-0.04     F
```

10-13

Table 10-3 (Continued). Cold Drawn Round Steel Wire (ISO 388)

TABLE 10-3.
COLD DRAWN ROUND STEEL WIRE (ISO 388)
BASIS, 1 IN = 25.4 MM
1 CUBIC METER STEEL = 7850 KG MASS

ORDER EXAMPLE:
LENGTH,STEEL WIRE 0.1 BS 4391 + STEEL QUALITY
PAGE NO. 3

THE NOMINAL SIZE IS NATIONAL STANDARD AS INDICATED
F=FIRST CHOICE,S=SECOND CHOICE,T=THIRD CHOICE,NUMBER=OTHER SIZE
* = COMMERCIAL SIZE

ISO	NOMINAL SIZE = D		SECTION AREA		MASS PER UNIT		U.S.A. ANSI B32.4	AUSTRAL AS 1153	JAPAN JIS G3521	EURO- NORM	FRANCE NF A47-411	U.K. BS 4391	GERMANY DIN 177	ITALY UNI 467
	MM	IN	MM**2	IN**2	KG/M	LB/FT								
F	1	0.039	0.785E+00	0.122E-02	0.617E-02	0.414E-02							F+-0.04	F
T	1.06	0.042	0.882E+00	0.137E-02	0.693E-02	0.466E-02						T	+-0.04	
	1.1	0.043	0.950E+00	0.147E-02	0.746E-02	0.501E-02							+-0.04	
S	1.12	0.044	0.985E+00	0.153E-02	0.773E-02	0.520E-02	S					S	F+-0.04	
T	1.18	0.046	0.109E+01	0.170E-02	0.858E-02	0.577E-02						T	+-0.04	
	1.2	0.047	0.113E+01	0.175E-02	0.888E-02	0.597E-02	F		F				F+-0.04	F
F	1.25	0.049	0.123E+01	0.190E-02	0.963E-02	0.647E-02						F	+-0.04	
T	1.32	0.052	0.137E+01	0.212E-02	0.107E-01	0.722E-02						T	+-0.04	1.3F
S	1.4	0.055	0.154E+01	0.239E-02	0.121E-01	0.812E-02	S*					S	F+-0.04	F
	1.5	0.059	0.177E+01	0.274E-02	0.139E-01	0.932E-02			F			T	+-0.06	F
F	1.6	0.063	0.201E+01	0.312E-02	0.158E-01	0.106E-01	F*	F	F			F	F+-0.06	F
T	1.7	0.067	0.227E+01	0.352E-02	0.178E-01	0.120E-01						T	+-0.06	F
S	1.8	0.071	0.254E+01	0.394E-02	0.200E-01	0.134E-01	S*		F			S	F+-0.06	F
	1.9	0.075	0.284E+01	0.439E-02	0.223E-01	0.150E-01						T	+-0.06	F
F	2	0.079	0.314E+01	0.487E-02	0.247E-01	0.166E-01	F*	F	F			F	F+-0.06	F
T	2.12	0.083	0.353E+01	0.547E-02	0.277E-01	0.186E-01						T	+-0.06	2.1F
S	2.2	0.087	0.380E+01	0.589E-02	0.298E-01	0.201E-01	S*					S	F+-0.06	F
	2.24	0.088	0.394E+01	0.611E-02	0.309E-01	0.208E-01						T	+-0.08	F
F	2.36	0.093	0.437E+01	0.678E-02	0.343E-01	0.231E-01			2.3F			F	F+-0.08	2.3F
T	2.5	0.098	0.491E+01	0.761E-02	0.385E-01	0.259E-01	F*	F				T	F+-0.08	F,2.4F
S	2.65	0.104	0.552E+01	0.855E-02	0.433E-01	0.291E-01	S		2.6F			S	+-0.08	2.6F
T	2.8	0.110	0.616E+01	0.954E-02	0.483E-01	0.325E-01	S*					T	F+-0.08	F,2.7F
F	3	0.118	0.707E+01	0.110E-01	0.555E-01	0.373E-01	F*	F	2.9F			F	F+-0.08	F
T	3.15	0.124	0.779E+01	0.121E-01	0.612E-01	0.411E-01						T	F+-0.08	
T	3.35	0.132	0.881E+01	0.137E-01	0.692E-01	0.465E-01	S*		3.2F			T	+-0.08	
	3.5	0.138	0.962E+01	0.149E-01	0.755E-01	0.508E-01			F			S	+-0.08	
G	3.55	0.140	0.990E+01	0.153E-01	0.777E-01	0.522E-01						T	+-0.08	
F	3.75	0.148	0.110E+02	0.171E-01	0.867E-01	0.583E-01	S						+-0.08	
F	4	0.157	0.126E+02	0.195E-01	0.986E-01	0.663E-01	F*	F			F	F	F+-0.10	F
	4.25	0.167	0.142E+02	0.220E-01	0.111E+00	0.748E-01							+-0.10	F
S	4.5	0.177	0.159E+02	0.247E-01	0.125E+00	0.839E-01	S				F	S	F+-0.10	F
F	4.75	0.187	0.177E+02	0.275E-01	0.139E+00	0.935E-01						T	+-0.10	F
F	5	0.197	0.196E+02	0.304E-01	0.154E+00	0.104E+00	F*	F			F	F	F+-0.10	F
T	5.3	0.209	0.221E+02	0.342E-01	0.173E+00	0.116E+00						T	+-0.10	F
	5.5	0.217	0.238E+02	0.368E-01	0.187E+00	0.125E+00	S*		F			S	F+-0.10	F
G	5.6	0.220	0.246E+02	0.382E-01	0.193E+00	0.130E+00							+-0.10	
T	6	0.236	0.283E+02	0.438E-01	0.222E+00	0.149E+00	F*	F			F	T	F+-0.10	F
F	6.3	0.248	0.312E+02	0.483E-01	0.245E+00	0.164E+00	F*					F	F+-0.15	F
	6.5	0.256	0.332E+02	0.514E-01	0.260E+00	0.174E+00	S*					T	+-0.15	F
T	6.7	0.264	0.353E+02	0.546E-01	0.277E+00	0.186E+00							+-0.15	F
	7	0.276	0.385E+02	0.597E-01	0.302E+00	0.203E+00	S*		F			T	+-0.15	F
S	7.1	0.280	0.396E+02	0.614E-01	0.311E+00	0.209E+00						S	F+-0.15	F

10-14 STEEL MATERIAL DATA

STEEL MATERIAL DATA

Table 10-3 (Continued). Cold Drawn Round Steel Wire (ISO 388)

TABLE 10-3.
COLD DRAWN ROUND STEEL WIRE (ISO 388)
BASIS; 1 IN. = 25.4 MM
1 CUBIC METER STEEL = 7850 KG MASS

ORDER EXAMPLE:
LENGTH,STEEL WIRE 0.1 BS 4391 + STEEL QUALITY
PAGE NO. 4

THE NOMINAL SIZE IS NATIONAL STANDARD AS INDICATED
F=FIRST CHOICE,S=SECOND CHOICE,T=THIRD CHOICE,NUMBER=OTHER SIZE
* = COMMERCIAL SIZE

I S O	NOMINAL SIZE = D MM / IN	SECTION AREA MM**2 / IN**2	MASS PER UNIT KG/M / LB/FT	U.S.A. ANSI B32.4	AUSTRAL AS 1153	JAPAN JIS G3521	EURO-NORM	FRANCE NF A47-411	U.K. BS 4391	GERMANY DIN 177	ITALY UNI 467
T	7.5 / 0.295	0.442E+02 / 0.685E-01	0.347E+00 / 0.233E+00	*					T	+-0.15	7.6F
F	8 / 0.315	0.503E+02 / 0.779E-01	0.395E+00 / 0.265E+00	F*	F			F	T	F+-0.15	
T	8.5 / 0.335	0.567E+02 / 0.880E-01	0.445E+00 / 0.299E+00						T	+-0.15	8.2F
S	9 / 0.354	0.636E+02 / 0.986E-01	0.499E+00 / 0.336E+00	S*	F	F		F	S	F+-0.15	8.8F
T	9.5 / 0.374	0.709E+02 / 0.110E+00	0.556E+00 / 0.374E+00						T	+-0.15	9.4F
F	10 / 0.394	0.785E+02 / 0.122E+00	0.617E+00 / 0.414E+00	F*	F	F		F	T	F+-0.20	F
T	10.6 / 0.417	0.882E+02 / 0.137E+00	0.693E+00 / 0.466E+00						T	+-0.20	
F	11 / 0.433	0.950E+02 / 0.147E+00	0.746E+00 / 0.501E+00	S*						+-0.20	
S	11.2 / 0.441	0.985E+02 / 0.153E+00	0.773E+00 / 0.520E+00		S			F	S	F+-0.20	
T	11.8 / 0.465	0.109E+03 / 0.170E+00	0.858E+00 / 0.577E+00						T	+-0.20	
	12 / 0.472	0.113E+03 / 0.175E+00	0.888E+00 / 0.597E+00	F*						+-0.20	
F	12.5 / 0.492	0.123E+03 / 0.190E+00	0.963E+00 / 0.647E+00		F	F		F	F	F+-0.20	
	13 / 0.512	0.133E+03 / 0.206E+00	0.104E+01 / 0.700E+00	S*					T	+-0.20	
	13.2 / 0.520	0.137E+03 / 0.212E+00	0.107E+01 / 0.722E+00					F		+-0.20	
S	14 / 0.551	0.154E+03 / 0.239E+00	0.121E+01 / 0.812E+00	F*	S			F	S	F+-0.20	
T	15 / 0.591	0.177E+03 / 0.274E+00	0.139E+01 / 0.932E+00	S*					T	+-0.20	
F	16 / 0.630	0.201E+03 / 0.312E+00	0.158E+01 / 0.106E+01	F*	F			F	F	F+-0.25	
	17 / 0.669	0.227E+03 / 0.352E+00	0.178E+01 / 0.120E+01	S*				F		+-0.25	
S	18 / 0.709	0.254E+03 / 0.394E+00	0.200E+01 / 0.134E+01	F*	S			F		F+-0.25	
T	19 / 0.748	0.284E+03 / 0.439E+00	0.223E+01 / 0.150E+01	F*				F		+-0.25	
F	20 / 0.787	0.314E+03 / 0.487E+00	0.247E+01 / 0.166E+01	F*	F			F		F+-0.25	
	21 / 0.827	0.346E+03 / 0.537E+00	0.272E+01 / 0.183E+01	S*				F		+-0.25	
T	21.2 / 0.835	0.353E+03 / 0.547E+00	0.277E+01 / 0.186E+01	F*						+-0.25	
	22 / 0.866	0.380E+03 / 0.589E+00	0.298E+01 / 0.201E+01	F*						+-0.25	
S	22.4 / 0.882	0.394E+03 / 0.611E+00	0.309E+01 / 0.208E+01	S*	S					+-0.25	
	23 / 0.906	0.415E+03 / 0.644E+00	0.326E+01 / 0.219E+01	S*						+-0.25	
T	23.6 / 0.929	0.437E+03 / 0.678E+00	0.343E+01 / 0.231E+01	S*						+-0.25	
	24 / 0.945	0.452E+03 / 0.701E+00	0.355E+01 / 0.239E+01	S*						+-0.25	
F	25 / 0.984	0.491E+03 / 0.761E+00	0.385E+01 / 0.259E+01	F*	F			F		+-0.25	

10-15

Table 10-4. Round Spring Steel Wire (ISO 388)

TABLE 10-4.
ROUND SPRING STEEL WIRE (ISO 388)
BASIS: 1 IN = 25.4 MM
1 CUBIC METER STEEL = 7850 KG MASS

ORDER EXAMPLE:
LENGTH,STEEL WIRE 0.0 OR DIN 2076+STEEL QUALITY

THE NOMINAL SIZE IS NATIONAL STANDARD AS INDICATED
F=FIRST CHOICE,S=SECOND CHOICE,T=THIRD CHOICE,NUMBER=OTHER SIZE
+ = COMMERCIAL SIZE

I S O	NOMINAL SIZE = D MM / IN	SECTION AREA MM**2 / IN**2	MASS PER UNIT KG/M / LB/FT	U.S.A. ANSI B32.4	AUSTRAL AS 1472	JAPAN JIS G3522	EURO- NORM	FRANCE NF A47-301	U.K. BS 5216	GERMANY DIN 2076	ITALY UNI 3823
F	0.16 0.0063	0.201E-01 0.312E-04	0.158E-03 0.106E-03	F						F +-0.005	
T	0.17 0.0067	0.227E-01 0.352E-04	0.178E-03 0.120E-03		F	F		S		+-0.005	
S	0.18 0.0071	0.254E-01 0.394E-04	0.200E-03 0.134E-03	S				F	T	F +-0.005	
T	0.19 0.0075	0.284E-01 0.439E-04	0.223E-03 0.150E-03					S		+-0.005	
F	0.2 0.0079	0.314E-01 0.487E-04	0.247E-03 0.166E-03	F		F		F		F +-0.01	
T	0.212 0.0083	0.353E-01 0.547E-04	0.277E-03 0.186E-03		F			0.21S		+-0.01	
S	0.22 0.0087	0.380E-01 0.589E-04	0.298E-03 0.201E-03	S				F		F +-0.01	
S	0.224 0.0088	0.394E-01 0.611E-04	0.309E-03 0.208E-03						S	+-0.01	
F	0.23 0.0091	0.415E-01 0.644E-04	0.326E-03 0.219E-03			F				+-0.01	
	0.236 0.0093	0.437E-01 0.678E-04	0.343E-03 0.231E-03							+-0.01	
F	0.25 0.0098	0.491E-01 0.761E-04	0.385E-03 0.259E-03	F		0.23F		F,0.24S	F	F +-0.01	
T	0.265 0.0104	0.552E-01 0.855E-04	0.433E-03 0.291E-03			0.26F				+-0.01	
S	0.28 0.0110	0.616E-01 0.954E-04	0.483E-03 0.325E-03	S				F,0.27S	S	F +-0.01	
T	0.3 0.0118	0.707E-01 0.110E-03	0.555E-03 0.373E-03			0.29F		F	F	S +-0.01	
F	0.315 0.0124	0.779E-01 0.121E-03	0.612E-03 0.411E-03		F				F	F +-0.01	
	0.32 0.0126	0.804E-01 0.125E-03	0.631E-03 0.424E-03							+-0.01	
T	0.335 0.0132	0.881E-01 0.137E-03	0.692E-03 0.465E-03					S	T	+-0.01	
	0.34 0.0134	0.908E-01 0.141E-03	0.713E-03 0.479E-03							+-0.01	
	0.35 0.0138	0.962E-01 0.149E-03	0.755E-03 0.508E-03			F				S +-0.01	
S	0.355 0.0140	0.990E-01 0.153E-03	0.777E-03 0.522E-03	S					S	+-0.01	
F	0.36 0.0142	0.102E+00 0.158E-03	0.799E-03 0.537E-03						T	F +-0.01	
T	0.375 0.0148	0.110E+00 0.171E-03	0.867E-03 0.583E-03							+-0.01	
	0.38 0.0150	0.113E+00 0.176E-03	0.890E-03 0.598E-03					S		S +-0.01	
F	0.4 0.0157	0.126E+00 0.195E-03	0.986E-03 0.663E-03	S		F		F	F	F +-0.01	
T	0.42 0.0165	0.139E+00 0.215E-03	0.109E-02 0.731E-03					S	T	+-0.01	
	0.425 0.0167	0.142E+00 0.220E-03	0.111E-02 0.748E-03		F					+-0.01	
	0.43 0.0169	0.145E+00 0.225E-03	0.114E-02 0.766E-03					F		S +-0.01	
S	0.45 0.0177	0.159E+00 0.247E-03	0.125E-02 0.839E-03	S				F		F +-0.01	
T	0.475 0.0187	0.177E+00 0.275E-03	0.139E-02 0.935E-03			F		0.47S	S	+-0.01	
	0.48 0.0189	0.181E+00 0.280E-03	0.142E-02 0.955E-03						T	F +-0.01	
F	0.5 0.0197	0.196E+00 0.304E-03	0.154E-02 0.104E-02	F				F	F	F +-0.01	
T	0.53 0.0209	0.221E+00 0.342E-03	0.173E-02 0.116E-02			F		F	T	+-0.01	
	0.55 0.0217	0.238E+00 0.368E-03	0.187E-02 0.125E-02							F +-0.01	
S	0.56 0.0220	0.246E+00 0.382E-03	0.193E-02 0.130E-02		S			F		S +-0.01	
T	0.6 0.0236	0.283E+00 0.438E-03	0.222E-02 0.149E-02					F	S	F +-0.01	
	0.63 0.0248	0.312E+00 0.483E-03	0.245E-02 0.164E-02	F		F		F	T	+-0.01	
	0.65 0.0256	0.332E+00 0.514E-03	0.260E-02 0.175E-02	S				F		S +-0.01	
T	0.67 0.0264	0.353E+00 0.546E-03	0.277E-02 0.186E-02					F		F +-0.01	
	0.71 0.0280	0.396E+00 0.614E-03	0.311E-02 0.209E-02		F					S +-0.01	
T	0.75 0.0295	0.442E+00 0.685E-03	0.347E-02 0.233E-02					F	T	S +-0.01	
F	0.8 0.0315	0.503E+00 0.779E-03	0.395E-02 0.265E-02	F		F		F		F +-0.01	

STEEL MATERIAL DATA

10-17

TABLE 10-4.
ROUND SPRING STEEL WIRE (ISO 388)
BASIS: 1 IN = 25.4 MM
1 CUBIC METER STEEL = 7850 KG MASS

Table 10-4 (Continued). Round Spring Steel Wire (ISO 388)

ORDER EXAMPLE:
LENGTH,STEEL WIRE 0.08 DIN 2076,STEEL QUALITY

PAGE NO. 2

THE NOMINAL SIZE IS NATIONAL STANDARD AS INDICATED
F=FIRST CHOICE,S=SECOND CHOICE,T=THIRD CHOICE,NUMBER=OTHER SIZE
* = COMMERCIAL SIZE

I S O	NOMINAL SIZE = D		SECTION AREA		MASS PER UNIT		U.S.A. ANSI B32.4	AUSTRAL AS 1472	JAPAN JIS G3522	EURO- NORM	FRANCE NF A47-301	U.K. BS 5216	GERMANY DIN 2076	ITALY UNI 3823
	MM	IN	MM**2	IN**2	KG/M	LB/FT								
T	0.85	0.033	0.567E+00	0.880E-03	0.445E-02	0.299E-02		S			S		S+-0.015	S
S	0.9	0.035	0.636E+00	0.986E-03	0.499E-02	0.336E-02					S		F+-0.015	F
O	0.95	0.037	0.709E+00	0.110E-02	0.556E-02	0.374E-02					S		S+-0.015	S
F	1	0.039	0.785E+00	0.122E-02	0.617E-02	0.414E-02	F		F		F		F+-0.015	F
	1.05	0.041	0.866E+00	0.134E-02	0.680E-02	0.457E-02							+-0.015	
	1.06	0.042	0.882E+00	0.137E-02	0.693E-02	0.466E-02		S			S		+-0.015	
T	1.1	0.043	0.950E+00	0.147E-02	0.746E-02	0.501E-02					F		F+-0.015	F
S	1.12	0.044	0.985E+00	0.153E-02	0.773E-02	0.520E-02	S					S	S+-0.015	S
	1.18	0.046	0.109E+01	0.170E-02	0.858E-02	0.577E-02						T	+-0.015	
	1.2	0.047	0.113E+01	0.175E-02	0.888E-02	0.597E-02	F				F		S+-0.015	F
F	1.25	0.049	0.123E+01	0.190E-02	0.963E-02	0.647E-02		F	F			T	F+-0.015	F
	1.3	0.051	0.133E+01	0.206E-02	0.104E-01	0.700E-02					S		S+-0.015	S
T	1.32	0.052	0.137E+01	0.212E-02	0.107E-01	0.722E-02	S					T	+-0.015	
	1.4	0.055	0.154E+01	0.239E-02	0.121E-01	0.812E-02		S			F	S	S+-0.02	F
	1.5	0.059	0.177E+01	0.274E-02	0.139E-01	0.932E-02			F		F	T	S+-0.02	F
	1.6	0.063	0.201E+01	0.312E-02	0.158E-01	0.106E-01	F*				F	F	F+-0.02	F
T	1.7	0.067	0.227E+01	0.352E-02	0.178E-01	0.120E-01					F	T	S+-0.02	S
S	1.8	0.071	0.254E+01	0.394E-02	0.200E-01	0.134E-01	S*	S			F	S	F+-0.02	F
T	1.9	0.075	0.284E+01	0.439E-02	0.223E-01	0.150E-01					F	T	S+-0.02	S
F	2	0.079	0.314E+01	0.487E-02	0.247E-01	0.166E-01	F*	F	F			F	F+-0.025	F
	2.1	0.083	0.346E+01	0.537E-02	0.272E-01	0.183E-01					F	T	S+-0.025	S
	2.12	0.083	0.353E+01	0.547E-02	0.277E-01	0.186E-01							F+-0.025	F
	2.2	0.087	0.380E+01	0.589E-02	0.299E-01	0.201E-01	S*						+-0.025	F
S	2.24	0.088	0.394E+01	0.611E-02	0.309E-01	0.208E-01		S			F	S	F+-0.025	F
	2.25	0.089	0.398E+01	0.616E-02	0.312E-01	0.210E-01							S+-0.025	F
	2.3	0.091	0.415E+01	0.644E-02	0.326E-01	0.219E-01					F		F+-0.025	F
T	2.36	0.093	0.437E+01	0.678E-02	0.343E-01	0.231E-01			F			T	F+-0.025	F
	2.4	0.094	0.452E+01	0.701E-02	0.355E-01	0.239E-01	S*				F		S+-0.025	F
F	2.5	0.098	0.491E+01	0.761E-02	0.385E-01	0.259E-01	F*				F		F+-0.025	F
	2.6	0.102	0.531E+01	0.823E-02	0.417E-01	0.281E-01						T	S+-0.025	F
T	2.65	0.104	0.552E+01	0.855E-02	0.433E-01	0.291E-01	S*				F/2.7F	S	F+-0.025	F
S	2.8	0.110	0.616E+01	0.954E-02	0.483E-01	0.325E-01	F*					T	S+-0.025	F
T	3	0.118	0.707E+01	0.110E-01	0.555E-01	0.373E-01	F*				F	T	F+-0.025	F
F	3.15	0.124	0.779E+01	0.121E-01	0.612E-01	0.411E-01			F		F	T	F+-0.025	F
T	3.2	0.126	0.804E+01	0.125E-01	0.631E-01	0.424E-01					F		+-0.025	F
	3.35	0.132	0.881E+01	0.137E-01	0.692E-01	0.465E-01						T	+-0.025	
	3.4	0.134	0.908E+01	0.141E-01	0.713E-01	0.479E-01					F		S+-0.03	F
	3.5	0.138	0.962E+01	0.149E-01	0.755E-01	0.508E-01	S*				S		+-0.03	
S	3.55	0.140	0.990E+01	0.153E-01	0.777E-01	0.522E-01		S				S	S+-0.03	F
	3.6	0.142	0.102E+02	0.158E-01	0.799E-01	0.537E-01							F+-0.03	
	3.75	0.148	0.110E+02	0.171E-01	0.867E-01	0.583E-01						T	+-0.03	
	3.8	0.150	0.113E+02	0.176E-01	0.890E-01	0.598E-01			F		F	F	S+-0.03	F
F	4	0.157	0.126E+02	0.195E-01	0.986E-01	0.663E-01	F*				F		F+-0.03	F

Table 10-4 (Continued). Round Spring Steel Wire (ISO 388)

TABLE 10-4.
ROUND SPRING STEEL WIRE (ISO 388)
BASIS: 1 IN = 25.4 MM
1 CUBIC METER STEEL = 7850 KG MASS

ORDER EXAMPLE:
LENGTH,STEEL WIRE,0.08,DIN 2076,STEEL QUALITY
PAGE NO. 3

THE NOMINAL SIZE IS NATIONAL STANDARD AS INDICATED
F=FIRST CHOICE, S=SECOND CHOICE, T=THIRD CHOICE, NUMBER=OTHER SIZE
* = COMMERCIAL SIZE

I S O	NOMINAL SIZE = D		SECTION AREA		MASS PER UNIT		U.S.A. ANSI B32.4	AUSTRAL AS 1472	JAPAN JIS G3522	EURO-NORM	FRANCE NF A47-301	U.K. BS 5216	GERMANY DIN 2076	ITALY UNI 3823
	MM	IN	MM**2	IN**2	KG/M	LB/FT								
T	4.25	0.167	0.142E+02	0.220E-01	0.111E+00	0.748E-01					4.2F		S*=0.03	
S	4.5	0.177	0.159E+02	0.247E-01	0.125E+00	0.839E-01	S*					S	F*=0.03	
T	4.75	0.187	0.177E+02	0.275E-01	0.139E+00	0.935E-01					4.7F	T	S*=0.03	
F	5	0.197	0.196E+02	0.304E-01	0.154E+00	0.104E+00	F*		F		F	F	F*=0.03	F
T	5.3	0.209	0.221E+02	0.342E-01	0.173E+00	0.116E+00						T	+=0.03	
S	5.5	0.217	0.238E+02	0.368E-01	0.187E+00	0.125E+00	S*						S*=0.03	
S	5.6	0.220	0.246E+02	0.382E-01	0.193E+00	0.130E+00			S		F	S	F*=0.03	
T	6	0.236	0.283E+02	0.438E-01	0.222E+00	0.149E+00	F*				F	F	S*=0.04	F
F	6.3	0.248	0.312E+02	0.483E-01	0.245E+00	0.164E+00			F		F	F	F*=0.04	F
T	6.5	0.256	0.332E+02	0.514E-01	0.260E+00	0.175E+00							S*=0.04	
T	6.7	0.264	0.353E+02	0.546E-01	0.277E+00	0.186E+00					F	T	F*=0.04	
F	7	0.276	0.385E+02	0.597E-01	0.302E+00	0.203E+00	F*				F	F	F*=0.04	F
S	7.1	0.280	0.396E+02	0.614E-01	0.311E+00	0.209E+00			S			S	+=0.04	
T	7.5	0.295	0.442E+02	0.685E-01	0.347E+00	0.233E+00					F	T	S*=0.04	
F	8	0.315	0.503E+02	0.779E-01	0.395E+00	0.265E+00	F*		F		F	F	F*=0.04	F
T	8.5	0.335	0.567E+02	0.880E-01	0.445E+00	0.299E+00						F	F*=0.04	
S	9	0.354	0.636E+02	0.986E-01	0.499E+00	0.336E+00	S*				F	S	F*=0.06	F
T	9.5	0.374	0.709E+02	0.110E+00	0.556E+00	0.374E+00			S			T	S*=0.06	
F	10	0.394	0.785E+02	0.122E+00	0.617E+00	0.414E+00	F*		F		F	F	F*=0.06	F
T	10.5	0.413	0.866E+02	0.134E+00	0.680E+00	0.457E+00						F	S*=0.06	
T	10.6	0.417	0.882E+02	0.137E+00	0.693E+00	0.466E+00						T	+=0.06	
S	11	0.433	0.950E+02	0.147E+00	0.746E+00	0.501E+00	S*		S			S	F*=0.08	F
S	11.2	0.441	0.985E+02	0.153E+00	0.773E+00	0.520E+00					F		S*=0.08	
T	11.8	0.465	0.109E+03	0.170E+00	0.858E+00	0.577E+00							+=0.08	
F	12	0.472	0.113E+03	0.175E+00	0.888E+00	0.597E+00	F*		F		F	F	F*=0.08	F
F	12.5	0.492	0.123E+03	0.190E+00	0.963E+00	0.647E+00	S*						S*=0.08	
T	13	0.512	0.133E+03	0.206E+00	0.104E+01	0.700E+00	F*				F		S*=0.1	
T	13.2	0.520	0.137E+03	0.212E+00	0.107E+01	0.722E+00	S*						+=0.1	
S	14	0.551	0.154E+03	0.239E+00	0.121E+01	0.812E+00	F*		S		F	S	S*=0.1	
T	15	0.591	0.177E+03	0.274E+00	0.139E+01	0.932E+00	S*					T	F*=0.1	
F	16	0.630	0.201E+03	0.312E+00	0.158E+01	0.106E+01	F*		F		F	F	S*=0.1	
T	17	0.669	0.227E+03	0.352E+00	0.178E+01	0.120E+01	S*					T	+=0.1	
S	18	0.709	0.254E+03	0.394E+00	0.200E+01	0.134E+01	F*		S		F	S	F*=0.1	
T	19	0.748	0.284E+03	0.439E+00	0.223E+01	0.150E+01	S*					T	S*=0.1	
F	20	0.787	0.314E+03	0.487E+00	0.247E+01	0.166E+01	F*		F		F	F	F*=0.1	
T	21	0.827	0.346E+03	0.537E+00	0.272E+01	0.183E+01	S*							
T	21.2	0.835	0.353E+03	0.547E+00	0.277E+01	0.186E+01	F*							
S	22	0.866	0.380E+03	0.589E+00	0.298E+01	0.201E+01	F*							
S	22.4	0.882	0.394E+03	0.611E+00	0.309E+01	0.208E+01	F*							
T	23	0.906	0.415E+03	0.644E+00	0.326E+01	0.219E+01	S*							
T	23.6	0.929	0.437E+03	0.678E+00	0.343E+01	0.231E+01								
F	24	0.945	0.452E+03	0.701E+00	0.355E+01	0.239E+01	S*							
F	25	0.984	0.491E+03	0.761E+00	0.385E+01	0.259E+01	F*		F					F

STEEL MATERIAL DATA

Table 10-5A. Hot Rolled Round Steel Bars (ISO 1035/I)

```
TABLE 10-5A. HOT ROLLED
ROUND STEEL BARS (ISO 1035/I)
BASIS, 1 IN = 25.4 MM
1 CUBIC METER STEEL = 7850 KG MASS
```

ORDER EXAMPLE:
LENGTH, ROUND STEEL 5 EURONORM 60 + ST QUALITY

THE NOMINAL SIZE IS NATIONAL STANDARD AS INDICATED
F=FIRST CHOICE, S=SECOND CHOICE, T=THIRD CHOICE, NUMBER=OTHER SIZE
* = COMMERCIAL SIZE

I S O	NOMINAL SIZE - D MM	IN	SECTION AREA MM**2	IN**2	MASS PER UNIT KG/M	LB/FT	U.S.A. ANSI B32.4	AUSTRAL AS 1027	JAPAN JIS G3191	EURO-NORM 60	FRANCE NF A45-003	U.K. BS 4229	GERMANY DIN 1013	ITALY UNI 6012
F	5	0.197	0.196E+02	0.304E-01	0.154E+00	0.104E+00	F			F	D+-0.4		F	F
F	5.5	0.217	0.238E+02	0.368E-01	0.187E+00	0.125E+00	S			S	D+-0.4		F	F
F	6	0.236	0.283E+02	0.438E-01	0.222E+00	0.149E+00	F	F		F	D+-0.4	S	F	F
F	6.5	0.256	0.332E+02	0.514E-01	0.260E+00	0.175E+00	S			S	D+-0.4		F	F
S	7	0.276	0.385E+02	0.597E-01	0.302E+00	0.203E+00	S			F	D+-0.4	S	F	F
F	7.5	0.295	0.442E+02	0.685E-01	0.347E+00	0.233E+00	S			F	D+-0.4	S	F	F
F	8	0.315	0.503E+02	0.779E-01	0.395E+00	0.265E+00	F	F		F	D+-0.4	F	F	F
F	8.5	0.335	0.567E+02	0.880E-01	0.445E+00	0.299E+00	F			F	D+-0.4	F	F	F
S	9	0.354	0.636E+02	0.986E-01	0.499E+00	0.336E+00	S			F	D+-0.4	S	F	F
F	9.5	0.374	0.709E+02	0.110E+00	0.556E+00	0.374E+00	F			F	D+-0.4		F	F
F	10	0.394	0.785E+02	0.122E+00	0.617E+00	0.414E+00	F	F	F	F	D+-0.4	F	F	F
F	10.5	0.413	0.866E+02	0.134E+00	0.680E+00	0.457E+00	F			F	D+-0.4		F	F
S	11	0.433	0.950E+02	0.147E+00	0.746E+00	0.501E+00	S		F	F	D+-0.4	S	F	F
F	11.5	0.453	0.104E+03	0.161E+00	0.815E+00	0.548E+00	F			F	D+-0.4		F	F
F	12	0.472	0.113E+03	0.175E+00	0.888E+00	0.597E+00	F		F	F	D+-0.4	F	F	F
F	12.5	0.492	0.123E+03	0.190E+00	0.963E+00	0.647E+00	F			F	D+-0.4		F	F
S	13	0.512	0.133E+03	0.206E+00	0.104E+01	0.700E+00	S*		F	F	D+-0.4	S	F	F
F	13.5	0.531	0.143E+03	0.222E+00	0.112E+01	0.755E+00	F			F	D+-0.4		F	F
F	14	0.551	0.154E+03	0.239E+00	0.121E+01	0.812E+00	F*		F	F	D+-0.4	F	F	F
F	14.5	0.571	0.165E+03	0.256E+00	0.130E+01	0.871E+00	F			F	D+-0.4		F	F
S	15	0.591	0.177E+03	0.274E+00	0.139E+01	0.932E+00	S*		F	F	D+-0.4	S	F	F
F	15.5	0.610	0.189E+03	0.292E+00	0.148E+01	0.994E+00	F			F	D+-0.4		F	F
F	16	0.630	0.201E+03	0.312E+00	0.158E+01	0.106E+01	F*		F	F	D+-0.5	F	F	F
F	16.5	0.650	0.214E+03	0.331E+00	0.168E+01	0.113E+01	F			F	D+-0.5		F	F
S	17	0.669	0.227E+03	0.352E+00	0.178E+01	0.120E+01	S*			S	D+-0.5	S	F	F
F	17.5	0.689	0.241E+03	0.373E+00	0.189E+01	0.127E+01	F			F	D+-0.5		F	F
F	18	0.709	0.254E+03	0.394E+00	0.200E+01	0.134E+01	F*		F	F	D+-0.5	F	F	F
F	18.5	0.728	0.269E+03	0.417E+00	0.211E+01	0.142E+01	F			F	D+-0.5		F	F
S	19	0.748	0.284E+03	0.439E+00	0.223E+01	0.150E+01	S*			F	D+-0.5	S	F	F
F	19.5	0.768	0.299E+03	0.463E+00	0.234E+01	0.158E+01	F			F	D+-0.5		F	F
F	20	0.787	0.314E+03	0.487E+00	0.247E+01	0.166E+01	F*		F	F	D+-0.5	F	F	F
F	20.5	0.807	0.330E+03	0.512E+00	0.259E+01	0.174E+01	F			F	D+-0.5		F	F
S	21	0.827	0.346E+03	0.537E+00	0.272E+01	0.183E+01	S*			F	D+-0.5	S	F	F
F	21.5	0.846	0.363E+03	0.563E+00	0.285E+01	0.192E+01	F			F	D+-0.5		F	F
F	22	0.866	0.380E+03	0.589E+00	0.298E+01	0.201E+01	F*		F	F	D+-0.5	F	F	F
F	22.5	0.886	0.398E+03	0.616E+00	0.312E+01	0.210E+01	F			F	D+-0.5		F	F
S	23	0.906	0.415E+03	0.644E+00	0.326E+01	0.219E+01	S*			S	D+-0.5	S	F	F
S	23.5	0.925	0.434E+03	0.672E+00	0.340E+01	0.229E+01	S			S	D+-0.5		F	F
S	24	0.945	0.452E+03	0.701E+00	0.355E+01	0.239E+01	S*		F	F	D+-0.5	S	F	F
S	24.5	0.965	0.471E+03	0.731E+00	0.370E+01	0.249E+01	S			S	D+-0.5		F	F
F	25	0.984	0.491E+03	0.761E+00	0.385E+01	0.259E+01	F*		F	F	D+-0.5	F	F	F
F	25.5	1.004	0.511E+03	0.792E+00	0.401E+01	0.269E+01	F			F	D+-0.6		F	F
S	26	1.024	0.531E+03	0.823E+00	0.417E+01	0.280E+01	S*			S	D+-0.6	S		

PAGE NO. 1

Table 10-5A (Continued). Hot Rolled Round Steel Bars (ISO 1035/I)

TABLE 10-5A. HOT ROLLED
ROUND STEEL BARS (ISO 1035/I)
BASIS: 1 IN = 25.4 MM
1 CUBIC METER STEEL = 7850 KG MASS

ORDER EXAMPLE:
LENGTH, ROUND STEEL 5 EURONORM 60 + ST QUALITY PAGE NO. 2

THE NOMINAL SIZE IS NATIONAL STANDARD AS INDICATED
F=FIRST CHOICE, S=SECOND CHOICE, T=THIRD CHOICE, NUMBER=OTHER SIZE
* = COMMERCIAL SIZE

	NOMINAL SIZE = D		SECTION AREA		MASS PER UNIT		U.S.A. ANSI B32.4	AUSTRAL AS 1027	JAPAN JIS G3191	EURO-NORM 60	FRANCE NF A45-003	U.K. BS 4229	GERMANY DIN 1013	ITALY UNI 6012
	MM	IN	MM**2	IN**2	KG/M	LB/FT								
S	26.5	1.043	0.552E+03	0.855E+00	0.433E+01	0.291E+01	*			D +-0.6			F	
	27	1.063	0.573E+03	0.887E+00	0.449E+01	0.302E+01	*			S D +-0.6	S		F	
	27.5	1.083	0.594E+03	0.921E+00	0.466E+01	0.313E+01	*			D +-0.6			F	
F	28	1.102	0.616E+03	0.954E+00	0.483E+01	0.325E+01	S*			F D +-0.6	F		F	
	28.5	1.122	0.638E+03	0.989E+00	0.501E+01	0.337E+01				D +-0.6			F	
	29	1.142	0.661E+03	0.102E+01	0.519E+01	0.348E+01		F		D +-0.6 T			F	
	29.5	1.161	0.683E+03	0.106E+01	0.537E+01	0.361E+01				D +-0.6			F	
S	30	1.181	0.707E+03	0.110E+01	0.555E+01	0.373E+01	F*		F	F D +-0.6	F		F	S
F	31	1.220	0.755E+03	0.117E+01	0.592E+01	0.398E+01				D +-0.6 T			F	
	32	1.260	0.804E+03	0.125E+01	0.631E+01	0.424E+01	S*	F		F D +-0.6	S		F	
S	33	1.299	0.855E+03	0.133E+01	0.671E+01	0.451E+01	*			D +-0.6 T			F	S
	34	1.339	0.908E+03	0.141E+01	0.713E+01	0.479E+01	*			S D +-0.6			F	
S	35	1.378	0.962E+03	0.149E+01	0.755E+01	0.508E+01	F*	F		F D +-0.6	F		F	
F	36	1.417	0.102E+04	0.158E+01	0.799E+01	0.537E+01	*		S	S D +-0.8	S		F	S
S	37	1.457	0.108E+04	0.167E+01	0.844E+01	0.567E+01	*			D +-0.8			F	S
	38	1.496	0.113E+04	0.176E+01	0.890E+01	0.598E+01	S*		F	F D +-0.8	S		F	F
	39	1.535	0.119E+04	0.185E+01	0.938E+01	0.630E+01				D +-0.8 T			F	
F	40	1.575	0.126E+04	0.195E+01	0.986E+01	0.663E+01	F*	F	F	F D +-0.8	F		F	F
	41	1.614	0.132E+04	0.205E+01	0.104E+02	0.696E+01				D +-0.8 T			F	
S	42	1.654	0.139E+04	0.215E+01	0.109E+02	0.731E+01	S*			F D +-0.8	F		F	S
	43	1.693	0.145E+04	0.225E+01	0.114E+02	0.766E+01				D +-0.8 T			F	
	44	1.732	0.152E+04	0.236E+01	0.119E+02	0.802E+01	*			F D +-0.8 S			F	
F	45	1.772	0.159E+04	0.247E+01	0.125E+02	0.839E+01	*		S	S D +-0.8	S		F	F
	46	1.811	0.166E+04	0.258E+01	0.130E+02	0.877E+01				D +-0.8 T			F	
S	47	1.850	0.173E+04	0.269E+01	0.136E+02	0.915E+01				F D +-0.8	F		F	
	48	1.890	0.181E+04	0.280E+01	0.142E+02	0.955E+01	S*	F		F D +-0.8	F		F	S
	49	1.929	0.189E+04	0.292E+01	0.148E+02	0.995E+01				D +-0.8 T			F	
F	50	1.969	0.196E+04	0.304E+01	0.154E+02	0.104E+02	F*		F	F D +-0.8	F		F	F
	51	2.008	0.204E+04	0.317E+01	0.160E+02	0.108E+02				D +-1			F	
S	52	2.047	0.212E+04	0.329E+01	0.167E+02	0.112E+02			S	S D +-1	S		F	S
S	53	2.087	0.221E+04	0.342E+01	0.173E+02	0.116E+02				D +-1 T			F	
	54	2.126	0.229E+04	0.355E+01	0.180E+02	0.121E+02				D +-1			F	
	55	2.165	0.238E+04	0.368E+01	0.187E+02	0.125E+02	S*			F D +-1	F		F	S
	56	2.205	0.246E+04	0.382E+01	0.193E+02	0.130E+02				D +-1			F	
S	57	2.244	0.255E+04	0.396E+01	0.200E+02	0.135E+02				S D +-1 T	S		F	S
	58	2.283	0.264E+04	0.410E+01	0.207E+02	0.139E+02				D +-1			F	
	59	2.323	0.273E+04	0.424E+01	0.215E+02	0.144E+02				D +-1			F	
S	60	2.362	0.283E+04	0.438E+01	0.222E+02	0.149E+02	F*	F	F	F D +-1	F		F	F
	61	2.402	0.292E+04	0.453E+01	0.229E+02	0.154E+02				D +-1			F	
	62	2.441	0.302E+04	0.468E+01	0.237E+02	0.159E+02				D +-1			F	
F	63	2.480	0.312E+04	0.483E+01	0.245E+02	0.164E+02				S D +-1	S		F	
	64	2.520	0.322E+04	0.499E+01	0.253E+02	0.170E+02				D +-1			F	
S	65	2.559	0.332E+04	0.514E+01	0.260E+02	0.175E+02	S*		F	F D +-1	F		F	S

STEEL MATERIAL DATA

Table 10-5A (Continued). Hot Rolled Round Steel Bars (ISO 1035/1)

```
TABLE 10-5A. HOT ROLLED                                          PAGE NO. 3
ROUND STEEL BARS (ISO 1035/1)               ORDER EXAMPLE:
BASIS; 1 IN = 25.4 MM                       LENGTH, ROUND STEEL 5 EURONORM 60 + ST QUALITY
1 CUBIC METER STEEL = 7850 KG MASS
```

THE NOMINAL SIZE IS NATIONAL STANDARD AS INDICATED
F=FIRST CHOICE, S=SECOND CHOICE, T=THIRD CHOICE, NUMBER=OTHER SIZE
* = COMMERCIAL SIZE

I S O	NOMINAL SIZE = D MM	IN	SECTION AREA MM**2	IN**2	MASS PER UNIT KG/M	LB/FT	U.S.A. ANSI B32.4	AUSTRAL AS 1027	JAPAN JIS G3191	EURO NORM 60	FRANCE NF A45-003	U.K. BS #229	GERMANY DIN 1013	ITALY UNI 6012
	66	2.598	0.342E+04	0.530E+01	0.269E+02	0.180F+02				D+-1			F	
	67	2.638	0.353E+04	0.546E+01	0.277E+02	0.186E+02				D+-1			F	
	68	2.677	0.363E+04	0.563E+01	0.285E+02	0.192E+02			S	D+-1	F	S	F	S
S	70	2.756	0.385E+04	0.597E+01	0.302E+02	0.203E+02	S*		F	D+-1	F	F	F	F
S	73	2.874	0.419E+04	0.649E+01	0.329E+02	0.221E+02				S D+-1	F	F	F	S
	75	2.953	0.442E+04	0.685E+01	0.347E+02	0.233E+02	S*		F	F D+-1	F	F	F	F
	78	3.071	0.478E+04	0.741E+01	0.375E+02	0.252E+02				S D+-1	T	F	F	S
F	80	3.150	0.503E+04	0.779E+01	0.395E+02	0.265E+02	F*		F	F D+-1	F	F	F	F
S	83	3.268	0.541E+04	0.839E+01	0.425E+02	0.285F+02							F	S
	85	3.346	0.567E+04	0.880E+01	0.445E+02	0.299E+02				S D+-1.3	S		F	F
S	88	3.465	0.608E+04	0.943E+01	0.477E+02	0.321E+02				F D+-1.3	F		F	S
F	90	3.543	0.636E+04	0.986E+01	0.499E+02	0.336E+02				F D+-1.3	F	S	F	F
S	95	3.740	0.709E+04	0.110E+02	0.556E+02	0.374E+02				S D+-1.3	F	S	F	F
F	100	3.937	0.785E+04	0.122E+02	0.617E+02	0.414E+02				F D+-1.3	F,98T	F	F	F
S	105	4.134	0.866E+04	0.134E+02	0.680E+02	0.457E+02				F D+-1.5	F	S	F	F
F	110	4.331	0.950E+04	0.147E+02	0.746E+02	0.501E+02	S*		F	F D+-1.5	F,108T	F	F	F
S	115	4.528	0.104E+05	0.161E+02	0.815E+02	0.548E+02				S D+-2	F	S	F	F
F	120	4.724	0.113E+05	0.175E+02	0.888E+02	0.597E+02	F*		F	F D+-2	F	F	F	F
F	125	4.921	0.123E+05	0.190E+02	0.963E+02	0.647E+02				F D+-2	F	S	F	F
S	130	5.118	0.133E+05	0.206E+02	0.104E+03	0.700E+02	S*			F D+-2	F	F	F	F
S	135	5.315	0.143E+05	0.222E+02	0.112E+03	0.755E+02				S D+-2	T	S	F	S
F	140	5.512	0.154E+05	0.239E+02	0.121E+03	0.812E+02	F*		F	F D+-2	T	F	F	F
S	145	5.709	0.165E+05	0.256E+02	0.130E+03	0.871E+02				S D+-2	T	S	F	S
S	150	5.906	0.177E+05	0.274E+02	0.139E+03	0.932E+02	S*			F D+-2	F	F	F	F
S	155	6.102	0.189E+05	0.292E+02	0.148E+03	0.995E+02				S D+-2	F	S	F	S
F	160	6.299	0.201E+05	0.312E+02	0.158E+03	0.106E+03	F*		F	F D+-2	F	F	F	F
	165	6.496	0.214E+05	0.331E+02	0.168E+03	0.113E+03				F D+-2.5	F	S	F	F
S	170	6.693	0.227E+05	0.352E+02	0.178E+03	0.120E+03	S*			S D+-2.5	F	F	F	F
	175	6.890	0.241E+05	0.373E+02	0.189E+03	0.127E+03				F D+-2.5	T	S	F	
F	180	7.087	0.254E+05	0.394E+02	0.200E+03	0.134E+03	F*		F	F D+-2.5	F	F	F	
	185	7.283	0.269E+05	0.416E+02	0.211E+03	0.142E+03				F D+-2.5	F	S	F	
S	190	7.480	0.284E+05	0.439E+02	0.223E+03	0.150E+03	S*			S D+-2.5	F	F	F	
	195	7.677	0.299E+05	0.463E+02	0.234E+03	0.158E+03				F D+-3	T	S	F	
F	200	7.874	0.314E+05	0.487E+02	0.247E+03	0.166E+03	F*		F	F D+-3	F	F	F	
	210	8.268	0.346E+05	0.537E+02	0.272E+03	0.183E+03	F*			S D+-3	F	S	F	
S	220	8.661	0.380E+05	0.589E+02	0.298E+03	0.201E+03	S*			S D+-3	F	F	F	
	230	9.055	0.415E+05	0.644E+02	0.326E+03	0.219E+03	*			D+-3	F	S	F	
	240	9.449	0.452E+05	0.701E+02	0.355E+03	0.239E+03	**			D+-3	F	S	F	
	250	9.843	0.491E+05	0.761E+02	0.385E+03	0.259E+03	F			D+-3	F	S	F	
	260	10.236	0.531E+05	0.823E+02	0.417E+03	0.280E+03					F	F	F	
	280	11.024	0.616E+05	0.954E+02	0.483E+03	0.325E+03	S					F,270S	F,270F	
	300	11.811	0.707E+05	0.110E+03	0.555E+03	0.373E+03	F					F,290S	F,290F	

10-21

Table 10-5B. Tolerances of Round, Square and Rectangular Hot Rolled Steel Bars (ISO 1035/IV)

Dimensions in millimeters

Nominal Size D	Tolerance + or −	Nominal Size A	Tolerance + or −	Nominal Size B	Tolerance + or −
⩾ 5 … 15	0.4	… 50	1	… 20	0.5
>15 … 25	0.5	>50 … 75	1.5	>20 … 40	1
>25 … 35	0.6	>75 … 100	2	>40 …	1.5
>35 … 50	0.8	>100 …	2% to a. maximum of 6		
>50 … 80	1				
>80 … 100	1.3				
>100 …	1.6%				

STEEL MATERIAL DATA

Table 10-6. Hot Rolled Round Steel Bars For Bolts And Rivets (EURONORM 65)

TABLE 10-6. HOT ROLLED ROUND STEEL BARS
FOR BOLTS AND RIVETS (EURONORM 65)
BASIS; 1 IN = 25.4 MM
1 CUBIC METER STEEL = 7850 KG MASS

ORDER EXAMPLE:
LENGTH,ROUND STEEL 7.8 EURONORM 65 + ST QUALIT

THE NOMINAL SIZE IS NATIONAL STANDARD AS INDICATED
F=FIRST CHOICE,S=SECOND CHOICE,T=THIRD CHOICE,NUMBER=OTHER SIZE
* = COMMERCIAL SIZE

EUR	NOMINAL SIZE − D		SECTION AREA		MASS PER UNIT		U.S.A. ANSI	AUSTRAL AS	JAPAN JIS G3104	EURO-NORM 65	FRANCE NF A45-075	U.K. BS	GERMANY DIN 59130	ITALY UNI 2850
	MM	IN	MM**2	IN**2	KG/M	LB/FT								
F	7.8	0.307	0.478E+02	0.741E-01	0.375E+00	0.252E+00				F+-0.15			F	7.75F
F	9.35	0.368	0.687E+02	0.106E+00	0.539E+00	0.362E+00			9F	F+-0.15			F	
F	9.75	0.384	0.747E+02	0.116E+00	0.586E+00	0.394E+00				F+-0.15 T			F	
F	10.9	0.429	0.933E+02	0.145E+00	0.733E+00	0.492E+00			10F	F+-0.15			10.5F	
F	11.75	0.463	0.108E+03	0.168E+00	0.851E+00	0.572E+00			11F	F+-0.20 S			F,11F	11.7F
F	12.5	0.492	0.123E+03	0.190E+00	0.963E+00	0.647E+00				F+-0.20			F	
F	13.5	0.531	0.143E+03	0.222E+00	0.112E+01	0.755E+00			13F	F+-0.20 T			F	13.7F
F	13.75	0.541	0.148E+03	0.230E+00	0.117E+01	0.783E+00				F+-0.20 S			F	
F	14.1	0.555	0.156E+03	0.242E+00	0.123E+01	0.824E+00				F+-0.20			14F	
F	15.5	0.610	0.189E+03	0.292E+00	0.148E+01	0.995E+00				F+-0.20 T			F	
F	15.7	0.618	0.194E+03	0.300E+00	0.152E+01	0.102E+01				F+-0.20 S			F	
F	17.4	0.685	0.238E+03	0.369E+00	0.187E+01	0.125E+01			16F	F+-0.20 17.5T			F	
F	17.7	0.697	0.246E+03	0.381E+00	0.193E+01	0.130E+01				F+-0.20			F	
F	18.4	0.724	0.266E+03	0.412E+00	0.209E+01	0.140E+01				F+-0.20 S			F	F
F	18.8	0.740	0.278E+03	0.430E+00	0.218E+01	0.146E+01				F+-0.20			18.5F	
F	19.4	0.764	0.296E+03	0.458E+00	0.232E+01	0.156E+01			19F	F+-0.20			F	
F	19.7	0.776	0.305E+03	0.472E+00	0.239E+01	0.161E+01				F+-0.20			F	F
F	20.3	0.799	0.324E+03	0.502E+00	0.254E+01	0.171E+01				F+-0.20 S			F	
F	21.3	0.839	0.356E+03	0.552E+00	0.280E+01	0.188E+01				F+-0.20 21.5S			F	
F	21.7	0.854	0.370E+03	0.573E+00	0.290E+01	0.195E+01				F+-0.20			F	
F	22.3	0.878	0.391E+03	0.605E+00	0.307E+01	0.206E+01			22F	F+-0.25			F	F
F	23.25	0.915	0.425E+03	0.658E+00	0.333E+01	0.224E+01				F+-0.25 22.5F			F	
F	23.65	0.931	0.439E+03	0.681E+00	0.345E+01	0.232E+01				F+-0.25 23.5S			F	
F	24.25	0.955	0.462E+03	0.716E+00	0.363E+01	0.244E+01				F+-0.25 23.75S			F	
F	25.1	0.988	0.495E+03	0.767E+00	0.388E+01	0.261E+01			25F	F+-0.25 24.75T			F	F
F	26.25	1.033	0.541E+03	0.839E+00	0.425E+01	0.285E+01				F+-0.25 22.5F			F	
F	26.65	1.049	0.558E+03	0.865E+00	0.438E+01	0.294E+01				F+-0.25 26.5F			F	
F	28.35	1.116	0.631E+03	0.978E+00	0.496E+01	0.333E+01			28F	F+-0.25			F	F
F	29.25	1.152	0.672E+03	0.104E+01	0.527E+01	0.354E+01				F+-0.25			F	29.65F
F	29.6	1.165	0.688E+03	0.107E+01	0.540E+01	0.363E+01				F+-0.25 T			F	
F	31.5	1.240	0.779E+03	0.121E+01	0.612E+01	0.411E+01				F+-0.30			F	32.5F
F	32.55	1.281	0.832E+03	0.129E+01	0.653E+01	0.439E+01			32F	F+-0.30			F	
F	34.6	1.362	0.940E+03	0.146E+01	0.738E+01	0.496E+01				F+-0.30 32.7T			F	35.5F
F	35.55	1.400	0.993E+03	0.154E+01	0.779E+01	0.524E+01			36F	F+-0.30			F	
F	37.8	1.488	0.112E+04	0.174E+01	0.881E+01	0.592E+01			38F	F+-0.30			F	38.5F
F	38.55	1.518	0.117E+04	0.181E+01	0.916E+01	0.616E+01				F+-0.30			F	
F	41.5	1.634	0.135E+04	0.210E+01	0.106E+02	0.714E+01			40F	F+-0.40			F	41.4F
F	44.15	1.738	0.153E+04	0.237E+01	0.120E+02	0.808E+01			42F	F+-0.40			F	44.4F
F	44.5	1.752	0.156E+04	0.241E+01	0.122E+02	0.820E+01			44F	F+-0.40			F	47.4F
F	47.5	1.870	0.177E+04	0.275E+01	0.139E+02	0.935E+01				F+-0.40			F	
F	50.5	1.988	0.200E+04	0.310E+01	0.157E+02	0.106E+02				F+-0.40			F	51.4F
F	51.5	2.028	0.208E+04	0.323E+01	0.164E+02	0.110E+02				F+-0.40			F	

Table 10-7. Bright Finish Round Steel Bars (DIN 668 Tolerance h 11)

TABLE 10-7. BRIGHT FINISH ROUND STEEL
BARS (DIN 668 TOLERANCE h 11)   PAGE NO. 1
BASIS: 1 IN = 25.4 MM
1 CUBIC METER STEEL = 7850 KG MASS

ORDER EXAMPLE:
LENGTH, ROUND STEEL 1.5 DIN 668 + STEEL QUALITY

THE NOMINAL SIZE IS NATIONAL STANDARD AS INDICATED
F=FIRST CHOICE, S=SECOND CHOICE, T=THIRD CHOICE, NUMBER=OTHER SIZE
* = COMMERCIAL SIZE

D I N	NOMINAL SIZE - D		SECTION AREA		MASS PER UNIT		U.S.A. ANSI B32.4	AUSTRAL AS 1027	JAPAN JIS G3123	EURO- NORM	FRANCE NF A47-11	U.K. BS *229	GERMANY DIN 668	ITALY UNI *68
	MM	IN	MM**2	IN**2	KG/M	LB/FT								
F	1	0.039	0.785E+00	0.122E-02	0.617E-02	0.414E-02	F						F+0-0.05	
S	1.1	0.043	0.950E+00	0.147E-02	0.746E-02	0.501E-02	S						S+0-0.05	
F	1.2	0.047	0.113E+01	0.175E-02	0.888E-02	0.597E-02	F						F+0-0.05	
S	1.3	0.051	0.133E+01	0.206E-02	0.104E-01	0.700E-02	S						S+0-0.05	
F	1.4	0.055	0.154E+01	0.239E-02	0.121E-01	0.812E-02	S						F+0-0.05	
S	1.5	0.059	0.177E+01	0.274E-02	0.139E-01	0.932E-02							S+0-0.05	
F	1.6	0.063	0.201E+01	0.312E-02	0.158E-01	0.106E-01	F						F+0-0.05	
S	1.7	0.067	0.227E+01	0.352E-02	0.178E-01	0.120E-01							S+0-0.05	
F	1.8	0.071	0.254E+01	0.394E-02	0.200E-01	0.134E-01	S						F+0-0.06	
S	1.9	0.075	0.284E+01	0.439E-02	0.223E-01	0.150E-01							S+0-0.06	
F	2.0	0.079	0.314E+01	0.487E-02	0.247E-01	0.166E-01	F						F+0-0.06	
S	2.1	0.083	0.346E+01	0.537E-02	0.272E-01	0.183E-01							S+0-0.06	
F	2.2	0.087	0.380E+01	0.589E-02	0.298E-01	0.201E-01	S						F+0-0.06	
S	2.3	0.091	0.415E+01	0.644E-02	0.326E-01	0.219E-01							S+0-0.06	
S	2.4	0.094	0.452E+01	0.701E-02	0.355E-01	0.239E-01	F						F+0-0.06	
F	2.5	0.098	0.491E+01	0.761E-02	0.385E-01	0.259E-01							F+0-0.06	
S	2.6	0.102	0.531E+01	0.823E-02	0.417E-01	0.280E-01	S						S+0-0.06	
F	2.7	0.106	0.573E+01	0.887E-02	0.449E-01	0.302E-01							S+0-0.06	
S	2.8	0.110	0.616E+01	0.954E-02	0.483E-01	0.325E-01	F						F+0-0.06	
F	2.9	0.114	0.661E+01	0.102E-01	0.519E-01	0.348E-01							S+0-0.06	
F	3	0.118	0.707E+01	0.110E-01	0.555E-01	0.373E-01	S						F+0-0.06 F IN	
F	3.5	0.136	0.962E+01	0.149E-01	0.755E-01	0.508E-01	F						F+0-0.075F 0.1	
F	4	0.157	0.126E+02	0.195E-01	0.986E-01	0.663E-01	S						F+0-0.075F INCR	
F	4.5	0.177	0.159E+02	0.247E-01	0.125E+00	0.839E-01					F		F+0-0.075F FROM	
F	5	0.197	0.196E+02	0.304E-01	0.154E+00	0.104E+00	F				F		F+0-0.075F 3	
S	5.5	0.217	0.238E+02	0.368E-01	0.187E+00	0.125E+00	S						S+0-0.075F THRU	
F	6	0.236	0.283E+02	0.438E-01	0.222E+00	0.149E+00	F				F		F+0-0.075F 6	
F	6.3	0.248	0.312E+02	0.483E-01	0.245E+00	0.164E+00	F				F		F+0-0.09 6.2F	
S	6.5	0.256	0.332E+02	0.514E-01	0.260E+00	0.175E+00	S						S+0-0.09 F	
F	7	0.276	0.385E+02	0.597E-01	0.302E+00	0.203E+00	S				F		F+0-0.09 F,6.8F	
S	7.5	0.295	0.442E+02	0.685E-01	0.347E+00	0.233E+00	S				F		S+0-0.09 F,7.2F	
F	8	0.315	0.503E+02	0.779E-01	0.395E+00	0.265E+00	F				F		F+0-0.09 F,7.8F	
F	8.5	0.335	0.567E+02	0.880E-01	0.445E+00	0.299E+00					F		S+0-0.09 F,8.2F	
S	9	0.354	0.636E+02	0.986E-01	0.499E+00	0.336E+00	S				F		F+0-0.09 F,8.8F	
S	9.5	0.374	0.709E+02	0.110E+00	0.556E+00	0.374E+00	S				F		S+0-0.09 F,9.2F	
F	10	0.394	0.785E+02	0.122E+00	0.617E+00	0.414E+00	F				F		F+0-0.09 F,9.8F	
S	10.5	0.413	0.866E+02	0.134E+00	0.680E+00	0.457E+00					F		S+0-0.11 F	
F	11	0.433	0.950E+02	0.147E+00	0.746E+00	0.501E+00	S						F+0-0.11 F	
S	11.5	0.453	0.104E+03	0.161E+00	0.815E+00	0.548E+00					F		S+0-0.11 F	
F	12	0.472	0.113E+03	0.175E+00	0.888E+00	0.597E+00	F						F+0-0.11 F	
S	12.5	0.492	0.123E+03	0.190E+00	0.963E+00	0.647E+00							S+0-0.11 F	
S	13	0.512	0.133E+03	0.206E+00	0.104E+01	0.700E+00	S						S+0-0.11 F	
S	13.5	0.531	0.143E+03	0.222E+00	0.112E+01	0.755E+00							S+0-0.11 F	

STEEL MATERIAL DATA

Table 10-7 (Continued). Bright Finish Round Steel Bars (DIN 668 Tolerance h 11)

TABLE 10-7. BRIGHT FINISH ROUND STEEL BARS (DIN 668 TOLERANCE h 11)
BASIS; 1 IN = 25.4 MM
1 CUBIC METER STEEL = 7850 KG MASS

PAGE NO. 2

ORDER EXAMPLE;
LENGTH, ROUND STEEL 1.5 DIN 668 + STEEL QUALITY

THE NOMINAL SIZE IS NATIONAL STANDARD AS INDICATED
F=FIRST CHOICE, S=SECOND CHOICE, T=THIRD CHOICE, NUMBER=OTHER SIZE
* = COMMERCIAL SIZE

DIN	NOMINAL SIZE = D		SECTION AREA		MASS PER UNIT		U.S.A. ANSI B32.4	AUSTRAL AS 1027	JAPAN JIS G3123	EURO- NORM	FRANCE NF A47-411	U.K. BS #229	GERMANY DIN 668	ITALY UNI #68
	MM	IN	MM**2	IN**2	KG/M	LB/FT								
F	14	0.551	0.154E+03	0.239E+00	0.121E+01	0.812E+00		F			F	F	F+0-0.11	F
S	14.5	0.571	0.165E+03	0.256E+00	0.130E+01	0.871E+00						S	S+0-0.11	F
S	15	0.591	0.177E+03	0.274E+00	0.139E+01	0.932E+00	S					S	S+0-0.11	F
S	15.5	0.610	0.189E+03	0.292E+00	0.148E+01	0.995E+00						S	S+0-0.11	F
F	16	0.630	0.201E+03	0.312E+00	0.158E+01	0.106E+01		F	F		F	F	F+0-0.11	F
S	16.5	0.650	0.214E+03	0.331E+00	0.168E+01	0.113E+01						S	S+0-0.11	F
S	17	0.669	0.227E+03	0.352E+00	0.178E+01	0.120E+01	S		F			S	S+0-0.11	F
S	17.5	0.689	0.241E+03	0.373E+00	0.189E+01	0.127E+01						S	+0-0.11	F
F	18	0.709	0.254E+03	0.394E+00	0.200E+01	0.134E+01	F	F	F		F	F	F+0-0.11	F
S	19	0.748	0.284E+03	0.439E+00	0.223E+01	0.150E+01	S		F			S	S+0-0.13	F
19.5	0.768	0.299E+03	0.463E+00	0.234E+01	0.158E+01							+0-0.13	F
F	20	0.787	0.314E+03	0.487E+00	0.247E+01	0.166E+01	F	F	F		F	F	F+0-0.13	F
20.5	0.807	0.330E+03	0.512E+00	0.259E+01	0.174E+01							+0-0.13	F
S	21	0.827	0.346E+03	0.537E+00	0.272E+01	0.183E+01	S					S	S+0-0.13	F
21.5	0.846	0.363E+03	0.563E+00	0.285E+01	0.192E+01							+0-0.13	F
F	22	0.866	0.380E+03	0.589E+00	0.298E+01	0.201E+01	F	F	F		F	F	F+0-0.13	F
	22.5	0.886	0.398E+03	0.616E+00	0.312E+01	0.210E+01							+0-0.13	F
S	23	0.906	0.415E+03	0.644E+00	0.326E+01	0.219E+01	S				S	S	S+0-0.13	F
	23.5	0.925	0.434E+03	0.672E+00	0.340E+01	0.229E+01							+0-0.13	F
F	24	0.945	0.452E+03	0.701E+00	0.355E+01	0.239E+01	F		F		S	F	+0-0.13	F
F	24.5	0.965	0.471E+03	0.731E+00	0.370E+01	0.249E+01						S	+0-0.13	F
	25	0.984	0.491E+03	0.761E+00	0.385E+01	0.259E+01	F				F	F	F+0-0.13	F
S	26	1.024	0.531E+03	0.823E+00	0.417E+01	0.280E+01					F	F	S+0-0.13	F
S	27	1.063	0.573E+03	0.887E+00	0.449E+01	0.302E+01		F			T	S	S+0-0.13	F
F	28	1.102	0.616E+03	0.954E+00	0.483E+01	0.325E+01	S	F	F		F	F	F+0-0.13	F
S	29	1.142	0.661E+03	0.102E+01	0.519E+01	0.348E+01	S				T	S	S+0-0.13	F
S	30	1.181	0.707E+03	0.110E+01	0.555E+01	0.373E+01	S	F	F		F	F	S+0-0.13	F
S	31	1.220	0.755E+03	0.117E+01	0.592E+01	0.398E+01					T	S	+0-0.16	F
F	32	1.260	0.804E+03	0.125E+01	0.631E+01	0.424E+01	S	F	F		S	F	F+0-0.16	F
S	33	1.299	0.855E+03	0.133E+01	0.671E+01	0.451E+01	S					S	S+0-0.16	F
S	34	1.339	0.908E+03	0.141E+01	0.713E+01	0.479E+01		F				F	S+0-0.16	F
S	35	1.378	0.962E+03	0.149E+01	0.755E+01	0.508E+01		F				F	S+0-0.16	F
F	36	1.417	0.102E+04	0.158E+01	0.799E+01	0.537E+01	S	F	F		S	F	F+0-0.16	F
S	38	1.496	0.113E+04	0.176E+01	0.890E+01	0.598E+01		F			F,37T	S	S+0-0.16	F
S	39	1.535	0.119E+04	0.185E+01	0.938E+01	0.630E+01	S	F			T	S	S+0-0.16	F
F	40	1.575	0.126E+04	0.195E+01	0.986E+01	0.663E+01	F	F	F		F	F	F+0-0.16	F
F	41	1.614	0.132E+04	0.205E+01	0.104E+02	0.696E+01					T	F	+0-0.16	F
	42	1.654	0.139E+04	0.215E+01	0.109E+02	0.731E+01	S				S	S	S+0-0.16	F
S	44	1.732	0.152E+04	0.236E+01	0.119E+02	0.802E+01					T,43F		S+0-0.16	F
F	45	1.772	0.159E+04	0.247E+01	0.125E+02	0.839E+01	F		F		S	F	F+0-0.16	F
	46	1.811	0.166E+04	0.258E+01	0.130E+02	0.877E+01		F			T	F	+0-0.16	F
S	48	1.890	0.181E+04	0.280E+01	0.142E+02	0.955E+01	S	F			F,47T	F	S+0-0.16	F
F	50	1.969	0.196E+04	0.304E+01	0.154E+02	0.104E+02	F	F	F		F	F	F+0-0.16	F

10-25

Table 10-7 (*Continued*). Bright Finish Round Steel Bars (DIN 668 Tolerance h 11)

```
TABLE 10-7.  BRIGHT FINISH ROUND STEEL                              ORDER EXAMPLE;
BARS (DIN 668 TOLERANCE h11)                                        LENGTH,ROUND STEEL 1.5 DIN 668 + STEEL QUALITY
BASIS; 1 IN = 25.4 MM
1 CUBIC METER STEEL = 7850 KG MASS
                                                                    PAGE NO. 3

                                                THE NOMINAL SIZE IS NATIONAL STANDARD AS INDICATED
                                                F=FIRST CHOICE,S=SECOND CHOICE,T=THIRD CHOICE,NUMBER=OTHER SIZE
                                                * = COMMERCIAL SIZE
                                                U.S.A.  AUSTRAL  JAPAN   EURO-   FRANCE   U.K.   GERMANY   ITALY
                                                ANSI    AS       JIS     NORM    NF       BS     DIN       UNI
D                                               B32.4   1027     G3123           A47-411  4229   668       468
I
N NOMINAL SIZE = D   SECTION AREA    MASS PER UNIT
  MM    IN           MM**2    IN**2  KG/M       LB/FT
S  52   2.047        0.212E+04 0.329E+01  0.167E+02 0.112E+02   S                                           S+0-0.19 F
S  55   2.165        0.238E+04 0.368E+01  0.187E+02 0.125E+02   F            F,54S                          S+0-0.19 F
F  56   2.205        0.246E+04 0.382E+01  0.193E+02 0.130E+02   F            F                              F+0-0.19 F
   58   2.283        0.264E+04 0.410E+01  0.207E+02 0.139E+02                                               S+0-0.19 F

F  60   2.362        0.283E+04 0.438E+01  0.222E+02 0.149E+02   F   F                                       S+0-0.19 F
F  63   2.480        0.312E+04 0.483E+01  0.245E+02 0.164E+02   F            64S                            F+0-0.19 F
S  65   2.559        0.332E+04 0.514E+01  0.260E+02 0.175E+02   S            F                              F+0-0.19 F
S  68   2.677        0.363E+04 0.563E+01  0.285E+02 0.192E+02                S                              S+0-0.19 F

F  70   2.756        0.385E+04 0.597E+01  0.302E+02 0.203E+02   S   F                                       F+0-0.19 F
S  75   2.953        0.442E+04 0.685E+01  0.347E+02 0.233E+02   S   F        F,72S                          S+0-0.19 F
F  80   3.150        0.503E+04 0.779E+01  0.395E+02 0.265E+02   F   F        F,76S                          F+0-0.19 F
S  85   3.346        0.567E+04 0.880E+01  0.445E+02 0.299E+02       F        F                              S+0-0.22 F

F  90   3.543        0.636E+04 0.986E+01  0.499E+02 0.336E+02   S   F                                       F+0-0.22 F
S  95   3.740        0.709E+04 0.110E+02  0.556E+02 0.374E+02       F                                       +0-0.22
F 100   3.937        0.785E+04 0.122E+02  0.617E+02 0.414E+02   F   F                                       S+0-0.22 F
F 105   4.134        0.866E+04 0.134E+02  0.680E+02 0.457E+02   F   F                                       F+0-0.22 F

S 110   4.331        0.950E+04 0.147E+02  0.746E+02 0.501E+02   S   F        F                              +0-0.22
S 115   4.528        0.104E+05 0.161E+02  0.815E+02 0.548E+02       F                                       +0-0.22
S 120   4.724        0.113E+05 0.175E+02  0.888E+02 0.597E+02   F   F        F                              S+0-0.22 F
F 125   4.921        0.123E+05 0.190E+02  0.963E+02 0.647E+02   F   F                                       F+0-0.25 F

S 130   5.118        0.133E+05 0.206E+02  0.104E+03 0.700E+02   S   F                                       S+0-0.25 F
F 140   5.512        0.154E+05 0.239E+02  0.121E+03 0.812E+02   F   F                                       F+0-0.25 F
F 150   5.906        0.177E+05 0.274E+02  0.139E+03 0.932E+02   F   F                                       S+0-0.25 F
F 160   6.299        0.201E+05 0.312E+02  0.158E+03 0.106E+03   F   F                                       S+0-0.25 F

  170   6.693        0.227E+05 0.352E+02  0.178E+03 0.120E+03   S            F                              +0-0.25
F 180   7.087        0.254E+05 0.394E+02  0.200E+03 0.134E+03   F            F                              +0-0.25
  190   7.480        0.284E+05 0.439E+02  0.223E+03 0.150E+03   S                                           +0-0.29
F 200   7.874        0.314E+05 0.487E+02  0.247E+03 0.166E+03   F            F                              F+0-0.29 F

  220   8.661        0.380E+05 0.589E+02  0.298E+03 0.201E+03   S                                           +0-0.29
  250   9.843        0.491E+05 0.761E+02  0.385E+03 0.259E+03   F                                           +0-0.29
  280  11.024        0.616E+05 0.954E+02  0.483E+03 0.325E+03   S                                           +0-0.32
  300  11.811        0.707E+05 0.110E+03  0.555E+03 0.373E+03   F                                           F+0-0.32 F

  320  12.598        0.804E+05 0.125E+03  0.631E+03 0.424E+03   S                                           +0-0.36
```

STEEL MATERIAL DATA

Table 10-8. Bright Finish Round Steel Bars (DIN 669 Tolerance h 9)

```
TABLE 10-8.  BRIGHT FINISH ROUND                                             ORDER EXAMPLE:
STEEL BARS (DIN 669 TOLERANCE h9)                                            LENGTH,ROUND STEEL 5 DIN 669 + STEEL QUALITY
BASIS; 1 IN = 25.4 MM
1 CUBIC METER STEEL = 7850 KG MASS

                                                        THE NOMINAL SIZE IS NATIONAL STANDARD AS INDICATED
                                                        F=FIRST CHOICE,S=SECOND CHOICE,T=THIRD CHOICE,NUMBER=OTHER SIZE
                                                        * = COMMERCIAL SIZE
                                                        U.S.A.   AUSTRAL  JAPAN   EURO-   FRANCE   U.K.   GERMANY  ITALY
                                                        ANSI     AS       JIS     NORM    NF       BS     DIN      UNI
                                                        B32.4    1027     G3123           A47-411  4229   669
D                                                                                                                PAGE NO. 1
I
N  NOMINAL SIZE = D   SECTION AREA    MASS PER UNIT
   MM       IN        MM**2    IN**2  KG/M       LB/FT
   3        0.118     0.707E+01 0.110E-01 0.555E-01 0.373E-01  F                                         +  0-0.030  F
   3.5      0.138     0.962E+01 0.149E-01 0.755E-01 0.508E-01  S                                         +  0-0.030  F
   4        0.157     0.126E+02 0.195E-01 0.986E-01 0.663E-01  F            F                            +  0-0.030  F
   4.5      0.177     0.159E+02 0.247E-01 0.125E+00 0.839E-01  S                                         +  0-0.030  F

F  5        0.197     0.196E+02 0.304E-01 0.154E+00 0.104E+00  F     F      F                            F+ 0-0.030  F
   5.5      0.217     0.238E+02 0.368E-01 0.187E+00 0.125E+00  S                                         +  0-0.030  F
F  6        0.236     0.283E+02 0.438E-01 0.222E+00 0.149E+00  F     F      F                            F+ 0-0.030  F
   6.5      0.256     0.332E+02 0.514E-01 0.260E+00 0.175E+00  S                                         +  0-0.036  F

F  7        0.276     0.385E+02 0.597E-01 0.302E+00 0.203E+00  S            F                            F+ 0-0.036  F
F  8        0.315     0.503E+02 0.779E-01 0.395E+00 0.265E+00  F     F      F                            F+ 0-0.036  F
F  9        0.354     0.636E+02 0.986E-01 0.499E+00 0.336E+00  S                                         F+ 0-0.036  F
F  10       0.394     0.785E+02 0.122E+00 0.617E+00 0.414E+00  F     F      F                            F+ 0-0.043  F

F  11       0.433     0.950E+02 0.147E+00 0.746E+00 0.501E+00  S            F                            F+ 0-0.043  F
F  12       0.472     0.113E+03 0.175E+00 0.888E+00 0.597E+00  F     F      F                            F+ 0-0.043  F
S  13       0.512     0.133E+03 0.206E+00 0.104E+01 0.700E+00  S            F                            S+ 0-0.043  F
F  14       0.551     0.154E+03 0.239E+00 0.121E+01 0.812E+00  F     F      F                            F+ 0-0.043  F

   14.5     0.571     0.165E+03 0.256E+00 0.130E+01 0.871E+00                                            +  0-0.043  F
S  15       0.591     0.177E+03 0.274E+00 0.139E+01 0.932E+00  S     S      F                            S+ 0-0.043  F
   15.5     0.610     0.189E+03 0.292E+00 0.148E+01 0.995E+00                                            +  0-0.043  F
F  16       0.630     0.201E+03 0.312E+00 0.158E+01 0.106E+01  F     F      F     F                      F+ 0-0.043  F

   16.5     0.650     0.214E+03 0.331E+00 0.168E+01 0.113E+01                                            +  0-0.043  F
   17       0.669     0.227E+03 0.352E+00 0.178E+01 0.120E+01  S     S      F                            +  0-0.043  F
   17.5     0.689     0.241E+03 0.373E+00 0.189E+01 0.127E+01                                            +  0-0.043  F
F  18       0.709     0.254E+03 0.394E+00 0.200E+01 0.134E+01  F     F      F                            F+ 0-0.043  F

   18.5     0.728     0.269E+03 0.417E+00 0.211E+01 0.142E+01                                            +  0-0.052  F
   19       0.748     0.284E+03 0.439E+00 0.222E+01 0.150E+01  S     S      F                            +  0-0.052  F
   19.5     0.768     0.299E+03 0.463E+00 0.234E+01 0.158E+01                                            +  0-0.052  F
F  20       0.787     0.314E+03 0.487E+00 0.247E+01 0.166E+01  F     F      F     F                      F+ 0-0.052  F

   20.5     0.807     0.330E+03 0.512E+00 0.259E+01 0.174E+01  S     S      F                            +  0-0.052  F
   21       0.827     0.346E+03 0.537E+00 0.272E+01 0.183E+01  F     F      F                            +  0-0.052  F
   21.5     0.846     0.363E+03 0.563E+00 0.285E+01 0.192E+01  S            F           S                +  0-0.052  F
F  22       0.866     0.380E+03 0.589E+00 0.298E+01 0.201E+01  F     F      F           S                F+ 0-0.052  F

   22.5     0.886     0.398E+03 0.616E+00 0.312E+01 0.210E+01  S     S      F                            +  0-0.052  F
   23       0.906     0.415E+03 0.644E+00 0.326E+01 0.219E+01  F     F      F                            F+ 0-0.052  F
   23.5     0.925     0.434E+03 0.672E+00 0.340E+01 0.229E+01  S            F           F                F+ 0-0.052  F
   24       0.945     0.452E+03 0.701E+00 0.355E+01 0.239E+01  F     F      F           F                F+ 0-0.052  F

   24.5     0.965     0.471E+03 0.731E+00 0.370E+01 0.249E+01  S     S      F           F                +  0-0.052  F
F  25       0.984     0.491E+03 0.761E+00 0.385E+01 0.259E+01  F     F      F           F      F         F+ 0-0.052  F
   26       1.024     0.531E+03 0.823E+00 0.417E+01 0.280E+01  S            F           F      F         F+ 0-0.052  F
   27       1.063     0.573E+03 0.887E+00 0.449E+01 0.302E+01                           T      F         +  0-0.052  F

F  28       1.102     0.616E+03 0.954E+00 0.483E+01 0.325E+01  S            F           F      F         F+ 0-0.052  F
   29       1.142     0.661E+03 0.102E+01 0.519E+01 0.348E+01                           T      F         +  0-0.052  F
S  30       1.181     0.707E+03 0.110E+01 0.555E+01 0.373E+01  F            F           F      F         S+ 0-0.052  F
```

10-27

Table 10-8 (Continued). Bright Finish Round Steel Bars (DIN 669 Tolerance h 9)

TABLE 10-8. BRIGHT FINISH ROUND STEEL BARS (DIN 669 TOLERANCE h 9)
BASIS; 1 IN = 25.4 MM
1 CUBIC METER STEEL = 7850 KG MASS

ORDER EXAMPLE:
LENGTH,ROUND STEEL 5 DIN 669 + STEEL QUALITY

PAGE NO. 2

THE NOMINAL SIZE IS NATIONAL STANDARD AS INDICATED
F=FIRST CHOICE,S=SECOND CHOICE,T=THIRD CHOICE,NUMBER=OTHER SIZE
* = COMMERCIAL SIZE

DIN	NOMINAL SIZE = D		SECTION AREA		MASS PER UNIT		U.S.A. ANSI B32.4	AUSTRAL AS 1027	JAPAN JIS G3123	EURO- NORM	FRANCE NF A47-411	U.K. BS 4229	GERMANY DIN 669	ITALY UNI 469
	MM	IN	MM**2	IN**2	KG/M	LB/FT								
	31	1.220	0.755E+03	0.117E+01	0.592E+01	0.398E+01					T	T	+0=0.062	
	32	1.260	0.804E+03	0.125E+01	0.631E+01	0.424E+01	S				S	F	+0=0.062	
	33	1.299	0.855E+03	0.133E+01	0.671E+01	0.451E+01						S	+0=0.062	
	34	1.339	0.908E+03	0.141E+01	0.713E+01	0.479E+01							+0=0.062	
S	35	1.378	0.962E+03	0.149E+01	0.755E+01	0.508E+01	F	F	F		F	F	S+0=0.062	F
F	36	1.417	0.102E+04	0.158E+01	0.799E+01	0.537E+01			F		S	F	F+0=0.062	F
S	38	1.496	0.113E+04	0.176E+01	0.890E+01	0.598E+01	S		F		F,37T	S	S+0=0.062	F
	39	1.535	0.119E+04	0.185E+01	0.938E+01	0.630E+01					T		+0=0.062	
F	40	1.575	0.126E+04	0.195E+01	0.986E+01	0.663E+01	F	F	F		F	F	F+0=0.062	F
	42	1.654	0.139E+04	0.215E+01	0.109E+02	0.731E+01	S		F		S,41T	F,41S	S+0=0.062	F
	44	1.732	0.152E+04	0.236E+01	0.119E+02	0.802E+01			F		T,43F	F	T+0=0.062	F
F	45	1.772	0.159E+04	0.247E+01	0.125E+02	0.839E+01	F	F	F		S	F	F+0=0.062	F
F	46	1.811	0.166E+04	0.258E+01	0.130E+02	0.877E+01					T		+0=0.062	F
	48	1.890	0.181E+04	0.280E+01	0.142E+02	0.955E+01	S	F	F		F,47T	F	S+0=0.062	F
F	50	1.969	0.196E+04	0.304E+01	0.154E+02	0.104E+02	F	F	F			F	F+0=0.062	F
	52	2.047	0.212E+04	0.329E+01	0.167E+02	0.112E+02	F					F	F+0=0.074	F
	55	2.165	0.238E+04	0.368E+01	0.187E+02	0.125E+02	S		F			F,54S	+0=0.074	F
F	56	2.205	0.246E+04	0.382E+01	0.193E+02	0.130E+02			F			F	F+0=0.074	F
	58	2.283	0.264E+04	0.410E+01	0.207E+02	0.139E+02			F			F	S+0=0.074	F
S	60	2.362	0.283E+04	0.438E+01	0.222E+02	0.149E+02	S	F	F			F	S+0=0.074	F
	63	2.480	0.312E+04	0.483E+01	0.245E+02	0.164E+02						F	F+0=0.074	F
S	65	2.559	0.332E+04	0.514E+01	0.260E+02	0.175E+02	S	F	F			F,64S	S+0=0.074	F
F	68	2.677	0.363E+04	0.563E+01	0.285E+02	0.192E+02						S	F+0=0.074	F
F	70	2.756	0.385E+04	0.597E+01	0.302E+02	0.203E+02			F			F	F+0=0.074	F
S	75	2.953	0.442E+04	0.685E+01	0.347E+02	0.233E+02	S		F			F,72S	S+0=0.087	F
F	80	3.150	0.503E+04	0.779E+01	0.395E+02	0.265E+02	S		F			F,76S	S+0=0.087	F
	85	3.346	0.567E+04	0.880E+01	0.445E+02	0.299E+02	S					F	F+0=0.087	F
	90	3.543	0.636E+04	0.986E+01	0.499E+02	0.336E+02	S					F	S+0=0.087	F
S	95	3.740	0.709E+04	0.110E+02	0.556E+02	0.374E+02			F			F	+0=0.087	F
F	100	3.937	0.785E+04	0.122E+02	0.617E+02	0.414E+02	F	F	F			F	S+0=0.087	F
	105	4.134	0.866E+04	0.134E+02	0.680E+02	0.457E+02	F					F	F+0=0.087	F
S	110	4.331	0.950E+04	0.147E+02	0.746E+02	0.501E+02			F			F	S+0=0.087	F
	115	4.528	0.104E+05	0.161E+02	0.815E+02	0.548E+02	F		F			F	F+0=0.087	F
F	120	4.724	0.113E+05	0.175E+02	0.888E+02	0.597E+02			F			F	S+0=0.087	F
S	125	4.921	0.123E+05	0.190E+02	0.963E+02	0.647E+02	F	F	F			F	F+0=0.100	F
	130	5.118	0.135E+05	0.206E+02	0.104E+03	0.700E+02			F			F	F+0=0.100	F
F	140	5.512	0.154E+05	0.239E+02	0.121E+03	0.812E+02	F	F	F			F	F+0=0.100	F
S	150	5.906	0.177E+05	0.274E+02	0.139E+03	0.932E+02	S		F			F	S+0=0.100	F
F	160	6.299	0.201E+05	0.312E+02	0.158E+03	0.106E+03	S		F			F	F+0=0.100	F
	170	6.693	0.227E+05	0.352E+02	0.178E+03	0.120E+03	S					F	F+0=0.100	F
F	180	7.087	0.254E+05	0.394E+02	0.200E+03	0.134E+03	F					F	F+0=0.100	F
	190	7.480	0.284E+05	0.439E+02	0.223E+03	0.150E+03	S						+0=0.100	F
F	200	7.874	0.314E+05	0.487E+02	0.247E+03	0.166E+03	F					F	F+0=0.115	F

STEEL MATERIAL DATA

Table 10-9. Ground or Polished Finish Round Steel Bars (DIN 59360 Tolerance h7)*

```
TABLE 10-9.  GROUND OR POLISHED FINISH ROUND            ORDER EXAMPLE:                         PAGE NR. 1
STEEL BARS (DIN 59360 TOLERANCE h7)                     LENGTH,ROUND STEEL 2 DIN 59360 + STEEL QUALITY
BASIS: 1 IN = 25.4 MM
1 CUBIC METER STEEL = 7850 KG MASS
                                                        THE NOMINAL SIZE IS NATIONAL STANDARD AS INDICATED
                                                        F=FIRST CHOICE,S=SECOND CHOICE,T=THIRD CHOICE,NUMBER=OTHER SIZE
                                                        * = COMMERCIAL SIZE
                                                        U.S.A.  AUSTRAL JAPAN  EURO-  FRANCE  U.K.  GERMANY ITALY
                                                        ANSI    AS      JIS    NORM   NF      BS    DIN     UNI
                                                        B32.4   1027    G3123         A47-411 4229  59360   5953
```

D I N	NOMINAL SIZE = D		SECTION AREA		MASS PER UNIT		U.S.A.	AUSTRAL	JAPAN	EURO-NORM	FRANCE	U.K.	GERMANY DIN 59360	ITALY UNI 5953
	MM	IN	MM**2	IN**2	KG/M	LB/FT								
F	2	0.079	0.314E+01	0.487E-02	0.247E-01	0.166E-01	F						F +0-0.010	
S	2.1	0.083	0.346E+01	0.537E-02	0.272E-01	0.183E-01							S +0-0.010	
F	2.2	0.087	0.380E+01	0.589E-02	0.298E-01	0.201E-01		F					F +0-0.010	
S	2.3	0.091	0.415E+01	0.644E-02	0.326E-01	0.219E-01	S						S +0-0.010	
B	2.4	0.094	0.452E+01	0.701E-02	0.355E-01	0.239E-01						S	S +0-0.010	
F	2.5	0.098	0.491E+01	0.761E-02	0.385E-01	0.259E-01	F					F	F +0-0.010	
S	2.6	0.102	0.531E+01	0.823E-02	0.417E-01	0.280E-01						S	S +0-0.010	
B	2.7	0.106	0.573E+01	0.887E-02	0.449E-01	0.302E-01						S	S +0-0.010	
F	2.8	0.110	0.616E+01	0.954E-02	0.483E-01	0.325E-01	S					F	F +0-0.010	
S	2.9	0.114	0.661E+01	0.102E-01	0.519E-01	0.348E-01						S	S +0-0.010	
F	3	0.118	0.707E+01	0.110E-01	0.555E-01	0.373E-01	F					F	F +0-0.012	F
	3.1	0.122	0.755E+01	0.117E-01	0.592E-01	0.398E-01						S	F +0-0.012	F
	3.2	0.126	0.804E+01	0.125E-01	0.631E-01	0.424E-01						F	F +0-0.012	F
	3.3	0.130	0.855E+01	0.133E-01	0.671E-01	0.451E-01						S	F +0-0.012	F
	3.4	0.134	0.908E+01	0.141E-01	0.713E-01	0.479E-01						F	F +0-0.012	F
F	3.5	0.138	0.962E+01	0.149E-01	0.755E-01	0.508E-01	S					F	F +0-0.012	F
	3.6	0.142	0.102E+02	0.158E-01	0.799E-01	0.537E-01						S	F +0-0.012	F
	3.7	0.146	0.108E+02	0.167E-01	0.844E-01	0.567E-01						F	F +0-0.012	F
	3.8	0.150	0.113E+02	0.176E-01	0.890E-01	0.598E-01						S	F +0-0.012	F
	3.9	0.154	0.119E+02	0.185E-01	0.938E-01	0.630E-01						S	F +0-0.012	F
F	4	0.157	0.126E+02	0.195E-01	0.986E-01	0.663E-01	F				F	F	F +0-0.012	F
	4.1	0.161	0.132E+02	0.205E-01	0.104E+00	0.696E-01						S	F +0-0.012	F
	4.2	0.165	0.139E+02	0.215E-01	0.109E+00	0.731E-01						S	F +0-0.012	F
	4.3	0.169	0.145E+02	0.225E-01	0.114E+00	0.766E-01						S	F +0-0.012	F
	4.4	0.173	0.152E+02	0.236E-01	0.119E+00	0.802E-01						S	F +0-0.012	F
F	4.5	0.177	0.159E+02	0.247E-01	0.125E+00	0.839E-01	S				F	F	F +0-0.012	F
	4.6	0.181	0.166E+02	0.258E-01	0.130E+00	0.877E-01						S	F +0-0.012	F
	4.7	0.185	0.173E+02	0.269E-01	0.136E+00	0.915E-01						S	F +0-0.012	F
	4.8	0.189	0.181E+02	0.280E-01	0.142E+00	0.955E-01	S					F	F +0-0.012	F
	4.9	0.193	0.189E+02	0.292E-01	0.148E+00	0.995E-01						S	F +0-0.012	F
F	5	0.197	0.196E+02	0.304E-01	0.154E+00	0.104E+00	F				F	F	F +0-0.012	F
	5.1	0.201	0.204E+02	0.317E-01	0.160E+00	0.108E+00						S	F +0-0.012	F
	5.2	0.205	0.212E+02	0.329E-01	0.167E+00	0.112E+00						S	S +0-0.012	F
	5.3	0.209	0.221E+02	0.342E-01	0.173E+00	0.116E+00						S	S +0-0.012	F
	5.4	0.213	0.229E+02	0.355E-01	0.180E+00	0.121E+00						S	S +0-0.012	F
	5.5	0.217	0.238E+02	0.368E-01	0.187E+00	0.125E+00	S					F	S +0-0.012	F
	5.6	0.220	0.246E+02	0.382E-01	0.193E+00	0.130E+00						S	S +0-0.012	F
	5.7	0.224	0.255E+02	0.396E-01	0.200E+00	0.135E+00						S	S +0-0.012	F
	5.8	0.228	0.264E+02	0.410E-01	0.207E+00	0.139E+00						S	F +0-0.012	F
	5.9	0.232	0.273E+02	0.424E-01	0.215E+00	0.144E+00						F	F +0-0.012	F
F	6	0.236	0.283E+02	0.438E-01	0.222E+00	0.149E+00	F				F	F	F +0-0.012	F
	6.1	0.240	0.292E+02	0.453E-01	0.229E+00	0.154E+00						S	F +0-0.015	F
	6.2	0.244	0.302E+02	0.468E-01	0.237E+00	0.159E+00						S	F +0-0.015	F
	6.3	0.248	0.312E+02	0.483E-01	0.245E+00	0.164E+00						F	F +0-0.015	F

*DIN 59361 covers steel bars to the same nominal sizes as shown for DIN 59360 above, except to ISO tolerance h6.

Table 10-9 (Continued). Ground or Polished Finish Round Steel Bars (DIN 59360 Tolerance h7)*

TABLE 10-9. GROUND OR POLISHED FINISH ROUND
STEEL BARS (DIN 59360 TOLERANCE h7)
BASIS: 1 IN = 25.4 MM
1 CUBIC METER STEEL = 7850 KG MASS

ORDER EXAMPLE:
LENGTH,ROUND STEEL 2 DIN 59360 + STEEL QUALITY

PAGE NO. 2

THE NOMINAL SIZE IS NATIONAL STANDARD AS INDICATED
F=FIRST CHOICE,S=SECOND CHOICE,T=THIRD CHOICE,NUMBER=OTHER SIZE
* = COMMERCIAL SIZE

DIN	NOMINAL SIZE = D		SECTION AREA		MASS PER UNIT		U.S.A. ANSI B32.4	AUSTRAL AS 1027	JAPAN JIS G3123	EURO- NORM	FRANCE NF A47-411	U.K. BS *229	GERMANY DIN 59360	ITALY UNI 5953
	MM	IN	MM**2	IN**2	KG/M	LB/FT								
S	6.4	0.252	0.322E+02	0.499E-01	0.253E+00	0.170E+00	S					S	+0-0.015	
	6.5	0.256	0.332E+02	0.514E-01	0.260E+00	0.175E+00					F	F	S+0-0.015	F
	6.6	0.260	0.342E+02	0.530E-01	0.269E+00	0.180E+00						S	+0-0.015	F
	6.7	0.264	0.353E+02	0.546E-01	0.277E+00	0.186E+00						S	+0-0.015	F
	6.8	0.268	0.363E+02	0.563E-01	0.285E+00	0.192E+00						F	+0-0.015	F
	6.9	0.272	0.374E+02	0.580E-01	0.294E+00	0.197E+00						F	+0-0.015	F
F	7	0.276	0.385E+02	0.597E-01	0.302E+00	0.203E+00						F	F+0-0.015	F
	7.1	0.280	0.396E+02	0.614E-01	0.311E+00	0.209E+00	S					S	+0-0.015	F
	7.2	0.283	0.407E+02	0.631E-01	0.320E+00	0.215E+00						F	+0-0.015	F
	7.3	0.287	0.419E+02	0.649E-01	0.329E+00	0.221E+00						F	+0-0.015	F
	7.4	0.291	0.430E+02	0.667E-01	0.338E+00	0.227E+00						F	+0-0.015	F
S	7.5	0.295	0.442E+02	0.685E-01	0.347E+00	0.233E+00						S	S+0-0.015	F
	7.6	0.299	0.454E+02	0.703E-01	0.356E+00	0.239E+00						F	+0-0.015	F
	7.7	0.303	0.466E+02	0.722E-01	0.366E+00	0.246E+00						F	+0-0.015	F
	7.8	0.307	0.478E+02	0.741E-01	0.375E+00	0.252E+00						F	+0-0.015	F
S	7.9	0.311	0.490E+02	0.760E-01	0.385E+00	0.259E+00						S	+0-0.015	F
F	8	0.315	0.503E+02	0.779E-01	0.395E+00	0.265E+00	F				F	F	F+0-0.015	F
	8.1	0.319	0.515E+02	0.799E-01	0.405E+00	0.272E+00						F	+0-0.015	F
	8.2	0.323	0.528E+02	0.819E-01	0.415E+00	0.279E+00						F	+0-0.015	F
	8.3	0.327	0.541E+02	0.839E-01	0.425E+00	0.285E+00						F	+0-0.015	F
	8.4	0.331	0.554E+02	0.859E-01	0.435E+00	0.292E+00	F					F	+0-0.015	F
	8.5	0.335	0.567E+02	0.880E-01	0.445E+00	0.299E+00						F	S+0-0.015	F
	8.6	0.339	0.581E+02	0.900E-01	0.456E+00	0.306E+00						F	+0-0.015	F
	8.7	0.343	0.594E+02	0.921E-01	0.467E+00	0.314E+00						F	+0-0.015	F
	8.8	0.346	0.608E+02	0.943E-01	0.477E+00	0.321E+00						F	+0-0.015	F
	8.9	0.350	0.622E+02	0.964E-01	0.488E+00	0.328E+00						F	+0-0.015	F
F	9	0.354	0.636E+02	0.986E-01	0.499E+00	0.336E+00	S				F	F	F+0-0.015	F
	9.1	0.358	0.650E+02	0.101E+00	0.511E+00	0.343E+00						S	+0-0.015	F
	9.2	0.362	0.665E+02	0.103E+00	0.522E+00	0.351E+00						S	+0-0.015	F
	9.3	0.366	0.679E+02	0.105E+00	0.533E+00	0.358E+00						F	+0-0.015	F
	9.4	0.370	0.694E+02	0.108E+00	0.545E+00	0.366E+00						S	+0-0.015	F
S	9.5	0.374	0.709E+02	0.110E+00	0.556E+00	0.374E+00						S	+0-0.015	F
	9.7	0.382	0.739E+02	0.115E+00	0.580E+00	0.390E+00						S	+0-0.015	F
	9.8	0.386	0.754E+02	0.117E+00	0.592E+00	0.398E+00						S	+0-0.015	F
F	10	0.394	0.785E+02	0.122E+00	0.617E+00	0.414E+00		F			F	F,9.6S	F+0-0.015	F
S	10.5	0.413	0.866E+02	0.134E+00	0.680E+00	0.457E+00			F			S,9.6S	S+0-0.015	F
F	11	0.433	0.950E+02	0.147E+00	0.746E+00	0.501E+00	S				F	F	F+0-0.018	F
S	11.5	0.453	0.104E+03	0.161E+00	0.815E+00	0.548E+00						F	+0-0.018	F
F	12	0.472	0.113E+03	0.175E+00	0.888E+00	0.597E+00	F				F	F	F+0-0.018	F
S	12.5	0.492	0.123E+03	0.190E+00	0.963E+00	0.647E+00						F	+0-0.018	F
S	13	0.512	0.133E+03	0.206E+00	0.104E+01	0.700E+00	S					F	S+0-0.018	F
S	13.5	0.531	0.143E+03	0.222E+00	0.112E+01	0.755E+00						F	S+0-0.018	F
F	14	0.551	0.154E+03	0.239E+00	0.121E+01	0.812E+00	F		F		F	F	F+0-0.018	F

*DIN 59361 covers steel bars to the same nominal sizes as shown for DIN 59360 above, except to ISO tolerance h6.

STEEL MATERIAL DATA

10-31

Table 10-9 (Continued). Ground or Polished Finish Round Steel Bars (DIN 59360 Tolerance h7)*

```
TABLE 10-9.  GROUND OR POLISHED FINISH ROUND                                PAGE NO. 3
STEEL BARS (DIN 59360 TOLERANCE h7)                    ORDER EXAMPLE:
BASIS; 1 IN = 25.4 MM                                  LENGTH,ROUND STEEL 2 DIN 59360 + STEEL QUALITY
1 CUBIC METER STEEL = 7850 KG MASS
```

                                                              THE NOMINAL SIZE IS NATIONAL STANDARD AS INDICATED
                                                              F=FIRST CHOICE,S=SECOND CHOICE,T=THIRD CHOICE,NUMBER=OTHER SIZE
                                                              * = COMMERCIAL SIZE

D I N	NOMINAL SIZE = D MM / IN	SECTION AREA MM**2 / IN**2	MASS PER UNIT KG/M / LB/FT	U.S.A. ANSI B32.4	AUSTRAL AS 1027	JAPAN JIS G3123	EURO- NORM	FRANCE NF A47-411	U.K. BS 4229	GERMANY DIN 59360	ITALY UNI 5953
S	14.5 / 0.571	0.165E+03 / 0.256E+00	0.130E+01 / 0.871E+00	S						S+0-0.018	F
S	15 / 0.591	0.177E+03 / 0.274E+00	0.139E+01 / 0.932E+00	S						S+0-0.018	F
S	15.5 / 0.610	0.189E+03 / 0.292E+00	0.148E+01 / 0.995E+00		F			F		S+0-0.018	F
F	16 / 0.630	0.201E+03 / 0.312E+00	0.158E+01 / 0.106E+01	F						F+0-0.018	F
S	16.5 / 0.650	0.214E+03 / 0.331E+00	0.168E+01 / 0.113E+01	S						S+0-0.018	F
F	17 / 0.669	0.227E+03 / 0.352E+00	0.178E+01 / 0.120E+01		F	F		F		F+0-0.018	F
F	18 / 0.709	0.254E+03 / 0.394E+00	0.200E+01 / 0.134E+01		F	F		F		F+0-0.018	F
S	19 / 0.748	0.284E+03 / 0.439E+00	0.223E+01 / 0.150E+01	S						S+0-0.021	F
F	20 / 0.787	0.314E+03 / 0.487E+00	0.247E+01 / 0.166E+01	F		F		F		F+0-0.021	F
S	21 / 0.827	0.346E+03 / 0.537E+00	0.272E+01 / 0.183E+01	S		F		F		S+0-0.021	F
F	22 / 0.866	0.380E+03 / 0.589E+00	0.298E+01 / 0.201E+01	F		F		S		F+0-0.021	F
S	23 / 0.906	0.415E+03 / 0.644E+00	0.326E+01 / 0.219E+01	S		F		F		S+0-0.021	F
S	24 / 0.945	0.452E+03 / 0.701E+00	0.355E+01 / 0.239E+01	S		F		S		S+0-0.021	F
F	25 / 0.984	0.491E+03 / 0.761E+00	0.385E+01 / 0.259E+01		F	F		F		F+0-0.021	F
S	26 / 1.024	0.531E+03 / 0.823E+00	0.417E+01 / 0.280E+01	S		F		F		S+0-0.021	F
S	27 / 1.063	0.573E+03 / 0.887E+00	0.449E+01 / 0.302E+01	S		F		T		S+0-0.021	F
F	28 / 1.102	0.616E+03 / 0.954E+00	0.483E+01 / 0.325E+01	F		F		T		F+0-0.021	F
S	29 / 1.142	0.661E+03 / 0.102E+01	0.519E+01 / 0.348E+01	S		F		T		S+0-0.021	F
F	30 / 1.181	0.707E+03 / 0.110E+01	0.555E+01 / 0.373E+01	F		F		F		F+0-0.021	F
F	32 / 1.260	0.804E+03 / 0.125E+01	0.631E+01 / 0.424E+01	F		F		S , 31T		F+0-0.025	F
S	33 / 1.299	0.855E+03 / 0.133E+01	0.671E+01 / 0.451E+01	S		F		F		S+0-0.025	F
S	34 / 1.339	0.908E+03 / 0.141E+01	0.713E+01 / 0.479E+01	S		F		S		S+0-0.025	F
F	35 / 1.378	0.962E+03 / 0.149E+01	0.755E+01 / 0.508E+01	F		F		F		F+0-0.025	F
F	36 / 1.417	0.102E+04 / 0.158E+01	0.799E+01 / 0.537E+01	F		F		S		F+0-0.025	F
S	38 / 1.496	0.113E+04 / 0.176E+01	0.890E+01 / 0.598E+01	S		F		F , 37T		S+0-0.025	F
S	39 / 1.535	0.119E+04 / 0.185E+01	0.938E+01 / 0.630E+01	S		F		S		S+0-0.025	F
F	40 / 1.575	0.126E+04 / 0.195E+01	0.986E+01 / 0.663E+01	F		F		F		F+0-0.025	F
S	42 / 1.654	0.139E+04 / 0.215E+01	0.109E+02 / 0.731E+01	S		F		S , 41T		S+0-0.025	F
S	44 / 1.732	0.152E+04 / 0.236E+01	0.119E+02 / 0.802E+01	S		F		F , 43F		S+0-0.025	F
F	45 / 1.772	0.159E+04 / 0.246E+01	0.125E+02 / 0.839E+01		F	F		T		F+0-0.025	F
F	46 / 1.811	0.166E+04 / 0.258E+01	0.130E+02 / 0.877E+01		F	F		S		F+0-0.025	F
S	48 / 1.890	0.181E+04 / 0.280E+01	0.142E+02 / 0.955E+01	S		F		F , 47T		S+0-0.025	F
F	50 / 1.969	0.196E+04 / 0.304E+01	0.154E+02 / 0.104E+02		F	F		F		F+0-0.030	F
S	52 / 2.047	0.212E+04 / 0.329E+01	0.167E+02 / 0.112E+02	S		F		F		S+0-0.030	F
F	55 / 2.165	0.238E+04 / 0.368E+01	0.187E+02 / 0.125E+02		F	F		F		F+0-0.030	F
F	56 / 2.205	0.246E+04 / 0.382E+01	0.193E+02 / 0.130E+02		F	F		F		F+0-0.030	F
S	58 / 2.283	0.264E+04 / 0.410E+01	0.207E+02 / 0.139E+02	S		F		F		S+0-0.030	F
F	60 / 2.362	0.283E+04 / 0.438E+01	0.222E+02 / 0.149E+02		F	F		F		F+0-0.030	F
F	63 / 2.480	0.312E+04 / 0.483E+01	0.245E+02 / 0.164E+02		F	F		F		F+0-0.030	F
S	65 / 2.559	0.332E+04 / 0.514E+01	0.260E+02 / 0.175E+02	S		F		F		S+0-0.030	F
F	70 / 2.756	0.385E+04 / 0.597E+01	0.302E+02 / 0.203E+02		F	F		F		F+0-0.030	F
S	75 / 2.953	0.442E+04 / 0.685E+01	0.347E+02 / 0.233E+02	S		F		F		S+0-0.030	F
F	80 / 3.150	0.503E+04 / 0.779E+01	0.395E+02 / 0.265E+02		F	F		F		F+0-0.030	F

*DIN 59361 covers steel bars to the same nominal sizes as shown for DIN 59360 above, except to ISO tolerance h6.

STEEL MATERIAL DATA

Table 10-10. Round Spring Steel Bars (JIS G4801)

TABLE 10-10. ROUND SPRING STEEL BARS
(JIS G4801)
BASIS: 1 IN = 25.4 MM
1 CUBIC METER STEEL = 7850 KG MASS

ORDER EXAMPLE:
LENGTH,ROUND STEEL 8 JIS G4801 + STEEL QUAL

THE NOMINAL SIZE IS NATIONAL STANDARD AS INDICATED
F=FIRST CHOICE, S=SECOND CHOICE, T=THIRD CHOICE, NUMBER=OTHER SIZE
* = COMMERCIAL SIZE

J I S	NOMINAL SIZE = D MM	IN	SECTION AREA MM**2	IN**2	MASS PER UNIT KG/M	LB/FT	U.S.A. ANSI B32.4	AUSTRAL AS 1027	JAPAN JIS G4801	EURO- NORM	FRANCE NF	U.K. BS	GERMANY DIN 2077	ITALY UNI 3871
F	6	0.236	0.283E+02	0.438E-01	0.222E+00	0.149E+00			F*=0.20				F	F
F	7	0.276	0.385E+02	0.597E-01	0.302E+00	0.203E+00	S ,6.5S	F	F*=0.20				F	F
F	8	0.315	0.503E+02	0.779E-01	0.395E+00	0.265E+00	F	F	F*=0.20				F	F
F	9	0.354	0.636E+02	0.986E-01	0.499E+00	0.336E+00	S		F*=0.20				F	F
F	10	0.394	0.785E+02	0.122E+00	0.617E+00	0.414E+00	F	F	F*=0.25				F	F
F	11	0.433	0.950E+02	0.147E+00	0.746E+00	0.501E+00	S		F*=0.25				S	F
F	12	0.472	0.113E+03	0.175E+00	0.888E+00	0.597E+00	F		F*=0.25				F	F
F	12.5	0.492	0.123E+03	0.190E+00	0.963E+00	0.647E+00			+=0.25					
S	13	0.512	0.133E+03	0.206E+00	0.104E+01	0.700E+00	S		S*=0.25				S	S
F	14	0.551	0.154E+03	0.239E+00	0.121E+01	0.812E+00		F	F*=0.25				S	F
S	15	0.591	0.177E+03	0.274E+00	0.139E+01	0.932E+00			S*=0.25				S	F
F	16	0.630	0.201E+03	0.312E+00	0.158E+01	0.106E+01	F		F*=0.30				F	F
S	17	0.669	0.227E+03	0.352E+00	0.178E+01	0.120E+01	S		S*=0.30				S	S
F	18	0.709	0.254E+03	0.394E+00	0.200E+01	0.134E+01	F		F*=0.30				S	F
S	19	0.748	0.284E+03	0.439E+00	0.223E+01	0.150E+01	S		S*=0.30				S	F
F	20	0.787	0.314E+03	0.487E+00	0.247E+01	0.166E+01	F		F*=0.30				F	F
S	21	0.827	0.346E+03	0.537E+00	0.272E+01	0.183E+01	S		S*=0.40				S	S
F	22	0.866	0.380E+03	0.589E+00	0.298E+01	0.201E+01	F		F*=0.40				F	F
	22.5	0.886	0.398E+03	0.616E+00	0.312E+01	0.210E+01			+=0.40					
	23	0.906	0.415E+03	0.644E+00	0.326E+01	0.219E+01	S		+=0.40					
S	24	0.945	0.452E+03	0.701E+00	0.355E+01	0.239E+01	S		S*=0.40				S	S
F	25	0.984	0.491E+03	0.761E+00	0.385E+01	0.259E+01	F		F*=0.40				F	F
S	26	1.024	0.531E+03	0.823E+00	0.417E+01	0.280E+01	S		S*=0.40				S	F
F	28	1.102	0.616E+03	0.954E+00	0.483E+01	0.325E+01	S	27F	F*=0.40				F	F
S	30	1.181	0.707E+03	0.110E+01	0.555E+01	0.373E+01	F	F	S*=0.50				S	S
F	32	1.260	0.804E+03	0.125E+01	0.631E+01	0.424E+01	F	F	F*=0.50				F	F
S	34	1.339	0.908E+03	0.141E+01	0.713E+01	0.479E+01	F	33F	S*=0.50				F	S
F	35	1.378	0.962E+03	0.149E+01	0.755E+01	0.508E+01	S		+=0.50					
F	36	1.417	0.102E+04	0.158E+01	0.799E+01	0.537E+01	F	F	F*=0.50				F	F
F	38	1.496	0.113E+04	0.176E+01	0.890E+01	0.598E+01	F	39F	F*=0.50				S	F
S	40	1.575	0.126E+04	0.195E+01	0.986E+01	0.663E+01	F	F	F*=0.50				F	S
F	42	1.654	0.139E+04	0.215E+01	0.109E+02	0.731E+01	S		S*=0.50				S	F
F	45	1.772	0.159E+04	0.247E+01	0.125E+02	0.839E+01	F	F	F*=0.50				F	F
F	48	1.890	0.181E+04	0.280E+01	0.142E+02	0.955E+01	F	F	F*=0.70				F	F
S	50	1.969	0.196E+04	0.304E+01	0.154E+02	0.104E+02	F	F	F*=0.70				F	F
F	53	2.087	0.221E+04	0.342E+01	0.173E+02	0.116E+02	F	F	S*=0.70				S	F
F	56	2.205	0.246E+04	0.382E+01	0.193E+02	0.130E+02 55S	F		F*=0.70				F	F
S	60	2.362	0.283E+04	0.438E+01	0.222E+02	0.149E+02	F		F*=0.70				F	F
F	63	2.480	0.312E+04	0.483E+01	0.245E+02	0.164E+02	F		F*=0.70				S	F
S	65	2.559	0.332E+04	0.514E+01	0.260E+02	0.175E+02	S		S*=0.70				F	S
F	70	2.756	0.385E+04	0.597E+01	0.302E+02	0.203E+02	S		F*=0.70				F	F
S	75	2.953	0.442E+04	0.685E+01	0.347E+02	0.233E+02	F		S*=1.00				S	S
F	80	3.150	0.503E+04	0.779E+01	0.395E+02	0.265E+02	F		F*=1.00				F	F

STEEL MATERIAL DATA

Table 10-11. Hot Rolled Hexagon Steel Bars (EURONORM 61)

```
TABLE 10-11. HOT ROLLED HEXAGON                                              ORDER EXAMPLE:
STEEL BARS (EURONORM 61)                                                     LENGTH,HEXAGON STEEL 9.5 BS 4229 + ST QUAL      PAGE NO. 1
BASIS: 1 IN = 25.4 MM
1 CUBIC METER STEEL = 7850 KG MASS

                                                            THE NOMINAL SIZE IS NATIONAL STANDARD AS INDICATED
                                                            F=FIRST CHOICE,S=SECOND CHOICE,T=THIRD CHOICE,NUMBER=OTHER SIZE
                                                            *=COMMERCIAL SIZE
 E                                                          U.S.A.  AUSTRAL JAPAN           FRANCE  U.K.    GERMANY ITALY
 U                                                          ANSI    AS      JIS     EURO-   NF      BS      DIN     UNI
 R  NOMINAL SIZE = D   SECTION AREA    MASS PER UNIT        B32.4   1027    G3191   NORM    A45-006 4229    1015    7061
    MM    IN           MM**2    IN**2    KG/M      LB/FT                            61

    3.2   0.126        0.887E+01 0.137E-01 0.696E-01 0.468E-01  F
    4     0.157        0.139E+02 0.215E-01 0.109E+00 0.731E-01  F
    5     0.197        0.217E+02 0.336E-01 0.170E+00 0.114E+00  F
    5.5   0.217        0.262E+02 0.406E-01 0.206E+00 0.138E+00  F

    7     0.276        0.424E+02 0.658E-01 0.333E+00 0.224E+00  F
    8     0.315        0.554E+02 0.859E-01 0.435E+00 0.292E+00  F       F
    9     0.354        0.701E+02 0.109E+00 0.551E+00 0.370E+00                                  S
    9.5   0.374        0.782E+02 0.121E+00 0.614E+00 0.412E+00                                  F

   10     0.394        0.866E+02 0.134E+00 0.680E+00 0.457E+00  F                               S       F
   10.5   0.413        0.955E+02 0.148E+00 0.750E+00 0.504E+00                                  F       F
   11     0.433        0.105E+03 0.162E+00 0.823E+00 0.553E+00                                  S       F
   11.5   0.453        0.115E+03 0.178E+00 0.899E+00 0.604E+00                                  S       F

   12     0.472        0.125E+03 0.193E+00 0.979E+00 0.658E+00  F                               F       S
   12.5   0.492        0.135E+03 0.210E+00 0.106E+01 0.714E+00                                  S       F
   13     0.512        0.146E+03 0.227E+00 0.115E+01 0.772E+00                                  S       S
   13.5   0.531        0.158E+03 0.245E+00 0.124E+01 0.833E+00  F                               S       S

   14     0.551        0.170E+03 0.263E+00 0.133E+01 0.895E+00          F=±0.25                 F       F
   14.5   0.571        0.182E+03 0.282E+00 0.143E+01 0.960E+00          F=±0.25                 F       F
   15     0.591        0.195E+03 0.302E+00 0.153E+01 0.103E+01          F=±0.25                 F       F
   15.5   0.610        0.208E+03 0.322E+00 0.163E+01 0.110E+01                                  F       F

   16     0.630        0.222E+03 0.344E+00 0.174E+01 0.117E+01  F       F=±0.3                  F       F
   16.5   0.650        0.236E+03 0.365E+00 0.185E+01 0.124E+01          F=±0.3          F       S       F
   17     0.669        0.250E+03 0.388E+00 0.196E+01 0.132E+01          F=±0.3                  F       F
   17.5   0.689        0.265E+03 0.411E+00 0.208E+01 0.140E+01          F=±0.3                  F       F

   18     0.709        0.281E+03 0.435E+00 0.220E+01 0.148E+01  F       F=±0.3                  F       F
   18.5   0.728        0.296E+03 0.459E+00 0.233E+01 0.156E+01          F=±0.3                  F       F
   19     0.748        0.313E+03 0.485E+00 0.245E+01 0.165E+01          F=±0.3                  S       F
   19.5   0.768        0.329E+03 0.510E+00 0.259E+01 0.174E+01          F=±0.3                  F       F

   20     0.787        0.346E+03 0.537E+00 0.272E+01 0.183E+01  F       F=±0.3                  F       F
   20.5   0.807        0.364E+03 0.564E+00 0.286E+01 0.192E+01          F=±0.3                  S       F
   21     0.827        0.382E+03 0.592E+00 0.300E+01 0.201E+01          F=±0.3           S      S       F
   21.5   0.846        0.400E+03 0.620E+00 0.314E+01 0.211E+01          F=±0.3                  F       F

   22     0.866        0.419E+03 0.650E+00 0.329E+01 0.221E+01  F       F=±0.3                  F       F
   22.5   0.886        0.438E+03 0.680E+00 0.344E+01 0.231E+01          F=±0.3                  F       F
   23     0.906        0.458E+03 0.710E+00 0.360E+01 0.242E+01          F=±0.3                  F       F
   23.5   0.925        0.478E+03 0.741E+00 0.375E+01 0.252E+01          F=±0.3                  F       F

   24     0.945        0.499E+03 0.773E+00 0.392E+01 0.263E+01  F       F=±0.3                  F       F
   24.5   0.965        0.520E+03 0.806E+00 0.408E+01 0.274E+01          F=±0.3                  S       F
   25     0.984        0.541E+03 0.839E+00 0.425E+01 0.286E+01          F=±0.3                  F       F
```

Table 10-11 (Continued). Hot Rolled Hexagon Steel Bars (EURONORM 61)

```
TABLE 10-11. HOT ROLLED HEXAGON                          ORDER EXAMPLE:
STEEL BARS (EURONORM 61)                                 LENGTH,HEXAGON STEEL 9.5 BS 4229 + ST QUAL
BASIS: 1 IN = 25.4 MM
1 CUBIC METER STEEL = 7850 KG MAS

                                                 THE NOMINAL SIZE IS NATIONAL STANDARD AS INDICATED
                                                 F=FIRST CHOICE,S=SECOND CHOICE,T=THIRD CHOICE,NUMBER=OTHER SIZE
                                                 * = COMMERCIAL SIZE
                                                 U.S.A.    AUSTRAL JAPAN   EURO-   FRANCE   U.K.    GERMANY ITALY
                                                 ANSI      AS      JIS     NORM    NF       BS      DIN     UNI
                                                 B32.4     1027    G3191   61      A45-006  4229    1015    7061
```

EUR	NOMINAL SIZE = D		SECTION AREA		MASS PER UNIT		U.S.A	AUSTRAL	JAPAN	EURO-NORM 61	FRANCE	U.K.	GERMANY	ITALY
	MM	IN	MM**2	IN**2	KG/M	LB/FT								
F	25.5	1.004	0.563E+03	0.873E+00	0.442E+01	0.297E+01				F+-0.4	S	S	F	F
	26	1.024	0.585E+03	0.907E+00	0.460E+01	0.309E+01				+-0.4				
	27	1.063	0.631E+03	0.979E+00	0.496E+01	0.333E+01				+-0.4				
F	27.5	1.083	0.655E+03	0.102E+01	0.514E+01	0.345E+01				F+-0.4				
	28	1.102	0.679E+03	0.105E+01	0.533E+01	0.358E+01				+-0.4				
	28.5	1.122	0.703E+03	0.109E+01	0.552E+01	0.371E+01				+-0.4				
	29	1.142	0.728E+03	0.113E+01	0.572E+01	0.384E+01				+-0.4				
F	30	1.181	0.779E+03	0.121E+01	0.612E+01	0.411E+01			F	F+-0.4	F	S	F	F
F	30.5	1.201	0.806E+03	0.125E+01	0.632E+01	0.425E+01				F+-0.4				
	31	1.220	0.832E+03	0.129E+01	0.653E+01	0.439E+01				+-0.4				
	31.5	1.240	0.859E+03	0.133E+01	0.675E+01	0.453E+01				F+-0.4				
	32	1.260	0.887E+03	0.137E+01	0.696E+01	0.468E+01				F+-0.4				
F	33	1.299	0.943E+03	0.146E+01	0.740E+01	0.497E+01			F	F+-0.4	F	S	F	F
	33.5	1.319	0.972E+03	0.151E+01	0.763E+01	0.513E+01				+-0.4				
	34	1.339	0.100E+04	0.155E+01	0.786E+01	0.528E+01				F+-0.4				
	35	1.378	0.106E+04	0.164E+01	0.833E+01	0.560E+01				+-0.4				
F	36	1.417	0.112E+04	0.174E+01	0.881E+01	0.592E+01			F	F+-0.6	F	S	F	F
	37	1.457	0.119E+04	0.184E+01	0.931E+01	0.625E+01				+-0.6				
	37.5	1.476	0.122E+04	0.189E+01	0.956E+01	0.642E+01				+-0.6				
	38	1.496	0.125E+04	0.194E+01	0.982E+01	0.660E+01				+-0.6				
F	39	1.535	0.132E+04	0.204E+01	0.103E+02	0.695E+01				F+-0.6		S	F	F
	39.5	1.555	0.135E+04	0.209E+01	0.106E+02	0.713E+01				+-0.6				
	40	1.575	0.139E+04	0.215E+01	0.109E+02	0.731E+01				F+-0.6	F	S	F	F
	41	1.614	0.146E+04	0.226E+01	0.114E+02	0.768E+01				+-0.6		F		
F	42	1.654	0.153E+04	0.237E+01	0.120E+02	0.806E+01			F	F+-0.6	F	S	F	F
F	42.5	1.673	0.156E+04	0.242E+01	0.123E+02	0.825E+01				F+-0.6		F		
	43	1.693	0.160E+04	0.248E+01	0.126E+02	0.845E+01				+-0.6		S		
	44	1.732	0.168E+04	0.260E+01	0.132E+02	0.884E+01				+-0.6		S		
	45	1.772	0.175E+04	0.272E+01	0.138E+02	0.925E+01				+-0.6		F		
	46	1.811	0.183E+04	0.284E+01	0.144E+02	0.967E+01				+-0.6		F		
	47	1.850	0.191E+04	0.297E+01	0.150E+02	0.101E+02				+-0.6		F		
F	47.5	1.870	0.195E+04	0.303E+01	0.153E+02	0.103E+02				+-0.6	S	S	F	F
	48	1.890	0.200E+04	0.309E+01	0.157E+02	0.105E+02				+-0.6		F		
	49	1.929	0.208E+04	0.322E+01	0.163E+02	0.110E+02				+-0.6		F		
	50	1.969	0.217E+04	0.336E+01	0.170E+02	0.114E+02				+-0.6	S	S	F	F
	51	2.008	0.225E+04	0.349E+01	0.177E+02	0.119E+02				+-0.8		F		
F	52	2.047	0.234E+04	0.363E+01	0.184E+02	0.124E+02			F	F+-0.8	F	F	F	F
	53	2.087	0.243E+04	0.377E+01	0.191E+02	0.128E+02				+-0.8		F		
	54	2.126	0.253E+04	0.391E+01	0.198E+02	0.133E+02				+-0.8		S		
	55	2.165	0.262E+04	0.406E+01	0.206E+02	0.138E+02	F*			+-0.8		F		

STEEL MATERIAL DATA

Table 10-11 (Continued). Hot Rolled Hexagon Steel Bars (EURONORM 61)

```
TABLE 10-11. HOT ROLLED HEXAGON                              ORDER EXAMPLE:
STEEL BARS (EURONORM 61)                          PAGE NO. 3   LENGTH,HEXAGON STEEL 9.5 BS 4229 + ST QUAL
BASIS, 1 IN = 25.4 MM
1 CUBIC METER STEEL = 7850 KG MASS

                                                  THE NOMINAL SIZE IS NATIONAL STANDARD AS INDICATED
                                                  F=FIRST CHOICE,S=SECOND CHOICE,T=THIRD CHOICE,NUMBER=OTHER SIZE
                                                  * = COMMERCIAL SIZE
                                                  U.S.A.  AUSTRAL JAPAN  EURO-  FRANCE  U.K.    GERMANY ITALY
                                                  ANSI    AS.     JIS    NORM   NF      BS      DIN     UNI
                                                  B32.4   1027    G3191  61     A45-006 4229    1015    7061
E
U  NOMINAL SIZE = D   SECTION AREA    MASS PER UNIT
R  MM     IN          MM**2   IN**2   KG/M      LB/FT
   56    2.205        0.272E+04  0.421E+01  0.213E+02  0.143E+02          +-0.8   S
F  57    2.244        0.281E+04  0.436E+01  0.221E+02  0.148E+02          F+-0.8  S       F
   58    2.283        0.291E+04  0.452E+01  0.229E+02  0.154E+02          +-0.8   F
   59    2.323        0.301E+04  0.467E+01  0.237E+02  0.159E+02          +-0.8   S       F

   60    2.362        0.312E+04  0.483E+01  0.245E+02  0.164E+02          +-0.8   F
   61    2.402        0.322E+04  0.499E+01  0.253E+02  0.170E+02          +-0.8   S
F  62    2.441        0.333E+04  0.516E+01  0.261E+02  0.176E+02          F+-0.8  F       S
   63    2.480        0.344E+04  0.533E+01  0.270E+02  0.181E+02          +-0.8   S       F

   64    2.520        0.355E+04  0.550E+01  0.278E+02  0.187E+02          +-0.8   S       S
   65    2.559        0.366E+04  0.567E+01  0.287E+02  0.193E+02 F        +-0.8   F
   66    2.598        0.377E+04  0.585E+01  0.296E+02  0.199E+02          +-0.8   S       S
F  67    2.638        0.389E+04  0.603E+01  0.305E+02  0.205E+02          F+-0.8  F

   68    2.677        0.400E+04  0.621E+01  0.314E+02  0.211E+02          +-0.8   F
   69    2.717        0.412E+04  0.639E+01  0.324E+02  0.217E+02          +-0.8   S       S
   70    2.756        0.424E+04  0.658E+01  0.333E+02  0.224E+02                S    F
   71    2.795        0.437E+04  0.677E+01  0.343E+02  0.230E+02          +-0.8   F

F  72    2.835        0.449E+04  0.696E+01  0.352E+02  0.237E+02          F+-0.8  S       S
   73    2.874        0.462E+04  0.715E+01  0.362E+02  0.243E+02          +-0.8   F
   74    2.913        0.474E+04  0.735E+01  0.372E+02  0.250E+02          +-0.8   S
   75    2.953        0.487E+04  0.755E+01  0.382E+02  0.257E+02 F        +-0.8   F

F  78    3.071        0.527E+04  0.817E+01  0.414E+02  0.278E+02          F+-0.8  F       S
   80    3.150        0.554E+04  0.859E+01  0.435E+02  0.292E+02          +-0.8   F
F  83    3.268        0.597E+04  0.925E+01  0.468E+02  0.315E+02          F+-0.8  S       S
   85    3.346        0.626E+04  0.970E+01  0.491E+02  0.330E+02 F                F

   88    3.465        0.671E+04  0.104E+02  0.526E+02  0.354E+02                        S     S
   90    3.543        0.701E+04  0.109E+02  0.551E+02  0.370E+02                        F     S
   93    3.661        0.749E+04  0.116E+02  0.588E+02  0.395E+02                        S     S
   95    3.740        0.782E+04  0.121E+02  0.614E+02  0.412E+02                        F

   98    3.858        0.832E+04  0.129E+02  0.653E+02  0.439E+02                        S
  100    3.937        0.866E+04  0.134E+02  0.680E+02  0.457E+02                        F     S
  105    4.134        0.955E+04  0.148E+02  0.750E+02  0.504E+02 F                      S
  115    4.528        0.115E+05  0.178E+02  0.899E+02  0.604E+02 F                            103S

  130    5.118        0.146E+05  0.227E+02  0.115E+03  0.772E+02 F
  145    5.709        0.182E+05  0.282E+02  0.143E+03  0.960E+02 F
```

Table 10-12. Bright Finish Hexagon Steel Bars (DIN 176 Tolerances h 11 And h 12)

TABLE 10-12. BRIGHT FINISH HEXAGON STEEL
BARS (DIN 176 TOLERANCES h 11 AND h 12)
BASIS; 1 IN = 25.4 MM
1 CUBIC METER STEEL = 7850 KG MASS

ORDER EXAMPLE:
LENGTH,HEXAGON STEEL 10 DIN 176 + STEEL QUAL

THE NOMINAL SIZE IS NATIONAL STANDARD AS INDICATED
F=FIRST CHOICE,S=SECOND CHOICE,T=THIRD CHOICE,NUMBER=OTHER SIZE
*=COMMERCIAL SIZE

DIN	NOMINAL SIZE = D MM	IN	SECTION AREA MM**2	IN**2	MASS PER UNIT KG/M	LB/FT	U.S.A. ANSI B32.4	AUSTRAL AS 1027	JAPAN JIS G3123	EURO-NORM	FRANCE NF A47-413	U.K. BS 4229	GERMANY DIN 176	ITALY UNI 470
F	3.2	0.126	0.887E+01	0.137E-01	0.696E-01	0.468E-01							F+0-.075	F
F	3.5	0.138	0.106E+02	0.164E-01	0.833E-01	0.560E-01							F+0-.075	F
F	4	0.157	0.139E+02	0.215E-01	0.109E+00	0.731E-01							F+0-.075	F
F	4.5	0.177	0.175E+02	0.272E-01	0.138E+00	0.925E-01							F+0-.075	F
F	5	0.197	0.217E+02	0.336E-01	0.170E+00	0.114E+00							F+0-.075	F
F	5.5	0.217	0.262E+02	0.406E-01	0.206E+00	0.138E+00							F+0-.075	F
F	6	0.236	0.312E+02	0.483E-01	0.245E+00	0.164E+00							F+0-.075	F
F	7	0.276	0.424E+02	0.658E-01	0.333E+00	0.224E+00							F+0-.090	F
F	8	0.315	0.554E+02	0.859E-01	0.435E+00	0.292E+00				S			F+0-.090	F
F	9	0.354	0.701E+02	0.109E+00	0.551E+00	0.370E+00							F+0-.090	F
F	10	0.394	0.866E+02	0.134E+00	0.680E+00	0.457E+00				S			F+0-.090	F
F	11	0.433	0.105E+03	0.162E+00	0.823E+00	0.553E+00				T			F+0-.110	F
F	12	0.472	0.125E+03	0.193E+00	0.979E+00	0.658E+00				T			F+0-.110	F
S	13	0.512	0.146E+03	0.227E+00	0.115E+01	0.772E+00				F			F+0-.110	F
F	14	0.551	0.170E+03	0.263E+00	0.133E+01	0.895E+00							F+0-.110	F
F	15	0.591	0.195E+03	0.302E+00	0.152E+01	0.103E+01							F+0-.110	F
S	16	0.630	0.222E+03	0.344E+00	0.174E+01	0.117E+01				F			F+0-.110	F
F	17	0.669	0.250E+03	0.388E+00	0.196E+01	0.132E+01							F+0-.110	F
F	18	0.709	0.281E+03	0.435E+00	0.220E+01	0.148E+01							+0-.110	F
F	19	0.748	0.313E+03	0.485E+00	0.245E+01	0.165E+01							F+0-.130	F
S	21	0.827	0.382E+03	0.592E+00	0.300E+01	0.201E+01				F			F+0-.130	F
S	22	0.866	0.419E+03	0.650E+00	0.329E+01	0.221E+01				F			F+0-.130	F,20F
F	24	0.945	0.499E+03	0.773E+00	0.392E+01	0.263E+01				F			F+0-.130	F,23F
F	27	1.063	0.631E+03	0.979E+00	0.496E+01	0.333E+01			F,26F	F	S,23F		F+0-.130	F,25F
S	30	1.181	0.779E+03	0.121E+01	0.612E+01	0.411E+01				F	F,26F		F+0-.130	F,26F
F	32	1.260	0.887E+03	0.137E+01	0.696E+01	0.468E+01				F	F,29F		F+0-.160	F,28F
F	36	1.417	0.112E+04	0.174E+01	0.881E+01	0.592E+01				F	F,35F		F+0-.160	F
F	38	1.496	0.125E+04	0.194E+01	0.982E+01	0.660E+01				S	F		F+0-.160	F
F	41	1.614	0.146E+04	0.226E+01	0.114E+02	0.768E+01				F	S,40F		F+0-.160	F
F	46	1.811	0.183E+04	0.284E+01	0.144E+02	0.967E+01				F	S,42S		F+0-.160	F
F	50	1.969	0.217E+04	0.336E+01	0.170E+02	0.114E+02	F			F	S		F+0-.160	F
F	55	2.165	0.262E+04	0.406E+01	0.206E+02	0.138E+02				F	S,54T		F+0-.190	F
F	60	2.362	0.312E+04	0.483E+01	0.245E+02	0.164E+02			F	F	S,58T		F+0-.190	F
F	65	2.559	0.366E+04	0.567E+01	0.287E+02	0.193E+02				F	T		F+0-.190	F
F	70	2.756	0.424E+04	0.658E+01	0.332E+02	0.224E+02				F			F+0-.300	F
F	75	2.953	0.487E+04	0.755E+01	0.382E+02	0.257E+02				F			F+0-.300	F
F	80	3.150	0.554E+04	0.859E+01	0.435E+02	0.292E+02				F			F+0-.300	F
F	85	3.346	0.626E+04	0.970E+01	0.491E+02	0.330E+02							F+0-.350	F
F	90	3.543	0.701E+04	0.109E+02	0.551E+02	0.370E+02							F+0-.350	F
F	95	3.740	0.782E+04	0.121E+02	0.614E+02	0.412E+02							F+0-.350	F
F	100	3.937	0.866E+04	0.134E+02	0.680E+02	0.457E+02							F+0-.350	F
F	105	4.134	0.955E+04	0.148E+02	0.750E+02	0.504E+02							+0-.350	F

STEEL MATERIAL DATA

Table 10-13. Hot Rolled Square Steel Bars (ISO 1035/II)

```
TABLE 10-13. HOT ROLLED SQUARE                                                    ORDER EXAMPLE:                      PAGE NO. 1
STEEL BARS (ISO 1035/II)                                                          LENGTH,SQUARE STEEL 5 EURONORM 59 + ST QUAL
BASIS; 1 IN = 25.4 MM
1 CUBIC METER STEEL = 7850 KG MASS
                                                    THE NOMINAL SIZE IS NATIONAL STANDARD AS INDICATED
                                                    F=FIRST CHOICE,S=SECOND CHOICE,T=THIRD CHOICE,NUMBER=OTHER SIZE
                                                    * = COMMERCIAL SIZE
I                                                   U.S.A.   AUSTRAL  JAPAN   EURO-   FRANCE   U.K.    GERMANY  ITALY
S                       SECTION AREA   MASS PER UNIT ANSI    AS       JIS     NORM    NF       BS      DIN UNI
O  NOMINAL SIZE = D                                 B32.4    1027     G3191   59      A45-004  4229    1014    6013
   MM    IN          MM**2    IN**2   KG/M    LB/FT
   3     0.118       0.900E+01 0.140E-01 0.706E-01 0.475E-01 F
F  4     0.157       0.160E+02 0.248E-01 0.126E+00 0.844E-01 F
F  5     0.197       0.250E+02 0.388E-01 0.196E+00 0.132E+00 F                            D*=0.4
   6     0.236       0.360E+02 0.558E-01 0.283E+00 0.190E+00 F                            D*=0.4 S

S  7     0.276       0.490E+02 0.760E-01 0.385E+00 0.258E+00 F                            D*=0.4 T       F     F
F  8     0.315       0.640E+02 0.992E-01 0.502E+00 0.338E+00 F                            D*=0.4 F       F     F
S  9     0.354       0.810E+02 0.126E+00 0.636E+00 0.427E+00 F                            D*=0.4 F       F     F
S  10    0.394       0.100E+03 0.155E+00 0.785E+00 0.527E+00 F                            D*=0.4         F     F

S  11    0.433       0.121E+03 0.188E+00 0.950E+00 0.638E+00 F                            D*=0.4 T       S     F
F  12    0.472       0.144E+03 0.223E+00 0.113E+01 0.760E+00 F                            D*=0.4 F       F     F
S  13    0.512       0.169E+03 0.262E+00 0.133E+01 0.891E+00 F                            D*=0.4 F       F     F
F  14    0.551       0.196E+03 0.304E+00 0.154E+01 0.103E+01 S*                           D*=0.4 F       F     F

S  15    0.591       0.225E+03 0.349E+00 0.177E+01 0.119E+01 F                            D*=0.4 S       F     F
F  16    0.630       0.256E+03 0.397E+00 0.201E+01 0.135E+01 F                            D*=0.5 F       F     F
F  17    0.669       0.289E+03 0.448E+00 0.227E+01 0.152E+01 F                            D*=0.5 S       F     F
F  18    0.709       0.324E+03 0.502E+00 0.254E+01 0.171E+01 S*                           D*=0.5 F       F     F

S  19    0.748       0.361E+03 0.560E+00 0.283E+01 0.190E+01 F                            D*=0.5 S       F     F
F  20    0.787       0.400E+03 0.620E+00 0.314E+01 0.211E+01 F*                           D*=0.5 S       F     F
F  21    0.827       0.441E+03 0.684E+00 0.346E+01 0.233E+01 F*                           D*=0.5 S       F     F
   22    0.866       0.484E+03 0.750E+00 0.380E+01 0.255E+01 S*                           D*=0.5 F       F     F,21.5F F

S  23    0.906       0.529E+03 0.820E+00 0.415E+01 0.279E+01 *                            D*=0.5 S       S     F
F  24    0.945       0.576E+03 0.893E+00 0.452E+01 0.304E+01 F*                           D*=0.5 F       F     F
L  25    0.984       0.625E+03 0.969E+00 0.491E+01 0.330E+01 F*                           D*=0.6 S       F     F
S  26    1.024       0.676E+03 0.105E+01 0.531E+01 0.357E+01 S*                           D*=0.6 F       S     F

F  28    1.102       0.784E+03 0.122E+01 0.615E+01 0.414E+01 S*                           D*=0.6 T,27S   F     F,26.5F F
S  30    1.181       0.900E+03 0.140E+01 0.706E+01 0.475E+01 F*                           D*=0.6 F       F     F,29F F
F  32    1.260       0.102E+04 0.159E+01 0.804E+01 0.540E+01 F                            D*=0.6 S       F     F,33S F
   34    1.339       0.116E+04 0.179E+01 0.907E+01 0.610E+01 F                            D*=0.6 T       S     F

S  35    1.378       0.122E+04 0.190E+01 0.962E+01 0.646E+01 S*                           F D*=0.6 S     F     F
S  36    1.417       0.130E+04 0.201E+01 0.102E+02 0.684E+01 S*                           F D*=0.8 S     F     S,.37F F
S  38    1.496       0.144E+04 0.224E+01 0.113E+02 0.762E+01 F*                           F D*=0.8 F     F     F
F  40    1.575       0.160E+04 0.248E+01 0.126E+02 0.844E+01 F*                           F D*=0.8 F     F     F

S  42    1.654       0.176E+04 0.273E+01 0.138E+02 0.931E+01 F*                           F D*=0.8 T     F     F,.43F F
F  45    1.772       0.202E+04 0.314E+01 0.159E+02 0.107E+02 S*                           F D*=0.8 F     F,.44S F,.43F F
F  48    1.890       0.230E+04 0.357E+01 0.181E+02 0.122E+02 S*                           F D*=0.8 F     S,.46S F,.47F F
F  50    1.969       0.250E+04 0.388E+01 0.196E+02 0.132E+02 F*                           F D*=0.8 F     F     F

   52    2.047       0.270E+04 0.419E+01 0.212E+02 0.143E+02 F*                           F D*=1.0 S     S     F
L  54    2.126       0.292E+04 0.452E+01 0.229E+02 0.154E+02 F*                           F D*=1.0 F     F     F
F  55    2.165       0.302E+04 0.469E+01 0.237E+02 0.160E+02 S*                           F D*=1.0 F     F     F
   56    2.205       0.314E+04 0.486E+01 0.246E+02 0.165E+02                              F D*=1.0 S     F     F
```

10-37

Table 10-13 (Continued). Hot Rolled Square Steel Bars (ISO 1035/II)

```
TABLE 10-13. HOT ROLLED SQUARE                                              ORDER EXAMPLE:
STEEL BARS (ISO 1035/II)                                                    LENGTH,SQUARE STEEL 5 EURONORM 59 * ST QUAL
BASIS; 1 IN = 25.4 MM                                                                                            PAGE NO. 2
1 CUBIC METER STEEL = 7850 KG MASS

                                                         THE NOMINAL SIZE IS NATIONAL STANDARD AS INDICATED
                                                         F=FIRST CHOICE,S=SECOND CHOICE,T=THIRD CHOICE,NUMBER=OTHER SIZE
                                                         * = COMMERCIAL SIZE
I                                                        U.S.A.  AUSTRAL JAPAN   EURO-  FRANCE  U.K.   GERMANY ITALY
S                                                        ANSI    AS      JIS     NORM   NF      BS     DIN     UNI
O NOMINAL SIZE = D   SECTION AREA      MASS PER UNIT     B32.4   1027    G3191   59     A45-004 4229   1014    6013
  MM      IN         MM**2    IN**2    KG/M      LB/FT
  58      2.283      0.336E+04 0.521E+01 0.264E+02 0.177E+02  F*                    D*=1.0         S
F 60      2.362      0.360E+04 0.558E+01 0.283E+02 0.190E+02          F      D*=1.0         S      57S   F
  64      2.520      0.410E+04 0.635E+01 0.322E+02 0.216E+02                 D*=1.0         S,62S  63F  F
  65      2.559      0.422E+04 0.655E+01 0.332E+02 0.223E+02                 D*=1.0         F            F

  68      2.677      0.462E+04 0.717E+01 0.363E+02 0.244E+02                                S,66S        F
S 70      2.756      0.490E+04 0.760E+01 0.385E+02 0.258E+02  S*     F       D*=1.0  S                   F
  74      2.913      0.548E+04 0.849E+01 0.430E+02 0.289E+02                 D*=1.0         S,72S  73S   F
  75      2.953      0.562E+04 0.872E+01 0.442E+02 0.297E+02                 D*=1.0         F            F

F 80      3.150      0.640E+04 0.992E+01 0.502E+02 0.338E+02  F*             D*=1.0  S      F            F
  85      3.346      0.722E+04 0.112E+02 0.567E+02 0.381E+02                 D*=1.3         S      F,83S F
S 90      3.543      0.810E+04 0.126E+02 0.636E+02 0.427E+02  S*             D*=1.3  S      F            F
  95      3.740      0.902E+04 0.140E+02 0.708E+02 0.476E+02                 D*=1.3         S      93S   F

F 100     3.937      0.100E+05 0.155E+02 0.785E+02 0.527E+02  F*             D*=1.3  S      F            F
  105     4.134      0.110E+05 0.171E+02 0.865E+02 0.582E+02                 D*=1.5         F      103S  F
  110     4.331      0.121E+05 0.188E+02 0.950E+02 0.638E+02                 D*=1.5  T      F            F
  115     4.528      0.132E+05 0.205E+02 0.104E+03 0.698E+02                 D*=1.5         S            F

S 120     4.724      0.144E+05 0.223E+02 0.113E+03 0.760E+02  F*             D*=1.5  S      F            F
  125     4.921      0.156E+05 0.242E+02 0.123E+03 0.824E+02                 D*=2.0         S            F
S 130     5.118      0.169E+05 0.262E+02 0.133E+03 0.891E+02                 D*=2.0  T      F            F
  140     5.512      0.196E+05 0.304E+02 0.154E+03 0.103E+03  S*             D*=2.0         F            F

  150     5.906      0.225E+05 0.349E+02 0.177E+03 0.119E+03                 D*=2.0  T
  160     6.299      0.256E+05 0.397E+02 0.201E+03 0.135E+03                 D*=2.0
  180     7.087      0.324E+05 0.502E+02 0.254E+03 0.171E+03                 D*=2.5
  200     7.874      0.400E+05 0.620E+02 0.314E+03 0.211E+03                 D*=2.5

  220     8.661      0.484E+05 0.750E+02 0.380E+03 0.255E+03                 D*=3.0
  250     9.843      0.625E+05 0.969E+02 0.491E+03 0.330E+03                 D*=3.0
  280    11.024      0.784E+05 0.122E+03 0.615E+03 0.414E+03                 D*=3.0
  300    11.811      0.900E+05 0.140E+03 0.706E+03 0.475E+03                 D*=3.0

  320    12.598      0.102E+06 0.159E+03 0.804E+03 0.540E+03                 D*=3.0
```

STEEL MATERIAL DATA

Table 10-14. Bright Finish Square Steel Bars (DIN 178 Tolerances h 11 and h 12)

TABLE 10-14. BRIGHT FINISH SQUARE STEEL
BARS (DIN 178 TOLERANCES h11 AND h12)
BASIS; 1 IN = 25.4 MM
1 CUBIC METER STEEL = 7850 KG MASS

ORDER EXAMPLE:
LENGTH,SQUARE STEEL 3 DIN178 + STEEL QUALITY

THE NOMINAL SIZE IS NATIONAL STANDARD AS INDICATED
F=FIRST CHOICE,S=SECOND CHOICE,T=THIRD CHOICE,NUMBER=OTHER SIZE
* = COMMERCIAL SIZE

DIN	NOMINAL SIZE = D		SECTION AREA		MASS PER UNIT		U.S.A. ANSI B32.4	AUSTRAL AS 1027	JAPAN JIS G3123	EURO- NORM	FRANCE NF A47-12	U.K. BS 4229	GERMANY DIN 178	ITALY UNI 472
	MM	IN	MM**2	IN**2	KG/M	LB/FT								
F	3	0.118	0.900E+01	0.140E-01	0.706E-01	0.475E-01	F						F+0-.060	F
F	3.5	0.138	0.122E+02	0.190E-01	0.962E-01	0.646E-01							F+0-.075	F
F	4	0.157	0.160E+02	0.248E-01	0.126E+00	0.844E-01							F+0-.075	F
F	4.5	0.177	0.202E+02	0.314E-01	0.159E+00	0.107E+00							F+0-.075	F
F	5	0.197	0.250E+02	0.388E-01	0.196E+00	0.132E+00	F		F				F+0-.075	F
F	5.5	0.217	0.302E+02	0.469E-01	0.237E+00	0.160E+00							F+0-.075	F
F	6	0.236	0.360E+02	0.558E-01	0.283E+00	0.190E+00	F						F+0-.075	F
F	7	0.276	0.490E+02	0.760E-01	0.385E+00	0.258E+00							F+0-.090	F
F	8	0.315	0.640E+02	0.992E-01	0.502E+00	0.338E+00	F						F+0-.090	F
F	9	0.354	0.810E+02	0.126E+00	0.636E+00	0.427E+00							F+0-.090	F
F	10	0.394	0.100E+03	0.155E+00	0.785E+00	0.527E+00	F				T		F+0-.090	F
F	11	0.433	0.121E+03	0.188E+00	0.950E+00	0.638E+00							F+0-.110	F
F	12	0.472	0.144E+03	0.223E+00	0.113E+01	0.760E+00	F		F		T		F+0-.110	F
F	13	0.512	0.169E+03	0.262E+00	0.133E+01	0.891E+00							F+0-.110	F
F	14	0.551	0.196E+03	0.304E+00	0.154E+01	0.103E+01	S				T		F+0-.110	F
S	15	0.591	0.225E+03	0.349E+00	0.177E+01	0.119E+01							F+0-.110	F
F	16	0.630	0.256E+03	0.397E+00	0.201E+01	0.135E+01	F		F		T		F+0-.110	F,21F
S	17	0.669	0.289E+03	0.448E+00	0.227E+01	0.152E+01					S	F	S+0-.110	F
S	18	0.709	0.324E+03	0.502E+00	0.254E+01	0.171E+01	S				S	F	F+0-.110	F,23F
S	19	0.748	0.361E+03	0.560E+00	0.283E+01	0.190E+01					F		S+0-.130	F
F	20	0.787	0.400E+03	0.620E+00	0.314E+01	0.211E+01	F		F		S	F	F+0-.130	F,26F
F	22	0.866	0.484E+03	0.750E+00	0.380E+01	0.255E+01	S				F	F	S+0-.130	F
S	24	0.945	0.576E+03	0.893E+00	0.452E+01	0.304E+01					S	F	S+0-.130	F
F	25	0.984	0.625E+03	0.969E+00	0.491E+01	0.330E+01	F		F		S		F+0-.130	F
S	27	1.063	0.729E+03	0.113E+01	0.572E+01	0.385E+01	S		F		26T		S+0-.130	F
F	28	1.102	0.784E+03	0.122E+01	0.615E+01	0.414E+01	F		F			F	F+0-.160	F
S	30	1.181	0.900E+03	0.140E+01	0.706E+01	0.475E+01	F		F		S	F	S+0-.160	F
F	32	1.260	0.102E+04	0.159E+01	0.804E+01	0.540E+01	S		F		T		F+0-.160	F
S	35	1.378	0.122E+04	0.190E+01	0.962E+01	0.646E+01	S		F		34S		S+0-.160	F
F	36	1.417	0.130E+04	0.201E+01	0.102E+02	0.684E+01	F		F		T,38F	F	F+0-.160	F
S	40	1.575	0.160E+04	0.248E+01	0.126E+02	0.844E+01	F		F		43F	F,41S	F+0-.160	41F
F	45	1.772	0.202E+04	0.314E+01	0.159E+02	0.107E+02	S		F				F+0-.160	F
F	50	1.969	0.250E+04	0.388E+01	0.196E+02	0.132E+02	F		F		48F	F,46S	F+0-.160	F,46F
S	55	2.165	0.302E+04	0.469E+01	0.237E+02	0.160E+02	F		F			F	S+0-.190	F
S	60	2.362	0.360E+04	0.558E+01	0.283E+02	0.190E+02	S		F			F	S+0-.190	F
F	63	2.480	0.397E+04	0.615E+01	0.312E+02	0.209E+02			F				F+0-.190	F
F	65	2.559	0.422E+04	0.655E+01	0.332E+02	0.223E+02			F			F	T+0-.190	F
T	70	2.756	0.490E+04	0.760E+01	0.385E+02	0.258E+02	S		F			F	F+0-.300	F
T	75	2.953	0.562E+04	0.872E+01	0.442E+02	0.297E+02			F				T+0-.300	F
F	80	3.150	0.640E+04	0.992E+01	0.502E+02	0.338E+02			F				F+0-.300	F
F	90	3.543	0.810E+04	0.126E+02	0.636E+02	0.427E+02	S					F,85S	+0-.300	F
F	100	3.937	0.100E+05	0.155E+02	0.785E+02	0.527E+02	F					F	F+0-.350	F
F	120	4.724	0.144E+05	0.223E+02	0.113E+03	0.760E+02	F,110S						+0-.350	F

Table 10-15. Hot Rolled Rectangular Steel Bars (ISO 1035/III)

```
TABLE 10-15. HOT ROLLED RECTANGULAR                          ORDER EXAMPLE:
STEEL BARS (ISO 1035/III)                                      LENGTH,FLAT STEEL 10X3 EURONORM 58 + ST QUAL
BASIS: 1 IN = 25.4 MM
1 CUBIC METER STEEL = 7850 KG MASS
```

```
                                                THE NOMINAL SIZE IS NATIONAL STANDARD AS INDICATED
                                                F=FIRST CHOICE,S=SECOND CHOICE,T=THIRD CHOICE,NUMBER=OTHER SIZE
                                                * = COMMERCIAL SIZE
                                                U.S.A.  AUSTRAL  JAPAN  EURO-  FRANCE   U.K.   GERMANY ITALY
I    NOMINAL SIZE = AXB   SECTION AREA   MASS PER UNIT    ANSI    AS     JIS   NORM    NF       BS     DIN   UNI
S                                                         B32.3   1256   G3194   58    A45-005  4229   1017  6014
6    MM        IN        MM**2    IN**2   KG/M   LB/FT
F 10 x  3    0.394X0.118  0.300E+02  0.465E-01  0.235E+00  0.158E+00  F              F  A+=1.0 S  F      F      F
F       4                 0.400E+02  0.620E-01  0.314E+00  0.211E+00  F              F  B+=0.5 S  F      F      F
F       5                 0.500E+02  0.775E-01  0.392E+00  0.264E+00  F              F  B+=0.5 T  F      F      F
F       6    X0.197       0.600E+02  0.930E-01  0.471E+00  0.316E+00  F              F  B+=0.5    F      F      F
F       7    X0.236       0.700E+02  0.109E+00  0.549E+00  0.369E+00  F              F  B+=0.5    F      F      F
F       8    X0.276

F 12 X  3    0.472X0.118  0.360E+02  0.558E-01  0.283E+00  0.190E+00  F              F  A+=1.0 S  F      F      F
F       4    X0.157       0.480E+02  0.744E-01  0.377E+00  0.253E+00  F              F  B+=0.5 S  F      F      F
F       5    X0.197       0.600E+02  0.930E-01  0.471E+00  0.316E+00  F              F  B+=0.5 S  F      F      F
F       6    X0.236       0.720E+02  0.112E+00  0.565E+00  0.380E+00  F              F  B+=0.5 S  F      F      F
F       7    X0.276       0.840E+02  0.130E+00  0.659E+00  0.443E+00  F              F  B+=0.5 S  F      F      F
F       8    X0.315       0.960E+02  0.149E+00  0.754E+00  0.506E+00  F              F  B+=0.5    F      F      F

F 14 X  3    0.551X0.118  0.420E+02  0.651E-01  0.330E+00  0.222E+00  F              F  A+=1.0 F  F      F      F
F       4    X0.157       0.560E+02  0.868E-01  0.440E+00  0.295E+00  F              F  B+=0.5 S  F      F      F
F       5    X0.197       0.700E+02  0.109E+00  0.549E+00  0.369E+00  F              F  B+=0.5 S  F      F      F
F       6    X0.236       0.840E+02  0.130E+00  0.659E+00  0.443E+00  F              F  B+=0.5 S  F      F      F
F       7    X0.276       0.980E+02  0.152E+00  0.769E+00  0.517E+00  F              F  B+=0.5    F      F      F
F       8    X0.315       0.112E+03  0.174E+00  0.879E+00  0.591E+00  F              F  B+=0.5    F      F      S

F 16 X  3    0.630X0.118  0.480E+02  0.744E-01  0.377E+00  0.253E+00  F              F  A+=1.0 F  F      F      F
F       4    X0.157       0.640E+02  0.992E-01  0.502E+00  0.338E+00  F              F  B+=0.5 F  F      F      F
F       5    X0.197       0.800E+02  0.124E+00  0.628E+00  0.422E+00  F              F  B+=0.5 F  F      F      F
F       6    X0.236       0.960E+02  0.149E+00  0.754E+00  0.506E+00  F   F    F     F  B+=0.5 F  F      F,6.5S F
F       7    X0.276       0.112E+03  0.174E+00  0.879E+00  0.591E+00  F              F  B+=0.5 F  F      F,9F   F
F       8    X0.315       0.128E+03  0.198E+00  0.100E+01  0.675E+00  F              F  B+=0.5 F  F      F,11S  F
F      10    X0.394       0.160E+03  0.248E+00  0.126E+01  0.844E+00  F              F  B+=0.5 S  F      F      F

F 18 X  3    0.709X0.118  0.540E+02  0.837E-01  0.424E+00  0.285E+00  F              F  A+=1.0 F  F      F      F
F       4    X0.157       0.720E+02  0.112E+00  0.565E+00  0.380E+00  F              F  B+=0.5 F  F      F      F
F       5    X0.197       0.900E+02  0.140E+00  0.706E+00  0.475E+00  F              F  B+=0.5 F  F      F      F
F       6    X0.236       0.108E+03  0.167E+00  0.848E+00  0.570E+00  F              F  B+=0.5 S  F      F      F
F       7    X0.276       0.126E+03  0.195E+00  0.989E+00  0.665E+00  F              F  B+=0.5 S  F      T,6.5T F
F       8    X0.315       0.144E+03  0.223E+00  0.113E+01  0.760E+00  F              F  B+=0.5 S  F      F,9S   F
F      10    X0.394       0.180E+03  0.279E+00  0.141E+01  0.949E+00  F              F  B+=0.5 S  F      F,11T  F

F 20 X  3    0.787X0.118  0.600E+02  0.930E-01  0.471E+00  0.316E+00  F              F  A+=1.0 F  F      F      F
F       4    X0.157       0.800E+02  0.124E+00  0.628E+00  0.422E+00  F              F  B+=0.5 F  F      F      F
F       5    X0.197       0.100E+03  0.155E+00  0.785E+00  0.527E+00  F              F  B+=0.5 F  F      F      F
F       6    X0.236       0.120E+03  0.186E+00  0.942E+00  0.633E+00  F              F  B+=0.5 F  F,6.5S F,6.5F F
F       7    X0.276       0.140E+03  0.217E+00  0.110E+01  0.738E+00  F              F  B+=0.5 F  F      F,9F   F
F       8    X0.315       0.160E+03  0.248E+00  0.126E+01  0.844E+00  F              F  B+=0.5 F  F      F      F
F      10    X0.394       0.200E+03  0.310E+00  0.157E+01  0.105E+01  F              F  B+=0.5 F  F      F      F
F      12    X0.472       0.240E+03  0.372E+00  0.188E+01  0.127E+01  F              F  A+=0.5 F  F      F      F
F      15    X0.591       0.300E+03  0.465E+00  0.235E+01  0.158E+01  14S            F  B+=0.5 14F F      F,13F  F
F      16    X0.630       0.320E+03  0.496E+00  0.251E+01  0.169E+01  F              F  B+=0.5 F  F      F      F
F      18    X0.709       0.360E+03  0.558E+00  0.283E+01  0.190E+01  S              F  B+=0.5    F      F      F

F 22 X  3    0.866E+00    0.660E+02  0.102E+00  0.518E+00  0.348E+00  F              F  A+=1.0 F  F      F      F
F       4    X0.157       0.880E+02  0.136E+00  0.691E+00  0.454E+00  F              F  B+=0.5 F  F      F      F
F       5    X0.197       0.110E+03  0.171E+00  0.863E+00  0.580E+00  F              F  B+=0.5 T  F      F      F
F       6    X0.236       0.132E+03  0.205E+00  0.104E+01  0.696E+00  F              F  A+=0.5 T  F      F      F
```

10-40   STEEL MATERIAL DATA   PAGE NO. 1

STEEL MATERIAL DATA

Table 10-15 (Continued). Hot Rolled Rectangular Steel Bars (ISO 1035/III)

```
TABLE 10-15. HOT ROLLED RECTANGULAR                                    ORDER EXAMPLE;
STEEL BARS (ISO 1035/III)                                              LENGTH,FLAT STEEL 10X3 EURONORM 58 + ST QUAL              PAGE NO. 2
BASIS, 1 IN = 25.4 MM
1 CUBIC METER STEEL = 7850 KG MASS

                                                              THE NOMINAL SIZE IS NATIONAL STANDARD AS INDICATED
                                                              F=FIRST CHOICE,S=SECOND CHOICE,T=THIRD CHOICE,NUMBER=OTHER SIZE
                                                              * = COMMERCIAL SIZE
                                                              U.S.A.   AUSTRAL  JAPAN    EURO-    FRANCE    U.K.    GERMANY  ITALY
I                                                             ANSI     AS       JIS      NORM     NF        BS      DIN      UNI
S  NOMINAL SIZE = AXB  SECTION AREA    MASS PER UNIT          B32.3    1256     G3194    58       A45-005   4229    1017     6014
O     MM      IN       MM**2    IN**2  KG/M       LB/FT

F  22 X  7   0.866X0.276  0.154E+03  0.239E+00  0.121E+01  0.812E+00                             A+-1.0           F,6.5S       F
F         8               0.176E+03  0.273E+00  0.138E+01  0.928E+00           F                 B+-0.5      F    F            F
F        10               0.220E+03  0.341E+00  0.173E+01  0.116E+01           F                 B+-0.5      F    F            F
F        12               0.264E+03  0.409E+00  0.207E+01  0.139E+01           F                 B+-0.5  T   F    F,11S        F
F        15               0.330E+03  0.512E+00  0.259E+01  0.174E+01           F                 B+-0.5      F    S,13F        F
F        18               0.396E+03  0.614E+00  0.311E+01  0.209E+01  16F      F                 B+-0.5      F    17S,14F      F

F  25 X  3   0.984X0.118  0.750E+02  0.116E+00  0.589E+00  0.396E+00                             A+-1.0      F    F            F
F         5               0.125E+03  0.194E+00  0.981E+00  0.659E+00           F                 B+-0.5      F    F            F
F         6               0.150E+03  0.233E+00  0.118E+01  0.791E+00           F     4.5F        B+-0.5      F    F,6.5F       F
F         7               0.175E+03  0.271E+00  0.137E+01  0.923E+00           F     F           B+-0.5  S,6.5T F  F            F
F         8               0.200E+03  0.310E+00  0.157E+01  0.105E+01           F     F           B+-0.5      F    F            F
F        10               0.250E+03  0.388E+00  0.196E+01  0.132E+01           F     9F          B+-0.5      F    F            F
F        12               0.300E+03  0.465E+00  0.235E+01  0.158E+01           F                 B+-0.5      F    S,13F        F
F        14               0.350E+03  0.543E+00  0.275E+01  0.185E+01  S        F                 B+-0.5      F    F            F
F        15               0.375E+03  0.581E+00  0.294E+01  0.198E+01           F                 B+-0.5      F    F            F
F        16               0.400E+03  0.620E+00  0.314E+01  0.211E+01           F                 B+-0.5      F    F            F
F        18               0.450E+03  0.698E+00  0.353E+01  0.237E+01  S        F                 B+-0.5      F    F            F

F  28 X  3   1.102X0.118  0.840E+02  0.130E+00  0.659E+00  0.443E+00                             A+-1.0      F                 F
F         4               0.112E+03  0.174E+00  0.879E+00  0.591E+00                             B+-0.5      F    S            F
F         5               0.140E+03  0.217E+00  0.110E+01  0.738E+00                             B+-0.5      F    S            F
F         6               0.168E+03  0.260E+00  0.132E+01  0.886E+00                             B+-0.5  S   F    S,6.5S       F
F         7               0.196E+03  0.304E+00  0.154E+01  0.103E+01                             B+-0.5      F    F            F
F         8               0.224E+03  0.347E+00  0.176E+01  0.118E+01                             B+-0.5      F    F            F
F        10               0.280E+03  0.434E+00  0.220E+01  0.148E+01                             B+-0.5      F    F            F
F        12               0.336E+03  0.521E+00  0.264E+01  0.177E+01                             B+-0.5      F    14F,13S      F
F        15               0.420E+03  0.651E+00  0.330E+01  0.222E+01                             B+-0.5      F    S,16S        F
F        18               0.504E+03  0.781E+00  0.396E+01  0.266E+01                             B+-0.5      F    F            F

F  30 X  3   1.181X0.118  0.900E+02  0.140E+00  0.706E+00  0.475E+00                             A+-1.0      F    F            F
F         4               0.120E+03  0.186E+00  0.942E+00  0.633E+00           F                 B+-0.5      F    F            F
F         5               0.150E+03  0.233E+00  0.118E+01  0.791E+00           F                 B+-0.5      F    F,6.5F       F
F         6               0.180E+03  0.279E+00  0.141E+01  0.949E+00           F                 B+-0.5  S,6.5T F  F,6.5F       F
F         7               0.210E+03  0.326E+00  0.165E+01  0.111E+01           F                 B+-0.5      F    F            F
F         8               0.240E+03  0.372E+00  0.188E+01  0.127E+01           F                 B+-0.5      F    F,9F         F
F        10               0.300E+03  0.465E+00  0.235E+01  0.158E+01           F                 B+-0.5  F,9S F    F            F
F        12               0.360E+03  0.558E+00  0.283E+01  0.190E+01           F                 B+-0.5      F    S,13F        F
F        14               0.420E+03  0.651E+00  0.330E+01  0.222E+01           F                 B+-0.5      F    F            F
F        15               0.450E+03  0.698E+00  0.353E+01  0.237E+01           F                 B+-0.5      F    S            F
F        16               0.480E+03  0.744E+00  0.377E+01  0.253E+01           F                 B+-0.5      F    S            F
F        18               0.540E+03  0.837E+00  0.424E+01  0.285E+01           S                 B+-0.5      F    S            F
F        20               0.600E+03  0.930E+00  0.471E+01  0.316E+01           F                 B+-0.5  T   F    S            F
F        25               0.750E+03  0.116E+01  0.589E+01  0.396E+01           F                 B+-1.0      F    F,22S        F

F  32 X  3   1.260X0.118  0.960E+02  0.149E+00  0.754E+00  0.506E+00                             A+-1.0      F    F            F
F         4               0.128E+03  0.198E+00  0.100E+01  0.675E+00           F     4.5F        B+-0.5  T   F    F            F
F         5               0.160E+03  0.248E+00  0.126E+01  0.844E+00           F     F           B+-0.5  T   F    S            F
F         6               0.192E+03  0.298E+00  0.151E+01  0.101E+01           F                 B+-0.5  T   F    6.5F         F
F         7               0.224E+03  0.347E+00  0.176E+01  0.118E+01           F                 B+-0.5  6.5T F    F            F
F         8               0.256E+03  0.397E+00  0.201E+01  0.135E+01           F                 B+-0.5  T   F    F            F
```

10-41

10-42   STEEL MATERIAL DATA

Table 10-15 (*Continued*). Hot Rolled Rectangular Steel Bars (ISO 1035/III)

```
TABLE 10-15. HOT ROLLED RECTANGULAR                              ORDER EXAMPLE;
STEEL BARS (ISO 1035/III)                                        LENGTH,FLAT STEEL 10X3 EURONORM 58 + ST QUAL
BASIS; 1 IN = 25.4 MM                                                                                   PAGE NO. 3
1 CUBIC METER STEEL = 7850 KG MASS

                                                  THE NOMINAL SIZE IS NATIONAL STANDARD AS INDICATED
                                                  F=FIRST CHOICE,S=SECOND CHOICE,T=THIRD CHOICE,NUMBER=OTHER SIZE
                                                  * = COMMERCIAL SIZE
I                                                 U.S.A.   AUSTRAL  JAPAN   EURO-   FRANCE   U.K.    GERMANY  ITALY
S                                                 ANSI     AS       JIS     NORM    NF       BS      DIN      UNI
O  NOMINAL SIZE = AXB  SECTION AREA  MASS PER UNIT B32.3   1256     G3194    58     A45-005  4229    1017     6014
   MM         IN        MM**2   IN**2    KG/M  LB/FT

F  32 X 10  1.260X0.394  0.320E+03  0.496E+00  0.251E+01  0.169E+01                        A+=1.0 T           S,13T  F
F       12          X0.472  0.384E+03  0.595E+00  0.301E+01  0.203E+01                 B+=0.5 T           S,14S  F
F       15          X0.591  0.480E+03  0.744E+00  0.377E+01  0.253E+01                 B+=0.5                 S       F
F       16          X0.630  0.512E+03  0.794E+00  0.402E+01  0.270E+01                 B+=0.5                 S       F
F       18          X0.709  0.576E+03  0.893E+00  0.452E+01  0.304E+01                 B+=0.5                 S       F
F       20          X0.787  0.640E+03  0.992E+00  0.502E+01  0.338E+01                 B+=0.5                 F,22S   F
F       25          X0.984  0.800E+03  0.124E+01  0.628E+01  0.422E+01                 B+=1.0

F  35 X  3  1.378X0.118  0.105E+03  0.163E+00  0.824E+00  0.554E+00                        A+=1.0                    F
F        4          X0.157  0.140E+03  0.217E+00  0.110E+01  0.738E+00                 B+=0.5                        F
F        5          X0.197  0.175E+03  0.271E+00  0.137E+01  0.923E+00                 B+=0.5                        F
F        6          X0.236  0.210E+03  0.326E+00  0.165E+01  0.111E+01                 B+=0.5        F,6.5F  F
F        7          X0.276  0.245E+03  0.380E+00  0.192E+01  0.129E+01                 B+=0.5                 F        F
F        8          X0.315  0.280E+03  0.434E+00  0.220E+01  0.148E+01                 B+=0.5                 F        F
F       10          X0.394  0.350E+03  0.543E+00  0.275E+01  0.185E+01                 B+=0.5                 S        F
F       12          X0.472  0.420E+03  0.651E+00  0.330E+01  0.222E+01                 A+=1.0                 F        F
F       14          X0.551  0.490E+03  0.760E+00  0.385E+01  0.258E+01                 B+=0.5                 S,13S   F
F       15          X0.591  0.525E+03  0.814E+00  0.412E+01  0.277E+01                 B+=0.5                 F        F
F       16          X0.630  0.560E+03  0.868E+00  0.440E+01  0.295E+01                 A+=1.0                 F        F
F       18          X0.709  0.630E+03  0.977E+00  0.495E+01  0.332E+01                 B+=0.5                 F        F
F       20          X0.787  0.700E+03  0.109E+01  0.549E+01  0.369E+01                 B+=0.5                 F        F
F       25          X0.984  0.875E+03  0.136E+01  0.687E+01  0.462E+01                 B+=1.0        22T              F
F       30          X1.181  0.105E+04  0.163E+01  0.824E+01  0.554E+01                 A+=1.0                 F,22S   F
F       32          X1.260  0.112E+04  0.174E+01  0.879E+01  0.591E+01                 B+=1.0                 F        F

F  40 X  3  1.575X0.118  0.120E+03  0.186E+00  0.942E+00  0.633E+00                        A+=1.0                    F
F        4          X0.157  0.160E+03  0.248E+00  0.126E+01  0.844E+00                 B+=0.5                        F
F        5          X0.197  0.200E+03  0.310E+00  0.157E+01  0.105E+01                 B+=0.5                        F
F        6          X0.236  0.240E+03  0.372E+00  0.188E+01  0.127E+01                 B+=0.5                        F
F        7          X0.276  0.280E+03  0.434E+00  0.220E+01  0.148E+01                 B+=0.5        F,6.5T          F
F        8          X0.315  0.320E+03  0.496E+00  0.251E+01  0.169E+01   38X4.5F      B+=0.5                        F
F       10          X0.394  0.400E+03  0.620E+00  0.314E+01  0.211E+01   38X6F        B+=0.5        F,6.5T  F,6.5S
F       12          X0.472  0.480E+03  0.744E+00  0.377E+01  0.253E+01   38X8F        B+=0.5                F,9F    F
F       14          X0.551  0.560E+03  0.868E+00  0.440E+01  0.295E+01   38X9F        B+=0.5        F,11T           F
F       15          X0.591  0.600E+03  0.930E+00  0.471E+01  0.316E+01   38X12F       B+=0.5                F,13S   F
F       16          X0.630  0.640E+03  0.992E+00  0.502E+01  0.338E+01                 A+=1.0                 F        F
F       18          X0.709  0.720E+03  0.112E+01  0.565E+01  0.380E+01   38X16F       B+=0.5                 F        F
F       20          X0.787  0.800E+03  0.124E+01  0.628E+01  0.422E+01   38X19F       B+=0.5                 F,22S   F
F       25          X0.984  0.100E+04  0.155E+01  0.785E+01  0.527E+01                 B+=1.0                 F        F
F       30          X1.181  0.120E+04  0.186E+01  0.942E+01  0.633E+01                 A+=1.0                 F        F
F       32          X1.260  0.128E+04  0.198E+01  0.100E+02  0.675E+01                 B+=1.0                 F        F

F  45 X  4  1.772X0.157  0.180E+03  0.279E+00  0.141E+01  0.949E+00   3F               A+=1.0 T          F,3F          F
F        5          X0.197  0.225E+03  0.349E+00  0.177E+01  0.119E+01        F        B+=0.5                        F
F        6          X0.236  0.270E+03  0.419E+00  0.212E+01  0.143E+01                 B+=0.5                        F
F        7          X0.276  0.315E+03  0.488E+00  0.247E+01  0.166E+01                 B+=0.5                        F
F        8          X0.315  0.360E+03  0.558E+00  0.283E+01  0.190E+01   44X4.5F      B+=0.5                F,6.5S   F
F       10          X0.394  0.450E+03  0.698E+00  0.353E+01  0.237E+01   44X6F        B+=0.5                        F
F       12          X0.472  0.540E+03  0.837E+00  0.424E+01  0.285E+01   44X8F        B+=0.5                 S        F
F       14          X0.551  0.630E+03  0.977E+00  0.495E+01  0.332E+01   44X9F        B+=0.5                 F        F
F       15          X0.591  0.675E+03  0.105E+01  0.530E+01  0.356E+01   44X12F       A+=1.0 T               F,13F   F
F       16          X0.630  0.720E+03  0.112E+01  0.565E+01  0.380E+01   44X16F  F    B+=-0.5 T              S        F
```

STEEL MATERIAL DATA

10-43

TABLE 10-15. HOT ROLLED RECTANGULAR
STEEL BARS (ISO 1035/III)
BASIS, 1 IN = 25.4 MM
1 CUBIC METER STEEL = 7850 KG MASS

ORDER EXAMPLE:
LENGTH,FLAT STEEL 10X3 EURONORM 58 + ST QUAL

Table 10-15 (Continued). Hot Rolled Rectangular Steel Bars (ISO 1035/III)

THE NOMINAL SIZE IS NATIONAL STANDARD AS INDICATED
F=FIRST CHOICE, S=SECOND CHOICE, T=THIRD CHOICE, NUMBER=OTHER SIZE
+ = COMMERCIAL SIZE

I S O	NOMINAL SIZE = AXB		SECTION AREA		MASS PER UNIT		U.S.A. ANSI B32.3	AUSTRAL AS 1256	JAPAN JIS G3194	EURO-NORM 58	FRANCE NF A45-005	U.K. BS 4229	GERMANY DIN 1017	ITALY UNI 6014
	MM	IN	MM**2	IN**2	KG/M	LB/FT								
F 45 X	18	1.772 X 0.709	0.810E+03	0.126E+01	0.636E+01	0.427E+01	T			A+=1.0 T		F	F	F
F	20	X0.787	0.900E+03	0.140E+01	0.706E+01	0.475E+01	S	F	44X19F	F B+=0.5 S	F	F	F,22F	F
F	25	X0.984	0.113E+04	0.174E+01	0.883E+01	0.593E+01	S	F		F B+=1.0 T	F	F	F	F
F	30	X1.181	0.135E+04	0.209E+01	0.106E+02	0.712E+01	S			F B+=1.0 T	F	F	F	F
F	32	X1.260	0.144E+04	0.223E+01	0.113E+02	0.760E+01	T			A+=1.0	F	F	F	F
F 50 X	3	1.969 X 0.118	0.150E+03	0.233E+00	0.118E+01	0.791E+00	F			F B+=1.0	F	F	F	F
F	4	X0.157	0.200E+03	0.310E+00	0.157E+01	0.105E+01	F			F B+=0.5	F	F	F	F
F	5	X0.197	0.250E+03	0.388E+00	0.196E+01	0.132E+01	F	F		F B+=0.5	F	F	F	F
F	6	X0.236	0.300E+03	0.465E+00	0.235E+01	0.158E+01	F	F	4.5F	F B+=0.5	F	F	F,6.5S	F
F	7	X0.276	0.350E+03	0.543E+00	0.275E+01	0.185E+01	F		F	F B+=0.5 S,6.5T	F	F	F	F
F	8	X0.315	0.400E+03	0.620E+00	0.314E+01	0.211E+01	F	F	F	F B+=0.5	F	F	F	F
F	9	X0.354	0.450E+03	0.698E+00	0.353E+01	0.238E+01	F		9F	F B+=0.5	F	F	F	F
F	10	X0.394	0.500E+03	0.775E+00	0.392E+01	0.264E+01	F	F	F	F B+=0.5	F	F	S,9S	F
F	12	X0.472	0.600E+03	0.930E+00	0.471E+01	0.316E+01	F	F	F	F B+=0.5 F,13T	F	F	S,13S	F
F	14	X0.551	0.700E+03	0.109E+01	0.549E+01	0.369E+01	S			F B+=0.5	F	F	S	F
F	15	X0.591	0.750E+03	0.116E+01	0.589E+01	0.396E+01	S			B+=0.5	F	F	S	F
F	16	X0.630	0.800E+03	0.124E+01	0.628E+01	0.422E+01	S			F B+=0.5	F	F	S	F
F	18	X0.709	0.900E+03	0.140E+01	0.706E+01	0.475E+01	S		19F	F B+=0.5	F	F	S	F
F	20	X0.787	0.100E+04	0.155E+01	0.785E+01	0.527E+01	S	F	F	F B+=1.0	F	F	S	F
F	22	X0.866	0.110E+04	0.171E+01	0.863E+01	0.580E+01	S			F B+=1.0	F	F		F
F	25	X0.984	0.125E+04	0.194E+01	0.981E+01	0.659E+01	S			B+=1.0	F	F	F	F
F	30	X1.181	0.150E+04	0.233E+01	0.118E+02	0.791E+01	S			B+=1.0	F	F	F	F
F	32	X1.260	0.160E+04	0.248E+01	0.126E+02	0.844E+01	S			B+=1.0	F	F	F	F
F	35	X1.378	0.175E+04	0.271E+01	0.137E+02	0.923E+01	S			B+=1.0	F	F	F	F
F	40	X1.575	0.200E+04	0.310E+01	0.157E+02	0.105E+02	S			B+=1.0	F	F	F	F
F 55 X	4	2.165 X 0.157	0.220E+03	0.341E+00	0.173E+01	0.116E+01		3F		F A+=1.0	F	F	F	F
F	5	X0.197	0.275E+03	0.426E+00	0.216E+01	0.145E+01		F		F B+=0.5 T	F	F	F	F
F	6	X0.236	0.330E+03	0.512E+00	0.259E+01	0.174E+01		F		F B+=0.5	F	F	6.5S	F
F	7	X0.276	0.385E+03	0.597E+00	0.302E+01	0.203E+01				F B+=0.5 T	F	F	F	F
F	8	X0.315	0.440E+03	0.682E+00	0.345E+01	0.232E+01		F		F B+=0.5 T	F	F	F	F
F	10	X0.394	0.550E+03	0.853E+00	0.432E+01	0.290E+01		F		F B+=0.5 T	F	F	F	F
F	12	X0.472	0.660E+03	0.102E+01	0.518E+01	0.348E+01				F B+=0.5 T	F	F	F,13F	F
F	14	X0.551	0.825E+03	0.128E+01	0.471E+01	0.316E+01				F B+=0.5 T	F	F	F,14S	F
F	15	X0.591	0.990E+03	0.153E+01	0.644E+01	0.435E+01		16F		F A+=0.5 T 16T	F	F	F,16S	F
F	18	X0.709	0.110E+04	0.171E+01	0.777E+01	0.522E+01				F B+=1.0	F	F	F	F
F	20	X0.787	0.138E+04	0.213E+01	0.863E+01	0.580E+01	S			F B+=1.0	F	F	S,22S	F
F	25	X0.984	0.165E+04	0.256E+01	0.108E+02	0.725E+01				F B+=1.0	F	F	F	F
F	30	X1.181	0.176E+04	0.273E+01	0.130E+02	0.870E+01	F,40F			F B+=1.0	F	F	F	F
F	32	X1.260			0.138E+02	0.928E+01				F B+=1.0	F	F	F	F
F 60 X	4	2.362 X 0.157	0.240E+03	0.372E+00	0.188E+01	0.127E+01	F			F A+=1.0	F	F	F	F
F	5	X0.197	0.300E+03	0.465E+00	0.235E+01	0.158E+01	F			F B+=0.5	F	F	F	F
F	6	X0.236	0.360E+03	0.558E+00	0.283E+01	0.190E+01	F			F B+=0.5	F	F	F,6.5S	F
F	7	X0.276	0.420E+03	0.651E+00	0.330E+01	0.222E+01	F			F B+=0.5 T,6.5T	F	F	F	F
F	8	X0.315	0.480E+03	0.744E+00	0.377E+01	0.253E+01	F			F B+=0.5	F	F	F	F
F	10	X0.394	0.600E+03	0.930E+00	0.471E+01	0.316E+01	F			F B+=0.5	F	F	F,9F	F
F	12	X0.472	0.720E+03	0.112E+01	0.565E+01	0.380E+01	F			F B+=0.5	F	F	F	F
F	14	X0.551	0.840E+03	0.130E+01	0.659E+01	0.443E+01	S			F B+=0.5	F	F	13F	F
F	15	X0.591	0.900E+03	0.140E+01	0.706E+01	0.475E+01	F			F B+=0.5	F	F	F	F
F	16	X0.630	0.960E+03	0.149E+01	0.754E+01	0.506E+01	F			F B+=0.5	F	F	F	F
F	18	X0.709	0.108E+04	0.167E+01	0.848E+01	0.570E+01	S			F B+=0.5	F	F	S	F
F	20	X0.787	0.120E+04	0.186E+01	0.942E+01	0.633E+01	F			F A+=0.5	F	F	8	F

Table 10-15 (Continued). Hot Rolled Rectangular Steel Bars (ISO 1035/III)

TABLE 10-15. HOT ROLLED RECTANGULAR
STEEL BARS (ISO 1035/III)
BASIS: 1 IN = 25.4 MM
1 CUBIC METER STEEL = 7850 KG MASS

ORDER EXAMPLE:
LENGTH,FLAT STEEL 10X3 EURONORM 58 + ST QUAL
PAGE NO. 5

THE NOMINAL SIZE IS NATIONAL STANDARD AS INDICATED
F=FIRST CHOICE, S=SECOND CHOICE, T=THIRD CHOICE, NUMBER=OTHER SIZE
* = COMMERCIAL SIZE

I S O	NOMINAL SIZE = A×B MM / IN	SECTION AREA MM**2 / IN**2	MASS PER UNIT KG/M / LB/FT	U.S.A. ANSI B32.3	AUSTRAL AS 1256	JAPAN JIS G3194	EURO- NORM 58	FRANCE NF A45-005	U.K. BS 4229	GERMANY DIN 1017	ITALY UNI 6014
F	60 × 25	2.362×0.984	0.150E+04 0.233E+01	0.118E+02 0.791E+01	F			F A+-1.0 S	F	F,22F	F
F	30	X1.181	0.180E+04 0.279E+01	0.141E+02 0.949E+01	F			B+-1.0	F	S	F
F	32	X1.260	0.192E+04 0.298E+01	0.151E+02 0.101E+02	S			B+-1.0 T	F	F	F
F	35	X1.378	0.210E+04 0.326E+01	0.165E+02 0.111E+02	F			B+-1.0 S	F	F	F
F	40	X1.575	0.240E+04 0.372E+01	0.188E+02 0.127E+02	F			B+-1.0 S	F	50T	F
F	45	X1.772	0.270E+04 0.419E+02	0.212E+02 0.142E+02	S			B+-1.5	F	F	F
F	65 × 4	2.559×0.157	0.260E+03 0.403E+00	0.204E+01 0.137E+01	3F			F A+-1.0 S	F	F,22F	F
F	5	X0.197	0.325E+03 0.504E+00	0.255E+01 0.171E+01	F			B+-0.5	F	S	F
F	6	X0.236	0.390E+03 0.605E+00	0.306E+01 0.206E+01	F			B+-0.5	F	F	F
F	7	X0.276	0.455E+03 0.705E+00	0.357E+01 0.240E+01	F			B+-0.5 T	F	6.5F	F
F	8	X0.315	0.520E+03 0.806E+00	0.408E+01 0.274E+01	F			B+-0.5 S	F	F,9S	F
F	10	X0.394	0.650E+03 0.101E+01	0.510E+01 0.343E+01	F			B+-0.5 S	F	F	F
F	12	X0.472	0.780E+03 0.121E+01	0.612E+01 0.411E+01	F		9F	B+-0.5 T	F	F	F
F	15	X0.591	0.975E+03 0.151E+01	0.765E+01 0.514E+01	F			B+-0.5 S	F	F,13S	F
F	16	X0.630	0.104E+04 0.161E+01	0.816E+01 0.549E+01	F			B+-0.5	F	S	F
F	18	X0.709	0.117E+04 0.181E+01	0.918E+01 0.617E+01	F			B+-0.5 T	F	F	F
F	20	X0.787	0.130E+04 0.202E+01	0.102E+02 0.686E+01	F		19F	B+-0.5 T	F	S,22S	F
F	25	X0.984	0.163E+04 0.252E+01	0.128E+02 0.857E+01	F		F,22F	B+-1.0 T	F	F	F
F	30	X1.181	0.195E+04 0.302E+01	0.153E+02 0.103E+02	F			B+-1.0	F	F	F
F	32	X1.260	0.208E+04 0.322E+01	0.163E+02 0.110E+02	F			B+-1.0	F	F	F
F	35	X1.378	0.228E+04 0.353E+01	0.179E+02 0.120E+02	F			B+-1.0	F	F	F
F	40	X1.575	0.260E+04 0.403E+01	0.204E+02 0.137E+02	F			B+-1.0	F	F	F
F	45	X1.772	0.293E+04 0.453E+01	0.230E+02 0.154E+02	F			B+-1.5	F	F	F
F	70 × 4	2.756×0.157	0.280E+03 0.434E+00	0.220E+01 0.148E+01	F			F A+-1.0 S	F	F,6.5S	F
F	5	X0.197	0.350E+03 0.543E+00	0.275E+01 0.185E+01	F			B+-0.5	F	F	F
F	6	X0.236	0.420E+03 0.651E+00	0.330E+01 0.222E+01	F			B+-0.5	F	F	F
F	7	X0.276	0.490E+03 0.760E+00	0.385E+01 0.258E+01	F			B+-0.5 T	F	F	F
F	8	X0.315	0.560E+03 0.868E+00	0.440E+01 0.295E+01	F			B+-0.5 F	F	F	F
F	10	X0.394	0.700E+03 0.109E+01	0.549E+01 0.369E+01	F			B+-0.5 F	F	F,13F	F
F	12	X0.472	0.840E+03 0.130E+01	0.659E+01 0.443E+01	F			B+-0.5 F	F	F	F
F	15	X0.591	0.105E+04 0.163E+01	0.824E+01 0.554E+01	F			B+-0.5 S,1AS	F	F	F
F	16	X0.630	0.112E+04 0.174E+01	0.879E+01 0.591E+01	F			B+-0.5 S	F	F,22S	F
F	18	X0.709	0.126E+04 0.195E+01	0.989E+01 0.665E+01	F			B+-0.5 S	F	S	F
F	20	X0.787	0.140E+04 0.217E+01	0.110E+02 0.738E+01	F			B+-0.5 S	F	F	F
F	25	X0.984	0.175E+04 0.271E+01	0.137E+02 0.923E+01	F			B+-1.0 S	F	F	F
F	30	X1.181	0.210E+04 0.326E+01	0.165E+02 0.111E+02	F			B+-1.0 S	F	F	F
F	32	X1.260	0.224E+04 0.347E+01	0.176E+02 0.118E+02	F			B+-1.0 T	F	F	F
F	35	X1.378	0.245E+04 0.380E+01	0.192E+02 0.129E+02	F			B+-1.0 T	F	F	F
F	40	X1.575	0.280E+04 0.434E+01	0.220E+02 0.148E+02	F			B+-1.0 T	F	F	F
F	45	X1.772	0.315E+04 0.488E+01	0.247E+02 0.166E+02	F			B+-1.0 T	F	F	F
F	50	X1.969	0.350E+04 0.543E+01	0.275E+02 0.185E+02	F			B+-1.5 T	F	F	F
F	75 × 4	2.953×0.157	0.300E+03 0.465E+00	0.235E+01 0.158E+01	F			F A+-1.0 T	F	F	F
F	5	X0.197	0.375E+03 0.581E+00	0.294E+01 0.198E+01	F			B+-0.5	F	S	F
F	6	X0.236	0.450E+03 0.698E+00	0.353E+01 0.237E+01	F			B+-0.5 T	F	6.5S	F
F	7	X0.276	0.525E+03 0.814E+00	0.412E+01 0.277E+01	F			B+-0.5 T	F	F	F
F	8	X0.315	0.600E+03 0.930E+00	0.471E+01 0.316E+01	F		9F	B+-0.5 T	F	F	F
F	10	X0.394	0.750E+03 0.116E+01	0.589E+01 0.396E+01	F		F	B+-0.5 T	F	F	F
F	12	X0.472	0.900E+03 0.140E+01	0.706E+01 0.475E+01	F			B+-0.5 T	F	F	F
F	15	X0.591	0.113E+04 0.174E+01	0.883E+01 0.593E+01	F			B+-0.5 T	F	F,13F	F

STEEL MATERIAL DATA

TABLE 10-15. HOT ROLLED RECTANGULAR
STEEL BARS (ISO 1035/III)
BASIS: 1 IN = 25.4 MM
1 CUBIC METER STEEL - 7850 KG MASS

ORDER EXAMPLE:
LENGTH,FLAT STEEL 10X3 EURONORM 58 + ST QUAL
PAGE NO. 6

THE NOMINAL SIZE IS NATIONAL STANDARD AS INDICATED
F=FIRST CHOICE,S=SECOND CHOICE,T=THIRD CHOICE,NUMBER=OTHER SIZE
* = COMMERCIAL SIZE

Table 10-15 (Continued). Hot Rolled Rectangular Steel Bars (ISO 1035/III)

NOMINAL SIZE = AxB		SECTION AREA		MASS PER UNIT		U.S.A. ANSI B32.3	AUSTRAL AS 1256	JAPAN JIS G3194	EUR- NORM 58	FRANCE NF A45-005	U.K. BS 4229	GERMANY DIN 1017	ITALY UNI 6014
MM	IN	MM**2	IN**2	KG/M	LB/FT								
75 x 16	2.953X0.630	0.120E+04	0.186E+01	0.942E+01	0.633E+01				A*=1.0 T			S	F
18	X0.709	0.135E+04	0.209E+01	0.106E+02	0.712E+01				B*=0.5 T		F		F
20	X0.787	0.150E+04	0.233E+01	0.118E+02	0.791E+01				F B*=1.0 T		F	F	F
25	X0.984	0.188E+04	0.291E+01	0.147E+02	0.989E+01				F B*=1.0 F		F	S	F
30	X1.181	0.225E+04	0.349E+01	0.177E+02	0.119E+02			F	B*=1.0 F		F	F	F
32	X1.260	0.240E+04	0.372E+01	0.188E+02	0.127E+02			F,22F	B*=1.0 F		F	F	F
35	X1.378	0.263E+04	0.407E+01	0.206E+02	0.138E+02				B*=1.0 F		F		F
40	X1.575	0.300E+04	0.465E+01	0.235E+02	0.158E+02				B*=1.0 F				F
45	X1.772	0.338E+04	0.523E+01	0.265E+02	0.178E+02				B*=1.5				F
50	X1.969	0.375E+04	0.581E+01	0.294E+02	0.198E+02				B*=1.5 T			60T	F
80 x 4	3.150X0.157	0.320E+03	0.496E+00	0.251E+01	0.169E+01				A*=1.5 F		F	F	F
5	X0.197	0.400E+03	0.620E+00	0.314E+01	0.211E+01				B*=0.5 F		F	F	F
6	X0.236	0.480E+03	0.744E+00	0.377E+01	0.253E+01				B*=0.5 T		F	F	F
7	X0.276	0.560E+03	0.868E+00	0.440E+01	0.295E+01				B*=0.5 F		F	F,6.5F	F
8	X0.315	0.640E+03	0.992E+00	0.502E+01	0.338E+01				B*=0.5 F		F	F	F
10	X0.394	0.800E+03	0.124E+01	0.628E+01	0.422E+01				B*=0.5 F		F	F	F
12	X0.472	0.960E+03	0.149E+01	0.754E+01	0.506E+01				B*=0.5 F		F	F,11S	F
14	X0.551	0.112E+04	0.174E+01	0.879E+01	0.591E+01				B*=0.5 S		F	13F	F
15	X0.591	0.120E+04	0.186E+01	0.942E+01	0.633E+01				B*=0.5 S			S	F
16	X0.630	0.128E+04	0.198E+01	0.100E+02	0.675E+01				B*=1.0 F		F	F	F
18	X0.709	0.144E+04	0.223E+01	0.113E+02	0.760E+01				B*=0.5 T		F	F	F
20	X0.787	0.160E+04	0.248E+01	0.126E+02	0.844E+01				B*=1.0 F		F	F	F
25	X0.984	0.200E+04	0.310E+01	0.157E+02	0.105E+02				B*=1.0 F,22T		F	S	F
30	X1.181	0.240E+04	0.372E+01	0.188E+02	0.127E+02				B*=1.0 F		F	S	F
32	X1.260	0.256E+04	0.397E+01	0.201E+02	0.135E+02				B*=1.0 F		F	F	F
35	X1.378	0.280E+04	0.434E+01	0.220E+02	0.148E+02				B*=1.0 F		F	T	F
40	X1.575	0.320E+04	0.496E+01	0.251E+02	0.169E+02				B*=1.5 F		F	S	F
45	X1.772	0.360E+04	0.558E+01	0.283E+02	0.190E+02				B*=1.5 F		F	F	F
50	X1.969	0.400E+04	0.620E+01	0.314E+02	0.211E+02				B*=1.5 T		F	T	F
60	X2.362	0.480E+04	0.744E+01	0.377E+02	0.253E+02				B*=1.5 F		F	T	F
90 x 5	3.543X0.197	0.450E+03	0.698E+00	0.353E+01	0.237E+01				B*=0.5 F		F	S	F
6	X0.236	0.540E+03	0.837E+00	0.424E+01	0.285E+01				B*=0.5 F		F	S	F
7	X0.276	0.630E+03	0.977E+00	0.495E+01	0.332E+01				B*=0.5 T		F	6.5F	F
8	X0.315	0.720E+03	0.112E+01	0.565E+01	0.380E+01				B*=0.5 F,9T		F	F	F
10	X0.394	0.900E+03	0.140E+01	0.706E+01	0.475E+01			F	B*=0.5 F,9F		F	F,9F	F
12	X0.472	0.108E+04	0.167E+01	0.848E+01	0.570E+01			9F	B*=0.5 F		F	F,11S	F
14	X0.551	0.126E+04	0.195E+01	0.989E+01	0.665E+01			F	B*=0.5 T		F	13S	F
15	X0.591	0.135E+04	0.209E+01	0.106E+02	0.712E+01				B*=0.5 T				F
16	X0.630	0.144E+04	0.223E+01	0.113E+02	0.760E+01				B*=0.5 T		F	F	F
18	X0.709	0.162E+04	0.251E+01	0.127E+02	0.855E+01				B*=0.5 S		F	F	F
20	X0.787	0.180E+04	0.279E+01	0.141E+02	0.949E+01			19F	B*=1.0 F		F	F	F
25	X0.984	0.225E+04	0.349E+01	0.177E+02	0.119E+02			F,22F	B*=1.0 F		F	F	F
30	X1.181	0.270E+04	0.419E+01	0.212E+02	0.142E+02				B*=1.0 F		F	F	F
32	X1.260	0.288E+04	0.446E+01	0.226E+02	0.152E+02				B*=1.0 F		F		F
35	X1.378	0.315E+04	0.488E+01	0.247E+02	0.166E+02				B*=1.0 F		F		F
40	X1.575	0.360E+04	0.558E+01	0.283E+02	0.190E+02				B*=1.0 F		F	F	F
45	X1.772	0.405E+04	0.628E+01	0.318E+02	0.214E+02				B*=1.0 T		F		F
50	X1.969	0.450E+04	0.698E+01	0.353E+02	0.237E+02				B*=1.5 F		F	S	F
60	X2.362	0.540E+04	0.837E+01	0.424E+02	0.285E+02				B*=1.5 F		F	S	F

10-45

STEEL MATERIAL DATA

Table 10-15 (Continued). Hot Rolled Rectangular Steel Bars (ISO 1035/III)

TABLE 10-15, HOT ROLLED RECTANGULAR
STEEL BARS (ISO 1035/III)
BASIS; 1 IN = 25.4 MM
1 CUBIC METER STEEL = 7850 KG MASS

ORDER EXAMPLE:
LENGTH,FLAT STEEL 10X3 EURONORM 58 + ST QUAL   PAGE NO. 7

THE NOMINAL SIZE IS NATIONAL STANDARD AS INDICATED
F=FIRST CHOICE,S=SECOND CHOICE,T=THIRD CHOICE,NUMBER=OTHER SIZE
* = COMMERCIAL SIZE

NOMINAL SIZE = AxB		SECTION AREA		MASS PER UNIT		U.S.A. ANSI B32.3	AUSTRAL AS 1256	JAPAN JIS G3194	EURO- NORM 58	FRANCE NF A45-005	U.K. BS 4229	GERMANY DIN 1017	ITALY UNI 6014
MM	IN	MM**2	IN**2	KG/M	LB/FT								
F100 x 5	3.937X0.197	0.500E+03	0.775E+00	0.392E+01	0.264E+01	F	F		F	A+-1.5 F	F	F	F
6	X0.236	0.600E+03	0.930E+00	0.471E+01	0.316E+01	F	F		F	B+-0.5 F	F	F	F
7	X0.276	0.700E+03	0.109E+01	0.549E+01	0.369E+01	F			F	B+-0.5 T	F	F	F
8	X0.315	0.800E+03	0.124E+01	0.628E+01	0.422E+01	F	F		F	B+-0.5 F	F	6.5F	F
10	X0.394	0.100E+04	0.155E+01	0.785E+01	0.527E+01	F	F	9F	F	B+-0.5 F	F	F	F
12	X0.472	0.120E+04	0.186E+01	0.942E+01	0.633E+01	F	F		F	B+-0.5 S	F	F,11S	F
14	X0.551	0.140E+04	0.217E+01	0.110E+02	0.738E+01	S			F	B+-0.5 S	F	F,13F	F
15	X0.591	0.150E+04	0.233E+01	0.118E+02	0.791E+01	F			F	B+-0.5 S	F		F
16	X0.630	0.160E+04	0.248E+01	0.126E+02	0.844E+01	S	F		F	B+-0.5 S	F		F
18	X0.709	0.180E+04	0.279E+01	0.141E+02	0.949E+01	S			F	B+-0.5 S	F		F
20	X0.787	0.200E+04	0.310E+01	0.157E+02	0.105E+02	F		19F	F	B+-1.0 S	F	S	F
25	X0.984	0.250E+04	0.388E+01	0.196E+02	0.132E+02	F	F	F,22F	F	B+-1.0 F	F	S	F
30	X1.181	0.300E+04	0.465E+01	0.235E+02	0.158E+02	F		28F	F	B+-1.0 F	F		F
32	X1.260	0.320E+04	0.496E+01	0.251E+02	0.169E+02	F			F	B+-1.0 F	F		F
35	X1.378	0.350E+04	0.543E+01	0.275E+02	0.185E+02	F		36F	F	B+-1.0 F	F		F
40	X1.575	0.400E+04	0.620E+01	0.314E+02	0.211E+02	F			F	B+-1.0 S	F		F
45	X1.772	0.450E+04	0.698E+01	0.353E+02	0.237E+02	S			F	B+-1.5 S	F		F
50	X1.969	0.500E+04	0.775E+01	0.392E+02	0.264E+02	F			F	B+-1.5 S	F		F
60	X2.362	0.600E+04	0.930E+01	0.471E+02	0.316E+02	F			F	B+-1.5 S	F		F
F110 x 5	4.331X0.197	0.550E+03	0.853E+00	0.432E+01	0.290E+01	S			F	A+-2.0 T	F		F
6	X0.236	0.660E+03	0.102E+01	0.518E+01	0.348E+01	S			F	B+-0.5 T	F		F
7	X0.276	0.770E+03	0.119E+01	0.604E+01	0.406E+01	S			F	R+-0.5 T	F		F
8	X0.315	0.880E+03	0.136E+01	0.691E+01	0.464E+01	S			F	B+-0.5 F	F		F
10	X0.394	0.110E+04	0.171E+01	0.863E+01	0.580E+01	S			F	B+-0.5 F	F	S,9F	F
12	X0.472	0.132E+04	0.205E+01	0.104E+02	0.696E+01	S			F	B+-0.5 F	F	F,11S	F
14	X0.551	0.154E+04	0.239E+01	0.121E+02	0.812E+01	T			F	B+-0.5 F	F	S,13S	F
15	X0.591	0.165E+04	0.256E+01	0.130E+02	0.870E+01	F			F	B+-0.5 T	F	S	F
16	X0.630	0.176E+04	0.273E+01	0.138E+02	0.928E+01	S			F	B+-0.5 S	F		F
18	X0.709	0.198E+04	0.307E+01	0.155E+02	0.104E+02	S			F	B+-0.5 S	F		F
20	X0.787	0.220E+04	0.341E+01	0.173E+02	0.116E+02	S			F	B+-1.0 S	F		F
25	X0.984	0.275E+04	0.426E+01	0.216E+02	0.145E+02	S			F	B+-1.0 F	F		F
30	X1.181	0.330E+04	0.512E+01	0.259E+02	0.174E+02	S			F	B+-1.0 T	F		F
32	X1.260	0.352E+04	0.546E+01	0.276E+02	0.186E+02	T			F	B+-1.0 S	F		F
35	X1.378	0.385E+04	0.597E+01	0.302E+02	0.203E+02	S			F	B+-1.0 T	F		F
40	X1.575	0.440E+04	0.682E+01	0.345E+02	0.232E+02	S			F	B+-1.0 S	F		F
45	X1.772	0.495E+04	0.767E+01	0.389E+02	0.261E+02	T			F	B+-1.5 S	F		F
50	X1.969	0.550E+04	0.853E+01	0.432E+02	0.290E+02	S			F	B+-1.5 S	F		F
60	X2.362	0.660E+04	0.102E+02	0.518E+02	0.348E+02	F			F	B+-1.5 S	F		F
F120 x 5	4.724X0.197	0.600E+03	0.930E+00	0.471E+01	0.316E+01	F	F		F	A+-2.0 T	F		F
6	X0.236	0.720E+03	0.112E+01	0.565E+01	0.380E+01	F	F		F	B+-0.5 F	F		F
7	X0.276	0.840E+03	0.130E+01	0.659E+01	0.443E+01	F			F	B+-0.5 F	F		F
8	X0.315	0.960E+03	0.149E+01	0.754E+01	0.506E+01	F	F		F	B+-0.5 F	F	F,11T	F
10	X0.394	0.120E+04	0.186E+01	0.942E+01	0.633E+01	F	F		F	B+-0.5 F	F	F	F
12	X0.472	0.144E+04	0.223E+01	0.113E+02	0.760E+01	F	F		F	B+-0.5 F	F	13S	F
14	X0.551	0.168E+04	0.260E+01	0.132E+02	0.886E+01	S			F	B+-0.5 S	F	F	F
15	X0.591	0.180E+04	0.279E+01	0.141E+02	0.945E+01	F			F	B+-0.5 S	F		F
16	X0.630	0.192E+04	0.298E+01	0.151E+02	0.101E+02	F	F		F	B+-0.5 S	F		F
18	X0.709	0.216E+04	0.335E+01	0.170E+02	0.114E+02	F			F	B+-0.5 T	F		F
20	X0.787	0.240E+04	0.372E+01	0.188E+02	0.127E+02	F	F		F	B+-0.5 F	F		F
25	X0.984	0.300E+04	0.465E+01	0.235E+02	0.158E+02	F			F	B+-1.0 F	F		S

STEEL MATERIAL DATA

Table 10-15 (Continued). Hot Rolled Rectangular Steel Bars (ISO 1035/III)

```
TABLE 10-15. HOT ROLLED RECTANGULAR                                  ORDER EXAMPLE;
STEEL BARS (ISO 1035/III)                                            LENGTH,FLAT STEEL 10X3 EURONORM 58 + ST QUAL
BASIS; 1 IN = 25.4 MM                                                                                    PAGE NO. 8
1 CUBIC METER STEEL = 7850 KG MASS

                                                     THE NOMINAL SIZE IS NATIONAL STANDARD AS INDICATED
                                                     F=FIRST CHOICE,S=SECOND CHOICE,T=THIRD CHOICE,NUMBER=OTHER SIZE
                                                     *=COMMERCIAL SIZE
                                                     U.S.A.   AUSTRAL JAPAN          EURO-   FRANCE   U.K.    GERMANY  ITALY
I                                                    ANSI     AS      JIS            NORM    NF       BS      DIN ●    UNI
S   NOMINAL SIZE = AXB   SECTION AREA   MASS PER UNIT B32.3   1256    G3194          58      A45-005  4229    1017     6014
O      MM    IN            MM**2   IN**2   KG/M   LB/FT

F120 x  30  4.724X1.181  0.360E+04 0.558E+01 0.283E+02 0.190E+02 F                              F              S         F
F       32           1.260  0.384E+04 0.595E+01 0.301E+02 0.203E+02 S                              F
F       35           1.378  0.420E+04 0.651E+01 0.330E+02 0.222E+02 F                     F        F              F
F       40           1.575  0.480E+04 0.744E+01 0.377E+02 0.253E+02 F          B+=1.0  T   F        F              F
F       45           1.772  0.540E+04 0.837E+01 0.424E+02 0.285E+02 S          B+=1.0  T   F        F              F
F       50           1.969  0.600E+04 0.930E+01 0.471E+02 0.316E+02 F          B+=1.5  T   F        F              S       F
F       60           2.362  0.720E+04 0.112E+02 0.565E+02 0.380E+02 F          B+=1.5  T   F        F              S       F

F130 x   6  5.118X0.236  0.780E+03 0.121E+01 0.612E+01 0.411E+01 F     F,5F    125X6F    A+=2.5 T  F              F
F        7           0.276  0.910E+03 0.141E+01 0.714E+01 0.480E+01                        A+=0.5     F           F
F        8           0.315  0.104E+04 0.161E+01 0.816E+01 0.549E+01 F           125X8F    B+=0.5 T  F              S       F
F       10           0.394  0.130E+04 0.202E+01 0.102E+02 0.686E+01 F           125X9F    B+=0.5 S  F              F,9F   F
F       12           0.472  0.156E+04 0.242E+01 0.123E+02 0.823E+01 F           125X12F   B+=0.5 S  F              F,11S  F
F       15           0.591  0.195E+04 0.302E+01 0.153E+02 0.103E+02 F                     B+=0.5 T  F              S,13S  F
F       16           0.630  0.208E+04 0.322E+01 0.163E+02 0.110E+02 F           125X16F   B+=0.5 T  F              S,15S  F
F       18           0.709  0.234E+04 0.363E+01 0.184E+02 0.123E+02 F           125X19F   B+=0.5 T  F              F
F       20           0.787  0.260E+04 0.403E+01 0.204E+02 0.137E+02 F           125X22F   B+=1.0 T  F              F
F       25           0.984  0.325E+04 0.504E+01 0.255E+02 0.171E+02 F           125X25F   B+=1.0 T  F              S       F
F       30           1.181  0.390E+04 0.605E+01 0.306E+02 0.206E+02 F           125X28F   B+=1.0 T  F              F
F       32           1.260  0.416E+04 0.645E+01 0.327E+02 0.219E+02 F                     B+=1.0    F              F
F       35           1.378  0.455E+04 0.705E+01 0.357E+02 0.240E+02 F           125X32F   B+=1.0    F              S       F
F       40           1.575  0.520E+04 0.806E+01 0.408E+02 0.274E+02 F                     B+=1.0    F              F
F       45           1.772  0.585E+04 0.907E+01 0.459E+02 0.309E+02 F                     B+=1.5    F              F
F       50           1.969  0.650E+04 0.101E+02 0.510E+02 0.343E+02 F           125X36F   B+=1.5    F              S       F
F       60           2.362  0.780E+04 0.121E+02 0.612E+02 0.411E+02 F                     B+=1.5    F              F

F140 x   7  5.512X0.276  0.980E+03 0.152E+01 0.769E+01 0.517E+01 F                        A+=2.5 6T F              F
F        8           0.315  0.112E+04 0.174E+01 0.879E+01 0.591E+01 F                     B+=0.5 S  F              S       F
F       10           0.394  0.140E+04 0.217E+01 0.110E+02 0.738E+01 F                     B+=0.5 T  F              S       F
F       12           0.472  0.168E+04 0.260E+01 0.132E+02 0.886E+01 F                     B+=0.5 T  F              S       F
F       14           0.551  0.196E+04 0.304E+01 0.154E+02 0.103E+02 F                     B+=0.5 T  F              S       F
F       15           0.591  0.210E+04 0.326E+01 0.165E+02 0.111E+02 F                     B+=0.5 T  F              F
F       16           0.630  0.224E+04 0.347E+01 0.176E+02 0.118E+02 F                     B+=0.5 T  F              S       F
F       18           0.709  0.252E+04 0.391E+01 0.198E+02 0.133E+02 S                     B+=0.5 T  F              F
F       20           0.787  0.280E+04 0.434E+01 0.220E+02 0.148E+02 F                     B+=1.0 T  F              S       F
F       25           0.984  0.350E+04 0.543E+01 0.275E+02 0.185E+02 S                     B+=1.0 T  F              F
F       30           1.181  0.420E+04 0.651E+01 0.330E+02 0.222E+02 F                     B+=1.0    F              S       F
F       32           1.260  0.448E+04 0.694E+01 0.352E+02 0.236E+02 F                     B+=1.0    F              S       F
F       35           1.378  0.490E+04 0.760E+01 0.385E+02 0.258E+02 S                     B+=1.5    F              S       F
F       40           1.575  0.560E+04 0.868E+01 0.440E+02 0.295E+02 F                     B+=1.5    F              T       F
F       45           1.772  0.630E+04 0.977E+01 0.495E+02 0.332E+02 F                     B+=1.5    F              F
F       50           1.969  0.700E+04 0.109E+02 0.549E+02 0.369E+02 S                     B+=1.5    F              S       F
F       60           2.362  0.840E+04 0.130E+02 0.659E+02 0.443E+02 T                     B+=1.5    F              S       F

F150 x   7  5.906X0.276  0.105E+04 0.163E+01 0.824E+01 0.554E+01 F     5F,6F              A+=2.5 5T,6S F           F
F        8           0.315  0.120E+04 0.186E+01 0.942E+01 0.633E+01 S                     B+=0.5 T  F              F
F       10           0.394  0.150E+04 0.233E+01 0.118E+02 0.791E+01 S                     B+=0.5 T  F              F,11S  F
F       12           0.472  0.180E+04 0.279E+01 0.141E+02 0.949E+01 S             9F      F B+=0.5 F,11T F         S,13F  F
F       14           0.551  0.210E+04 0.326E+01 0.165E+02 0.111E+02 T                     B+=0.5 T  F              F
F       15           0.591  0.225E+04 0.349E+01 0.177E+02 0.119E+02 F                     B+=0.5 T  F              F
F       16           0.630  0.240E+04 0.372E+01 0.188E+02 0.127E+02 S                     B+=0.5 T  F              F
F       18           0.709  0.270E+04 0.419E+01 0.212E+02 0.142E+02 T             F       B+=0.5    F              F
```

10-47

10-48 STEEL MATERIAL DATA

Table 10-15 (Continued). Hot Rolled Rectangular Steel Bars (ISO 1035/III)

TABLE 10-15. HOT ROLLED RECTANGULAR STEEL BARS (ISO 1035/III)
BASIS: 1 IN = 25.4 MM
1 CUBIC METER STEEL = 7850 KG MASS

ORDER EXAMPLE:
LENGTH=FLAT STEEL 10X3 EURONORM 58 + ST QUAL
PAGE NO. 9

THE NOMINAL SIZE IS NATIONAL STANDARD AS INDICATED
F=FIRST CHOICE, S=SECOND CHOICE, T=THIRD CHOICE, NUMBER=OTHER SIZE
* = COMMERCIAL SIZE

NOMINAL SIZE = AxB		SECTION AREA		MASS PER UNIT		U.S.A. ANSI B32.3	AUSTRAL AS 1256	JAPAN JIS G3194	EURO- NORM 58	FRANCE NF A45-005	U.K. BS 4229	GERMANY DIN 1017	ITALY UNI 6014
MM	IN	MM**2	IN**2	KG/M	LB/FT								
F150 x 20	5.906x0.787	0.300E+04	0.465E+01	0.235E+02	0.158E+02	S		19F	F A+-2.5	F	F	S	F
25	x0.984	0.375E+04	0.581E+01	0.294E+02	0.198E+02	S	F	F,22F	F B+-1.0	T	F	S	
30	x1.181	0.450E+04	0.698E+01	0.353E+02	0.237E+02	T	F	28F	F B+-1.0	T	F	S	
32	x1.260	0.480E+04	0.744E+01	0.377E+02	0.253E+02	T			B+-1.0		F		
35	x1.378	0.525E+04	0.814E+01	0.412E+02	0.277E+02	S			B+-1.0	T	F		
40	x1.575	0.600E+04	0.930E+01	0.471E+02	0.316E+02	S	F	36F	F B+-1.5		F	S	F
45	x1.772	0.675E+04	0.105E+02	0.530E+02	0.356E+02	T			B+-1.5		F		
50	x1.969	0.750E+04	0.116E+02	0.589E+02	0.396E+02	S	F		F B+-1.5		F	S	F
60	x2.362	0.900E+04	0.140E+02	0.706E+02	0.475E+02	S			B+-1.5		F		
F160 x 7	6.299x0.276	0.112E+04	0.174E+01	0.879E+01	0.591E+01	F			A+-2.5		F		
8	x0.315	0.128E+04	0.198E+01	0.100E+02	0.675E+01	F			B+-0.5		F		
10	x0.394	0.160E+04	0.248E+01	0.126E+02	0.844E+01	F			B+-0.5		F		
12	x0.472	0.192E+04	0.298E+01	0.151E+02	0.101E+02	F			B+-0.5		F		
14	x0.551	0.224E+04	0.347E+01	0.176E+02	0.118E+02	F			B+-0.5		F		
15	x0.591	0.240E+04	0.372E+01	0.188E+02	0.127E+02	F			B+-0.5		F		
16	x0.630	0.256E+04	0.397E+01	0.201E+02	0.135E+02	F			B+-0.5		F		
18	x0.709	0.288E+04	0.446E+01	0.226E+02	0.152E+02	F			B+-0.5		F		
20	x0.787	0.320E+04	0.496E+01	0.251E+02	0.169E+02	F			B+-0.5		F		
25	x0.984	0.400E+04	0.620E+01	0.314E+02	0.211E+02	F			B+-1.0		F		
30	x1.181	0.480E+04	0.744E+01	0.377E+02	0.253E+02	S			B+-1.0		F		
32	x1.260	0.512E+04	0.794E+01	0.402E+02	0.270E+02	S			B+-1.0		F		
35	x1.378	0.560E+04	0.868E+01	0.440E+02	0.295E+02	S			B+-1.0		F		
40	x1.575	0.640E+04	0.992E+01	0.502E+02	0.338E+02	S			B+-1.5		F		
45	x1.772	0.720E+04	0.112E+02	0.565E+02	0.380E+02	F			B+-1.5		F		
50	x1.969	0.800E+04	0.124E+02	0.628E+02	0.422E+02	F			B+-1.5		F		
60	x2.362	0.960E+04	0.149E+02	0.754E+02	0.506E+02	F			B+-1.5		F		
F180 x 7	7.087x0.276	0.126E+04	0.195E+01	0.989E+01	0.665E+01	F	5F,6F		A+-2.5		F		
8	x0.315	0.144E+04	0.223E+01	0.113E+02	0.760E+01	F	F		B+-0.5		F		
10	x0.394	0.180E+04	0.279E+01	0.141E+02	0.949E+01	F	F	9F	F B+-0.5		F		
12	x0.472	0.216E+04	0.335E+01	0.170E+02	0.114E+02	F	F		B+-0.5		F		
14	x0.551	0.252E+04	0.391E+01	0.198E+02	0.133E+02	F			B+-0.5		F		
15	x0.591	0.270E+04	0.419E+01	0.212E+02	0.142E+02	F			B+-0.5		F		
16	x0.630	0.288E+04	0.446E+01	0.226E+02	0.152E+02	F			B+-0.5		F		
18	x0.709	0.324E+04	0.502E+01	0.254E+02	0.171E+02	S		19F	B+-0.5		F		
20	x0.787	0.360E+04	0.558E+01	0.283E+02	0.190E+02	F		F,22F	B+-0.5		F		
25	x0.984	0.450E+04	0.698E+01	0.353E+02	0.237E+02	F		28F	B+-1.0		F		
30	x1.181	0.540E+04	0.837E+01	0.424E+02	0.285E+02	F			B+-1.0		F		
32	x1.260	0.576E+04	0.893E+01	0.452E+02	0.304E+02	F			B+-1.0		F		
35	x1.378	0.630E+04	0.977E+01	0.495E+02	0.332E+02	F			B+-1.0		F		
40	x1.575	0.720E+04	0.112E+02	0.565E+02	0.380E+02	F		36F	B+-1.0		F		
45	x1.772	0.810E+04	0.126E+02	0.636E+02	0.427E+02	F			B+-1.5		F		
50	x1.969	0.900E+04	0.140E+02	0.706E+02	0.475E+02	F			B+-1.5		F		
60	x2.362	0.108E+05	0.167E+02	0.848E+02	0.570E+02	F			B+-1.5		F		
F200 x 10	7.874x0.394	0.200E+04	0.310E+01	0.157E+02	0.105E+02	F	F,5F,6F	9F	F A+-2.5		F		F,8F
12	x0.472	0.240E+04	0.372E+01	0.188E+02	0.127E+02	F	F,8F		B+-0.5		F		F
14	x0.551	0.280E+04	0.434E+01	0.220E+02	0.148E+02	F			B+-0.5		F		
15	x0.591	0.300E+04	0.465E+01	0.235E+02	0.158E+02	F			B+-0.5		F		
16	x0.630	0.320E+04	0.496E+01	0.251E+02	0.169E+02	F		F	B+-0.5		F		
18	x0.709	0.360E+04	0.558E+01	0.283E+02	0.190E+02	S		F	B+-0.5		F		

STEEL MATERIAL DATA

Table 10-15 (Continued). Hot Rolled Rectangular Steel Bars (ISO 1035/III)

TABLE 10-15. HOT ROLLED RECTANGULAR STEEL BARS (ISO 1035/III)
BASIS, 1 IN.= 25.4 MM
1 CUBIC METER STEEL = 7850 KG MASS

ORDER EXAMPLE:
LENGTH,FLAT STEEL 10X3 EURONORM 58 + ST QUAL   PAGE NO. 1b

THE NOMINAL SIZE IS NATIONAL STANDARD AS INDICATED
F=FIRST CHOICE,S=SECOND CHOICE,T=THIRD CHOICE,NUMBER=OTHER SIZE
* = COMMERCIAL SIZE

I S O	NOMINAL SIZE = AXB MM	IN	SECTION AREA MM**2	IN**2	MASS PER UNIT KG/M	LB/FT	U.S.A. ANSI B32.3	AUSTRAL AS 1256	JAPAN JIS G3194	EURO- NORM 58	FRANCE NF A45-005	U.K. BS 4229	GERMANY DIN 1017	ITALY UNI 6014
F	200 X 20	7.874X0.787	0.400E+04	0.620E+01	0.314E+02	0.211E+02	F	F	19F	A+=2.5		F		
F	25	X0.984	0.500E+04	0.775E+01	0.392E+02	0.264E+02	F	F	F,22F	B+=1.0		F		
F	30	X1.181	0.600E+04	0.930E+01	0.471E+02	0.316E+02	F	F	28F	B+=1.0		F		
F	32	X1.260	0.640E+04	0.992E+01	0.502E+02	0.338E+02	S	F	F	B+=1.0		F		
F	35	X1.378	0.700E+04	0.105E+02	0.549E+02	0.369E+02	F			B+=1.0		F		
F	40	X1.575	0.800E+04	0.124E+02	0.628E+02	0.422E+02	F	F	36F	B+=1.0		F		
F	45	X1.772	0.900E+04	0.140E+02	0.706E+02	0.475E+02	S			B+=1.5		F		
F	50	X1.969	0.100E+05	0.155E+02	0.785E+02	0.527E+02	F	F		B+=1.5		F		
F	60	X2.362	0.120E+05	0.186E+02	0.942E+02	0.633E+02	F			B+=1.5		F		
F 250 X 10	9.843X0.394	0.250E+04	0.388E+01	0.196E+02	0.132E+02	F	F,5F,6F	9F	A+=2.5		F			
F	12	X0.472	0.300E+04	0.465E+01	0.235E+02	0.158E+02	F	F,8F	F	B+=0.5		F		
F	14	X0.551	0.350E+04	0.543E+01	0.275E+02	0.185E+02	S			B+=0.5		F		
F	15	X0.591	0.375E+04	0.581E+01	0.294E+02	0.198E+02	F			B+=0.5		F		
F	16	X0.630	0.400E+04	0.620E+01	0.314E+02	0.211E+02	F	F	F	B+=0.5		F		
F	18	X0.709	0.450E+04	0.698E+01	0.353E+02	0.237E+02	S			B+=0.5		F		
F	20	X0.787	0.500E+04	0.775E+01	0.392E+02	0.264E+02	F	F	19F	B+=1.0		F		
F	25	X0.984	0.625E+04	0.969E+01	0.491E+02	0.330E+02	F	F	F,22F	B+=1.0		F		
F	30	X1.181	0.750E+04	0.116E+02	0.589E+02	0.396E+02	F	F	28F	B+=1.0		F		
F	32	X1.260	0.800E+04	0.124E+02	0.628E+02	0.422E+02	S	F	F	B+=1.0		F		
F	35	X1.378	0.875E+04	0.136E+02	0.687E+02	0.462E+02	F			B+=1.0		F		
F	40	X1.575	0.100E+05	0.155E+02	0.785E+02	0.527E+02	F	F	36F	B+=1.0		F		
F	45	X1.772	0.113E+05	0.174E+02	0.883E+02	0.593E+02	S			B+=1.5		F		
F	50	X1.969	0.125E+05	0.194E+02	0.981E+02	0.659E+02	F			B+=1.5		F		
F	60	X2.362	0.150E+05	0.233E+02	0.118E+03	0.791E+02	F			B+=1.5		F		
300 X 6	11.811X0.236	0.180E+04	0.279E+01	0.141E+02	0.949E+01	F	F,5F		A+=2.5		F			
	8	X0.315	0.240E+04	0.372E+01	0.188E+02	0.127E+02	F			B+=0.5		F		
	10	X0.394	0.300E+04	0.465E+01	0.235E+02	0.158E+02	S			B+=0.5		F		
	12	X0.472	0.360E+04	0.558E+01	0.283E+02	0.190E+02	F			B+=0.5		F		
	14	X0.551	0.420E+04	0.651E+01	0.330E+02	0.222E+02	S			B+=0.5		F		
	15	X0.591	0.450E+04	0.698E+01	0.353E+02	0.237E+02	F			B+=0.5		F		
	16	X0.630	0.480E+04	0.744E+01	0.377E+02	0.253E+02	S			B+=0.5		F		
	18	X0.709	0.540E+04	0.837E+01	0.424E+02	0.285E+02	S			B+=0.5		F		
	20	X0.787	0.600E+04	0.930E+01	0.471E+02	0.316E+02	F	F	19F	B+=1.0		F		
	25	X0.984	0.750E+04	0.116E+02	0.589E+02	0.396E+02	F	F	F,22F	B+=1.0		F		
	30	X1.181	0.900E+04	0.140E+02	0.706E+02	0.475E+02	F	F	28F	B+=1.0		F		
	32	X1.260	0.960E+04	0.149E+02	0.754E+02	0.506E+02	S	F	F	B+=1.0		F		
	35	X1.378	0.105E+05	0.163E+02	0.824E+02	0.555E+02	F			B+=1.0		F		
	40	X1.575	0.120E+05	0.186E+02	0.942E+02	0.633E+02	F	F	36F	B+=1.0		F		
	45	X1.772	0.135E+05	0.209E+02	0.106E+03	0.712E+02	S			B+=1.5		F		
	50	X1.969	0.150E+05	0.233E+02	0.118E+03	0.791E+02	F			B+=1.5		F		
	60	X2.362	0.180E+05	0.279E+02	0.141E+03	0.949E+02	F			B+=1.5		F		
F 400 X 15	15.748X0.591	0.600E+04	0.930E+01	0.471E+02	0.316E+02	F			A+=2.5		F			
F	20	X0.787	0.800E+04	0.124E+02	0.628E+02	0.422E+02	F			B+=0.5		F		
F	25	X0.984	0.100E+05	0.155E+02	0.785E+02	0.527E+02	F			B+=1.0		F		
F	30	X1.181	0.120E+05	0.186E+02	0.942E+02	0.633E+02	F			B+=1.0		F		
F	32	X1.260	0.128E+05	0.198E+02	0.100E+03	0.675E+02	S			B+=1.0		F		
F	40	X1.575	0.160E+05	0.248E+02	0.126E+03	0.844E+02	F			B+=1.5		F		
F	50	X1.969	0.200E+05	0.310E+02	0.157E+03	0.105E+03	F			B+=1.5		F		
F	60	X2.362	0.240E+05	0.372E+02	0.188E+03	0.127E+03	F			B+=1.5		F		

Table 10-16. Bright Finish Rectangular Steel Bars with Sharp Corners (DIN 174)

```
TABLE 10-16. BRIGHT FINISH RECTANGULAR                                    ORDER EXAMPLE;
STEEL BARS WITH SHARP CORNERS (DIN 174)                                   LENGTH,FLAT STEEL 5X2 DIN 174 + STEEL QUALITY        PAGE NO. 1
BASIS; 1 IN = 25.4 MM
1 CUBIC METER STEEL = 7850 KG MASS

                                                                          THE NOMINAL SIZE IS NATIONAL STANDARD AS INDICATED
                                                                          F=FIRST CHOICE,S=SECOND CHOICE,T=THIRD CHOICE,NUMBER=OTHER SIZE
                                                                          * = COMMERCIAL SIZE
D                                                                         U.S.A.  AUSTRAL  JAPAN   EURO-   FRANCE  U.K.   GERMANY  ITALY
I                         SECTION AREA    MASS PER UNIT                   ANSI    AS       JIS     NORM    NF      BS     DIN      UNI
N   NOMINAL SIZE = AXB                                                    B32.3            G3123                   4229   174      757
    MM       IN       MM**2    IN**2      KG/M      LB/FT

F    5 X 2   0.197X0.079  0.100E+02 0.155E-01  0.785E-01 0.527E-01                                                A+0-0.075 F
F    2.5     X0.098       0.125E+02 0.194E-01  0.981E-01 0.659E-01                                                B+0-0.060 F
F    3       X0.118       0.150E+02 0.233E-01  0.118E+00 0.791E-01                                                B+0-0.060 F

F    6 X 2   0.236X0.079  0.120E+02 0.186E-01  0.942E-01 0.633E-01                                                A+0-0.075 F
F    2.5     X0.098       0.150E+02 0.233E-01  0.118E+00 0.791E-01                                                B+0-0.060 F
F    3       X0.118       0.180E+02 0.279E-01  0.141E+00 0.949E-01                                                B+0-0.060 F
F    4       X0.157       0.240E+02 0.372E-01  0.188E+00 0.127E+00                                                B+0-0.075 F

S    8 X 1.5 0.315X0.059  0.120E+02 0.186E-01  0.942E-01 0.633E-01                                                B+0-0.090 F
F    1.6     X0.063       0.128E+02 0.198E-01  0.100E+00 0.675E-01                                                B+0-0.060 F
F    2       X0.079       0.160E+02 0.248E-01  0.126E+00 0.844E-01                                                B+0-0.060 F
F    2.5     X0.098       0.200E+02 0.310E-01  0.157E+00 0.105E+00                                                B+0-0.060 F
F    3       X0.118       0.240E+02 0.372E-01  0.188E+00 0.127E+00                                                B+0-0.060 F
F    4       X0.157       0.320E+02 0.496E-01  0.251E+00 0.169E+00                                                B+0-0.075 F
F    5       X0.197       0.400E+02 0.620E-01  0.314E+00 0.211E+00                                                B+0-0.075 F
F    6       X0.236       0.480E+02 0.744E-01  0.377E+00 0.253E+00                                                B+0-0.075 F

S   10 X 1.5 0.394X0.059  0.150E+02 0.233E-01  0.118E+00 0.791E-01  1.4S                                          A+0-0.090 F
F    1.6     X0.063       0.160E+02 0.248E-01  0.126E+00 0.844E-01  F                                             B+0-0.060 F
F    2       X0.079       0.200E+02 0.310E-01  0.157E+00 0.105E+00  F                                             B+0-0.060 F
F    2.5     X0.098       0.250E+02 0.388E-01  0.196E+00 0.132E+00  F                                             B+0-0.060 F
F    3       X0.118       0.300E+02 0.465E-01  0.235E+00 0.158E+00  F                                             B+0-0.060 F
F    4       X0.157       0.400E+02 0.620E-01  0.314E+00 0.211E+00  F                                             B+0-0.075 F
F    5       X0.197       0.500E+02 0.775E-01  0.392E+00 0.264E+00  F                                             B+0-0.075 F
F    6       X0.236       0.600E+02 0.930E-01  0.471E+00 0.316E+00  F                                             B+0-0.075 F,8F

S   12 X 1.5 0.472X0.059  0.180E+02 0.279E-01  0.141E+00 0.949E-01  1.4S                                          A+0-0.110 F
F    1.6     X0.063       0.192E+02 0.298E-01  0.151E+00 0.101E+00  F                                             B+0-0.060 F
F    2       X0.079       0.240E+02 0.372E-01  0.188E+00 0.127E+00  F                                             B+0-0.060 F
F    2.5     X0.098       0.300E+02 0.465E-01  0.235E+00 0.158E+00  F                                             B+0-0.060 F
F    3       X0.118       0.360E+02 0.558E-01  0.283E+00 0.190E+00  F                                             B+0-0.060 F
F    4       X0.157       0.480E+02 0.744E-01  0.377E+00 0.253E+00  F                                             B+0-0.075 F
F    5       X0.197       0.600E+02 0.930E-01  0.471E+00 0.316E+00  F                                             B+0-0.075 F
F    6       X0.236       0.720E+02 0.112E+00  0.565E+00 0.380E+00  F                                             B+0-0.075 F
F    8       X0.315       0.960E+02 0.149E+00  0.754E+00 0.506E+00  F                                    10F      B+0-0.090 F

S   14 X 1.5 0.551X0.059  0.210E+02 0.326E-01  0.165E+00 0.111E+00                                                A+0-0.110 F
F    1.6     X0.063       0.224E+02 0.347E-01  0.176E+00 0.118E+00                                                B+0-0.060 F
F    2       X0.079       0.280E+02 0.434E-01  0.220E+00 0.148E+00                                                B+0-0.060 F
F    2.5     X0.098       0.350E+02 0.543E-01  0.275E+00 0.185E+00                                                B+0-0.060 F
F    3       X0.118       0.420E+02 0.651E-01  0.330E+00 0.222E+00                                                B+0-0.060 F
F    4       X0.157       0.560E+02 0.868E-01  0.440E+00 0.295E+00                                                B+0-0.075 F
F    5       X0.197       0.700E+02 0.109E+00  0.549E+00 0.369E+00                                                B+0-0.075 F
F    6       X0.236       0.840E+02 0.130E+00  0.659E+00 0.443E+00                                                B+0-0.075 F
F    8       X0.315       0.112E+03 0.174E+00  0.879E+00 0.591E+00                           9F                   B+0-0.090 F,10F

T   15 X 1.5 0.591X0.059  0.225E+02 0.349E-01  0.177E+00 0.119E+00                                                A+0-0.110 F
S    1.6     X0.063       0.240E+02 0.372E-01  0.188E+00 0.127E+00                                                B+0-0.060 F
S    2       X0.079       0.300E+02 0.465E-01  0.235E+00 0.158E+00                                                B+0-0.060 S
S    2.5     X0.098       0.375E+02 0.581E-01  0.294E+00 0.198E+00                                                B+0-0.060 S
S    3       X0.118       0.450E+02 0.698E-01  0.353E+00 0.237E+00                                                B+0-0.060 S
```

STEEL MATERIAL DATA 10-51

Table 10-16 (*Continued*). Bright Finish Rectangular Steel Bars with Sharp Corners (DIN 174)

```
TABLE 10-16, BRIGHT FINISH RECTANGULAR                             ORDER EXAMPLE:
STEEL BARS WITH SHARP CORNERS (DIN 174)                            LENGTH,FLAT STEEL 5X2 DIN 174 + STEEL QUALITY    PAGE NO. 2
BASIS: 1 IN = 25.4 MM
1 CUBIC METER STEEL = 7850 KG MASS
                                                                   THE NOMINAL SIZE IS NATIONAL STANDARD AS INDICATED
                                                                   F=FIRST CHOICE,S=SECOND CHOICE,T=THIRD CHOICE,NUMBER=OTHER SIZE
                                                                   * = COMMERCIAL SIZE
D                                                                  U.S.A.  AUSTRAL  JAPAN   EURO-  FRANCE  U.K.   GERMANY  ITALY
I   NOMINAL SIZE = AXB    SECTION AREA    MASS PER UNIT            ANSI    AS       JIS     NORM   NF      BS     DIN      UNI
N       IN             MM         IN**2    KG/M        LB/FT       B32.3            G3123                 4229   174      757
    MM

S  15 x 4   0.591X0.157  0.600E+02  0.930E-01  0.471E+00  0.316E+00                                              A+0-0.110  S
S       5      X0.197    0.750E+02  0.116E+00  0.589E+00  0.396E+00                                              B+0-0.075  S
S       6      X0.236    0.900E+02  0.140E+00  0.706E+00  0.475E+00                                              B+0-0.090  S
S       8      X0.315    0.120E+03  0.186E+00  0.942E+00  0.633E+00                                              B+0-0.090  S
S      10      X0.394    0.150E+03  0.233E+00  0.118E+01  0.791E+00                                              B+0-0.090  S

S  16 x 1.5  0.630X0.059  0.240E+02  0.372E-01  0.188E+00  0.127E+00  1.4S                                       A+0-0.110
F      1.6     X0.063     0.256E+02  0.397E-01  0.201E+00  0.135E+00                                             B+0-0.060  F
F       2      X0.079     0.320E+02  0.496E-01  0.251E+00  0.169E+00                                             B+0-0.060  F
F       2.5    X0.098     0.400E+02  0.620E-01  0.314E+00  0.211E+00                                             B+0-0.060  F
F       3      X0.118     0.480E+02  0.744E-01  0.377E+00  0.253E+00                                             B+0-0.060  F
F       4      X0.157     0.640E+02  0.992E-01  0.502E+00  0.338E+00                                             B+0-0.075  F
F       5      X0.197     0.800E+02  0.124E+00  0.628E+00  0.422E+00                                             B+0-0.075  F
F       6      X0.236     0.960E+02  0.149E+00  0.754E+00  0.506E+00                                             B+0-0.075  F
F       8      X0.315     0.128E+03  0.198E+00  0.100E+01  0.675E+00                                             B+0-0.090  F
F      10      X0.394     0.160E+03  0.248E+00  0.126E+01  0.844E+00                          9F                 B+0-0.090  F,12F

S  18 x 1.5  0.709X0.059  0.270E+02  0.419E-01  0.212E+00  0.142E+00                                             A+0-0.110
F      1.6     X0.063     0.288E+02  0.446E-01  0.226E+00  0.152E+00                                             B+0-0.060  F
F       2      X0.079     0.360E+02  0.558E-01  0.283E+00  0.190E+00                                             B+0-0.060  F
F       2.5    X0.098     0.450E+02  0.698E-01  0.353E+00  0.237E+00                                             B+0-0.060  F
F       3      X0.118     0.540E+02  0.837E-01  0.424E+00  0.285E+00                                             B+0-0.060  F
F       4      X0.157     0.720E+02  0.112E+00  0.565E+00  0.380E+00                                             B+0-0.075  F
F       5      X0.197     0.900E+02  0.140E+00  0.706E+00  0.475E+00   F                                         B+0-0.075  F
F       6      X0.236     0.108E+03  0.167E+00  0.848E+00  0.570E+00   F                                         B+0-0.075  F
F       8      X0.315     0.144E+03  0.223E+00  0.113E+01  0.760E+00   F                                         B+0-0.090  F
F      10      X0.394     0.180E+03  0.279E+00  0.141E+01  0.949E+00                                             B+0-0.090  F
F      12      X0.472     0.216E+03  0.335E+00  0.170E+01  0.114E+01                          9F                 B+0-0.110  F

S  20 x 1.5  0.787X0.059  0.300E+02  0.465E-01  0.235E+00  0.158E+00  1.4S                                       A+0-0.130
F      1.6     X0.063     0.320E+02  0.496E-01  0.251E+00  0.169E+00                                             B+0-0.060  F
F       2      X0.079     0.400E+02  0.620E-01  0.314E+00  0.211E+00                                             B+0-0.060  F
F       2.5    X0.098     0.500E+02  0.775E-01  0.392E+00  0.264E+00                                             B+0-0.060  F
F       3      X0.118     0.600E+02  0.930E-01  0.471E+00  0.316E+00                                             B+0-0.060  F
F       4      X0.157     0.800E+02  0.124E+00  0.628E+00  0.422E+00                                             B+0-0.075  F
F       5      X0.197     0.100E+03  0.155E+00  0.785E+00  0.527E+00                                             B+0-0.075  F
F       6      X0.236     0.120E+03  0.186E+00  0.942E+00  0.633E+00                                             B+0-0.075  F
F       8      X0.315     0.160E+03  0.248E+00  0.126E+01  0.844E+00                                             B+0-0.090  F
F      10      X0.394     0.200E+03  0.310E+00  0.157E+01  0.105E+01                                             B+0-0.090  F
F      12      X0.472     0.240E+03  0.372E+00  0.188E+01  0.127E+01   F                                         B+0-0.110  F
F      15      X0.591     0.300E+03  0.465E+00  0.235E+01  0.158E+01  1.4S                                       B+0-0.110  F
F      16      X0.630     0.320E+03  0.496E+00  0.251E+01  0.169E+01   F                                         B+0-0.110  F

F  22 x 2   0.866X0.079  0.440E+02  0.682E-01  0.345E+00  0.232E+00                                              A+0-0.130
F       3      X0.118     0.660E+02  0.102E+00  0.518E+00  0.348E+00                                             B+0-0.060  F,2.5F
F       4      X0.157     0.880E+02  0.136E+00  0.691E+00  0.464E+00                                             B+0-0.075  F
F       5      X0.197     0.110E+03  0.171E+00  0.863E+00  0.580E+00                                             B+0-0.075  F
F       6      X0.236     0.132E+03  0.205E+00  0.104E+01  0.696E+00                                             B+0-0.090  F
F       8      X0.315     0.176E+03  0.273E+00  0.138E+01  0.928E+00                                             B+0-0.090  F
F      10      X0.394     0.220E+03  0.341E+00  0.173E+01  0.116E+01   9F                                   F,16F B+0-0.090  F
F      12      X0.472     0.264E+03  0.409E+00  0.207E+01  0.139E+01   F,16F                                F,20F B+0-0.110  F,15F
```

STEEL MATERIAL DATA

Table 10-16 (*Continued*). **Bright Finish Rectangular Steel Bars with Sharp Corners (DIN 174)**

```
TABLE 10-16. BRIGHT FINISH RECTANGULAR                          ORDER EXAMPLE:
STEEL BARS WITH SHARP CORNERS (DIN 174)                         LENGTH,FLAT STEEL 5X2 DIN 174 + STEEL QUALITY
BASIS: 1 IN = 25.4 MM                                                                                    PAGE NO. 3
1 CUBIC METER STEEL = 7850 KG MASS

                                                    THE NOMINAL SIZE IS NATIONAL STANDARD AS INDICATED
                                                    F=FIRST CHOICE,S=SECOND CHOICE,T=THIRD CHOICE,NUMBER=OTHER SIZE
                                                    * = COMMERCIAL SIZE
D                                                   U.S.A.  AUSTRAL  JAPAN   EURO-   FRANCE  U.K.    GERMANY  ITALY
I  NOMINAL SIZE = AXB   SECTION AREA   MASS PER UNIT ANSI   AS       JIS     NORM    NF      BS      DIN      UNI
N     MM          IN      MM**2   IN**2    KG/M    LB/FT   B32.3             G3123                   4229    174      757
```

	Nominal	Section Area	Mass per unit	ANSI	AS	JIS	EURO	NF	BS	DIN	UNI
F	25 x 2 0.984X0.079	0.500E+02 0.775E-01	0.392E+00 0.264E+00							A+0-0.130	F
F	2.5 X0.098	0.625E+02 0.969E-01	0.491E+00 0.330E+00							B+0-0.060	F
F	3 X0.118	0.750E+02 0.116E+00	0.589E+00 0.396E+00							B+0-0.060	F
F	4 X0.157	0.100E+03 0.155E+00	0.785E+00 0.527E+00							B+0-0.075	F
F	5 X0.197	0.125E+03 0.194E+00	0.981E+00 0.659E+00			F				B+0-0.075	F
F	6 X0.236	0.150E+03 0.233E+00	0.118E+01 0.791E+00			F				B+0-0.075	F
F	8 X0.315	0.200E+03 0.310E+00	0.157E+01 0.105E+01			F				B+0-0.090	F
F	10 X0.394	0.250E+03 0.388E+00	0.196E+01 0.132E+01			9F				B+0-0.090	F
S	12 X0.472	0.300E+03 0.465E+00	0.235E+01 0.158E+01			F				B+0-0.110	F
S	15 X0.591	0.375E+03 0.581E+00	0.294E+01 0.198E+01	14S						B+0-0.110	F
F	16 X0.630	0.400E+03 0.620E+00	0.314E+01 0.211E+01							B+0-0.110	F
F	20 X0.787	0.500E+03 0.775E+00	0.392E+01 0.264E+01			19F				B+0-0.130	F,18F
F	28 x 2 1.102X0.079	0.560E+02 0.868E-01	0.440E+00 0.295E+00							A+0-0.130	
F	3 X0.118	0.840E+02 0.130E+00	0.659E+00 0.443E+00							B+0-0.060	F
F	4 X0.157	0.112E+03 0.174E+00	0.879E+00 0.591E+00							B+0-0.075	F
F	5 X0.197	0.140E+03 0.217E+00	0.110E+01 0.738E+00			F				B+0-0.075	F
F	6 X0.236	0.168E+03 0.260E+00	0.132E+01 0.886E+00			F				B+0-0.075	F
F	8 X0.315	0.224E+03 0.347E+00	0.176E+01 0.118E+01			F				B+0-0.090	F
F	10 X0.394	0.280E+03 0.434E+00	0.220E+01 0.148E+01			F				B+0-0.090	F
F	12 X0.472	0.336E+03 0.521E+00	0.264E+01 0.177E+01			F				B+0-0.110	F
S	15 X0.591	0.420E+03 0.651E+00	0.330E+01 0.222E+01							B+0-0.110	F
F	16 X0.630	0.448E+03 0.694E+00	0.352E+01 0.236E+01							B+0-0.110	F
F	20 X0.787	0.560E+03 0.868E+00	0.440E+01 0.295E+01							B+0-0.130	F,18F
F	30 x 2 1.181X0.079	0.600E+02 0.930E-01	0.471E+00 0.316E+00							A+0-0.130	
S	2.5 X0.098	0.750E+02 0.116E+00	0.589E+00 0.396E+00							B+0-0.060	
S	3 X0.118	0.900E+02 0.140E+00	0.706E+00 0.475E+00							B+0-0.060	
S	4 X0.157	0.120E+03 0.186E+00	0.942E+00 0.633E+00							B+0-0.075	
S	5 X0.197	0.150E+03 0.233E+00	0.118E+01 0.791E+00			F				B+0-0.075	
S	6 X0.236	0.180E+03 0.279E+00	0.141E+01 0.949E+00			F				B+0-0.075	
S	8 X0.315	0.240E+03 0.372E+00	0.188E+01 0.127E+01			F				B+0-0.090	F
S	10 X0.394	0.300E+03 0.465E+00	0.235E+01 0.158E+01			F				B+0-0.090	F
S	12 X0.472	0.360E+03 0.558E+00	0.283E+01 0.190E+01			F				B+0-0.110	F
T	15 X0.591	0.450E+03 0.698E+00	0.353E+01 0.237E+01	14S						B+0-0.110	F
S	16 X0.630	0.480E+03 0.744E+00	0.377E+01 0.253E+01							B+0-0.110	F
S	20 X0.787	0.600E+03 0.930E+00	0.471E+01 0.316E+01			F				B+0-0.130	F,18F
T	25 X0.984	0.750E+03 0.116E+01	0.589E+01 0.396E+01							B+0-0.130	
F	32 x 2 1.260X0.079	0.640E+02 0.992E-01	0.502E+00 0.338E+00							A+0-0.160	
F	2.5 X0.098	0.800E+02 0.124E+00	0.628E+00 0.422E+00							B+0-0.060	
F	3 X0.118	0.960E+02 0.149E+00	0.754E+00 0.506E+00							B+0-0.060	
F	4 X0.157	0.128E+03 0.198E+00	0.100E+01 0.675E+00							B+0-0.075	
F	5 X0.197	0.160E+03 0.248E+00	0.126E+01 0.844E+00			F				B+0-0.075	F
F	6 X0.236	0.192E+03 0.298E+00	0.151E+01 0.101E+01			F				B+0-0.075	F
F	8 X0.315	0.256E+03 0.397E+00	0.201E+01 0.135E+01			F				B+0-0.090	F
F	10 X0.394	0.320E+03 0.496E+00	0.251E+01 0.169E+01			9F				B+0-0.090	F
S	12 X0.472	0.384E+03 0.595E+00	0.301E+01 0.203E+01			F				B+0-0.110	F
S	15 X0.591	0.480E+03 0.744E+00	0.377E+01 0.253E+01			F				B+0-0.110	F
F	16 X0.630	0.512E+03 0.794E+00	0.402E+01 0.270E+01							B+0-0.110	F
F	20 X0.787	0.640E+03 0.992E+00	0.502E+01 0.338E+01			19F				B+0-0.130	F
F	25 X0.984	0.800E+03 0.124E+01	0.628E+01 0.422E+01			22F				B+0-0.130	

10-52

STEEL MATERIAL DATA

Table 10-16 (*Continued*). **Bright Finish Rectangular Steel Bars with Sharp Corners (DIN 174)**

```
TABLE 10-16. BRIGHT FINISH RECTANGULAR              ORDER EXAMPLE:
STEEL BARS WITH SHARP CORNERS (DIN 174)               LENGTH,FLAT STEEL 5X2 DIN 174 + STEEL QUALITY
BASIS; 1 IN = 25.4 MM                                                                PAGE NO. 4
CUBIC METER STEEL = 7850 KG MASS

                                                    THE NOMINAL SIZE IS NATIONAL STANDARD AS INDICATED
                                                    F=FIRST CHOICE,S=SECOND CHOICE,T=THIRD CHOICE,NUMBER=OTHER SIZE
                                                    * = COMMERCIAL SIZE
NOMINAL SIZE = AXB  SECTION AREA    MASS PER UNIT   U.S.A.  AUSTRAL JAPAN   EURO-  FRANCE  U.K.   GERMANY ITALY
  MM      IN          MM**2   IN**2   KG/M   LB/FT   ANSI    AS      JIS    NORM    NF     BS     DIN     UNI
                                                    B32.3           G3123                 4229   174     757

35 x 2   1.378x0.079 0.700E+02 0.109E+00 0.549E+00 0.369E+00                                     A+0=0.160
     2.5         0.098 0.875E+02 0.136E+00 0.687E+00 0.462E+00  F                                B+0=0.060
     3           0.118 0.105E+03 0.163E+00 0.824E+00 0.554E+00  F                                B+0=0.060 F
     4           0.157 0.140E+03 0.217E+00 0.110E+01 0.738E+00  F                                B+0=0.075 F
     5           0.197 0.175E+03 0.271E+00 0.137E+01 0.923E+00  F                                B+0=0.075 F
     6           0.236 0.210E+03 0.326E+00 0.165E+01 0.111E+01  F                                A+0=0.075 F
     8           0.315 0.280E+03 0.434E+00 0.220E+01 0.148E+01  F                                B+0=0.090 F
    10           0.394 0.350E+03 0.543E+00 0.275E+01 0.185E+01  F                                B+0=0.090 F
    12           0.472 0.420E+03 0.651E+00 0.330E+01 0.222E+01  F                                B+0=0.110 F
    15           0.591 0.525E+03 0.814E+00 0.412E+01 0.277E+01  F                                B+0=0.110 F
    20           0.787 0.700E+03 0.109E+01 0.549E+01 0.369E+01 14S                               B+0=0.130 F,1S
    25           0.984 0.875E+03 0.136E+01 0.687E+01 0.462E+01  F                                B+0=0.130

36 x 2   1.417x0.079 0.720E+02 0.112E+00 0.565E+00 0.380E+00                                     A+0=0.160
     2.5         0.098 0.900E+02 0.140E+00 0.706E+00 0.475E+00                                   B+0=0.060 F
     3           0.118 0.108E+03 0.167E+00 0.848E+00 0.570E+00                                   B+0=0.060 F
     4           0.157 0.144E+03 0.223E+00 0.113E+01 0.760E+00                                   B+0=0.075 F
     5           0.197 0.180E+03 0.279E+00 0.141E+01 0.949E+00                                   B+0=0.075 F
     6           0.236 0.216E+03 0.335E+00 0.170E+01 0.114E+01                                   B+0=0.075 F
     8           0.315 0.288E+03 0.446E+00 0.226E+01 0.152E+01                                   B+0=0.090 F
    10           0.394 0.360E+03 0.558E+00 0.283E+01 0.190E+01                                   B+0=0.090 F
    12           0.472 0.432E+03 0.670E+00 0.339E+01 0.228E+01                                   B+0=0.110 F
    15           0.591 0.540E+03 0.837E+00 0.424E+01 0.285E+01                                   B+0=0.110 F
    16           0.630 0.576E+03 0.893E+00 0.452E+01 0.304E+01                                   B+0=0.110 F
    20           0.787 0.720E+03 0.112E+01 0.565E+01 0.380E+01                                   B+0=0.130 F
    25           0.984 0.900E+03 0.140E+01 0.706E+01 0.475E+01                                   B+0=0.130

40 x 2   1.575x0.079 0.800E+02 0.124E+00 0.628E+00 0.422E+00                                     A+0=0.160
     3           0.118 0.120E+03 0.186E+00 0.942E+00 0.633E+00                                   B+0=0.060 F
     4           0.157 0.160E+03 0.248E+00 0.126E+01 0.844E+00                                   B+0=0.060 F
     5           0.197 0.200E+03 0.310E+00 0.157E+01 0.105E+01  F                                B+0=0.075 F
     6           0.236 0.240E+03 0.372E+00 0.188E+01 0.126E+01  F                                B+0=0.075 F
     8           0.315 0.320E+03 0.496E+00 0.251E+01 0.169E+01  F                                B+0=0.090 F
    10           0.394 0.400E+03 0.620E+00 0.314E+01 0.211E+01  F                                B+0=0.090 F
    12           0.472 0.480E+03 0.744E+00 0.377E+01 0.253E+01  F                                B+0=0.110 F
    15           0.591 0.600E+03 0.930E+00 0.471E+01 0.316E+01 14S                               B+0=0.110 F
    16           0.630 0.640E+03 0.992E+00 0.502E+01 0.338E+01  F                                B+0=0.110 F
    20           0.787 0.800E+03 0.124E+01 0.628E+01 0.422E+01  F                                B+0=0.130 F,1S
    25           0.984 0.100E+04 0.155E+01 0.785E+01 0.527E+01  F                                B+0=0.130 F
    30           1.181 0.120E+04 0.186E+01 0.942E+01 0.633E+01  F                                B+0=0.130 F
    32           1.260 0.128E+04 0.198E+01 0.100E+02 0.675E+01  S                                B+0=0.250 F

45 x 2   1.772x0.079 0.900E+02 0.140E+00 0.706E+00 0.475E+00                                     A+0=0.160
     3           0.118 0.135E+03 0.209E+00 0.106E+01 0.712E+00  S                                B+0=0.060 F
     4           0.157 0.180E+03 0.279E+00 0.141E+01 0.949E+00  S                                B+0=0.075 F
     5           0.197 0.225E+03 0.349E+00 0.177E+01 0.119E+01  S                                B+0=0.075 F
     6           0.236 0.270E+03 0.419E+00 0.212E+01 0.142E+01  S                                B+0=0.075 F
     8           0.315 0.360E+03 0.558E+00 0.283E+01 0.190E+01  S                                B+0=0.090 F
    10           0.394 0.450E+03 0.698E+00 0.353E+01 0.237E+01  S                                B+0=0.090 F
    12           0.472 0.540E+03 0.837E+00 0.424E+01 0.285E+01  S                                H+0=0.110 F
```

10-54

STEEL MATERIAL DATA

Table 10-16 (Continued). Bright Finish Rectangular Steel Bars with Sharp Corners (DIN 174)

```
TABLE 10-16. BRIGHT FINISH RECTANGULAR                              ORDER EXAMPLE:
STEEL BARS WITH SHARP CORNERS (DIN 174)                             LENGTH,FLAT STEEL 5X2 DIN 174 + STEEL QUALITY
BASIS: 1 IN = 25.4 MM                                                                                    PAGE NO. 5
1 CUBIC METER STEEL = 7850 KG MASS

                                                    THE NOMINAL SIZE IS NATIONAL STANDARD AS INDICATED
                                                    F=FIRST CHOICE,S=SECOND CHOICE,T=THIRD CHOICE,NUMBER=OTHER SIZE
                                                    * = COMMERCIAL SIZE
D                                                   U.S.A.  AUSTRAL JAPAN   EURO-   FRANCE U.K.    GERMANY ITALY
I                                                   ANSI    AS      JIS     NORM    NF     BS      DIN     UNI
N  NOMINAL SIZE = AXB    SECTION AREA   MASS PER UNIT B32.3         G3123                  4229    174     757
   MM        IN          MM**2   IN**2  KG/M   LB/FT
```

	Nominal	AxB (IN)	Section MM²	Section IN²	KG/M	LB/FT	ANSI B32.3	JIS G3123	BS 4229	DIN 174	UNI 757
F	45 x16	1.772x0.630	0.720E+03	0.112E+01	0.565E+01	0.380E+01	S			A+0=0.160	
F	20	X0.787	0.900E+03	0.140E+01	0.706E+01	0.475E+01	S		F	B+0=0.130	F,18F
F	25	X0.984	0.113E+04	0.174E+01	0.883E+01	0.593E+01	S		F	B+0=0.130	F
T	30	X1.181	0.135E+04	0.209E+01	0.106E+02	0.712E+01	S		F	B+0=0.130	F
F	32	X1.260	0.144E+04	0.223E+01	0.113E+02	0.760E+01	T	4OF		B+0=0.250	
F	50 x 2	1.969x0.079	0.100E+03	0.155E+00	0.785E+00	0.527E+00	F			A+0=0.160	
F	3	X0.118	0.150E+03	0.233E+00	0.118E+01	0.791E+00	F		F	B+0=0.060	F
F	4	X0.157	0.200E+03	0.310E+00	0.157E+01	0.105E+01	F			B+0=0.075	
F	5	X0.197	0.250E+03	0.388E+00	0.196E+01	0.132E+01	F		F	B+0=0.075	F
F	6	X0.236	0.300E+03	0.465E+00	0.235E+01	0.158E+01	F		F	B+0=0.075	F
F	8	X0.315	0.400E+03	0.620E+00	0.314E+01	0.211E+01	F		F	B+0=0.090	F
F	10	X0.394	0.500E+03	0.775E+00	0.392E+01	0.264E+01	F		F	B+0=0.090	F
F	12	X0.472	0.600E+03	0.930E+00	0.471E+01	0.316E+01	F		F	B+0=0.090	F
F	15	X0.591	0.750E+03	0.116E+01	0.589E+01	0.396E+01	F	9F	F	B+0=0.110	F
F	16	X0.630	0.800E+03	0.124E+01	0.628E+01	0.422E+01	14S		F	B+0=0.110	F
F	20	X0.787	0.100E+04	0.155E+01	0.785E+01	0.527E+01	F	F	F	B+0=0.110	F,18F
F	25	X0.984	0.125E+04	0.194E+01	0.981E+01	0.659E+01	F	19F	F	B+0=0.130	F
S	30	X1.181	0.150E+04	0.233E+01	0.118E+02	0.791E+01	F	F,22F	F	B+0=0.130	F
F	32	X1.260	0.160E+04	0.248E+01	0.126E+02	0.844E+01	S	4OF		B+0=0.250	
S	55 x 3	2.165x0.118	0.165E+03	0.256E+00	0.130E+01	0.870E+00				A+0=0.190	
S	3	X0.118	0.165E+03	0.256E+00	0.130E+01	0.870E+00				B+0=0.060	
S	4	X0.157	0.220E+03	0.341E+00	0.173E+01	0.116E+01				B+0=0.075	
S	5	X0.197	0.275E+03	0.426E+00	0.216E+01	0.145E+01				B+0=0.075	F,6F
S	8	X0.315	0.440E+03	0.682E+00	0.345E+01	0.232E+01				B+0=0.090	F
S	10	X0.394	0.550E+03	0.853E+00	0.432E+01	0.290E+01				B+0=0.090	F
S	12	X0.472	0.660E+03	0.102E+01	0.518E+01	0.348E+01			F	B+0=0.110	F
T	15	X0.591	0.825E+03	0.128E+01	0.648E+01	0.435E+01			F	B+0=0.110	F
S	16	X0.630	0.880E+03	0.136E+01	0.691E+01	0.464E+01			F	B+0=0.110	F
S	20	X0.787	0.110E+04	0.171E+01	0.863E+01	0.580E+01			F	B+0=0.130	F,18F
S	25	X0.984	0.138E+04	0.213E+01	0.108E+02	0.725E+01			F	B+0=0.130	F
F	30	X1.181	0.165E+04	0.256E+01	0.130E+02	0.870E+01			F	B+0=0.130	F
F	32	X1.260	0.176E+04	0.273E+01	0.138E+02	0.928E+01		4OF		B+0=0.250	
F	56 x 3	2.205x0.118	0.168E+03	0.260E+00	0.132E+01	0.886E+00				A+0=0.190	
F	3	X0.118	0.168E+03	0.260E+00	0.132E+01	0.886E+00				B+0=0.060	
F	4	X0.157	0.224E+03	0.347E+00	0.176E+01	0.118E+01				B+0=0.075	
F	5	X0.197	0.280E+03	0.434E+00	0.220E+01	0.148E+01				B+0=0.075	
F	8	X0.315	0.448E+03	0.694E+00	0.352E+01	0.236E+01				B+0=0.090	
F	10	X0.394	0.560E+03	0.868E+00	0.440E+01	0.295E+01				B+0=0.090	
F	12	X0.472	0.672E+03	0.104E+01	0.528E+01	0.354E+01				B+0=0.110	
F	15	X0.591	0.840E+03	0.130E+01	0.659E+01	0.443E+01				B+0=0.110	
F	16	X0.630	0.896E+03	0.139E+01	0.703E+01	0.473E+01				B+0=0.110	
F	20	X0.787	0.112E+04	0.174E+01	0.879E+01	0.591E+01				B+0=0.130	
F	25	X0.984	0.140E+04	0.217E+01	0.110E+02	0.738E+01				B+0=0.130	
F	32	X1.260	0.179E+04	0.278E+01	0.141E+02	0.945E+01				B+0=0.250	

STEEL MATERIAL DATA

Table 10-16 (Continued). Bright Finish Rectangular Steel Bars with Sharp Corners (DIN 174)

TABLE 10-16. BRIGHT FINISH RECTANGULAR
STEEL BARS WITH SHARP CORNERS (DIN 174)
BASIS: 1 IN = 25.4 MM
1 CUBIC METER STEEL = 7850 KG MASS

ORDER EXAMPLE:
LENGTH,FLAT STEEL 5X2 DIN 174 + STEEL QUALITY
PAGE NO. 6

THE NOMINAL SIZE IS NATIONAL STANDARD AS INDICATED
F=FIRST CHOICE,S=SECOND CHOICE,T=THIRD CHOICE,NUMBER=OTHER SIZE
* = COMMERCIAL SIZE

D I N	NOMINAL SIZE = AxB		SECTION AREA		MASS PER UNIT		U.S.A. ANSI B32.3	AUSTRAL AS	JAPAN JIS G3123	EURO- NORM	FRANCE NF	U.K. BS 4229	GERMANY DIN 174	ITALY UNI 757
	MM	IN	MM**2	IN**2	KG/M	LB/FT								
S	60 x 3	2.362X0.118	0.180E+03	0.279E+00	0.141E+01	0.949E+00	F						A+0-0.190	
S	4	X0.157	0.180E+03	0.279E+00	0.141E+01	0.949E+00	F						B+0-0.060	
S	5	X0.197	0.240E+03	0.372E+00	0.188E+01	0.127E+01	F						B+0-0.075	
S	6	X0.236	0.300E+03	0.465E+00	0.235E+01	0.158E+01	F						B+0-0.075	F
S	8	X0.315	0.360E+03	0.558E+00	0.283E+01	0.190E+01	F						B+0-0.090	F
S	10	X0.394	0.480E+03	0.744E+00	0.377E+01	0.253E+01	F						B+0-0.090	F
S	12	X0.472	0.600E+03	0.930E+00	0.471E+01	0.316E+01	F						B+0-0.090	F
S	15	X0.591	0.720E+03	0.112E+01	0.565E+01	0.380E+01	14S						B+0-0.110	F
T	16	X0.630	0.900E+03	0.140E+01	0.706E+01	0.475E+01	F						B+0-0.110	
S	20	X0.787	0.960E+03	0.149E+01	0.754E+01	0.506E+01	F						B+0-0.130	F,18F
S	25	X0.984	0.120E+04	0.186E+01	0.942E+01	0.633E+01	F						B+0-0.130	F
T	30	X1.181	0.150E+04	0.233E+01	0.118E+02	0.791E+01	F						B+0-0.130	
S	40	X1.575	0.180E+04	0.279E+01	0.141E+02	0.949E+01	F						B+0-0.250	
			0.240E+04	0.372E+01	0.188E+02	0.127E+02	F							
F	63 x 3	2.480X0.118	0.189E+03	0.293E+00	0.148E+01	0.997E+00							A+0-0.190	
F	3	X0.118	0.189E+03	0.293E+00	0.148E+01	0.997E+00							B+0-0.060	
F	5	X0.157	0.252E+03	0.391E+00	0.198E+01	0.133E+01							B+0-0.075	
F	5	X0.197	0.315E+03	0.488E+00	0.247E+01	0.166E+01							B+0-0.075	F
F	6	X0.236	0.378E+03	0.586E+00	0.297E+01	0.199E+01							B+0-0.090	F
F	8	X0.315	0.504E+03	0.781E+00	0.396E+01	0.266E+01							B+0-0.090	F
F	10	X0.394	0.630E+03	0.977E+00	0.495E+01	0.332E+01							B+0-0.110	F
F	12	X0.472	0.756E+03	0.117E+01	0.593E+01	0.399E+01							B+0-0.110	F
F	15	X0.591	0.945E+03	0.146E+01	0.742E+01	0.498E+01							B+0-0.110	F
F	16	X0.630	0.101E+04	0.156E+01	0.791E+01	0.532E+01							B+0-0.130	F
F	20	X0.787	0.126E+04	0.195E+01	0.989E+01	0.665E+01							B+0-0.130	
F	25	X0.984	0.158E+04	0.244E+01	0.124E+02	0.831E+01							B+0-0.250	
F	32	X1.260	0.202E+04	0.312E+01	0.158E+02	0.106E+02							B+0-0.250	
S	40	X1.575	0.252E+04	0.391E+01	0.198E+02	0.133E+02								
S	65 x 4	2.559X0.157	0.260E+03	0.403E+00	0.204E+01	0.137E+01	3F						A+0-0.190	
S	5	X0.197	0.325E+03	0.504E+00	0.255E+01	0.171E+01	F						B+0-0.075	F
S	6	X0.236	0.390E+03	0.605E+00	0.306E+01	0.206E+01							B+0-0.075	F
	8	X0.315	0.520E+03	0.806E+00	0.408E+01	0.274E+01							B+0-0.090	F
	10	X0.394	0.650E+03	0.101E+01	0.510E+01	0.343E+01	F						B+0-0.090	F
	12	X0.472	0.780E+03	0.121E+01	0.612E+01	0.411E+01							B+0-0.110	F
	15	X0.591	0.975E+03	0.151E+01	0.765E+01	0.514E+01							B+0-0.110	F
	16	X0.630	0.104E+04	0.161E+01	0.816E+01	0.549E+01	F,30F						B+0-0.110	F
	20	X0.787	0.130E+04	0.202E+01	0.124E+02	0.686E+01	F,40F						B+0-0.130	F,18F
	25	X0.984	0.163E+04	0.252E+01	0.128E+02	0.857E+01	F,50F						B+0-0.130	F
F	70 x 4	2.756X0.157	0.280E+03	0.434E+00	0.220E+01	0.148E+01							A+0-0.190	
F	5	X0.197	0.350E+03	0.543E+00	0.275E+01	0.185E+01							B+0-0.075	F
F	6	X0.236	0.420E+03	0.651E+00	0.330E+01	0.222E+01							B+0-0.075	F
F	8	X0.315	0.560E+03	0.868E+00	0.440E+01	0.295E+01							B+0-0.090	F
F	10	X0.394	0.700E+03	0.109E+01	0.549E+01	0.369E+01							B+0-0.090	F
F	12	X0.472	0.840E+03	0.130E+01	0.659E+01	0.443E+01							B+0-0.110	F
S	15	X0.591	0.105E+04	0.163E+01	0.824E+01	0.554E+01							B+0-0.110	F
F	16	X0.630	0.112E+04	0.174E+01	0.879E+01	0.591E+01							B+0-0.110	

10-55

Table 10-16 (Continued). Bright Finish Rectangular Steel Bars with Sharp Corners (DIN 174)

```
TABLE 10-16. BRIGHT FINISH RECTANGULAR                                    ORDER EXAMPLE;
STEEL BARS WITH SHARP CORNERS (DIN 174)                                   LENGTH,FLAT STEEL 5X2 DIN 174 + STEEL QUALITY
BASIS; 1 IN = 25.4 MM                                                                                              PAGE NO. 7
1 CUBIC METER STEEL = 7850 KG MASS

                                                            THE NOMINAL SIZE IS NATIONAL STANDARD AS INDICATED
                                                            F=FIRST CHOICE,S=SECOND CHOICE,T=THIRD CHOICE,NUMBER=OTHER SIZE
                                                            *= COMMERCIAL SIZE
D                                                           U.S.A.  AUSTRAL  JAPAN    EURO-    FRANCE  U.K.    GERMANY  ITALY
I                                                           ANSI    AS       JIS      NORM     NF      BS      DIN      UNI
N   NOMINAL SIZE = AXB   SECTION AREA    MASS PER UNIT      B32.3            G3123                    4229     174      757
    MM         IN        MM**2    IN**2   KG/M     LB/FT

F 70 X20    2.756X0.787  0.140E+04 0.217E+01 0.110E+02 0.738E+01                                              A+0-0.190  18F
F      20                0.140E+04 0.217E+01 0.110E+02 0.738E+01                                              B+0-0.130  F
F      25                0.175E+04 0.271E+01 0.137E+02 0.923E+01                                              B+0-0.130  F
S      30                0.210E+04 0.326E+01 0.165E+02 0.111E+02                                              B+0-0.130  F
F      40                0.280E+04 0.434E+01 0.220E+02 0.148E+02                                              B+0-0.250

F 80 X 5    3.150X0.197  0.400E+03 0.620E+00 0.314E+01 0.211E+01    F,3F                                      A+0-0.190
F      6                 0.480E+03 0.744E+00 0.377E+01 0.253E+01    F                                         B+0-0.075
F      8                 0.640E+03 0.992E+00 0.502E+01 0.338E+01    F                                         B+0-0.090
F     10                 0.800E+03 0.124E+01 0.628E+01 0.422E+01    F                                         B+0-0.090
F     12                 0.960E+03 0.149E+01 0.754E+01 0.506E+01    F                                         B+0-0.090
F     15                 0.120E+04 0.186E+01 0.942E+01 0.633E+01    F                                         B+0-0.110
S     16                 0.128E+04 0.198E+01 0.100E+02 0.675E+01    F                                         B+0-0.110
F     20                 0.160E+04 0.248E+01 0.126E+02 0.844E+01    F                                         B+0-0.110
F     25                 0.200E+04 0.310E+01 0.157E+02 0.105E+02    F                                         B+0-0.130
S     30                 0.240E+04 0.372E+01 0.188E+02 0.127E+02    F                                         B+0-0.130
S     40                 0.320E+04 0.496E+01 0.251E+02 0.169E+02    F                                         B+0-0.130
S     50                 0.400E+04 0.620E+01 0.314E+02 0.211E+02    F                                         B+0-0.250

F 90 X 5    3.543X0.197  0.450E+03 0.698E+00 0.353E+01 0.237E+01    S                                         A+0-0.220
S      6                 0.540E+03 0.837E+00 0.424E+01 0.285E+01    S                                         B+0-0.075
F      8                 0.720E+03 0.112E+01 0.565E+01 0.380E+01    S                                         B+0-0.090
S     10                 0.900E+03 0.140E+01 0.706E+01 0.475E+01    S                                         B+0-0.090
F     12                 0.108E+04 0.167E+01 0.848E+01 0.570E+01    S                                         B+0-0.110
S     15                 0.135E+04 0.209E+01 0.106E+02 0.712E+01    S                                         B+0-0.110
F     16                 0.144E+04 0.223E+01 0.113E+02 0.760E+01  14T                                         B+0-0.110
F     20                 0.180E+04 0.279E+01 0.141E+02 0.949E+01    S                                         B+0-0.110
F     25                 0.225E+04 0.349E+01 0.177E+02 0.119E+02    S                                         B+0-0.130
F     30                 0.270E+04 0.419E+01 0.212E+02 0.142E+02    S                                         B+0-0.130
S     32                 0.288E+04 0.446E+01 0.226E+02 0.152E+02    T                                         B+0-0.130
S     40                 0.360E+04 0.558E+01 0.283E+02 0.190E+02    S    F,50F                                B+0-0.250

F100 X 5    3.937X0.197  0.500E+03 0.775E+00 0.392E+01 0.264E+01    F                                         A+0-0.220
F      6                 0.600E+03 0.930E+00 0.471E+01 0.316E+01    F                                         B+0-0.075
S      8                 0.800E+03 0.124E+01 0.628E+01 0.422E+01    F                                         B+0-0.090
F     10                 0.100E+04 0.155E+01 0.785E+01 0.527E+01    F                                         B+0-0.090
F     12                 0.120E+04 0.186E+01 0.942E+01 0.633E+01    F                                         B+0-0.110
S     15                 0.150E+04 0.233E+01 0.118E+02 0.791E+01    F                                         B+0-0.110
F     16                 0.160E+04 0.248E+01 0.126E+02 0.844E+01  14S                                         B+0-0.110
F     20                 0.200E+04 0.310E+01 0.157E+02 0.105E+02    F                                         B+0-0.130
F     25                 0.250E+04 0.387E+01 0.196E+02 0.132E+02    F    F,25F                                B+0-0.130
F     30                 0.300E+04 0.465E+01 0.235E+02 0.158E+02    F                                         B+0-0.130
S     40                 0.400E+04 0.620E+01 0.314E+02 0.211E+02    F                                         B+0-0.250
S     50                 0.500E+04 0.775E+01 0.392E+02 0.264E+02    F                                         B+0-0.250
```

STEEL MATERIAL DATA

Table 10-17. Bright Steel for Parallel Keys-Square Section (DIN 6880)

```
TABLE 10-17. BRIGHT STEEL FOR PARALLEL
KEYS - SQUARE SECTION (DIN 6880)
BASIS; 1 IN = 25.4 MM
1 CUBIC METER STEEL = 7850 KG MASS
```

ORDER EXAMPLE;
LENGTH,KEY STEEL 6X6 DIN 6880 + STEEL QUALITY

THE NOMINAL SIZE IS NATIONAL STANDARD AS INDICATED
F=FIRST CHOICE,S=SECOND CHOICE,T=THIRD CHOICE,NUMBER=OTHER SIZE
* = COMMERCIAL SIZE

	U.S.A. ANSI	AUSTRAL AS	JAPAN JIS	EURO- NORM	FRANCE NF	U.K. BS 4235	GERMANY DIN 6880	ITALY UNI

D I N	NOMINAL SIZE = AXB		SECTION AREA		MASS PER UNIT			
	MM	IN	MM**2	IN**2	KG/M	LB/FT		
S	2 x 2	0.079X0.079	0.400E+01	0.620E-02	0.314E-01	0.211E-01	S A+0-0.025	
	2	X0.079	0.400E+01	0.620E-02	0.314E-01	0.211E-01	B+0-0.025	
S	3 x 3	0.118X0.118	0.900E+01	0.140E-01	0.706E-01	0.475E-01	S A+0-0.025	
	3	X0.118	0.900E+01	0.140E-01	0.706E-01	0.475E-01	B+0-0.025	
S	4 x 4	0.157X0.157	0.160E+02	0.248E-01	0.126E+00	0.844E-01	S A+0-0.030	
	4	X0.157	0.160E+02	0.248E-01	0.126E+00	0.844E-01	B+0-0.030	
F	5 x 5	0.197X0.197	0.250E+02	0.388E-01	0.196E+00	0.132E+00	F A+0-0.030	
	5	X0.197	0.250E+02	0.388E-01	0.196E+00	0.132E+00	B+0-0.030	
F	6 x 6	0.236X0.236	0.360E+02	0.558E-01	0.283E+00	0.190E+00	F A+0-0.030	
	6	X0.236	0.360E+02	0.558E-01	0.283E+00	0.190E+00	B+0-0.030	
F	7 x 7	0.276X0.276	0.490E+02	0.760E-01	0.385E+00	0.258E+00	F A+0-0.036	
	7	X0.276	0.490E+02	0.760E-01	0.385E+00	0.258E+00	B+0-0.036	
F	8 x 8	0.315X0.315	0.640E+02	0.992E-01	0.502E+00	0.338E+00	F A+0-0.036	
	8	X0.315	0.640E+02	0.992E-01	0.502E+00	0.338E+00	B+0-0.036	
F	10 x10	0.394X0.394	0.100E+03	0.155E+00	0.785E+00	0.527E+00	F A+0-0.036	
	10	X0.394	0.100E+03	0.155E+00	0.785E+00	0.527E+00	B+0-0.036	
F	12 x12	0.472X0.472	0.144E+03	0.223E+00	0.113E+01	0.760E+00	F A+0-0.043	
	12	X0.472	0.144E+03	0.223E+00	0.113E+01	0.760E+00	B+0-0.043	
F	14 x14	0.551X0.551	0.196E+03	0.304E+00	0.154E+01	0.103E+01	F A+0-0.043	
	14	X0.551	0.196E+03	0.304E+00	0.154E+01	0.103E+01	B+0-0.043	
F	16 x16	0.630X0.630	0.256E+03	0.397E+00	0.201E+01	0.135E+01	F A+0-0.043	
	16	X0.630	0.256E+03	0.397E+00	0.201E+01	0.135E+01	B+0-0.043	
F	18 x18	0.709X0.709	0.324E+03	0.502E+00	0.254E+01	0.171E+01	F A+0-0.043	
	18	X0.709	0.324E+03	0.502E+00	0.254E+01	0.171E+01	B+0-0.043	
F	20 x20	0.787X0.787	0.400E+03	0.620E+00	0.314E+01	0.211E+01	F A+0-0.052	
	20	X0.787	0.400E+03	0.620E+00	0.314E+01	0.211E+01	B+0-0.052	
F	22 x22	0.866X0.866	0.484E+03	0.750E+00	0.380E+01	0.255E+01	F A+0-0.052	
	22	X0.866	0.484E+03	0.750E+00	0.380E+01	0.255E+01	B+0-0.052	

STEEL MATERIAL DATA

Table 10-18. Bright Steel for Parallel Keys-Rectangular Section (DIN 6880)

```
TABLE 10-18. BRIGHT STEEL FOR PARALLEL                                   ORDER EXAMPLE:
KEYS - RECTANGULAR SECTION (DIN 6880)                                    LENGTH,KEY STEEL 12X8 DIN 6880 + STEEL QUALITY
BASIS; 1 IN = 25.4 MM
1 CUBIC METER STEEL = 7850 KG MASS
                                                                         THE NOMINAL SIZE IS NATIONAL STANDARD AS INDICATED
D                                                                        F=FIRST CHOICE,S=SECOND CHOICE,T=THIRD CHOICE,NUMBER=OTHER SIZE
I                                                                        * = COMMERCIAL SIZE
N  NOMINAL SIZE = AXB   SECTION AREA       MASS PER UNIT                 U.S.A.  AUSTRAL JAPAN   EURO-   FRANCE  U.K.    GERMANY ITALY
   MM         IN        MM**2     IN**2    KG/M       LB/FT              ANSI    AS      JIS     NORM    NF      BS      DIN     UNI
                                                                                                                 4235    6880

S   8 x  7   0.315x0.276 0.560E+02 0.868E-01 0.440E+00 0.295E+00                                                         S A+0-0.036
                   x0.276 0.560E+02 0.868E-01 0.440E+00 0.295E+00                                                          B+0-0.090

S  10 x  8   0.394x0.315 0.800E+02 0.124E+00 0.628E+00 0.422E+00                                                         S A+0-0.036
                   x0.315 0.800E+02 0.124E+00 0.628E+00 0.422E+00                                                          B+0-0.036

F  12 x  8   0.472x0.315 0.960E+02 0.149E+00 0.754E+00 0.506E+00                                                         F A+0-0.043
                   x0.315 0.960E+02 0.149E+00 0.754E+00 0.506E+00                                                          B+0-0.036

S  12 x 10   0.472x0.394 0.120E+03 0.186E+00 0.942E+00 0.633E+00                                                         S A+0-0.043
                   x0.394 0.120E+03 0.186E+00 0.942E+00 0.633E+00                                                          B+0-0.036

F  14 x  9   0.551x0.354 0.126E+03 0.195E+00 0.989E+00 0.665E+00                                                         F A+0-0.043
                   x0.354 0.126E+03 0.195E+00 0.989E+00 0.665E+00                                                          B+0-0.090

F  16 x 10   0.630x0.394 0.160E+03 0.248E+00 0.126E+01 0.844E+00                                                         F A+0-0.043
                   x0.394 0.160E+03 0.248E+00 0.126E+01 0.844E+00                                                          B+0-0.090

F  18 x 11   0.709x0.433 0.198E+03 0.307E+00 0.155E+01 0.104E+01                                                         F A+0-0.043
                   x0.433 0.198E+03 0.307E+00 0.155E+01 0.104E+01                                                          B+0-0.110

F  20 x 12   0.787x0.472 0.240E+03 0.372E+00 0.188E+01 0.127E+01                                                         F A+0-0.052
                   x0.472 0.240E+03 0.372E+00 0.188E+01 0.127E+01                                                          B+0-0.110

F  22 x 14   0.866x0.551 0.308E+03 0.477E+00 0.242E+01 0.162E+01                                                         F A+0-0.052
                   x0.551 0.308E+03 0.477E+00 0.242E+01 0.162E+01                                                          B+0-0.110

F  25 x 14   0.984x0.551 0.350E+03 0.543E+00 0.275E+01 0.185E+01                                                         F A+0-0.052
                   x0.551 0.350E+03 0.543E+00 0.275E+01 0.185E+01                                                          B+0-0.110

S  25 x 22   0.984x0.866 0.550E+03 0.853E+00 0.432E+01 0.290E+01                                                         S A+0-0.052
                   x0.866 0.550E+03 0.853E+00 0.432E+01 0.290E+01                                                          B+0-0.130

F  28 x 16   1.102x0.630 0.448E+03 0.694E+00 0.352E+01 0.236E+01                                                         F A+0-0.052
                   x0.630 0.448E+03 0.694E+00 0.352E+01 0.236E+01                                                          B+0-0.110

S  28 x 25   1.102x0.984 0.700E+03 0.109E+01 0.549E+01 0.369E+01                                                         S A+0-0.052
                   x0.984 0.700E+03 0.109E+01 0.549E+01 0.369E+01                                                          B+0-0.130

F  32 x 18   1.260x0.709 0.576E+03 0.893E+00 0.452E+01 0.304E+01                                                         F A+0-0.062
                   x0.709 0.576E+03 0.893E+00 0.452E+01 0.304E+01                                                          B+0-0.110

S  32 x 30   1.260x1.181 0.960E+03 0.149E+01 0.754E+01 0.506E+01                                                         S A+0-0.062
                   x1.181 0.960E+03 0.149E+01 0.754E+01 0.506E+01                                                          B+0-0.130
```

PAGE NO. 1

10-58

STEEL MATERIAL DATA

10-59

Table 10-18 (Continued). Bright Steel for Parallel Keys-Rectangular Section (DIN 6880)

```
TABLE 10-18. BRIGHT STEEL FOR PARALLEL                           ORDER EXAMPLE:
KEYS - RECTANGULAR SECTION (DIN 6880)                            LENGTH,KEY STEEL 12X8 DIN 6880 + STEEL QUALITY
BASIS: 1 IN = 25.4 MM                                                                                        PAGE NO. 2
1 CUBIC METER STEEL = 7850 KG MASS

                                                THE NOMINAL SIZE IS NATIONAL STANDARD AS INDICATED
                                                F=FIRST CHOICE,S=SECOND CHOICE,T=THIRD CHOICE,NUMBER=OTHER SIZE
                                                * = COMMERCIAL SIZE
                                                U.S.A.  AUSTRAL  JAPAN   EURO-   FRANCE   U.K.     GERMANY  ITALY
                                                ANSI    AS       JIS     NORM    NF       BS       DIN      UNI
                                                                                          4235     6880
D
I                                    MASS PER UNIT
N  NOMINAL SIZE = AXB   SECTION AREA
   MM    IN             MM**2    IN**2      KG/M       LB/FT

F  36 x20  1.417x0.787  0.720E+03  0.112E+01  0.565E+01  0.380E+01                                         F  A+0-0.062
      20   X0.787       0.720E+03  0.112E+01  0.565E+01  0.380E+01                                            B+0-0.130

S  36 x3*  1.417x1.339  0.122E+04  0.190E+01  0.961E+01  0.646E+01                                         S  A+0-0.062
      3*   X1.339       0.122E+04  0.190E+01  0.961E+01  0.646E+01                                            B+0-0.160

F  40 x22  1.575x0.866  0.880E+03  0.136E+01  0.691E+01  0.464E+01                                         F  A+0-0.062
      22   X0.866       0.880E+03  0.136E+01  0.691E+01  0.464E+01                                            B+0-0.130

S  40 x38  1.575x1.496  0.152E+04  0.236E+01  0.119E+02  0.802E+01                                         S  A+0-0.062
      38   X1.496       0.152E+04  0.236E+01  0.119E+02  0.802E+01                                            B+0-0.160

S  45 x25  1.772x0.984  0.113E+04  0.174E+01  0.883E+01  0.593E+01                                         S  A+0-0.062
      25   X0.984       0.113E+04  0.174E+01  0.883E+01  0.593E+01                                            B+0-0.130

S  45 x43  1.772x1.693  0.194E+04  0.300E+01  0.152E+02  0.102E+02                                         S  A+0-0.062
      43   X1.693       0.194E+04  0.300E+01  0.152E+02  0.102E+02                                            B+0-0.160

S  50 x28  1.969x1.102  0.140E+04  0.217E+01  0.110E+02  0.738E+01                                         S  A+0-0.062
      28   X1.102       0.140E+04  0.217E+01  0.110E+02  0.738E+01                                            B+0-0.130

S  50 x48  1.969x1.890  0.240E+04  0.372E+01  0.188E+02  0.127E+02                                         S  A+0-0.062
      48   X1.890       0.240E+04  0.372E+01  0.188E+02  0.127E+02                                            B+0-0.160

S  56 x32  2.205x1.260  0.179E+04  0.278E+01  0.141E+02  0.945E+01                                         S  A+0-0.074
      32   X1.260       0.179E+04  0.278E+01  0.141E+02  0.945E+01                                            B+0-0.160

S  63 x32  2.480x1.260  0.202E+04  0.312E+01  0.158E+02  0.106E+02                                         S  A+0-0.074
      32   X1.260       0.202E+04  0.312E+01  0.158E+02  0.106E+02                                            B+0-0.160

S  70 x36  2.756x1.417  0.252E+04  0.391E+01  0.198E+02  0.133E+02                                         S  A+0-0.074
      36   X1.417       0.252E+04  0.391E+01  0.198E+02  0.133E+02                                            B+0-0.160

S  80 x40  3.150x1.575  0.320E+04  0.496E+01  0.251E+02  0.169E+02                                         S  A+0-0.074
      40   X1.575       0.320E+04  0.496E+01  0.251E+02  0.169E+02                                            B+0-0.160

T  90 x45  3.543x1.772  0.405E+04  0.628E+01  0.318E+02  0.214E+02                                         T  A+0-0.087
      45   X1.772       0.405E+04  0.628E+01  0.318E+02  0.214E+02                                            B+0-0.160

S 100 x50  3.937x1.969  0.500E+04  0.775E+01  0.392E+02  0.264E+02                                         S  A+0-0.087
      50   X1.969       0.500E+04  0.775E+01  0.392E+02  0.264E+02                                            B+0-0.160
```

Table 10-19. Bright Steel for Parallel Keys-Flat Section (DIN 6880)

TABLE 10-19. BRIGHT STEEL FOR PARALLEL
KEYS-FLAT SECTION (DIN 6880)
BASIS: 1 IN = 25.4 MM
1 CUBIC METER STEEL = 7850 KG MASS

ORDER EXAMPLE:
LENGTH,KEY STEEL 7X4 DIN 6880 + STEEL QUALITY

THE NOMINAL SIZE IS NATIONAL STANDARD AS INDICATED
F=FIRST CHOICE,S=SECOND CHOICE,T=THIRD CHOICE,NUMBER=OTHER SIZE
* = COMMERCIAL SIZE

D I N	NOMINAL SIZE = AXB MM / IN	SECTION AREA MM**2 / IN**2	MASS PER UNIT KG/M / LB/FT	U.S.A. ANSI	AUSTRAL AS	JAPAN JIS	EURO- NORM	FRANCE NF	U.K. BS	GERMANY DIN 6880	ITALY UNI
S	5 x 3	0.197×0.118 / 0.118	0.150E+02 / 0.233E-01	0.118E+00 / 0.791E-01							S A+0-0.030 / B+0-0.060
S	6 x 4	0.236×0.157 / 0.157	0.240E+02 / 0.372E-01	0.188E+00 / 0.127E+00							S A+0-0.030 / B+0-0.075
F	7 x 4	0.276×0.157 / 0.157	0.280E+02 / 0.434E-01	0.220E+00 / 0.148E+00							F A+0-0.036 / B+0-0.030
F	8 x 5	0.315×0.197 / 0.197	0.400E+02 / 0.620E-01	0.314E+00 / 0.211E+00							F A+0-0.036 / B+0-0.030
F	10 x 6	0.394×0.236 / 0.236	0.600E+02 / 0.930E-01	0.471E+00 / 0.316E+00							F A+0-0.036 / B+0-0.030
F	12 x 6	0.472×0.236 / 0.236	0.720E+02 / 0.112E+00	0.565E+00 / 0.380E+00							F A+0-0.043 / B+0-0.075
F	14 x 6	0.551×0.236 / 0.236	0.840E+02 / 0.130E+00	0.659E+00 / 0.443E+00							F A+0-0.043 / B+0-0.075
F	16 x 7	0.630×0.276 / 0.276	0.112E+03 / 0.174E+00	0.879E+00 / 0.591E+00							F A+0-0.043 / B+0-0.090
F	18 x 7	0.709×0.276 / 0.276	0.126E+03 / 0.195E+00	0.989E+00 / 0.665E+00							F A+0-0.043 / B+0-0.090
F	20 x 8	0.787×0.315 / 0.315	0.160E+03 / 0.248E+00	0.126E+01 / 0.844E+00							F A+0-0.052 / B+0-0.090
F	22 x 9	0.866×0.354 / 0.354	0.198E+03 / 0.307E+00	0.155E+01 / 0.104E+01							F A+0-0.052 / B+0-0.090
F	25 x 9	0.984×0.354 / 0.354	0.225E+03 / 0.349E+00	0.177E+01 / 0.119E+01							F A+0-0.052 / B+0-0.090
F	28 x 10	1.102×0.394 / 0.394	0.280E+03 / 0.434E+00	0.220E+01 / 0.148E+01							F A+0-0.052 / B+0-0.090
F	32 x 11	1.260×0.433 / 0.433	0.352E+03 / 0.546E+00	0.276E+01 / 0.186E+01							F A+0-0.062 / B+0-0.110
F	36 x 12	1.417×0.472 / 0.472	0.432E+03 / 0.670E+00	0.339E+01 / 0.228E+01							F A+0-0.062 / B+0-0.110
S	40 x 14	1.575×0.551 / 0.551	0.560E+03 / 0.868E+00	0.440E+01 / 0.295E+01							S A+0-0.062 / B+0-0.110

STEEL MATERIAL DATA

Table 10-20. Hot Rolled Steel Bars for Woodruff Keys-Half Rounds (EURONORM 66)

```
TABLE 10-20. HOT ROLLED STEEL BARS FOR WOODRUFF KEYS,
HALF ROUNDS (EURONORM 66)
BASIS; 1 IN. = 25.4 MM
1 CUBIC METER STEEL = 7850 KG MASS
```

ORDER EXAMPLE;
LENGTH,HALF ROUNDS 16X8 EURONORM 66,+ST QUALITY

THE NOMINAL SIZE IS NATIONAL STANDARD AS INDICATED
F=FIRST CHOICE,S=SECOND CHOICE,T=THIRD CHOICE,NUMBER=OTHER SIZE
* = COMMERCIAL SIZE

E U R	NOMINAL SIZE = AXB MM IN	SECTION AREA MM**2 IN**2	MASS PER UNIT KG/M LB/FT	U.S.A. ANSI	AUSTRAL AS	JAPAN JIS	EURO- NORM 66	FRANCE NF	U.K. BS	GERMANY DIN 1018	ITALY UNI
F	16 x 8	0.630X0.315 0.101E+03 0.156E+00	0.789E+00 0.530E+00				F A+=1 B+=0.5			F	
F	20 x10	0.787X0.394 0.157E+03 0.243E+00	0.123E+01 0.829E+00				F A+=1 B+=0.5			F	
	26 x13	1.024X0.512 0.265E+03 0.411E+00	0.208E+01 0.140E+01				A+=1 B+=0.5			F	
	30 x15	1.181X0.591 0.353E+03 0.548E+00	0.277E+01 0.186E+01				A+=1 B+=0.6			F	
F	60 x30	2.362X1.181 0.141E+04 0.219E+01	0.111E+02 0.746E+01				F A+=1.2 B+=1.2			F	
F	75 x37.5	2.953X1.476 0.221E+04 0.342E+01	0.173E+02 0.117E+02				F A+=1.5 B+=1.5			F	

Table 10-21. Hot Rolled Steel Bars for Woodruff Keys-Half Ovals (EURONORM 66)

TABLE 10-21. HOT ROLLED STEEL BARS FOR WOODRUFF KEYS
HALF OVALS (EURONORM 66)
BASIS, 1 IN = 25.4 MM
1 CUBIC METER STEEL = 7850 KG MASS

ORDER EXAMPLE:
LENGTH,HALF OVALS 16X3 EURONORM 66 + ST QUAL

THE NOMINAL SIZE IS NATIONAL STANDARD AS INDICATED
F=FIRST CHOICE,S=SECOND CHOICE,T=THIRD CHOICE,NUMBER=OTHER SIZE
* = COMMERCIAL SIZE

EUR	NOMINAL SIZE = AXB			SECTION AREA			MASS PER UNIT		U.S.A. ANSI	AUSTRAL AS	JAPAN JIS	EURO-NORM 66	FRANCE NF	U.K. BS	GERMANY DIN 1018	ITALY UNI
	A	B	C	A	B	C	KG/M	LB/FT								
	14 × 4		16.2	0.551 ×	0.157	0.640	0.311E+00	0.209E+00				A*=1 B*=0.5				
F	16 × 3		24.3	0.630 ×	0.118	0.958	0.258E+00	0.173E+00				F A*=1 B*=0.5			F	F
	16 × 3.5		21.8	0.630 ×	0.138	0.858	0.304E+00	0.204E+00				A*=1 B*=0.5			F	F
	18 × 3.2		28.5	0.709 ×	0.126	1.123	0.309E+00	0.208E+00				A*=1 B*=0.5			F	F
	20 × 6.5		21.9	0.787 ×	0.256	0.862	0.735E+00	0.494E+00				F A*=1 B*=0.5			F	F
F	25 × 8		27.5	0.984 ×	0.315	1.084	0.113E+01	0.758E+00				A*=1 B*=0.5			F	F
	28 × 6		38.7	1.102 ×	0.236	1.522	0.911E+00	0.612E+00				F A*=1 B*=0.5			F	F
F	33 × 8		42.0	1.299 ×	0.315	1.655	0.144E+01	0.971E+00				A*=1 B*=0.5			F	F
F	35 × 10		40.6	1.378 ×	0.394	1.599	0.195E+01	0.131E+01				A*=1 B*=0.5			F	F
	40 × 10		50.0	1.575 ×	0.394	1.969	0.219E+01	0.147E+01				A*=1 B*=0.5			F	F
	50 × 12		64.1	1.969 ×	0.472	2.523	0.328E+01	0.220E+01				A*=1 B*=0.5			F	F
	75 × 18		96.1	2.953 ×	0.709	3.784	0.738E+01	0.496E+01				A*=1.5 B*=0.6			F	F
	100 × 25		125.0	3.937 ×	0.984	4.921	0.137E+02	0.922E+01				A*=1.5 B*=0.8			F	F

STEEL MATERIAL DATA

Table 10-22A. Seamless (SL) and Welded (WE) Steel Tubes for General Use (ISO 336, 134, 221, 2937)

TABLE 10-22A. SEAMLESS(SL) AND WELDED(WE) STEEL TUBES
FOR GENERAL USE (ISO 336,134,221,2937)
BASIS; 1 IN = 25.4 MM,1 CUBIC METER
STEEL = 7850 KG MASS

ORDER EXAMPLE:
LENGTH,SEAMLESS TUBE 10.2x1.6 DIN2448 + ST QUA
PAGE NO. 1

THE NOMINAL SIZE IS NATIONAL STANDARD AS INDICATED
F=FIRST CHOICE,S=SECOND CHOICE,T=THIRD CHOICE,NUMBER=OTHER SIZE
L=LIGHT,N=NORMAL,H=HEAVY SERIES

ISO	MILLIMETERS A	B	C	INCHES A	B	C	MASS PER UNIT KG/M	LB/FT SL WE	U.S.A. ANSI B36.1	U.S.A. ANSI B36.1	AUSTRAL AS 1238	AUSTRAL AS 1238	JAPAN JIS G3452,4,5	JAPAN JIS G3452,4,5	FRANCE NF A49-110	FRANCE NF A49-140	U.K. BS 1775	U.K. BS 1775	GERMANY DIN 2448	GERMANY DIN 2458	ITALY UNI 4991	ITALY UNI 7091
F 10.2	0.5	9.2	0.020	0.402	0.363		0.120E+00	0.804E-01	ISO C=0.6 0.8 1 1.2 1.4 1.8													
	1.6	7.0	0.063		0.276		0.339E+00	0.228E+00	FF	L	FF		FF						FF	N,L	FF	
	2.0	6.2	0.079		0.245		0.404E+00	0.272E+00	FF		FF		FF						FF		FF	
	2.3	5.6	0.091		0.221		0.448E+00	0.301E+00	FF		FF		FF								F	
N	2.6	5.0	0.102		0.197		0.487E+00	0.327E+00	FF	N	FF		2.4FF							N		N
F 12.0	0.5	11.0	0.020	0.472	0.433		0.142E+00	0.953E-01	ISO C=0.6 0.8 1 1.2 1.4 1.8 2 2.3 2.6 2.9 3.2													
	1.6	8.8	0.063		0.346		0.410E+00	0.276E+00	FF		FF								F		F	N
F 13.5	0.5	12.5	0.020	0.531	0.492		0.160E+00	0.108E+00	ISO C=0.6 0.8 1 1.2 1.4 2.6 3.2													
	1.6	10.3	0.063		0.405		0.470E+00	0.316E+00	FF										F		FF	
	1.8	9.9	0.071		0.389		0.519E+00	0.349E+00	FF	L									FF	N,L	FF	N
	2.0	9.5	0.079		0.374		0.567E+00	0.381E+00	FF				2.2FF						FF		FF	
	2.3	8.9	0.091		0.350		0.635E+00	0.427E+00	FF	L			FF		F	L			FF		F	
	2.9	7.7	0.114		0.303		0.758E+00	0.509E+00	FF	N			FF		FF	N			F			N
F 16.0	0.5	15.0	0.020	0.630	0.591		0.191E+00	0.128E+00	ISO C=0.6 0.8 1 1.2 1.4 1.6 2 2.3 2.6 2.9 3.2 3.6 4													
	1.8	12.4	0.071		0.488		0.630E+00	0.424E+00	FF										F	N,L		
F 17.2	0.5	16.2	0.020	0.677	0.638		0.206E+00	0.138E+00	ISO C=0.6 0.8 1 1.2 1.4 2.6 3.6 4.5													
	1.6	14.0	0.063		0.551		0.616E+00	0.414E+00	FF										F		FF	
	1.8	13.6	0.071		0.535		0.684E+00	0.459E+00	FF										FF	N,L	FF	N
	2.0	13.2	0.079		0.520		0.750E+00	0.504E+00	FF										FF		FF	
	2.3	12.6	0.091		0.496		0.845E+00	0.568E+00	FF	L			FF		F	L			FF		F	
	2.9	11.4	0.114		0.449		0.102E+01	0.687E+00	FF	N			FF		FF	N			FF		F	
	3.2	10.8	0.126		0.425		0.110E+01	0.742E+00	FF				FF		FF	H			F			
F 19.0	0.5	18.0	0.020	0.750	0.711		0.228E+00	0.153E+00	ISO C=0.6 0.8 1 1.2 1.4 1.6 1.8 2 2.3 2.6 2.9 3.2 3.6 4 4.5													
F 20.0	0.5	19.0	0.020	0.787	0.748		0.240E+00	0.162E+00	ISO C=0.6 0.8 1 1.2 1.4 1.6 1.8 2 2.3 2.6 2.9 3.2 3.6 4 4.5													
	1.6	16.8	0.063		0.661		0.726E+00	0.488E+00	FF										F		FF	
	2.0	16.0	0.079		0.630		0.888E+00	0.597E+00	FF										FF	N,L	FF	N
F 21.3	0.5	20.3	0.020	0.840	0.801		0.256E+00	0.172E+00	ISO C=0.6 0.8 1 1.2 1.4 1.6 4 4.5 5.4													
	1.8	17.7	0.071		0.698		0.866E+00	0.582E+00	FF										FF		FF	
	2.0	17.3	0.079		0.683		0.952E+00	0.640E+00	FF										FF		FF	N
	2.3	16.7	0.091		0.659		0.108E+01	0.724E+00	FF										FF		FF	
	2.6	16.1	0.102		0.635		0.120E+01	0.806E+00	FF										FF		FF	
	2.9	15.5	0.114		0.612		0.132E+01	0.884E+00	FF				FF				F		FF		F	
	3.2	14.9	0.126		0.588		0.143E+01	0.960E+00	FF				FF						FF		F	
	3.6	14.1	0.142		0.557		0.157E+01	0.106E+01	FF				FF		F	H			F			
	5.0	11.3	0.197		0.446		0.201E+01	0.135E+01	4.8FF				FF		FF	H			F			
N	7.1	7.1	0.280		0.281		0.249E+01	0.167E+01	7.5FF	H	FF											
F 25.0	0.5	24.0	0.020	0.984	0.945		0.302E+00	0.203E+00	ISO C=0.6 0.8 1 1.2 1.4 1.8 2.3 2.6 2.9 3.2 3.6 4 4.5 5.4 5.6													
	1.6	21.8	0.063		0.858		0.923E+00	0.620E+00	FF				FF						FF	N,L	FF	
	2.0	21.0	0.079		0.827		0.113E+01	0.762E+00	FF				FF						FF		FF	N
	6.3	12.4	0.248		0.488		0.291E+01	0.195E+01	ISO C=5.9										F			

10-64   **STEEL MATERIAL DATA**

Table 10-22A (Continued). Seamless (SL) and Welded (WE) Steel Tubes for General Use (ISO 336, 134, 221, 2937)

```
TABLE 10-22A, SEAMLESS(SL) AND WELDED(WE) STEEL TUBES                           ORDER EXAMPLE;
FOR GENERAL USE (ISO 336,134,221,2937)                                          LENGTH,SEAMLESS TUBE 10*2X1.6 DIN2448 + ST QUA
BASIS; 1 IN = 25.4 MM, 1 CUBIC METER                                            PAGE NO. 2
STEEL = 7850 KG MASS
```

```
                                                 THE NOMINAL SIZE IS NATIONAL STANDARD AS INDICATED
                                                 F=FIRST CHOICE,S=SECOND CHOICE,T=THIRD CHOICE,NUMBER=OTHER SIZE
                                                 L=LIGHT, N=NORMAL,H=HEAVY SERIES
I            MILLIMETERS     INCHES    MASS PER UNIT     U.S.A.    AUSTRAL  JAPAN           FRANCE  U.K.         GERMANY ITALY
S            A    B    C     A    B    C     KG/M    LB/FT  SL     ANSI   AS    JIS              NF    BS           DIN    UNI
0                                                            WE    B36.1  1238  G3452.1.5        A49-110 1775       2448   4991
                                                                   B36.1  1238  G3452.4.5        A49-140 1775       2458   7091

F   26.9   25.9  0.5   0.020  1.059  0.326E+00   0.219E+00   ISO C=0.6  0.8  1.1.2  1.4  1.6  3.6  4.5  5  5.4  5.9  6.3  7.1
        1.8 23.3       0.071  0.917  0.111E+01   0.749E+00         FF                                    F      FF              FF
L       2.0 22.9       0.079  0.902  0.123E+01   0.825E+00         FF                                    FF     FF              FF
        2.3 22.3       0.091  0.878  0.140E+01   0.938E+00         FF                            F   L   N          FF     N
N       2.6 21.7       0.102  0.854  0.156E+01   0.105E+01         FF                            FF  L              FF
        2.9 21.1       0.114  0.831  0.172E+01   0.115E+01         FF  L                         FF  L   N          FF     N
        3.2 20.5       0.126  0.807  0.187E+01   0.126E+01   FF  L                               FF                 FF
        3.6 19.7       0.142  0.776  0.207E+01   0.139E+01   FF  N                                                  FF
        4.0 18.9       0.157  0.744  0.226E+01   0.152E+01   FF  N                                                  F
        5.6 15.7       0.220  0.618  0.294E+01   0.198E+01   FF  H                                                  F
        8.0 10.9       0.315  0.429  0.373E+01   0.251E+01

S   30.0   29.0  0.5   0.020  1.142  0.364E+00   0.244E+00   ISO C=0.6  0.8  1.1.2  1.4  1.8  2.9  3.2  3.6  4  4.5  5  5.4  5.6  5.9  6.3
        1.6 26.8       0.063  1.055  0.112E+01   0.753E+00                                                                   FF
L       2.0 26.0       0.079  1.024  0.138E+01   0.928E+00                                                                   FF         FF
        2.3 25.4       0.091  1.000  0.157E+01   0.106E+01                                       FF  L   N                   FF         N
N       2.6 24.8       0.102  0.976  0.176E+01   0.118E+01                                       FF  L                       F
        8.0 14.0       0.315  0.551  0.434E+01   0.292E+01                                       F                           F

    31.8   30.8  0.5   0.020  1.211  0.386E+00   0.259E+00   ISO C=0.6  0.8  1.1.2  1.4  1.6  1.8
        2.0 27.8       0.079  1.093  0.147E+01   0.988E+00
L       2.6 26.6       0.102  1.045  0.187E+01   0.126E+01
        8.0 15.8       0.315  0.620  0.470E+01   0.316E+01   ISO C=7.1

F   33.7   32.7  0.5   0.020  1.288  0.409E+00   0.275E+00   ISO C=0.6  0.8  1.1.2  1.4  1.6  3.6  4  4.5  5  5.4  5.6  5.9  6.3  7.1
        2.0 29.7       0.079  1.170  0.156E+01   0.105E+01         FF                                                        FF
L       2.3 29.1       0.091  1.146  0.178E+01   0.120E+01         FF                                    F                   FF
        2.6 28.5       0.102  1.122  0.199E+01   0.134E+01         FF                            FF  L   N                   FF         N
N       2.9 27.9       0.114  1.099  0.220E+01   0.148E+01         FF  L                         FF  L                       FF
        3.2 27.3       0.126  1.075  0.241E+01   0.162E+01         FF  N                         FF  L   N                   FF         N
        3.6 26.5       0.142  1.044  0.267E+01   0.180E+01   3.4FF  L  FF  N                     FF                          FF
        4.0 25.7       0.157  1.012  0.293E+01   0.197E+01   FF  N                                                           FF
        4.5 24.7       0.177  0.973  0.324E+01   0.218E+01   FF  N                                                           F
        5.0 23.7       0.197  0.933  0.354E+01   0.238E+01   FF  H                                                           F
        6.3 21.1       0.248  0.831  0.426E+01   0.286E+01                                                                   F
        8.8 16.1       0.346  0.634  0.540E+01   0.363E+01   ISO C=5.4  5.6  5.9  7.1  8

S   38.0   37.0  0.5   0.020  1.461  0.462E+00   0.311E+00   ISO C=0.6  0.8  1.1.2  1.4  1.6  1.8
        1.6 34.8       0.063  1.343  0.178E+01   0.119E+01
L       2.3 33.4       0.091  1.319  0.205E+01   0.136E+01                                                                   FF
        2.6 32.8       0.102  1.295  0.227E+01   0.153E+01                                       FF  L   N                   FF         N
        10.0 18.0      0.394  0.713  0.691E+01   0.464E+01

F   42.4   41.4  0.5   0.020  1.630  0.517E+00   0.347E+00   ISO C=0.6  0.8  1.1.2  1.4  1.6  3.6  4  4.5  5  5.4  5.6  5.9  6.3  7.1  8
        1.8 38.4       0.071  1.512  0.195E+01   0.134E+01         FF                                    F                   FF
L       2.3 37.8       0.091  1.488  0.222E+01   0.153E+01         FF                                                        FF
        2.6 37.2       0.102  1.464  0.255E+01   0.171E+01         FF                            FF  L   N                   FF         N
        2.9 36.6       0.114  1.441  0.288E+01   0.190E+01         FF                            FF  L                       FF
N       3.2 36.0       0.126  1.417  0.309E+01   0.208E+01         FF                            FF  L   N                   FF         N
        3.6 35.2       0.142  1.386  0.344E+01   0.231E+01   FF  L                               FF  H                       FF
        4.0 34.4       0.157  1.354  0.379E+01   0.255E+01   FF  N                                                           FF
```

STEEL MATERIAL DATA

10-65

Table 10-22A (Continued). Seamless (SL) and Welded (WE) Steel Tubes for General Use (ISO 336, 134, 221, 2937)

TABLE 10-22A. SEAMLESS(SL) AND WELDED(WE) STEEL TUBES
FOR GENERAL USE (ISO 336,134,221,2937)
BASIS; 1 IN = 25.4 MM, 1 CUBIC METER
STEEL = 7850 KG MASS

PAGE NO. 3

ORDER EXAMPLE;
LENGTH,SEAMLESS TUBE 10.2X1.6 DIN 2448 + ST QUA

THE NOMINAL SIZE IS NATIONAL STANDARD AS INDICATED
F=FIRST CHOICE, S=SECOND CHOICE, T=THIRD CHOICE, NUMBER=OTHER SIZE
L=LIGHT, N=NORMAL, H=HEAVY SERIES

	MILLIMETERS		INCHES			MASS PER UNIT		U.S.A. ANSI B36.1	AUSTRAL AS 1238	JAPAN JIS G3452,4,5	FRANCE NF A49-110	U.K. BS 1775	GERMANY DIN 2448	ITALY UNI 4991
I S O	A	B C	A	C	B	KG/M	LB/FT				A49-140	1775	2458	7091

	42.4	4.5	33.4	1.669	0.177	1.315	0.421E+01	0.283E+01	FF	FF	FF			FF	FF
F		5.0	32.4		0.197	1.275	0.461E+01	0.310E+01	FF N	FF	FF			FF	FF
		6.3	29.8		0.248	1.173	0.561E+01	0.377E+01	FF	FF			5.4F	F	F
		10.0	22.4		0.394	0.882	0.799E+01	0.537E+01	FF H	FF					
		11.0	20.4		0.433	0.803	0.852E+01	0.572E+01							
	44.5	0.5	43.5	1.750	0.020	1.711	0.543E+00	0.365E+00	ISO C=5.4 5.6 5.9 7.1 8 8.8						
S		2.0	40.5		0.079	1.593	0.210E+01	0.141E+01	ISO C=0.6 0.8 1 1.2 1.4 1.6 1.8						
		2.3	39.9		0.091	1.569	0.239E+01	0.161E+01		FF	F			F	FF
L		2.9	38.7		0.114	1.545	0.269E+01	0.181E+01			FF N			FF N	FF N
N		12.5	19.5		0.492	0.766	0.986E+01	0.663E+01			8.8 10 11		8 8.8 10 11		
	48.3	0.5	47.3	1.900	0.020	1.861	0.589E+00	0.396E+00	ISO C=2.9 3.2 3.6 4 4.5 5 5.4 5.6 5.9 6.3 7.1 8						
F		2.3	43.7		0.091	1.719	0.261E+01	0.175E+01		FF				F L	FF
L		2.9	42.5		0.114	1.695	0.293E+01	0.197E+01		FF				FF N	FF N
N		3.2	41.9		0.126	1.672	0.325E+01	0.218E+01		FF				FF	FF
		3.6	41.1		0.142	1.648	0.356E+01	0.239E+01	FF L	FF L		FF L N		FF	FF
		4.0	40.3		0.157	1.617	0.397E+01	0.267E+01				FF H		FF	FF
		4.5	39.3		0.177	1.585	0.437E+01	0.294E+01	FF	FF N				FF	FF
		5.0	38.3		0.197	1.546	0.486E+01	0.327E+01	FF N	FF				FF	FF
		5.6	37.1		0.220	1.506	0.534E+01	0.359E+01	FF	FF				FF	FF
		7.1	34.1		0.280	1.459	0.590E+01	0.396E+01	FF H	FF				F	F
		10.0	28.3		0.394	1.341	0.721E+01	0.485E+01		FF			5.9F	F	F
		12.5	23.3		0.492	1.113	0.945E+01	0.635E+01	ISO C=5.4 5.6 6.3 8 8.8 11						
						0.916	0.110E+02	0.742E+01							
	51.0	0.5	50.0	2.000	0.020	1.961	0.623E+00	0.418E+00	ISO C=0.6 0.8 1 1.2 1.4 1.6 1.8 2						
F		2.3	46.4		0.091	1.819	0.276E+01	0.186E+01		FF			FF	F L	S
L		2.6	45.8		0.102	1.795	0.310E+01	0.209E+01		FF				FF N	S N
N		14.2	22.6		0.559	0.882	0.129E+02	0.866E+01					8 8.8 10 11		12.5
	54.0	0.5	53.0	2.125	0.020	2.086	0.660E+00	0.443E+00	ISO C=0.6 0.8 1 1.2 1.4 1.8						
S		1.6	50.8		0.063	1.999	0.207E+01	0.139E+01		FF				FF	FF
		2.0	50.0		0.079	1.968	0.256E+01	0.172E+01		FF				FF	FF
		2.3	49.4		0.091	1.944	0.293E+01	0.197E+01		FF				FF	FF
		2.6	48.8		0.102	1.920	0.330E+01	0.221E+01		FF				FF	FF
		2.9	48.2		0.114	1.897	0.365E+01	0.246E+01		FF				FF	FF N
		14.2	25.6		0.559	1.007	0.139E+02	0.937E+01					8 8.8 10 11		12.5
	57.0	0.5	56.0	2.250	0.020	2.211	0.697E+00	0.468E+00	ISO C=3.2 3.6 4 4.5 5 5.4 5.6 5.9 6.3 7.1 8						
S		1.6	53.8		0.063	2.124	0.219E+01	0.147E+01		FF				F	FF
		2.0	53.0		0.079	2.093	0.271E+01	0.182E+01		FF				F L	FF
L		2.3	52.4		0.091	2.069	0.310E+01	0.208E+01		FF				FF L	FF
		2.6	51.8		0.102	2.045	0.349E+01	0.234E+01		FF				FF	FF
		2.9	51.2		0.114	2.022	0.387E+01	0.260E+01		FF				FF N	FF N
		16.0	25.0		0.630	0.990	0.162E+02	0.109E+02					8 8.8 10 11		12.5 14.2
	60.3	0.5	59.3	2.375	0.020	2.336	0.737E+00	0.495E+00	ISO C=0.6 0.8 1 1.2 1.4 1.6 1.8 2 2.6						
F		2.3	55.7		0.091	2.194	0.221E+01		2.1FF					F L	FF
L		2.9	54.5		0.114	2.147	0.411E+01	0.276E+01	FF	FF			F	FF	FF
N		3.2	53.9		0.126	2.123	0.451E+01	0.303E+01	FF	FF			F	FF	FF N

Table 10-22A (Continued). Seamless (SL) and Welded (WE) Steel Tubes for General Use (ISO 336, 134, 221, 2937)

TABLE 10-22A. SEAMLESS(SL) AND WELDED(WE) STEEL TUBES
FOR GENERAL USE (ISO 336,134,221,2937)
BASIS: 1 IN = 25.4 MM, 1 CUBIC METER
STEEL = 7850 KG MASS

ORDER EXAMPLE:
LENGTH,SEAMLESS TUBE 10.2X1.6 DIN2448 + ST QUA
PAGE NO. 4

THE NOMINAL SIZE IS NATIONAL STANDARD AS INDICATED
F=FIRST CHOICE,S=SECOND CHOICE,T=THIRD CHOICE,NUMBER=OTHER SIZE
L=LIGHT,N=NORMAL,H=HEAVY SERIES

	MILLIMETERS			INCHES			MASS PER UNIT		U.S.A. ANSI B36.1	AUSTRAL AS 1238	JAPAN JIS G3452,4,5	FRANCE NF A49-110 A49-140	U.K. BS 1775 1775	GERMANY DIN 2448 2458	ITALY UNI 7091
ISO	A	B	C	A	C	B	KG/M SL	LB/FT WE							

(Table data for nominal sizes 60.3, 63.5, 70.0, 73.0, 76.1 mm with columns for thickness, mass per unit, and national standard designations omitted for brevity)

STEEL MATERIAL DATA
10-66

STEEL MATERIAL DATA

Table 10-22A (Continued). Seamless (SL) and Welded (WE) Steel Tubes for General Use (ISO 336, 134, 221, 2937)

TABLE 10-22A. SEAMLESS(SL) AND WELDED(WE) STEEL TUBES
FOR GENERAL USE (ISO 336,134,221,2937)
BASIS; 1 IN = 25.4 MM, 1 CUBIC METER
STEEL = 7850 KG MASS

ORDER EXAMPLE;
LENGTH,SEAMLESS TUBE 10.2x1.6 DIN2448 + ST QUA
PAGE NO. 5

THE NOMINAL SIZE IS NATIONAL STANDARD AS INDICATED
F=FIRST CHOICE,S=SECOND CHOICE,T=THIRD CHOICE,NUMBER=OTHER SIZE
L=LIGHT,N=NORMAL,H=HEAVY SERIES

	MILLIMETERS			INCHES			MASS PER UNIT		U.S.A. ANSI B36.1	AUSTRAL AS 1238	JAPAN JIS G3452,4,5	FRANCE NF A49-110 A49-140	U.K. BS 1775 1775	GERMANY DIN 2448 2458	ITALY UNI 4991 7091
I S O	A	B	C	A	B	C	KG/M	LB/FT SL WE	B36.1	1238	G3452,4,5				

	82.5	0.5	81.5	3.250	0.020	3.211	0.101E+01	0.679E+00	ISO	C=0.6	C=0.8	1	1.2	1.4	1.6	1.8	2	2.3	2.9	3.6	4	4.5	5
		2.6	77.3		0.102	3.045	0.512E+01	0.344E+01														F	L	S
		3.2	76.1		0.126	2.998	0.626E+01	0.421E+01													FF	N	S	N
		22.2	38.1		0.874	1.502	0.330E+02	0.222E+02											14.2	16	17.5	20
F	88.9	0.5	87.9	3.500	0.020	3.461	0.109E+01	0.732E+00	ISO	C=0.6	0.8	1	1.2	1.4	1.6	1.8	2	2.3	2.6				
		2.0	84.9		0.079	3.343	0.429E+01	0.288E+01	FF														
		2.9	83.1		0.114	3.272	0.615E+01	0.413E+01	FF												F	F	
		3.2	82.5		0.126	3.248	0.676E+01	0.454E+01	FF												FF	FF	
		3.6	81.7		0.142	3.217	0.757E+01	0.509E+01	FF												FF	FF	
L		4.0	80.9		0.157	3.185	0.838E+01	0.563E+01	FF	L								F	L		FF	FF	N
		4.5	79.9		0.177	3.146	0.937E+01	0.629E+01	FF									FF	N		FF	FF	
N		5.0	78.9		0.197	3.106	0.103E+02	0.695E+01	FF	N								FF	H		FF	FF	
		5.4	78.1		0.213	3.075	0.111E+02	0.747E+01	4.8FF												FF	FF	
		5.6	77.7		0.220	3.059	0.115E+02	0.773E+01	FF	L								F			FF	FF	
		6.3	76.3		0.248	3.004	0.128E+02	0.862E+01	FF									F			FF	FF	
		7.1	74.7		0.280	2.941	0.143E+02	0.962E+01	7.6FF												FF	FF	
		8.0	72.9		0.315	2.870	0.160E+02	0.107E+02	6.6FF												FF	FF	
		11.0	66.9		0.433	2.634	0.211E+02	0.142E+02	7.6FF	N											FF	FF	
		16.0	56.9		0.630	2.240	0.288E+02	0.193E+02	15.2FF	H											FF	FF	
		25.0	38.9		0.984	1.531	0.394E+02	0.265E+02	ISO	C=5.9	8.8	10	12.5	14.2	17.5	20	22.2						
F	101.6	0.5	100.6	4.000	0.020	3.961	0.125E+01	0.838E+00	ISO	C=0.6	0.8	1	1.2	1.4	1.6	1.8	2.3	2.6					
		2.0	97.6		0.079	3.843	0.491E+01	0.330E+01	FF														
		2.9	95.8		0.114	3.772	0.706E+01	0.474E+01	FF											F	F		
		3.2	95.2		0.126	3.748	0.777E+01	0.522E+01	FF											FF	FF		
		3.6	94.4		0.142	3.717	0.870E+01	0.585E+01	FF											FF	FF		
L		4.0	93.6		0.157	3.685	0.963E+01	0.647E+01	FF	L							F	L		FF	FF		
		4.5	92.6		0.177	3.646	0.108E+02	0.724E+01	FF									SS	N		FF	FF		
N		5.0	91.6		0.197	3.606	0.119E+02	0.800E+01	FF	N							SS	H		FF	FF		
		5.6	90.4		0.220	3.559	0.131E+02	0.891E+01	4.8FF												FF	FF		
		6.3	89.0		0.248	3.504	0.148E+02	0.995E+01	FF L												FF	FF		
		7.1	87.4		0.280	3.441	0.165E+02	0.111E+02	FF												FF	FF		
		8.0	85.6		0.315	3.370	0.185E+02	0.124E+02	FF N												FF	FF		
		28.0	45.6		1.102	1.795	0.508E+02	0.342E+02	ISO	C=5.4	5.9	8.8	10	11	12.5	14.2	16	17.5	20	22.2	25		
S	108x0	0.5	107.0	4.250	0.020	4.211	0.133E+01	0.891E+00	ISO	C=0.6	0.8	1	1.2	1.4	1.6	1.8	2.3	4	4.5	5.4	5.6	5.9	6.3	7.1
		2.0	104.0		0.079	4.093	0.523E+01	0.351E+01	FF													F	F	
		2.9	102.2		0.114	4.045	0.676E+01	0.454E+01	FF													FF	FF	
		3.2	101.6		0.126	4.022	0.752E+01	0.505E+01	FF													FF	FF	
		3.6	100.8		0.142	3.998	0.827E+01	0.556E+01	FF										F	L		FF	FF	
		4.5	99.0		0.177	3.967	0.922E+01	0.623E+01	FF													FF	FF	
		30.0	48.0		1.181	1.888	0.577E+02	0.388E+02	ISO	C=8	8.8	10	11	12.5	14.2	16	17.5	20	22.2	25	28		
F	114.3	0.5	113.3	4.500	0.020	4.461	0.140E+01	0.943E+00	ISO	C=0.6	0.8	1	1.2	1.4	1.6	1.8	2.3	2.6					
		2.0	110.3		0.079	4.343	0.554E+01	0.372E+01	FF													F	F	
		2.6	109.1		0.102	4.295	0.716E+01	0.481E+01	FF													FF	FF	
		2.9	108.5		0.114	4.272	0.797E+01	0.535E+01	FF													FF	FF	
L		3.2	107.9		0.126	4.248	0.877E+01	0.589E+01	FF										F	L	F	FF	FF	
N		3.6	107.1		0.142	4.217	0.983E+01	0.660E+01	FF										FF	N		FF	FF	N
		4.0	106.3		0.157	4.185	0.109E+02	0.731E+01	FF													FF	FF	

STEEL MATERIAL DATA

Table 10-22A (Continued). Seamless (SL) and Welded (WE) Steel Tubes for General Use (ISO 336, 134, 221, 2937)

TABLE 10-22A. SEAMLESS(SL) AND WELDED(WE) STEEL TUBES.
FOR GENERAL USE (ISO 336,134,221,2937)
BASIS; 1 IN = 25.4 MM, 1 CUBIC METER
STEEL = 7850 KG MASS

ORDER EXAMPLE:
LENGTH, SEAMLESS TUBE 10.2X1.6 DIN2448 + ST QUA
PAGE NO. 6

THE NOMINAL SIZE IS NATIONAL STANDARD AS INDICATED
F=FIRST CHOICE, S=SECOND CHOICE, T=THIRD CHOICE, NUMBER=OTHER SIZE
L=LIGHT, N=NORMAL, H=HEAVY SERIES

ISO	MILLIMETERS A C B	INCHES A	C	B	MASS PER UNIT KG/M	LB/FT	U.S.A. ANSI B36.1	AUSTRAL AS 1238	JAPAN JIS G3452,4,5	FRANCE NF A49-110	U.K. BS 1775	GERMANY DIN 2448	ITALY UNI 4991
							B36.1	1238	G3452,4,5	A49-140	1775	2458	7091
F 114.3	4.5 105.3	4.500	0.177	4.146	0.122E+02	0.819E+01	FF	FF	FF		FF	FF	FF
	5.0 104.3		0.197	4.106	0.135E+02	0.906E+01	4.8FF					FF	FF
	5.4 103.5		0.213	4.075	0.145E+02	0.975E+01	5.2FF					FF	FF
	5.6 103.1		0.220	4.059	0.150E+02	0.101E+02	FF				FF	FF	FF
	5.9 102.5		0.232	4.035	0.158E+02	0.106E+02	FF L	FF N				FF	FF
	6.3 101.7		0.248	4.004	0.168E+02	0.113E+02	FF					FF	FF
	7.1 100.1		0.280	3.941	0.188E+02	0.126E+02	FF		FF			FF	FF
	8.8 96.7		0.346	3.807	0.229E+02	0.154E+02	FF N					FF	FF
	11.0 92.3		0.433	3.634	0.280E+02	0.188E+02	FF					FF	FF
	14.2 85.9		0.559	3.382	0.350E+02	0.236E+02	13.5FF						FF
	17.5 79.3		0.689	3.122	0.418E+02	0.281E+02	17.1FF H						FF
	32.0 50.3		1.260	1.980	0.649E+02	0.436E+02	ISO C=10 12.5 16 20 22.2 25 28 30						
127.0	0.6 125.8	5.000	0.024	4.953	0.187E+01	0.126E+01	ISO C=0.8 1 1.2 1.4 1.6 1.8 2.3 2.6 2.9 3.6 4.5 5.4 5.6 5.9						
	3.2 120.6		0.126	4.748	0.977E+01	0.657E+01						FL S	N
	4.0 119.0		0.157	4.685	0.121E+02	0.815E+01						FF N	
	36.0 55.0		1.417	2.165	0.808E+02	0.543E+02	ISO C=6.3 7.1 8 8.8 10 11 12.5 14.2 16 17.5 20 22.2 25 28 30 32						
S 133.0	0.6 131.8	5.250	0.024	5.203	0.196E+01	0.132E+01	ISO C=0.8 1 1.2 1.4 1.6 1.8 2.3 2.6 2.9 3.2 4.5 5.4 5.6 5.9						
L	3.6 125.8		0.142	4.967	0.115E+02	0.772E+01						FF L	FF N
N	4.0 125.0		0.157	4.935	0.127E+02	0.855E+01						FF	FF
	36.0 61.0		1.417	2.415	0.861E+02	0.579E+02	ISO C=6.3 7.1 8 8.8 10 11 12.5 14.2 16 17.5 20 22.2 25 28 30 32						
F 139.7	0.6 138.5	5.500	0.024	5.453	0.206E+01	0.138E+01	ISO C=0.8 1 1.2 1.4 1.6 1.8 2.3 2.6 2.9 3.2 5.9 7.1 8.8						
L	3.6 132.5		0.142	5.217	0.121E+02	0.812E+01						FF L	FF N
N	4.0 131.7		0.157	5.185	0.134E+02	0.900E+01						FF	FF
	4.5 130.7		0.177	5.146	0.150E+02	0.101E+02	FF	FF				FF	FF
	5.0 129.7		0.197	5.106	0.166E+02	0.112E+02	FF					FF	FF
	5.4 128.9		0.213	5.075	0.179E+02	0.120E+02	FF					FF	FF
	5.6 128.5		0.220	5.059	0.185E+02	0.124E+02						FF	FF
	6.3 127.1		0.248	5.004	0.207E+02	0.139E+02	FF N		FF			FF	FF
	8.0 123.7		0.315	4.870	0.260E+02	0.175E+02	FF		6.6FF			FF	FF
	10.0 119.7		0.394	4.713	0.320E+02	0.215E+02			9.5FF		9.5F	FF	FF
	40.0 59.7		1.575	2.350	0.984E+02	0.661E+02	ISO C=11 12.5 14.2 16 17.5 20 22.2 25 28 30 32 36						
141.3	0.6 140.1	5.563	0.024	5.516	0.208E+01	0.140E+01	ISO C=0.8 1 1.2 1.4 1.6 1.8 2.3 2.6 2.9 3.6 5.4 5.9						
	2.0 137.3		0.079	5.406	0.687E+01	0.462E+01	FF						
	3.2 134.9		0.126	5.311	0.109E+02	0.732E+01	FF						
	4.0 133.3		0.157	5.248	0.135E+02	0.910E+01	FF						
	4.5 132.3		0.177	5.209	0.152E+02	0.102E+02	4.8FF						
	5.0 131.3		0.197	5.169	0.168E+02	0.113E+02	FF						
	5.4 130.5		0.213	5.138	0.181E+02	0.122E+02	FF						
	6.3 128.7		0.248	5.067	0.210E+02	0.141E+02	6.6FF L	FF N					
	7.1 127.1		0.280	5.004	0.235E+02	0.158E+02	FF,8FF						
	8.8 123.7		0.346	4.870	0.288E+02	0.193E+02	FF,8FF	FF,8FF					
	10.0 121.3		0.394	4.776	0.324E+02	0.218E+02	9.5FF N						
	12.5 116.3		0.492	4.579	0.397E+02	0.267E+02	FF						
	16.0 109.3		0.630	4.303	0.494E+02	0.332E+02	19FF H						
	20.0 101.3		0.787	3.988	0.598E+02	0.402E+02	FF						
	40.0 61.3		1.575	2.413	0.999E+02	0.671E+02	ISO C=11 14.2 17.5 22.2 25 28 30 32 36						

10-68

STEEL MATERIAL DATA

10-69

Table 10-23. Hollow Steel Bars for Machining (ISO 2938)

```
TABLE 10-23.  HOLLOW STEEL BARS FOR                                    ORDER EXAMPLE:
MACHINING (ISO 2938)                                                     LENGTH,HOLLOW BAR, D63XID32 ISO 2938 GRADE 1        PAGE NO. 1
BASIS; 1 IN = 25.4 MM
1 CUBIC METER STEEL = 7850 KG MASS

                                                       THE NOMINAL SIZE IS NATIONAL STANDARD AS INDICATED
                                                       F=FIRST CHOICE,S=SECOND CHOICE,T=THIRD CHOICE,NUMBER=OTHER SIZE
                                                       * = COMMERCIAL SIZE
I                                                      U.S.A.  AUSTRAL  JAPAN           FRANCE   U.K.       GERMANY  ITALY
S                                                      ANSI    AS       JIS      ISO    NF       BS        DIN      UNI
O   MILLIMETERS     INCHES         MASS PER UNIT                                 2938   A49-312
    A    B   C    A     B     C      KG/M      LB/FT
```

	A	B	C	A	B	C	KG/M	LB/FT	ANSI	AS	JIS	ISO 2938	NF A49-312	BS	DIN	UNI
	32	3.5	25	1.260	0.138	0.984	0.246E+01	0.165E+01				A+1-0		F		
T		6	20		0.236	0.787	0.385E+01	0.259E+01				C=5%		F		
T		8	16		0.315	0.630	0.474E+01	0.318E+01				C=5%		F		
	36	4	28	1.417	0.157	1.102	0.316E+01	0.212E+01				A+1-0		F		
F		5.5	25		0.217	0.984	0.414E+01	0.278E+01				C=5%		F		
T		8	20		0.315	0.787	0.552E+01	0.371E+01				C=5%		F		
T		10	16		0.394	0.630	0.641E+01	0.431E+01				C=5%		F		
	40	4	32	1.575	0.157	1.260	0.355E+01	0.239E+01				A+1-0		F		
F		6	28		0.236	1.102	0.503E+01	0.338E+01				C=5%		F		
T		7.5	25		0.295	0.984	0.601E+01	0.404E+01				C=5%		F		
T		10	20		0.394	0.787	0.740E+01	0.497E+01				C=5%		F		
	45	4.5	36	1.772	0.177	1.417	0.449E+01	0.302E+01				A+1-0		F		
F		6.5	32		0.256	1.260	0.617E+01	0.415E+01				C=5%		F		
T		8.5	28		0.335	1.102	0.765E+01	0.514E+01				C=5%		F		
T		12.5	20		0.492	0.787	0.100E+02	0.673E+01				C=5%		F		
	50	5	40	1.969	0.197	1.575	0.555E+01	0.373E+01				A+1-0		F		
F		7	36		0.276	1.417	0.742E+01	0.499E+01				C=5%		F		
T		9	32		0.354	1.260	0.910E+01	0.611E+01				C=5%		F		
T		12.5	25		0.492	0.984	0.116E+02	0.777E+01				C=5%		F		
	56	5.5	45	2.205	0.217	1.772	0.685E+01	0.460E+01				A+2-0		F		
F		8	40		0.315	1.575	0.947E+01	0.636E+01				C=5%		F		
T		10	36		0.394	1.417	0.113E+02	0.762E+01				C=5%		F		
T		14	28		0.551	1.102	0.145E+02	0.974E+01				C=5%		F		
	63	5	53	2.480	0.197	2.087	0.715E+01	0.481E+01				A+2-0		F		
F		6.5	50		0.256	1.969	0.906E+01	0.609E+01				C=5%		F		
S		9	45		0.354	1.772	0.120E+02	0.805E+01				C=5%		F		
T		11.5	40		0.453	1.575	0.146E+02	0.981E+01				C=5%		F		
T		13.5	36		0.531	1.417	0.165E+02	0.111E+02				C=5%		F		
T		15.5	32		0.610	1.260	0.182E+02	0.122E+02				C=5%		F		
	71	5.5	60	2.795	0.217	2.362	0.888E+01	0.597E+01				A+2-0		F		
F		7.5	56		0.295	2.205	0.117E+02	0.789E+01				C=5%		F		
S		10.5	50		0.413	1.969	0.157E+02	0.105E+02				C=5%		F		
T		13	45		0.512	1.772	0.186E+02	0.125E+02				C=5%		F		
T		17.5	36		0.689	1.417	0.231E+02	0.155E+02				C=5%		F		
	75	6	63	2.953	0.236	2.480	0.102E+02	0.686E+01				A+2-0		F		
F		7.5	60		0.295	2.362	0.125E+02	0.839E+01				C=5%		F		
S		9.5	56		0.374	2.205	0.153E+02	0.103E+02				C=5%		F		
T		12.5	50		0.492	1.969	0.193E+02	0.129E+02				C=5%		F		
T		15	45		0.591	1.772	0.222E+02	0.149E+02				C=5%		F		
T		17.5	40		0.689	1.575	0.248E+02	0.167E+02				C=5%		F		

Table 10-23 (Continued). Hollow Steel Bars for Machining (ISO 2938)

```
TABLE 10-23. HOLLOW STEEL BARS FOR                                    ORDER EXAMPLE:
MACHINING (ISO 2938)                                                  LENGTH,HOLLOW BAR D63XID32 ISO 2938 GRADE I   PAGE NO. 2
BASIS; 1 IN = 25.4 MM
1 CUBIC METER STEEL = 7850 KG MASS

                                                     THE NOMINAL SIZE IS NATIONAL STANDARD AS INDICATED
                                                     F=FIRST CHOICE,S=SECOND CHOICE,T=THIRD CHOICE,NUMBER=OTHER SIZE
                                                     *.= COMMERCIAL SIZE
I                                                    U.S.A.  AUSTRAL JAPAN   FRANCE  U.K.   GERMANY ITALY
S                                                    ANSI    AS      JIS     NF      BS     DIN     UNI
O   MILLIMETERS    INCHES       MASS PER UNIT                                A49-312
    A    B   C    A      B    C        KG/M      LB/FT                ISO
                                                                      2938
```

	A	B	C	A	B	C	KG/M	LB/FT	ISO 2938	ANSI	AS	JIS	NF A49-312	BS	DIN	UNI
F	80	6.5	67	3.150	0.256	2.638	0.118E+02	0.79E+01	A+2%-0				F			
S		8.5	63		0.335	2.480	0.150E+02	0.101E+02	C-5%				F			
T		12	56		0.472	2.205	0.201E+02	0.135E+02	C-5%				F			
T		15	50		0.591	1.969	0.240E+02	0.162E+02	A+2%-0				F			
T		17.5	45		0.689	1.772	0.270E+02	0.181E+02	C-5%				F			
T		20	40		0.787	1.575	0.296E+02	0.199E+02	C-5%				F			
F	85	7.5	70	3.346	0.295	2.756	0.143E+02	0.963E+01	A+2%-0				F			
S		9	67		0.354	2.638	0.169E+02	0.113E+02	C-5%				F			
T		12	61		0.472	2.402	0.216E+02	0.145E+02	C-5%				F			
T		15	55		0.591	2.165	0.259E+02	0.174E+02	C-5%				F			
T		17.5	50		0.689	1.969	0.291E+02	0.196E+02	C-5%				F			
T		20	45		0.787	1.772	0.321E+02	0.215E+02	C-5%				F			
F	90	7.5	75	3.543	0.295	2.953	0.153E+02	0.103E+02	A+2%-0				F			
S		9.5	71		0.374	2.795	0.189E+02	0.127E+02	C-5%				F			
T		11.5	67		0.453	2.638	0.223E+02	0.150E+02	C-5%				F			
T		13.5	63		0.531	2.480	0.255E+02	0.171E+02	C-5%				F			
T		17	56		0.669	2.205	0.306E+02	0.206E+02	C-5%				F			
T		20	50		0.787	1.969	0.345E+02	0.232E+02	C-5%				F			
F	95	7.5	80	3.740	0.295	3.150	0.162E+02	0.109E+02	A+2%-0				F			
		10	75		0.394	2.953	0.210E+02	0.141E+02	C-5%				F			
		12	71		0.472	2.795	0.246E+02	0.165E+02	C-5%				F			
S		13	69		0.512	2.717	0.263E+02	0.177E+02	C-5%				F			
		14	67		0.551	2.638	0.280E+02	0.188E+02	C-5%				F			
T		16	63		0.630	2.480	0.312E+02	0.209E+02	C-5%				F			
T		18	59		0.709	2.323	0.342E+02	0.230E+02	C-5%				F			
T		19.5	56		0.768	2.205	0.363E+02	0.244E+02	C-5%				F			
T		22.5	50		0.886	1.969	0.402E+02	0.270E+02	C-5%				F			
F	100	7.5	85	3.937	0.295	3.346	0.171E+02	0.115E+02	A+2%-0				F			
		10	80		0.394	3.150	0.222E+02	0.149E+02	C-5%				F			
S		12.5	75		0.492	2.953	0.270E+02	0.181E+02	C-5%				F			
T		14.5	71		0.571	2.795	0.306E+02	0.205E+02	C-5%				F			
T		18.5	63		0.728	2.480	0.372E+02	0.250E+02	C-5%				F			
T		22	56		0.866	2.205	0.423E+02	0.284E+02	C-5%				F			
F	106	8	90	4.173	0.315	3.543	0.193E+02	0.130E+02	A+2%-0				F			
S		10.5	85		0.413	3.346	0.247E+02	0.166E+02	C-5%				F			
		13	80		0.512	3.150	0.298E+02	0.200E+02	C-5%				F			
		15.5	75		0.610	2.953	0.346E+02	0.232E+02	C-5%				F			
T		17.5	71		0.689	2.795	0.383E+02	0.257E+02	C-5%				F			
T		21.5	63		0.846	2.480	0.448E+02	0.301E+02	C-5%				F			
T		25	56		0.984	2.205	0.499E+02	0.336E+02	C-5%				F			

STEEL MATERIAL DATA

Table 10-23 (Continued). Hollow Steel Bars for Machining (ISO 2938)

TABLE 10-23. HOLLOW STEEL BARS FOR MACHINING (ISO 2938)
BASIS; 1 IN = 25.4 MM
1 CUBIC METER STEEL = 7850 KG MASS

ORDER EXAMPLE:
LENGTH,HOLLOW BAR D63XID32 ISO 2938 GRADE 1

THE NOMINAL SIZE IS NATIONAL STANDARD AS INDICATED
F=FIRST CHOICE,S=SECOND CHOICE,T=THIRD CHOICE,NUMBER=OTHER SIZE
* = COMMERCIAL SIZE

	U.S.A.	AUSTRAL	JAPAN		FRANCE	U.K.	GERMANY	ITALY
	ANSI	AS	JIS	ISO 2938	NF A49-312	BS	DIN	UNI

I S O	MILLIMETERS A B C	INCHES A B C	MASS PER UNIT KG/M LB/FT							
112	8.5 95	4.409 0.335 3.740	0.217E+02 0.146E+02	A+2%-0	F
F	11 90	0.433 3.543	0.274E+02 0.184E+02	C-5%	F
S	13.5 85	0.531 3.346	0.328E+02 0.220E+02	C-5%	F
T	16 80	0.630 3.150	0.379E+02 0.255E+02	C-5%	F
T	20.5 71	0.807 2.795	0.463E+02 0.311E+02	C-5%	F
T	24.5 63	0.965 2.480	0.529E+02 0.355E+02	C-5%	F
118	9 100	4.646 0.354 3.937	0.242E+02 0.163E+02	A+2%-0	F
F	11.5 95	0.453 3.740	0.302E+02 0.203E+02	C-5%	F
S	14 90	0.551 3.543	0.359E+02 0.241E+02	C-5%	F
T	16.5 85	0.650 3.346	0.413E+02 0.278E+02	C-5%	F
T	19 80	0.748 3.150	0.464E+02 0.312E+02	C-5%	F
T	23.5 71	0.925 2.795	0.548E+02 0.368E+02	C-5%	F
T	27.5 63	1.083 2.480	0.614E+02 0.412E+02	C-5%	F
125	9.5 106	4.921 0.374 4.173	0.271E+02 0.182E+02	A+2%-0	F
F	12.5 100	0.492 3.937	0.347E+02 0.233E+02	C-5%	F
S	15 95	0.591 3.740	0.407E+02 0.273E+02	C-5%	F
T	17.5 90	0.689 3.543	0.464E+02 0.312E+02	C-5%	F
T	22.5 80	0.886 3.150	0.569E+02 0.382E+02	C-5%	F
T	27 71	1.063 2.795	0.653E+02 0.438E+02	C-5%	F
132	10 112	5.197 0.394 4.409	0.301E+02 0.202E+02	A+2%-0	F
F	13 106	0.512 4.173	0.382E+02 0.256E+02	C-5%	F
S	17 98	0.669 3.858	0.482E+02 0.324E+02	C-5%	F
T	21 90	0.827 3.543	0.575E+02 0.386E+02	C-5%	F
T	25 82	0.984 3.228	0.660E+02 0.443E+02	C-5%	F
T	26 80	1.024 3.150	0.680E+02 0.457E+02	C-5%	F
T	30.5 71	1.201 2.795	0.763E+02 0.513E+02	C-5%	F
140	11 118	5.512 0.433 4.646	0.350E+02 0.235E+02	A+2%-0	F
F	14 112	0.551 4.409	0.435E+02 0.292E+02	C-5%	F
S	17 106	0.669 4.173	0.516E+02 0.347E+02	C-5%	F
T	20 100	0.787 3.937	0.592E+02 0.398E+02	C-5%	F
T	25 90	0.984 3.543	0.709E+02 0.476E+02	C-5%	F
T	30 80	1.181 3.150	0.814E+02 0.547E+02	C-5%	F
150	9 132	5.906 0.354 5.197	0.313E+02 0.210E+02	A+2%-0	F
F	12.5 125	0.492 4.921	0.424E+02 0.285E+02	C-5%	F
S	16 118	0.630 4.646	0.529E+02 0.355E+02	C-5%	F
T	22 106	0.866 4.173	0.694E+02 0.467E+02	C-5%	F
T	27.5 95	1.083 3.740	0.831E+02 0.558E+02	C-5%	F
T	35 80	1.378 3.150	0.993E+02 0.667E+02	C-5%	F
160	12 136	6.299 0.472 5.354	0.438E+02 0.294E+02	A+2%-0	F
F	14 132	0.551 5.197	0.504E+02 0.339E+02	C-5%	F
S	19 122	0.748 4.803	0.661E+02 0.444E+02	C-5%	F
T	24 112	0.945 4.409	0.805E+02 0.541E+02	C-5%	F
T	30 100	1.181 3.937	0.962E+03 0.646E+02	C-5%	F
T	35 90	1.378 3.543	0.108E+03 0.725E+02	C-5%	F

10-71

Table 10-23 (Continued). Hollow Steel Bars for Machining (ISO 2938)

TABLE 10-23. HOLLOW STEEL BARS FOR
MACHINING (ISO 2938)
BASIS: 1 IN = 25.4 MM
1 CUBIC METER STEEL = 7850 KG MASS

ORDER EXAMPLE;
LENGTH,HOLLOW BAR D63XID32 ISO 2938 GRADE 1   PAGE NO. 4

THE NOMINAL SIZE IS NATIONAL STANDARD AS INDICATED
F=FIRST CHOICE,S=SECOND CHOICE,T=THIRD CHOICE,NUMBER=OTHER SIZE
* = COMMERCIAL SIZE

I S O	MILLIMETERS A B C	INCHES A B C	MASS PER UNIT KG/M LB/FT	U.S.A. ANSI	AUSTRAL AS	JAPAN JIS	ISO 2938	FRANCE NF A49-312	U.K. BS	GERMANY DIN	ITALY UNI
170	12.5 145	6.693 0.492 5.709	0.886E+02 0.326E+02				A+2%=0	F			
F	15 140	0.591 5.512	0.573E+02 0.385E+02				C=5%	F			
S	20 130	0.787 5.118	0.740E+02 0.497E+02				C=5%	F			
T	26 118	1.024 4.646	0.923E+02 0.620E+02				C=5%	F			
T	30 110	1.181 4.331	0.104E+03 0.696E+02				C=5%	F			
T	35 100	1.378 3.937	0.117E+03 0.783E+02				C=5%	F			
180	12.5 155	7.087 0.492 6.102	0.516E+02 0.347E+02				A+2%=0	F			
F	15 150	0.591 5.906	0.610E+02 0.410E+02				C=5%	F			
S	20 140	0.787 5.512	0.789E+02 0.530E+02				C=5%	F			
T	27.5 125	1.083 4.921	0.103E+03 0.695E+02				C=5%	F			
T	34 112	1.339 4.409	0.122E+03 0.823E+02				C=5%	F			
T	40 100	1.575 3.937	0.138E+03 0.928E+02				C=5%	F			
190	12.5 165	7.480 0.492 6.496	0.547E+02 0.368E+02				A+2%=0	F			
F	15 160	0.591 6.299	0.647E+02 0.435E+02				C=5%	F			
S	20 150	0.787 5.906	0.838E+02 0.563E+02				C=5%	F			
T	22 146	0.866 5.748	0.911E+02 0.612E+02				C=5%	F			
T	29 132	1.142 5.197	0.115E+03 0.774E+02				C=5%	F			
T	36 118	1.417 4.646	0.137E+03 0.919E+02				C=5%	F			
T	42 106	1.654 4.173	0.153E+03 0.103E+03				C=5%	F			
F 200	20 160	7.874 0.787 6.299	0.888E+02 0.597E+02				A+2%=0	F			
T	30 140	1.181 5.512	0.126E+03 0.845E+02				C=5%	F			
T	44 112	1.732 4.409	0.169E+03 0.114E+03				C=5%	F			
F 212	21 170	8.346 0.827 6.693	0.989E+02 0.665E+02				A+2%=0	F			
T	31 150	1.220 5.906	0.138E+03 0.930E+02				C=5%	F			
T	43.5 125	1.713 4.921	0.181E+03 0.121E+03				C=5%	F			
F 224	22 180	8.819 0.866 7.087	0.110E+03 0.736E+02				A+2%=0	F			
T	32 160	1.260 6.299	0.152E+03 0.102E+03				C=5%	F			
T	46 132	1.811 5.197	0.202E+03 0.136E+03				C=5%	F			
F 236	23 190	9.291 0.906 7.480	0.121E+03 0.812E+02				A+2%=0	F			
T	33 170	1.299 6.693	0.165E+03 0.111E+03				C=5%	F			
T	48 140	1.890 5.512	0.223E+03 0.150E+03				C=5%	F			
F 250	25 200	9.843 0.984 7.874	0.139E+03 0.932E+02				A+2%=0	F			
T	35 180	1.378 7.087	0.186E+03 0.125E+03				C=5%	F			
T	50 150	1.969 5.906	0.247E+03 0.166E+03				C=5%	F			

STEEL MATERIAL DATA

Table 10-24. Seamless (SL) and Welded (WE) Precision Steel Tubes (ISO 560, 3304, 3305)

```
TABLE 10-24. SEAMLESS (SL) AND WELDED (WE) PRECISION STEEL                    ORDER EXAMPLE:
TUBES (ISO 560,3304,3305)                                                       LENGTH,STEEL TUBE 4X0.5 DIN2393 + STEEL QUAL
BASIS: 1 IN = 25.4 MM
1 CUBIC METER STEEL = 7850 KG MASS                                                                                PAGE NO. 1

                                                             THE NOMINAL SIZE IS NATIONAL STANDARD AS INDICATED
                                                             F=FIRST CHOICE,S=SECOND CHOICE,T=THIRD CHOICE,NUMBER=OTHER SIZE
                                                             * = COMMERCIAL SIZE
```

I S O	MILLIMETERS B	C	INCHES A	C	B	MASS PER UNIT KG/M	LB/FT	U.S.A. ANSI B32.5 SL	U.S.A. ANSI B32.5 WE	AUSTRAL AS	JAPAN JIS	ISO 3304 3305	FRANCE NF A49-310	U.K. BS	GERMANY DIN 2391 2393	ITALY UNI 2898 5921
F	4	0.5 3	0.157	0.020	0.118	0.433E-01	0.290E-01	FF	FF			A+-0.1			FF FF	F F
		0.6 2.8		0.024	0.110	0.503E-01	0.338E-01	FF	FF			B+-0.3			FF FF	F F
S		0.8 2.4		0.031	0.094	0.631E-01	0.424E-01	FF	FF			B+-0.3	S		FF FF	F F
F		1 2		0.039	0.079	0.740E-01	0.497E-01	FF	FF			B+-0.3	F		FF FF	F F
F	5	0.5 4	0.197	0.020	0.157	0.555E-01	0.373E-01	FF	FF			A+-0.1			FF FF	F F
		0.6 3.8		0.024	0.150	0.651E-01	0.437E-01	FF	FF			B+-0.3			FF FF	F F
S		0.8 3.4		0.031	0.134	0.829E-01	0.557E-01	FF	FF			B+-0.3	S		FF FF	F S
F		1 3		0.039	0.118	0.986E-01	0.663E-01	FF	FF			B+-0.3	F		FF FF	F F
F	6	0.5 5	0.236	0.020	0.197	0.678E-01	0.456E-01	FF	FF			A+-0.1			FF FF	F F
		0.6 4.8		0.024	0.189	0.799E-01	0.537E-01	FF	FF			B+-0.25			FF FF	F F
S		0.8 4.4		0.031	0.173	0.103E+00	0.689E-01	FF	FF			B+-0.25	S		FF FF	F S
F		1 4		0.039	0.157	0.123E+00	0.829E-01	FF	FF			B+-0.25	F		FF FF	F F
F		1.2 3.6		0.047	0.142	0.142E+00	0.955E-01	FF	FF			B+-0.3	S		FF FF	F S
S		1.5 3		0.059	0.118	0.166E+00	0.112E+00	FF	FF			B+-0.3			FF FF		
F	8	0.5 7	0.315	0.020	0.276	0.925E-01	0.621E-01	FF	FF			A+-0.1			FF FF	FF F
		0.6 6.8		0.024	0.268	0.109E+00	0.736E-01	FF	FF			B+-0.2			FF FF	SF F
S		0.8 6.4		0.031	0.252	0.142E+00	0.955E-01	FF	FF			B+-0.2	S		FF FF	SF F
F		1 6		0.039	0.236	0.173E+00	0.116E+00	FF	FF			B+-0.2	F		FF FF	F F
S		1.2 5.6		0.047	0.220	0.201E+00	0.135E+00	FF	FF			B+-0.3	S		FF FF	SF F
F		1.5 5		0.059	0.197	0.240E+00	0.162E+00	FF	FF			B+-0.3	F		FF FF	F F
S		1.8 4.4		0.071	0.173	0.275E+00	0.185E+00	FF	FF			B+-0.35	S		FF FF	SF F
F		2 4		0.079	0.157	0.296E+00	0.199E+00	FF	FF			B+-0.35	F		FF FF	F F
S		2.2 3.6		0.087	0.142	0.315E+00	0.211E+00	FF	FF			B+-0.4	S		FF FF	SF F
F		2.5 3		0.098	0.118	0.339E+00	0.228E+00	FF	FF			B+-0.4	F		FF FF	F F
F	10	0.5 9	0.394	0.020	0.354	0.117E+00	0.787E-01	FF	FF			A+-0.1			FF FF	FF F
S		0.6 8.8		0.024	0.346	0.139E+00	0.935E-01	FF	FF			B+-0.15	S		FF FF	SF F
F		0.8 8.4		0.031	0.331	0.182E+00	0.122E+00	FF	FF			B+-0.15	F		FF FF	F F
S		1 8		0.039	0.315	0.222E+00	0.149E+00	FF	FF			B+-0.2	S		FF FF	SF F
F		1.2 7.6		0.047	0.299	0.260E+00	0.175E+00	FF	FF			B+-0.25	F		FF FF	F F
S		1.5 7		0.059	0.276	0.314E+00	0.211E+00	FF	FF			B+-0.25	S		FF FF	SF F
F		1.8 6.4		0.071	0.252	0.364E+00	0.245E+00	FF	FF			B+-0.3	F		FF FF	F F
S		2 6		0.079	0.236	0.395E+00	0.265E+00	FF	FF			B+-0.3	S		FF FF	SF F
F		2.2 5.6		0.087	0.220	0.423E+00	0.284E+00	FF	FF			B+-0.35	F		FF FF	F F
F		2.5 5		0.098	0.197	0.462E+00	0.311E+00	FF	FF			B+-0.35	F		FF FF	F F
F	12	0.5 11	0.472	0.020	0.433	0.142E+00	0.953E-01	FF	FF			A+-0.1			FF FF	F F
S		0.6 10.8		0.024	0.425	0.169E+00	0.113E+00	FF	FF			B+-0.15	S		FF FF	SF F
F		0.8 10.4		0.031	0.409	0.221E+00	0.148E+00	FF	FF			B+-0.15	F		FF FF	F F
S		1 10		0.039	0.394	0.271E+00	0.18E+00	FF	FF			B+-0.2	S		FF FF	SF F
F		1.2 9.6		0.047	0.378	0.320E+00	0.215E+00	FF	FF			B+-0.2	F		FF FF	F F
S		1.5 9		0.059	0.354	0.388E+00	0.261E+00	FF	FF			B+-0.25	S		FF FF	SF F
F		1.8 8.4		0.071	0.331	0.453E+00	0.304E+00	FF	FF			B+-0.25	F		FF FF	F F
S		2 8		0.079	0.315	0.493E+00	0.331E+00	FF	FF			B+-0.3	S		FF FF	SF F
F		2.2 7.6		0.087	0.299	0.532E+00	0.357E+00	FF	FF			B+-0.3	F		FF FF	F F
S		2.5 7		0.098	0.276	0.586E+00	0.394E+00	FF	FF			B+-0.35	S		FF FF	SF F
F		2.8 6.4		0.110	0.252	0.635E+00	0.427E+00	FF	FF			B+-0.4	S		FF FF	SF F
F		3 6		0.118	0.236	0.666E+00	0.447E+00	FF	FF			B+-0.4	F		FF FF	F F

Table 10-24 (Continued). Seamless (SL) and Welded (WE) Precision Steel Tubes (ISO 560, 3304, 3305)

TABLE 10-24. SEAMLESS (SL) AND WELDED (WE) PRECISION STEEL
TUBES (ISO 560,3304,3305)
BASIS 1 IN = 25.4 MM
1 CUBIC METER STEEL = 7850 KG MASS

ORDER EXAMPLE:
LENGTH,STEEL TUBE 4X0.5 DIN2393 + STEEL QUAL

PAGE NO. 2

THE NOMINAL SIZE IS NATIONAL STANDARD AS INDICATED
F=FIRST CHOICE,S=SECOND CHOICE,T=THIRD CHOICE,NUMBER=OTHER SIZE
* = COMMERCIAL SIZE

I S O	MILLIMETERS A B C	INCHES A B C	MASS PER UNIT KG/M LB/FT	U.S.A. ANSI B32.5 SL	B32.5 WE	AUSTRAL AS	JAPAN JIS	ISO 3304 3305	FRANCE NF A49-310	U.K. BS	GERMANY DIN 2391 2393	ITALY UNI 2898 5921
F 14 0.5 13 0.551 0.020 0.512	0.166E+00 0.112E+00	FF	FF			A*+0.1	F		FF	FF	F
F 0.6 12.8 0.024 0.504	0.198E+00 0.133E+00	FF	FF			B*+0.1	F		FF	FF	F
S 0.8 12.4 0.031 0.488	0.260E+00 0.175E+00	FF	FF			B*+0.1	S		FF	FF	SF
F 1 12 0.039 0.472	0.321E+00 0.215E+00	FF	FF			B*+0.15 F			FF	FF	SF
S 1.2 11.6 0.047 0.457	0.379E+00 0.255E+00	FF	FF			B*+0.15 F			FF	FF	FF
F 1.5 11 0.059 0.433	0.462E+00 0.311E+00	FF	FF			B*+0.15 F			FF	FF	FF
S 1.8 10.4 0.071 0.409	0.542E+00 0.364E+00	FF	FF			B*+0.2 S			FF	FF	F
F 2 10 0.079 0.394	0.592E+00 0.398E+00	FF	FF			B*+0.2 F			FF	FF	S
S 2.2 9.6 0.087 0.378	0.640E+00 0.430E+00	FF	FF			B*+0.25 S			FF	FF	S
F 2.5 9 0.098 0.354	0.709E+00 0.476E+00	FF	FF			B*+0.25 F			FF	FF	S
S 2.8 8.4 0.110 0.331	0.773E+00 0.520E+00	FF	FF			B*+0.3 S			FF	FF	S
F 3 8 0.118 0.315	0.814E+00 0.547E+00	FF	FF			B*+0.3 F			FF	FF	F
F 15 0.5 14 0.591 0.020 0.551	0.179E+00 0.120E+00	FF	FF			A*+0.1 F			FF	FF	F
F 0.6 13.8 0.024 0.543	0.213E+00 0.143E+00	FF	FF			B*+0.1 F			FF	FF	F
S 0.8 13.4 0.031 0.528	0.280E+00 0.188E+00	FF	FF			B*+0.1 S			FF	FF	S
F 1 13 0.039 0.512	0.345E+00 0.232E+00	FF	FF			B*+0.1 F			FF	FF	S
S 1.2 12.6 0.047 0.496	0.408E+00 0.274E+00	FF	FF			B*+0.15 S			FF	FF	S
F 1.5 12 0.059 0.472	0.499E+00 0.336E+00	FF	FF			B*+0.15 F			FF	FF	F
S 1.8 11.4 0.071 0.449	0.586E+00 0.394E+00	FF	FF			B*+0.2 S			FF	FF	F
F 2 11 0.079 0.433	0.641E+00 0.431E+00	FF	FF			B*+0.2 F			FF	FF	S
S 2.2 10.6 0.087 0.417	0.694E+00 0.467E+00	FF	FF			B*+0.25 S			FF	FF	S
F 2.5 10 0.098 0.394	0.771E+00 0.518E+00	FF	FF			B*+0.25 F			FF	FF	S
S 2.8 9.4 0.110 0.370	0.842E+00 0.566E+00	FF	FF			B*+0.3 S			FF	FF	S
F 3 9 0.118 0.354	0.888E+00 0.597E+00	FF	FF			B*+0.3 F			FF	FF	F
F 16 0.5 15 0.630 0.020 0.591	0.191E+00 0.128E+00	FF	FF			A*+0.1 F			FF	FF	F
F 0.6 14.8 0.024 0.583	0.228E+00 0.153E+00	FF	FF			B*+0.1 F			FF	FF	F
S 0.8 14.4 0.031 0.567	0.300E+00 0.202E+00	FF	FF			B*+0.1 S			FF	FF	SF
F 1 14 0.039 0.551	0.370E+00 0.249E+00	FF	FF			B*+0.1 F			FF	FF	SF
S 1.2 13.6 0.047 0.535	0.438E+00 0.294E+00	FF	FF			B*+0.1 S			FF	FF	FF
F 1.5 13 0.059 0.512	0.536E+00 0.360E+00	FF	FF			B*+0.1 F			FF	FF	FF
S 1.8 12.4 0.071 0.488	0.630E+00 0.424E+00	FF	FF			B*+0.15 S			FF	FF	F
F 2 12 0.079 0.472	0.691E+00 0.464E+00	FF	FF			B*+0.15 F			FF	FF	S
S 2.2 11.6 0.087 0.457	0.749E+00 0.503E+00	FF	FF			B*+0.2 S			FF	FF	S
F 2.5 11 0.098 0.433	0.832E+00 0.559E+00	FF	FF			B*+0.2 F			FF	FF	S
S 2.8 10.4 0.110 0.409	0.911E+00 0.612E+00	FF	FF			B*+0.3 S			FF	FF	S
F 3 10 0.118 0.394	0.962E+00 0.646E+00	FF	FF			B*+0.3 F			FF	FF	F
S 3.5 9 0.138 0.354	0.108E+01 0.725E+00	FF	FF			B*+0.35 S			FF	FF	SF
F 4 8 0.157 0.315	0.118E+01 0.795E+00	FF	FF			B*+0.35 F			FF	FF	FF
F 18 0.5 17 0.709 0.020 0.669	0.216E+00 0.145E+00	FF	FF			A*+0.1 F			FF	FF	F
F 0.6 16.8 0.024 0.661	0.257E+00 0.173E+00	FF	FF			B*+0.1 F			FF	FF	F
S 0.8 16.4 0.031 0.646	0.339E+00 0.228E+00	FF	FF			B*+0.1 S			FF	FF	SF
F 1 16 0.039 0.630	0.419E+00 0.282E+00	FF	FF			B*+0.1 F			FF	FF	SF
S 1.2 15.6 0.047 0.614	0.497E+00 0.334E+00	FF	FF			B*+0.1 S			FF	FF	FF
F 1.5 15 0.059 0.591	0.610E+00 0.410E+00	FF	FF			B*+0.1 F			FF	FF	FF
S 1.8 14.4 0.071 0.567	0.719E+00 0.483E+00	FF	FF			B*+0.15 S			FF	FF	F
F 2 14 0.079 0.551	0.789E+00 0.530E+00	FF	FF			B*+0.1 F			FF	FF	F
S 2.2 13.6 0.087 0.535	0.857E+00 0.576E+00	FF	FF			B*+0.2 S			FF	FF	S
F 2.5 13 0.098 0.512	0.955E+00 0.642E+00	FF	FF			B*+0.2 F			FF	FF	F

STEEL MATERIAL DATA

10-75

Table 10-24 (Continued). Seamless (SL) and Welded (WE) Precision Steel Tubes (ISO 560, 3304, 3305)

TABLE 10-24. SEAMLESS (SL) AND WELDED (WE) PRECISION STEEL
TUBES (ISO 560,3304,3305)
BASIS; 1 IN = 25.4 MM
1 CUBIC METER STEEL = 7850 KG MASS

ORDER EXAMPLE;
LENGTH,STEEL TUBE 4X0.5 DIN2393 + STEEL QUAL

PAGE NO. 3

THE NOMINAL SIZE IS NATIONAL STANDARD AS INDICATED
F=FIRST CHOICE,S=SECOND CHOICE,T=THIRD CHOICE,NUMBER=OTHER SIZE
* = COMMERCIAL SIZE

I S O		MILLIMETERS			INCHES			MASS PER UNIT		U.S.A. ANSI B32.5	AUSTRAL AS B32.5	JAPAN JIS	ISO 3304 3305	FRANCE NF A49-310	U.K. BS	GERMANY DIN 2391 2393	ITALY UNI 2898 5921
	A	B	C		A	B	C	KG/M SL	LB/FT WE								
S	18	12.4	2.8		0.709	0.488	0.110	0.105E+01	0.705E+00	FF	FF		B+-0.2	S		FF	S
F		12	3			0.472	0.118	0.111E+01	0.746E+00	FF	FF		B+-0.2	F		FF	F
S		11	3.5			0.433	0.138	0.125E+01	0.841E+00	FF	FF		B+-0.35	S		FF	S
F		10	4			0.394	0.157	0.138E+01	0.928E+00	FF	FF		B+-0.35	F		FF	F
F	20	19	0.5		0.787	0.748	0.020	0.240E+00	0.162E+00	FF	FF		A+-0.1			FF	F
S		18.8	0.6			0.740	0.024	0.287E+00	0.193E+00	FF	FF		B+-0.1			FF	F
F		18.4	0.8			0.724	0.031	0.379E+00	0.255E+00	FF	FF		B+-0.1	S		FF	S
F		18	1			0.709	0.039	0.469E+00	0.315E+00	FF	FF		B+-0.1	F		FF	FF
F		17.6	1.2			0.693	0.047	0.556E+00	0.374E+00	FF	FF		B+-0.1	F		FF	SF
F		17	1.5			0.669	0.059	0.684E+00	0.460E+00	FF	FF		B+-0.1	F		FF	FF
F		16.4	1.8			0.646	0.071	0.808E+00	0.543E+00	FF	FF		B+-0.1	S		FF	S
F		16	2			0.630	0.079	0.888E+00	0.597E+00	FF	FF		B+-0.1	F		FF	FF
F		15.6	2.2			0.614	0.087	0.966E+00	0.649E+00	FF	FF		B+-0.1	F		FF	FF
F		15	2.5			0.591	0.098	0.108E+01	0.725E+00	FF	FF		B+-0.15	F		FF	FF
F		14.4	2.8			0.567	0.110	0.119E+01	0.798E+00	FF	FF		B+-0.15	S		FF	F
S		14	3			0.551	0.118	0.126E+01	0.845E+00	FF	FF		B+-0.2	F		FF	F
F		13	3.5			0.512	0.138	0.142E+01	0.957E+00	FF	FF		B+-0.3	S		FF	F
F		12	4			0.472	0.157	0.158E+01	0.106E+01	FF	FF		B+-0.35	F		FF	F
F		11	4.5			0.433	0.177	0.172E+01	0.116E+01	FF	FF		B+-0.35	S		FF	F
F		10	5			0.394	0.197	0.185E+01	0.124E+01	FF	FF		B+-0.35	F		FF	F
F	22	21	0.5		0.866	0.827	0.020	0.265E+00	0.178E+00	FF	FF		A+-0.1			FF	F
S		20.8	0.6			0.819	0.024	0.317E+00	0.213E+00	FF	FF		B+-0.1			FF	F
F		20.4	0.8			0.803	0.031	0.418E+00	0.281E+00	FF	FF		B+-0.1	S		FF	S
F		20	1			0.787	0.039	0.518E+00	0.348E+00	FF	FF		B+-0.1	F		FF	FF
F		19.6	1.2			0.772	0.047	0.616E+00	0.414E+00	FF	FF		B+-0.1	F		FF	FF
F		19	1.5			0.748	0.059	0.758E+00	0.510E+00	FF	FF		B+-0.1	F		FF	SF
F		18.4	1.8			0.724	0.071	0.897E+00	0.603E+00	FF	FF		B+-0.1	F		FF	FF
F		18	2			0.709	0.079	0.986E+00	0.663E+00	FF	FF		B+-0.1	F		FF	FF
F		17.6	2.2			0.693	0.087	0.107E+01	0.722E+00	FF	FF		B+-0.1	F		FF	FF
F		17	2.5			0.669	0.098	0.120E+01	0.808E+00	FF	FF		B+-0.15	F		FF	FF
F		16.4	2.8			0.646	0.110	0.133E+01	0.891E+00	FF	FF		B+-0.15	F		FF	FF
F		16	3			0.630	0.118	0.141E+01	0.945E+00	FF	FF		B+-0.2	S		FF	F
F		15	3.5			0.591	0.138	0.160E+01	0.107E+01	FF	FF		B+-0.2	F		FF	F
F		14	4			0.551	0.157	0.178E+01	0.119E+01	FF	FF		B+-0.3	F		FF	F
F		13	4.5			0.512	0.177	0.194E+01	0.131E+01	FF	FF		B+-0.35	S		FF	F
F		12	5			0.472	0.197	0.210E+01	0.141E+01	FF	FF		B+-0.35	F		FF	F
F	25	24	0.5		0.984	0.945	0.020	0.302E+00	0.203E+00	FF	FF		A+-0.1			FF	F
S		23.8	0.6			0.937	0.024	0.361E+00	0.243E+00	FF	FF		B+-0.1			FF	F
F		23.4	0.8			0.921	0.031	0.477E+00	0.321E+00	FF	FF		B+-0.1	S		FF	S
F		23	1			0.906	0.039	0.592E+00	0.398E+00	FF	FF		B+-0.1	F		FF	SF
F		22.6	1.2			0.890	0.047	0.704E+00	0.473E+00	FF	FF		B+-0.1	F		FF	FF
S		22	1.5			0.866	0.059	0.869E+00	0.584E+00	FF	FF		B+-0.1	F		FF	FF
F		21.4	1.8			0.843	0.071	0.103E+01	0.692E+00	FF	FF		B+-0.1	F		FF	FF
F		21	2			0.827	0.079	0.113E+01	0.762E+00	FF	FF		B+-0.1	F		FF	FF
F		20.6	2.2			0.811	0.087	0.124E+01	0.831E+00	FF	FF		B+-0.1	F		FF	FF
F		20	2.5			0.787	0.098	0.139E+01	0.932E+00	FF	FF		B+-0.15	F		FF	FF
F		19.4	2.8			0.764	0.110	0.153E+01	0.103E+01	FF	FF		B+-0.15	S		FF	F
F		19	3			0.748	0.118	0.163E+01	0.109E+01	FF	FF		B+-0.15	F		FF	F

10-76   STEEL MATERIAL DATA

Table 10-24 (Continued). Seamless (SL) and Welded (WE) Precision Steel Tubes (ISO 560, 3304, 3305)

TABLE 10-24. SEAMLESS (SL) AND WELDED (WE) PRECISION STEEL
TUBES (ISO 560,3304,3305) PAGE NO. 4
BASIS; 1 IN = 25.4 MM
1 CUBIC METER STEEL = 7850 KG MASS

ORDER EXAMPLE;
LENGTH,STEEL TUBE 4X0.5, DIN2393 + STEEL QUAL

THE NOMINAL SIZE IS NATIONAL STANDARD AS INDICATED
F=FIRST CHOICE, S=SECOND CHOICE, T=THIRD CHOICE, NUMBER=OTHER SIZE
* = COMMERCIAL SIZE

I S O	MILLIMETERS A B	C	INCHES A	B	C	MASS PER UNIT KG/M	LB/FT	U.S.A. ANSI B32.5 SL	B32.5 WE	AUSTRAL AS	JAPAN JIS	ISO 3304 3305	FRANCE NF A49-310	U.K. BS	GERMANY DIN 2391 2393	ITALY UNI 2898 5921
S	25 3.5	18	0.984	0.709	0.138	0.186E+01	0.125E+01	FF	FF			B+-0.15	S		FF	FF
F	4	17		0.669	0.157	0.207E+01	0.139E+01	FF	FF			B+-0.2	F		FF	F
S	4.5	16		0.630	0.177	0.228E+01	0.153E+01	FF	FF			B+-0.2	S		FF	F
F	5	15		0.591	0.197	0.247E+01	0.166E+01	FF	FF			B+-0.3	F		FF	F
F	28 0.5	27	1.102	1.063	0.020	0.339E+00	0.228E+00	FF	FF			A+-0.1			FF	F
S	0.6	26.8		1.055	0.024	0.405E+00	0.272E+00	FF	FF			B+-0.1			FF	F
S	0.8	26.4		1.039	0.031	0.537E+00	0.361E+00	FF	FF			B+-0.1			FF	F
F	1	26		1.024	0.039	0.666E+00	0.447E+00	FF	FF			B+-0.1			FF	F
S	1.2	25.6		1.008	0.047	0.793E+00	0.533E+00	FF	FF			B+-0.1			FF	F
F	1.5	25		0.984	0.059	0.980E+00	0.659E+00	FF	FF			B+-0.1			FF	SF
S	1.8	24.4		0.961	0.071	0.116E+01	0.782E+00	FF	FF			B+-0.1			FF	FF
F	2	24		0.945	0.079	0.128E+01	0.862E+00	FF	FF			B+-0.1	F		FF	F
S	2.2	23.6		0.929	0.087	0.140E+01	0.941E+00	FF	FF			B+-0.1	F		FF	FF
F	2.5	23		0.906	0.098	0.157E+01	0.106E+01	FF	FF			B+-0.1	S		FF	S
S	2.8	22.4		0.882	0.110	0.174E+01	0.117E+01	FF	FF			B+-0.1	F		FF	FF
F	3	22		0.866	0.118	0.185E+01	0.124E+01	FF	FF			B+-0.15			FF	SF
S	3.5	21		0.827	0.138	0.211E+01	0.142E+01	FF	FF			B+-0.15			FF	FF
F	4	20		0.787	0.157	0.237E+01	0.159E+01	FF	FF			B+-0.15	F		FF	FF
S	4.5	19		0.748	0.177	0.264E+01	0.175E+01	FF	FF			B+-0.15	F		FF	FF
F	5	18		0.709	0.197	0.284E+01	0.191E+01	FF	FF			B+-0.15	F		FF	FF
S	5.5	17		0.669	0.217	0.305E+01	0.205E+01	FF	FF			B+-0.2	F		FF	SF
F	6	16		0.630	0.236	0.326E+01	0.219E+01	FF	FF			B+-0.3	F		FF	S
F	30 0.5	29	1.181	1.142	0.020	0.364E+00	0.244E+00	FF	FF			A+-0.1			FF	F
S	0.6	28.8		1.134	0.024	0.435E+00	0.292E+00	FF	FF			B+-0.1			FF	F
S	0.8	28.4		1.118	0.031	0.576E+00	0.387E+00	FF	FF			B+-0.1			FF	F
F	1	28		1.102	0.039	0.715E+00	0.481E+00	FF	FF			B+-0.1	S		FF	SF
S	1.2	27.6		1.087	0.047	0.852E+00	0.573E+00	FF	FF			B+-0.1	F		FF	FF
F	1.5	27		1.063	0.059	0.105E+01	0.708E+00	FF	FF			B+-0.1	F		FF	FF
S	1.8	26.4		1.039	0.071	0.125E+01	0.841E+00	FF	FF			B+-0.1	F		FF	FF
F	2	26		1.024	0.079	0.138E+01	0.928E+00	FF	FF			B+-0.1	F		FF	FF
S	2.2	25.6		1.008	0.087	0.151E+01	0.101E+01	FF	FF			B+-0.1	F		FF	FF
F	2.5	25		0.984	0.098	0.170E+01	0.114E+01	FF	FF			B+-0.1	S		FF	FF
S	2.8	24.4		0.961	0.110	0.188E+01	0.126E+01	FF	FF			B+-0.1	F		FF	FF
F	3	24		0.945	0.118	0.200E+01	0.134E+01	FF	FF			B+-0.15	F		FF	FF
S	3.5	23		0.906	0.138	0.229E+01	0.154E+01	FF	FF			B+-0.15			FF	FF
F	4	22		0.866	0.157	0.256E+01	0.172E+01	FF	FF			B+-0.15	F		FF	SF
S	4.5	21		0.827	0.177	0.283E+01	0.190E+01	FF	FF			B+-0.15	F		FF	FF
F	5	20		0.787	0.197	0.308E+01	0.207E+01	FF	FF			B+-0.15	F		FF	FF
S	5.5	19		0.748	0.217	0.332E+01	0.223E+01	FF	FF			B+-0.3	F		FF	F
F	32 0.5	31	1.260	1.220	0.020	0.388E+00	0.261E+00	FF	FF			A+-0.15			FF	FF
S	0.6	30.8		1.213	0.024	0.465E+00	0.312E+00	FF	FF			B+-0.15			FF	FF
S	0.8	30.4		1.197	0.031	0.616E+00	0.414E+00	FF	FF			B+-0.15			FF	FF
F	1	30		1.181	0.039	0.765E+00	0.514E+00	FF	FF			B+-0.15	F		FF	FF
S	1.2	29.6		1.165	0.047	0.911E+00	0.612E+00	FF	FF			B+-0.15	F		FF	SF
F	1.5	29		1.142	0.059	0.113E+01	0.758E+00	FF	FF			B+-0.15	F		F	F
S	1.8	28.4		1.118	0.071	0.134E+01	0.901E+00	FF	FF			B+-0.15	F		FF	FF
F	2	28		1.102	0.079	0.148E+01	0.994E+00	FF	FF			B+-0.15	F		FF	FF
S	2.2	27.6		1.087	0.087	0.162E+01	0.109E+01	FF	FF			B+-0.15	S		FF	S

STEEL MATERIAL DATA

Table 10-24 (Continued). Seamless (SL) and Welded (WE) Precision Steel Tubes (ISO 560, 3304, 3305)

TABLE 10-24. SEAMLESS (SL) AND WELDED (WE) PRECISION STEEL
TUBES (ISO 560,3304,3305)
BASIS: 1 IN = 25.4 MM
1 CUBIC METER STEEL = 7850 KG MASS

ORDER EXAMPLE:
LENGTH,STEEL TUBE 4X0.5 DIN2393 + STEEL QUAL

PAGE NO. 5

THE NOMINAL SIZE IS NATIONAL STANDARD AS INDICATED
F=FIRST CHOICE,S=SECOND CHOICE,T=THIRD CHOICE,NUMBER=OTHER SIZE
* = COMMERCIAL SIZE

I S O	MILLIMETERS			INCHES			MASS PER UNIT		U.S.A. ANSI B32.5	AUSTRAL AS B32.5	JAPAN JIS	ISO 3304 3305	FRANCE NF A49-310	U.K. BS	GERMANY DIN 2391 2393	ITALY UNI 2898 5921
	A	B	C	A	B	C	KG/M SL	LB/FT WE								
F	32	2.5	27	1.260	0.098	1.063	0.182E+01	0.122E+01	FF			B+-0.15	F		FF	F
S		2.8	26.4		0.110	1.039	0.202E+01	0.135E+01	FF			B+-0.15	F		FF	F
F		3	26		0.118	1.024	0.215E+01	0.144E+01	FF			B+-0.15	S		FF	S
S		3.5	25		0.138	0.984	0.246E+01	0.165E+01	FF			B+-0.15	F		FF	F
S		4	24		0.157	0.945	0.276E+01	0.186E+01	FF			B+-0.15	S		FF	S
F		4.5	23		0.177	0.906	0.305E+01	0.205E+01	FF			B+-0.15	F		F	F
S		5	22		0.197	0.866	0.333E+01	0.224E+01	FF			B+-0.15	S		FF	S
F		5.5	21		0.217	0.827	0.359E+01	0.242E+01	FF			B+-0.3	F		F	F
F		6	20		0.236	0.787	0.385E+01	0.259E+01	FF			B+-0.3	F		F	F
F	35	0.5	34	1.378	0.020	1.339	0.425E+00	0.286E+00	FF			A+-0.15	F		FF	F
F		0.6	33.8		0.024	1.331	0.509E+00	0.342E+00	FF			B+-0.15	S		FF	S
S		0.8	33.4		0.031	1.315	0.675E+00	0.453E+00	FF			B+-0.15	S		FF	S
F		1	33		0.039	1.299	0.838E+00	0.563E+00	FF			B+-0.15	F		FF	F
S		1.2	32.6		0.047	1.283	0.100E+01	0.672E+00	FF			B+-0.15	F		FF	F
F		1.5	32		0.059	1.260	0.124E+01	0.833E+00	FF			B+-0.15	F		FF	F
S		1.8	31.4		0.071	1.236	0.147E+01	0.990E+00	FF			B+-0.15	F		FF	SF
F		2	31		0.079	1.220	0.163E+01	0.109E+01	FF			B+-0.15	F		FF	FF
F		2.2	30.6		0.087	1.205	0.178E+01	0.120E+01	FF			B+-0.15	F		FF	F
S		2.5	30		0.098	1.181	0.200E+01	0.135E+01	FF			B+-0.15	F		FF	S
F		2.8	29.4		0.110	1.157	0.222E+01	0.149E+01	FF			B+-0.15	F		FF	F
F		3	29		0.118	1.142	0.237E+01	0.159E+01	FF			B+-0.15	F		FF	F
S		3.5	28		0.138	1.102	0.272E+01	0.183E+01	FF			B+-0.15	F		FF	F
F		4	27		0.157	1.063	0.306E+01	0.205E+01	FF			B+-0.15	F		FF	F
S		4.5	26		0.177	1.024	0.338E+01	0.227E+01	FF			B+-0.15	F		FF	F
S		5	25		0.197	0.984	0.370E+01	0.249E+01	FF			B+-0.15	F		FF	F
S		5.5	24		0.217	0.945	0.400E+01	0.269E+01	FF			B+-0.2	F		FF	F
F		6	23		0.236	0.906	0.429E+01	0.288E+01	FF			B+-0.2	S		FF	F
S		7	21		0.276	0.827	0.483E+01	0.325E+01	FF			B+-0.2	F		FF	F
F		8	19		0.315	0.748	0.533E+01	0.358E+01	FF			B+-0.2	S		FF	F
F	38	0.5	37	1.496	0.020	1.457	0.462E+00	0.311E+00	FF			A+-0.15	F		FF	F
S		0.6	36.8		0.024	1.449	0.553E+00	0.372E+00	FF			B+-0.15	S		FF	S
F		0.8	36.4		0.031	1.433	0.734E+00	0.493E+00	FF			B+-0.15	S		FF	S
F		1	36		0.039	1.417	0.912E+00	0.613E+00	FF			B+-0.15	F		FF	F
F		1.2	35.6		0.047	1.402	0.109E+01	0.732E+00	FF			B+-0.15	F		FF	F
F		1.5	35		0.059	1.378	0.135E+01	0.907E+00	FF			B+-0.15	F		FF	SF
F		1.8	34.4		0.071	1.354	0.161E+01	0.108E+01	FF			B+-0.15	F		FF	FF
F		2	34		0.079	1.339	0.178E+01	0.119E+01	FF			B+-0.15	F		FF	F
S		2.2	33.6		0.087	1.323	0.194E+01	0.131E+01	FF			B+-0.15	F		FF	S
F		2.5	33		0.098	1.299	0.219E+01	0.147E+01	FF			B+-0.15	F		FF	F
F		2.8	32.4		0.110	1.276	0.243E+01	0.163E+01	FF			B+-0.15	F		FF	F
S		3	32		0.118	1.260	0.259E+01	0.174E+01	FF			B+-0.15	F		FF	F
F		3.5	31		0.138	1.220	0.298E+01	0.200E+01	FF			B+-0.15	F		FF	F
F		4	30		0.157	1.181	0.335E+01	0.225E+01	FF			B+-0.15	F		FF	SF
S		4.5	29		0.177	1.142	0.372E+01	0.250E+01	FF			B+-0.15	F		FF	FF
F		5	28		0.197	1.102	0.407E+01	0.273E+01	FF			B+-0.15	F		FF	S
S		5.5	27		0.217	1.063	0.441E+01	0.296E+01	FF			B+-0.15	F		FF	F
F		6	26		0.236	1.024	0.474E+01	0.318E+01	FF			B+-0.15	S		FF	S
S		7	24		0.276	0.945	0.535E+01	0.360E+01	FF			B+-0.2	S		FF	F
F		8	22		0.315	0.866	0.592E+01	0.398E+01	F			B+-0.25	F		F	F

Table 10-24 (Continued). Seamless (SL) and Welded (WE) Precision Steel Tubes (ISO 560, 3304, 3305)

TABLE 10-24. SEAMLESS (SL) AND WELDED (WE) PRECISION STEEL
TUBES (ISO 560,3304,3305)
BASIS; 1 IN = 25.4 MM
1 CUBIC METER STEEL = 7850 KG MASS

ORDER EXAMPLE;
LENGTH,STEEL TUBE 4X0.5 DIN2393 + STEEL QUAL
PAGE NO. 6

THE NOMINAL SIZE IS NATIONAL STANDARD AS INDICATED
F=FIRST CHOICE,S=SECOND CHOICE,T=THIRD CHOICE,NUMBER=OTHER SIZE
* = COMMERCIAL SIZE

I S B	MILLIMETERS A C B	INCHES A C	B	MASS PER UNIT KG/M LB/FT	SL WE	U.S.A. ANSI B32.5	AUSTRAL AS B32.5	JAPAN JIS	ISO 3304 3305	FRANCE NF A49-310	U.K. BS	GERMANY DIN 2391 2393	ITALY UNI 2898 5921
F	40 0.5 39	1.575 0.020	1.535	0.487E+00 0.327E+00	FF				A+-0.15			FF	F
S	0.6 38.8	0.024	1.528	0.583E+00 0.392E+00	FF				B+-0.15			FF	F
F	0.8 38.4	0.031	1.512	0.773E+00 0.520E+00	FF				B+-0.15	S		FF	S
SF	1 38	0.039	1.496	0.962E+00 0.646E+00	FF				B+-0.15	S		FF	SF
F	1.2 37.6	0.047	1.480	0.115E+01 0.772E+00	FF				B+-0.15	F		FF	SF
F	1.5 37	0.059	1.457	0.142E+01 0.957E+00	FF				B+-0.15	S		FF	FF
F	1.8 36.4	0.071	1.433	0.170E+01 0.114E+01	FF				B+-0.15	F		FF	F
SF	2 36	0.079	1.417	0.187E+01 0.126E+01	FF				B+-0.15	S		FF	S
F	2.2 35.6	0.087	1.402	0.205E+01 0.138E+01	FF				B+-0.15	F		FF	F
SF	2.5 35	0.098	1.378	0.231E+01 0.155E+01	FF				B+-0.15	S		FF	S
F	2.8 34.4	0.110	1.354	0.257E+01 0.173E+01	FF				B+-0.15	F		FF	F
SF	3 34	0.118	1.339	0.274E+01 0.184E+01	FF				B+-0.15	F		FF	SF
F	3.5 33	0.138	1.299	0.315E+01 0.212E+01	FF				B+-0.15	S		FF	S
S	4 32	0.157	1.260	0.355E+01 0.239E+01	FF				B+-0.15	F		FF	F
F	4.5 31	0.177	1.220	0.394E+01 0.265E+01	FF				B+-0.15	S		FF	S
SF	5 30	0.197	1.181	0.432E+01 0.290E+01	FF				B+-0.15	F		FF	S
F	5.5 29	0.217	1.142	0.468E+01 0.314E+01	FF				B+-0.15	S		FF	S
SF	6 28	0.236	1.102	0.503E+01 0.338E+01	FF				B+-0.15	F		FF	S
F	7 26	0.276	1.024	0.570E+01 0.383E+01	FF				B+-0.2	S		FF	F
F	8 24	0.315	0.945	0.631E+01 0.424E+01	FF				B+-0.25	F		FF	L
F	42 1 40	1.654 0.039	1.575	0.101E+01 0.679E+00	FF				A+-0.2	F		FF	F
SF	1.2 39.6	0.047	1.559	0.121E+01 0.811E+00	FF				B+-0.2	S		FF	SF
F	1.5 39	0.059	1.535	0.150E+01 0.101E+01	FF				B+-0.2	F		FF	F
SF	1.8 38.4	0.071	1.512	0.178E+01 0.120E+01	FF				B+-0.2	S		FF	S
F	2 38	0.079	1.496	0.197E+01 0.133E+01	FF				B+-0.2	F		FF	F
SF	2.2 37.6	0.087	1.480	0.216E+01 0.145E+01	FF				B+-0.2	S		FF	F
F	2.5 37	0.098	1.457	0.244E+01 0.164E+01	FF				B+-0.2	F		FF	S
SF	2.8 36.4	0.110	1.433	0.271E+01 0.182E+01	FF				B+-0.2	S		FF	S
F	3 36	0.118	1.417	0.289E+01 0.194E+01	FF				B+-0.2	F		FF	S
SF	3.5 35	0.138	1.378	0.332E+01 0.223E+01	FF				B+-0.2	F		FF	SF
F	4 34	0.157	1.339	0.375E+01 0.252E+01	FF				B+-0.2	S		FF	S
SF	4.5 33	0.177	1.299	0.416E+01 0.280E+01	FF				B+-0.2	F		FF	S
F	5 32	0.197	1.260	0.456E+01 0.307E+01	FF				B+-0.2	S		FF	S
SF	5.5 31	0.217	1.220	0.495E+01 0.333E+01	FF				B+-0.2	F		FF	S
F	6 30	0.236	1.181	0.533E+01 0.358E+01	FF				B+-0.2	S		FF	S
F	7 28	0.276	1.102	0.604E+01 0.406E+01	FF				B+-0.2	F		FF	F
F	8 26	0.315	1.024	0.671E+01 0.451E+01	FF				B+-0.2	S		FF	S
F	9 24	0.354	0.945	0.732E+01 0.492E+01	FF				B+-0.2	F		FF	F
F	45 1 43	1.772 0.039	1.693	0.109E+01 0.729E+00	FF				A+-0.2	F		FF	F
SF	1.2 42.6	0.047	1.677	0.130E+01 0.871E+00	FF				B+-0.2	S		FF	SF
F	1.5 42	0.059	1.654	0.161E+01 0.108E+01	FF				B+-0.2	F		FF	F
SF	1.8 41.4	0.071	1.630	0.192E+01 0.129E+01	FF				B+-0.2	S		FF	S
F	2 41	0.079	1.614	0.212E+01 0.143E+01	FF				B+-0.2	F		FF	F
SF	2.2 40.6	0.087	1.598	0.232E+01 0.156E+01	FF				B+-0.2	S		FF	F
F	2.5 40	0.098	1.575	0.262E+01 0.176E+01	FF				B+-0.2	F		FF	F
SF	2.8 39.4	0.110	1.551	0.291E+01 0.196E+01	FF				B+-0.2	S		FF	F
F	3 39	0.118	1.535	0.311E+01 0.209E+01	FF				B+-0.2	F		FF	F
S	3.5 38	0.138	1.496	0.358E+01 0.241E+01	FF				B+-0.2	S		FF	S

STEEL MATERIAL DATA

Table 10-24 (Continued). Seamless (SL) and Welded (WE) Precision Steel Tubes (ISO 560, 3304, 3305)

TABLE 10-24. SEAMLESS (SL) AND WELDED (WE) PRECISION STEEL TUBES (ISO 560, 3304, 3305)
BASIS: 1 IN = 25.4 MM
1 CUBIC METER STEEL = 7850 KG MASS

ORDER EXAMPLE:
LENGTH, STEEL TUBE 4X0.5 DIN2393 + STEEL QUAL
PAGE NO. 7

THE NOMINAL SIZE IS NATIONAL STANDARD AS INDICATED
F=FIRST CHOICE, S=SECOND CHOICE, T=THIRD CHOICE, NUMBER=OTHER SIZE
* = COMMERCIAL SIZE

I S O	MILLIMETERS A B C	INCHES A B C	MASS PER UNIT KG/M LB/FT	U.S.A. ANSI B32.5	AUSTRAL AS	JAPAN JIS	ISO 3304 3305	FRANCE NF A49-310	U.K. BS	GERMANY DIN 2391 2393	ITALY UNI 2898 5921
F	45 4 37	1.772 0.157 1.457	0.404E+01 0.272E+01	FF			B+-0.2	F		FF	F
S	4.5 36.5	0.177 1.417	0.449E+01 0.302E+01	FF			B+-0.2	S		FF	S
S	5 36	0.197 1.378	0.493E+01 0.331E+01	FF			B+-0.2	F		FF	F
F	5.5 34	0.217 1.339	0.536E+01 0.360E+01	FF			B+-0.2	F		FF	S
S	6 33	0.236 1.299	0.577E+01 0.388E+01	FF			B+-0.2	F		FF	F
S	7 31	0.276 1.220	0.656E+01 0.441E+01	FF			B+-0.2	S		FF	S
S	8 29	0.315 1.142	0.730E+01 0.491E+01	FF			B+-0.2	F		FF	F
S	9 27	0.354 1.063	0.799E+01 0.537E+01	FF			B+-0.2	S		FF	F
F	10 25	0.394 0.984	0.863E+01 0.580E+01	FF			B+-0.25	F		FF	
F	48 1 46	1.890 0.039 1.811	0.116E+01 0.779E+00				A+-0.2	F		FF	SF
S	1.2 45.6	0.047 1.795	0.138E+01 0.931E+00				B+-0.2	S		FF	FF
S	1.5 45	0.059 1.772	0.172E+01 0.116E+01				B+-0.2	F		FF	SF
S	1.8 44.4	0.071 1.748	0.205E+01 0.138E+01				B+-0.2	S		FF	FF
S	2 44	0.079 1.732	0.227E+01 0.152E+01				B+-0.2	F		FF	FF
F	2.2 43.6	0.087 1.717	0.248E+01 0.167E+01				B+-0.2	F		FF	FF
S	2.5 43	0.098 1.693	0.281E+01 0.189E+01				B+-0.2	S		FF	SF
S	2.8 42.4	0.110 1.669	0.312E+01 0.210E+01				B+-0.2	F		FF	FF
S	3 42	0.118 1.654	0.333E+01 0.224E+01				B+-0.2	F		FF	SF
F	3.5 41	0.138 1.614	0.384E+01 0.258E+01				B+-0.2	F		FF	FF
S	4 40	0.157 1.575	0.434E+01 0.292E+01				B+-0.2	F		FF	SF
S	4.5 39	0.177 1.535	0.483E+01 0.324E+01				B+-0.2	F		FF	FF
S	5 38	0.197 1.496	0.530E+01 0.356E+01				B+-0.2	S		FF	SF
S	5.5 37	0.217 1.457	0.576E+01 0.387E+01				B+-0.2	F		FF	FF
S	6 36	0.236 1.417	0.621E+01 0.418E+01				B+-0.2	F		FF	SF
S	7 34	0.276 1.339	0.708E+01 0.476E+01				B+-0.2	S		FF	FF
S	8 32	0.315 1.260	0.789E+01 0.530E+01				B+-0.2	F		FF	SF
F	9 30	0.354 1.181	0.866E+01 0.582E+01				B+-0.2	F		FF	FF
F	10 28	0.394 1.102	0.937E+01 0.630E+01				B+-0.2	F		FF	SF
F	50 1 48	1.969 0.039 1.890	0.121E+01 0.812E+00	FF			A+-0.2	F		FF	F
S	1.2 47.6	0.047 1.874	0.144E+01 0.970E+00	FF			B+-0.2	S		FF	SF
S	1.5 47	0.059 1.850	0.179E+01 0.121E+01	FF			B+-0.2	F		FF	FF
S	1.8 46.4	0.071 1.827	0.214E+01 0.144E+01	FF			B+-0.2	S		FF	SF
S	2 46	0.079 1.811	0.237E+01 0.159E+01	FF			B+-0.2	F		FF	SF
F	2.2 45.6	0.087 1.795	0.259E+01 0.174E+01	FF			B+-0.2	F		FF	SF
S	2.5 45	0.098 1.772	0.293E+01 0.197E+01	FF			B+-0.2	S		FF	FF
S	2.8 44.4	0.110 1.748	0.326E+01 0.219E+01	FF			B+-0.2	F		FF	SF
S	3 44	0.118 1.732	0.348E+01 0.234E+01	FF			B+-0.2	F		FF	SF
F	3.5 43	0.138 1.693	0.401E+01 0.270E+01	FF			B+-0.2	F		FF	FF
S	4 42	0.157 1.654	0.454E+01 0.305E+01	FF			B+-0.2	F		FF	SF
S	4.5 41	0.177 1.614	0.505E+01 0.339E+01	FF			B+-0.2	S		FF	SF
S	5 40	0.197 1.575	0.555E+01 0.373E+01	FF			B+-0.2	F		FF	SF
S	5.5 39	0.217 1.535	0.604E+01 0.406E+01	FF			B+-0.2	F		FF	SF
S	6 38	0.236 1.496	0.651E+01 0.437E+01	FF			B+-0.2	F		FF	SF
S	7 36	0.276 1.417	0.742E+01 0.499E+01	FF			B+-0.2	F		FF	SF
S	8 34	0.315 1.339	0.829E+01 0.557E+01	FF			B+-0.2	F		FF	SF
S	9 32	0.354 1.260	0.910E+01 0.611E+01	FF			B+-0.2	F		FF	SF
F	10 30	0.394 1.181	0.986E+01 0.663E+01	FF			B+-0.2	F		FF	

Table 10-24 (Continued). Seamless (SL) and Welded (WE) Precision Steel Tubes (ISO 560, 3304, 3305)

TABLE 10-24. SEAMLESS (SL) AND WELDED (WE) PRECISION STEEL
TUBES (ISO 560,3304,3305)
BASIS: 1 IN = 25.4 MM
1 CUBIC METER STEEL = 7850 KG MASS

ORDER EXAMPLE:
LENGTH,STEEL TUBE 4X0.5 DIN2393 + STEEL QUAL

PAGE NO. 8

THE NOMINAL SIZE IS NATIONAL STANDARD AS INDICATED
F=FIRST CHOICE,S=SECOND CHOICE,T=THIRD CHOICE,NUMBER=OTHER SIZE
* = COMMERCIAL SIZE

I S O	MILLIMETERS A B C	INCHES A B C	MASS PER UNIT KG/M LB/FT	U.S.A. ANSI B32.5 SL	AUSTRAL AS B32.5 WE	JAPAN JIS	ISO 3304 3305	FRANCE NF A49-310	U.K. BS	GERMANY DIN 2391 2393	ITALY UNI 2898 5921
F 55 1 53	2.165 0.039 2.087	0.133E+01 0.895E+00	FF			A+-0.25 1.2S			FF	
F 5 45	0.197 1.772	0.617E+01 0.414E+01	FF			B+-0.25 F			FF	
F 10 35	0.394 1.378	0.111E+02 0.746E+01	FF			A+-0.25 F			F	
F 11 33	0.433 1.299	0.119E+02 0.802E+01	FF			A+-0.25 F				
F 12.5 30	0.492 1.181	0.131E+02 0.880E+01	12FF			B+-0.25 F				
F 60 1 58	2.362 0.039 2.283	0.146E+01 0.978E+00	FF			A+-0.25 1.2S			FF	
S 5.5 49	0.217 1.929	0.739E+01 0.497E+01	FF			B+-0.25 S			FF	
F 10 40	0.394 1.575	0.123E+02 0.829E+01	FF			B+-0.25 F			F	
F 11 38	0.433 1.496	0.133E+02 0.893E+01	FF			B+-0.25 F				
F 12.5 35	0.492 1.378	0.146E+02 0.984E+01	12FF			B+-0.25 F				
F 63 1 61	2.480 0.039 2.402	0.153E+01 0.103E+01	FF			A+-0.3 1.5F			FF	F
S 5.5 52	0.217 2.047	0.780E+01 0.524E+01	FF			B+-0.3 S			FF	S
F 10 43	0.394 1.693	0.131E+02 0.878E+01	FF			B+-0.3 F			F	
F 11 41	0.433 1.614	0.141E+02 0.948E+01				B+-0.3 F				
F 12.5 38	0.492 1.496	0.156E+02 0.105E+02				B+-0.3 F				
F 70 1 68	2.756 0.039 2.677	0.170E+01 0.114E+01	FF			A+-0.3 1.5F			FF	F
S 5.5 59	0.217 2.323	0.875E+01 0.588E+01	FF			B+-0.3 S			FF	S
F 10 50	0.394 1.969	0.148E+02 0.994E+01	FF			B+-0.3 F			F	
F 11 48	0.433 1.890	0.160E+02 0.108E+02	FF			B+-0.3 F				
F 12.5 45	0.492 1.772	0.177E+02 0.119E+02	12FF			B+-0.3 F				
F 80 1 78	3.150 0.039 3.071	0.195E+01 0.131E+01	FF			A+-0.35 1.8S			FF	F
S 6 68	0.236 2.677	0.109E+02 0.736E+01	FF			B+-0.35 F			FF	F
F 10 60	0.394 2.362	0.173E+02 0.116E+02	FF			B+-0.35 F			F	
F 11 58	0.433 2.283	0.187E+02 0.126E+02	FF			B+-0.35 F				
F 12.5 55	0.492 2.165	0.208E+02 0.140E+02	12FF			B+-0.35 F				
F 90 1.5 87	3.543 0.059 3.425	0.327E+01 0.220E+01	FF			A+-0.4 1.8S			FF	F
F 6 78	0.236 3.071	0.124E+02 0.835E+01	FF			B+-0.4 F			FF	F
F 10 70	0.394 2.756	0.197E+02 0.133E+02	FF			B+-0.4 F			F	
F 11 68	0.433 2.677	0.214E+02 0.144E+02	FF			B+-0.4 F				
F 12.5 65	0.492 2.559	0.239E+02 0.161E+02	12FF			B+-0.4 F				
F 100 1.5 97	3.937 0.059 3.819	0.364E+01 0.245E+01	FF			A+-0.45 2F			FF	F
F 7 86	0.276 3.386	0.161E+02 0.108E+02	FF			B+-0.45 S			FF	S
F 10 80	0.394 3.150	0.222E+02 0.149E+02	FF			B+-0.45 F			F	
F 11 78	0.433 3.071	0.241E+02 0.162E+02	FF			B+-0.45 F				
F 12.5 75	0.492 2.953	0.270E+02 0.181E+02	12FF			B+-0.45 F				
F 110 2 106	4.331 0.079 4.173	0.533E+01 0.358E+01	FF			A+-0.5 2.2S			FF	F
F 7 96	0.276 3.780	0.178E+02 0.119E+02	FF			B+-0.5 S			FF	S
F 10 90	0.394 3.543	0.247E+02 0.166E+02	FF			B+-0.5 F			F	
F 120 2 116	4.724 0.079 4.567	0.582E+01 0.391E+01	FF			A+-0.5 2.5F			FF	F
F 10 100	0.394 3.937	0.271E+02 0.182E+02	FF			B+-0.5 F			FF	F

STEEL MATERIAL DATA

Table 10-25. Seamless (SL) and Welded (WE) Stainless Steel Tubes (ISO 1127, 5252)

TABLE 10-25. SEAMLESS(SL) AND WELDED(WE) STAINLESS
STEEL TUBES (ISO 1127,5252)
BASIS; 1 IN = 25.4 MM, 1 CUBIC METER
AUSTENITIC STEEL = 7967.75 KG MASS

ORDER EXAMPLE:
LENGTH,STAINLESS TUBE 25X2 DIN2462 + STEEL QUA
PAGE NO. 1

THE NOMINAL SIZE IS NATIONAL STANDARD AS INDICATED
F=FIRST CHOICE,S=SECOND CHOICE,T=THIRD CHOICE,NUMBER=OTHER SIZE
* = COMMERCIAL SIZE

	MILLIMETERS		INCHES				MASS PER UNIT		U.S.A. ANSI B36.19	AUSTRAL AS	JAPAN JIS G3459 G3463	ISO 1127 D2,T3	FRANCE NF	U.K. BS	GERMANY DIN 2462 2463	ITALY UNI 6904
I S O	A	B	A	C	B	KG/M	LB/FT	SL WE								
F	10.2	0.5 9.2	0.402	0.020	0.363	0.121E+00	0.816E-01				A+-0.5			F	F
F		0.8 8.6		0.031	0.339	0.188E+00	0.126E+00				C+-10%			FF	F
F		1.0 8.2		0.039	0.323	0.230E+00	0.155E+00				C+-10%			FF	F
F		1.2 7.8		0.047	0.308	0.270E+00	0.182E+00			FF	C+-10%			FF	
F		1.6 7.0		0.063	0.276	0.344E+00	0.231E+00			FF	C+-10%			FF	
F		2.0 6.2		0.079	0.245	0.411E+00	0.276E+00				C+-10%			FF	
F		2.3 5.6		0.091	0.221	0.455E+00	0.306E+00				C+-10%			FF	
F		2.6 5.0		0.102	0.197	0.495E+00	0.332E+00				C+-10%			FF	
F	13.5	0.5 12.5	0.531	0.020	0.492	0.163E+00	0.109E+00				A+-0.5			F	F
F		0.8 11.9		0.031	0.468	0.254E+00	0.171E+00				C+-10%			FF	F
F		1.0 11.5		0.039	0.452	0.313E+00	0.210E+00				C+-10%			FF	FF
F		1.2 11.1		0.047	0.437	0.369E+00	0.248E+00				C+-10%			FF	FF
F		1.6 10.3		0.063	0.405	0.477E+00	0.320E+00			FF	C+-10%			FF	FF
F		2.0 9.5		0.079	0.374	0.576E+00	0.387E+00				C+-10%			FF	FF
F		2.3 8.9		0.091	0.350	0.645E+00	0.433E+00				C+-10%			FF	FF
F		2.6 8.3		0.102	0.326	0.709E+00	0.477E+00				C+-10%			FF	FF
F		2.9 7.7		0.114	0.303	0.769E+00	0.517E+00				C+-10%			FF	FF
F	16.0	0.5 15.0	0.630	0.020	0.591	0.194E+00	0.130E+00				A+-0.5			F	F
F		0.8 14.4		0.031	0.567	0.304E+00	0.205E+00				C+-10%			FF	FF
F		1.0 14.0		0.039	0.551	0.375E+00	0.252E+00				C+-10%			FF	FF
F		1.2 13.6		0.047	0.536	0.445E+00	0.299E+00				C+-10%			FF	FF
F		1.6 12.8		0.063	0.504	0.577E+00	0.388E+00			FF	C+-10%			FF	FF
F		2.0 12.0		0.079	0.473	0.701E+00	0.471E+00				C+-10%			FF	FF
F		2.3 11.4		0.091	0.449	0.789E+00	0.530E+00				C+-10%			FF	FF
F		2.6 10.8		0.102	0.425	0.872E+00	0.588E+00				C+-10%			FF	FF
F		3.2 9.6		0.126	0.378	0.103E+01	0.689E+00				C+-10%			FF	FF
F		3.6 8.8		0.142	0.347	0.112E+01	0.751E+00				C+-10%			FF	FF
F		4.0 8.0		0.157	0.315	0.120E+01	0.807E+00				C+-10%			FF	FF
F	17.2	0.5 16.2	0.677	0.020	0.638	0.209E+00	0.140E+00				A+-0.5			F,O.8F	F
F		0.8 15.6		0.031	0.598	0.406E+00	0.272E+00				C+-10%			FF	FF
F		1.0 15.2		0.039	0.583	0.481E+00	0.323E+00				C+-10%			FF	FF
F		1.2 14.8		0.047	0.551	0.520E+00	0.420E+00				C+-10%			FF	FF
F		1.6 14.0		0.063	0.520	0.625E+00	0.420E+00			FF	C+-10%			FF	FF
F		2.0 13.2		0.079	0.496	0.761E+00	0.511E+00				C+-10%			FF	FF
F		2.3 12.6		0.091	0.472	0.858E+00	0.576E+00			FF	C+-10%			FF	FF
F		2.6 12.0		0.102	0.449	0.950E+00	0.639E+00				C+-10%			FF	FF
F		2.9 11.4		0.114	0.425	0.104E+01	0.698E+00				C+-10%			FF	FF
F		3.2 10.8		0.126	0.402	0.112E+01	0.754E+00				C+-10%			FF	FF
F		3.6 10.0		0.142	0.394	0.123E+01	0.824E+00				C+-10%			FF	FF
F		4.0 9.2		0.157	0.362	0.132E+01	0.888E+00				C+-10%			FF	FF
F	21.3	0.5 20.3	0.840	0.020	0.801	0.260E+00	0.175E+00				A+-0.5			F	F
F		0.8 19.7		0.031	0.777	0.411E+00	0.276E+00				C+-10%			FF	FF
F		1.0 19.3		0.039	0.761	0.508E+00	0.341E+00				C+-10%			FF	FF
F		1.2 18.9		0.047	0.746	0.604E+00	0.406E+00				C+-10%			FF	FF
F		1.6 18.1		0.063	0.714	0.789E+00	0.530E+00			FF	C+-10%			FF	FF
F		2.0 17.3		0.079	0.683	0.966E+00	0.649E+00			FF	C+-10%			FF	FF

10-81

Table 10-25 (Continued). Seamless (SL) and Welded (WE) Stainless Steel Tubes (ISO 1127, 5252)

TABLE 10-25. SEAMLESS(SL) AND WELDED(WE) STAINLESS
STEEL TUBES (ISO 1127,5252)
BASIS; 1 IN = 25.4 MM, 1 CUBIC METER
AUSTENITIC STEEL = 7967.75 KG MASS

PAGE NR. 2

ORDER EXAMPLE:
LENGTH,STAINLESS TUBE 25x2 DIN2462 + STEEL QUA

THE NOMINAL SIZE IS NATIONAL STANDARD AS INDICATED
F=FIRST CHOICE,S=SECOND CHOICE,T=THIRD CHOICE,NUMBER=OTHER SIZE
* = COMMERCIAL SIZE

MILLIMETERS			INCHES			MASS PER UNIT		U.S.A. ANSI B36.19 SL	U.S.A. ANSI B36.19 WE	AUSTRAL AS	JAPAN JIS G3459	JAPAN JIS G3463	ISO 1127	ISO D2,T3	FRANCE NF	U.K. BS	GERMANY DIN 2462	GERMANY DIN 2463	ITALY UNI 6904
A	B	C	A	B	C	KG/M	LB/FT												
21.3	16.7	2.3	0.840	0.659	0.091	0.109E+01	0.735E+00						A+-0.5				F	F	F
	16.1	2.6		0.635	0.102	0.122E+01	0.818E+00				FF		C+-10%				F	F	F
	15.5	2.9		0.612	0.114	0.134E+01	0.898E+00	FF			FF		C+-10%				F	F	F
	14.9	3.2		0.588	0.126	0.145E+01	0.974E+00				FF		C+-10%				F	F	F
	14.1	3.6		0.557	0.142	0.160E+01	0.107E+01	FF	FF		FF		C+-10%				F	F	F
	13.3	4.0		0.525	0.157	0.173E+01	0.116E+01						C+-10%				F	F	F
	12.3	4.5		0.486	0.177	0.189E+01	0.127E+01				FF		C+-10%				F	F	F
	11.3	5.0		0.446	0.197	0.204E+01	0.137E+01						C+-10%				F	F	F
	10.1	5.6		0.399	0.220	0.220E+01	0.148E+01						C+-10%				F	F	F
25.0	24.0	0.5	0.984	0.945	0.020	0.307E+00	0.206E+00						A+-0.5				F	F	F
	23.4	0.8		0.921	0.031	0.485E+00	0.326E+00						C+-10%				F	F	F
	23.0	1.0		0.905	0.039	0.601E+00	0.404E+00						C+-10%				F	F	F
	22.6	1.2		0.890	0.047	0.715E+00	0.480E+00						C+-10%				F	F	F
	21.8	1.6		0.858	0.063	0.937E+00	0.630E+00				FF		C+-10%				F	F	F
	21.0	2.0		0.827	0.079	0.115E+01	0.774E+00				FF		C+-10%				F	F	F
	20.4	2.3		0.803	0.091	0.131E+01	0.878E+00				FF		C+-10%				F	F	F
	19.8	2.6		0.779	0.102	0.146E+01	0.980E+00				FF		C+-10%				F	F	F
	19.2	2.9		0.756	0.114	0.160E+01	0.108E+01						C+-10%				F	F	F
	18.6	3.2		0.732	0.126	0.175E+01	0.117E+01				FF		C+-10%				F	F	F
	17.8	3.6		0.701	0.142	0.193E+01	0.130E+01				FF		C+-10%				F	F	F
	17.0	4.0		0.669	0.157	0.210E+01	0.141E+01						C+-10%				F	F	F
	16.0	4.5		0.630	0.177	0.231E+01	0.155E+01						C+-10%				F	F	F
	15.0	5.0		0.590	0.197	0.250E+01	0.168E+01						C+-10%				F	F	F
	13.8	5.6		0.543	0.220	0.272E+01	0.183E+01						C+-10%				F	F	F
	12.4	6.3		0.488	0.248	0.295E+01	0.198E+01						C+-10%				F	F	F
26.9	25.3	0.8	1.059	0.996	0.031	0.523E+00	0.351E+00						A+-0.5				F	F	F
	24.9	1.0		0.980	0.039	0.648E+00	0.436E+00				FF		C+-10%				F	F	F
	24.5	1.2		0.965	0.047	0.772E+00	0.519E+00				FF		C+-10%				F	F	F
	23.7	1.6		0.933	0.063	0.101E+01	0.681E+00	FF	FF		FF		C+-10%				F	F	F
	22.9	2.0		0.902	0.079	0.125E+01	0.838E+00	FF	FF		FF		C+-10%				F	F	F
	22.3	2.3		0.878	0.091	0.142E+01	0.952E+00				FF		C+-10%				F	F	F
	21.7	2.6		0.854	0.102	0.158E+01	0.106E+01				FF		C+-10%				F	F	F
	21.1	2.9		0.831	0.114	0.174E+01	0.117E+01						C+-10%				F	F	F
	20.5	3.2		0.807	0.126	0.190E+01	0.128E+01	FF			FF		C+-10%				F	F	F
	19.7	3.6		0.776	0.142	0.210E+01	0.141E+01				FF		C+-10%				F	F	F
	18.9	4.0		0.744	0.157	0.229E+01	0.154E+01	FF					C+-10%				F	F	F
	17.9	4.5		0.705	0.177	0.252E+01	0.170E+01						C+-10%				F	F	F
	16.9	5.0		0.665	0.197	0.274E+01	0.184E+01						C+-10%				F	F	F
	15.7	5.6		0.618	0.220	0.299E+01	0.201E+01						C+-10%				F	F	F
	14.3	6.3		0.563	0.248	0.325E+01	0.218E+01						C+-10%				F	F	F
	12.7	7.1		0.500	0.280	0.352E+01	0.236E+01						C+-10%				F	F	F
30.0	28.4	0.8	1.181	1.118	0.031	0.585E+00	0.393E+00						A+-0.5				F	F	F
	28.0	1.0		1.102	0.039	0.726E+00	0.488E+00						C+-10%				F	F	F
	27.6	1.2		1.087	0.047	0.865E+00	0.581E+00						C+-10%				F	F	F
	26.8	1.6		1.055	0.063	0.114E+01	0.764E+00						C+-10%				F	F	F
	26.0	2.0		1.024	0.079	0.140E+01	0.942E+00						C+-10%				F	F	F
	25.4	2.3		1.000	0.091	0.159E+01	0.107E+01						C+-10%				F	F	F
	24.8	2.6		0.976	0.102	0.178E+01	0.120E+01						C+-10%				F	F	F

STEEL MATERIAL DATA

Table 10-25 (Continued). Seamless (SL) and Welded (WE) Stainless Steel Tubes (ISO 1127, 5252)

TABLE 10-25. SEAMLESS(SL) AND WELDED(WE) STAINLESS
STEEL TUBES (ISO 1127,5252)
BASIS; 1 IN = 25.4 MM, 1 CUBIC METER
AUSTENITIC STEEL = 7967.75 KG MASS

ORDER EXAMPLE:
LENGTH/STAINLESS TUBE 25x2 DIN2462- + STEEL QUA
PAGE NO. 3

THE NOMINAL SIZE IS NATIONAL STANDARD AS INDICATED
F=FIRST CHOICE, S=SECOND CHOICE, T=THIRD CHOICE, NUMBER=OTHER SIZE
* = COMMERCIAL SIZE

ISO	MILLIMETERS A	B	C	INCHES A	B	C	MASS PER UNIT KG/M LB/FT SL/WE	U.S.A. ANSI B36.19 B36.19	AUSTRAL AS	JAPAN JIS G3459 G3463	ISO 1127 D2,T3	FRANCE NF	U.K. BS	GERMANY DIN 2462 2463	ITALY UNI 6904
F	30.0	2.9	24.2	1.181	0.114	0.953	0.197E+01 0.132E+01				A+-0.5				
F		3.2	23.6		0.126	0.929	0.215E+01 0.144E+01				C+-10%			F	F
F		3.6	22.8		0.142	0.898	0.238E+01 0.160E+01				C+-10%			F	F
F		4.0	22.0		0.157	0.866	0.260E+01 0.175E+01				C+-10%			F	F
F		4.5	21.0		0.177	0.827	0.287E+01 0.193E+01				C+-10%			F	F
F		5.0	20.0		0.197	0.787	0.313E+01 0.210E+01				C+-10%			F	F
F		5.6	18.8		0.220	0.740	0.342E+01 0.230E+01				C+-10%			F	F
F		6.3	17.4		0.248	0.685	0.374E+01 0.251E+01				C+-10%			F	F
F		7.1	15.8		0.280	0.622	0.407E+01 0.273E+01				C+-10%			F	F
F	31.8	0.8	30.2	1.250	0.031	1.187	0.621E+00 0.417E+00				A+-0.5				
F		1.0	29.8		0.039	1.171	0.771E+00 0.518E+00				C+-10%			FF	F
F		1.2	29.4		0.047	1.156	0.919E+00 0.618E+00				C+-10%			FF	F
F		1.6	28.6		0.063	1.124	0.121E+01 0.813E+00				C+-10%			FF	F
F		2.0	27.8		0.079	1.093	0.149E+01 0.100E+01		FF		C+-10%			FF	F
F		2.3	27.2		0.091	1.069	0.170E+01 0.114E+01		FF		C+-10%			FF	F
F		2.6	26.6		0.102	1.045	0.190E+01 0.128E+01		FF		C+-10%			FF	F
F		2.9	26.0		0.114	1.022	0.210E+01 0.141E+01		FF		C+-10%			FF	F
F		3.2	25.4		0.126	0.998	0.229E+01 0.154E+01		FF		C+-10%			FF	F
F		3.6	24.6		0.142	0.967	0.254E+01 0.171E+01		FF		C+-10%			FF	F
F		4.0	23.8		0.157	0.935	0.278E+01 0.187E+01		FF		C+-10%			FF	F
F		4.5	22.8		0.177	0.896	0.308E+01 0.207E+01				C+-10%			FF	F
F		5.6	20.6		0.220	0.809	0.367E+01 0.247E+01				C+-10%			FF	F
F		6.3	19.2		0.248	0.754	0.402E+01 0.270E+01				C+-10%			FF	F
F		7.1	17.6		0.280	0.691	0.439E+01 0.295E+01				C+-10%			FF	F
F	33.7	0.8	32.1	1.327	0.031	1.264	0.659E+00 0.443E+00				A+-0.5				
F		1.0	31.7		0.039	1.248	0.819E+00 0.550E+00				C+-10%			FF	F
F		1.2	31.3		0.047	1.233	0.976E+00 0.656E+00				C+-10%			FF	F
F		1.6	30.5		0.063	1.201	0.129E+01 0.864E+00	FF			C+-10%			FF	F
F		2.0	29.7		0.079	1.170	0.159E+01 0.107E+01		FF		C+-10%			FF	F
F		2.3	29.1		0.091	1.146	0.181E+01 0.121E+01		FF		C+-10%			FF	F
F		2.6	28.5		0.102	1.122	0.202E+01 0.136E+01		FF		C+-10%			FF	F
F		2.9	27.9		0.114	1.099	0.224E+01 0.150E+01	FF	FF		C+-10%			FF	F
F		3.2	27.3		0.126	1.075	0.244E+01 0.164E+01	FF	FF		C+-10%			FF	F
F		3.6	26.5		0.142	1.044	0.271E+01 0.182E+01	FF	FF		C+-10%			FF	F
F		4.0	25.7		0.157	1.012	0.297E+01 0.200E+01		FF		C+-10%			FF	F
F		4.5	24.7		0.177	0.973	0.329E+01 0.221E+01		FF		C+-10%			FF	F
F		5.0	23.7		0.197	0.933	0.359E+01 0.241E+01		FF		C+-10%			FF	F
F		5.6	22.5		0.220	0.886	0.394E+01 0.265E+01		FF		C+-10%			FF	F
F		6.3	21.1		0.248	0.831	0.432E+01 0.290E+01				C+-10%			FF	F
F		7.1	19.5		0.280	0.768	0.473E+01 0.318E+01				C+-10%			FF	F
F		8.0	17.7		0.315	0.697	0.515E+01 0.346E+01				C+-10%			FF	F
F	38.0	0.8	36.4	1.500	0.031	1.437	0.745E+00 0.501E+00				A+-0.5				
F		1.0	36.0		0.039	1.421	0.926E+00 0.622E+00				C+-10%			FF	F
F		1.2	35.6		0.047	1.406	0.111E+01 0.743E+00				C+-10%			FF	F
F		1.6	34.8		0.063	1.374	0.146E+01 0.980E+00				C+-10%			FF	F
F		2.0	34.0		0.079	1.343	0.180E+01 0.121E+01		FF		C+-10%			FF	F
F		2.3	33.4		0.091	1.319	0.206E+01 0.138E+01		FF		C+-10%			FF	F
F		2.6	32.8		0.102	1.295	0.230E+01 0.155E+01		FF		C+-10%			FF	F

Table 10-25 (Continued). Seamless (SL) and Welded (WE) Stainless Steel Tubes (ISO 1127, 5252)

TABLE 10-25. SEAMLESS(SL) AND WELDED(WE) STAINLESS
STEEL TUBES (ISO 1127,5252)
BASIS; 1 IN = 25.4 MM, 1 CUBIC METER
AUSTENITIC STEEL = 7967.75 KG MASS

ORDER EXAMPLE;
LENGTH, STAINLESS TUBE 25X2 DIN2462 + STEEL QUA
PAGE NO. 4

THE NOMINAL SIZE IS NATIONAL STANDARD AS INDICATED
F=FIRST CHOICE, S=SECOND CHOICE, T=THIRD CHOICE, NUMBER=OTHER SIZE
*=COMMERCIAL SIZE

ISO	MILLIMETERS A B C	INCHES A B C	MASS PER UNIT KG/M LB/FT	SL WE	U.S.A. ANSI B36.19	AUSTRAL AS B36.19	JAPAN JIS G3459 G3463	ISO 1127 D2,T3	FRANCE NF	U.K. BS	GERMANY DIN 2462 2463	ITALY UNI 6904
F	38.0 2.9 32.2	1.500 0.114 1.272	0.255E+01 0.171E+01					A+-0.5				
F	3.2 31.6	0.126 1.248	0.279E+01 0.187E+01					C+-10%			FF	FF
F	3.6 30.8	0.142 1.217	0.310E+01 0.208E+01					C+-10%			FF	F
F	4.0 30.0	0.157 1.185	0.340E+01 0.229E+01					C+-10%			FF	F
F	4.5 29.0	0.177 1.146	0.377E+01 0.254E+01				FF	C+-10%			FF	F
F	5.0 28.0	0.197 1.106	0.413E+01 0.278E+01				FF	C+-10%			FF	F
F	5.6 26.8	0.220 1.059	0.454E+01 0.305E+01				FF	C+-10%			FF	F
F	6.3 25.4	0.248 1.004	0.500E+01 0.336E+01				FF	C+-10%			FF	F
F	7.1 23.8	0.280 0.941	0.549E+01 0.369E+01				FF	C+-10%			FF	F
F	8.0 22.0	0.315 0.870	0.601E+01 0.404E+01				FF	C+-10%			FF	F
F	8.8 20.4	0.346 0.807	0.643E+01 0.432E+01					C+-10%			F	F
F	10.0 18.0	0.394 0.713	0.701E+01 0.471E+01					C+-10%			F	F
F	42.4 1.0 40.4	1.669 0.039 1.590	0.104E+01 0.696E+00					A+-0.5				
F	1.6 39.2	0.063 1.543	0.163E+01 0.110E+01					C+-10%			FF	FF/1.2FF
F	2.0 38.4	0.079 1.512	0.202E+01 0.136E+01					C+-10%			FF	F
F	2.3 37.8	0.091 1.488	0.231E+01 0.155E+01					C+-10%			FF	F
F	2.6 37.2	0.102 1.464	0.259E+01 0.174E+01			FF		C+-10%			FF	F
F	2.9 36.6	0.114 1.441	0.287E+01 0.193E+01					C+-10%			FF	F
F	3.2 36.0	0.126 1.417	0.314E+01 0.211E+01			FF		C+-10%			FF	F
F	3.6 35.2	0.142 1.386	0.350E+01 0.235E+01					C+-10%			FF	F
F	4.0 34.4	0.157 1.354	0.384E+01 0.258E+01			FF		C+-10%			FF	F
F	4.5 33.4	0.177 1.315	0.427E+01 0.287E+01					C+-10%			FF	F
F	5.0 32.4	0.197 1.275	0.468E+01 0.315E+01					C+-10%			FF	F
F	5.6 31.2	0.220 1.228	0.516E+01 0.347E+01					C+-10%			FF	F
F	6.3 29.8	0.248 1.173	0.569E+01 0.383E+01			6FF		C+-10%			FF	F
F	7.1 28.2	0.280 1.110	0.627E+01 0.422E+01					C+-10%			FF	F
F	8.0 26.4	0.315 1.039	0.689E+01 0.463E+01					C+-10%			FF	F
F	8.8 24.8	0.346 0.976	0.740E+01 0.497E+01					C+-10%			FF	F
F	10.0 22.4	0.394 0.882	0.811E+01 0.545E+01					C+-10%			F	F
F	11.0 20.4	0.433 0.803	0.865E+01 0.581E+01					C+-10%			F	F
F	44.5 1.0 42.5	1.750 0.039 1.671	0.109E+01 0.732E+00					A+-0.5				
F	1.2 42.1	0.047 1.656	0.130E+01 0.874E+00					C+-10%			FF	FF
F	1.6 41.3	0.063 1.624	0.172E+01 0.115E+01					C+-10%			FF	F
F	2.0 40.5	0.079 1.593	0.213E+01 0.143E+01					C+-10%			FF	F
F	2.3 39.9	0.091 1.569	0.243E+01 0.163E+01					C+-10%			FF	F
F	2.6 39.3	0.102 1.545	0.273E+01 0.183E+01			FF		C+-10%			FF	F
F	2.9 38.7	0.114 1.522	0.302E+01 0.203E+01			FF		C+-10%			FF	F
F	3.2 38.1	0.126 1.498	0.331E+01 0.222E+01			FF		C+-10%			FF	F
F	3.6 37.3	0.142 1.467	0.369E+01 0.248E+01			FF		C+-10%			FF	F
F	4.0 36.5	0.157 1.435	0.406E+01 0.272E+01			FF		C+-10%			FF	F
F	4.5 35.5	0.177 1.396	0.451E+01 0.303E+01			FF		C+-10%			FF	F
F	5.0 34.5	0.197 1.356	0.494E+01 0.332E+01			FF		C+-10%			FF	F
F	5.6 33.3	0.220 1.309	0.545E+01 0.366E+01			FF		C+-10%			FF	F
F	6.3 31.9	0.248 1.254	0.602E+01 0.405E+01			6.5FF		C+-10%			FF	F
F	7.1 30.3	0.280 1.191	0.665E+01 0.447E+01			FF		C+-10%			FF	F
F	8.0 28.5	0.315 1.120	0.731E+01 0.491E+01			FF		C+-10%			FF	F
F	8.8 26.9	0.346 1.057	0.786E+01 0.528E+01					C+-10%			F	F
F	10.0 24.5	0.394 0.963	0.864E+01 0.580E+01					C+-10%			F	F
F	11.0 22.5	0.433 0.884	0.922E+01 0.620E+01					C+-10%			F	F

/ STEEL MATERIAL DATA 10-85

Table 10-25 (Continued). Seamless (SL) and Welded (WE) Stainless Steel Tubes (ISO 1127, 5252)

TABLE 10-25. SEAMLESS(SL) AND WELDED(WE) STAINLESS
STEEL TUBES (ISO 1127,5252)
BASIS; 1 IN. = 25.4 MM, 1 CUBIC METER
AUSTENITIC STEEL = 7967.75 KG MASS

ORDER EXAMPLE;
LENGTH,STAINLESS TUBE 25X2 DIN2462 + STEEL QUA

PAGE NO. 5

THE NOMINAL SIZE IS NATIONAL STANDARD AS INDICATED
F=FIRST CHOICE,S=SECOND CHOICE,T=THIRD CHOICE,NUMBER=OTHER SIZE
* = COMMERCIAL SIZE

I S O	MILLIMETERS A C B	INCHES A C B	MASS PER UNIT KG/M LB/FT	U.S.A. ANSI B36.19 B36.19	AUSTRAL AS	JAPAN JIS G3459 G3463	ISO 1127 D2,T3	FRANCE NF	U.K. BS	GERMANY DIN 2462 2463	ITALY UNI 6904	SL WE
	46.3 1.0 46.3	1.900 0.039 1.821	0.118E+01 0.796E+00				A+-0.5					FF
	45.9 1.2 45.9	0.047 1.806	0.141E+01 0.951E+00				C+-10%			FF	FF	FF
	45.1 1.6 45.1	0.063 1.774	0.187E+01 0.126E+01				C+-10%			FF	FF	FF
	44.3 2.0 44.3	0.079 1.743	0.232E+01 0.156E+01				C+-10%			FF	FF	FF
	43.7 2.3 43.7	0.091 1.719	0.265E+01 0.178E+01			FF	C+-10%			FF	FF	FF
	43.1 2.6 43.1	0.102 1.695	0.297E+01 0.200E+01			FF	C+-10%			FF	FF	FF
	42.5 2.9 42.5	0.114 1.672	0.330E+01 0.221E+01	FF		FF	C+-10%			FF	FF	FF
	41.9 3.2 41.9	0.126 1.648	0.361E+01 0.243E+01			FF	C+-10%			FF	FF	FF
	41.1 3.6 41.1	0.142 1.617	0.403E+01 0.271E+01	FF		FF	C+-10%			FF	FF	FF
	40.3 4.0 40.3	0.157 1.585	0.444E+01 0.298E+01			FF	C+-10%			FF	FF	FF
	39.3 4.5 39.3	0.177 1.546	0.493E+01 0.332E+01			FF	C+-10%			FF	FF	FF
	38.3 5.0 38.3	0.197 1.506	0.542E+01 0.364E+01	FF		FF	C+-10%			FF	FF	FF
	37.1 5.6 37.1	0.220 1.459	0.599E+01 0.402E+01			FF	C+-10%			FF	FF	FF
	35.7 6.3 35.7	0.248 1.404	0.662E+01 0.445E+01			6.5FF	C+-10%			FF	FF	FF
	34.1 7.1 34.1	0.280 1.341	0.732E+01 0.492E+01			FF	C+-10%			FF	FF	FF
	32.3 8.0 32.3	0.315 1.270	0.807E+01 0.542E+01				C+-10%			FF	FF	FF
	30.7 8.8 30.7	0.346 1.207	0.870E+01 0.585E+01				C+-10%			FF	FF	FF
	28.3 10.0 28.3	0.394 1.113	0.959E+01 0.644E+01				C+-10%			FF	FF	FF
	26.3 11.0 26.3	0.433 1.034	0.103E+02 0.690E+01				C+-10%			FF	FF	FF
51.0	49.0 1.0 49.0	2.000 0.039 1.921	0.125E+01 0.841E+00				A+-1%			FF	FF	SS
	48.8 1.2 48.8	0.047 1.906	0.150E+01 0.101E+01				C+-10%			FF	FF	SS
	47.8 1.6 47.8	0.063 1.874	0.198E+01 0.133E+01				C+-10%			FF	FF	SS
	47.0 2.0 47.0	0.079 1.843	0.245E+01 0.165E+01			FF	C+-10%			FF	FF	SS
	46.4 2.3 46.4	0.091 1.819	0.280E+01 0.188E+01			FF	C+-10%			FF	FF	SS
	45.8 2.6 45.8	0.102 1.795	0.315E+01 0.212E+01			FF	C+-10%			FF	FF	SS
	45.2 2.9 45.2	0.114 1.772	0.349E+01 0.235E+01			FF	C+-10%			FF	FF	SS
	44.6 3.2 44.6	0.126 1.748	0.383E+01 0.257E+01			FF	C+-10%			FF	FF	SS
	43.8 3.6 43.8	0.142 1.717	0.427E+01 0.287E+01			FF	C+-10%			FF	FF	SS
	43.0 4.0 43.0	0.157 1.685	0.471E+01 0.316E+01			FF	C+-10%			FF	FF	SS
	42.0 4.5 42.0	0.177 1.646	0.524E+01 0.352E+01			FF	C+-10%			FF	FF	SS
	41.0 5.0 41.0	0.197 1.606	0.576E+01 0.387E+01			FF	C+-10%			FF	FF	SS
	39.8 5.6 39.8	0.220 1.559	0.636E+01 0.428E+01			FF	C+-10%			FF	FF	SS
	38.4 6.3 38.4	0.248 1.504	0.705E+01 0.474E+01			6.5FF	C+-10%			FF	FF	SS
	36.8 7.1 36.8	0.280 1.441	0.780E+01 0.524E+01			FF	C+-10%			FF	FF	SS
	35.0 8.0 35.0	0.315 1.370	0.861E+01 0.579E+01			FF	C+-10%			FF	FF	SS
	33.4 8.8 33.4	0.346 1.307	0.930E+01 0.625E+01			9.5FF	C+-10%			FF	FF	SS
	31.0 10.0 31.0	0.394 1.213	0.103E+02 0.690E+01			12.5FF	C+-10%			FF	FF	SS
	29.0 11.0 29.0	0.433 1.134	0.110E+02 0.740E+01			FF	C+-10%			FF	FF	SS
54.0	52.0 1.0 52.0	2.125 0.039 2.046	0.133E+01 0.891E+00				A+-1%			SS	SS	FF
	50.0 2.0 50.0	0.079 1.968	0.260E+01 0.175E+01			FF	C+-10%			SS	SS	FF
	46.0 4.0 46.0	0.157 1.810	0.501E+01 0.336E+01			FF	C+-10%			SS	SS	FF
	32.0 11.0 32.0	0.433 1.259	0.118E+02 0.796E+01			FF	C+-10%			FF	FF	FF
	29.0 12.5 29.0	0.492 1.141	0.130E+02 0.873E+01			FF	C+-10%			FF	FF	FF
57.0	55.0 1.0 55.0	2.250 0.039 2.171	0.140E+01 0.942E+00				A+-1%			FF	FF	FF
	53.0 2.0 53.0	0.079 2.093	0.275E+01 0.185E+01			FF	C+-10%			FF	FF	FF
	45.8 5.6 45.8	0.220 1.809	0.721E+01 0.484E+01			FF	C+-10%			FF	FF	FF
	32.0 12.5 32.0	0.492 1.266	0.139E+02 0.936E+01			FF	C+-10%			FF	FF	FF

Table 10-25 (Continued). Seamless (SL) and Welded (WE) Stainless Steel Tubes (ISO 1127, 5252)

TABLE 10-25. SEAMLESS(SL) AND WELDED(WE) STAINLESS
STEEL TUBES (ISO 1127,5252)   PAGE NO. 6
BASIS; 1 IN = 25.4 MM, 1 CUBIC METER   ORDER EXAMPLE;
AUSTENITIC STEEL = 7967.75 KG MASS   LENGTH;STAINLESS TUBE 25X2 DIN2462 + STEEL QUA

```
                                                    THE NOMINAL SIZE IS NATIONAL STANDARD AS INDICATED
                                                    F=FIRST CHOICE,S=SECOND CHOICE,T=THIRD CHOICE,NUMBER=OTHER SIZE
                                                    * = COMMERCIAL SIZE
I                                                   U.S.A.  AUSTRAL JAPAN           FRANCE U.K.   GERMANY ITALY
S                                                   ANSI    AS      JIS     ISO     NF     BS    DIN     UNI
O                                                   B36.19          G3459   1127           2462  6904
      MILLIMETERS   INCHES    MASS PER UNIT         B36.19          G3463   D2,13          2463
      A    C    B   A    C    B    KG/M    LB/FT SL
                                                 WE
```

A	C	B	A	C	B	KG/M	LB/FT	SL/WE	ANSI B36.19	AS	JIS G3459 G3463	ISO 1127 D2,13	NF	BS	DIN 2462 2463	UNI 6904
60.3	1.0	58.3	2.375	0.039	2.296	0.148E+01	0.997E+00	SL				A+-1%			FF	
	1.6	57.1		0.063	2.249	0.235E+01	0.158E+01	WE	FF		FF	C+-10%			FF	F
	2.0	56.3		0.079	2.218	0.292E+01	0.196E+01		FF		FF	C+-10%			FF	F
	2.9	54.5		0.114	2.147	0.417E+01	0.280E+01		FF		FF	C+-10%			FF	F
	4.9	50.5		0.193	1.989	0.680E+01	0.457E+01		FF		FF	C+-10%			FF	F
	5.6	49.1		0.220	1.934	0.767E+01	0.515E+01		FF		FF	C+-10%			FF	F
	6.3	47.7		0.248	1.879	0.852E+01	0.572E+01				6.5FF	C+-10%			FF	F
	12.5	35.3		0.492	1.391	0.150E+02	0.101E+02				FF	C+-10%			FF	F
63.5	1.0	61.5	2.500	0.039	2.421	0.156E+01	0.105E+01					A+-1%			FF	2S
	2.3	58.9		0.091	2.319	0.352E+01	0.237E+01				FF	C+-10%			FF	S
	6.3	50.9		0.248	2.004	0.902E+01	0.606E+01				6.5FF	C+-10%			FF	S
	12.5	38.5		0.492	1.516	0.160E+02	0.107E+02				FF	C+-10%			F	S
70.0	1.0	68.0	2.750	0.039	2.671	0.173E+01	0.116E+01					A+-1%			FF	2F
	1.6	66.8		0.063	2.624	0.274E+01	0.184E+01				FF	C+-10%			FF	F
	2.3	65.4		0.091	2.569	0.390E+01	0.262E+01				FF	C+-10%			FF	F
	6.3	57.4		0.248	2.254	0.100E+02	0.675E+01				6.5FF	C+-10%			FF	F
	12.5	45.0		0.492	1.766	0.180E+02	0.121E+02				FF	C+-10%			FF	F
76.1	1.0	74.1	3.000	0.039	2.921	0.188E+01	0.126E+01					A+-1%			FF	
	1.6	72.9		0.063	2.874	0.298E+01	0.200E+01				FF	C+-10%			FF	F
	2.3	71.5		0.091	2.819	0.425E+01	0.286E+01				FF	C+-10%			FF	F
	6.3	63.5		0.248	2.504	0.110E+02	0.740E+01				6.5FF	C+-10%			FF	F
	12.5	51.1		0.492	2.016	0.199E+02	0.134E+02				FF	C+-10%			FF	F
88.9	1.0	86.9	3.500	0.039	3.421	0.220E+01	0.148E+01					A+-1%			FF	
	1.6	85.7		0.063	3.374	0.350E+01	0.235E+01				FF	C+-10%			FF	F
	2.0	84.9		0.079	3.343	0.435E+01	0.292E+01				FF	C+-10%			FF	F
	2.9	83.1		0.114	3.272	0.624E+01	0.420E+01				FF	C+-10%			FF	F
	3.2	82.5		0.126	3.248	0.686E+01	0.461E+01		FF		FF	C+-10%			FF	F
	5.6	77.7		0.220	3.059	0.117E+02	0.785E+01		FF		FF	C+-10%			FF	2,3F
	8.0	72.9		0.315	2.870	0.162E+02	0.109E+02				7.6FF	C+-10%			FF	F
	12.5	63.9		0.492	2.516	0.239E+02	0.161E+02				FF	C+-10%			F	F
101.6	1.0	99.6	4.000	0.039	3.921	0.252E+01	0.169E+01					A+-1%			FF	
	2.0	97.6		0.079	3.843	0.499E+01	0.335E+01				FF	C+-10%			FF	F
	2.9	95.8		0.114	3.772	0.716E+01	0.481E+01				FF	C+-10%			FF	F
	5.6	90.4		0.220	3.559	0.135E+02	0.904E+01				FF	C+-10%			FF	F
	8.0	85.6		0.315	3.370	0.187E+02	0.126E+02				FF	C+-10%			FF	F
	10.0	81.6		0.394	3.213	0.229E+02	0.154E+02				9.5FF	C+-10%			FF	F
	12.5	76.6		0.492	3.016	0.279E+02	0.187E+02				FF	C+-10%			FF	F
	20.0	61.6		0.787	2.425	0.409E+02	0.275E+02				FF	C+-10%			F	F
114.3	1.2	111.9	4.500	0.047	4.406	0.340E+01	0.228E+01					A+-1%			FF	
	2.0	110.3		0.079	4.343	0.562E+01	0.378E+01				FF	C+-10%			FF	F
	2.9	108.5		0.114	4.272	0.809E+01	0.543E+01				6FF	C+-10%			FF	F
	6.3	101.7		0.248	4.004	0.170E+02	0.114E+02				6FF	C+-10%			FF	F
	8.8	96.7		0.346	3.807	0.232E+02	0.156E+02				FF	C+-10%			FF	F
	12.5	89.3		0.492	3.516	0.319E+02	0.214E+02				FF	C+-10%			FF	F
	20.0	74.3		0.787	2.925	0.472E+02	0.317E+02				FF	C+-10%			FF	F

STEEL MATERIAL DATA

STEEL MATERIAL DATA

Table 10-26. Hot Rolled Equal Leg Angles with Round Corners (ISO 657/I)

TABLE 10-26. HOT ROLLED EQUAL LEG ANGLES
WITH ROUND CORNERS (ISO 657/I)
BASIS: 1 IN = 25.4 MM

ORDER EXAMPLE:
LENGTH,L25X3 EURONORM 56 + STEEL QUALITY

THE NOMINAL SIZE IS NATIONAL STANDARD AS INDICATED
F=FIRST CHOICE,S=SECOND CHOICE,T=THIRD CHOICE,NUMBER=OTHER SIZE
* = COMMERCIAL SIZE

ISO	MILLIMETERS A x A x C	R1	INCHES A x A x C	R1	U.S.A. ANSI	AUSTRAL AS	JAPAN JIS G3192	EURO-NORM 56	FRANCE NF A45-009	U.K. BS 4848/4	GERMANY DIN 1028	ITALY UNI 5783
F	L 20 x 20 x 3.0	3.5	0.787 x 0.787 x 0.118	0.14				F A+=1	F,2T	S	F	F
F	4.0	3.5	0.157	0.14				F C+=0.5	2.5T	S	F	F
F	L 25 x 25 x 3.0	3.5	0.984 x 0.984 x 0.118	0.14				F A+=1	F	F	F	F
F	4.0	3.5	0.157	0.14				F C+=0.5	T	F	F	F
F	5.0	3.5	0.197	0.14				F C+=0.5	2.5S	F	F	F
F	L 30 x 30 x 3.0	5.0	1.181 x 1.181 x 0.118	0.20				F A+=1	F	F	F	F
F	4.0	5.0	0.157	0.20				F C+=0.5	S	F	F	F
F	5.0	5.0	0.197	0.20				F C+=0.5	T	F	F	F
F	L 35 x 35 x 3.0	5.0	1.378 x 1.378 x 0.118	0.20				F A+=1	S	S	F	F
F	4.0	5.0	0.157	0.20				F C+=0.5	S	S	F	F
F	5.0	5.0	0.197	0.20				F C+=0.5	3.5F	S	F 6F	F
F	L 40 x 40 x 3.0	6.0	1.575 x 1.575 x 0.118	0.24				S A+=1	S	S	F	F
F	4.0	6.0	0.157	0.24				F C+=0.5	S	F	F	F
F	5.0	6.0	0.197	0.24				F C+=0.5	T	F	F	F
F	6.0	6.0	0.236	0.24				F C+=0.5		F	F	F
F	L 45 x 45 x 3.0	7.0	1.772 x 1.772 x 0.118	0.28			F	F C+=0.5	S	S	F	F
F	4.0	7.0	0.157	0.28				F C+=0.5	S	F	F	F
F	4.5	7.0	0.177	0.28				F C+=0.5	T	F	F	F
F	5.0	7.0	0.197	0.28				F C+=0.5		F	F	F
F	6.0	7.0	0.236	0.28				F C+=0.5		F	F,7F	F
F	L 50 x 50 x 3.0	7.0	1.969 x 1.969 x 0.118	0.28			F	S A+=1	S	S	F	F
F	4.0	7.0	0.157	0.28				F C+=0.5	S	S	F	F
F	5.0	7.0	0.197	0.28				F C+=0.5	F	F	F	F
F	6.0	7.0	0.236	0.28				F C+=0.5	T	F	F,9F	F
F	L 60 x 60 x 4.0	8.0	2.362 x 2.362 x 0.157	0.31			F	S A+=1.5	S	S	F	F
F	5.0	8.0	0.197	0.31			F	F C+=0.75	S	S	F	F
F	6.0	8.0	0.236	0.31				F C+=0.75	F	F	F	F
F	8.0	8.0	0.315	0.31				F C+=0.75	T	F	F	F
F	10.0	8.0	0.394	0.31				F C+=0.75		F	F	F
F	L 65 x 65 x 6.0	9.0	2.559 x 2.559 x 0.236	0.35			F	S A+=1.5	S	S	F,7F,9F	
F	8.0	9.0	0.315	0.35			F	F C+=0.75		F	F,11F	
F	L 70 x 70 x 5.0	9.0	2.756 x 2.756 x 0.197	0.35				A+=1.5	S		F	F
F	6.0	9.0	0.236	0.35				F C+=0.75	S	S	F	F
F	7.0	9.0	0.276	0.35				F C+=0.75	F	S	F	F
F	8.0	9.0	0.315	0.35				F C+=0.75	S	F	F	F
F	9.0	9.0	0.354	0.35				F C+=0.75	T	F	11F	F
F	10.0	9.0	0.394	0.35				F C+=0.75		F		
F	L 75 x 75 x 6.0	10.0	2.953 x 2.953 x 0.236	0.39			F	A+=1.5	S		F,7F	
F	8.0	10.0	0.315	0.39			9F	C+=0.75			F,7F	
F	12.0	10.0	0.472	0.39			F	C+=0.75			F,10F	

PAGE NO. 1

10-87

Table 10-26 (Continued). Hot Rolled Equal Leg Angles with Round Corners (ISO 657/I)

TABLE 10-26. HOT ROLLED EQUAL LEG ANGLES
WITH ROUND CORNERS (ISO 657/I)
BASIS: 1 IN = 25.4 MM

ORDER EXAMPLE;
LENGTH,L25X3 EURONORM 56 + STEEL QUALITY   PAGE NO. 2

THE NOMINAL SIZE IS NATIONAL STANDARD AS INDICATED
F=FIRST CHOICE,S=SECOND CHOICE,T=THIRD CHOICE,NUMBER=OTHER SIZE
*=COMMERCIAL SIZE

ISO	MILLIMETERS A x A x C	R1	INCHES A x A x C	R1	U.S.A. ANSI	AUSTRAL AS	JAPAN JIS G3192	EURO- NORM 56	FRANCE NF A45-009	U.K. BS 4848/4	GERMANY DIN 1028	ITALY UNI 5783
F	L 80 x 80 x 6.0	10.0	3.150 x 3.150 x 0.236	0.39				S A*=1.5	S		7F	F
F	8.0	10.0	0.315	0.39			F	F C*=0.75	F,7S	F	F	F
F	10.0	10.0	0.394	0.39				F C*=0.75	T F	F	F	F
F	12.0	10.0	0.472	0.39				F C*=0.75	S,5S	F	F	F
F	14.0	10.0	0.551	0.39				C*=0.75			F	F
F	L 90 x 90 x 6.0	11.0	3.543 x 3.543 x 0.236	0.43				S A*=1.5	S			F
F	8.0	11.0	0.315	0.43			7F	F C*=0.75	S,7S	F	F	F
F	9.0	11.0	0.354	0.43				S C*=0.75	F		F	S
F	10.0	11.0	0.394	0.43			F	F C*=0.75	F		F	F
F	12.0	11.0	0.472	0.43				C*=0.75			11F	F
F	13.0	11.0	0.512	0.43			F	C*=0.75			F,16F	F
F	L100 x 100 x 6.5	12.0	3.937 x 3.937 x 0.256	0.47				S A*=1.5	S			F
F	8.0	12.0	0.315	0.47			7F	F C*=0.75	S,7S	F	F	F
F	10.0	12.0	0.394	0.47			F	F C*=0.75	F,9S	F	F	F
F	12.0	12.0	0.472	0.47				F C*=0.75	S		F	F
F	14.0	12.0	0.551	0.47			13F	F C*=0.75			F	F
F	15.0	12.0	0.591	0.47				C*=0.75				
F	16.0	12.0	0.630	0.47				C*=0.75				
	20.0	12.0	0.787	0.47				C*=0.75				
F	L120 x 120 x 8.0	13.0	4.724 x 4.724 x 0.315	0.51				S A*=2	S			F
F	10.0	13.0	0.394	0.51				F C*=1	F		F	F
F	11.0	13.0	0.433	0.51				C*=1	S			
F	12.0	13.0	0.472	0.51				C*=1	S		F,13F	F
	15.0	13.0	0.591	0.51				F C*=1	S,14S	F	F	F
F	L130 x 130 x 12.0	14.0	5.118 x 5.118 x 0.472	0.55			F	A*=2				
F	14.0	14.0	0.551	0.55			9F	C*=1				
F	16.0	14.0	0.630	0.55			15F	C*=1				
F	L150 x 150 x 10.0	16.0	5.906 x 5.906 x 0.394	0.63			F	S A*=2	S		F	F
F	12.0	16.0	0.472	0.63			F	F C*=1	F		F,16F	F
F	14.0	16.0	0.551	0.63				F C*=1	F		F	F
F	15.0	16.0	0.591	0.63				F C*=1	F		F,20F	F
F	18.0	16.0	0.709	0.63			19F	C*=1	F		F	F
F	L180 x 180 x 15.0	18.0	7.087 x 7.087 x 0.591	0.71			175x12F	F A*=3			16F	F
F	18.0	18.0	0.709	0.71			175x15F	F C*=1.25	S		F	F
F	20.0	18.0	0.787	0.71				C*=1.25	S		F	F
F	L200 x 200 x 16.0	18.0	7.874 x 7.874 x 0.630	0.71			15F	F A*=3			F,22F	F
F	18.0	18.0	0.709	0.71				F C*=1.25	S		F	F
F	20.0	18.0	0.787	0.71				F C*=1.25	S		F	F
F	24.0	18.0	0.945	0.71			25F	F C*=1.25	T		F	F
F	L250 x 250 x 25.0	24.0	9.843 x 9.843 x 0.984	0.94			F	A*=3			F,28F	F
F	35.0	24.0	1.378	0.94			F	C*=1.25			F	F

STEEL MATERIAL DATA

Table 10-27. Hot Rolled Equal Leg Angles with Sharp Corners (DIN 1022)

TABLE 10-27. HOT ROLLED EQUAL LEG ANGLES
WITH SHARP CORNERS (DIN 1022)
BASIS; 1 IN = 25.4 MM

ORDER EXAMPLE:
LENGTH,LS20X3 DIN 1022+STEEL QUALITY

THE NOMINAL SIZE IS NATIONAL STANDARD AS INDICATED
F=FIRST CHOICE,S=SECOND CHOICE,T=THIRD CHOICE,NUMBER=OTHER SIZE
*.= COMMERCIAL SIZE

D I N	MILLIMETERS A × A × C	R1	INCHES A × A × C	R1	U.S.A. ANSI	AUSTRAL AS	JAPAN JIS	EURO- NORM	FRANCE NF	U.K. BS	GERMANY DIN 1022	ITALY UNI
F	LS 20 × 20 × 3.0	0.0	0.787 × 0.787 × 0.118	0.00					F		F A+=1.0	
F	LS 20 × 20 × 4.0	0.0	0.787 × 0.787 × 0.157	0.00							F C+=0.5	
F	LS 25 × 25 × 3.0	0.0	0.984 × 0.984 × 0.118	0.00					F		F A+=1.0	
F	LS 25 × 25 × 4.0	0.0	0.984 × 0.984 × 0.157	0.00							F C+=0.5	
F	LS 30 × 30 × 3.0	0.0	1.181 × 1.181 × 0.118	0.00					F		F A+=1.0	
F	LS 30 × 30 × 4.0	0.0	1.181 × 1.181 × 0.157	0.00							F C+=0.5	
F	LS 35 × 35 × 4.0	0.0	1.378 × 1.378 × 0.157	0.00							F A+=1.0	
F	LS 35 × 35 × 4.0	0.0	1.378 × 1.378 × 0.157	0.00							F C+=0.5	
F	LS 40 × 40 × 4.0	0.0	1.575 × 1.575 × 0.157	0.00							F A+=1.0	
F	LS 40 × 40 × 5.0	0.0	1.575 × 1.575 × 0.197	0.00							F C+=0.5	
F	LS 45 × 45 × 5.0	0.0	1.772 × 1.772 × 0.197	0.00							F A+=1.0	
F	LS 45 × 45 × 5.0	0.0	1.772 × 1.772 × 0.197	0.00							F C+=0.5	
F	LS 50 × 50 × 5.0	0.0	1.969 × 1.969 × 0.197	0.00							F A+=1.0	
F	LS 50 × 50 × 5.0	0.0	1.969 × 1.969 × 0.197	0.00							F C+=0.5	

Table 10-28. Bright Steel Equal Leg Angles with Sharp Corners (DIN 59370)

TABLE 10-28. BRIGHT STEEL EQUAL LEG ANGLES
WITH SHARP CORNERS (DIN 59370)
BASIS; 1 IN = 25.4 MM

ORDER EXAMPLE;
LENGTH,ANGLE S10X1 DIN 59370+STEEL QUALITY

THE NOMINAL SIZE IS NATIONAL STANDARD AS INDICATED
F=FIRST CHOICE,S=SECOND CHOICE,T=THIRD CHOICE,NUMBER=OTHER SIZE
*= COMMERCIAL SIZE

		U.S.A.	AUSTRAL	JAPAN	EURO-	FRANCE	U.K.		GERMANY	ITALY
		ANSI	AS	JIS	NORM	NF	BS		DIN 59370	UNI

DIN	MILLIMETERS A x A x C	R1	INCHES A x A x C	R1							
F S	10 x 10 x 1.0	0.5	0.394 x 0.394 x 0.039	0.02						F A*=0.10	
F	x 2.0	0.5	0.079	0.02						F C*=0.10	
F	x 3.0	0.5	0.118	0.02						F C*=0.10	
F S	12 x 12 x 2.0	0.5	0.472 x 0.472 x 0.079	0.02						F A*=0.10	
F	x 3.0	0.5	0.118	0.02						F C*=0.10	
F	x 4.0	0.5	0.157	0.02						F C*=0.10	
F S	15 x 15 x 2.0	0.5	0.591 x 0.591 x 0.079	0.02						F A*=0.10	
F	x 3.0	0.5	0.118	0.02						F C*=0.10	
F	x 4.0	0.5	0.157	0.02						F C*=0.10	
F S	18 x 18 x 2.0	0.8	0.709 x 0.709 x 0.079	0.03						F A*=0.15	
F	x 3.0	0.8	0.118	0.03						F C*=0.10	
F	x 4.0	0.8	0.157	0.03						F C*=0.15	
F S	20 x 20 x 2.0	0.8	0.787 x 0.787 x 0.079	0.03						F A*=0.15	
F	x 3.0	0.8	0.118	0.03						F C*=0.10	
F	x 4.0	0.8	0.157	0.03						F C*=0.15	
F S	25 x 25 x 3.0	0.8	0.984 x 0.984 x 0.118	0.03						F A*=0.15	
F	x 4.0	0.8	0.157	0.03						F C*=0.10	
F	x 5.0	0.8	0.197	0.03						F C*=0.15	
F S	30 x 30 x 3.0	0.8	1.181 x 1.181 x 0.118	0.03						F A*=0.15	
F	x 4.0	0.8	0.157	0.03						F C*=0.10	
F	x 5.0	0.8	0.197	0.03						F C*=0.15	
F S	35 x 35 x 3.0	1.0	1.378 x 1.378 x 0.118	0.04						F A*=0.20	
F	x 4.0	1.0	0.157	0.04						F C*=0.10	
F	x 5.0	1.0	0.197	0.04						F C*=0.15	
F	x 6.0	1.0	0.236	0.04						F C*=0.15	
F S	40 x 40 x 4.0	1.0	1.575 x 1.575 x 0.157	0.04						F A*=0.20	
F	x 5.0	1.0	0.197	0.04						F C*=0.15	
F S	45 x 45 x 4.0	1.0	1.772 x 1.772 x 0.157	0.04						F A*=0.20	
F	x 5.0	1.0	0.197	0.04						F C*=0.15	
F S	50 x 50 x 5.0	1.0	1.969 x 1.969 x 0.197	0.04						F A*=0.20	
F	x 6.0	1.0	0.236	0.04						F C*=0.15	
F S	60 x 60 x 5.0	1.0	2.362 x 2.362 x 0.197	0.04						F A*=0.20	
F	x 6.0	1.0	0.236	0.04						F C*=0.15	

STEEL MATERIAL DATA

Table 10-29. Hot Rolled Unequal Leg Angles with Round Corners (ISO 657/II)

TABLE 10-29. HOT ROLLED UNEQUAL LEG ANGLES
WITH ROUND CORNERS (ISO 657/II)
BASIS; 1 IN = 25.4 MM

ORDER EXAMPLE:
LENGTH,L30X20X3 EURONORM 57 + STEEL QUALITY
PAGE NO. 1

THE NOMINAL SIZE IS NATIONAL STANDARD AS INDICATED
F=FIRST CHOICE,S=SECOND CHOICE,T=THIRD CHOICE,NUMBER=OTHER SIZE
* = COMMERCIAL SIZE

ISO	MILLIMETERS A × B × C	R1	INCHES A × B × C	R1	U.S.A. ANSI	AUSTRAL AS	JAPAN JIS G3192	EURO-NORM 57	FRANCE NF A45-010	U.K. BS 4848/4	GERMANY DIN 1029	ITALY UNI 5784
F	L 30 × 20 × 3.0	4.0	1.181 × 0.787 × 0.118	0.16				F A+=1.0 F	F	F		F
F	L 30 × 20 × 4.0	4.0	1.181 × 0.787 × 0.157	0.16				F B+=1.0 F	F	F		F
F	L 30 × 20 × 5.0	4.0	1.181 × 0.787 × 0.197	0.16				F C+=0.5 F	F	F		F
F	L 40 × 20 × 3.0	4.0	1.575 × 0.787 × 0.118	0.16				F A+=1.0 T	F	F		F
F	L 40 × 20 × 4.0	4.0	1.575 × 0.787 × 0.157	0.16				F B+=1.0 T	F	F		F
F	L 40 × 20 × 5.0	4.0	1.575 × 0.787 × 0.197	0.16				F C+=0.5				F
F	L 40 × 25 × 4.0	4.0	1.575 × 0.984 × 0.157	0.16				F A+=1.0 F				F
F	L 40 × 25 × 4.0	4.0	1.575 × 0.984 × 0.157	0.16				F B+=1.0 F				F
F	L 40 × 25 × 5.0	4.0	1.575 × 0.984 × 0.197	0.16				F C+=0.5	F			F
F	L 45 × 30 × 4.0	4.0	1.772 × 1.181 × 0.157	0.16				F A+=1.0 F	F	F	F,3F	F
F	L 45 × 30 × 4.0	4.0	1.772 × 1.181 × 0.157	0.16				F B+=1.0 F	F	F	F,3F	F
F	L 45 × 30 × 5.0	4.0	1.772 × 1.181 × 0.197	0.16				F C+=0.5 S		F		F
S	L 50 × 30 × 4.0	5.0	1.969 × 1.181 × 0.157	0.20				S A+=1.0 T	F	F	F	S
S	L 50 × 30 × 5.0	5.0	1.969 × 1.181 × 0.197	0.20				S B+=1.0 F	F	F		S
S	L 50 × 30 × 6.0	5.0	1.969 × 1.181 × 0.236	0.20				S C+=0.5				S
F	L 60 × 30 × 5.0	6.0	2.362 × 1.181 × 0.197	0.24				F A+=1.0 F	F	F		F
F	L 60 × 30 × 5.0	6.0	2.362 × 1.181 × 0.197	0.24				F B+=1.0 F	F	F		F
F	L 60 × 30 × 6.0	6.0	2.362 × 1.181 × 0.236	0.24				F C+=.75	F		7F	F
F	L 60 × 40 × 5.0	6.0	2.362 × 1.575 × 0.197	0.24				F A+=1.5 F	F	F	F	F
F	L 60 × 40 × 6.0	6.0	2.362 × 1.575 × 0.236	0.24				F B+=1.5 F	F	F	F	F
F	L 60 × 40 × 7.0	6.0	2.362 × 1.575 × 0.276	0.24				F C+=.75 T	F	F		F
F	L 65 × 50 × 5.0	6.0	2.559 × 1.969 × 0.197	0.24				F A+=1.5 T				F
F	L 65 × 50 × 6.0	6.0	2.559 × 1.969 × 0.236	0.24				F B+=1.5 F				F
F	L 65 × 50 × 7.0	6.0	2.559 × 1.969 × 0.276	0.24				F C+=.75	F		9F	F
F	L 65 × 50 × 8.0	6.0	2.559 × 1.969 × 0.315	0.24				F				F
S	L 70 × 50 × 5.0	7.0	2.756 × 1.969 × 0.197	0.28				S A+=1.5 T	F	F		S
S	L 70 × 50 × 6.0	7.0	2.756 × 1.969 × 0.236	0.28				S B+=1.5 F	F	F		S
S	L 70 × 50 × 7.0	7.0	2.756 × 1.969 × 0.276	0.28				S C+=.75 T	F	F		S
S	L 70 × 50 × 8.0	7.0	2.756 × 1.969 × 0.315	0.28								S
F	L 75 × 50 × 5.0	7.0	2.953 × 1.969 × 0.197	0.28				F A+=1.5 F	F	F		F
F	L 75 × 50 × 6.0	7.0	2.953 × 1.969 × 0.236	0.28				F B+=1.5 F	F	F		F
F	L 75 × 50 × 7.0	7.0	2.953 × 1.969 × 0.276	0.28				F C+=.75	F	F		F
F	L 75 × 50 × 8.0	7.0	2.953 × 1.969 × 0.315	0.28							9F	F
F	L 80 × 40 × 5.0	7.0	3.150 × 1.575 × 0.197	0.28				F A+=1.5 F		F		F
F	L 80 × 40 × 6.0	7.0	3.150 × 1.575 × 0.236	0.28				F B+=1.5 T		F		F
F	L 80 × 40 × 7.0	7.0	3.150 × 1.575 × 0.276	0.28				F C+=.75	F	F		F
F	L 80 × 40 × 8.0	7.0	3.150 × 1.575 × 0.315	0.28								F
F	L 80 × 60 × 6.0	8.0	3.150 × 2.362 × 0.236	0.31				F A+=1.5 F		F		F
F	L 80 × 60 × 7.0	8.0	3.150 × 2.362 × 0.276	0.31				F B+=1.5 F		F		F
F	L 80 × 60 × 8.0	8.0	3.150 × 2.362 × 0.315	0.31				F C+=.75 S				F

Table 10-29 (Continued). Hot Rolled Unequal Leg Angles with Round Corners (ISO 657/II)

TABLE 10-29. HOT ROLLED UNEQUAL LEG ANGLES
WITH ROUND CORNERS (ISO 657/II)
BASIS; 1 IN = 25.4 MM

PAGE NO. 2

ORDER EXAMPLE;
LENGTH,L30X20X3 EURONORM 57 + STEEL QUALITY

THE NOMINAL SIZE IS NATIONAL STANDARD AS INDICATED
F=FIRST CHOICE,S=SECOND CHOICE,T=THIRD CHOICE,NUMBER=OTHER SIZE
*=COMMERCIAL SIZE

ISO	MILLIMETERS A x A x C	R1	INCHES A x A x C	R1	U.S.A. ANSI	AUSTRAL AS	JAPAN JIS G3192	EURO-NORM 57	FRANCE NF A45-010	U.K. BS 4848/4	GERMANY DIN 1029	ITALY UNI 5784
S	L 90 × 65 × 6.0	8.0	3.543 × 2.559 × 0.236	0.31				S A+=1.5				S
S	7.0	8.0	0.276	0.31				S B+=1.5				S
S	8.0	8.0	0.315	0.31				S C+=.75				S
S	10.0	8.0	0.394	0.31				S				S
F	L100 × 50 × 6.0	9.0	3.937 × 1.969 × 0.236	0.35				F A+=1.5 T				F
F	7.0	9.0	0.276	0.35				F B+=1.5 T		F		F
F	8.0	9.0	0.315	0.35				F C+=.75 T				F
F	10.0	9.0	0.394	0.35				F				F
F	L100 × 65 × 7.0	10.0	3.937 × 2.559 × 0.276	0.39				F A+=1.5			9F	F
F	8.0	10.0	0.315	0.39				F B+=1.5		F	11F	F
F	10.0	10.0	0.394	0.39				F C+=.75		F		F
F	L100 × 75 × 8.0	10.0	3.937 × 2.953 × 0.315	0.39				F A+=1.5			7F	F
F	10.0	10.0	0.394	0.39				F B+=1.5			9F	F
F	12.0	10.0	0.472	0.39				F C+=.75			11F	F
F	L120 × 80 × 8.0	11.0	4.724 × 3.150 × 0.315	0.43				F A+=2.0 T				F
F	10.0	11.0	0.394	0.43				F B+=2.0 F				F
F	12.0	11.0	0.472	0.43				F C+=1.0 S			F,14F	F
S	L125 × 75 × 8.0	11.0	4.921 × 2.953 × 0.315	0.43		7F		S A+=2.0				S
S	10.0	11.0	0.394	0.43		F		S B+=2.0				S
S	12.0	11.0	0.472	0.43		13F		S C+=1.0				S
F	L135 × 65 × 8.0	11.0	5.315 × 2.559 × 0.315	0.43				F A+=2.0				F
F	10.0	11.0	0.394	0.43				F B+=2.0				F
F	12.0	11.0	0.472	0.43				F C+=1.0				F
S	L150 × 75 × 9.0	11.0	5.906 × 2.953 × 0.354	0.43				F A+=2.0				
S	10.0	11.0	0.394	0.43				F B+=2.0		F		
S	12.0	11.0	0.472	0.43				F C+=1.0				
S	15.0	11.0	0.591	0.43				F				
F	L150 × 90 × 10.0	12.0	5.906 × 3.543 × 0.394	0.47		9F		F A+=2.0 S				F
F	12.0	12.0	0.472	0.47		F		F B+=2.0 T,11S				F
F	15.0	12.0	0.591	0.47				F C+=1.0 13T				F
F	L200 × 100 × 10.0	15.0	7.874 × 3.937 × 0.394	0.59				F A+=3.0				F
F	12.0	15.0	0.472	0.59				F B+=3.0				F
F	15.0	15.0	0.591	0.59				F C+=1.25				F
F	L200 × 150 × 12.0	15.0	7.874 × 5.906 × 0.472	0.59				F A+=3.0				F
F	15.0	15.0	0.591	0.59				F B+=3.0		F	11F	F
F	18.0	15.0	0.709	0.59				F C+=1.25				F
F	L250 × 90 × 10.0	15.0	9.843 × 3.543 × 0.394	0.59				A+=4.0				F
F	12.0	15.0	0.472	0.59				B+=4.0		F		F
F	14.0	15.0	0.551	0.59				C+=1.25				F
F	16.0	15.0	0.630	0.59								F

STEEL MATERIAL DATA

Table 10-30. Hot Rolled Large Unequal Leg Angles with Round Corners (JIS G3192)

TABLE 10-30. HOT ROLLED LARGE UNEQUAL LEG
ANGLES WITH ROUND CORNERS (JIS G3192)
BASIS: 1 IN = 25.4 MM

ORDER EXAMPLE:
LENGTH,L200X90 JIS G3192 + STEEL QUALITY

THE NOMINAL SIZE IS NATIONAL STANDARD AS INDICATED
F=FIRST CHOICE,S=SECOND CHOICE,T=THIRD CHOICE,NUMBER=OTHER SIZE
* = COMMERCIAL SIZE

NOMINAL SIZE MILLIMETERS						INCHES						U.S.A. ANSI	AUSTRAL AS	JAPAN JIS G3192-71	EURO- NORM	FRANCE NF	U.K. BS	GERMANY DIN	ITALY UNI
A	B	C	D	R1	R2	A	B	C	D	R1	R2								
L200 x	90	9.0	14.0	7.0	7.874x	3.543	0.35	0.55	0.28					F A+=4 B+=2 C+=0.8 D+=1					
L250 x*	90	10.0	15.0	17.0	8.5	9.843x	3.543	0.39	0.59	0.67	0.33			F A+=4 B+=2 C+=1 D+=1					
L250 x	90	12.0	16.0	17.0	8.5	9.843x	3.543	0.47	0.63	0.67	0.33			F A+=4 B+=2 C+=1 D+=1.2					
L300 x	90	11.0	16.0	19.0	9.5	11.811x	3.543	0.43	0.63	0.75	0.37			F A+=4 B+=2 C+=1 D+=1.2					
L300 x	90	13.0	17.0	19.0	9.5	11.811x	3.543	0.51	0.67	0.75	0.37			F A+=4 B+=2 C+=1 D+=1.2					
L400 x	100	13.0	18.0	24.0	12.0	15.748x	3.937	0.51	0.71	0.94	0.47			F A+=4 B+=3 C+=1 D+=1.2					

Table 10-31. Hot Rolled T-Steel with Round Corners (EURONORM 55)

TABLE 10-31. HOT ROLLED T-STEEL WITH ROUND
CORNERS (EURONORM 55)
BASIS: 1 IN = 25.4 MM

ORDER EXAMPLE:
LENGTH,T20X20 EURONORM 55 + STEEL QUALITY

THE NOMINAL SIZE IS NATIONAL STANDARD AS INDICATED
F=FIRST CHOICE,S=SECOND CHOICE,T=THIRD CHOICE,NUMBER=OTHER SIZE
* = COMMERCIAL SIZE

NOMINAL SIZE						INCHES				EURO- NORM 55	U.S.A. ANSI	AUSTRAL AS	JAPAN JIS	FRANCE NF A45-008	U.K. BS	GERMANY DIN 1024	ITALY UNI 5785
MILLIMETERS																	
A	B	C	R1	R2		A	B	C	R1 R2								
T20X20	20 × 20	3.0 3.0	1.5 1.5	1.0 1.0		0.787 ×	0.787	0.12 0.12	0.06 0.04 0.06 0.04	F A,B+-1 F C+-0.5				T T		F F F F	F F
T25X25	25 × 25	3.5 3.5	2.0 2.0	1.0 1.0		0.984 ×	0.984	0.14 0.14	0.08 0.04 0.08 0.04	F A,B+-1 F C+-0.5				S S		F F F F	F F
T30X30	30 × 30	4.0 4.0	2.0 2.0	1.0 1.0		1.181 ×	1.181	0.16 0.16	0.08 0.04 0.08 0.04	F A,B+-1 F C+-0.5				F F		F F F F	F F
T35X35	35 × 35	4.5 4.5	2.5 2.5	1.0 1.0		1.378 ×	1.378	0.18 0.18	0.10 0.04 0.10 0.04	F A,B+-1 F C+-0.5				F F		F F F F	F F
T40X40	40 × 40	5.0 5.0	2.5 2.5	1.0 1.0		1.575 ×	1.575	0.20 0.20	0.10 0.04 0.10 0.04	F A,B+-1 F C+-0.5				F F		F F F F	F F
T45X45	45 × 45	5.5 5.5	3.0 3.0	1.5 1.5		1.772 ×	1.772	0.22 0.22	0.12 0.06 0.12 0.06	F A,B+-1 F C+-0.5				T T		F F F F	F F
T50X50	50 × 50	6.0 6.0	3.0 3.0	1.5 1.5		1.969 ×	1.969	0.24 0.24	0.12 0.06 0.12 0.06	F A,B+-1 F C+-0.5				F F		F F F F	F F
T60X60	60 × 60	7.0 7.0	3.5 3.5	2.0 2.0		2.362 ×	2.362	0.28 0.28	0.14 0.08 0.14 0.08	F A,B+-1.5 F C+-0.75				F F		F F F F	F F
T70X70	70 × 70	8.0 8.0	4.0 4.0	2.0 2.0		2.756 ×	2.756	0.31 0.31	0.16 0.08 0.16 0.08	F A,B+-1.5 F C+-0.75				F F		F F F F	F F
T80X80	80 × 80	9.0 9.0	4.5 4.5	2.0 2.0		3.150 ×	3.150	0.35 0.35	0.18 0.08 0.18 0.08	F A,B+-1.5 F C+-0.75				T T		F F F F	F F
T100X	100 × 100	11.0 11.0	5.5 5.5	3.0 3.0		3.937 ×	3.937	0.43 0.43	0.22 0.12 0.22 0.12	F A,B+-1.5 F C+-0.75				T T		F F F F	F F
T120X	120 × 120	13.0 13.0	6.5 6.5	3.0 3.0		4.724 ×	4.724	0.51 0.51	0.26 0.12 0.26 0.12	F A,B+-2 F C+-1						F F F F	F F
T140X	140 × 140	15.0 15.0	7.5 7.5	4.0 4.0		5.512 ×	5.512	0.59 0.59	0.30 0.16 0.30 0.16	A,B+-2 C+-1						F F	F F

STEEL MATERIAL DATA

Table 10-32. Hot Rolled T-Steel with Round Corners and Wide Base (DIN 1024)

TABLE 10-32. HOT ROLLED T-STEEL WITH ROUND
CORNERS AND WIDE BASE (DIN 1024)
BASIS; 1 IN = 25.4 MM

ORDER EXAMPLE;
LENGTH,TB 30 DIN 1024 + STEEL QUALITY

THE NOMINAL SIZE IS NATIONAL STANDARD AS INDICATED
F=FIRST CHOICE,S=SECOND CHOICE,T=THIRD CHOICE,NUMBER=OTHER SIZE
* = COMMERCIAL SIZE

NOMINAL SIZE MILLIMETERS						INCHES					U.S.A. ANSI	AUSTRAL AS	JAPAN JIS	EURO- NORM	FRANCE NF	U.K. BS	GERMANY DIN DIN1024	ITALY UNI
	A	B	C	R1	R2	A	B	C	R1	R2								
TB 30	60 ×	30	5.5 5.5	3.0 3.0	1.5 1.5	2.362 ×	1.181	0.22 0.22	0.12 0.12	0.06 0.06					A,B+-1.5 C+-0.75		F	F
TB 35	70 ×	35	6.0 6.0	3.0 3.0	1.5 1.5	2.756 ×	1.378	0.24 0.24	0.12 0.12	0.06 0.06					A,B+-1.5 C+-0.75		F	F
TB 40	80 ×	40	7.0 7.0	3.5 3.5	2.0 2.0	3.150 ×	1.575	0.28 0.28	0.14 0.14	0.08 0.08					A,B+-1.5 C+-0.75		F	F
TB 50	100 ×	50	8.5 8.5	4.0 4.0	2.0 2.0	3.937 ×	1.969	0.33 0.33	0.16 0.16	0.08 0.08					A,B+-1.5 C+-0.75		F	F
TB 60	120 ×	60	10.0 10.0	5.0 5.0	2.5 2.5	4.724 ×	2.362	0.39 0.39	0.20 0.20	0.10 0.10					A,B+-2 C+-1		F	F

10-96 STEEL MATERIAL DATA

Table 10-33. Hot Rolled T-Steel with Sharp Corners (DIN 59051)

TABLE 10-33. HOT ROLLED T-STEEL WITH SHARP
CORNERS (DIN 59051)
BASIS: 1 IN = 25.4 MM

ORDER EXAMPLE:
LENGTH,TPS 20 DIN 59051 + STEEL QUALITY

THE NOMINAL SIZE IS NATIONAL STANDARD AS INDICATED
F=FIRST CHOICE, S=SECOND CHOICE, T=THIRD CHOICE, NUMBER=OTHER SIZE
* = COMMERCIAL SIZE

NOMINAL SIZE						INCHES					U.S.A. ANSI	AUSTRAL AS	JAPAN JIS	EURO-NORM	FRANCE NF	U.K. BS	GERMANY DIN 59051	ITALY UNI 5681
	MILLIMETERS A B C			R1	R2	A	B	C	R1	R2								
TPS 20	20 × 20	3.0 4.0		0.0 0.0	0.0 0.0	0.787 ×	0.787	0.12 0.16	0.00 0.00	0.00 0.00				A,B+-1 C+-0.5	F		F	F
TPS 25	25 × 25	3.5 4.5		0.0 0.0	0.0 0.0	0.984 ×	0.984	0.14 0.18	0.00 0.00	0.00 0.00				A,B+-1 C+-0.5	F		F	F
TPS 30	30 × 30	4.0 5.0		0.0 0.0	0.0 0.0	1.181 ×	1.181	0.16 0.20	0.00 0.00	0.00 0.00				A,B+-1 C+-0.5	F		F	F
TPS 35	35 × 35	4.5 5.5		0.0 0.0	0.0 0.0	1.378 ×	1.378	0.18 0.22	0.00 0.00	0.00 0.00				A,B+-1 C+-0.5	F		F	F
TPS 40	40 × 40	5.0 6.0		0.0 0.0	0.0 0.0	1.575 ×	1.575	0.20 0.24	0.00 0.00	0.00 0.00				A,B+-1 C+-0.5	F		F	F
TPS 45	45 × 45	6.5 6.5		0.0 0.0	0.0 0.0	1.772 ×	1.772	0.26 0.26	0.00 0.00	0.00 0.00				A,B+-1 C+-0.5			F	F
TPS 50	50 × 50	7.0 7.0		0.0 0.0	0.0 0.0	1.969 ×	1.969	0.28 0.28	0.00 0.00	0.00 0.00				A,B+-1 C+-0.5			F	F
TPS 60	60 × 60	8.0 8.0		0.0 0.0	0.0 0.0	2.362 ×	2.362	0.31 0.31	0.00 0.00	0.00 0.00				A,B+-1.5 C+-0.75			F	F
TPS 70	70 × 70	9.0 9.0		0.0 0.0	0.0 0.0	2.756 ×	2.756	0.35 0.35	0.00 0.00	0.00 0.00				A,B+-1.5 C+-0.75			F	F
TPS 80	80 × 80	10.0 10.0		0.0 0.0	0.0 0.0	3.150 ×	3.150	0.39 0.39	0.00 0.00	0.00 0.00				A,B+-1.5 C+-0.75			F	F
TPS 100	100 × 100	11.0 11.0		0.0 0.0	0.0 0.0	3.937 ×	3.937	0.43 0.43	0.00 0.00	0.00 0.00				A,B+-1.5 C+-0.75			F	F

STEEL MATERIAL DATA

Table 10-34. Hot Rolled Z-Steel with Round Corners (DIN 1027)

TABLE 10-34. HOT ROLLED Z-STEEL WITH ROUND CORNERS (DIN 1027)
BASIS: 1 IN = 25.4 MM

ORDER EXAMPLE:
LENGTH, Z 30 DIN 1027 + STEEL QUALITY

THE NOMINAL SIZE IS NATIONAL STANDARD AS INDICATED
F=FIRST CHOICE, S=SECOND CHOICE, T=THIRD CHOICE, NUMBER=OTHER SIZE
* = COMMERCIAL SIZE

Nominal Size						Inches						U.S.A. ANSI	AUSTRAL AS	JAPAN JIS	EURO-NORM	FRANCE NF	U.K. BS	GERMANY DIN 1027	ITALY UNI
Millimeters A	B	C	R1	R2		A	B	C	R1	R2									
Z 30	30 × 38	4.0 4.0 4.0	4.5 4.5 4.5	2.5 2.5 2.5		1.181 ×	1.496	0.16 0.16 0.16	0.18 0.18 0.18	0.10 0.10 0.10						F		F A,R+-1 C+-0.5 R1+-0.5	
Z 40	40 × 40	4.5 4.5 4.5	5.0 5.0 5.0	2.5 2.5 2.5		1.575 ×	1.575	0.18 0.18 0.18	0.20 0.20 0.20	0.10 0.10 0.10						F		F A,R+-1 C+-0.5 R1+-0.5	
Z 50	50 × 43	5.0 5.0 5.0	5.5 5.5 5.5	3.0 3.0 3.0		1.969 ×	1.693	0.20 0.20 0.20	0.22 0.22 0.22	0.12 0.12 0.12						F		F A,R+-1 C+-0.5 R1+-0.5	
Z 60	60 × 45	5.0 5.0 5.0	6.0 6.0 6.0	3.0 3.0 3.0		2.362 ×	1.772	0.20 0.20 0.20	0.24 0.24 0.24	0.12 0.12 0.12						F		F A+-1.5 R+-1 C+-0.5 R1+-0.5	
Z 80	80 × 50	6.0 6.0 6.0	7.0 7.0 7.0	3.5 3.5 3.5		3.150 ×	1.969	0.24 0.24 0.24	0.28 0.28 0.28	0.14 0.14 0.14						F		F A+-1.5 B+-1 C+-0.5 R1+-0.5	
Z 100	100 × 55	6.5 6.5 6.5	8.0 8.0 8.0	4.0 4.0 4.0		3.937 ×	2.165	0.26 0.26 0.26	0.31 0.31 0.31	0.16 0.16 0.16						F		F A+-1.5 B+-1.5 C+-0.75 R1+-0.75	
Z 120	120 × 60	7.0 7.0 7.0	9.0 9.0 9.0	4.5 4.5 4.5		4.724 ×	2.362	0.28 0.28 0.28	0.35 0.35 0.35	0.18 0.18 0.18						F		F A+-2 B+-1.5 C+-0.75 R1+-0.75	
Z 140	140 × 65	8.0 8.0 8.0	10.0 10.0 10.0	5.0 5.0 5.0		5.512 ×	2.559	0.31 0.31 0.31	0.39 0.39 0.39	0.20 0.20 0.20						F		F A+-2 B+-1.5 C+-0.75 R1+-0.75	
Z 160	160 × 70	8.5 8.5 8.5	11.0 11.0 11.0	5.5 5.5 5.5		6.299 ×	2.756	0.33 0.33 0.33	0.43 0.43 0.43	0.22 0.22 0.22								F A+-4 B+-1.5 C+-0.75 R1+-0.75	
Z 180	180 × 75	9.5 9.5 9.5	12.0 12.0 12.0	6.0 6.0 6.0		7.087 ×	2.953	0.37 0.37 0.37	0.47 0.47 0.47	0.24 0.24 0.24								S A+-4 R+-1.5 C+-0.75 R1+-0.75	
Z 200	200 × 80	10.0 10.0 10.0	13.0 13.0 13.0	6.5 6.5 6.5		7.874 ×	3.150	0.39 0.39 0.39	0.51 0.51 0.51	0.26 0.26 0.26								S A+-4 R+-1.5 C+-0.75 R1+-0.75	

Table 10-35. Hot Rolled Small U-Steel with Round Corners (EURONORM 54)

TABLE 10-35. HOT ROLLED SMALL U-STEEL WITH
ROUND CORNERS (EURONORM 54)
BASIS; 1 IN = 25.4 MM

ORDER EXAMPLE;
LENGTH,U30X15 EURONORM 54 + STEEL QUALITY N

THE NOMINAL SIZE IS NATIONAL STANDARD AS INDICATED
F=FIRST CHOICE,S=SECOND CHOICE,T=THIRD CHOICE,NUMBER=OTHER SIZE
* = COMMERCIAL SIZE

NOMINAL SIZE MILLIMETERS						INCHES						U.S.A. ANSI	AUSTRAL AS	JAPAN JIS	EURO- NORM 54	FRANCE NF A45-007	U.K. BS	GERMANY DIN 1026	ITALY UNI 5786
A	B	C	D	R1	R2	A	B	C	D	R1	R2								
U 30 x 15	4.0	4.5	4.5	2.0	1.181x	0.591	0.16	0.18	0.18	0.08				F	F A,B+-1.5 C+-0.5 D-0.5	F	F	F
U 40 x 20	5.0	5.5	5.0	2.5	1.575x	0.787	0.20	0.22	0.20	0.10				F	F A,B+-1.5 C+-0.5 D-0.5	F	F	F
U 50 x 25	5.0	6.0	6.0	3.0	1.969x	0.984	0.20	0.24	0.24	0.12				F	F A,B+-1.5 C+-0.5 D-0.5	F	F	F
U 60 x 30	6.0	6.0	6.0	3.0	2.362x	1.181	0.24	0.24	0.24	0.12				F	F A,B+-1.5 C+-0.5 D-0.5	F	F	F
U 30 x 33	5.0	7.0	7.0	3.5	1.181x	1.299	0.20	0.28	0.28	0.14				F	F A,B+-1.5 C+-0.5 D-0.5	F	F	F
U 40 x 35	5.0	7.0	7.0	3.5	1.575x	1.378	0.20	0.28	0.28	0.14				F	F A,B+-1.5 C+-0.5 D-0.5	F	F	F
U 50 x 38	5.0	7.0	7.0	3.5	1.969x	1.496	0.20	0.28	0.28	0.14				F	F A,B+-1.5 C+-0.5 D-0.5	F	F	F
U 65 x 42	5.5	7.5	7.5	4.0	2.559x	1.654	0.22	0.30	0.30	0.16				F	F A,B+-1.5 C+-0.5 D-0.5	F	F	F

STEEL MATERIAL DATA

Table 10-36. Hot Rolled UPN-Steel with Round Corners (UNI 5680)

TABLE 10-36. HOT ROLLED UPN-STEEL WITH
ROUND CORNERS (UNI 5680)
BASIS: 1 IN = 25.4 MM

ORDER EXAMPLE:
LENGTH,UPN80X45 NF A45-202 + STEEL QUALITY

THE NOMINAL SIZE IS NATIONAL STANDARD AS INDICATED
F=FIRST CHOICE,S=SECOND CHOICE,T=THIRD CHOICE,NUMBER=OTHER SIZE
* = COMMERCIAL SIZE

NOMINAL SIZE MILLIMETERS						INCHES						U.S.A. ANSI	AUSTRAL AS	JAPAN JIS	EURO-NORM 24	FRANCE NF A45-202	U.K. BS	GERMANY DIN 1026	ITALY UNI 5680
A	B	C	D	R1	R2	A	B	C	D	R1	R2								
80 × 45	6.0	8.0	8.0	4.0	3.150×	1.772	0.24	0.31	0.31	0.16				A+=2 B+=1.5 C+=0.5 D=0.5	F	F	F	F
100 × 50	6.0	8.5	8.5	4.5	3.937×	1.969	0.24	0.33	0.33	0.18				A+=2 B+=1.5 C+=0.5 D=0.5	F	F	F	F
120 × 55	7.0	9.0	9.0	4.5	4.724×	2.165	0.28	0.35	0.35	0.18				A+=2 B+=1.5 C+=0.5 D=0.5	F	F	F	F
140 × 60	7.0	10.0	10.0	5.0	5.512×	2.362	0.28	0.39	0.39	0.20				A+=2 B+=1.5 C+=0.5 D=0.5	F	F	F	F
160 × 65	7.5	10.5	10.5	5.5	6.299×	2.559	0.30	0.41	0.41	0.22				A+=2 B+=1.5 C+=0.5 D=1	F	F	F	F
180 × 70	8.0	11.0	11.0	5.5	7.087×	2.756	0.31	4.33	4.33	0.22				A+=2 B+=1.5 C+=0.5 D=1	F	F	F	F
200 × 75	8.5	11.5	11.5	6.0	7.874×	2.953	0.33	0.45	0.45	0.24				A+=2 B+=1.5 C+=0.5 D=1	F	F	F	F
220 × 80	9.0	12.5	12.5	6.5	8.661×	3.150	0.35	0.49	0.49	0.26				A+=3 B+=2 C+=0.5 D=1	F	F	F	F
240 × 85	9.5	13.0	13.0	6.5	9.449×	3.346	0.37	0.51	0.51	0.26				A+=3 B+=2 C+=0.5 D=1	F	F	F	F
260 × 90	10.0	14.0	14.0	7.0	10.236×	3.543	0.39	0.55	0.55	0.28				A+=3 B+=2 C+=0.5 D=1	F	F	F	F

Table 10-36 (Continued). Hot Rolled UPN-Steel with Round Corners (UNI 5680)

TABLE 10-36. HOT ROLLED UPN-STEEL WITH
ROUND CORNERS (UNI 5680)
BASIS; 1 IN = 25.4 MM

ORDER EXAMPLE:
LENGTH,UPN80X45 NF A45-202 + STEEL QUALITY
PAGE N0. 2

THE NOMINAL SIZE IS NATIONAL STANDARD AS INDICATED
F=FIRST CHOICE,S=SECOND CHOICE,T=THIRD CHOICE,NUMBER=OTHER SIZE
* = COMMERCIAL SIZE

NOMINAL SIZE															U.S.A. ANSI	AUSTRAL AS	JAPAN JIS	EURO-NORM 24	FRANCE NF A45-202	U.K. BS	GERMANY DIN 1026	ITALY UNI 5680
MILLIMETERS						INCHES																
A	B	C	D	R1	R2	A	B	C	D	R1	R2											
280 × 95	10.0	15.0	15.0	7.5	11.024×	3.740	0.39	0.59	0.59	0.30									F		F	
300 × 100	10.0	16.0	16.0	8.0	11.811×	3.937	0.39	0.63	0.63	0.31					A+=3 B+=2 C+=0.5 D=1				F		F	
320 × 100	14.0	17.5	17.5	8.8	12.598×	3.937	0.55	0.69	0.69	0.35					A+=3 B+=2 C+=0.5 D=1						F	
350 × 100	14.0	16.0	16.0	8.0	13.780×	3.937	0.55	0.63	0.63	0.31					A+=3 B+=2 C+=0.7 D=1.5				F			
380 × 102	13.5	16.0	16.0	8.0	14.961×	4.016	0.53	0.63	0.63	0.31					A+=3 B+=2.5 C+=0.7 D=1.5						F	
400 × 110	14.0	18.0	18.0	9.0	15.748×	4.331	0.55	0.71	0.71	0.35					A+=3 B+=2.5 C+=0.7 D=1.5						F	

STEEL MATERIAL DATA

Table 10-37. Hot Rolled IPN-Beams with Round Corners (UNI 5679)

TABLE 10-37, HOT ROLLED IPN-BEAMS WITH
ROUND CORNERS (UNI 5679)
BASIS; 1 IN = 25.4 MM

ORDER EXAMPLE;
LENGTH,IPN80 UNI 5679 + STEEL QUALITY
PAGE NO. 1

THE NOMINAL SIZE IS NATIONAL STANDARD AS INDICATED
F=FIRST CHOICE,S=SECOND CHOICE,T=THIRD CHOICE,NUMBER=OTHER SIZE
* = COMMERCIAL SIZE

	NOMINAL SIZE MILLIMETERS A B C D R1	INCHES A B C D R1	U.S.A. ANSI	AUSTRAL AS	JAPAN JIS	EURO- NORM 24	FRANCE NF A45-209	U.K. BS	GERMANY DIN 1025	ITALY UNI 5679
IPN080	80 42 3.9 5.9 2.3	3.150 1.654 0.15 0.23 0.09				A+=2 F B+=1.5 C+=0.5 D=0.5	F	F	F	F
IPN100	100 50 4.5 6.8 2.7	3.937 1.969 0.18 0.27 0.11				A+=2 F B+=1.5 C+=0.5 D=0.5	F	F	F	F
IPN120	120 58 5.1 7.7 3.1	4.724 2.283 0.20 0.30 0.12				A+=2 F B+=1.5 C+=0.5 D=0.5	F	F	F	F
IPN140	140 66 5.7 8.6 3.4	5.512 2.598 0.22 0.34 0.13				A+=2 F B+=1.5 C+=0.5 D=0.5	F	F	F	F
IPN160	160 74 6.3 9.5 3.8	6.299 2.913 0.25 0.37 0.15				A+=2 F B+=1.5 C+=0.5 D=1	F	F	F	F
IPN180	180 82 6.9 10.4 4.1	7.087 3.228 0.27 0.41 0.16				A+=2 F B+=2 C+=0.5 D=1	F	F	F	F
IPN200	200 90 7.5 11.3 4.5	7.874 3.543 0.30 0.44 0.18				A+=2 F B+=2 C+=0.5 D=1	F	F	F	F
IPN220	220 98 8.1 12.2 4.9	8.661 3.858 0.32 0.48 0.19				A+=3 F B+=2 C+=0.5 D=1	F	F	F	F
IPN240	240 106 8.7 13.1 5.2	9.449 4.173 0.34 0.52 0.20				A+=3 F B+=2.5 C+=0.5 D=1	F	F	F	F
IPN260	260 13 9.4 14.1 5.6	10.236 0.512 0.37 0.56 0.22				A+=3 F B+=2.5 C+=0.5 D=1	F	F	F	F

Table 10-37 (Continued). Hot Rolled IPN-Beams with Round Corners (UNI 5679)

```
TABLE 10-37, HOT ROLLED IPN-BEAMS WITH                      ORDER EXAMPLE;
ROUND CORNERS (UNI 5679)                                    LENGTH,IPN80 UNI 5679 + STEEL QUALITY    PAGE NO. 2
BASIS; 1 IN = 25.4 MM

        NOMINAL SIZE                                  THE NOMINAL SIZE IS NATIONAL STANDARD AS INDICATED
                                                      F=FIRST CHOICE,S=SECOND CHOICE,T=THIRD CHOICE,NUMBER=OTHER SIZE
        MILLIMETERS         INCHES                    * = COMMERCIAL SIZE
        A   B    C    D   R1    A      B    C    D   R1   U.S.A.  AUSTRAL JAPAN  EURO-   FRANCE   U.K.  GERMANY ITALY
                                                          ANSI    AS      JIS    NORM    NF       BS    DIN     UNI
                                                                                 24      A45-209        1025    5679

IPN280  280 119 10.1 15.2 6.1  11.024  4.685 0.40 0.60 0.24                             A*=3  F                F    F
                                                                                        B*=2.5
                                                                                        C*=0.5
                                                                                        D=1

IPN300  300 125 10.8 16.2 6.5  11.811  4.921 0.43 0.64 0.26                             A*=3  F                F    F
                                                                                        B*=2.5
                                                                                        C*=0.5
                                                                                        D=1

IPN320  320 131 11.5 17.3 6.9  12.598  5.157 0.45 0.68 0.27                             A*=3  F                F    F
                                                                                        B*=3
                                                                                        C*=0.6
                                                                                        D=1.5

IPN340  340 137 12.5 18.3 7.3  13.386  5.394 0.49 0.72 0.29                             A*=3  F                F    F
                                                                                        B*=3
                                                                                        C*=0.6
                                                                                        D=1.5

IPN360  360 143 13.0 19.5 7.8  14.173  5.630 0.51 0.77 0.31                             A*=3  F                F    F
                                                                                        B*=3
                                                                                        C*=0.6
                                                                                        D=1.5

IPN400  400 155 14.4 21.6 8.6  15.748  6.102 0.57 0.85 0.34                             A*=3  F                F    F
                                                                                        B*=3
                                                                                        C*=0.7
                                                                                        D=1.5

IPN450  450 170 16.2 24.3 9.7  17.717  6.693 0.64 0.96 0.38                             A*=4  F                F    F
                                                                                        B*=3
                                                                                        C*=0.8
                                                                                        D=1.5

IPN500  500 185 18.0 27.0 10.8 19.685  7.283 0.71 1.06 0.43                             A*=4  F                F    F
                                                                                        B*=3
                                                                                        C*=0.9
                                                                                        D=1.5

IPN600  600 215 21.6 32.4 13.0 23.622  8.465 0.85 1.28 0.51                             A*=4  F                F    F
                                                                                        B*=3
                                                                                        C*=1.1
                                                                                        D=1.5
```

STEEL MATERIAL DATA

Table 10-38. Hot Rolled IPE-Beams with Sharp Corners (EURONORM 19, 44)

TABLE 10-38. HOT ROLLED IPE-BEAMS WITH
SHARP CORNERS (EURONORM 19,44)
BASIS; 1 IN = 25.4 MM

ORDER EXAMPLE;
LENGTH,IPE80 EURONORM 19,44 + STEEL QUALITY   PAGE NO. 1

THE NOMINAL SIZE IS NATIONAL STANDARD AS INDICATED
F=FIRST CHOICE,S=SECOND CHOICE,T=THIRD CHOICE,NUMBER=OTHER SIZE
* = COMMERCIAL SIZE

NOMINAL SIZE											EURO-	FRANCE	U.K.	GERMANY	ITALY
	MILLIMETERS					INCHES					U.S.A. AUSTRAL JAPAN	NORM	NF	BS	DIN	UNI
	A	B	C	D	R1	A	B	C	D	R1	ANSI AS JIS	19,44	A45-205		1025	5398
IPE080	80	46	3.8	5.2	5.0	3.150	1.811	0.15	0.20	0.20		A+=2 B+=2 C+=0.5 D+=1	F	F	F	F
IPE100	100	55	4.1	5.7	7.0	3.937	2.165	0.16	0.22	0.28		A+=2 B+=2 C+=0.5 D+=1	F	F	F	F
IPE120	120	64	4.4	6.3	7.0	4.724	2.520	0.17	0.25	0.28		A+=2 B+=2 C+=0.5 D+=1	F	F	F	F
IPE140	140	73	4.7	6.9	7.0	5.512	2.874	0.19	0.27	0.28		A+3=2 B+3=2 C+=0.75 D+=1.5	F	F	F	F
IPE160	160	82	5.0	7.4	9.0	6.299	3.228	0.20	0.29	0.35		A+3=2 B+3=2 C+=0.75 D+=1.5	F	F	F	F
IPE180	180	91	5.3	8.0	9.0	7.087	3.583	0.21	0.31	0.35		A+3=2 B+3=2 C+=0.75 D+=1.5	F	F	F	F
IPE200	200	100	5.6	8.5	12.0	7.874	3.937	0.22	0.33	0.47		A+=3 B+=3 C+=0.75 D+=1.5	F	F	F	F
IPE220	220	110	5.9	9.2	12.0	8.661	4.331	0.23	0.36	0.47		A+=3 B+=3 C+=0.75 D+=1.5	F	F	F	F
IPE240	240	120	6.2	9.8	15.0	9.449	4.724	0.24	0.39	0.59		A+=3 B+=3 C+=0.75 D+=1.5	F	F	F	F
IPE270	270	135	6.6	10.2	15.0	10.630	5.315	0.26	0.40	0.59		A+=3 B+=3 C+=0.75 D+=1.5	F	F	F	F

10-103

Table 10-38 (Continued). Hot Rolled IPE-Beams with Sharp Corners (EURONORM 19, 44)

TABLE 10-38. HOT ROLLED IPE-BEAMS WITH
SHARP CORNERS (EURONORM 19,44)
BASIS; 1 IN = 25.4 MM

ORDER EXAMPLE:
LENGTH,IPE80 EURONORM 19,44 + STEEL QUALITY
PAGE NO. 2

THE NOMINAL SIZE IS NATIONAL STANDARD AS INDICATED
F=FIRST CHOICE,S=SECOND CHOICE,T=THIRD CHOICE,NUMBER=OTHER SIZE
* = COMMERCIAL SIZE

NOMINAL SIZE											U.S.A. ANSI	AUSTRAL AS	JAPAN JIS	EURO- NORM 19,44	FRANCE NF A45-205	U.K. BS	GERMANY DIN 1025	ITALY UNI 5398
	MILLIMETERS A	B	C	D	R1	INCHES A	B	C	D	R1								
IPE300	300	150	7.1	10.7	15.0	11.811	5.906	0.28	0.42	0.59				A*=3 B*=3 C*=1 D*=2	F		F	F
IPE330	330	160	7.5	11.5	18.0	12.992	6.299	0.30	0.45	0.71				A*=3 B*=3 C*=1 D*=2	F		F	F
IPE360	360	170	8.0	12.7	18.0	14.173	6.693	0.31	0.50	0.71				A*=3 B*=3 C*=1 D*=2	F		F	F
IPE400	400	180	8.6	13.5	21.0	15.748	7.087	0.34	0.53	0.83				A*=3 B*=4 C*=1 D*=2	F		F	F
IPE450	450	190	9.4	14.6	21.0	17.717	7.480	0.37	0.57	0.83				A*=4 B*=4 C*=1 D*=2	F		F	F
IPE500	500	200	10.2	16.0	21.0	19.685	7.874	0.40	0.63	0.83				A*=4 B*=4 C*=1 D*=2	F		F	F
IPE550	550	210	11.1	17.2	24.0	21.654	8.268	0.44	0.68	0.94				A*=5 B*=4 C*=1 D*=2	F		F	F
IPE600	600	220	12.0	19.0	24.0	23.622	8.661	0.47	0.75	0.94				A*=5 B*=4 C*=1 D*=2	F		F	F

STEEL MATERIAL DATA

Table 10-39. Hot Rolled Wide Flange HE-Beams with Round Corners (EURONORM 53)

TABLE 10-39. HOT ROLLED WIDE FLANGE HE-BEAMS
WITH ROUND CORNERS (EURONORM 53)
BASIS; 1 IN = 25.4 MM

ORDER EXAMPLE:
LENGTH,HE100A EURONORM 53 + STEEL QUALITY

PAGE NO. 1

THE NOMINAL SIZE IS NATIONAL STANDARD AS INDICATED
F=FIRST CHOICE,S=SECOND CHOICE,T=THIRD CHOICE,NUMBER=OTHER SIZE
*= COMMERCIAL SIZE

	NOMINAL SIZE MILLIMETERS					INCHES					EURO- NORM 53	U.S.A. ANSI	AUSTRAL AS	JAPAN JIS	FRANCE NF A45-201	U.K. BS	GERMANY DIN 1025	ITALY UNI 5397
	A	B	C	D	R1	A	B	C	D	R1								
HE 100A	96	100	5.0	8.0	12.0	3.780	3.937	0.20	0.31	0.47	F				F A+4=2 B+=3 C+=1 D+=1.5	F	F	F
HE 100B	100	100	6.0	10.0	12.0	3.937	3.937	0.24	0.39	0.47	F				F A+4=2 B+=3 C+=1 D+=1.5	F	F	F
HE 100M	120	106	12.0	20.0	12.0	4.724	4.173	0.47	0.79	0.47	F				F A+4=2 B+=3 C+=1 D+=2	F	F	F
HE 120A	114	120	5.0	8.0	12.0	4.488	4.724	0.20	0.31	0.47	F				F A+4=2 B+=3 C+=1 D+=1.5	F	F	F
HE 120B	120	120	6.5	11.0	12.0	4.724	4.724	0.26	0.43	0.47	F				F A+4=2 B+=3 C+=1 D+=1.5	F	F	F
HE 120M	140	126	12.5	21.0	12.0	5.512	4.961	0.49	0.83	0.47	F				F A+4=2 B+=3 C+=1 D+=2	F	F	F
HE 140A	133	140	5.5	8.5	12.0	5.236	5.512	0.22	0.33	0.47	F				F A+4=2 B+=3 C+=1 D+=1.5	F	F	F
HE 140B	140	140	7.0	12.0	12.0	5.512	5.512	0.28	0.47	0.47	F				F A+4=2 B+=3 C+=1 D+=1.5	F	F	F
HE 140M	160	146	13.0	22.0	12.0	6.299	5.748	0.51	0.87	0.47	F				F A+4=2 B+=3 C+=1 D+=2	F	F	F
HE 160A	152	160	6.0	9.0	15.0	5.984	6.299	0.24	0.35	0.59	F				F A+4=2 B+=3 C+=1 D+=1.5	F	F	F

Table 10-39 (Continued). Hot Rolled Wide Flange HE-Beams with Round Corners (EURONORM 53)

TABLE 10-39. HOT ROLLED WIDE FLANGE HE-BEAMS
WITH ROUND CORNERS (EURONORM 53)   PAGE NO. 2
BASIS; 1 IN = 25.4 MM

ORDER EXAMPLE;
LENGTH,HE100A EURONORM 53 + STEEL QUALITY

THE NOMINAL SIZE IS NATIONAL STANDARD AS INDICATED
F=FIRST CHOICE,S=SECOND CHOICE,T=THIRD CHOICE,NUMBER=OTHER SIZE
*= COMMERCIAL SIZE

NOMINAL SIZE					INCHES					U.S.A. ANSI	AUSTRAL AS	JAPAN JIS	EURO- NORM 53	FRANCE NF A45-201	U.K. BS	GERMANY DIN 1025	ITALY UNI 5397
MILLIMETERS																	
A	B	C	D	R1	A	B	C	D	R1								
HE 160B	160	160	8.0	13.0	15.0	6.299	6.299	0.31	0.51	0.59				F A+4=2 B+=3 C+=1 D+=1.5	F		F	F
HE 160M	180	166	14.0	23.0	15.0	7.087	6.535	0.55	0.91	0.59				F A,B+=3 C+=1 D+=2	F		F	F
HE 180A	171	180	6.0	9.5	15.0	6.732	7.087	0.24	0.37	0.59				F A,B+=3 C+=1 D+=1.5	F		F	F
HE 180B	180	180	8.5	14.0	15.0	7.087	7.087	0.33	0.55	0.59				F A,B+=3 C+=1 D+=1.5	F		F	F
HE 180M	200	186	14.5	24.0	15.0	7.874	7.323	0.57	0.94	0.59				F A,B+=3 C+=1 D+=2	F		F	F
HE 200A	190	200	6.5	10.0	18.0	7.480	7.874	0.26	0.39	0.71				F A,B+=3 C+=1 D+=1.5	F		F	F
HE 200B	200	200	9.0	15.0	18.0	7.874	7.874	0.35	0.59	0.71				F A,B+=3 C+=1 D+=1.5	F		F	F
HE 200M	220	206	15.0	25.0	18.0	8.661	8.110	0.59	0.98	0.71				F A,B+=3 C+=1 D+=2	F		F	F
HE 220A	210	220	7.0	11.0	18.0	8.268	8.661	0.28	0.43	0.71				F A,B+=3 C+=1 D+=1.5	F		F	F
HE 220B	220	220	9.5	16.0	18.0	8.661	8.661	0.37	0.63	0.71				F A,B+=3 C+=1 D+=1.5	F		F	F
HE 220M	240	226	15.5	26.0	18.0	9.449	8.898	0.61	1.02	0.71				F A,B+=3 C+=1 D+=2	F		F	F
HE 240A	230	240	7.5	12.0	21.0	9.055	9.449	0.30	0.47	0.83				F A,B+=3 C+=1 D+=1.5	F		F	F
HE 240B	240	240	10.0	17.0	21.0	9.449	9.449	0.39	0.67	0.83				F A,B+=3 C+=1 D+=2	F		F	F

STEEL MATERIAL DATA

Table 10-39 (Continued). Hot Rolled Wide Flange HE-Beams with Round Corners (EURONORM 53)

TABLE 10-39. HOT ROLLED WIDE FLANGE HE-BEAMS
WITH ROUND CORNERS (EURONORM 53)   PAGE NO. 3
BASIS; 1 IN = 25.4 MM

ORDER EXAMPLE:
LENGTH,HE100A EURNORM 53 + STEEL QUALITY

THE NOMINAL SIZE IS NATIONAL STANDARD AS INDICATED
F=FIRST CHOICE,S=SECOND CHOICE,T=THIRD CHOICE,NUMBER=OTHER SIZE
* = COMMERCIAL SIZE

NOMINAL SIZE												U.S.A. ANSI	AUSTRAL AS	JAPAN JIS	EURO- NORM 53	FRANCE NF A45-201	U.K. BS	GERMANY DIN 1025	ITALY UNI 5397
MILLIMETERS					INCHES														
A	B	C	D	R1	A	B	C	D	R1										

HE 240A 270 248 18.0 7.5 21.0 10.630 9.764 0.71 0.30 0.83   F A,B+=3 F   F   F
                                                                 C+=1
                                                                 D+=2

HE 260A 250 260 7.5 12.5 24.0 9.843 10.236 0.30 0.49 0.94   F A,B+=3 F   F   F
                                                                 C+=1
                                                                 D+=2

HE 260B 260 260 10.0 17.5 24.0 10.236 10.236 0.39 0.69 0.94   F A,B+=3 F   F   F
                                                                 C+=1
                                                                 D+=2

HE 260M 268 260 18.0 23.5 24.0 11.417 10.551 0.71 0.93 0.94   F A,B+=3 F   F   F
                                                                 C+=1.5
                                                                 D+=2

HE 280A 270 280 8.0 13.0 24.0 10.630 11.024 0.31 0.51 0.94   F A,B+=3 F   F   F
                                                                 C+=1.5
                                                                 D+=2

HE 280B 280 280 10.5 18.0 24.0 11.024 11.024 0.41 0.71 0.94   F A,B+=3 F   F   F
                                                                 C+=1.5
                                                                 D+=2

HE 280M 288 280 18.5 33.0 24.0 12.205 11.339 0.73 1.30 0.94   F A,B+=3 F   F   F
                                                                 C+=1.5
                                                                 D+=2

HE 300A 290 300 8.5 14.0 27.0 11.417 11.811 0.33 0.55 1.06   F A,B+=3 F   F   F
                                                                 C+=1.5
                                                                 D+=2

HE 300B 300 300 11.0 19.0 27.0 11.811 11.811 0.43 0.75 1.06   F A,B+=3 F   F   F
                                                                 C+=1.5
                                                                 D+=2

HE 300C 320 305 16.0 29.0 27.0 12.598 12.008 0.63 1.14 1.06   F A,B+=3 F   F   F
                                                                 C+=1.5
                                                                 D+=2.5

HE 300M 340 310 21.0 39.0 27.0 13.386 12.205 0.83 1.54 1.06   F A+=4   F   F   F
                                                                 B+=3
                                                                 C+=1.5
                                                                 D+=2.5

HE 320A 310 300 9.0 15.5 27.0 12.205 11.811 0.35 0.61 1.06   F A,B+=3 F   F   F
                                                                 C+=1.5
                                                                 D+=2

HE 320B 320 300 11.5 20.5 27.0 12.598 11.811 0.45 0.81 1.06   F A,B+=3 F   F   F
                                                                 C+=1.5
                                                                 D+=2

Table 10-39 (Continued). Hot Rolled Wide Flange HE-Beams with Round Corners (EURONORM 53)

```
TABLE 10-39. HOT ROLLED WIDE FLANGE HE-BEAMS                                    ORDER EXAMPLE:
WITH ROUND CORNERS (EURONORM 53)                                                LENGTH,HE100A EURONORM 53 + STEEL QUALITY    PAGE NO. 4
BASIS; 1 IN = 25.4 MM

    NOMINAL SIZE
                                                        THE NOMINAL SIZE IS NATIONAL STANDARD AS INDICATED
    MILLIMETERS          INCHES                         F=FIRST CHOICE,S=SECOND CHOICE,T=THIRD CHOICE,NUMBER=OTHER SIZE
    A    B    C    D    R1   A       B       C     D    R1      * = COMMERCIAL SIZE
                                                                U.S.A.  AUSTRAL  JAPAN  EURO-  FRANCE  U.K.   GERMANY  ITALY
                                                                ANSI    AS       JIS    NORM   NF      BS     DIN      UNI
                                                                                        53     A45-201         1025     5397
```

	A	B	C	D	R1	A	B	C	D	R1			
HE 320M 359	309	21.0	40.0	27.0	14.134	12.165	0.83	1.57	1.06	F A+-4 B+-3 C+-1.5 D+-2.5	F	F
HE 340A 330	300	9.5	16.5	27.0	12.992	11.811	0.37	0.65	1.06	F A,B+-3 F C+-1.5 D+-2	F	F
HE 340B 340	300	12.0	21.5	27.0	13.386	11.811	0.47	0.85	1.06	F A,B+-3 F C+-1.5 D+-2	F	F
HE 340M 377	309	21.0	40.0	27.0	14.843	12.165	0.83	1.57	1.06	F A+-4 B+-3 C+-1.5 D+-2.5	F	F
HE 360A 350	300	10.0	17.5	27.0	13.780	11.811	0.39	0.69	1.06	F A,B+-3 F C+-1.5 D+-2	F	F
HE 360B 360	300	12.5	22.5	27.0	14.173	11.811	0.49	0.89	1.06	F A,B+-3 F C+-1.5	F	F
HE 360M 395	308	21.0	40.0	27.0	15.551	12.126	0.83	1.57	1.06	F A+-4 B+-3 C+-1.5 D+-2.5	F	F
HE 400A 390	300	11.0	19.0	27.0	15.354	11.811	0.43	0.75	1.06	F A,B+-3 F C+-1.5 D+-2	F	F
HE 400B 400	300	13.5	24.0	27.0	15.748	11.811	0.53	0.94	1.06	F A,B+-3 F C+-1.5	F	F
HE 400M 432	307	21.0	40.0	27.0	17.008	12.087	0.83	1.57	1.06	F A+-4 B+-3 C+-1.5 D+-2.5	F	F
HE 450A 440	300	11.5	21.0	27.0	17.323	11.811	0.45	0.83	1.06	F A+-4 B+-3 C+-1.5 D+-2	F	F

STEEL MATERIAL DATA

10-109

Table 10-39 (Continued). Hot Rolled Wide Flange HE-Beams with Round Corners (EURONORM 53)

TABLE 10-39. HOT ROLLED WIDE FLANGE HE-BEAMS
WITH ROUND CORNERS (EURONORM 53)
BASIS: 1 IN = 25.4 MM

PAGE NO. 5

ORDER EXAMPLE:
LENGTH,HE100A EURONORM 53 + STEEL QUALITY

THE NOMINAL SIZE IS NATIONAL STANDARD AS INDICATED
F=FIRST CHOICE,S=SECOND CHOICE,T=THIRD CHOICE,NUMBER=OTHER SIZE
* = COMMERCIAL SIZE

NOMINAL SIZE	MILLIMETERS					INCHES					U.S.A. ANSI	AUSTRAL AS	JAPAN JIS	EURO- NORM 53	FRANCE NF A45-201	U.K. BS	GERMANY DIN 1025	ITALY UNI 5397
	A	B	C	D	R1	A	B	C	D	R1								
HE 450B 450	300	14.0	26.0	27.0	17.717	11.811	0.55	1.02	1.06				F A+=4 B+=3 C+=1.5 D+=2	F	F	F	F
HE 450M 478	307	21.0	40.0	27.0	18.819	12.087	0.83	1.57	1.06				F A+=5 B+=3 C+=1.5 D+=2.5	F	F	F	F
HE 500A 490	300	12.0	23.0	27.0	19.291	11.811	0.47	0.91	1.06				F A+=4 B+=3 C+=1.5 D+=2	F	F	F	F
HE 500B 500	300	14.5	28.0	27.0	19.685	11.811	0.57	1.10	1.06				F A+=4 B+=3 C+=1.5 D+=2	F	F	F	F
HE 500M 524	306	21.0	40.0	27.0	20.630	12.047	0.83	1.57	1.06				F A+=5 B+=3 C+=1.5 D+=3	F	F	F	F
HE 550A 540	300	12.5	24.0	27.0	21.260	11.811	0.49	0.94	1.06				F A+=5 B+=3 C+=1.5 D+=2	F	F	F	F
HE 550B 550	300	15.0	29.0	27.0	21.654	11.811	0.59	1.14	1.06				F A+=5 B+=3 C+=1.5 D+=2	F	F	F	F
HE 550M 572	306	21.0	40.0	27.0	22.520	12.047	0.83	1.57	1.06				F A+8=6 B+=3 C+=1.5 D+=3	F	F	F	F
HE 600A 590	300	13.0	25.0	27.0	23.228	11.811	0.51	0.98	1.06				F A+=5 B+=3 C+=1.5 D+=2	F	F	F	F
HE 600B 600	300	15.5	30.0	27.0	23.622	11.811	0.61	1.18	1.06				F A+=5 B+=3 C+=1.5 D+=2	F	F	F	F

Table 10-39 (Continued). Hot Rolled Wide Flange HE-Beams with Round Corners (EURONORM 53)

TABLE 10-39. HOT ROLLED WIDE FLANGE HE-BEAMS
WITH ROUND CORNERS (EURONORM 53)
BASIS; 1 IN = 25.4 MM

ORDER EXAMPLE:
LENGTH,HE100A EURONORM 53 + STEEL QUALITY PAGE NO. 6

THE NOMINAL SIZE IS NATIONAL STANDARD AS INDICATED
F=FIRST CHOICE,S=SECOND CHOICE,T=THIRD CHOICE,NUMBER=OTHER SIZE
* = COMMERCIAL SIZE

NOMINAL SIZE MILLIMETERS						INCHES						EURO-NORM 53	U.S.A. ANSI	AUSTRAL AS	JAPAN JIS	FRANCE NF A45-201	U.K. BS	GERMANY DIN 1025	ITALY UNI 5397
A	B	C	D	R1		A	B	C	D	R1									
HE 600M 620	305	21.0	40.0	27.0	24.409	12.008	0.83	1.57	1.06	F A+=8-6 B+=3 C+=1.5 D+=3	F			F		F	F
HE 650A 640	300	13.5	26.0	27.0	25.197	11.811	0.53	1.02	1.06	F A+=5 B+=3 C+=1.5 D+=3				F		F	
HE 650B 650	300	16.0	31.0	27.0	25.591	11.811	0.63	1.22	1.06	F A+=5 B+=3 C+=1.5 D+=2				F		F	
HE 650M 668	305	21.0	40.0	27.0	26.299	12.008	0.83	1.57	1.06	F A+=8-6 B+=3 C+=1.5 D+=3				F		F	
HE 700A 690	300	14.5	27.0	27.0	27.165	11.811	0.57	1.06	1.06	F A+=5 B+=3 C+=1.5 D+=2				F		F	
HE 700B 700	300	17.0	32.0	27.0	27.559	11.811	0.67	1.26	1.06	F A+=5 B+=3 C+=1.5 D+=2				F		F	
HE 700M 716	304	21.0	40.0	27.0	28.189	11.969	0.83	1.57	1.06	F A+=8-6 B+=3 C+=1.5 D+=3				F		F	
HE 800A 790	300	15.0	28.0	30.0	31.102	11.811	0.59	1.10	1.18	F A+=5 B+=3 C+=2 D+=2				F		F	
HE 800B 800	300	17.5	33.0	30.0	31.496	11.811	0.69	1.30	1.18	F A+=5 B+=3 C+=2 D+=2				F		F	
HE 800M 814	303	21.0	40.0	30.0	32.047	11.929	0.83	1.57	1.18	F A+=8-6 B+=3 C+=2 D+=3				F		F	

STEEL MATERIAL DATA

Table 10-39 (Continued). Hot Rolled Wide Flange HE-Beams with Round Corners (EURONORM 53)

TABLE 10-39. HOT ROLLED WIDE FLANGE HE-BEAMS
WITH ROUND CORNERS (EURONORM 53)
BASIS; 1 IN = 25.4 MM

ORDER EXAMPLE:
LENGTH,HE100A EURONORM 53 + STEEL QUALITY
PAGE NO. 7

THE NOMINAL SIZE IS NATIONAL STANDARD AS INDICATED
F=FIRST CHOICE,S=SECOND CHOICE,T=THIRD CHOICE,NUMBER=OTHER SIZE
* = COMMERCIAL SIZE

NOMINAL SIZE	MILLIMETERS					INCHES					U.S.A. ANSI	AUSTRAL AS	JAPAN JIS	EURO- NORM 53	FRANCE NF A45-201	U.K. BS	GERMANY DIN 1025	ITALY UNI 5397
	A	B	C	D	R1	A	B	C	D	R1								
HE 900A 890	300	16.0	30.0	30.0		35.039	11.811	0.63	1.18	1.18				F A+=5 B+=3 C+=2 D+=2	F	F	F	F
HE 900B 900	300	18.5	35.0	30.0		35.433	11.811	0.73	1.38	1.18				F A+=5 B+=3 C+=2 D+=2	F		F	F
HE 900M 910	302	21.0	40.0	30.0		35.827	11.890	0.83	1.57	1.18				F A+=8-6 B+=3 C+=2 D+=3	F		F	F
HE1000A 990	300	16.5	31.0	30.0		38.976	11.811	0.65	1.22	1.18				F A+=5 B+=3 C+=2 D+=2	F		F	F
HE1000B1000	300	19.0	36.0	30.0		39.370	11.811	0.75	1.42	1.18				F A+=5 B+=3 C+=2 D+=2	F		F	F
HE1000M1008	302	21.0	40.0	30.0		39.685	11.890	0.83	1.57	1.18				F A+=8-6 B+=3 C+=2 D+=3	F		F	F

Table 10-40. Hot-Finished Structural Hollow Square Sections (ISO 657/XIV)

TABLE 10-40 HOT-FINISHED STRUCTURAL
HOLLOW SQUARE SECTIONS (ISO 657/XIV)
BASIS; 1 IN = 25.4 MM

ORDER EXAMPLE:
LENGTH/SQUARE TUBE 50X50X4 ISO657/14 + ST QUAL

THE NOMINAL SIZE IS NATIONAL STANDARD AS INDICATED
F=FIRST CHOICE,S=SECOND CHOICE,T=THIRD CHOICE,NUMBER=OTHER SIZE
* = COMMERCIAL SIZE

ISO	MILLIMETERS A X A X C	R1	INCHES A X A X C	R1	U.S.A. ANSI B32.5	AUSTRAL AS	JAPAN JIS G3466	ISO 657/XIV	FRANCE NF A49.652	U.K. BS 4848/2	GERMANY DIN 59410	ITALY UNI
F	20 X 20 X 2.0	4.0	0.787 X 0.787 X 0.079	0.16	F			A+-0.5	S			
F	20 X 20 X 2.6	5.2	0.787 X 0.787 X 0.102	0.20	2.5F			C-0.4	F			
F	30 X 30 X 2.0	4.0	1.181 X 1.181 X 0.079	0.16	F			A+-0.5	F			
F	30 X 30 X 2.6	5.2	1.181 X 1.181 X 0.102	0.20	2.5F			C-0.4				
F	30 X 30 X 3.2	6.4	1.181 X 1.181 X 0.126	0.25	3F			C-0.4				
F	40 X 40 X 2.6	5.2	1.575 X 1.575 X 0.102	0.20	2.5F			A+-0.5	2F		2.9F	
F	40 X 40 X 3.2	6.4	1.575 X 1.575 X 0.126	0.25	3F			C-0.4		F	F	
F	40 X 40 X 4.0	8.0	1.575 X 1.575 X 0.157	0.31	F			C-12.5%				
F	50 X 50 X 3.2	6.4	1.969 X 1.969 X 0.126	0.25	3F		F	A+-0.5	2F	F	2.9F	
F	50 X 50 X 4.0	8.0	1.969 X 1.969 X 0.157	0.31	F			C-12.5%			F	
F	50 X 50 X 5.0	10.0	1.969 X 1.969 X 0.197	0.39	F			C-12.5%				
F	60 X 60 X 3.2	6.4	2.362 X 2.362 X 0.126	0.25	3F			A+-0.5	2F	F	2.9F	
F	60 X 60 X 4.0	8.0	2.362 X 2.362 X 0.157	0.31	F			C-12.5%			F	
F	60 X 60 X 5.0	10.0	2.362 X 2.362 X 0.197	0.39	F			C-12.5%				
F	70 X 70 X 3.2	6.4	2.756 X 2.756 X 0.126	0.25	3F			A+-1%		F		
F	70 X 70 X 3.6	7.2	2.756 X 2.756 X 0.142	0.28	3.5F		F	C-12.5%				
F	70 X 70 X 4.0	8.0	2.756 X 2.756 X 0.157	0.31	F			C-12.5%				
F	70 X 70 X 5.0	10.0	2.756 X 2.756 X 0.197	0.39	F			C-12.5%				
F	80 X 80 X 3.2	6.4	3.150 X 3.150 X 0.126	0.25	3F			A+-1%		F		
F	80 X 80 X 3.6	7.2	3.150 X 3.150 X 0.142	0.28	3.5F			C-12.5%				
F	80 X 80 X 4.0	8.0	3.150 X 3.150 X 0.157	0.31	F			C-12.5%				
F	80 X 80 X 5.0	10.0	3.150 X 3.150 X 0.197	0.39	F			C-12.5%				
F	80 X 80 X 6.3	12.6	3.150 X 3.150 X 0.248	0.50	6F			C-12.5%				
F	90 X 90 X 3.2	6.4	3.543 X 3.543 X 0.126	0.25	3F			A+-1%		F	4.5F	
F	90 X 90 X 3.6	7.2	3.543 X 3.543 X 0.142	0.28	3.5F			C-12.5%			5.6F	
F	90 X 90 X 4.0	8.0	3.543 X 3.543 X 0.157	0.31	F			C-12.5%				
F	90 X 90 X 5.0	10.0	3.543 X 3.543 X 0.197	0.39	F			C-12.5%				
F	90 X 90 X 6.3	12.6	3.543 X 3.543 X 0.248	0.50	6F			C-12.5%				
F	90 X 90 X 8.0	16.0	3.543 X 3.543 X 0.315	0.63	F			C-12.5%				
F	100 X 100 X 3.2	6.4	3.937 X 3.937 X 0.126	0.25	3F			A+-1%		F	4.5F	
F	100 X 100 X 4.0	8.0	3.937 X 3.937 X 0.157	0.31	F			C-12.5%			5.6F	
F	100 X 100 X 5.0	10.0	3.937 X 3.937 X 0.197	0.39	F		F	C-12.5%				
F	100 X 100 X 6.3	12.6	3.937 X 3.937 X 0.248	0.50	6F		F	C-12.5%				
F	100 X 100 X 8.0	16.0	3.937 X 3.937 X 0.315	0.63	F		4.5F	C-12.5%				
F	100 X 100 X 10.0	20.0	3.937 X 3.937 X 0.394	0.79	F			C-12.5%				
F	120 X 120 X 3.2	6.4	4.724 X 4.724 X 0.126	0.25	3F			A+-1%		F	4.5F	
F	120 X 120 X 5.0	10.0	4.724 X 4.724 X 0.197	0.39	F			C-12.5%			5.6F	
F	120 X 120 X 6.3	12.6	4.724 X 4.724 X 0.248	0.50	6F			C-12.5%			F	
F	120 X 120 X 8.0	16.0	4.724 X 4.724 X 0.315	0.63	F			C-12.5%				
F	120 X 120 X 10.0	20.0	4.724 X 4.724 X 0.394	0.79	F			C-12.5%				

Notes:
1. R1 is the maximum corner radius which is specified in ISO 657/14 standard to R1 = 2C up to sizes A ≤ 100 or A x B ≤ 100 x 150 mm. The corner radii for larger sizes is according to BS 4848/2.

2. Cold-Finished

Square and rectangular tubing are specified in ISO 4019 standard to the same, with a few exceptions, outside dimensions as shown here and with wall thicknesses of 1.2 1.6 2 2.6 3.2 4 5 6.3 7.1 8 10 12.5 mm.

STEEL MATERIAL DATA

Table 10-40 (Continued). Hot-Finished Structural Hollow Square Sections (ISO 657/XIV)

```
TABLE 10-40  HOT-FINISHED STRUCTURAL                          PAGE NO. 2
HOLLOW SQUARE SECTIONS (ISO 657/XIV)
BASIS: 1 IN = 25.4 MM

                                    ORDER EXAMPLE:
                                    LENGTH, SQUARE TUBE 50X50X4 ISO657/14 + ST QUAL

                                    THE NOMINAL SIZE IS NATIONAL STANDARD AS INDICATED
                                    F=FIRST CHOICE, S=SECOND CHOICE, T=THIRD CHOICE, NUMBER=OTHER SIZE
                                    * = COMMERCIAL SIZE
                                    U.S.A.  AUSTRAL  JAPAN              FRANCE   U.K.    GERMANY  ITALY
                                    ANSI    AS       JIS       ISO      NF       BS      DIN      UNI
                                    B32.5            G3466     657/XIV  A49-652  4848/2  59410
```

ISO	MILLIMETERS A × A × C	R1	INCHES A × A × C	R1	ANSI	ISO	NF	BS	DIN	UNI
F	140 × 140 × 3.6	7.2	5.512 × 5.512	0.28		A+-1%				
F	× 5.0	10.0		0.39		C-12.5%			5.6F	
F	× 6.3	12.6		0.50		C-12.5%			7.1F	
F	× 8.0	16.0		0.63		C-12.5%			8.8F	
F	× 10.0	20.0		0.79		C-12.5%				
F	150 × 150 × 4.0	8.0	5.906 × 5.906	0.31	4.5F	A+-1%				
F	× 5.0	10.0		0.39	F	C-12.5%		F		
F	× 6.3	12.6		0.50	6F	C-12.5%		F		
F	× 8.0	16.0		0.63	F	C-12.5%		F		
F	× 10.0	20.0		0.79	F	C-12.5%		F		
F	× 12.5	25.0		0.98	12F	C-12.5%		F		
F	× 16.0	32.0		1.26	F	C-12.5%		F		
F	160 × 160 × 4.0	8.0	6.299 × 6.299	0.31	F	A+-1%				
F	× 5.0	10.0		0.39	F	C-12.5%				
F	× 6.3	12.6		0.50	F	C-12.5%				
F	× 8.0	16.0		0.63	F	C-12.5%				
F	× 10.0	20.0		0.79	F	C-12.5%				
F	× 12.5	25.0		0.98	F	C-12.5%				
F	× 16.0	32.0		1.26	F	C-12.5%				
F	180 × 180 × 4.0	8.0	7.087 × 7.087	0.31	F	A+-1%		F	F	
F	× 5.0	10.0		0.39	F	C-12.5%		F	F	
F	× 6.3	12.6		0.50	6F	C-12.5%		F	F	
F	× 8.0	16.0		0.63	F	C-12.5%		F	F	
F	× 10.0	20.0		0.79	F	C-12.5%		F	F	
F	× 12.5	25.0		0.98	12F	C-12.5%		F	F	
F	× 16.0	32.0		1.26	F	C-12.5%		F	F	
F	200 × 200 × 5.0	10.0	7.874 × 7.874	0.39	F	A+-1%		F	F	
F	× 6.3	12.6		0.50	6F	C-12.5%		F	F	
F	× 8.0	16.0		0.63	F	C-12.5%		F	F	
F	× 10.0	20.0		0.79	F	C-12.5%		F	F	
F	× 12.5	25.0		0.98	12F	C-12.5%		F	F	
F	× 16.0	32.0		1.26	F	C-12.5%		F	F	
F	220 × 220 × 5.0	10.0	8.661 × 8.661	0.39	F	A+-1%		F	F	
F	× 6.3	12.6		0.50	6F	C-12.5%		F	F	
F	× 8.0	16.0		0.63	F	C-12.5%		F	F	
F	× 10.0	20.0		0.79	F	C-12.5%		F	F	
F	× 12.5	25.0		0.98	12F	C-12.5%		F	F	
F	× 16.0	32.0		1.26	F	C-12.5%		F	F	
F	250 × 250 × 5.9	11.8	9.843 × 9.843	0.46	6F	A+-1%		F	F	
F	× 6.3	12.6		0.50	F	C-12.5%		F	F	
F	× 8.0	16.0		0.63	F	C-12.5%		F	F	
F	× 10.0	20.0		0.79	F	C-12.5%		F	F	
F	× 12.5	25.0		0.98	12F	C-12.5%		F	F	
F	× 16.0	32.0		1.26	F	C-12.5%		F	F	
F	× 20.0	40.0		1.57		C-12.5%		F	F	

Notes:
1. R1 is the maximum corner radius which is specified in ISO 657/14 standard to R1 = 2C up to sizes A ≤ 100 or A × B ≤ 100 × 150 mm. The corner radii for larger sizes is according to BS 4848/2.
2. Cold-Finished
   Square and rectangular tubing are specified in ISO 4019 standard to the same, with a few exceptions, outside dimensions as shown here and with wall thicknesses of 1.2 1.6 2 2.6 3.2 4 5 6.3 7.1 8 10 12.5 mm.

STEEL MATERIAL DATA

Table 10-40 (Continued). Hot-Finished Structural Hollow Square Sections (ISO 657/XIV)

TABLE 10-40 HOT-FINISHED STRUCTURAL
HOLLOW SQUARE SECTIONS (ISO 657/XIV)
BASIS; 1 IN = 25.4 MM

PAGE NO. 3

ORDER EXAMPLE;
LENGTH,SQUARE TUBE 50X50X4 ISO657/I4 + ST QUAL

THE NOMINAL SIZE IS NATIONAL STANDARD AS INDICATED
F=FIRST CHOICE,S=SECOND CHOICE,T=THIRD CHOICE,NUMBER=OTHER SIZE
* = COMMERCIAL SIZE

ISO	MILLIMETERS A X A X C	R1	INCHES A X A X C	R1	U.S.A. ANSI B32.5	AUSTRAL AS	JAPAN JIS G3466	ISO 657/XIV	FRANCE NF A49-652	U.K. BS 4848/2	GERMANY DIN 59410	ITALY UNI
F	260 X 260 X 5.9	11.8	10.236 X10.236 X 0.232	0.46				A+-1%				
F	6.3	12.6	0.248	0.50				C-12.5%				
F	8.0	16.0	0.315	0.63				C-12.5%				7.1F
F	10.0	20.0	0.394	0.79				C-12.5%				8.8F
F	12.5	25.0	0.492	0.98				C-12.5%				11F
F	16.0	32.0	0.630	1.26				C-12.5%				
F	20.0	40.0	0.787	1.57				C-12.5%				
F	300 X 300 X 7.1	14.2	11.811 X11.811 X 0.280	0.56	7F			A+-1%				
F	8.0	16.0	0.315	0.63	F			C-12.5%				
F	10.0	20.0	0.394	0.79	F		6F	C-12.5%				
F	12.5	25.0	0.492	0.98	12F			C-12.5%		F		
F	16.0	32.0	0.630	1.26	F			C-12.5%		F		
F	20.0	40.0	0.787	1.57	F			C-12.5%				
F	25.0	50.0	0.984	1.97	F			C-12.5%				
F	350 X 350 X 8.0	16.0	13.780 X13.780 X 0.315	0.63	F			A+-1%				
F	10.0	20.0	0.394	0.79	F			C-12.5%		F		
F	12.5	25.0	0.492	0.98	12F			C-12.5%		F		
F	16.0	32.0	0.630	1.26	F			C-12.5%		F		
F	20.0	40.0	0.787	1.57	F			C-12.5%				
F	25.0	50.0	0.984	1.97	F			C-12.5%				
F	400 X 400 X 10.0	20.0	15.748 X15.748 X 0.394	0.79	F			A+-1%				
F	12.5	25.0	0.492	0.98	12F			C-12.5%		F		F
F	16.0	32.0	0.630	1.26	F			C-12.5%		F		F
F	20.0	40.0	0.787	1.57	F			C-12.5%				
F	25.0	50.0	0.984	1.97	F			C-12.5%				

Notes:

1. R1 is the maximum corner radius which is specified in ISO 657/I4 standard to R1 = 2C up to sizes A≤100 or A x B≤100 x 150 mm. The corner radii for larger sizes is according to BS 4848/2.

2. **Cold-Finished**

   Square and rectangular tubing are specified in ISO 4019 standard to the same, with a few exceptions, outside dimensions as shown here and with wall thicknesses of 1.2 1.6 2 2.6 3.2 4 5 6.3 7.1 8 10 12.5 mm.

STEEL MATERIAL DATA

Table 10-41. Hot-Finished Structural Hollow Rectangular Sections (ISO 657/XIV)

```
TABLE 10-41  HOT-FINISHED STRUCTURAL                                         PAGE NO. 1
HOLLOW RECTANGULAR SECTIONS (ISO 657/XIV)       ORDER EXAMPLE:
BASIS: 1 IN = 25.4 MM                           LENGTH,RECT TUBE 80X40X4 ISO657/14 + ST QUAL
```

B THE NOMINAL SIZE IS NATIONAL STANDARD AS INDICATED
F=FIRST CHOICE, S=SECOND CHOICE, T=THIRD CHOICE, NUMBER=OTHER SIZE
* = COMMERCIAL SIZE

ISO	MILLIMETERS A × B × C	R1	INCHES A × B	C	R1	U.S.A. ANSI B32.5	AUSTRAL AS	JAPAN JIS G3466	ISO 657/XIV	FRANCE NF A49-652	U.K. BS 4848/2	GERMANY DIN 59410	ITALY UNI
F	50 × 30 × 2.6	5.2	1.969 × 1.181	0.102	0.20	2.5F			A,B+-0.5 C=-0.4 C-12.5%	2F F	F F	2.9F F	
F	3.2	6.4		0.126	0.25	3F							
F	4.0	8.0		0.157	0.31	F							
F	60 × 40 × 3.2	6.4	2.362 × 1.575	0.126	0.25	3F			A+-1% B+-0.5 C-12.5%	2S 2.5S	F F	2.9F F	
F	3.6	6.4		0.142	0.25	3.5F							
F	4.0	8.0		0.157	0.31	F							
F	5.0	10.0		0.197	0.39	F							
F	70 × 40 × 3.2	6.4	2.756 × 1.575	0.126	0.25	3F			A+-1% B+-0.5 C-12.5%		F F	2.9F F	
F	4.0	6.4		0.157	0.25	F							
F	4.0	8.0		0.157	0.31	F							
F	5.0	10.0		0.197	0.39	F							
F	80 × 40 × 3.2	6.4	3.150 × 1.575	0.126	0.25	3F			A+-1% B+-0.5 C-12.5%	2F 2.5S	F F	2.9F F	
F	3.6	6.4		0.142	0.25	3.5F							
F	4.0	8.0		0.157	0.31	F							
F	5.0	10.0		0.197	0.39	F							
F	90 × 50 × 3.2	6.4	3.543 × 1.969	0.126	0.25	3F			A,B+-1% C-12.5%		F F	F F	
F	3.6	7.2		0.142	0.28	3.5F	F						
F	4.0	8.0		0.157	0.31	F							
F	5.0	10.0		0.197	0.39	F							
F	100 × 50 × 3.2	6.4	3.937 × 1.969	0.126	0.25	3F			A,B+-1% C-12.5%		F F	4.5F 5.6F	
F	3.6	7.2		0.142	0.28	3.5F							
F	4.0	8.0		0.157	0.31	F							
F	5.0	10.0		0.197	0.39	F							
F	100 × 60 × 3.2	6.4	3.937 × 2.362	0.126	0.25	3F			A,B+-1% C-12.5%		F F	F F	
F	3.6	7.2		0.142	0.28	3.5F							
F	4.0	8.0		0.157	0.31	F							
F	5.0	10.0		0.197	0.39	F							
F	6.3	12.6		0.248	0.50	6F							
F	120 × 60 × 3.2	6.4	4.724 × 2.362	0.126	0.25	3F			A,B+-1% C-12.5%		F F	4.5F 5.6F	
F	3.6	7.2		0.142	0.28	3.5F							
F	4.0	8.0		0.157	0.31	F							
F	5.0	10.0		0.197	0.39	F							
F	6.3	12.6		0.248	0.50	6F							
F	8.0	16.0		0.315	0.63	F							
F	120 × 80 × 3.2	6.4	4.724 × 3.150	0.126	0.25	3F			A,B+-1% C-12.5%		F F	F F	
F	4.0	8.0		0.157	0.31	F							
F	5.0	10.0		0.197	0.39	F							
F	6.3	12.6		0.248	0.50	6F							
F	8.0	16.0		0.315	0.63	F							
F	10.0	20.0		0.394	0.79	F							

Notes:

1. R1 is the maximum corner radius which is specified in ISO 657/14 standard to R1 = 2C up to sizes A ≤ 100 or A × B ≤ 100 × 150 mm. The corner radii for larger sizes is according to BS 4848/2.

2. Cold-Finished

   Square and rectangular tubing are specified in ISO 4019 standard to the same, with a few exceptions, outside dimensions as shown here and with wall thicknesses of 1.2 1.6 2 2.6 3.2 4 5 6.3
   7.1 8 10 12.5 mm.

Table 10-41 (Continued). Hot-Finished Structural Hollow Rectangular Sections (ISO 657/XIV)

TABLE 10-41 HOT-FINISHED STRUCTURAL
HOLLOW RECTANGULAR SECTIONS (ISO 657/XIV)
BASIS: 1 IN = 25.4 MM

ORDER EXAMPLE:
LENGTH, RECT TUBE 80×40×4 ISO657/14 + ST QUAL
PAGE NO. 2

THE NOMINAL SIZE IS NATIONAL STANDARD AS INDICATED
F=FIRST CHOICE, S=SECOND CHOICE, T=THIRD CHOICE, NUMBER=OTHER SIZE
* = COMMERCIAL SIZE

ISO	MILLIMETERS A × B × C	R1	INCHES A × B	× C	R1	U.S.A. ANSI B32.5	AUSTRAL AS	JAPAN JIS G3466	ISO 657/XIV	FRANCE NF A49-652	U.K. BS 4848/2	GERMANY DIN 59410	ITALY UNI
F	140 × 80 × 3.2	6.4	5.512 × 3.150	0.126	0.25				A,B+-1%	C=12.5%			
F	4.0	8.0		0.157	0.31				C=12.5%				
F	5.0	10.0		0.197	0.39				C=12.5%				
F	6.3	12.6		0.248	0.50				C=12.5%				
F	8.0	16.0		0.315	0.63				C=12.5%				
F	10.0	20.0		0.394	0.79				C=12.5%				
F	150 × 100 × 3.2	6.4	5.906 × 3.937	0.126	0.25	3F			A,B+-1%	C=12.5%			
F	4.0	8.0		0.157	0.31	F		4.5F	C=12.5%				
F	5.0	10.0		0.197	0.39				C=12.5%				
F	6.3	12.6		0.248	0.50	6F		6F	C=12.5%	F			
F	8.0	16.0		0.315	0.63				C=12.5%				
F	10.0	20.0		0.394	0.79				C=12.5%				
F	160 × 80 × 3.2	6.4	6.299 × 3.150	0.126	0.25				A,B+-1%	C=12.5%			
F	4.0	8.0		0.157	0.31				C=12.5%				
F	5.0	10.0		0.197	0.39				C=12.5%				
F	6.3	12.6		0.248	0.50				C=12.5%				
F	8.0	16.0		0.315	0.63				C=12.5%				
F	10.0	20.0		0.394	0.79				C=12.5%				
F	180 × 100 × 3.6	7.2	7.087 × 3.937	0.142	0.28	3.5F			A,B+-1%	C=12.5%			
F	5.0	10.0		0.197	0.39	F			C=12.5%				
F	6.3	12.6		0.248	0.50	6F			C=12.5%	F			
F	8.0	16.0		0.315	0.63	F			C=12.5%				
F	10.0	20.0		0.394	0.79	F			C=12.5%				
F	200 × 100 × 4.0	8.0	7.874 × 3.937	0.157	0.31	F		4.5F	A,B+-1%	C=12.5%			
F	5.0	10.0		0.197	0.39	F		6F	C=12.5%				
F	6.3	12.6		0.248	0.50	6F			C=12.5%	F			
F	8.0	16.0		0.315	0.63	F			C=12.5%	F			
F	10.0	20.0		0.394	0.79	F			C=12.5%	F			
F	12.5	25.0		0.492	0.98	12F			C=12.5%	F			
F	16.0	32.0		0.630	1.26	F			C=12.5%				
F	200 × 120 × 4.0	8.0	7.874 × 4.724	0.157	0.31	F			A,B+-1%	C=12.5%			5.6F
F	5.0	10.0		0.197	0.39	F			C=12.5%				7.1F
F	6.3	12.6		0.248	0.50	6F			C=12.5%	F			8.8F
F	8.0	16.0		0.315	0.63	F			C=12.5%	F			
F	10.0	20.0		0.394	0.79	F			C=12.5%	F			
F	12.5	25.0		0.492	0.98	12F			C=12.5%	F			
F	16.0	32.0		0.630	1.26	F			C=12.5%				
F	220 × 140 × 4.0	8.0	8.661 × 5.512	0.157	0.31				A,B+-1%	C=12.5%			F
F	5.0	10.0		0.197	0.39				C=12.5%				F
F	6.3	12.6		0.248	0.50				C=12.5%				F
F	8.0	16.0		0.315	0.63				C=12.5%				
F	10.0	20.0		0.394	0.79				C=12.5%				
F	12.5	25.0		0.492	0.98				C=12.5%				
F	16.0	32.0		0.630	1.26				C=12.5%				

Notes:

1. R1 is the maximum corner radius which is specified in ISO 657/14 standard to R1 = 2C up to sizes A ≤ 100 or A × B ≤ 100 × 150 mm. The corner radii for larger

2. **Cold-Finished**

   Square and rectangular tubing are specified in ISO 4019 standard to the same, with a few exceptions, outside dimensions as shown here and with wall thicknesses of 1.2 1.6 2 2.6 3.2 4 5 6.3

STEEL MATERIAL DATA

Table 10-41 (Continued). Hot-Finished Structural Hollow Rectangular Sections (ISO 657/XIV)

```
TABLE 10-41 HOT-FINISHED STRUCTURAL                                              PAGE NO. 3
HOLLOW RECTANGULAR SECTIONS (ISO 657/XIV)
BASIS; 1 IN = 25.4 MM

                                                   ORDER EXAMPLE;
                                                   LENGTH,RECT TUBE 80X40X4 ISO657/14 + ST QUAL

                                              THE NOMINAL SIZE IS NATIONAL STANDARD AS INDICATED
                                              F=FIRST CHOICE,S=SECOND CHOICE,T=THIRD CHOICE,NUMBER=OTHER SIZE
                                              * = COMMERCIAL SIZE
                                              U.S.A.   AUSTRAL  JAPAN                        U.K.               GERMANY  ITALY
                                              ANSI     AS       JIS              ISO         BS                 DIN      UNI
                                              B32.5             G3466            657/XIV     4848/2             59410
                                                                        FRANCE
                                                                        NF A49-652

  I
  S                        INCHES
  O  MILLIMETERS
     A  X  B  X  C   R1    A   X    B    X   C      R1
F  250 X 150 X  5.0  10.0  9.843 X 5.906    0.197   0.39    F               A,B+-1%
F              6.3   12.6                   0.248   0.50    6F              C=12.5%
F              8.0   16.0                   0.315   0.63    F               C=12.5%
F             10.0   20.0                   0.394   0.79    F               C=12.5%
F             12.5   25.0                   0.492   0.98    12F             C=12.5%
F             16.0   32.0                   0.630   1.26    F               C=12.5%

F  300 X 200 X  5.9  11.8  11.811 X 7.874   0.232   0.46    F               A,B+-1%
F              6.3   12.6                   0.248   0.50    6F              C=12.5%
F              8.0   16.0                   0.315   0.63    F               C=12.5%
F             10.0   20.0                   0.394   0.79    F               C=12.5%
F             12.5   25.0                   0.492   0.98    12F             C=12.5%
F             16.0   32.0                   0.630   1.26    F               C=12.5%
F             20.0   40.0                   0.787   1.57    F               C=12.5%

F  400 X 200 X  7.1  14.2  15.748 X 7.874   0.280   0.56    7F              A,B+-1%
F              8.0   16.0                   0.315   0.63    F               C=12.5%
F             10.0   20.0                   0.394   0.79    F               C=12.5%
F             12.5   25.0                   0.492   0.98    12F             C=12.5%
F             16.0   32.0                   0.630   1.26    F               C=12.5%
F             20.0   40.0                   0.787   1.57    F               C=12.5%
F             25.0   50.0                   0.984   1.97    F               C=12.5%

F  450 X 250 X  8.0  16.0  17.717 X 9.843   0.315   0.63    F               A,B+-1%
F             10.0   20.0                   0.394   0.79    F               C=12.5%
F             12.5   25.0                   0.492   0.98    F               C=12.5%
F             16.0   32.0                   0.630   1.26    F               C=12.5%
F             20.0   40.0                   0.787   1.57    F F             C=12.5%
F             25.0   50.0                   0.984   1.97    F               C=12.5%

F  500 X 300 X 10.0  20.0  19.685 X11.811   0.394   0.79    F               A,B+-1%
F             12.5   25.0                   0.492   0.98    F               C=12.5%
F             16.0   32.0                   0.630   1.26    F               C=12.5%
F             20.0   40.0                   0.787   1.57    F F             C=12.5%
F             25.0   50.0                   0.984   1.97    F               C=12.5%
```

Notes:

1. R1 is the maximum corner radius which is specified in ISO 657/14 standard to R1 = 2C up to sizes A ≤ 100 or A x B ≤ 100 x 150 mm. The corner radii for larger sizes is according to BS 4848/2.

2. Cold-Finished

   Square and rectangular tubing are specified in ISO 4019 standard to the same, with a few exceptions, outside dimensions as shown here and with wall thicknesses of 1.2 1.6 2 2.6 3.2 4 5 6.3 7.1 8 10 12.5 mm.

Table 10-42A. Hot Rolled Low Carbon Steel Plate, Sheet, and Strip (ISO 3573, 630, 1052, 3575)

		1	2	3	4	5	6	7
		\multicolumn{4}{c}{Ductile Grades}	\multicolumn{3}{c}{Structural Grades}		
Quality		Commercial	Drawing	Deep Drawing	Deep Drawing Special Killed	Commercial		
(1) (2) Physical Properties	Tensile Rm (MPa) Yield Re (MPa) Elong A (%) Rockwell HR30T	25-	-430 28-	-370 28-	-390	330-440 195- 22-	360-470 210- 20-	410-520 230- 17-
(3) Chemical Composition (ISC)	Carbon (%) Manganese (%) Silicon (%) Phosphor (%) Sulphur (%)	C= -0.15 Mn= -0.60 P= -0.05 S= -0.05	C= -0.12 Mn= -0.50 P= -0.04 S= -0.04	C= -0.10 Mn= -0.45 P= -0.03 S= -0.03	C= -0.08 Mn= -0.45 P= -0.03 S= -0.03	C= -0.15 P= -0.05 S= -0.05	C= -0.17 P= -0.05 S= -0.05	C= -0.20 P= -0.05 S= -0.05
ISC		ISC 3573 HR 1	ISC 3573 HR 2	ISC 3573 HR 3	ISC 3573 HR 4	ISC 630 Fe 33	ISC 630 Fe 37-A Fe 37-B Fe 37 C Fe 37 D	ISC 630 Fe 42 A Fe 42 B Fe 42 C Fe 42 D
EURONORM		EU 46 Fe KC1	EU 46 FeKC2	EU 46 FeK11 FeK12 FeK2	EU 46 FeK3 FeK4	EU 47 Fe 34B1, 2, 3 Fe34D2, 3 Fe33-C	EU 47 Fe37A Fe37B1, 2, 3 Fe37C2, 3 Fe37D2, 3	EU 47 Fe42A Fe42B1, 2, 3 Fe42C2, 3 Fe42D2, 3
NORTH AMERICA (5)		ASTM - A 569 (AISI-CQ)	ASTM - A 621 (AISI-DQ)		ASTM - A 622 (AISI-DQ Spec. Killed)	ASTM - A 570 Grade A	ASTM - A 570 Grade B	ASTM - A 570 Grade C
JAPAN		JIS G 3131 SPHC		JIS G 3131 SPHD	JIS G 3131 SPHE	JIS G 3101 SS34		JIS G 3101 SS41
GERMANY		DIN 1624 St 0, St1	DIN 1624 St2	DIN 1624 St3	DIN 1624 St4		DIN 1623 TUSt37 WUSt37-2 USt37-2 RSt37-2	DIN 1623 USt42-2 RSt42-2
FRANCE		NF A36- 301 0C	NF A36- 301 1C	NF A36- 301 2C	NF A36- 301 3C	NF A35-501 A34 A33	NF A35-501 E24 (A37)	NF A35-501 E26 (A42) E30 (A47)
UNITED KINGDOM (BS 1449)		HR15, HS15 HR4, HS4	HR14, HS14	HR3, HS3	HR2, HS2 HR1, HS1		Grade 34/20	Grade 37/23
ITALY		UNI 5867 FeP10	UNI 5867 FeP11	UNI 5867 FeP12	UNI 5867 FeP13	UNI 7070 Fe34A Fe34B Fe34C	UNI 7070 Fe37A Fe37B Fe37C Fe37D	UNI 7070 Fe42A Fe42B Fe42C Fe42D
AUSTRALIA		AS 1594 HRC			AS 1594 HRD	AS 1594 HR 200 AS 1405 (180)	AS 1405 210	AS 1594 HR240

NOTES:
1. 1 MPa = 1 N/mm² ≈ 0.1 kgf/mm² · 1000 psi = 6.894757 MPa ≈ 7 MPa.
2. Rm, Re and A5 are ISO symbols for tensile strength, yield strength and elongation.
3. Unless otherwise noted steels fall within chemical composition limits shown for that column.
4. 1 oz/ft² = 305 g/m²
5. Structrual steels in each column may differ in tensile strength and/or chemistry but are equivalent in terms of yield strength.

STEEL MATERIAL DATA

Table 10-42A (*Continued*). Hot Rolled Low Carbon Steel Plate, Sheet, and Strip (ISO 3573, 630, 1052, 3575)

8	9	10	11	12	13	14	15	16
				\multicolumn{5}{c}{Hot-Dip Zinc Coated Sheets (0.25-5 mm Thick) (4)}
				Commercial	Lock Forming	Drawing	Deep Drawing	Deep Drawing Special Killed
510-610 250- 17-	490-620 290- 15-	590-750 330- 11-	690-870 360- 7-			-430 24-	-410 26-	-410 29-
C= -0.20 P= -0.05 S= -0.05	C≈0.30 P= -0.06 S= -0.05	C≈0.40 P= -0.06 S= -0.05	C≈0.50 P= -0.06 S= -0.05	C= -0.15 Mn= -0.60 P= -0.05 S= -0.05	C= -0.12 Mn= -0.60 P= -0.04 S= -0.04	C= -0.12 Mn= -0.50 P= -0.04 S= -0.04	C= -0.10 Mn= -0.45 P= -0.03 S= -0.03	C= -0.08 Mn= -0.45 P= -0.03 S= -0.03
ISC 630 Fe 52-B Fe 52-C Fe 52-D	ISC 1052 Fe 50-1 Fe 50-2	ISC 1052 Fe 60-1 Fe 60-2	ISC 1052 Fe 70-2	ISC 3575 Z1, ZF1 ISC 3575 Hot Dipped Coatings Zinc: Z700 (700 g/m2 total both sides) Z600, Z450, Z350, Z275, Z200, Z100, Z001 (No min.) Zinc-Iron ZF180, ZF100, ZF001 (No min.)	ISC 3575 Z1A, ZF1A	ISC 3575 Z2, ZF2	ISC 3575 Z3, ZF3	ISC 3575 Z4, ZF4
EU 47 Fe52C2,3 Fe52D2,3	EU 47 Fe 50-1 Fe 50-2	EU 47 Fe 60-1 Fe 60-2	EU 47 Fe 70-1 Fe 70-2	EU 46 Fe K01		EU 46 Fe K02	EU 46 Fe K11 Fe K12 Fe K2	EU 46 Fe K3 Fe K4
ASTM- A 570 Grade D	ASTM- A 570 Grade E	AISI 1040	AISI 1050	ASTM A526 ASTM A525 ZINC COATING CLASSES (IN g/m²) 720, 640, 565, 500, 430, 350 1.25 COMMERCIAL = 275, LIGHT COMMERCIAL = 183	ASTM A527	ASTM A528		ASTM- A642
	JIS G 3101 SS50	JIS G 3101 SS55		JIS G 3302 SPG 1, SPG 3C SPG 2C	JIS G 3302 SPG 2S SPG 3S	JIS G 3302 SPG 2L SPG 3L	Zinc Coatings No1 = 183 g/m² No2 = 244 g/m² No3 = 381 g/m²	JIS G 3302 SPG 2D
DIN 1623 St52-3	DIN 1623 St 50-2	DIN 1623 St 60-2	DIN 1623 St 70-2	DIN 1624 St 0 DIN 50 961 Zinc Coatings: gal Zn 3, gal Zn 5, gal Zn 8, gal Zn 12, gal Zn 25, gal Zn 40 (No indicates min. zinc coating thickness in micrometres)	DIN 1624 St 1	DIN 1624 St 2	DIN 1624 St 3	DIN 1624 St 4
NF A35-501 E36 (A52)	NF A35-501 A50	NF A35-501 A60	NF A35-501 A70	NF A36-321 Class I NFA36-320 Zinc Coating: T.P.G. =400g/m², T.C.G. =450 g/m² Zinc Coating	NF A36-321 Class II	NF A36-321 Class III	NF A36-321 Class IV (Aging)	NF A36-321 Class V (non-aging)
Grade 43/25	Grade 43/28	Grade 50/35	Grade 46/40	BS 2989 Hot Dipped Zinc Coatings: 1A (610-763 g/m²), 1B (549-610 g/m²), 1C (458-549 g/m²) 2A (275-428 g/m²), 2B (214-365 g/m²)	HR15, HS15 HR14, HS14, HR4, HS4	HR3, HS3	HR2, HS2, HR1, HS1
UNI 7070 Fe52B Fe52C Fe52D	UNI 7070 Fe 50	UNI 7070 Fe 60	UNI 7070 Fe 70	UNI 5753 Fe Z0 Hot Dipped Zinc Coating No 1 = 198g/m² No 2 = 275g/m²	UNI 5753 Fe Z1	UNI 5753 Fe Z2 No 3 = 350 g/m² No 4 = 504 g/m²	UNI 5753 Fe Z3	
HR340	AS 1594 HR 280			AS 1397 GC Hot dipped zinc coatings:		AS 1397 GD Z100, Z200, Z300, Z430, Z550, ZF 100, ZF 200		

Table 10-42B. Cold Reduced Low Carbon Steel Sheets and Strips (ISO 3574, 1111)

	1	2	3	4	5	6	7
	\multicolumn{4}{c}{Cold Rolled}	\multicolumn{3}{c}{(4) Cold Reduced Tinplate and Blackplate}		
Quality	Commercial	Drawing	Deep Drawing	Deep Drawing Special Killed	Soft		
(1) (2) Physical Properties — Tensile Rm (MPa), Yield Re (MPa), Elong A (%), Rockwell HR30T	30-	-370 34-	-350 36-	-340	-52	48-56	54-61
(3) Chemical Composition (ISC) — Carbon (%), Manganese (%), Silicon (%), Phosphor (%), Sulphur (%)	C= -0.15 Mn= -0.60 P= -0.05 S= -0.05	C= -0.12 Mn= -0.50 P= -0.04 S= -0.04	C= -0.10 Mn= -0.45 P= -0.04 S= -0.03	C= -0.08 Mn= -0.45 P= -0.03 S= -0.03			
ISC	ISC 3574 CR 1	ISC 3574 CR 2	ISC 3574 CR 3	ISC 3574 CR 4	ISC 1111 T 50	ISC 1111 T 52	ISC 1111 T 57
					ISC 1111 Tin Coatings Hot Dipped Tinplate: H 12/12 (24 g/m^2), H 14/14, Electrolytic Tinplate: E 2.8/2.8, E 5.6/5.6, E 8. El. Tinplate: D 8.4/2.8, D 11.2/2.8, D 11.2/5.6,		
EUROCNORM	EU 32 FePC1	EU 32 FePC2		EU 32 FePC3	EU 77 A	EU 77 B	
					EU 77 Tin Coatings Hot Dipped Tinplate: F 24 (24 g/m^2), F 30 Electrolytic Tinplate: E 1 (5.6 g/m^2), E 2 (11.2 g Diff. Coated El Tinplate: E 3-1 (16.8 & 5.6 g/m^2) One Sided Tinplate: E 1-0 (5.6 & 0 g/m^2), E 2-0		
NORTH AMERICA	ASTM A366 CQ (AISI CR CQ)	ASTM A619 DR (AISI CR DQ)		ASTM A620 DQ	ASTM A623 T-1	ASTM A623 T-2	ASTM A623 T-3
					AISI Hot Dipped Tinplate (g/m^2): Common Cokes = 19, Standard Cokes = 24 Best Cokes = 27, Kanners Sp Cokes = 31 1A Charcoal = 40, 2A Charcoal = 52 ASTM A624 Electrolytic Tinplate (g/m^2): #10 = 1.1/1.1, #25 = 2.8/2.8, #50 = 5.6/5.6, #75 = #75/25 = 8.4/2.8, #100/25 = 11.2/2.8, #100/50 = 11		
JAPAN	JIS G 3141 SPCC		JIS G 3141 SPCD	JIS G 3141 SPCE	JIS G 3303 T-1	JIS G 3303 T-2	JIS G 3303 T-3
					Blackplate Grade SPB Standard Tin Coatings Conform to ISC 1111 Hot Dipped Tinplate: SPTH #110 (H12/12), #125, # Electrolytic Tinplate: SPTE #25 (E2.8/2.8), #50, SPTE-D #50/25 (D5.6/2.8),		
GERMANY	DIN 1623 St 12	DIN 1623 St 13		DIN 1623 St 14	DIN 1616 A	DIN 1616 B	
	03 = standard surface finish, 05 = best finish				Finish Groups W3, W2, W1 (highest quality) Coatings according to EU77 Standard		
FRANCE	NF A36-401 TC	NF A36-401 XE ZE		NF A36-401 XES ZES	NF A36-150 A	NF A36-150 B	
	X = standard surface finish, Z = best finish				Coatings according to EU77 standard		
UNITED KINGDOM		BS 1449 CR4, CS4	BS 1449 CR3, CS3	BS 1449 CR2, CS2 CR1, CS1	BS 2920 T1A, T1B	BS 2920 T2	BS 2920 T3
					Coatings same as ISO 1111 standard		
ITALY	UNI 5866 FeP00	UNI 5866 Fe P01	UNI 5866 Fe P02	UNI 5866 FePC3 Fe P04	UNI 5755 A	UNI 5755 B	
					Coatings according to EU 77 standard		
AUSTRALIA	AS 1595-I CRC		AS 1595-I CRD	AS 1595-I CRE	AS 1517 T1	AS 1517 T2	AS 1517 T3
					Hot dipped: H25, H30, H35; Electrolytic: ED2 Elec. Diff: D05/02, D10/02, D10/05, D15/02...10		

NOTES:
1. 1 MPa = 1 N/mm² ≈ 0.1 kgf/mm² · 1000 psi = 6.894757 MPa ≈ 7 MPa
2. Rm, Re and A5 are ISO symbols for tensile strength, yield strength and elongation.
3. Unless otherwise noted steels fall within chemical composition limits shown for that column.
4. 1 lb/base box = 22.46 g/m²
5. 1 oz/ft² = 305 g/m²

STEEL MATERIAL DATA

Table 10-42B (Continued). Cold Reduced Low Carbon Steel Sheets and Strips (ISO 3574, 1111)

8	9	10	11	12	13	14
(0.15-0.49 mm Thick)		\multicolumn{5}{c}{Cold Reduced Sheet and Strip to Hardness (Temper) Classifications (0.4-5 mm thick)}
	Hard	Annealed	Commercial	Quarter Hard (1/4 hard)	Half Hard (1/2 hard)	Full Hard
57-68	66-73					
ISC 1111 T 61, T 65 H 15/15, H 17/17 4/8.4, E 11.2/11.2 D 15.1/5.6	ISC 1111 T 70					
EU 77 C /m²), E 3 (16.8 g/m²), E4 (22.4g/), E 4-1, E 4-2	EU 77 D					
ASTM A623 T-4, T-5 8.4/8.4, #100 =11. 2/5.6, #135/25 =	ASTM A623 T-6 2/11.2, #50/25 15.1/2.8	AISI DQ	AISI CQ	AISI CR PQ 1/4 hard ASTM A109 No. 3 Temper	AISI CR PQ 1/2 hard ASTM A109 No. 2 Temper	AISI CR PQ full hard ASTM A109 No. 1 Temper
JIS G 3303 T-4, T-5 135, #150 #75, #100 #75/25, #75/50, #	JIS G 3303 T-6 100/25, etc.	JIS G 3141 A D = Dull finish, B = Bright finish	JIS G 3141 S, 8	JIS G 3141 4	JIS G 3141 2	JIS G 3141 1
DIN 1616 C	DIN 1616 D	DIN 1624 G GD = Dull Finish, GBK = Bright Finish, RP, RPG = Best Bright Finish	DIN 1624 LG, K32	DIN 1624 K40	DIN 1624 K50	DIN 1624 K60, K70
NF A36-150 C	NF A36-150 D		NF A FO	NF A FOT1	NF A FOT2	NF A FOT4
BS 2920 T4, T4CA, T5CA	BS 2920 T6CA	BS 1449 A P, M = Matt Finish, BR = Bright Finish, PL, MF, SF = Best Bright Finish	BS 1449 SP	BS 1449 QH	BS 1449 HH	BS 1449 H
UNI 5755 C	UNI 5755 D					
AS 1517 T4, T5 T4CA, T5CA , E05, E10, E15, E20, E25, E27 D20/02···15, D25/02···20, D27/02···20	AS 1517 T6, T6CA	AS 1595-II Temper 6 No 1 = Dull finish, No 2 = Bright Finish, No 3 = Best Bright Finish	AS 1595-II Temper 4 Temper 5	AS 1595-II Temper 3	AS 1595-II Temper 2	AS 1595-II Temper 1

Table 10-42C. Carbon Steels for Wires and Rods (EURONORM 16)

(1) (2) (3)	1	2	3	4	5	6	7	8
PHYSICAL PROPERTIES (4) (5) Tensile Rm (MPa) Yield Re (MPa) Elong A5 (%)	COMMERCIAL		LOW CARBON					
Chemical Composition (in %) (to EU 16) Carbon Silicon Manganese Phosphorous Sulphur Nitrogen	C≤0.06 Mn=0.25-0.50 P≤0.035 S≤0.035 N≤0.007	C≤0.08 Mn=0.25-0.50 All EURONORM Prefixes 3 N≤0.007	C≤0.10 Mn=0.25-0.60 Designation N≤0.007	C=0.08-0.13 Mn=0.30-0.60 N≤0.007	C=0.13-0.18 Si≤0.35 Mn=0.30-0.60 N≤0.007	C=0.18-0.23 Si≤0.35 Mn=0.40-0.70 P≤0.060 S≤0.050 N≤0.007	C=0.23-0.28 Si=0.15-0.35 All All EURONORM Prefixes 1 N≤0.008	C=0.28-0.33 All All EURONORM Designation All
ISO								
EURONORM 16	3CD5	1CD6 3CD6	1CD8 2CD8 3CD8	1CD10 3CD10	1CD15 3CD15	2CD20 3CD20	2CD25 3CD25	2CD30 3CD30
North America (AISI)	1005	1006	1008	1010	1015	1020	1025	1030
	ASTM A227 Hard Drawn CL I (1570-1800 MPa); CL II (1810-2040 MPa) - ASTM A229 Oil Tempered CL I (1620-1830 MPa);							
Japan		JIS G3505 SWRM6	JIS G3505 SWRM8	JIS G3505 SWRM10 SWRM12	JIS G3505 SWRM15 SWRM17	JIS G3505 SWRM20 SWRM22	JIS G3506 SWRH27	JIS G3506 SWRH32
	JIS G3521 Hard Drawn Steel Wire to the following 3 classes: SWA (1270-1470 MPa); SWB (1470-1720 MPa); SWC (1720-1960)							
Germany (DIN 17140)	D6-2 D5-1	D8-2 D7-1	D9-1	D12-2	D15-2 Hardness 1	D20-2 Hardness 2	D26-2 Hardness 2½	
	DIN 17223 Round Spring Steel Wire - Quality Specification 6 Classes: A (1420 - 1710 MPa); B (1720 - 2010 MPa);							
France (NF A35-051)	FM5-2 FM5-3	FM6-2 FM6-3	FM8-1 FM8-3	FM10-2 FM10-3	FM15-1 FM15-3	FM18-1 FM18-3 FM20-1 FM20-3	FM26-1 FM26-3	FM32-3
	NF A47-301 Round Drawn Steel Wire for Springs: Class B1 (1760-2000 MPa); Class C1 (2000-2200 MPa)							
United Kingdom (BS 970)	015A03	030A04 040A04 050A04		040A10 050A10 060A10	040A15 050A15 060A15	040A20 050A20	060A25 060A27	060A30 060A32
	BS 5216 Patented Cold Drawn Steel Spring Wire - Codes NS, HS, Grade 1 (1370-1570 MPa); Grade 2 (1570-1770 MPa); BS 2803 Oil Hardened and Tempered Steel Wire for Springs - Grades I, II and III (1622-1776 MPa)							
Italy				UNI 5331 C10	UNI 5332 C16	UNI 5332 C20		UNI 5332 C30
	UNI 3823 Round Steel Wire for Springs - Types I, II, III, IV - Class A (1270-1470 MPa); B (1470-1720 MPa);							
Australia (DR 72134) (Grade AS 1135)		S 1006	S 1008 K 1008	S 1010 K 1010	S 1015 K 1012	S 1020 K 1020	S 1025 K 1026	S 1030 K 1030
	AS 1472 Carbon Steel Spring Wire - Soft Drawn (C = 0.65-0.75%); Oil Hardened (C = 0.55-0.85%) Class 1 (1620-1830 MPa);							

NOTES:
1. 1 MPa = 1 N/mm² ≈ 0.1 kgf/mm² · 1000 psi = 6.894757 MPa ≈ 7 MPa
2. Rm, Re and A5 are ISO symbols for tensile strength, yield strength and elongation.
3. Unless otherwise noted steels fall within chemical composition limits shown for that column.
4. Physical properties are shown for reference only and may not correspond to qualities listed in that column.
5. Tensile strength values shown for finished wire products are valid for 2 mm diameters.

STEEL MATERIAL DATA

Table 10-42C (*Continued*). Carbon Steels for Wires and Rods (EURONORM 16)

9	10	11	12	13	14	15	16	17	18
MEDIUM CARBON			HIGH CARBON AND STRENGTH						
C=0.33-0.39 $P \leq 0.050$ $S \leq 0.045$	C=0.38-0.43 All EURONORM Designation Prefixes 2	C=0.43-0.48	C=0.48-0.53	C=0.53-0.58	C=0.58-0.63	C=0.63-0.68	C=0.73-0.78	C=0.83-0.88	C=0.93-0.98
2CD35 3CD35	2CD40 3CD40	2CD45 3CD45	2CD50 3CD50	2CD55 3CD55	2CD60 3CD60	2CD65 3CD65	3CD75 (2CD70) (3CD70)	3CD85 (3CD80)	3CD95 (3CD90)
1035	1040	1045	1050	1055	1060	1065	1070	1080	1086

CL II (1830-2040) - ASTM A228 Music Wire (1940-2150 MPa)   ASTM A230 Oil Tempered Valve Spring Quality (1650-1790 MPa)

| JIS G3506 SWRH37 | JIS G3506 SWRH42A SWRH42B | JIS G3506 SWRH47A SWRH47B | JIS G3506 SWRH52A SWRH52B | JIS G3506 SWRH57A SWRH57B | JIS G3506 SWRH62A SWRH62B | JIS G3506 SWRH67A SWRH67B | JIS G3506 SWRH72A SWRH72B SWRH77A | JIS G3506 SWRH82A SWRH82B | |

MPa) - JIS G3522 Piano Wires Class SWPA (1810-2010 MPa); SWPB (2010-2210 MPa); SWPV (1720-1860 MPa)

| D35-2 Hardness 3 | | D45-2 Hardness 4 | | D55-2 Hardness 5 | | D65-2 Hardness 6 | D75-2 Hardness 7 | D85-2 Hardness 8 | D95-2 Hardness 9 |

C (2020-2210 MPa);   II (2110 - 2350 MPa);   FD (1620 - 1760 MPa);   VD (1520 - 1620 MPa)

| FM36-3 | FM38-3 FM40-3 | FM42-3 FM46-3 | FM50-3 FM52-3 | FM56-3 FM58-3 | FM60-3 FM62-3 | FM66-3 FM68-3 | FM72-3 MF76-3 FM78-3 | FM80-3 FM82-3 FM86-3 | |
| 060A35 060A37 | 060A40 060A42 | 060A47 | 060A52 | 060A57 | 060A62 | 060A67 | 060A72 070A72 | 060A86 060A83 | 060A99 060A96 |

Grade 3 (1770-1970 MPa) - Codes ND, HD Grade 2 (1370-1770 MPa); Grade 3 (1770-1970 MPa) - Code M Grade 4 (1970-2120 MPa); Grade 5 (2120-2270 MPa)

| UNI 5332 C35 | UNI 3823 Type I | UNI 3823 Type I | UNI 3823 Type I | UNI 3823 Type II | UNI 3823 Type II | UNI 3823 Type III | UNI 3823 Type III | UNI 3823 Type IV | UNI 3823 Type IV |

C (1720-1960 MPa); D (1960-2210 MPa); E ($\geq$ 2210 MPa)

| S1035 K1035 | S1040 K1039 | S1045 K1045 | | | | | | | |

Hard Drawn (C=0.45-0.85%) - Range #1 (1600-1840 MPa); #2 (1810-2040 MPa); #3 (2010-2270 MPa)

Table 10-42D. Steel Types Used for Tubing (ISO 3304, 3305, 3306, 2937, 2938)

		1	2	3	4	5	6	7	8
	Physical Properties	\multicolumn{5}{c|}{Seamless and Welded Precision Tubes}	\multicolumn{3}{c}{Plain End As-Welded Precision Tubes}			
1	Condition Tensile Rm (MPa) Yield ReL(MPa) Elongation A(%)	BK 400- - 6-	BK 420- - 6-	BK 450- - 6-	BK 520- - 5-	BK 600- - 4-	KM 320- - 10-	KM 330- - 8-	KM 400- - 7-
2	Condition Tensile Rm(MPa) Yield ReL(MPa) Elongation A(%)	BKW 350- - 10-	BKW 370- - 10-	BKW 400- - 9-	BKW 450- - 8-	BKW 550- - 7-	GKM, GZF 270- - 27-	GKM, GZF 320- - 27-	GKM, GZF 340- - 26-
3	Condition Tensile Rm(MPa) Yield ReL(MPa) Elongation A(%)	GBK, GZF 270- - 27-	GBK, GZF 320- - 27-	GBK, GZF 340- - 26-	GBK, GZF 400- - 24-	GBK, GZF 480- - 23-	NKM, NZF 280- 155- 25-	NKM, NZF 320- 195- 25-	NKM, NZF 360- 215- 24-
4	Condition Tensile Rm(MPa) Yield ReL(MPa) Elongation A(%)	NBK, NZF 280- 155- 25-	NBK, NZF 320- 195- 25-	NBK, NZF 360- 215- 24-	NBK, NZF 410- 235- 22-	NBK, NZF 490- 285- 21-			
	Chemical Composition (ISO) Carbon % Silicon % Manganese % Phosphorus % Sulphur %	$C \leq 0.13$ - $Mn \leq 0.6$ $P \leq 0.05$ $S \leq 0.05$	$C \leq 0.16$ - $Mn \leq 0.7$ $P \leq 0.05$ $S \leq 0.05$	$C \leq 0.17$ $Si \leq 0.35$ $Mn \leq 0.8$ $P \leq 0.05$ $S \leq 0.05$	$C \leq 0.21$ $Si \leq 0.35$ $Mn \leq 1.2$ $P \leq 0.05$ $S \leq 0.05$	$C \leq 0.23$ $Si \leq 0.35$ $Mn \leq 1.5$ $P \leq 0.05$ $S \leq 0.05$	$C \leq 0.13$ - $Mn \leq 0.6$ $P \leq 0.05$ $S \leq 0.05$	$C \leq 0.16$ - $Mn \leq 0.7$ $P \leq 0.05$ $S \leq 0.05$	$C \leq 0.17$ $Si \leq 0.35$ $Mn \leq 0.8$ $P \leq 0.05$ $S \leq 0.05$
ISO		ISO 3304 ISO 3305 R28	ISO 3304 ISO 3305 R33	ISO 3304 ISO 3305 R37	ISO 3304 ISO 3305 R42	ISO 3304 ISO 3305 R50	ISO 3306 R28	ISO 3306 R33	ISO 3306 R37
NORTH AMERICA		ASTM A519 and A512 MT 1010 MD, MDSA NORM-MD-SR	ASTM A519 and A512 MT 1015 MD, MDSA NORM-MD-SR	ASTM A519 and A512 MT 1020 MD, MDSA NORM-MD-SR	ASTM A519 and A512 MTX 1015 MD, MDSA NORM-MD-SR	ASTM A519 and A512 MTX 1020 MD, MDSA NORM-MD-SR	ASTM A512 MT 1010 MF NORM-MD-SA	ASTM A512 MT 1015 MD, MDSA NORM-MD-SA	ASTM A512 MTX 1015 MD, MDSA NORM-MD-SA
JAPAN							JIS G3445 STKM 11A		
GERMANY		DIN 2391 St35	DIN 2393 St34-2	DIN 2393 St37-2	DIN 2391 St45 DIN 2393 St42-2	DIN 2391 St52, St55 DIN 2393 St52-3	DIN 1626 St33	DIN 2394 SSt2	DIN 2394 SSt2 DIN 1626 St 34
FRANCE		Pr A49-341 TS 28-a	Pr A49-341 TS 34-a	NF A49-310 TU 37-b Pr A49-341 TS 37-a	Pr A49-341 TS 42-a	NF A49-310 TU 52-b Pr A49-341 TS 47-a			
UNITED KINGDOM		BS 1775 CDS-11 CEW-11 BS 980 CDS-1 CEW-1 CEW-2	BS 1775 CDS-13	BS 1775 CDS-16 CEW-16 BS 980 CDS-2 CDS-3 ERW-2	BS 1775 CDS-20 CEW-24 CEW-23	BS 1775 CEW-28 BS 980 CDS-9	BS 980 ERW-1 BS 1775 ERW-11	BS-1775 HFW-13 ERW-13	BS-1775 HFW-16 ERW-16
ITALY			UNI 2897 Fe 35-1	UNI 2897 Fe 35-2	UNI 2897 Fe 45-2	UNI 2897 Fe 52-2			UNI 7091 Fe 34
AUSTRALIA (AS 1450)		CDS 170 ERW 170 HFS 170	CDS 200 ERW 200 CDS 250 ERW 250 HFS 200	CDS 300 ERW 300	CDS 350 ERW 350	CDS 370 CDS 540 CDS 430 ERW 380	CEW 170 ERW 170 EFW 170	CEW 200 ERW 200 EFW 200	CEW 250 ERW 250 EFW 250

NOTES:
1. I MPa = 1 N/mm² ≈ 0.1 kgf/mm² · 1000 psi = 6.894757 MPa ≈ 7 MPa.
2. Rm, Re and A are ISO symbols for tensile strength, yield strength and elongation.
3. Unless otherwise noted steels fall within chemical composition limits shown for that column.

STEEL MATERIAL DATA

Table 10-42D (*Continued*). Steel Types Used for Tubing (ISO 3304, 3305, 3306, 2937, 2938)

9	10	11	12	13	14	15	16	17
\multicolumn{7}{c	}{Mechanical Application Tubes - Seamless}	\multicolumn{2}{c	}{Hollow Steel Bars - Seamless}	
KM 460- - 6-	KM 520- - 5-	320-440 195- 25-	360-480 215- 24-	410-530 235- 22-	490-610 285- 21-	540-660 275- 20-	Hot Finished 490-610 335- 21-	Hot Finished 490-640 275- 21-
GKM, GZF 400- - 24-	GKM, GZF 480- - 23-						Normalized 490-610 345- 21-	Normalized 490-610 275- 21-
NKM,NZF 410- 235- 22-	NKM, NZF 490- 285- 21-							
C ≤ 0.21 Si ≤ 0.35 Mn ≤ 1.2 P ≤ 0.05 S ≤ 0.05	C ≤ 0.23 Si ≤ 0.35 Mn ≤ 1.5 P ≤ 0.05 S ≤ 0.05	C ≤ 0.16 - Mn=0.5-0.7 P ≤ 0.05 S ≤ 0.05	C ≤ 0.17 Si ≤ 0.35 Mn=0.4-0.8 P ≤ 0.045 S ≤ 0.045	C ≤ 0.21 Si ≤ 0.35 Mn=0.4-1.2 P ≤ 0.045 S ≤ 0.045	C ≤ 0.23 Si≤ 0.35 Mn=0.8-1.5 P ≤ 0.045 S ≤ 0.045	C=0.32-0.39 Si=0.15-0.4 Mn=0.5-0.8 P ≤ 0.035 S ≤ 0.035	C ≤ 0.2 Si ≤ 0.5 Mn ≤ 1.6 P ≤ 0.045 S ≤ 0.045	C=0.32-0.39 Si=0.15-0.4 Mn=0.5-0.8 P ≤ 0.035 S ≤ 0.035
ISO 3306 R42	ISO 3306 R50	ISO 2937 T51	ISO 2937 T54	ISO 2937 T59	ISO 2937 T518	ISO 2937 C35	ISO 2938 1	ISO 2938 2
ASTM A512 2 MT 1020 MD-MDSA NORM-MD-SA	ASTM A512 MTX 1020 MD-MDSA NORM-MD-SA	ASTM A519 MT 1010 N	ASTM A519 MT 1015 N	ASTM A519 MT 1020 N	ASTM A519 MTX 1020 N	ASTM A519 1045 A	ASTM A618 I II	ASTM A519 1035 A OR N
	JIS G3444 STK 50 STK 55	JIS G3444 STK 30		JIS G3445 STKM 12A STKM 12B STKM 12C	JIS G3445 STKM 13A STKM 13B STKM 13C	JIS G3445 STKM 17A STKM 17C	JIS G3445 STKM 18A STKM 18B STKM 18C	JIS G3445 STKM 16A STKM 16C
DIN 1626 St 37 DIN 2393 St 37-2	DIN 1626 St 42 St 52 DIN 2393 St 42 St 52	DIN 1629 St 00	DIN 17175 St 35.8 DIN 1629 St 35.4	DIN 17175 St 45.8 DIN 1629 St 45.4	DIN 1629 St 52.4	DIN 1629 St 55.4	DIN 1629 St 52.4	DIN 1629 St 55.4
			NF A49-311 TU 37-b		NF A49-311 TU 52-b	NF A49-311 TU XC35 TU 56-b	Pr A49-312 TU52-b	Pr A49-312 TU XC 35
BS 1775 ERW-20	BS 1775 HFW-23 ERW-23	BS 980 CDS-1 BS 1775 CDS-13	BS 980 CDS-2 CDS-3 BS 1775 CDS-16	BS 980 CDS-9 BS 4360 43C 43D 43E	BS 980 CDS-10 BS 4360 50B 50C 50D	BS 980 CDS-6 BS 4360 55C 55E	BS 1775 HFS-20	BS 980 ERW-3
UNI 7091 Fe 37	UNI 7091 Fe 42 Fe 52		UNI 663 Fe 35-1	UNI 663 Fe 45-1	UNI 6403 C20	UNI 663 Fe 55-1		
CEW 350 ERW 350 EFW 350	CEW 300 ERW 300 EFW 300		CDS 200	CDS 250	CDS 300	CDS 430	HFS 350	HFS 300

Nomenclature—North America
MD — Mandrel Drawn
MDSA — Mandrel Drawn – Soft Annealed
NORM-MD-SA — Normalized – Mandrel Drawn, Stress Relieved
HR — Hot Rolled   N — Normalized   A — Annealed

Nomenclature — ISO
BK — Cold Finished/Hard
BWK — Cold Finished/Soft
GBK — Annealed with Controlled Atmosphere
NBK — Normalized with Controlled Atmosphere
NZF — Normalized and Pickled

KM — As Welded and Sized
GKM — Annealed with Controlled Atmosphere
GZF — Annealed and Pickled
NKM — Normalized with Controlled Atmosphere

Table 10-43. Steel for Structural and General Engineering Purposes (ISO 630 and 1052)

	(1)(2)	1	2	3	4	5	6	7	8
PHYS. PROP.	Tensile Rm(MPa) Yield Re(MPa) Elong. A5 (%)	-510	360-440 230- 26-	410-490 250- 23-	430-510 270- 23-	490-610 350- 22-	490-610 290- 20-	590-710 330- 15-	690-830 360- 11-
CHEMICAL COMPOSITION (ISO)	CARBON		C=0.20%max	C=0.25%max	C=0.25%max	C=0.22%max			
ISO 630 and 1052		ISO 630 Fe 33	ISO 630 Fe 37-A Fe 37-B Fe 37-C Fe 37-D	ISO 630 Fe 42-A Fe 42-B Fe 42-C Fe 42-D	ISO 630 Fe 44-A Fe 44-B Fe 44-C Fe 44-D	ISO 630 Fe 52-B Fe 52-C Fe 52-D	ISO 1052 Fe 50-1 Fe 50-2	ISO 1052 Fe 60-1 Fe 60-2	ISO 1052 Fe 70-2
EURONORM 25		Fe 310	Fe 360-A Fe 360-BFU Fe 360-BFN Fe 360-C Fe 360-D		Fe 430-A Fe 430-B Fe 430-C Fe 430-D	Fe 510-B Fe 510-C Fe 510-D Fe 510-DD	Fe 490-1 Fe 490-2	Fe 590-2	Fe 690-2
NORTH AMERICA 3)		ASTM A284 Grades A, B, C ASTM A570 Grades A, B ASTM A283 Grades A, B, C	ASTM A284 Grade D ASTM A570 Grade C ASTM A283 Grade D	ASTM A36 ASTM A570 Grade D	ASTM A570 Grade E ASTM A572 Grade 42	ASTM A572 Grade 50	ASTM A572 Grade 42	ASTM A572 Grade 50	ASTM A572 Grade 55
JAPAN (JIS G 3101)		SS34		SS41		SS50	SS55		
GERMANY (DIN 17100)		St33-1 St33-2 USt34-1 or RSt34-1 USt34-2 or RSt34-2	USt37-1 RSt37-1 USt37-2 RSt37-2 St37-3	USt42-1 RSt42-1 USt42-2 RSt42-2 St42-3	RSt46-2 St46-3	St52-3	St50-1 St50-2	St60-1 St60-2	St70-2
FRANCE (NF A35-501)		A33 A33-2 A34-1 A34-2	E24-1 E24-2 E24-3 E24-4 (A37)	E26-1 E26-2 E26-3 E26-4 (A42)	E30-2 E30-3 E30-4 (A47)	E36-2 E36-3 E36-4 (A52)	A50-1 A50-2	A60-1 A60-2	A70-2
UNITED KINGDOM		BS1449 Grade 34/20	BS1449 Grade 37/23 BS 4360 Grades 40 A, B, C	BS1449 Grade 43/25 BS 4360 Grades 40D,E	BS4360 Grade 43 D,E	BS4360 Grade 50 A,B,C,D	BS4360 Grade 50 A,B,C,D	BS4360 Grade 55 C,E	
ITALY (UNI 7070)		Fe 33	Fe37A Fe37C Fe37B Fe37D	Fe42A Fe42C Fe42B Fe42D	Fe44A Fe44B Fe44C Fe44D	Fe52B Fe52C Fe 52D	Fe50	Fe60	Fe70
AUSTRALIA		AS 1405 Grade 310	AS 1405 Grade 370	AS 1204 Grade 250	AS1204 Grade 300	AS 1204 Grade 350 AS 1205 Grade 350	AS 1204 Grade 400	AS 1204 Grade 500 AS 1205 Grade 500	

NOTES:
1. 1 MPa = 1 N/mm² ≈ 0.1 kgf/mm² · 1000 psi = 6.894757 MPa ≈ 7 MPa.
2. Rm, Re and A5 are ISO symbols for tensile strength, yield strength and elongation.
3. Unless otherwise noted steels fall within chemical composition limits shown for that column.

STEEL MATERIAL DATA

(Continued from page 10-6)

A third type of plain-end (as-welded and sized) precision steel tube is covered in ISO 3306, and the nominal diameters are, to a large extent, the same as those shown in Table 10-24.

Table 10-25. Seamless (SL) and Welded (WE) Stainless Steel Tubes (ISO 1127, 5252)—The International and national standards for stainless steel tubes have been based on customary inch standards, and nominal diameters from 10.2 to 114.3 mm have been listed and compared worldwide in Table 10-25.

Tolerances

Table 10-22B (ISO 5252), on page 10-6, gives the ISO tolerance classes for tubular products.

Table 10-22A—Outside Diameter: According to ISO D_2: ± 1% with a minimum of ± 0.5 mm (ISO 2937).

Wall Thickness

a. $\frac{C}{A} \leqslant 3\%$ according to ISO T_1: ± 15%

b. $3 < \frac{C}{A} \leqslant 10\%$ according to ISO T_2: ± 12.5%

c. $\frac{C}{A} > 10\%$ $A \leqslant 168.3$ according to ISO T_2: ± 12.5%
$A > 168.3$ according to ISO T_3: ± 10%

When C is the specified thickness in millimeters and A is the specified outside diameter in mm.

Table 10-23—For hollow steel bars ISO standard tolerances for outside diameters and wall thicknesses are given in the column marked ISO 2938 in Table 10-23.

Table 10-24—The ISO standard tolerances for outside and inside precision steel tube diameters are as shown in columns marked ISO 3304 and ISO 3305 in Table 10-24. The ISO allowances for the wall thicknesses are T_3: ± 10% and a minimum of ± 0.12 mm.

Table 10-25—Stainless steel tubes of the austenitic type are specified in the German standards for ISO tolerance Classes D_2 and T_3 as shown in Table 10-25.

Material

Steel tubes are generally supplied in steel qualities listed in Table 10-42D for regular steel and in Tables 10-49 and 10-50 for stainless steels. Stainless seamless steel tubes for pressure applications are specified in ISO 2605 and for regular steel qualities in ISO 2604, Parts 2 and 3.

Qualities and Coatings

Steel tubes are supplied with a number of different finishes as specified in Table 10-42D. Both the ISO and the North American finish symbols are explained in the footnote of the above table.

Order Examples

Steel tubular products in metric standard sizes are specified with the name of the product, national standards number, tolerance, steel designation, and any special requirements. Typical examples are as follows:

Example 1—2000 m welded precision steel tubes in steel R37, as per ISO 3305, annealed in controlled atmosphere (GBK) outside diameter 25 mm, thickness 2 mm.

Example 2—Hollow bar according to ISO 2938, steel Grade 1, normalized, D63 mm, ID 32 mm, 18.2 Mg, in random lengths.

Example 3—10000 m tubes according to ISO 2937, steel Grade TS1, 60.3 mm D, 5 mm thick.

European Steel Section Standards

Most of the steel section types shown in Tables 10-26 through 10-41 are not covered in any ISO standards. They are for the most part used in Europe and standardized in the various EURONORM publications quoted in each table. An extensive cooperation in standardization of sizes as well as tolerances is being accomplished throughout Europe. Steel material types used for the above sections are for the most part from Table 10-43.

WORLD STEEL DESIGNATION CROSS REFERENCES*

Description of World Designation Systems for Steel

Tables 10-42 through 10-54 compare the steel types chemically defined in an ISO standard to the equivalent steel types in the national standards of the major industrial countries. The first tables, 10-42A, B, C, and D, are arranged according to product types. Table 10-43 covers steel types classified according to strength properties.

Tables 10-44 through 10-51—These are grouped according to chemistry and special process features. Tables 10-52 through 10-54 cover steel and iron qualities in castings.

As a guide for specifying, corresponding national standards numbers are listed for each type of steel in the tables. In cases where further details about a type of steel are required, the international or national standard should be studied.

International Material Cross References

Table 10-42A—Hot-rolled low carbon steel plate, sheet, and strip in ductile grades, structural grades, and zinc-coated qualities are given in a worldwide comparison in this table.

Table 10-42B—Cold-reduced low carbon steel sheets and strip in commercial and drawing qualities, tin plate, black plate, and cold-reduced hardness classifications are shown.**

---

*See World Ferrous Material Index at the end of this section.

**The text is continued on page 10-137.

Table 10-44. Carbon Steels (ISO 683: Part 1 and 4)

(1) (2)	1	2	3	4	5	6	7
PHYSICAL PROPERTIES (HEAT TREATED) — Tensile R_m(MPa) / Yield R_e(MPa) / Elong. A_5(%)					540-690 / 360- / 19-	580-730 / 390- / 18-	620-760 / 420- / 17-
CHEMICAL COMPOSITION (ISO, AISI) — Carbon / Manganese / Silicon / Phosphor / Sulphur (3)	C=0.08% max / Mn=0.25-0.40% / Si=0.15-0.40% all / P=0.050% max and S=0.050% max for ISO types without suffix / P=0.035% max for ISO types eb, e and ea / S=0.030-0.050% (type eb), S=0.035% max (type e), S=0.020-0.035% (type ea)	C=0.08-0.13% / Mn=0.30-0.60%	C=0.13-0.18% / Mn=0.30-0.60%	C=0.18-0.23% / Mn=0.30-0.60%	C=0.22-0.29% / Mn=0.40-0.70%	C=0.27-0.34% / Mn=0.50-0.80%	C=0.32-0.39% / Mn=0.50-0.80%
ISO 683/1 and 4					C25 / C25eb / C25e / C25ea	C30 / C30eb / C30e / C30ea	C35 / C35eb / C35e / C35ea
EURONORM 83					1C25 / 2C25 / 3C25		1C35 / 2C35 / 3C35
NORTH AMERICA (AISI, SAE)	1006	1010	1015	1020	1025	1030	1035
JAPAN (JIS G 4051)		S10C	S15C	S20C	S25C	S30C	S35C
GERMANY (DIN 17200)		DIN 17210 / C10	DIN 17210 / C15	C22 / Ck22			C35 / Ck35 / Cm35
FRANCE (NF A 35-551)		CC10 / XC10	XC12 / XC18	CC20	XC25	XC32	CC35
UNITED KINGDOM (BS 970)	030A04 / 040A04	040A10 / 050A10	040A15 / 050A15	040A20 / 050A20	060A25 / 060A27	060A30 / 060A32	060A35 / 060A37
ITALY		UNI 5331 / C10	UNI 5332 / C16	UNI 5332 / C20		UNI 5332 / C30	UNI 5332 / C35 / [UNI 5333] C33
AUSTRALIA (DR72135 Grade AS1135)	S1006 / K1008	S1010 / K1010	S1015 / K1012	S1020 / K1020	S1025 / K1026	S1030 / K1030	S1035 / K1035

NOTES:
1. 1 MPa = 1 N/mm² ≈ 0.1 kgf/mm² · 1000 psi = 6.894757 MPa ≈ 7 MPa
2. R_m, R_e and A_5 are ISO symbols for tensile strength, yield strength and elongation.
3. Steels in each column may differ in tensile strength and chemistry but are equivalent in terms of yield strength.

… STEEL MATERIAL DATA

Table 10-44 (*Continued*). Carbon Steels (ISO 683: Part 1 and 4)

8	9	10	11	12	13	14	15	16
660-800 450- 16-	700-840 480- 14-	720-880 510- 13-	780-930 540- 12-	830-980 570- 11-				
C=0.37-0.44% Mn=0.50-0.80%	C=0.42-0.50% Mn=0.50-0.80%	C=0.47-0.55% Mn=0.60-0.90%	C=0.52-0.60% Mn=0.60-0.90%	C=0.57-0.65% Mn=0.60-0.90%	C=0.60-0.70% Mn=0.60-0.90%	C=0.65-0.75% Mn=0.60-0.90%	C=0.75-0.88% Mn=0.60-0.90%	C=0.85-0.98% Mn=0.60-0.90%
C40 C40eb C40e C40ea	C45 C45eb C45e C45ea	C50 C50eb C50e C50ea	C55 C55eb C55e C55ea	C60 C60eb C60e C60ea				
	1C45 2C45 3C45		1C55 2C55 3C55	1C60 2C60 3C60				
1040	1045	1050	1055	1060	1065	1070	1080	1090
S40C	S45C	S50C	S55C	S58C				
	C45 Ck45 Cm45		C55 Ck55 Cm55	C60 Ck60 Cm60				
XC38 XC42	CC45 XC48		XC55		XC65	XC70	XC80	
060A40 060A42	060A47 080A47	080A52 060A52	080A57 060A57	080A62 060A62	080A67 060A67	080A72 060A72	080A78 080A83 070A78 060A78	080A86 060A96
UNI 5332 C40 UNI 5333] C38]	UNI 5332 C45 UNI 5333] C43]	UNI 5332 C50 UNI 5333] C48]		UNI 5332 C60	UNI 3545 C60	UNI 3545 C70	UNI 3545 C75	UNI 3545 C90 C100
S1040 K1039	S1045 K1045	S1050 K1050	K1055	S1058 K1060	K1065	S1070 K1070	XK1082	

Table 10-45. Alloy Direct Hardening Steels (ISO 683: Part 2, 4, 5, 6, 7, 8)

(1)	(2)	1	2	3	4	5	6
PHYSICAL PROPERTIES (HEAT TREATED) (5)	Tensile Rm(MPa) Yield Re (MPa) Elong. A5 (%)	880-1080 690- 12-	980-1180 790- 11-	1080-1270 880- 10-	780-930 590- 13-	1080-1270 880- 10-	880-1080 690- 12-
CHEMICAL COMPOSITION (ISO)	(3) Carbon Silicon Mangan. Chrom. Molyb. Nickel	C=0.22-0.29% Si=0.15-0.40% Mn=0.50-0.80% Cr=0.90-1.20% Mo=0.15-0.30%	C=0.30-0.37% Si=0.15-0.40% Mn=0.50-0.80% Cr=0.90-1.20% Mo=0.15-0.30%	C=0.38-0.45% Si=0.15-0.40% Mn=0.50-1.00% Cr=0.90-1.20% Mo=0.15-0.30%	C=0.25-0.32% Si=0.15-0.40% Mn=1.30-1.65%	C=0.28-0.35% Si=0.15-0.40% Mn=0.40-0.70% Cr=2.80-3.30% Mo=0.30-0.50% Ni=0.30 max	C=0.30-0.37% Si=0.15-0.40% Mn=0.60-0.90% Cr=0.90-1.20%
ISO R683	(4)	Part 2 and 4 Grade 1	Part 2 and 4 Grade 2,2a,2b	Part 2 and 4 Grade 3,3a,3b	Part 5 Grade 1,1a,1b	Part 6 Grade 1	Part 7 Grade 1,1a,1b
EURONORM 83		A25CrMo4 B25CrMo4	34CrMo4	42CrMo4	28Mn6	32CrMo12	34Cr4
NORTH AMERICA (AISI,SAE)	1	4130	4135 4137	4140 4142	1527 1526		5132 5130
JAPAN		JIS G 4105 SCM 2	JIS G 4105 SCM 3	JIS G 4105 SCM 4	JIS G 4106 SMn1		JIS G 4104 SCr2
GERMANY (DIN 17200)		25CrMo4	34CrMo4 34CrMoS4	43CrMo4 42CrMoS4	28Mn6	32CrMo12	34Cr4 34CrS4
FRANCE (NF A35-551)		25CD4	35CD4 30CD4	42CD4	35M5 20M5		32C4
UNITED KINGDOM (BS 970)			708A37	708M40 708A42 709M40	150M28		530A32
ITALY		UNI 5332 25CrMo4 30CrMo4	UNI 5333 38CrMo4 UNI 5332 35CrMo4	UNI 5332 40CrMo4			UNI 5332 35CrMo5
AUSTRALIA (AS 1444)		4130		4140	DR72134 Grade AS 1135 XK1325 XK1330		5132

NOTES:
1. 1 MPa = 1N/mm² ≈ 0.1 kgf/mm² · 1000 psi = 6.894757 MPa ≈ 7 MPa
2. Rm, Re and A5 are ISO symbols for tensile strength, yield strength and elongation.
3. Unless otherwise noted steels fall within chemical composition limits shown for that column.
4. ISO classification of steel is in its fifth draft and it is similar to the Euronorm designation.
5. Steel within each column may differ sightly in chemistry, but compositions are essentially equivalent for similar response to heat treatment.
6. Physical properties are shown for reference only.

STEEL MATERIAL DATA

Table 10-45 (*Continued*). Alloy Direct Hardening Steels (ISO 683: Part 2, 4, 5, 6, 7, 8)

7	8	9	10	11	12	13
930-1130 740- 11-	980-1240 780- 11-	1030-1230 830- 10-	1030-1230 830- 10-	1180-1370 980- 9-	1230-1420 1030- 9-	1230-1420 1030- 9-
C=0.30-0.41% Si=0.15-0.40% Mn=0.60-0.90% Cr=0.90-1.20%	C=0.38-0.45% Si=0.15-0.40% Mn=0.60-0.90% Cr=0.90-1.20%	C=0.37-0.44% Si=0.15-0.40% Mn=0.70-1.0% Cr=0.40-0.60% Mo=0.15-0.30% Ni=0.40-0.70%	C=0.36-0.43% Si=0.15-0.40% Mn=0.50-0.80% Cr=0.60-0.90% Mo=0.15-0.30% Ni=0.70-1.0%	C=0.32-0.39% Si=0.15-0.40% Mn=0.50-0.80% Cr=1.30-1.70% Mo=0.15-0.30% Ni=1.30-1.70%	C=0.26-0.33% Si=0.15-0.40% Mn=0.30-0.60% Cr=1.80-2.20% Mo=0.30-0.50% Ni=1.80-2.20%	C=0.30-0.37% Si=0.15-0.40% Mn=0.30-0.60% Cr=1.60-2.00% Mo=0.25-0.45% Ni=3.70-4.20%
Part 7 Grade 2,2a,2b	Part 7 Grade 3,3a,3b	Part 8 Grade 1,1a,1b	Part 8 Grade 2,2a,2b	Part 8 Grade 3,3a,3b	Part 8 Grade 5,5a,5b	Part 8 Grade 6,6a,6b
37Cr4	41Cr4	40NiCrMo2	39NiCrMo3	35CrNiMo6	30CrNiMo8	34NiCrMo16
5135	5140 5145	8740 8637 8640 8642	(9840)	4340 Cr=0.70-0.90%		
JIS G 4104 SCr4	JIS G 4104 SCr	JIS G 4103 SNCM 6	JIS G 4103 SNCM 8 Ni=1.6-2.0%			
37Cr4 37Cr S4	41Cr4 41Cr S4			34CrNiMo6	30CrNiMo8	
38C4	42C4	40NCD3	35NCD6	35NCD6	30CND8	35NCD16
530A36		945A40	(640A35)	816M40 817M40	823M30	835M30
UNI 5333 36CrMn4	UNI 5332 40Cr4	UNI 5332 40NiCrMo2	UNI 5332 38NiCrMo4 UNI 5333 40NiCrMo4	UNI 5332 40NiCrMo7		UNI 5332 35NiCrMo15
	5140 5145	8740 8640		4340 Cr=0.70-0.90%		

Table 10-46. Free Cutting Steels (ISO 683: Part 9)

		1	2	3	4	5	6	7
		\multicolumn{5}{c	}{NON-HARDENING TYPES}		CASE HARDENING
PHYSICAL PROPERTIES (HEAT TREATED)	Tensile Rm(MPa) (1)(2) Yield Re(MPa) Elong. As(%)	490-780 390- 8-	510-800 410- 7-	510-800 410- 7-	540-830 430- 7-	540-830 430- 7-	490-780 390- 8-	490-780 390- 8-
CHEMICAL COMPOSITION	Carbon (3) Silicon Mangan. Phosph. Sulph. Lead	C=0.13% max Si=0.05% max Mn=0.60-1.20% P=0.11% max S=0.18-0.25%	C=0.14% max Si=0.05% max Mn=0.90-1.30% P=0.11% max S=0.24-0.32%	C=0.14% max Si=0.05% max Mn=0.90-1.30% P=0.11% max S=0.24-0.32% Pb=0.15-0.35%	C=0.15% max Si=0.05% max Mn=1.0-1.50% P=0.11% max S=0.30-0.40%	C=0.15% max Si=0.05% max Mn=1.0-1.50% P=0.11% max S=0.30-0.40% Pb=0.15-0.35%	C=0.07-0.13% Si=0.15-0.40% Mn=0.50-0.90% P=0.06% max S=0.15-0.25%	C=0.07-0.13% Si=0.15-0.40% Mn=0.50-0.90% P=0.06% max S=0.15-0.25% Pb=0.15-0.35%
ISO R683/9	(4)	Grade 1	Grade 2	Grade 2Pb	Grade 3	Grade 3Pb	Grade 4	Grade 4Pb
EURONORM 87		10S22	11SMn28	11SMnPb28	12SMn35	12SMnPb35	10S20	10SPb20
NORTH AMERICA (AISI or SAE J403f)		1211	1212	12L13	1214	12L14	1108	11L08
JAPAN (JIS G4804)		SUM21	SUM22	SUM22L		SUM24L	SUM12	SUM23L
GERMANY (DIN 1651)		9S20	9SMn28	9SMnPb28	9SMn36	9SMnPb36	10S20	10SPb20
FRANCE			NF A35-561 S250	NF A35-561 S250Pb	NF A35-561 S300	NF A35-561 S300Pb	NF A35-562 10F1	
UNITED KINGDOM (BS 970)		220M07	230M07		240M07		045M10	
ITALY (UNI 4838)		10S20	9SMn23	9SMnPb23			10S22	
AUSTRALIA (DR 72134 Grade AS1135)			XS1112 S1214	S12L14	CS1100	XS11L12		

NOTES:
1. I MPa = 1 N/mm² ≈ 0.1 kgf/mm² · 1000 psi = 6.894757 MPa ≈ 7 MPa
2. Rm, Re and A5 are ISO symbols for tensile strength, yield strength and elongation.
3. Unless otherwise noted steels fall within chemical composition limits shown for that column.
4. ISO classification of steel is in its fifth draft and it is similar to the Euronorm designation.
5. Physical properties are shown for reference only.

STEEL MATERIAL DATA

Table 10-46 (*Continued*). **Free Cutting Steel (ISO 683: Part 9)**

8	9	10	11	12	13	14
TYPES			HARDENING TYPES			
510-800 410- 7-	540-830 410- 7-	570-760 390- 14-	620-810 420- 14-	740-930 510- 12-	650-840 450- 11-	660-780 360- 9-
C=0.09-0.15% Si=0.15-0.40% Mn=0.90-1.20% P=0.06% max S=0.15-0.25%	C=0.14-0.20% Si=0.15-0.40% Mn=0.50-0.90% P=0.06 % max S=0.15-0.25%	C=0.32-0.39% Si=0.15-0.40% Mn=0.50-0.90% P=0.06 % max S=0.15-0.25%	C=0.32-0.39% Si=0.15-0.40% Mn=0.90-1.20% P=0.06 % max S=0.15-0.25%	C=0.32-0.39% Si=0.15-0.40% Mn=1.30-1.65% P=0.06 % max S=0.15-0.25%	C=0.42-0.50% Si=0.15-0.40% Mn=0.50-0.90% P=0.06 % max S=0.15-0.25%	C=0.57-0.65% Si=0.10-0.40% Mn=0.50-0.90% P=0.06 % max S=0.15-0.25%
Grade 5	Grade 6	Grade 7	Grade 8	Grade 9	Grade 10	
22SMn20	17S20	35S20	35SMn20		45S20	60S20
1117	1115	1138	1140	1137 1141 11L41 Pb=0.15-0.35%	1146	
	SUM32			SUM41	SUM43	
		35S20			45S20	60S20
NF A35-562 13MF4	NF A35-562 18MF5		NF A35-562 35MF6		NF A35-562 45MF6	
212M14	Mn=1.20-1.60 S = 0.10-0.18 214M15		212M36		212M44	
		35SMn 10		216M36		
XS1115 S=0.08-0.13%		K1138 S=0.08-0.13%		K1137 S=0.08-0.13%	K1146 S=0.08-0.13%	

Table 10-47. Nitriding and Case Hardening Steels (ISO 683: Part 10 and 11)

		1	2	3	4	5	6	7	8
		\multicolumn{8}{c}{NITRIDING STEELS}							
PHYSICAL PROPERTIES (HEAT TREATED)	Tensile Rm(MPa) (1)(2) Yield Re(MPa) Elong. A5(%)	1080-1270 880- 10-	1270-1470 1080- 8-	780-930 590- 14-	930-1130 740- 12-	490-830 290- 13-	590-930 340- 12-	640-980 390- 10-	830-1180 540- 10-
CHEMICAL COMPOSITION (ISO)	Carbon (3) Silicon Mangan. Chrom. Molybd. Nickel Alumin.	C=0.28-0.35% Si=0.15-0.40% Mn=0.40-0.70% Cr=2.80-3.30% Mo=0.30-0.50% Ni=0.30 max	C=0.35-0.42% Si=0.15-0.40% Mn=0.40-0.70% Cr=3.00-3.50% Mo=0.80-1.10%	C=0.30-0.37% Si=0.20-0.50% Mn=0.50-0.80% Cr=1.00-1.30% Mo=0.15-0.25% Al=0.80-1.20%	C=0.38-0.45% Si=0.20-0.50% Mn=0.50-0.80% Cr=1.50-1.80% Mo=0.25-0.40% Al=0.80-1.20%	C=0.07-0.13% Si=0.15-0.40% Mn=0.30-0.60%	C=0.12-0.18% all Mn=0.30-0.60%	C=0.12-0.18% Mn=0.60-0.90%	C=0.17-0.23% Mn=0.60-0.90% Cr=0.70-1.00%
ISO R683/	(4)	Part 10 Grade 1	Part 10 Grade 2	Part 10 Grade 3	Part 10 Grade 4	Part 11 Grades 1,1a	Part 11 Grades 2,2a	Part 11 Grade 3	Part 11 Grade 4
EURONORM		EU 85 31CrMo12	EU 85 39CrMoV13	EU 85 34CrAlMo5	EU 85 41CrAlMo7	EU 84 2C10 3C10	EU 84 2C15 3C15		EU 84 15Cr2
NORTH MAERICA (AISI, SAE)				ASTM A355 Class D	ASTM A355 Class A (135 Mod)	1010	1015	1016	5120
JAPAN					JIS G 4202 SACM1	JIS G 4051 S9CK	JIS G 4051 S15CK	JIS G 4051 S20CK	JIS G 4104 SCr22
GERMANY		DIN17211 31CrMo12	DIN17211 39CrMoV139	DIN17211 34CrAlMo5	DIN17211 41CrAlMo7	DIN17210 C10 Ck10	DIN17210 C15 Ck15 Cm15		DIN17210 15Cr3
FRANCE (NF A35-551)				30CAD6-12	40CAD6-12	CC10 XC10	XC18	CC20	16MC5
UNITED KINGDOM (BS 970)			897M39	905M31 Cr=1.40-1.80%	905M39	040A10	040A15 523A14	080A15	527A19
ITALY		UNI 6120 30CrMo12			UNI 6120 38CrAlMo7	UNI 5331 C10	UNI 5331 C16		
AUSTRALIA (DR72134 Grade 1135)						S1010 K1010	S1015	S1016 K1016	

NOTES:
1. 1 MPa = 1 N/mm² ≈ 0.1 kgf/mm² · 1000 psi = 6.894757 MPa ≈ 7 MPa
2. Rm, Re and A5 are ISO symbols for tensile strength, yield strength and elongation.
3. Unless otherwise noted steels fall within chemical composition limits shown for that column.
4. ISO classification of steel is in its fifth draft and it is similar to the Euronorm designation.
5. Physical properties are shown for reference only.

STEEL MATERIAL DATA

Table 10-47 (*Continued*). Nitriding and Case Hardening Steels (ISO 683: Part 10 and 11)

9	10	11	12	13	14	15	16	17	18
\multicolumn{10}{c}{CASE HARDENING STEELS}
930-1270 640- 9-	1030-1370 690- 8-	930-1270 640- 9-	980-1320 640- 8-	1030-1370 690- 8-	980-1320 640- 8-	980-1320 640- 8-	1080-1420 640- 8-	1130-1470 780- 8-	1270-1620 8880- 7-
C=0.13-0.19% Mn=1.00-1.30% Cr=0.80-1.10%	C=0.15-0.21% Mn=0.60-0.90% Cr=0.85-1.15% Mo=0.15-0.25%	C=0.17-0.23% Mn=0.60-0.90% Cr=0.30-0.50% Mo=0.40-0.50%	C=0.12-0.18% Mn=0.60-0.90% Cr=0.80-1.10% Ni=1.30-1.70%	C=0.11-0.17% Mn=0.35-0.65% Cr=1.40-1.70% Ni=1.30-1.70%	C=0.10-0.16% Mn=0.35-0.65% Cr=0.60-0.90% Ni=2.75-3.25%	C=0.17-0.23% Mn=0.60-0.90% Cr=0.35-0.65% Mo=0.15-0.24% Ni=0.40-0.70%	C=0.14-0.20% Mn=0.60-0.90% Cr=0.80-1.10% Mo=0.15-0.25% Ni=1.20-1.60%	C=0.11-0.17% Mn=0.30-0.60% Cr=0.80-1.10% Mo=0.20-0.30% Ni=3.00-3.50%	C=0.12-0.18% Mn=0.25-0.55% Cr=1.1-1.4% Mo=0.20-0.30% Ni=3.80-4.30%
Part 11 Grades 5,5a	Part 11 Grade 7	Part 11 Grades 8,8a	Part 11 Grade 9	Part 11 Grade 10	Part 11 Grade 11	Part 11 Grade 12	Part 11 Grade 13	Part 11 Grade 14	Part 11 Grade 15
EU 84 16MnCr5	EU 84 18CrMo4	EU 84 20MoCr4	EU84 15NiCr6	EU 84 14CrNi6	EU 84 13NiCr12	EU 84 20NiCrMo2	EU 84 17NiCrMo5	EU 84 14NiCrMo13	
5115		4118 Mo=0.08-0.15%	4718			8620 8617 8615	4320	9310 Mo=0.08-0.15%	JIS G 4103 SNCM 25 Cr=0.70-1.00%
JIS G 4104 SCr21	JIS G 4105 SCM22 Cr=0.90-1.20%	JIS G 4105 SCM24 Cr=0.90-1.20%	JIS G 4102 SNC1 C=0.32-0.40%		JIS G 4102 SNC22	JIS G 4103 SNCM21			
DIN17210 16MnCr5 16MnCrS5		DIN17210 20MoCr4 20MoCrS4		DIN17210 15CrNi6					(16NCD13)
20MC5	18CD4		16NC6		14NC11	20NCD2	18NCD6		
		637H17				635A14	637A16	655A12	659A15
		UNI 5331 16CrNi4			UNI 5331 16NiCr11	UNI 5331 20NiCrMo2	UNI 5331 18NiCrMo5	UNI 5331 16NiCrMo12	
	AS1444 4130 C=0.27-0.33%					AS1444 8620			

Table 10-48. Flame and Induction Hardening Steels (ISO 683: Part 12)

		1	2	3	4	5	6	7	8	9	10	11
PHYSICAL PROPERTIES (HEAT TREATED)	(1) (2) Tensile Rm(MPa) Yield Re(MPa) Elong. A5(%)	620-760 420- 17-	660-800 450- 16-	700-840 480- 14-	740-880 510- 13-	740-880 510- 12-	880-1080 640- 12-	930-1130 740- 11-	980-1180 780- 11-	1080-1270 880- 10-	1030-1230 830- 10-	1030-1230 830- 10-
CHEMICAL COMPOSITION (ISO)	(3) Carbon Silicon Mangan. Chrom. Molybd. Nickel	C=0.33-0.39% Si=0.15-0.40% Mn=0.50-0.80%	C=0.38-0.44% all Mn=0.50-0.80%	C=0.43-0.49% Mn=0.50-0.80%	C=0.48-0.55% Mn=0.60-0.90%	C=0.50-0.57% Mn=0.40-0.70%	C=0.42-0.48% Mn=0.50-0.80% Cr=0.40-0.60%	C=0.34-0.40% Mn=0.60-0.90% Cr=0.90-1.20%	C=0.38-0.44% Mn=0.60-0.90% Cr=0.90-1.20%	C=0.38-0.44% Mn=0.50-0.80% Cr=0.90-1.20% Mo=0.15-0.30%	C=0.38-0.44% Mn=0.70-1.00% Cr=0.40-0.60% Mo=0.15-0.30% Ni=0.40-0.70%	C=0.37-0.43% Mn=0.50-0.80% Cr=0.60-0.90% Mo=0.15-0.30% Ni=0.70-1.00%
ISO 683/12	(4)	Grade 1	Grade 2	Grade 3	Grade 4	Grade 5	Grade 6	Grade 7	Grade 8	Grade 9	Grade 10	Grade 11
EURONORM 86		C36		C46		C53	45Cr2	38Cr4		41CrMo4		40NiCrMo3
NORTH AMERICA (AISI)		1035	1040	1045	1050	1055	5145	5135	5140	4140	8640 8740	
JAPAN		JIS G 4051 S35C	JIS G 4051 S40C	JIS G 4051 S45C	JIS G 4051 S50C	JIS G 4051 S55C		JIS G 4104 SCr 3	JIS G 4104 SCr4	JIS G 4105 SCM4		
GERMANY (DIN 17212)		Cf35		Cf45		Cf53	45Cr2	38Cr4	42Cr4	41CrMo4		
FRANCE (NF A 35-551)		CC35 XC38	XC42	CC45	XC48	XC55	42C2	38C4	42C4	42CD4		40NCD3
UNITED KINGDOM (BS 970)		060A35 080A35	060A40 080A40	060A47 080A47	060A52 080A52	060A57 080A57	630M40 Cr=0.90-1.20	530M36	530A40 630M40	708A42 708M40 709M40	945A40 945M38 Mn=1.20-1.60 Cr=0.60-0.90	
ITALY (UNI 5333)		C35	C40	C45	C50			36CrMn 6	40Cr4			40NiCrMo4
AUSTRALIA		DR72134 Grade AS1135 S1035 K1035	DR72134 Grade AS1135 S1040	DR72134 Grade AS1135 S1045 K1045	DR72134 Grade AS1135 S1050 K1050	DR72134 Grade AS1135 S1058 K1055		DR73044 Grade AS1444 5140	Dr73044 Grade AS1444 5145			Dr73044 Grade AS1444 8640

NOTES:
1. 1 MPa = 1 N/mm² ≈ 0.1 kgf/mm² · 1000 psi = 6.894757 MPa ≈ 7 MPa
2. Rm, Re and A5 are ISO symbols for tensile strength, yield strength and elongation.
3. Unless otherwise noted steels fall within chemical composition limits shown for that column.
4. ISO classification of steel is in its fifth draft and it is similar to the Euronorm designation.
5. Steel within each column may differ slightly in chemistry, but compositions are essentially equivalent for similar response to heat treatment.
6. Physical properties are shown for reference only.

STEEL MATERIAL DATA

(Continued from page 10-127)

Table 10-42C—This table gives an international comparison of carbon steels for wire manufacturing and finished spring steel wire qualities.

Table 10-42D—Steel types are shown which are used in tubular products such as seamless and welded precision, and seamless mechanical application, as well as for heavy welded tubes.

Table 10-43—Steel grades for structural and general engineering purposes are shown. These steel types are extensively used for numerous steel products in Europe and throughout the world.

Table 10-44—The carbon steels listed, having carbon contents from C = 0.08% maximum to C = 0.98%, are very popular throughout the world. The various designations and national standards for carbon steels are shown.

Table 10-45—Alloy steels of the direct hardening types are standardized in ISO 683—Parts 2, 4, 5, 6, 7 and 8. A worldwide comparison is given in the table.

Table 10-46—The free cutting steels specified exhibit the special qualities necessary in machining for chip-breaking.

Table 10-47—The steel types, particularly suited for nitriding and case-hardening processes, are compared internationally.

Table 10-48—Steels used in flame and induction hardening processes are standardized in ISO 683 Parts 10 and 11. These and other corresponding national standards are shown in the table.

Table 10-49—Ferritic and martensitic stainless steel types are covered in the ISO 683 Part 13 standard. A worldwide comparison of designations is shown.

Table 10-50—Austenitic stainless steels from the major industrial countries are shown and compared.

Table 10-51—Alloy spring steels are described in the ISO 683, Part 14 standard. A world comparison of standard steel types is given.

Table 10-52—Cast irons with laminated, spheroidal and nodular graphite are internationally compared.

Table 10-53—This table lists malleable iron and steel for casting, comparing the various national and international standards.

Table 10-54—Austenitic cast iron designations and qualities are given.

Table 10-49. Stainless Steels-Ferritic and Martensitic (ISO 683: Part 13)

			1	2	3	4	5	6
			\multicolumn{6}{c|}{FERRITIC}					
PHYSICAL PROPERTIES (HEAT TREATED)	Tensile Rm(MPa) Yield Re(MPa) Elong. A5(%)	(1)(2)	440-640 250- 20-	410-610 250- 20-	440-640 250- 18-	440-640 250- 15-	440-640 250- 18-	440-640 250- 18-
CHEMICAL COMPOSITION (ISO)	Carbon Silicon Mangan. Chrom. Molybd. Nickel	(3)	C= -0.08% Si=1.0 max Mn= -1.0% Cr=11.5-14.0% Ni= -0.50%	C= -0.08% all Mn= -1.0% Cr=11.5-14.0% Ni= -0.50% Al=0.10-0.30%	C= -0.10% Mn= -1.0% Cr=16.0-18.0% Ni= -0.50%	C= -0.12% Mn= -1.5% Cr=16.0-18.0% Mo= -0.6% Ni= -0.50%	C= -0.10% Mn= -1.0% Cr=16.0-18.0% Ni= -0.50% Ti=0.5-0.8%	C= -0.10% Mn= -1.0% Cr=16.0-18.0% Mo=0.9-1.3%
ISO 683/13		(4)	Grade 1	Grade 2	Grade 8	Grade 8a	Grade 8b	Grade 9c
EURONORM 88			X6Cr13	X6CrAl13	X8Cr17	X10CrS17	X8CrTi17	X8CrMo17
NORTH AMERICA (AISI)			403	405	430	430F		436
JAPAN (JIS G 4303)			SUS403	SUS405	SUS430	SUS430F SUS429		SUS434
GERMANY (5) (DIN 17440)			X7Cr13	X7CrAl13	X8Cr17		X8CrTi17	X6CrMo17
FRANCE (NF A 35-578)			Z6C13	Z6CA13	Z8C17			Z8CD17.01
UNITED KINGDOM (BS 970)			403S17	BS PD 6290 405S17	430S15			BS PD 6290 432S19
ITALY (UNI 6900)			X6 Cr13	X6CrAl13	X8Cr17	X10CrS17		X8CrMo17
AUSTRALIA (AS 1444)					430	430F		

NOTES:
1. 1 MPa = 1N/mm² ≈ 0.1 kgf/mm² · 1000 psi = 6.894757 MPa ≈ 7 MPa
2. Rm, Re and A5 are ISO symbols for tensile strength, yield strength and elongation.
3. Unless otherwise noted steels fall within chemical composition limits shown for that column.
4. ISO classification of steel is in its fifth draft and it is similar to the Euronorm designation.
5. Steel types shown in brackets ( ) are not covered in DIN 17440 yet, but are shown in published specifications.

STEEL MATERIAL DATA

Table 10-49 (*Continued*). Stainless Steels-Ferritic and Martensitic (ISO 683: Part 13)

7	8	9	10	11	12	13	14	15
\multicolumn{9}{c}{MARTENSITIC}							
590-780 410- 16-	640-830 440- 12-	690-880 490- 14-	830-1030 640- 10-	880-1130 690- 9-	780-980 590- 11-			
C=0.09-0.15% Mn= -1.0% Cr=11.5-14.0% Ni= -1.0%	C-0.08-0.15% Mn= -1.5% Cr=12.0-14.0% Mo= -0.6% Ni= -1.0%	C=0.16-0.25% Mn= -1.0% Cr=12.0-14.0% Ni= -1.0%	C=0.10-0.20% Mn= -1.0% Cr=15.0-18.0% Ni=1.5-3.0%	C=0.17-0.25% Mn= -1.0% Cr=16.0-18.0% Ni=1.5-3.0%	C=0.26-0.35% Mn= -1.0% Cr=12.0-14.0% Ni= -1.0%	C=0.36-0.45% Mn= -1.0% Cr=12.5-14.5% Ni= -1.0%	C=0.42-0.50% Mn= -1.0% Cr=12.5-14.5% Ni= -1.0%	C=0.95-1.20% Mn= -1.0% Cr=16.0-18.0% Mo= -0.75% Ni= -0.5%
Grade 3	Grade 7	Grade 4	Grade 9	Grade 9b	Grade 5	Grade 6	Grade 6a	Grade A-1b
X12Cr13	X12CrS13	X20Cr13		X21CrNi17	X30Cr13	X40Cr13	X45Cr13	X105CrMo17
410	416 416Se	420 420F	431		420 FSe			440C
SUS410 SUS403	SUS416 SUS410J1	SUS420J1	SUS431		SUS420J2 SUS420F			SUS440C SUS440F
X10Cr13	(X12CrS13)	X20Cr13		X22CiNi17	(X30Cr13)		X40Cr13	(X105CrMo17)
Z12C13 Z10C13	Z12CF13	Z20C13			Z30C13		Z40C14	
410S21	416S41 416S21	420S37 420S29 416S37 416S29	431S29 441S29 441S49	431S29 C=0.15%	420S45			
X12Cr13	X12CrS13	X20Cr13	X14CrNi19	X16CrNi16	X30Cr13	X40Cr14		
410	416	420	431					

Table 10-50. Stainless Steels—Austenitic (ISO 683: Part 13, ISO 2604, 2605, 2607)

		1	2	3	4	5	6	7	8
PHYSICAL PROPERTIES (HEAT TREATED)	(1) (2) Tensile Rm(MPa) Yield Re(MPa) Elong. A5(%)	440-640 180- 40-	490-690 210- 35-	490-690 210- 35-	490-690 200- 40-	490-690 210- 40-	490-690 210- 35-	490-690 180- 40-	590-780 220- --
CHEMICAL COMPOSITION (ISO)	(3) Carbon Silicon Mangan. Chrom. Molybd. Nickel	C= -0.03% Si=1.0 max Mn=2.0 max Cr=17.0-19.0% Ni=9.0-12.0%	C= -0.08% all Cr=17.0-19.0% Ni=9.0-12.0% Ti=0.4-0.8%	C= -0.08% all except as noted Cr=17.0-19.0% Ni=9.0-12.0% Nb=0.4-1.0%	C= -0.07% Cr=17.0-19.0% Ni=8.0-11.0%	C= -0.12% Cr=17.0-19.0% Ni=8.0-10.0%	C= -0.12% Cr=17.0-19.0% Mo= -0.60% Ni=8.0-10.0%	C= 0.10% Cr=17.0-19.0% Ni=11.0-13.0%	C= -0.15% Cr=16.0-18.0% Ni=6.0-8.0%
ISO 683/13 ISO 2604 ISO 2605 ISO 2607	(4)	Grade 10 F46 TS46 P46	Grade 15 F53 TS53 P53	Grade 16 F50, F51 TS50 P50	Grade 11 F47 TS47 P47	Grade 12 F48 TS48 P48	Grade 17	Grade 13	Grade 14
EURONORM 88		X3CrNi18 10	X6CrNiTi18 10	X6CrNiNb18 10	X5CrNi18 9	X10CrNi18 9	X10CrNiS18 9	X8CrNi18 12	X12CrNi17 7
NORTH AMERICA (AISI)		304L	321	347	304	302	303	305	301
JAPAN (JIS G 4303)		SUS304L	SUS321	SUS347	SUS304	SUS302	SUS303 SUS303Se	SUS305	SUS301
GERMANY (DIN 17440)	(5)	X2CrNi18 9	X10CrNiTi18 9	X10CrNiNb18 9	X5CrNi18 9	(X12CrNi18 8)	X12CrNiS18 8	X5CrNi19 11	DIN 17224 X12CrNi17 7
FRANCE (NF A 35-602)		Z2CN18.10	Z6CNT18.11	Z6CNNb18.11	Z6CN18.09	Z10CN18.09	Z10CNF18.09	Z8CN18.12	Z12CN17.08
UNITED KINGDOM (BS 970)		304S12	321S12	347S17	304S15	302S25	303S21	BS PD 6290 305S19	BS PD 6290 301S01
ITALY (UNI 6900)		X2CrNi18 11	X6CrNiTi18 11 X8CrNiTi1811	X6CrNiNb18 11 X8CrNiNb1811	X5CrNi18 10	X10CrNi18 09	X10CrNiS18 9	X8CrNi18 12	X12CrNi17 07
AUSTRALIA (AS 1444)		304L	321		304	302	303		

NOTES:
1. I MPa = 1 N/mm² ≈ 0.1 kgf/mm² · 1000 psi = 6.894757MPa ≈ 7 MPa
2. Rm, Re and A5 are ISO symbols for tensile strength, yield strength and elongation.
3. Unless otherwise noted steels fall within chemical composition limits shown for that column.
4. ISO classification of steel is in its fifth draft and it is similar to the Euronorm designation.
5. Steel within each column may differ sightly in chemistry, but compositions are essentially equivalent for similar response to heat treatment.
6. Physical properties are shown for reference only.

STEEL MATERIAL DATA

10-141

Table 10-50 (Continued). Stainless Steels—Austenitic (ISO 683: Part 13, ISO 2604, 2605, 2607)

9	10	11	12	13	14	15	16	17	18
440-640 200- 40-	490-690 210- 40-	490-690 220- 35-	490-690 220- 35-	440-640 200- 40-	490-690 210- 40-	490-690 220- 35-	490-690 220- 35-	490-690 200- 35-	640-630 300- 40-
C= -0.03% Cr=16.0-18.5% Mo=2.0-2.5% Ni=11.0-14.0%	C= 0.07% Cr=16.0-18.5% Mo=2.0-2.5% Ni=10.5-14.0%	C= -0.08% Cr=16.0-18.5% Mo=2.0-2.5% Ni=10.5-14.0% Ti=0.4-0.8%	C= -0.08% Cr=16.0-18.5% Mo=2.0-2.5% Ni=10.5-14.0% Nb=0.8-1.0%	C= -0.03% Cr=16.0-18.5% Mo=2.5-3.0% Ni=11.5-14.5%	C= 0.07% Cr=16.0-18.5% Mo=2.5-3.0% Ni=11.0-14.5%	C= -0.08% Cr=16.0-18.5% Mo=2.5-3.0% Ni=11.0-14.5% Ti=0.4-0.8%	C= -0.08% Cr=16.0-18.5% Mo=2.5-3.0% Ni=11.0-14.5% Nb=0.8-1.0%	C= -0.03% Cr=17.5-19.5% Mo=3.0-4.0% Ni=14.0-17.0%	C= -0.15% Mn=7.5-10.5% Cr=17.0-19.0% Ni=4.0-6.0% N=0.05-0.25%
Grade 19 TS57 P57	Grade 20 TS60 P60	Grade 21	Grade 23	Grade 19a F59 TS58 P58	Grade 20a F62 TS61 P61	Grade 21a F66	Grade 23a	Grade 24	Grade A-3
X3CrNi- Mo17 12 2	X6CrNiMo- 17 12 2	X6CrNiMo- Ti17 12 2	X6CrNiMo- Nb17 12 2	X3CrNi- Mo17 13 3	X6CrNiMo- Ti17 13 3	X6CrNiMo- Ti17 13 3	X6CrNiMo- Nb17 13 3	X3CrNi- Mo18 16 4	
316L	316			316L	316			317	202
SUS316L	SUS316			SUS316L	SUS316			SUS317	SUS202
X2CrNiMo18 10	X5CrNiMo18 10	X10CrNiMo- Ti18 10	X10CrNiMo- Nb18 10	X2CrNi- Mo18 12	X5CrNi- Mo18 12	(X10CrNiMo- Ti18 12)	(X10CrNiMo- Nb18 12)	X2CrNiMo18 16	(X8CrMnNi18 9)
Z2CND17.12	Z6CND17.11	Z8CNDT17.12	Z6CNDNb17.12	Z2CND17.13	Z6CND17.13	Z8CNDT17.13	Z6CNDNb17.13	Z2CND19.15	
316S12	316S16	320S17	318S17	316S12	316S16	320S17	BS PD 6290 318S17	317S12	
X2CrNiMo17 13	X5CrNiMo17 13	X6CrNiMoTi- 17 12	X6CrNiMoNb- 17 12	X2CrNiMo 17 13	X5CrNiMo 17 13	X6CrNiMoTi- 17 13	X6CrNiMo- Nb 17 13	X2CrNiMo 18 16	
316L	316			316L	316			317	

Table 10-51. Spring Steels (ISO 683: Part 14)

		1	2	3	4	5	6
PHYSICAL PROPERTIES (HEAT TREATED)	(1) (2) Tensile Rm(MPa) Yield Re(MPa) Elong A5(%)	1180- 880- 6-	1180- 880- 6-	1270- 1080- 6-	1320- 1130- 6-	1320- 1130- 6-	1370- 1180- 5-
CHEMICAL COMPOSITION (ISO)	(3) Carbon Silicon Mangan. Chrom. Molybd. Vanadium Phosphor Sulphur Boron	C=0.72-0.85% Si=0.15-0.40% Mn=0.50-0.80% P= -0.050% S= -0.050%	C=0.72-0.85% Si=0.15-0.40% Mn=0.50-0.80% P= -0.035% S= -0.035%	C=0.43-0.50% Si=1.50-2.00% Mn=0.50-0.80% P= -0.040% S= -0.040%	C=0.47-0.55% Si=1.50-2.00% Mn=0.50-0.80% P= -0.040% S= -0.040%	C=0.52-0.60% Si=1.50-2.00% Mn=0.60-0.90% P= -0.040% S= -0.040%	C=0.57-0.64% Si=1.70-2.20% Mn=0.70-1.00% P= -0.040% S= -0.040%
ISO 683	Part 14 (4)	Grade 1	Grade 2	Grade 3	Grade 4	Grade 5	Grade 6
EURONORM 89		EU 16 2CD 70	EU 16 3CD 75	45Si7	50Si7	55Si7	60Si7
NORTH AMERICA (AISI, SAE)		1074 1080				9255	9260
JAPAN (JIS G4801)		SUP3				SUP6	SUP7
GERMANY (DIN 17221)		DIN 17140 D75-2		46Si7	51Si7	55Si7	
FRANCE (NF A35-551)		XC80		45S7		55S7	
UNITED KINGDOM (BS 970)		060A78 070A78			250A53	250A58	250A61
ITALY (UNI 3545)			C75		50Si7	55Si8	
AUSTRALIA (AS 1444)		XK1082				9255	9260

NOTES:
1. 1 MPa = 1 N/mm² ≈ 0.1 kgf/mm² · 1000 psi = 6.894757 MPa ≈ 7 MPa
2. Rm, Re and A5 are ISO symbols for tensile strength, yield strength and elongation.
3. Unless otherwise noted steels fall within chemical composition limits shown for that column.
4. ISO classification of steel is in its fifth draft and it is similar to the Euronorm designation.

STEEL MATERIAL DATA

Table 10-51 *(Continued)*. **Spring Steels (ISO 683: Part 14)**

7	8	9	10	11	12	13	14
1370- 1180- 5-	1370- 1180- 6-	1370- 1180- 5-	1370- 1180- 6-	1370- 1180- 6-	1370- 1180- 6-	1370- 1180- 6-	1370- 1180- 6-
C=0.57-0.64% Si=1.70-2.20% Mn=0.70-1.00% Cr=0.25-0.40% P= -0.040% S= -0.040%	C=0.52-0.59% Si=0.15-0.40% Mn=0.70-1.00% Cr=0.60-0.90% P= -0.035% S= -0.035%	C=0.56-0.64% Si=0.15-0.40% Mn=0.70-1.00% Cr=0.60-0.90% P= -0.035% S= -0.035%	C=0.56-0.64% Si=0.15-0.40% Mn=0.70-1.00% Cr=0.60-0.90% P= 0.035% S= -0.035% B= -0.0005%	C=0.42-0.50% Si=1.30-1.70% Mn=0.50-0.80% Cr=0.50-0.75% Mo=0.15-0.30% P= -0.035% S= -0.035%	C=0.56-0.64% Si=0.15-0.40% Mn=0.70-1.00% Cr=0.70-0.90% Mo=0.25-0.35% P= -0.035% S= -0.035%	C=0.48-0.55% Si=0.15-0.40% Mn=0.70-1.00% Cr=0.90-1.20% V=0.10-0.20% P= -0.035% S= -0.035%	C=0.48-0.56% Si=0.15-0.40% Mn=0.70-1.00% Cr=0.90-1.20% Mo=0.15-0.25% V=0.07-0.12% P= -0.035% S= -0.035%
Grade 7	Grade 8	Grade 9	Grade 10	Grade 11	Grade 12	Grade 13	Grade 14
60SiCr8	55Cr3			45SiCrMo6		50CrV4	51CrMoV4
					4161		
	5155	5160	51B60			6150	
	SUP9		SUP11			SUP10	
60SiCr7	55Cr3					50CrV4	51CrMoV4
60SC7				45SCD6		50CV4	
		527A60				735A50	
60SiCr8						50CrV4	
	5155	5160				6150	

Table 10-52. Cast Iron with Laminated, Spheroidal and Nodular Graphite (ISO 185, 1083)

		1	2	3	4	5	6
	(1) (2)	\multicolumn{6}{c}{LAMINATED (GREY) CAST IRON}
PHYSICAL PROPERTIES	Tensile Rm(MPa) Proof Rpo.2(MPa) Elong. A5(%)	100-	150-	200-	250-	290-	340-
ISO		ISO 185 Grade 10	ISO 185 Grade 15	ISO 185 Grade 20	ISO 185 Grade 25	ISO 185 Grade 30	ISO 185 Grade 35
EURONORM							
NORTH AMERICA		ASTM A159 and SAE J431b G1800	ASTM A159 and SAE J431b G2500	ASTM A159 and SAE J431b G3000	ASTM A159 and SAE J431b G3500	ASTM A159 and SAE J431b G4000	ASTM A48 Class 50
JAPAN		JIS G5501 FC-10	JIS G5501 FC-15	JIS G5501 FC-20	JIS G5501 FC-25	JIS G5501 FC-30	JIS G5501 FC-35
GERMANY		DIN 1691 GG-10	DIN1691 GG-15	DIN 1691 GG-20	DIN 1691 GG-25	DIN 1691 GG-30	DIN 1691 GG-35
FRANCE		NF A32.101 Ft10	NF A32.101 Ft15	NF A32.101 Ft20	NF A32.101 Ft25	NF A32.101 Ft30	NF A32.101 Ft35
UNITED KINGDOM		BS 1452 Grade 10	BS1452 Grade 12	BS1452 Grade 14	BS1452 Grade 17	BS1452 Grade 20	BS1452 Grade 23
ITALY		UNI5007 G10	UNI5007 G15	UNI5007 G20	UNI5007 G25	UNI5007 G30	UNI5007 G35
AUSTRALIA		AS G8 Grade 10	AS G8 Grade 12	AS G8 Grade 14	AS G8 Grade 17	AS G8 Grade 20	AS G8 Grade 23

1. I MPa = 1 N/mm² ≈ 0.1 kgf/mm² · 1000 psi = 6.894757 MPa ≈ 7 MPa
2. Rm, Re and Rpo.2 are ISO symbols for tensile strength, proof strength and elongation

STEEL MATERIAL DATA

Table 10-52 (*Continued*). **Cast Iron with Laminated, Spheroidal and Nodular Graphite (ISO 185, 1083)**

7	8	9	10	11	12	13
		SPHEROIDAL OR NODULAR CAST IRON				
390–	370– 230– 17–	410– 270– 12–	490– 340– 7–	590– 390– 2–	690– 440– 2–	800– 2–
ISO 185 Grade 40	ISO 1083 Grade 38-17	ISO 1083 Grade 42-12	ISO 1083 Grade 50-7	ISO 1083 Grade 60-2	ISO 1083 Grade 70-2	
ASTM A48 Class 55 Class 60	ASTM A536 Gr 60-40-18 SAE J 434 b D 4018	ASTM A536 Gr 60-45-12 SAE J434b D 4512	ASTM A536 Gr 80-55-06 SAE J434b D 5506		ASTM A536 Gr 100-70-03 SAE J434b D7003	ASTM A536 Gr 120-90-02 SAE J434b DQ&T
		JIS G5502 FCD-40	JIS G5502 FCD-50	JIS G5502 FCD-60	JIS G5502 FCD-70	
DIN 1691 GG-40		DIN 1693 GGG-40	DIN 1693 GGG-50	DIN 1693 GGG-60	DIN 1693 GGG-70	DIN 1693 GGG-80
NF A32.101 Ft 40	NF A32.301 FGS 38/15	NF A32.301 FGS42/12	NF A32.301 FGS50/7	NF A32.301 FGS60/2	NF A32.301 FGS70/2	
BS1452 Grade 26	BS2789 370/17	BS2789 420/12	BS2789 500/7	BS2789 600/3	BS2789 700/2	BS 2789 800/2
		UNI 4544 GS42/10	UNI 4544 GS50/5	UNI 4544 GS55/2		
AS G8 Grade 26	AS G9 370/17	AS G9 420/12	AS G9 500/7	AS G9 600/3	AS G9 700/2	AS G9 800/2

Table 10-53. Mallable Iron and Steel Castings (ISO 942, 943, 944)

		1	2	3	4	5	6	7	8
	(1) (2)	BLACKHEART			WHITEHEART	PEARLITIC			
PHYSICAL PROPERTIES (ISO/ASTM)	Tensile Rm(MPa) Yield Re(MPa) Elong. A5(%)	290- 190- 6-	310- 190- 10-	340- 210- 12-	310- 4-	410- 270- 10-	440- 270- 7-	490- 310- 5-	540- 350- 4-
ISO		ISO 943 Grade C	ISO 943 Grade B	ISO 943 Grade A	ISO 942 Grade B Grade A		ISO 944 Grade E	ISO 944 Grade D	ISO 944 Grade C
NORTH AMERICA	ASTM		ASTM A47 Gr 32510	ASTM A47 Gr 35018		ASTM A220 Gr 40010	ASTM A220 Gr 45006 Gr 45008	ASTM A220 Gr 50005	ASTM A220 Gr 60004
	SAE					SAE J158 Gr M3210	SAE J158 Gr M4504	SAE J158 Gr M5003	SAE J158 Gr M5503
JAPAN		JIS G5702 FCMB28	JIS G5702 FCMB32	JIS G5702 FCMB35 FCMB37	JIS G5703 FCMW34 FCMW38		JIS G5704 FCMP45	JIS G5704 FCMP50	JIS G5704 FCMP55
GERMANY				DIN1692 GTS-35	DIN1692 GTW-35 GTW-40 GTW-55		DIN1692 GTS-45		DIN1692 GTS-55
FRANCE			NF A32.702 MN32-8	NF A32.702 MN35-10 MN38-18	NF A32.701 MB 40-10 MB 35-7			NF A32.703 MP50-5	
UNITED KINGDOM		BS310 Gr B290/6	BS310 Gr B310/10	BS310 Gr B340/12	BS309 Gr W 410/4 Gr W 340/3		BS3333 Gr P440/7	BS3333 Gr P510/4	BS3333 Gr P540/5
ITALY				UNI 3779 GMN35 GMN37	UNI 3779 GMB35		UNI 3779 GMN 45	UNI 3779 GMN 50	UNI 3779 GMN 55
AUSTRALIA		AS G12 Gr B18/6	AS G12 Gr B20/10	AS G12 Gr B22/14	AS G11 Gr W24/18 Gr W22/14	AS G14 PM63-40-09	AS G14 PM68-45-07	AS G14 PM70-48-05	AS G14 PM80-60-03 PM75-50-07

NOTES:
1. I MPa = 1 N/mm² ≈ 0.1 kgf/mm² · 1000 psi = 6.894757 MPa ≈ 7 MPa
2. Rm, Re and A5 are ISO symbols for tensile strength, yield strength and elongation.
3. Unless otherwise noted steels fall within chemical composition limits shown for that column.

STEEL MATERIAL DATA

Table 10-53 (*Continued*). Mallable Iron and Steel Castings (ISO 942, 943, 944)

9	10	11	12	13	14	15	16	17	18
		\multicolumn{8}{c}{STEEL CASTINGS}
640- 420- 3-	690- 540- 2-	400-550 200- 25-	450-600 230- 22-	520-670 260- 18-	570-720 300- 15-	620- 410- 20-	720- 590- 17-	830- 650- 14-	1030- 860- 9-
ISO 944 Grade B	ISO 944 Grade A	ISO 3755 Grade 20-40	ISO 3755 Grade 23-45	ISO 3755 Grade 26-52	ISO 3755 Grade 30-57				
ASTM A220 Gr 80002 Gr 70003 SAE J158 Gr M7002	ASTM A220 Gr 90001 SAE J158 Gr M8501	ASTM A27 Gr 60-30 SAE J435c Gr 0025	ASTM A27 Gr 65-35 SAE J435c Gr 0030	ASTM A27 Gr 70-40 SAE J435c Gr 0050A	ASTM A148 Gr 80-50 Gr 80-40 SAE J435c Gr 080	ASTM A148 Gr 90-60 SAE J435c Gr 090	ASTM A148 Gr 105-85 SAE J435c Gr 0105	ASTM A148 Gr 120-95 SAE J435c Gr 0120	ASTM A148 Gr 150-125 Gr 175-145 SAE J435c Gr 0150 Gr 0175
JIS G5704 FCMP60	JIS G5704 FCMP70	JIS G5101 SC42	JIS G5101 SC46	JIS G5101 SC49	JIS G5111 SCMnCr2A	JIS G5111 SCMnCr2B SCMnCr3A	JIS G5111 SCMnM3B SCMnCrM2B SCCrM1B	JIS G5111 SCMnCrM3B	JIS G5111 SCNCrM2B
DIN1692 GTS-65	DIN1692 GTS-70	DIN1681 GS-38 GS-38.3	DIN1681 GS-45 GS-45.3	DIN1681 GS-52 GS-52.3	DIN1681 GS-60 GS-60.3	DIN1681 GS-62 GS-62.3	DIN1681 GS-70		
NF A32.703 MP60-3	NF A32.703 MP70-2	NF A32.051 E20-40-M	NF A32.051 E23-45-M	NF A32.051 E26-52-M	NF A32.051 E30-57-M				
BS3333 Gr P570/3	BS3333 Gr P690/2	BS592 Gr A	BS592 Gr B	BS592 Gr C	BS1456 Gr A	BS1456 Gr B1	BS1456 Gr B2	BS1458 Gr A	BS1458 Gr B Gr C
UNI 3779 GMN 65	UNI 3779 GMN 70	UNI3158 FeG38 FeG38VR FeG00	UNI3158 FeG45 FeG45VR	UNI3158 FeG52 FeG52VR					
AS G14 PM90-70-02	AS G14 PM100-80-02 PM105-85-02	AS G22 Gr 0025 Gr 0022	AS G22 Gr 0030	AS G22 Gr 0050A	AS G22 Gr 080	AS G22 Gr 090 Gr 00503	AS G22 Gr 0105	AS G22 Gr 0120	AS G22 Gr 0150 Gr 0175

Table 10-54. Austenitic Cast Iron (ISO 2892)

	1	2	3	4	5	6	7	8
				SPHEROIDAL GRAPHITE (DUCTILE)				
PHYSICAL PROPERTIES (1)(2) Tensile Rm(MPa) / Proof Rp0.2(MPa) / Elong. A (%)	390- 210- 15-	370- 210- 7-	390- 210- 7-	370- 210- 10-	370- 210- 20-	440- 170- 25-	370- 210- 13-	370- 210- 7-
CHEMICAL COMPOSITION (ISO) (3) Carbon / Silicon / Mangan. / Nickel / Chrom. / Copper / Phosphor.	C= -3.0% Si=2.0-3.0% Mn=6.0-7.0% Ni=12.0-14.0% Cr= -0.2% Cu=0.5% max P=0.08% max	C= -3.0% Si=1.5-3.0% Mn=0.5-1.5% Ni=18.0-22.0% Cr=1.0-2.5% all all	C= -3.0% Si=1.5-3.0% Mn=0.5-1.5% Ni=18.0-22.0% Cr=2.5-3.5%	C= -3.0% Si=4.5-5.5% Mn=0.5-1.5% Ni=18.0-22.0% Cr=1.0-2.5%	C= -3.0% Si=1.0-3.0% Mn=1.5-2.5% Ni=21.0-24.0% Cr= -0.5%	C= -2.6% Si=1.5-2.5% Mn=4.0-4.5% Ni=22.0-24.0% Cr= -0.2%	C= -2.6% Si=1.5-3.0% Mn=0.5-1.5% Ni=28.0-32.0% Cr=1.0-1.5%	C= -2.6% Si=1.5-3.0% Mn=0.5-1.5% Ni=28.0-32.0% Cr=2.5-3.5%
ISO 2892	S-NiMn13 7	S-NiCr20 2	S-NiCr20 3	S-NiSiCr-20 5 2	S-Ni22	S-NiMn23 4	S-NiCr30 1	S-NiCr30 3
EURONORM								
NORTH AMERICA		ASTM A439 Type D-2	ASTM A439 Type D-2B		ASTM A439 Type D-2C		ASTM A439 Type D-3A	ASTM A439 Type D-3
JAPAN								
GERMANY (DIN 1694)	GGG-NiMn13 7	GGG-NiCr20 2	GGG-NiCr20 3	GGG-NiSiCr20 4 2	GGG-Ni 22	GGG-NiMn23 4	GGG-NiCr30 1	GGG-NiCr30 3
FRANCE (NF A32-301)	S-NM13 7	S-NC20 2	S-NC20 3	S-NSC20 5 2	S-N22	S-NM23 4	S-NC30 1	S-NC30 3
UNITED KINGDOM (BS 3468)		AUS202 Grade A	AUS202 Grade B	AUS204	AUS203			AUS205
ITALY								
AUSTRALIA (AS G15)		AUS 202 Grade A	AUS 202 Grade B	AUS 204	AUS 203			AUS 205

NOTES:
1. MPa = 1 N/mm² ≈ 0.1 kgf/mm² · 1000 psi = 6.894757 MPa ≈ 7 MPa
2. Rm, Re and A5 are ISO symbols for tensile strength, yield strength and elongation.
3. Unless otherwise noted steels fall within chemical composition limits shown for that column.

STEEL MATERIAL DATA

10-149

Table 10-54 (Continued). Austenitic Cast Iron (ISO 2892)

9	10	11	12	13	14	15	16	17	18
			LAMINATED GRAPHITE (GREY)						
390– 240– –	370– 210– 20–	370– 210– 7–	170–	190–	170–	190–	190–	170–	120–
C= -2.6% Si=5.0-6.0% Mn=0.5-1.5% Mn=28.0-32.0% Cr=4.5-5.5%	C= -2.4% Si=1.5-3.0% Mn=0.5-1.5% Ni=34.0-36.0% Cr= -0.2%	C= -2.4% Si=1.5-3.0% Mn=0.5-1.5% Ni=34.0-36.0% Cr=2.0-3.0%	C= -3.0% Si=1.0-2.8% Mn=0.5-1.5% Ni=13.5-17.5% Cr=1.0-2.5% Cu=5.5-7.5%	C= -3.0% Si=1.0-2.8% Mn=0.5-1.5% Ni=13.5-17.5% Cr=2.5-3.5% Cu=5.5-7.5%	C= -3.0% Si=1.0-2.8% Mn=0.5-1.5% Ni=18.0-22.0% Cr=1.0-2.5% Cu= -0.5%	C= -3.0% Si=1.0-2.8% Mn=0.5-1.5% Ni=18.0-22.0% Cr=2.5-3.5% Cu= -0.5%	C= -2.5% Si=1.0-2.0% Mn=0.5-1.5% Ni=28.0-32.0% Cr=2.5-3.5% Cu= -0.5%	C= -2.5% Si=5.0-6.0% Mn=0.5-1.5% Ni=29.0-32.0% Cr=4.5-5.5% Cu= -0.5%	C= -2.4% Si=1.0-2.0% Mn=0.5-1.5% Ni=34.0-36.0% Cr= -0.2% Cu= -0.5%
S-NiSiCr- 30 5 5	S-Ni35	S-NiCr35 3	L-NiCu- Cr15 6 2	L-NiCu- Cr15 6 3	L-NiCr- 20 2	L-NiCr. 20 3	L-NiCr- 30 3	L-NiSi- Cr30 5 5	L-Ni35
ASTM A439 Type D-4	ASTM A439 Type D-5	ASTM A439 Type D-5B	ASTM A436 Type 1	ASTM A436 Type 1b	ASTM A436 Type 2	ASTM A436 Type 2b	ASTM A436 Type 3	ASTM A436 Type 4	ASTM A436 Type 5
GGG- NiSiCr30 5 5	GGG- Ni35	GGG- NiCr35 3	GGL- NiCuCr15 6 2	GGL-NiCu Cr15 6 3	GGL- NiCr20 2	GGL- NiCr20 3	GGL- NiCr30 3	GGL-Ni SiCr30 5 5	GGL- Ni35
S-NSC30 5 5	S-N35	S-NC35 3	L-NUC15 6 2	L-NUC15 6 3	L-NC20 2	L-NC20 3	L-NC30 3	L-NSC30 5 5	L-N35
			AUS101 Grade A	AUS101 Grade B	AUS102 Grade A	AUS102 Grade B	AUS105		
			AUS 101 Grade A	AUS 101 Grade B	AUS 102 Grade A	AUS 102 Grade B	AUS 105		

(*The World Ferrous Materials Index follows on page 10-150.*)

STEEL MATERIAL DATA

WORLD FERROUS MATERIALS INDEX

DESIGNATION	STANDARD	TABLE	COL
GRADE 10	ISO 185		
GRADE 15	ISO 185		
GRADE 20	ISO 185		
GRADE 25	ISO 185		
GRADE 30	ISO 185		
GRADE 35	ISO 185		
GRADE 40	ISO 185		
Fe 33	ISO 630		
Fe 33	ISO 630		
Fe37-A	ISO 630		
Fe37-A	ISO 630		
Fe37-B	ISO 630		
Fe37-B	ISO 630		
Fe37-C	ISO 630		
Fe37-C	ISO 630		
Fe37-D	ISO 630		
Fe37-D	ISO 630		
Fe42-A	ISO 630		
Fe42-A	ISO 630		
Fe42-B	ISO 630		
Fe42-B	ISO 630		
Fe42-C	ISO 630		
Fe42-C	ISO 630		
Fe42-D	ISO 630		
Fe42-D	ISO 630		
Fe44-A	ISO 630		
Fe44-B	ISO 630		
Fe44-C	ISO 630		
Fe44-D	ISO 630		
Fe52-B	ISO 630		
Fe52-B	ISO 630		
Fe52-C	ISO 630		
Fe52-C	ISO 630		
Fe52-D	ISO 630		
Fe52-D	ISO 630		
C25	ISO 683/1 AND 4		
C25	ISO 683/1 AND 4		
C25	ISO 683/1 AND 4		
C25	ISO 683/1 AND 4		
C30	ISO 683/1 AND 4		
C30	ISO 683/1 AND 4		
C30	ISO 683/1 AND 4		
C30	ISO 683/1 AND 4		
C35	ISO 683/1 AND 4		
C35	ISO 683/1 AND 4		
C35	ISO 683/1 AND 4		
C35	ISO 683/1 AND 4		
C40	ISO 683/1 AND 4		
C40	ISO 683/1 AND 4		
C40	ISO 683/1 AND 4		

TABLE	COL	DESIGNATION	STANDARD	TABLE	COL
10-52	1	GRADE 3A	ISO 683/8	10-45	11
10-52	2	GRADE 3B	ISO 683/8	10-45	11
10-52	3	GRADE 5	ISO 683/8	10-45	12
10-52	4	GRADE 5A	ISO 683/8	10-45	12
10-52	5	GRADE 5B	ISO 683/8	10-45	13
10-52	6	GRADE 6	ISO 683/8	10-45	13
10-52	7	GRADE 6A	ISO 683/8	10-45	13
10-42A	5	GRADE 6B	ISO 683/8	10-45	13
10-43	1	GRADE 1	ISO 683/9	10-46	1
10-42A	6	GRADE 2	ISO 683/9	10-46	2
10-43	2	GRADE 2P	ISO 683/9	10-46	3
10-42A	6	GRADE 3	ISO 683/9	10-46	4
10-43	2	GRADE 3P	ISO 683/9	10-46	5
10-42A	6	GRADE 4	ISO 683/9	10-46	6
10-43	2	GRADE 4P	ISO 683/9	10-46	7
10-42A	6	GRADE 5	ISO 683/9	10-46	8
10-43	2	GRADE 6	ISO 683/9	10-46	9
10-42A	7	GRADE 7	ISO 683/9	10-46	10
10-43	3	GRADE 8	ISO 683/9	10-46	11
10-42A	7	GRADE 9	ISO 683/9	10-46	12
10-43	3	GRADE 10	ISO 683/9	10-46	13
10-42A	7	GRADE 1	ISO 683/10	10-47	1
10-43	3	GRADE 2	ISO 683/10	10-47	2
10-42A	7	GRADE 3	ISO 683/10	10-47	3
10-43	3	GRADE 4	ISO 683/10	10-47	4
10-43	4	GRADE 1A	ISO 683/11	10-47	5
10-43	4	GRADE 2	ISO 683/11	10-47	5
10-43	4	GRADE 2A	ISO 683/11	10-47	6
10-43	4	GRADE 3	ISO 683/11	10-47	7
10-42A	8	GRADE 4	ISO 683/11	10-47	8
10-43	5	GRADE 5	ISO 683/11	10-47	9
10-42A	8	GRADE 5A	ISO 683/11	10-47	10
10-43	5	GRADE 7	ISO 683/11	10-47	11
10-42A	8	GRADE 8	ISO 683/11	10-47	12
10-43	5	GRADE 8A	ISO 683/11	10-47	13
10-44	5	GRADE 9	ISO 683/11	10-47	14
10-44	5	GRADE 10	ISO 683/11	10-47	15
10-44	5	GRADE 11	ISO 683/11	10-47	16
10-44	6	GRADE 12	ISO 683/11	10-47	17
10-44	6	GRADE 13	ISO 683/11	10-47	18
10-44	6	GRADE 14	ISO 683/11	10-47	1
10-44	6	GRADE 15	ISO 683/11	10-48	2
10-44	7	GRADE 1	ISO 683/12	10-48	3
10-44	7	GRADE 2	ISO 683/12	10-48	4
10-44	7	GRADE 3	ISO 683/12	10-48	5
10-44	7	GRADE 4	ISO 683/12	10-48	6
10-44	8	GRADE 5	ISO 683/12	10-48	7
10-44	8	GRADE 6	ISO 683/12	10-48	8
10-44	8	GRADE 7	ISO 683/12	10-48	9
		GRADE 8	ISO 683/12	10-48	10
		GRADE 9	ISO 683/12	10-48	11
		GRADE 10	ISO 683/12		
		GRADE 11	ISO 683/12		

STEEL MATERIAL DATA

10-151

WORLD FERROUS MATERIALS INDEX (Continued)

Material	Standard	Page	Col
C40	ISO 683/1 AND 4	10-44	8
C45	ISO 683/1 AND 4	10-44	9
C45	ISO 683/1 AND 4	10-44	9
C45	ISO 683/1 AND 4	10-44	9
C50	ISO 683/1 AND 4	10-44	10
C50	ISO 683/1 AND 4	10-44	10
C50	ISO 683/1 AND 4	10-44	10
C55	ISO 683/1 AND 4	10-44	11
C55	ISO 683/1 AND 4	10-44	11
C55	ISO 683/1 AND 4	10-44	11
C60	ISO 683/1 AND 4	10-44	12
C60	ISO 683/1 AND 4	10-44	12
C60	ISO 683/1 AND 4	10-44	12
GRADE 1	ISO 683/2 AND 4	10-45	1
GRADE 2	ISO 683/2 AND 4	10-45	2
GRADE 2A	ISO 683/2 AND 4	10-45	2
GRADE 3	ISO 683/2 AND 4	10-45	3
GRADE 3A	ISO 683/2 AND 4	10-45	3
GRADE 3B	ISO 683/2 AND 4	10-45	3
GRADE 1	ISO 683/5	10-45	4
GRADE 1A	ISO 683/5	10-45	4
GRADE 1B	ISO 683/5	10-45	5
GRADE 1	ISO 683/6	10-45	6
GRADE 1	ISO 683/7	10-45	6
GRADE 1A	ISO 683/7	10-45	6
GRADE 1B	ISO 683/7	10-45	7
GRADE 2	ISO 683/7	10-45	7
GRADE 2A	ISO 683/7 AND 4	10-45	7
GRADE 2B	ISO 683/7	10-45	8
GRADE 3	ISO 683/7	10-45	8
GRADE 3A	ISO 683/7	10-45	8
GRADE 3B	ISO 683/8	10-45	9
GRADE 1A	ISO 683/8	10-45	9
GRADE 1B	ISO 683/8	10-45	10
GRADE 2	ISO 683/8	10-45	10
GRADE 2A	ISO 683/8	10-45	10
GRADE 2B	ISO 683/8	10-45	11
GRADE 3	ISO 683/8	10-45	11

Material	Standard	Page	Col
GRADE A-1B	ISO 683/13	10-49	15
GRADE A-3	ISO 683/13	10-49	18
GRADE 1	ISO 683/13	10-49	1
GRADE 2	ISO 683/13	10-49	2
GRADE 3	ISO 683/13	10-49	7
GRADE 4	ISO 683/13	10-49	9
GRADE 5	ISO 683/13	10-49	12
GRADE 6A	ISO 683/13	10-49	13
GRADE 7	ISO 683/13	10-49	14
GRADE 8	ISO 683/13	10-49	8
GRADE 8A	ISO 683/13	10-49	3
GRADE 8B	ISO 683/13	10-49	4
GRADE 9	ISO 683/13	10-49	5
GRADE 9B	ISO 683/13	10-49	10
GRADE 9C	ISO 683/13	10-49	11
GRADE 10	ISO 683/13	10-49	6
GRADE 11	ISO 683/13	10-50	1
GRADE 12	ISO 683/13	10-50	4
GRADE 13	ISO 683/13	10-50	5
GRADE 14	ISO 683/13	10-50	7
GRADE 15	ISO 683/13	10-50	8
GRADE 16	ISO 683/13	10-50	2
GRADE 17	ISO 683/13	10-50	3
GRADE 19	ISO 683/13	10-50	6
GRADE 19A	ISO 683/13	10-50	9
GRADE 20	ISO 683/13	10-50	13
GRADE 20A	ISO 683/13	10-50	10
GRADE 21	ISO 683/13	10-50	14
GRADE 21A	ISO 683/13	10-50	11
GRADE 23	ISO 683/13	10-50	15
GRADE 23A	ISO 683/13	10-50	12
GRADE 24	ISO 683/13	10-50	16
GRADE 1	ISO 683/14	10-50	17
GRADE 2	ISO 683/14	10-51	1
GRADE 3	ISO 683/14	10-51	2
GRADE 4	ISO 683/14	10-51	3
GRADE 5	ISO 683/14	10-51	4
GRADE 6	ISO 683/14	10-51	5
GRADE 7	ISO 683/14	10-51	6
GRADE 8	ISO 683/14	10-51	7
GRADE 9	ISO 683/14	10-51	8
GRADE 10	ISO 683/14	10-51	9
GRADE 11	ISO 683/14	10-51	10
	ISO 683/14	10-51	11

STEEL MATERIAL DATA

WORLD FERROUS MATERIALS INDEX (Continued)

DESIGNATION	STANDARD	TABLE	COL
GRADE 12	ISO 683/14	10-51	12
GRADE 13	ISO 683/14	10-51	13
GRADE 14	ISO 683/14	10-51	14
GRADE A	ISO 942	10-53	4
GRADE B	ISO 942	10-53	4
GRADE A	ISO 943	10-53	3
GRADE B	ISO 943	10-53	3
GRADE C	ISO 943	10-53	2
GRADE A	ISO 943	10-53	1
GRADE B	ISO 944	10-53	10
GRADE C	ISO 944	10-53	9
GRADE D	ISO 944	10-53	8
GRADE E	ISO 944	10-53	7
Fe 50-1	ISO 1052	10-53	6
Fe 50-1	ISO 1052	10-42A	9
Fe 50-2	ISO 1052	10-53	6
Fe 50-2	ISO 1052	10-42A	9
Fe 60-1	ISO 1052	10-53	6
Fe 60-1	ISO 1052	10-42A	10
Fe 60-2	ISO 1052	10-53	7
Fe 60-2	ISO 1052	10-42A	10
Fe 70-1	ISO 1052	10-43	11
Fe 70-2	ISO 1052	10-43	8
GRADE 38-17	ISO 1083	10-52	8
GRADE 42-12	ISO 1083	10-52	9
GRADE 50-7	ISO 1083	10-52	10
GRADE 60-2	ISO 1083	10-52	11
GRADE 70-2	ISO 1083	10-52	12
T50	ISO 1111	10-42B	5
T52	ISO 1111	10-42B	6
T57	ISO 1111	10-42B	7
T61	ISO 1111	10-42B	8
T65	ISO 1111	10-42B	8
T70	ISO 1111	10-42B	9
F46	ISO 2604	10-50	1
F47	ISO 2604	10-50	4
F48	ISO 2604	10-50	5
F50	ISO 2604	10-50	3
F51	ISO 2604	10-50	3
F53	ISO 2604	10-50	2
F59	ISO 2604	10-50	13
F62	ISO 2604	10-50	14
F66	ISO 2604	10-50	15
TS46	ISO 2605	10-50	1
TS47	ISO 2605	10-50	4
TS48	ISO 2605	10-50	4
TS50	ISO 2605	10-50	3
TS53	ISO 2605	10-50	3
TS57	ISO 2605	10-50	2
TS58	ISO 2605	10-50	9
TS60	ISO 2605	10-50	13
TS61	ISO 2605	10-50	14

DESIGNATION	STANDARD	TABLE	COL
R37	ISO 3306	10-42D	8
R42	ISO 3306	10-42D	9
R50	ISO 3306	10-42D	10
HR1	ISO 3573	10-42A	1
HR2	ISO 3573	10-42A	2
HR3	ISO 3573	10-42A	3
HR4	ISO 3573	10-42A	4
CR1	ISO 3574	10-42B	1
CR2	ISO 3574	10-42B	2
CR3	ISO 3574	10-42B	3
CR4	ISO 3574	10-42B	4
Z1	ISO 3575	10-42A	12
Z1A	ISO 3575	10-42A	13
Z2	ISO 3575	10-42A	14
Z3	ISO 3575	10-42A	15
Z4	ISO 3575	10-42A	16
ZF1	ISO 3575	10-42A	12
ZF1A	ISO 3575	10-42A	13
ZF2	ISO 3575	10-42A	14
ZF3	ISO 3575	10-42A	15
ZF4	ISO 3575	10-42A	16
GRADE 20-40	ISO 3755	10-53	11
GRADE 23-45	ISO 3755	10-53	12
GRADE 26-52	ISO 3755	10-53	13
GRADE 30-57	ISO 3755	10-53	14

EUROPEAN EURONORM FERROUS MATERIALS INDEX

DESIGNATION	STANDARD	TABLE	COL
A25C M 4	EURONORM 83	10-45	1
B25C M 4	EURONORM 83	10-45	1
C36	EURONORM 86	10-48	1
C46	EURONORM 86	10-48	3
C53	EURONORM 86	10-48	5
FeK01	EURONORM 46	10-42A	1
FeK01	EURONORM 46	10-42A	12
FeK02	EURONORM 46	10-42A	2
FeK02	EURONORM 46	10-42A	14
FeK2	EURONORM 46	10-42A	3
FeK2	EURONORM 46	10-42A	15
FeK3	EURONORM 46	10-42A	4
FeK4	EURONORM 46	10-42A	16
FeK4	EURONORM 46	10-42A	4
FeK11	EURONORM 46	10-42A	16
FeK11	EURONORM 46	10-42A	3
FeK12	EURONORM 46	10-42A	15
FePO1	EURONORM 32	10-42B	1
FePO2	EURONORM 32	10-42B	2

STEEL MATERIAL DATA

WORLD FERROUS MATERIALS INDEX (Continued)

P46	ISO 2607		FeP03	EURONORM 32	10-42B	4
P47	ISO 2607	1	Fe33-0	EURONORM 47	10-50	5
P48	ISO 2607	4	Fe34B1	EURONORM 47	10-42A	5
P50	ISO 2607	5	Fe34B2	EURONORM 47	10-42A	5
P53	ISO 2607	3	Fe34B3	EURONORM 47	10-42A	5
P57	ISO 2607	2	Fe34C2	EURONORM 47	10-42A	5
P58	ISO 2607	9	Fe34C3	EURONORM 47	10-42A	5
P60	ISO 2607	13	Fe37A	EURONORM 47	10-42A	6
P61	ISO 2607	10	Fe37B1	EURONORM 47	10-42A	6
L-Ni 35	ISO 2892	14	Fe37B2	EURONORM 47	10-42A	6
L-Ni Cr 20 2	ISO 2892	18	Fe37B3	EURONORM 47	10-42A	6
L-Ni Cr 20 3	ISO 2892	14	Fe37C2	EURONORM 47	10-42A	6
L-Ni Cr 30 3	ISO 2892	15	Fe37C3	EURONORM 47	10-42A	6
L-Ni CuCr 15 6-2	ISO 2892	16	Fe37D2	EURONORM 47	10-42A	6
L-Ni CuCr 15 6 3	ISO 2892	12	Fe37D3	EURONORM 47	10-42A	6
L-Ni SiCr 30 5 5	ISO 2892	13	Fe42A	EURONORM 47	10-42A	7
S-Ni 22	ISO 2892	17	Fe42B1	EURONORM 47	10-42A	7
S-Ni 35	ISO 2892	5	Fe42B3	EURONORM 47	10-42A	7
S-Ni Cr 20 2	ISO 2892	10	Fe42C2	EURONORM 47	10-42A	7
S-Ni Cr 20 3	ISO 2892	2	Fe42C3	EURONORM 47	10-42A	7
S-Ni Cr 30 1	ISO 2892	3	Fe42D2	EURONORM 47	10-42A	7
S-Ni Cr 30 3	ISO 2892	7	Fe42D3	EURONORM 47	10-42A	7
S-Ni Cr 35 3	ISO 2892	8	Fe50-1	EURONORM 47	10-42A	9
S-Ni Mn 13 7	ISO 2892	11	Fe50-2	EURONORM 47	10-42A	9
S-Ni Mn 23 4	ISO 2892	1	Fe52C2	EURONORM 47	10-42A	8
S-Ni SiCr 20 5 2	ISO 2892	6	Fe52C3	EURONORM 47	10-42A	8
S-Ni SiCr 30 5 5	ISO 2892	9	Fe52C2	EURONORM 47	10-42A	8
C35	ISO 2937	15	Fe52D3	EURONORM 47	10-42A	8
TS1	ISO 2937	11	Fe60-1	EURONORM 47	10-42A	10
TS4	ISO 2937	12	Fe60-2	EURONORM 47	10-42A	11
TS9	ISO 2937	13	Fe70-1	EURONORM 47	10-42A	11
TS18	ISO 2937	14	Fe70-2	EURONORM 47	10-42A	11
GRADE 1	ISO 2938	16	Fe31C	EURONORM 25	10-43	1
GRADE 2	ISO 2938	17	Fe36C-A	EURONORM 25	10-43	2
R28	ISO 3304	1	Fe36C-BFN	EURONORM 25	10-43	2
R33	ISO 3304	2	Fe36C-BFL	EURONORM 25	10-43	2
R37	ISO 3304	3	Fe36C-C	EURONORM 25	10-43	2
R42	ISO 3304	4	Fe36C-C	EURONORM 25	10-43	2
R50	ISO 3304	5	Fe43C-A	EURONORM 25	10-43	4
R28	ISO 3305	1	Fe43C-B	EURONORM 25	10-43	4
R33	ISO 3305	2	Fe43C-C	EURONORM 25	10-43	4
R37	ISO 3305	3	Fe43C-C	EURONORM 25	10-43	4
R42	ISO 3305	4	Fe49C-1	EURONORM 25	10-43	6
R50	ISO 3305	5	Fe49C-2	EURONORM 25	10-43	6
R28	ISO 3306	6	Fe490-2	EURONORM 25	10-43	6
R33	ISO 3306	7	Fe510-B	EURONORM 25	10-43	5

WORLD FERROUS MATERIALS INDEX (Continued)

DESIGNATION	STANDARD	TABLE	COL
Fe510-C	EURONORM 25	10-43	5
Fe510-D	EURONORM 25	10-43	5
Fe510-DD	EURONORM 25	10-43	5
Fe590-2	EURONORM 25	10-43	7
Fe690-2	EURONORM 25	10-43	6
GRADE A	EURONORM 77	10-42B	6
GRADE B	EURONORM 77	10-42B	7
GRADE C	EURONORM 77	10-42B	8
GRADE D	EURONORM 77	10-42B	9
X3CrNi18 10	EURONORM 88	10-50	1
X3CrNiMo17 12 2	EURONORM 88	10-50	9
X3CrNiMo17 13 3	EURONORM 88	10-50	13
X3CrNiMo18 16 4	EURONORM 88	10-50	17
X5CrNi18 9	EURONORM 88	10-50	1
X6Cr13	EURONORM 88	10-49	1
X6CrAl13	EURONORM 88	10-49	2
X6CrNiTi18 10	EURONORM 88	10-50	2
X6CrNiNb18 10	EURONORM 88	10-50	3
X6CrNiMo17 12 2	EURONORM 88	10-50	10
X6CrNiMoTi17 12 2	EURONORM 88	10-50	11
X6CrNiMoNb17 12 2	EURONORM 88	10-50	12
X6CrNiMoTi17 13 3	EURONORM 88	10-50	14
X6CrNiMoTi17 13 3	EURONORM 88	10-50	15
X6CrNiMoNb17 13 3	EURONORM 88	10-50	16
X8Cr17	EURONORM 88	10-49	3
X8CrMo17	EURONORM 88	10-49	6
X8CrNi18 12	EURONORM 88	10-50	7
X8CrTi17	EURONORM 88	10-49	5
X10CrNi18 9	EURONORM 88	10-50	5
X10CrNiS18 9	EURONORM 88	10-50	6
X10CrS17	EURONORM 88	10-49	4
X12Cr13	EURONORM 88	10-49	7
X12CrNi17 7	EURONORM 88	10-50	8
X12CrS13	EURONORM 88	10-49	8
X20Cr13	EURONORM 88	10-49	9
X21CrNi17	EURONORM 88	10-49	11
X30Cr13	EURONORM 88	10-49	12
X40Cr13	EURONORM 88	10-49	13
X45Cr13	EURONORM 88	10-49	14
X105CrMo17	EURONORM 88	10-49	15
1C25	EURONORM 83	10-44	7
1C35	EURONORM 83	10-44	9
1C45	EURONORM 83	10-44	11
1C55	EURONORM 83	10-44	12
1C60	EURONORM 16	10-42C	2
1CD6	EURONORM 16	10-42C	3
1CD8	EURONORM 16	10-42C	4
1CD10	EURONORM 16	10-42C	5
1CD15	EURONORM 16	10-42C	5
2C10	EURONORM 84	10-47	5

DESIGNATION	STANDARD	TABLE	COL
10S20	EURONORM 87	10-46	6
10S22	EURONORM 87	10-46	1
10SPb 2C	EURONORM 87	10-46	7
11SMn28	EURONORM 87	10-46	2
11SMnPb28	EURONORM 87	10-46	3
12SMn35	EURONORM 87	10-46	4
12SMnPb35	EURONORM 87	10-46	5
13NiCr12	EURONORM 84	10-47	14
14CrNi6	EURONORM 84	10-47	13
14NiCrMo13	EURONORM 84	10-47	17
15Cr2	EURONORM 84	10-47	8
15NiCr6	EURONORM 84	10-47	12
16MnCr5	EURONORM 84	10-47	9
17NiCrMo5	EURONORM 84	10-47	16
17S2C	EURONORM 87	10-46	9
18CrMo4	EURONORM 84	10-47	10
20MoCr4	EURONORM 84	10-47	11
20NiCrMo2	EURONORM 84	10-47	15
22SMn2C	EURONORM 87	10-46	8
28Mn6	EURONORM 83	10-45	4
30CrNiMo8	EURONORM 83	10-45	12
31CrMo12	EURONORM 83	10-45	14
32CrMo12	EURONORM 83	10-45	15
34Cr4	EURONORM 85	10-45	5
34CrAlMo5	EURONORM 85	10-47	6
34CrMo4	EURONORM 83	10-45	2
34NiCrMo16	EURONORM 83	10-45	13
35CrNiMo6	EURONORM 83	10-45	11
35S20	EURONORM 87	10-46	10
35SMn2C	EURONORM 87	10-46	11
37Cr4	EURONORM 83	10-45	7
38Cr4	EURONORM 86	10-48	7
38CrMoV13	EURONORM 85	10-47	2
39NiCrMo3	EURONORM 83	10-45	10
40NiCrMo2	EURONORM 83	10-45	9
40NiCrMo3	EURONORM 83	10-45	11
41Cr4	EURONORM 86	10-48	8
41CrAlMo7	EURONORM 85	10-47	4
41CrMo4	EURONORM 86	10-48	9
42CrMo4	EURONORM 83	10-45	3
45Cr2	EURONORM 86	10-48	6
45Si7	EURONORM 87	10-46	13
45S20	EURONORM 89	10-51	3
45SiCrMo6	EURONORM 89	10-51	11
50CrV4	EURONORM 87	10-46	13
50Si7	EURONORM 89	10-51	13
51CrMoV4	EURONORM 89	10-51	14
55Cr3	EURONORM 89	10-51	8
55Si7	EURONORM 89	10-51	5
60S20	EURONORM 87	10-46	14

STEEL MATERIAL DATA

10-155

WORLD FERROUS MATERIALS INDEX (Continued)

2C15	EURONORM 84	10-47	6	
2C25	EURONORM 83	10-44	5	
2C35	EURONORM 83	10-44	7	
2C45	EURONORM 83	10-44	9	
2C55	EURONORM 83	10-44	11	
2C60	EURONORM 83	10-44	12	
2CD8	EURONORM 16	10-42C	3	
2CD20	EURONORM 16	10-42C	6	
2CD25	EURONORM 16	10-42C	7	
2CD30	EURONORM 16	10-42C	8	
2CD35	EURONORM 16	10-42C	9	
2CD40	EURONORM 16	10-42C	10	
2CD45	EURONORM 16	10-42C	11	
2CD50	EURONORM 16	10-42C	12	
2CD55	EURONORM 16	10-42C	13	
2CD60	EURONORM 16	10-42C	14	
2CD65	EURONORM 16	10-42C	15	
2CD70	EURONORM 16	10-42C	16	
2CD70	EURONORM 16	10-51	1	60Si7	EURONORM 89	10-51	6
3C10	EURONORM 84	10-47	5	60SiCr8	EURONORM 89	10-51	7
3C25	EURONORM 83	10-44	6				
3C35	EURONORM 83	10-44	5	UNITED STATES AISI FERROUS MATERIALS INDEX			
3C45	EURONORM 83	10-44	7				
3C55	EURONORM 83	10-44	9	DESIGNATION	STANDARD	TABLE	COL
3C60	EURONORM 83	10-44	11				
3CD5	EURONORM 16	10-42C	12	CQ	AISI	10-42A	1
3CD6	EURONORM 16	10-42C	1	CQ	AISI	10-42B	11
3CD8	EURONORM 16	10-42C	2	CR CG	AISI	10-42B	1
3CD10	EURONORM 16	10-42C	3	CR DG	AISI	10-42B	2
3CD15	EURONORM 16	10-42C	4	CR PG FULL HARD	AISI	10-42B	14
3CD20	EURONORM 16	10-42C	5	CR PG 1/2 HARD	AISI	10-42B	13
3CD25	EURONORM 16	10-42C	6	CR PG 1/4 HARD	AISI	10-42B	12
3CD30	EURONORM 16	10-42C	7	DQ	AISI	10-42A	2
3CD35	EURONORM 16	10-42C	8	DQ	AISI	10-42B	10
3CD40	EURONORM 16	10-42C	9	DQ-SK	AISI	10-42A	4
3CD45	EURONORM 16	10-42C	10	11L08	AISI	10-46	7
3CD50	EURONORM 16	10-42C	11	11L41	AISI	10-46	12
3CD55	EURONORM 16	10-42C	12	12L14	AISI	10-46	5
3CD60	EURONORM 16	10-42C	13	12L13	AISI	10-46	3
3CD65	EURONORM 16	10-42C	14	51B6C	AISI	10-51	10
3CD70	EURONORM 16	10-42C	15	202	AISI	10-50	18
3CD75	EURONORM 16	10-42C	16	301	AISI	10-50	8
3CD80	EURONORM 16	10-51	2	302	AISI	10-50	5
3CD85	EURONORM 16	10-42C	17	303	AISI	10-50	6
3CD90	EURONORM 16	10-42C	18	304	AISI	10-50	4
3CD95	EURONORM 16	10-42C	18	304L	AISI	10-50	1
				305	AISI	10-50	7
				316	AISI	10-50	10
				316L	AISI	10-50	14
				316L	AISI	10-50	9
				316L	AISI	10-50	13
				317	AISI	10-50	17
				321	AISI	10-50	2
				347	AISI	10-50	3
				403	AISI	10-49	1
				405	AISI	10-49	2
				410	AISI	10-49	7
				416	AISI	10-49	8
				416Se	AISI	10-49	8
				420	AISI	10-49	9
				420F	AISI	10-49	9
				420FSe	AISI	10-49	12
				430	AISI	10-49	3
				430F	AISI	10-49	4
				431	AISI	10-49	10

WORLD FERROUS MATERIALS INDEX (Continued)

DESIGNATION	STANDARD	TABLE	COL
436	AISI	10-49	6
440C	AISI	10-49	15
1005	AISI	10-42C	1
1006	AISI	10-42C	2
1006	AISI	10-44	1
1008	AISI	10-42C	3
1010	AISI	10-42C	4
1010	AISI	10-44	2
1010	AISI	10-47	5
1015	AISI	10-42C	5
1015	AISI	10-44	3
1015	AISI	10-47	6
1016	AISI	10-47	7
1020	AISI	10-42C	6
1020	AISI	10-44	4
1025	AISI	10-42C	7
1025	AISI	10-44	5
1030	AISI	10-42C	8
1030	AISI	10-44	6
1035	AISI	10-42C	9
1035	AISI	10-44	7
1040	AISI	10-48	1
1040	AISI	10-42A	10
1040	AISI	10-42C	10
1040	AISI	10-44	8
1045	AISI	10-42C	11
1045	AISI	10-44	9
1045	AISI	10-48	3
1050	AISI	10-42A	11
1050	AISI	10-42C	12
1050	AISI	10-44	10
1050	AISI	10-48	4
1055	AISI	10-42C	13
1055	AISI	10-44	11
1055	AISI	10-48	5
1060	AISI	10-42C	14
1065	AISI	10-42C	15
1065	AISI	10-44	13
1070	AISI	10-42C	16
1070	AISI	10-44	14
1074	AISI	10-51	1
1080	AISI	10-51	1
1080	AISI	10-42C	17
1080	AISI	10-44	15
1086	AISI	10-42C	18
1090	AISI	10-44	16
1108	AISI	10-46	6
1115	AISI	10-46	9
1117	AISI	10-46	8
1137	AISI	10-46	12
1138	AISI	10-46	10

UNITED STATES ASTM FERROUS MATERIALS INDEX

DESIGNATION	STANDARD	TABLE	COL
A36	ASTM A36	10-43	3
A526	ASTM A526	10-42A	12
A527	ASTM A527	10-42A	13
A528	ASTM A528	10-42A	14
A569	ASTM A569	10-42A	1
A621	ASTM A621	10-42A	2
A622	ASTM A622	10-42A	4
A642	ASTM A642	10-42A	16
CLASS A	ASTM A355	10-47	4
CLASS C	ASTM A355	10-47	3
CLASS I	ASTM A227	10-42C	2
CLASS I	ASTM A229	10-42C	7
CLASS II	ASTM A227	10-42C	4
CLASS II	ASTM A229	10-42C	9
CLASS 50	ASTM A48	10-52	6
CLASS 55	ASTM A48	10-52	7
CLASS 60	ASTM A48	10-52	7
CQ	ASTM A366	10-42B	1
DQ	ASTM A620	10-42B	4
DR	ASTM A619	10-42B	2
G1800	ASTM A159	10-52	1
G2500	ASTM A159	10-52	2
G3000	ASTM A159	10-52	3
G3500	ASTM A159	10-52	4
G4000	ASTM A159	10-52	5
GR 60-30	ASTM A27	10-53	11
GR 60-40-18	ASTM A536	10-53	9
GR 60-45-12	ASTM A536	10-52	10
GR 65-35	ASTM A27	10-53	12
GR 70-40	ASTM A27	10-53	13
GR 80-40	ASTM A148	10-53	14
GR 80-50	ASTM A148	10-53	14
GR 80-55-06	ASTM A536	10-53	11
GR 90-60	ASTM A148	10-53	15
GR 100-70-03	ASTM A536	10-52	12
GR 105-85	ASTM A148	10-53	16
GR 120-90-02	ASTM A536	10-52	13
GR 120-95	ASTM A148	10-53	17
GR 150-125	ASTM A148	10-53	18
GR 175-145	ASTM A148	10-53	18
GR 32510	ASTM A47	10-53	2
GR 35018	ASTM A47	10-53	3
GR 40010	ASTM A220	10-53	5
GR 45006	ASTM A220	10-53	6
GR 45008	ASTM A220	10-53	6
GR 50005	ASTM A220	10-53	7
GR 60004	ASTM A220	10-53	8
GR 70003	ASTM A220	10-53	9
GR 80002	ASTM A220	10-53	9

STEEL MATERIAL DATA

WORLD FERROUS MATERIALS INDEX (Continued)

AISI	1140	10-46	11		ASTM	A220	GR 900C1	10-53	10
AISI	1141	10-46	12		ASTM	A283	GRADE A	10-43	1
AISI	1146	10-46	13		ASTM	A284	GRADE A	10-43	5
AISI	1211	10-46	1		ASTM	A570	GRADE A	10-42A	1
AISI	1212	10-46	2		ASTM	A283	GRADE B	10-43	1
AISI	1214	10-46	4		ASTM	A570	GRADE B	10-43	6
AISI	1526	10-45	4		ASTM	A284	GRADE B	10-42A	1
AISI	1527	10-45	11		ASTM	A570	GRADE B	10-43	1
AISI	4118	10-47	1		ASTM	A283	GRADE C	10-43	7
AISI	4130	10-45	2		ASTM	A284	GRADE C	10-42A	2
AISI	4135	10-45	2		ASTM	A570	GRADE C	10-43	2
AISI	4137	10-45	3		ASTM	A283	GRADE D	10-43	8
AISI	4140	10-48	9		ASTM	A284	GRADE D	10-42A	3
AISI	4142	10-45	3		ASTM	A570	GRADE D	10-43	9
AISI	4161	10-51	12		ASTM	A570	GRADE E	10-43	4
AISI	4320	10-47	16		ASTM	A572	GRADE 42	10-43	4
AISI	4340	10-45	11		ASTM	A572	GRADE 42	10-43	6
AISI	4718	10-47	12		ASTM	A572	GRADE 50	10-43	5
AISI	5115	10-47	9		ASTM	A572	GRADE 55	10-43	7
AISI	5120	10-47	8		ASTM	A618	I	10-43	8
AISI	5130	10-45	6		ASTM	A512	MT1010	10-42D	16
AISI	5132	10-45	6		ASTM	A512	MT1010	10-42D	16
AISI	5135	10-45	7		ASTM	A519	MT1010	10-42D	1
AISI	5140	10-48	8		ASTM	A512	MT1010	10-42D	6
AISI	5145	10-48	8		ASTM	A519	MT1015	10-42D	11
AISI	5145	10-45	6		ASTM	A512	MT1015	10-42D	2
AISI	5155	10-51	8		ASTM	A519	MT1015	10-42D	7
AISI	5160	10-51	9		ASTM	A512	MT1020	10-42D	12
AISI	6150	10-51	13		ASTM	A519	MT1020	10-42D	4
AISI	8615	10-47	15		ASTM	A512	MT1020	10-42D	9
AISI	8617	10-47	15		ASTM	A519	MTX1015	10-42D	13
AISI	8620	10-45	9		ASTM	A512	MTX1015	10-42D	3
AISI	8637	10-48	10		ASTM	A519	MTX1020	10-42D	8
AISI	8640	10-45	9		ASTM	A512	MTX1020	10-42D	5
AISI	8642	10-45	9		ASTM	A519	MTX1020	10-42D	10
AISI	8740	10-45	10		ASTM	A512	MTX1020	10-42D	5
AISI	9255	10-48	10		ASTM	A519	MTX1020	10-42D	14
AISI	9260	10-51	6						
AISI	9310	10-47	17						
AISI	9840	10-45	10						

10-157

STEEL MATERIAL DATA

WORLD FERROUS MATERIALS INDEX (Continued)

DESIGNATION	STANDARD	TABLE	COL
MUSIC WIRE	ASTM A228	10-42C	11
NO 1 TEMPER	ASTM A109	10-42B	14
NO 2 TEMPER	ASTM A109	10-42B	13
NO 3 TEMPER	ASTM A109	10-42B	12
T-1	ASTM A623	10-42B	5
T-2	ASTM A623	10-42B	6
T-3	ASTM A623	10-42B	7
T-4	ASTM A623	10-42B	8
T-5	ASTM A623	10-42B	8
T-6	ASTM A623	10-42B	9
TYPE D-2	ASTM A439	10-54	2
TYPE D-2B	ASTM A439	10-54	3
TYPE D-2C	ASTM A439	10-54	5
TYPE D-3	ASTM A439	10-54	8
TYPE D-3A	ASTM A439	10-54	7
TYPE D-4	ASTM A439	10-54	9
TYPE D-5	ASTM A439	10-54	10
TYPE D-5B	ASTM A439	10-54	11
TYPE 1	ASTM A436	10-54	12
TYPE 1B	ASTM A436	10-54	13
TYPE 2	ASTM A436	10-54	14
TYPE 2B	ASTM A436	10-54	15
TYPE 3	ASTM A436	10-54	16
TYPE 4	ASTM A436	10-54	17
TYPE 5	ASTM A436	10-54	18
VALVE SPRING	ASTM A230	10-42C	13
1035	ASTM A519	10-42D	17
1045	ASTM A519	10-42D	15

UNITED STATES SAE FERROUS MATERIALS INDEX

DESIGNATION	STANDARD	TABLE	COL
D4018	SAE J434B	10-52	8
D4512	SAE J434B	10-52	9
D5506	SAE J434B	10-52	10
D7003	SAE J434B	10-52	12
DQ&T	SAE J434B	10-52	13
G1800	SAE J431B	10-52	1
G2500	SAE J431B	10-52	2
G3000	SAE J431B	10-52	3
G3500	SAE J431B	10-52	4
G4000	SAE J431B	10-52	5
GR M3210	SAE J158	10-53	5
GR M4504	SAE J158	10-53	6

DESIGNATION	STANDARD	TABLE	COL
S40C	JIS G4051	10-44	8
S40C	JIS G4051	10-48	2
S45C	JIS G4051	10-44	9
S45C	JIS G4051	10-48	3
S50C	JIS G4051	10-44	10
S50C	JIS G4051	10-48	4
S55C	JIS G4051	10-44	11
S55C	JIS G4051	10-48	5
S58C	JIS G4051	10-44	12
SACM1	JIS G4202	10-44	11
SC42	JIS G5101	10-47	11
SC46	JIS G5101	10-53	12
SC49	JIS G5101	10-53	13
SCCrM1B	JIS G5111	10-53	16
SCM 2	JIS G4105	10-45	1
SCM 3	JIS G4105	10-45	2
SCM 4	JIS G4105	10-45	3
SCM22	JIS G4105	10-48	9
SCM22	JIS G4105	10-47	10
SCM24	JIS G4105	10-47	11
SCMnCr2A	JIS G5111	10-53	14
SCMnCr2B	JIS G5111	10-53	15
SCMnCr3A	JIS G5111	10-53	15
SCMnM3B	JIS G5111	10-53	16
SCMnCr2B	JIS G5111	10-53	16
SCMnCr3B	JIS G5111	10-53	17
SCNCrM2B	JIS G5111	10-53	18
SCr2	JIS G4104	10-45	8
SCr2	JIS G4104	10-48	6
SCr3	JIS G4104	10-45	7
SCr4	JIS G4104	10-48	8
SCr4	JIS G4104	10-47	8
SCr21	JIS G4104	10-48	9
SCr22	JIS G4106	10-47	8
SMn1	JIS G4102	10-45	4
SNC1	JIS G4102	10-47	12
SNC22	JIS G4102	10-47	14
SNCM 6	JIS G4103	10-45	9
SNCM 8	JIS G4103	10-47	10
SNCM21	JIS G4103	10-45	15
SNCM25	JIS G4103	10-47	18
SPCC	JIS G3141	10-42B	1
SPCD	JIS G3141	10-42B	3
SPCE	JIS G3141	10-42B	4
SPG1	JIS G3302	10-42A	12
SPG2C	JIS G3302	10-42A	12

STEEL MATERIAL DATA

WORLD FERROUS MATERIALS INDEX (Continued)

GR M5003	SAE J158	10-53	7
GR M5503	SAE J158	10-53	8
GR M7002	SAE J158	10-53	9
GR M8501	SAE J158	10-53	10
GR 0025	SAE J435C	10-53	11
GR 0030	SAE J435C	10-53	12
GR 0050A	SAE J435C	10-53	13
GR 080	SAE J435C	10-53	14
GR 090	SAE J435C	10-53	15
GR 0105	SAE J435C	10-53	16
GR 0120	SAE J435C	10-53	17
GR 0150	SAE J435C	10-53	18
GR 0175	SAE J435C	10-53	18

JAPANESE JIS FERROUS MATERIALS INDEX

DESIGNATION	STANDARD	TABLE	COL
A		10-42B	10
FC-10	JIS G3141	10-52	1
FC-15	JIS G5501	10-52	2
FC-20	JIS G5501	10-52	3
FC-25	JIS G5501	10-52	4
FC-30	JIS G5501	10-52	5
FC-35	JIS G5501	10-52	6
FCD-40	JIS G5501	10-52	9
FCD-50	JIS G5502	10-52	10
FCD-60	JIS G5502	10-52	11
FCD-70	JIS G5502	10-52	12
FCMB28	JIS G5702	10-53	1
FCMB32	JIS G5702	10-53	2
FCMB35	JIS G5702	10-53	3
FCMB37	JIS G5702	10-53	3
FCMP45	JIS G5704	10-53	6
FCMP50	JIS G5704	10-53	7
FCMP55	JIS G5704	10-53	8
FCMP60	JIS G5704	10-53	9
FCMP70	JIS G5704	10-53	10
FCMW34	JIS G5703	10-53	4
FCMW38	JIS G5703	10-53	4
S	JIS G3141	10-42B	11
S9CK	JIS G4051	10-47	5
S10C	JIS G4051	10-44	2
S15C	JIS G4051	10-44	3
S15CK	JIS G4051	10-47	6
S20C	JIS G4051	10-44	4
S20CK	JIS G4051	10-47	7
S25C	JIS G4051	10-44	5
S30C	JIS G4051	10-44	6
S35C	JIS G4051	10-44	7
S35C	JIS G4051	10-48	1

SPG2D	JIS G3302	10-42A	16
SPG2L	JIS G3302	10-42A	14
SPG2S	JIS G3302	10-42A	13
SPG3C	JIS G3302	10-42A	12
SPG3L	JIS G3302	10-42A	14
SPG3S	JIS G3302	10-42A	13
SPHC	JIS G3131	10-42A	1
SPHD	JIS G3131	10-42A	3
SPHE	JIS G3131	10-42A	4
SS34	JIS G3101	10-42A	5
SS34	JIS G3101	10-43	1
SS41	JIS G3101	10-42A	7
SS41	JIS G3101	10-43	2
SS50	JIS G3101	10-42A	9
SS50	JIS G3101	10-43	6
SS55	JIS G3101	10-42A	10
SS55	JIS G3101	10-43	7
STK30	JIS G3445	10-42D	11
STK50	JIS G3444	10-42D	10
STK55	JIS G3445	10-42D	10
STKM11A	JIS G3445	10-42D	6
STKM12A	JIS G3445	10-42D	13
STKM12B	JIS G3445	10-42D	13
STKM12C	JIS G3445	10-42D	13
STKM13A	JIS G3445	10-42D	14
STKM13B	JIS G3445	10-42D	14
STKM13C	JIS G3445	10-42D	14
STKM16A	JIS G3445	10-42D	17
STKM16C	JIS G3445	10-42D	17
STKM17A	JIS G3445	10-42D	15
STKM17C	JIS G3445	10-42D	15
STKM18A	JIS G3445	10-42D	16
STKM18B	JIS G3445	10-42D	16
STKM18C	JIS G3445	10-42D	16
SUM12	JIS G4804	10-46	6
SUM21	JIS G4804	10-46	1
SUM22	JIS G4804	10-46	2
SUM22L	JIS G4804	10-46	3
SUM23L	JIS G4804	10-46	7
SUM24L	JIS G4804	10-46	4
SUM32	JIS G4804	10-46	5
SUM41	JIS G4804	10-46	9
SUM43	JIS G4804	10-46	12
SUP3	JIS G4801	10-46	13
SUP6	JIS G4801	10-51	1
SUP7	JIS G4801	10-51	5
SUP9	JIS G4801	10-51	6
SUP10	JIS G4801	10-51	8
SUP11	JIS G4801	10-51	13
SUS2C2	JIS G4303	10-50	10
SUS3C1	JIS G4303	10-50	18
SUS3C2	JIS G4303	10-50	5

WORLD FERROUS MATERIALS INDEX (Continued)

DESIGNATION	STANDARD	TABLE	COL
SUS303	JIS G4303	10-50	6
SUS303se	JIS G4303	10-50	6
SUS304	JIS G4303	10-50	4
SUS304L	JIS G4303	10-50	1
SUS305	JIS G4303	10-50	7
SUS316	JIS G4303	10-50	10
SUS316L	JIS G4303	10-50	14
SUS316L	JIS G4303	10-50	9
SUS317	JIS G4303	10-50	13
SUS321	JIS G4303	10-50	17
SUS347	JIS G4303	10-50	2
SUS403	JIS G4303	10-49	1
SUS405	JIS G4303	10-49	7
SUS410	JIS G4303	10-49	2
SUS416	JIS G4303	10-49	7
SUS420F	JIS G4303	10-49	8
SUS420J1	JIS G4303	10-49	8
SUS420J2	JIS G4303	10-49	12
SUS429	JIS G4303	10-49	4
SUS430	JIS G4303	10-49	3
SUS430F	JIS G4303	10-49	4
SUS431	JIS G4303	10-49	10
SUS434	JIS G4303	10-49	6
SUS440C	JIS G4303	10-49	15
SUS440F	JIS G4303	10-49	15
SWA	JIS G3521	10-42C	5
SWB	JIS G3521	10-42C	6
SWC	JIS G3521	10-42C	8
SWPA	JIS G3522	10-42C	11
SWPB	JIS G3522	10-42C	13
SWPV 27	JIS G3522	10-42C	14
SWRH 32	JIS G3506	10-42C	7
SWRH 37	JIS G3506	10-42C	8
SWRH 42A	JIS G3506	10-42C	9
SWRH 42B	JIS G3506	10-42C	10
SWRH 47A	JIS G3506	10-42C	10
SWRH 47B	JIS G3506	10-42C	11
SWRH 52A	JIS G3506	10-42C	11
SWRH 52B	JIS G3506	10-42C	12
SWRH 57A	JIS G3506	10-42C	12
SWRH 57B	JIS G3506	10-42C	13
SWRH 62A	JIS G3506	10-42C	13
SWRH 62B	JIS G3506	10-42C	14
SWRH 67A	JIS G3506	10-42C	15

DESIGNATION	STANDARD	TABLE	COL
Ck35	DIN 17200	10-44	7
Ck45	DIN 17200	10-44	9
Ck55	DIN 17200	10-44	11
Ck60	DIN 17200	10-44	12
Cm15	DIN 17210	10-47	6
Cm35	DIN 17200	10-44	7
Cm45	DIN 17200	10-44	9
Cm55	DIN 17200	10-44	11
Cm60	DIN 17200	10-44	12
D	DIN 1616	10-42B	9
D5-1	DIN 17140	10-42C	1
D6-2	DIN 17140	10-42C	2
D7-1	DIN 17140	10-42C	2
D8-2	DIN 17140	10-42C	3
D9-1	DIN 17140	10-42C	4
D12-2	DIN 17140	10-42C	4
D15-2	DIN 17140	10-42C	5
D20-2	DIN 17140	10-42C	6
D26-2	DIN 17140	10-42C	7
D35-2	DIN 17140	10-42C	8
D45-2	DIN 17140	10-42C	9
D55-2	DIN 17140	10-42C	11
D65-2	DIN 17140	10-42C	13
D75-2	DIN 17140	10-42C	15
D75-2	DIN 17140	10-42C	16
D85-2	DIN 17140	10-51	1
D95-2	DIN 17140	10-42C	17
FD	DIN 17223	10-42C	18
G	DIN 1624	10-42B	10
GG-10	DIN 1691	10-52	1
GG-15	DIN 1691	10-52	2
GG-2C	DIN 1691	10-52	3
GG-25	DIN 1691	10-52	4
GG-3C	DIN 1691	10-52	5
GG-35	DIN 1691	10-52	6
GG-4C	DIN 1691	10-52	7
GGG-40	DIN 1693	10-52	9
GGG-50	DIN 1693	10-52	10
GGG-60	DIN 1693	10-52	11
GGG-70	DIN 1693	10-52	12
GGG-80	DIN 1693	10-52	13
GGG-NiMn13 7	DIN 1694	10-54	1
GGG-NiCr20 2	DIN 1694	10-54	2
GGG-NiCr20 3	DIN 1694	10-54	3
GGG-NiSiCr20 4 2	DIN 1694	10-54	4
GGG-Ni22	DIN 1694	10-54	5
GGG-NiMn23 4	DIN 1694	10-54	6
GGG-NiCr30 1	DIN 1694	10-54	7

STEEL MATERIAL DATA

WORLD FERROUS MATERIALS INDEX (*Continued*)

SWRH 67B	JIS G3506	10-42C	15	GGG-NiCr30 3	DIN 1694	10-54	8
SWRH 72A	JIS G3506	10-42C	16	GGG-NiSiCr30 5 5	DIN 1694	10-54	9
SWRH 72B	JIS G3506	10-42C	16	GGG-Ni35	DIN 1694	10-54	10
SWRH 77A	JIS G3506	10-42C	16	GGL-NiCuCr15 6 2	DIN 1694	10-54	11
SWRH 82A	JIS G3506	10-42C	17	GGL-NiCuCr15 6 3	DIN 1694	10-54	12
SWRH 82B	JIS G3506	10-42C	17	GGL-NiCr20 2	DIN 1694	10-54	13
SWRM 6	JIS G3505	10-42C	2	GGL-NiCr20 3	DIN 1694	10-54	14
SWRM 8	JIS G3505	10-42C	3	GGL-NiCr30 3	DIN 1694	10-54	15
SWRM 10	JIS G3505	10-42C	4	GGL-NiSiCr30 5 5	DIN 1694	10-54	16
SWRM 12	JIS G3505	10-42C	5	GGL-Ni35	DIN 1694	10-54	17
SWRM 15	JIS G3505	10-42C	5	GS-38	DIN 1681	10-54	18
SWRM 17	JIS G3505	10-42C	6	GS-38.3	DIN 1681	10-53	11
SWRM 20	JIS G3505	10-42C	6	GS-45	DIN 1681	10-53	11
SWRM 22	JIS G3505	10-42B	5	GS-45.3	DIN 1681	10-53	12
T-1	JIS G3303	10-42B	6	GS-52	DIN 1681	10-53	13
T-2	JIS G3303	10-42B	7	GS-52.3	DIN 1681	10-53	13
T-3	JIS G3303	10-42B	8	GS-60	DIN 1681	10-53	14
T-4	JIS G3303	10-42B	8	GS-60.3	DIN 1681	10-53	14
T-5	JIS G3303	10-42B	14	GS-62	DIN 1681	10-53	15
1 TEMPER	JIS G3141	10-42B	13	GS-62.3	DIN 1681	10-53	15
2 TEMPER	JIS G3141	10-42B	12	GS-70	DIN 1681	10-53	16
4 TEMPER	JIS G3141	10-42B	11	GTS-35	DIN 1692	10-53	3
8 TEMPER	JIS G3141			GTS-45	DIN 1692	10-53	6
				GTS-55	DIN 1692	10-53	8
				GTS-65	DIN 1692	10-53	9
				GTS-70	DIN 1692	10-53	10
				GTW-35	DIN 1692	10-53	4
				GTW-40	DIN 1692	10-53	4

GERMAN DIN FERROUS MATERIALS INDEX

DESIGNATION	STANDARD	TABLE	COL				
A	DIN 1616	10-42B	6	GTW-55	DIN 1692	10-53	4
A	DIN 17223	10-42C	5	II	DIN 17223	10-42C	10
B	DIN 1616	10-42B	7	K32	DIN 1624	10-42B	11
B	DIN 17223	10-42C	7	K40	DIN 1624	10-42B	12
C	DIN 1616	10-42B	8	K50	DIN 1624	10-42B	13
C	DIN 17223	10-42C	9	K60	DIN 1624	10-42B	14
C10	DIN 17210	10-44	2	K70	DIN 1624	10-42B	14
C10	DIN 17210	10-47	5	LG	DIN 1624	10-42B	11
C15	DIN 17210	10-47	3	RSt34-1	DIN 17100	10-43	1
C15	DIN 17210	10-44	6	RSt34-2	DIN 17100	10-43	1
C22	DIN 17200	10-44	4	RSt37-1	DIN 17100	10-43	2
C35	DIN 17200	10-44	7	RSt37-2	DIN 17100	10-43	2
C45	DIN 17200	10-44	9	RSt37-2	DIN 1623	10-42A	6
C55	DIN 17200	10-44	11	RSt42-1	DIN 17100	10-43	3
C60	DIN 17200	10-44	12	RSt42-2	DIN 1623	10-42A	7
Cf53	DIN 17212	10-47	5	RSt42-2	DIN 17100	10-43	3
Cf45	DIN 17212	10-48	3	RSt46-2	DIN 17100	10-43	4
Cf35	DIN 17212	10-48	1	StcO	DIN 1629	10-42D	11
Ck10	DIN 17210	10-47	5	StO	DIN 1624	10-42A	1
Ck15	DIN 17210	10-47	6	StO	DIN 1624	10-42A	12
Ck22	DIN 17200	10-44	4	St1	DIN 1624	10-42A	1

STEEL MATERIAL DATA

WORLD FERROUS MATERIALS INDEX (Continued)

DESIGNATION	STANDARD	TABLE	COL
St1	DIN 1624	10-42A	13
St2	DIN 1624	10-42A	2
St2	DIN 1624	10-42A	14
St3	DIN 1624	10-42A	3
St3	DIN 1624	10-42A	15
St4	DIN 1624	10-42A	4
St4	DIN 1624	10-42A	16
St12	DIN 1623	10-42B	1
St13	DIN 1623	10-42B	2
St14	DIN 1623	10-42B	4
St33	DIN 1626	10-42D	6
St33-1	DIN 17100	10-43	1
St33-2	DIN 17100	10-43	1
St34	DIN 1626	10-42D	8
St34-2	DIN 2393	10-42D	2
St35	DIN 2391	10-42D	1
St35.4	DIN 1629	10-42D	12
St35.8	DIN 17175	10-42D	9
St37	DIN 1626	10-42D	3
St37-2	DIN 2393	10-42D	9
St37-2	DIN 2393	10-43	2
St37-3	DIN 17100	10-42D	10
St42	DIN 1626	10-42D	10
St42-2	DIN 2393	10-43	3
St42-2	DIN 2393	10-42D	4
St42-3	DIN 17100	10-42D	4
St45	DIN 2391	10-42D	13
St45.4	DIN 1629	10-42D	13
St45.8	DIN 17175	10-42D	4
St46-3	DIN 17100	10-46	6
St50-2	DIN 17100	10-43	6
St50-2	DIN 17100	10-43	6
St52	DIN 2393	10-42D	10
St52	DIN 1626	10-42D	10
St52	DIN 2393	10-42D	5
St52-3	DIN 2391	10-42D	8
St52-3	DIN 1623	10-42A	5
St52-3	DIN 2393	10-42D	5
St52-3	DIN 17100	10-43	5
St52.4	DIN 1629	10-42D	14
St55	DIN 2391	10-42D	16
St55.4	DIN 1629	10-42D	15
St60-1	DIN 17100	10-42D	17
St60-2	DIN 1623	10-42A	10
St60-2	DIN 17100	10-43	7
St70-2	DIN 1623	10-42A	11
St70-2	DIN 17100	10-43	8
SSt2	DIN 2394	10-42D	7

DESIGNATION	STANDARD	TABLE	COL
15CrNi6	DIN 17210	10-47	13
16MnCr5	DIN 17210	10-47	9
16MnCrS5	DIN 17210	10-47	9
20MoCr4	DIN 17210	10-47	11
20MoCrS4	DIN 17210	10-47	11
25CrMo4	DIN 17200	10-45	1
28Mn6	DIN 17200	10-45	4
30CrNiMo8	DIN 17200	10-45	12
31CrMo12	DIN 17211	10-47	1
32CrMo12	DIN 17211	10-47	5
34Cr4	DIN 17200	10-45	6
34CrAlMo5	DIN 17211	10-47	3
34CrMo4	DIN 17200	10-45	2
34CrMoS4	DIN 17200	10-45	2
34CrNiMo6	DIN 17200	10-45	11
34CrS4	DIN 17200	10-45	6
35S20	DIN 1651	10-46	10
37Cr4	DIN 17200	10-45	7
37CrS4	DIN 17200	10-45	7
38Cr4	DIN 17212	10-48	7
39CrMoV139	DIN 17211	10-47	2
41Cr4	DIN 17200	10-45	8
41CrAlMo7	DIN 17211	10-47	4
41CrMo4	DIN 17212	10-48	9
41CrS4	DIN 17200	10-45	8
42Cr4	DIN 17212	10-48	8
42CrMoS4	DIN 17200	10-45	3
43CrMo4	DIN 17200	10-45	3
45Cr2	DIN 17212	10-48	6
45S20	DIN 1651	10-46	13
46Si7	DIN 17221	10-51	3
50CrV4	DIN 17221	10-51	13
51CrMoV4	DIN 17221	10-51	14
51Si7	DIN 17221	10-51	2
55Cr3	DIN 17221	10-51	8
55Si7	DIN 17221	10-51	5
60S20	DIN 1651	10-46	14
60SiCr7	DIN 17221	10-51	7

FRENCH NF FERROUS MATERIALS INDEX

DESIGNATION	STANDARD	TABLE	COL
A	NF A36-150	10-42B	6
A33	NF A35-501	10-42A	5
A33	NF A35-501	10-43	1
A33-2	NF A35-501	10-43	1
A34	NF A35-501	10-42A	5

STEEL MATERIAL DATA 10-163

WORLD FERROUS MATERIALS INDEX (Continued)

TUSt37	DIN 1623		A34-1	NF A35-501	10-43	1
USt34-1	DIN 17100	6	A34-2	NF A35-501	10-43	1
USt34-2	DIN 17100	1	A37	NF A35-501	10-42A	6
USt37-2	DIN 1623	6	A37	NF A35-501	10-43	2
USt37-1	DIN 17100	2	A42	NF A35-501	10-42A	7
USt37-2	DIN 17100	2	A42	NF A35-501	10-43	3
USt42-1	DIN 17100	3	A47	NF A35-501	10-42A	7
USt42-2	DIN 1623	7	A50	NF A35-501	10-43	4
USt42-2	DIN 17100	3	A50-1	NF A35-501	10-42A	9
VD	DIN 17223	13	A50-2	NF A35-501	10-43	6
WUSt37-2	DIN 1623		A52	NF A35-501	10-42A	8
X2CrNi18 9	DIN 17440	1	A52	NF A35-501	10-43	5
X2CrNiMo18 10	DIN 17440	9	A60	NF A35-501	10-42A	10
X2CrNiMo18 12	DIN 17440	13	A60-1	NF A35-501	10-43	7
X2CrNiMo18 16	DIN 17440	17	A60-2	NF A35-501	10-42A	11
X5CrNi18 9	DIN 17440	4	A70	NF A35-501	10-43	8
X5CrNi19 11	DIN 17440	7	A70-2	NF A35-501	10-42A	7
X5CrNiMo18 10	DIN 17440	10	B	NF A36-150	10-42B	4
X5CrNiMo18 12	DIN 17440	14	B1	NF A36-150	10-42C	8
X6CrMo17	DIN 17440	6	C	NF A47-301	10-42B	6
X7Cr13	DIN 17440	1	C1	NF A47-301	10-42C	12
X7CrA113	DIN 17440	2	C30C13	NF A35-578	10-49	2
X8Cr17	DIN 17440	3	CC10	NF A35-551	10-44	5
X8CrTi17	DIN 17440	5	CC10	NF A35-551	10-47	4
X8CrMnNi18 9	DIN 17440	18	CC20	NF A35-551	10-44	7
X10Cr13	DIN 17440	7	CC20	NF A35-551	10-47	7
X10CrNiMoNb18 10	DIN 17440	12	CC35	NF A35-551	10-44	1
X10CrNiMoNb18 12	DIN 17440	16	CC35	NF A35-551	10-48	9
X10CrNiMoTi18 12	DIN 17440	15	CC45	NF A35-551	10-44	3
X10CrNiMoTi18 10	DIN 17440	11	CC45	NF A35-551	10-48	12
X10CrNiTi18 9	DIN 17440	3	CLASS I	NF A36-321	10-42A	13
X10CrNiNb18 9	DIN 17440	8	CLASS II	NF A36-321	10-42A	14
X12Crs13	DIN 17440	8	CLASS III	NF A36-321	10-42A	15
X12CrNi17 7	DIN 17224	6	CLASS IV	NF A36-321	10-42A	16
X12CrNiS18 8	DIN 17440	5	CLASS V	NF A36-150	10-42B	9
X12CrNi18 8	DIN 17440	9	D	NF A32-051	10-53	11
X20Cr13	DIN 17440	11	E20-40-M	NF A32-051	10-53	12
X22CrNi17	DIN 17440	12	E23-45-M	NF A35-501	10-42A	6
X30Cr13	DIN 17440	14	E24	NF A35-501	10-43	2
X40Cr13	DIN 17440	15	E24-1	NF A35-501	10-43	2
X105CrMo17	DIN 1651	1	E24-2	NF A35-501	10-43	2
9S20	DIN 1651	2	E24-3	NF A35-501	10-42A	7
9SMn28	DIN 1651	4	E24-4	NF A35-501	10-43	3
9SMn36	DIN 1651	3	E26	NF A35-501	10-43	3
9SMnPb28	DIN 1651	5	E26-1	NF A35-501	10-43	3
9SMnPb36	DIN 1651	6	E26-2	NF A35-501	10-43	3
10S20	DIN 1651	7	E26-3	NF A35-501	10-43	3
10SPb20	DIN 17210	8	E26-4	NF A35-501	10-43	3
15Cr3						

STEEL MATERIAL DATA

WORLD FERROUS MATERIALS INDEX (Continued)

DESIGNATION	STANDARD	TABLE	COL	DESIGNATION	STANDARD	TABLE	COL
E26-52-M	NF A32-051	10-53	13	TU37-B	NF A49-311	10-42D	12
E30	NF A35-501	10-42A	7	TU37-B	NF A49-310	10-42D	3
E30-2	NF A35-501	10-43	4	TU52-B	NF A49-310	10-42D	5
E30-3	NF A35-501	10-43	4	TU52-B	NF A49-311	10-42D	14
E30-4	NF A35-501	10-43	4	TU52-B	PR A49-312	10-42D	16
E30-57-M	NF A32-051	10-53	14	TU56-B	NF A49-311	10-42D	15
E36	NF A35-501	10-42A	8	TUXC35	NF A49-311	10-42D	15
E36-2	NF A35-501	10-43	5	TUXC35	PR A49-312	10-42D	17
E36-3	NF A35-501	10-43	5	XC10	NF A35-551	10-44	2
E36-4	NF A35-501	10-43	5	XC10	NF A35-551	10-47	5
F0	NF	10-42B	11	XC12	NF A35-551	10-44	3
FOT1	NF	10-42B	12	XC18	NF A35-551	10-44	3
FOT2	NF	10-42B	13	XC18	NF A35-551	10-47	6
FOT3	NF	10-42B	14	XC25	NF A35-551	10-44	5
FGS38/15	NF A32-301	10-52	8	XC32	NF A35-551	10-44	6
FGS42/12	NF A32-301	10-52	9	XC38	NF A35-551	10-44	8
FGS50/7	NF A32-301	10-52	10	XC38	NF A35-551	10-48	1
FGS60/2	NF A32-301	10-52	11	XC42	NF A35-551	10-44	8
FGS70/2	NF A32-301	10-52	12	XC42	NF A35-551	10-48	2
FM5-2	NF A35-051	10-42C	1	XC48	NF A35-551	10-44	9
FM5-3	NF A35-051	10-42C	1	XC48	NF A35-551	10-48	4
FM6-2	NF A35-051	10-42C	2	XC55	NF A35-551	10-44	11
FM6-3	NF A35-051	10-42C	2	XC55	NF A35-551	10-48	5
FM8-1	NF A35-051	10-42C	3	XC65	NF A35-551	10-44	13
FM8-3	NF A35-051	10-42C	3	XC70	NF A35-551	10-44	14
FM10-2	NF A35-051	10-42C	4	XC80	NF A35-551	10-44	15
FM10-3	NF A35-051	10-42C	4	XC80	NF A35-551	10-51	1
FM15-1	NF A35-051	10-42C	5	XE	NF A36-401	10-42B	2
FM15-3	NF A35-051	10-42C	5	XES	NF A36-401	10-42B	4
FM18-1	NF A35-051	10-42C	6	Z2CN18.1C	NF A35-602	10-50	1
FM18-3	NF A35-051	10-42C	6	Z2CNC17.12	NF A35-602	10-50	9
FM20-1	NF A35-051	10-42C	6	Z2CNC17.13	NF A35-602	10-50	13
FM20-3	NF A35-051	10-42C	6	Z2CNC19.5	NF A35-602	10-50	17
FM26-3	NF A35-051	10-42C	7	Z6C13	NF A35-578	10-50	1
FM32-3	NF A35-051	10-42C	8	Z6CA13	NF A35-578	10-49	2
FM36-3	NF A35-051	10-42C	9	Z6CN18.09	NF A35-602	10-50	4
FM38-3	NF A35-051	10-42C	10	Z6CNC17.11	NF A35-602	10-50	10
FM40-3	NF A35-051	10-42C	10	Z6CNC17.12	NF A35-602	10-50	14
FM42-3	NF A35-051	10-42C	11	Z6CNCNb17.12	NF A35-602	10-50	12
FM46-3	NF A35-051	10-42C	11	Z6CNCNb17.13	NF A35-602	10-50	16
FM50-3	NF A35-051	10-42C	11	Z6CNNb18.11	NF A35-602	10-50	2
FM52-3	NF A35-051	10-42C	12	Z6CNT18.11	NF A35-602	10-49	3
FM56-3	NF A35-051	10-42C	12	Z8C17	NF A35-578	10-50	10
FM58-3	NF A35-051	10-42C	13	Z8CD17.01	NF A35-578	10-49	6
FM60-3	NF A35-051	10-42C	13	Z8CN18.12	NF A35-602	10-50	14
FM62-3	NF A35-051	10-42C	14	Z8CNDT17.12	NF A35-602	10-50	11
FM66-3	NF A35-051	10-42C	14	Z8CNDT17.13	NF A35-602	10-50	15
		10-42C	15	Z10C13	NF A35-578	10-49	7

STEEL MATERIAL DATA

10-165

WORLD FERROUS MATERIALS INDEX (Continued)

FM68-3	NF A35-051	15		Z10CN18.09	NF A35-602	10-50	5
FM72-3	NF A35-051	16		Z10CNF18.09	NF A35-602	10-50	6
FM76-3	NF A35-051	16		Z12C13	NF A35-578	10-49	7
FM78-3	NF A35-051	17		Z12CF13	NF A35-602	10-49	8
FM80-3	NF A35-051	17		Z12CN17.08	NF A35-578	10-50	8
FM82-3	NF A35-051	17		Z20C13	NF A35-578	10-49	9
FM86-3	NF A35-051			Z40C14	NF A36-401	10-49	14
Ft10	NF A32-101	1		ZE	NF A36-401	10-42B	2
Ft15	NF A32-101	2		ZES	NF A36-301	10-42B	1
Ft20	NF A32-101	3		0C	NF A36-301	10-42A	2
Ft25	NF A32-101	4		1C	NF A36-301	10-42A	3
Ft30	NF A32-101	5		2C	NF A36-301	10-42A	4
Ft35	NF A32-101	6		3C	NF A35-562	10-46	6
Ft40	NF A32-101	7		10F1	NF A35-562	10-46	8
L-N35	NF A32-301	18		13MF4	NF A35-551	10-47	14
L-NC20 2	NF A32-301	14		14NC11	NF A35-551	10-47	8
L-NC20 3	NF A32-301	15		16MC5	NF A35-551	10-47	12
L-NC30 3	NF A32-301	16		16NC6	NF A35-551	10-47	18
L-NC30 5 5	NF A32-301	17		16NCC13	NF A35-551	10-47	10
L-NUC15 6 2	NF A32-301	12		18CD4	NF A35-562	10-46	9
L-NUC15 6 3	NF A32-301	13		18MF5	NF A35-551	10-47	16
MB35-7	NF A32-701	4		18NCD6	NF A35-551	10-45	4
MB40-10	NF A32-701	4		20M5	NF A35-551	10-47	9
MN32-8	NF A32-702	2		20MC5	NF A35-551	10-47	15
MN35-10	NF A32-702	3		20NCC2	NF A35-551	10-45	1
MN38-18	NF A32-702	3		25CD4	NF A35-551	10-47	3
MP50-5	NF A32-703	7		30CAC6-12	NF A35-551	10-45	2
MP60-3	NF A32-703	9		30CD4	NF A35-551	10-45	12
MP70-2	NF A32-703	10		30CND8	NF A35-551	10-45	6
S-N22	NF A32-301	5		32C4	NF A35-551	10-45	2
S-N35	NF A32-301	10		35CD4	NF A35-551	10-45	4
S-NC20 2	NF A32-301	2		35M5	NF A35-562	10-46	11
S-NC20 3	NF A32-301	3		35MF6	NF A35-551	10-46	10
S-NC30 1	NF A32-301	7		35NCD6	NF A35-551	10-45	11
S-NC30 3	NF A32-301	8		35NCD6	NF A35-551	10-45	13
S-NC35 3	NF A32-301	11		35NCD16	NF A35-551	10-45	7
S-NM13 7	NF A32-301	1		38C4	NF A35-551	10-45	7
S-NM23 4	NF A32-301	6		38C4	NF A35-551	10-48	4
S-NSC20 5 2	NF A32-301	9		40CAC6-12	NF A35-551	10-47	9
S-NSC30 5 5	NF A32-301	4		40NCC3	NF A35-551	10-45	11
S250	NF A35-561	2		40NCC3	NF A35-551	10-48	6
S250P	NF A35-561	3		42C2	NF A35-551	10-45	8
S300	NF A35-561	4		42C4	NF A35-551	10-48	8
S300P	NF A35-561	5		42CD4	NF A35-551	10-45	3
TC	NF A36-401	1		42CD4	NF A35-551	10-48	9
TS28-A	NF PR A49-341	1		45MF6	NF A35-562	10-46	13
TS34-A	NF PR A49-341	2		45S7	NF A35-551	10-51	3
TS37-A	NF PR A49-341	3		45SCD6	NF A35-551	10-51	11
TS42-A	NF PR A49-341	4		50CV4	NF A35-551	10-51	13
TS47-A	NF PR A49-341	5					

WORLD FERROUS MATERIALS INDEX (Continued)

DESIGNATION	STANDARD	TABLE	COL
55S7	NF A35-551	10-51	5
60SC7	NF A35-551	10-51	7

BRITISH BS FERROUS MATERIALS INDEX

DESIGNATION	STANDARD	TABLE	COL
A	BS 1449	10-42B	10
AUS101 GRADE A	BS 3468	10-54	12
AUS101 GRADE B	BS 3468	10-54	13
AUS102 GRADE A	BS 3468	10-54	14
AUS102 GRADE B	BS 3468	10-54	15
AUS105	BS 3468	10-54	16
AUS202 GRADE A	BS 3468	10-54	2
AUS202 GRADE B	BS 3468	10-54	3
AUS203	BS 3468	10-54	5
AUS204	BS 3468	10-54	4
AUS205	BS 3468	10-54	8
CDS-1	BS 980	10-42D	1
CDS-1	BS 980	10-42D	11
CDS-2	BS 980	10-42D	3
CDS-2	BS 980	10-42D	12
CDS-3	BS 980	10-42D	3
CDS-3	BS 980	10-42D	12
CDS-6	BS 980	10-42D	15
CDS-9	BS 980	10-42D	13
CDS-10	BS 980	10-42D	14
CDS-11	BS 980	10-42D	1
CDS-13	BS 1775	10-42D	2
CDS-13	BS 1775	10-42D	11
CDS-16	BS 1775	10-42D	3
CDS-16	BS 1775	10-42D	12
CDS-20	BS 1775	10-42D	4
CEW-1	BS 980	10-42B	1
CEW-2	BS 980	10-42B	1
CEW-16	BS 1775	10-42B	3
CEW-23	BS 1775	10-42B	4
CEW-24	BS 1775	10-42B	5
CEW-28	BS 1775	10-42B	4
CR1	BS 1449	10-42B	4
CR2	BS 1449	10-42B	4
CR3	BS 1449	10-42B	3
CR4	BS 1449	10-42B	2
CS1	BS 1449	10-42B	4
CS2	BS 1449	10-42B	4
CS3	BS 1449	10-42B	3
CS4	BS 1449	10-42B	2
ERW-1	BS 980	10-42D	6
ERW-2	BS 980	10-42D	3

DESIGNATION	STANDARD	TABLE	COL
GRADE 43D	BS 4360	10-43	4
GRADE 43E	BS 4360	10-43	4
GRADE 46/40	BS 1449	10-42A	11
GRADE 50/35	BS 1449	10-42A	10
GRADE 50A	BS 4360	10-43	5
GRADE 50A	BS 4360	10-43	6
GRADE 50B	BS 4360	10-43	5
GRADE 50B	BS 4360	10-43	6
GRADE 50C	BS 4360	10-43	5
GRADE 50C	BS 4360	10-43	6
GRADE 50C	BS 4360	10-43	5
GRADE 55C	BS 4360	10-43	7
GRADE 55E	BS 4360	10-43	7
H	BS 1449	10-42B	14
HD GRADE 2	BS 5216	10-42C	12
HD GRADE 3	BS 5216	10-42C	13
HFS-20	BS 1775	10-42D	16
HFW-13	BS 1775	10-42D	7
HFW-16	BS 1775	10-42D	8
HFW-23	BS 1775	10-42D	10
HH	BS 1449	10-42B	13
HR1	BS 1449	10-42A	4
HR1	BS 1449	10-42A	16
HR2	BS 1449	10-42A	4
HR2	BS 1449	10-42A	16
HR3	BS 1449	10-42A	3
HR3	BS 1449	10-42A	15
HR4	BS 1449	10-42A	2
HR14	BS 1449	10-42A	14
HR14	BS 1449	10-42A	2
HR15	BS 1449	10-42A	14
HR15	BS 1449	10-42A	1
HS GRADE 1	BS 1449	10-42A	12
HS GRADE 1	BS 5216	10-42C	4
HS GRADE 2	BS 5216	10-42C	16
HS GRADE 3	BS 5216	10-42C	15
HS1	BS 1449	10-42A	4
HS1	BS 1449	10-42A	16
HS2	BS 1449	10-42A	3
HS2	BS 1449	10-42A	15
HS3	BS 1449	10-42A	2
HS3	BS 1449	10-42A	14
HS4	BS 1449	10-42A	2
HS4	BS 1449	10-42A	14
HS14	BS 1449	10-42A	14
HS14	BS 1449	10-42A	1
HS15	BS 1449	10-42A	14
HS15	BS 1449	10-42A	12
M GRADE 4	BS 5216	10-42C	16
M GRADE 5	BS 5216	10-42C	17

STEEL MATERIAL DATA

WORLD FERROUS MATERIALS INDEX (Continued)

ERW-3	BS 980	10-42D 17	ND GRADE 2	BS 5216	10-42C	12
ERW-11	BS 1775	10-42D 6	ND GRADE 3	BS 5216	10-42C	13
ERW-13	BS 1775	10-42D 7	NS GRADE 1	BS 5216	10-42C	5
ERW-16	BS 1775	10-42D 8	NS GRADE 2	BS 5216	10-42C	7
ERW-20	BS 1775	10-42D 9	NS GRADE 3	BS 5216	10-42C	9
ERW-23	BS 1775	10-42D 10	QH	BS 1449	10-42B	12
GR A	BS 592	10-53 11	SP	BS 1449	10-42B	11
GR A	BS 1456	10-53 14	T1A	BS 2920	10-42B	5
GR A	BS 1458	10-53 17	T1B	BS 2920	10-42B	5
GR B	BS 592	10-53 12	T2	BS 2920	10-42B	6
GR B	BS 1458	10-53 18	T3	BS 2920	10-42B	7
GR C	BS 592	10-53 13	T4	BS 2920	10-42B	8
GR C	BS 1456	10-53 18	T4CA	BS 2920	10-42B	8
GR B1	BS 1456	10-53 15	T5CA	BS 2920	10-42B	9
GR B2	BS 1456	10-53 16	T6CA	BS 970	10-42C	1
GR B290/6	BS 310	10-53 1	015AC3	BS 970	10-42C	1
GR B310/10	BS 310	10-53 2	030AC4	BS 970	10-44	1
GR B340/12	BS 310	10-53 3	040AC4	BS 970	10-42C	2
GR P440/7	BS 3333	10-53 6	040CAC4	BS 970	10-44	1
GR P510/4	BS 3333	10-53 7	040A10	BS 970	10-44	4
GR P540/5	BS 3333	10-53 8	040A10	BS 970	10-44	2
GR P570/3	BS 3333	10-53 9	040A15	BS 970	10-47	5
GR P690/2	BS 309	10-53 10	040A15	BS 970	10-44	5
GR W340/3	BS 2803	10-53 4	040CA15	BS 970	10-44	3
GR W410/4	BS 2803	10-42C 5	040A20	BS 970	10-47	6
GRADE I	BS 2803	10-42C 5	040A20	BS 970	10-42C	6
GRADE II	BS 1452	10-42C 6	43C	BS 970	10-44	4
GRADE III	BS 1452	10-52 2	43D	BS 4360	10-42D	13
GRADE 10	BS 1452	10-52 2	43E	BS 4360	10-42D	13
GRADE 12	BS 1452	10-52 3	045M10	BS 970	10-46	6
GRADE 14	BS 1452	10-52 4	050AC4	BS 970	10-42C	2
GRADE 17	BS 1452	10-52 5	050A10	BS 970	10-42C	4
GRADE 20	BS 1452	10-52 6	050A15	BS 970	10-44	2
GRADE 23	BS 1452	10-52 7	050A15	BS 970	10-42C	5
GRADE 26	BS 1449	10-42A 6	050A20	BS 970	10-42C	3
GRADE 34/20	BS 1449	10-43 1	50B	BS 970	10-42C	6
GRADE 37/23	BS 1449	10-43 7	50C	BS 4360	10-44	4
GRADE 37/23	BS 1449	10-43 2	50D	BS 970	10-46	14
GRADE 40A	BS 4360	10-43 2	55C	BS 970	10-42C	14
GRADE 40B	BS 4360	10-43 2	55E	BS 4360	10-42D	14
GRADE 40C	BS 4360	10-43 3	060A10	BS 4360	10-42D	15
GRADE 40D	BS 4360	10-43 3	060A15	BS 970	10-42D	15
GRADE 40E	BS 1449	10-42A 8	060A25	BS 970	10-42C	5
GRADE 43/25	BS 1449	10-43 3		BS 970	10-42C	7
GRADE 43/25	BS 1449	10-42A 9				
GRADE 43/28						

WORLD FERROUS MATERIALS INDEX (Continued)

DESIGNATION	STANDARD	TABLE	COL
060A25	BS 970	10-44	5
060A27	BS 970	10-42C	7
060A27	BS 970	10-42C	5
060A30	BS 970	10-42C	8
060A30	BS 970	10-44	6
060A32	BS 970	10-42C	8
060A32	BS 970	10-42C	6
060A35	BS 970	10-42C	9
060A35	BS 970	10-44	7
060A35	BS 970	10-42C	1
060A37	BS 970	10-42C	9
060A37	BS 970	10-44	7
060A40	BS 970	10-42C	10
060A40	BS 970	10-48	8
060A40	BS 970	10-48	2
060A42	BS 970	10-42C	10
060A42	BS 970	10-44	8
060A47	BS 970	10-42C	11
060A47	BS 970	10-44	9
060A47	BS 970	10-48	3
060A52	BS 970	10-42C	12
060A52	BS 970	10-44	10
060A52	BS 970	10-48	4
060A57	BS 970	10-42C	13
060A57	BS 970	10-44	11
060A57	BS 970	10-48	5
060A62	BS 970	10-42C	14
060A62	BS 970	10-44	12
060A67	BS 970	10-42C	15
060A67	BS 970	10-44	13
060A72	BS 970	10-42C	16
060A72	BS 970	10-42C	16
060A72	BS 970	10-44	14
060A78	BS 970	10-44	15
060A78	BS 970	10-51	1
060A83	BS 970	10-42C	17
060A86	BS 970	10-42C	17
060A96	BS 970	10-42C	18
060A96	BS 970	10-44	16
060A99	BS 970	10-42C	18
070A72	BS 970	10-42C	16
070A78	BS 970	10-44	15
080A15	BS 970	10-51	1
080A35	BS 970	10-47	7
080A40	BS 970	10-48	1
080A47	BS 970	10-44	9
080A52	BS 970	10-48	3
080A52	BS 970	10-44	10
080A57	BS 970	10-44	11
080A57	BS 970	10-48	5

DESIGNATION	STANDARD	TABLE	COL
430S15	BS 970	10-49	3
431S29	BS 970	10-49	10
431S29	BS PD6290	10-49	11
432S19	BS 970	10-49	6
441S29	BS 970	10-49	10
441S29	BS 970	10-49	10
500/7	BS 2789	10-52	6
523A14	BS 970	10-47	6
527A19	BS 970	10-47	8
527A60	BS 970	10-51	9
530A32	BS 970	10-45	6
530A36	BS 970	10-45	7
530A36	BS 970	10-48	7
530A40	BS 970	10-48	8
530M40	BS 970	10-48	8
600/3	BS 2789	10-52	11
630M40	BS 970	10-48	6
635A14	BS 970	10-47	15
637H17	BS 970	10-47	12
637A16	BS 970	10-47	16
640A35	BS 970	10-45	10
655A12	BS 970	10-47	17
659A15	BS 970	10-47	18
700/2	BS 2789	10-52	12
708A37	BS 970	10-45	2
708M40	BS 970	10-45	3
708M40	BS 970	10-48	9
708A42	BS 970	10-45	3
708A42	BS 970	10-48	9
709M40	BS 970	10-45	3
709M40	BS 970	10-48	9
735A50	BS 970	10-51	13
800/2	BS 2789	10-52	13
816M40	BS 970	10-45	11
817M40	BS 970	10-45	11
823M30	BS 970	10-45	12
835M30	BS 970	10-45	13
897M39	BS 970	10-47	2
905M31	BS 970	10-47	3
905M39	BS 970	10-47	4
945A40	BS 970	10-45	9
945A40	BS 970	10-48	10
945M38	BS 970	10-48	10

ITALIAN UNI FERROUS MATERIALS INDEX

DESIGNATION	STANDARD	TABLE	COL
A	UNI 5755	10-42B	6
B	UNI 5755	10-42B	7

STEEL MATERIAL DATA 10-169

WORLD FERROUS MATERIALS INDEX (Continued)

080A62	BS 970	10-44	12	C	UNI 5755	10-42B	8
080A67	BS 970	10-44	13	C10	UNI 5331	10-42C	4
080A72	BS 970	10-44	14	C10	UNI 5331	10-44	2
080A78	BS 970	10-44	15	C10	UNI 5331	10-47	5
080A83	BS 970	10-44	15	C16	UNI 5332	10-42C	5
080A86	BS 970	10-44	16	C16	UNI 5332	10-44	3
150M28	BS 970	10-45	4	C16	UNI 5331	10-47	6
212M14	BS 970	10-46	8	C20	UNI 5332	10-42C	6
212M36	BS 970	10-46	11	C20	UNI 6403	10-42D	14
212M44	BS 970	10-46	13	C30	UNI 5332	10-44	4
214M15	BS 970	10-46	9	C30	UNI 5332	10-42C	8
216M36	BS 970	10-46	12	C33	UNI 5332	10-44	6
220M07	BS 970	10-46	1	C35	UNI 5333	10-42C	7
230M07	BS 970	10-46	2	C35	UNI 5332	10-44	9
240M07	BS 970	10-46	4	C35	UNI 5333	10-48	1
250A53	BS 970	10-51	4	C38	UNI 5332	10-44	8
250A58	BS 970	10-51	5	C40	UNI 5333	10-44	8
250A61	BS 970	10-51	6	C40	UNI 5332	10-42C	2
301S01	BS PD6290	10-50	8	C43	UNI 5333	10-48	9
302S25	BS 970	10-50	5	C45	UNI 5333	10-44	3
303S21	BS 970	10-50	6	C48	UNI 5332	10-44	10
304S15	BS 970	10-50	1	C50	UNI 5333	10-48	10
305S19	BS PD6290	10-50	7	C50	UNI 5333	10-44	4
316S12	BS 970	10-50	9	C60	UNI 5332	10-44	12
316S16	BS 970	10-50	13	C60	UNI 5333	10-44	13
316S16	BS 970	10-50	10	C70	UNI 3545	10-44	14
317S12	BS 970	10-50	14	C75	UNI 3545	10-44	15
318S17	BS 970	10-50	17	C90	UNI 3545	10-51	2
320S17	BS 970	10-50	12	C100	UNI 3545	10-44	16
320S17	BS PD6290	10-50	16	CLASS A	UNI 3823	10-42C	5
321S12	BS 970	10-50	11	CLASS B	UNI 3823	10-42C	7
347S17	BS 970	10-50	15	CLASS C	UNI 3823	10-42C	9
370/17	BS 2789	10-50	3	CLASS D	UNI 3823	10-42C	10
403S17	BS 970	10-52	8	CLASS E	UNI 3823	10-42C	11
405S17	BS PD6290	10-49	1	D	UNI 5755	10-42B	9
410S21	BS 970	10-49	2	Fe33	UNI 7070	10-43	1
416S21	BS 970	10-49	7	Fe34	UNI 7091	10-42D	8
416S29	BS 970	10-49	8	Fe34A	UNI 7070	10-42D	5
416S37	BS 970	10-49	9	Fe34B	UNI 7070	10-42A	5
416S41	BS 970	10-49	9	Fe34C	UNI 7070	10-42C	5
420/12	BS 2789	10-52	8	Fe35-1	UNI 663	10-42A	12
420S29	BS 970	10-49	9	Fe35-1	UNI 2897	10-42D	2
420S37	BS 970	10-49	9	Fe35-2	UNI 2897	10-42D	3
420S45	BS 970	10-49	12	Fe37	UNI 7091	10-42D	9

WORLD FERROUS MATERIALS INDEX (Continued)

DESIGNATION	STANDARD	TABLE	COL	DESIGNATION	STANDARD	TABLE	COL
Fe37A	UNI 7070	10-42A	6	x6CrNiMoNb17 13	UNI 6900	10-50	16
Fe37A	UNI 7070	10-43	2	x8Cr17	UNI 6900	10-49	3
Fe37B	UNI 7070	10-42A	6	x8CrMo17	UNI 6900	10-49	6
Fe37B	UNI 7070	10-43	2	x8CrNi18 12	UNI 6900	10-50	7
Fe37C	UNI 7070	10-42A	6	x10Crs17	UNI 6900	10-49	4
Fe37C	UNI 7070	10-43	2	x10CrNi18 09	UNI 6900	10-50	5
Fe37D	UNI 7070	10-42A	6	x10CrNiS18 9	UNI 6900	10-50	6
Fe37D	UNI 7070	10-43	2	x12Cr13	UNI 6900	10-49	7
Fe42	UNI 7091	10-42D	10	x12CrNi17 07	UNI 6900	10-50	8
Fe42A	UNI 7070	10-42A	3	x12CrS13	UNI 6900	10-49	8
Fe42B	UNI 7070	10-43	7	x14CrNi19	UNI 6900	10-49	10
Fe42B	UNI 7070	10-42A	3	x16CrNi16	UNI 6900	10-49	11
Fe42C	UNI 7070	10-43	7	x20Cr13	UNI 6900	10-49	9
Fe42C	UNI 7070	10-42A	3	x30Cr13	UNI 6900	10-49	12
Fe42C	UNI 7070	10-43	3	x40Cr14	UNI 6900	10-49	13
Fe42D	UNI 7070	10-42A	7	9SMn23	UNI 4838	10-46	2
Fe42D	UNI 7070	10-43	3	9SMnPb23	UNI 4838	10-46	3
Fe44A	UNI 7070	10-43	4	10S20	UNI 4838	10-46	1
Fe44B	UNI 7070	10-43	4	10S22	UNI 4838	10-46	6
Fe44C	UNI 7070	10-43	4	16Cri4	UNI 5331	10-47	12
Fe44D	UNI 7070	10-43	4	16NiCr11	UNI 5331	10-47	14
Fe45-1	UNI 663	10-42D	13	16NiCrMo12	UNI 5331	10-47	17
Fe45-2	UNI 2897	10-42D	4	18NiCrMo5	UNI 5331	10-47	16
Fe50	UNI 7070	10-42A	9	20NiCrMo2	UNI 5331	10-47	15
Fe50	UNI 7070	10-43	6	25CrMo4	UNI 5332	10-45	1
Fe52	UNI 7091	10-42D	10	30CrMo4	UNI 5332	10-45	1
Fe52-2	UNI 2897	10-42D	5	30CrMo12	UNI 6120	10-47	1
Fe52B	UNI 7070	10-42A	8	35CrMo4	UNI 5332	10-45	2
Fe52B	UNI 7070	10-43	5	35NiCrMo15	UNI 5332	10-45	13
Fe52C	UNI 7070	10-42A	8	35SMn1C	UNI 4838	10-46	10
Fe52C	UNI 7070	10-43	5	36CrMn4	UNI 5333	10-46	7
Fe52D	UNI 7070	10-42A	8	36CrMn6	UNI 5333	10-48	4
Fe52D	UNI 7070	10-43	5	38CrA1Mo7	UNI 6120	10-47	2
Fe55-1	UNI 663	10-42D	15	38CrMo4	UNI 5333	10-45	8
Fe60	UNI 7070	10-42A	10	38NiCrMo4	UNI 5332	10-45	8
Fe60	UNI 7070	10-43	7	40Cr4	UNI 5332	10-45	3
Fe70	UNI 7070	10-42A	11	40Cr4	UNI 5333	10-48	3
Fe70	UNI 7070	10-43	8	40CrMo4	UNI 5332	10-45	9
FeG00	UNI 3158	10-53	11	40NiCrMo2	UNI 5333	10-45	10
FeG38	UNI 3158	10-53	11	40NiCrMo4	UNI 5333	10-48	11
FeG38VR	UNI 3158	10-53	11	40NiCrMo4	UNI 5333	10-45	11
FeG45	UNI 3158	10-53	12	40NiCrMo7	UNI 5332	10-45	13
FeG45VR	UNI 3158	10-53	12	50CrV4	UNI 3545	10-51	4
FeG52	UNI 3158	10-53	13	50Si7	UNI 3545	10-51	5
FeG52VR	UNI 3158	10-53	13	55Si8	UNI 3545	10-51	7
FeP00	UNI 5866	10-42B	1	60SiCr8	UNI 3545	10-51	
FeP01	UNI 5866	10-42B	2				
FeP02	UNI 5866	10-42B	3				

10-170   STEEL MATERIAL DATA

STEEL MATERIAL DATA

10-171

WORLD FERROUS MATERIALS INDEX (Continued)

AUSTRALIAN AS FERROUS MATERIALS INDEX

			DESIGNATION	STANDARD	TABLE	COL
FeP03	UNI 5866	4	AUS101 GRADE A	AS G15	10-54	12
FeP04	UNI 5866	4	AUS101 GRADE B	AS G15	10-54	13
FeP10	UNI 5867	1	AUS102 GRADE A	AS G15	10-54	14
FeP11	UNI 5867	2	AUS102 GRADE B	AS G15	10-54	15
FeP12	UNI 5867	3	AUS105	AS G15	10-54	16
FeP13	UNI 5867	4	AUS202 GRADE A	AS G15	10-54	2
FeZ0	UNI 5753	12	AUS202 GRADE B	AS G15	10-54	3
FeZ1	UNI 5753	13	AUS203	AS G15	10-54	5
FeZ2	UNI 5753	14	AUS204	AS G15	10-54	4
FeZ3	UNI 5753	15	AUS205	AS G15	10-54	8
G10	UNI 5007	1	CDS170	AS 1450	10-42D	1
G15	UNI 5007	2	CDS200	AS 1450	10-42D	2
G20	UNI 5007	3	CDS200	AS 1450	10-42D	12
G25	UNI 5007	4	CDS250	AS 1450	10-42D	2
G30	UNI 5007	5	CDS250	AS 1450	10-42D	13
G35	UNI 5007	6	CDS300	AS 1450	10-42D	3
GMB35	UNI 3779	4	CDS300	AS 1450	10-42D	14
GMN35	UNI 3779	3	CDS350	AS 1450	10-42D	4
GMN37	UNI 3779	3	CDS370	AS 1450	10-42D	5
GMN45	UNI 3779	6	CDS430	AS 1450	10-42D	5
GMN50	UNI 3779	7	CDS430	AS 1450	10-42D	15
GMN55	UNI 3779	8	CDS540	AS 1450	10-42D	5
GMN65	UNI 3779	9	CEW170	AS 1450	10-42D	6
GMN70	UNI 3779	10	CEW200	AS 1450	10-42D	7
GS42/10	UNI 4544	9	CEW250	AS 1450	10-42D	8
GS50/5	UNI 4544	10	CEW300	AS 1450	10-42D	10
GS55/2	UNI 4544	11	CEW350	AS 1450	10-42D	9
TYPE I	UNI 3823	10	CRC	AS 1595-I	10-42B	1
TYPE I	UNI 3823	11	CRD	AS 1595-I	10-42B	3
TYPE II	UNI 3823	12	CRE	AS 1595-I	10-42B	4
TYPE III	UNI 3823	13	CS1100	AS DR 72134 GR AS113	10-46	
TYPE III	UNI 3823	14	EFW170	AS 1450	10-42D	6
TYPE III	UNI 3823	15	EFW200	AS 1450	10-42D	7
TYPE III	UNI 3823	16	EFW250	AS 1450	10-42D	8
TYPE IV	UNI 3823	17	EFW300	AS 1450	10-42D	10
TYPE IV	UNI 3823	18	EFW350	AS 1450	10-42D	9
X2CrNi18 11	UNI 6900	1	ERW170	AS 1450	10-42D	6
X2CrNiMo17 13	UNI 6900	9	ERW200	AS 1450	10-42D	7
X2CrNiMo17 13	UNI 6900	13	ERW250	AS 1450	10-42D	8
X2CrNiMo18 16	UNI 6900	17	ERW300	AS 1450	10-42D	10
X5CrNi18 10	UNI 6900	4	ERW350	AS 1450	10-42D	9
X5CrNiMo17 13	UNI 6900	10				
X5CrNiMo17 13	UNI 6900	14				
X6CrAl13	UNI 6900	1				
X6CrNiTi18 11	UNI 6900	2				
X6CrNiNb18 11	UNI 6900	3				
X6CrNiMoTi17 12	UNI 6900	11				
X6CrNiMoNb17 12	UNI 6900	12				
X6CrNiMoTi17 13	UNI 6900	15				

STEEL MATERIAL DATA

WORLD FERROUS MATERIALS INDEX (Continued)

DESIGNATION	STANDARD	TABLE	COL
ERW380	AS 1450	10-42D	5
GC 12	AS 1397	10-42A	12
GD 14	AS 1397	10-42A	14
GR B18/6	AS G12	10-53	1
GR B20/10	AS G12	10-53	2
GR B22/14	AS G12	10-53	3
GR W22/14	AS G11	10-53	4
GR W24/18	AS G11	10-53	5
GR 0022	AS G11	10-53	11
GR 0025	AS G22	10-53	11
GR 0030	AS G22	10-53	12
GR 0050A	AS G22	10-53	13
GR 080	AS G22	10-53	14
GR J90	AS G22	10-53	15
GR 0105	AS G22	10-53	16
GR 0120	AS G22	10-53	17
GR 0150	AS G22	10-53	18
GR C175	AS G22	10-53	18
GR 00503	AS G22	10-53	15
GRADE 10	AS G8	10-52	1
GRADE 12	AS G8	10-52	2
GRADE 14	AS G8	10-52	3
GRADE 17	AS G8	10-52	4
GRADE 20	AS G8	10-52	5
GRADE 23	AS G8	10-52	6
GRADE 26	AS G8	10-52	7
GRADE 250	AS 1204	10-43	3
GRADE 300	AS 1405	10-43	4
GRADE 310	AS 1204	10-43	1
GRADE 350	AS 1205	10-43	5
GRADE 370	AS 1405	10-43	5
GRADE 400	AS 1204	10-43	2
GRADE 500	AS 1205	10-43	7
GRADE 500	AS 1204	10-43	7
HD RANGE 1	AS 1472	10-42C	11
HD RANGE 2	AS 1472	10-42C	12
HD RANGE 3	AS 1472	10-42C	14
HFS1/0	AS 1450	10-42D	1
HFS200	AS 1450	10-42D	2
HFS300	AS 1450	10-42D	17
HFS350	AS 1450	10-42D	16
HR200	AS 1594	10-42A	5
HR240	AS 1594	10-42A	7
HR280	AS 1594	10-42A	9
HR340	AS 1594	10-42A	8
HRC	AS 1594	10-42A	1
HRD	AS 1594	10-42A	4

DESIGNATION	STANDARD	TABLE	COL
S101C	AS DR72134 GR AS1135	10-47	5
S1015	AS DR72134 GR AS1135	10-42C	5
S1015	AS DR72135 GR AS1135	10-44	3
S1015	AS DR72134 GR AS1135	10-47	6
S1016	AS DR72134 GR AS1135	10-47	7
S102C	AS DR72134 GR AS1135	10-42C	6
S102C	AS DR72135 GR AS1135	10-44	4
S1025	AS DR72134 GR AS1135	10-42C	7
S1025	AS DR72135 GR AS1135	10-44	5
S1030	AS DR72134 GR AS1135	10-42C	8
S1030	AS DR72135 GR AS1135	10-44	6
S1035	AS DR72134 GR AS1135	10-42C	9
S1035	AS DR72135 GR AS1135	10-44	7
S104C	AS DR72134 GR AS1135	10-48	1
S104C	AS DR72134 GR AS1135	10-42C	10
S104C	AS DR72135 GR AS1135	10-44	8
S1045	AS DR72134 GR AS1135	10-48	2
S1045	AS DR72134 GR AS1135	10-42C	11
S1045	AS DR72135 GR AS1135	10-44	9
S1050	AS DR72134 GR AS1135	10-48	3
S105C	AS DR72135 GR AS1135	10-44	10
S1058	AS DR72134 GR AS1135	10-48	4
S1058	AS DR72134 GR AS1135	10-42C	12
S1070	AS DR72134 GR AS1135	10-48	5
S1214	AS DR72134 GR AS1135	10-46	14
SD	AS 1472	10-42C	3
SNG 27/12	AS G9	10-52	9
SNG 32/7	AS G9	10-52	10
SNG 37/2	AS G9	10-52	10
SNG 42/2	AS G9	10-52	11
SNG 47/2	AS G9	10-52	12
T1	AS 1517	10-42B	5
T2	AS 1517	10-42B	6
T3	AS 1517	10-42B	7
T4	AS 1517	10-42B	8
T4CA	AS 1517	10-42B	8
T5	AS 1517	10-42B	8
T5CA	AS 1517	10-42B	8
T6	AS 1517	10-42B	9
T6CA	AS 1517	10-42B	9
XK1082	AS DR72135 GR AS1135	10-44	15
XK1082	AS 1444	10-51	1
XK1325	AS DR72134 GR AS1135	10-45	4
XK1330	AS DR72134 GR AS1135	10-45	4
XS11L12	AS DR72134 GR AS1135	10-46	5
XS1112	AS DR72134 GR AS1135	10-46	2
XS1115	AS DR72134 GR AS1135	10-46	8

STEEL MATERIAL DATA

10-173

WORLD FERROUS MATERIALS INDEX (Continued)

K1008	AS DR72134 GR AS1135	10-42C	3	1 TEMPER	AS 1595-II	10-42B	14
K1008	AS DR72135 GR AS1135	10-44	1	2 TEMPER	AS 1595-II	10-42B	13
K1010	AS DR72135 GR AS1135	10-42C	4	3 TEMPER	AS 1595-II	10-42B	12
K1010	AS DR72135 GR AS1135	10-44	2	4 TEMPER	AS 1595-II	10-42B	11
K1010	AS DR72134 GR AS1135	10-47	5	5 TEMPER	AS 1595-II	10-42B	11
K1012	AS DR72134 GR AS1135	10-42C	5	6 TEMPER	AS 1595-II	10-42B	10
K1012	AS DR72134 GR AS1135	10-44	3	180	AS 1405	10-42A	5
K1016	AS DR72134 GR AS1135	10-42C	7	210	AS 1405	10-42A	6
K1020	AS DR72135 GR AS1135	10-42C	6	302	AS 1444	10-50	5
K1020	AS DR72134 GR AS1135	10-47	4	303	AS 1444	10-50	6
K1022	AS DR72134 GR AS1135	10-42C	8	304	AS 1444	10-50	4
K1026	AS DR72135 GR AS1135	10-42C	7	304L	AS 1444	10-50	1
K1026	AS DR72135 GR AS1135	10-44	5	316	AS 1444	10-50	10
K1030	AS DR72134 GR AS1135	10-42C	8	316	AS 1444	10-50	14
K1030	AS DR72134 GR AS1135	10-44	6	316L	AS 1444	10-50	13
K1035	AS DR72135 GR AS1135	10-42C	9	316L	AS 1444	10-50	17
K1035	AS DR72135 GR AS1135	10-44	7	317	AS 1444	10-50	2
K1039	AS DR72134 GR AS1135	10-48	1	321	AS G9	10-52	8
K1039	AS DR72134 GR AS1135	10-42C	10	370/17	AS 1444	10-49	7
K1045	AS DR72135 GR AS1135	10-42C	11	410	AS 1444	10-49	8
K1045	AS DR72135 GR AS1135	10-44	9	416	AS 1444	10-49	9
K1050	AS DR72134 GR AS1135	10-48	3	420	AS G9	10-52	9
K1050	AS DR72134 GR AS1135	10-44	10	420/12	AS 1444	10-49	3
K1055	AS DR72134 GR AS1135	10-48	4	430	AS 1444	10-49	4
K1055	AS DR72135 GR AS1135	10-44	11	430F	AS 1444	10-49	10
K1060	AS DR72135 GR AS1135	10-48	5	431	AS 1444	10-49	10
K1060	AS DR72135 GR AS1135	10-44	12	500/7	AS G9	10-52	11
K1065	AS DR72135 GR AS1135	10-44	13	600/3	AS G9	10-52	12
K1070	AS DR72134 GR AS1135	10-44	14	700/2	AS G9	10-52	13
K1137	AS DR72134 GR AS1135	10-44	12	800/2	AS G9	10-52	1
K1138	AS DR72134 GR AS1135	10-46	10	4130	AS 1444	10-45	10
K1146	AS DR72135 GR AS1135	10-46	13	4130	AS 1444	10-47	3
L	AS 1517	10-42B	1	4140	AS 1444	10-45	11
MC	AS 1517	10-42B	2	4340	AS 1444	10-45	6
MR	AS 1472	10-42C	3	5132	AS 1444	10-45	8
OH CLASS I	AS G14	10-53	5	5140	AS DR73044 GR AS1444	10-48	7
PM63-40-09	AS G14	10-53	5	5145	AS 1444	10-45	8
PM68-45-07	AS G14	10-53	6	5145	AS DR73044 GR AS1444	10-45	8
PM70-48-05	AS G14	10-53	7	5145	AS 1444	10-48	8
PM75-50-07	AS G14	10-53	8	5155	AS 1444	10-51	8
PM80-60-03	AS G14	10-53	9	5160	AS 1444	10-51	9
PM90-70-02	AS G14	10-53	9	6150	AS 1444	10-51	13
PM100-80-02	AS G14	10-53	10	8620	AS 1444	10-51	15
PM105-85-02	AS DR72134 GR AS1135	10-46	10	8640	AS 1444	10-47	9
S12L14	AS DR72135 GR AS1135	10-42C	3	8640	AS DR73044 GR AS1444	10-45	9
S1006	AS DR72134 GR AS1135	10-42C	2	9255	AS 1444	10-48	11
S1006	AS DR72134 GR AS1135	10-42C	1	9260	AS 1444	10-45	9
S1008	AS DR72134 GR AS1135	10-42C	3		AS 1444	10-51	5
S1010	AS DR72134 GR AS1135	10-42C	4				6
S1010	AS DR72135 GR AS1135	10-44	2				

FOREIGN STEEL SUPPLIERS

Country	Agencies Serving the U.S.	National
Australia	The Broken Hill Proprietary Company Ltd. 100 Park Avenue Room 3200 New York, NY 10017 212/679-1300	The Broken Hill Proprietary Company Ltd. 140 William Street Melbourne, Victoria 3000, Australia
Japan	Japan Iron & Steel Exporter Association 60 East 42nd Street New York, NY 10017 212/697-6864	Japan Iron & Steel Exporters Association 3-16, Nihonbashi-kayabacho, Chuo-ku, Tokyo, Japan
France	Usinor Sales Corp. 600 3rd Avenue New York, NY 10016 212/697-2532	OTUA 129 Avenue Charles de Gaulle 92200 Neuilly-Sur-Seine France
U.K.	British Steel Corp., Inc. 601 Jefferson Houston Texas 77002 713/224-6441	BSC Head Office P. O. Box 142 151 Gower Street London WCIE 6BB, England
Italy	Siderius 1345 Avenue of the Americas New York, NY 10019 212/489-7470	Finsider Viale Castro Pretorio 122 00185 Roma Italy
Germany	American Saar Steel Corp. 41 East 42nd Street New York, NY 10017	Roechlingsche Eisen & Stahlwerke GmbH Voelklingen/Saar West Germany
	Ameropean Industries Inc. 295 Treadwell Street Hamden, CT 06514	Isolation BmbH Mannheim West Germany
	Cosid Inc. 30 East 42nd Street New York, NY 10017	Saarlaendische Stahlexport- und Handelsgesellschaft GmbH Sulzbach, West Germany
	Daub Steel & Metals 230 Park Avenue New York, NY 10017	Stahlwerk Unna Mueller & Co. KG, Unna/Westf, West Germany
	Fagersta Steels Inc. 2 Henderson Drive West Caldwell, NY 07006	Stahlwerk Westig GmbH Unna/Westf West Germany
	Ferrostaal Overseas Corp. 17 Battery Place New York, NY 10004	
Germany	Ferrostaal Pacific Corp. 231 Sansome Street San Francisco, CA 94104	Ferrostaal AG, Essen West Germany
	Hoesch America Inc. 225 Peachtree Street N.E. Atlanta, GA 30303	Hoesch Handel AG. Dortmund Hoesch Huettenwerke, AG Dortmund Hoesch Werke Hohenlimburg-Schwerte AG. Schwerte West Germany
	Kloeckner Inc. 1270 Ave. of the Americas New York, NY 10020	Kloeckner-Werke AG, Duisburg, Kloeckner & Co. Duisburg, West Germany
	Krupp International Inc. 350 Executive Blvd. Elmsford, NY 10523	Friek Krupp GmbH Essen West Germany
	Mannesmann Pipe & Steel Corp. 9501 W. Devon Avenue Rosemont, Ill. 60018	Mannesmann Export GmbH Duesseldorf West Germany
	Marathon Specialty Steels Inc. 350 5th Avenue New York, NY 10001	Deutsche Edelstahlwerke AG Krefeld, West Germany
	Thyssen Steel Corp. 350 5th Avenue New York, NY 10001	Aug Thyssen Huette AG Duisburg-Hamborn Phoenix-Rheinrohr AG, Dusseldorf, West Germany
	Walzstahl-Exportburo* GMBH 4 Dusseldorf 1 Postfach 8420 West Germany	Verein Deutscher* Eisenhutten-leute 4 Dusseldorf 1 Postfach 8209 West Germany
	German American Chamber of Commerce, Inc. 666 Fifth Avenue New York, NY 10019 212/582-7788	

RELATED ISO STANDARDS (TC 17, TC 5, TC 107)

ISO TC 17**	Steel
ISO/R 79-1968	Brinell Hardness test for steel, 2nd Edition
ISO/R 80-1968	Rockwell hardness test (B and C scales) for steel, 2nd Edition
ISO/R 81-1967	Vickers hardness test for steel (load 5 to 100 kgf), 2nd Edition
ISO 82-1974	Steel – Tensile testing
ISO/R 83-1959	Charpy impact test (U-notch) for steel

*German Steel Export and Trade Associations.
**Now also listed under ISO/TC 164-mechanical testing of metals.

STEEL MATERIAL DATA

ISO/R 84-1959	Izod impact test for steel
ISO/R 85-1959	Bend test for steel
ISO 86-1974	Steel – Tensile testing of sheet and strip less than 3 mm and not less than 0.5 mm thick
ISO/R 87-1959	Simple bend testing of steel sheet and strip less than 3 mm thick
ISO/R 88-1959	Reverse bend testing of steel sheet and strip less than 3 mm thick
ISO 89-1974	Steel – Tensile testing of wire
ISO 136-1972	Steel – Simple torsion testing of wire
ISO 144-1973	Steel – Reverse bend testing of wire
ISO/R 145-1960	Wrapping test for steel wire
ISO/R 146-1968	Verification of Vickers hardness testing machines, 2nd Edition
ISO/R 147-1960	Load calibration of testing machines for tensile testing of steel
ISO/R 148-1960	Beam impact test (V-notch) for steel
ISO/R 149-1960	Modified Erichsen cupping test for steel sheet and strip
ISO/R 156-1967	Verification of Brinell hardness testing machines, 2nd Edition
ISO/R 165-1960	Flanging test on steel tubes
ISO/R 166-1960	Drift expanding test on steel tubes
ISO/R 167-1960	Bend test on steel tubes
ISO/R 202-1961	Flattening test on steel tubes
ISO/R 203-1961	Interrupted creep testing of steel at elevated temperatures (load and temperature interrupted)
ISO/R 204-1961	Non-interrupted creep testing of steel at elevated temperatures
ISO 205-1976	Determination of proof stress and proving test for steel at elevated temperatures
ISO/R 206-1961	Creep stress rupture testing of steel at elevated temperatures
ISO/R 373-1964	General principles for fatigue testing of metals
ISO/R 374-1964	Ring expanding test on steel tubes
ISO 375-1974	Steel – Tensile testing of tubes
ISO/R 376-1964	Calibration of elastic proving devices
ISO/R 377-1964	Selection and preparation of samples and test pieces for wrought steel
ISO/R 404-1964	General technical delivery requirements for steel
ISO/R 409-1964	Tables of Vickers hardness values (HV) for metallic materials
ISO/R 410-1964	Tables of Brinell hardness values (HB) for use in tests made on flat surfaces
ISO/R 437-1965	Chemical analysis of steels–Determination of total carbon (Gravimetric method after combustion in a stream of oxygen)
ISO/R 439-1969	Chemical analysis of steel and cast iron–Determination of total silicon (Gravimetric method), 2nd Edition
ISO/R 442-1965	Verification of pendulum impact testing machines for testing steels
ISO/R 629-1967	Chemical analysis of steels–Determination of manganese (Spectrophotometric method)
ISO/R 630-1967	Structural steels
ISO/R 640-1967	Calibration of standardized blocks to be used for Vickers hardness testing machines
ISO/R 642-1967	Hardenability test by end quenching steel (Jominy test)
ISO/R 643-1967	Micrographic determination of the austenitic grain size of steels
ISO/R 657/I-1968	Dimensions of hot-rolled steel sections–Part 1 : Equal-leg angles–Metric series–Dimensions and sectional properties
ISO/R 657/II-1968	Dimensions of hot-rolled steel sections–Part 2 : Unequal-leg angles–Metric series–Dimensions and sectional properties
ISO/R 657/III-1969	Dimensions of hot-rolled steel sections–Equal-leg angles–Inch series –Dimensions and sectional properties
ISO/R 657/IV-1969	Dimensions of hot-rolled steel sections–Unequal-leg angles–Inch series–Dimensions and sectional properties
ISO 657/V-1976	Hot-rolled steel sections–Part V Equal-leg angles and unequal-leg angles–Tolerances for metric and inch series.
ISO/R 657/VII-1969	Dimensions of hot-rolled steel sections–Parallel flange I-beams–Inch series–Dimensions and sectional properties
ISO/R 657/IX-1969	Dimensions of hot-rolled steel sections–Parallel flange column sections–Inch series–Dimensions and sectional properties
ISO/R 671-1968	Chemical analysis of steel and cast iron–Determination of sulphur (Method after combustion in current of oxygen and titration with sodium borate)

ISO/R 674-1968	Calibration of standardized blocks to be used for Rockwell B and C hardness scale testing machines
ISO/R 683/I-1968	Heat-treated steels, alloy steels and free-cutting steels—Part 1 : Quenched and tempered unalloyed steels
ISO/R 683/II-1968	Heat-treated steels, alloy steels and free-cutting steels—Part 2 : Wrought quenched and tempered steels with 1% chromium and 0,2% molybdenum
ISO/R 683/III-1970	Heat-treated steels, alloy steels and free-cutting steels—Part 3 : Wrought quenched and tempered unalloyed steels with controlled sulphur content
ISO/R 683/IV-1970	Heat-treated steels, alloy steels and free-cutting steels—Part 4 : Wrought quenched and tempered steels with 1% chromium and 0,2% molybdenum and controlled sulphur content
ISO/R 683/V-1970	Heat-treated steels, alloy steels and free-cutting steels—Part 5 : Wrought quenched and tempered manganese steels
ISO/R 683/VI-1970	Heat-treated steels, alloy steels and free-cutting steels—Part 6 : Wrought quenched and tempered steels with 3% chromium and 0,5% molybdenum
ISO/R 683/VII-1970	Heat-treated steels, alloy steels and free-cutting steels—Part 7 : Wrought quenched and tempered chromium steels
ISO/R 683/VIII-1970	Heat-treated steels, alloy steels and free-cutting steels—Part 8 : Wrought quenched and tempered chromium-nickel-molybdenum steels
ISO/R 683/IX-1970	Heat-treated steels, alloy steels and free-cutting steels—Part 9 : Wrought free-cutting steels
ISO 683/X-1975	Heat-treated steels, alloy steels and free-cutting steels—Part 10 : Wrought nitriding steels
ISO/R 683/XI-1970	Heat-treated steels, alloy steels and free-cutting steels—Part 11 : Wrought case hardening steels
ISO 683/XII-1972	Heat-treated steels, alloy steels and free-cutting steels—Part 12 : Flame and induction hardening steels
ISO 683/XIII-1974	Heat-treated steels, alloy steels and free-cutting steels—Part 13 : Wrought stainless steels
ISO 683/XIV-1973	Heat-treated steels, alloy steels and free-cutting steels—Part 14 : Steels for hot-formed and heat-treated springs
ISO 683/XV-1976	Steels—Part 15 : Valve steels for internal combustion engines
ISO 683/XVI-1976	Steels—Part 16 : Precipitation hardening stainless steels
ISO 683/XVII-1976	Steel—Part 17: Ball and roller bearing steels
ISO 683/XVIII-1976	Steels—Part 18 : Wrought unalloyed steels
ISO/R 716-1968	Verification of Rockwell B and C scale hardness testing machines
ISO/R 726-1968	Calibration of standardized blocks to be used for Brinell hardness testing machines
ISO/R 783-1968	Mechanical testing of steel at elevated temperatures—Determination of lower yield stress and proof stress and proving test
ISO/R 1005/I-1969	Railway rolling stock material—Tyres for trailer stock
ISO/R 1005/II-1969	Railway rolling stock material—Rough tyres for trailer stock: Dimensions and tolerances
ISO/R 1005/III-1969	Railway rolling stock material—Axles for trailer stock
ISO/R 1005/IV-1969	Railway rolling stock material—Rolled or forged wheel centres for tyred wheels for trailer stock
ISO/R 1005/V-1969	Railway rolling stock material—Cast wheel centres in non-alloy steel for tyred wheels for trailer stock
ISO/R 1005/VI-1969	Railway rolling stock material—Solid wheels for trailer stock
ISO/R 1005/VII-1969	Railway rolling stock material—Wheel sets for trailer stock
ISO/R 1024-1969	Rockwell superficial hardness test (N and T scales) for steel
ISO/R 1035/I-1969	Dimensions of hot-rolled steel bars—Round bars—Metric series
ISO/R 1035/II-1969	Dimensions of hot-rolled steel bars—Square bars—Metric series
ISO/R 1035/III-1969	Dimensions of hot-rolled steel bars—Flat bars—Metric series
ISO 1035/IV-1976	Hot-rolled steel bars—Part IV: Tolerances of round, square and flat bars—Metric series.

STEEL MATERIAL DATA

ISO/R 1052-1969	Steels for general engineering purposes
ISO/R 1079-1969	Verification of Rockwell superficial N and T scale hardness testing machines
ISO 1099-1975	Axial load fatigue testing
ISO/R 1111-1969	Cold-reduced tinplate and cold-reduced black plate—Part 1 - Sheet
ISO 1111/II-1976	Cold reduced tinplate and cold reduced blackplate—Part II : Coil for subsequent cutting into sheet form
ISO 1143-1975	Rotating bar bending fatigue testing
ISO 1352-1977	Steel—Torsional stress fatigue test
ISO/R 1355-1970	Calibration of standardized blocks to be used for Rockwell superficial N and T scale hardness testing machines
ISO 2566/I-1973	Steel—Conversion of elongation values—Part I - Carbon and low alloy steels
ISO 2573-1977	Determination of K-values of a tensile testing system
ISO 2604.1-1975	Steels for pressure purposes—Part 1 : Forgings—Quality requirements
ISO 2604.2-1975	Steels for pressure purposes—Part 2 : Wrought seamless tubes
ISO 2604.3-1975	Steels for pressure purposes—Part 3 : ERW and induction welded tubes
ISO 2604.4-1975	Steels for pressure purposes—Part 4 : Plates
DIS 2604/5	Longitudinally welded austenitic stainless steel tubes
DIS 2604/6	Submerged arc welded steel tubes
ISO 2605/1-1976	Steel products for pressure purposes—Elevated temperature properties
ISO 2605/2-1976	Steel products for pressure purposes—Elevated temperature properties
ISO 2639-1973	Steel—Determination and verification of the effective depth of carburized and hardened cases
ISO 2732-1973	Steels and cast iron—Determination of phosphorus—Spectrophotometric method
ISO 3573-1976	Hot-rolled carbon steel sheet
ISO 3574-1976	Cold reduced carbon steel sheet
ISO 3575-1976	Continuous hot-dip zinc coated carbon steel
ISO 3576-1976	Hot-rolled carbon steel sheet for production of cold reduced products
ISO 3651/1-1976	Resistance to intergranular corrosion of stainless steels—Part 1
ISO 3651/2-1976	Resistance to intergranular corrosion of stainless steels—Part 2
ISO 3754-1976	Steel-depth of hardening after flame or induction hardening
ISO 3755-1976	Grades of cast steels for general engineering purposes
ISO 3763-1976	Methods for assessing non-metallic inclusions in wrought steels
ISO 3785-1976	Steel-identification of test piece axes
ISO 3798-1976	Tinplate and blackplate-packaging requirements
ISO 3887-1976	Steel non-alloy and low alloy—depth of decarburization
DIS 4829	Chemical analysis of steel and cast iron—Determination of silicon—Spectrophotometric method using the reduced silicomolybdic complex plex
DTR 4830/1	Chemical analysis of ferrous materials—Determination of low carbon contents in steel—Monometric low-pressure method
DTR 4830/2	Chemical analysis of ferrous materials—Determination of low carbon contents in steel—Volumetric method after combustion in an electric resistance
DTR 4830/3	Chemical analysis of ferrous materials—Determination of low carbon contents in steel—Conductimetric measurement after combustion
DTR 4830/4	Chemical analysis of ferrous materials—Determination of low carbon contents in steels and cast iron—Measurement by coulometric titration after combustion
DIS 4941	Chemical analysis of steel and cast iron—Determination of molybdenum—Photometric method
DIS 4945	Chemical analysis of steel—Determination of nitrogen—Spectrophotometric method
DIS 4948/1	Classification of steels—Part 1 : Based on chemical composition
DIS 4950/1	High yield strength steel flat products for structural purposes

DIS 4950/2	High yield strength flat steel products—Part 2: High yield strength flat steel products supplied in the normalized or controlled conditions
DIS 4951	High yield strength steel sections for structural purposes
DIS 4965	Dynamic force calibration of axial load fatigue testing machines by means of a strain gage technique
DIS 4967	Micrographic determination of the content of non-metallic inclusions in steels by means of standard diagrams
DIS 4970/2	High yield strength flat steel products—Part 2—High yield strength flat steel products supplied in the normalized or controlled condition
DIS 4995	Hot rolled steel sheet of structural quality
DIS 4996	Hot rolled steel sheet of high yield stress structural quality
DIS 4997	Cold reduced steel sheet of structural quality
ISO 4998-1977	Continuous hot-dip zinc coated steel sheet of structural quality
DIS 4999	Continuous hot-dip terne lead alloy coated cold reduced carbon steel sheet
ISO TC 5	**Metal Pipes and Fittings**
ISO/R 7-1954	Pipe threads for gas list tubes and screwed fittings where pressure-tight joints are made on the threads—1/8 inch to 6 inches
ISO/R 13-1955	Cast iron pipes, special castings and cast iron parts for pressure main lines
ISO/R 49-1957	Malleable cast iron pipe fittings screwed in accordance with International Standard ISO 7
ISO 50-1977	Steel sockets screwed in accordance with ISO 7—Minimum lengths
ISO 64-1974	Steel tubes—Outside diameters
ISO 65-1973	Steel tubes suitable for screwing in accordance with International Standard ISO 7
ISO 134-1973	Plain and steel tubes for general purposes
ISO 221-1976	Steel tubes—Wall thickness—2nd Edition
ISO/R 228-1961	Pipe threads where pressure-tight joints are not made on the threads (1/8 inch to 6 inches)
ISO 274-1975	Copper tubes of circular sections—Dimensions
ISO/R 285-1962	Steel tubes—Butt welding bends (90° and 180°)
ISO 336-1976	Plain end steel tubes, welded or seamless—General table of dimensions and masses per unit length
ISO/R 531-1966	Cast iron sanitary pipes and fittings for waste water and ventilation
ISO 559-1976	Steel pipes for gas, water and sewage, welded or seamless
ISO 560-1975	Plain end precision steel tubes, seamless and welded—Dimensions and masses per unit length
ISO 657/14-1977	Hot finished steel structural hollow sections—Part 14
ISO 1127-1977	Stainless steel tubes—Dimensions, tolerances and conventional masses per unit length
ISO/R 1128-1969	Steel tubes—Butt welding bends 5 D (90° and 180°)
ISO 1129-1977	Boiler tubes—Dimensions, tolerances and conventional masses per unit length
ISO 1179-1973	Pipe connections for plain end steel and other metal tubes in industrial application
ISO/R 2016-1971	Capillary solder fittings for copper tubes—Dimensions of sockets and male ends
ISO/R 2037-1971	Pipes and fittings—Stainless steel tubes for the food industry
ISO 2084-1974	Pipeline flanges for general use—Metric series—Mating dimensions
ISO 2441-1975	Pipeline flanges—Shapes, dimensions of pressure-tight surfaces
ISO 2531-1974	Ductile iron pipes fittings and accessories for pressure pipe-lines
ISO 2546-1973	Seamless plain end tubes made from unalloyed steel and without quality requirements
ISO 2547-1973	Welded plain end tubes made from unalloyed steel and without quality requirements
ISO 2851-1973	Metal pipes and fittings—Stainless steel bends and tees for the food industry

STEEL MATERIAL DATA

ISO 2852-1974	Metal pipes and fittings–Stainless steel clamp liners with gaskets for the food industry	ISO 1461-1973	Metallic coatings–Hot dip galvanized coatings on fabricated ferrous products–Requirements
ISO 2853-1976	Stainless steel screwed couplings for the food industry	ISO 1462-1973	Metallic coatings–Coatings other than those anodic to the basis metal–Accelerated corrosion tests–Method for evaluation of the results
ISO 2937-1974	Plain and seamless steel tubes for mechanical application		
ISO 2938-1974	Hollow steel bars for machining	ISO 1463-1973	Metallic and oxide coatings–Measurement of thickness by microscopical examination of cross-sections
DIS 3151	Steel tubes for facade scaffolds		
ISO 3304-1975	Plain end seamless precision steel tubes–Technical conditions for delivery		
		ISO 2063-1973	Metallic coatings–Protection of iron and steel against corrosion–Metal spraying of zinc and aluminum
ISO 3305-1975	Welded precision steel tubes–Delivery		
ISO 3306-1975	Plain end as-welded precision steel tubes–Delivery	ISO 2064-1973	Metallic and other non-organic coatings–Definitions and conventions concerning the measurement of the thickness
ISO TR 3311-1974	Seamless steel tubes for pipelines–Preferred sizes		
ISO 3418-1975	Steel tubes–Butt-welding bends without quality requirements	ISO 2079-1973	Surface treatment and metallic coatings–General classification of terms
ISO 3419-1975	Steel tubes–Butt-welding bends with quality requirements	ISO 2080-1973	Metallic coatings–Electroplating and related processes–Vocabulary
DIS 3545	Symbols to be used in specifications for steel tubes	ISO 2081-1973	Metallic coatings–Electroplated coatings of zinc on iron or steel
DIS 4019	Dimensions of cold finished steel–Structural hollow sections	ISO 2082-1973	Metallic coatings–Electroplated coatings of cadmium or iron or steel
DIS 4054	Couplers for scaffolds made of steel tubes	ISO 2093-1973	Metallic coatings–Electroplated coatings of tin
ISO 5252-1977	Steel tubes–System of tolerances	ISO 2177-1972	Metallic coatings–Measurement of coating thickness–Coulometric method by anodic dissolution
DIS 5256	Wrapping and protection of underground pipelines using bitumen and coal tar		
		ISO 2178-1972	Non-magnetic metallic and vitreous or porcelain enamel coatings on magnetic basis metals–Measurement of coating thickness–Magnetic method
ISO TC 107	**Metallic and Other Non-Organic Coatings**		
ISO 1456-1974	Metallic coatings – Electroplated coatings of nickel plus chromium		
		ISO 2179-1972	Electroplated coatings of tin-nickel alloy
ISO 1457-1974	Metallic coatings – Electroplated coatings of copper plus nickel pluu chromium on iron or steel	ISO 2360-1972	Non-conductive coatings on non-magnetic basis metals–Measurement of coating thickness–Eddy current method
ISO 1458-1974	Metallic coatings – Electroplated coatings of nickel		
ISO 1459-1973	Metallic coatings–Protection against corrosion by hot dip galvanizing–Guiding principles	ISO 2361-1972	Electrodeposited nickel coatings on magnetic and non-magnetic substrates–Measurement of coating thickness–Magnetic method
ISO 1460-1973	Metallic coatings–Hot dip galvanized coatings on ferrous materials–Determination of the mass per unit area–Gravimetric method		
		ISO 2722-1973	Vitreous and porcelain enamels–Determination of resistance to citric acid at room temperature

ISO 2723-1973	Vitreous and porcelain enamels for sheet steel—Production of specimens for testing
ISO 2724-1973	Vitreous and porcelain enamels for cast iron—Production of specimens for testing
ISO 2733-1973	Vitreous and porcelain enamels—Apparatus for testing with acid and neutral liquids and their vapours
ISO 2734-1973	Vitreous and porcelain enamels—Apparatus for testing with alkaline liquids
ISO 2742-1973	Vitreous and porcelain enamels—Determination of resistance to boiling citric acid
ISO 2743-1973	Vitreous and porcelain enamels—Determination of resistance to boiling hydrochloric acid
ISO 2744-1973	Vitreous and porcelain enamels—Determination of resistance to boiling water and water vapour
ISO 2745-1973	Vitreous and porcelain enamels—Determination of resistance to hot sodium hydroxide
ISO 2746-1973	Vitreous and porcelain enamels—Enamelled articles for service under highly corrosive conditions—High voltage test
ISO 2747-1973	Vitreous and porcelain enamels—Enamelled cooking utensils—Determination of resistance to thermal shock
ISO 2819/I-1973	Metallic coatings on metallic substrates—Review of methods available for testing adhesion—Part I : Electrodeposited and chemically deposited coating
ISO 3497-1976	Metallic coatings—Measurement of thickness by X-ray spectrometry
DIS 3543	Measurement of coating thicknesses—Beta backscatter principle
DIS 3613	Test methods for chromate treatments on zinc and cadmium
ISO 3768-1976	Neutral salt spray test
ISO 3769-1976	Acetic acid salt spray test
ISO 3770-1976	Copper accelerated acetic acid salt spray test—Cass test
ISO 3868-1976	Measurement of coating thicknesses
ISO 3882-1976	Measurement of thickness of metallic or non-organic coatings
DIS 3892	Coating mass of conversion coatings on metallic materials
DIS 4519	Standard method of sampling procedures for inspection of electrodeposited metallic and related finishes
DIS 4520	Chromate treatment of coatings of zinc and cadmium
DIS 4521	Electroplated coatings of silver engineering purposes
DIS 4522	Electrodesited coatings of silver for decorative purposes
DIS 4523	Electroplated coatings of gold alloy for engineering purposes
DIS 4524	Electroplated coatings of gold and gold alloy for decorative purposes
DIS 4536	Metallic and other non-organic coatings—Corrosion tests
DIS 4538	Metallic and other non-organic coatings—Corrosion tests—Thioacetamide corrosion test
DIS 4540	Rating of electroplated test specimens subjected to corrosion tests
DIS 4541	Corrosion testing by the corrodkote test

foreign alloys/general information

AA Wrought Alloys and Similar Foreign Alloys

The following Table 1.2 lists the AA wrought alloys and the corresponding foreign alloys that are covered by standards as used in Canada, France, Germany, Great Britain, Italy, Spain and Switzerland and in the Recommendations of the International Organization for Standardization (ISO).

The Table includes only those foreign alloys that are essentially equivalent in composition to the corresponding AA alloys, but whose composition limits are not necessarily exactly the same as their AA counterparts. Standards are subject to change and the actual issue of the specification or standard currently in effect should be consulted for full information.

TABLE 1.2 Designations for AA and Similar Foreign Alloys

AA Alloy	Canada CSA	France NF	Germany DIN Werkstoff-Nr.	Great Britain BS	Great Britain DTD	Italy UNI	Spain UNE	Switzerland VSM	ISO
EC		A5/L	E-Al99.5 3.0257	1E				Al99.5E	
1100	990C	A45							Al99.0Cu
2011	CB60		AlCuBiPb 3.1655						
2014	CS41N	A–U4SG	AlCuSiMn 3.1255			P-AlCu4.4-SiMnMg	L–313		AlCu4SiMg
Alclad 2014	CS41N Alclad					P-AlCu4.4-SiMnMg placc.			
2017	CM41	A–U4G	AlCuMg1 3.1325	H14 5L.37 L.87	150A	P-AlCu4-MgMn		Al3.5Cu-0.5Mg	AlCu4MgSi
2018	CN42								
2024	CG42	A–U4G1	AlCuMg2 3.1355	L.97 L.98	5090	P-AlCu4.5-MgMn	L–314	Al4Cu1.2-Mg	AlCu4Mg1
Alclad 2024	CG42 Alclad				5100	P-AlCu4.5-MgMn placc.			
2025	CS41P								
2117	CG30	A–U2G	AlCuMg0.5 3.1305	L.86		P-AlCu2.5-MgSi			AlCu2Mg
2218		A–U4N		6L.25			L–315	Al-Cu-Ni	
2618		A–U2GN			717 724 731A 745 5014 5084				
3003	MC10	A–M1							AlMn1Cu
3004		A–M1G							
4032	SG121	A–S12UN			324A	P-AlSi12-MgCuNi			
4043	S5		AlSi5 3.2245	N21					
5005		A–G0.6							AlMg1
5050		A–G1		3L.44		P-AlMg1.5		Al–1.5Mg	AlMg1.5
5052	GR20			2L.55 2L.56 L.80 L.81		P-AlMg2.5			AlMg2.5
5056	GM50R			N6 2L.58					AlMg5
5083	GM41		AlMg4.5Mn 3.3547	N8					AlMg4.5Mn

CSA: Canadian Standards Association
NF: Normes Francaises
DIN: Deutsche Industrie-Norm
BS: British Standard
DTD: Directorate of Technical Development
UNI: Unificazione Nazionale Italiana
UNE: Una Norma Espanola
VSM: Verein Schweizerischer Maschinenindustrieller
ISO: International Organization for Standardization

11 Non-Ferrous Material

General

Only copper and aluminum products are covered in this section. Major industrial countries such as Japan, Germany, France, and Italy have produced non-ferrous material dimensioned to metric modules for a number of years, and the material standards[1] for non-ferrous sizes reflect the available commercial sizes in these countries.

Recently the American National Standards Institute issued American National Standards B32.3- and B32.4-1974 for preferred metric sizes for flat, round, square, and hexagon metal products. The preferred metric sizes were the result of the informational input of representatives, as well as the experience of other countries where the metric system has been in use and where it is being undertaken.

Description of Tables

The non-ferrous material dimension tables given in this section list standard metric sizes produced in major countries already on the metric system. Linear dimensions, section areas, and mass per length of area are shown both in SI units and in customary inch units in the left-hand portions of each table. The left most column in the tables, which are headed in vertical fashion by ISO, ANS, DIN or BS, indicates the preference rating in the appropriate standard from which the metric sizes are chosen. In the right-hand portions of the tables, national standards for the applicable products are listed with their specified preference rating.

Tolerances shown in the right-hand portions of the tables are in millimeters, and are listed by column under a heading noting the original standard. Most of the tolerances are specified within the ISO system of limits and fits which is covered in national product standards throughout the world.

All conversions and calculations were processed by a computer, and the data shown are rounded off to the nearest number in each case. The computer exponential "E" format (see Section 10) was selected in order to cover a wide range of steel sizes and at the same time present a three significant digit accuracy for all numbers shown. The maximum error range is from 0.5 percent to 0.05 percent. The standard density for copper used is 8900 kilogram mass per cubic meter, and for aluminum 2700 kilogram mass per cubic meter. The density is not constant, and varies with added alloying materials.

World Standards for Non-Ferrous Plate and Sheet

Table 11-1. Cold Rolled Copper and Copper Alloy Sheets (ANSI B32.3)–The nominal thicknesses in the range from 0.1 to 10 millimeters have been based on the ANSI B32.3 standard, and the tolerances shown are based on the German DIN 1751 standard for cold rolled copper sheets.

Table 11-2. Cold Rolled Aluminum and Aluminum Alloy Sheets (ANSI B32.3)–Aluminum sheets are shown in sizes ranging from 0.25 to 15 millimeter thicknesses with preference ratings as specified in ANSI B32.3 standard. The tolerances shown are for aluminum sheets of Material Group I in the DIN 1783 standard, which is applicable for most of the aluminum sheet products.

Table 11-3. Cold Rolled Copper and Copper Alloy Plates (ANSI B32.3)–Copper plates in thicknesses from 3 to 50 millimeters are shown in Table 11-3, and the selection of sizes has been based on the ANSI B32.3 standard. The tolerances listed are those specified in the Japanese Standard JIS H3111 for oxygen-free copper sheets and plates.

Table 11-4. Hot Rolled Aluminum and Aluminum Alloy Plates (ANSI B32.3)–Hot-rolled aluminum plates in thicknesses from 5 to 100 millimeters have been selected from the ANSI B32.3 standard, and some tolerances from the DIN 59600 standard for sheets and rounds made of hot rolled aluminum are also shown.

World Non-Ferrous Wire Standards

Table 11-5. Copper Wires for Electrical Winding (DIN 46461)–The wire diameters in ranges from 0.04 to 8 millimeters and the tolerances shown in Table 11-5 have been based on the German Standard DIN 46461. The above

[1] For information about the term "standard" as used in this book, please see page 1-2.

standard is an agreement with the International Standard IEC 182-4 1971 for basic dimensions of winding wires and the ISO R388 diameter of wire recommendation.

Table 11-6. Aluminum Wire Sizes for Electrical Winding (BS 4391)—Aluminum wire diameters in sizes ranging from 0.1 to 10 millimeters have been based on the British Standard BS 4391. The tolerances shown are those from the DIN 46420 standard.

World Non-Ferrous Bar Standards

Table 11-7. Round Copper Bar Sizes (DIN 1756-ISO Tolerance h11)—Round copper bars in sizes ranging from 3 to 80 millimeters with tolerances based on the German Standard DIN 1756 are shown in Table 11-7.

Table 11-8. Round Aluminum Bar Sizes (DIN 1798-ISO Tolerance h11)—The aluminum bar diameters shown in Table 11-8 range from 3 to 100 millimeters, and the tolerances and nominal sizes are based on DIN 1798 standard for round aluminum bars.

Table 11-9. Square Copper Bar Sizes (DIN 1761-ISO Tolerance h11)—Square copper bars in sizes ranging from 2 to 100 millimeters are shown in Table 11-9, and the tolerances and sizes up to 60 millimeters are specified in DIN 1761 for square drawn copper bars.

Table 11-10. Square Aluminum Bar Sizes (DIN 1796-ISO Tolerance h11)—Aluminum bars with square cross sections are shown for sizes ranging from 3 to 200 millimeters in Table 11-10. The nominal sizes and tolerances listed in the above tables have been based on data from DIN 1796 on square drawn aluminum bars.

Table 11-11. Hexagon Copper Bar Sizes (DIN 1763-ISO Tolerance h11)—Copper bars with hexagonal cross sections in sizes ranging from 3 to 115 millimeters are shown in Table 11-11. Widths across flats and tolerances listed are specified in the DIN 1763 standard for hexagon drawn copper bars. Products with ISO limits and fits tolerances shown are available in countries already on the metric system.

Table 11-12. Hexagon Aluminum Bar Sizes (DIN 1797-ISO Tolerance h11)—Hexagon aluminum bars in sizes ranging from 3 to 105 millimeters are compared worldwide in Table 11-12. The German Standard DIN 1797 has been the basis for sizes and tolerances listed.

Table 11-13. Rectangular Copper Bar Sizes (DIN 1759)—National standards for rectangular copper bars in sizes ranging from 5 × 2 to 150 × 40 millimeters have been compared in Table 11-13. The DIN 1759 standard has provided the basis for nominal sizes and tolerances shown. The German standard tolerances listed are applicable for material in Group I as defined in DIN 1759, which covers most copper and copper alloy types.

Table 11-14. Rectangular Aluminum Bar Sizes (DIN 1769)—The German standard DIN 1769 covers drawn aluminum and aluminum alloy bars and the tolerances shown are for material group I (group II has larger tolerances). See the standard for the aluminum alloys belonging to each group.

World Non-Ferrous Tube Standards

Table 11-15. Copper Tubing for General Purposes (ISO/R274)—The ISO/R274* recommendation specifies dimensions for copper tubes of circular sections, and copper tubes in sizes ranging from 3 to 80 millimeters are listed and compared with national standards in major countries on the metric system. The deviation shown in Table 11-15 is for seamless copper tubes, and it is an average tolerance applicable to the outside tube diameter.

Table 11-16. Aluminum Tubing for General Purposes (DIN 1795)—Seamless aluminum tubes for general purposes in sizes ranging from 3 to 60 millimeter diameters are compared worldwide in Table 11-16. The tolerances for the outside diameters shown in "A" in the table are specified in DIN 1795 as a mean value.

---

*The ISO/R274 recommendation is now approved as an ISO 274-1975 standard and it specifies slightly different preference ratings and tolerances than shown in Table 11-15.

NON-FERROUS MATERIAL

Table 11-1. Cold Rolled Copper and Copper Alloy Sheets (ANSI B32.3)

```
TABLE 11-1. COLD ROLLED COPPER AND
COPPER ALLOY SHEETS (ANSI B32.3)
BASIS: 1 IN = 25.4 MM
1 CUBIC METER COPPER = 8900 KG MASS
```

ORDER EXAMPLE:
SIZE,COPPER SHEETS 0.40 BS 2870 + MATERIAL

THE NOMINAL SIZE IS NATIONAL STANDARD AS INDICATED
F=FIRST CHOICE,S=SECOND CHOICE,T=THIRD CHOICE,NUMBER=OTHER SIZE
* = COMMERCIAL SIZE

A N S	NOMINAL SIZE = D MM IN	MASS PER UNIT KG/M**2 LB/FT**2	U.S.A. ANSI B32.3	JAPAN JIS H3111	GERMANY DIN 1751	FRANCE NF	U.K. BS 2870	ITALY UNI 3233	AUSTRAL AS 1566
F	0.1 0.0039	0.890E+00 0.182E+00	F						F
F	0.11 0.0043	0.979E+00 0.201E+00			F+-0.02				S
F	0.12 0.0047	0.107E+01 0.219E+00	F		+-0.02				F
S	0.14 0.0055	0.125E+01 0.255E+00	S		F+-0.02				S
F	0.16 0.0063	0.142E+01 0.292E+00	F	0.15F	F+-0.02			0.15F	F
S	0.18 0.0071	0.160E+01 0.328E+00	F		F+-0.03				S
F	0.2 0.0079	0.178E+01 0.365E+00	S	F	F+-0.03			F	F
S	0.22 0.0087	0.196E+01 0.401E+00	F		F+-0.03				S
F	0.25 0.0098	0.222E+01 0.456E+00	F	F	F+-0.03		F		F
S	0.28 0.0110	0.249E+01 0.510E+00	S		+-0.03				S
F	0.3 0.0118	0.267E+01 0.547E+00	F	F	F+-0.03		F		F
S	0.35 0.0138	0.311E+01 0.638E+00	F		+-0.03		S		S
F	0.4 0.0157	0.356E+01 0.729E+00	F	F	F+-0.04		F		F
S	0.45 0.0177	0.400E+01 0.820E+00	S		F+-0.04				S
F	0.5 0.0197	0.445E+01 0.911E+00	F	F	F+-0.04		F		F
S	0.55 0.0217	0.489E+01 0.100E+01	F		+-0.04				S
F	0.6 0.0236	0.534E+01 0.109E+01	S,0.65S	F	F+-0.04		F		F
S	0.7 0.0276	0.623E+01 0.128E+01	S	F	F+-0.04		S		S
F	0.8 0.0315	0.712E+01 0.146E+01	F	F	F+-0.05		S		F
S	0.9 0.0354	0.801E+01 0.164E+01	F		F+-0.05				S
F	1 0.0394	0.890E+01 0.182E+01	F	F	F+-0.05		F		F
S	1.1 0.0433	0.979E+01 0.201E+01	S		F+-0.06				S
F	1.2 0.0472	0.107E+02 0.219E+01	F	F	F+-0.06		S		F
S	1.4 0.0551	0.125E+02 0.255E+01	F		F+-0.06				S
	1.5 0.0591	0.133E+02 0.273E+01	F	F	F+-0.07		F		F
F	1.6 0.0630	0.142E+02 0.292E+01	S	F	F+-0.07				S
S	1.8 0.0709	0.160E+02 0.328E+01	F		F+-0.07				F
F	2 0.0787	0.178E+02 0.365E+01	F	F	F+-0.08		F		F
S	2.2 0.0866	0.196E+02 0.401E+01	S	2.3F	F+-0.08		S		S
F	2.5 0.0984	0.222E+02 0.456E+01	F	F	F+-0.09				F
S	2.8 0.1102	0.249E+02 0.510E+01	F		F+-0.09				S
F	3 0.1181	0.267E+02 0.547E+01	F		F+-0.1		F		F
S	3.2 0.1260	0.285E+02 0.583E+01	F	F	F+-0.1				S
F	3.5 0.1378	0.311E+02 0.638E+01	F	3.8S F	F+-0.11		F		F
F	4 0.1575	0.356E+02 0.729E+01	F	F	F+-0.12				S
S	4.5 0.1772	0.400E+02 0.820E+01	F	4.2S F	F+-0.13		F		F
F	5 0.1969	0.445E+02 0.911E+01	F	4.8S F	F+-0.13				F
S	5.5 0.2165	0.489E+02 0.100E+02	S						S
F	6 0.2362	0.534E+02 0.109E+02	F						F
F	7 0.2756	0.623E+02 0.128E+02	F						S
F	8 0.3150	0.712E+02 0.146E+02	F						F
F	9 0.3543	0.801E+02 0.164E+02	F	F					S
F10	0.3937	0.890E+02 0.182E+02		F			F		F

Table 11-2. Cold Rolled Aluminum and Aluminum Alloy Sheets (ANSI B32.3)

TABLE 11-2. COLD ROLLED ALUMINUM AND
ALUMINUM ALLOY SHEETS (ANSI B32.3)
BASIS; 1 IN = 25.4 MM
1 CUBIC METER ALUMINUM = 2700 KG MASS

ORDER EXAMPLE;
SIZE, ALUMINUM SHEETS 1.2 UNI 4196 + MATERIAL

THE NOMINAL SIZE IS NATIONAL STANDARD AS INDICATED
F=FIRST CHOICE,S=SECOND CHOICE,T=THIRD CHOICE,NUMBER=OTHER SIZE
* = COMMERCIAL SIZE

A/N/S	NOMINAL SIZE = D		MASS PER UNIT		U.S.A. ANSI B32.3	JAPAN JIS	GERMANY DIN 1783	FRANCE NF	U.K. BS DD5	ITALY UNI 4196	AUSTRAL AS 1123
	MM	IN	KG/M**2	LB/FT**2							
F	0.25	0.0098	0.675E+00	0.138E+00	F				F		F
S	0.28	0.0110	0.756E+00	0.155E+00							S
F	0.3	0.0118	0.810E+00	0.166E+00	F			F			F
S	0.35	0.0138	0.945E+00	0.194E+00	S						S
F	0.4	0.0157	0.108E+01	0.221E+00	F		F +- 0.02		F	F	F
S	0.45	0.0177	0.121E+01	0.249E+00	S						S
F	0.5	0.0197	0.135E+01	0.277E+00	F		F +- 0.03		F	F	F
S	0.55	0.0217	0.148E+01	0.304E+00	S						S
F	0.6	0.0236	0.162E+01	0.332E+00	F		F +- 0.03		F	F	F
S	0.7	0.0276	0.189E+01	0.387E+00	S	, 0.65S	F +- 0.03		S	F	S
F	0.8	0.0315	0.216E+01	0.442E+00	F				S	F	F
S	0.9	0.0354	0.243E+01	0.498E+00	S						S
F	1	0.0394	0.270E+01	0.553E+00	F		F +- 0.04		F	F	F
S	1.1	0.0433	0.297E+01	0.608E+00	S						S
F	1.2	0.0472	0.324E+01	0.664E+00	F		F +- 0.04		F	F	F
S	1.4	0.0551	0.378E+01	0.774E+00	S						S
F	1.5	0.0591	0.405E+01	0.830E+00	F		F +- 0.04		F	F	F
F	1.6	0.0630	0.432E+01	0.885E+00	S		F +- 0.05				S
F	1.8	0.0709	0.486E+01	0.995E+00	F		F +- 0.05				F
F	2	0.0787	0.540E+01	0.111E+01	F						F
S	2.2	0.0866	0.594E+01	0.122E+01	S		F +- 0.06				S
F	2.5	0.0984	0.675E+01	0.138E+01	F						F
S	2.8	0.1102	0.756E+01	0.155E+01	S		F +- 0.07				S
F	3	0.1181	0.810E+01	0.166E+01	F						F
F	3.5	0.1378	0.945E+01	0.194E+01	F	, 3.2S	F +- 0.08				S
F	4	0.1575	0.108E+02	0.221E+01	F	, 3.8S	F +- 0.09				F
F	4.5	0.1772	0.121E+02	0.249E+01	F	, 4.2S					S
F	5	0.1969	0.135E+02	0.277E+01	F	, 4.8S	F +- 0.1				F
S	5.5	0.2165	0.148E+02	0.304E+01	S						S
F	6	0.2362	0.162E+02	0.332E+01	F		F		F		F
F	7	0.2756	0.189E+02	0.387E+01	F		F				S
F	8	0.3150	0.216E+02	0.442E+01	F						F
S	9	0.3543	0.243E+02	0.498E+01	S						S
F	10	0.3937	0.270E+02	0.553E+01	F	, 11S	F		F		F
F	12	0.4724	0.324E+02	0.664E+01	F		F				
	15	0.5906	0.405E+02	0.830E+01	F						

NON-FERROUS MATERIAL

Table 11-3. Cold Rolled Copper and Copper Alloy Plates (ANSI B32.3)

TABLE 11-3. COLD ROLLED COPPER AND
COPPER ALLOY PLATES (ANSI B32.3)
BASIS: 1 IN = 25.4 MM
1 CUBIC METER COPPER = 8900 KG MASS

ORDER EXAMPLE:
SIZE,COPPER PLATE 16 BS 2875 + MATERIAL

THE NOMINAL SIZE IS NATIONAL STANDARD AS INDICATED
F=FIRST CHOICE, S=SECOND CHOICE, T=THIRD CHOICE, NUMBER=OTHER SIZE
* = COMMERCIAL SIZE

A N S	NOMINAL SIZE = D		MASS PER UNIT		U.S.A. ANSI B32.3	JAPAN JIS H3111	GERMANY DIN 1751	FRANCE NF	U.K. BS 2875	ITALY UNI 3233	AUSTRAL AS 1566
	MM	IN	KG/M**2	LB/FT**2							
F	3	0.118	0.267E+02	0.547E+01		F+=0.09	F				
S	3.2	0.126	0.285E+02	0.583E+01	S	F+=0.09	F				
F	3.5	0.138	0.311E+02	0.638E+01		F+=0.11	F				
S	3.8	0.150	0.338E+02	0.693E+01	S	F+=0.11	F				
F	4	0.157	0.356E+02	0.729E+01	F	F+=0.11	F			F	F
S	4.2	0.165	0.374E+02	0.766E+01	S	F+=0.11	F				
F	4.5	0.177	0.400E+02	0.820E+01	F	F+=0.11	F				
S	4.8	0.189	0.427E+02	0.875E+01	S	F+=0.11	F				
F	5	0.197	0.445E+02	0.911E+01	F	F+=0.11	F		F	F	F
F	5.5	0.217	0.489E+02	0.100E+02	F	F+=0.14	F			F	F
F	6	0.236	0.534E+02	0.109E+02	F	F+=0.14	F		F	F	F
F	7	0.276	0.623E+02	0.128E+02	F	F+=0.14	F				
F	8	0.315	0.712E+02	0.146E+02	F	F+=0.14	F		F	F	F
S	9	0.354	0.801E+02	0.164E+02	S	F+=0.2	F				
F	10	0.394	0.890E+02	0.182E+02	F	F+=0.2	F		F	F	F
S	11	0.433	0.979E+02	0.201E+02	S	F+=0.2	F				
F	12	0.472	0.107E+03	0.219E+02	F	F+=0.2	F		F	F	F
S	14	0.551	0.125E+03	0.255E+02	S	F+=0.25	F		S		
F	15	0.591	0.133E+03	0.273E+02	F	F+=0.25	F		S	F	F
F	16	0.630	0.142E+03	0.292E+02	F	F+=0.25	F		F		
S	18	0.709	0.160E+03	0.328E+02	S	F+=0.25	F		S	F	F
F	20	0.787	0.178E+03	0.365E+02	F	F+=0.25	F		S	F	F
S	22	0.866	0.196E+03	0.401E+02	S	+=1.4%			S	F	F
F	25	0.984	0.222E+03	0.456E+02	F	+=1.4%			F	F	F
S	28	1.102	0.249E+03	0.510E+02	S	+=1.4%			S	F	F
F	30	1.181	0.267E+03	0.547E+02	F	+=1.4%			F	F	F
F	32	1.260	0.285E+03	0.583E+02	F				S		
F	35	1.378	0.311E+03	0.638E+02	F						
S	38	1.496	0.338E+03	0.693E+02	S						36F
F	40	1.575	0.356E+03	0.729E+02	F				S	F	F
S	45	1.772	0.400E+03	0.820E+02	S				F	F	F
F	50	1.969	0.445E+03	0.911E+02	F					F	F

Table 11-4. Hot Rolled Aluminum and Aluminum Alloy Plates (ANSI B32.3)

TABLE 11-4. HOT ROLLED ALUMINUM AND
ALUMINUM ALLOY PLATES (ANSI B32.3)
BASIS; 1 IN = 25.4 MM
1 CUBIC METER ALUMINUM = 2700 KG MASS

ORDER EXAMPLE;
SIZE,ALUMINUM PLATES 20 AS 1184 + MATERIAL

THE NOMINAL SIZE IS NATIONAL STANDARD AS INDICATED
F=FIRST CHOICE,S=SECOND CHOICE,T=THIRD CHOICE,NUMBER=OTHER SIZE
* = COMMERCIAL SIZE

A N S	NOMINAL SIZE = D		MASS PER UNIT		U.S.A. ANSI B32.3	JAPAN JIS	GERMANY DIN 59600	FRANCE NF	U.K. BS 1470	ITALY UNI 4196	AUSTRAL AS 1184
	MM	IN	KG/M**2	LB/FT**2							
F	5	0.197	0.135E+02	0.277E+01	F	F,S,5S	F+=0.18		F	F	S
F	6	0.236	0.162E+02	0.332E+01	F,S,5S		F+=0.2		F	F	F
F	7	0.276	0.189E+02	0.387E+01	F				F		S
F	8	0.315	0.216E+02	0.442E+01	F		F+=0.21		F		F
S	9	0.354	0.243E+02	0.498E+01	S				F		S
F	10	0.394	0.270E+02	0.553E+01	S		F+=0.24		F	F	F
S	11	0.433	0.297E+02	0.608E+01	S				F		S
F	12	0.472	0.324E+02	0.664E+01	F		F+=0.27		F		F
S	14	0.551	0.378E+02	0.774E+01	S				S		S
F	15	0.591	0.405E+02	0.830E+01			F+=0.32		F		F
F	16	0.630	0.432E+02	0.885E+01	F				S		F
S	18	0.709	0.486E+02	0.995E+01	S				S		S
F	20	0.787	0.540E+02	0.111E+02	F		F+=0.4		F		F
F	22	0.866	0.594E+02	0.122E+02	F				S		F
F	25	0.984	0.675E+02	0.138E+02	F		F+=0.47		F		F
S	28	1.102	0.756E+02	0.155E+02	S				F		S
F	30	1.181	0.810E+02	0.166E+02	F				F	F	F
S	32	1.260	0.864E+02	0.177E+02	F		F+=0.55		F		
F	35	1.378	0.945E+02	0.194E+02	F				S		36S
F	40	1.575	0.108E+03	0.221E+02	F		F+=0.7		F		F
S	45	1.772	0.121E+03	0.249E+02					S		S
F	50	1.969	0.135E+03	0.277E+02	F				F		F
F	55	2.165	0.148E+03	0.304E+02	F				S		F
F	60	2.362	0.162E+03	0.332E+02	F				F		F
S	70	2.756	0.189E+03	0.387E+02	S				S		S
F	75	2.953	0.202E+03	0.415E+02					S		S
F	80	3.150	0.216E+03	0.442E+02	S				F	F	F
S	90	3.543	0.243E+03	0.498E+02					S		S
F	100	3.937	0.270E+03	0.553E+02	F				F		F

NON-FERROUS MATERIAL

Table 11-5. Copper Wires for Electrical Winding (DIN 46461)

TABLE 11-5. COPPER WIRES
FOR ELECTRICAL WINDING (DIN 46461)
BASIS: 1 IN = 25.4 MM
1 CUBIC METER COPPER = 8900 KG MASS

ORDER EXAMPLE:
LENGTH,COPPER WIRE 0.1 DIN 46461 + MATERIAL
PAGE NO. 1

THE NOMINAL SIZE IS NATIONAL STANDARD AS INDICATED
F=FIRST CHOICE, S=SECOND CHOICE, T=THIRD CHOICE, NUMBER=OTHER SIZE
* = COMMERCIAL SIZE

D I N	NOMINAL SIZE = D MM / IN	SECTION AREA MM**2 / IN**2	MASS PER UNIT KG/M / LB/FT	U.S.A. ANSI B32.4	JAPAN JIS H3503	GERMANY DIN 46461	FRANCE NF C31-111	U.K. BS 2873	ITALY UNI 3605	AUSTRAL AS 1573
F	0.04 / 0.0016	0.126E-02 / 0.195E-05	0.112E-04 / 0.752E-05			F+-8%				
F	0.045 / 0.0018	0.159E-02 / 0.247E-05	0.142E-04 / 0.951E-05			F+-8%		F		
F	0.05 / 0.0020	0.196E-02 / 0.304E-05	0.175E-04 / 0.117E-04			F+-8%	F	F		
F	0.056 / 0.0022	0.246E-02 / 0.382E-05	0.219E-04 / 0.147E-04	0.055S		F+-8%		S		
S	0.06 / 0.0024	0.283E-02 / 0.436E-05	0.252E-04 / 0.169E-04			S+-8%		T		
F	0.063 / 0.0025	0.312E-02 / 0.483E-05	0.277E-04 / 0.186E-04	0.065S		F+-8%		F		
F	0.065 / 0.0026	0.385E-02 / 0.597E-05	0.343E-04 / 0.230E-04			F+-8%		F		
F	0.071 / 0.0028	0.396E-02 / 0.614E-05	0.352E-04 / 0.237E-04			F+-8%		S		
F	0.08 / 0.0031	0.503E-02 / 0.779E-05	0.447E-04 / 0.301E-04			F+-8%	F	F		
F	0.09 / 0.0035	0.636E-02 / 0.986E-05	0.566E-04 / 0.380E-04			F+-8%		S		
F	0.1 / 0.0039	0.785E-02 / 0.122E-04	0.699E-04 / 0.470E-04			F+-8%	F	F	F	
S	0.11 / 0.0043	0.950E-02 / 0.147E-04	0.846E-04 / 0.568E-04			S+-8%		S	F	
F	0.112 / 0.0044	0.985E-02 / 0.153E-04	0.877E-04 / 0.589E-04			F+-7%				
F	0.12 / 0.0047	0.113E-01 / 0.175E-04	0.101E-03 / 0.676E-04			S+-7%	F	F	F	
S	0.125 / 0.0049	0.123E-01 / 0.190E-04	0.109E-03 / 0.734E-04			F+-7%				
F	0.13 / 0.0051	0.133E-01 / 0.206E-04	0.118E-03 / 0.794E-04			S+-7%		S		S
F	0.14 / 0.0055	0.154E-01 / 0.239E-04	0.137E-03 / 0.921E-04	S		F+-7%		S		F
S	0.15 / 0.0059	0.177E-01 / 0.274E-04	0.157E-03 / 0.106E-03			S+-7%	F			F
F	0.16 / 0.0063	0.201E-01 / 0.312E-04	0.179E-03 / 0.120E-03			F+-7%		F		S
F	0.18 / 0.0071	0.254E-01 / 0.394E-04	0.226E-03 / 0.152E-03	S		F+-7%		S		
F	0.2 / 0.0079	0.314E-01 / 0.487E-04	0.280E-03 / 0.188E-03	S		F+-6%	F	S	F	
S	0.22 / 0.0087	0.380E-01 / 0.589E-04	0.338E-03 / 0.227E-03	S		S+-6%		F	F	S
F	0.224 / 0.0088	0.394E-01 / 0.611E-04	0.351E-03 / 0.236E-03			F+-6%			F	
F	0.25 / 0.0098	0.491E-01 / 0.761E-04	0.437E-03 / 0.294E-03	F		F+-6%		S	F	S
F	0.28 / 0.0110	0.616E-01 / 0.954E-04	0.548E-03 / 0.368E-03	S		S+-6%	F	S	F	
S	0.3 / 0.0118	0.707E-01 / 0.110E-03	0.629E-03 / 0.423E-03	F		F+-6%		F	F	S
F	0.315 / 0.0124	0.779E-01 / 0.121E-03	0.694E-03 / 0.466E-03			F+-5%		S	F	
F	0.32 / 0.0126	0.804E-01 / 0.125E-03	0.716E-03 / 0.481E-03			S+-5%			F	
S	0.35 / 0.0138	0.962E-01 / 0.149E-03	0.856E-03 / 0.575E-03	S		S+-5%		S	F 0.38F	S
F	0.355 / 0.0140	0.990E-01 / 0.153E-03	0.881E-03 / 0.592E-03			F+-5%			F	F
F	0.4 / 0.0157	0.126E+00 / 0.195E-03	0.112E-02 / 0.752E-03	F		F+-5%	F	F		S
F	0.45 / 0.0177	0.159E+00 / 0.247E-03	0.142E-02 / 0.951E-03	S		F+-5%		S	F 0.42F	S
F	0.5 / 0.0197	0.196E+00 / 0.304E-03	0.175E-02 / 0.117E-02	F		F+-5%	F	F	F 0.48F	F
S	0.55 / 0.0217	0.238E+00 / 0.368E-03	0.211E-02 / 0.142E-02			S+-4%			F 0.52F	F
F	0.56 / 0.0220	0.246E+00 / 0.382E-03	0.219E-02 / 0.147E-02	S		F+-4%		F	0.58F	S
F	0.6 / 0.0236	0.283E+00 / 0.438E-03	0.252E-02 / 0.169E-02	F		F+-4%		S		
F	0.63 / 0.0248	0.312E+00 / 0.483E-03	0.277E-02 / 0.186E-02			F+-4%		F	F	
F	0.65 / 0.0256	0.332E+00 / 0.514E-03	0.295E-02 / 0.198E-02	S		S+-4%		S	F	
S	0.7 / 0.0276	0.385E+00 / 0.597E-03	0.343E-02 / 0.230E-02	S		S+-4%			F	
F	0.71 / 0.0280	0.396E+00 / 0.614E-03	0.352E-02 / 0.237E-02			F+-4%		S	F	
F	0.75 / 0.0295	0.442E+00 / 0.685E-03	0.393E-02 / 0.264E-02			F+-4%	F	T	F	
F	0.8 / 0.0315	0.503E+00 / 0.779E-03	0.447E-02 / 0.301E-02	F		F+-4%		F	F	
F	0.85 / 0.0335	0.567E+00 / 0.880E-03	0.505E-02 / 0.339E-02			F+-4%		T	F	

11-7

Table 11-5 (Continued). Copper Wires for Electrical Winding (DIN 46461)

```
TABLE 11-5. COPPER WIRES                                           ORDER EXAMPLE:
FOR ELECTRICAL WINDING (DIN 46461)                                 LENGTH*COPPER WIRE 0.1 DIN 46461 + MATERIAL
BASIS: 1 IN = 25.4 MM                                                                                    PAGE NO. 2
1 CUBIC METER COPPER = 8900 KG MASS
```

                                                                  THE NOMINAL SIZE IS NATIONAL STANDARD AS INDICATED
                                                                  F=FIRST CHOICE, S=SECOND CHOICE, T=THIRD CHOICE, NUMBER=OTHER SIZE
                                                                  * = COMMERCIAL SIZE

D I N	NOMINAL SIZE = D		SECTION AREA		MASS PER UNIT		U.S.A. ANSI B32.4	JAPAN JIS	GERMANY DIN 46461	FRANCE NF C31-111	U.K. BS 2873	ITALY UNI 3605	AUSTRAL AS 1153
	MM	IN	MM**2	IN**2	KG/M	LB/FT							
F	0.9	0.0354	0.636E+00	0.986E-03	0.566E-02	0.380E-02			F++4%	F	F	F	F
F	0.95	0.0374	0.709E+00	0.110E-02	0.631E-02	0.424E-02			F++4%	F	T	F	S
F	1.0	0.0394	0.785E+00	0.122E-02	0.699E-02	0.470E-02	S		F++4%	F	T	F	
F	1.06	0.0417	0.882E+00	0.137E-02	0.785E-02	0.528E-02			F++4%	F	T	F	
S	1.1	0.0433	0.950E+00	0.147E-02	0.846E-02	0.568E-02	S		S++4%	F		F	
F	1.12	0.0441	0.985E+00	0.153E-02	0.877E-02	0.589E-02			F++4%	F	S	F	S
F	1.18	0.0465	0.109E+01	0.170E-02	0.973E-02	0.654E-02			F++4%	F	T	F	
S	1.2	0.0472	0.113E+01	0.175E-02	0.101E-01	0.676E-02			S++4%	F		F	
F	1.25	0.0492	0.123E+01	0.190E-02	0.109E-01	0.734E-02			F++4%	F	T	1.3F	F
F	1.32	0.0520	0.137E+01	0.212E-02	0.122E-01	0.818E-02			++4%	F	S	F	S
F	1.4	0.0551	0.154E+01	0.239E-02	0.137E-01	0.921E-02			++4%	F	T	F	F
F	1.5	0.0591	0.177E+01	0.274E-02	0.157E-01	0.106E-01			++4%	F	T	F	
F	1.6	0.0630	0.201E+01	0.312E-02	0.179E-01	0.120E-01	F		F++4%	F	T	F	F
F	1.7	0.0669	0.227E+01	0.352E-02	0.202E-01	0.136E-01			++4%	F	F	F	S
S	1.8	0.0709	0.254E+01	0.394E-02	0.226E-01	0.152E-01	S		S++4%	F	S	F	F
F	1.9	0.0748	0.284E+01	0.439E-02	0.252E-01	0.170E-01			++4%	F	T	F	
F	2	0.0787	0.314E+01	0.487E-02	0.280E-01	0.188E-01	F		F++4%	F	S	F	F
S	2.12	0.0835	0.353E+01	0.547E-02	0.314E-01	0.211E-01			S++4%	F	F	2.1F	S
S	2.2	0.0866	0.380E+01	0.589E-02	0.338E-01	0.227E-01			S++4%	F	T	F	
F	2.24	0.0882	0.394E+01	0.611E-02	0.351E-01	0.236E-01			F++4%	F	S	F	
F	2.36	0.0929	0.437E+01	0.678E-02	0.389E-01	0.262E-01	S		F++4%	F	T	2.3F	
F	2.5	0.0984	0.491E+01	0.761E-02	0.437E-01	0.294E-01			F++4%	F	F	F.2+4F	
F	2.65	0.1043	0.552E+01	0.855E-02	0.491E-01	0.330E-01	S		F++4%	F	T	2.6F	F
F	2.8	0.1102	0.616E+01	0.954E-02	0.548E-01	0.368E-01			F++4%	F	S	F	S
F	3	0.1181	0.707E+01	0.110E-01	0.629E-01	0.423E-01	F		F++4%	F	F	F	
F	3.15	0.1240	0.779E+01	0.121E-01	0.694E-01	0.466E-01			F++4%	F		F	F
S	3.2	0.1260	0.804E+01	0.125E-01	0.716E-01	0.481E-01			S++4%	F	S	F	
F	3.35	0.1319	0.881E+01	0.137E-01	0.784E-01	0.527E-01			F++4%	F	T	F	
S	3.5	0.1378	0.962E+01	0.149E-01	0.856E-01	0.575E-01	S		S++4%	F	S	3.8F	
S	3.55	0.1398	0.990E+01	0.153E-01	0.881E-01	0.592E-01			F++4%	F	T	F	F
F	3.75	0.1476	0.110E+02	0.171E-01	0.983E-01	0.661E-01			F++4%	F	F	F	
F	4	0.1575	0.126E+02	0.195E-01	0.112E+00	0.752E-01	F		F++4%	F	T	F	
F	4.25	0.1673	0.142E+02	0.220E-01	0.126E+00	0.848E-01			F++4%	F	T	4.2F	
F	4.5	0.1772	0.159E+02	0.247E-01	0.142E+00	0.951E-01	S		F++4%	F	S	F	S
F	4.75	0.1870	0.177E+02	0.275E-01	0.158E+00	0.106E+00			F++4%	F	T	4.8F	F
F	5	0.1969	0.196E+02	0.304E-01	0.175E+00	0.117E+00			F++4%	F	F	F	
F	5.3	0.2087	0.221E+02	0.342E-01	0.196E+00	0.132E+00			F++4%	F	T	5.2F	
S	5.5	0.2165	0.238E+02	0.368E-01	0.211E+00	0.142E+00	S		S++4%	F	T	F	S
F	5.6	0.2205	0.244E+02	0.382E-01	0.219E+00	0.147E+00			F++4%	F	S	5.8F	F
F	6	0.2362	0.283E+02	0.438E-01	0.252E+00	0.169E+00			F++4%	F	T	F	
F	6.3	0.2480	0.312E+02	0.483E-01	0.277E+00	0.186E+00	6.5S		F++4%	F	F	6.5F	F
F	7.1	0.2795	0.396E+02	0.614E-01	0.352E+00	0.237E+00	7S		++4%	F	F	7F	S
F	8	0.3150	0.503E+02	0.779E-01	0.447E+00	0.301E+00	F		++4%	F	S	F	F

NON-FERROUS MATERIAL

Table 11-6. Aluminum Wire Sizes for Electrical Winding (BS 4391)

```
TABLE 11-6. ALUMINUM WIRE SIZES                                    ORDER EXAMPLE;
FOR ELECTRICAL WINDING (BS 4391)                                   LENGTH,ALUMINUM WIRE 0.4 DIN 46420 + MATERIAL
BASIS: 1 IN = 25.4 MM                                              PAGE NO. 1
1 CUBIC METER ALUMINUM = 2700 KG MASS

                                                    THE NOMINAL SIZE IS NATIONAL STANDARD AS INDICATED
                                                    F=FIRST CHOICE,S=SECOND CHOICE,T=THIRD CHOICE,NUMBER=OTHER SIZE
                                                    * = COMMERCIAL SIZE
                                                    U.S.A.   JAPAN   GERMANY   FRANCE   U.K.   ITALY    AUSTRAL
B                                                   ANSI     JIS     DIN       NF       BS     UNI      AS
S     NOMINAL SIZE = D    SECTION AREA  MASS PER UNIT        B32.4            46420              4391    3817     1153
      MM        IN        MM**2  IN**2   KG/M      LB/FT
F    0.1      0.0039     0.785E-02  0.122E-04  0.212E-04  0.142E-04           F+=0.02                   F        F
S    0.112    0.0044     0.985E-02  0.153E-04  0.266E-04  0.179E-04   0.11S   F+=0.02                   S        F
F    0.125    0.0049     0.123E-01  0.190E-04  0.331E-04  0.223E-04   0.12F   F+=0.02                   F        S
S    0.14     0.0055     0.154E-01  0.239E-04  0.416E-04  0.279E-04   S       F+=0.02                   S        F

F    0.16     0.0063     0.201E-01  0.312E-04  0.543E-04  0.365E-04   F       F+=0.02           F       F        F
S    0.18     0.0071     0.254E-01  0.394E-04  0.687E-04  0.462E-04   S                                 S        S
F    0.2      0.0079     0.314E-01  0.487E-04  0.848E-04  0.570E-04   F       F+=0.02           F       F        F
S    0.22     0.0087     0.380E-01  0.589E-04  0.103E-03  0.690E-04   S                                          

S    0.224    0.0088     0.394E-01  0.611E-04  0.106E-03  0.715E-04           F+=0.02                            S
F    0.25     0.0098     0.491E-01  0.761E-04  0.133E-03  0.891E-04   F       F+=0.025          F       F        F
S    0.28     0.0110     0.616E-01  0.954E-04  0.166E-03  0.112E-03   S       F+=0.025                  F        S
F    0.3      0.0118     0.707E-01  0.110E-03  0.191E-03  0.128E-03   F       +=0.025           F                

F    0.315    0.0124     0.779E-01  0.121E-03  0.210E-03  0.141E-03           F+=0.025                  0.32F    F
S    0.35     0.0138     0.962E-01  0.149E-03  0.260E-03  0.175E-03   F       F+=0.025          F                
S    0.355    0.0140     0.990E-01  0.153E-03  0.267E-03  0.180E-03           F+=0.025                  F        S
F    0.4      0.0157     0.126E+00  0.195E-03  0.339E-03  0.228E-03   F       F+=0.03           F       F,0.38F  F

S    0.45     0.0177     0.159E+00  0.247E-03  0.429E-03  0.289E-03   S       F+=0.03                   F,0.42F  S
F    0.5      0.0197     0.196E+00  0.304E-03  0.530E-03  0.356E-03   F       F+=0.03           F       F,0.48F  F
S    0.55     0.0217     0.238E+00  0.368E-03  0.641E-03  0.431E-03   S       +=0.03                    F,0.52F  F
S    0.56     0.0220     0.246E+00  0.382E-03  0.665E-03  0.447E-03           +=0.03                    0.58F    S

F    0.6      0.0236     0.283E+00  0.438E-03  0.763E-03  0.513E-03   F       +=0.03            F       F        F
F    0.63     0.0248     0.312E+00  0.483E-03  0.842E-03  0.566E-03           F+=0.03                   F        F
S    0.65     0.0256     0.332E+00  0.514E-03  0.896E-03  0.602E-03   S       0.03                      F        
F    0.7      0.0276     0.385E+00  0.597E-03  0.104E-02  0.698E-03   S       F+=0.03                            

S    0.71     0.0280     0.396E+00  0.614E-03  0.107E-02  0.718E-03           F+=0.04                   F        S
T    0.75     0.0295     0.442E+00  0.685E-03  0.119E-02  0.802E-03   F       F+=0.04                   F        T
F    0.8      0.0315     0.503E+00  0.779E-03  0.136E-02  0.912E-03   F       F+=0.04           F       F        F
T    0.85     0.0335     0.567E+00  0.880E-03  0.153E-02  0.103E-02           F+=0.04                   F        T

S    0.9      0.0354     0.636E+00  0.986E-03  0.172E-02  0.115E-02   S       F+=0.04                   F        S
F    0.95     0.0374     0.709E+00  0.110E-02  0.191E-02  0.129E-02   F       F+=0.04                   F        F
F    1.0      0.0394     0.785E+00  0.122E-02  0.212E-02  0.142E-02   F       F+=0.04           F       F        F
T    1.06     0.0417     0.882E+00  0.137E-02  0.238E-02  0.160E-02   F       F+=0.05                   F        T

S    1.1      0.0433     0.950E+00  0.147E-02  0.257E-02  0.172E-02   S       F+=0.05                   1.15F    S
S    1.12     0.0441     0.985E+00  0.153E-02  0.266E-02  0.179E-02   S       F+=0.05                   S        T
T    1.18     0.0465     0.109E+01  0.170E-02  0.295E-02  0.198E-02   T       F+=0.05                   F        T
F    1.2      0.0472     0.113E+01  0.175E-02  0.305E-02  0.205E-02           +=0.05                             

F    1.25     0.0492     0.123E+01  0.190E-02  0.331E-02  0.223E-02   F       F+=0.05                   1.3F     F
T    1.32     0.0520     0.137E+01  0.212E-02  0.369E-02  0.248E-02   T       F+=0.05                   T        T
S    1.4      0.0551     0.154E+01  0.239E-02  0.416E-02  0.279E-02   S       F+=0.05                   F,1.35F  S
T    1.5      0.0591     0.177E+01  0.274E-02  0.477E-02  0.321E-02           F+=0.05                   F        T

F    1.6      0.0630     0.201E+01  0.312E-02  0.543E-02  0.365E-02   F       F+=0.05                   F,1.55F  F
T    1.7      0.0669     0.227E+01  0.352E-02  0.613E-02  0.412E-02           F+=0.05                   F        T
```

Table 11-6 (Continued). Aluminum Wire Sizes for Electrical Winding (BS 4391)

TABLE 11-6. ALUMINUM WIRE SIZES
FOR ELECTRICAL WINDING (BS 4391)
BASIS; 1 IN = 25.4 MM
1 CUBIC METER ALUMINUM = 2700 KG MASS

ORDER EXAMPLE:
LENGTH,ALUMINUM WIRE 0.4 DIN 46420 + MATERIAL
PAGE NO. 2

THE NOMINAL SIZE IS NATIONAL STANDARD AS INDICATED
F=FIRST CHOICE,S=SECOND CHOICE,T=THIRD CHOICE,NUMBER=OTHER SIZE
* = COMMERCIAL SIZE

B S	NOMINAL SIZE = D MM / IN	SECTION AREA MM**2 / IN**2	MASS PER UNIT KG/M / LB/FT	U.S.A. ANSI B32.4	JAPAN JIS	GERMANY DIN 46420	FRANCE NF	U.K. BS 4391	ITALY UNI 3817	AUSTRAL AS 1153
S	1.8 / 0.071	0.254E+01 / 0.394E-02	0.687E-02 / 0.462E-02	S		F+-0.05		S	F,1.75F	S
T	1.9 / 0.075	0.284E+01 / 0.439E-02	0.766E-02 / 0.514E-02			F+-0.05		T	F	T
F	2 / 0.079	0.314E+01 / 0.487E-02	0.848E-02 / 0.570E-02	F		F+-0.05		F	F,1.95F	F
T	2.12 / 0.083	0.353E+01 / 0.547E-02	0.953E-02 / 0.640E-02			F+-0.05		T	2.1F	T
S	2.24 / 0.088	0.394E+01 / 0.611E-02	0.106E-01 / 0.715E-02			F+-0.05		S	F	S
	2.2 / 0.087	0.380E+01 / 0.589E-02	0.103E-01 / 0.690E-02	S		+-0.05			F	
	2.3 / 0.091	0.415E+01 / 0.644E-02	0.112E-01 / 0.754E-02			+-0.06			F	
T	2.36 / 0.093	0.437E+01 / 0.678E-02	0.118E-01 / 0.794E-02			F+-0.06		T	2.35F	T
F	2.5 / 0.098	0.491E+01 / 0.761E-02	0.133E-01 / 0.891E-02	F		F+-0.06		F	F,2.4F	F
T	2.65 / 0.104	0.552E+01 / 0.855E-02	0.149E-01 / 0.100E-01			F+-0.06		T	2.6F	T
S	2.8 / 0.110	0.615E+01 / 0.954E-02	0.166E-01 / 0.112E-01	S		F+-0.06		S	F,2.7F	S
T	3 / 0.118	0.707E+01 / 0.110E-01	0.191E-01 / 0.128E-01			F+-0.06		T	F,2.9F	T
F	3.15 / 0.124	0.779E+01 / 0.121E-01	0.210E-01 / 0.141E-01	F		F+-0.08		F	3.2F	F
T	3.35 / 0.132	0.881E+01 / 0.137E-01	0.238E-01 / 0.160E-01			F+-0.08		T	3.4F	T
	3.5 / 0.138	0.962E+01 / 0.149E-01	0.260E-01 / 0.175E-01			+-0.08			F	
S	3.55 / 0.140	0.990E+01 / 0.153E-01	0.267E-01 / 0.180E-01	S		F+-0.08		S	3.6F	S
T	3.75 / 0.148	0.110E+02 / 0.171E-01	0.298E-01 / 0.200E-01			F+-0.08		T	3.8F	T
F	4 / 0.157	0.126E+02 / 0.195E-01	0.339E-01 / 0.228E-01	F		F+-0.08		F	F,3.85F	F
T	4.25 / 0.167	0.142E+02 / 0.220E-01	0.383E-01 / 0.257E-01			F+-0.08		T	4.2F	T
S	4.5 / 0.177	0.159E+02 / 0.247E-01	0.429E-01 / 0.289E-01			F+-0.08		S	F	S
T	4.75 / 0.187	0.177E+02 / 0.275E-01	0.478E-01 / 0.322E-01			F+-0.08		T	4.8F	T
F	5 / 0.197	0.196E+02 / 0.304E-01	0.530E-01 / 0.356E-01	F		F+-0.08		F	F	F
T	5.3 / 0.209	0.221E+02 / 0.342E-01	0.596E-01 / 0.400E-01			F+-0.08		T	5.2F	T
	5.5 / 0.217	0.238E+02 / 0.368E-01	0.641E-01 / 0.431E-01			+-0.08			F	
S	5.6 / 0.220	0.246E+02 / 0.382E-01	0.665E-01 / 0.447E-01			F+-0.08		S	F	S
T	5.8 / 0.228	0.264E+02 / 0.410E-01	0.713E-01 / 0.479E-01			F+-0.08		T	F	T
	6 / 0.236	0.283E+02 / 0.438E-01	0.763E-01 / 0.513E-01			F+-0.08			F	
F	6.3 / 0.248	0.312E+02 / 0.483E-01	0.842E-01 / 0.566E-01	F		F+-0.1		F	F	F
S	6.5 / 0.256	0.332E+02 / 0.514E-01	0.896E-01 / 0.602E-01	S		+-0.1			F	
T	6.7 / 0.264	0.353E+02 / 0.546E-01	0.952E-01 / 0.640E-01			+-0.1		T	F	T
	7 / 0.276	0.385E+02 / 0.597E-01	0.104E+00 / 0.698E-01			F+-0.1			F	
S	7.1 / 0.280	0.396E+02 / 0.614E-01	0.107E+00 / 0.718E-01			+-0.1		S	F	S
T	7.5 / 0.295	0.442E+02 / 0.685E-01	0.119E+00 / 0.802E-01			F+-0.1		T	F	T
F	8 / 0.315	0.503E+02 / 0.779E-01	0.136E+00 / 0.912E-01	F		F+-0.1		F	F	F
T	8.5 / 0.335	0.567E+02 / 0.880E-01	0.153E+00 / 0.103E+00			+-0.1		T	F	T
S	9 / 0.354	0.636E+02 / 0.986E-01	0.172E+00 / 0.115E+00	S		+-0.1		S	F	S
T	9.5 / 0.374	0.709E+02 / 0.110E+00	0.191E+00 / 0.129E+00			+-0.1		T	F	T
F	10 / 0.394	0.785E+02 / 0.122E+00	0.212E+00 / 0.142E+00	F		F+-0.1		F	F	F

NON-FERROUS MATERIAL

Table 11-7. Round Copper Bar Sizes (DIN 1756-ISO Tolerance h11)

```
TABLE 11-7. ROUND COPPER BAR SIZES
(DIN 1756-ISO TOLERANCE h11)
BASIS; 1 IN = 25.4 MM
1 CUBIC METER COPPER = 8900 KG MASS

ORDER EXAMPLE;
   LENGTH,ROUND 10 UNI 3606 + MATERIAL

THE NOMINAL SIZE IS NATIONAL STANDARD AS INDICATED
F=FIRST CHOICE,S=SECOND CHOICE,T=THIRD CHOICE,NUMBER=OTHER SIZE
* = COMMERCIAL SIZE
```

D I N	NOMINAL SIZE = D		SECTION AREA		MASS PER UNIT		U.S.A. ANSI B32.4	JAPAN JIS H3403	GERMANY DIN 1756	FRANCE NF	U.K. BS 4229	ITALY UNI 3606	AUSTRAL AS 1027
	MM	IN	MM**2	IN**2	KG/M	LB/FT							
F	3	0.118	0.707E+01	0.110E-01	0.629E-01	0.423E-01	F		F+0-0.06			F	F
S	3.5	0.138	0.962E+01	0.149E-01	0.856E-01	0.575E-01	S		S+0-0.08			F	F
F	4	0.157	0.126E+02	0.195E-01	0.112E+00	0.752E-01	F		F+0-0.08		F	F	F
S	4.5	0.177	0.159E+02	0.247E-01	0.142E+00	0.951E-01	S		S+0-0.08			F	F
F	5	0.197	0.196E+02	0.304E-01	0.175E+00	0.117E+00	F		F+0-0.08		F	F	F
S	5.5	0.217	0.238E+02	0.368E-01	0.211E+00	0.142E+00	S		S+0-0.08			F	F
F	6	0.236	0.283E+02	0.438E-01	0.252E+00	0.169E+00	F		F+0-0.08		F	F	F
S	6.5	0.256	0.332E+02	0.514E-01	0.295E+00	0.198E+00	S		S+0-0.09			S	F
S	7	0.276	0.385E+02	0.597E-01	0.343E+00	0.230E+00	S		S+0-0.09		F	F	F
F	8	0.315	0.503E+02	0.779E-01	0.447E+00	0.301E+00	F		F+0-0.09		F	F	F
S	9	0.354	0.636E+02	0.986E-01	0.566E+00	0.380E+00	S		S+0-0.09		F	F	F
F	10	0.394	0.785E+02	0.122E+00	0.699E+00	0.470E+00	F		F+0-0.09		F	F	F
S	11	0.433	0.950E+02	0.147E+00	0.846E+00	0.568E+00	S		S+0-0.11		F	F	F
F	12	0.472	0.113E+03	0.175E+00	0.101E+01	0.676E+00	F		F+0-0.11		F	F	F
F	14	0.551	0.154E+03	0.239E+00	0.137E+01	0.921E+00	F		F+0-0.11		F	F	F
S	15	0.591	0.177E+03	0.274E+00	0.157E+01	0.106E+01	S		+0-0.11			F	F
F	16	0.630	0.201E+03	0.312E+00	0.179E+01	0.120E+01	F		F+0-0.11		F	F	F
F	18	0.709	0.254E+03	0.394E+00	0.226E+01	0.152E+01	F,17S		S+0-0.13		F	F	F
F	20	0.787	0.314E+03	0.487E+00	0.280E+01	0.188E+01	F,19S		F+0-0.13		F	F	F
S	22	0.866	0.380E+03	0.589E+00	0.338E+01	0.227E+01	F,21S		S+0-0.13		F	F	F
F	24	0.945	0.452E+03	0.701E+00	0.403E+01	0.271E+01	F		+0-0.13		F	F	F
	25	0.984	0.491E+03	0.761E+00	0.437E+01	0.294E+01			F+0-0.13		F	F	F
F	26	1.024	0.531E+03	0.823E+00	0.473E+01	0.318E+01			F+0-0.13		F	F	F
S	28	1.102	0.616E+03	0.954E+00	0.548E+01	0.368E+01	S,23S		S+0-0.13		F	F	27F
F	30	1.181	0.707E+03	0.110E+01	0.629E+01	0.423E+01	F		+0-0.13		F	F	F
F	32	1.260	0.804E+03	0.125E+01	0.716E+01	0.481E+01	F		F+0-0.16		F	F	F
F	34	1.339	0.908E+03	0.141E+01	0.808E+01	0.543E+01	F		+0-0.16		F	F	F
S	36	1.417	0.102E+04	0.158E+01	0.906E+01	0.609E+01	35F		S+0-0.16		35F	F	33F
F	38	1.496	0.113E+04	0.176E+01	0.101E+02	0.678E+01	S		+0-0.16		F	F	F
F	40	1.575	0.126E+04	0.195E+01	0.112E+02	0.752E+01	F		F+0-0.16		F	F	F
F	42	1.654	0.139E+04	0.215E+01	0.123E+02	0.822E+01	F		F+0-0.16		F	F	F
S	45	1.772	0.159E+04	0.247E+01	0.142E+02	0.951E+01	F		S+0-0.16		F	F	39F
F	48	1.890	0.181E+04	0.280E+01	0.161E+02	0.108E+02	S		+0-0.16		F	F	F
	50	1.969	0.196E+04	0.304E+01	0.175E+02	0.117E+02	F		F+0-0.16		F	F	F
F	52	2.047	0.212E+04	0.329E+01	0.189E+02	0.125E+02	F		F+0-0.19		F	F	F
	55	2.165	0.238E+04	0.368E+01	0.211E+02	0.142E+02	S		+0-0.19		F	F	F
S	56	2.205	0.246E+04	0.382E+01	0.219E+02	0.147E+02	S		S+0-0.19		F	F	F
S	60	2.362	0.283E+04	0.438E+01	0.252E+02	0.169E+02	S		S+0-0.19		F	F	F
F	63	2.480	0.312E+04	0.483E+01	0.277E+02	0.186E+02	F		F+0-0.19		F	F	F
	65	2.559	0.332E+04	0.514E+01	0.295E+02	0.198E+02	S		+0-0.19				
S	70	2.756	0.385E+04	0.597E+01	0.343E+02	0.230E+02	S		S+0-0.19		F	F	F
S	75	2.953	0.442E+04	0.685E+01	0.393E+02	0.264E+02	S		S+0-0.19		F	F	F
F	80	3.150	0.503E+04	0.779E+01	0.447E+02	0.301E+02	F		F+0-0.19		F		F

Table 11-8. Round Aluminum Bar Sizes (DIN 1798-ISO Tolerance h11)

```
TABLE 11-8.  ROUND ALUMINUM BAR SIZES                                    ORDER EXAMPLE:
(DIN 1798-ISO TOLERANCE h11)                                             LENGTH,ROUND 20 DIN 1798 + AL QUALITY
BASIS: 1 IN = 25.4 MM
1 CUBIC METER ALUMINUM = 2700 KG MASS

                                                                         THE NOMINAL SIZE IS NATIONAL STANDARD AS INDICATED
                                                                         F=FIRST CHOICE,S=SECOND CHOICE,T=THIRD CHOICE,NUMBER=OTHER SIZE
                                                                         * = COMMERCIAL SIZE
                                                                         U.S.A.   JAPAN  GERMANY  FRANCE    U.K.     ITALY   AUSTRAL
                                                                         ANSI     JIS    DIN      NF        BS       UNI     AS
                                                                         B32.4           1798     A50-731   4229     3818    1027
```

D I N	NOMINAL SIZE = D		SECTION AREA		MASS PER UNIT		U.S.A. ANSI B32.4	JAPAN JIS	GERMANY DIN 1798	FRANCE NF A50-731	U.K. BS 4229	ITALY UNI 3818	AUSTRAL AS 1027
	MM	IN	MM**2	IN**2	K/GM	LB/FT							
F	3	0.118	0.707E+01	0.110E-01	0.191E-01	0.128E-01	F		F+0-0.06	F		F	
S	3.5	0.138	0.962E+01	0.149E-01	0.260E-01	0.175E-01	S		S+0-0.08				
F	4	0.157	0.126E+02	0.195E-01	0.339E-01	0.228E-01	F		F+0-0.08	F		F	
S	4.5	0.177	0.159E+02	0.247E-01	0.429E-01	0.289E-01	S		S+0-0.08				
F	5	0.197	0.196E+02	0.304E-01	0.530E-01	0.356E-01	F		F+0-0.08	F		F	
S	5.5	0.217	0.238E+02	0.368E-01	0.641E-01	0.431E-01	S		S+0-0.08				
F	6	0.236	0.283E+02	0.438E-01	0.763E-01	0.513E-01	F		F+0-0.08	F		F	
S	6.5	0.256	0.332E+02	0.514E-01	0.896E-01	0.602E-01	S		S+0-0.09				
S	7	0.276	0.385E+02	0.597E-01	0.104E+00	0.698E-01	S		S+0-0.09	F		F	
F	8	0.315	0.503E+02	0.779E-01	0.136E+00	0.912E-01	F		F+0-0.09	F		F	
S	9	0.354	0.636E+02	0.986E-01	0.172E+00	0.115E+00	S		S+0-0.09	F		F	
F	10	0.394	0.785E+02	0.122E+00	0.212E+00	0.142E+00	F		F+0-0.09	F		F	
S	11	0.433	0.950E+02	0.147E+00	0.257E+00	0.172E+00	S		S+0-0.11	F		F	
F	12	0.472	0.113E+03	0.175E+00	0.305E+00	0.205E+00	F		F+0-0.11	F		F	
F	13	0.512	0.133E+03	0.206E+00	0.358E+00	0.241E+00	F		F+0-0.11	F		F	
S	14	0.551	0.154E+03	0.239E+00	0.416E+00	0.279E+00	S		S+0-0.11	6		6	
F	15	0.591	0.177E+03	0.274E+00	0.477E+00	0.321E+00	S		S+0-0.11	F		F,17F	
S	16	0.630	0.201E+03	0.312E+00	0.543E+00	0.365E+00	F		F+0-0.11	F		F	
F	18	0.709	0.254E+03	0.394E+00	0.687E+00	0.462E+00	F,17S		F+0-0.13	F		F,19F	
S	20	0.787	0.314E+03	0.487E+00	0.848E+00	0.570E+00	F,19S		F+0-0.13	F		F	
S	22	0.866	0.380E+03	0.589E+00	0.103E+01	0.690E+00	F,21S		F+0-0.13	F		F,21F	
F	24	0.945	0.452E+03	0.701E+00	0.122E+01	0.821E+00	S,23S		S+0-0.13	F		F,23F	
S	25	0.984	0.491E+03	0.761E+00	0.133E+01	0.891E+00	F		F+0-0.13	F		F,26F	
S	28	1.102	0.616E+03	0.954E+00	0.166E+01	0.112E+01	S,26S		S+0-0.13	F,26F		F,27F	27F
S	30	1.181	0.707E+03	0.110E+01	0.191E+01	0.128E+01	F		F+0-0.13	F		F	F
S	32	1.260	0.804E+03	0.125E+01	0.217E+01	0.146E+01	F		F+0-0.16	F		F,33F	F,33F
F	36	1.417	0.102E+04	0.158E+01	0.275E+01	0.185E+01	35F		S+0-0.16	F,34F	35F	F,35F	
S	38	1.496	0.113E+04	0.176E+01	0.306E+01	0.206E+01	S		S+0-0.16	F		F	
F	40	1.575	0.126E+04	0.195E+01	0.339E+01	0.228E+01	F		F+0-0.16	F		F	39F
F	42	1.654	0.139E+04	0.215E+01	0.374E+01	0.251E+01	F		F+0-0.16	F		F,39F	F
S	45	1.772	0.159E+04	0.247E+01	0.429E+01	0.289E+01	F		S+0-0.16	F		F	F
F	48	1.890	0.181E+04	0.280E+01	0.489E+01	0.328E+01	S		+0-0.16	F		F	F
F	50	1.969	0.196E+04	0.304E+01	0.530E+01	0.356E+01	F		F+0-0.16	F		F	F
S	55	2.165	0.238E+04	0.368E+01	0.641E+01	0.431E+01	F		F+0-0.20	F		F,52F	
S	56	2.205	0.246E+04	0.382E+01	0.665E+01	0.447E+01	S		S+0-0.19	F		F	
S	60	2.362	0.283E+04	0.438E+01	0.763E+01	0.513E+01	F		+0-0.20	F		F,58F	
F	63	2.480	0.312E+04	0.483E+01	0.842E+01	0.566E+01	F		F+0-0.19	F		F	F
S	65	2.559	0.332E+04	0.514E+01	0.896E+01	0.602E+01	F		+0-0.20	F		F	F
S	70	2.756	0.385E+04	0.597E+01	0.104E+02	0.698E+01	S		S+0-0.30	F		F,67F	F
F	75	2.953	0.442E+04	0.685E+01	0.119E+02	0.802E+01	S		+0-0.30	F		F,67F	F
F	80	3.150	0.503E+04	0.779E+01	0.136E+02	0.912E+01	F		F+0-0.30	F		F	F
S	90	3.543	0.636E+04	0.986E+01	0.172E+02	0.115E+02	S		S+0-0.35	F		F,85F	F
F	100	3.937	0.785E+04	0.122E+02	0.212E+02	0.142E+02	F		F+0-0.35	F		F,95F	F

NON-FERROUS MATERIAL

Table 11-9. Square Copper Bar Sizes (DIN 1761-ISO Tolerance h11)

```
TABLE 11-9. SQUARE COPPER BAR SIZES
(DIN 1761-ISO TOLERANCE h11)
BASIS; 1 IN = 25.4 MM
1 CUBIC METER COPPER = 8900 KG MASS
```

ORDER EXAMPLE:
LENGTH,SQUARE 6 BS 4229 + COPPER QUALITY

THE NOMINAL SIZE IS NATIONAL STANDARD AS INDICATED
F=FIRST CHOICE, S=SECOND CHOICE, T=THIRD CHOICE, NUMBER=OTHER SIZE
* = COMMERCIAL SIZE

D I N	NOMINAL SIZE = D MM / IN	SECTION AREA MM**2 / IN**2	MASS PER UNIT KG/M / LB/FT	U.S.A. ANSI B32.4	JAPAN JIS	GERMANY DIN 1761	FRANCE NF	U.K. BS 4229	ITALY UNI 3217	AUSTRAL AS 1027
F	2 0.079	0.400E+01 0.620E-02	0.356E-01 0.239E-01			F +0 -0.06				
F	2.2 0.087	0.484E+01 0.750E-02	0.431E-01 0.289E-01			F +0 -0.06				
F	2.5 0.098	0.625E+01 0.969E-02	0.556E-01 0.374E-01			F +0 -0.06				
F	2.8 0.110	0.784E+01 0.122E-01	0.698E-01 0.469E-01			F +0 -0.06				
F	3 0.118	0.900E+01 0.140E-01	0.801E-01 0.538E-01			F +0 -0.06		F		
F	3.5 0.138	0.122E+02 0.190E-01	0.109E+00 0.733E-01			F +0 -0.08				
F	4 0.157	0.160E+02 0.248E-01	0.142E+00 0.957E-01			F +0 -0.08		F	F	
F	4.5 0.177	0.202E+02 0.314E-01	0.180E+00 0.121E+00			F +0 -0.08				
F	5 0.197	0.250E+02 0.388E-01	0.222E+00 0.150E+00			F +0 -0.08		F	F	
F	5.5 0.217	0.302E+02 0.469E-01	0.269E+00 0.181E+00			F +0 -0.08				
F	6 0.236	0.360E+02 0.558E-01	0.320E+00 0.215E+00			F +0 -0.08		F	F	
F	7 0.276	0.490E+02 0.760E-01	0.436E+00 0.293E+00			F +0 -0.09				
F	8 0.315	0.640E+02 0.992E-01	0.570E+00 0.383E+00			F +0 -0.09		F	F	
F	9 0.354	0.810E+02 0.126E+00	0.721E+00 0.484E+00			F +0 -0.09				
F	10 0.394	0.100E+03 0.155E+00	0.890E+00 0.598E+00			F +0 -0.09		F	F	
F	11 0.433	0.121E+03 0.188E+00	0.108E+01 0.724E+00			F +0 -0.11				
F	12 0.472	0.144E+03 0.223E+00	0.128E+01 0.861E+00			F +0 -0.11		F	F	
F	13 0.512	0.169E+03 0.262E+00	0.150E+01 0.101E+01			F +0 -0.11				
F	14 0.551	0.196E+03 0.304E+00	0.174E+01 0.117E+01			F +0 -0.11		F	F	
F	15 0.591	0.225E+03 0.349E+00	0.200E+01 0.135E+01			F +0 -0.11				
F	16 0.630	0.256E+03 0.397E+00	0.228E+01 0.153E+01			F +0 -0.11		F	F	
F	17 0.669	0.289E+03 0.448E+00	0.257E+01 0.173E+01			F +0 -0.11				
F	18 0.709	0.324E+03 0.502E+00	0.288E+01 0.194E+01		S	F +0 -0.11		F	F	
F	19 0.748	0.361E+03 0.560E+00	0.321E+01 0.216E+01			F +0 -0.13				
F	20 0.787	0.400E+03 0.620E+00	0.356E+01 0.239E+01			F +0 -0.13		F	F	
F	21 0.827	0.441E+03 0.684E+00	0.392E+01 0.264E+01			F +0 -0.13				
F	22 0.866	0.484E+03 0.750E+00	0.431E+01 0.289E+01		S	F +0 -0.13		F	F	
F	24 0.945	0.576E+03 0.893E+00	0.513E+01 0.344E+01			F +0 -0.13				
F	25 0.984	0.625E+03 0.969E+00	0.556E+01 0.374E+01			F +0 -0.13		F	F	
F	27 1.063	0.729E+03 0.113E+01	0.649E+01 0.436E+01			F +0 -0.13		F	F	
F	30 1.181	0.900E+03 0.140E+01	0.801E+01 0.538E+01			F +0 -0.16		F	F	
F	32 1.260	0.102E+04 0.159E+01	0.911E+01 0.612E+01			F +0 -0.16		F	F	
F	36 1.417	0.130E+04 0.201E+01	0.115E+02 0.775E+01	35S		F +0 -0.16		F	F	35F
F	40 1.575	0.160E+04 0.248E+01	0.142E+02 0.957E+01			F +0 -0.16		F		
F	41 1.614	0.168E+04 0.261E+01	0.150E+02 0.101E+02			F +0 -0.16		F		
F	46 1.811	0.212E+04 0.328E+01	0.188E+02 0.127E+02	45S		F +0 -0.16		F		45F
F	50 1.969	0.250E+04 0.388E+01	0.222E+02 0.150E+02			F +0 -0.16		F	F	
F	55 2.165	0.302E+04 0.469E+01	0.269E+02 0.181E+02		S	F +0 -0.19		F	F	
F	60 2.362	0.360E+04 0.558E+01	0.320E+02 0.215E+02		S	F +0 -0.19		F	F	
F	70 2.756	0.490E+04 0.760E+01	0.436E+02 0.293E+02		S	F +0 -0.19		F		
	80 3.150	0.640E+04 0.992E+01	0.570E+02 0.383E+02			+0 -0.19		F	F	
	90 3.543	0.810E+04 0.126E+02	0.721E+02 0.484E+02		S	+0 -0.22		F	F	
	100 3.937	0.100E+05 0.155E+02	0.890E+02 0.598E+02			+0 -0.22		F		

Table 11-10. Square Aluminum Bar Sizes (DIN 1796-ISO Tolerance h11)

```
TABLE 11-10. SQUARE ALUMINUM BAR SIZES              ORDER EXAMPLE;
(DIN 1796-ISO TOLERANCE h11)                        LENGTH,SQUARE 8 AS 1027 + MATERIAL
BASIS; 1 IN. = 25.4 MM
1 CUBIC METER ALUMINUM = 2700 KG MASS
```

THE NOMINAL SIZE IS NATIONAL STANDARD AS INDICATED
F=FIRST CHOICE,S=SECOND CHOICE,T=THIRD CHOICE,NUMBER=OTHER SIZE
* = COMMERCIAL SIZE

	NOMINAL SIZE = D		SECTION AREA		MASS PER UNIT		U.S.A. ANSI B32.4	JAPAN JIS	GERMANY DIN 1796	FRANCE NF A50-732	U.K. BS 4229	ITALY UNI 3821	AUSTRAL AS 1027
	MM	IN	MM**2	IN**2	KG/M	LB/FT							
F	3	0.118	0.900E+01	0.140E-01	0.243E-01	0.163E-01	F		F+0-0.06	F			
F	3.5	0.138	0.122E+02	0.190E-01	0.331E-01	0.222E-01			F+0-0.08				
F	4	0.157	0.160E+02	0.248E-01	0.432E-01	0.290E-01	F		F+0-0.08	F			
F	4.5	0.177	0.202E+02	0.314E-01	0.547E-01	0.367E-01			F+0-0.08				
F	5	0.197	0.250E+02	0.388E-01	0.675E-01	0.454E-01	F		F+0-0.08	F			
F	5.5	0.217	0.302E+02	0.469E-01	0.817E-01	0.549E-01			F+0-0.08				
F	6	0.236	0.360E+02	0.558E-01	0.972E-01	0.653E-01	F		F+0-0.08	F			
F	7	0.276	0.490E+02	0.760E-01	0.132E+00	0.889E-01			F+0-0.09				
F	8	0.315	0.640E+02	0.992E-01	0.173E+00	0.116E+00	F		F+0-0.09	F		F	
F	9	0.354	0.810E+02	0.126E+00	0.219E+00	0.147E+00			F+0-0.09				
F	10	0.394	0.100E+03	0.155E+00	0.270E+00	0.181E+00	F		F+0-0.09	F		F	
F	11	0.433	0.121E+03	0.188E+00	0.327E+00	0.220E+00			F+0-0.11				
F	12	0.472	0.144E+03	0.223E+00	0.389E+00	0.261E+00	F		F+0-0.11	F		F	
F	13	0.512	0.169E+03	0.262E+00	0.456E+00	0.307E+00			F+0-0.11				
F	14	0.551	0.196E+03	0.304E+00	0.529E+00	0.356E+00	S		F+0-0.11			F	
F	15	0.591	0.225E+03	0.349E+00	0.607E+00	0.408E+00			F+0-0.11			F	
F	16	0.630	0.256E+03	0.397E+00	0.691E+00	0.464E+00	F		F+0-0.11	F		F	
F	17	0.669	0.289E+03	0.448E+00	0.780E+00	0.524E+00			F+0-0.11				
F	18	0.709	0.324E+03	0.502E+00	0.875E+00	0.588E+00			F+0-0.11			F	
F	19	0.748	0.361E+03	0.560E+00	0.975E+00	0.655E+00			F+0-0.13				
F	20	0.787	0.400E+03	0.620E+00	0.108E+01	0.726E+00	F		F+0-0.13	F		F	
F	22	0.866	0.484E+03	0.750E+00	0.131E+01	0.878E+00	S		F+0-0.13				
F	24	0.945	0.576E+03	0.893E+00	0.156E+01	0.105E+01			F+0-0.13			F	
F	25	0.984	0.625E+03	0.969E+00	0.169E+01	0.113E+01	F		F+0-0.13	F		F	
F	27	1.063	0.729E+03	0.113E+01	0.197E+01	0.132E+01			F+0-0.13				
F	28	1.102	0.784E+03	0.122E+01	0.212E+01	0.142E+01	S		F+0-0.13				
F	30	1.181	0.900E+03	0.140E+01	0.243E+01	0.163E+01	S		F+0-0.13	F	F,26F	F	
F	32	1.260	0.102E+04	0.159E+01	0.276E+01	0.186E+01			F+0-0.16			F	
F	35	1.378	0.122E+04	0.190E+01	0.331E+01	0.222E+01	S		F+0-0.16				
F	36	1.417	0.130E+04	0.201E+01	0.350E+01	0.235E+01	S		F+0-0.16			F	
F	40	1.575	0.160E+04	0.248E+01	0.432E+01	0.290E+01	S		F+0-0.16			F	
F	41	1.614	0.168E+04	0.261E+01	0.454E+01	0.305E+01	S		F+0-0.16				
F	45	1.772	0.202E+04	0.314E+01	0.547E+01	0.367E+01			F+0-0.16			F	
F	46	1.811	0.212E+04	0.328E+01	0.571E+01	0.384E+01	S		F+0-0.16			F	
F	50	1.969	0.250E+04	0.388E+01	0.675E+01	0.454E+01	F		F+0-0.16	F		F	
F	55	2.165	0.302E+04	0.469E+01	0.817E+01	0.549E+01	S		F+0-0.19				
F	60	2.362	0.360E+04	0.558E+01	0.972E+01	0.653E+01	F		F+0-0.19	F,75F	F,65F	F	
	70	2.756	0.490E+04	0.760E+01	0.132E+02	0.889E+01	S		F+0-0.19	F,75F	F,75F		
	80	3.150	0.640E+04	0.992E+01	0.173E+02	0.116E+02	S		F+0-0.19	F,85F	F,85F		
	100	3.937	0.100E+05	0.155E+02	0.270E+02	0.181E+02	F,90S		+0-0.22	F,90F	F,90F	95F	
	120	4.724	0.144E+05	0.223E+02	0.389E+02	0.261E+02	F,110S		+0-0.22	F,110F			
	160	6.299	0.256E+05	0.397E+02	0.691E+02	0.464E+02	F,140S		+0-0.25	F,140F			
	200	7.874	0.400E+05	0.620E+02	0.108E+03	0.726E+02	F,180S		+0-0.29	F,180F			

NON-FERROUS MATERIAL

NON-FERROUS MATERIAL

Table 11-11. Hexagon Copper Bar Sizes (DIN 1763-ISO Tolerance h11)

TABLE 11-11. HEXAGON COPPER BAR SIZES
(DIN 1763-ISO TOLERANCE h11)
BASIS; 1 IN = 25.4 MM
1 CUBIC METER COPPER = 8900 KG MASS

ORDER EXAMPLE:
LENGTH,HEXAGON 14 BS 4229 + MATERIAL

THE NOMINAL SIZE IS NATIONAL STANDARD AS INDICATED
F=FIRST CHOICE,S=SECOND CHOICE,T=THIRD CHOICE,NUMBER=OTHER SIZE
* = COMMERCIAL SIZE

DIN	NOMINAL SIZE = D		SECTION AREA		MASS PER UNIT		U.S.A. ANSI B32.4	JAPAN JIS	GERMANY DIN 1763	FRANCE NF	U.K. BS 4229	ITALY UNI 3216	AUSTRAL AS 1027
	MM	IN	MM**2	IN**2	KG/M	LB/FT							
F	3	0.118	0.779E+01	0.121E-01	0.694E-01	0.466E-01			F +0-0.06				
F	3.2	0.126	0.887E+01	0.137E-01	0.789E-01	0.530E-01			+0-0.08		F		F
F	3.5	0.138	0.106E+02	0.164E-01	0.944E-01	0.634E-01			F +0-0.08				
F	4	0.157	0.139E+02	0.215E-01	0.123E+00	0.829E-01	F		F +0-0.08		F		F
F	4.5	0.177	0.175E+02	0.272E-01	0.156E+00	0.105E+00			F +0-0.08				
F	5	0.197	0.217E+02	0.336E-01	0.193E+00	0.129E+00			F +0-0.08		F		F
F	5.5	0.217	0.262E+02	0.406E-01	0.233E+00	0.157E+00			F +0-0.08				
F	6	0.236	0.312E+02	0.483E-01	0.277E+00	0.186E+00			F +0-0.08		F	F	F
F	7	0.276	0.424E+02	0.658E-01	0.378E+00	0.254E+00	F		F +0-0.09				
F	8	0.315	0.554E+02	0.859E-01	0.493E+00	0.331E+00			F +0-0.09		F	F	F
F	9	0.354	0.701E+02	0.109E+00	0.624E+00	0.420E+00			F +0-0.09				
F	10	0.394	0.866E+02	0.134E+00	0.771E+00	0.518E+00	F		F +0-0.09		F	F	F
F	11	0.433	0.105E+03	0.162E+00	0.933E+00	0.627E+00			F +0-0.11				
F	12	0.472	0.125E+03	0.193E+00	0.111E+01	0.746E+00			F +0-0.11		F	F	F
F	13	0.512	0.146E+03	0.227E+00	0.130E+01	0.875E+00			F +0-0.11				
F	14	0.551	0.170E+03	0.263E+00	0.151E+01	0.102E+01			F +0-0.11		F	F	F
F	15	0.591	0.195E+03	0.302E+00	0.173E+01	0.117E+01	F		+0-0.11				
F	17	0.669	0.250E+03	0.388E+00	0.223E+01	0.150E+01			F +0-0.11		F	F	F
F	18	0.709	0.281E+03	0.435E+00	0.250E+01	0.168E+01			F +0-0.11				
F	19	0.748	0.313E+03	0.485E+00	0.278E+01	0.187E+01			F +0-0.13		F		F
F	21	0.827	0.382E+03	0.592E+00	0.340E+01	0.228E+01	F		F +0-0.13				
F	22	0.866	0.419E+03	0.650E+00	0.373E+01	0.251E+01			F +0-0.13		F	F	F
F	24	0.945	0.499E+03	0.773E+00	0.444E+01	0.298E+01			F +0-0.13		F		F
F	27	1.063	0.631E+03	0.979E+00	0.562E+01	0.378E+01			F +0-0.13		F		F
F	30	1.181	0.779E+03	0.121E+01	0.694E+01	0.466E+01	F		F +0-0.13		F	F	F
F	32	1.260	0.887E+03	0.137E+01	0.789E+01	0.530E+01			F +0-0.16		F		F
F	36	1.417	0.112E+04	0.174E+01	0.999E+01	0.671E+01			F +0-0.16		F		F
F	41	1.614	0.146E+04	0.226E+01	0.130E+02	0.871E+01			F +0-0.16		F		F
F	46	1.811	0.183E+04	0.284E+01	0.163E+02	0.110E+02	F		F +0-0.16		F		F
F	50	1.969	0.217E+04	0.336E+01	0.193E+02	0.129E+02			F +0-0.16		F		F
F	55	2.165	0.262E+04	0.406E+01	0.233E+02	0.157E+02			F +0-0.19		F		F
F	60	2.362	0.312E+04	0.483E+01	0.277E+02	0.186E+02			F +0-0.19		F		F
F	65	2.559	0.366E+04	0.567E+01	0.326E+02	0.219E+02	F		+0-0.19		F		F
F	70	2.756	0.424E+04	0.658E+01	0.378E+02	0.254E+02			+0-0.19		F		F
F	75	2.953	0.487E+04	0.755E+01	0.434E+02	0.291E+02			+0-0.19		F		F
F	80	3.150	0.554E+04	0.859E+01	0.493E+02	0.331E+02			+0-0.19		F		F
F	85	3.346	0.626E+04	0.970E+01	0.557E+02	0.374E+02	F		+0-0.22		F		F
F	90	3.543	0.701E+04	0.109E+02	0.624E+02	0.420E+02			+0-0.22		F		F
F	95	3.740	0.782E+04	0.121E+02	0.696E+02	0.467E+02			+0-0.22		F		F
F	100	3.937	0.866E+04	0.134E+02	0.771E+02	0.518E+02			+0-0.22		F		F
	105	4.134	0.955E+04	0.148E+02	0.850E+02	0.571E+02	F		+0-0.22				
	115	4.528	0.115E+05	0.178E+02	0.102E+03	0.685E+02	F		+0-0.22				

11-15

Table 11-12. Hexagon Aluminum Bar Sizes (DIN 1797-ISO Tolerance h11)

```
TABLE 11-12. HEXAGON ALUMINUM BAR SIZES
(DIN 1797-ISO TOLERANCE 11)
BASIS: 1 IN = 25.4 MM
1 CUBIC METER ALUMINUM = 2700 KG MASS

                                                    ORDER EXAMPLE;
                                                    LENGTH,HEXAGON 4 NF A50-733 + MATERIAL

                                   THE NOMINAL SIZE IS NATIONAL STANDARD AS INDICATED
                                   F=FIRST CHOICE,S=SECOND CHOICE,T=THIRD CHOICE,NUMBER=OTHER SIZE
                                   * = COMMERCIAL SIZE
```

D NOMINAL SIZE = D		SECTION AREA		MASS PER UNIT		U.S.A. ANSI B32.4	JAPAN JIS	GERMANY DIN 1797	FRANCE NF A50-733	U.K. BS 4229	ITALY UNI 3820	AUSTRAL AS 1027
MM	IN	MM**2	IN**2	KG/M	LB/FT							
F 3	0.118	0.779E+01	0.121E-01	0.210E-01	0.141E-01			F	F+0-0.06 F			
F 3.2	0.126	0.887E+01	0.137E-01	0.239E-01	0.161E-01				+0-0.08 F			
F 3.5	0.138	0.106E+02	0.164E-01	0.286E-01	0.192E-01			F	F+0-0.08			
F 4	0.157	0.139E+02	0.215E-01	0.374E-01	0.251E-01	F		F	F+0-0.08 F			
F 4.5	0.177	0.175E+02	0.272E-01	0.473E-01	0.318E-01			F	F+0-0.08			
F 5	0.197	0.217E+02	0.336E-01	0.585E-01	0.393E-01			F	F+0-0.08 F	F		F
F 5.5	0.217	0.262E+02	0.406E-01	0.707E-01	0.475E-01			F	F+0-0.08 F	F		F
F 6	0.236	0.312E+02	0.483E-01	0.842E-01	0.566E-01			F	F+0-0.08 F			
F 7	0.276	0.424E+02	0.658E-01	0.115E+00	0.770E-01			F	F+0-0.09 F			
F 8	0.315	0.554E+02	0.859E-01	0.150E+00	0.101E+00			F	F+0-0.09 F	F		F
F 9	0.354	0.701E+02	0.109E+00	0.189E+00	0.127E+00			F	F+0-0.09 F			
F 10	0.394	0.866E+02	0.13E+00	0.234E+00	0.157E+00			F	F+0-0.09 F	F		F
F 11	0.433	0.105E+03	0.162E+00	0.283E+00	0.190E+00			F	F+0-0.11 F			
F 12	0.472	0.125E+03	0.193E+00	0.337E+00	0.226E+00			F	F+0-0.11 F	F		F
F 13	0.512	0.146E+03	0.227E+00	0.395E+00	0.266E+00			F	F+0-0.11 F			
F 14	0.551	0.170E+03	0.263E+00	0.458E+00	0.308E+00			F	F+0-0.11 F			
F 15	0.591	0.195E+03	0.302E+00	0.526E+00	0.354E+00			F	F+0-0.11 F			
F 16	0.630	0.222E+03	0.344E+00	0.599E+00	0.402E+00			F	F+0-0.11 F	F		F
F 17	0.669	0.250E+03	0.388E+00	0.676E+00	0.454E+00			F	F+0-0.11 F			
F 18	0.709	0.281E+03	0.435E+00	0.758E+00	0.509E+00			F	F+0-0.11 F			
F 19	0.748	0.313E+03	0.485E+00	0.844E+00	0.567E+00			F	F+0-0.13 F,23F	F		F
F 20	0.787	0.346E+03	0.537E+00	0.935E+00	0.628E+00			F	+0-0.13 F			
F 21	0.827	0.382E+03	0.592E+00	0.103E+01	0.693E+00			F	F+0-0.13 F			
F 22	0.866	0.419E+03	0.650E+00	0.113E+01	0.760E+00			F	F+0-0.13 F			
F 24	0.945	0.499E+03	0.773E+00	0.135E+01	0.905E+00			F	F+0-0.13 F,26F	F		F
F 25	0.984	0.541E+03	0.839E+00	0.146E+01	0.982E+00			F	+0-0.13 F			
F 27	1.063	0.631E+03	0.979E+00	0.170E+01	0.115E+01			F	F+0-0.16 F,29F	F		F
F 28	1.102	0.679E+03	0.105E+01	0.183E+01	0.123E+01			F	+0-0.16 F			
F 30	1.181	0.779E+03	0.121E+01	0.210E+01	0.141E+01	F		F	F+0-0.16 40F	F		F
F 32	1.260	0.887E+03	0.137E+01	0.239E+01	0.161E+01			F	F+0-0.16 F	F		F
F 36	1.417	0.112E+04	0.174E+01	0.303E+01	0.204E+01			F	F+0-0.16 F	F		F
F 41	1.614	0.146E+04	0.226E+01	0.393E+01	0.264E+01			F	F+0-0.16 40F			
F 46	1.811	0.183E+04	0.284E+01	0.495E+01	0.332E+01			F	F+0-0.16 F,45F	F		F
F 50	1.969	0.217E+04	0.336E+01	0.585E+01	0.393E+01			F	F+0-0.16 F	F		F
F 55	2.165	0.262E+04	0.406E+01	0.707E+01	0.475E+01			F	F+0-0.19 F	F		F
F 60	2.362	0.312E+04	0.483E+01	0.842E+01	0.566E+01			F	F+0-0.19 F			
F 65	2.559	0.366E+04	0.567E+01	0.988E+01	0.664E+01			F	F+0-0.19 F	F		F
F 75	2.953	0.487E+04	0.755E+01	0.132E+02	0.884E+01			F	F+0-0.19 F	F	F,70F	F
F 85	3.346	0.626E+04	0.970E+01	0.169E+02	0.114E+02			F	+0-0.22		F,80F	F
F 95	3.740	0.782E+04	0.121E+02	0.211E+02	0.142E+02			F	+0-0.22		F,90F	F
100	3.937	0.866E+04	0.134E+02	0.234E+02	0.157E+02				+0-0.22			F
105	4.134	0.955E+04	0.148E+02	0.258E+02	0.173E+02				+0-0.22			

11-16 NON-FERROUS MATERIAL

NON-FERROUS MATERIAL

Table 11-13. Rectangular Copper Bar Sizes (DIN 1759)

```
TABLE 11-13. RECTANGULAR COPPER BAR SIZES                                                    ORDER EXAMPLE:                    PAGE NO. 1
(DIN 1759)                                                                                   LENGTH,RECTANGULAR 8 X 2 UNI 3607 + MATERIAL
BASIS; 1 IN = 25.4 MM
1 CUBIC METER COPPER = 8900 KG MASS

                                                               THE NOMINAL SIZE IS NATIONAL STANDARD AS INDICATED
                                                               F=FIRST CHOICE,S=SECOND CHOICE,T=THIRD CHOICE,NUMBER=OTHER SIZE
                                                               * = COMMERCIAL SIZE
 D                                                             U.S.A.   JAPAN   GERMANY  FRANCE   U.K.    ITALY    AUSTRAL
 I                                                             ANSI     JIS     DIN      NF       BS      UNI      AS
 N    NOMINAL SIZE = AXB    SECTION AREA    MASS PER UNIT      B32.3            1759                     4229     3607    1256
         IN           MM**2    IN**2        KG/M      LB/FT

F   5 X 2   0.197X0.079  0.100E+02  0.155E-01  0.890E-01  0.598E-01                       F A+-0.08        F         F
F       3                0.150E+02  0.233E-01  0.133E+00  0.897E-01                       F B+-0.05        F  F,2.5F  F
F       4                0.200E+02  0.310E-01  0.178E+00  0.120E+00                       F B+-0.07        F         F

F   6 X 2   0.236X0.079  0.120E+02  0.186E-01  0.107E+00  0.718E-01                       F A+-0.08        F         F
F       3                0.180E+02  0.279E-01  0.160E+00  0.108E+00                       F B+-0.05        F  F,2.5F  F
F       4                0.240E+02  0.372E-01  0.214E+00  0.144E+00                       F B+-0.07        F         F
F       5                0.300E+02  0.465E-01  0.267E+00  0.179E+00                       F B+-0.07        F         F

F   8 X 2   0.315X0.079  0.160E+02  0.248E-01  0.142E+00  0.957E-01                       F A+-0.08        F         F
F       3                0.240E+02  0.372E-01  0.214E+00  0.144E+00                       F B+-0.05        F  F,2.5F  F
F       4                0.320E+02  0.496E-01  0.285E+00  0.191E+00                       F B+-0.07        F         F
F       5                0.400E+02  0.620E-01  0.356E+00  0.239E+00                       F B+-0.07        F         F
F       6                0.480E+02  0.744E-01  0.427E+00  0.287E+00                       F B+-0.07        F         F

F  10 X 2   0.394X0.079  0.200E+02  0.310E-01  0.178E+00  0.120E+00 F                     F A+-0.08  F,1.6F F  F,1.6F F
F       3                0.300E+02  0.465E-01  0.267E+00  0.179E+00 F                     F B+-0.05  F      F         F
F       4                0.400E+02  0.620E-01  0.356E+00  0.239E+00 F                     F B+-0.07  F      F         F
F       5                0.500E+02  0.775E-01  0.445E+00  0.299E+00 F                     F B+-0.07  F      F         F
F       6                0.600E+02  0.930E-01  0.534E+00  0.359E+00 F                     F B+-0.07  F      F         F
F       8                0.800E+02  0.124E+00  0.712E+00  0.478E+00 F                     F B+-0.08  F      F         F

F  12 X 2   0.472X0.079  0.240E+02  0.372E-01  0.214E+00  0.144E+00 F                     F A+-0.1   F,1.6F F  F,1.6F F
F       3                0.360E+02  0.558E-01  0.320E+00  0.215E+00 F                     F B+-0.05  F      F         F
F       4                0.480E+02  0.744E-01  0.427E+00  0.287E+00 F                     F B+-0.07  F      F  F,2.5F F
F       5                0.600E+02  0.930E-01  0.534E+00  0.359E+00 F                     F B+-0.07  F      F         F
F       6                0.720E+02  0.112E+00  0.641E+00  0.431E+00 F                     F B+-0.07  F      F         F
F       8                0.960E+02  0.149E+00  0.854E+00  0.574E+00 F                     F B+-0.09  F      F         F
F      10                0.120E+03  0.186E+00  0.107E+01  0.718E+00 F                     F B+-0.09  F      F         F

F  15 X 2   0.591X0.079  0.300E+02  0.465E-01  0.267E+00  0.179E+00 F                     F A+-0.1          F         F
F       3                0.450E+02  0.698E-01  0.400E+00  0.269E+00 F                     F B+-0.05         F         F
F       4                0.600E+02  0.930E-01  0.534E+00  0.359E+00 F                     F B+-0.07         F         F
F       5                0.750E+02  0.116E+00  0.667E+00  0.449E+00 F                     F B+-0.07         F         F
F       6                0.900E+02  0.140E+00  0.801E+00  0.538E+00 F                     F B+-0.07         F         F
F       8                0.120E+03  0.186E+00  0.107E+01  0.718E+00 F                     F B+-0.09         F         F
F      10                0.150E+03  0.233E+00  0.133E+01  0.897E+00 F                     F B+-0.09         F         F
F      12                0.180E+03  0.279E+00  0.160E+01  0.108E+01 F                     F B+-0.1          F         F

F  16 X 1.6 0.630X0.063  0.256E+02  0.397E-01  0.228E+00  0.153E+00 F                                             2F     F
F       3                0.480E+02  0.744E-01  0.427E+00  0.287E+00 F                                             F,2.5F F
F       4                0.640E+02  0.992E-01  0.570E+00  0.383E+00 F                                             F      F
F       5                0.960E+02  0.149E+00  0.854E+00  0.574E+00 F                                             F,5F   F
F       6                0.120E+03  0.198E+00  0.114E+01  0.766E+00 F                                             F      F
F       8                0.160E+03  0.248E+00  0.142E+01  0.957E+00 F                                             F      F
F      10                0.192E+03  0.298E+00  0.171E+01  0.115E+01 F                                             F      F

F  18 X 2   0.709X0.079  0.360E+02  0.558E-01  0.320E+00  0.215E+00                       F A+-0.1
F       3                0.540E+02  0.837E-01  0.481E+00  0.323E+00                       F B+-0.05
F       4                0.720E+02  0.112E+00  0.641E+00  0.431E+00                       F B+-0.07
F       5                0.900E+02  0.140E+00  0.801E+00  0.538E+00                       F B+-0.07
F       6                0.108E+03  0.167E+00  0.961E+00  0.646E+00                       F B+-0.07
```

Table 11-13 (Continued). Rectangular Copper Bar Sizes (DIN 1759)

TABLE 11-13. RECTANGULAR COPPER BAR SIZES
(DIN 1759)
BASIS; 1 IN = 25.4 MM
1 CUBIC METER COPPER = 8900 KG MASS

ORDER EXAMPLE;
LENGTH,RECTANGULAR 8 X 2 UNI 3607 + MATERIAL

PAGE NO. 2

THE NOMINAL SIZE IS NATIONAL STANDARD AS INDICATED
F=FIRST CHOICE,S=SECOND CHOICE,T=THIRD CHOICE,NUMBER=OTHER SIZE
* = COMMERCIAL SIZE

DIN	NOMINAL SIZE = A×B		SECTION AREA		MASS PER UNIT		U.S.A. ANSI B32.3.	JAPAN JIS	GERMANY DIN 1759	FRANCE NF	U.K. BS 4229	ITALY UNI 3607	AUSTRAL AS 1256
	MM	IN	MM**2	IN**2	KG/M	LB/FT							
F	18 × 8	0.709×0.315	0.144E+03	0.223E+00	0.128E+01	0.861E+00			F A+−0.1				
F	10	X0.394	0.180E+03	0.279E+00	0.160E+01	0.108E+01			F B+−0.09				
F	12	X0.472	0.216E+03	0.335E+00	0.192E+01	0.129E+01			F B+−0.1				
F	15	X0.591	0.270E+03	0.419E+00	0.240E+01	0.161E+01			F B+−0.1				
F	20 × 2	0.787×0.079	0.400E+02	0.620E−01	0.356E+00	0.239E+00			F A+−0.15				
F	3	X0.118	0.600E+02	0.930E−01	0.534E+00	0.359E+00			F B+−0.05			F,2.5F	
F	4	X0.157	0.800E+02	0.124E+00	0.712E+00	0.478E+00			F B+−0.07			F	
F	5	X0.197	0.100E+03	0.155E+00	0.890E+00	0.598E+00			F B+−0.07			F	
F	6	X0.236	0.120E+03	0.186E+00	0.107E+01	0.718E+00			F B+−0.07			F	
F	8	X0.315	0.160E+03	0.248E+00	0.142E+01	0.957E+00			F B+−0.09			F	
F	10	X0.394	0.200E+03	0.310E+00	0.178E+01	0.120E+01			F B+−0.09			F	
F	12	X0.472	0.240E+03	0.372E+00	0.214E+01	0.144E+01			F B+−0.1			F	
F	15	X0.591	0.300E+03	0.465E+00	0.267E+01	0.179E+01			F B+−0.1			F	
F	16	X0.630	0.320E+03	0.496E+00	0.285E+01	0.191E+01			F B+−0.1			F	
F	18	X0.709	0.360E+03	0.558E+00	0.320E+01	0.215E+01			F B+−0.1			F	
F	25 × 2	0.984×0.079	0.500E+02	0.775E−01	0.445E+00	0.299E+00			F A+−0.15			F	
F	3	X0.118	0.750E+02	0.116E+00	0.667E+00	0.449E+00			F B+−0.05			F,2.5F	
F	4	X0.157	0.100E+03	0.155E+00	0.890E+00	0.598E+00			F B+−0.07			F	
F	5	X0.197	0.125E+03	0.194E+00	0.111E+01	0.748E+00			F B+−0.07			F	
F	6	X0.236	0.150E+03	0.233E+00	0.133E+01	0.897E+00			F B+−0.07			F	
F	8	X0.315	0.200E+03	0.310E+00	0.178E+01	0.120E+01			F B+−0.09			F	
F	10	X0.394	0.250E+03	0.388E+00	0.222E+01	0.150E+01			F B+−0.09			F	
F	12	X0.472	0.300E+03	0.465E+00	0.267E+01	0.179E+01			F B+−0.1			F	
F	15	X0.591	0.375E+03	0.581E+00	0.334E+01	0.224E+01			F B+−0.1			F	
F	16	X0.630	0.400E+03	0.620E+00	0.356E+01	0.239E+01			F B+−0.1			F	
F	18	X0.709	0.450E+03	0.698E+00	0.400E+01	0.269E+01			F B+−0.1			F	
F	20	X0.787	0.500E+03	0.775E+00	0.445E+01	0.299E+01			F B+−0.15			F	
F	30 × 3	1.181×0.118	0.900E+02	0.140E+00	0.801E+00	0.538E+00			F A+−0.15			F	32×2F
F	3	X0.118	0.900E+02	0.140E+00	0.801E+00	0.538E+00			F B+−0.05				32×3F
F	4	X0.157	0.120E+03	0.186E+00	0.107E+01	0.718E+00			F B+−0.07			F	32×4F
F	5	X0.197	0.150E+03	0.233E+00	0.133E+01	0.897E+00			F B+−0.07			F	32×5F
F	6	X0.236	0.180E+03	0.279E+00	0.160E+01	0.108E+01			F B+−0.07			F	32×6F
F	8	X0.315	0.240E+03	0.372E+00	0.214E+01	0.144E+01			F B+−0.09			F	32×8F
F	10	X0.394	0.300E+03	0.465E+00	0.267E+01	0.179E+01			F B+−0.09			F	32×10F
F	12	X0.472	0.360E+03	0.558E+00	0.320E+01	0.215E+01			F B+−0.1			F	32×12F
F	15	X0.591	0.450E+03	0.698E+00	0.400E+01	0.269E+01			F B+−0.1			F	
F	16	X0.630	0.480E+03	0.744E+00	0.427E+01	0.287E+01			F B+−0.1			F	
F	18	X0.709	0.540E+03	0.837E+00	0.481E+01	0.323E+01			F B+−0.1			F	
F	20	X0.787	0.600E+03	0.930E+00	0.534E+01	0.359E+01			S B+−0.1			F	
F	25	X0.984	0.750E+03	0.116E+01	0.667E+01	0.449E+01			F B+−0.15			F	
F	40 × 3	1.575×0.118	0.120E+03	0.186E+00	0.107E+01	0.718E+00			F A+−0.2			F	2F
F	4	X0.157	0.160E+03	0.248E+00	0.142E+01	0.957E+00			F B+−0.07			F	F,2.5F
F	5	X0.197	0.200E+03	0.310E+00	0.178E+01	0.120E+01			F B+−0.09			F	F
F	6	X0.236	0.240E+03	0.372E+00	0.214E+01	0.144E+01			F B+−0.09			F	F
F	8	X0.315	0.320E+03	0.496E+00	0.285E+01	0.191E+01			F B+−0.1			F	F
F	10	X0.394	0.400E+03	0.620E+00	0.356E+01	0.239E+01			F B+−0.1			F	F

NON-FERROUS MATERIAL

Table 11-13 (Continued). Rectangular Copper Bar Sizes (DIN 1759)

```
TABLE 11-13. RECTANGULAR COPPER BAR SIZES                              ORDER EXAMPLE:
(DIN 1759)                                                             LENGTH,RECTANGULAR 8 X 2 UNI 3607 + MATERIAL
BASIS; 1 IN = 25.4 MM
1 CUBIC METER COPPER = 8900 KG MASS
                                                                       PAGE NO. 3
                                                                       THE NOMINAL SIZE IS NATIONAL STANDARD AS INDICATED
                                                                       F=FIRST CHOICE,S=SECOND CHOICE,T=THIRD CHOICE,NUMBER=OTHER SIZE
                                                                       * = COMMERCIAL SIZE
D I
N    NOMINAL SIZE = AXB   SECTION AREA    MASS PER UNIT   U.S.A.  JAPAN  GERMANY  FRANCE  U.K.   ITALY   AUSTRAL
     MM         IN        MM**2   IN**2   KG/M    LB/FT   ANSI   JIS    DIN      NF      BS      UNI     AS
                                                          B32.3          1759             4229    3607    1256

F  40 X12  1.575X0.472  0.480E+03 0.744E+00  0.427E+01 0.287E+01  F    F    F  A+-0.2    F       F       F
F     12   X0.472       0.480E+03 0.744E+00  0.427E+01 0.287E+01  F    F    F  B+-0.12   F       F       F
F     15   X0.591       0.600E+03 0.930E+00  0.534E+01 0.359E+01       F    F  B+-0.12                   F
F     16   X0.630       0.640E+03 0.992E+00  0.570E+01 0.383E+01       F    F  B+-0.12                   F
F     18   X0.709       0.720E+03 0.112E+01  0.641E+01 0.431E+01  S    F    F  B+-0.15                   F
F     20   X0.787       0.800E+03 0.124E+01  0.712E+01 0.478E+01       F    F  B+-0.15                   F
F     25   X0.984       0.100E+04 0.155E+01  0.890E+01 0.598E+01       F    F  B+-0.15                   F
F     30   X1.181       0.120E+04 0.186E+01  0.107E+02 0.718E+01       F    F  B+-0.15                   F
F     35   X1.378       0.140E+04 0.217E+01  0.125E+02 0.837E+01       F    F  B+-0.2                    F

F  50 X 3  1.969X0.118  0.150E+03 0.233E+00  0.133E+01 0.897E+00  F    F    F  A+-0.2    F       2F      F
F      3   X0.118       0.150E+03 0.233E+00  0.133E+01 0.897E+00  F    F    F  B+-0.07           F,2.5F  F
F      4   X0.157       0.200E+03 0.310E+00  0.178E+01 0.120E+01  F    F    F  B+-0.09                   F
F      5   X0.197       0.250E+03 0.388E+00  0.222E+01 0.150E+01  F    F    F  B+-0.09                   F
F      6   X0.236       0.300E+03 0.465E+00  0.267E+01 0.179E+01  F    F    F  B+-0.09           F       F
F      8   X0.315       0.400E+03 0.620E+00  0.356E+01 0.239E+01  F    F    F  B+-0.1            F       F
F     10   X0.394       0.500E+03 0.775E+00  0.445E+01 0.299E+01  F    F    F  B+-0.12           F       F
F     12   X0.472       0.600E+03 0.930E+00  0.534E+01 0.359E+01  F    F    F  B+-0.12           F       F
F     15   X0.591       0.750E+03 0.116E+01  0.667E+01 0.449E+01  F    F    F  B+-0.12                   F
F     16   X0.630       0.800E+03 0.124E+01  0.712E+01 0.478E+01  S    F    F  B+-0.12                   F
F     18   X0.709       0.900E+03 0.140E+01  0.801E+01 0.538E+01  S    F    F  B+-0.12                   F
F     20   X0.787       0.100E+04 0.155E+01  0.890E+01 0.598E+01  F    F    F  B+-0.12                   F
F     25   X0.984       0.125E+04 0.194E+01  0.111E+02 0.748E+01  F    F    F  F+-0.12                   F
F     30   X1.181       0.150E+04 0.233E+01  0.133E+02 0.897E+01  F    F    F  F+-0.15                   F
F     35   X1.378       0.175E+04 0.271E+01  0.156E+02 0.105E+02  F    F    F  B+-0.2                    F
F     40   X1.575       0.200E+04 0.310E+01  0.178E+02 0.120E+02  F    F    F  B+-0.2                    F

F  60 X 3  2.362X0.118  0.180E+03 0.279E+00  0.160E+01 0.108E+01  F    F    F  A+-0.25                   F
F      3   X0.118       0.180E+03 0.279E+00  0.160E+01 0.108E+01  F    F    F  B+-0.09                   F
F      4   X0.157       0.240E+03 0.372E+00  0.214E+01 0.144E+01  F    F    F  B+-0.11           63X4F   F
F      5   X0.197       0.300E+03 0.465E+00  0.267E+01 0.179E+01  F    F    F  B+-0.11           63X5F   F
F      6   X0.236       0.360E+03 0.558E+00  0.320E+01 0.215E+01  F    F    F  B+-0.11           63X6F   F
F      8   X0.315       0.480E+03 0.744E+00  0.427E+01 0.287E+01  F    F    F  B+-0.11           63X8F   F
F     10   X0.394       0.600E+03 0.930E+00  0.534E+01 0.359E+01  F    F    F  B+-0.12           63X10F  F
F     12   X0.472       0.720E+03 0.112E+01  0.641E+01 0.431E+01  F    F    F  B+-0.12           63X12F  F
F     15   X0.591       0.900E+03 0.140E+01  0.801E+01 0.538E+01  F    F    F  B+-0.15                   F
F     16   X0.630       0.960E+03 0.149E+01  0.854E+01 0.574E+01  S    F    F  B+-0.15           63X16F  F
F     18   X0.709       0.108E+04 0.167E+01  0.961E+01 0.646E+01  F    F    F  B+-0.15                   F
F     20   X0.787       0.120E+04 0.186E+01  0.107E+02 0.718E+01  F    F    F  B+-0.2                    F
F     25   X0.984       0.150E+04 0.233E+01  0.133E+02 0.897E+01  F    F    F  B+-0.2                    F
F     30   X1.181       0.180E+04 0.279E+01  0.160E+02 0.108E+02  F    F    F  B+-0.25                   F
F     35   X1.378       0.210E+04 0.326E+01  0.187E+02 0.126E+02  F    F    F  B+-0.25          F,50F    F,50F
F     40   X1.575       0.240E+04 0.372E+01  0.214E+02 0.144E+02  F    F    F  B+-0.25                   F

F  80 X 4  3.150X0.157  0.320E+03 0.496E+00  0.285E+01 0.191E+01  F    F    F  A+-0.25                   F
F      5   X0.197       0.400E+03 0.620E+00  0.356E+01 0.239E+01  F    F    F  B+-0.11                   F
F      6   X0.236       0.480E+03 0.744E+00  0.427E+01 0.287E+01  F    F    F  B+-0.11                   F
F      8   X0.315       0.640E+03 0.992E+00  0.570E+01 0.383E+01  F    F    F  B+-0.12                   F
F     10   X0.394       0.800E+03 0.124E+01  0.712E+01 0.478E+01  F    F    F  B+-0.15                   F
F     12   X0.472       0.960E+03 0.149E+01  0.854E+01 0.574E+01  F    F    F  B+-0.15                   F
F     15   X0.591       0.120E+04 0.186E+01  0.107E+02 0.718E+01  F    F    F  B+-0.15                   F
F     16   X0.630       0.128E+04 0.198E+01  0.114E+02 0.766E+01  F    F    F  B+-0.15                   F
```

11-19

Table 11-13 (Continued). Rectangular Copper Bar Sizes (DIN 1759)

TABLE 11-13. RECTANGULAR COPPER BAR SIZES
(DIN 1759)
BASIS: 1 IN = 25.4 MM
1 CUBIC METER COPPER = 8900 KG MASS

ORDER EXAMPLE:
LENGTH,RECTANGULAR 8 X 2 UNI 3607 PAGE NO. 4
+ MATERIAL

THE NOMINAL SIZE IS NATIONAL STANDARD AS INDICATED
F=FIRST CHOICE,S=SECOND CHOICE,T=THIRD CHOICE,NUMBER=OTHER SIZE
* = COMMERCIAL SIZE

NOMINAL SIZE = AxB		SECTION AREA		MASS PER UNIT		U.S.A. ANSI B32.3	JAPAN JIS	GERMANY DIN 1759	FRANCE NF	U.K. BS 4229	ITALY UNI 3607	AUSTRAL AS 1256
MM	IN	MM**2	IN**2	KG/M	LB/FT							
F 80 x18	3.150x0.709	0.144E+04	0.223E+01	0.128E+02	0.861E+01	S		A+-0.25				
F 18	x0.709	0.144E+04	0.223E+01	0.128E+02	0.861E+01	S		B+-0.15				
F 20	x0.787	0.160E+04	0.248E+01	0.178E+02	0.957E+01	F		B+-0.2	F			
F 25	x0.984	0.200E+04	0.310E+01	0.178E+02	0.120E+02	F		B+-0.2	F			
F 30	x1.181	0.240E+04	0.372E+01	0.214E+02	0.144E+02	F		B+-0.2	F			
F 35	x1.378	0.280E+04	0.434E+01	0.249E+02	0.167E+02	F		B+-0.25	50F		50F	
F 40	x1.575	0.320E+04	0.496E+01	0.285E+02	0.191E+02	F		B+-0.25	F,60F		F,60F	
F100 x 5	3.937x0.197	0.500E+03	0.775E+00	0.445E+01	0.299E+01	F		A+-0.3	4F		F,4F	4F
F 6	x0.236	0.600E+03	0.930E+00	0.534E+01	0.359E+01	F		B+-0.12	F		F	F
F 8	x0.315	0.800E+03	0.124E+01	0.712E+01	0.478E+01	F		B+-0.15	F		F	
F 10	x0.394	0.100E+04	0.155E+01	0.890E+01	0.598E+01	F		B+-0.15	F		F	
F 12	x0.472	0.120E+04	0.186E+01	0.107E+02	0.718E+01	F		B+-0.18	F		F	
F 15	x0.591	0.150E+04	0.233E+01	0.133E+02	0.897E+01	F		B+-0.18	F		F	
F 16	x0.630	0.160E+04	0.248E+01	0.142E+02	0.957E+01	F		B+-0.18			F	
F 18	x0.709	0.180E+04	0.279E+01	0.160E+02	0.108E+02	S		B+-0.18	F			
F 20	x0.787	0.200E+04	0.310E+01	0.178E+02	0.120E+02	F		B+-0.23	F		F	
F 25	x0.984	0.250E+04	0.388E+01	0.222E+02	0.150E+02	F		B+-0.23	F		F	
F 30	x1.181	0.300E+04	0.465E+01	0.267E+02	0.179E+02	F		B+-0.23	F,50F		F,50F	
F 35	x1.378	0.350E+04	0.543E+01	0.311E+02	0.209E+02	F		B+-0.3	60F		60F	
F 40	x1.575	0.400E+04	0.620E+01	0.356E+02	0.239E+02	F		B+-0.3	F,80F		F,80F	
F120 x 6	4.724x0.236	0.720E+03	0.112E+01	0.641E+01	0.431E+01	F		A+-0.3			125X6F	
F 6	x0.236	0.720E+03	0.112E+01	0.641E+01	0.431E+01	F		B+-0.12				
F 8	x0.315	0.960E+03	0.149E+01	0.854E+01	0.574E+01	F		B+-0.15			125X8F	
F 10	x0.394	0.120E+04	0.186E+01	0.107E+02	0.718E+01	F		B+-0.15			125X10F	
F 12	x0.472	0.144E+04	0.223E+01	0.128E+02	0.861E+01	F		B+-0.18			125X12F	
F 15	x0.591	0.180E+04	0.279E+01	0.160E+02	0.108E+02	F		B+-0.18			125X16F	
F 18	x0.709	0.216E+04	0.335E+01	0.192E+02	0.129E+02	S		B+-0.18				
F 20	x0.787	0.240E+04	0.372E+01	0.214E+02	0.144E+02	S		B+-0.23				
F 25	x0.984	0.300E+04	0.465E+01	0.267E+02	0.179E+02	F		B+-0.23				
F 30	x1.181	0.360E+04	0.558E+01	0.320E+02	0.215E+02	F		B+-0.23				
F 35	x1.378	0.420E+04	0.651E+01	0.374E+02	0.251E+02	F		B+-0.3				
F 40	x1.575	0.480E+04	0.744E+01	0.427E+02	0.287E+02	F		B+-0.3				
F140 x10	5.512x0.394	0.140E+04	0.217E+01	0.125E+02	0.837E+01	F		A+-0.4				
F 10	x0.394	0.140E+04	0.217E+01	0.125E+02	0.837E+01	F		B+-0.18				
F 15	x0.591	0.210E+04	0.326E+01	0.187E+02	0.126E+02	F		B+-0.18				
F 20	x0.787	0.280E+04	0.434E+01	0.249E+02	0.167E+02	F		B+-0.25				
F 25	x0.984	0.350E+04	0.543E+01	0.311E+02	0.209E+02	F		B+-0.25				
F 30	x1.181	0.420E+04	0.651E+01	0.374E+02	0.251E+02	F		B+-0.25				
F 35	x1.378	0.490E+04	0.760E+01	0.436E+02	0.293E+02	F		B+-0.4				
F 40	x1.575	0.560E+04	0.868E+01	0.498E+02	0.335E+02	F		B+-0.4				
F150 x10	5.906x0.394	0.150E+04	0.233E+01	0.133E+02	0.897E+01	S		A+-0.4			160X8F	
F 0	x0.000	0.000E+00	0.000E+00	0.000E+00	0.000E+00	S		B+-0.18			160X10F	
F 15	x0.591	0.225E+04	0.349E+01	0.135E+02	0.135E+02	S		B+-0.2			160X12F	
F 20	x0.787	0.300E+04	0.465E+01	0.267E+02	0.179E+02	S		B+-0.25			160X16F	
F 25	x0.984	0.375E+04	0.581E+01	0.33E+02	0.224E+02	S		B+-0.25				
F 30	x1.181	0.450E+04	0.698E+01	0.400E+02	0.269E+02	S		B+-0.25				
F 35	x1.378	0.525E+04	0.814E+01	0.467E+02	0.314E+02	S		B+-0.4				
F 40	x1.575	0.600E+04	0.930E+01	0.534E+02	0.359E+02	S		B+-0.4				

NON-FERROUS MATERIAL

Table 11-14. Drawn Rectangular Aluminum Bar Sizes (DIN 1769)

```
TABLE 11-14.  DRAWN RECTANGULAR                                    ORDER EXAMPLE;
ALUMINUM BAR SIZES (DIN 1769)                                      LENGTH,RECTANGULAR 10 X 3 AS 1256 + MATERIAL
BASIS; 1 IN = 25.4 MM
1 CUBIC METER ALUMINUM = 2700 KG MASS

                                                  THE NOMINAL SIZE IS NATIONAL STANDARD AS INDICATED
                                                  F=FIRST CHOICE,S=SECOND CHOICE,T=THIRD CHOICE,NUMBER=OTHER SIZE
                                                  * = COMMERCIAL SIZE
D                              MASS PER UNIT      U.S.A.  JAPAN  GERMANY  FRANCE    U.K.   ITALY    AUSTRAL
I  NOMINAL SIZE = AXB  SECTION AREA                ANSI    JIS     DIN      NF       BS     UNI       AS
N    MM          IN      MM**2    IN**2    KG/M      LB/FT      B32.3            1769   A50-734   4229    3822      1256

F   5 x  2   0.197X0.079  0.100E+02  0.155E-01  0.270E-01  0.181E-01                      F A+-0.08 F                     F
F       3    X0.118       0.150E+02  0.233E-01  0.405E-01  0.272E-01                      F B+-0.06 F       F             F
F       4    X0.157       0.200E+02  0.310E-01  0.540E-01  0.363E-01                      F B+-0.06 F       F             F

F   6 x  2   0.236X0.079  0.120E+02  0.186E-01  0.324E-01  0.218E-01                      F A+-0.08 F                     F
F       3    X0.118       0.180E+02  0.279E-01  0.486E-01  0.327E-01                      F B+-0.06 F       F             F
F       4    X0.157       0.240E+02  0.372E-01  0.648E-01  0.435E-01                      F B+-0.06 F       F             F
F       5    X0.197       0.300E+02  0.465E-01  0.810E-01  0.544E-01                      F B+-0.06 F       F             F

F   8 x  2   0.315X0.079  0.160E+02  0.248E-01  0.432E-01  0.290E-01                      F A+-0.08 F                     F
F       3    X0.118       0.240E+02  0.372E-01  0.648E-01  0.435E-01                      F B+-0.06 F       F             F
F       4    X0.157       0.320E+02  0.496E-01  0.864E-01  0.581E-01                      F B+-0.06 F       F             F
F       5    X0.197       0.400E+02  0.620E-01  0.108E+00  0.726E-01                      F B+-0.06 F       F             F
F       6    X0.236       0.480E+02  0.744E-01  0.130E+00  0.871E-01                      F B+-0.06 F       F             F

F  10 x  3   0.394X0.079  0.200E+02  0.310E-01  0.540E-01  0.363E-01                      F A+-0.08 F                     F
F       3    X0.118       0.300E+02  0.465E-01  0.810E-01  0.544E-01                      F B+-0.06 F       F             F
F       4    X0.157       0.400E+02  0.620E-01  0.108E+00  0.726E-01                      F B+-0.06 F       F             F
F       5    X0.197       0.500E+02  0.775E-01  0.135E+00  0.907E-01                      F B+-0.06 F       F             F
F       6    X0.236       0.600E+02  0.930E-01  0.162E+00  0.109E+00                      F B+-0.06 F       F             F
F       8    X0.315       0.800E+02  0.124E+00  0.216E+00  0.145E+00                      F B+-0.08 F       F             F

F  12 x  2   0.472X0.079  0.240E+02  0.372E-01  0.648E-01  0.435E-01                      F A+-0.1  F                     F
F       3    X0.118       0.360E+02  0.558E-01  0.972E-01  0.653E-01                      F B+-0.06 F       F             F
F       4    X0.157       0.480E+02  0.744E-01  0.130E+00  0.871E-01                      F B+-0.06 F       F             F
F       5    X0.197       0.600E+02  0.930E-01  0.162E+00  0.109E+00                      F B+-0.06 F       F             F
F       6    X0.236       0.720E+02  0.112E+00  0.194E+00  0.131E+00                      F B+-0.06 F       F             F
F       8    X0.315       0.960E+02  0.149E+00  0.259E+00  0.174E+00                      F B+-0.06 F       F             F
F      10    X0.394       0.120E+03  0.186E+00  0.324E+00  0.218E+00                      F B+-0.08 F       F             F

F  15 x  2   0.591X0.079  0.300E+02  0.465E-01  0.810E-01  0.544E-01                      F A+-0.1  F                     F
F       3    X0.118       0.450E+02  0.698E-01  0.121E+00  0.816E-01                      F B+-0.06 F       F             F
F       4    X0.157       0.600E+02  0.930E-01  0.162E+00  0.109E+00                 2.5F F B+-0.06 F       F             F
F       5    X0.197       0.750E+02  0.116E+00  0.202E+00  0.136E+00                      F B+-0.06 F       F             F
F       6    X0.236       0.900E+02  0.140E+00  0.243E+00  0.163E+00                    F F B+-0.06 F       F             F
F       8    X0.315       0.126E+03  0.149E+00  0.174E+00 ...  ... ...
F      10    X0.394       0.150E+03  0.233E+00  0.405E+00  0.272E+00                    F F B+-0.08 F       F             F
F      12    X0.472       0.180E+03  0.279E+00  0.486E+00  0.327E+00                      F B+-0.1  F       F             F

F  16 x 2.5  0.630X0.098  0.400E+02  0.620E-01  0.108E+00  0.726E-01                      S             S       2F        S
F       3    X0.118       0.480E+02  0.744E-01  0.130E+00  0.871E-01                      S             S       F         S
F       4    X0.157       0.640E+02  0.992E-01  0.173E+00  0.116E+00                      S             S       F         S
F       5    X0.197       0.800E+02  0.124E+00  0.216E+00  0.145E+00                      S             S       F         F
F       6    X0.236       0.960E+02  0.149E+00  0.259E+00  0.174E+00                      F             S       F         F
F       8    X0.315       0.128E+03  0.198E+00  0.346E+00  0.232E+00                      S             S       F         S
F      10    X0.394       0.160E+03  0.248E+00  0.432E+00  0.290E+00                      F                     F         F

F  18 x  2   0.709X0.079  0.360E+02  0.558E-01  0.972E-01  0.653E-01                      F A+-0.1  F                     F
F       3    X0.118       0.540E+02  0.837E-01  0.146E+00  0.980E-01                      F B+-0.06 F       F             F
F       5    X0.157       0.900E+02  0.112E+00  0.194E+00  0.131E+00                      F B+-0.06 F       F             F
F       6    X0.236       0.108E+03  0.167E+00  0.292E+00  0.196E+00                      F B+-0.06 F       F             F
```

Table 11-14 (Continued). Drawn Rectangular Aluminum Bar Sizes (DIN 1769)

TABLE 11-14. DRAWN RECTANGULAR ALUMINUM BAR SIZES (DIN 1769)
BASIS; 1 IN = 25.4 MM
1 CUBIC METER ALUMINUM = 2700 KG MASS

ORDER EXAMPLE:
LENGTH,RECTANGULAR 10 X 3 AS 1256 + MATERIAL

PAGE NO. 2

THE NOMINAL SIZE IS NATIONAL STANDARD AS INDICATED
F=FIRST CHOICE,S=SECOND CHOICE,T=THIRD CHOICE,NUMBER=OTHER SIZE
* = COMMERCIAL SIZE

D I N	NOMINAL SIZE = A×B		SECTION AREA		MASS PER UNIT		U.S.A. ANSI B32.3	JAPAN JIS	GERMANY DIN 1769	FRANCE NF A50-734	U.K. BS 4229	ITALY UNI 3822	AUSTRAL AS 1256
MM	IN		MM**2	IN**2	KG/M	LB/FT							
F 18 × 8	0.709	X0.315	0.144E+03	0.223E+00	0.389E+00	0.261E+00			F A+-0.1	F		F	
F 10		X0.394	0.180E+03	0.279E+00	0.486E+00	0.327E+00			F B+-0.08	F		F	
F 12		X0.472	0.216E+03	0.335E+00	0.583E+00	0.392E+00			F B+-0.1	F		F	
F 15		X0.591	0.270E+03	0.419E+00	0.729E+00	0.490E+00			F B+-0.1	F		F	
F 20 × 2	0.787	X0.079	0.400E+02	0.620E-01	0.108E+00	0.726E-01			F A+-0.15 F			F	
F 3		X0.118	0.600E+02	0.930E-01	0.162E+00	0.109E+00			F A+-0.06 F			F	
F 4		X0.157	0.800E+02	0.124E+00	0.216E+00	0.145E+00			F B+-0.06 F			F	
F 5		X0.197	0.100E+03	0.155E+00	0.270E+00	0.181E+00			F B+-0.06 F			F	
F 6		X0.236	0.120E+03	0.186E+00	0.324E+00	0.218E+00			F B+-0.06 F			F	
F 8		X0.315	0.160E+03	0.248E+00	0.432E+00	0.290E+00			F B+-0.08 F			F	
F 10		X0.394	0.200E+03	0.310E+00	0.540E+00	0.363E+00			F B+-0.08 F			F	
F 12		X0.472	0.240E+03	0.372E+00	0.648E+00	0.435E+00			F B+-0.1 F			F	
F 15		X0.591	0.300E+03	0.465E+00	0.810E+00	0.544E+00			F B+-0.1 F		16F	F	
F 18		X0.709	0.360E+03	0.558E+00	0.972E+00	0.653E+00			S B+-0.1 S				
F 25 × 2	0.984	X0.079	0.500E+02	0.775E-01	0.135E+00	0.907E-01			F A+-0.15 F		2.5F	F	
F 3		X0.118	0.750E+02	0.116E+00	0.202E+00	0.136E+00			F B+-0.06 F			F	
F 4		X0.157	0.100E+03	0.155E+00	0.270E+00	0.181E+00			F B+-0.06 F			F	
F 5		X0.197	0.125E+03	0.194E+00	0.337E+00	0.227E+00			F B+-0.06 F			F	
F 6		X0.236	0.150E+03	0.233E+00	0.405E+00	0.272E+00			F B+-0.06 F			F	
F 8		X0.315	0.200E+03	0.310E+00	0.540E+00	0.363E+00			F B+-0.08 F			F	
F 10		X0.394	0.250E+03	0.388E+00	0.675E+00	0.454E+00			F B+-0.08 F		2.5F	F	
F 12		X0.472	0.300E+03	0.465E+00	0.810E+00	0.544E+00			F B+-0.1 F			F	
F 15		X0.591	0.375E+03	0.581E+00	0.101E+01	0.680E+00			F B+-0.1 F		16F	F	
F 18		X0.709	0.450E+03	0.698E+00	0.121E+01	0.816E+00			S B+-0.1 S				
F 20		X0.787	0.500E+03	0.775E+00	0.135E+01	0.907E+00			F B+-0.1 F				
F 30 × 3	1.181	X0.118	0.900E+02	0.140E+00	0.243E+00	0.163E+00			F A+-0.15 F			F	32X3F
F 4		X0.157	0.120E+03	0.186E+00	0.324E+00	0.218E+00			F B+-0.06 F			F	32X4F
F 5		X0.197	0.150E+03	0.233E+00	0.405E+00	0.272E+00			F B+-0.06 F			F	
F 6		X0.236	0.180E+03	0.279E+00	0.486E+00	0.327E+00			F B+-0.06 F			F	32X6F
F 8		X0.315	0.240E+03	0.372E+00	0.648E+00	0.435E+00			F B+-0.08 F			F	
F 10		X0.394	0.300E+03	0.465E+00	0.810E+00	0.544E+00			F B+-0.08 F			F	32X10F
F 12		X0.472	0.360E+03	0.558E+00	0.972E+00	0.653E+00			F B+-0.1 F			F	32X12S
F 15		X0.591	0.450E+03	0.698E+00	0.121E+01	0.816E+00			F B+-0.1 F		16F	F	32X16S
F 18		X0.709	0.540E+03	0.837E+00	0.146E+01	0.980E+00			S B+-0.1 S				
F 20		X0.787	0.600E+03	0.930E+00	0.162E+01	0.109E+01			F B+-0.1 F				32X20S
F 25		X0.984	0.750E+03	0.116E+01	0.202E+01	0.136E+01			F B+-0.15 F				
F 40 × 3	1.575	X0.118	0.120E+03	0.186E+00	0.324E+00	0.218E+00			F A+-0.2 F			F	
F 4		X0.157	0.160E+03	0.248E+00	0.432E+00	0.290E+00			F B+-0.08 F			F	
F 5		X0.197	0.200E+03	0.310E+00	0.540E+00	0.363E+00			F B+-0.08 F			F	
F 6		X0.236	0.240E+03	0.372E+00	0.648E+00	0.435E+00			F B+-0.08 F			F	
F 8		X0.315	0.320E+03	0.496E+00	0.864E+00	0.581E+00			F B+-0.1 F			F	
F 10		X0.394	0.400E+03	0.620E+00	0.108E+01	0.726E+00			F B+-0.1 F			F	
F 12		X0.472	0.480E+03	0.744E+00	0.130E+01	0.871E+00			F B+-0.12 F			F	
F 15		X0.591	0.600E+03	0.930E+00	0.162E+01	0.109E+01			F B+-0.12 F		16F	F	
F 18		X0.709	0.720E+03	0.112E+01	0.194E+01	0.131E+01			S B+-0.12 S				
F 20		X0.787	0.800E+03	0.124E+01	0.216E+01	0.145E+01			F B+-0.15 F				
F 25		X0.984	0.100E+04	0.155E+01	0.270E+01	0.181E+01			F B+-0.15 F				
F 30		X1.181	0.120E+04	0.186E+01	0.324E+01	0.218E+01			F B+-0.15 F				

NON-FERROUS MATERIAL

Table 11-14 (Continued). Drawn Rectangular Aluminum Bar Sizes (DIN 1769)

```
TABLE 11-14.  DRAWN RECTANGULAR                                                    ORDER EXAMPLE:
ALUMINUM BAR SIZES (DIN 1769)                                                      LENGTH,RECTANGULAR 10 X 3 AS 1256 + MATERIAL
BASIS; 1 IN = 25.4 MM                                                              PAGE No. 3
1 CUBIC METER ALUMINUM = 2700 KG MASS
```

THE NOMINAL SIZE IS NATIONAL STANDARD AS INDICATED
F=FIRST CHOICE, S=SECOND CHOICE, T=THIRD CHOICE, NUMBER=OTHER SIZE
* = COMMERCIAL SIZE

NOMINAL SIZE = A×B		SECTION AREA		MASS PER UNIT		U.S.A. ANSI B32.3	JAPAN JIS	GERMANY DIN 1769	FRANCE NF A50-734	U.K. BS 4229	ITALY UNI 3822	AUSTRAL AS 1256
MM	IN	MM**2	IN**2	KG/M	LB/FT							
F 40 ×35	1.575×1.378	0.140E+04	0.217E+01	0.378E+01	0.254E+01	F		F A=0.2	F			F
F 35	X1.378	0.140E+04	0.217E+01	0.378E+01	0.254E+01	F		F B=0.2	F			F
F 50 × 3	1.969×0.118	0.150E+03	0.233E+00	0.405E+00	0.272E+00	F		F A=0.2	F			F
F 4	X0.157	0.200E+03	0.310E+00	0.540E+00	0.363E+00	F		F B=0.08	F			F
F 5	X0.197	0.250E+03	0.388E+00	0.675E+00	0.454E+00	F		F B=0.08	F			F
F 6	X0.236	0.300E+03	0.465E+00	0.810E+00	0.544E+00	F		F B=0.08	F			F
F 8	X0.315	0.400E+03	0.620E+00	0.108E+01	0.726E+00	F		F B=0.1	F			F
F 10	X0.394	0.500E+03	0.775E+00	0.135E+01	0.907E+00	F		F B=0.12	F			F
F 12	X0.472	0.600E+03	0.930E+00	0.162E+01	0.109E+01	F		F B=0.12	F			F
F 15	X0.591	0.750E+03	0.116E+01	0.202E+01	0.136E+01	16F		F B=0.12	F			F
F 18	X0.709	0.900E+03	0.140E+01	0.243E+01	0.163E+01	S		F B=0.15	F			F 16S
F 20	X0.787	0.100E+04	0.155E+01	0.270E+01	0.181E+01	F		F B=0.15	F			F
F 25	X0.984	0.125E+04	0.194E+01	0.337E+01	0.227E+01	F		F B=0.15	F			F
F 30	X1.181	0.150E+04	0.233E+01	0.405E+01	0.272E+01	F		F B=0.15	F			F
F 35	X1.378	0.175E+04	0.271E+01	0.472E+01	0.318E+01	F		F B=0.2	F			F
F 40	X1.575	0.200E+04	0.310E+01	0.540E+01	0.363E+01	F		F B=0.2	F			F
F 60 × 3	2.362×0.118	0.180E+03	0.279E+00	0.486E+00	0.327E+00	F		F A=0.25	F		F	F
F 4	X0.157	0.240E+03	0.372E+00	0.648E+00	0.435E+00	F		F B=0.1	F		F	S
F 5	X0.197	0.300E+03	0.465E+00	0.810E+00	0.544E+00	F		F B=0.1	F		F	F
F 6	X0.236	0.360E+03	0.558E+00	0.972E+00	0.653E+00	F		F B=0.1	F		F	F
F 8	X0.315	0.480E+03	0.744E+00	0.130E+01	0.871E+00	F		F B=0.12	F			F
F 10	X0.394	0.600E+03	0.930E+00	0.162E+01	0.109E+01	F		F B=0.12	F			F S
F 12	X0.472	0.720E+03	0.112E+01	0.194E+01	0.131E+01	F		F B=0.12	F			F
F 15	X0.591	0.900E+03	0.140E+01	0.243E+01	0.163E+01	16F		F B=0.12	F			F 16S
F 18	X0.709	0.108E+04	0.167E+01	0.292E+01	0.196E+01	S		F B=0.15	F			F
F 20	X0.787	0.120E+04	0.186E+01	0.324E+01	0.218E+01	F		F B=0.15	F			S
F 25	X0.984	0.150E+04	0.233E+01	0.405E+01	0.272E+01	F		F B=0.15	F			S
F 30	X1.181	0.180E+04	0.279E+01	0.486E+01	0.327E+01	F		F B=0.15	F			F
F 35	X1.378	0.210E+04	0.326E+01	0.567E+01	0.381E+01	F		F B=0.2	F			F
F 40	X1.575	0.240E+04	0.372E+01	0.648E+01	0.435E+01	F		F B=0.2	F			F
F 80 × 4	3.150×0.157	0.320E+03	0.496E+00	0.864E+00	0.581E+00	F		F A=0.25	3F		F	F
F 5	X0.197	0.400E+03	0.620E+00	0.108E+01	0.726E+00	F		F B=0.1	F		F	F
F 6	X0.236	0.480E+03	0.744E+00	0.130E+01	0.871E+00	F		F B=0.1	F		F	F
F 8	X0.315	0.640E+03	0.992E+00	0.173E+01	0.116E+01	F		F B=0.1	F		F	F
F 10	X0.394	0.800E+03	0.124E+01	0.216E+01	0.145E+01	F		F B=0.12	F			F
F 12	X0.472	0.960E+03	0.149E+01	0.259E+01	0.174E+01	F		F B=0.12	F			F
F 15	X0.591	0.120E+04	0.186E+01	0.324E+01	0.218E+01	16F		F B=0.12	F			F 16F
F 18	X0.709	0.144E+04	0.223E+01	0.389E+01	0.261E+01	S		F B=0.15	F			F
F 20	X0.787	0.160E+04	0.248E+01	0.432E+01	0.290E+01	F		F B=0.15	F			F
F 25	X0.984	0.200E+04	0.310E+01	0.540E+01	0.363E+01	F		F B=0.15	F			F
F 30	X1.181	0.240E+04	0.372E+01	0.648E+01	0.435E+01	F		F B=0.15	F			F
F 35	X1.378	0.280E+04	0.434E+01	0.756E+01	0.508E+01	F		F B=0.2	F			F
F 40	X1.575	0.320E+04	0.496E+01	0.864E+01	0.581E+01	F		F B=0.2	F			F
F100 × 5	3.937×0.197	0.500E+03	0.775E+00	0.135E+01	0.907E+00	F		F A=0.28	3F		F	F 4F
F 6	X0.236	0.600E+03	0.930E+00	0.162E+01	0.109E+01	F		F B=0.12	F		F	F
F 8	X0.315	0.800E+03	0.124E+01	0.216E+01	0.145E+01	F		F B=0.12	F		F	F
F 10	X0.394	0.100E+04	0.155E+01	0.270E+01	0.181E+01	F		F B=0.12	F			F
F 12	X0.472	0.120E+04	0.186E+01	0.324E+01	0.218E+01	F		F B=0.15	F			F

Table 11-14 (Continued). Drawn Rectangular Aluminum Bar Sizes (DIN 1769)

```
TABLE 11-14.  DRAWN RECTANGULAR                                    ORDER EXAMPLE:
ALUMINUM BAR SIZES (DIN 1769)                                      LENGTH,RECTANGULAR 10 X 3 AS 1256 + MATERIAL
BASIS: 1 IN = 25.4 MM                                              PAGE NO. 4
1 CUBIC METER ALUMINUM = 2700 KG MASS

                                                                   THE NOMINAL SIZE IS NATIONAL STANDARD AS INDICATED
                                                                   F=FIRST CHOICE,S=SECOND CHOICE,T=THIRD CHOICE,NUMBER=OTHER SIZE
                                                                   * = COMMERCIAL SIZE
   D                                                               U.S.A.    JAPAN   GERMANY  FRANCE    U.K.    ITALY   AUSTRAL
   I  NOMINAL SIZE = AxB   SECTION AREA    MASS PER UNIT           ANSI      JIS     DIN      NF        BS      UNI     AS
   N    MM        IN         MM**2   IN**2      KG/M     LB/FT     B32.3             1769     A50-734   4229    3822    1256

F  F100 x15  3.937x0.591  0.150E+04  0.233E+01  0.405E+01  0.272E+01  16F           F  A +-0.28             16F    F       16S
F       18        x0.709  0.180E+04  0.279E+01  0.486E+01  0.327E+01  S             F  B +-0.15                            S
F       20        x0.787  0.200E+04  0.310E+01  0.540E+01  0.363E+01  F             F  B +-0.2                             F
F       25        x0.984  0.250E+04  0.388E+01  0.675E+01  0.454E+01  F             F  B +-0.2
F       30        x1.181  0.300E+04  0.465E+01  0.810E+01  0.544E+01  F             F  B +-0.2
F       35        x1.378  0.350E+04  0.543E+01  0.945E+01  0.635E+01  F             F  B +-0.25
F       40        x1.575  0.400E+04  0.620E+01  0.108E+02  0.726E+01  F             F  B +-0.25

F  F120 x 6   .724x0.236  0.720E+03  0.112E+01  0.194E+01  0.131E+01  F             F  A +-0.28             F      F
F       8         x0.315  0.960E+03  0.149E+01  0.259E+01  0.174E+01  F             F  B +-0.12
F       10        x0.394  0.120E+04  0.186E+01  0.324E+01  0.218E+01  F             F  B +-0.12
F       12        x0.472  0.144E+04  0.223E+01  0.389E+01  0.261E+01  F             F  B +-0.15
F       15        x0.591  0.180E+04  0.279E+01  0.486E+01  0.327E+01  16F           F  B +-0.15             16F
F       18        x0.709  0.216E+04  0.335E+01  0.583E+01  0.392E+01  S             F  B +-0.15                     F
F       20        x0.787  0.240E+04  0.372E+01  0.648E+01  0.435E+01  F             F  B +-0.2
F       25        x0.984  0.300E+04  0.465E+01  0.810E+01  0.544E+01  F             F  B +-0.2
F       30        x1.181  0.360E+04  0.558E+01  0.972E+01  0.653E+01  F             F  B +-0.2
F       35        x1.378  0.420E+04  0.651E+01  0.113E+02  0.762E+01  F             F  B +-0.25
F       40        x1.575  0.480E+04  0.744E+01  0.130E+02  0.871E+01  F             F  B +-0.25

F  F140 x 8  5.512x0.315  0.112E+04  0.174E+01  0.302E+01  0.203E+01  F             F  A +-0.32
F       10        x0.394  0.140E+04  0.217E+01  0.378E+01  0.254E+01  F             F  B +-0.12
F       12        x0.472  0.168E+04  0.260E+01  0.454E+01  0.305E+01  F             F  B +-0.15
F       15        x0.591  0.210E+04  0.326E+01  0.567E+01  0.381E+01  16F           F  B +-0.15
F       18        x0.709  0.252E+04  0.391E+01  0.680E+01  0.457E+01  S             F  B +-0.15
F       20        x0.787  0.280E+04  0.434E+01  0.756E+01  0.508E+01  F             F  B +-0.2
F       25        x0.984  0.350E+04  0.543E+01  0.945E+01  0.635E+01  F             F  B +-0.2
F       30        x1.181  0.420E+04  0.651E+01  0.113E+02  0.762E+01  F             F  B +-0.2
F       35        x1.378  0.490E+04  0.760E+01  0.132E+02  0.889E+01  F             F  B +-0.3
F       40        x1.575  0.560E+04  0.868E+01  0.151E+02  0.102E+02  F             F  B +-0.3

F  F160 x 8  6.299x0.315  0.128E+04  0.198E+01  0.346E+01  0.232E+01  F             F  A +-0.32             6F     F       6F
F       10        x0.394  0.160E+04  0.248E+01  0.432E+01  0.290E+01  F             F  B +-0.12             F      F       F
F       12        x0.472  0.192E+04  0.298E+01  0.518E+01  0.348E+01  F             F  B +-0.15
F       15        x0.591  0.240E+04  0.372E+01  0.648E+01  0.435E+01  16F           F  B +-0.15             16F            16S
F       18        x0.709  0.288E+04  0.446E+01  0.778E+01  0.523E+01  S             F  B +-0.15                            S
F       20        x0.787  0.320E+04  0.496E+01  0.864E+01  0.581E+01  F             F  B +-0.2                             F
F       25        x0.984  0.400E+04  0.620E+01  0.108E+02  0.726E+01  F             F  B +-0.2
F       30        x1.181  0.480E+04  0.744E+01  0.130E+02  0.871E+01  F             F  B +-0.2
F       35        x1.378  0.560E+04  0.868E+01  0.151E+02  0.102E+02  F             F  B +-0.3
F       40        x1.575  0.640E+04  0.992E+01  0.173E+02  0.116E+02  F             F  B +-0.3

F  F180 x10  7.087x0.394  0.180E+04  0.279E+01  0.486E+01  0.327E+01  F             F  A +-0.35
F       10        x0.394  0.180E+04  0.279E+01  0.486E+01  0.327E+01  F             F  B +-0.15
F       12        x0.472  0.216E+04  0.335E+01  0.583E+01  0.392E+01  F             F  B +-0.2
F       15        x0.591  0.270E+04  0.419E+01  0.729E+01  0.490E+01  16F           F  B +-0.2
F       18        x0.709  0.324E+04  0.502E+01  0.875E+01  0.588E+01  S             F  B +-0.2
F       20        x0.787  0.360E+04  0.558E+01  0.972E+01  0.653E+01  F             F  B +-0.25
F       25        x0.984  0.450E+04  0.698E+01  0.121E+02  0.816E+01  F             F  B +-0.25
F       30        x1.181  0.540E+04  0.837E+01  0.146E+02  0.980E+01  F             F  B +-0.25
F       35        x1.378  0.630E+04  0.977E+01  0.170E+02  0.114E+02  F             F  B +-0.35
F       40        x1.575  0.720E+04  0.112E+02  0.194E+02  0.131E+02  F             F  B +-0.35
```

NON-FERROUS MATERIAL

Table 11-15. Copper Tubing for General Purposes (ISO 274)

TABLE 11-15. COPPER TUBING
FOR GENERAL PURPOSES (ISO 274)
BASIS: 1 IN = 25.4 MM
1 CUBIC METER COPPER = 8900 KG MASS

ORDER EXAMPLE:
LENGTH,TUBE R X 1 DIN 1754 + MATERIAL

PAGE NO. 1

THE NOMINAL SIZE IS NATIONAL STANDARD AS INDICATED
F=FIRST CHOICE,S=SECOND CHOICE,T=THIRD CHOICE,NUMBER=OTHER SIZE
* = COMMERCIAL SIZE

		MILLIMETERS		INCHES				MASS PER UNIT		U.S.A. ANSI B32.5	JAPAN JIS H3611	GERMANY DIN 1754	FRANCE NF A51-103	U.K. BS 2871	ITALY UNI 1455	AUSTRAL AS 1572
		A	C	B	A	C	B	KG/M	LB/FT							
F		3	0.5	2	0.118	0.020	0.079	0.350E-01	0.235E-01	F		F A+-0.05	F		F	F
F			0.6	1.8		0.024	0.071	0.403E-01	0.271E-01	F			F		F	F
F			0.8	1.4		0.031	0.055	0.492E-01	0.331E-01	F						
	S		1	1		0.039	0.039	0.559E-01	0.376E-01	F						
F		4	0.5	3	0.157	0.020	0.118	0.489E-01	0.329E-01	F		F A+-0.05 F			F	F
F			0.6	2.8		0.024	0.110	0.570E-01	0.383E-01	F					F	F
F			0.8	2.4		0.031	0.094	0.716E-01	0.481E-01	F					F	F
	S		1	2		0.039	0.079	0.839E-01	0.564E-01	F						
F		5	0.5	4	0.197	0.020	0.157	0.629E-01	0.423E-01	F		F A+-0.05 F			F	F
F			0.6	3.8		0.024	0.150	0.738E-01	0.496E-01	F					F	F
F			0.8	3.4		0.031	0.134	0.939E-01	0.631E-01	F					F	F
	S		1	3		0.039	0.118	0.112E+00	0.752E-01	F						
F		6	0.5	5	0.236	0.020	0.197	0.769E-01	0.517E-01	F		F A+-0.05 F		F	F	F
F			0.6	4.8		0.024	0.189	0.906E-01	0.609E-01	F					F	F
F			0.8	4.4		0.031	0.173	0.116E+00	0.782E-01	F					F	F
	S		1	4		0.039	0.157	0.140E+00	0.939E-01	F						
F		8	0.5	7	0.315	0.020	0.276	0.105E+00	0.705E-01	F		F A+-0.05 F		F	F	F
F			0.6	6.8		0.024	0.268	0.124E+00	0.834E-01	F					F	F
F			0.8	6.4		0.031	0.252	0.161E+00	0.108E+00	F					F	F
F			1	6		0.039	0.236	0.196E+00	0.132E+00	F					F	F
F			1.2	5.6		0.047	0.220	0.228E+00	0.153E+00	F					F	F
	S		1.5	5		0.059	0.197	0.273E+00	0.183E+00	F						
F		10	0.5	9	0.394	0.020	0.354	0.133E+00	0.892E-01	F		F A+-0.06 F		F	F	F
F			0.6	8.8		0.024	0.346	0.158E+00	0.106E+00	F					F	F
F			0.8	8.4		0.031	0.331	0.206E+00	0.138E+00	F					F	F
F			1	8		0.039	0.315	0.252E+00	0.169E+00	F					F	F
F			1.2	7.6		0.047	0.299	0.295E+00	0.198E+00	F					F	F
F			1.5	7		0.059	0.276	0.356E+00	0.240E+00	F				1.6F	F	F
	S		2	6		0.079	0.236	0.447E+00	0.301E+00	F						
F		12	0.5	11	0.472	0.020	0.433	0.161E+00	0.108E+00	F		F A+-0.08 F		F	F	F
F			0.6	10.8		0.024	0.425	0.191E+00	0.129E+00	F					F	F
F			0.8	10.4		0.031	0.409	0.251E+00	0.168E+00	F					F	F
F			1	10		0.039	0.394	0.308E+00	0.207E+00	F					F	F
F			1.2	9.6		0.047	0.378	0.362E+00	0.243E+00	F					F	F
F			1.5	9		0.059	0.354	0.440E+00	0.296E+00	F				1.6F	F	F
	S		2	8		0.079	0.315	0.559E+00	0.376E+00	F						
F		14	0.5	13	0.551	0.020	0.512	0.189E+00	0.127E+00	F		F A+-0.08 F		F	F	F
S			0.6	12.8		0.024	0.504	0.225E+00	0.151E+00	F					F	F
F			0.8	12.4		0.031	0.488	0.295E+00	0.198E+00	F					F	F
F			1	12		0.039	0.472	0.363E+00	0.244E+00	F					F	F
F			1.2	11.6		0.047	0.457	0.429E+00	0.289E+00	F					F	F
F			1.5	11		0.059	0.433	0.524E+00	0.352E+00	F				1.6F	F	F
	S		2	10		0.079	0.394	0.671E+00	0.451E+00	F						

Table 11-15 (Continued). Copper Tubing for General Purposes (ISO 274)

TABLE 11-15. COPPER TUBING
FOR GENERAL PURPOSES (ISO 274)
BASIS; 1 IN = 25.4 MM
1 CUBIC METER COPPER = 8900 KG MASS

ORDER EXAMPLE:
LENGTH,TUBE 8 X 1 DIN 1754 + MATERIAL PAGE NO. 2

THE NOMINAL SIZE IS NATIONAL STANDARD AS INDICATED
F=FIRST CHOICE,S=SECOND CHOICE,T=THIRD CHOICE,NUMBER=OTHER SIZE
* = COMMERCIAL SIZE

		MILLIMETERS			INCHES				MASS PER UNIT		U.S.A. ANSI B32.5	JAPAN JIS H3611	GERMANY DIN 1754	FRANCE NF A51-103	U.K. BS 2871	ITALY UNI 1455	AUSTRAL AS 1572
I	S	O	A	B	C	A	B	C	B	KG/M	LB/FT							
F		15	0.5	14	0.020	0.591	0.551	0.020		0.203E+00	0.136E+00	F		F A+-0.08		F		
S			0.6	13.8	0.024		0.543	0.024		0.242E+00	0.162E+00	F						
F			0.8	13.4	0.031		0.528	0.031		0.318E+00	0.213E+00	F		F		0.7F	S	
F			1	13	0.039		0.512	0.039		0.391E+00	0.263E+00	F		F		F	S	
F			1.2	12.6	0.047		0.496	0.047		0.463E+00	0.311E+00	F		F			S	
F			1.5	12	0.059		0.472	0.059		0.566E+00	0.380E+00	F		F			S	
F			2	11	0.079		0.433	0.079		0.727E+00	0.488E+00	F		F				
F			2.5	10	0.098		0.394	0.098		0.874E+00	0.587E+00	F		F				
S		16	0.5	15	0.020	0.630	0.591	0.020		0.217E+00	0.146E+00	F		F				
F			0.6	14.8	0.024		0.583	0.024		0.258E+00	0.174E+00	F						
F			0.8	14.4	0.031		0.567	0.031		0.340E+00	0.228E+00	F		F A+-0.08		F		
F			1	14	0.039		0.551	0.039		0.419E+00	0.282E+00	F		F		F		
F			1.2	13.6	0.047		0.535	0.047		0.497E+00	0.334E+00	F		F		F		
F			1.5	13	0.059		0.512	0.059		0.608E+00	0.409E+00	F		F		1.6F	F	
F			2	12	0.079		0.472	0.079		0.783E+00	0.526E+00	F		F			F	
S			2.5	11	0.098		0.433	0.098		0.944E+00	0.634E+00	F		F			F	
F		18	0.8	16.4	0.031	0.709	0.646	0.031		0.385E+00	0.259E+00	F		F A+-0.08		F		
F			1	16	0.039		0.630	0.039		0.475E+00	0.319E+00	F		F		F		
F			1.2	15.6	0.047		0.614	0.047		0.564E+00	0.379E+00	F		F		F		
F			1.5	15	0.059		0.591	0.059		0.692E+00	0.465E+00	F		F		1.6F	F	
F			2	14	0.079		0.551	0.079		0.895E+00	0.601E+00	F		F			F	
F			2.5	13	0.098		0.512	0.098		0.108E+01	0.728E+00	F		F			F	
F			3	12	0.118		0.472	0.118		0.126E+01	0.845E+00	F		F			F	
S			3.5	11	0.138		0.433	0.138		0.142E+01	0.958E+00	F		F			F	
F			4	10	0.157		0.394	0.157		0.157E+01	0.105E+01	F		F				
S			4.5	9	0.177		0.354	0.177		0.170E+01	0.114E+01	F		F				
S			5	8	0.197		0.315	0.197		0.182E+01	0.122E+01	F		F				
F		20	0.8	18.4	0.031	0.787	0.724	0.031		0.429E+00	0.289E+00	F		F A+-0.12	F,0.5F	F		
F			1	18	0.039		0.709	0.039		0.531E+00	0.357E+00	F		F		F		
F			1.2	17.6	0.047		0.693	0.047		0.631E+00	0.424E+00	F		F		F		
F			1.5	17	0.059		0.669	0.059		0.776E+00	0.521E+00	F		F		F 1.6F	F	
F			2	16	0.079		0.630	0.079		0.101E+01	0.676E+00	F		F			F	
F			2.5	15	0.098		0.591	0.098		0.122E+01	0.822E+00	F		F			F	
F			3	14	0.118		0.551	0.118		0.143E+01	0.958E+00	F		F			F	
S			3.5	13	0.138		0.512	0.138		0.161E+01	0.109E+01	F		F			F	
F			4	12	0.157		0.472	0.157		0.181E+01	0.122E+01	F		F				
S			4.5	11	0.177		0.433	0.177		0.195E+01	0.131E+01	F		F				
S			5	10	0.197		0.394	0.197		0.210E+01	0.141E+01	F		F				
F		22	1	20	0.039	0.866	0.787	0.039		0.587E+00	0.395E+00	F		F A+-0.12	F	0.9F	F	
F			1.2	19.6	0.047		0.772	0.047		0.698E+00	0.469E+00	F		F		F	F	
F			1.5	19	0.059		0.748	0.059		0.860E+00	0.578E+00	F		F			F	
F			2	18	0.079		0.709	0.079		0.112E+01	0.752E+00	F		F			F	
F			2.5	17	0.098		0.669	0.098		0.136E+01	0.916E+00	F		F			F	
S			3	16	0.118		0.630	0.118		0.159E+01	0.107E+01	F		F			F	
F			3.5	15	0.138		0.591	0.138		0.181E+01	0.122E+01	F		F			F	
F			4	14	0.157		0.551	0.157		0.201E+01	0.135E+01	F		F			F	
S			4.5	13	0.177		0.512	0.177		0.220E+01	0.148E+01	F		F			F	
S			5	12	0.197		0.472	0.197		0.238E+01	0.160E+01	F		F			F	

NON-FERROUS MATERIAL

Table 11-15 (Continued). Copper Tubing for General Purposes (ISO 274)

```
TABLE 11-15. COPPER TUBING                                                PAGE NO. 3
FOR GENERAL PURPOSES (ISO 274)
BASIS; 1 IN = 25.4 MM                                ORDER EXAMPLE;
1 CUBIC METER COPPER = 8900 KG MASS                  LENGTH,TUBE 8 X 1 DIN 1754 + MATERIAL

                                          THE NOMINAL SIZE IS NATIONAL STANDARD AS INDICATED
                                          F=FIRST CHOICE,S=SECOND CHOICE,T=THIRD CHOICE,NUMBER=OTHER SIZE
                                          * = COMMERCIAL SIZE
       MILLIMETERS      INCHES    MASS PER UNIT    U.S.A.  JAPAN    GERMANY   FRANCE     U.K.    ITALY   AUSTRAL
I       A    B    C    A    B     C    KG/M    LB/FT    ANSI    JIS      DIN      NF         BS      UNI     AS
S                                                       B32.5   H3611    1754    A51-103    2871    1455    1572
O
F  25   1    23       0.984  0.039  0.906   0.671E+00  0.451E+00   F              F  A+-0.12  F,0.5F          F       F
F       1.2  22.6            0.047  0.890   0.799E+00  0.537E+00   F              F                           F       F
F       1.5  22              0.059  0.866   0.986E+00  0.662E+00   F              F         1.6F              F       F
F       2    21              0.079  0.827   0.129E+01  0.864E+00   F              F                           F       F
S       2.5  20              0.098  0.787   0.157E+01  0.106E+01   F              F                           F       F
S       3    19              0.118  0.748   0.185E+01  0.124E+01   F              F                           F       F
S       3.5  18              0.138  0.709   0.210E+01  0.141E+01   F              F                           F       F
S       4    17              0.157  0.669   0.235E+01  0.158E+01   F              F                           F       F
S       4.5  16              0.177  0.630   0.258E+01  0.173E+01   F              F                           F       F
S       5    15              0.197  0.591   0.280E+01  0.188E+01   F              F                           F       F

F  28   1    26       1.102  0.039  1.024   0.755E+00  0.507E+00   F              F  A+-0.12  F            0.6F        F
F       1.2  25.6            0.047  1.008   0.899E+00  0.604E+00   F              F                  F,0.9F F
F       1.5  25              0.059  0.984   0.111E+01  0.747E+00   F              F         1.6F             F        F
F       2    24              0.079  0.945   0.145E+01  0.977E+00   F              F                          F        F
S       2.5  23              0.098  0.906   0.178E+01  0.120E+01   F              F                          F        F
S       3    22              0.118  0.866   0.210E+01  0.141E+01   F              F                          F        F
S       3.5  21              0.138  0.827   0.240E+01  0.161E+01   F              F                                   F
S       4    20              0.157  0.787   0.268E+01  0.180E+01   F              F                                   F
S       4.5  19              0.177  0.748   0.296E+01  0.199E+01   F              F                                   F
S       5    18              0.197  0.709   0.322E+01  0.216E+01   F              F                                   F

F  30   1    28       1.181  0.039  1.102   0.811E+00  0.545E+00   F              F  A+-0.12  F,0.5F                  F
F       1.2  27.6            0.047  1.087   0.966E+00  0.649E+00   F              F                                   F
F       1.5  27              0.059  1.063   0.120E+01  0.803E+00   F              F         1.6F                      F
F       2    26              0.079  1.024   0.157E+01  0.105E+01   F              F                                   F
S       2.5  25              0.098  0.984   0.192E+01  0.129E+01   F              F         F,1.6F                    F
S       3    24              0.118  0.945   0.226E+01  0.152E+01   F              F                                   F
S       3.5  23              0.138  0.906   0.259E+01  0.174E+01   F              F                                   F
S       4    22              0.157  0.866   0.291E+01  0.195E+01   F              F                  F                F
S       4.5  21              0.177  0.827   0.321E+01  0.216E+01   F              F                                   F
S       5    20              0.197  0.787   0.350E+01  0.235E+01   F              F                                   F

F  32   1    30       1.260  0.039  1.181   0.867E+00  0.582E+00   F              F  A+-0.15  F                       F
F       1.2  29.6            0.047  1.165   0.103E+01  0.694E+00   F              F                                   F
F       1.5  29              0.059  1.142   0.128E+01  0.860E+00   F              F                                   F
F       2    28              0.079  1.102   0.168E+01  0.113E+01   F              F                                   F
S       2.5  27              0.098  1.063   0.206E+01  0.139E+01   F              F                                   F
S       3    26              0.118  1.024   0.243E+01  0.163E+01   F              F                                   F
S       3.5  25              0.138  0.984   0.279E+01  0.187E+01   F              F                                   F
S       4    24              0.157  0.945   0.313E+01  0.210E+01   F              F                                   F
S       4.5  23              0.177  0.906   0.346E+01  0.233E+01   F              F                                   F
S       5    22              0.197  0.866   0.377E+01  0.254E+01   F              F                                   F

F  35   1    33       1.378  0.039  1.299   0.951E+00  0.639E+00   F              F  A+-0.15  F            0.7F       F
F       1.2  32.6            0.047  1.283   0.113E+01  0.762E+00   F              F                        F          F
F       1.5  32              0.059  1.260   0.141E+01  0.944E+00   F              F                                   F
F       2    31              0.079  1.220   0.185E+01  0.124E+01   F              F                                   F
S       2.5  30              0.098  1.181   0.227E+01  0.153E+01   F              F                                   F
S       3    29              0.118  1.142   0.268E+01  0.180E+01   F              F                                   F
S       3.5  28              0.138  1.102   0.308E+01  0.207E+01   F              F                                   F
S       4    27              0.157  1.063   0.347E+01  0.233E+01   F              F                                   F
```

Table 11-15 (Continued). Copper Tubing for General Purposes (ISO 274)

```
TABLE 11-15. COPPER TUBING                                                    ORDER EXAMPLE:
FOR GENERAL PURPOSES (ISO 274)                                                LENGTH,TUBE 8 X 1 DIN 1754 + MATERIAL
BASIS: 1 IN = 25.4 MM                                                                                        PAGE NO. 4
1 CUBIC METER COPPER = 8900 KG MASS
                                                                              THE NOMINAL SIZE IS NATIONAL STANDARD AS INDICATED
                                                                              F=FIRST CHOICE,S=SECOND CHOICE,T=THIRD CHOICE,NUMBER=OTHER SIZE
                                                                              * = COMMERCIAL SIZE
 I                                                                            U.S.A.  JAPAN   GERMANY  FRANCE    U.K.    ITALY    AUSTRAL
 S    MILLIMETERS    INCHES             MASS PER UNIT                         ANSI    JIS     DIN      NF        BS      UNI      AS
 0    A    B    C    A       B     C      KG/M          LB/FT                 B32.5   H3611   1754     A51-103   2871    1455     1572

 S   35   26   4.5  1.378   1.024  0.177  0.384E+01     0.258E+01              F                                                 
 S        25        1.378   0.984  0.197  0.419E+01     0.282E+01              F                                                 

 F   38   35.6 1.2  1.496   1.402  0.047  0.123E+01     0.830E+00              F       F       F A+-0.15  F               F
 F        35   1.5          1.378  0.059  0.153E+01     0.103E+01              F       F       F         F,1.6F           F
 F        34   2            1.339  0.079  0.201E+01     0.135E+01              F       F       F         F                F
 F        33   2.5          1.299  0.098  0.248E+01     0.167E+01              F       F       F         F                F
 S        32   3            1.260  0.118  0.294E+01     0.197E+01              F                                                 
 S        31   3.5          1.220  0.138  0.338E+01     0.227E+01              F                         F                       
 S        30   4            1.181  0.157  0.380E+01     0.256E+01              F                                                 
 S        29   4.5          1.142  0.177  0.422E+01     0.283E+01              F                                                 
 S        28   5            1.102  0.197  0.461E+01     0.310E+01              F                                                 

 F   40   38   1    1.575   1.496  0.039  0.109E+01     0.733E+00              F       F       F A+-0.15  F               F
 F        37.6 1.2          1.480  0.047  0.130E+01     0.875E+00              F                                                 
 F        37   1.5          1.457  0.059  0.161E+01     0.109E+01              F       F       F         F,1.6F           F
 F        36   2            1.417  0.079  0.212E+01     0.143E+01              F       F       F         F                F
 F        35   2.5          1.378  0.098  0.262E+01     0.176E+01              F       F       F         F                F  S
 S        34   3            1.339  0.118  0.310E+01     0.209E+01              F                                                 
 S        33   3.5          1.299  0.138  0.357E+01     0.240E+01              F                                                 
 S        32   4            1.260  0.157  0.403E+01     0.271E+01              F                                                 
 S        31   4.5          1.220  0.177  0.447E+01     0.300E+01              F                                                 
 S        30   5            1.181  0.197  0.489E+01     0.329E+01              F                                                 

 F   42   39   1.5  1.654   1.535  0.059  0.170E+01     0.114E+01              F                 F A+-0.15    F                  
 F        38   2            1.496  0.079  0.224E+01     0.150E+01              F                                                 
 F        37   2.5          1.457  0.098  0.276E+01     0.186E+01              F                                                 
 F        36   3            1.417  0.118  0.327E+01     0.220E+01              F                                                 
 S        35   3.5          1.378  0.138  0.377E+01     0.253E+01              F                                                 
 S        34   4            1.339  0.157  0.425E+01     0.286E+01              F                                                 
 S        33   4.5          1.299  0.177  0.472E+01     0.317E+01              F                                                 S
 S        32   5            1.260  0.197  0.517E+01     0.348E+01              F                                                 S

 F   50   47   1.5  1.969   1.850  0.059  0.203E+01     0.137E+01              F       F       F A+-0.15  1.2F             F
 F        46   2            1.811  0.079  0.268E+01     0.180E+01              F       F       F         F,1.6F           F
 F        45   2.5          1.772  0.098  0.332E+01     0.223E+01              F       F       F         F                F  S
 F        44   3            1.732  0.118  0.394E+01     0.265E+01              F       F       F         F                F
 S        43   3.5          1.693  0.138  0.455E+01     0.306E+01              F                                                 
 S        42   4            1.654  0.157  0.514E+01     0.346E+01              F                                                 
 S        41   4.5          1.614  0.177  0.572E+01     0.385E+01              F                                                 S
 S        40   5            1.575  0.197  0.629E+01     0.423E+01              F                                                 S
```

NON-FERROUS MATERIAL

Table 11-16. Aluminum Tubing for General Purposes (DIN 1795)

```
TABLE 11-16. ALUMINUM TUBING                                    ORDER EXAMPLE:
FOR GENERAL PURPOSES (DIN 1795)                                 LENGTH,TUBE 10 X 1.5 DIN 1795 + MATERIAL
BASIS; 1 IN = 25.4 MM
1 CUBIC METER ALUMINUM = 2700 KG MASS
```

THE NOMINAL SIZE IS NATIONAL STANDARD AS INDICATED
F=FIRST CHOICE,S=SECOND CHOICE,T=THIRD CHOICE,NUMBER=OTHER SIZE
* = COMMERCIAL SIZE

D I N	MILLIMETERS A B C	INCHES A B C	MASS PER UNIT KG/M LB/FT	U.S.A. ANSI B32.5	JAPAN JIS	GERMANY DIN 1795	FRANCE NF A50-737	U.K. BS	ITALY UNI	AUSTRAL AS
F	3 0.5 2 0.020	0.118 0.079 0.020	0.106E-01 0.712E-02	F		F A+-0.04	F			
F	0.8 1.4 0.031	0.055 0.031	0.149E-01 0.100E-01	F		F	F			
F	1 1 0.039	0.039 0.039	0.170E-01 0.114E-01	F		F				
F	4 0.5 3 0.020	0.157 0.118 0.020	0.148E-01 0.997E-02	F		F A+-0.04	F			
F	0.8 2.4 0.031	0.094 0.031	0.217E-01 0.146E-01	F		F	F			
F	1 2 0.039	0.079 0.039	0.254E-01 0.171E-01	F		F				
F	5 0.5 4 0.020	0.197 0.157 0.020	0.191E-01 0.128E-01	F		F A+-0.04	F			
F	0.8 3.4 0.031	0.134 0.031	0.285E-01 0.192E-01	F		F	F			
F	1 3 0.039	0.118 0.039	0.339E-01 0.228E-01	F		F				
F	6 0.5 5 0.020	0.236 0.197 0.020	0.233E-01 0.157E-01	F		F A+-0.04	F			
F	0.8 4.4 0.031	0.173 0.031	0.353E-01 0.237E-01	F		F	F			
F	1 4 0.039	0.157 0.039	0.424E-01 0.285E-01	F		F		F,1.2F		
F	8 0.5 7 0.020	0.315 0.276 0.020	0.318E-01 0.214E-01	F		F A+-0.04	F			
F	0.8 6.4 0.031	0.252 0.031	0.489E-01 0.328E-01	F		F	F			
F	1 6 0.039	0.236 0.039	0.594E-01 0.399E-01	F		F	F			
F	1.5 5 0.059	0.197 0.059	0.827E-01 0.556E-01	F		F	F	F,1.2F		
F	2 4 0.079	0.157 0.079	0.102E+00 0.684E-01	F		F	F			
F	10 0.5 9 0.020	0.394 0.354 0.020	0.403E-01 0.271E-01	F		F A+-0.04	F			
F	0.8 8.4 0.031	0.331 0.031	0.624E-01 0.420E-01	F		F	F			
F	1 8 0.039	0.315 0.039	0.763E-01 0.513E-01	F		F	F			
F	1.5 7 0.059	0.276 0.059	0.108E+00 0.727E-01	F		F	F	F,1.2F		
F	2 6 0.079	0.236 0.079	0.136E+00 0.912E-01	F		F	F			
F	2.5 5 0.098	0.197 0.098	0.159E+00 0.107E+00	F		F	F			
F	3 4 0.118	0.157 0.118	0.178E+00 0.120E+00	F		F	F			
F	12 0.5 11 0.020	0.472 0.433 0.020	0.488E-01 0.328E-01	F		F A+-0.05	F			
F	0.8 10.4 0.031	0.409 0.031	0.760E-01 0.511E-01	F		F	F			
F	1 10 0.039	0.394 0.039	0.933E-01 0.627E-01	F		F	F			
F	1.5 9 0.059	0.354 0.059	0.134E+00 0.898E-01	F		F	F	F,1.2F		
F	2 8 0.079	0.315 0.079	0.170E+00 0.114E+00	F		F	F	F		
F	2.5 7 0.098	0.276 0.098	0.201E+00 0.135E+00	F		F	F			
F	3 6 0.118	0.236 0.118	0.229E+00 0.154E+00	F		F	F			
F	3.5 5 0.138	0.197 0.138	0.252E+00 0.170E+00	F		F	F			
F	4 4 0.157	0.157 0.157	0.271E+00 0.182E+00	F		F	F			
F	14 0.5 13 0.020	0.551 0.512 0.020	0.573E-01 0.385E-01	F		F A+-0.05	F		S	
F	0.8 12.4 0.031	0.488 0.031	0.896E-01 0.602E-01	F		F	F		S	
F	1 12 0.039	0.472 0.039	0.110E+00 0.741E-01	F		F	F		S	
F	1.5 11 0.059	0.433 0.059	0.159E+00 0.107E+00	F		F	F		S,1.2S	
F	2 10 0.079	0.394 0.079	0.204E+00 0.137E+00	F		F	F		S	
F	2.5 9 0.098	0.354 0.098	0.244E+00 0.164E+00	F		F	F			
F	3 8 0.118	0.315 0.118	0.280E+00 0.188E+00	F		F	F			
F	3.5 7 0.138	0.276 0.138	0.312E+00 0.209E+00	F		F	F			
F	4 6 0.157	0.236 0.157	0.339E+00 0.228E+00	F		F	F			
F	15 0.5 14 0.020	0.591 0.551 0.020	0.615E-01 0.413E-01	F		F A+-0.05	F			
F	0.8 13.4 0.031	0.528 0.031	0.964E-01 0.648E-01	F		F	F			

11-29

Table 11-16 (Continued). Aluminum Tubing for General Purposes (DIN 1795)

TABLE 11-16. ALUMINUM TUBING
FOR GENERAL PURPOSES (DIN 1795)
BASIS: 1 IN = 25.4 MM
1 CUBIC METER ALUMINUM = 2700 KG MASS

ORDER EXAMPLE:
LENGTH,TUBE 10 x 1.5 DIN 1795 + MATERIAL   PAGE NO. 2

THE NOMINAL SIZE IS NATIONAL STANDARD AS INDICATED
F=FIRST CHOICE, S=SECOND CHOICE, T=THIRD CHOICE, NUMBER=OTHER SIZE
* = COMMERCIAL SIZE

DIN	MILLIMETERS		INCHES			MASS PER UNIT		U.S.A. ANSI B32.5	JAPAN JIS	GERMANY DIN 1795	FRANCE NF. A50-737	U.K. BS	ITALY UNI	AUSTRAL AS
A	B	C	A	B	C	KG/M	LB/FT							
15	1	13	0.591	0.039	0.512	0.119E+00	0.798E-01	F	F	F A+-0.05	F	F		
	1.5	12		0.059	0.472	0.172E+00	0.115E+00	F	F	F	F	F		
	2	11		0.079	0.433	0.221E+00	0.148E+00	F	F	F	F	F		
	2.5	10		0.098	0.394	0.265E+00	0.178E+00	F	F	F	F	F		
	3	9		0.118	0.354	0.305E+00	0.205E+00	F	F	F	F	F		
	3.5	8		0.138	0.315	0.341E+00	0.229E+00	F	F	F	F	F		
	4	7		0.157	0.276	0.373E+00	0.251E+00	F	F	F	F	F		
16	0.8	14.4	0.630	0.031	0.567	0.103E+00	0.693E-01	F	F	F A+-0.05	F,0.5F			
	1	14		0.039	0.551	0.127E+00	0.855E-01	F	F	F	F	F		
	1.5	13		0.059	0.512	0.184E+00	0.124E+00	F	F	F	F,1.2F			
	2	12		0.079	0.472	0.238E+00	0.160E+00	F	F	F	F			
	2.5	11		0.098	0.433	0.286E+00	0.192E+00	F	F	F	F			
	3	10		0.118	0.394	0.331E+00	0.222E+00	F	F	F	F			
	3.5	9		0.138	0.354	0.371E+00	0.249E+00	F	F	F	F			
	4	8		0.157	0.315	0.407E+00	0.274E+00	F	F	F	F			
18	0.8	16.4	0.709	0.031	0.646	0.117E+00	0.784E-01	F	F	F A+-0.05	S,0.5S			
	1	16		0.039	0.630	0.144E+00	0.969E-01	F	F	F	S			
	1.5	15		0.059	0.591	0.210E+00	0.141E+00	F	F	F	S,1.2F			
	2	14		0.079	0.551	0.271E+00	0.182E+00	F	F	F	S			
	2.5	13		0.098	0.472	0.382E+00	0.256E+00	F	F	F	2.5S			
	3	12		0.118	0.472	0.382E+00	0.256E+00	F	F	F				
	4	10		0.157	0.394	0.475E+00	0.319E+00	F	F	F				
	5	8		0.197	0.315	0.551E+00	0.370E+00	F	F	F				
	6	6		0.236	0.236	0.611E+00	0.410E+00	F	F	F				
20	0.8	18.4	0.787	0.031	0.724	0.130E+00	0.875E-01	F	F	F A+-0.05	F,0.5F			
	1	18		0.039	0.709	0.161E+00	0.108E+00	F	F	F	F,1.2F			
	1.5	17		0.059	0.669	0.235E+00	0.158E+00	F	F	F	F			
	2	16		0.079	0.630	0.305E+00	0.205E+00	F	F	F	F			
	2.5	15		0.098	0.591	0.373E+00	0.251E+00	F	F	F	F,2.5F			
	3	14		0.118	0.551	0.433E+00	0.291E+00	F	F	F	F			
	4	12		0.157	0.472	0.543E+00	0.365E+00	F	F	F				
	5	10		0.197	0.394	0.636E+00	0.427E+00	F	F	F				
	6	8		0.236	0.315	0.713E+00	0.479E+00	F	F	F				
22	1	20	0.866	0.039	0.787	0.178E+00	0.120E+00	F	F	F A+-0.05	S,0.5S			
	1.5	19		0.059	0.748	0.261E+00	0.175E+00	F	F	F	S,0.8S			
	2	18		0.079	0.709	0.339E+00	0.228E+00	F	F	F	S,1.2S			
	3	16		0.118	0.630	0.483E+00	0.325E+00	F	F	F	S,2.5S			
	4	14		0.157	0.551	0.611E+00	0.410E+00	F	F	F				
	5	12		0.197	0.472	0.721E+00	0.484E+00	F	F	F				
	6	10		0.236	0.394	0.814E+00	0.547E+00	F	F	F				
25	1	23	0.984	0.039	0.906	0.204E+00	0.137E+00	F	F	F A+-0.05	F,0.5F			
	1.5	22		0.059	0.866	0.299E+00	0.201E+00	F	F	F	F,0.8F			
	2	21		0.079	0.827	0.390E+00	0.262E+00	F	F	F	F,1.2F			
	2.5	20		0.098	0.787	0.477E+00	0.321E+00	F	F	F	F			
	3	19		0.118	0.748	0.560E+00	0.376E+00	F	F	F	F			
	4	17		0.157	0.669	0.713E+00	0.479E+00	F	F	F	F			
	5	15		0.197	0.591	0.848E+00	0.570E+00	F	F	F	F			
	6	13		0.236	0.512	0.967E+00	0.650E+00	F	F	F	F			

Table 11-16 (Continued). Aluminum Tubing for General Purposes (DIN 1795)

TABLE 11-16. ALUMINUM TUBING
FOR GENERAL PURPOSES (DIN 1795)
BASIS: 1 IN = 25.4 MM
1 CUBIC METER ALUMINUM = 2700 KG MASS

ORDER EXAMPLE:
LENGTH,TUBE 10 X 1.5 DIN 1795 + MATERIAL   PAGE NO. 3

THE NOMINAL SIZE IS NATIONAL STANDARD AS INDICATED
F=FIRST CHOICE,S=SECOND CHOICE,T=THIRD CHOICE,NUMBER=OTHER SIZE
* = COMMERCIAL SIZE

DIN	MILLIMETERS			INCHES			MASS PER UNIT		U.S.A. ANSI B32.5	JAPAN JIS	GERMANY DIN 1795	FRANCE NF A50-737	U.K. BS	ITALY UNI	AUSTRAL AS
A	B	C		A	B	C	KG/M	LB/FT							
28	26	1.5		1.102	1.024	0.039	0.229E+00	0.154E+00	F	F	F A+-0.05	S,0.8S	F	F	F
	25	1.5			0.984	0.059	0.337E+00	0.227E+00	F	F	F	S,1.2S	F	F	F
	24	2			0.945	0.079	0.441E+00	0.299E+00	F	F	F	S	F	F	F
	23	2.5			0.906	0.098	0.541E+00	0.363E+00	F	F	F	S	F	F	F
	22	3			0.866	0.118	0.636E+00	0.427E+00	F	F	F	S	F	F	F
	20	4			0.787	0.157	0.814E+00	0.547E+00	F	F	F		F	F	F
	18	5			0.709	0.197	0.975E+00	0.656E+00	F	F	F		F	F	F
	16	6			0.630	0.236	0.112E+01	0.752E+00	F	F	F		F	F	F
	12	8			0.472	0.315	0.136E+01	0.91E+00	F	F	F		F	F	F
30	28	1		1.181	1.102	0.039	0.246E+00	0.165E+00	F	F	F		F	F	F
	27	1.5			1.063	0.059	0.363E+00	0.244E+00	F	F	F A+-0.05	S,0.8S	F	F	F
	26	2			1.024	0.079	0.475E+00	0.319E+00	F	F	F	S,1.2S	F	F	F
	25	2.5			0.984	0.098	0.583E+00	0.392E+00	F	F	F	S	F	F	F
	24	3			0.945	0.118	0.687E+00	0.462E+00	F	F	F	S	F	F	F
	22	4			0.866	0.157	0.882E+00	0.593E+00	F	F	F		F	F	F
	20	5			0.787	0.197	0.106E+01	0.712E+00	F	F	F		F	F	F
	18	6			0.709	0.236	0.122E+01	0.821E+00	F	F	F		F	F	F
	14	8			0.551	0.315	0.149E+01	0.100E+01	F	F	F		F	F	F
32	30	1		1.260	1.181	0.039	0.263E+00	0.177E+00	F	F	F		F	F	F
	29	1.5			1.142	0.059	0.388E+00	0.261E+00	F	F	F A+-0.06	F,0.8F	F	F	F
	28	2			1.102	0.079	0.509E+00	0.342E+00	F	F	F	F,1.2F	F	F	F
	27	2.5			1.063	0.098	0.626E+00	0.420E+00	F	F	F	F	F	F	F
	26	3			1.024	0.118	0.738E+00	0.496E+00	F	F	F	F	F	F	F
	24	4			0.945	0.157	0.950E+00	0.638E+00	F	F	F	F	F	F	F
	22	5			0.866	0.197	0.115E+01	0.769E+00	F	F	F	F	F	F	F
	20	6			0.787	0.236	0.132E+01	0.889E+00	F	F	F	F	F	F	F
	16	8			0.630	0.315	0.163E+01	0.109E+01	F	F	F	F	F	F	F
35	33	1		1.378	1.299	0.039	0.288E+00	0.194E+00	F	F	F		F	F	F
	32	1.5			1.260	0.059	0.426E+00	0.286E+00	F	F	F A+-0.06	36S	F	F	F
	31	2			1.220	0.079	0.560E+00	0.376E+00	F	F	F		F	F	F
	30	2.5			1.181	0.098	0.689E+00	0.463E+00	F	F	F		F	F	F
	29	3			1.142	0.118	0.814E+00	0.547E+00	F	F	F		F	F	F
	27	4			1.063	0.157	0.105E+01	0.707E+00	F	F	F		F	F	F
	25	5			0.984	0.197	0.127E+01	0.855E+00	F	F	F		F	F	F
	23	6			0.906	0.236	0.148E+01	0.992E+00	F	F	F		F	F	F
	19	8			0.748	0.315	0.183E+01	0.123E+01	F	F	F		F	F	F
38	36	1		1.496	1.417	0.039	0.314E+00	0.211E+00	F	F	F A+-0.06	S,0.8S	F	F	F
	35	1.5			1.378	0.059	0.464E+00	0.312E+00	F	F	F	S,1.2S	F	F	F
	34	2			1.339	0.079	0.611E+00	0.410E+00	F	F	F	S	F	F	F
	33	2.5			1.299	0.098	0.753E+00	0.506E+00	F	F	F	S	F	F	F
	32	3			1.260	0.118	0.891E+00	0.598E+00	F	F	F	S	F	F	F
	30	4			1.181	0.157	0.115E+01	0.775E+00	F	F	F		F	F	F
	28	5			1.102	0.197	0.140E+01	0.940E+00	F	F	F		F	F	F
	26	6			1.024	0.236	0.163E+01	0.109E+01	F	F	F		F	F	F
	22	8			0.866	0.315	0.204E+01	0.137E+01	F	F	F		F	F	F

Table 11-16 (Continued). Aluminum Tubing for General Purposes (DIN 1795)

TABLE 11-16. ALUMINUM TUBING
FOR GENERAL PURPOSES (DIN 1795)
BASIS; 1 IN = 25.4 MM
1 CUBIC METER ALUMINUM = 2700 KG MASS

ORDER EXAMPLE;
LENGTH,TUBE 10 X 1.5 DIN 1795 + MATERIAL

PAGE NO. 4

THE NOMINAL SIZE IS NATIONAL STANDARD AS INDICATED
F=FIRST CHOICE,S=SECOND CHOICE,T=THIRD CHOICE,NUMBER=OTHER SIZE
* = COMMERCIAL SIZE

D I N	MILLIMETERS A C B	INCHES A C B	MASS PER UNIT KG/M LB/FT	U.S.A. ANSI B32.5	JAPAN JIS	GERMANY DIN 1795	FRANCE NF A50-737	U.K. BS	ITALY UNI	AUSTRAL AS
F 40	1 38	1.575 0.039 1.496	0.331E+00 0.222E+00	F	F	F A+-0.06	F,0.8F	F	F	F
F	1.5 37	0.059 1.457	0.490E+00 0.329E+00	F	F	F	F,1.2F	F	F	F
F	2 36	0.079 1.417	0.645E+00 0.433E+00	F	F	F	F	F	F	F
F	3 34	0.118 1.339	0.942E+00 0.633E+00	F	F	F	F	F	F	F
F	4 32	0.157 1.260	0.122E+01 0.821E+00	F	F	F	F	F	F	F
F	5 30	0.197 1.181	0.148E+01 0.997E+00	F	F	F	F	F	F	F
F	6 28	0.236 1.102	0.173E+01 0.116E+01	F	F	F	F	F	F	F
F	8 24	0.315 0.945	0.217E+01 0.146E+01	F	F	F	F	F	F	F
F	10 20	0.394 0.787	0.254E+01 0.171E+01	F	F	F	F	F	F	F
F 42	1.5 39	1.654 0.059 1.535	0.515E+00 0.346E+00	F	F	F A+-0.06	F	F	F	F
F	2 38	0.079 1.496	0.679E+00 0.456E+00	F	F	F	F	F	F	F
F	2.5 37	0.098 1.457	0.838E+00 0.563E+00	F	F	F	F	F	F	F
F 50	1 48	1.969 0.039 1.890	0.416E+00 0.279E+00	F	F	F A+-0.06	F,1.2F	F	F	F
F	2 46	0.079 1.811	0.814E+00 0.547E+00	F	F	F	F,1.5F	F	F	F
F	3 44	0.118 1.732	0.120E+01 0.804E+00	F	F	F	F,2.5F	F	F	F
F	4 42	0.157 1.654	0.156E+01 0.105E+01	F	F	F	F	F	F	F
F	5 40	0.197 1.575	0.191E+01 0.128E+01	F	F	F	F	F	F	F
F	6 38	0.236 1.496	0.224E+01 0.150E+01	F	F	F	F	F	F	F
F	8 34	0.315 1.339	0.285E+01 0.192E+01	F	F	F	F	F	F	F
F	10 30	0.394 1.181	0.339E+01 0.228E+01	F	F	F	F	F	F	F
F	12 26	0.472 1.024	0.387E+01 0.260E+01	F	F	F	F	F	F	F
F	16 18	0.630 0.709	0.461E+01 0.310E+01	F	F	F	F	F	F	F
F 55	1 53	2.165 0.039 2.087	0.458E+00 0.308E+00	F	F	F A+-0.08	F	F	F	F
F	2 51	0.079 2.008	0.899E+00 0.604E+00	F	F	F	F	F	F	F
F	3 49	0.118 1.929	0.132E+01 0.889E+00	F	F	F	F	F	F	F
F	4 47	0.157 1.850	0.173E+01 0.116E+01	F	F	F	F	F	F	F
F	5 45	0.197 1.772	0.212E+01 0.142E+01	F	F	F	F	F	F	F
F	6 43	0.236 1.693	0.249E+01 0.168E+01	F	F	F	F	F	F	F
F	8 39	0.315 1.535	0.319E+01 0.214E+01	F	F	F	F	F	F	F
F	10 35	0.394 1.378	0.382E+01 0.256E+01	F	F	F	F	F	F	F
F	12 31	0.472 1.220	0.438E+01 0.294E+01	F	F	F	F	F	F	F
F	16 23	0.630 0.906	0.529E+01 0.356E+01	F	F	F	F	F	F	F
F 60	1 58	2.362 0.039 2.283	0.500E+00 0.336E+00	F	F	F A+-0.08	F	F	F	F
F	2 56	0.079 2.205	0.984E+00 0.661E+00	F	F	F	F	F	F	F
F	3 54	0.118 2.126	0.145E+01 0.975E+00	F	F	F	F	F	F	F
F	4 52	0.157 2.047	0.190E+01 0.128E+01	F	F	F	F	F	F	F
F	5 50	0.197 1.969	0.233E+01 0.157E+01	F	F	F	F	F	F	F
F	6 48	0.236 1.890	0.275E+01 0.185E+01	F	F	F	F	F	F	F
F	8 44	0.315 1.732	0.353E+01 0.237E+01	F	F	F	F	F	F	F
F	10 40	0.394 1.575	0.424E+01 0.285E+01	F	F	F	F	F	F	F
F	12 36	0.472 1.417	0.489E+01 0.328E+01	F	F	F	F	F	F	F
F	16 28	0.630 1.102	0.597E+01 0.401E+01	F	F	F	F	F	F	F

NON-FERROUS MATERIAL

RELATED ISO STANDARDS AND PUBLICATIONS (TC 18, TC 26, TC 79)

Zinc and Zinc Alloys (TC 18)

ISO/R 301-1963	Zinc alloys ingots
ISO 713-1975	Zinc—Determination of lead and cadmium contents—Polarographic method
ISO 714-1975	Zinc—Determination of iron content—Photometric method
ISO 715-1975	Zinc—Determination of lead content—Polarographic method
ISO/R 752-1968	Zinc ingots
ISO 1053-1975	Zinc—Determination of copper content—Spectrophotometric method
ISO 1054-1975	Zinc—Determination of cadmium content—Polarographic method
ISO 1055-1975	Zinc and zinc alloys—Determination of iron content—Spectrophotometric method
ISO 1169-1975	Zinc alloys—Determination of aluminum content—Volumetric method
ISO 1570-1975	Zinc and zinc alloys—Determination of tin content—Spectrophotometric method
ISO 1976-1975	Zinc alloys—Determination of copper content—Electrolytic method
ISO 2576-1972	Chemical analysis of zinc alloys—Polarographic determination of lead and cadmium in zinc alloys containing copper
ISO 2741-1973	Zinc alloys—Complexometric determination of magnesium
ISO 3750-1976	Zinc alloys—Determination of magnesium content—Atomic absorption method
ISO 3751-1976	Zinc ingots—Selection and preparation of samples for chemical analysis
ISO 3752-1976	Zinc alloys ingots—Selection and preparation of samples for chemical analysis
ISO 3815-1976	Zinc and zinc alloys—Spectrographic analysis
ISO 3816-1976	Zinc ingots—Selection and preparation of samples for spectrographic analysis
ISO 3817-1976	Zinc alloy ingots—Selection and preparation of samples for spectrographic analysis

Copper and Copper Alloys (TC 26)

ISO/R 196-1961	Method of mercurous nitrate test for copper and copper alloys
ISO/TR 197/I-1976	Copper and copper alloys—Terms and definitions—Part I : Materials
ISO/TR 197/II -1976	Copper and copper alloys—Terms and definitions—Part II : Refinery shapes
ISO/TR 197/III -1976	Copper and copper alloys—Terms and definitions—Part III : Wrought products
ISO/TR 197/IV -1976	Copper and copper alloys—Terms and definitions—Part IV : Castings
ISO 426/I-1973	Wrought copper-zinc alloys—Chemical composition and form of wrought products—Part 1 : Non-leaded, special and high tensile alloys
ISO 426/II-1973	Wrought copper-zinc alloys—Chemical composition and forms of wrought products—Part II : Leaded alloys
ISO 427-1973	Wrought copper-tin alloys—Chemical composition and forms of wrought products
ISO 428-1973	Wrought copper-aluminum alloys—Chemical composition and forms of wrought products
ISO 429-1973	Wrought copper-nickel alloys—Chemical composition and forms of wrought products
ISO 430-1973	Wrought copper-nickel-zinc alloys—Chemical composition and forms of wrought products
ISO 431-1972	Electrolytic tough pitch copper—Refinery shapes
ISO/R 1187-1971	Special wrought copper alloys
ISO/R 1190/I -1971	Copper and copper alloys—Code of designation—Part I : Designation of materials
ISO/R 1190/II -1971	Copper and copper alloys—Code of designation—Part II : Designation of tempers
ISO/R 1336-1971	Wrought alloyed coppers
ISO/R 1337-1971	Wrought coppers
ISO/R 1338-1971	Cast copper alloys
ISO/R 1428-1971	Fire-refined high-conductivity tough pitch copper—Refinery shapes
ISO/R 1429-1971	Fire-refined tough pitch copper—Refinery shapes

ISO/R 1430-1971	Phosphorus-deoxidized copper—Refinery shapes
ISO 1553-1976	Unalloyed copper containing not less than 99.90% of copper—Determination of copper content—Electrolytic method
ISO 1554-1976	Wrought and cast copper alloys—Determination of copper content—Electrolytic method
ISO 1634-1974	Wrought copper and copper alloys—Rolled flat products (plate, sheet, strip)—Mechanical properties
ISO 1635-1974	Wrought copper and copper alloys—Round tubes for general purposes—Mechanical properties
ISO 1637-1974	Wrought copper and copper alloys—Solid products supplied in straight lengths—Mechanical properties
ISO 1638-1974	Wrought copper and copper alloys—Drawn solid products supplied in coils or on reels—Mechanical properties
ISO 1639-1974	Wrought copper alloys—Extruded sections—Mechanical properties
ISO 1640-1974	Wrought copper alloys—Forgings—Mechanical properties
ISO 1810-1976	Copper alloys—Determination of nickel (low contents)—Dimethylglyoxime spectrophotometric method
ISO/R 1811-1971	Chemical analysis of copper and copper alloys—Sampling of copper refinery shapes
ISO 1812-1976	Copper alloys—Determination of iron content—1.10-Phenanthroline spectrophotometric method
ISO 2311-1972	Electrolytic cathode copper
ISO 2543-1973	Copper and copper alloys—Determination of manganese—Spectrophotometric method
ISO 2624-1973	Copper and copper alloys—Estimation of average grain size
ISO 2626-1973	Copper—Hydrogen embrittlement test
ISO 3110-1975	Copper alloys—Determination of aluminum as alloying element—Volumetric method
ISO 3111-1975	Copper alloys—Determination of tin as alloying element—Volumetric method
ISO 3112-1975	Copper and copper alloys—Determination of lead—Extracting titration method
ISO 3220-1975	Copper and copper alloys—Determination of arsenic—Photometric method
ISO DIS 3486	Cold rolled flat products delivered in straight length
ISO DIS 3487	Cold rolled flat products in coils or on reels/strip/
ISO DIS 3488	Wrought copper & alloys—Extruded round, square & hexagonal bars
ISO DIS 3489	Copper & alloys—Drawn round bars
ISO DIS 3490	Copper & alloys—Drawn hexagonal bars
ISO DIS 3491	Drawn square copper and copper alloys—Drawn square bars
ISO DIS 3492	Copper and alloys—Drawn round wire
ISO DIS 4746	Oxygen-free copper—Scale adhesion test

Light Metals and their Alloys (TC 79)

ISO/R 114-1959	Composition of 99.8 unalloyed magnesium ingots
ISO/R 115-1968	Classification and composition of unalloyed aluminum ingots for remelting
ISO/R 121-1971	Composition of magnesium-aluminum-zinc alloy castings and mechanical properties of sand cast reference test bars
ISO/R 122-1959	Composition of magnesium-aluminum-zinc alloy ingots for casting purposes
ISO/R 164-1960	Composition of aluminum alloy castings
ISO/R 207-1961	Composition of 99.95 unalloyed magnesium ingots
ISO/R 208-1961	Composition of aluminum alloy castings (complement to R 164)
ISO/R 209-1971	Composition of wrought products of aluminum and aluminum alloys—Chemical composition (per cent)
ISO/R 503-1966	Composition of wrought magnesium-aluminum-zinc alloys
ISO 791-1973	Magnesium alloys—Determination of aluminum—8-hydroxyquinoline gravimetric method

NON-FERROUS MATERIAL

ISO 792-1973	Magnesium and magnesium alloys—Determination of iron—Orthophenanthroline photometric method
ISO 793-1973	Aluminum and aluminum alloys—Determination of iron—Orthophenanthroline photometric method
ISO 794-1976	Magnesium and magnesium alloys—Determination of copper content—Oxalyldihydrazide photometric method
ISO 795-1976	Aluminum and aluminum alloys—Determination of copper content—Oxalyldihydrazide photometric method
ISO 796-1973	Aluminum alloys—Determination of copper—Electrolytic method
ISO 797-1973	Aluminum and aluminum alloys—Determination of silicon—Gravimetric method
ISO/R 807-1968	Chemical analysis of magnesium and magnesium alloys—Polarographic determination of zinc (zinc content between 0.1 and 0.4%)
ISO 808-1973	Aluminum and aluminum alloys—Determination of silicon—Spectrophotometric methods with the reduced silicomolybdic complex
ISO 809-1973	Magnesium and magnesium alloys—Determination of manganese—Periodate photometric method (manganese content between 0.01 and 0.8%)
ISO 810-1973	Magnesium and magnesium alloys—Determination of manganese—Periodate photometric method (manganese content less than 0.01%)
ISO/R 826-1968	Mechanical property limits for rolled products of aluminum and aluminum alloys
ISO/R 827-1968 Add 1-1968	Mechanical property limits for extruded products of aluminum and aluminum alloys
ISO/R 828-1968	Mechanical property limits for rivet stock of aluminum and aluminum alloys
ISO/R 829-1968	Mechanical property limits for aluminum alloy forgings
ISO 886-1973	Aluminum and aluminun alloys—Determination of manganese—Photometric method (manganese content between 0.005 and 1.5%)
ISO/R 1118-1969	Chemical analysis of aluminum and aluminum alloys—Determination of titanium (spectrophotometric method with chromotropic acid)
ISO 1178-1976	Magnesium alloys—Determination of soluble zirconium—Alizarin sulphonate photometric method
ISO 1783-1973	Magnesium alloys—Determination of zinc—Volumetric method
ISO 1784-1976	Aluminum alloys—Determination of zinc—EDTA titrimetric method
ISO 1975-1973	Magnesium and magnesium alloys—Determination of silicon—Spectrophotometric method with the reduced silicomolybdic complex
ISO 2085-1976	Anodizing of aluminum and its alloys—Check of continuity of thin anodic oxide coatings—Copper sulphate test
ISO/R 2092-1971	Light metals and their alloys—Code of designation
ISO 2106-1976	Anodizing of aluminum and its alloys—Determination of mass per unit area of anodic oxide coatings—Gravimetric method
ISO/R 2107-1971	Light metals and their alloys—Temper designations
ISO 2119-1972	Magnesium-zinc-zirconium alloy castings—Chemical composition
ISO 2128-1976	Anodizing of aluminum and its alloys—Determination of thickness of anodic oxide coatings—Non-destructive measurement by split-beam microscope
ISO 2135-1976	Anodizing of aluminum and its alloys—Accelerated test of lightfastness of colored anodic oxide coatings
ISO DTR 2136	Wrought aluminum and aluminum alloys—Rolled products—Mechanical properties
ISO/R 2142-1971	Wrought aluminum and aluminum alloys—Selection of specimens and test pieces
ISO/R 2143-1971	Surface treatment of metals—Anodisation of aluminum and its alloys—Estimation of the loss of absorptive power by colorant drop test with prior acid treatment
ISO/R 2147-1971	Aluminum alloys—Sand cast test pieces—Mechanical properties

ISO 2297-1973	Chemical analysis of aluminum and its alloys—Complexometric determination of mangesium
ISO 2353-1972	Magnesium and its alloys—Determination of manganese in magnesium alloys containing zirconium, rare earths, thorium and silver—Periodate photometric method
ISO 2354-1976	Magnesium alloys—Determination of insoluble zirconium—Alizarin sulphonate photometric method
ISO 2355-1972	Chemical analysis of magnesium and its alloys—Determination of rare earths—Gravimetric method
ISO 2376-1972	Anodization (anodic oxidation) of aluminum and its alloys—Insulation check by measurement of breakdown potential
ISO 2377-1972	Magnesium alloy sand castings—Reference test bar
ISO 2378-1972	Aluminum alloy chill castings—Reference test bar
ISO 2379-1972	Aluminum alloy sand castings—Reference test bar
ISO 2637-1973	Aluminum and its alloys—Determination of zinc—Atomic absorption method
ISO 2767-1973	Surface treatments of metals—Anodic oxidation of aluminum and its alloys—Specular reflectance at 45°—Total reflectance—Image clarity
ISO DTR 2778	Wrought aluminum and aluminun alloys—Drawn tubes—Mechanical properties
ISO 2779-1973	Aluminum machining alloys—Chemical composition and mechanical properties of alloys A1-Cu6 Bi Pb and A1-Cu4 Pb Mg
ISO 2931-1975	Anodizing of aluminum and its alloys—Assessment of quality of sealed anodic oxide coatings by measurement of admittance or impedance
ISO 2932-1973	Anodizing of aluminum and its alloys—Assessment of sealing quality by measurement of the loss of mass after immersion in acid solution
ISO 3115-1974	Magnesium - zinc - zirconium alloy sand castings—Mechanical properties
ISO 3116-1974	Wrought magnesium-aluminum-zinc alloys—Mechanical properties
ISO DTR 3134/1	Light metals and their alloys—Terms and definitions—Part I : Materials
ISO DTR 3134/2	Light metals and their alloys—Terms and definitions—Part II : Unwrought products
ISO DTR 3134/3	Light metals and their alloys—Terms and definitions—Part III : Wrought products
ISO DTR 3134/4	Light metals and their alloys—Terms and definitions—Part IV : Castings
ISO 3210-1974	Anodizing of aluminum and its alloys—Assessment of sealing quality by measurement of the loss of mass after immersion in phosphoric-chromic acid solution
ISO 3211-1974	Anodizing of aluminum and its alloys—Assessment of resistance of anodic coatings to cracking by deformation
ISO 3255-1974	Magnesium and magnesium alloys—Determination of aluminum—Chromazurol S photometric method
ISO DIS 3256	Aluminum and alloys—Determination of magnesium
ISO DIS 3335	Extruded solid profiles in aluminum —Magnesium alloy Al Zn4, 5-MgL 7020—Chemical composition and mechanical properties
ISO DIS 3843	Test for lightfastness of colored anodic oxide coatings
ISO 3978-1976	Aluminum and aluminum alloys—Determination of chromium—Spectrophotometric method using diphenylcarbazide, after extraction
ISO DIS 3979	Aluminum and alloys—Determination of nickel
ISO DIS 3980	Aluminum alloys—Determination of copper
ISO DIS 3981	Aluminum alloys—Determination of nickel
ISO DIS 3982	Aluminum alloys—Determination of chromium
ISO DIS 4058	Chemical analysis of magnesium and its alloys—Determination of nickel

NON-FERROUS MATERIAL

Other Sources for National Non-Ferrous Material Standards

U.S.A.

Metals Handbook
Volume 1: Properties and selection of metals
Volume 2: Heat treating, cleaning and finishing
Volume 3: Machining
Volume 4: Forming
Available from: American Society of Metals, Metals Park, Ohio, U.S.A. 44703

Standards Handbook for Copper and Copper Alloy Wrought Mill Products
Available from: Copper Development Association
405 Lexington Ave.
New York, N.Y. 10017

France

AFNOR Summary of French Standards for Light Metals (in French)
Available from: AFNOR, Tour Europe,
CEDEX 7
92080 Paris La Defense
France

U. K.

Metallic Materials Specification Handbook by Robert B. Ross
Available from: Halsted Press
A Division of John Wiley & Sons, Inc.
605 Third Ave.
New York, N.Y. 10016

Germany

DIN Pocketbook 26–Standards for Copper and Copper Alloys (in German)
DIN Pocketbook 27–Standards for Aluminum and Light Metals (in German)
Available from: Beuth-Vertrieb GMBH
Burggrafenstrasse 4-7
1000 Berlin 30
W. Germany

12 Bearings

Introduction

Roller bearings have been subject to extensive world standardization efforts. The most popular bearing types in countries already using the metric system are the ball and cylindrical roller bearing types. The secretariat for the Technical Committee ISO/TC4 Rolling Bearings is held by Sweden. Its first recommendation R15 for radial bearings boundary dimensions was issued in 1955, and the second edition of this important standard[1] received worldwide approval in 1968. The largest ball and cylindrical roller bearing manufacturing companies in the world have adhered to the ISO recommendation, so that today a number of bearing types, produced by different manufacturing companies, can be interchanged.

The ANSI is representing the United States position on an international level, and the Anti-Friction Bearing Manufacturing Association, Inc., 60 East 42nd Street, New York, New York 10017, has provided ANSI with North American bearing standards through its AFBMA standards series numbers 1 through 20 available from the above organization. The ANSI B3.14 (AFBMA #20) standard describes metric ball and roller bearings conforming to basic boundary plans, which completely covers the following ISO recommendations: R15/1, R119, R201, 464, R465, and R1038.

Covered with certain exceptions in the ANSI B3.14 standard are:

R104—Except for Section 2, Double Direction Thrust Bearings
R300/1—Except for Group III, Tapered Roller Bearings
R492—Except for Class 6

World standardization of other bearing types, such as needle and tapered roller bearings, has had a slow start because of a commitment to the customary inch measuring system. In recent years this has changed, and diligent participation at the ISO level by leading North American producers of these bearing types has led to the introduction of American bearing design concept to the world.

[1] For information about the term "standard" as used in this book, please see page 1-2.

The secretariat for ISO/TC123-Plain Bearings is held by the Soviet Union, where several standards have been issued on the subject, as shown on page 12-16. Self-aligning bushings would fall in the category of TC123. However, the Russians stated that they were not interested in this matter. Self-aligning bushings will, therefore, be covered by TC4/SC7; and the Subcommittee 7 is handled by Germany. The draft proposals shown in Table 12-13 were presented to TC4 in 1972, and a draft international standard on self-aligning bushings will be issued soon.

BALL AND CYLINDRICAL ROLLER BEARINGS

General

National standards and industry manufacturing practices for ball and cylindrical roller bearings conform mostly to the ISO worldwide recommendations. The purpose of this section is to provide references to important national standards on bearings, and to facilitate worldwide specification data to be used for the selection and procurement of bearings.

Boundary Dimensions for Ball and Cylindrical Roller Bearings

Bearing boundary dimensions are important, and ball and cylindrical roller bearings of various internal designs and of the same dimension series are interchangeable. Boundary dimensions for radial roller bearings are specified in ISO/R15: Part I and Part II for diameter Series 7. Thrust roller bearings with flat seats have their boundary dimensions specified in ISO/R104.

The ANSI B3.14 (AFBMA #20) standard agrees with the ISO recommendations except for double direction thrust bearings in Section 2 of ISO/R104.

The corner radii or nominal chamfers on the bearings are specified for the various dimension series in ISO/R15, and the ISO 582 describes the chamfer dimension limits and maximum shaft and housing fillet radius for metric series bearings. The ISO boundary plan lists r as a nominal bearing chamfer whereas ANSI boundary plan lists $r_{a\ Max}$ as

Table 12-1. Bearing Chamfers (ISO 582)

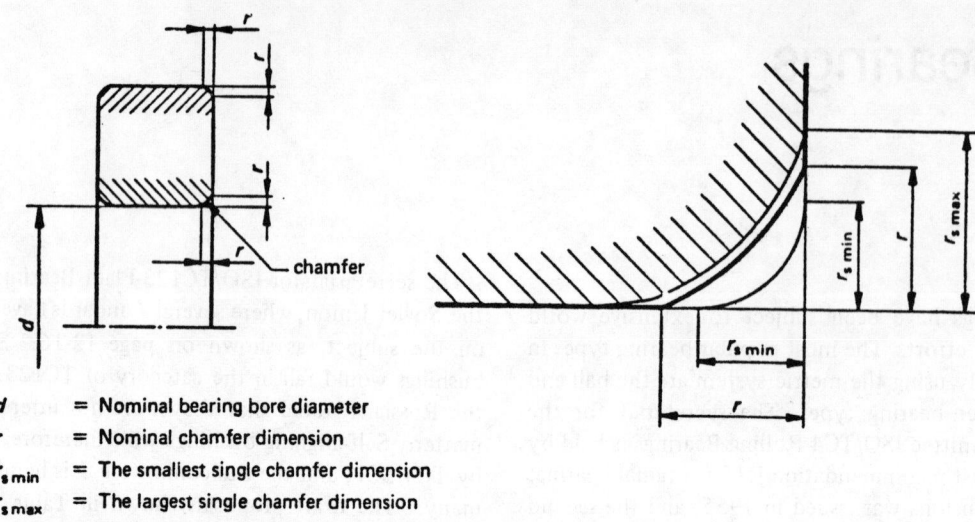

- d = Nominal bearing bore diameter
- r = Nominal chamfer dimension
- $r_{s\,min}$ = The smallest single chamfer dimension
- $r_{s\,max}$ = The largest single chamfer dimension
- $r_{as\,max}$ = The largest single shaft and housing fillet radius

NOTE – In the axial direction the $r_{s\,max}$ values given in the table may be slightly exceeded because the NOT-GO tolerance limit for the diameters of the bore and outside surface does not necessarily apply for a distance of up to twice r from the ring side face.

Values in millimetres

Nominal chamfer dimension r	Nominal bearing bore diameter d Over	Incl.	Chamfer dimension limits $r_{s\,max}$ Radial[1] bearings	$r_{s\,max}$ Thrust bearings	$r_{s\,min}$ All bearings	Shaft and housing fillet radius $r_{as\,max}$
0.1	–	–	0.1	0.1	0.05	0.05
0.15	–	–	0.16	0.16	0.08	0.08
0.2	–	–	0.2	0.2	0.1	0.1
0.3	–	–	0.3	0.3	0.15	0.15
0.4	–	–	0.5	0.5	0.2	0.2
0.5	–	40	0.6	0.8	0.3	0.3
	40	–	0.8			
1	–	40	1	1.5	0.6	0.6
	40	–	1.3			
1.5	–	50	1.6	2	1	1
	50	–	1.9			
2	–	120	2	2.5	1	1
	120	–	2.5			
2.5	–	120	2.5	3	1.5	1.5
	120	–	3			
3	–	220	3.5	4	2	2
	220	–	3.8			
3.5	–	280	4	4.5	2	2
	280	–	4.5			
4	–	280	5	5.5	3	2.5
	280	–	5.5			
5	–	–	6.5	6.5	4	3
6	–	–	8	8	5	4
8	–	–	10	10	6	5
10	–	–	12.5	12.5	7.5	6
12	–	–	15	15	9.5	8
15	–	–	18	18	12	10
18	–	–	21	21	15	12
22	–	–	25	25	19	15

1) Including tapered roller bearings, for which the limits apply to the chamfer r on the back face but not to the chamfer r_1 on the front face of the rings.

BEARINGS

Table 12-2. World Bearing Tolerance Classes

ISO R492 R199	U.S.A. ANSI B3.14	Japan JIS B1514	Germany DIN 620	France NF E22-321	U.K. BS 292	Italy UNI 4505 4506	Australia
Normal Class 0	ABEC1 RBEC1	0	Normal P0	0	0	0	
Class 6	(ABEC3)	6	P6	6	6	6	
Class 5	ABEC5 RBEC5	5	P5	5	5	5	
Class 4	ABEC7 ABEC9	4	P4	4	4	4	

NOTES:
1. ISO/R492 is applicable to radial roller bearing tolerances and ISO/R199 to thrust roller bearings.
2. The former ANSI DESIGNATION ABEC3 is rarely used. Tolerance class ABEC9 is not defined in ISO standards yet.
3. For other tolerances and special requirements such as internal bearing clearance or heat stabilization temperatures, see national standards referred to in Table 12-2.

the maximum fillet radius that the bearing chamfer will clear. The maximum fillet radius $r_{a\,Max}$ specified for bearings in the American standards is equal to or less than $r_{s\,Min}$ as shown in Table 12-1. Although a worldwide agreement on this subject is not foreseen in the near future, the countries outside North America are following the ISO practice.

Tolerancing

Tolerance limits for dimensions and runouts are covered in ISO/R492 for radial roller bearings and in ISO/R199 for thrust roller bearings with flat seats. A normal class (Class 0) and Classes 6, 5, and 4 listed in increasing order of precision are defined in the ISO standards. How these tolerance classes relate to other national standards is shown in Table 12-2.

The ANSI B3.14 standard defines four tolerance classes for ball bearings as follows: ABEC1, ABEC5, ABEC7, and ABEC9.

For roller bearings there are two tolerance classes, RBEC1 and RBEC5. The differences existing between the ANSI and ISO values are minor and generally too small to be of practical significance or to jeopardize interchangeability.

Identification Codes

The identification codes for bearings are specified in national standards shown in Table 12-3.

Table 12-3. World Bearing Identification Code Standards

	ISO/ R300
U.S.A.	ANSI B3.14
JAPAN	JIS B1513
GERMANY	DIN 623
FRANCE	NF E22-395
U.K.	BS 292
ITALY	UNI 5417
AUSTRALIA	AS

The ANSI B3.14 standard conforms to the ISO/R300 recommendations and the DIN standard has a different identification code. However, a cross reference to the ISO/R300 system is shown in the German standard.

The ISO identification codes have been used in the bearing tables in this section, and a brief description of the designation system is as follows:

The ISO Identification code system has a basic number and supplementary numbers as shown in Table 12-4. The basic bearing number consists of a bore dimension in millimeters, a type code when the first letter indicates class of bearing (B = radial ball bearings; R = cylindrical roller bearings; S = self-aligning roller bearings; T = thrust ball and roller bearings), and finally the dimension code the diameter and width (or height) of bearings.

The basic bearing code numbers are shown in the tables for ball and roller bearings. For a complete bearing code designation system, refer to ANSI B3.14 (AFBMA #20) or ISO/R300. The description of how to use the ISO-ANSI identification code system has been kept to a minimum here, since future changes to the above system are expected.

Popular Ball and Cylindrical Roller Bearing Types

The most popular deep grove, angular contact, double row, self-aligning and thrust ball bearings are shown in Table 12-5. ISO and national standards for each bearing type as well as their names in English (E), German (G), French (F), and Italian (I) are also given in the table.

Deep-Groove Ball Bearings—The single-row, deep-groove ball bearings of the ISO dimension series 10, 02, and 03 shown in Table 12-5 are generally available from 10 to 100 mm and larger shaft diameters. These ball bearings can be supplied with single or double shields or seals and with grooves for retaining rings. Ball bearings of this type are well-suited for light to medium radial loads at high speeds. The largest number of bearings sold are of type BC 02.

Angular Contact Ball Bearings—Ball bearings of the single row, angular contact types of the dimension series 02 and 03, with shaft diameters from 10 to 110 mm, are shown in Table 12-5. Bearings are generally available for sizes ranging from 25 to 180 mm shaft diameters. The angular contact bearings are suited for light to medium radial loads and for light axial loads in one direction at high speeds.

Double Row Ball Bearings—Double row ball bearings of the ISO dimension series 32 and 33 and shaft diameters from 10(15) to 110 mm are shown in Table 12-5. The bearings listed in the table are generally available, and they are well-suited for light to medium radial loads and light thrust loads in both directions at high speeds.

Self-Aligning Ball Bearings—Self-aligning ball bearings of the ISO dimension series 02 and 22 having nominal shaft diameters, as shown in Table 12-5, are commercially available. The self-aligning ball bearings are used when some shaft deflection is expected. However, this type of bearing is less popular than the self-aligning roller bearings shown in Table 12-6. Self-aligning ball bearings might be perferred over self-aligning roller bearings when applications are for small-shaft diameter sizes with height axial and radial loads and high speeds.

Self-Aligning Taper Roller Thrust Bearings—Self-aligning taper roller thrust bearings of the dimension series 94 and with bore diameters from 60 to 140 mm are listed in Table 12-5. These types of bearings are well-suited for heavy thrust loads and some shaft deflections at low to medium speeds.

Ball Thrust Bearings—Ball thrust bearings of the dimension series 11 having nominal bore diameters from 30 to 200 mm are listed in Table 12-5. For proper functioning, particularly at high speeds, a minimum thrust load is required. Thrust ball bearings are suitable for light to medium axial loads and for high speeds.

Cylindrical Roller Bearings—Single-row cylindrical roller bearings with nominal shaft diameters from 25 to 170 mm in dimension series 02; and shaft sizes from 20 to 160 mm in dimension series 03 are shown in Table 12-6. Bearings of this type are well-suited for heavy radial loads and cannot transmit any axial forces. The bearings are recommended for use at moderate speeds when no shaft deflection is expected.

Self-Aligning Roller Bearings—Spherical roller bearings in dimension series 22 with bore sizes ranging from 25 to 180 mm, and in dimension series 30 and 31 with shaft diameters from 110 to 480 mm are shown in Table 12-6. The self-aligning roller bearings are available with cylindrical or 1:12 tapered bores to the nominal dimensions shown. This type of bearing is well-suited for heavy radial loads and moderate axial loads at medium rotating speeds. Bearings will perform satisfactorily with some shaft deflections. The most popular bearing type is the SD 22.

Table 12-4. Schematic Arrangement of a Complete ISO/R300 Code Number

Section 1			Section 2	Section 3	Section 4	Section 5
Basic Number			Supplementary Number			
Bore	Type	Outside diameter and width (or height)	Modification of design	Internal fit and tolerances	Lubricants and preservatives	Special requirements
0000	AAA	00	AAAA	00	A	000

BEARINGS

Table 12-5. World Ball and Thrust Bearing Standards

Deep Groove Ball Bearings - Single Row (Dimension Series 10)

- ISO/R15
- U.S.A. ANSI B3.14
- JAPAN JIS B1521
- GERMANY DIN 625
- FRANCE NF E22-300
- U.K. BS 292
- ITALY UNI 4473
- AUSTRAL AS

Names: (E) DEEP GROOVE BALL BEARINGS - SINGLE ROW; (G) RILLENKUGELLAGER; (F) ROULEMENTS A BILLES; (I) CUSCINETTI RADIALI RIGIDI AD UNA CORONA DI SFERE

BEARING TYPE BC					DIMENSION SERIES 10				
DESIG. (1)	d	D	B	r(2)	DESIG.	d	D	B	r
10 BC10	10	26	8	0,5	60 BC10	60	95	18	2
12 BC10	12	28	8	0,5	65 BC10	65	100	18	2
15 BC10	15	32	9	0,5	70 BC10	70	110	20	2
17 BC10	17	35	10	0,5	75 BC10	75	115	20	2
20 BC10	20	42	12	1	80 BC10	80	125	22	2
25 BC10	25	47	12	1	85 BC10	85	130	22	2
30 BC10	30	55	13	1,5	90 BC10	90	140	24	2,5
35 BC10	35	62	14	1,5	95 BC10	95	145	24	2,5
40 BC10	40	68	15	1,5	100 BC10	100	150	24	2,5
45 BC10	45	75	16	1,5	105 BC10	105	160	26	3
50 BC10	50	80	16	1,5	110 BC10	110	170	28	3
55 BC10	55	90	18	2	120 BC10	120	180	28	3

Deep Groove Ball Bearings - Single Row (Dimension Series 02)

- ISO/R15
- U.S.A. ANSI B3.14
- JAPAN JIS B1521
- GERMANY DIN 625
- FRANCE NF E22-300
- U.K. BS 292
- ITALY UNI 4203
- AUSTRAL AS

BEARING TYPE BC					DIMENSION SERIES 02				
DESIG.	d	D	B	r	DESIG.	d	D	B	r
10 BC02	10	30	9	1	45 BC02	45	85	19	2
12 BC02	12	32	10	1	50 BC02	50	90	20	2
15 BC02	15	35	11	1	55 BC02	55	100	21	2,5
17 BC02	17	40	12	1	60 BC02	60	110	22	2,5
20 BC02	20	47	14	1,5	65 BC02	65	120	23	2,5
22 BC02	22	50	14	1,5	70 BC02	70	125	24	2,5
25 BC02	25	52	15	1,5	75 BC02	75	130	25	2,5
28 BC02	28	58	16	1,5	80 BC02	80	140	26	3
30 BC02	30	62	16	1,5	85 BC02	85	150	28	3
32 BC02	32	65	17	1,5	90 BC02	90	160	30	3
35 BC02	35	72	17	2	95 BC02	95	170	32	3,5
40 BC02	40	80	18	2	100 BC02	100	180	34	3,5

Deep Groove Ball Bearings - Single Row (Dimension Series 03)

- ISO/R15
- U.S.A. ANSI B3.14
- JAPAN JIS B1521
- GERMANY DIN 625
- FRANCE NF E22-300
- U.K. BS 292
- ITALY UNI 4204
- AUSTRAL AS

BEARING TYPE BC					DIMENSION SERIES 03				
DESIG. (1)	d	D	B	r(2)	DESIG.	d	D	B	r
10 BC03	10	35	11	1	45 BC03	45	100	25	2,5
12 BC03	12	37	12	1,5	50 BC03	50	110	27	3
15 BC03	15	42	13	1,5	55 BC03	55	120	29	3
17 BC03	17	47	14	1,5	60 BC03	60	130	31	3,5
20 BC03	20	52	15	2	65 BC03	65	140	33	3,5
22 BC03	22	56	16	2	70 BC03	70	150	35	3,5
25 BC03	25	62	17	2	75 BC03	75	160	37	3,5
28 BC03	28	68	18	2	80 BC03	80	170	39	3,5
30 BC03	30	72	19	2	85 BC03	85	180	41	4
32 BC03	32	75	20	2	90 BC03	90	190	43	4
35 BC03	35	80	21	2,5	95 BC03	95	200	45	4
40 BC03	40	90	23	2,5	100 BC03	100	215	47	4

Angular Contact Ball Bearings - Single Row (Dimension Series 02)

- ISO/R15
- U.S.A. ANSI B3.14
- JAPAN JIS B1522
- GERMANY DIN 628
- FRANCE NF E22-300
- U.K. BS 292
- ITALY UNI 6336
- AUSTRAL AS

Names: (E) ANGULAR CONTACT BALL BEARINGS - SINGLE ROW; (G) SCHRAEGKUGELLAGER-EINREIHIG; (F) ROULEMENTS A BILLES; (I) CUSCINETTI RADIALI OBLIQUI AD UNA CORONA DI SFERE CON ANGOLO 32-45°

BEARING TYPE BT					DIMENSION SERIES 02				
DESIG.	d	D	B	r	DESIG.	d	D	B	r
10 BT02	10	30	9	1	55 BT02	55	100	21	2,5
12 BT02	12	32	10	1	60 BT02	60	110	22	2,5
15 BT02	15	35	11	1	65 BT02	65	120	23	2,5
17 BT02	17	40	12	1	70 BT02	70	125	24	2,5
20 BT02	20	47	14	1,5	75 BT02	75	130	25	2,5
25 BT02	25	52	15	1,5	80 BT02	80	140	26	3
30 BT02	30	62	16	1,5	85 BT02	85	150	28	3
35 BT02	35	72	17	2	90 BT02	90	160	30	3
40 BT02	40	80	18	2	95 BT02	95	170	32	3,5
45 BT02	45	85	19	2	100 BT02	100	180	34	3,5
50 BT02	50	90	20	2	105 BT02	105	190	36	3,5
					110 BT02	110	200	38	3,5

Angular Contact Ball Bearings - Single Row (Dimension Series 03)

- ISO/R15
- U.S.A. ANSI B3.14
- JAPAN JIS B1522
- GERMANY DIN 628
- FRANCE NF E22-300
- U.K. BS 292
- ITALY UNI
- AUSTRAL AS

BEARING TYPE BT					DIMENSION SERIES 03				
DESIG.	d	D	B	r	DESIG.	d	D	B	r
10 BT03	10	35	11	1	60 BT03	60	130	31	3,5
12 BT03	12	37	12	1,5	65 BT03	65	140	33	3,5
15 BT03	15	42	13	1,5	70 BT03	70	150	35	3,5
17 BT03	17	47	14	1,5	75 BT03	75	160	37	3,5
20 BT03	20	52	15	2	80 BT03	80	170	39	3,5
25 BT03	25	62	17	2	85 BT03	85	180	41	4
30 BT03	30	72	19	2	90 BT03	90	190	43	4
35 BT03	35	80	21	2,5	95 BT03	95	200	45	4
40 BT03	40	90	23	2,5	100 BT03	100	215	47	4
45 BT03	45	100	25	2,5	105 BT03	105	225	49	4
50 BT03	50	110	27	3	110 BT03	110	240	50	4
55 BT03	55	120	29	3					

Figures Courtesy of SKF.

NOTES:
1. Bearing codes are according to ISO/R 300 and ANSI B3.14 standards.
2. r = r nominal bearing chamfer. The values listed are according to ISO/R 15 and a comparison with the housing and shaft maximum fillet radius specified in ANSI B3.14 is shown in Table 12-1.

Table 12-5 (Continued). World Ball and Thrust Bearing Standards

STANDARD	FIGURE	NAME	DIMENSIONS in Millimeters
ISO/R15 U.S.A. ANSI B3.14 JAPAN JIS B1521 GERMANY DIN 628 FRANCE NF E22-300 U.K. BS ITALY UNI 4211 AUSTRAL AS		(E) DOUBLE ROW BALL BEARINGS (G) SCHRAEGKUGELLAGER- ZWEIREIHIG (F) ROULEMENTS A BILLES (I) CUSCINETTI RADIALI OBLIQUI A DUE CORONE DI SFERE	**BEARING TYPE BE — DIMENSION SERIES 32** DESIG. d D B r / DESIG. d D B r 10 BE32 10 30 14,3 1 / 55 BE32 55 100 33,3 2,5 12 BE32 12 32 15,9 1 / 60 BE32 60 110 36,5 2,5 15 BE32 15 35 15,9 1 / 65 BE32 65 120 38,1 2,5 17 BE32 17 40 17,5 1 / 70 BE32 70 125 39,7 2,5 20 BE32 20 47 20,6 1,5 / 75 BE32 75 130 41,3 2,5 25 BE32 25 52 20,6 1,5 / 80 BE32 80 140 44,4 3 30 BE32 30 62 23,8 1,5 / 85 BE32 85 150 49,2 3 35 BE32 35 72 27,0 2 / 90 BE32 90 160 52,4 3 40 BE32 40 80 30,2 2 / 95 BE32 95 170 55,6 3,5 45 BE32 45 85 30,2 2 / 100 BE32 100 180 60,3 3,5 50 BE32 50 90 30,2 2 / 105 BE32 105 190 65,1 3,5 110 BE32 110 200 69,8 3,5
ISO/R15 U.S.A. ANSI B3.14 JAPAN JIS B1521 GERMANY DIN 628 FRANCE NF E22-300 U.K. BS ITALY UNI 4212 AUSTRAL AS		(E) DOUBLE ROW BALL BEARINGS (G) SCHRAEGKUGELLAGER- ZWEIREIHIG (F) ROULEMENTS A BILLES (I) CUSCINETTI RADIALI OBLIQUI A DUE CORONE DI SFERE	**BEARING TYPE BE — DIMENSION SERIES 33** DESIG. d D B r / DESIG. d D B r 15 BE33 15 42 19,0 1,5 / 60 BE33 60 130 54,0 3,5 17 BE33 17 47 22,2 1,5 / 65 BE33 65 140 58,7 3,5 20 BE33 20 52 22,2 2 / 70 BE33 70 150 63,5 3,5 25 BE33 25 62 25,4 2 / 75 BE33 75 160 68,3 3,5 30 BE33 30 72 30,2 2 / 80 BE33 80 170 68,3 3,5 35 BE33 35 80 34,9 2,5 / 85 BE33 85 180 73,0 4 40 BE33 40 90 36,5 2,5 / 90 BE33 90 190 73,0 4 45 BE33 45 100 39,7 2,5 / 95 BE33 95 200 77,8 4 50 BE33 50 110 44,4 3 / 100 BE33 100 215 82,6 4 55 BE33 55 120 49,2 3 / 105 BE33 105 225 87,3 4 110 BE33 110 240 92,1 4
ISO/R15 U.S.A. ANSI B3.14 JAPAN JIS B1523 GERMANY DIN 630 FRANCE NF E22-300 U.K. BS 292 ITALY UNI 4477,79 AUSTRAL AS	CYLINDRICAL BORE / TAPERED BORE (Taper 1 to 12)	(E) SELF-ALIGNING BALL BEARINGS (G) RADIAL PENDELKUGEL-LAGER (F) ROULEMENTS A SURFACE EXTERIEURE SPHERIQUE (I) CUSCINETTI RADIALI ORIENTABILI A DUE CORONE DI SFERE	**BEARING TYPE BS DIM. SERIES 02 / DIM. SERIES 22** DESIG. d D B r / DESIG. d D B r 10 BS02 10 30 9 1 / 17 BS22 17 40 16 1 12 BS02 12 32 10 1 / 20 BS22 20 47 18 1,5 15 BS02 15 35 11 1 / 25 BS22 25 52 18 1,5 17 BS02 17 40 12 1 / 30 BS22 30 62 20 1,5 20 BS02 20 47 14 1,5 / 35 BS22 35 72 23 2 25 BS02 25 52 15 1,5 / 40 BS22 40 80 23 2 30 BS02 30 62 16 1,5 / 45 BS22 45 85 23 2 35 BS02 35 72 17 2 / 50 BS22 50 90 23 2 40 BS02 40 80 18 2 / 55 BS22 55 100 25 2,5 45 BS02 45 85 19 2 / 60 BS22 60 110 28 2,5 50 BS02 50 90 20 2 / 65 BS22 65 120 31 2,5 55 BS02 55 100 21 2,5 / 70 BS22 70 125 31 2,5
(3) ISO U.S.A. ANSI JAPAN JIS B1539 GERMANY DIN 728 FRANCE NF U.K. BS 3134 ITALY UNI 6043 AUSTRAL AS	Courtesy of DIN.	(E) SELF-ALIGNING TAPER ROLLER THRUST BEARING (G) AXIAL-PENDELROLLENLAGER - LAGERREIHE 294 (F) ROULEMENTS A ROULEAUX CONIQUES (I) CUSCINETTI ASSIALI ORIENTABILI A RULLI	**BEARING TYPE TS — DIMENSION SERIES 94** DESIG. dw Dg H Dw dg Hw Hg A r 60 TS94 60 130 42 123 89 15 20 38 2.5 65 TS94 65 140 45 133 96 16 21 42 3 70 TS94 70 150 48 142 103 17 23 44 3 75 TS94 75 160 51 152 109 18 24 47 3 80 TS94 80 170 54 160 117 19 26 50 3.5 85 TS94 85 180 58 170 125 21 28 54 3.5 90 TS94 90 190 60 180 132 22 29 56 3.5 100 TS94 100 210 67 200 146 24 32 62 4 110 TS94 110 230 73 220 162 26 35 69 4 120 TS94 120 250 78 236 174 29 37 74 5 130 TS94 130 270 85 255 189 31 41 81 5 140 TS94 140 280 85 268 199 31 41 86 5
ISO/R U.S.A. ANSI B3.14 JAPAN JIS B1532 GERMANY DIN 711 FRANCE NF E22-300 U.K. BS 292 ITALY UNI 4493 AUSTRAL AS		(E) BALL THRUST BEARINGS (G) AXIAL-RILLENKUGEL-LAGER (F) ROULEMENTS A BILLES-AXIALES (I) CUSCINETTI ASSIALI A SFERE A SEMPLICE EFFETTO	**BEARING TYPE TA — DIMENSION SERIES 11** DESIG. d d2 D H r / DESIG. d d2 D H r 30 TA11 30 32 47 11 1 / 90 TA11 90 92 120 22 1,5 35 TA11 35 37 52 12 1 / 100 TA11 100 102 135 25 1,5 40 TA11 40 42 60 13 1 / 110 TA11 110 112 145 25 1,5 45 TA11 45 47 65 14 1 / 120 TA11 120 122 155 25 1,5 50 TA11 50 52 70 14 1 / 130 TA11 130 132 170 30 1,5 55 TA11 55 57 78 16 1 / 140 TA11 140 142 180 31 1,5 60 TA11 60 62 85 17 1,5 / 150 TA11 150 152 190 31 1,5 65 TA11 65 67 90 18 1,5 / 160 TA11 160 162 200 31 1,5 70 TA11 70 72 95 18 1,5 / 170 TA11 170 172 215 34 2 75 TA11 75 77 100 19 1,5 / 180 TA11 180 183 225 34 2 80 TA11 80 82 105 19 1,5 / 190 TA11 190 193 240 37 2 85 TA11 85 87 110 19 1,5 / 200 TA11 200 203 250 37 2

NOTES: *Figures Courtesy of SKF, DIN.*
1. Bearing codes are according to ISO/R 300 and ANSI B3.14 standards.
2. r = r nominal bearing chamfer. The values listed are according to ISO/R 15 and a comparison with the housing and shaft maximum fillet radius specified in ANSI B3.14 is shown in Table 12-1.

BEARINGS

Table 12-6. World Cylindrical and Spherical Roller Bearing Standards

STANDARD	FIGURE	NAME	DIMENSIONS (In Millimeters)
ISO/R15 U.S.A. ANSI B3.14 JAPAN JIS B1533 GERMANY DIN 635 FRANCE NF E22-300 U.K. BS 292 ITALY UNI 4481 AUSTRAL AS		(E) CYLINDRICAL ROLLER BEARINGS - SINGLE ROW (G) RADIAL-ROLLENLAGER (TONNENLAGER) (F) ROULEMENTS A ROULEAUX (I) CUSCINETTI RADIALI ORIENTABILI AD UNA CORONA DI RULLI	**BEARING TYPE RN — DIMENSION SERIES 02** DESIG.(1) d D B r(2) — DESIG. d D B r 25 RN02 25 52 15 1.5 — 85 RN02 85 150 28 3 30 RN02 30 62 16 1.5 — 90 RN02 90 160 30 3 35 RN02 35 72 17 2 — 95 RN02 95 170 32 3.5 40 RN02 40 80 18 2 — 100 RN02 100 180 34 3.5 45 RN02 45 85 19 2 — 105 RN02 105 190 36 3.5 50 RN02 50 90 20 2 — 110 RN02 110 200 38 3.5 55 RN02 55 100 21 2.5 — 120 RN02 120 215 40 3.5 60 RN02 60 110 22 2.5 — 130 RN02 130 230 40 4 65 RN02 65 120 23 2.5 — 140 RN02 140 250 42 4 70 RN02 70 125 24 2.5 — 150 RN02 150 270 45 4 75 RN02 75 130 25 2.5 — 160 RN02 160 290 48 4 80 RN02 80 140 26 3 — 170 RN02 170 310 52 5
ISO/R15 U.S.A. ANSI B3.14 JAPAN JIS B1533 GERMANY DIN 635 FRANCE NF E22-300 U.K. BS 292 ITALY UNI 4482 AUSTRAL AS		(E) CYLINDRICAL ROLLER BEARINGS - SINGLE ROW (G) RADIAL-ROLLENLAGER (TONNENLAGER) (F) ROULEMENTS A ROULEAUX (I) CUSCINETTI RADIALI ORIENTABILI AD UNA CORONA DI RULLI	**BEARING TYPE RN — DIMENSION SERIES 03** DESIG. d D B r — DESIG. d D B r 20 RN03 20 52 15 2 — 80 RN03 80 170 39 3.5 25 RN03 25 62 17 2 — 85 RN03 85 180 41 4 30 RN03 30 72 19 2 — 90 RN03 90 190 43 4 35 RN03 35 80 21 2.5 — 95 RN03 95 200 45 4 40 RN03 40 90 23 2.5 — 100 RN03 100 215 47 4 45 RN03 45 100 25 2.5 — 105 RN03 105 225 49 4 50 RN03 50 110 27 3 — 110 RN03 110 240 50 4 55 RN03 55 120 29 3 — 120 RN03 120 260 55 4 60 RN03 60 130 31 3.5 — 130 RN03 130 280 58 5 65 RN03 65 140 33 3.5 — 140 RN03 140 300 62 5 70 RN03 70 150 35 3.5 — 150 RN03 150 320 65 5 75 RN03 75 160 37 3.5 — 160 RN03 160 340 68 5
ISO/R15 U.S.A. ANSI B3.14 JAPAN JIS B1535 GERMANY DIN 635 FRANCE NF E22-314 U.K. BS ITALY UNI 4489 AUSTRAL AS	CYLINDRICAL BORE / TAPERED BORE (Taper 1 to 12)	(E) SPHERICAL ROLLER BEARINGS - SELF-ALIGNING (G) RADIAL-ROLLENLAGER (PENDELROLLENLAGER) (F) ROULEMENTS A ROULEAUX (SPHERIQUE) (I) CUSCINETTI RADIALI ORIENTABILI A DUE CORONE DI RULLI	**BEARING TYPE SD, SC, SL — DIMENSION SERIES 22** DESIG. d D B r — DESIG. d D B r 25 SD22 25 52 18 1.5 — 85 SD22 85 150 36 3 30 SD22 30 62 20 1.5 — 90 SD22 90 160 40 3 35 SD22 35 72 23 2 — 95 SD22 95 170 43 3.5 40 SD22 40 80 23 2 — 100 SD22 100 180 46 3.5 45 SD22 45 85 23 2 — 110 SD22 110 200 53 3.5 50 SD22 50 90 23 2 — 120 SD22 120 215 58 3.5 55 SD22 55 100 25 2.5 — 130 SD22 130 230 64 4 60 SD22 60 110 28 2.5 — 140 SD22 140 250 68 4 65 SD22 65 120 31 2.5 — 150 SD22 150 270 73 4 70 SD22 70 125 31 2.5 — 160 SD22 160 290 80 4 75 SD22 75 130 31 2.5 — 170 SD22 170 310 86 5 80 SD22 80 140 33 3 — 180 SD22 180 320 86 5
ISO/R15 U.S.A. ANSI B3.14 JAPAN JIS B1535 GERMANY DIN 635 FRANCE NF E22-314 U.K. BS ITALY UNI 4484 AUSTRAL AS	CYLINDRICAL BORE / TAPERED BORE (Taper 1 to 12)	(E) SPHERICAL ROLLER BEARINGS - SELF-ALIGNING (G) RADIAL-ROLLENLAGER (PENDELROLLENLAGER) (F) ROULEMENTS A ROULEAUX (SPHERIQUE) (I) CUSCINETTI RADIALI ORIENTABILI A DUE CORONE DI RULLI	**BEARING TYPE SD, SC, SL — DIMENSION SERIES 30** DESIG. d D B r — DESIG. d D B r 110 SD30 110 170 45 3 — 260 SD30 260 400 104 5 120 SD30 120 180 46 3 — 280 SD30 280 420 106 5 130 SD30 130 200 52 3 — 300 SD30 300 460 118 5 140 SD30 140 210 53 3 — 320 SD30 320 480 121 5 150 SD30 150 225 56 3.5 — 340 SD30 340 520 133 6 160 SD30 160 240 60 3.5 — 360 SD30 360 540 134 6 170 SD30 170 260 67 3.5 — 380 SD30 380 560 135 6 180 SD30 180 280 74 3.5 — 400 SD30 400 600 148 6 190 SD30 190 290 75 3.5 — 420 SD30 420 620 150 6 200 SD30 200 310 82 3.5 — 440 SD30 440 650 157 8 220 SD30 220 340 90 4 — 460 SD30 460 680 163 8 240 SD30 240 360 92 4 — 480 SD30 480 700 165 8
ISO/R15 U.S.A. ANSI B3.14 JAPAN JIS B1535 GERMANY DIN 635 FRANCE NF E22-314 U.K. BS ITALY UNI 4485 AUSTRAL AS	CYLINDRICAL BORE / TAPERED BORE (Taper 1 to 12)	(E) SPHERICAL ROLLER BEARINGS - SELF-ALIGNING (G) RADIAL-ROLLENLAGER (PENDELROLLENLAGER) (F) ROULEMENTS A ROULEAUX (SPHERIQUE) (I) CUSCINETTI RADIALI ORIENTABILI A DUE CORONE DI RULLI	**BEARING TYPE SD, SC, SL — DIMENSION SERIES 31** DESIG. d D B r — DESIG. d D B r 110 SD31 110 180 56 3 — 260 SD31 260 440 144 5 120 SD31 120 200 62 3 — 280 SD31 280 460 146 6 130 SD31 130 210 64 3 — 300 SD31 300 500 160 6 140 SD31 140 225 68 3.5 — 320 SD31 320 540 176 6 150 SD31 150 250 80 3.5 — 340 SD31 340 580 190 6 160 SD31 160 270 86 3.5 — 360 SD31 360 600 192 6 170 SD31 170 280 86 3.5 — 380 SD31 380 620 194 6 180 SD31 180 300 96 4 — 400 SD31 400 650 200 8 190 SD31 190 320 104 4 — 420 SD31 420 700 224 8 200 SD31 200 340 112 4 — 440 SD31 440 720 226 8 220 SD31 220 370 120 5 — 460 SD31 460 760 240 10 240 SD31 240 400 128 5 — 480 SD31 480 790 248 10

Figures Courtesy of SKF.

NOTES:
1. Bearing codes are according to ISO/R 300 and ANSI B3.14 standards.
2. r = r nominal bearing chamfer. The values listed are according to ISO/R 15 and a comparison with the housing and shaft maximum fillet radius specified in ANSI B3.14 is shown in Table 12-1.

NEEDLE ROLLER BEARINGS

General

The world needle roller bearing standards shown in Table 12-9 are the results of intensive work by ISO delegates from North America and other leading industrial countries. Most of these standards have been recently issued. In one case, American industry practice for bearing designs deviates from that of the European. Therefore, before a specific bearing is selected from Table 12-9, the text for each bearing type should be read carefully.

Drawn Cup Needle Roller Bearings

Boundary dimensions and tolerances for drawn cup needle roller bearings are specified in the ISO 3245 standard. These bearings are of small radial section, and therefore useful where space is limited. The needle bearings are normally retained in the housing by press fit, and no retaining shoulders are required. The use of split housings is not recommended.

The radial load capacity is moderate to high for bearings with hardened and ground shaft or inner ring. However, no thrust load can be transmitted. The bearings should not be used where dynamic overloads can be expected. Drawn-cup needle roller bearings are generally available in two types, one with a full complement of needle rollers and the other with caged rollers. Full complement types are best suited for moderate speeds and caged types perform well at high speeds.

The boundary plans shown in Table 12-9 do not conform to the ISO R15 recommendation. The dimensions listed for drawn-cup needle roller bearings are the preferred sizes chosen from two large tables of standard sizes, and the selections are for diameter series 2 based on information supplied by a large producer of these bearing types. The preferred sizes of series 1 are listed in ISO 3245. For other bearings than shown, see the ISO or national standards listed.

Machined Rings Needle Roller Bearings

Two types of machined rings needle roller bearings, named the *American* and *European*, are shown in Table 12-9. The American type is proportioned for the heavy-duty applications prevalent in North America, and is specified in the AFBMA #18.1 (ANSI B3.18) standards and covered in an ISO draft proposal (#4/SC5 N160). The European type has found widespread usage within the continental European countries, and is specified in the ISO 1206 and 3097 standards.

A brief description of the performance of the two bearing types is as follows: The machined rings needle-roller bearings are of moderately small radial section, and they are, therefore, useful where space is limited. The bearings should be retained in the housing by shoulders or rings and may be used in split housings. The radial load capacity for these bearings is moderate to high, but no thrust load can be transmitted. These bearing types may be used where the possibility of some dynamic overload exists, and they perform quite well at high speeds.

The main differences between the American and the European types are in the basic proportions. For the same shaft diameter, the American type has a larger outside diameter and width, thus providing more radial capacity.

From a comparison of four typical ISO and AFBMA bearings, the basic dynamic load capacity was from *34%* to *99%* higher for the North American type, and the projected bearing life was extended by from *165%* to *900%*.

Boundary Dimensions and Tolerances

American Type—The basic dimensions for machined ring bearings are shown in Table 12-9 for nominal inside diameters from 15 to 300 mm. The tolerances are given in Table 12-7.

Inner bearing rings in bore diameters from 10 to 280 mm are specified in proposed revision to ANSI B3.18 standard, and bore diameters from 10 to 130 mm are shown in Table 12-9. Tolerances for needle roller bearing inner rings are listed in Table 12-8.

European Type—Basic dimensions for machined-type needle roller bearings are shown in the ISO 1206 standard for dimension series 48 (ISO/R15 Part I) with 110 to 360 mm shaft diameters and for dimension series 49 with 5 to 160 mm inside diameters. Tolerances for these bearings are covered in the ISO 3097 standard. The boundary dimensions for the series 49 bearings are shown in Table 12-9 in nominal shaft diameters from 5 to 160 mm for bearings without inner rings and from 5 to 140 mm for complete needle bearings.

Caged and Loose Needle Rollers for Radial and Thrust Bearings

Radial Needle Roller and Cage Assemblies—Needle roller and cage assemblies are supplied without auxiliary inner and outer rings. Housing and shafts should be hardened and ground to serve as raceways, and the space requirements are kept to a minimum. The assembly provides high radial and no thrust load capacity, and is well suited for high-speed applications. Radial needle cage assemblies are specified in the ISO 3030 standard for shaft sizes from 5 to 60 mm in diameter series IC, and for shaft sizes from 10 to 100 mm in diameter series 2C, 3C, and 4C. Nominal dimensions for radial needle cage assemblies are shown in Table 12-9.

BEARINGS

Axial or Thrust Needle Roller and Cage Assemblies—Axial needle roller and cage assemblies may be used in conjunction with hardened and ground thrust washers or may operate against machine elements of proper hardness and finish. Generally, thrust needle roller and cage assemblies are used with radial bearings. The needle cage assemblies must be piloted to maintain concentricity with shaft centerline and the needle rollers have high speed and axial load capacity when properly assembled. The basic dimensions for thrust needle roller and cage assemblies are specified in the ISO 3031 standard and shown in Table 12-9.

Loose Needle Rollers—A complement of needle rollers used with hardened and ground shaft and housing as raceways provides the maximum radial load capacity in a minimum space. The loose rollers will operate at moderate speeds. Loose needle rollers are specified in ISO 3096 standard, and the basic dimensions are shown in Table 12-9.

TAPERED ROLLER BEARINGS

General

World standardization of metric tapered roller bearings has been attempted by ISO since 1963 when the first part of standard number ISO/R355 was issued. Eight parts of the above standard have been published to date.

A new ISO Standard for Metric Tapered Roller Bearings has now been published. This new standard will replace ISO/R355 in total.

Boundary Dimensions for Tapered Roller Bearings

Table 12-11 shows the boundary dimensions for the metric tapered roller bearings taken from the new ISO 355. The bearing corners have been specified as minimum

Table 12-7. Tolerance Limits for Needle Roller Bearings, With Cage, Machined Ring, Without Inner Ring*

BASIC OUTSIDE DIAMETER, D		ALLOWABLE DEVIATION FROM D OF SINGLE MEAN DIAMETER, D_{mp}		BASIC BORE DIAMETER UNDER NEEDLE ROLLERS F_w		ALLOWABLE DEVIATION FROM F_w		ALLOWABLE DEVIATION FROM WIDTH B	
mm		ISO R492 NORM TOL CLASS		mm		F7				
OVER	INCL	HIGH	LOW	OVER	INCL	LOW	HIGH	HIGH	LOW
18	30	+0	-10	10	10	+16	+39	+0	-120
30	50	+0	-12	18	30	+20	+41	+0	-120
50	80	+0	-15	30	50	+25	+50	+0	-120
80	120	+0	-20	50	60	+30	+60	+0	-120
120	180	+0	-25	60	80	+30	+60	+0	-150
180	250	+0	-30	80	90	+36	+71	+0	-150
250	315	+0	-35	90	120	+36	+71	+0	-200
315	400	+0	-40	120	140	+43	+83	+0	-200
				140	180	+43	+83	+0	-250
				180	200	+50	+96	+0	-250
				200	250	+50	+96	+0	-300
				250	260	+56	+108	+0	-300
				260	315	+56	+108	+0	-350

*Deviations are in micrometers.

Courtesy of Torrington.

Table 12-8. Tolerance Limits for Needle Roller Bearing Inner Rings*

BASIC BORE DIAMETER d		ALLOWABLE DEVIATION FROM d OF SINGLE MEAN DIAMETER, d_{mp}		ALLOWABLE DEVIATION FROM OUTSIDE DIAMETER D OF SINGLE MEAN DIAMETER, D_{mp}		ALLOWABLE DEVIATION FROM WIDTH B	
mm		ISO R492 NORM TOL CLASS					
OVER	INCL	HIGH	LOW	HIGH	LOW	HIGH	LOW
25	10	+0	-8	-10	-21	+0	-120
10	18	+0	-8	-12	-25	+0	-120
18	30	+0	-10	-17	-33	+0	-120
30	40	+0	-12	-21	-37	+0	-120
40	50	+0	-12	-26	-45	+0	-120
50	80	+0	-15	-34	-56	+0	-150
80	100	+0	-20	-43	-65	+0	-200
100	120	+0	-20	-47	-72	+0	-200
120	140	+0	-25	-56	-81	+0	-250
140	180	+0	-25	-68	-97	+0	-250
180	250	+0	-30	-82	-114	+0	-300
250	280	+0	-35	-94	-126	+0	-300

*Deviations are in micrometers. *Courtesy of Torrington.*

to insure that fillets can be designed that will not interfere with the bearing corners.

Tolerancing

Tolerances for metric tapered roller bearings are covered in ISO/R577/1 for normal tolerances, ISO/R577/2 for tolerance classes 5 and 6, ISO 577/3 for tolerance class 4, and ISO 2349 for subunit normal tolerances and tolerance class 6.

Series Designation

The ISO Standard specifies a three-symbol dimension series designation in order of contact angle series designation, diameter series designation, and width series designation. The definitions for these three designations are shown in Table 12-10. The dimension series designation for each standard metric tapered roller bearing is shown in the boundary dimension tabulation, Table 12-11.

Bearings with similar geometrical characteristics are referred to the same dimension series. Each dimension series is designated by a combination of three symbols; the first is a numeric value for the contact angle range (angle series), the second is an alphanumeric value for the outside diameter to bore diameter relationship range (diameter series), and the third is an alphanumeric value for the width to section height relationship range (width series).

Tapered Roller Bearing Selection

Bearing boundary dimensions, as seen in Table 12-11, have been grouped by contact angle series designation and then listed within each group in ascending order of (1) bore size, (2) outside diameter, and (3) width. Figure 12-1 shows the geometric definition for the boundary dimension tabulation symbols used in Table 12-11. These symbols denote basic dimensions, except for chamfers which are shown as minimums.

The bearing rating may be calculated using the formulas shown in ISO 281 (ANSI B3.16 or AFBMA 11). It is recommended that the manufacturer be contacted for ratings and availability of specific bearings.

A typical bearing selection could be based on the dimension series designation as follows:[1]

---

[1] These example dimension series designations are bearings that were part of the metric tapered roller bearing standards proposal of a major tapered roller bearing manufacturer and are now included in the ISO 355 standard.

BEARINGS

Table 12-9. World Needle Roller Bearing Standards

STANDARD	FIGURE	NAME	\multicolumn{7}{c}{Dimensions in millimeters}

STANDARD	NAME	TYPES			DIAMETER SERIES 1 (1)		
ISO 3245 U.S.A. ANSI B3.18 JAPAN JIS GERMANY DIN 618 FRANCE NF E22-372 U.K. BS ITALY UNI AUSTRAL AS	(E) DRAWN CUP NEEDLE ROLLER BEARINGS (SERIES 1) (G) NADELLAGER MIT KAEFIG-NADELHUELSEN (F) ROULEMENTS A AIGUILLES-DOUILLES A AIGUILLES (I) CUSCINETTI RADIALI A RULLINI SENZA ANELLO INTERNO	F_W	D	WIDTH B	F_W	D	WIDTH B
		4	8	8	20	26	12 16
		5	9	9	22	28	12 16
		6	10	9	25	32	16 20
		7	11	9	28	35	16 20
		8	12	10	30	37	16 20
		9	13	10	35	42	16 20
		10	14	10	40	47	16 20
		12	16	10	45	52	16 20
		14	20	12 16	50	58	20 24
		16	22	12 16	55	63	20 24
		18	24	12 16			

TOLERANCES — Deviations in micrometres

F_W mm		$\Delta F_{wmin} = F_{wmin} - F_W$	
OVER	INCL	HIGH	LOW
3	6	+28	+10
6	10	+31	+13
10	18	+34	+16
18	30	+41	+20
30	50	+50	+25
50	70	+60	+30

The minimum roller complement bore diameter is checked with the cup mounted in a ring gage. The diameter of which is equal to the low limit of N6.

NAME	TYPES			DIAMETER SERIES 2		
(E) DRAWN CUP NEEDLE ROLLER BEARINGS (SERIES 2) (G) NADELLAGER MIT KAEFIG-NADELHUELSEN (F) ROULEMENTS A AIGUILLES-DOUILLES A AIGUILLES (I) CUSCINETTI RADIALI A RULLINI SENZE ANELLO INTERNO	F_W	D	WIDTH B	F_W	D	WIDTH B
	8	14	10 14	22	30	14 20
	9	15	10 16	25	35	18 28
	10	16	10 16	28	38	18 28
	12	18	12 16	30	40	18 28
	14	22	14 18	32	42	18 28
	15	23	14 20	35	45	20 28
	16	24	14 20	38	48	20 28
	17	25	14 20	40	50	20 28
	18	26	14 20	42	52	20 28
	20	28	14 20	45	55	20 28

STANDARD	NAME	TYPE NA			DIMENSION SERIES 21, 51 and 81 (2)					
AMERICAN TYPE ISO U.S.A. ANSI B 3.18 JAPAN JIS GERMANY DIN FRANCE NF U.K. BS ITALY UNI AUSTRAL AS	(E) MACHINED RING NEEDLE ROLLER BEARINGS WITH CAGE (G) NADELLAGER MIT KAEFIG OHNE INNENRING (F) ROULEMENTS A AIGUILLES-SANS BAGUE INTERIEURE (I) CUSCINETTI RADIALI A RULLINI SENZE ANELLO INTERNO	F_W	D	WIDTH B	F_W	D	WIDTH B	F_W	D	WIDTH B
		15	28	18 21 25	70	95	34 41 48	150	185	48 56 65
		17	30	18 21 25	75	100	34 41 48	160	195	48 56 65
		20	33	18 21 25	80	105	34 41 48	170	210	56 65 75
		25	38	18 21 25	85	110	34 41 48	180	220	56 65 75
		30	45	21 25 29	90	120	41 48 56	190	230	56 65 75
		35	50	21 25 29	95	125	41 48 56	200	240	56 65 75
		40	55	21 25 29	100	130	41 48 56	210	260	56 65 75
		45	60	21 25 29	105	135	41 48 56	220	270	71 82 95
		50	70	29 34 38	110	140	41 48 56	240	290	71 82 95
		55	75	29 34 38	120	150	48 56 65	260	310	71 82 95
		60	80	29 34 38	130	165	48 56 65	280	330	71 82 95
		65	85	29 34 38	140	175	48 56 65	300	350	71 82 95

FOR TOLERANCES SEE TABLE 12-7

STANDARD	NAME	TYPE NR			DIMENSION SERIES 21, 51 and 81					
AMERICAN TYPE ISO U.S.A. ANSI B3.18 JAPAN JIS GERMANY DIN FRANCE NF U.K. BS ITALY UNI AUSTRAL AS	(E) MACHINED INNER RINGS FOR NEEDLE ROLLER BEARINGS (G) NADELLAGER INNENRING (F) ROULEMENTS A AIGUILLES-BAGUE INTERIEURE (I) ANELLO INTERNO CUSCINETTI RADIALI A RULLINI	d	D	WIDTH B	d	D	WIDTH B			
		10	15	18.5 21.5 25.5	60	70	34.5 41.5 48.5			
		12	17	18.5 21.5 25.5	65	75	34.5 41.5 48.5			
		15	20	18.5 21.5 25.5	70	80	34.5 41.5 48.5			
		17	25	18.5 21.5 25.5	75	85	34.5 41.5 48.5			
		20	25	18.5 21.5 25.5	80	90	41.5 48.5 56.5			
		25	30	21.5 25.5 29.5	85	95	41.5 48.5 56.5			
		30	35	21.5 25.5 29.5	90	100	41.5 48.5 56.5			
		35	40	21.5 25.5 29.5	95	105	41.5 48.5 56.5			
		40	45	21.5 25.5 29.5	100	110	41.5 48.5 56.5			
		40	50	29.5 34.5 38.5	105	120	48.5 56.5 65.5			
		45	55	29.5 34.5 38.5	110	130	48.5 56.5 65.5			
		50	60	29.5 34.5 38.5	120	140	48.5 56.5 65.5			
		55	65	29.5 34.5 38.5						

FOR TOLERANCES SEE TABLE 12-8

STANDARD	NAME	TYPES			DIMENSION SERIES 49 (3)					
EUROPEAN TYPE ISO 1206,3097 U.S.A. ANSI JAPAN JIS B1536 GERMANY DIN 617 FRANCE NF E22-370 U.K. BS ITALY UNI 3081 AUSTRAL AS	(E) MACHINED RING NEEDLE ROLLER BEARINGS (G) NADELLAGER MIT KAEFIG OHNE INNENRING (F) ROULEMENTS A AIGUILLES-SANS BAGUE INTERIEURE (I) CUSCINETTI RADIALI A RULLINI SENZA ANELLO INTERNO	F_W	D	B	F_W	D	B	F_W	D	B
		5	11*	10	22	30	13	52	68	22	85	105	30
		6	12*	10	25	37	17	22	90	110	30
		7	13	10	28	39	17	58	72	22	95	115*	30
		8	15	10	30	42	17	60	75*	22	100	120	35
		9	17	10	32	45	17	63	80	25	105	125	35
		10	19	11	35	47	17	65	82*	25	110	130	35
		12	20	11	40	52	20	68	85	25	145	140	40
		14	22	13	42	55	20	70	88	25	125	150	40
		16	24	13	45	58*	20	72	90	25	135	165	43
		18	26*	13	48	62	22	75	95*	30	150	180	50
		20	28	13	50	65*	22	80	100	30	160	190	50

TOLERANCES SEE ISO 3097 STANDARD

NOTES:
1. Diameter series specified in ISO R15 standard.
2. Dimension series 21 = width series 2 and diameter series 1 (ISO R15).
3. Dimension series 49 is applicable except for those dimensions marked with an asterisk (*).

Table 12-9. (Continued) World Needle Roller Bearing Standards

STANDARD	FIGURE	NAME	Dimensions in millimeters
EUROPEAN TYPE ISO 1206 3097 U.S.A. ANSI JAPAN JIS B1536 GERMANY DIN 617 FRANCE NF U.K. BS ITALY UNI 6341 AUSTRAL AS	TOLERANCES See ISO 3097 standard	(E) Complete Needle Roller Bearings With Machined Rings (G) NADELLAGER MIT KAEFIG-MASSREIHE 49 (F) ROULEMENTS A AIGUILLES (I) CUSCINETTI RADIALI RIGIDI A RULLINI CON ANELLO INTERNO	TYPES DIMENSION SERIES 49 d / D / B / d / D / B / d / D / B 5 / 13 / 10 / 25 / 42 / 17 / 70 / 100 / 30 6 / 15 / 10 / 28 / 45 / 17 / 75 / 105 / 30 7 / 17 / 10 / 30 / 47 / 17 / 80 / 110 / 30 8 / 19 / 11 / 32 / 52 / 20 / 85 / 120 / 35 9 / 20 / 11 / 35 / 55 / 20 / 90 / 125 / 35 10 / 22 / 13 / 40 / 62 / 22 / 95 / 130 / 35 12 / 24 / 13 / 45 / 68 / 22 / 100 / 140 / 40 15 / 28 / 13 / 50 / 72 / 22 / 110 / 150 / 40 17 / 30 / 13 / 55 / 80 / 25 / 120 / 165 / 45 20 / 37 / 17 / 60 / 85 / 25 / 130 / 180 / 50 22 / 39 / 17 / 65 / 90 / 25 / 140 / 190 / 50
ISO 3030 U.S.A. ANSI JAPAN JIS GERMANY DIN 5405 FRANCE NF U.K. BS ITALY UNI AUSTRAL AS	TOLERANCES Tolerances for the cage width B_c -0.2 -0.55^{mm}	(E) RADIAL NEEDLE ROLLER AND CAGE ASSEMBLIES (IC) (G) RADIAL-NADEL-KRAENZE MIT KAEFIG.(IC) (F) ROULEMENTS A AIGUILLES-CAGES A RADIALES (I) CUSCINETTI RADIALI A RULLINI	TYPE NM DIAMETER SERIES 1C F_W / E_W / B_C / F_W / E_W / B_C / F_W / E_W / B_C 5 / 8 / 8 / -- / 20 / 24 / 10 / 13 6 / 9 / 8 / 10 / 22 / 26 / 10 / 13 7 / 10 / 8 / 10 / 25 / 29 / 10 / 13 8 / 11 / 10 / 13 / 28 / 33 / 13 / 17 9 / 12 / 10 / 13 / 30 / 35 / 13 / 17 10 / 13 / 10 / 13 / 32 / 37 / 13 / 17 12 / 15 / 10 / 13 / 35 / 40 / 13 / 17 14 / 18 / 10 / 13 / 40 / 45 / 17 / 27 15 / 19 / 10 / 13 / 45 / 50 / 17 / 27 16 / 20 / 10 / 13 / 50 / 55 / 20 / 27 17 / 21 / 10 / 13 / 55 / 61 / 20 / 30 18 / 22 / 10 / 13 / 60 / 65 / 20 / 30
ISO 3030 U.S.A. ANSI B3.18 JAPAN JIS GERMANY DIN 5405 FRANCE NF U.K. BS ITALY UNI AUSTRAL AS	TOLERANCES Tolerances for the cage width B_c -0.2 -0.55^{mm}	(E) RADIAL NEEDLE ROLLER AND CAGE ASSEMBLIES (2C, 3C, 4C) (G) RADIAL-NADEL-KRAENZE MIT KAEFIG (2C, 3C, 4C) (F) ROULEMENTS A AIGUILLES-CAGES A RADIALES (I) CUSCINETTI RADIALI A RULLINI	TYPE NM DIAMETER SERIES 2C, 3C and 4C F_W / E_W / B_C / F_W / E_W / B_C / F_W / E_W / B_C 10 / 14 / 13 / 25 / 31 / 17 / 60 / 68 / 25 12 / 16 / 13 / 26 / 34 / 17 / 66 / 73 / 30 14 / 20 / 17 / 30 / 37 / 20 / 70 / 78 / 30 15 / 21 / 17 / 32 / 39 / 20 / 75 / 83 / 30 16 / 22 / 17 / 35 / 42 / 20 / 80 / 88 / 30 17 / 23 / 17 / 40 / 48 / 25 / 85 / 93 / 30 18 / 24 / 17 / 45 / 53 / 25 / 90 / 98 / 30 20 / 26 / 17 / 50 / 58 / 25 / 95 / 103 / 30 22 / 28 / 17 / 55 / 63 / 25 / 100 / 108 / 30
ISO 3031 U.S.A. ANSI JAPAN JIS GERMANY DIN 5405 FRANCE NF E22-374 U.K. BS ITALY UNI AUSTRAL AS	TOLERANCES (1) Tolerance for cage outside diameter D_c c12 Tolerance for cage bore diameter D_{c1} E11	(E) Thrust Needle Roller And Cage Assemblies (G) AXIAL-NADEL-KRAENZE MIT KAEFIG (F) ROULEMENTS A AIGUILLES-CAGES A AXIALES (I) CUSCINETTI ASSIALI A RULLINI	D_{c1} / D_c / D_W / D_{c1} / D_c / D_W 15 / 28 / 2 / 50 / 70 / 3 16 / 29 / 2 / 55 / 78 / 3 17 / 30 / 2 / 60 / 85 / 3 18 / 31 / 2 / 65 / 90 / 3 20 / 35 / 2 / 70 / 95 / 4 25 / 42 / 2 / 75 / 100 / 4 30 / 47 / 2 / 80 / 105 / 4 35 / 52 / 2 / 85 / 110 / 4 40 / 60 / 3 / 90 / 120 / 4 45 / 65 / 3 / 100 / 135 / 4
ISO 3096 U.S.A. ANSI JAPAN JIS GERMANY DIN 5402 FRANCE NF E22-383 U.K. BS ITALY UNI AUSTRAL AS	ROUNDED END FLAT END (2) Radius limits for rounded end needle rollers $R_{min} = \frac{D_W}{2}$ and $R_{max} = \frac{L_W}{2}$	(E) Needle Rollers (G) NADEL ROLLEN (F) AIGUILLES (I) RULLINI	L_W values: 5.8, 6.8, 7.8, 9.8, 11.8, 13.8, 15.8, 17.8, 19.8, 21.8, 23.8, 25.8, 27.8, 29.8, 34.8, 39.8, 49.8 D_W: 1, 1.5, 2, 2.5, 3, 3.5, 4, 5

NOTES:
1. For tolerance limits see Section 6.
2. The radius r = 0.1 mm for all needle rollers shown.

BEARINGS

General Purpose Application (60 mm Bore)
   Dimension Series Designations: 2CC, 2CD, 2ED
Combination General Purpose and Heavy-Pinion Applications (60 mm Bore)
   Dimension Series Designation: 4FE
Pinion Applications (60 mm Bore)
   Dimension Series Designation: 5DD, 5ED
High-Speed Applications (60 mm Bore)
   Dimension Series Designation: 4CB
Steep Angle High Thrust Applications (60 mm Bore)
   Dimension Series Designation: 7FC

PLAIN BEARINGS AND SELF-ALIGNING BUSHINGS

General

The ISO technical committee TC 123 has developed several Draft International Standards (DIS) for plain bearings. Other national standards referred to in Table 12-12 do not necessarily agree with the ISO drafts yet, and they are shown for reference purposes only. The worldwide standardizing of self-aligning bushings has been assigned to TC4; and subcommittee SC7 has proposed an existing German standard for a new ISO proposal on this subject.

Plain Cylindrical Bearings with or Without Flange Made from Sintered Material

Dimensions and tolerances for plain bearings made from sintered material are specified in size range from 1 to 60 mm in ISO 2795 and shown in sizes from 3 to 60 mm in Table 12-12. It is envisaged that as far as possible the same outside diameters will be recommended for all types of plain bearings. The outside diameters for thin wall series are shown in columns marked D1, the normal series in columns marked D2, and the preferred plain bearing lengths are shown under L. A further reduction of the preferred nominal diameters (d) in the ISO standard is recommended by this text (see Preferred Numbers, Table 4-5).

Plain Spherical Bearings Made from Sintered Material

Dimensions and tolerances for spherical bearings made from sintered material are covered in the ISO 2795 standard in sizes from 1 to 20 mm nominal diameters. The largest sizes are shown in Table 12-12.

Plain Wrapped Bearings (Bushes)

Dimensions and tolerances for plain wrapped bearings in nominal housing diameters from 6 to 150 mm are specified in the ISO 3547 standard, and the recommended dimensions for bearing housing sizes from 12 to 60 mm are shown in Table 12-12. Other details concerning tolerances, chamfers, and lubrication grooves are shown in the ISO standard.

Plain Thin-Walled Half Bearings

The ISO 3548 draft lays down the main dimensions and tolerances for thin-walled half bearings for use in reciprocating machinery in the housing size range from 20 to 500 mm. The housing and inside diameters for sizes from 25 to 120 mm are shown in Table 12-12. For recommended tolerances, notches, and lubrication grooves, see ISO 3548.

Self-Aligning Bushings

The ISO draft proposal for self-aligning bushings is based on the DIN 648 standard, and it is an extension of existing European aircraft standards.

The following series are covered in the proposal and shown in Table 12-13.
   Series E—Basic series (nominal diameters from 4 to 300 mm)
   Series G—Heavy series (nominal diameters from 4 to 280 mm)

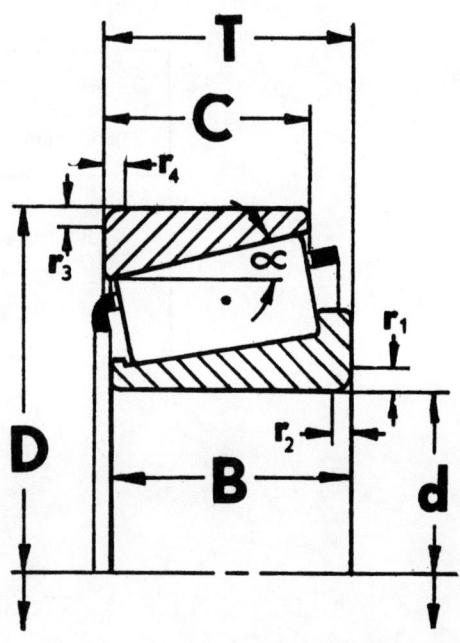

Courtesy of Timken.

Fig. 12-1. Symbols for the dimensions used in Table 12-11: α = bearing contact angle, nominal; d = bearing bore diameter, nominal; D = bearing outside diameter, nominal; T = bearing width, nominal; B = cone width, nominal; C = cup width, nominal; r_1 = cone back face chamfer height; r_2 = cone back face chamfer width; r_3 = cup back face chamfer height; r_4 = cup back face chamfer width; r_1 smin = r_2 smin = cone back face smallest single chamfer height and width; r_3 smin = r_4 smin = cup back face smallest single chamfer height and width.

Table 12-10. Angle, Diameter, and Width Series Designations (ISO 355)

Angle Series Designation	α Over	α Incl.
1	Reserved For	Future Use
2	10°–00′	13°–52′
3	13°–52′	15°–59′
4	15°–59′	18°–55′
5	18°–55′	23°–00′
6	23°–00′	27°–00′
7	27°–00′	30°–00′

Diameter Series Designation	$\frac{T}{(D-d)^{0.95}}$ Over	$\frac{T}{(D-d)^{0.95}}$ Incl.
A	Reserved For	Future Use
B	3.40	3.80
C	3.80	4.40
D	4.40	4.70
E	4.70	5.00
F	5.00	5.60
G	5.60	7.00

Width Series Designation	$\frac{D}{d^{0.77}}$ Over	$\frac{D}{d^{0.77}}$ Incl.
A	Reserved For	Future Use
B	0.50	0.68
C	0.68	0.80
D	0.80	0.88
E	0.88	1.00

BEARINGS

Table 12-11. World Tapered Roller Bearing Standards

STANDARD		NAME
(1) ISO 355		(E) TAPERED ROLLER BRG - GENERAL PURPOSE APPLICATION
U.S.A.	ANSI	
JAPAN	JIS B1534	(G) KEGELROLLENLAGER - LAGERREIHE 2
GERMANY	DIN 720	
FRANCE	NF R95-506	(F) ROULEMENTS A ROULEAUX CONIQUES - SERIE 2
U.K.	BS 3134	
ITALY	UNI 6042	(I) CUSCINETTI RADIALI A RULLI CONICI - SERIE 2
AUSTRAL	AS	

ANGLE SERIES DESIGNATION (2)

Desig.	d	D	B	C	T	$r_1 \, \& \, r_2$	$r_3 \, \& \, r_4$
15 2FB	15	42	13	11	14.25	1	1
17 2DB	17	40	12	11	13.25	1	1
17 2DD	17	40	16	14	17.25	1	1
17 2FB	17	47	14	12	15.25	1	1
17 2FD	17	47	19	16	20.25	1	1
20 2BD	20	37	12	9	12	0.3	0.3
20 2DC	20	45	17.5	13.5	17	1	1
20 2DB	20	47	14	12	15.25	1	1
20 2DD	20	47	18	15	19.25	1	1
20 2ED	20	50	22	18.5	22	2	1.5
20 2FB	20	52	15	13	16.25	1.5	1.5
20 2FD	20	52	21	18	22.25	1.5	1.5
22 2BC	22	40	12	9	12	0.3	0.3
22 2CC	22	47	17.5	13.5	17	1	1
22 2ED	22	52	22	18.5	22	2	1.5
25 2BD	25	42	12	9	12	0.3	0.3
25 2CE	25	47	14	17	17	0.6	0.6
25 2CC	25	50	17.5	13.5	17	1.5	1
25 2CD	25	52	18	16	19.25	1	1
25 2DE	25	52	22	18	22	1	1
25 2EE	25	58	26	21	26	2	1.5
25 2FB	25	62	17	15	18.25	1.5	1.5
25 2FD	25	62	24	20	25.25	1.5	1.5
28 2BD	28	45	12	9	12	0.3	0.3
28 2CD	28	55	19.5	15.5	19	1.5	1.5
28 2DE	28	58	24	19	24	1	1
28 2ED	28	65	27	22	27	2	2
30 2BD	30	47	12	9	12	0.3	0.3
30 2CE	30	55	20	16	20	1	1
30 2CD	30	58	19.5	15.5	19	1.5	1.5
30 2DE	30	62	25	19.5	25	1	1
30 2EE	30	68	29	24	29	2	2

ANGLE SERIES DESIGNATION (2) (CONT.)

Desig.	d	D	B	C	T	$r_1 \, \& \, r_2$	$r_3 \, \& \, r_4$
30 2FB	30	72	19	16	20.75	1.5	1.5
30 2FD	30	72	27	23	28.75	1.5	1.5
32 2BD	32	52	15	10	14	0.6	0.6
32 2CD	32	62	21	17	21	1.5	1.5
32 2DE	32	65	26	20.5	26	1	1
32 2ED	32	72	29	24	29	2	2
35 2BD	35	55	14	11.5	14	0.6	0.6
35 2CE	35	62	21	17	21	1	1
35 2DD	35	68	23	18.5	23	2	2
35 2DE	35	72	28	22	28	1.5	1.5
35 2EE	35	78	32.5	27	33	2.5	2
35 2FB	35	80	21	18	22.75	2	1.5
35 2FE	35	80	31	25	32.75	2	1.5
40 2BC	40	62	15	12	15	0.6	0.6
40 2BE	40	68	22	18	22	1	1
40 2CD	40	75	24	19.5	24	2	2
40 2CE	40	75	26	20.5	26	1.5	1.5
40 2DE	40	80	32	25	32	1.5	1.5
40 2EE	40	85	32.5	28	33	2.5	2
40 2FB	40	90	23	20	25.25	2	1.5
40 2FD	40	90	33	27	35.25	2	1.5
45 2BC	45	68	15	12	15	0.6	0.6
45 2CE	45	75	24	19	24	1	1
45 2CD	45	80	24	19.5	24	2	2
45 2ED	45	95	35	30	36	2.5	2.5
45 2FB	45	100	25	22	27.25	2	1.5
45 2FD	45	100	36	30	38.25	2	1.5
50 2BC	50	72	15	12	15	0.6	0.6
50 2CE	50	80	24	19	24	1	1
50 2CD	50	85	24	19.5	24	2	2
50 2ED	50	100	35	30	36	2.5	2.5
50 2FB	50	110	27	23	29.25	2.5	2
50 2FD	50	110	40	33	42.25	2.5	2
55 2BC	55	80	17	14	17	1	1
55 2CC	55	85	18.5	14	18	2	2
55 2CE	55	90	27	21	27	1.5	1.5
55 2CD	55	95	27	21.5	27	2	2
55 2ED	55	110	39	32	39	2.5	2.5
55 2FB	55	120	29	25	31.5	2.5	2
55 2FD	55	120	43	35	45.5	2.5	2
60 2BC	60	85	17	14	17	1	1
60 2CC	60	90	18.5	14	18	2	2
60 2CE	60	95	27	21	27	1.5	1.5
60 2CD	60	100	27	21.5	27	2	2
60 2EE	60	115	39	33	40	2.5	2.5
60 2FB	60	130	31	26	33.5	3	2.5
60 2FD	60	130	46	37	48.5	3	2.5
65 2BC	65	90	17	14	17	1	1
65 2CC	65	100	22	17.5	22	2	2
65 2CE	65	100	27	21	27	1.5	1.5

STANDARD		NAME
(1) ISO 355		(E) TAPERED ROLLER BEARING- GENERAL PURPOSE APPLICATION
U.S.A.	ANSI	
JAPAN	JIS B1534	(G) KEGELROLLENLAGER - LAGERREIHE 3
GERMANY	DIN 720	
FRANCE	NF R95-506	(F) ROULEMENTS A ROULEAUX CONIQUES - SERIE 3
U.K.	BS 3134	
ITALY	UNI 6042	(I) CUSCINETTI RADIALI A RULLI CONICI - SERIE 3
AUSTRAL	AS	

ANGLE SERIES DESIGNATION (3)

Desig	d	D	B	C	T	$r_1 \, \& \, r_2$	$r_3 \, \& \, r_4$
20 3CC	20	42	15	12	15	0.6	0.6
22 3CC	22	44	15	11.5	15	0.6	0.6
25 3CC	25	52	15	13	16.25	1	1
30 3DB	30	62	16	14	17.25	1	1
30 3DC	30	62	20	17	21.25	1	1
32 3DB	32	65	17	15	18.25	1	1
35 3DB	35	72	17	15	18.25	1.5	1.5
35 3DC	35	72	23	19	24.25	1.5	1.5
40 3CD	40	68	19	14.5	19	1	1
40 3DB	40	80	18	16	19.75	1.5	1.5
40 3DC	40	80	23	19	24.75	1.5	1.5
45 3CC	45	75	20	15.5	20	1	1
45 3CE	45	80	26	20.5	26	1.5	1.5
45 3DB	45	85	19	16	20.75	1.5	1.5

ANGLE SERIES DESIGNATION (3) (CONT.)

Desig.	d	D	B	C	T	$r_1 \, \& \, r_2$	$r_3 \, \& \, r_4$
45 3DC	45	85	23	19	24.75	1.5	1.5
45 3DE	45	85	32	25	32	1.5	1.5
50 3CC	50	80	20	15.5	20	1	1
50 3CE	50	85	26	20	26	1.5	1.5
50 3DB	50	90	20	17	21.75	1.5	1.5
50 3DC	50	90	23	19	24.75	1.5	1.5
50 3DE	50	90	32	25	32	1.5	1.5
55 3CC	55	90	23	17.5	23	1.5	1.5
55 3CE	55	95	30	23	30	1.5	1.5
55 3DB	55	100	21	18	22.75	2	1.5
55 3DC	55	100	25	21	26.75	2	1.5
55 3DE	55	100	35	27	35	2	1.5
60 3CE	60	100	30	23	30	1.5	1.5
60 3DB	60	110	22	19	23.75	2	1.5
60 3EC	60	110	28	24	29.75	2	1.5
60 3EE	60	110	38	29	38	2	1.5
65 3DE	65	110	34	26.5	34	1.5	1.5
65 3EB	65	120	23	20	24.75	2	1.5
65 3EC	65	120	31	27	32.75	2	1.5
65 3EE	65	120	41	32	41	2	1.5
65 3FE	65	135	51	43	52	5	3
70 3DE	70	120	37	29	37	1.5	1.5
70 3EB	70	125	24	21	26.25	2	1.5
70 3EC	70	125	31	27	33.25	2	1.5
70 3EE	70	125	41	32	41	2	1.5
75 3DE	75	125	37	29	37	1.5	1.5
75 3EE	75	130	41	31	41	2	1.5
75 3FE	75	145	51	43	52	5	3
80 3CC	80	125	29	22	29	1.5	1.5
80 3DE	80	130	37	29	37	2	1.5
80 3EB	80	140	26	22	28.25	2.5	2
80 3EC	80	140	33	28	35.25	2.5	2

Courtesy of Timken.

NOTE:
The tapered roller bearings shown in this table are those included in the ISO 355 standard. The national standards listed do not agree with the ISO standard yet.

Table 12-11 (Continued). World Tapered Roller Bearings Standards

STANDARD	FIGURE	NAME	DIMENSIONS In Millimeters
(1) ISO 355 U.S.A. ANSI JAPAN JIS B1534 GERMANY DIN 720 FRANCE NF R95-506 U.K. BS 3134 ITALY UNI AUSTRAL AS		(E) TAPERED ROLLER BEARING-SPECIAL APPLICATION (G) KEGELROLLENLAGER - LAGERREIHE 4 (F) ROULEMENTS A ROULEAUX CONIQUES - SERIE 4 (I) CUSCINETTI RADIALI A RULLI CONICI - SERIE 4	ANGLE SERIES DESIGNATION (4) Desig. d D B C T $r_1 \& r_2$ $r_3 \& r_4$ 20 4DB 20 45 14 10 14 1 1 22 4CB 22 47 14 10 14 1 1 25 4CC 25 47 15 11.5 15 0.6 0.6 25 4CB 25 50 14 10 14 1 1 28 4CC 28 52 16 12 16 1 1 28 4CB 28 55 14.5 11 15 1 1 30 4CC 30 55 17 13 17 1 1 30 4CB 30 60 16.5 12.5 17 1 1 32 4CC 32 58 17 13 17 1 1 32 4DB 32 65 17.5 13.5 18 1 1 35 4CC 35 62 18 14 18 1 1 35 4DB 35 70 18 14 19 1 1 40 4DB 40 75 18 14 19 1 1 45 4DB 45 85 20 15.5 21 2 2 50 4DB 50 90 20 15.5 21 2 2 50 4FE 50 105 40 34 41 4 2.5 55 4CB 55 95 20 15.5 21 2 2 55 4FE 55 115 42 37 44 5 2.5 60 4CC 60 95 23 17.5 23 1.5 1.5 60 4CB 60 100 20 15.5 21 2 2 60 4FE 60 125 46 40 48 5 2.5 65 4CC 65 100 23 17.5 23 1.5 1.5 65 4CB 65 105 20 15.5 21 2 2 70 4CB 70 110 20 15.5 21 2 2 70 4CC 70 110 25 19 25 1.5 1.5 70 4FE 70 140 51 43 52 5 3 75 4CB 75 115 20 15.5 21 2 2 75 4CC 75 115 25 19 25 1.5 1.5 75 4DB 75 130 25 22 27.25 2 1.5 75 4DC 75 130 31 27 33.25 2 1.5
(1) ISO 355 U.S.A. ANSI JAPAN JIS B1534 GERMANY DIN 720 FRANCE NF R95-506 U.K. BS 3134 ITALY UNI AUSTRAL AS		(E) TAPERED ROLLER BEARING-PINION APPLICATION (G) KEGELROLLENLAGER - LAGERREIHE 5 (F) ROULEMENTS A ROULEAUX CONIQUES - SERIE 5 (I) CUSCINETTI RADIALI A RULLI CONICI - SERIE 5	ANGLE SERIES DESIGNATION (5) Desig. d D B C T $r_1 \& r_2$ $r_3 \& r_4$ 20 5DD 20 47 18 15 19.25 1 1 25 5CD 25 52 18 15 19.25 1 1 28 5DD 28 58 19 16 20.25 1 1 30 5DC 30 62 20 17 21.25 1 1 30 5FD 30 72 27 23 28.75 1.5 1.5 32 5DC 32 65 21.5 17 22 1 1 32 5FD 32 75 28 23 29.75 1.5 1.5 35 5DC 35 72 23 19 24.25 1.5 1.5 35 5FE 35 80 31 25 32.75 2 1.5 40 5DC 40 80 23 19 24.75 1.5 1.5 40 5DD 40 80 26.5 21.5 27 4 2 40 5FD 40 90 33 27 35.25 2 1.5 45 5DC 45 85 23 19 24.75 1.5 1.5 45 5ED 45 90 31 26 32 4 2 45 5FD 45 100 36 30 38.25 2 1.5 50 5DC 50 90 23 18 24.75 1.5 1.5 50 5ED 50 100 34.5 29 36 4 2 50 5FD 50 110 40 33 42.25 2.5 2 55 5DD 55 100 28.5 24 30 4 2.5 55 5ED 55 105 34.5 29 36 4 2.5 55 5FD 55 120 43 35 45.5 2.5 2 60 5DD 60 110 32 27 34 4 2.5 60 5ED 60 115 38 31 39 4 2.5 60 5FD 60 130 46 37 48.5 3 2.5 65 5DD 65 115 32 27 34 4 2.5 65 5ED 65 120 38 31 39 4 2.5 65 5GD 65 140 48 39 51 3 2.5 70 5DD 70 125 34.5 30 37 4 2.5 70 5ED 70 130 40 34 42 4 2.5 70 5GD 70 150 51 42 54 3 2.5
(1,2) ISO 355 U.S.A. ANSI JAPAN JIS GERMANY DIN FRANCE NF U.K. BS ITALY UNI AUSTRAL AS		(E) TAPERED ROLLER BEARING-STEEP ANGLE HIGH THRUST APPLICATIONS (G) KEGELROLLENLAGER - LAGERREIHE 7 (F) ROULEMENTS A ROULEAUX CONIQUES - SERIE 7 (I) CUSCINETTI ASSIALI A RULLI CONICI - SERIE 7	ANGLE SERIES DESIGNATION (7) Desig. d D B C T $r_1 \& r_2$ $r_3 \& r_4$ 25 7FB 25 62 17 13 18.25 1.5 1.5 30 7FB 30 72 19 14 20.75 1.5 1.5 35 7FB 35 80 21 15 22.75 2 1.5 40 7FB 40 90 23 17 25.25 2 1.5 45 7FC 45 95 26.5 20 29 2.5 2.5 45 7FB 45 100 25 18 27.25 2 1.5 50 7FC 50 105 29 22 32 3 3 50 7FB 50 110 27 19 29.25 2.5 2 55 7FC 55 115 31 23.5 34 3 3 55 7FB 55 120 29 21 31.5 2.5 2 60 7FC 60 125 33.5 26 37 3 3 60 7FB 60 130 31 22 33.5 3 2.5 65 7FC 65 130 33.5 26 37 3 3 65 7GB 65 140 33 23 36 3 2.5

Courtesy of Timken.

NOTES:
1. The tapered roller bearings shown in this table are those included in the ISO 355 standard. The national standards listed do not agree with the ISO standard yet.
2. Self-aligning tapered roller thrust bearings to existing metric standards are shown in Table 12-5.

BEARINGS

Table 12-12. World Plain Bearing Standards

STANDARD	FIGURE	NAME	Dimensions in millimeters
ISO 2795 U.S.A. ANSI JAPAN JIS GERMANY DIN 1850 FRANCE NF U.K. BS 4480 ITALY UNI AUSTRAL AS	L, d, D1, D2 D1 = Normal Series D2 = Thin Series TOLERANCES (2) Housing: H7 Fitted bore size: H7 Length: j_s13 Insertion pin: m5 Concentricity (full indicator movement): IT9	(E) Plain cylindrical bearings made from sintered material (G) Buchsen fuer Gleitlager aus Sinter Metall (F) PALIERS LISSES-COUSSINETS FRITTÉS (I) CUSCINETTI CILINDRICI SEMPLICI	d D1 D2 L(1) d D1 D2 L(1) 3 6 5 3 4 22 28 27 15 20 25 30 4 8 7 3 4 6 25 32 30 20 25 30 35 5 9 8 4 5 8 28 36 33 20 25 30 40 6 10 9 4 6 10 30 38 35 20 25 30 40 7 11 10 5 8 10 32 40 38 20 25 30 40 8 12 11 6 8 12 35 45 41 25 35 40 50 9 14 12 6 10 14 38 48 44 25 35 45 55 10 16 14 8 10 16 40 50 46 30 40 50 60 12 18 16 8 12 20 42 52 48 30 40 50 60 14 20 18 10 14 20 45 55 51 35 45 55 65 15 21 19 10 15 25 48 58 55 35 50 70 16 22 20 12 16 25 50 60 58 35 50 70 18 24 22 12 18 30 55 65 63 40 55 70 20 26 25 15 20 25 30 60 72 68 50 60 70
ISO 2795 U.S.A. ANSI JAPAN JIS GERMANY DIN 1850 FRANCE NF U.K. BS 4480 ITALY UNI AUSTRAL AS	D1, D2, L, T, d TOLERANCES Housing: H7 Fitted bore size: H7 Flange diameter, thickness and length: j_s13 Insertion pin: m5 Concentricity (full indicator movement): IT9	(E) Plain flanged bearings made from sintered material (G) Buchsen mit Bund fuer Gleitlager aus Sinter Metall (F) PALIERS LISSES-COUSSINETS FRITTÉS (I) CUSCINETTI SEMPLICI CON FLANGIA	d D1 D2 L T d D1 D2 L T 3 6 9 4 1.5 22 28 34 15 20 25 30 3 4 8 12 3 4 6 2 25 32 39 20 25 30 3.5 5 9 13 4 5 8 2 28 36 44 20 25 30 4 6 10 14 4 6 10 2 30 38 46 20 25 30 4 7 11 15 5 8 10 2 32 40 48 20 25 30 4 8 12 16 6 8 12 2 35 45 55 25 35 40 5 9 14 19 6 10 14 2.5 38 48 58 25 35 45 5 10 16 22 8 10 16 3 40 50 60 30 40 50 5 12 18 24 8 12 20 3 42 52 62 30 40 50 5 14 20 26 10 14 20 3 45 55 65 35 45 55 5 15 21 27 10 15 25 3 48 58 68 35 50 5 16 22 28 12 16 25 3 50 60 70 35 50 5 18 24 30 12 18 30 3 55 65 75 40 55 5 20 26 32 15 20 25 30 3 60 72 84 50 60 6
ISO 2795 U.S.A. ANSI JAPAN JIS GERMANY DIN FRANCE NF U.K. BS 4480 ITALY UNI AUSTRAL AS	Spherical D, d, L TOLERANCES (3)(4) Bore: H7 Spherical diameter: h11 Length: j_s13	(E) Plain spherical bearings made from sintered material (G) BUCHSEN FUER GLEITLAGER-KUGEL (F) PALIERS LISSES-COUSSINETS FRITTÉS-SPHERIQUE (I) CUSCINETTI SPERICI	d D L d D L 2 5 3 9 18 12 2.5 6 4 10 22 14 3 8 6 12 22 15 4 10 8 14 24 17 5 12 9 15 27 20 6 14 10 16 28 20 7 16 11 18 30 20 8 16 11 20 36 25
ISO 3548 U.S.A. ANSI JAPAN JIS GERMANY DIN FRANCE NF U.K. BS ITALY UNI AUSTRAL AS	L, d, D	(E) Plain thin-walled half bearings (G) HALBRUND BUCHSEN FUER GLEITLAGER (F) PALIERS LISSES-DEMI-COUSSINETS MINCES (I) META-CUSCINETTI FINI	D d D d 25 22 21.5 56 52.5 52 51 26 23 22.5 60 56.5 56 55 28 25 24.5 63 59.5 59 58 30 27 26.5 67 63 62 61 32 29 28.5 28 71 67 66 65 34 31 30.5 30 75 71 70 69 36 33 32.5 32 80 76 75 74 38 35 34.5 34 85 81 80 79 40 36.5 35 90 85 84 83 42 38.5 38 37 95 90 89 88 45 41.5 41 40 100 95 94 93 48 44.5 44 43 105 100 99 98 50 46.5 46 45 110 105 104 103 53 49.5 49 48 120 115 114 113
ISO 3547 U.S.A. ANSI JAPAN JIS GERMANY DIN 1498 FRANCE NF U.K. BS 1131 ITALY UNI AUSTRAL AS	L, D, d TOLERANCES Width L ± 0.25	(E) Plain wrapped (bushes) bearings (G) EINSPANN BUCHSEN FUER LAGERUNGEN (F) PALIERS LISSES-BAGUES ROULEES (I) CUSCINETTI SEMPLICI	D d L D d L 12 10.5 10 10 15 20 28 25 24 20 25 30 40 13 11.5 11 10 15 20 30 27 26 20 25 30 40 14 12.5 12 10 15 20 32 29 28 20 25 30 40 15 13 12 10 15 20 34 31 30 20 25 30 40 16 14 13 15 20 25 36 33 32 25 40 50 17 15 14 15 20 25 38 35 34 25 40 50 18 16 15 15 20 25 40 37 36 25 40 50 19 17 16 15 20 25 42 39 38 25 40 50 20 18 17 15 20 25 45 42 41 40 25 40 50 21 19 18 15 20 25 30 48 45 44 43 25 40 50 22 20 19 15 20 25 30 50 47 46 45 25 40 60 24 22 21 15 20 25 30 53 50 49 48 25 40 60 25 23 22 15 20 25 30 56 52 51 50 25 40 60 26 23 22 20 25 30 40 60 56 55 54 30 50 70

NOTES:
1. For the thin series, from diameter 20 mm (included), the last value for length is not applicable.
2. For tolerance limits, see Section 6.
3. Tolerance for housing diameter should normally be H10 but this depends on the method of assembly. Where an easier fit is preferred for lighter self-alignment, G10 is suggested.
4. A cylindrical surface is permissible on the sphere at the center of the bearing length. The diameter of the resulting cylinder shall be not less than 75% of the spherical diameter.

Table 12-13. World Self-Aligning Bushing Standards

STANDARD	NAME/FIGURE	TABLE DIMENSIONS IN MILLIMETERS
ISO 6124 U.S.A. ANSI JAPAN JIS GERMANY DIN 648 FRANCE NF U.K. BS ITALY UNI AUSTRALIA AS	(E) SPHERICAL PLAIN BEARINGS (BASIC SERIES E) (G) GELENKLAGER (MASSREIHE E) (F) SPHERIQUE PALIERS LISSES (SERIE E) (I) CUSCINETTI SFERICI (SERIE E) FOR TOLERANCES SEE ISO 6125 STANDARD	**BASIC SERIES E** d / D / B / C / d_1 min / r_1 min / r_2 min / α approx 4 / 12 / 5 / 3 / 6 / 0.3 / 0.3 / 16° 5 / 14 / 6 / 4 / 7 / 0.3 / 0.3 / 13° 6 / 14 / 6 / 4 / 8 / 0.3 / 0.3 / 13° 8 / 16 / 8 / 5 / 10 / 0.3 / 0.3 / 15° 10 / 19 / 9 / 6 / 13 / 0.6 / 0.3 / 12° 12 / 22 / 10 / 7 / 15 / 0.6 / 0.6 / 10° 14 / 26 / 12 / 9 / 18 / 0.6 / 0.6 / 8° 15 / 26 / 12 / 9 / 18 / 0.6 / 0.6 / 8° 16 / 30 / 14 / 10 / 20 / 0.6 / 0.6 / 10° 17 / 30 / 14 / 10 / 20 / 0.6 / 0.6 / 10° 20 / 35 / 16 / 12 / 24 / 0.6 / 0.6 / 9° 24 / 42 / 20 / 16 / 29 / 0.6 / 0.6 / 7° 25 / 42 / 20 / 16 / 29 / 0.6 / 0.6 / 7° 30 / 47 / 22 / 18 / 34 / 0.6 / 0.6 / 6° 35 / 55 / 25 / 20 / 39 / 0.6 / 1.0 / 6° 40 / 62 / 28 / 22 / 45 / 0.6 / 1.0 / 7° 45 / 68 / 32 / 25 / 50 / 0.6 / 1.0 / 7° 50 / 75 / 35 / 28 / 55 / 0.6 / 1.0 / 6° 60 / 90 / 44 / 36 / 66 / 1.0 / 1.0 / 6° 70 / 105 / 49 / 40 / 77 / 1.0 / 1.0 / 6° 80 / 120 / 55 / 45 / 88 / 1.0 / 1.0 / 6° 90 / 130 / 60 / 50 / 98 / 1.0 / 1.0 / 5° 100 / 150 / 70 / 55 / 109 / 1.0 / 1.0 / 7° 110 / 160 / 70 / 55 / 120 / 1.0 / 1.0 / 6° 120 / 180 / 85 / 70 / 130 / 1.0 / 1.0 / 6° 140 / 210 / 90 / 70 / 150 / 1.0 / 1.0 / 7° 160 / 230 / 105 / 80 / 170 / 1.0 / 1.0 / 8° 180 / 260 / 105 / 80 / 192 / 1.1 / 1.1 / 6° 200 / 290 / 130 / 100 / 212 / 1.1 / 1.1 / 7° 220 / 320 / 135 / 100 / 238 / 1.1 / 1.1 / 7° 240 / 340 / 140 / 100 / 265 / 1.1 / 1.1 / 8° 260 / 370 / 150 / 110 / 285 / 1.1 / 1.1 / 7° 280 / 400 / 155 / 120 / 310 / 1.1 / 1.1 / 6° 300 / 430 / 165 / 120 / 330 / 1.1 / 1.1 / 7°
ISO 6124 U.S.A. ANSI JAPAN JIS GERMANY DIN 648 FRANCE NF U.K. BS ITALY UNI AUSTRALIA AS	(E) SPHERICAL PLAIN BEARINGS (HEAVY SERIES G) (G) GELENKLAGER (MASSREIHE G) (F) SPHERIQUE PALIERS LISSES (SERIE G) (I) CUSCINETTI SFERICI (SERIE G) FOR TOLERANCES SEE ISO 6125 STANDARD	**HEAVY SERIES G** d / D / B / C / d_1 min / r_1 min / r_2 min / α approx 4 / 14 / 7 / 4 / 7 / 0.3 / 0.3 / 20° 5 / 16 / 9 / 5 / 8 / 0.3 / 0.3 / 21° 6 / 16 / 9 / 5 / 9 / 0.3 / 0.3 / 21° 8 / 19 / 11 / 6 / 11 / 0.3 / 0.3 / 21° 10 / 22 / 12 / 7 / 13 / 0.6 / 0.6 / 18° 12 / 26 / 15 / 9 / 16 / 0.6 / 0.6 / 18° 14 / 30 / 16 / 10 / 19 / 0.6 / 0.6 / 16° 15 / 30 / 16 / 10 / 19 / 0.6 / 0.6 / 16° 16 / 35 / 20 / 12 / 21 / 0.6 / 0.6 / 19° 17 / 35 / 20 / 12 / 21 / 0.6 / 0.6 / 19° 20 / 42 / 25 / 16 / 24 / 0.6 / 0.6 / 17° 24 / 47 / 28 / 18 / 29 / 0.6 / 0.6 / 17° 25 / 47 / 28 / 18 / 29 / 0.6 / 0.6 / 17° 30 / 55 / 32 / 20 / 34 / 0.6 / 1.0 / 17° 35 / 62 / 35 / 22 / 39 / 0.6 / 1.0 / 16° 40 / 68 / 40 / 25 / 44 / 0.6 / 1.0 / 17° 45 / 75 / 43 / 28 / 50 / 0.6 / 1.0 / 15° 50 / 90 / 56 / 36 / 57 / 0.6 / 1.0 / 17° 60 / 105 / 63 / 40 / 67 / 1.0 / 1.0 / 17° 70 / 120 / 70 / 45 / 77 / 1.0 / 1.0 / 16° 80 / 130 / 75 / 50 / 87 / 1.0 / 1.0 / 14° 90 / 150 / 85 / 55 / 98 / 1.0 / 1.0 / 15° 100 / 160 / 85 / 55 / 110 / 1.0 / 1.0 / 14° 110 / 180 / 100 / 70 / 122 / 1.0 / 1.0 / 12° 120 / 210 / 115 / 70 / 132 / 1.0 / 1.0 / 16° 140 / 230 / 130 / 80 / 151 / 1.0 / 1.0 / 16° 160 / 260 / 135 / 80 / 176 / 1.0 / 1.1 / 16° 180 / 290 / 155 / 100 / 196 / 1.1 / 1.1 / 14° 200 / 320 / 165 / 100 / 220 / 1.1 / 1.1 / 15° 220 / 340 / 175 / 100 / 243 / 1.1 / 1.1 / 16° 240 / 370 / 190 / 110 / 263 / 1.1 / 1.1 / 15° 260 / 400 / 205 / 120 / 283 / 1.1 / 1.1 / 15° 280 / 430 / 210 / 120 / 310 / 1.1 / 1.1 / 15°

BEARINGS

RELATED ISO STANDARDS

Roller Bearings (TC4)

ISO/R 15/I-1968	Rolling bearings—Radial bearings—Boundary dimensions—General plan—Diameter series 8, 9, 0, 1, 2, 3, and 4, 2nd Edition
ISO/R 15/II-1970	Rolling bearings—Radial bearings—Boundary dimensions—General plan—Part 2: Diameter series 7
ISO/R 76-1958	Ball and roller bearings—Methods of evaluating static load ratings
ISO/R 104-1966	Rolling bearings—Thrust bearings with flat seats—Boundary dimensions, 2nd Edition
ISO/R 113-1969	Rolling bearings—Accessories, 2nd Edition
ISO/R 199/I-1961	Rolling bearings—Thrust ball bearings with flat seats—Normal tolerances
ISO/R 199/II-1968	Rolling bearings—Thrust ball bearings with flat seats—Tolerances—Part 2: Tolerances classes 6, 5, and 4
ISO/R 200-1961	Rolling bearings—Internal clearance in unloaded bearings—Definitions
ISO/R 201-1961	Rolling bearings—Radial internal clearance in unloaded radial groove-type ball bearings with cylindrical bore—Values
ISO/R 246-1962	Rolling bearings—Cylindrical roller bearings—Separate thrust collars—Boundary dimensions
ISO 281/I-1977	Rolling bearings—Methods of evaluating dynamic load ratings
ISO/R 300/I-1963**	ISO identification code for rolling bearings—Group I: Radial ball and roller bearings—Group II: Thrust ball and roller bearings—Group III: Tapered roller bearings, metric series
ISO/R 300/II-1965**	ISO identification code for rolling bearings—Group IV: Tapered roller bearings, inch series
ISO/R 300/III**-1968	ISO identification code for rolling bearings—Group V: Air frame bearings.
ISO/R 355/I-1963*	Rolling bearings—Tapered roller bearings boundary dimensions
ISO/R 355/II-1965*	Rolling bearings—Tapered roller bearings boundary dimensions
ISO/R 355/III*-1967	Rolling bearings—Tapered roller bearings boundary dimensions—Metric series—Diameter series 9 and 0
ISO/R 355/IV*-1968	Rolling bearings—Tapered roller bearings boundary dimensions—Sub-units—Inch series
ISO/R 355/V-1969*	Rolling bearings—Tapered roller bearings boundary dimensions—Sub-units—Metric series
ISO/R 355/VI* 1970	Rolling bearings—Tapered roller bearings boundary dimensions—Metric series—Dimensions series 31 and 32
ISO/R 355/VII* 1970	Rolling bearings—Tapered roller bearings boundary dimensions—Sub-units—Metric series—Dimensions series 30, 31, and 32
ISO 355/VIII*-1973	Rolling bearings—Tapered roller bearings—Boundary dimensions—Sub-units—Metric series—Dimension series 29 and 13
ISO 464-1976	Rolling bearings—Bearings with locating snap ring—Dimensions
ISO/R 465-1965	Rolling bearings—Double-row self-aligning rolling bearings—Radial internal clearance
ISO/R 492-1966	Rolling bearings—Radial bearings—Tolerances
ISO/R 533-1966	Rolling bearings—Double row cylindrical roller bearings, type RD with tapered bore 1:12—Tolerance class 5—Special requirements
ISO/R 577/I-1967	Rolling bearings—Tapered roller bearings—Metric series—Tolerances—Part 1: Normal tolerances
ISO/R 577/II-1968	Rolling bearings—Tapered roller bearings—Metric series—Tolerances—Part 2: Tolerance classes 6 and 5
ISO/R 577/III-1973	Rolling bearings—Tapered roller bearings—Metric series—Tolerances—Tolerance Class 4
ISO 578-1973	Rolling bearings—Tapered roller bearings—Inch series—Tolerance class 4 (normal tolerance) 3, 0 and 00
ISO 582-1972	Rolling bearings—Metric series bearings—Chamfer dimension limits and maximum shaft and housing fillet radius
ISO/R 1002-1969	Rolling bearings—Airframe bearings—introduction, general, boundary dimensions tolerances, permissible load

*Replaced by the ISO 355-1977 standard
**ISO recommendations withdrawn

ISO/R 1038-1969	Rolling bearings—Cylindrical roller bearings—Radial internal clearance
ISO 1123-1976	Rolling bearings—Tapered roller bearings—Inch series—Chamfer dimension limits
ISO/R 1132-1969	Rolling bearings—Tolerances—Definitions
ISO 1160-1976	Rolling bearings—Rolling bearings for railway axle-boxes—Acceptance inspection
ISO 1206-1976	Rolling bearings—Needle roller bearings—Boundary dimensions—Metric series—Dimension series 48 and 49
ISO/R 1224-1971	Rolling bearings—Instrument precision bearings
ISO/R 1646-1970	Rolling bearings—Double-row self-aligning ball bearings—Radial internal clearance
ISO/R 1648-1971	Rolling bearings—Radial bearings with shields or seals—Outside diameter tolerances—Normal tolerance class and tolerance class 6
ISO 2264-1972	Rolling bearings—Bearings with spherical outside surface and extended inner ring width
ISO 2265-1972	Rolling bearings—Bearings with locating snap ring—Inner diameter of snap ring
ISO 2316-1973	Rolling bearings—Tapered roller bearing—Boundary dimensions—Subunits—Metric series—Outer rings with flange
ISO 2349-1973	Rolling bearings—Tapered roller bearings—Subunits—Tolerances—Metric series—Normal tolerance class and tolerance class 6—Inch series—Tolerance class 4 normal tolerance class
ISO 2982-1972	Rolling bearings—Accessories—Locknuts, narrow series and lockwashers with straight inner tab
ISO 2983-1975	Roller Bearings—Locknuts, wide series and lockwashers with bent inner tabs
ISO 3030-1974	Radial needle roller and cage assemblies—Boundary dimensions and tolerances
ISO 3031-1974	Thrust needle roller and cage assemblies—Boundary dimensions and tolerances
ISO 3096-1974	Needle roller bearings—Needle rollers—Dimensions
ISO 3097-1974	Radial needle roller bearings—Dimension series 48 and 49—Metric series—Normal tolerance class
ISO 3145-1974	Rolling bearings—Bearings with spherical outside surface and extended inner ring width—Eccentric locking collars
ISO 3228-1977	Rolling bearings—Bearings with spherical outside surface and extended inner ring width—Cast and pressed housings
ISO 3245-1974	Needle roller bearings—Drawn cup without inner ring—Metric series—Boundary dimensions and tolerances
ISO 3290-1975	Roller bearings—Bearing parts—Balls for rolling bearings
ISO DPR 5593	Vocabulary—Bearings and parts
ISO DPR 5753	Rolling bearings—Radial internal clearance
ISO DPR 6124	Rolling bearings—Spherical plain radial bearings, joint type—Boundary dimensions, dimension series E and G
ISO DPR 6125	Rolling bearings—Spherical plain radial bearings, joint type—Tolerances
ISO DPR 6126	Rolling bearings—Spherical plain bearings, rod ends—Boundary dimensions and tolerances—Dimensions series E and JK
ISO DPR 6193	Roller bearings, needle rollers—Tolerances

Plain Bearings (TC 123)

ISO 2795-1975	Plain bearings made from sintered material—Dimension and tolerances
ISO 3547-1976	Plain bearings—Wrapped bushes—Dimensions and tolerances
ISO DIS 3548	Plain bearings—Thin-walled half bearings—Dimensions and tolerances
ISO DPR 4378	Plain bearings—Terms, definitions and classification
ISO DIS 4379	Plain bearings—Solid bushes made from copper alloys—Dimensions and tolerances
ISO DPR 4380	Method of calculation of hydrodynamic thrust bearings

BEARINGS

ISO DPR 4381	Lead and tin-based alloys for plain bearings	ISO DPR 4385	Testing of the bond strength of metallic compound bearings—Destructive testing
ISO DPR 4382	Copper alloys for plain bearings		
ISO DPR 4383	Materials of wrapped bushes for plain bearings	ISO DPR 4386	Testing of the bond strength of metallic compound bearings—Nondestructive testing
ISO DPR 4384	Testing of antifriction metal		

13 Mechanical Power Transmission Systems

General

The purpose of this section is to describe world standards[1] for the most commonly used power source in industry—the electric motor—and some of the important power-transmission components such as transmission chains and sprockets, endless belt drives, metric module gearing, splines and serrations, keys and keyways. Some standards for these items have been based on the metric measuring system, others on the customary inch system. The main differences, if any, will be pointed out in this section to help facilitate worldwide interchangeability of parts. A number of ISO and IEC standards have been issued on the subject, and some of the most important recommendations as well as the ISO technical committee organizing the technical work behind each standard, are listed at the end of the section. Note that the technical committees are designated by the letters TC followed by the appropriate committee number.

Information presented in ISO and IEC standards and draft proposals have been included, but since they are subject to change, care should be exercised with some of the material presented in the draft proposals.

ELECTRIC MOTORS

Introduction

Material presented in this chapter is intended to give a brief introduction into standards related to the usage of electric motors, ISO has published several important standards on this subject, and IEC is active in developing standards for rotating electrical machines through its IEC-TC-2 working groups and committees.

Rotating Speeds

North America uses alternating electrical current with 60 Hz frequency, but most other parts of the world use 50 Hz power supply. The synchronous rotating speed for induction motors is calculated by using the following formula:

$$n = \frac{f \cdot 120 \text{ (RPM)}}{p} = \frac{\text{frequency} \cdot 120 \text{ (RPM)}}{\text{number of poles}} \quad (13\text{-}1)$$

Electric motors (other than dc) rotate with speeds depending on the frequency of the power supply, and the most commonly used output speeds produced for industry are shown in Table 13-1.

Table 13-1. Nominal Synchronous Output Speeds (RPM) for Motors Operating on a 50 Hz or 60 Hz ac Supply

POLES	SPEED (RPM) 50 Hz	SPEED (RPM) 60 Hz
2	3000	3600
4	1500	1800
6	1000	1200
8	750	900
10	600	720

Power Output Ratings

The nominal power outputs in kilowatts (kW) are specified in the IEC 72 standard and shown in Table 13-2 for output ratings from 0.06 to 250 kW.

Preferred output powers for larger electric motors in a size range from 280 to 1000 kW are specified in the IEC 72 standard.

FOOT-MOUNTED ELECTRIC MOTORS (IEC 72)

Designation of Foot-mounted Motors

Foot-mounted machines may be designated by the frame number immediately followed by the diameter of the shaft extension. When the frame number does not end

---

[1] For information about the term "standard" as used in this book, please see page 1-2.

Table 13-2. Nominal Power Outputs for Electric Motors (IEC 72)

kW Primary series	kW Secondary series [2]	hp [1]
0.06		1/12
0.09		1/8
0.12		1/6
0.18		1/4
0.25		1/3
0.37		1/2
0.55		3/4
0.75		1
1.1		1.5
1.5		2
	1.8	
2.2		3
	3	
3.7		5
	4	
5.5		7.5
	6.3	
7.5		10
	10	
11		15
	13	
15		20
	17	
18.5		25
	20	
22		30
	25	
30		40
	33	
37		50
	40	
45		60
	50	
55		75
	63	
75		100
	80	
90		125
	100	
110		150
	125	
132		175
150		200
160		220
185		250
200		270
220		300
250		350

NOTES:
1. 1 hp = 746 W.
2. To be used as intermediate values only in cases of special need.

MECHANICAL POWER TRANSMISSION SYSTEMS

with a letter, frame number and shaft diameter are separated by a dash.

Examples: 12M 28
80 − 7

When a foot-mounted machine is also provided with a flange at the driven end, the flange number may be added immediately after the shaft diameter.

Frame numbers for the various basic size ac motors are shown in Table 13-3.

Shaft Height for Electric Motors

The ISO 496 standard specifies shaft heights for driving and driven machines in all possible sizes, since it is based on the preferred numbering series R5, R10, R20, and R40 (R = Renard, see Table 4-1 in Section 4). The number of standard heights in the IEC 72 standard is limited to the standard heights in the R20 series except for the shaft height, 132 mm, taken from the R40 series and substituting the shaft heights 125 and 140 mm. The only rounded-off preferred number in the ISO 496 and IEC 72 standards is the 224 mm, replaced by the 225 mm shaft height.

The IEC 72 standard shaft heights with tolerances for electric motors are shown in Table 13-3, in the 2nd and 3rd columns, under the "H" heading.

Foot Base Dimensions and Mounting Bolt Sizes

The IEC 72 standard specifies the basic foot-mounting dimensions in millimeters and the recommended coarse-threaded metric fasteners required to anchor down the motor; the metric values are shown in Table 13-3. For inch conversion and a more complete description of the rotating electrical machines, see the IEC 72 standard.

FLANGE-MOUNTED ELECTRIC MOTORS (IEC 72)

Designation of Flange-Mounted Motors

Machines having only flange mounting may be designated by the diameter of the shaft extension and immediately followed by the flange number.

Example: 28F215

Note the different position of shaft diameter compared with foot-mounted motor designations.

Basic Dimensions for Flange-Mounted Electric Motors

Detail dimensions for flanges and mounting holes for electric motors are specified in the IEC 72 standard, and the most important sizes and tolerances for flange numbers from F55 to F1080 are shown in Table 13-4. Position of holes in mounting the flange is as shown in Table 13-4 for flanges having 4 holes, with a 45° mounting angle. For flanges with 8 holes, the mounting angle would be 22.5°. When a flange-mounted machine also has feet, the holes in the flange are spaced similarly to the above while the flange clamping surface is perpendicular to the mounting plane of the feet. Metric fasteners recommended are interchangeable with the North American preferred fasteners.

Cylindrical Shaft End Details

Parallel shaft end dimensions for foot-and flange-mounted electric motors are specified in IEC 72; the most important sizes are listed in Table 13-5. Cylindrical and conical shaft ends for driving and driven machines are also specified in the ISO 775 standard. The selected shaft sizes in the IEC 72 standard are the same as the long series specified in ISO 775. The ISO standard has many intermediate shaft diameters and a short cylindrical shaft series as well as several types of tapered shaft ends, with or without keys.

POWER TRANSMISSION CHAINS AND SPROCKETS

Most of the national and international standards on power transmission chains are soft-converted, customary inch standards. A difference between the European and the American roller chains of types specified in ISO/R 606 and 1275 exists, and transmission chains of the two types are not interchangeable. Both types are, however, covered in the ISO standards, and the American types are produced and included in several national standards in Europe and throughout the world.

A brief summary of the most important standards for chains is shown in Table 13-6. A selection guide for chains, indicating the ISO designations and the most important mechanical data for each type, is also presented in the table.

Additional Information Relating to ISO Standards

ISO/R 606: The North American standard ANSI B29.1 has been modified to be more closely aligned with the ISO /R 606 standard.

ISO 1275: Extended (double) pitch precision roller chains are shown in Table 13-6, wirh curved side plates. This type is also supplied with straight waist side plates and/or conveyor-chain attachment plates.

ISO/R 1395: Short pitch transmission bush (rollerless) chains are standardized in ANSI B29.1 and larger pitches in ANSI B29.12.

Table 13-3. Basic Dimensions for Foot Mounted Electric Motors (IEC 72)

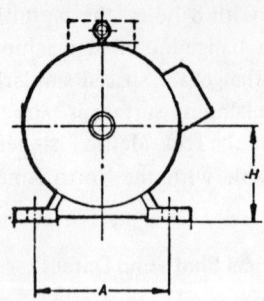
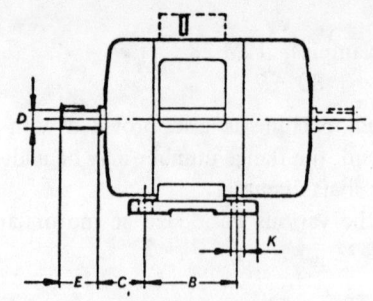

Frame number [1]	H Nominal mm	H Maximum deviation mm	A mm	B mm	C mm	K [2] Nominal mm	K Tolerance [3] μm	K Tolerance [3] μm	Bolt or screw
56	56	−0.5	90	71	36	5.8	+300	0	M5
63	63	−0.5	100	80	40	7	+360	0	M6
71	71	−0.5	112	90	45	7	+360	0	M6
80	80	−0.5	125	100	50	10	+360	0	M8
90 S	90	−0.5	140	100	56	10	+360	0	M8
90 L	90	−0.5	140	125	56	10	+360	0	M8
100 S	100	−0.5	160	112	63	12	+430	0	M10
100 L	100	−0.5	160	140	63	12	+430	0	M10
112 S	112	−0.5	190	114	70	12	+430	0	M10
112 M	112	−0.5	190	140	70	12	+430	0	M10
(112 L)	112	−0.5	190	159	70	12	+430	0	M10
132 S	132	−0.5	216	140	89	12	+430	0	M10
132 M	132	−0.5	216	178	89	12	+430	0	M10
(132 L)	132	−0.5	216	203	89	12	+430	0	M10
160 S	160	−0.5	254	178	108	15	+430	0	M12
160 M	160	−0.5	254	210	108	15	+430	0	M12
160 L	160	−0.5	254	254	108	15	+430	0	M12
180 S	180	−0.5	279	203	121	15	+430	0	M12
180 M	180	−0.5	279	241	121	15	+430	0	M12
180 L	180	−0.5	279	279	121	15	+430	0	M12
200 S	200	−0.5	318	228	133	19	+520	0	M16
200 M	200	−0.5	318	267	133	19	+520	0	M16
200 L	200	−0.5	318	305	133	19	+520	0	M16
225 S	225	−0.5	356	286	149	19	+520	0	M16
225 M	225	−0.5	356	311	149	19	+520	0	M16
(225 L)	225	−0.5	356	356	149	19	+520	0	M16
250 S	250	−0.5	406	311	168	24	+520	0	M20
250 M	250	−0.5	406	349	168	24	+520	0	M20
(250 L)	250	−0.5	406	406	168	24	+520	0	M20
280 S	280	−1	457	368	190	24	+520	0	M20
280 M	280	−1	457	419	190	24	+520	0	M20
(280 L)	280	−1	457	457	190	24	+520	0	M20
315 S	315	−1	508	406	216	28	+520	0	M24
315 M	315	−1	508	457	216	28	+520	0	M24
(315 L)	315	−1	508	508	216	28	+520	0	M24
355 S	355	−1	610	500	254	28	+520	0	M24
355 M	355	−1	610	560	254	28	+520	0	M24
355 L	355	−1	610	630	254	28	+520	0	M24
400 S	400	−1	686	560	280	35	+620	0	M30
400 M	400	−1	686	630	280	35	+620	0	M30
400 L	400	−1	686	710	280	35	+620	0	M30

NOTES:
1. Frame numbers within brackets should be regarded as non-preferred for A.C. induction machines.
2. Open-ended slots are not permitted.
3. These tolerances are those given in coarse series H14 according to ISO Recommendation R273.

MECHANICAL POWER TRANSMISSION SYSTEMS

Table 13-4. Basic Dimensions for Flange-Mounted Electric Motors (IEC 72)

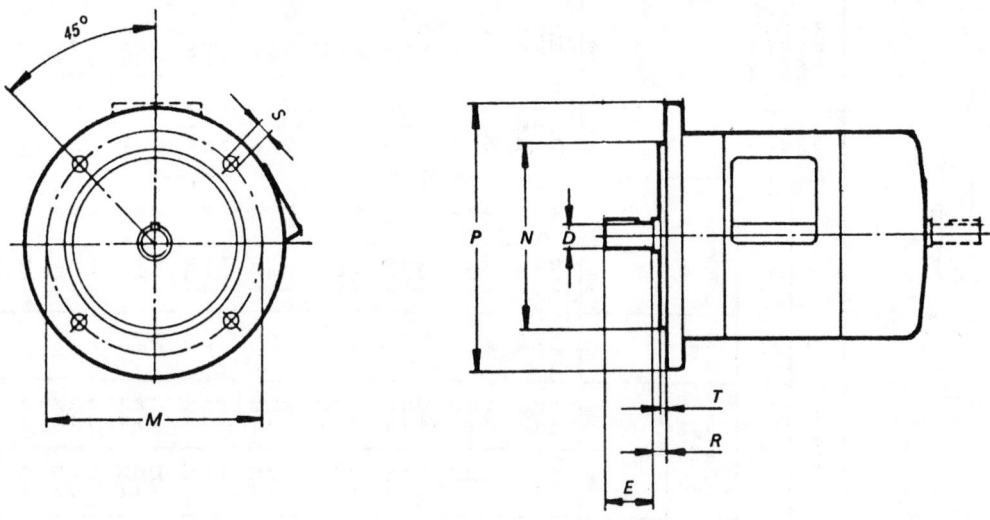

Flange number	M	N Nominal	N Tolerance ISO		P [1]	R	Number of holes	S Clearance hole [2] Nominal	S Tolerance [3]		Thread [2]	T Maximum
	mm	mm	μm	μm	mm	mm		mm	μm	μm		mm
F55	55	40	j6	+11	−5	70	0	4	5.8	+300	0	M5	2.5
F65	65	50	j6	+11	−5	80	0	4	5.8	+300	0	M5	2.5
F75	75	60	j6	+12	−7	90	0	4	5.8	+300	0	M5	2.5
F85	85	70	j6	+12	−7	105	0	4	7	+360	0	M6	2.5
F100	100	80	j6	+12	−7	120	0	4	7	+360	0	M6	3
F115	115	95	j6	+13	−9	140	0	4	10	+360	0	M8	3
F130	130	110	j6	+13	−9	160	0	4	10	+360	0	M8	3.5
F165	165	130	j6	+14	−11	200	0	4	12	+430	0	M10	3.5
F215	215	180	j6	+14	−11	250	0	4	15	+430	0	M12	4
F265	265	230	j6	+16	−13	300	0	4	15	+430	0	M12	4
F300	300	250	j6	+16	−13	350	0	4	19	+520	0	M16	5
F350	350	300	j6	+16	−16	400	0	4	19	+520	0	M16	5
F400	400	350	j6	+18	−18	450	0	8	19	+520	0	M16	5
F500	500	450	j6	+20	−20	550	0	8	19	+520	0	M16	5
F600	600	550	js6	+22	−22	660	0	8	24	+520	0	M20	6
F740	740	680	js6	+25	−25	800	0	8	24	+520	0	M20	6
F940	940	880	js6	+28	−28	1000| 0	8	28	+520	0	M24	6
F1080	1080	1000	js6	+28	−28	1150	0	8	28	+520	0	M24	6

[1] The external outline of mounting flanges up to and including F300 may be other than circular. Dimension P may deviate from that given in the table only on the minus side.
[2] When the flange has clearance holes (Designation I), screws with the thread specified should be used. When the flange is tapped (Designation II), the thread size should be as shown. It is recommended that the clearance holes in the mating part should be according to the table.
[3] These tolerances are those given in coarse series H14 according to ISO Recommendation R273.

MECHANICAL POWER TRANSMISSION SYSTEMS

Table 13.5. Cylindrical Shaft End Details (IEC 72)

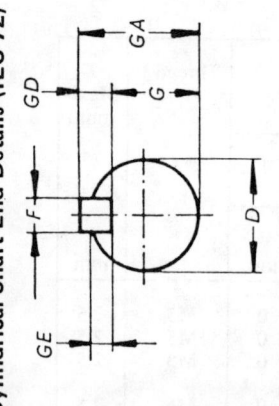

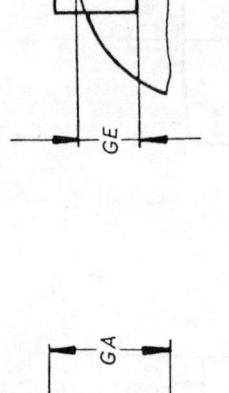

D [1]			E [5]		F		Key			GD			F		Keyway			GE		GA Nominal [4]	Greatest torque for continuous duty on a.c. motors [5]
Nominal	Designation ISO	Tolerance			Nominal	Tolerance h9	Nominal	Designation ISO	Tolerance		Nominal	Tolerance N9 [5]		Tolerance P9 [5]		Nominal	Tolerance			
mm		μm	μm	mm	mm	μm	mm		μm	μm	mm	μm	μm	μm	μm	mm	μm	μm	mm	Nm
7 9 11	j6 j6 j6	+7 +7 +8	−2 −2 −3	16 20 23	2 3 4	0 0 0	−25 −25 −30	2 3 4	h9 h9 h9	0 0 0	−25 −25 −30	2 3 4	−4 −4 0	−29 −29 −30	−6 −6 −12	−31 −31 −42	1.2 1.8 2.5	+100 +100 +100	0 0 0	7.8 10.2 12.5	0.25 0.63 1.25
14 16 18	j6 j6 j6	+8 +8 +8	−3 −3 −3	30 40 40	5 5 6	0 0 0	−30 −30 −30	5 5 6	h9 h9 h9	0 0 0	−30 −30 −30	5 5 6	0 0 0	−30 −30 −30	−12 −12 −12	−42 −42 −42	3 3 3.5	+100 +100 +100	0 0 0	16 18 20.5	2.8 4.5 7.1
19 22 24	j6 j6 j6	+9 +9 +9	−4 −4 −4	40 50 50	6 6 8	0 0 0	−30 −30 −36	6 6 8	h9 h9 h9	0 0 0	−30 −30 −36	6 6 8	0 0 0	−30 −30 −36	−15 −15 −15	−51 −51 −51	3.5 3.5 4	+100 +100 +200	0 0 0	21.5 24.5 27	8.25 14 18
28 32 38	j6 k6 k6	+9 +18 +18	−4 +2 +2	60 80 80	8 10 10	0 0 0	−36 −36 −36	8 10 10	h9 h11 h11	0 0 0	−36 −90 −90	8 10 10	0 0 0	−36 −36 −36	−15 −15 −15	−51 −51 −51	4 5 5	+200 +200 +200	0 0 0	31 35 41	31.5 50 90
42 48 55	k6 k6 m6	+18 +18 +30	+2 +2 +11	110 110 110	12 14 16	0 0 0	−43 −43 −43	12 14 16	h11 h11 h11	0 0 0	−90 −90 −110	12 14 16	0 0 0	−43 −43 −43	−18 −18 −18	−61 −61 −61	5 5.5 6	+200 +200 +200	0 0 0	45 51.5 59	125 200 355
60 65 70	m6 m6 m6	+30 +30 +30	+11 +11 +11	140 140 140	18 18 20	0 0 0	−43 −43 −52	18 18 20	h11 h11 h11	0 0 0	−110 −110 −110	18 18 20	0 0 0	−43 −43 −52	−18 −18 −22	−61 −61 −74	7 7 7.5	+200 +200 +200	0 0 0	64 69 74.5	450 630 800
75 80 85	m6 m6 m6	+30 +30 +30	+11 +11 +11	140 170 170	20 22 22	0 0 0	−52 −52 −52	20 22 22	h11 h11 h11	0 0 0	−110 −110 −110	20 22 22	0 0 0	−52 −52 −52	−22 −22 −22	−74 −74 −74	7.5 9 9	+200 +200 +200	0 0 0	79.5 85 90	1 000 1 250 1 600
90 95 100	m6 m6 m6	+35 +35 +35	+13 +13 +13	170 170 210	25 25 28	0 0 0	−52 −52 −52	25 25 28	h11 h11 h11	0 0 0	−110 −110 −110	25 25 28	0 0 0	−52 −52 −52	−22 −22 −22	−74 −74 −74	9 9 10	+200 +200 +200	0 0 0	95 100 106	1 900 2 360 2 800
110	m6	+35	+13	210	28	0	−52	28	h11	0	−110	28	0	−52	−22	−74	10	+200	0	116	4 000

NOTES:
1. For diameters up to 25 mm, a shoulder of 0.5 mm is considered sufficient.
2. In cases where the service conditions are well defined, shaft extensions might also be selected in accordance with existing ISO Recommendations.
3. The keyway tolerance N9 applies for normal keys and P9 for fitted keys
4. Tolerances for GA can be calculated from values of the other dimensions given in the table. See also ISO Recommendation in preparation.
5. The torque values are chosen from series R 40. In cases where the service conditions are well defined, torque values might also be selected in accordance with existing ISO Recommendations.

MECHANICAL POWER TRANSMISSION SYSTEMS

Table 13-6. World Power Transmission Chain Standards

NATIONAL STANDARD	FIGURE	NAME	RATINGS (1)(2)	DATA
TYPE A (B) ISO 606 (606) U.S.A. ANSI B29.1 JAPAN JIS B1802 (B1801) GERMANY DIN 8188 (8187) FRANCE NF E26-102 U.K. BS (228) ITALY UNI 2578 (2579) AUSTRAL AS 1532 (1532)	TYPES: A (AMERICAN) B (EUROPEAN) ISO CHAIN NO: 05B,06B,08A,08B,10A,10B, 12A,12B,16A,16B,20A,20B, 24A,24B,28A,28B,32A,32B, 40A,40B,48A,48B,56B,64B, 72B EXAMPLE: 05 - 05/16 inch pitch	(E) SINGLE PITCH TRANS-MISSION PRECISION ROLLER CHAINS (G) ROLLENKETTEN (F) TRANSMISSION PAR CHAINES A PAS COURTS (I) CATENE A RULLI	PITCH: Type A 12.7-76.2 mm Type B 8 - 114.3 mm STRENGTH: Type A 14 -500 kN Type B 4.4-900 kN Proof Load 33% MAX SPEED: 8000 RPM MAX POWER: 360 kW (SINGLE STRAND) 1200 kW (FOUR STRANDS)	USAGE: Power transmission SPROCKET: Machined (see ISO 606). Number of teeth range from 9 to 150. Preferred: 17,19,21,23,25, 38,57,76,95,114 LUBRICATION: Required REMARKS: Available in single, double or triple widths. Types A and B not interchangeable.
TYPE A (B) ISO 1275 (1275) U.S.A. ANSI B29.4 CONV B29.3 TRAN JAPAN JIS GERMANY DIN 8181 (8181) FRANCE NF E26-103 U.K. BS (4687) ITALY UNI AUSTRAL AS	TYPES: A (AMERICAN) B (EUROPEAN) ISO CHAIN NO: 208A,208B,210A,210B,212A 212B,216A,216B,220A,220B 224A,224B,228A,228B EXAMPLE: 208 - 2 DOUBLE PITCH 08 08/16 INCH PITCH	(E) EXTENDED (DOUBLE) PITCH PRECISION ROLLER CHAINS (G) LANGGLIEDRIGE ROLLENKETTEN (F) TRANSMISSION PAR CHAINES A PAS LONGS (I) CATENE A RULLI-A PASSO LUNGO	PITCH: Type A 25.4- 76.2 mm Type B 25.4-101.6 mm STRENGTH: Type A 13.8-124.6 kN Type B 17.8-169 kN Proof Load 33% MAX SPEED: 1300 RPM MAX POWER: 69 kW	USAGE: Power transmission and material handling (less speed and power than ISO 606 chains) SPROCKET: Single or double cut (see ISO 1275). Number of teeth range from 5 to 75. Preferred: 7,9,10,11,13,19 27,38,57 LUBRICATION: Optional REMARKS: Usually used in single widths. Can be used with conveyor attachment plates. Types A and B not interchangeable.
ISO 1395 U.S.A. ANSI B29.12 JAPAN JIS GERMANY DIN 8164 FRANCE NF E26-104 U.K. BS ITALY UNI AUSTRAL AS	ISO CHAIN NO: .04C .06C EXAMPLE: 1.04C,1.06C SINGLE WIDTH 2.04C,2.06C DOUBLE WIDTH 3.04C,3.06C TRIPLE WIDTH	(E) SHORT PITCH TRANS-MISSION PRECISION BUSH (ROLLERLESS) CHAINS (G) BUCHSENKETTE (F) TRANSMISSION PAR CHAINES A DOUILLES (I) CATENE A BUSSOLE	PITCH: 6.35 and 9.525 mm STRENGTH: 10.5 and 23.7 kN Proof Load 33% MAX SPEED: MAX POWER:	USAGE: Power transmission and material handling SPROCKET: Machines (see ISO 1395). Number of teeth range from 9 to 150. Preferred: 17,19,21,23,25,38, 57,76,95 and 114 LUBRICATION: REMARKS: Used in single, double and triple widths. Can have conveyor attachment plates.
ISO 487 U.S.A. ANSI JAPAN JIS B1801 GERMANY DIN 8189 FRANCE NF E26-105 U.K. BS 2947 ITALY UNI AUSTRAL AS	ISO CHAIN NO: S32, S42, S45, S52 S55, S62, S77, S88	(E) STEEL ROLLER CHAINS (G) ROLLENKETTEN FUER LANDMASCHINEN (F) TRANSMISSION PAR CHAINES, TYPE S (I) CATENE AGRICOLE	PITCH: 29.21-66.27 mm STRENGTH: 8-44 kN Proof Load 33% MAX SPEED: MAX POWER:	USAGE: Agriculture, building, material handling SPROCKET: Cast or machine finished teeth. Material cast iron ISO 185 grade 15 (see Table 10-52). Number of teeth range from 9 to 34. Preferred: 9,11,13,15,17,18, 27,30 LUBRICATION: Optional REMARKS: Usually used in single widths. Can be used with conveyor attachment plates (types K1=extended straight link; M1=bent link plate)
ISO 1977 U.S.A. ANSI JAPAN JIS GERMANY DIN 8165 FRANCE NF U.K. BS 4116 ITALY UNI AUSTRAL AS	ISO CHAIN NO: TYPE M M20,M28,M40,M56 M80,M112,M160 M224,M315,M450 M630,M900 TYPE MC MC28,MC56,MC112, MC224	(E) CONVEYOR CHAINS (METRIC SERIES) (G) STAHLGELENK-KETTE FUER STETIG-FOERDERER (F) CHAINES CONVOYEURS (I) CATENE TRANSPORTO	PITCH: Type M 40-1000 mm Type MC 63- 500 mm STRENGTH: Type M 20-900 kN Type MC 28-224 kN MAX SPEED: Type M Type MC MAX POWER: Type M Type MC	USAGE: Material handling SPROCKET: Specified in ISO 1977 Part 2. Material cast iron ISO 185 grade 15 (see Table 10-52). Number of teeth range from 8 to 32. Preferred 8,10,12,16 and 24. LUBRICATION: Optional REMARKS: Used in single widths. Can be used with conveyor attachment plates (types K1, K2, and K3)

NOTES:
1. Conversion factors: 1 mm = 0.0394 IN; 1kN = 225LB (F); 1 m/s = 200 FT/MIN; 1kW = 1.34 HP.
2. Ratings based on ISO Standards.

Table 13-6 (Continued). World Power Transmission Chain Standards

NATIONAL STANDARD	FIGURE	NAME	RATINGS (1)(2)	DATA
ISO U.S.A. ANSI 29.2 JAPAN JIS GERMANY DIN 8190 FRANCE NF U.K. BS ITALY UNI AUSTRAL AS		(E) INVERTED TOOTH (SILENT) CHAINS (G) ZAHNKETTEN MIT INNENFUEHRUNG (F) (I)	PITCH: STRENGTH: MAX SPEED: MAX POWER:	USAGE: SPROCKET: LUBRICATION: REMARKS:
ISO 3512 U.S.A. ANSI B29.10 JAPAN JIS GERMANY DIN 8182 (3) FRANCE NF U.K. BS ITALY UNI AUSTRAL AS	ISO CHAIN NO: 2010,2512,2814,3315, 3618,4020,4824,5628 DESIGNATION EXAMPLE: 2010: 20- 20/8 INCH PITCH 10-10/16 INCH BEARING PIN DIA.	(E) HEAVY DUTY CRANKED LINK TRANSMISSION CHAINS (G) ROLLENKETTEN MIT GEKROEPFTES GLIED (F) CHAINES DE TRANSMISSION A MAILLO NS COUDES DE HAUTE RESISTANCE (I) CATENE PER TRANS-MISSIONI PESANTI	PITCH: 63.5-177.8 mm STRENGTH: 262-2068 kN MAX SPEED: 600 RPM MAX POWER: 315 kW	USAGE: Power transmission (heavy duty and dirty environment) SPROCKET: Number of teeth range from 7 to 24. LUBRICATION: Desirable at all speeds and generally used at high speeds REMARKS:
ISO DIS 4347 U.S.A. ANSI B29.8 JAPAN JIS GERMANY DIN 8152 FARNCE NF U.K. BS ITALY UNI AUSTRAL AS	TYPES: LH (ANSI B29.8 SERIES SIDE PLATES) LL (ISO 606B SERIES SIDE PLATES) ISO CHAIN NO: LH TYPES LH08,LH10,LH12,LH16,LH20,LH24,LH28,LH32 ALL WITH LACING SUFFIX 23,34 or 46. LL TYPES: LL08,LL10,LL12,LL16,LL20,LL24,LL28,LL32,LL40,LL48 ALL WITH LACING SUFFIX 22,44 or 66	(E) LEAF CHAIN (G) LASCHENKETTE-STAHLGELENKKETTE (F) CHAINES DE LEVAGE A MAILLES JOIN TES, CHAPES ET TOURTEAUX DE RENNO (I) CATENE FLEYER TIPO AL	PITCH: Type LH 12.7-50.8 mm Type LL 12.7-76.2 mm STRENGTH: Type LH 22.2-578.3 kN Type LL 17.8-1201 kN MAX SPEED: SLOW MAX POWER: RATED ONLY IN TENSION	USAGE: Lifting purposes SHEAVES: Min. diameter = 5 x pitch LUBRICATION: Desirable REMARKS: Cranked links not available
ISO DIS 4348 U.S.A. ANSI B29.17 JAPAN JIS GERMANY DIN 8153 FRANCE NF E26-107 U.K. BS ITALY UNI AUSTRAL AS	ISO CHAIN NO: 24A26 (A=Austenitic stainless steel) 24C26 (C=Carbon or alloy steel)	(E) HINGE TYPE FLAT TOP CONVEYOR CHAINS (G) SCHARNIERBAND-KETTE (F) TRANSMISSION PAR CHAINES-A MAILLES JOINTIVES (I) CATENE PER TRANSP. INCERN. CON PIASTRA PIATTA	PITCH: 38.1 mm STRENGTH: 24A26 8 kN 24C26 10 kN MAX SPEED: MAX POWER:	USAGE: Conveyors for bulk material, containers packages, or small parts. SPROCKET: Number of teeth from 12 to 41. Type A - Radius tooth form; Type B - Straight tooth form LUBRICATION: Not required REMARKS: Odd number of "actual" teeth are preferred in order to extend wear life of sprockets

NOTES:
1. Conversion factors: 1 mm = 0.0394 IN; 1kN = 225 LB (F); 1 m/s = 200 FT/MIN; 1kW = 1.34 HP.
2. Ratings based on ISO Standards.
3. DIN 8182 is cancelled.

MECHANICAL POWER TRANSMISSION SYSTEMS

Table 13-7. World Endless-Belt Standards

STANDARD	NAME	TECHNICAL DATA (All Dimensions In Millimeters)
ISO R459, R460 U.S.A. RMA/MPTA IP/22 JAPAN GERMANY DIN 7753 (BLATT 1) FRANCE NF T47-117 U.K. B.S. 3790 ITALY AUSTRAL AS B243	(E) NARROW INDUSTRIAL V-BELTS (G) ENDLOSE SCHMALKEILRIEMEN FÜR DEN MASCHINENBAU (F) COURROIES TRAPEZOIDALES E'TROITES (I)	(1) Type (2) l_p W T kW D_{min} e h_{min} α (4) 9N 9 8 1-30 67 10.3 8 36 38 40 42 15N (3) 15 13 20-200 180 17.5 13.7 38 40 42 25N 25 23 150-1000 315 28.6 22.6 38 40 42 SPZ 8.5 9.5 8 1-30 63 12 9 34 38 SPA 11 13 10 20-40 90 15 11 34 38 SPB 14 16 13 30-200 140 19 14 34 38 SPC 19 22 18 100-1000 224 26 19 34 38
ISO R253, R434, R608 U.S.A. JAPAN JIS K6323 GERMANY DIN 2215 FRANCE NF T47-106 U.K. B.S. 1440 ITALY UNI 5265/5266 AUSTRAL AS 1215	(E) CLASSICAL INDUSTRIAL V-BELTS (SECTIONS Y & Z) (G) ENDLOSE KEILRIEMEN (F) COURROIES TRAPEZOIDALES CLASSIQUES (I) TRASMISSIONE a CINGHIE TRAPEZOIDALI PER APPLICAZIONI INDUSTRIALI	Type l_p W T kW D_{min} e h_{min} α (4) Y 5.3 6.5 4 0-1 20 8 4.7 32 36 Z 8.5 10 6 0-2 50 12 7 34 38
ISO R52, R608 U.S.A. RMA/MPTA IP-20 JAPAN JIS K6323 GERMANY DIN 2215 FRANCE NF T47-106 U.K. B.S. 1440 ITALY UNI 5265/5266 AUSTRAL AS 1215	(E) CLASSICAL INDUSTRIAL V-BELTS (SECTIONS A, B, C, D AND E) (G) ENDLOSE KEILRIEMEN (F) COURROIES TRAPEZOIDALES CLASSIQUES (I) TRASMISSIONE a CINGHIE TRAPEZOIDALI PER APPLICAZIONI INDUSTRIALI	Type l_p W T kW D_{min} e h_{min} α (4) A 11 13 8 1-8 75 15 8.7 34 38 B 14 17 11 6-20 125 19 10.8 34 38 C 19 22 14 15-75 200 25.5 14.3 36 38 D 27 32 19 50-600 355 37 19.9 36 38 E 32 38 25 500-1000 500 44.5 23.4 36 38
ISO 5290 U.S.A. RMA/MPTA IP-22 JAPAN GERMANY FRANCE U.K. ITALY AUSTRAL	(E) NARROW JOINED INDUSTRIAL V-BELTS (G) (F) (I)	Type l_p W T kW D_{min} e P α (4) 9J 9 10 1-30 67 10.3 8.9 36 38 40 42 15J (3) 15 16 20-200 180 17.5 15.2 38 40 42 20J 20 20 150-1000 265 24.4 20.3 38 40 42 25J 25 25 150-1000 315 28.6 25.4 38 40 42
ISO 2790 U.S.A. SAE J636b JAPAN GERMANY DIN 7753 (BLATT 3) FRANCE PR R155-03 U.K. BS AU150 ITALY AUSTRAL	(E) NARROW AUTOMOTIVE V-BELT DRIVES (G) ENDLOSE SCHMALKEILRIEMEN FÜR DEN KRAFTFAHRZEUGBAU (F) COURROIES TRAPEZOIDALES POUR SERVITUDES DE MOTEURS DE VEHICULES AUTOMOBILES (I)	Type l_p W T kW D_{min} e P_{min} α AV10 8.5 10 8 -- 67 -- 11 36 AV13 11 13 10 -- 90 -- 13.75 36

NOTES:
1. l_p = pitch width; W = belt width; T = belt height; KW = approx. belt power range (1 KW = 1.34 HP); D_{min} = Approx. minimum sheave diameter; e = groove center distance; h = groove depth from pitch line; p = groove depth from outside of sheave; α = Sheave groove angle.
2. Types 9N, 15N, 25N are standard belts in North America and Australia. These types are also designated 3V, 5V, 8V, respectively.
3. These belts are not designated by pitch width.
4. Groove angle in degrees (°) depends on sheave diameter.

13-9

Table 13-7 (Continued). World Endless-Belt Standards

STANDARD	FIGURE	NAME	TECHNICAL DATA (All Dimensions in Millimeters)
ISO 3410 USA ASAE S211.3 JAPAN GERMANY FRANCE U.K. B.S. 3733 ITALY AUSTRAL	(W, l_p, T, D_{min})	(E) VARIABLE SPEED AGRICULTURAL V-BELTS (G) (F) COURROIES TRAPEZOIDALES SANS FINDE VARIATEURS POUR MACHINES AGRICOLES (I)	(1) Type l_p W T kW D_{min} e h_{min} α HI 23.6 25.4 12.7 -- 170 -- 13 26 HJ 30.0 31.8 15.1 -- 210 -- 16 26 HK 35.5 38.1 17.5 -- 255 -- 19 26 HL 41.4 44.5 19.8 -- 300 -- 22 26 HM 47.5 50.8 22.2 -- 340 -- 25 26
ISO 5296,5294 USA RMA/MPTA IP-24 JAPAN JIS B9078 GERMANY FRANCE U.K. B.S. 4548 ITALY AUSTRAL	(PITCH, h_s, S, h_t, D_{min}, β)	(E) SYNCHRONOUS BELTS (G) (F) COURROIES SYNCHRONE (I)	(1) (2) Type Pitch S kW D_{min} h_t h_s β XL 5.080 2.57 0-1 16 1.27 2.3 50 L 9.525 4.65 0-3 30 1.91 3.6 40 H 12.700 6.12 2-35 57 2.29 4.3 40 XH 22.225 12.57 20-75 141 6.35 11.2 40 XXH 31.750 19.05 50-150 202 9.53 15.7 40
ISO USA RMA MPTA IP-26 JAPAN GERMANY FRANCE U.K. ITALY AUSTRAL	(T, h, e, D_{min})	(E) V-RIBBED BELTS (G) (F) (I)	Type e T kW D_{min} h α H 1.60 3 0-2 20 1.34 40 J 2.34 4 1-7 33 2.15 40 K 3.56 7 5-15 55 3.64 40 L 4.70 10 10-25 75 5.40 40 M 9.40 17 20-200 180 11.18 40

NOTES: (1) l_p = pitch width; W = belt width; T = belt height;
kW = Approx. belt power range (1 kW = 1.34 HP);
D_{min} = Approx. minimum sheave diameter;
e = groove center distance; h = groove depth from pitch line;
P = groove depth from outside of sheave; α = sheave groove angle, degrees

(2) S = belt tooth width at root; h_t = belt tooth height;
h_s = belt height; β = belt tooth angle, degrees

MECHANICAL POWER TRANSMISSION SYSTEMS

ISO 487-1976   This ISO standard covers steel roller chains for agricultural machinery and it has no corresponding ANSI standard. (This type of roller chain is not produced in North America.)

ISO 1977-1976   This ISO standard applies to metric-dimensioned bush, plain and flanged roller chains of both solid and hollow bearing pin types designed for general conveying and mechanical handling duties. It has limited usage and acceptance, and no corresponding American standard. The nearest ANSI type is the B29.15 standard for heavy-duty roller conveyer chains.

ISO 3512-1976   This international standard is the ISO version of ANSI B29.10, specifying heavy-duty offset sidebar power transmission roller chains and separate teeth.

ISO/DIS 4347   A similar ISO standard to the ANSI B29.8 on leaf chains is pending processing by ISO, and is soon to be published.

ISO/DIS 4348   An ISO version of the ANSI B29.17 standard on flat top conveyor chains is soon to be published.

ENDLESS-BELT DRIVES

Endless-belt drives are used to transmit power on a wide variety of types and sizes of machinery. The ability to transmit a wide range of power efficiently and economically is one of the main advantages of endless-belt drives. They also allow for versatility in the positioning of the driveR or driveN unit and offer an excellent means of achieving large speed ratios.

Basic Types of Endless-Belt Drives

V-Belt Drives—These are the most popular of the endless-belt drives. They are manufactured in a wide variety of cross sections and types, to handle general power transmission requirements on industrial and agricultural machinery, automobiles, and commercial vehicles.

There are International Standards (ISO) applying to most types of V-belt drives and many of the industrialized nations have their national standards as well. A typical V-belt cross section is shown in Fig. 13-1.

V-belts are specified by a cross-section designation which also denotes belt type. Nominal belt top width and nominal belt thickness are also often specified.

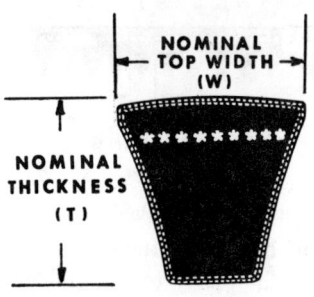

Courtesy of Gates Rubber Company

Fig. 13-1. Typical V-Belt Cross Section

Synchronous Belt Drives—While synchronous belt drives are suitable for general power transmission, they are usually used on applications where an exact speed ratio must be maintained between the driveR and driveN units. Typical applications are automotive cam-shaft drives and machine-tool drives where indexing is critical.

This type of endless belt was originally developed in the United States, where standards have been in existence for some time. International Standards have been developed now by the ISO/TC 41.

The sketch, Fig. 13-2, shows the method by which a synchronous belt transmits power as the teeth of the belt mate with teeth on the pulley.

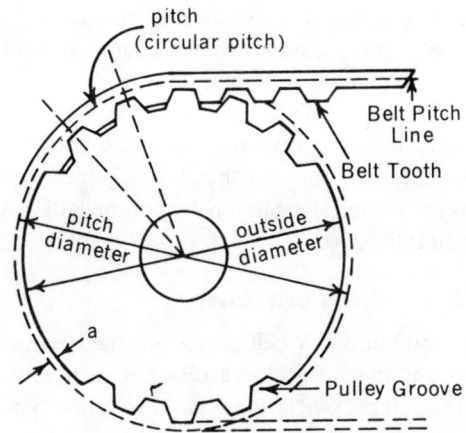

Courtesy of the Rubber Manufacturers Association

Fig. 13-2. Synchronous belt drive

Special Belt Drives—The most popular type of endless belt in the special-belt category, is the V-ribbed belt. A cross-sectional sketch of this type of belt is shown in Fig. 13-3.

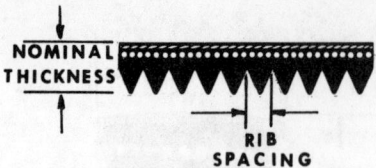

Courtesy of the Gates Rubber Company
Fig. 13-3. V-ribbed belt

Small V-ribs on the bottom of the belt mate with corresponding grooves in the pulley to help guide the belt and to transmit power. Since the belt ribs fill the grooves completely, it does not have the full wedging characteristics of a V-belt.

The thickness of a V-ribbed belt is small in comparison to its width, thus it has the capability of operating satisfactorily on small pulley diameters.

Narrow Joined, Industrial V-Belt Drives

For most drives, multiple V-belts will operate trouble-free for a long period of time. Sometimes, however, V-belts can become unstable and tend to turn over or come off in certain industrial applications—especially those with high pulsating or shock loads. By joining two or more belts with a tie-band, a more stable operation will result. The basic area of application for narrow joined industrial V-belts is the same as for narrow industrial V-belts except for the added advantage of more stability.

The narrow joined V-belt was developed in the United States and was designed to operate in the 9N, 15N, and 25N sheave grooves. The freedom of interchangeability between belts and sheave grooves, of approximately the same dimensions, does not apply to joined type belts. If the belt does not fit the sheave grooves in a precise manner the sheave-groove land may interfere with the belt tie-band. Table 13-7 shows a sketch of a joined type belt as well as the technical data on belt and groove dimensions as shown in ISO 5290.

Narrow Automotive V-Belt Drives

Narrow automotive V-belt drives are used to transmit power to automotive engine accessories such as cooling fans, water pumps, generators, etc. The narrow type belts are used on passenger cars as well as on small- to medium-sized trucks. Large trucks and earthmoving machinery often use the larger conventional cross-section automotive belts. Narrow automotive V-belts are usually used as single belts although they are sometimes used in matched sets of 2, or more.

North America as well as the European countries have used the same basic belts for several years. Complete interchangeability is attainable among the narrow automotive V-belts built in all industrialized countries of the world. Table 13-7 gives the various standards that apply to narrow automotive V-belts as well as the belt and groove dimensions.

Variable-Speed Agricultural V-Belt Drives

Variable-speed agricultural V-belts are used on a variety of farm machinery where variable driven speeds are desirable. The traction drive on a combine is a very popular example. Variable-speed V-belts provide a very economical and practical method of obtaining infinite speed variation. This allows farm machinery to operate at the optimum speed for increased efficiency.

Variable-speed agricultural V-belts were developed in North America and standardized by the American Society of Agricultural Engineers (ASAE). The five cross-sections standardized by ASAE are used universally around the world. An ISO standard is currently in the Draft Standard phase which standardizes the ASAE cross sections on an international level.

Table 13-7 gives the technical data on belt and sheave-groove dimensions. Many variable-speed drives use one variable sheave along with a regular fixed-pitched sheave. Technical data can be applied to both types. The pitch width (1_p) of the variable sheave is measured with the sheave fully closed. The groove depth from the pitch line is measured with the sheave fully open.

Classical Agricultural V-Belt Drives

Classical agricultural V-belts are used for a wide variety of power transmission applications on farm machinery where variable speed is not required. Agricultural V-belts provide an economical method of transmitting power with the added advantages of having shock-absorbing characteristics.

In most industrial countries, the V-belt and sheave-groove dimensions recommended for classical agricultural V-belts are essentially the same as those recommended for classical industrial V-belts. Therefore, the belts produced by the various industrial nations of the world are usually interchangeable. Since the belt and sheave groove dimensions are essentially the same as those listed in Table 13-7 for classical industrial V-belts—A, B, C, D, and E cross-sections—they will not be repeated for agricultural classical V-belts.

Synchronous-Belt Drives

Synchronous belts can be used for most types of power transmission but, as the name implies, are especially designed for applications where the drive*N* RPM must be synchronized with the drive*R* RPM. Synchronous belts are

MECHANICAL POWER TRANSMISSION SYSTEMS

available in several sizes which can handle lightly loaded applications such as film projectors, to heavily loaded applications such as large machine tools.

Synchronous-belt drives were developed and standardized in the USA. Their usage has spread to all industrialized nations of the world—most of which have adopted the USA dimensions. To allow interchangeability all dimensions pertaining to the teeth on the belt and pulley must be maintained very closely. International standards have been developed based on the USA dimensions to assure worldwide interchangeability.

Table 13-7 also gives technical data on the basic dimensions of the synchronous belt tooth as well as the approximate kW range for each of the cross sections.

V-Ribbed Belt Drives

Usually, the V-Ribbed type belt is used in the general industrial power transmission area. It is not, however, restricted to this area and is sometimes used on automotive accessory drives as well as drives on agricultural machinery and appliances.

This belt has the capability of operating over small diameter pulleys but generally requires higher tensions than a V-belt because it does not incorporate the full wedging principle; it is available in a wide variety of number of ribs per belt.

This type of belt is used primarily in North America. Any use in Europe or other countries of the world would be based on the standards established in the USA. There are no ISO standards on this belt nor are there any immediate plans to develop any.

The belt cross-section is specified by the spacing between the ribs. This is shown as dimension *e* in Table 13-7. Other pertinent dimensions of the belt are also shown in Table 13-7. The USA Standard, published jointly by the Rubber Manufacturers Association and the Mechanical Power Transmission Association, gives complete specifications for manufacturing the pulley grooves.

MECHANICAL POWER TRANSMISSION SYSTEMS— GEARING[1]

Gearing is one of the subjects singled out for attention because of its key importance in machinery, power systems, transportation equipment, and instrumentation. Gears range in size from tiny, delicate watch and instrument gearing to massive rolling-mill and earthmoving-equipment gears; they are used throughout almost every industry and are a basic component permeating all sorts of designs. This, in turn, has made them a standard interchangeable component that must reflect design standards; be compatible with entire systems of which they are a part, and conform to the prevalent units of measure. Further, in the interest of worldwide product markets and the need for easy and efficient maintenance and field repairs, it is advantageous to have gears conform to international standards; preferably one basic design standard and units of measure.

Present Status of Gear Standards

Throughout the world major producing countries have been at variance regarding gear standards ever since gearing took a definitive form in the mid-18th century. Not only has there been a major split into the metric and inch worlds, that voids interchangeability, but also each of the two groups has had several versions, resulting in the non-interchangeability of many metric gears; and also the non-interchangeability of American and British inch gears.

The USA is, at present, still very heavily committed to the *"inch diametral pitch"* gear system. Only very recently are metric module gears beginning to appear in selected industries. Of course, the goal is to increase the rate of metrification with ultimate metric take-over in the not-too-distant future.

Europe and Japan are completely committed to the *"metric module."* The several differences that existed in past years have been almost entirely eliminated, and all countries are using the ISO metric gear standards, or very near equivalents.

Britain is in the midst of converting to the metric system, and consequently, presently is heavily committed to a dual systems situation. This will prevail until the process of conversion is completed and there is complete replacement of inch gear machinery and products in the field, which can be a long-drawn-out process depending upon wear-out.

The developing countries are mostly in the advantageous situation of coming-on-board the technological era just when the metric gear system has been designated as the preferred and ultimate system. These countries are following good advice and are developing their gear standards around the international ISO recommendations. The only exceptions are isolated cases of remaining with the inch diametral pitch due to inherited equipment and standards from "inch nations" that either were in control of those countries in the past or introduced inch gearing during early years of economic assistance for development. These cases are rapidly phasing out.

---

[1] This section on gearing has been contributed by Prof. George W. Michalec of the Stevens Institute of Technology, Hoboken, N.J., and Dr. Frank Bushsbaum, V.P., Designatronics Inc., New Hyde Park, N.Y.

The Future of Gear Standards

In the foreseeable future, measured only in tens of years, the world will be entirely standardized on metric module gears per ISO standards. This will spread the advantages of interchangeable gearing throughout the world. There will be less duplication of machinery and tooling, with resultant economies. Further, field maintenance and spare-part procurement will be both improved and less expensive.

The Gear Standards' Transition Period

Metrification of various engineering subjects, such as gearing, involves not only a change in measure scales and introduction of SI units, but also, in many cases, additional specific considerations regarding parameters and different design standards. This is of concern only to those switching to metric gearing during this era of change. Once the transition is accomplished, all designers and gear users will be adept with metric gears and SI units. The advantages of converting to metric gears are the already well-stated and known features of metric measures and SI units. The disadvantages for gearing relate mostly to the loss of empirical design formulas that are peculiar to the inch system. However, this is only a problem for the transition period. New empirical design formulas will develop as application of "metric module" gears and familiarity with the SI units increases.

Symbols

Gear symbols have never had worldwide uniformity, both within the inch and metric worlds, let alone between the two systems. The ISO has made recommendations for gear symbols and notation in ISO 701-1976. Unfortunately, this standard has not had complete acceptance and implementation. Further, American nomenclature and symbol practice is in wide variance. Since the USA is still in the early phase of transition and is living with a dual system and since the gear field is still exclusively inch, this writing conforms to the previous customary American nomenclature and symbol practices.

METRIC GEARING

The Metric Module

Metric gearing is not only based upon different units of length measure, but also involves its own unique design standard. Historically, metric gears began with a different approach to tooth-proportion standardization, and this constitutes a major obstacle to simple adoption of the metric system by the American gear industry. In the inch system diametral pitch was created as a convenient means for relating pitch diameters to center distance. Thus, diametral pitch is defined as:

$$P = \frac{N}{D} = \left[\begin{array}{c}\text{Number of teeth per inch} \\ \text{of pitch diameter}\end{array}\right] \quad (13\text{-}1)$$

Where: N = Number of teeth
D = Pitch diameter
P = Diametrical pitch

From this relationship there are particular integer-values of diametral pitch that yield integer values for center distance, in inches. Thus 8, 16, 32, and 64 diametral pitches, to mention only some, provide tooth numbers which can result in center distances that are even inches and/or convenient fractions of an inch. In the metric system the module is analogous to pitch, and is defined as:

$$m = \frac{D}{N} \left[\begin{array}{c}\text{Amount of pitch diameter} \\ \text{per tooth, in millimeters}\end{array}\right] \quad (13\text{-}2)$$

This defines the module as analogous to the reciprocal of diametral pitch. However, the module is a dimension (length of pitch diameter per tooth); whereas diametral pitch is the number of teeth to a unit length of pitch diameter. Again convenient center distances in metric measure are obtained by choosing integer module values and selected fractional values.

One consequence is that each system, inch diametral pitch and metric module, have adopted preferred standard values which are non-interchangeable. (Table 13-8 lists commonly used pitches in both systems. Their correspondence is best portrayed in the circular tooth thickness measure).

Note too, that the term "diametral pitch" belongs to the inch system. In the metric system the nearest analogous pitch is termed "module," and the word "pitch" is reserved for tooth spacing along the pitch circle. In the inch system, the tooth spacing measure is more accurately called "circular" pitch.

Metric Gear-Tooth Proportions

Unfortunately, the metric module was developed in a number of versions, each differing in minor ways. The German module, defined by the DIN standard, is widely used throughout Europe. However, the Japanese have their own version defined in JIS standards. Also, the Soviets follow their own metric module standard. The deviations are neither as staggering as the inch vs. metric system nor do they cause insurmountable problems. The various metric standards are akin to the several American diametral pitch standards which differ only in regard to dedendum values and root radii.

MECHANICAL POWER TRANSMISSION SYSTEMS

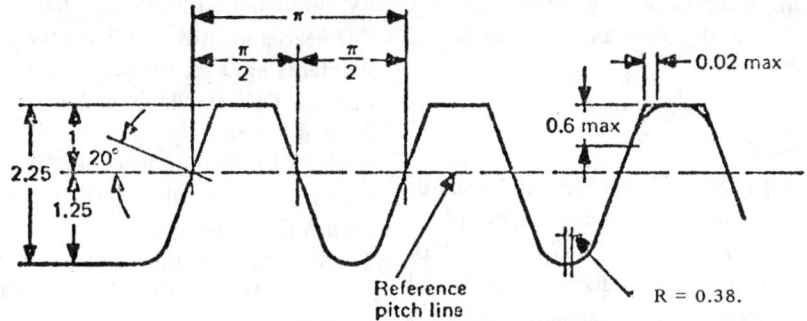

Fig. 13-4. Representation of the profile of module (ISO 53) $m = 1$ (or of diametral pitch $P = 1$)

Fortunately, in time for metrication in the USA, there is a new unified module standard sponsored and promoted by the International Standards Organization (ISO). This is not only a metric module standard but conforms to the new SI units system in all respects. It is the one and only system that is to be adopted in the USA in the interest of simplifying and standardizing metrication. Currently, France; Germany; Great Britain; Japan, and other major metric countries are shifting to this international standard. Only this one metric standard is advocated as the basis for the USA's metric gearing.

ISO Metric Module Tooth Proportions

ISO metric gearing, defined in terms of a basic rack, is applicable to cylindrical gears of the spur and helical types with involute teeth. ISO 53 Standard (1st issue, 1957) provides tooth proportions. Because the world contains considerable inch gearing, this standard has a dual purpose in that it also lists data for inch diametral pitch gears.

Tooth proportions for metric gears are given in terms of the basic rack, Fig. 13-4 Dimensions, in millimeters, are normalized for module, $m = 1$. Corresponding values for other modules are obtained by multiplying each dimension by the value of the specific module, m. The standard applies to cylindrical gears of the spur and helical varieties. Note that the same basic rack defines inch diametral pitch gears, with the data in the figure normalized for diametral pitch 1.

Tooth Form   The rack is straight-sided, full-depth; forming the basis of full-depth interchangeable gears.
Pressure Angle   20°. This conforms to worldwide acceptance of 20° as the most versatile pressure angle.
Addendum   This is equal to the module, m; which conforms to the American practice of $1/P$.
Dedendum   A value of $1.250\ m$ is used, which corresponds to recent American practice, in ANSI 6.1-1968 and AGMA 201.02.
Root Radius   This is slightly greater than current American standards specify.
Tip Radius   A maximum tip rounding is specified. This is a deviation from American standards which do not specify rounding. However, as a maximum or limit value this does not prevent American technology from specifying a tip radius as near zero as possible.

Note that the basic rack for metric gears and the basic rack for American inch diametral pitch gears are essentially identical. (See ANSI B6.1-1968.) For metric gears specific size dimensions are obtained by multiplying the given basic rack tooth dimensions by m (the module).[1] Gears conforming to diametral pitch American standards have tooth proportions sized by dividing the basic rack dimensions by the specific diametral pitch (P).

The ISO metric gear standard will permit wide interchangeability of products. Minor discrepancies can arise within individual national standards due to tip and root radii variations. However, a dedendum of $1.250\ m$ will permit interchangeability with all DIN and JIS gears. A major step to wide interchangeability will be limiting

---

[1] To avoid any possible confusion in this section, the symbol in italicized type, m, is used to designate "module" values in formulas, etc. However, care must be taken—particularly when using other publications—that the module symbol is not confused with that for metric linear measure, the meter (m).

module choice to a practical number of preferred modules as standard values within the standard. (See preferred modules in Table 13-9.)

American Gear Standards

American gearing has been guided by several standards affording the designer a choice of pressure angles (14½°, 20°, and 25°) and dedendum choices ($1.157/p$ and $1.250/p$). In particular, fine-pitch gears (20 diametral pitch; and finer) have followed a special design standard involving a dedendum of $(1.200 + .002)/P$. This means American gear designers and users, in changing to metric standards, must make additional accommodations if previously using one of these ANSI or AGMA standards. For details of these other standards and also tooth proportions of the basic rack, see footnote[1].

American gear standards have been generated and sponsored by two organizations. The AGMA (American Gear Manufacturers Association) has been the chief generator and ANSI (formally ASA) sponsors many of the AGMA standards including the most basic. A complete list of current AGMA standards can be obtained from AGMA headquarters (1330 Massachusetts Ave. N.W., Washington, D.C. 20005). Basic and key American standards are listed at the end of this section.

An important point to remember is that currently, all ANSI and AGMA standards are committed to inch gearing. There is no information about metric gearing in any of these standards. However, there is now an AGMA committee studying metrication, and presumably there will be recommendations and standards in the future.

AGMA 390.03 includes gear quality classifications and much detail regarding gear errors and quality parameters. Tolerance limits for quality grades are given, which makes this information somewhat parallel to that given in ISO 1328, although the figures are not identical. (Further details about gear accuracy, errors and their propagation can be found in Michalec,[2] Chapters 4 and 5).

Japan JIS Standards

Gearing in Japan is guided by the Japanese Industrial Standards Committee through its JIS standards. These are based upon the metric module and essentially conform to the ISO metric module system. The key JIS gear standards are also listed at the end of this section.

German DIN Gear Standards

In formulation and usage for a long time are the German Industrial (DIN) Standards which have had wide popularity throughout Europe and have strongly influenced the ISO gear standards. Utilizing the metric module, the DIN standards have set the stage for worldwide metric gearing. Current DIN standards and the accepted ISO gear standards differ in only minor details. The key DIN gear standards in English are listed at the end of this section, and they specify a detailed metric module gear calculation.

French Gear Standards

The French standards on gears, which are listed at the end of this section, are similar to the ISO standards listed in brackets.

British Standards (B.S.)

Great Britain's early industrial development gave rise to a collection of British Standards (B.S.). Originally, and until the 1960's, the B.S. gear standards were entirely wedded to the inch diametral pitch system. However, upon Britain's implementation of an earnest program to convert to the metric system, B.S. standards based upon the metric module have been generated. Thus, B.S. standards currently cover both systems. The key standards for both systems are listed at the end of this Chapter. The terms and definitions are in accordance with the ISO/R 1122 and the notations have been derived from ISO/R701. The grades and elemental tolerances are generally in accordance with agreement reached within the ISO.

It should be noted that B.S. and American inch diametral gears are not interchangeable. The basic racks are very nearly the same, but differences in some details prevent complete compatibility of gears.

The major differences between B.S. metric gearing and ISO gear standards are as follows:

1. The total depth is permitted to vary between 2.25 to 2.40. This added clearance allows use of different manufacturing techniques.
2. The root radius is limited to be within 0.25 to 0.39.

Italian Gear Standards

A series of UNI standards have been published that correspond to the ISO recommendations.

Australian Gear Standards

The standards listed for Australia are based on the customary inch system.

Interchangeability of Metric Module and Inch Diametral Gears

Limitations—If Equations (13-1) and (13-2) are solved for pitch diameter and these values are set equal to one another by introduction of the conversion factor 25.4, then there is the relationship:

$$P \times m = 25.4 \qquad (13-3)$$

---

[1] G. W. Michalec, *Precision Gearing: Theory and Practice*, (New York: John Wiley and Sons, Inc.) 1966, pp. 13-14.
[2] *Ibid:* Michalec.

MECHANICAL POWER TRANSMISSION SYSTEMS

This shows that inch diametral pitch and the metric module are related by the decimal factor 25.4. It is obvious that conversion results in decimal values, often awkward numbers, for one or the other measure. This means convenient values in one system will not be nice, convenient values in the other. The result is that each system, inch diametral pitch and metric module, have adopted preferred standard values which are non-interchangeable. Table 13-8 lists the commonly used pitches/modules of both systems with preferred values in boldface type. Corresponding equivalent values are given, but these are of no help since odd valued pitches and modules are usually not tooled in gear manufacturing shops.

Converting Practices Between the Metric and Inch Systems

Table 13-8 indicates that exact conversion and replacement of inch gearing with metric gearing is impossible, assuming using special tooling is not permitted. The best that can be done is to shift to the nearest standard module when converting from the inch system. Keep in mind, however, that preferred module sizes exist in different countries (see Table 13-9). The degree of non-correspondence is best measured by the circular pitch and the circular-tooth thickness values.

The consequence is that metrication of gearing requires complete new design regarding gear dimensions and center distance. This also involves new gear cutting tools. But there is special exception in the case of helical gear meshes. Referring to the equation for center distance based upon the pitch diameters:

$$C = \frac{N_1 + N_2}{2 P_n \cos \psi} \quad (13\text{-}4)$$

Replacing the normal diametral pitch P_n with the nearest standard module will yield an inexact equivalent center distance, C. However, the helix angle, ψ, can be adjusted slightly to bring the center distance into proper exact value. This procedure is practical as long as the helical gears are produced by hobbing and the setting of the helix angle is an open option.

Preferred Modules and National Practices

The generalized list of preferred modules identified in Table 13-8 is narrowed to a smaller number by ISO as shown in the column marked ISO 54, in Table 13-9. Most metric gear countries have developed national standards which have guided practice to settle upon a limited number of preferred modules. The preferred modules vary among the major metric countries. An over-all view of national preferences for selected countries is given in Table 13-9.

Kinematics of Gearing—Gear Geometry and Definitions

The basic geometry of a meshed spur gear pair is shown in Fig. 13-5 including key symbols. Similar terms apply to helical gearing with the introduction of a helix angle, ψ. Formulas for the various parameters are given in Table 13-10. Geometry of bevel gearing is presented in Fig. 13-6. Force vectors associated with the transmission of torque for spur, helical, and bevel gearing are defined in Fig. 13-7. Magnitudes of forces are computed as follows:

General terms

F^n = Normal tooth force between teeth
F^t = Tangenital force in plane of rotation
F^b = Component of force directed towards gear center
F^a = Component of force in axial direction
ϕ = Pressure angle

For spur gearing:

$$F^n = \frac{F^t}{\cos \phi} \quad (13\text{-}5)$$

$$F^b = F^t \tan \phi = F^n \sin \phi \quad (13\text{-}6)$$

For helical gearing:

$$F^n = \frac{F^t}{\cos \psi \cos \phi} \quad (13\text{-}7)$$

$$F^b = \frac{F^t \sin \phi_n}{\cos \psi \cos \phi} = F^n \sin \phi_n \quad (13\text{-}8)$$

$$F^a = F^n \cos \phi_n \sin \phi \quad (13\text{-}9)$$

Where: $\tan \phi = \dfrac{\tan \phi_n}{\cos \psi}$ \quad (13-10)

ϕ_n = Normal pressure angle
ψ = Helix angle

For bevel gearing:

$$F^n = \frac{F^t}{\cos \phi} \quad (13\text{-}11)$$

$$F^b = F^t \sin \phi \sin \gamma \quad (13\text{-}12)$$

$$F^a = F^t \sin \phi \cos \gamma \quad (13\text{-}13)$$

Where: γ = Bevel gear pitch angle

Kinematic Gearing Formulas

Many kinematic equations are universal, that is independent of the specific units of length and therefore can

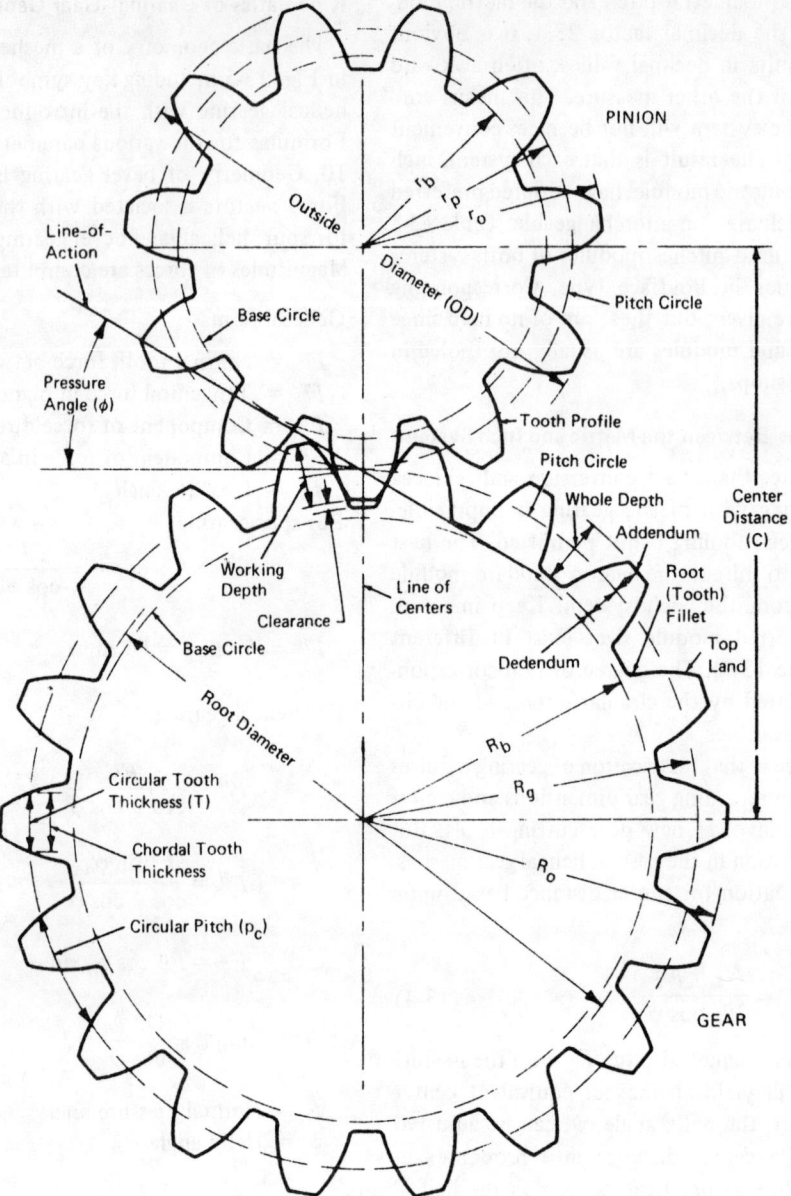

Courtesy of Stock Drive Products

Fig. 13-5.

be used with inch or metric units. However, units must be consistent throughout; i.e., all measures must be in millimeters (or whatever other unit chosen). An example of a universal units equation is:

$$m_p = \frac{\sqrt{_1R^2_o - {_1R^2_b}} + \sqrt{_2R^2_o - {_2R^2_b}} - C \sin \phi}{P_c \cos \phi} \quad (13\text{-}14)$$

In this equation m_p is contact ratio and the outside radius (R_o), base circle radius (R_b), center distance (C), and circular pitch (P_c) must be consistent units. For metric usage all dimensions are in millimeters; and for American inch use, all dimensions must be in inches. This equation, when applied to standard gears, can be manipulated to yield:

$$m_p = \frac{\sqrt{(N_1 + 2)^2 - N^2_1 \cos^2 \phi} + \sqrt{(N_2 + 2)^2 - N^2_2 \cos^2 \phi} - (N_1 + N_2 \sin \phi)}{2\pi \cos \phi} \quad (13\text{-}15)$$

MECHANICAL POWER TRANSMISSION SYSTEMS

Table 13-8. Metric/American Gear Equivalents
(Courtesy of Stock Drive Products)

Diametral Pitch, P	Module m	Circular Pitch inches	Circular Pitch millimeters	Circular Tooth Thickness inches	Circular Tooth Thickness millimeters	Addendum inches	Addendum millimeters
0.5000	50.8080	6.2832	159.594	3.1416	79.809	2.0000	50.808
0.5080	**50**	6.1840	157.080	3.0921	78.540	1.9685	50.000
0.5640	45	5.5660	141.372	2.7850	70.686	1.7730	45.000
0.6050	42	5.1950	131.947	2.5964	65.973	1.6529	42.000
0.6510	39	4.8240	122.522	2.4129	61.261	1.5361	39.000
0.7060	36	4.4530	113.097	2.2249	56.549	1.4164	36.000
0.7500	33.8667	4.1888	106.396	2.0943	53.198	1.3333	33.867
0.7700	33	4.0820	103.673	2.0400	51.836	1.2987	33.000
0.8470	30	3.7110	94.248	1.8545	47.124	1.1806	30.000
0.9410	27	3.3390	84.823	1.6693	42.412	1.0627	27.000
1	25.4000	3.1416	79.800	1.5708	39.898	1.0000	25.400
1.0580	24	2.9685	75.398	1.4847	37.699	0.9452	24.000
1.1550	22	2.7210	69.115	1.3600	34.558	0.8658	22.000
1.2700	**20**	2.4737	62.832	1.2368	31.416	0.7874	20.000
1.4111	18	2.2263	56.548	1.1132	28.274	0.7087	18.000
1.5000	16.9333	2.0944	53.198	1.0472	26.599	0.6667	16.933
1.5875	**16**	1.9790	50.267	0.9894	25.133	0.6299	16.000
1.8143	14	1.7316	43.983	0.8658	21.991	0.5512	14.000
2	12.7000	1.5708	39.898	0.7854	19.949	0.5000	12.700
2.1167	**12**	1.4842	37.699	0.7420	18.850	0.4724	12.000
2.5000	10.1600	1.2566	31.918	0.6283	15.959	0.4000	10.160
2.5400	**10**	1.2368	31.415	0.6184	15.708	0.3937	10.000
2.8222	9	1.1132	28.275	0.5565	14.137	0.3543	9.000
3	8.4667	1.0472	26.599	0.5235	13.300	0.3333	8.467
3.1416	8.0851	1.0000	25.400	0.5000	12.700	0.3183	8.085
3.1750	**8**	0.9895	25.133	0.4948	12.566	0.3150	8.000
3.5000	7.2571	0.8976	22.799	0.4488	11.399	0.2857	7.257
3.6286	7	0.8658	21.991	0.4329	10.996	0.2756	7.000
3.9078	6.5000	0.8039	20.420	0.4020	10.210	0.2559	6.500
4	6.3500	0.7854	19.949	0.3927	9.975	0.2500	6.350
4.2333	**6**	0.7421	18.850	0.3710	9.425	0.2362	6.000
4.6182	5.5000	0.6803	17.279	0.3401	8.639	0.2165	5.500
5	5.0801	0.6283	15.959	0.3142	7.979	0.2000	5.080
5.0802	**5**	0.6184	15.707	0.3092	7.854	0.1968	5.000
5.3474	4.75	0.5875	14.923	0.2938	7.461	0.1870	4.750
5.6444	4.5	0.5566	14.133	0.2783	7.069	0.1772	4.500
6	4.2333	0.5236	13.299	0.2618	6.650	0.1667	4.233
6.3500	**4**	0.4947	12.555	0.2473	6.283	0.1575	4.000
6.7733	3.75	0.4638	11.781	0.2319	5.890	0.1476	3.750
7	3.6286	0.4488	11.399	0.2244	5.700	0.1429	3.629
7.2571	3.5	0.4329	10.996	0.2164	5.498	0.1378	3.500
7.8154	3.25	0.4020	10.211	0.2010	5.105	0.1279	3.250
8	3.1750	0.3927	9.974	0.1964	4.989	0.1250	3.175
8.4667	3	0.3711	9.426	0.1855	4.713	0.1181	3.000
9	2.8222	0.3491	8.867	0.1745	4.432	0.1111	2.822
9.2364	2.75	0.3401	8.639	0.1700	4.319	0.1082	2.750
10	2.5400	0.3142	7.981	0.1571	3.990	0.1000	2.540
10.1600	**2.50**	0.3092	7.854	0.1546	3.927	0.0984	2.500

NOTE: Bold face modules and diametral pitches designates preferred values.

MECHANICAL POWER TRANSMISSION SYSTEMS

Table 13-9. World Standard Modules and Diametral Pitches of Cylindrical Gears

Module m (mm)	DIAMETRAL PITCH P 1/in	25.4/P mm	ISO 54	JAPAN JIS B 1701	GERMANY DIN 780	FRANCE NF E 23-011	U.K. BS 4582	ITALY UNI 6586	AUSTRALIA AS DR 74059
0.05					F				
0.055					S				
0.06					F				
0.07					S		S		
0.08					F		F,0.075T		
0.09					S		S,0.085T		
0.1				F	F		F,0.095T		
0.11					S		S,0.105T		
0.12					F		F,0.115T		
0.14					S		S,0.13T		
0.16				0.15S	F		F,0.15T		
0.18					S		S,0.17T		
0.20				F	F		F,0.19T		
0.22					S		S,0.21T		
0.25				S	F		F,0.24T		
0.28					S		S,0.26T		
0.3				F	F		F		
0.35				S	S		S,0.32T		
0.4				F	F		F,0.38T		
0.45				S	S		S,0.42T		
0.5				F	F	F	F,0.48T	F	
0.55				S	S	S	S,0.52T		
0.6				F	F	F	F,0.58T		
0.65				T	S		T		
0.7				S	F	S	F		
0.75				S	S		T	S	
0.8				F	F	F	F		
0.85					S		T		
0.9				S	F	S	F		
0.95					S		T		
1.			F	F	F	F	F BS 436	F	F
1.125			S		S	S	S	S	S
1.25	20	1.27000	F	F	F	F	F	F	F
1.375	18	1.41111	S		S	S	S	S	S
1.5	16	1.58750	F	F	F	F	F	F	F
1.75	14	1.81429	S	S	S	S	S	S	S
2	12	2.11667	F	F	F	F	F	F	F
2.25	11	2.30909	S	S	S	S	S	S	S
2.5	10	2.54000	F	F	F	F	F	F	F
2.75	9	2.82222	S	S	S	S	S	S	S
3	8	3.17500	F	F	F	F	F	F	F
3.25			T	T	T			T	T
3.5	7	3.62857	S	S	S	S	S	S	S
3.75			T	T	T			T	T
4	6	4.23333	F	F	F	F	F	F	F
4.5	5.5	4.61818	S	S	S	S	S	S	S
5	5	5.0800	F	F	F	F	F	F	F
5.5	4.5	5.64444	S	S	S	S	S	S	S
6	4	6.35000	F	F	F	F	F	F	F
6.5			T	T	T			T	T
7	3.5	7.25714	S	S	S	S	S	S	S
8	3	8.46667	F	F	F	F	F	F	F
9	2.75	9.23636	S	S	S	S	S	S	S
10	2.5	10.16000	F	F	F	F	F	F	F
11	2.25	11.28889	S	S	S	S	S	S	S
12	2	12.70000	F	F	F	F	F	F	F
14	1.75	14.51429	S	S	S	S	S	S	S
16	1.5	16.93333	F	F	F	F	F	F	F
18			S	S	S	S	S	S	S
20	1.25	20.32000	F	F	F	F	F	F	F
22			S	S	S	S	S	S	S
25	1	25.40000	F	F	F	F	F	F	F
28	0.875	29.02857	S	S	S	S	S	S	S
32	0.75	33.86667	F	F	F	F	F	F	F
36			S	S	S	S	S	S	S
40	0.625	40.64000	F	F	F	F	F	F	F
45			S	S	S	S	S	S	S
50	0.5	50.80000	F	F	F,60F	F	F	F	F

NOTE: F = First Choice, S = Second Choice, T = Third Choice, Number = other module. Preference rating listed in ISO column refers to both modules and diametral pitches. All other national standards have metric modules only.

MECHANICAL POWER TRANSMISSION SYSTEMS

In this form the equation is completely free of units, and clearly shows it is equally applicable to metric and inch gears. However, in the form of Equation 13-15, this is a restricted and limited contact ratio equation in that it is applicable to only standard metric gears whose addendums are module (m), and standard inch gears with addendum values $1/P$.

A large number of equations are independent of the measuring units system and with metrification there is no problem. Most of the kinematic design equations that appear in American gear text books, and are associated with inch-system gears, are suitable for use with metric gear dimensions providing a proper substitution of module (m) is made for pitch.

For equations involving diametral pitch:

$$P \text{ is replaced by } \frac{25.4}{m}$$

Then recalling:

$$P \times p_c = \pi$$

For equations involving circular pitch:

$$p_c \text{ is replaced by } \frac{\pi}{25.4} m$$

NOTE: When converting between metric module and the inch diametral pitch the conversion factor and relationship can be remembered from the simple product of the two pitch measures, as previously given by Equation 13-3:

$$m \times P = 25.4$$

By this means all kinematic equations involving pitch parameters can be utilized for calculations. However, by the above conversion substitutions, equation results are still inch measurements. Thus, this is a way to adapt the metric module to the inch system of kinematic design equations.

Basic kinematic and geometric design equations for spur gears in both metric module and inch diametral pitch forms are given in Table 13-10. These equations show the essence of using the module versus inch diametral pitch. There are many other equations which are identical in both systems. Equations 9 and 15 of Table 13-10 are examples. Others are:
1. Over-pins measurement formula
2. Relationship of tooth thicknesses at different radii from the gear center
3. Long and short addendum equations
4. Profile-shifted gear design equations: i.e., enlarged gear teeth; non-standard center distances; etc.

*One newton (N) = 1/9.806650 kilogram force or approximately 102 gram force.
or 1 N = 1/4.448222 pound force or approximately 4 ounces force.

Gear Strength and Durability Rating Formulas

SI Units—Users of metric gearing, predominately in Europe and Japan, have developed strength and durability formulas that utilize metric units. Ultimately, when some time in the relatively near future SI units become universal and exclusively used, these formulations will need to be slightly modified to conform to the basic SI units. However, present metric based formulas are essentially in conformance except for scale (cm^2 vs m^2, etc.) and will be readily converted and directly usable with SI units. Thus, there will be no difficulty associated with any of the metric based formulas now in use.

Conversion of Classical Strength Formulas for Metric Usage

Lewis Formula—The oldest and most classic strength formula is that of Wilbur Lewis, presented in 1892. Now bearing his name, the Lewis Equation formulates beam strength as:

$$W_b = \frac{FSY}{P} \qquad (13\text{-}16)$$

Where: F = Gear-face width, inches
P = Diametral pitch
S = Stress, lbs/in^2
Y = Lewis tooth form factor
W_b = Transmitted tooth force, lbs.

Lewis developed this equation using inch units and its application in America has been on that basis. However, since Y is a dimensionless factor, the equation is equally suitable for use with metric units. In metric countries the Lewis equation has had equally wide use. For use with metric units it is more convenient to introduce the Lewis factor based on circular pitch, y, where the two Lewis factors are related by:

$$y = \frac{Y}{\pi} \qquad (13\text{-}17)$$

Then the metric formulation is (expressed in SI units):

$$W'_b = \pi m F' S' y \qquad (13\text{-}18)$$

and for metric units:

$W^1{}_b$ = Transmitted tooth force, in newtons (N)*
F^1 = Face width, in millimeters
m = Module, in millimeters
y = Lewis factor, dimensionless
S^1 = Stress, megapascals (MPa)**

**One megapascal = 1 MPa = 1 N/mm^2 = 1/9.806650 kilogram per square millimeter or approximately 0.1 kgf/mm^2.
1000 psi = 6.894757 MPa ≈ 7 MPa.

Table 13-10. Basic Spur Gear Design Equations

	TO OBTAIN	FROM KNOWN	METRIC FORMULA (all dimensions millimeters)	INCH DIAMETRAL PITCH FORMULA (all dimensions inches)
1.	Diametral Pitch	module	—	$P_d = \dfrac{25.4}{m}$
2.	Module	diametral pitch	$m = \dfrac{25.4}{P_d}$	—
3.	Circular pitch	module; diametral pitch	$P_c = m\pi$	$P_c = \dfrac{\pi}{P_d}$
4.	Pitch diameter	module; diametral pitch	$D = mN$	$D = \dfrac{N}{P_d}$
5.	No. of teeth	"	$N = \dfrac{D}{m}$	$N = DP_d$
6.	Addendum	"	$a = m$	$a = \dfrac{1}{P_d}$
7.	Dedendum	"	$b = 1.25m$	$b = \dfrac{1.25}{P_d}$
8.	Base pitch	"	$P_b = m\pi \cos\phi$	$P_b = \dfrac{\pi}{P_d}\cos\phi$
9.	Base circle diameter	pitch diameter	$D_b = D\cos\phi$	$D_b = D\cos\phi$
10.	Outside diameter	module; diametral pitch	$D_o = D + 2m$	$D_o = D + \dfrac{2}{P_d}$
11.	Root diameter	"	$D_r = D - 2.5m$	$D_r = D - \dfrac{2.5}{P_d}$
12.	Tooth thickness at standard pitch radius	"	$T_{Std.} = \dfrac{\pi}{2}m$	$T_{Std.} = \dfrac{\pi}{2P_d}$
13.	Center distance	"	$C = \dfrac{m(N_1 + N_2)}{2}$	$C = \dfrac{N_1 + N_2}{2P_d}$
14.	Contact ratio	module and metric dimensions	$m_p = \dfrac{\sqrt{{}_1R_o^2 - {}_1R_b^2} + \sqrt{{}_2R_o^2 - {}_2R_b^2} - C\sin\phi}{m\pi\cos\phi}$	$m_p = \dfrac{\sqrt{{}_1R_o^2 - {}_1R_b^2} + \sqrt{{}_2R_o^2 - {}_2R_b^2} - C\sin\phi}{\dfrac{\pi}{P_d}\cos\phi}$
15.	Min. number teeth for no undercutting	Pressure angle	$N_c = \dfrac{2}{\sin^2\phi}$	$N_c = \dfrac{2}{\sin^2\phi}$
16.	Backlash, (angular)	Change in center distance (metric; inch dimensions)	$_aB = \dfrac{4\Delta C \tan\phi}{D}$ (radians)	$_aB = \dfrac{4\Delta C \tan\phi}{D}$ (radians)

MECHANICAL POWER TRANSMISSION SYSTEMS

Associated with inch gearing the same value of Lewis factor, y, can be used. However, this is limited to metric gears that have addendums and dedendums proportioned in the same values as the American gear standards for which the y factor applies. Fortunately, the metric ISO gear standard and American inch gears are compatible in this regard. Nevertheless, this may not be true of all metric gears (non-ISO standard) and one must carefully watch for this possible difference, then use the proper Lewis factor, y, if a difference exists.

Buckingham Equations—Of equal notoriety are Buckingham's formulas for predicting gear mesh dynamic loading and durability. These classical equations are:

Dynamic loading:

$$W_d = \frac{.05V(FC+W)}{.05V + \sqrt{FC+W}} + W \qquad (13\text{-}20)$$

Where: W_d = Total dynamic load
W = Applied load
V = Pitchline velocity, ft per min.
F = Face width of gears, inches
C = Deformation factor

Wear formula (surface durability):

$$W_w = DFKQ \qquad (13\text{-}21)$$

Where: W_w = Equivalent static load for wear
D = Pitch diameter of pinion, inches
K = Load stress factor
Q = Ratio factor

Although the original equations were refined by Buckingham and others over many years of use, all versions are intimately related to the inch and other customary units. This is due to the introduction of constants and factors, such as C and K, and the disregard for consistency of units. Neither equation can fulfill a dimensional test, and therefore, are arbitrary arrangements of dimensioned parameters. Thus, they are largely empirically assembled parameter equations to give desired results. Units cannot be arbitrarily converted to metric measures without making appropriate changes to the empirically set constants and various factors. The result is that these equations and many similar strength and durability formulas are inappropriate when using metric gearing and SI units.

AGMA Rating Formulas—In the United States wide use is made of the AGMA rating formulas, which are based upon the inch system. For spur gears the basic formulas are:

Strength:

Tooth bending stress:

$$S_t = \frac{W_t K_o}{K_v} \frac{P_d}{F} \frac{K_s K_m}{J} \qquad (13\text{-}22)$$

Allowable stress:

$$S_t = \frac{K_L}{K_R K_T} S_{at} \qquad (13\text{-}23)$$

Surface Durability:

Surface stress:

$$S_c = C_P \sqrt{\frac{W_t C_o}{C_v} \frac{C_s}{dF} \frac{C_m C_f}{I}} \qquad (13\text{-}24)$$

Allowable stress

$$S_c = \frac{C_L C_H}{C_T C_R} S_{ac} \qquad (13\text{-}25)$$

Terms are defined in Table 13-11.

Table 13-11. Definitions of Terms in AGMA Rating Formulas

Term	Strength	Durability
Load:		
Transmitted Load	W_t	W_t
Dynamic Factor	K_v	C_v
Overload Factor	K_o	C_o
Size:		
Pinion Pitch Diameter	...	d
Net Face Width	F	F
Transverse Diametral Pitch	P_d	...
Size Factor	K_s	C_s
Stress Distribution:		
Load Distribution Factor	K_m	C_m
Geometry Factor	J	I
Surface Condition Factor	...	C_f
Stress:		
Calculated Stress	S_t	S_c
Allowable Stress	S_{at}	S_{ac}
Elastic Coefficient	...	C_p
Hardness-Ratio Factor	...	C_H
Life Factor	K_L	C_L
Temperature Factor	K_T	C_T
Factor of Safety	K_R	C_R

These formulas are directly related to the inch system through specific inch unit terms, the various factors, and the fact that the equations are not dimensionally consistent. Therefore, they cannot be arbitrarily used for metric gearing. These formulas are usable provided all metric gear

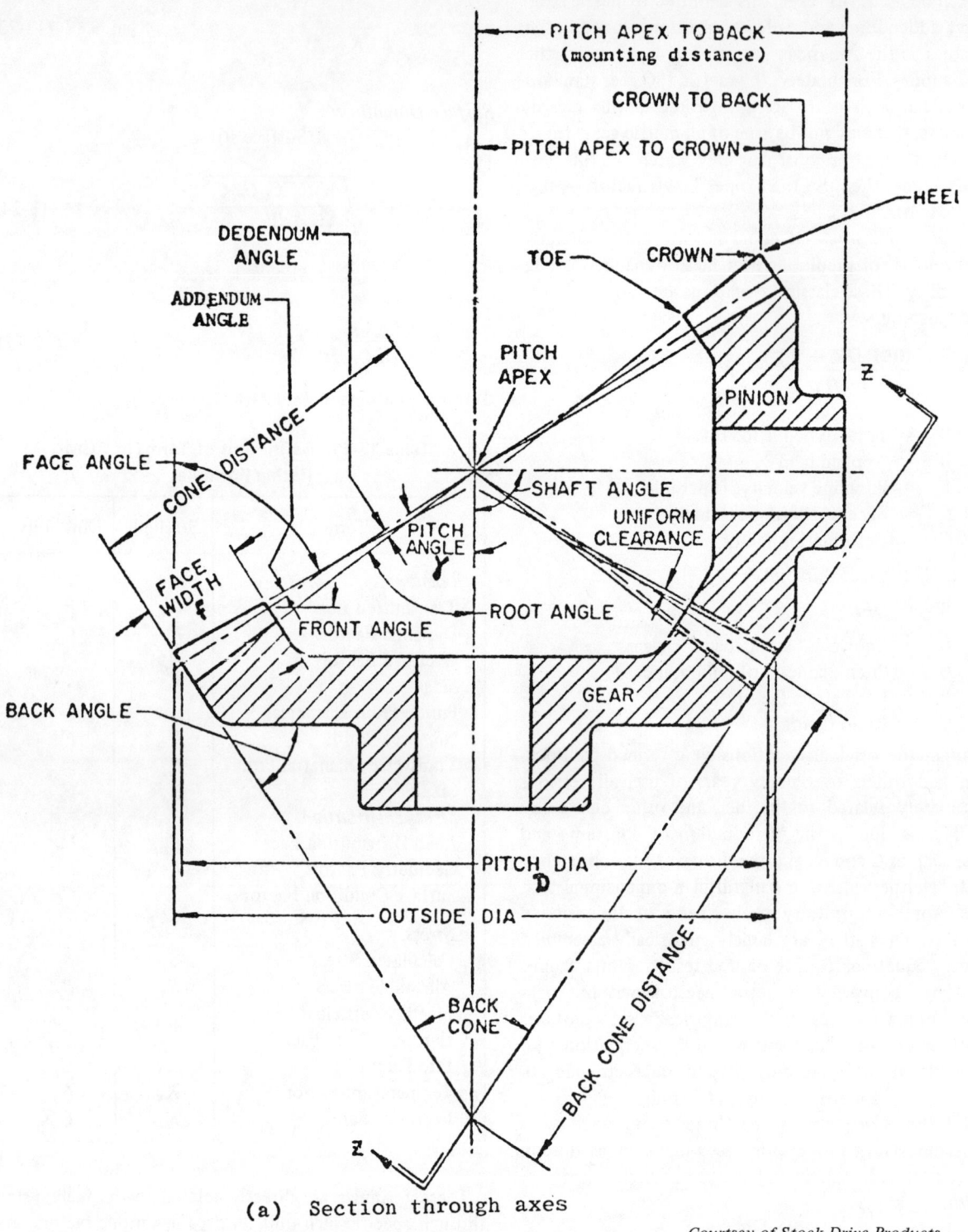

(a) Section through axes

Courtsey of Stock Drive Products

Fig. 13-6. Basic gear geometry, nomenclature and symbols for bevel gearing.

MECHANICAL POWER TRANSMISSION SYSTEMS

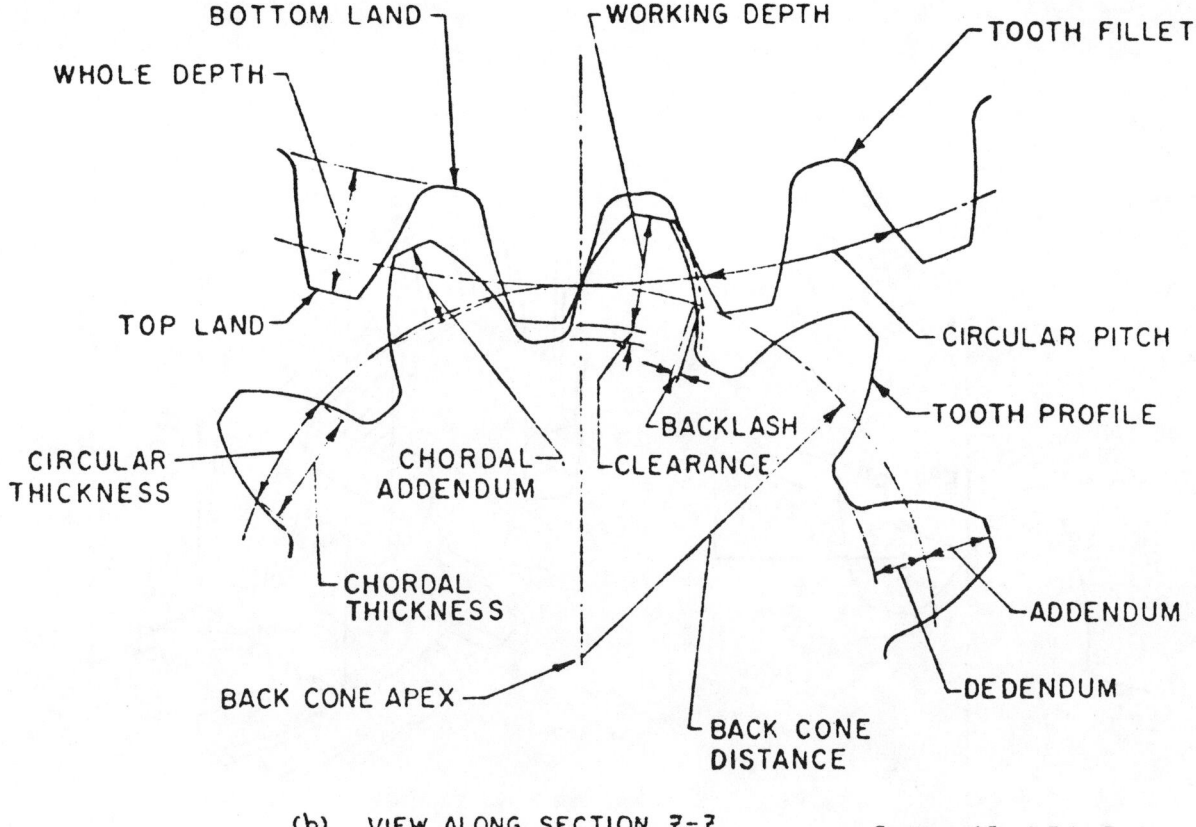

(b) VIEW ALONG SECTION Z-Z

Courtesy of Stock Drive Products

Fig. 13-6 (*Continued*). Basic gear geometry, nomenclature and symbols for bevel gearing.

dimensions are converted to customary units (to inches, diametral pitch, etc.). However, such an intermediary step reduces the utility and value of the formulas, since many of the factors come from graphs and tables which are also in customary units, requiring further intermediary conversions. An example is the velocity factor, K_v. This factor comes from graph plots of pitchline velocities, in feet per minute units.

Until these formulas and backup data are all converted to metric, usage of the formulas will be somewhat awkward. To assist, Equations (13-22) and (13-24) are modified to permit application to metric gears using preferred SI units. Equation (13-26) needs no conversion factor as long as the indicated metric SI units are consistently used. This is because the introduction of face width in millimeters and the metric module cancel each other's conversion factors. Equation (13-27) requires the 0.083 factor to accommodate conversion of $W'_t d'$ and F' within the radical.

$$S'_t = \frac{W'_t K_o}{K_v} \frac{1}{mF'} \frac{K_s K_m}{J} \qquad (13\text{-}26)$$

$$S'_c = 0.083\, C_P \sqrt{\frac{W'_t C_o}{C_v} \frac{C_s}{d'F'} \frac{C_m C_f}{I}} \qquad (13\text{-}27)$$

Where: S'_t = Bending stress, megapascals (1 MPa = 1 N/mm²)*
S'_c = Surface stress, megapascals
W'_t = Transmitted tooth load, newtons (N)*
m = Module, millimeters
F' = Face width, millimeters
d' = Pitch diameter, millimeters

The values of J and I will be, as now presented in AGMA literature, for inch gears, provided the metric gears have addendums and dedendums in the same proportion.

---
*For notes on the megapascal and newton, see page 13-21.

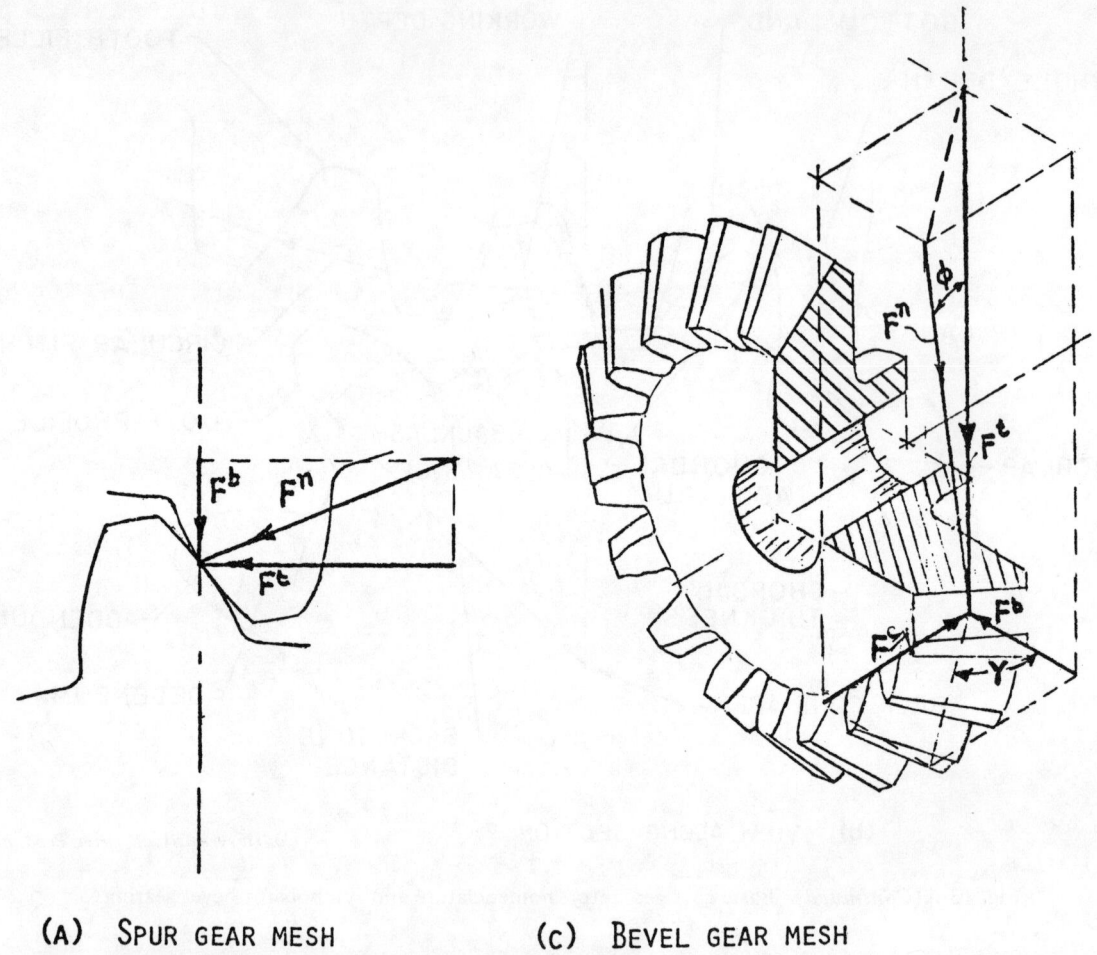

(A) Spur gear mesh

(C) Bevel gear mesh

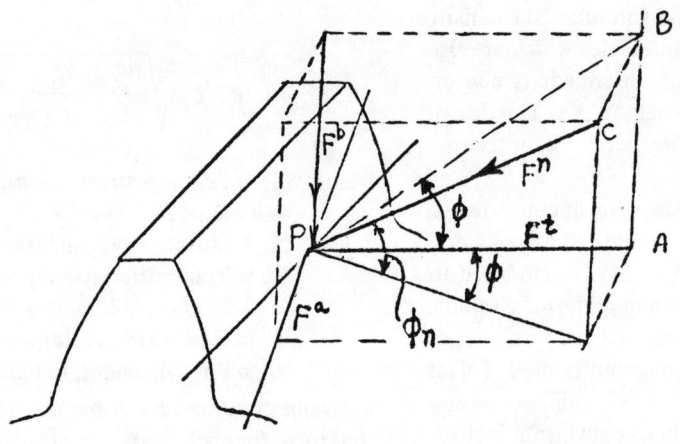

(B) Helical gear mesh

Courtesy of Stock Drive Products

Fig. 13-7. Force vectors in meshed gears.

MECHANICAL POWER TRANSMISSION SYSTEMS

Factors such as K_v and K_s must be obtained from inch-based values in the literature by converting the associated metric values that define the design conditions. Other miscellaneous factors, such as K_o, K_R, and K_L are independent of the units system and are directly applicable as listed in AGMA standards.

REFERENCES

1. D. W. Dudley. *Gear Handbook* (New York: McGraw-Hill Book Co.) 1962
2. Erik Oberg and Franklin D. Jones. *Machinery's Handbook*, 20th ed. (New York: Industrial Press, Inc.) 1975, pp. 735-990.
3. G. W. Michalec. "*Precision Gearing: Theory and Practice.*" (New York: John Wiley and Sons, Inc.) 1966, chs. 4 and 5.
4. *Sterling Instruments Catalog 80*, Designatronics Inc., New Hyde Park, N.Y. 11040

SPLINES

Introduction

Splines are used to transmit power from the power source to a driven machine. The use of splines is specially well-suited for applications that require axial flexibility such as: engine driveshafts, transmission output, etc. The most important types of splines are the straight spline, and the involute spline and serrations shown in Fig. 13-8.

Straight Spline

Metric dimensioned straight-sided splines for cylindrical shafts is specified in ISO/R14, and the sizes recommended have been in use for a number of years by countries already on the metric system. The ISO recommendation shows nominal sizes only, and manufacturing tolerances must be based on existing inch straight-sided spline practices or on national standards. (Listed at the end of this chapter.)

The values shown in Table 13-12 for d, D, and B are the nominal dimensions common to shaft and hub. The profile of a splined shaft or hub is designated by stating in order: the number of splines N, the minor diameter d, and the outside diameter D, all in millimeters as follows: 6 × 23 × 26.

Involute Splines

At the present time, there is no internationally recognized involute spline and serration standard. Incidentally, the American National Standards Institute (ANSI) has eliminated the term "serration" from involute standard terminology. In some foreign countries, the term is still used to describe involute splines with a nominal pressure angle of 45°. The task of establishing an international spline standard comes under the jurisdiction of the International Standards Organization Technical Committee (ISO TC32). The secretariat of this committee at the present time is held by France (AFNOR).

North American Spline Standard (ANSI-B92) The ANSI-B92 spline standard has been in effect in the United States for many years as ASA B5.15. It was most recently updated in 1970, and the designation was then changed. This standard features selections in pressure angles of 30, 37.5, and 45 degrees. Numbers of teeth from 6 to 60 in successive integers are tabularized except in finer pitches of 45-degree pressure angle where the range is 6 to 100. Nominal sizes range from 1/8-inch pitch diameter to 24-inch pitch diameter. Flat root or fillet root splines are offered in the 30° pressure angle while only fillet root types are offered in 37.5 and 45° pressure angles. Centering between mating parts is accomplished by tooth side fit. Major diameter centering is offered in 30° pressure angle only. The 1970 publication of this standard incorporated several additional changes which improved it from the standpoints of versatility and usability.

Summary of the Major Changes

1. Spline tables of 37.5° pressure angle with approximately the same range in diametral pitches in the 30° splines listed in the standard.
2. Three additional classes of tolerance ranges were added. The additional quality classes were designated Classes 4, 6, and 7. The original tolerance level in the ASA B5.15 was designated Class 5. The new Classes 4, 6, and 7 are 71%, 140%, and 200% of Class 5 values, respectively.
3. Greater clearance between the form diameter and the root diameter of the external part is allowed, to provide for more versatile use of standard shaper cutters. It is a feature of the standard that all external splines of a given pitch may be cut with a single standard shaper cutter.

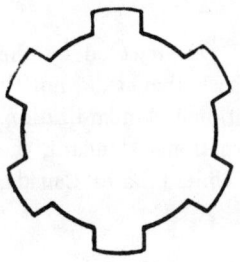 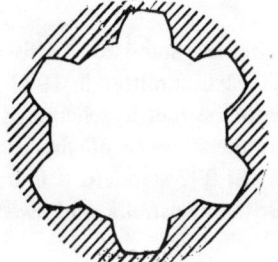

STRAIGHT SPLINE   INVOLUTE SPLINE

Fig. 13-8. Straight-sided and involute splines.

Table 13-12. Straight-Sided Splines for Cylindrical Shafts* (ISO 14)

	ISO	14
U.S.A.	ANSI	
JAPAN	JIS	B1601
GERMANY	DIN	5461
FRANCE	NF	
U.K.	BS	
ITALY	UNI	219-225
AUSTRALIA	AS	

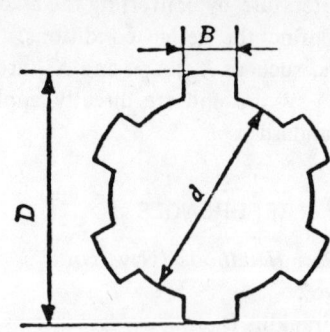

d	Light series Designation	N	D	B	Medium Series Designation	N	D	B
11					6 × 11 × 14	6	14	3
13					6 × 13 × 16	6	16	3.5
16					6 × 16 × 20	6	20	4
18					6 × 18 × 22	6	22	5
21					6 × 21 × 25	6	25	5
23	6 × 23 × 26	6	26	6	6 × 23 × 28	6	28	6
26	6 × 26 × 30	6	30	6	6 × 26 × 32	6	32	6
28	6 × 28 × 32	6	32	7	6 × 28 × 34	6	34	7
32	8 × 32 × 36	8	36	6	8 × 32 × 38	8	38	6
36	8 × 36 × 40	8	40	7	8 × 36 × 42	8	42	7
42	8 × 42 × 46	8	46	8	8 × 42 × 48	8	48	8
46	8 × 46 × 50	8	50	9	8 × 46 × 54	8	54	9
52	8 × 52 × 58	8	58	10	8 × 52 × 60	8	60	10
56	8 × 56 × 62	8	62	10	8 × 56 × 65	8	65	10
62	8 × 62 × 68	8	68	12	8 × 62 × 72	8	72	12
72	10 × 72 × 78	10	78	12	10 × 72 × 82	10	82	12
82	10 × 82 × 88	10	88	12	10 × 82 × 92	10	92	12
92	10 × 92 × 98	10	98	14	10 × 92 × 102	10	102	14
102	10 × 102 × 108	10	108	16	10 × 102 × 112	10	112	16
112	10 × 112 × 120	10	120	18	10 × 112 × 125	10	125	18

*Dimensions in millimeters

The basic tooth proportions as a ratio of diametral pitch were retained.

This basic standard had gained widespread use throughout the industrial nations prior to 1970; therefore, the dimensions were converted to metric equivalents by the ANSI-B92 Committee and published as a metric version of the B92-1970 Standard. This document was submitted to ISO TC32 in 1973 as a proposal for an International Standard, on the premise that it had been accepted as an inch-based standard on a worldwide basis and it would then maintain continuity of interchangeability as an International Standard. This proposal was rejected by the TC32 Committee in 1973 on the basis that it did not fit into the metric scheme of international standardization. In the absence of any other international standard, the ANSI-B92 standard is in use in the United States, Canada, Britain, Australia, and Sweden.

German Spline Standard (DIN 5480)—The other standard in contention for the international standard is DIN 5480 which is being proposed by the German delegation to ISO TC 32.

MECHANICAL POWER TRANSMISSION SYSTEMS

This standard is derived from metric modules selected from ISO 54 standard (see Table 13-9) ranging from 0.6-10, and provides for a basic pressure angle of 30°. The preferred modules are as follows: 0.8, 1.25, 2, 3, 5, and 8. Pressure angles of 37.5° and 45° are not provided in this standard.

The DIN 5480 standard has 14 sheets, and it includes tolerances and nominal sizes. The German standards are available in English or German from ANSI.

Basic rack profile proportions are constant throughout the range of metric modules. These proportions are similar, but not exactly equal to the proportions of the rack profile for the ANSI-B92 standard. The DIN 5480 standard provides for centering of the mating parts by the "side fit" method, but provides the option of centering by fit on the major or minor diameters.

The application of radial displacement of the profile is utilized for the purpose of obtaining major diameters of the internal parts equal to the bores of standard metric ball bearings. This is done to guarantee the largest possible transverse sectional area of the male member next to a standard ball bearing. The bore of the ball bearing is to slide freely over the major diameter of the male spline. This standard has been in use as a National Standard in Germany for many years and has gained popularity in the Eastern European countries in recent years. It has not gained wide acceptance in the Western nations, however, because of the added complications in tooling caused by the profile shift. It seems that a reasonable solution for the International Standard may be reached by compromising the basic ideas of both the B92 standard and the DIN 5480 standard. This is possible because the tooth proportions of the B92 standard and the DIN 5480 standard are similar; the gaging techniques are identical.

It is possible to arrive at a compromise in rack proportions and basic modules. Then, nominal splines at each integer number of teeth can be tabularized together with those numbers which require the profile shift to account for ball-bearing bores. Those numbers which have had profile shift applied should be highlighted in some manner to show that they are particularly suited for use adjacent to ball bearings. This type of proposal is in the planning stages and may be offered to TC32.

KEYS AND KEYWAYS

Introduction

The data presented in this chapter is based on the following ISO standards: ISO/R773, R774, 2491, and 2492. These recommendations include standards for square, rectangular, and flat section parallel and tapered keys with their corresponding keyways. The above standards are generally accepted and included in national standards in the major industrial countries of the world already on the metric system. Metric national standards for Woodruff keys do exist; however, a world standard on this type has not been published as yet (see Section 10, Tables 10-20 and 10-21, and ISO DIS 3912).

Material

The ISO standards listed above, recommend the use of a key steel having a tensile strength of not less than 590 MPa (86000 psi) in the finished condition. Keys are generally produced from ISO steel Fe 60 (see Table 10-43).

Dimensions and Tolerances

Dimensions and tolerances for keys and keyway sections are shown in Tables 13-14 through 13-21 and standard lengths, in Table 13-22. Tolerance on length is not specified in the ISO standards. DIN 6885 specifies the tolerances in lengths shown in Table 13-13. ISO and national standards are also listed for each type.

Table 13-13. Tolerances for Key Lengths (DIN 6885)

Key length		Tolerance	
from	to	key	keyway
10	28	−0.2	+0.2
32	80	−0.3	+0.3
90	250	−0.5	+0.5

Dimensions in millimeters

Fits

Three classes of standard fits are shown in Tables 13-15 and 13-19, and they are called "free," "normal," and "close" fits. The above fits have been used in Europe with satisfaction for many years. A brief description of the fits are as follows:

Free: Where the hub is required to slide over the key when in use.

Normal: Where the key is to be inserted in the keyway with minimum fitting, as is required for mass production assembly.

Close: Where an accurate fit of key is required. In this class, fitting will be required under maximum material conditions and if it is required to obtain these conditions, some selection of components may be necessary.

Table 13-14. Rectangular or Square Parallel Keys. (ISO 773)

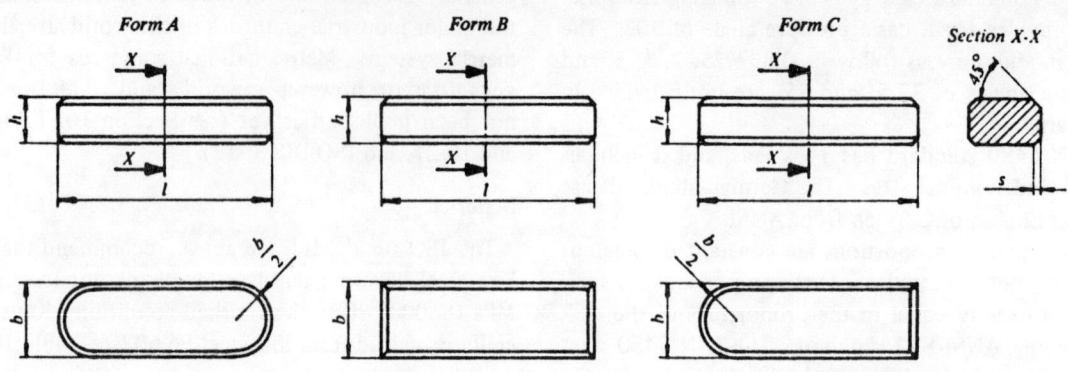

Dimensions in millimetres

Width b nominal	Tolerance h9	Thickness h nominal	Tolerance[1]	Chamfer s min.	max.	Range of lengths l [2] from	to
2	0	2	0	0.16	0.25	6	20
3	-0.025	3	-0.025	0.16	0.25	6	36
4		4		0.16	0.25	8	45
5	0	5	0	0.25	0.40	10	56
6	-0.030	6	-0.030	0.25	0.40	14	70
8	0	7		0.25	0.40	18	90
10	-0.036	8		0.40	0.60	22	110
12		8	0	0.40	0.60	28	140
14	0	9	-0.090	0.40	0.60	36	160
16	-0.043	10		0.40	0.60	45	180
18		11		0.40	0.60	50	200
20		12		0.60	0.80	56	220
22	0	14	0	0.60	0.80	63	250
25	-0.052	14	-0.110	0.60	0.80	70	280
28		16		0.60	0.80	80	320
32		18		0.60	0.80	90	360
36		20		1.00	1.20	100	400
40	0	22	0	1.00	1.20	–	–
45	-0.062	25	-0.130	1.00	1.20	–	–
50		28		1.00	1.20	–	–
56		32		1.60	2.00	–	–
63	0	32		1.60	2.00	–	–
70	-0.074	36	0	1.60	2.00	–	–
80		40	-0.160	2.50	3.00	–	–
90	0	45		2.50	3.00	–	–
100	-0.087	50		2.50	3.00	–	–

NOTES:
1. Tolerance on thickness h of the key: square section — h9; rectangular section — h11 (see Section 6, Table 6-26).
2. Preferred lengths are shown in Table 13-22.

MECHANICAL POWER TRANSMISSION SYSTEMS

Table 13-15. Keyways for Rectangular or Square Parallel Keys (ISO 773)

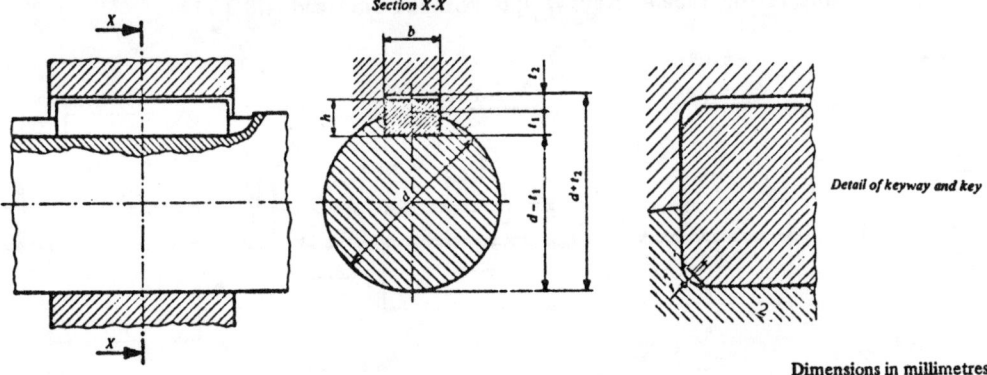

Detail of keyway and key

Dimensions in millimetres

Shaft Diameter d over	to	Key[1] Section $b \times h$	Keyway Width b nominal	Free keys Shaft H9	Free keys Hub D10	Normal keys Shaft N9	Normal keys Hub J_s9	Close keys Shaft and hub P9	Depth Shaft t_1 nom.	Tol.	Depth Hub t_2 nom.	Tol.	Radius r max.	min.
6	8	2 × 2	2	+0.025	+0.060	−0.004	+0.0125	−0.006	1.2		1		0.16	0.08
8	10	3 × 3	3	0	+0.020	−0.029	−0.0125	−0.031	1.8		1.4		0.16	0.08
10	12	4 × 4	4						2.5	+0.1 / 0	1.8	+0.1 / 0	0.16	0.08
12	17	5 × 5	5	+0.030	+0.078	0	+0.0150	−0.012	3.0		2.3		0.25	0.16
17	22	6 × 6	6	0	+0.030	−0.030	−0.0150	−0.042	3.5		2.8		0.25	0.16
22	30	8 × 7	8	+0.036	+0.098	0	+0.0180	−0.015	4.0		3.3		0.25	0.16
30	38	10 × 8	10	0	+0.040	−0.036	−0.0180	−0.051	5.0		3.3		0.40	0.25
38	44	12 × 8	12						5.0		3.3		0.40	0.25
44	50	14 × 9	14	+0.043	+0.120	0	+0.0215	−0.018	5.5		3.8		0.40	0.25
50	58	16 × 10	16	0	+0.050	−0.043	−0.0215	−0.061	6.0		4.3		0.40	0.25
58	65	18 × 11	18						7.0	+0.2 / 0	4.4	+0.2 / 0	0.40	0.25
65	75	20 × 12	20						7.5		4.9		0.60	0.40
75	85	22 × 14	22	+0.052	+0.149	0	+0.0260	−0.022	9.0		5.4		0.60	0.40
85	95	25 × 14	25	0	+0.065	−0.052	−0.0260	−0.074	9.0		5.4		0.60	0.40
95	110	28 × 16	28						10.0		6.4		0.60	0.40
110	130	32 × 18	32						11.0		7.4		0.60	0.40
130	150	36 × 20	36						12.0		8.4		1.00	0.70
150	170	40 × 22	40	+0.062	+0.180	0	+0.0310	−0.026	13.0		9.4		1.00	0.70
170	200	45 × 25	45	0	+0.080	−0.062	−0.0310	−0.088	15.0		10.4		1.00	0.70
200	230	50 × 28	50						17.0		11.4		1.00	0.70
230	260	56 × 32	56						20.0	+0.3 / 0	12.4	+0.3 / 0	1.60	1.20
260	290	63 × 32	63	+0.074	+0.220	0	+0.0370	−0.032	20.0		12.4		1.60	1.20
290	330	70 × 36	70	0	+0.100	−0.074	−0.0370	−0.106	22.0		14.4		1.60	1.20
330	380	80 × 40	80						25.0		15.4		2.50	2.00
380	440	90 × 45	90	+0.087	+0.260	0	+0.0435	−0.037	28.0		17.4		2.50	2.00
440	500	100 × 50	100	0	+0.120	−0.087	−0.0435	−0.124	31.0		19.5		2.50	2.00

NOTES:
1. The relation between the diameter and the section of key applies to normal use.
2. The depth of keyways in shafts and hubs should be obtained by direct measurement or by measuring the dimensions $(d - t_1)$ and $(d + t_2)$.

Table 13-16. Tapered Keys With or Without Gib Head (ISO 774)

Dimensions in millimetres

Width b nominal	Tolerance h9	Thickness h nominal	Tolerance [1]	Chamfer s min.	max.	Range of lengths l [2] from	to	Gib head h_1 nominal
2	0	2	0	0.16	0.25	6	20	—
3	− 0.025	3	− 0.025	0.16	0.25	6	36	—
4		4		0.16	0.25	8	45	7
5	0	5	0	0.25	0.40	10	56	8
6	− 0.030	6	− 0.030	0.25	0.40	14	70	10
8	0	7		0.25	0.40	18	90	11
10	− 0.036	8		0.40	0.60	22	110	12
12		8	0	0.40	0.60	28	140	12
14	0	9	− 0.090	0.40	0.60	36	160	14
16	− 0.043	10		0.40	0.60	45	180	16
18		11		0.40	0.60	50	200	18
20		12		0.60	0.80	56	220	20
22	0	14	0	0.60	0.80	63	250	22
25	− 0.052	14	− 0.110	0.60	0.80	70	280	22
28		16		0.60	0.80	80	320	25
32		18		0.60	0.80	90	360	28
36	0	20		1.00	1.20	100	400	32
40	− 0.062	22	0	1.00	1.20	—	—	36
45		25	− 0.130	1.00	1.20	—	—	40
50		28		1.00	1.20	—	—	45
56		32		1.60	2.00	—	—	50
63	0	32		1.60	2.00	—	—	50
70	− 0.074	36	0	1.60	2.00	—	—	56
80		40	− 0.160	2.50	3.00	—	—	63
90	0	45		2.50	3.00	—	—	70
100	− 0.087	50		2.50	3.00	—	—	80

NOTES:
1. Tolerance on thickness h of the key: square section — h9; rectangular section — h11 (see Table 6-26).
2. Preferred lengths are shown in Table 13-22.

MECHANICAL POWER TRANSMISSION SYSTEMS

Table 13-17. Keyways for Tapered Keys With or Without Gib Head (ISO 774)

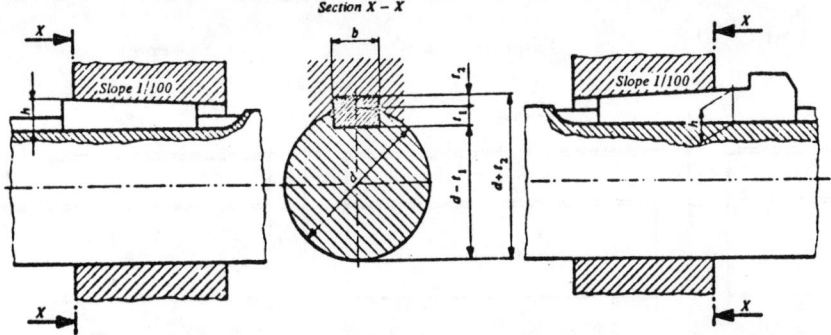

Dimensions in millimetres

Shaft Diameter d over	to	Key[1] Section $b \times h$	Keyway Width Shaft and hub b nominal	Tolerance D 10	Depth[2] Shaft t_1 nominal	Tolerance	Hub t_2[3] nominal	Tolerance	Radius r max.	min.
6	8	2 × 2	2	+0.060 +0.020	1.2	+0.1 0	0.5	+0.1 0	0.16	0.08
8	10	3 × 3	3		1.8		0.9		0.16	0.08
10	12	4 × 4	4		2.5		1.2		0.16	0.08
12	17	5 × 5	5	+0.078 +0.030	3.0		1.7		0.25	0.16
17	22	6 × 6	6		3.5		2.2		0.25	0.16
22	30	8 × 7	8	+0.098 +0.040	4.0		2.4		0.25	0.16
30	38	10 × 8	10		5.0		2.4		0.40	0.25
38	44	12 × 8	12		5.0		2.4		0.40	0.25
44	50	14 × 9	14	+0.120 +0.050	5.5		2.9		0.40	0.25
50	58	16 × 10	16		6.0	+0.2 0	3.4	+0.2 0	0.40	0.25
58	65	18 × 11	18		7.0		3.4		0.40	0.25
65	75	20 × 12	20		7.5		3.9		0.60	0.40
75	85	22 × 14	22	+0.149 +0.065	9.0		4.4		0.60	0.40
85	95	25 × 14	25		9.0		4.4		0.60	0.40
95	110	28 × 16	28		10.0		5.4		0.60	0.40
110	130	32 × 18	32		11.0		6.4		0.60	0.40
130	150	36 × 20	36	+0.180 +0.080	12.0		7.1		1.00	0.70
150	170	40 × 22	40		13.0		8.1		1.00	0.70
170	200	45 × 25	45		15.0		9.1		1.00	0.70
200	230	50 × 28	50		17.0		10.1		1.00	0.70
230	260	56 × 32	56		20.0	+0.3 0	11.1	+0.3 0	1.60	1.20
260	290	63 × 32	63	+0.220 +0.100	20.0		11.1		1.60	1.20
290	330	70 × 36	70		22.0		13.1		1.60	1.20
330	380	80 × 40	80		25.0		14.1		2.50	2.00
380	440	90 × 45	90	+0.260 +0.120	28.0		16.1		2.50	2.00
440	500	100 × 50	100		31.0		18.1		2.50	2.00

NOTES:
1. The relation between the diameter of shaft and the section of key applies to normal use.
2. The depth of keyways in shafts and hubs should be obtained by direct measurement or by measuring the dimensions $(d - t_1)$ and $(d + t_2)$.
3. The depth t_2 should be measured at the end of the hub at the side where the key enters.

Table 13-18. Thin Parallel Keys (ISO 2491)

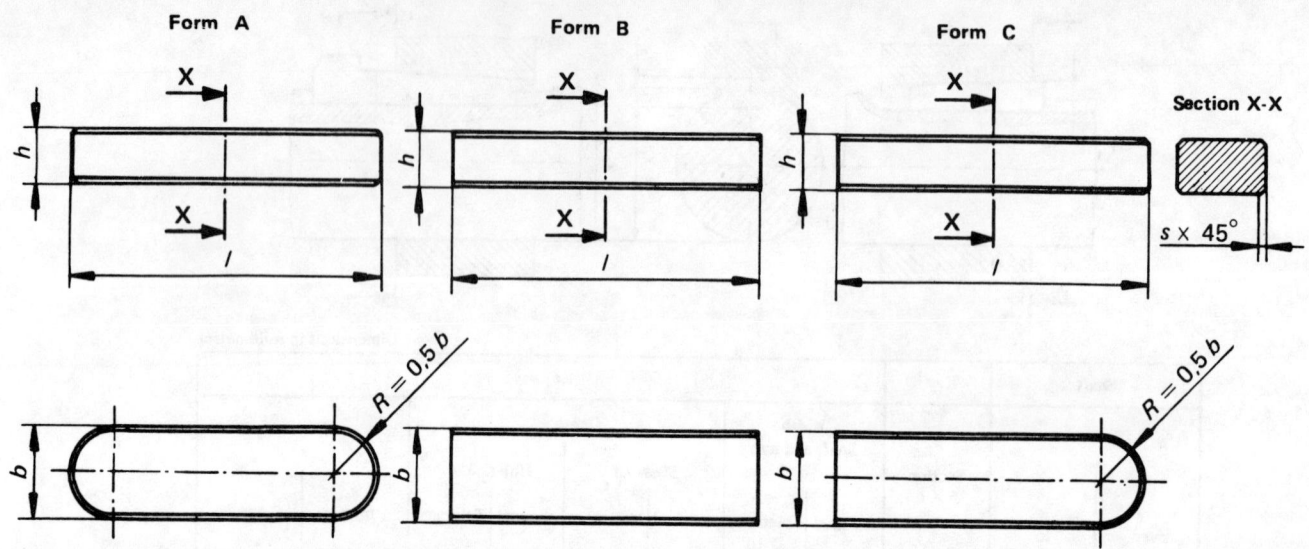

Values in millimetres

Width b		Thickness h		Chamfer[1] s		Length[2] l	
nominal	tolerance h9	nominal	tolerance h11	min.	max.	Range from	to
5	0 −0,030	3	0 −0,060	0,25	0,40	10	56
6		4		0,25	0,40	14	70
8	0 −0,036	5		0,25	0,40	18	90
10		6	0 −0,075	0,40	0,60	22	110
12		6		0,40	0,60	28	140
14	0 −0,043	6		0,40	0,60	36	160
16		7		0,40	0,60	45	180
18		7		0,40	0,60	50	200
20		8	0 −0,090	0,60	0,80	56	220
22	0 −0,052	9		0,60	0,80	63	250
25		9		0,60	0,80	70	280
28		10		0,60	0,80	80	320
32	0 −0,062	11	0 −0,110	0,60	0,80	90	360
36		12		1,00	1,20	100	400

NOTES:
1. Only the longitudinal edges and those of the rounded ends shall be chamfered; the other edges are merely broken.
2. Preferred lengths are shown in Table 13-22.

MECHANICAL POWER TRANSMISSION SYSTEMS

Table 13-19. Keyways for Thin Parallel Keys (ISO 2491)

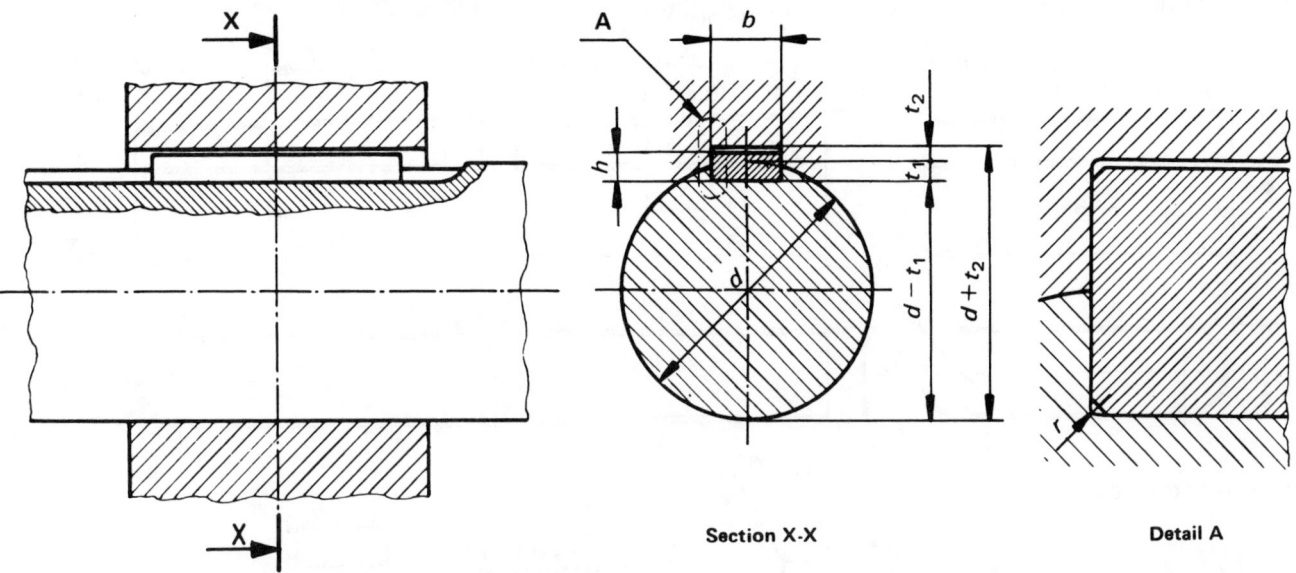

Section X-X   Detail A

Values in millimetres

Shaft Diameter d over	to	Key Section b × h	Keyway Width b nominal	Free Shaft H9	Free Hub D10	Normal Shaft N9	Normal Hub J$_s$9	Close Shaft and hub P9	Depth Shaft t_1 nom.	tol.	Hub t_2 nom.	tol.	Radius r max.	min.
12	17	5 × 3	5	+0,030	+0,078	0	±0,015	−0,012	1,8		1,4		0,25	0,16
17	22	6 × 4	6	0	+0,030	−0,030		−0,042	2,5		1,8		0,25	0,16
22	30	8 × 5	8	+0,036	+0,098	0	±0,018	−0,015	3	+0,1	2,3	+0,1	0,25	0,16
30	38	10 × 6	10	0	+0,040	−0,036		−0,051	3,5	0	2,8	0	0,40	0,25
38	44	12 × 6	12						3,5		2,8		0,40	0,25
44	50	14 × 6	14	+0,043	+0,120	0	±0,0215	−0,018	3,5		2,8		0,40	0,25
50	58	16 × 7	16	0	+0,050	−0,043		−0,061	4		3,3		0,40	0,25
58	65	18 × 7	18						4		3,3		0,40	0,25
65	75	20 × 8	20						5		3,3		0,60	0,40
75	85	22 × 9	22	+0,052	+0,149	0	±0,026	−0,022	5,5	+0,2	3,8	+0,2	0,60	0,40
85	95	25 × 9	25	0	+0,065	−0,052		−0,074	5,5	0	3,8	0	0,60	0,40
95	110	28 × 10	28						6		4,3		0,60	0,40
110	130	32 × 11	32	+0,062	+0,180	0	±0,031	−0,026	7		4,4		0,60	0,40
130	150	36 × 12	36	0	+0,080	−0,062		−0,088	7,5		4,9		1,00	0,70

NOTES:
1. The relation between the diameter of the shaft and the section of the key must be strictly respected.
2. The depth of keyways in shafts and hubs are obtained by direct measurement or by measuring the dimensions $(d - t_1)$ and $(d + t_2)$.

Table 13-20. Thin Taper Keys With or Without Gib Head (ISO 2492)

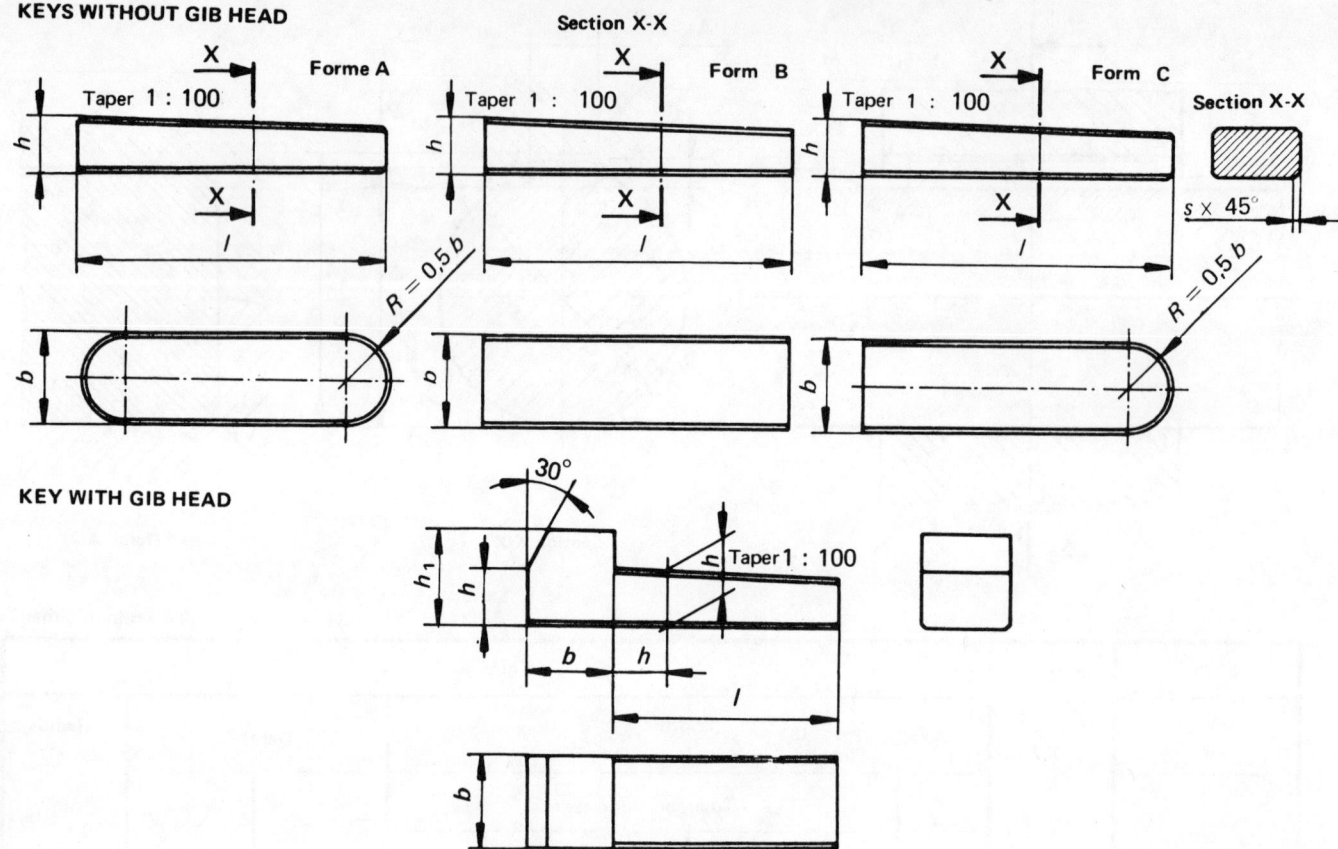

Values in millimetres

Width b nominal	Width b tolerance h9 [3]	Thickness h nominal	Thickness h tolerance h11 [3]	Chamfer [1] s min.	Chamfer [1] s max.	Length [2] l Range from	Length [2] l Range to	Gib head h_1 nom.
8	0 −0,036	5	0 −0,075	0,25	0,40	20	70	8
10	0 −0,036	6	0 −0,075	0,40	0,60	25	90	10
12	0 −0,036	6	0 −0,075	0,40	0,60	32	125	10
14	0 −0,043	6	0 −0,075	0,40	0,60	36	140	10
16	0 −0,043	7		0,40	0,60	45	180	11
18	0 −0,043	7		0,40	0,60	50	200	11
20	0 −0,052	8	0 −0,090	0,60	0,80	56	220	12
22	0 −0,052	9	0 −0,090	0,60	0,80	63	250	14
25	0 −0,052	9	0 −0,090	0,60	0,80	70	280	14
28	0 −0,052	10	0 −0,090	0,60	0,80	80	320	16
32	0 −0,062	11	0 −0,110	0,60	0,80	90	360	18
36	0 −0,062	12	0 −0,110	1,00	1,20	100	400	20
40	0 −0,062	14	0 −0,110	1,00	1,20	125	400	22
45	0 −0,062	16	0 −0,110	1,00	1,20	140	400	25
50	0 −0,062	18	0 −0,110	1,00	1,20	160	400	28

NOTES:
1. Only the longitudinal edges and those of the rounded ends shall be chamfered; the other edges shall be merely broken.
2. For key lengths see Table 13-22.
3. The tolerances h9 and h11 apply only to the dimensions of the section of the key.

MECHANICAL POWER TRANSMISSION SYSTEMS

Table 13-21. Keyways for Thin Taper Keys With or Without Gib Head (ISO 2492)

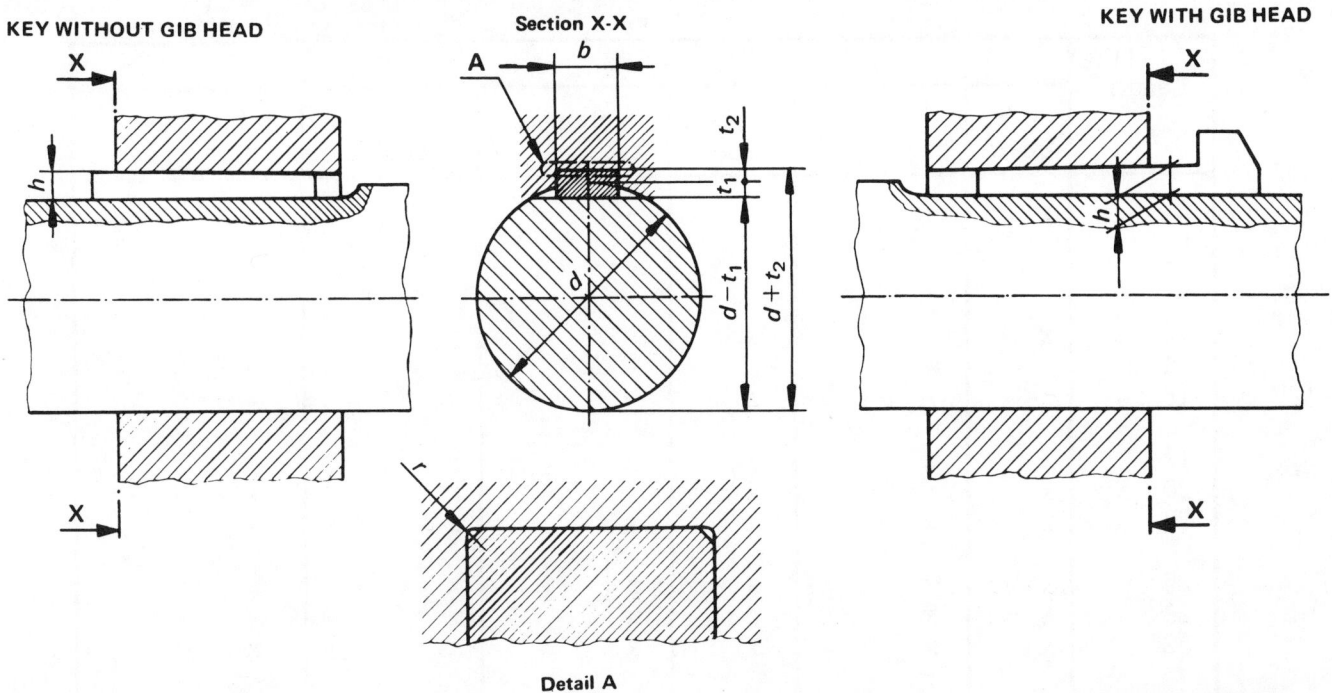

Values in millimetres

Shaft Diameter d		Key[1) Section $b \times h$	Keyway (hub) Width b		Depth[2) t_2		Radius r		Flat[3) (shaft) Height[2) t_1	
over	to		nom.	tol. D10	nom.	tol.	max.	min.	nom.	tol.
22	30	8 × 5	8	+ 0,098	1,7		0,25	0,16	3	
30	38	10 × 6	10	+ 0,040	2,2	+ 0,1	0,40	0,25	3,5	+ 0,1
38	44	12 × 6	12		2,2	0	0,40	0,25	3,5	0
44	50	14 × 6	14	+ 0,120	2,2		0,40	0,25	3,5	
50	58	16 × 7	16	+ 0,050	2,4		0,40	0,25	4	
58	65	18 × 7	18		2,4		0,40	0,25	4	
65	75	20 × 8	20		2,4		0,60	0,40	5	
75	85	22 × 9	22	+ 0,149	2,9		0,60	0,40	5,5	
85	95	25 × 9	25	+ 0,065	2,9	+ 0,2	0,60	0,40	5,5	+ 0,2
95	110	28 × 10	28		3,4	0	0,60	0,40	6	0
110	130	32 × 11	32		3,4		0,60	0,40	7	
130	150	36 × 12	36		3,9		1,00	0,70	7,5	
150	170	40 × 14	40	+ 0,180	4,4		1,00	0,70	9	
170	200	45 × 16	45	+ 0,080	5,4		1,00	0,70	10	
200	230	50 × 18	50		6,4		1,00	0,70	11	

NOTES:
1. The relation between the diameter of the shaft and the section of the key must be strictly respected.
2. The depth of the keyway in the hub and the height of the flat on the shaft should be obtained by direct measurement or by measuring the dimensions $(d - t_1)$ and $(d + t_2)$.
3. Subject to agreement between customer and manufacturer, the flat on the shaft may be replaced by a keyway with the same width (including tolerances) as that of the keyway in the hub and with a depth equal (including tolerances) to the height of the flat.

Table 13-22. Preferred Lengths of Keys

Dimensions in millimetres

(1) (2) Length	square	rectangular	square taper	rectangular taper	thin rectangular	thin taper rectangular
6	x	-	x	-	-	-
8	x	-	x	-	-	-
10	x	-	x	-	x	-
12	x	-	x	-	x	-
14	x	-	x	-	-	-
16	x	-	x	-	x	-
18	x	x	x	x	x	-
20	x	x	x	x	x	x
22	x	x	x	x	x	x
25	x	x	x	x	x	x
28	x	x	x	x	x	x
32	x	x	x	x	x	x
36	x	x	x	x	x	x
40	x	x	x	x	x	x
45	x	x	x	x	x	x
50	x	x	x	x	x	x
56	x	x	x	x	x	x
63	x	x	x	x	x	x
70	x	x	x	x	x	x
80	-	x	-	x	x	x
90	-	x	-	x	x	x
100	-	x	-	x	x	x
110	-	x	-	x	x	x
125	-	x	-	x	x	x
140	-	x	-	x	x	x
160	-	x	-	x	x	x
180	-	x	-	x	x	x
200	-	x	-	x	x	x
220	-	x	-	x	x	x
250	-	x	-	x	x	x
280	-	x	-	x	x	x
320	-	x	-	x	x	x
360	-	x	-	x	x	x
400	-	x	-	x	x	x

NOTES:
1. The lengths are those shown as standard in ISO/R 773, R 774, 2491, 2492.
2. The nominal lengths for keys shown are specified in national standards in countries already on the metric system.

MECHANICAL POWER TRANSMISSION SYSTEMS

RELATED ISO (TC 14) AND IEC STANDARDS FOR ELECTRIC MOTORS

ISO 496-1973	Shaft heights for driving and driven machines
ISO/R773-1969	Rectangular or square parallel keys and their corresponding keyways
ISO/R774-1969	Taper keys with or without gib head and their corresponding keyways
ISO R775-1969	Cylindrical and 1/10 conical shaft ends
ISO 2491-1974	Thin parallel keys and their corresponding keyways
ISO 2492-1974	Thin taper keys with or without gib head and their corresponding keyways
ISO 3117-1977	Tangential keying
ISO 3912-1977	Woodruff keys and keyways
ISO DPR 4328	Centre holes
ISO DPR 4329	Shaft ends with woodruff keys
ISO DPR 4455	Fixing method for parallel keys
ISO DPR 4456	Blending profiles of shouldered shafts
ISO DPR 4457	Fitments bores for shaft ends
ISO DPR 4458	Retaining washers for shaft ends
ISO DPR 4459	Length of keys for shaft ends
ISO DPR 4460	Mounted half couplings
ISO DPR 4461	Forged on flanges
IEC 34-1	Rotating electrical machines, Part 1: rating and performance
IEC 34-2	Rotating electrical machines, Part 2: methods for determining losses and efficiency of rotating electrical machinery from tests (excluding machines for traction vehicles)
IEC 34-3	Rotating electrical machines, Part 3: ratings and characteristics of three-phase, 50 Hz turbine-type machines
IEC 34-4	Recommendations for rotating electrical machinery (excluding machines for traction vehicles), Part 4: Methods for determining synchronous machine quantities from tests
IEC 34-5	Rotating electrical machines, Part 5: degrees of protection of enclosures for rotating machinery
IEC 34-6	Rotating electrical machines, Part 6: methods of cooling rotating machinery
IEC 34-7	Rotating electrical machines, Part 7: symbols for types of construction and mounting arrangements of rotating electrical machinery
IEC 34-8	Rotating electrical machines, Part 8: terminal markings and direction of rotation of rotating machines
IEC 34-9	Rotating electrical machines, Part 9: noise limits
IEC 38	Standard voltages
IEC 59	Standard current ratings
IEC 72	Dimensions and output ratings for foot-mounted electrical machines with frame numbers 355 to 1000.
IEC 72A	First supplement
IEC 117-1 through 16	Recommended graphical symbols
IEC 182-1	Basic dimensions of winding wires, Part 1: diameters of conductors for round winding wires
IEC 182-2	Basic dimensions of winding wires, Part 2- maximum overall diameters of enamelled round winding wires, including supplement 182-2A
IEC 182-2A	First supplement to publication 182-2 (1964), sold separately
IEC 182-3	Part 3: Dimensions of conductors of insulated rectangular copper winding wires
IEC 182-4	Basic dimensions of winding wires, Part 4: diameters of conductors for round resistance wires

ISO (TC 100) ROLLER CHAIN STANDARDS

ISO 487-1976	Steel roller chains, types S 32 to S 88, with their associated chain wheels
ISO/R606-1967	Short pitch transmission precision roller chains and chain wheels
ISO 1275-1972	Extended pitch precision roller chains and chain wheels for transmission and conveyors
ISO 1395-1977	Short pitch transmission precision bush chains and chain wheels
ISO 1977/1-1976	Conveyor chains, attachments and chain wheels, Part 1: Chains (metric series)
ISO 1977/2-1974	Conveyor chains, attachments and chain wheels, Part 2: chain wheels (metric series)
ISO 1977/3-1974	Conveyor chains, attachments and chain wheels, Part 3: attachments (metric series)
ISO 3512-1976	Heavy duty cranked link transmission chains

ISO 4347-1977	Leaf chains, clevises, and sheaves
ISO DIS 4348	Hinge-type flat top conveyor chains and sprocket teeth
ISO DPR 6039	Light duty chains, attachments and chain wheels for agricultural conveyors

INTERNATIONAL (ISO TC 41) V-BELT STANDARDS

ISO/R52-1957	Grooved pulleys for V-belts—Groove sections A, B, C, D, E
ISO/R253-1962	Grooved pulleys for V-belts—Groove sections Y and Z
ISO/R255-1962	Geometrical inspection of grooves of pulleys for V-belts
ISO/R256-1962	Section checking of V-belts
ISO/R434-1965	Lengths of Y-section V-belts (1p = 53. mm or 0.21 in.)
ISO/R459-1965	Grooved pulleys for narrow V-belts—Groove sections SPZ, SPA, SPB
ISO/R460-1965	Lengths of narrow V-belts—Sections SPZ, SPA, SPB
ISO/R608-1967	Lengths of classical V-belts—Sections Z, A, B, C, D, E
ISO/R1081-1969	Terms and definitions relating to drives using V-belts and grooved pulleys
ISO 1604-1976	Endless wide V-belts for industrial speed-changers and groove profiles for corresponding pulleys
ISO 1813-1976	Antistatic endless V-belts (Sections Y, Z, A, B, C, D, E)—Electrical conductivity—Characteristic and method of test
ISO 2790-1974	Narrow V-belt drives for the automotive industry—Dimensions
ISO 3410-1976	Endless variable speed V-belts for agricultural machinery
ISO DIS 5287	Bending test on belts for automobiles
ISO DPR 5288	Synchronous belt drives—Nomenclature
ISO DIS 5289	Hexagonal belts for agricultural machinery
ISO DIS 5290	Narrow twin belts
ISO DPR 5291	Standard twin belts
ISO DPR 5292	Transmissible powers of V-belts
ISO DPR 5294	Pulleys for synchronous belt drives
ISO DPR 5295	Transmissibles powers of synchronous belt drives
ISO DIS 5296	Synchronous belt drives—Belts

NATIONAL AND INTERNATIONAL (ISO TC 60) GEAR STANDARDS

INTERNATIONAL

ISO 53-1974	Basic rack of cylindrical gears for general and heavy engineering usage
ISO 54-1977	Modules and diametral pitches of cylindrical gears for general engineering and for heavy engineering, 2nd edition
ISO/R467-1966	Preferred modules and diametral pitches of cylindrical gears for general engineering
ISO 677-1976	Basic rack of straight bevel gears for general engineering and heavy engineering
ISO 678-1976	Modules and diametral pitches of straight bevel gears for general engineering and heavy engineering
ISO 701-1976	International gear notation—Symbols for geometrical data
ISO/R1122-1969	Glossary of gears—Geometrical definitions
ISO 1328-1975	Accuracy of parallel involute gears
ISO 1340-1976	Cylindrical gears—Information to be given to the manufacturer by the producer
ISO 1341-1976	Straight bevel gears—Information to be given to the manufacturer by the purchaser in order to obtain the gear required
ISO 2490-1975	Nominal dimension of single start gear hobs with axial keyway
ISO DPR 4467	Exterior parallel cylincrical gears—Division of teeth—Part I—Reducing gears for general engineering

GEAR DESIGN DOCUMENTS UNDER DEVELOPMENT IN ISO TC 60 WORKING GROUP 6

ISO/TC60 WG6-165E	Strength of spur and helical gear teeth
ISO/TC60 WG6-166E	Durability of spur and helical gear teeth
ISO/TC60 WG6-167E	Units for gear rating
ISO/TC60 WG6-168E	Symbols for calculating cylindrical gears

MECHANICAL POWER TRANSMISSION SYSTEMS

NORTH AMERICA—Basic Inch ANSI and AGMA Gear Standards

ANSI No.	AGMA No.	Title
B6.12-1964	110.03	Gear-tooth wear and failure
	112.04	Gear nomenclature—Terms, definitions, symbols and abbreviations
	115.01	Reference information—Basic gear geometry
	116.01	Glossary—Terms used in gearing
	150.03	Application classification for spur, helical, herringbone, and bevel gear motors
	151.02	Application classification for helical, herringbone, and spiral bevel gear speed reducers
B6.1-1968	201.02	Tooth proportions for coarse-pitch involute spur gears
	202.03	System for zero[1] bevel gears
	203.03	Fine-pitch on-center face gears for 20-degree involute spur pinions
B6.7-1967	207.05	Tooth proportions for fine-pitch involute spur and helical gears
B6.13-1965	208.02	System for straight bevel gears
	209.03	System for spiral bevel gears
	211.02	Surface durability (pitting) of helical and herringbone gear teeth
	215.01	Information sheet for surface durability (pitting) of spur, helical, herringbone, and bevel gear teeth
	221.02	Rating the strength of helical and herringbone gear teeth
	225.01	Information sheet for strength of spur, helical, herringbone and bevel gear teeth
	239.01	Measuring methods and practices manual for control of spur, helical, and herringbone gears
	241.02	Specification for general industrial gear materials—Steel (drawn, rolled, and forged)
	244.02	Nodular iron gear materials
	250.02	Lubrication of industrial enclosed gearing, includes appendix sheet 250.02A, typical manufacturer's oils meeting standard AGMA 250.02
	255.02	Bolting (allowable tensile stress) for gear drives
	260.01	Shafting—Allowable torsional and bending stresses
	265.01	Bearings—Allowable loads and speeds
	271.03	Ratios for helical, herringbone, and combination spiral bevel gear speed reducers
	290.02	Marking for enclosed gear drives
	291.01	Information sheet—Reducer assembly designations
	295.03	Specification for measurement of sound on high-speed helical gears
B6.9-1962	374.03	Design for fine pitch worm gearing
	390.02	Gear classification manual
	420.03	Practice for helical and herringbone gear speed reducers and increasers
	421.06	Practice for high-speed helical and herringbone gear units

JAPAN[1]

Standard Number	Title
JIS B 0121-1961	Letter symbols for gears
JIS B 1701-1963	Involute gear tooth profile
JIS B 1701-1960	Accuracy for spur and helical gears
JIS B 1703-1968	Backlash for spur and helical gears
JIS B 1751-1971	Master cylindrical gears
JIS B 1753-1971	Measuring method of noise of gears

NOTE:
1. Standards available in English from ANSI, 1430 Broadway, New York, N.Y. 10018, or Business Section of Japanese Standards Association, 1-24, Akasaka 4-chome, Minato-ku, Tokyo, 107 Japan.

GERMANY[2]

Standard Number	Title
DIN 780*	Module series for gears (modules for spur, bevel, and worm-gears) (corresponds to ISO 54)
DIN 867*	Basic rack for spur gears with involute teeth for general engineering (approximates to ISO 53)

DIN 869	Sheet 2—Gears—Guides to order bevel gears
DIN 3960*	Gear tooth systems[1]
	Sheet 1—Definitions and symbols for geometrical data
	Sheet 2—Geometrical data for spur gears (cylindrical) and transmissions (cylindrical)
	Sheet 3—Geometrical data for spur gears and transmissions with involute gears
DIN 3961*	Tolerances for spur gears according to DIN 867
DIN 3962*	Tolerances for spur gear tooth systems according to DIN 867
	Sheet 1—Tolerances for individual errors—Module to 0.6 mm
	Sheet 2—Tolerances for individual errors—Module above 0.6 to 1.6 mm
	Sheet 3—Tolerances for individual errors—Module above 1.6 to 4 mm
	Sheet 4—Tolerances for individual errors—Module above 4 to 10 mm
DIN 3963*	Tolerances for spur gears according to DIN 867—Permissible tolerance on tooth alignment, limits of tolerance on total composite error, and tooth thickness allowances
DIN 3964*	Center distance allowances
DIN 3966*	Specification of spur gears in drawing practice
DIN 3967*	Tolerances for spur gears according to DIN 867—Permissible tolerance on tooth alignment, limits of tolerances on total composite error, and tolerance on tooth widths
DIN 3971*	Specification factors and errors relating to bevel gears—Basic terms and definitions
DIN 3975	Geometrical data and errors on cylindrical worm gears—Basic terms and definitions
DIN 3976	Cylindrical worm gears, dimensions, coordination of center distance and transmission ratio in worm gear drives
DIN 3990	Load calculations for spur and bevel gears
	Sheet 1—Basic information and formulas
	Sheet 2—Tooth form factor y_F
	Sheet 3—Load distribution factor Y
	Sheet 4—Support factor q_L, spur load Distribution factor K_F for tooth base and K for flank
	Sheet 5—Flank form factor Z_H
	Sheet 6—Material factor Z_M
	Sheet 7—Pinion engagement factor Z_B, gear engagement factor Z_D, profile overlap
	Sheet 8—Overlap factor Z
	Sheet 9—Guides for the strength of material
	Sheet 10—Helix angle factor Y
DIN 3992	Spread of center distance for spur gears
DIN 3994	Spread of center distance for spur gears with 05—gears—Introduction
DIN 3995	Straight spur gears with 05—Gears Sheets 1–8

NOTE:
1. Standards marked with an asterisk (*) are in English translation available from BEUTH-VERTRIEB GmbH, Burggrafenstrasse 4-7, 1 BERLIN 30, W. Germany, or ANSI, 1430 Broadway, New York, N.Y. 10018.

FRANCE

Standard Number	Title
NF E 23-001/1972	Glossary of gears (Similar to ISO 1122)
NF E 23-002/1972	Glossary of worm gears
NF E 23-005/1965	Gearing—Symbols (Similar to ISO 701)
NF E 23-006/1967	Tolerances for spur gears with involute teeth (Similar to ISO 1328)
NF E 23-011/1972	Cylindrical gears for general and heavy engineering—Basic rack and modules (Similar to ISO 467 and ISO 53)
NF L 32-611/1955	Calculating spur gears to NF L 32-610
NF E 23-012/1972	Cylindrical gears—Information to be given to the manufacturer by the producer

UNITED KINGDOM[2]

Standard Number	Title
BS 235: 1972	Gears for electric traction—Spur and helical gears for the drive between electrical motor and axle
BS 436:	Spur and helical gears
Part 1/1967	Basic rack form, pitches and accuracy (diametrical pitch series)
Part 2/1970	Basic rack form, modules and accuracy (1 to 50 metric modules) (Agrees with ISO 53, 54, and 701)
BS 545/1949	Bevel gears (machine cut)

NOTE:
2. Standards available from ANSI, 1430 Broadway, New York, N.Y. 10018; or BSI, 2 Park Street, London, W1A 2BS, England.

MECHANICAL POWER TRANSMISSION SYSTEMS

BS 978:	Fine pitch gears
Part 1/1968	Involute spur and helical gears
Part 2/1952	Cycloidal type gears
PD 3376/1959	Addendum 1 to BS 978: Part 2: Double circular arc type gears
Part 3/1952	Bevel gears
Part 5/1965	Hobs and cutters
BS 1807:	Gears for turbines and similar drives
Part 1/1952	Accuracy
Part 2/1958	Tooth form and pitches
BS 3696/1963	Master gears
BS 4517/1969	Dimensions of spur and helical geared motor units (metric series)
BS 4582:	Fine pitch gears, involute spur and helical gears module from 0.07 to 1 mm) (according to ISO 701 and 1122 standards)
Part 1/1970	
PD 6457/1970	Addendum modification to involute, spur, and helical gears (Agrees with ISO 701 and 1122 standards)

ITALY

Standard Number	Title
UNI 3521/1954	Gearing—Module series
UNI 3522/1954	Gearing—Basic rack
UNI 4430/1960	Spur gears—Order information for straight and bevel gears
UNI 4760/1961	Gearing—Glossary and geometrical definitions
UNI 6586/1969	Modules and diametral pitches of cylindrical and straight bevel gears for general engineering and heavy engineering (correspond to ISO 54 and 678)
UNI 6587/1969	Basic rack of cylindrical gears for general engineering (corresponds to ISO 53)
UNI 6588/1969	Basic rack of straight bevel gears for general engineering and heavy engineering (corresponds to ISO 677)
UNI 6773/1970	International gear notation—Symbols for geometrical data (corresponds to ISO 701)

AUSTRALIA

Standard Number	Title
AS B 62/1965	Bevel gears
AS B 66/1969	Worm gears (inch series)
AS B 214/1966	Geometrical dimensions for worm gears—Units
AS B 217/1966	Glossary for gearing
AS 1637	International gear notation symbols for geometric data (similar to ISO 701)

RELATED NATIONAL AND INTERNATIONAL (ISO TC 32) STANDARDS FOR SPLINES[1]

ISO/R14-1955	Straight-sided splines (for cylindrical shafts)—Nominal dimensions in millimeters
ANSI B92.1	Involute splines and inspection (metric version)
DIN 5461, 2, 3, 4, 5*	Spline shaft connections with straight flanks (Introduction, light, medium, and heavy series—Tolerances
DIN 5480 Sheet 1-14*	Involute spline (30° pressure angle with modules from 0.6 to 10)
DIN 5481 Sheet 1*	Internal and external serrations

ISO (TC 16) STANDARDS FOR KEYS AND KEYWAYS

ISO/R773-1969	Rectangular or square parallel keys and their corresponding keyways
ISO/R774-1969	Taper keys with or without gib head and their corresponding keyways
ISO 2491-1974	Thin parallel keys and their corresponding keyways
ISO 2492-1974	Thin taper keys with or without gib head and their corresponding keyways
ISO 3117-1977	Tangential keying
ISO 3912-1977	Woodruff keys and keyways

NOTE:
1. Standards marked with an asterisk (*) are in English translation available from BEUTH-VERTRIEB GmbH, Burggrafenstrasse 4-7, 1 BERLIN 30, W. Germany, or ANSI, 1430 Broadway, New York, N.Y. 10018.

14 Fluid Power Systems and Components

General

Fluid power systems are those that transmit and control power through the use of a pressurized fluid (liquid or gas) within an enclosed circuit. Fluid power includes the technologies of hydraulics, pneumatics, and fluid logic. The world standards[1] presented in this section deal, for the most part, with hydraulic fluid power systems and components.

The preparation of the general outline and the related ISO standards and drafts have been supplied to the author by Mr. J. I. Morgan, Executive Vice President, National Fluid Power Association (NFPA), Milwaukee, Wisconsin 53222. NFPA is a manufacturer's association comprising more than 175 of the leading fluid power producers in the United States. With the authority granted by the American National Standards Institute (ANSI), NFPA administers the secretariat of the International Organization for Standardization's (ISO) technical committee for fluid power, ISO/TC 131, and TC 131's advisory panel. NFPA also administers the secretariat of two subcommittees and eight working groups for ISO/TC 131; and provides the secretariat for all United States Technical Advisory Group (USA TAG) committees to ISO/TC 131. Some liaison members are as follows: Farm and Industrial Equipment Institute, Fluid Power Society, National Fluid Power Association, Rubber Manufacturers' Association, and Society of Automotive Engineers.

The ISO standards and drafts presented in this section conform, with some exceptions, to the national standards and industry practices in the major industrial countries outside the United States.

Selected SI Units for Fluid Power

The recommended SI units and letter symbols for fluid power systems are standardized in the ISO draft DCS 131/AP N31. These are in agreement with the material presented in Section 2 of this text, except for the unit of pressure, bar—with one bar (10^5 N/m^2 = 10^5 Pa) being approximately equal to 1 kg (force)/cm² (see Section 2, page 2-7). The bar has been widely used throughout Europe for pressure measurements. The author recommends the use of the unit pascal (1 Pa = 1 N/m^2) and the practical unit kilopascal (kPa) in the United States. The pascal pressure unit is suggested for use in Europe now for theoretical hydraulic pressure calculations, and it should be used where new pressure gages are installed or calibrated. One megapascal equals 10 bar and 1 kilopascal equals 10 millibar. To bring practical units for volumetric flow in line with theoretical calculations, the cubic meter per second (m^3/s) or liter per second (ℓ/s) is recommended for use instead of the cubic meter and liter per minute (simplifications of power calculations).

Nominal Pressures

The ISO 2944 recommended pressures for fluid power are shown in Table 14-1.

Table 14-1. Nominal Pressures—Gage Pressures in Bar (1 bar = 100 kPa) (ISO 2944)

0,01	0,10	1,0	10	100	1 000
(0,012 5)	(0,125)	(1,25)	(12,5)	(125)	
0,016	0,16	1,6	16	160	
(0,02)	(0,2)	(2,0)	(20)	200	
0,025	0,25	2,5	25	250	
(0,031 5)	(0,315)	(3,15)	(31,5)	315	
0,04	0,4	4,0	40	400	
(0,05)	(0,5)	(5,0)	(50)	500	
0,063	0,63	6,3	63	630	
(0,08)	(0,8)	(8,0)	(80)	800	

NOTES:
1. Non-preferred values are in parentheses.
2. 1 bar = 100 kPa (kilopascal) = 14.5 psi
3. The decimal marker used in the ISO standard and in this table is a comma " , ".

PUMPS AND MOTORS

Mounting Flanges and Shafts

The following material is based on ISO 3019/I—Hydraulic fluid power, positive displacement pumps and

---

[1] For information about the term "standard" as used in this book, please see page 1-2.

Table 14-2. Two-Bolt Mounting Flange (ISO 3019/1)

Dimensions in millimeters

Identification code	Pilot dimensions				Flange dimensions					
	A 0 −0,05	W 0 −0,5	X min.	Y max.	B	J	K	D +0,3 −0,1	T	R
50−2	50,80	6,4	—	0,8	64	14	82	10,3	102	10
82−2	82,55	6,4	—	0,8	95	18	106	11,1	130	12
101−2	101,60	9,7	51	1,5	120	25	146	14,3	174	14
127−2	127,00	12,7	64	1,5	148	31	181	17,5	213	16
152−2	152,40	12,7	70	1,5	200	40	229	20,6	267	19
165−2	165,10	15,9	70	2,3	270	55	318	27,0	368	25
177−2	177,80	15,9	70	2,3	300	60	350	27,0	400	25

NOTES:
1. Tolerances : 1-place dimensions ± 0,5.
2. Slots instead of holes : optional.
3. Geometric tolerancing symbols shown in Section 3.
4. Surface Texture symbols shown in Section 5.
5. The decimal marker used in the ISO standard and in this table is a comma " , "

motors, dimensions and identification code for mounting flanges and shaft ends.

Part I: Inch series shown in metric units. Pumps are components which convert rotary mechanical power and fluid power. Motors are components which convert fluid power into rotary mechanical power.

Tables 14-2 through 14-7 specify sizes, dimensions, and an identification code for positive displacement, hydraulic fluid power, pump and motor mounting flanges of the following types: two-bolt flanges (see Table 14-2); and four-bolt flanges (see Table 14-3).

Also specified are sizes, dimensions, and an identification code for positive displacement hydraulic fluid power pump and motor shaft ends of the following types: straight shafts without thread (see Table 14-4); straight shafts with thread (see Table 14-5); tapered shafts with thread (see Table 14-6); and 30° involute spline (see Table 14-7).

In addition these tables provide: a minimum number of flange and shaft sizes; composite dimension reference and identification codes for pumps and motors; simplified dimensional interchangeability with regard to flanges and shafts; and preferred sizes and dimensions for new designs.

The identification codes recommended used for pumps and motors are shown in Tables 14-2 through 14-7, and a brief description of the codes will now be given.

Tables 14-2 and 14-3. The number preceding the dash (-) is an approximation, in millimeters, of the mounting flange diameter. The number following the dash (-) states the number of mounting bolt holes in the flange.

FLUID POWER SYSTEMS AND COMPONENTS

Table 14-3. Four-Bolt Mounting Flange (ISO 3019/1)

First Angle Projection

Dimensions in millimetres

Identifica-tion code	Pilot dimensions				Flange dimensions		
	A 0 −0,05	W 0 −0,51	X min.	Y max.	R	D +0,3 −0,1	S
101 − 4	101,60	9,7	51	1,5	14	14,3	89,8
127 − 4	127,00	12,7	64	1,5	14	14,3	114,5
152 − 4	152,40	12,7	70	1,5	19	20,6	161,6
165 − 4	165,10	15,9	70	2,3	19	20,6	224,5
177 − 4	177,80	15,9	70	2,3	25	27,0	247,5

NOTES:
1. Tolerances: 1-place dimensions ± 0,5.
2. Slots instead of holes: optional.
3. Geometric tolerancing symbols shown in Section 3.
4. Surface Texture symbols shown in Section 5.
5. The decimal marker used in the ISO standard and in this table is a comma ",".

Table 14-4. Straight Shaft Ends Without Thread (ISO 3019/1)

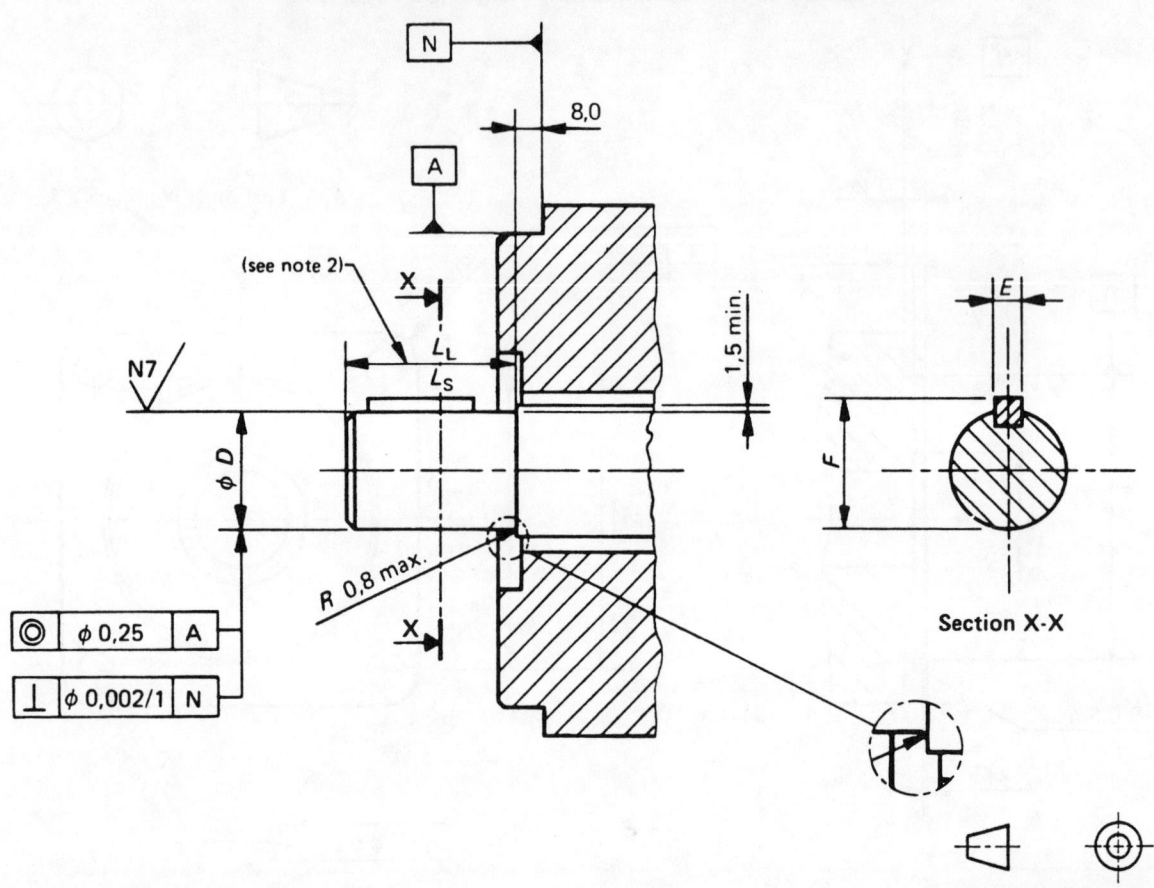

Dimensions in millimetres

Identification code	D max.	D min.	E (Key width) +0,03 0	F ± 0,13	L_L	L_S
13 – 1	12,70	12,67	3,18	14,07	–	19
16 – 1	15,88	15,85	3,97	17,60	51	24
22 – 1	22,23	22,20	6,35	24,90	63	33
25 – 1	25,40	25,35	6,35	28,10	70	38
32 – 1	31,75	31,70	7,94	35,20	76	48
38 – 1	38,10	38,05	9,53	42,27	83	54
44 – 1	44,45	44,40	11,11	49,30	92	67

NOTES:

1. Tolerances: 1-place dimensions ± 0,5.
2. L_L is an optional long length shaft.
3. Geometric tolerancing symbols shown in Section 3.
4. Surface Texture symbols shown in Section 5.
5. The decimal marker used in the ISO standard and in this table is a comma ",".

FLUID POWER SYSTEMS AND COMPONENTS

Table 14-5. Straight Shaft Ends with Thread (ISO 3019/1)

First Angle Projection

Section X-X

Dimensions in millimetres

Identification code	C + 0,13 − 0,08	D max.	D min.	D_T (Note 2)	E (Key width) + 0,03 0	F ± 0,13	L_C	L_S	L_T
13 – 2	2,4	12,70	12,67	3/8 – 24	3,18	14,07	29	19	14,25
16 – 2	3,2	15,88	15,85	1/2 – 20	3,97	17,60	34	24	18,25
22 – 2	4,0	22,23	22,20	5/8 – 18	6,35	24,90	48	33	23,00
25 – 2	4,0	25,40	25,35	3/4 – 16	6,35	28,10	52	38	27,00
32 – 2	4,0	31,75	31,70	1 – 12	7,94	35,20	67	48	31,00
38 – 2	4,0	38,10	38,05	1 1/8 – 12	9,53	42,27	73	54	34,90
44 – 2	4,0	44,45	44,40	1 1/4 – 12	11,11	49,30	89	67	39,70

NOTES:
1. Tolerances: 1-place dimensions ± 0,5.
2. Threads in accordance with ISO/R 725.
3. Geometric tolerancing symbols shown in Section 3.
4. Surface Texture symbols shown in Section 5.
5. The decimal marker used in the ISO standard and in this table is a comma ",".

Table 14-6. Tapered Shaft Ends with Thread (ISO 3019/1)

Dimensions in millimetres

Identification code	C +0,13 −0,08	D max.	D min.	D_T (Note 3)	E (Key width) +0,03 0	L_{CT}	L_{ST}	L_T	Z min.	Z max.
13 – 3	2,0	12,70	12,67	5/16 – 32	3,18	25	17	12,70	1,63	1,37
16 – 3	3,2	15,88	15,85	1/2 – 20	3,97	28	17	18,26	2,13	1,88
22 – 3	4,0	22,23	22,20	5/8 – 18	6,35	43	28	23,01	3,33	3,07
25 – 3	4,0	25,43	25,37	3/4 – 16	6,35	49	35	26,97	3,33	3,07
32 – 3	4,0	31,78	31,72	1 – 12	7,94	49	35	30,96	4,11	3,86
38 – 3	4,0	38,13	38,07	1 1/8 – 12	9,53	62	47	34,92	4,93	4,67
44 – 3	4,0	44,48	44,42	1 1/4 – 12	11,11	71	54	39,67	5,72	5,46
50 – 3	4,0	50,83	50,77	1 1/4 – 12	12,70	90	73	39,67	6,50	6,25

NOTES:
1. Tolerances: 1-place dimensions ± 0,5.
2. Dimension Z is normal to the key and at the larger taper.
3. Geometric tolerancing symbols shown in Section 3.
4. Surface Texture symbols shown in Section 5.
5. The decimal marker used in the ISO standard and in this table is a comma ",".

FLUID POWER SYSTEMS AND COMPONENTS

Table 14-7. —30° Involute Spline Shaft Ends (ISO 3019/1)

Dimensions in millimetres

Identification code	Spline	L_A min.	L_B	L_S	U min.
13 – 4	9T 20/40 DP	5,1	1,5	19	9,40
16 – 4	9T 16/32 DP	7,6	1,5	24	11,81
22 – 4	13T 16/32 DP	10,2	1,5	33	18,16
25 – 4	15T 16/32 DP	12,7	1,5	38	21,34
32 – 4	14T 12/24 DP	15,2	2,0	48	26,42
38 – 4	17T 12/24 DP	17,8	2,0	54	32,77
44 – 4	13T 8/16 DP	20,3	3,0	67	36,63
50 – 4	15T 8/16 DP	25,4	3,0	80	42,95

NOTES:
1. Tolerances: 1-place dimensions ± 0,5.
2. For spline dimensions, see ISO/TC 32 (Secr. 71) 143, 30° involute spline, Class 5, flat root, side fit.
3. Geometric tolerancing symbols shown in Section 3.
4. Undercut defined by dimension U is optional.
5. The decimal marker used in the ISO standard in this table is a comma ",".

Tables 14-4 through 14-7. The number preceding the dash (-) is an approximation, in millimeters, of the major diameter of the shaft. The number following the dash (-) is arbitrarily assigned as follows:

-1, straight shafts without thread
-2, straight shafts with thread
-3, tapered shafts with thread
-4, 30° involute spline.

Displacement Series

The recommended displacement values for pumps and motors shown in Table 14-8 are based on the ISO 3662. This International Standard establishes the geometric displacements (V) of hydraulic fluid power pumps and motors having rotating or oscillating drives. Refer to the Table 14-8.

Table 14-8. Nominal Values for Geometric Displacements (V) Geometric displacement in mℓ/r (milliliter/revolution) (ISO 3662)

0,1	1	10	100	1 000
		(11,2)	(112)	(1 120)
	1,25	12,5	125	1 250
		(14)	(140)	(1 400)
0,16	1,6	16	160	1 600
		(18)	(180)	(1 800)
	2	20	200	2 000
		(22,4)	(224)	(2 240)
0,25	2,5	25	250	2 500
		(28)	(280)	(2 800)
	3,15	31,5	315	3 150
		(35,5)	(355)	(3 550)
0,4	4	40	400	4 000
		(45)	(450)	(4 500)
	5	50	500	5 000
		(56)	(560)	(5 600)
0,63	6,3	63	630	6 300
		(71)	(710)	(7 100)
	8	80	800	8 000
		(90)	(900)	(9 000)

NOTES:
1. The nominal geometric displacement is the displacement assigned to hydraulic fluid power pumps and motors for the purpose of convenient designation.
2. Specify values of displacements in excess of 9 000 mℓ/rev with R 20 numbers, R 10 is the preferred series (see Section 4, Table 4-1.).
3. Values printed in parentheses are nonpreferred values.
4. The decimal marker used in the ISO standard and in this table is a comma " , ".

The volumes shown are also applicable to variable displacement units; in such cases, the values refer to the maximum displacement.

The geometric displacements shown in Table 14-8 may be used as guidelines for the design of positive displacement hydraulic fluid power pumps and motors. They may also be used to derive other basic design criteria and normal ratings.

CYLINDERS

Nominal Pressure Ratings

The recommended nominal pressures used for hydraulic and pneumatic fluid power cylinders are specified in ISO 3322. See Table 14-9.

Table 14-9. Nominal Pressures for Cylinders (ISO 3322)

Bar	6.3 10 16 25 40 63 100 160 250 400

NOTES:
1. 1 bar = 100 kPa (kilopascals) = 14.5 psi.
2. Stated values are given as gauge pressure.
3. Any other values required should be selected from ISO 2944.

Bore and Piston Rod Diameters

A metric series of cylinder bore and piston rod diameters for applications to hydraulic and pneumatic fluid power cylinders are specified in ISO 3320. The recommended values are shown in Tables 14-10 and 14-11.

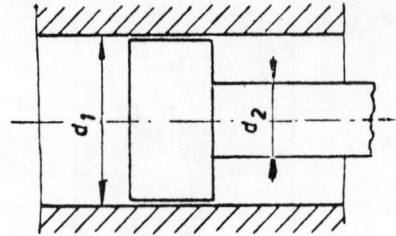

Fig. 14-1. Bore and piston rod diameters (ISO 3320),

where

d_1 is cylinder bore;
d_2 is piston rod diameters.

Piston Rod Thread

The piston rod thread is a thread with which the piston rod is attached to any components outside the cylinder.

The recommended piston rod thread sizes for hydraulic and pneumatic cylinders are specified in ISO 4395. See Table 14-12.

FLUID POWER SYSTEMS AND COMPONENTS

Table 14-10. Cylinder Bore (ISO 3320)

d_1	8	10	12	16	20	25	32	40	50
	63	80	100	125	160	200	250	320	400

NOTE:
An extension upwards of the diameter ranges may, if required, be made using the R 10 Series of preferred numbers (see Table 4-1).

Table 14-11. Piston Rod Diameters (ISO 3320)

d_2	4	5	6	8	10	12	14	16	18	20	22	25
	28	32	36	40	45	50	56	63	70	80	90	100
	110	125	140	160	180	200	220	250	280	320	360	

NOTE:
An extension upwards of the diameter ranges may, if required, be made using the R20 series of preferred numbers (see Section 4).

Table 14-12. Piston Rod Threads (ISO 4395)

*M3 × 0.35	M18 × 1.5	*M48 × 2	*M140 × 4
*M4 × 0.5	*M20 × 1.5	M56 × 2	*M160 × 4
*M5 × 0.5	(*M22 × 1.5)	*M64 × 3	M180 × 4
*M6 × 0.75	M24 × 2	M72 × 3	*M200 × 4
*M8 × 1	(*M27 × 2)	*M80 × 3	*M220 × 4
*M10 × 1.25	M30 × 2	M90 × 3	M250 × 6
*M12 × 1.25	(*M33 × 2)	*M100 × 3	*M280 × 6
(*M14 × 1.5)	M36 × 2	M110 × 3	
*M16 × 1.5	*M42 × 2	M125 × 4	

NOTES:
1. *Nominal size for female and shouldered male piston rod thread (see Fig. 14-2).
2. Second choice sizes are shown in parentheses ( ).
3. Metric thread details are shown in Table 8-1.

Cylinder Tube Sizes

The following discussion applies to steel hydraulic and pneumatic cylinder tubes with inside diameters from 25 mm up to 400 mm. The bores of the tubes have been specially finished with or without metal removal.

A range of preferred thicknesses is specified for each of the metric bore sizes designated as standards for hydraulic and pneumatic cylinders. Two tables of thicknesses are included to cover both cold-finished and hot-finished steel tubes.

Cylinder Barrels Made from Cold-Finished Tubes. Preferred sizes appropriate for cylinders made from cold-finished steel tubes are listed in Table 14-13 by inside diameter, wall thickness, and outside diameter. The inside diameter sizes, from 25 mm to 200 mm, are selected from those in ISO 3320. The tube dimensions may be specified by *either* inside diameter and wall thickness *or* inside diameter and outside diameter.

Cylinder Barrels Made from Hot-Finished Tubes. Preferred sizes appropriate for cylinders made from hot-finished steel tubes are listed in Table 14-14 by inside diameter and outside diameter. The inside diameter sizes, from 63 mm to 400 mm, are selected from those in ISO 3320.

The outside diameter sizes of hot-finished tubes are selected from ISO 64: 1974. The tube dimensions for hot-finished tubes can be specified *only* by inside diameter and outside diameter.

Tolerances. The tolerances on inside diameter, outside diameter and thickness shall be those included in "Specification for Steel Hydraulic and Pneumatic Cylinder Tubes." ISO 4394.

Cylinder Tube Specifications

Scope. The ISO 4394 standard covers round steel tubes, of seamless or welded type, in the hot or cold worked condition, with plain ends. The bores are specially finished with or without metal removal. The tubes are dimensionally defined by: (1) inside diameter and thickness, or (2) inside diameter and outside diameter.

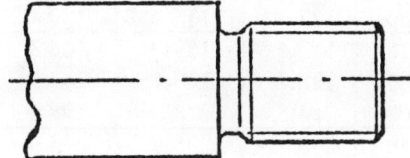 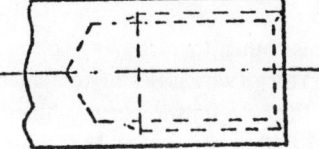

Fig. 14-2. Female and shouldered male piston rod threads (For thread lengths see ISO 4395).

Table 14-13. Preferred Sizes for Cylinder Barrels made from
Cold Finished Steel Tubes (ISO 4394)

Bore mm	Wall Thickness (mm)											
	1.5	2.0	2.5	3.0	3.5	5.0	6.0	7.5	10.0	12.5	15.0	20.0
	Outside Diameters (mm)											
25	28			31		35		40				
32	35	36		38		42		47				
40			45	46		50		55				
50			55	56		60		65	70	75		
63			68	69		73	75	78	83	88		
80			85	86		90	92	95	100	105	110	
100			105	106		110	112	115	120	125	130	
125					132	135	137	140	145	150	155	165
160					167	170		175	180	185	190	200
200						210		215	220	225	230	240

This standard defines the mechanical properties, dimensional tolerances, surface finishes and technical delivery conditions. A recommended designation for purchase orders is also shown.

The tubes covered by the ISO 4394 are intended for use as barrels in a wide variety of hydraulic and pneumatic cylinders. Note: Steel should not be considered as the only material suitable for such applications.

Mechanical Properties and Chemical Composition. The mechanical properties specified in ISO 4394 are shown in Tables 14-15 and 14-16.

The sulphur and phosphor content of steel must be below 0.05% each, and the carbon content held below 0.25%.

Surface Finish. Bore surface finish values are specified in R_a (Center-line-average—see Section 5), and the commercial classes are shown in Table 14-17.

Tolerances.

INSIDE DIAMETER—Five classes of tolerances are recognized: H8, H9, H11, H12, and H13 in accordance with Section 6. Tolerances include ovality and taper.

Tolerances H8 and H9 can normally be supplied only if the ratio of the tube's inside diameter to its thickness is less than 20 : 1. Tolerance H11 can normally be supplied only if the ratio of the tube's inside diameter to its thickness is less than 25 : 1.

Special tolerances, other than shown, can be specified.

OUTSIDE DIAMETER—Two classes of tolerances are recognized: Class 1, for cold finished or machined tubes, as shown in Table 14-18; class 2, for hot finished tubes. Class 2 tolerances shall be ± 1% of the nominal outside diameter (minimum of ± 0.5 mm). Tolerances include ovality and taper.

THICKNESS AND ECCENTRICITY—If the inside diameter and thickness of the tube are specified, then the thickness measured at any cross section along the tube length must not vary from the nominal thickness by more than ± 10%. This figure includes eccentricity.

Order Example. To order a seamless tube, 5 m long with an outside diameter 60 mm, inside diameter 50 mm (see Table 14-13), steel grade HP 5 (Table 14-16), surface finish d (see Table 14-17), inside diameter tolerance H9, specify the tube as follows:

5 m seamless tube 60×50 ISO 4394
Steel Grade HP5, Finish d, Tolerance H9

Table 14-14. Preferred Sizes for Cylinder Barrels made from Hot Finished Steel Tubes (ISO 4394)

Bore mm	Outside Diameter (mm)			
63	76.1	82.5	88.9	101.6
80	101.6	108	114.3	127
100	127	133	139.7	152.4
125	152.4	159	168.3	177.8
160	193.7	219.1	244.5	—
200	244.5	273	298.5	—
250	273	298.5	323.9	355.6
320	355.6	368	406.4	419
400	419	457	508	559

FLUID POWER SYSTEMS AND COMPONENTS

Table 14-15. Tubes with Lower Tensile Properties and Greater Ductility (ISO 4394)

Steel Grade	R_m minimum (1) MPa	(2) BHN	R_{eL} or $R_{p\,(0.2)}$ minimum, $a \leqslant 10$ mm (4) MPa	$10\text{ mm} < a \leqslant 20\text{ mm}$ MPa	$20\text{ mm} < a \leqslant 50\text{ mm}$ MPa	A (3) %
HP 1	360	102	235	225	215	24
HP 2	490	140	335	310	285	21
HP 3	550	163	460	450	420	17

See notes to Table 14-16.

Table 14-16. Tubes with Higher Tensile Properties (ISO 4394)

Steel Grade	R_m minimum (1) MPa	(2) BHN	$R_e L$ or R_p (0.2) minimum MPa	A (3) %
HP 4	450	126	380	10
HP 5	550	163	440	10
HP 6	640	190	540	10

NOTES:
1. 1 MPa = 1 N/mm² ≈ 0.1 kgf/mm².
   1000 psi = 6.894757 MPa ≈ 7 MPa
2. R_m, R_{eL} and R_p (0.2) are ISO symbols for tensile strength, lower yield stress and 0.2% proof stress.
3. A = minimum elongation (%) on 5.65 $\sqrt{S_0}$ (S_0 = cross section area of gage length)
4. a = nominal thickness of tube in mm

CONDUCTORS

Threads for Ports and Fitting Ends

The ISO 6149 standard for metric port and fitting end dimensions specifies O-ring of three other types or seals. The standard thread sizes are as follows:

M5 × 0.8, M8 × 1, M10 × 1, M12 × 1.5, M14 × 1.5, M16 × 1.5, M18 × 1.5, M22 × 1.5, M27 × 2, M33 × 2, M42 × 2, M50 × 2, M60 × 2

See the ISO 6149 standard for port, boss, and seal details.

Line Tubing Sizes

The line tubing sizes recommended used for fluid conductors are specified in ISO 4397 as shown in Table 14-19.

The tube sizes in Table 14-19 have been in use in Europe for many years. (For details concerning the ISO and national standards for seamless and welded metric tubes,

Table 14-17. Classes of Surface Finish (Ra) (ISO 4394)

Classes					
a	b	c	d	e	f
0.125	0.2	0.4	0.8	1.6	3.2	micrometers
5	8	16	32	63	125	microinches
–	N4	N5	N6	N7	N8	ISO 1302 roughness no.

refer to Table 10-24.) The just mentioned hydraulic metric tube sizes are marketed in the United States.

Hose Sizes

The recommended series of inside diameters of hoses made by rubber or plastics are specified in ISO 4397, as shown in Table 14-20.

Nominal Pressure Ratings for Conductors

The nominal pressure ratings to be used for connectors and associated components are given in ISO 4399. See Table 14-21.

SEALING DEVICES

O-Ring Sizes and Tolerances

Scope. The ISO/DIS 3601 standard for O-rings will be composed of three parts, as follows:

Part I: Inside diameters, cross sections, tolerances and size code identification for an "O" series of metric O-rings
Part II: Design criteria for standard applications
Part III: Quality acceptance criteria

The inside diameters for O-rings shown in Table 14-22 are based on the preferred number series (see Section 4

Table 14-18. Tolerances on Outside Diameters (ISO 4394)

Outside Diameters		Tolerances
Over	Up to and Including	
mm	mm	mm
	30	±0.10
30	40	±0.15
40	50	±0.20
50	60	±0.25
60	70	±0.30
70	80	±0.35
80	90	±0.40
90	100	±0.45
100	120	±0.50
120	140	±0.65
140	150	±0.75
150	160	±0.80
160	170	±0.85
170	180	±0.90
180	190	±0.95
190	200	±1.0
200	210	±1.05
210	220	±1.10
220	230	±1.15
230	240	±1.20

Table 14-19. Line Tubing Sizes (ISO 4397)

Outside diameters (mm)	4 5 6 8 10 12 16 20 25 32 40 50

for the Renard Series) and the range of sizes was checked against production records. The order of preference, when choosing inside diameters for O-Rings, should be as follows: I.D. to R5 (first choice), R10 (second choice), R20 (third choice), R40 (fourth choice) and R80 (fifth choice). See Tables 4-1 and 4-2. It is recommended that inside diameters for special size O-Rings, not shown in Table 14-22, be chosen from the Renard series of preferred numbers.

Identification Code (ISO/DIS 3601 /I)

The ISO recommends the use of a size code consisting of eight digits where the first three digits represent the section diameter (d_2). See Table 14-22. The next five digits would be the internal diameter (d_1) expressed in millimeters.

EXAMPLES:

Size code	d_2		d_1
18000355	= 1.80	X	3.55
26503450	= 2.65	X	34.5
35505000	= 3.55	X	50
53023000	= 5.30	X	230
70046200	= 7	X	462

NOTE: Nonstandard sizes could use this same format without confusing them with standard sizes.

Rotary Shaft Lip Seal Sizes

The basic dimensions for lip seals are specified in ISO 6194/I standard, and the recommended values are shown in Table 14-23. Seals to the shown basic dimensions and with various lip designs are supplied in countries already using the metric system.

Table 14-20. Hose Inside Diameters (ISO 4397)

Hose I. D. (mm)	3.2 5 6.3 8 10 12.5 16 19* 20 25 31.5 38* 40 50 51*

*for hydraulic purposes

Table 14-21. Nominal Pressures for Conductors in Bar (1 bar = 100 kPa = 10^5 N/m^2) (ISO 4399)

Nominal Pressures in bar
2.5 6.3 10 16 25 40 100 160 (200) 250 (315) 400 630

NOTE:
Nominal pressures shown in brackets are second choice.

Table 14-22. Inside Diameters, Cross Sections, and Tolerances for O-Rings (ISO 3601/I)

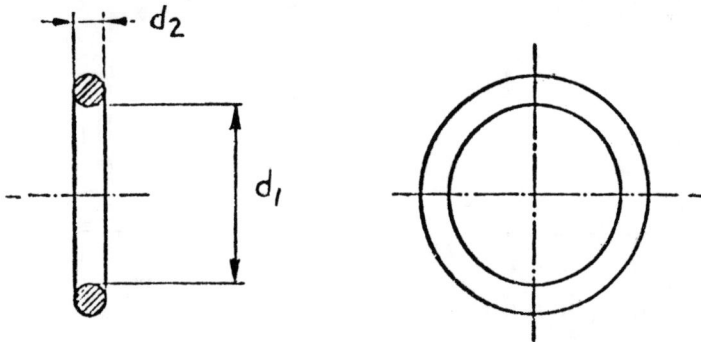

Dimensions in millimeters

d_1	tol. ±	d_2 and tol. ± 1,80 ± 0,08	2,65 ± 0,09	3,55 ± 0,10	5,30 ± 0,13	7,00 ± 0,15	d_1	tol. ±	1,80 ± 0,08	2,65 ± 0,09	3,55 ± 0,10	5,30 ± 0,13	7,00 ± 0,15	d_1	tol. ±	1,80 ± 0,08	2,65 ± 0,09	3,55 ± 0,10	5,30 ± 0,13	7,00 ± 0,15
1,80	0,13	x					40,0	0,38			x	x		170	1,34			x	x	
2,00	0,13	x					41,2	0,39			x	x		175	1,38			x	x	
2,24	0,13	x					42,5	0,40			x	x		180	1,41			x	x	
2,50	0,13	x					43,7	0,41			x	x		185	1,44			x	x	
2,80	0,14	x					45,0	0,42			x	x		190	1,48			x	x	
3,15	0,14	x					46,2	0,43			x	x		195	1,51			x	x	
3,55	0,14	x					47,5	0,44			x	x		200	1,55			x	x	
4,00	0,14	x					48,7	0,45			x	x		206	1,59				x	x
4,50	0,14	x					50,0	0,46			x	x		212	1,63				x	x
5,00	0,15	x					51,5	0,47			x	x		218	1,67				x	x
5,30	0,15	x					53,0	0,48			x	x		224	1,71				x	x
5,60	0,15	x					54,5	0,50			x	x		230	1,75				x	x
6,00	0,15	x					56,0	0,51			x	x		236	1,79				x	x
6,30	0,15	x					58,0	0,52			x	x		243	1,83				x	x
6,70	0,16	x					60,0	0,54			x	x		250	1,88				x	x
7,10	0,16	x					61,5	0,55			x	x		258	1,93				x	x
7,50	0,16	x					63,0	0,56			x	x		265	1,98				x	x
8,00	0,16	x					65,0	0,58			x	x		272	2,02				x	x
8,50	0,16	x					67,0	0,59			x	x		280	2,08				x	x
9,00	0,17	x					69,0	0,61			x	x		290	2,14				x	x
9,50	0,17	x					71,0	0,63			x	x		300	2,21				x	x
10,0	0,17	x					73,0	0,64			x	x		307	2,25				x	x
10,6	0,18	x					75,0	0,66			x	x		315	2,30				x	x
11,2	0,18	x					77,5	0,67			x	x		325	2,37				x	x
11,8	0,19	x					80,0	0,69			x	x		335	2,43				x	x
12,5	0,19	x					82,5	0,71			x	x		345	2,49				x	x
13,2	0,19	x					85,0	0,73			x	x		355	2,56				x	x
14,0	0,19	x	x				87,5	0,75			x	x		365	2,62				x	x
15,0	0,20	x	x				90,0	0,77			x	x		375	2,68				x	x
16,0	0,20	x	x				92,5	0,79			x	x		387	2,76				x	x
17,0	0,21	x	x				95,0	0,81			x	x		400	2,84				x	x
18,0	0,21		x	x			97,5	0,83			x	x		412	2,91					x
19,0	0,22		x	x			100	0,84			x	x		425	2,99					x
20,0	0,22		x	x			103	0,87			x	x		437	3,07					x
21,2	0,23		x	x			106	0,89			x	x		450	3,15					x
22,4	0,24		x	x			109	0,91			x	x		462	3,22					x
23,6	0,24		x	x			112	0,93			x	x		475	3,30					x
25,0	0,25		x	x			115	0,95			x	x		487	3,37					x
26,5	0,26		x	x			118	0,97			x	x		500	3,45					x
28,0	0,28		x	x			122	1,00			x	x		515	3,54					x
30,0	0,29		x	x			125	1,03			x	x		530	3,63					x
31,5	0,31		x	x			128	1,05			x	x		545	3,72					x
32,5	0,32		x	x			132	1,08			x	x		560	3,81					x
33,5	0,32		x	x			136	1,10			x	x		580	3,93					x
34,5	0,33		x	x			140	1,13			x	x		600	4,05					x
35,5	0,34		x	x			145	1,17			x	x		615	4,13					x
36,5	0,35		x	x			150	1,20			x	x		630	4,22					x
37,5	0,36		x	x			155	1,24			x	x		650	4,34					x
38,7	0,37		x	x			160	1,27			x	x		670	4,46					x
							165	1,31			x	x								

Table 14-23. Basic Dimensions for Lip Seals (ISO 6194/I)

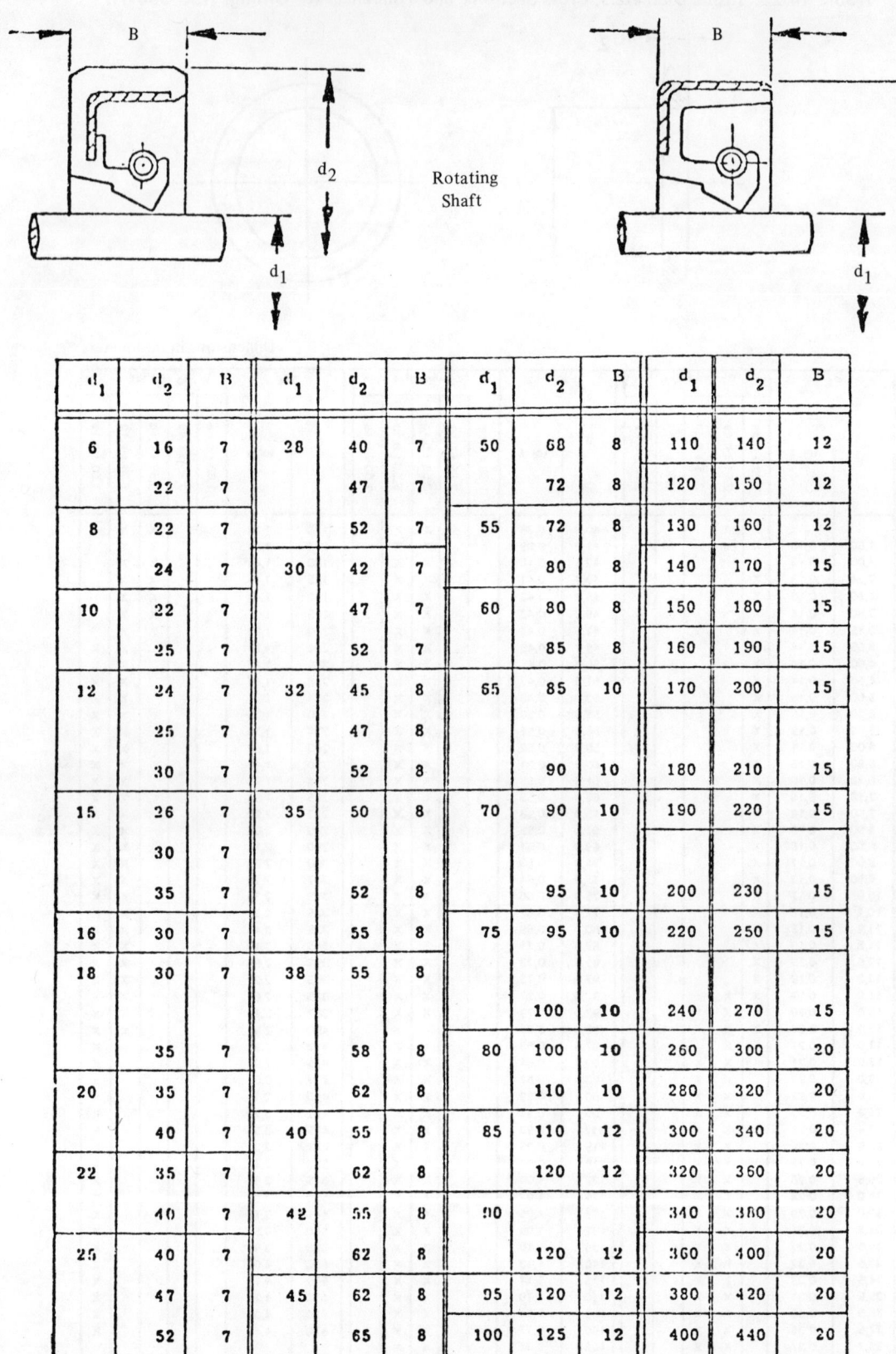

d_1	d_2	B	d_1	d_2	B	d_1	d_2	B	d_1	d_2	B
6	16	7	28	40	7	50	68	8	110	140	12
	22	7		47	7		72	8	120	150	12
8	22	7		52	7	55	72	8	130	160	12
	24	7	30	42	7		80	8	140	170	15
10	22	7		47	7	60	80	8	150	180	15
	25	7		52	7		85	8	160	190	15
12	24	7	32	45	8	65	85	10	170	200	15
	25	7		47	8						
	30	7		52	8		90	10	180	210	15
15	26	7	35	50	8	70	90	10	190	220	15
	30	7									
	35	7		52	8		95	10	200	230	15
16	30	7		55	8	75	95	10	220	250	15
18	30	7	38	55	8						
							100	10	240	270	15
	35	7		58	8	80	100	10	260	300	20
20	35	7		62	8		110	10	280	320	20
	40	7	40	55	8	85	110	12	300	340	20
22	35	7		62	8		120	12	320	360	20
	40	7	42	55	8	90			340	380	20
25	40	7		62	8		120	12	360	400	20
	47	7	45	62	8	95	120	12	380	420	20
	52	7		65	8	100	125	12	400	440	20

FLUID POWER SYSTEMS AND COMPONENTS

Table 14.24. Piston Seal Housing Dimensions for Hydraulic Cylinders (ISO 5597)
(Dimensions in millimeters)

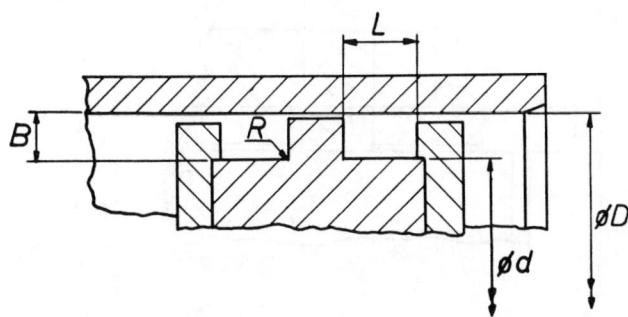

ϕD	B	$\phi D - 2B$ (ϕd)	L_1	L_2	L_3	R max
16	4	8	5	6.3	–	0.3
20	4	12	5	6.3	–	0.3
25	4	17	5	6.3	–	0.3
	5	15	6.3	8	16	0.3
32	4	24	5	6.3	–	0.3
	5	22	6.3	8	16	0.3
40	4	32	5	6.3	–	0.3
	5	30	6.3	8	16	0.3
50	5	40	6.3	8	16	0.3
	7.5	35	9.5	12.5	25	0.4
63	5	53	6.3	8	16	0.3
	7.5	48	9.5	12.5	25	0.4
80	7.5	65	9.5	12.5	25	0.4
	10	60	12.5	16	32	0.6
100	7.5	85	9.5	12.5	25	0.4
	10	80	12.5	16	32	0.6
125	10	105	12.5	16	32	0.6
	12.5	100	16	20	40	0.8
160	10	140	12.5	16	32	0.6
	12.5	135	16	20	40	0.8
200	12.5	175	16	20	40	0.8
	15	170	20	25	50	0.8
250	12.5	225	16	20	40	0.8
	15	220	20	25	50	0.8
320	15	290	20	25	50	0.8
400	20	360	25	32	63	1

Table 14-25. Rod Seal Housing Dimensions for Hydraulic Cylinders (ISO 5597)
(Dimensions in millimeters)

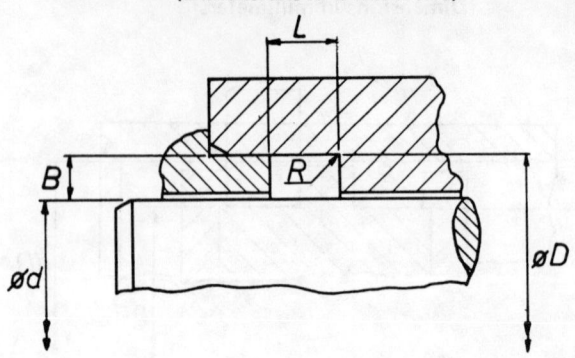

φd	B	d + 2B (φD)	L_1	L_2	L_3	R max
6	4	14	5	6.3	14.5	0.3
8	4	16	5	6.3	14.5	0.3
10	4 5	18 20	5	6.3 8	14.5 16	0.3 0.3
12	4 5	20 22	5	6.3 8	14.5 16	0.3 0.3
14	4 5	22 24	5	6.3 8	14.5 16	0.3 0.3
16	4 5	24 26	5	6.3 8	14.5 16	0.3 0.3
18	4 8	26 28	5	6.3 8	14.5 16	0.3 0.3
20	4 5	28 30	5	6.3 8	14.5 16	0.3 0.3
22	4 5	30 32	5	6.3 8	14.5 16	0.3 0.3
25	4 5	33 35	5	6.3 8	14.5 16	0.3 0.3
28	5 7.5	38 43	6.3	8 12.5	16 25	0.3 0.4
32	5 7.5	42 47	6.3	8 12.5	16 25	0.3 0.4
36	5 7.5	46 51	6.3	8 12.5	16 25	0.3 0.4
40	5 7.5	50 55	6.3	8 12.5	16 25	0.3 0.4
45	5 7.5	55 60	6.3	8 12.5	16 25	0.3 0.4
50	5 7.5	60 65	6.3	8 12.5	16 25	0.3 0.4

FLUID POWER SYSTEMS AND COMPONENTS

Table 14-25 (*Continued*). Rod Seal Housing Dimensions for Hydraulic Cylinders (ISO 5597) (Dimensions in millimeters)

φd	B	d + 2B (φD)	L_1	L_2	L_3	R max
56	7.5	71	9.5	12.5	25	0.4
	10	76		16	32	0.6
63	7.5	78	9.5	12.5	25	0.4
	10	83		16	32	0.6
70	7.5	85	9.5	12.5	25	0.4
	10	90		16	32	0.6
80	7.5	95	9.5	12.5	25	0.4
	10	100		16	32	0.6
90	7.5	105	9.5	12.5	25	0.4
	10	110		16	32	0.6
100	10	120	12.5	16	32	0.6
	12.5	125		20	40	0.8
110	10	130	12.5	16	32	0.6
	12.5	135		20	40	0.8
125	10	145	12.5	16	32	0.6
	12.5	150		20	40	0.8
140	10	160	12.5	16	32	0.6
	12.5	165		20	40	0.8
160	12.5	185	16	20	40	0.8
	15	190		25	50	0.8
180	12.5	205	16	20	40	0.8
	15	210		25	50	0.8
200	12.5	225	16	20	40	0.8
	15	230		25	50	0.8
220	15	250	20	25	50	0.8
250	15	280	20	25	50	0.8
280	15	310	20	25	50	0.8
320	20	360	25	32	63	1
360	20	400	25	32	63	1

Piston and Rod Seal Housing Dimensions

Housing dimensions for piston and rod seals are published in ISO 5597 standard, and the basic sizes are shown in Table 14-24 piston seals and Table 14-25 rod seals. The dimensions shown conform to industrial practices in countries already using the metric system.

RELATED ISO STANDARDS (ISO TC 131, TC 45)

ISO 1307-1975	Rubber hose-bore sizes, test pressures, and tolerances on length
ISO 1404-1970	Industrial air hoses
ISO 1436-1972	Wire-reinforced, rubber-covered hydraulic hoses
ISO/R 1939-1970	Pneumatic cylinders—Cylinder bores and parts
ISO/R 1941-1970	Flat seal for hydraulic couplings
ISO/R 1943-1970	Coupling threads for hydraulic or pneumatic piping (pipe threads)
ISO/R 1944-1970	Pipe coupling for hydraulic piping (pipe threads)
ISO/R 2091-1971	Hydraulic cylinders—Internal diameters and piston rod diameters—Metric series
ISO DR 2318	Air hoses
ISO 2941-1974	Hydraulic fluid power—Filter elements—Verification of collapse/burst resistance
ISO 2942-1974	Hydraulic fluid power—Filter elements—Determination of fabrication integrity
ISO 2943-1974	Hydraulic fluid power—Filter elements—Verification of material compatibility with fluids
ISO 2944-1974	Fluid power systems and components—Nominal pressures
ISO 3019/I-1975	Hydraulic fluid power—Positive displacement pumps and motors—Dimensions and identification code for mounting flanges and shaft ends. Part I: Inch series shown in metric units
ISO 3320-1975	Fluid power systems and components—Cylinders—Cylinder bores and piston rod diameters—Metric series
ISO 3321-1975	Fluid power systems and components—Cylinders—Cylinder bores and piston rod diameters—Inch series
ISO 3322-1975	Fluid power systems and components—Cylinders—Nominal pressures
ISO/DIS 3601	Fluid systems—O-Rings—Part I: Inside diameters, cross sections, tolerances and size code identification for a series of metric O-rings
ISO 3662-1976	Hydraulic fluid power-Pumps and motors—Geometric displacements
ISO 3722-1976	Hydraulic fluid power—Fluid sample containers—Qualifying and controlling cleaning methods
ISO 3723-1976	Hydraulic fluid power—method for end load test
ISO 3724-1976	Hydraulic fluid power—Filter elements—Verification of flow fatigue characteristics
DIS 3938	Hydraulic fluid power—Contamination analysis data
ISO 3939-1977	Multiple lip packing sets—Measuring stack heights
DIS 3968	Filter elements—Pressure drop versus flow characteristics
ISO DIS 4021	Particulate contamination analysis—Extraction of fluid samples
ISO DPR 4391	Hydraulic fluid power—pumps, motors and integral transmissions—Parameter definitions and symbols
ISO DPR 4392	Hydraulic fluid power—Motors—Starting conditions and operation at extremely low speeds
ISO DIS 4393	Fluid power systems and components—Cylinders—Basic series of piston strokes
ISO DIS 4394	Cylinder barrels—Requirements for steel tubes with specially finished bores
ISO DIS 4395	Cylinders—Piston rod thread dimensions and types
ISO DPR 4396	Fluid power cylinders—Piston rod threads—Preferred sizes, dimensions and types
ISO DIS 4397	Connectors and associated components—Outside diameters of tubes and inside diameter of hoses
ISO DPR 4398	Fluid power systems and components—Hoses—Series of inside diameters
ISO 4399-1977	Connectors and associated components—Nominal pressures

FLUID POWER SYSTEMS AND COMPONENTS

ISO DPR 4400	Fluid power systems and components—Electrical plug connectors—Control and regulation assemblies	ISO DPR 5596	Fluid power systems and components—Gas-loaded accumulator with separator—Characteristic quantities—Identification
ISO DPR 4401	Hydraulic fluid power—4 port directional valves—Mounting surfaces	ISO DP 5597	Housings for seals for reciprocating applications—Dimensions and tolerances
ISO 4402-1977	Calibration of liquid automatic particle count instruments	ISO DP 5598	Fluid power—Vocabulary
ISO DPR 4403	Hydraulic fluid power—Mineral oils—Schedule of required data	ISO DIS 5599/I	Pneumatic fluid power—5 port directional control valve—Mounting surfaces
ISO DPR 4404	Hydraulic fluid power—Fire resistant fluids—Determination of the anti-corrosive power		
ISO DPR 4405	Hydraulic fluid power—Determination of fluid contamination—Gravimetric method	ISO DP 5599/III	Pneumatic fluid power—5 port directional control valve—Optional electrical connectors
ISO DPR 4406	Hydraulic fluid power—Fluids—Solid contaminant code	ISO DPR 5781	Hydraulic fluid power—Two port valves—Pressure control valves—Check valves—Mounting surfaces
ISO DPR 4407	Hydraulic fluid power—Determining fluid contamination counting method under transmitted light	ISO DPR 5782	Pneumatic fluid power—Compressed air filters—Product standard
ISO DPR 4408	Hydraulic fluid power—Determining fluid contamination counting method under incident light	ISO DPR 5783	Hydraulic fluid power—Mounting surfaces—Codification
ISO DPR 4409	Hydraulic fluid power—Pumps, motors and integral transmissions—Method of testing positive displacement	ISO DPR 6020/1	Hydraulic fluid power—Cylinders—Mounting dimensions (160 bar)—Medium series—Part 1
ISO DPR 4410	Pneumatic fluid power systems—Industrial air pressure regulators—Test procedures and data presentation	ISO DPR 6020/2	Hydraulic fluid power—Cylinders—Mounting dimensions (160 bar)—Compact series—Part 2
		ISO DPR 6022	Hydraulic fluid power—Cylinders—Mounting dimensions (250 bar)
ISO DPR 4411	Hydraulic fluid power—Valves—Method of determining the pressure differential flow characteristics	ISO DPR 6052	Pneumatic fluid power—Code of identification of mounting surfaces for directional control and regulation valves
DIS 4412	Pumps—Airborne noise levels		
ISO DIS 4413	Application of hydraulic power equipment to transmission and control	ISO DPR 6071	Fluid for hydraulic transmission—Fire resistant fluids—Classification and designation
ISO DPR 4414	Fluid power systems—Industrial pneumatic equipment—Guide for detailed specifications and designs—Part 4: Circuit controls—Part 5: Actuators linear and rotary—Part 6: Air motors—Part 7: Compressors—Part 8:	ISO DPR 6072	Compatibility of hydraulic fluids with sealing material
		ISO DPR 6073	Prediction of bulk moduli for petroleum oils
		ISO DPR 6074	Fluids for hydraulic transmissions—Mineral oils—Classification
		ISO DPR 6075	Characteristics of mineral oils—Part 1: Standard test
ISO DPR 4572	Hydraulic fluid power—Filters—Multi-pass method for evaluating filtration performance	ISO DPR 6076	Compatibility of hydraulic fluids with elastomeric materials
ISO DPR 5594	Fluid power systems and components—Port and fitting end	ISO DPR 6099	Fluid power systems and components—Hydraulic and pneumatic cylinders—Identification code for mounting dimensions and for cylinder mountings working documents
ISO DPR 5595	Fluid power systems and components—Gas-loaded accumulator with separator—Volumes—Pressures		

ISO DP 6149	Metric Port—Dimensions and design
ISO DPR 6150	Pneumatic fluid power—Quick action couplers—Plug dimensions and tolerances
ISO DPR 6162	Fluid power systems and components—Flanged tube, pipe and hose connections—Dimensions and design considerations of four-bolt split flanges
ISO DPR 6163	Hydraulic fluid power—Square and round flanges—Pressure stages from 100 bar up to 400 bar
ISO DPR 6164	Hydraulic fluid power—Square welded collar flanges (250 bar and 400 bar)—Dimensions
ISO DP 6194/I	Fluid systems—Rotary shaft lip type seals—Part 1: Nominal dimensions and tolerances
ISO DPR 6195	Housing dimensions for wiper rings, and reciprocating rod applications for fluid power cylinders

15 Electrical Components

General

World standards[1] for electrical components for automotive use will be discussed in this section. Progress towards international interchangeability of this type of equipment has been slow, principally hampered by the lack of worldwide acceptance of the metric system of measure. The increase in recent years in the exchange of automobiles produced in many countries has created the need for world standards in this area, and the lack of accepted ISO and IEC standards has become more obvious. North America's position on international standardization in this field could be labeled as indifferent, and this has strengthened the influence other countries, already on the metric system, have had throughout the world.

MAJOR STANDARD-FORMING ORGANIZATIONS

ISO: The technical committees within the ISO on Road Vehicles (TC22) and Agricultural Tractors and Machinery (TC23) have the Secretariats assigned to FRANCE (AFNOR). ISO/TC22 have the following subcommittees working on electrical components:

SC 1 Ignition Equipment (organized by Germany-DNA)
SC 3 Electrical Connections (DNA)
SC 8 Lighting and Signalling (AFNOR)

IEC: The International Electrotechnical Commission (IEC) is active in writing world standards for electrical components in automotive use through its many technical committees and subcommittees, and they are as follows:

IEC/TC20	Cables
IEC/TC23	Electrical Accessories
SC23B	Plugs, Socket Outlets, and Switches
SC23C	Worldwide Plug and Socket Outlet Systems
SC23B	Lampholders
SC23F	Connecting Devices
IEC/TC32	Fuses
SC32B	Low-Voltage Fuses
IEC/TC34	Lamps and Related Equipment
SC34A	Lamps
SC34B	Lamp Caps and Holders
IEC/TC35	Primary Cells and Batteries

SAE: The Society of Automotive Engineers has published a series of customary-inch module standards for electrical equipment, 38 of which are listed in this section. Some of the SAE standards have inch-module fasteners and standard components; these are not readily adoptable by countries already on the metric system. Active participation by SAE in either ISO and IEC is only possible through ANSI.

Worldwide Acceptance of Standards

During 1974 the writer conducted a survey among leading producers of this type of equipment worldwide, including 17 of the major SAE standard electrical components. The presence of a dominant SAE role in other countries was established, however, an increasing trend in focusing new standards on the subject of existing metric standards in ISO, IEC, and in the Continental European countries was evident. It is the writer's opinion, that the worldwide North American influence in this area will diminish rapidly, unless we soon can present some fully metric standards in the area of electrical components for automotive use.

RELATED INTERNATIONAL, NATIONAL, AND SOCIETY STANDARDS

ISO/R 303-1963	Lighting and signalling for motor vehicles and trailers
ISO 512-1974	Sound signalling devices on motor vehicles—Acoustic standards and technical specifications
ISO 1185-1975	Electrical connections between prime movers and towed vehicles with 24 V electrical equipment for commercial international traffic
ISO-1724-1975	Electrical connections for vehicles with 6 or 12 V electrical equipment particularly for private motor-cars and light trailers or caravans
ISO 1919-1976	Automobiles—Spark plug M 14 X 1.25 with flat seating

---

[1] For information about the term "standard" as used in this book, please see page 1-2.

ISO 2344-1976	Automobiles—Spark plugs M 14 × 1.25 with conical seating
ISO 2345-1976	Automobiles—Spark plugs M 18 × 1.5 with conical seating
ISO 2346-1976	Automobiles—Compact spark plugs M 14 × 1.25 with flat seating
ISO 2347-1976	Automobiles—Compact spark plugs M 14 × 1.25 with conical seating
ISO 2416-1972	Automobiles—Load distribution for private cars
ISO 2542-1972	Internal combustion engines—Spark plug ignition—Terminology
ISO 2704-1976	Automobiles—Spark plugs M 10 × 1 with flat seating
ISO 2705-1976	Automobiles—Spark plugs M 12 × 1.25 with flat seating

IEC

IEC 95-1*	Lead-acid starter batteries, Part 1: General requirements and methods of test
IEC 95-2*	Lead-acid starter batteries, Part 2: Dimensions of batteries
IEC 95-3*	Lead-acid starter batteries, Part 3: Dimensions and marking of terminals
IEC 199	Dimensions of lead-acid motor scooter batteries
IEC 391	Marking of insulated conductors

*See also Battery Council International (BCI) *Battery Replacement Data Book* available from BCI, 1801 Murchison Dr., Burlingame, California 94010

United States (S.A.E.)

*J930a—	Storage Batteries for Construction & Industrial Machinery (1) & (2)
*J542b—	Starting Motor Mountings (1)
J543b—	Starting Motor Pinions & Ring Gears (1)
J545b—	Generator Mountings (1)
*J544b—	Starting Motor & Generator Curves (1)
J546 —	Magneto Mountings (1)
J548b—	Spark Plugs (1)
*J549a—	Preignition Rating of Spark Plugs (1)
*J550b—	Spark Plug Installation—Torque Requirement & Gasket Dimensioning (1)
J554 —	Electric Fuses (1)
*J553c—	Circuit Breakers (1)
J258 —	Circuit Breakers—Internal Mounted—Automatic Reset (1)
J259 —	Ignition Switch (1)
*J156 —	Fusible Links (1)
*J858a—	Electrical Terminals—Blade Type (1)
J928a—	Electrical Terminals—Pin & Receptacle Type (1)
J561b—	Electrical Terminals—Eyelet & Spade Type (1)
*J112a—	Electric Windshield Wiper Switch (1)
*J234 —	Electric Windshield Washer Switch (1)
*J235 —	Electric Blower Motor Switch (1)
*J759b—	Lighting Identification Code (3)
J567c—	Lamp Bulb Retention System (3)
J850 —	Connectors & Plugs (3)
J568 —	Sockets Receiving Prefocus Base Lamps (3)
J571c—	Dimensional Specs for Sealed Beam Headlamp Units (3)
J573d—	Lamp Bulbs & Sealed Units (3)
*J760 —	Dimensional Specs for General Service Sealed Lighting Units (3)
J575e—	Tests for Motor Vehicle Lighting Devices and Components (3)
J579a—	Sealed Beam Headlamp Units for Motor Vehicles (3)
J580a—	Sealed Beam Headlamp (3)
J585d—	Tail Lamps (Rear Position Lights) (3)
J586c—	Stop Lamps (3)
*J249 —	Mechanical Stop Lamp Switch (3)
J588e—	Turn Signal Lamps (3)
*J186 —	Supplemental High Mounted Stop & Turn Signal Lamps (3)
J589b—	Turn Signal Switch (3)
*J914 —	Side Turn Signal Lamps (3)
*J945 —	Vehicular Hazard Warning Signal Flasher (3)

*Recommended Practice

Prepared by: (1) ELECTRICAL EQUIPMENT COMMITTEE
(2) CIMTC
(3) LIGHTING COMMITTEE

NOTE:
S.A.E. Publications available from: Society of Automotive Engineers, Inc., 400 Commonwealth Drive, Warrendale, Pa. 15096

Japan

JIS D 5102-1960	Spark plugs with resistor for automobiles (for water-cooled engine)
JIS D 5103-1969	Glow plugs for automobiles
JIS D 5121-1972	Ignition coils for automobiles
JIS D 5202-1967	Mounting dimensions of distributors for automobiles
JIS D 5303-1969	Lead-acid traction batteries

ELECTRICAL COMPONENTS

JIS D 5500-1969	Lighting and signalling equipment for automobiles
JIS D 5504-1971	Sealed beam head lamp units for motor vehicles
JIS D 5701-1968	Electric horns for automobiles
JIS D 5706-1968	Horn relays for automobiles
JIS D 5707-1970	Turn signal flashers for automobiles
JIS D 5708-1970	Hazard warning signal flashers for automobiles
JIS D 5710-1969	Wiper blades and wiper arms for automobiles
JIS D 5711-1972	Red fuses for motor vehicles
JIS D 5712-1973	Warning buzzers for automobiles
JIS D 5801-1972	Hydraulic stop lamp switches for automobiles
JIS D 5802-1972	Dimmer switches for automobiles
JIS D 5803-1972	Warning lamp switches of lubricating oil pressure for automobiles
JIS D 5808-1972	Mechanical stop lamp switches for automobiles

Germany

DIN 72577	Electrical Connections for Motor Vehicles and Trailers, Seven Conductors for 6V and 12V equipment (Sheet 1-Socket; Sheet 2-Plug; Sheet 3-Wiring Diagram)
DIN 72781:Sheet 2	Windshield Wipers for Motor Vehicles; Wiper Motors Definitions, Testing

Facta-Handbuch Normen für den Kraftfahrzeugbau-1974 is a comprehensive German handbook with most electrical standards for automotive use. The book and other German standards are available from BEUTH-VERTRIEB GmbH, Burggrafenstr 4-7, 1 Berlin, W. Germany

Italy

Etas-Kompass book is a publication with Italian automotive electrical supplier information available from Etas-Kompass S.p.A., Via Mantegna, 6, 20154 Milano, Italy

16 Tires, Rims, and Valves

National and Regional Standardization[1]

Standards for tires and the closely related products, rims and valves, generally fall into two broad categories. One deals with size designations, dimensional standards, and capacity ratings to provide interchangeability among products made by a wide variety of manufacturers. The other deals with performance standards which provide a basis for minimum acceptable levels of safety, or possibly, for quality differentiation among similar products.

The first of these two categories has been the subject of organized industry activity since the founding in the United States, in 1903, of what is now known as the Tire and Rim Association (TRA). Initially, this organization confined its activity to the standardization of rim dimensions that related to tire fit and mounting. Later, the standardization of tire size designations, dimensions and load ratings, as well as certain key dimensions of valves for use with tires and inner tubes, were added to the work of this organization.

In other areas of the world, industry-wide standardization came about more slowly. In some countries, national standards were established by industry organizations with a scope very similar to that of the U.S. Tire and Rim Association.

The most important of these national organizations outside of the United States was probably the Society of Motor Manufacturers and Traders (S.M.M.T.) in the United Kingdom, which established a committee of technical personnel from among the tire and rim manufacturers of that country. The S.M.M.T. standards for tires and rims were widely adopted, particularly in the British Commonwealth countries, but since they were somewhat different from the American Tire and Rim Association standards (particularly in relation to tires), true international standardization did not exist.

On the European continent some national industry-wide organizations did develop for purposes similar to those of the Tire and Rim Association and the S.M.M.T. The most active of these was probably the Wirtschaftsverband der deutschen Kautchukindustrie (W. d. K.), in Germany. The Commissione Tecnica Di Unificazione Nell'Autoveicol (CUNA), in Italy, also had some activity related to publication of tire standards, and the Scandinavian Tire and Rim Organization (STRO) did some work on tire standardization related directly to the Scandinavian area.

In the Eastern European countries, tire and rim standards were established and published by governmental transportation agencies, particularly in Russia, Czechoslovakia, and Poland.

Throughout the world, outside of the United States and Europe, some national industry technical organizations were established to parallel (or merely to publish) on a local basis, tire and rim standards established by the S.M.M.T. or the Tire and Rim Association. Among such organizations were the Tire and Rim Association of Australia; The Japan Automobile Tire Manufacturers Association (JATMA); The South African Tire and Rim Association; The Tire and Rim Association of Brazil; etc. For the most part, these organizations were established and functioned when a significant tire manufacturing and vehicle assembly industry existed in a country. The amount of such activity outside of North America and Europe was therefore quite small prior to 1946.

After the cessation of hostilities in Europe in 1945, the tire and vehicle industries were rebuilt and as a result of the breaking down of some of the long-standing barriers between countries of that continent, a need for tire and rim dimension standards relating to interchangeability on an international basis became obvious. With the prodding of a few far-sighted individuals within the European tire and rim companies, the European Tire and Rim Technical Organization (E.T.R.T.O.) was formed and was officially established in 1964.

---

[1] For information about the term "standard" as used in this book, please see page 1-2.

Since that time, work has progressed to the point that an annual ETRTO *Data Book* (similar to the Tire and Rim Association's *Yearbook*) is published regularly. National organizations in the Western European countries have essentially stopped individual standardization work related to tires and rims and where a need for a national standard exists, the ETRTO standards are adopted.

The Canadian tire and rim industries have for many years been so closely related to those of the United States that they have considered the Tire and Rim Association standards completely adequate for their own needs. If a special need arises within Canada, this country can bring such a need to the attention of the appropriate TRA Subcommittee for action through liaison performed by a permanent Canadian representative named by the TRA. In practice, because of the similarity of vehicles and operating conditions, such a special situation almost never arises.

As the volume of tire and vehicle production outside of North America and Western Europe has grown considerably in the last 25 years, the need for local tire and rim standards has also grown. For the most part, such standards have been adopted directly from either the TRA or the ETRTO (or its predecessor European national organizations) depending usually on where the particular tire size or vehicle was originally established.

The result of this situation has been some overlap and resulting nonstandardization of some tire sizes. As long as European passenger cars were almost all smaller than those designed in the U.S., the amount of overlap in tires of this type was not a great problem. However, tires for trucks, farm equipment, aircraft, motorcycles, and industrial equipment tend to be similar in size no matter in what country of the world they are made, so the need of worldwide standards for tires of these types becomes obvious as soon as the desire for international trade exists.

World Standardization

This brings us to the activation of a technical committee covering tires, rims, and valves within the International Organization for Standardization (ISO). This organization and its predecessor were essentially European in make-up in their earlier days and although they had long ago designated a Technical Committee for tires, rims, and valves (TC 31), it had not been active, primarily due to the fact that the United States, with its dominant position in the industry, was not a participant.

However, in 1967, the United States and some of the important European tire producing countries that had previously been reluctant, agreed to become active in TC 31 and an organization meeting was held in Washington, D.C., in December 1968, following the designation of the USA as TC 31 Secretariat by ISO.

The following countries are now members of ISO/TC 31:

Participating Members	Observer Members
Bulgaria	Australia
Canada	Belgium
Czechoslovakia	Brazil
France	Chile
Federal Republic of Germany	Columbia
Israel	Ethiopia
Italy	Finland
Japan	Hungary
Netherlands	India
Poland	Iran
Republic of South Africa	Ireland
Spain	Mexico
Sweden	Pakistan
United Kingdom	Peru
U.S.S.R.	Portugal
U.S.A.	Romania
	Switzerland
	Thailand
	Turkey
	Yugoslavia

In the organization meeting of TC 31, the following committee scope was agreed upon:

"Standardization of classifications, size designations, dimensions, and ratings of tires, rims, and valves."

Further, two Working Groups were established to lay the foundation for future Committee work. WG#1 covered terminology and definitions, and WG#2 covered units to be used in TC 31.

Since the organization meeting, five additional Plenary Sessions of TC 31 have taken place and Subcommittees covering various tire types have been established to carry on the detailed work of drafting standards. (Because these Subcommittees were originally started as Working Groups, the numbering starts with 3.) The TC 31 Subcommittees and their Secretariats are as follows:

SC 3	Passenger Car Tires and Rims	France
SC 4	Truck Tires and Rims	Italy
SC 5	Agricultural Tires and Rims	Germany
SC 6	Off-The-Road Tires and Rims	Canada
SC 7	Industrial Tires and Rims	U.K.
SC 8	Aircraft Tires and Rims	U.S.A.
SC 9	Valves and Tubes	France
SC 10	Cycle, Moped, and Motorcycle Tires and Rims	France

While no complete ISO standards relating to tires, rims, and valves have as yet been established by TC 31, the work is partially completed in all Subcommittees and some parts of the standards are already agreed to.

TIRES, RIMS, AND VALVES

When ISO standards are available, they will be a valuable new source for standards covering tires, rims, and valves to be designed and manufactured in the future, since they will have been accepted by essentially all the countries of the world with substantial capacity for manufacturing these items.

Units of Measure for Tires

The subject of units used in the important standardizing bodies previously discussed in this chapter should now be considered.

Because of their origin, many of the standards in the tire, rim, and valve industries were initially published in inch-pound units. Where nations used metric units, the inch-pound units developed in the United States and the United Kingdom were frequently converted to whatever local metric units were normally being used, with the inevitable confusion caused by nonuniform rules for conversion and rounding.

Where such standards were initiated in metric countries, the particular metric units in use were, of course, those adopted. Thus, the units used for tire pressure were at one time "atmospheres" (see Section 2, page 2-7) in Germany, "kilograms (force) per square centimeter" in France, and "pounds per square inch" in the United Kingdom, the United States, and many other industrialized countries of the world which had not made the decision to convert to a metric system.

At present, the Tire and Rim Association publishes its *Yearbook* mostly in inch-pound units, althought it has for some years included a table of conversion factors to metric units.

The S.M.M.T. publications which were formerly widely used in many countries were also in inch-pound units. In years past they also used the British hundredweight unit.

The ETRTO standards and the Japanese Industrial Standard covering dimensions and load ratings of tires (JIS D4202), are published in metric (but not necessarily SI) units.

ISO requires that its standards be published using SI units. However, they do not forbid the inclusion of other units. Because the transition to a wholly SI world will no doubt take many years, TC 31 has decided to publish certain other metric and inch-pound units in addition, in order to avoid differences that could arise from converting and rounding in individual countries that want to use ISO standards for tires, rims, and valves but for legal or practical reasons need to issue them using non-SI units.

As a result of the work of TC 31/Working Group #2 and the subsequent decisions of TC 31, the following units will be shown in the initial ISO standards covering tires, rims, and valves:

Item	*SI Unit*	*Other Units*
Tire and Rim Dimensions	mm	(inch)
Loads (mass)	kg	(pound)
Pressure	kPa	(bar) (pounds per square inch)
Velocity	...	(km per hour) (mile per hour) (knot)
Distance	km or m	(mile)
Plane Angle	...	(degree)

While the radian is the official unit for angular measurement, the degree is acceptable for plane angles according to ISO Standard 1000.

Standard 1000 also permits the use of the bar for pressure. This unit is the subject of much discussion in many ISO Committees. It is convenient for those European countries which have used either atmospheres or kilograms (force) per square centimeter for tire pressure since all three of these units are the same, within the accuracy needed for use in the tire industry, and thus, existing gages and tables can continue to be used when ISO standards are adopted.

The use of the kilogram (mass) for tire load capacities is harder to rationalize. The tire's performance is generally directly related to the force applied to it (usually because of the weight of the vehicle it carries). However, consultation with ISO vehicle committees has revealed that they will consistently be referring to vehicle mass, rather than weight, and therefore will be using the kilogram (SI base unit for mass) as the appropriate unit. Therefore, it would be impractical to use newtons for tire load capacities when vehicle manufacturers will be dealing with vehicle mass in terms of kilograms.

The knot will be used only in relation to aircraft tires. This unit has been adopted because aircraft landing and take-off speeds are normally quoted in knots.

At the present time the extent of work accomplished in the various ISO/TC 31 Subcommittees is limited to agreement on tire size designations and certain other markings needed to properly identify a tire as well as dimensions of some existing tires and rims and some tolerances on such dimensions.

Tire Designations

Dimensions of tires and rims already in existence will essentially not be changed except for those cases where a tire (or rim) with a given designation has different dimensions in different existing standards. In those cases, either a compromise is reached with a tolerance which incorporates existing tires (or rims), or, if the differences are too wide to cover with reasonable tolerances, a choice is made on the basis of either the dominant usage or some engineering logic.

The first significant agreement reached in ISO/TC 31 related to the format of tire designations to be used for all future tires (except aircraft tires). This format consists of two parts which will be separated from each other, both on the sidewall marking of the tire and in printed matter. The first part contains elements relating to tire dimensions and construction; the second part consists of a service description (for example: load rating and speed category).

The system can best be illustrated by an example:

$$165/80 \text{ R } 15 \quad \text{S}/82$$
$$(1)\,(2)\,(3)\,(4) \quad (5)\,(6)$$

Each element in this example has a meaning as follows:
(1) The nominal cross-sectional width of the tire; the example indicates a width of 165 mm but this is not necessarily meant to be the exact width of the tire. It could be that for larger size tires, such as rear tractor or earthmover tires, the number in this position in the size designation may approximate the tire cross-section width in inches. This choice will be up to each Subcommittee as long as the number has some reference to the tire cross-section. Although this number is essentially dimensionless, a larger number will always indicate a larger cross-section width; thus it does provide the necessary differentiation among different tire sizes.
(2) The nominal "aspect ratio" of the tire cross-section multiplied by 100. The aspect ratio is the ratio of the cross section height above the bead seat diameter to the cross section width. It has been agreed that for all future sizes, aspect ratios will be in multiples of 5.
(3) A code letter indicating the basic type of tire construction. So far, TC 31 has established the following construction codes:

$$\text{D} = \text{diagonal (bias)}$$
$$\text{B} = \text{bias belted}$$
$$\text{R} = \text{radial}$$

(4) A numerical code referring to the bead seat diameter of the rim. Since the life of rims is quite lengthy (at least as long as vehicle life, in most cases) consideration must be given to making any ISO tire standards compatible with usage on existing rims over a period of perhaps 20 or more years. Rim bead seat diameter is particularly critical since if the tire and rim dimensions are not matched at this point there is a danger of violent rupture of the tire during inflation, with the consequent safety hazard.

In all countries of the world rim contours that were originally developed in inch units have become standard by virtue of usage. These dimensions will be adopted as ISO standards for rims and the dimensions to be published in metric units will merely be soft conversions from present inch units. Since it would be confusing (and of no real value) to change existing rim nomenclature when these rims are established as ISO standards, the nominal bead seat diameters now expressed in inches will continue to be expressed in those same numbers. While they are essentially dimensionless in the size designation, they do provide the necessary differentiation among rims of different diameter since a larger code number will always mean a larger rim diameter.

TC 31 takes the view that changes in existing nomenclature should not be made if no useful purpose is served. However, it is obvious to the TC 31 members that future design work should be carried on in metric modules. Therefore, in the case of the rim diameter portion of the tire size designation, it has been agreed that the nominal rim diameter will be expressed in millimeters when a new tire or rim is designed that will not be compatible from a rim-fit standpoint with existing rims or tires.

(5) The first part of the service description of the tire consists of a code letter which indicates the speed category of the tire. These code letters will be assigned by TC 31 to various speeds and it is planned that any given code letter will mean the same speed on any tire to which it will be applied. Code letters so far agreed to by TC 31 cover the range from 80 to 200 km/h in increments of 10 km/h, using the letters F through U, in their alphabetical order. The letters H, I and O are not used because of possible confusion with numbers.

(6) The second part of the service description consists of an index number indicating the basic load capacity of the tire. The Load Index number corresponds to a load capacity established in a table derived directly from a Renard R80 series with Load Index 0 being equal to 45 kg. This Load Index Table (16-1) is shown at the end of this chapter.

If TC 31 should determine that the "basic" maximum load carrying capacity of a given tire should be, for example, 475 kg, the tire would be assigned a Load Index of 82. The Load Index Table is capable of being extended indefinitely and so can be used to cover the load capacities of all existing or future tires.

The adoption of the Load Index system permits adjustment in the actual load capacity of a tire or group of tires when specific service conditions require some variation from the "basic" load capacity. The Load Index also avoids some confusion in cases where certain national safety regulations require marking of the tire load capacity on the sidewall, in pounds.

Use of the service description (items 5 and 6) will be at the discretion of the various TC 31 Subcommittees, depending on their individual needs.

TIRES, RIMS, AND VALVES

In addition to the tire markings, or designations, thus far explained, it has been agreed that tires designed to be operated without inner tubes will be marked "TUBELESS," in all countries.

Further, if a Subcommittee decides that it needs identification to distinguish between tires of identical size designation which are designed for entirely different types of vehicles, a prefix code letter placed before the size designation could be assigned by TC 31. The prefix letter "P" has been chosen for passenger car tires, but its use has been made optional.

From Soft to Hard Conversion in Tire Standards

It is probably obvious at this point that neither tires nor rims could be classified as "metric parts" regardless of what units are the basis for the different parts of the size designation. The dimensions of tires and rims can be expressed just as easily in metric as in inch units. Initially at least, the function of TC 31 may be principally to publish tire, rim, and valve dimensional data in SI units for whatever use various countries or industries might want to make of them.

This first step is followed almost immediately, however, by the natural decision to design new products in metric modules, TC 31/SC 3 has, for example, already decided that for future tires of any particular aspect ratio, the increment between tire cross-section widths will be 10 mm. The use of logical increments in metric units, in the long run will be a helpful step toward the complete conversion to "metric thinking."

One function of ISO is to stimulate national standardization bodies to establish identical standards for products. Once ISO standards are established, it is expected that such national bodies will begin to use those standards, perhaps even before they are published by ISO. In the United States, the Passenger Car Tire Subcommittee of the Tire and Rim Association has, in fact, already established several new tire sizes based entirely on the ISO agreements and from the beginning has designed these new tire sizes in metric units.

Tire Inflation Pressures

While this may indeed indicate progress toward real conversion to a metric system, it has brought to light some interesting problems. As an example, the matter of maximum inflation pressure for this new tire size must be considered.

In setting up load/inflation tables in SI units this TRA Subcommittee has determined that increments of 20 kPa are logical for tire pressures. (Existing tables are in increments of 2 psi for TRA passenger car tires.) For the particular tire size under consideration 280 kPa is the logical choice for maximum inflation pressure (based on the empirical rules which have been developed from experience over the years.)

However, the United States Federal Motor Vehicle Safety Standard covering tires for passenger cars originally stated that the maximum pressure for such tires may be *only 32, 36, or 40* psi. Since 280 kPa does not convert exactly into any one of these three numbers, a dilemma arose. Upon petition, by the Rubber Manufacturer Association, the government regulation was amended. This turned out to be a ponderous process that took over one year to complete

The resolution of these problems may be a long way off, unless the governments allowing or requiring inch/pound units take some forceful action to accelerate the conversion. Due to the disruptions that would ensue, and economic factors, this may not be a politically popular move.

Perhaps a corollary of this ISO activity will be increased use of Renard Series (see Section 4, "Preferred Numbers") in establishing new tire lines. While Renard Series are not really "metric" they are certainly closely related to "metric thinking" and during the transition from inch/pound to metric units, the opportunity is provided to adopt them, if a need exists for providing a logical basis for setting up increments between sizes or capacities.

Dimensional Standards for Tires

In concluding this discussion of worldwide standards relating to interchangeability, it could be said that there is very little that could be identified as "metric" in the field of tires, rims, and valves. Although ETRTO uses metric units in their publications, they are frequently merely a conversion from the inch/pound units from which they were derived. Even in those elements of ETRTO publications which were not originated in the inch/pound system, the metric units used often do not signify that the increments between steps in tire sizes, load capacities, inflation pressures, etc., were based on "metric" logic. Therefore, it is difficult to say whether current tire standards are either based on an inch/pound system of units or on a metric system. It is probably closer to the truth to say that the current standards are haphazard and not based on any system of units.

The rules of ISO procedure and publication will no doubt give some impetus to converting the tire, rim, and valve standards of the future to a logical metric system.

Table 16-1. Tire Load Index Table (ISO 4209/I)

LI	kg	LI	kg	LI	kg	LI	kg	LI	kg	LI	kg	LI	kg
0	45	40	140	80	450	120	1400	160	4500	200	14 000	240	45 000
1	46.2	41	145	81	462	121	1450	161	4625	201	14 500	241	46 250
2	47.5	42	150	82	475	122	1500	162	4750	202	15 000	242	47 500
3	48.7	43	155	83	487	123	1550	163	4875	203	15 500	243	48 750
4	50	44	160	84	500	124	1600	164	5000	204	16 000	244	50 000
5	51.5	45	165	85	515	125	1650	165	5150	205	16 500	245	51 500
6	53	46	170	85	530	126	1700	166	5300	206	17 000	246	53 000
7	54.5	47	175	87	545	127	1750	167	5450	207	17 500	247	54 500
8	56	48	180	88	560	128	1800	168	5600	208	18 000	248	56 000
9	58	49	185	89	580	129	1850	169	5800	209	18 500	249	58 000
10	60	50	190	90	600	130	1900	170	6000	210	19 000	250	60 000
11	61.5	51	195	91	615	131	1950	171	6150	211	19 500	251	61 500
12	63	52	200	92	630	132	2000	172	6300	212	20 000	252	63 000
13	65	53	206	93	650	133	2060	173	6500	213	20 600	253	65 000
14	67	54	212	94	670	134	2120	174	6700	214	21 200	254	67 000
15	69	55	218	95	690	135	2160	175	6900	215	21 800	255	69 000
16	71	56	224	96	710	710	2240	176	7100	216	22 400	256	71 000
17	73	57	230	97	730	137	2300	177	7360	217	23 000	257	73 000
18	75	58	236	98	750	138	2360	178	7500	218	23 600	258	75 000
19	77.5	59	243	99	775	139	2430	179	7750	219	24 300	259	77 500
20	80	60	250	100	800	140	2500	180	8000	220	25 000	260	80 000
21	82.5	61	257	101	825	141	2575	181	8250	221	25 750	261	82 500
22	85	62	265	102	850	142	2650	182	8500	222	26 500	262	85 000
23	87.5	63	272	103	875	143	2725	183	8750	223	27 250	263	87 500
24	90	64	280	104	900	144	2800	184	9000	224	28 000	264	90 000
25	92.5	65	290	105	925	145	2900	185	9250	225	29 000	265	92 500
26	95	66	300	106	950	146	3000	186	9500	226	30 000	266	95 000
27	97.5	67	307	107	975	147	3075	187	9750	227	30 750	267	97 500
28	100	68	315	108	1000	148	3150	188	10000	228	31 500	268	100 000
29	103	69	325	109	1030	149	3250	189	10300	229	32 500	269	103 000
30	106	70	335	110	1060	150	3350	190	10600	230	33 500	270	106 000
31	109	71	345	111	1090	151	3450	191	10900	231	34 500	271	109 000
32	112	72	355	112	1120	152	3550	192	11200	232	35 500	272	112 000
33	115	73	365	113	1150	153	3650	193	11500	233	36 500	273	115 000
34	118	74	375	114	1180	154	3750	194	11800	234	37 500	274	118 000
35	121	75	387	115	1215	155	3875	195	12150	235	38 750	275	121 000
36	125	76	400	116	1250	156	4000	196	12500	236	40 000	276	125 000
37	128	77	412	117	1285	157	4125	197	12850	237	41 250	277	128 500
38	132	78	425	118	1320	158	4250	198	13200	238	42 500	278	132 000
39	136	79	437	119	1360	159	4375	199	13500	239	43 750	279	136 000

TIRES, RIMS, AND VALVES

Dimensional standards for future sizes of passenger car, truck, and industrial tires and current sizes of agricultural, industrial, and aircraft tires and certain aspects of valve dimensions have now been agreed to by TC 31. Based on those agreements, published ISO standards for these products may be forthcoming in 1977. After that, the pace of conversion to metric standards in the worldwide tire, rim, and valve industries should certainly quicken.

Tire Performance and Safety Standards

We should now discuss briefly the status of that category of tire, rim, and valve standards which relate to performance. This field is relatively new, having been opened up as a part of the general interest in vehicle and highway safety on the part of government and consumer interest groups which began in earnest about 13 years ago.

Some early attempts to establish such standards for passenger car tires, and later, truck tires, were made by the U.S. Rubber Manufacturers Association and the Society of Automotive Engineers. When the United States Congress passed the Motor Vehicle and Traffic Safety Act in 1966, these standards became very significant since enforcement was placed in the hands of the United States Government.

Federal Motor Vehicle Safety Standard 109, covering new tires for passenger cars, became effective in the United States on January 1, 1968. Subsequently, FMVSS 110 covering passenger car tire selection and rims, FMVSS 117, covering retreaded passenger car tires; FMVSS 119 covering new truck and other highway tires, and FMVSS 120 covering tire selection and rims for truck and other highway tires have also become effective.

Except for the newly adopted ISO tire sizes, all U.S. Motor Vehicle Standards are written using inch/pound units and in fact require that certain labeling information be stated in inch/pound units. Until the federal government in the United States takes some positive action to endorse the use of metric units, it appears that the U.S. motor vehicle safety standards will be a significant obstacle to metrication in the tire, rim, and valve industries.

Following the lead of the U.S., safety standards covering passenger-car tires and rims have been established in Canada, Australia, Venezuela, and Columbia. The Canadian tire and rim safety standards are identical to the U.S. standards and are also numbered 109 and 110. Their counterparts in Australia are identified as Australian Design Rules 23 and 24. The Venezuelan standards are known as Covenin standards and there are several numbered standards covering different aspects of both passenger-car and truck tires. The Columbian tire standard is known as E-15, and was issued by the Institute Columbiano de Normas Technicas (INCONTECT).

In Japan, the Japanese Automobile Tire Manufacturers Association has established safety standards for tires which are intended for voluntary compliance by tire manufacturers.

For several years the Economic Commission for Europe (ECE - a division of the United Nations) has been working on the establishment of safety standards for vehicles and their components. A regulation relative to passenger-car tires (Reg. No. 30) has now been developed and has been approved by United Nations headquarters. Implementation is now at the discretion of any of the ECE countries who are signatories to the Convention relating to motor vehicle and highway safety. The U.K. has announced that regulation No. 30 became effective October 1, 1977.

The ECE Committee which developed this passenger-car tire regulation has now turned its attention to developing a parallel Regulation covering commercial-vehicle tires.

A Directive covering passenger-car tire performance is also being prepared within a committee of the European Common Market (EEC). It is anticipated that this Directive will have technical content identical to the ECE Regulation, but may have somewhat different labeling and enforcement provisions.

U.S., Canadian, and Australian tire safety standards are presently published using inch/pound units. From a practical standpoint this affects only the marking of the tire load capacity since everything else in these standards is either dimensionless or can be easily converted during the design or manufacturing process from one system of units to the other.

Since Australia is officially involved with a 10-year conversion to the metric system, as a matter of national policy, certain forces within Australia are pressing to require metric terminology in their tire safety standard, but there is considerable resistance to this since tire manufacturers throughout the world are already required by law in the U.S. and Canada to mark passenger-car tire load capacities in pounds. For the present, at least, it can be said that Australian Design Rule #23 is in the inch/pound system.

The Venezuelan standard uses metric (but not necessarily SI) units. The Columbian E-15 standard does not yet include numerical data. The JATMA standards use metric (but not necessarily SI) units.

The ECE passenger-car tire Regulation and its companion EEC Directive are being established in metric (although not necessarily SI) units. The system of units used in this case is relatively insignificant since the Regulation covers only those tire markings which are essentially dimensionless, and a high-speed test whose conditions, although stated in metric terms, can easily be converted to other units by testing laboratories.

When ISO standards for tires are finally established, it is probable that they will be incorporated, where applicable, into ECE and EEC Regulations and Directives.

The EEC has stated that in the future only SI (as contrasted with other metric or non-metric units) units will be accepted within its domain. It has yet to be determined whether this decision will be strictly adhered to or will be modified.

The Philippines, New Zealand, India, Thailand, Kenya, and Tanzania are also in the process of developing government tire standards. In each case, the units used are presently those currently in general use in each of the countries. Metric units are not SI units in all cases.

Conclusion

The worldwide standardization of tires, rims, and valves to facilitate the interchangeability of parts may become a reality several years hence. However, safety regulations and other non-tariff trade barriers may unfortunately be enforced in some countries before we have internationally accepted tire, rim, and valve standards in this respect.

RELATED ISO STANDARDS AND ORGANIZATIONS

ISO (TC 22)

ISO 3006-1974	Road vehicles — Passenger car wheels—Test methods
ISO DIS 3911	Wheels/rims—Nomenclature, designation, marking and units of measure

ISO (TC 31)

ISO 3324/I-1976	Aircraft tires and rims—Part 1 Specification
ISO DIS 3324/II	Aircraft tires test methods
ISO/DIS 3739/I	Industrial tires and rims—Current ranges
ISO/DIS 3877/I	Vocabulary—Terms relating to tires
ISO/DIS 3877/II	Vocabulary —Terms relating to valves
ISO/DIS 3877/III	Vocabulary—Terms relating to tubes
ISO 4000-1977	Passenger car tires and rims (future series)
ISO 4209/I-1977	Truck-bus tires and rims
ISO 4223/1-1977	Definitions of some terms used in the tire industry
ISO/DP 4249	Tires and rims for motorcycles—14 to 21-inch rim diameter
ISO/DP 4250	Off-the-road tires and rims
ISO/DP 4251	Tires and rims for agricultural tractors and machines
ISO/DP 4570	Tire valve threads
ISO/DP 5751	Motorcycle tires (future series)
ISO 5775-1977	Bicycle tires and rims
ISO/DP 5995/I	Moped tires and rims

Organizations

ETRTO	European Tyre and Rim Technical Organisation Avenue Brugmann, 32 1060 Bruxelles Belgique
JATMA	The Japan Automobile Tire Manufacturers Association, Inc. 9th Floor, Toranomon Building No. 15, Shiba Toranoman Minato-Ku, Tokyo, Japan
TRA	The Tire and Rim Association, Inc. 3200 West Market Street Akron, Ohio 44313 U.S.A.

17 Metal Cutting Tools

Introduction

Standards[1] for small tools used in removing material from workpieces are being developed worldwide by ISO Technical Committee TC 29. Metric standards for various types of cutting tools are very important, since they tie in with most other metric standards of parts and components in the mechanical design field. The use of metric threaded fasteners, for example, might require metric drill sizes for clearance and tapping holes and a metric cutting tool for the counterbores.

The metric cutting tool standards presented in this section should make it possible to find the most important dimensions, and also many references to other national standards on the subject that might be useful.

TWIST DRILLS

World standards for parallel shank and Morse taperedshank twist drills have been in use by countries already on the metric system for a number of years. The metric diameters for twist drills have been selected from an arithmetic number series where the difference between one size and the next smaller size remains constant. A selection of nominal diameters, based on the preferred number series, would reduce the number of standard sizes substantially and a preferred number diameter series has been under consideration by TC 29.

Limits of Tolerance on Diameter

The tolerance on diameter, as measured across the lands at the outer corners, is h8 as specified in the ISO System of Limits and Fits (see Section 6). The values for this tolerance are given in Table 17-1.

Tolerance on Length

Each flute and overall length may vary between the values specified for the range in question and those specified in the general tables for the ranges immediately above and below.

In the case of taper shank drills, if the next larger or smaller overall length is associated with a different taper shank from that of the length in question, then the permissible upper or lower limit must be that of the next larger or smaller overall length minus the difference between the lengths of the taper shanks concerned.

Table 17-1. Tolerance Limits for Twist Drills*

| Diameter || Tolerance (h8) ||
Over	To	High (+)	Low (−)
...	1	0	0.014
1	3	0	0.014
3	6	0	0.018
6	10	0	0.022
10	18	0	0.027
18	30	0	0.033
30	50	0	0.039
50	80	0	0.046
80	120	0	0.054

*Dimensions in millimeters.

Parallel-Shank Twist Drills

Parallel-shank twist drills are generally made of high-speed steel.

Designation of a twist drill with long parallel shank (see Table 17-4), 30 mm diameter and made of high-speed steel, is as follows: Long Twist Drill 30 BS 328, Part I.

Jobber Series

Parallel-shank twist drills of the short type (jobber series) are standardized in ISO 235: Part I. Nominal diameters and lengths, as well as other national standards, are shown in Table 17-2. The national standards shown conform to the ISO Standard.

---

[1] For information about the term "standard" as used in this book, please see page 1-2.

Table 17-2. Parallel Shank Twist Drills (Jobber Series) (ISO 235/I)

U.S.A.	ISO	235/I
U.S.A.	ANSI	
JAPAN	JIS	
GERMANY	DIN	338
FRANCE	NF	E66-073
U.K.	BS	328/I
ITALY	UNI	5620
AUSTRALIA	AS	

Recess optional

d	l_1	l	d	l_1	l	d	l_1	l	d	l_1	l	d	l_1	l
0,20	2,5	19	1,35	18	40	3,50	39	70	7,30	69	109	11,10	94	142
0,22	2,5	19	1,40	18	40	3,60	39	70	7,40	69	109	11,20	94	142
0,25	2,5	19	1,45	18	40	3,70	39	70	7,50	69	109	11,30	94	142
0,28	3	19	1,50	20	43	3,80	43	75	7,60	75	117	11,40	94	142
0,30	3	19	1,55	20	43	3,90	43	75	7,70	75	117	11,50	94	142
0,32	3	19	1,60	20	43	4,00	43	75	7,80	75	117	11,60	94	142
0,35	4	19	1,65	20	43	4,10	43	75	7,90	75	117	11,70	94	142
0,38	4	19	1,70	20	43	4,20	43	75	8,00	75	117	11,80	94	142
0,40	4	19	1,75	22	46	4,30	47	80	8,10	75	117	11,90	94	142
0,42	5	20	1,80	22	46	4,40	47	80	8,20	75	117	12,00	101	151
0,45	5	20	1,85	22	46	4,50	47	80	8,30	75	117	12,10	101	151
0,48	5	20	1,90	22	46	4,60	47	80	8,40	75	117	12,20	101	151
0,50	6	22	1,95	24	49	4,70	52	86	8,50	81	125	12,30	101	151
0,52	6	22	2,00	24	49	4,80	52	86	8,60	81	125	12,40	101	151
0,55	6	22	2,05	24	49	4,90	52	86	8,70	81	125	12,50	101	151
0,58	7	24	2,10	24	49	5,00	52	86	8,80	81	125	12,60	101	151
0,60	7	24	2,15	24	49	5,10	52	86	8,90	81	125	12,70	101	151
0,62	8	26	2,20	27	53	5,20	52	86	9,00	81	125	12,80	101	151
0,65	8	26	2,25	27	53	5,30	52	86	9,10	81	125	12,90	101	151
0,68	8	26	2,30	27	53	5,40	57	93	9,20	81	125	13,00	101	151
0,70	9	28	2,35	27	53	5,50	57	93	9,30	81	125	13,10	101	151
0,72	9	28	2,40	27	53	5,60	57	93	9,40	81	125	13,20	101	151
0,75	9	28	2,45	27	53	5,70	57	93	9,50	81	125	13,30	101	151
0,78	10	30	2,50	30	57	5,80	57	93	9,60	81	125	13,40	108	160
0,80	10	30	2,55	30	57	5,90	57	93	9,70	81	125	13,50	108	160
0,82	10	30	2,60	30	57	6,00	57	93	9,80	81	125	13,60	108	160
0,85	10	30	2,65	30	57	6,10	57	93	9,90	87	133	13,70	108	160
0,88	10	30	2,70	30	57	6,20	63	101	10,00	87	133	13,80	108	160
0,90	11	32	2,75	30	57	6,30	63	101	10,10	87	133	13,90	108	160
0,92	11	32	2,80	33	61	6,40	63	101	10,20	87	133	14,00	108	160
0,95	11	32	2,85	33	61	6,50	63	101	10,30	87	133	14,25	108	160
0,98	11	32	2,90	33	61	6,60	63	101	10,40	87	133	14,50	114	169
1,00	12	34	2,95	33	61	6,70	63	101	10,50	87	133	14,75	114	169
1,05	12	34	3,00	33	61	6,80	63	101	10,60	87	133	15,00	114	169
1,10	14	36	3,10	36	65	6,90	69	109	10,70	94	142	15,25	114	169
1,15	14	36	3,20	36	65	7,00	69	109	10,80	94	142	15,50	120	178
1,20	14	36	3,30	36	65	7,10	69	109	10,90	94	142	15,75	120	178
1,25	16	38	3,40	39	70	7,20	69	109	11,00	94	142	16,00	120	178
1,30	16	38												

Note: 1. The tolerance on diameter of measured near the point is h8 as shown in Table 17-1 or 6-26.
2. The decimal marker used in the ISO standard and in this table is a comma ",".

METAL CUTTING TOOLS

Table 17-3. Parallel Shank Twist Drill (Stub Series) (ISO 235/I)

U.S.A.	ISO 235/I
U.S.A.	ANSI
JAPAN	JIS
GERMANY	DIN 1897
FRANCE	NF E66-061
U.K.	BS 328/I
ITALY	UNI 5621
AUSTRALIA	AS

Dimensions in millimeters

d	l_1	l	d	l_1	l	d	l_1	l	d	l_1	l
0,50	3	20	9,50	40	84	18,50			27,50		
0,80	5	24	9,80			18,75	64	127	27,75	81	162
1,00	6	26	10,00			19,00			28,00		
1,20	8	30	10,20	43	89	19,25			28,25		
1,50	9	32	10,50			19,50	66	131	28,50		
1,80	11	36	10,80			19,75			28,75		
2,00	12	38	11,00			20,00			29,00		
2,20	13	40	11,20	47	95	20,25			29,25	84	168
2,50	14	43	11,50			20,50	68	136	29,50		
2,80	16	46	11,80			20,75			29,75		
3,00			12,00			21,00			30,00		
3,20	18	49	12,20			21,25			30,25		
3,50	20	52	12,50	51	102	21,50			30,50		
3,80			12,80			21,75	70	141	30,75	87	174
4,00	22	55	13,00			22,00			31,00		
4,20			13,20			22,25			31,25		
4,50	24	58	13,50			22,50			31,50		
4,80			13,80	54	107	22,75			31,75		
5,00	26	62	14,00			23,00	72	146	32,00		
5,20			14,25			23,25			32,50	90	180
5,50			14,50	56	111	23,50			33,00		
5,80	28	66	14,75			23,75			33,50		
6,00			15,00			24,00			34,00		
6,20			15,25			24,25			34,50		
6,50	31	70	15,50	58	115	24,50	75	151	35,00	93	186
6,80			15,75			24,75			35,50		
7,00			16,00			25,00			36,00		
7,20	34	74	16,25			25,25			36,50	96	193
7,50			16,50	60	119	25,50			37,00		
7,80			16,75			25,75	78	156	37,50		
8,00	37	79	17,00			26,00			38,00		
8,20			17,25			26,25			38,50		
8,50			17,50	62	123	26,50			39,00	100	200
8,80			17,75			26,75			39,50		
9,00	40	84	18,00			27,00	81	162	40,00		
9,20			18,25	64	127	27,25					

NOTE: 1. The tolerance on diameter d measured near the point is h8 as shown in Table 17-1 or 6-26
2. The decimal marker used in the ISO standard and in this table is a comma ",".

METAL CUTTING TOOLS

Stub Series

Extra short (stub series) parallel-shank twist drills are covered in ISO 235: Part I, for nominal diameters ranging from 0.5 to 40 mm. The standard diameters and lengths in the range from 0.5 to 40 mm are shown in Table 17-3, and the dimensions conform to those in other national standards shown.

Long Series

Long series parallel-shank twist drills are specified in ISO 494 Standard in sizes ranging from 1 to 31.5 mm, nominal diameters. Linear dimensions are shown in Table 17-4; they are the same as in other national standards shown.

Extra-Long Series

The extra-long parallel-shank twist drill series is covered in ISO 3292 Standard in the size range from 2 to 14 mm. Nominal dimensions for extra-long twist drills are shown in Table 17-5A, and the sizes conform to other national standards shown.

Carbide-Tipped Type

Parallel-shank twist drills with carbide tips are standardized in ISO DIS 3440, in sizes from 3 to 20 mm. National standards for carbide-tipped drills of several types, and nominal dimensions are shown in Table 17-6A.

Morse Taper-Shank Twist Drills—Standard Shank, Oversize Shank, and Core Drills

Morse taper-shank twist drills with standard shank, oversize shank, and the core-drill type are specified in ISO 235: Part 1 and Part 2 (core drills) standards in nominal diameters from 3 to 100 mm. National standards and nominal dimensions for the three types of drills are shown in Table 17-7.

Extra-Long Series

Morse taper twist drills of the extra-long series are covered in ISO 3291 Standard in size range from 6 to 50 mm. Nominal dimensions and other national standards for the above type drills are shown in Table 17-5B.

Carbide-Tip Series

The ISO DIS 3441 Standard specifies Morse taper twist drills with carbide tips, in sizes from 8 to 50 mm. National standards and basic dimensions are shown in Table 17-6B.

COMBINED DRILLS AND COUNTERSINKS

Center Drills, Types A, B, and R

The following three types of center drills and holes are covered in the ISO Standard as shown below:

ISO 866 - Type A. Combined drills and countersinks (center drills) for center holes without protecting chamfers.

ISO 2540- Type B. Combined drills and countersinks (center drills) for center holes with protecting chamfers.

ISO 2541- Type R. Combined drills and countersinks (center drills) for center holes with radius form.

The ISO recommended drill sizes are specified in the national standards as shown, except for the nominal diameter 0.63 mm, which is not a normal size in the DIN 332 and 333 Standards.

The three center drill types are shown in Table 17-8, and their associated recommended holes, in Table 17-9.

Countersinks for Angles 60°, 90°, and 120° with Parallel or Morse Taper Shanks

The ISO international Standards 3293 and 3294 describe countersinks for 60°, 90°, and 120° angles. This cutting tool is used in a preformed hole with larger diameter than the diameter d_2 shown in Tables 17-10 and 17-11. The ISO Standards cover eight nominal sizes ranging from 16 to 80 mm with Morse taper shanks, and six nominal sizes ranging from 8 to 25 mm, with parallel shanks. The nominal dimensions for the above countersinks are shown in Table 17-10 (Morse taper shanks) and Table 17-11 (parallel shanks).

REAMERS

Reamers are generally used to produce a precision round hole with parallel or tapered walls in steel or metal, after the hole is formed. The amount of material removed with reamers is small, and hand reamers and light machine reamers are frequently used. Therefore, a lead must be provided, and most types have a 45° chamfer angle.

ISO and national standards for metric reamers are shown in Tables 17-17 to 17-19. Some national standards deviate substantially from the ISO recommendations, and in those cases, a reference only (Ref. only) indication follows the national standards number.

Tolerance on Cutting Diameter

The tolerance on the cutting diameter, measured immediately behind the chamfer or taper lead, is the ISO tolerance m6, as described in Section 6 and shown in Table 17-12.

METAL CUTTING TOOLS

Table 17-4. Parallel Shank Twist Drills (Long Series) (ISO 494)

	ISO	494
U.S.A.	ANSI	B94.11
JAPAN	JIS	
GERMANY	DIN	340
FRANCE	NF	E66-075
U.K.	BS	328/I
ITALY	UNI	5619,
AUSTRALIA	AS	

Dimensions in millimeters

d	l_1	l	d	l_1	l	d	l_1	l	d	l_1	l
1,00	33	56	3,60	73	112	6,20	97	148	8,80	115	175
1,10	37	60	3,70			6,30			8,90		
1,20	41	65	3,80	78	119	6,40			9,00		
1,30			3,90			6,50			9,10		
1,40	45	70	4,00			6,60			9,20		
1,50			4,10			6,70			9,30		
1,60	50	76	4,20			6,80	102	156	9,40		
1,70			4,30			6,90			9,50		
1,80	53	80	4,40			7,00			9,60		
1,90			4,50	82	126	7,10			9,70		
2,00	56	85	4,60			7,20			9,80		
2,10			4,70			7,30			9,90		
2,20	59	90	4,80			7,40			10,00	121	184
2,30			4,90			7,50			10,10		
2,40	62	95	5,00	87	132	7,60			10,20		
2,50			5,10			7,70			10,30		
2,60			5,20			7,80			10,40		
2,70			5,30			7,90			10,50		
2,80	66	100	5,40			8,00	109	165	10,60		
2,90			5,50			8,10			10,70		
3,00			5,60			8,20			10,80		
3,10			5,70	91	139	8,30			10,90		
3,20	69	106	5,80			8,40			11,00	128	195
3,30			5,90			8,50			11,10		
3,40	73	112	6,00	97	148	8,60	115	175	11,20		
3,50			6,10			8,70			11,30		

NOTE: 1. The tolerance on diameter d measured near the point is h8 as shown in Table 17-1 or 6-26.
        2. The decimal marker used in the ISO standard and in this table is a comma " , ".

Table 17-4 (Continued). Parallel Shank Twist Drills (Long Series) (ISO 494)

d	l_1	l	d	l_1	l	d	l_1	l	d	l_1	l
11,40	128	195	13,90	140	214	20,00	166	254	26,25	190	290
11,50			14,00			20,25			26,50		
11,60			14,25	144	220	20,50	171	261	26,75	195	298
11,70			14,50			20,75			27,00		
11,80			14,75			21,00			27,25		
11,90	134	205	15,00	149	227	21,25	176	268	27,50		
12,00			15,25			21,50			27,75		
12,10			15,50			21,75			28,00		
12,20			15,75			22,00			28,25	201	307
12,30			16,00			22,25			28,50		
12,40			16,25	154	235	22,50	180	275	28,75		
12,50			16,50			22,75			29,00		
12,60			16,75			23,00			29,25		
12,70			17,00			23,25			29,50		
12,80			17,25	158	241	23,50	185	282	29,75		
12,90			17,50			23,75			30,00	207	316
13,00			17,75			24,00			30,25		
13,10			18,00			24,25			30,50		
13,20			18,25			24,50			30,75		
13,30	140	214	18,50	162	247	24,75			31,00		
13,40			18,75			25,00			31,25		
13,50			19,00			25,25	190	290	31,50		
13,60			19,25			25,50					
13,70			19,50	166	254	25,75					
13,80			19,75			26,00					

Special tolerances in the cutting diameters for reamers are standardized in the ISO 522 Standard.

It is not practicable to standardize reamer limits to suit each of the grades of holes provided for in the ISO system of limits and fits, and the reamer limits adopted have been chosen as likely to meet the most general demand of users working to the ISO system of Limits and Fits Standard. The reamers are intended to produce H8 holes, and by selection, will also be suitable for H7 holes.

Tolerance on Length

For hand reamers, parallel machine reamers, machine chucking reamers, and machine jig reamers, each flute and overall length may vary between the values specified for the range in question, and those specified in the appropriate general tables for the ranges immediately above and below that range.

In the case of taper-shank reamers, if the next larger, or smaller, overall length is associated with a different taper shank from that of the length in question, then the permissible upper or lower limit shall be that of the next larger or smaller overall length, plus or minus the difference between the lengths of the taper shanks concerned.

Tolerance on Parallel Shank Diameter

The tolerance on the diameter for parallel shanks is h9 for hand reamers and h9 for machine (chucking) and Morse finishing reamers. The limits are shown in Table 17-12 and the diameter and squares are shown in Table 17-13.

METAL CUTTING TOOLS

Table 17-5A. Extra-Long Twist Drills with Parallel Shanks (ISO 3292)

	ISO	3292
U.S.A.	ANSI	
JAPAN	JIS	
GERMANY	DIN	1869
FRANCE	NF	E66-075
U.K.	BS	
ITALY	UNI	
AUSTRALIA	AS	

Dimensions in millimeters

Preferred diameters d h8	$l = 125$ $l_1 = 80$	$l = 160$ $l_1 = 100$	$l = 200$ $l_1 = 125$	$l = 250$ $l_1 = 160$	$l = 315$ $l_1 = 200$	$l = 400$ $l_1 = 250$
2,0	X	X				
2,5	X	X				
3,0		X	X			
3,5		X	X	X		
4,0		X	X	X	X	
4,5		X	X	X	X	
5,0			X	X	X	X
5,5			X	X	X	X
6,0			X	X	X	X
6,5			X	X	X	X
7,0			X	X	X	X
7,5			X	X	X	X
8,0				X	X	X
8,5				X	X	X
9,0				X	X	X
9,5				X	X	X
10,0				X	X	X
10,5				X	X	X
11,0				X	X	X
11,5				X	X	X
12,0				X	X	X
12,5				X	X	X
13,0				X	X	X
13,5				X	X	X
14,0				X	X	X
Ranges of diameters	2,0 to 2,65	2,0 to 4,75	Over 2,65 to 7,5	Over 3,35 to 14,0	Over 3,75 to 14,0	Over 4,75 to 14,0

NOTE 1. The decimal marker used in the ISO standard and in this table is a comma ",".

Table 17-5B. Extra-Long Twist Drills with Morse Taper Shanks (ISO 3291)

	ISO	3291
U.S.A.	ANSI	
JAPAN	JIS	
GERMANY	DIN	1870
FRANCE	NF	E66-076
U.K.	BS	
ITALY	UNI	5622
AUSTRALIA	AS	

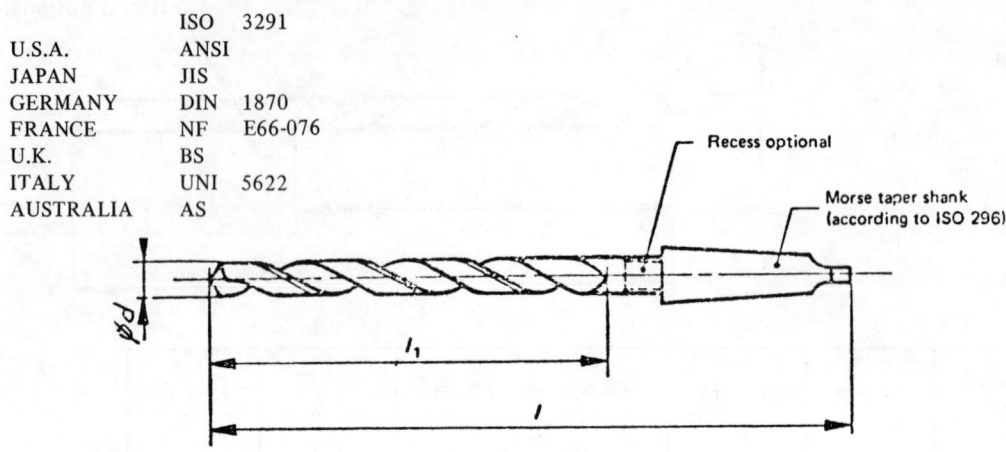

Dimensions in millimeters

Preferred diameters d h8	$l = 200$ $l_1 = 125$	$l = 250$ $l_1 = 160$	$l = 315$ $l_1 = 200$	$l = 400$ $l_1 = 250$	$l = 500$ $l_1 = 315$	$l = 630$ $l_1 = 400$	$l = 710$ $l_1 = 450$	Morse taper shank No.
6,0	X	X	X					
6,5	X	X	X					
7,0	X	X	X					
7,5	X	X	X					
8,0	X	X	X					
8,5	X	X	X					1
9,0	X	X	X					
9,5	X	X	X					
10,0	X	X	X	X				
10,5	X	X	X	X				
11,0	X	X	X	X				
11,5	X	X	X	X				
12,0		X	X	X				
12,5		X	X	X				
13,0		X	X	X				1 or 2*
13,5		X	X	X				
14,0		X	X	X				
14,5		X	X	X	X			
15,0		X	X	X	X			
15,5		X	X	X	X			
16,0		X	X	X	X			2
16,5		X	X	X	X			
17,0		X	X	X	X			
17,5			X	X	X			
Ranges of diameters	6,0 to 11,8	6,0 to 17,0	6,0 to 23,02	Over 9,5 to 40,0	Over 14,0 to 50,0	Over 23,02 to 50,0	Over 40,0 to 50,0	

NOTES:
1. Drills in the diameter range 12 to 14 mm may have either Morse taper shank size No. 1 or 2. Member bodies will have to choose one or the other for inclusion in their national standards.
2. The decimal marker used in the ISO standard and in this table is a comma " , ".

METAL CUTTING TOOLS

Table 17-5B (Continued). Extra-Long Twist Drills with Morse Taper Shanks (ISO 3291)

Dimensions in millimeters

Preferred diameters d h8	$l = 200$ $l_1 = 125$	$l = 250$ $l_1 = 160$	$l = 315$ $l_1 = 200$	$l = 400$ $l_1 = 250$	$l = 500$ $l_1 = 315$	$l = 630$ $l_1 = 400$	$l = 710$ $l_1 = 450$	Morse taper shank No.
18,0			X	X	X			2
18,5			X	X	X			
19,0			X	X	X			
19,5			X	X	X			
20,0			X	X	X			
20,5			X	X	X			
21,0			X	X	X			
21,5			X	X	X			
22,0			X	X	X			
22,5			X	X	X			
23,0			X	X	X			
23,5				X	X	X		3
24,0				X	X	X		
24,5				X	X	X		
25,0				X	X	X		
25,5				X	X	X		
26,0				X	X	X		
26,5				X	X	X		
27,0				X	X	X		
27,5				X	X	X		
28,0				X	X	X		
28,5				X	X	X		
29,0				X	X	X		
29,5				X	X	X		
30				X	X	X		
31				X	X	X		
32				X	X	X		4
33				X	X	X		
34				X	X	X		
35				X	X	X		
36				X	X	X		
37				X	X	X		
38				X	X	X		
39				X	X	X		
40				X	X	X		
41					X	X	X	
42					X	X	X	
43					X	X	X	
44					X	X	X	
45					X	X	X	
46					X	X	X	
47					X	X	X	
48					X	X	X	
49					X	X	X	
50					X	X	X	
Ranges of diameters	6,0 to 11,8	6,0 to 17,0	6,0 to 23,02	Over 9,5 to 40,0	Over 14,0 to 50,0	Over 23,02 to 50,0	Over 40,0 to 50,0	

Table 17-6A. Carbide-Tipped Twist Drills with Parallel Shanks (ISO 3440)

ISO	3440
U.S.A. ANSI	
JAPAN JIS	
GERMANY DIN	8037 (8038 plastics)
FRANCE NF	
U.K. BS	
ITALY UNI	
AUSTRALIA AS	

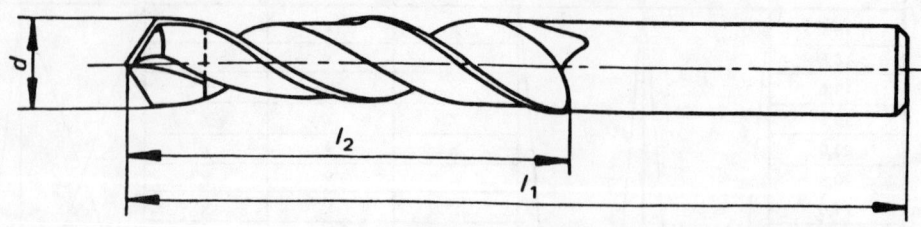

The point angle for metalworking shall be 118° and in the working of plastics shall be 85°

Dimensions in millimeters

Preferred diameters d	Diameter range d h8 over	Diameter range d h8 up to	l_1	l_2
3	2,5	3	50	20
3,5 / 4	3	4	56	25
4,5 / 5	4	5	63	28
5,5 / 6 / 6,5	5	6,5	71	32
7 / 7,5 / 8	6,5	8	80	40
8,5 / 9 / 9,5	8	9,5	85	45
10 / 10,5 / 11	9,5	11	95	50
11,5 / 12 / 13	11	13	106	56
14 / 15	13	15	118	63
16 / 17	15	17	132	71
18 / 19	17	19	150	80
20	19	20	160	90

NOTE 1. The decimal marker used in the ISO standard and in this table is a comma ",".

METAL CUTTING TOOLS

17-11

Table 17-6B. Carbide-Tipped Twist Drills with Morse Taper Shanks (ISO 3441)

U.S.A.	ISO 3441
	ANSI
JAPAN	JIS B4110
GERMANY	DIN 8041
FRANCE	NF
U.K.	BS
ITALY	UNI
AUSTRALIA	AS

The point angle for metalworking shall be 118° and in the working of plastics shall be 85°

Dimensions in millimeters

Preferred diameters d	Diameter range d h8 over	Diameter range d h8 up to	l_1	l_2	Morse taper No.
8	—	8	130	40	1
8,5 9 9,5	8	9,5	135	45	1
10 10,5 11	9,5	11	140	50	1
11,5 12 13	11	13	146	56	
14 15	13	15	168	63	
16 17	15	17	175	71	2
18 19	17	19	185	80	
20 21 22	19	22	215	90	3
23 24 25	22	25	225	100	3

Preferred diameters d	Diameter range d h8 over	Diameter range d h8 up to	l_1	l_2	Morse taper No.
26 27 28	25	28	262	112	
29 30 31 32	28	32	275	125	
33 34 35 36	32	36	290	140	4
37 38 39 40	36	40	310	160	
41 42 43 44 45	40	45	330	180	
46 47 48 49 50	45	50	350	200	

NOTE 1. The decimal marker used in the ISO standard and in this table is a comma ",".

Designation

Reamers are designated with their proper name and the nominal diameter. The national standards number is also included in a complete specification as follows:

Hand Reamer 16 BS 122: Part 2.

Hand Reamers and Long-Fluted Machine Reamers

The ISO 236 Standard covers hand reamers and long-fluted machine reamers in nominal diameters from 1.5 to 71 mm.

Hand reamers have virtually parallel cutting edges with taper and bevel lead, with the shank of the nominal di-

Table 17-7. Morse Taper Shank Twist Drills and Core Drills (ISO 235/I)

	ISO	235/I
U.S.A.	ANSI	
JAPAN	JIS	
GERMANY	DIN	345, 343
FRANCE	NF	E66-071
U.K.	BS	328/I
ITALY	UNI	
AUSTRALIA	AS	

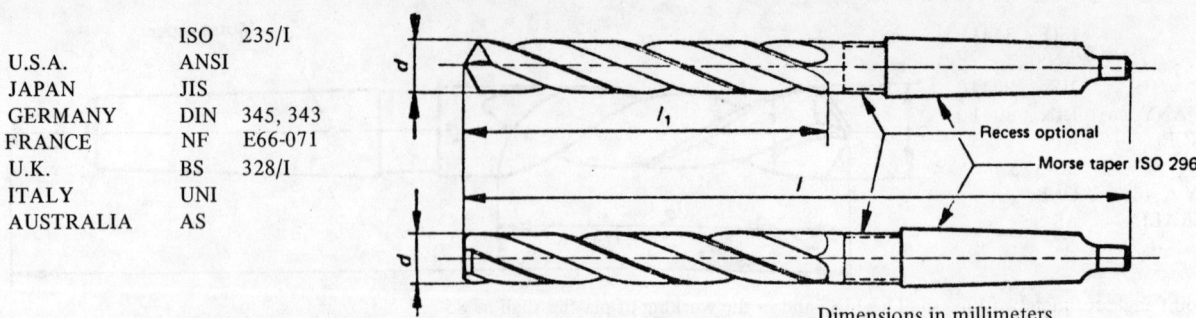

Dimensions in millimeters

d	l_1	Standard shank l	M.T.	Oversize shank l	M.T.	d	l_1	Standard shank l	M.T.	Oversize shank l	M.T.
3,00	33	114				12,00					
3,20	36	117				12,20					
3,50	39	120				12,50	101	182		199	
3,80						12,80					
4,00	43	124				13,00			1		2
4,20						13,20					
4,50	47	128				13,50					
4,80						13,80	108	189		206	
5,00	52	133				14,00					
5,20						14,25					
5,50						14,50	114	212			
5,80	57	138				14,75					
6,00						15,00					
6,20						15,25					
6,50	63	144				15,50	120	218			
6,80						15,75					
7,00						16,00					
7,20	69	150				16,25					
7,50			1			16,50	125	223			
7,80						16,75					
8,00	75	156				17,00					
8,20						17,25					
8,50						17,50	130	228	2		
8,80						17,75					
9,00	81	162				18,00					
9,20						18,25					
9,50						18,50	135	233		256	
9,80						18,75					
10,00	87	168				19,00					
10,20						19,25					
10,50						19,50	140	238		261	3
10,80						19,75					
11,00						20,00					
11,20	94	175				20,25					
11,50						20,50	145	243		266	
11,80						20,75					
						21,00					

NOTES:
1. The tolerance on diameter d measured near the point is h8 as shown in Table 17-1 or 6-26.
2. Twist drills have standard or oversize shanks and core drills have standard shanks only.
3. Morse taper shanks are according to ISO 296. See Table 17-18 for taper details and national standard references.

METAL CUTTING TOOLS

Table 17-7 (Continued). Morse Taper Shank Twist Drills and Core Drills (ISO 235/I)

Dimensions in millimeters

d	l_1	Standard shank l	M.T.	Oversize shank l	M.T.	d	l	Standard shank l	M.T.	Oversize shank l	M.T.	d	l_1	Standard shank l	M.T.	Oversize shank l	M.T.
21,25						32,50						58					
21,50						33,00	185	334				59	235	422			
21,75	150	248		271		33,50						60					
22,00			2		3	34,00						61					
22,25						34,50	190	339				62	240	427			
22,50						35,00						63					
22,75		253		276		35,50						64					
23,00	155					36,00						65		245	432		499	
23,25		276				36,50	195	344				66					
23,50						37,00						67			5		
23,75						37,50						68					
24,00						38,00						69		250	437		504	6
24,25	160	281				38,50						70					
24,50						39,00	200	349				71					
24,75						39,50						72					
25,00						40,00						73		255	442		509	
25,25						40,50						74					
25,50						41,00						75					
25,75						41,50	205	354	4	392		76		447		514	
26,00	165	286				42,00						77					
26,25						42,50						78	260	514			
26,50						43,00						79					
26,75						43,50						80					
27,00						44,00	210	359		397		81					
27,25	170	291		319		44,50						82					
27,50			3			45,00						83	265	519			
27,75						45,50					5	84					
28,00						46,00						85					
28,25						46,50	215	364		402		86					
28,50						47,00						87					
28,75						47,50						88	270	524			
29,00	175	296		324		48,00						89			6		
29,25						48,50						90					
29,50					4	49,00	220	369		407		91					
29,75						49,50						92					
30,00						50,00						93	275	529			
30,25						50,50		374		412		94					
30,50						51	225					95					
30,75	180	301		329		52		412				96					
31,00						53						97					
31,25						54			5			98	280	534			
31,50						55	230	417				99					
31,75		306		334		56						100					
32,00	185	334	4			57	235	422									

Table 17-8. Combined Drills and Countersinks (Center drills) (ISO 866, 2540, 2541)

TYPE	A	B	R
U.S.A.	ISO	866	2540	2541
JAPAN	ANSI	B94.11		
GERMANY	JIS	333	320	333
FRANCE	DIN	E66-051	E66-051	E66-051
U.K.	NF	328/II	328/II	328/II
ITALY	BS	3223	3223	
AUSTRALIA	UNI	1913	1913	1913
	AS			

TYPE A

P R E F (1)	(2)	d k12	d₁ h9	l max-min	L max-min
S		0.5		1–0.8	
S		0.63		1.2–0.9	21–19
S		0.8	3.15	1.5–1.1	
F		1		1.9–1.3	33.5–29.5
S		1.25		2.2–1.6	
F		1.6	4	2.8–2	37.5–33.5
F		2.	5	3.3–2.5	42–38
F		2.5	6.3	4.1–3.1	47–43
F		3.15	8	4.9–3.9	52–48
F		4	10	6.2–5	59–53
S		5	12.5	7.5–6.3	66–60
F		6.3	16	9.2–8	74–68
S		8	20	11.5–10.1	83–77
F		10	25	14.2–12.8	103–97

TYPE B

d₁ h9	d₂ k12	l max-min	L max-min
4	2.12	1.9–1.3	37.5–33.5
5	2.65	2.2–1.6	42–38
6.3	3.35	2.8–2	47–43
8	4.25	3.3–2.5	52–48
10	5.3	4.1–3.1	59–53
11.2	6.7	4.9–3.9	63–57
14	8.5	6.2–5	70–64
18	10.6	7.5–6.3	78–72
20	13.2	9.2–8	83–77
25	17	11.5–10.1	103–97
31.5	21.2	14.2–12.8	128–122

TYPE R

d₁ h9	r max-min	l approx.	L max-min
3.15	3.15–2.5	3	33.5–29.5
3.15	4–3.15	3.3	33.5–29.5
4	5–4	4.25	37.5–33.5
5	6.3–5	5.4	42–38
6.3	8–6.3	6.7	47–43
8	10–8	8.5	52–48
10	12.5–10	10.6	59–53
12.5	16–12.5	13.2	66–60
16	20–16	17	74–68
20	25–20	21.1	83–77
25	31.5–25	26.8	103–97

NOTES:
1. Preference ratings are based on ISO recommendations; F = first choice; S = second choice.
2. The ISO tolerances k12 and h9 are specified in Section 6.(Table 6-1 and 2).
3. Designation; Center drill, type B4/14 ISO 2540.

METAL CUTTING TOOLS

Table 17-9. Recommended Dimensions for Center Holes (ISO 866, 2540, 2541)

	Type	A	B	R
	ISO	866	2540	2541
U.S.A.	ANSI	B94.11		
JAPAN	JIS			
GERMANY	DIN	332	332	332
FRANCE	NF	E66-051	E66-051	E66-051
U.K.	BS	328/II	328/II	328/II
ITALY	UNI	3222	3222	
AUSTRALIA	AS	1913	1913	1913

Type A center hole
ISO 866

Type B center hole
ISO 2540

Type R center hole
ISO 2541

PREF (1)	d	TYPE A l_{min}	TYPE A D	TYPE B l_{min}	TYPE B D	TYPE R D
S	0.5	0.8	1.06			
S	0.63	0.9	1.32			
S	0.8	1.1	1.7			
F	1	1.3	2.12	1.3	3.15	2.12
S	1.25	1.6	2.65	1.6	4	2.65
F	1.6	2	3.35	2	5	3.35
F	2	2.5	4.25	2.5	6.3	4.25
F	2.5	3.1	5.3	3.1	8	5.3
F	3.15	3.9	6.7	3.9	10	6.7
F	4	5	8.5	5	12.5	8.5
S	5	6.3	10.6	6.3	16	10.6
F	6.3	8	13.2	8	18	13.2
S	8	10.1	17	10.1	22.4	17
F	10	12.8	21.2	12.8	28	21.2

NOTE:
Preference ratings are based on ISO standards; F = first choice; S = second choice.

17-16

METAL CUTTING TOOLS

ameter of the cutting edges, and with a square on the end. Basic dimensions for reamers of this type are shown in Table 17-14.

The taper lead is approximately 1 degree, and the length of the taper is 1 1/2 times the diameter of the reamer, or 20 mm, whichever is smaller. Long taper lead reamers, similar in all respects to parallel hand reamers except that the cutting edges are tapered, must have 4/5 of the flute length tapered and 1/5 parallel. Nominal sizes under 16 mm diameter have a taper of approximately 1 in 96, and sizes 16 mm, and above, have a taper of approximately 1 in 64.

Parallel machine reamers are designated long, fluted machine reamers with Morse taper shanks in ISO 236. Nominal sizes from 7 to 71 mm are shown in Table 17-15.

Machine Chucking Reamers with Parallel or Morse Taper Shanks

Machine chucking reamers with virtually parallel cutting edges are standardized in ISO 521, and reamers with Morse taper shanks are shown in Table 17-16 in sizes from 5.5 to 50 mm nominal diameters. Reamers of this type are used in lathe tailstocks or in turrets.

Also specified in ISO 521 are parallel reamers designated machine chucking reamers with parallel shanks. Reamers in sizes from 1.4 to 20 mm cutting diameters are shown in Table 17-17.

Morse Taper Reamers

Reamers to produce self-holding taper sockets of small metric types, numbers 4 and 6, and customary Morse taper

Table 17-10. Countersinks for 60°, 90°, and 120° Angles with Morse Taper Shanks (ISO 3293)

	ISO	3293
U.S.A.	ANSI	
JAPAN	JIS	
GERMANY	DIN	334 (60°), 335 (90°), 347 (120°)
FRANCE	NF	E66-249
U.K.	BS	
ITALY	UNI	6848
AUSTRALIA	AS	

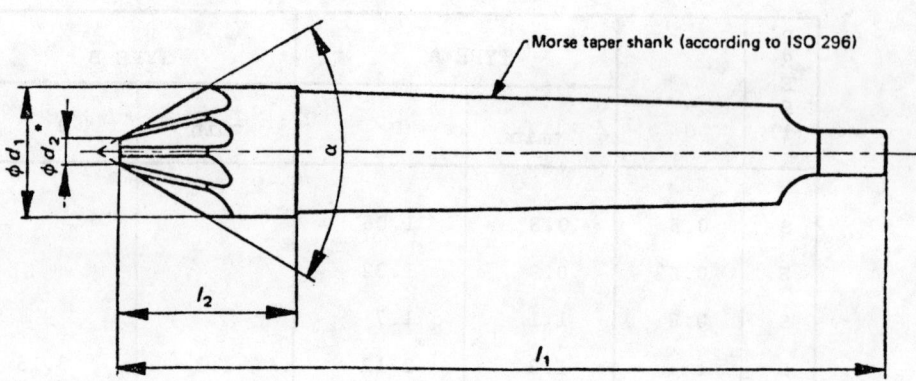

$\alpha = 60°, 90°$ or $120°$ inclusive (tolerance : $_{-1°}^{0}$)

Dimensions in millimeters

Nominal size d_1	Small diameter d_2	Overall length l_1 ($\alpha = 60°$)	Overall length l_1 ($\alpha = 90°$ & 120°)	Body length l_2 ($\alpha = 60°$)	Body length l_2 ($\alpha = 90°$ & 120°)	Morse taper shank No.
16	3,2	97	93	24	20	1
20	4	120	116	28	24	2
25	7	125	121	33	29	2
31,5	9	132	124	40	32	2
40	12,5	160	150	45	35	3
50	16	165	153	50	38	3
63	20	200	185	58	43	4
80	25	215	196	73	54	4

Front end design optional.

METAL CUTTING TOOLS

Table 17-11. Countersinks for 60°, 90°, and 120° Angles with Parallel Shanks (ISO 3294)

	ISO	3294
U.S.A.	ANSI	
JAPAN	JIS	
GERMANY	DIN	334 (60°) 335(90°) 347 (120°)
FRANCE	NF	E66-250
U.K.	BS	
ITALY	UNI	6847
AUSTRALIA	AS	

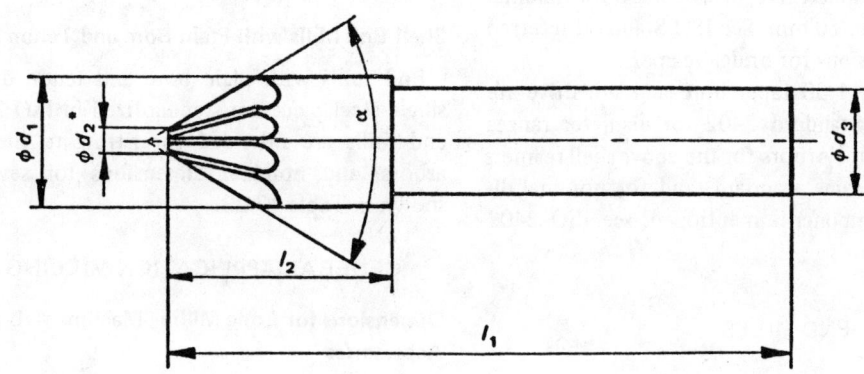

$\alpha = 60°, 90°$ or $120°$ inclusive (tolerance: $^{+0}_{-1°}$)

Dimensions in millimeters

Nominal size d_1	Small diameter d_2*	Overall length l_1 $\alpha = 60°$	Overall length l_1 $\alpha = 90°$ & $120°$	Body length l_2 $\alpha = 60°$	Body length l_2 $\alpha = 90°$ & $120°$	Shank diameter d_3 h9
8	1,6	48	44	16	12	8
10	2	50	46	18	14	8
12,5	2,5	52	48	20	16	8
16	3,2	60	56	24	20	10
20	4	64	60	28	24	10
25	7	69	65	33	29	10

*Front end design optional.

numbers 0 to 6, are defined in the ISO 2250 Standard. The basic dimensions for reamers with parallel shanks and Morse taper shanks are shown in Table 17-18. The diameters and squares of parallel shanks are according to ISO 237 (see Table 17-13). The self-holding taper for tool shanks conforms to ISO 296 (see Table 17-18.)

Hand and Machine Pin Reamers

ISO international Standards 3465, 3466, and 3467 specify reamers with a 1:50 taper intended to produce holes for taper pins manufactured to ISO 2339, in size ranges from 0.6 to 50 mm nominal diameters. These reamers, according to their shank drive, are grouped into three categories as follows:

- ISO 3465 - Hand taper pin reamers (0.6 to 50 mm)
- ISO 3466 - Machine taper pin reamers with parallel shanks (2 to 12 mm)
- ISO 3467 - Machine taper pin reamers with Morse taper shanks (5 to 50 mm)

The hand taper pin reamer standard ISO 3465 covers the total range of hole sizes for available tapered pins to ISO 2339, and the nominal sizes are shown in Table 17-19.

Other Reamers

Machine bridge reamers used to produce 1:10 tapered holes for rivets are standardized in ISO 2238 for nominal sizes, ranging from 6 to 50 mm. See ISO Standard referred to for nominal dimensions for bridge reamers.

Shell Reamers with 1:30 taper bore and slot drive are covered in the ISO Standard 2402 for diameter ranges from 9.9 to 101.6 mm. Arbors for the above shell reamers are described in the same standard, and for nominal dimensions for the components mentioned, see ISO 2402 Standard.

END MILLS

End Mills with Parallel or Morse Taper Shanks—Standard and Long Series

The nominal diameters to the metric end mills shown in Table 17-20 have been chosen based on the preferred number series R40 (Renard 40—See Section 4), on ISO Standard 523, and on current practices in countries already on the metric system. End mills with parallel shanks, in sizes from 2 to 38 mm, are covered in ISO/R 1641 Recommendations. The dimensions are shown in Table 17-20 for standard and long series types. Morse taper-shank end mills in nominal sizes from 6.3 to 63 mm, are covered in the same ISO Standard, and dimensions are shown in Table 17-21, for both standard and long end mills.

End Mills with 7/24 Taper Shanks—Standard and Long Series

The ISO 2324 Standard specifies nominal dimensions for end mills with 7/24 taper shanks, in sizes from 25 to 80 mm, nominal diameters. The basic dimensions for this type of end mill are shown in Table 17-22.

Shell End Mills with Plain Bore and Tenon Drive

End mills with plain bore and tenon drive made of a single steel piece are standardized in ISO 2586. The shell end mills are intended for fitting to the end of cutter arbors, and nominal dimensions for seven cutters are shown in Table 17-23.

GENERAL APPLICATION MILLING CUTTERS

Dimensions for Long Milling-Machine Arbors and Accessories

Tables 17-24 through 31 show the international and national standards and dimensions for milling cutting arbors and accessories.

The dimensions shown in the tables are based on ISO 297, ISO 839 and ISO 240 standards, and provide interchangeability for milling cutters to other ISO standards.

Table 17-12. Tolerance Limits for Reamers (ISO 522)

Nominal Diameter		Cutting Diameter m6		Shaft Diameter (hand reamers) f8*		Shaft Diameter (machine reamers) h9	
Over	to	High+	Low+	High−	Low−	High+	Low−
1	3	0.002	0.009	0.007	0.021	0	0.025
3	6	0.004	0.012	0.010	0.028	0	0.030
6	10	0.006	0.015	0.013	0.035	0	0.036
10	18	0.007	0.018	0.016	0.043	0	0.043
18	30	0.008	0.021	0.020	0.053	0	0.052
30	50	0.009	0.025	0.025	0.064	0	0.062
50	80	0.011	0.030	0.030	0.076		
80	120	0.013	0.035	0.036	0.090		

Dimensions in millimeters

NOTES:
Tolerances are the same as specified in Section 6 – ISO System of Limits and Fits.
*The ISO 236/1 standard specifies the shank diameter tolerance h9 for hand reamers. The f8 tolerance is specified in the BS 122/2 standard.

METAL CUTTING TOOLS

Table 17-13. Squares for Parallel Shank Reamers (ISO 237)*

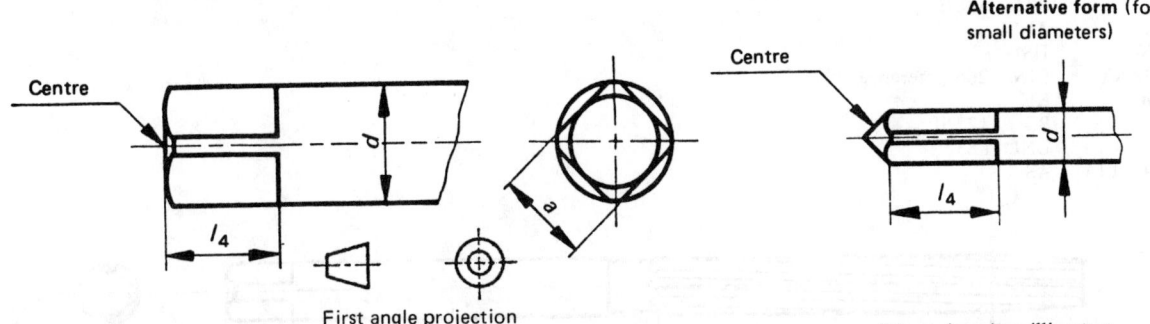

First angle projection

Dimensions in millimeters

Shank diameter* d over	incl.	Driving square a	l_4	Preferred shank diameter	Shank diameter* d over	incl.	Driving square a	l_4	Preferred shank diameter
1,06	1,18	0,90		1,12	10,60	11,80	9,00	12	11,20
1,18	1,32	1,00		1,25	11,80	13,20	10,00	13	12,50
1,32	1,50	1,12		1,40	13,20	15,00	11,20	14	14,00
1,50	1,70	1,25	4	1,60	15,00	17,00	12,50	16	16,00
1,70	1,90	1,40		1,80	17,00	19,00	14,00	18	18,00
1,90	2,12	1,60		2,00	19,00	21,20	16,00	20	20,00
2,12	2,36	1,80		2,24	21,20	23,60	18,00	22	22,40
2,36	2,65	2,00		2,50	23,60	26,50	20,00	24	25,00
2,65	3,00	2,24		2,80	26,50	30,00	22,40	26	28,00
3,00	3,35	2,50	5	3,15	30,00	33,50	25,00	28	31,50
3,35	3,75	2,80		3,55	33,50	37,50	28,00	31	35,50
3,75	4,25	3,15	6	4,00	37,50	42,50	31,50	34	40,00
4,25	4,75	3,55		4,50	42,50	47,50	35,50	38	45,00
4,75	5,30	4,00	7	5,00	47,50	53,00	40,00	42	50,00
5,30	6,00	4,50		5,60	53,00	60,00	45,00	46	56,00
6,00	6,70	5,00	8	6,30	60,00	67,00	50,00	51	63,00
6,70	7,50	5,60		7,10	67,00	75,00	56,00	56	71,00
7,50	8,50	6,30	9	8,00	75,00	85,00	63,00	62	80,00
8,50	9,50	7,10	10	9,00	85,00	95,00	71,00	68	90,00
9,50	10,60	8,00	11	10,00	95,00	106,00	80,00	75	100,00

NOTES:
1. From a number of possible diameters in a particular step, choose the value nearest to the preferred value (see the last column of the table).
2. The decimal marker used in the ISO standard and in this table is a comma ",".

Table 17-14. Hand Reamers with Parallel Shanks (ISO 236/I)

U.S.A.	ISO	236/I
	ANSI	
JAPAN	JIS	
GERMANY	DIN	206 (reference only)
FRANCE	NF	
U.K.	BS	122/II
ITALY	UNI	6852
AUSTRALIA	AS	

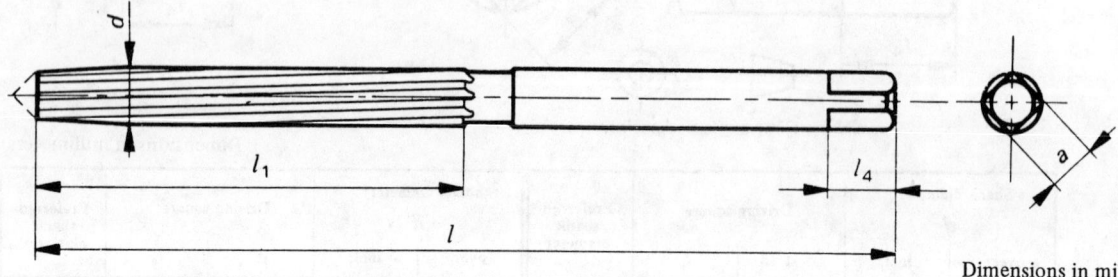

Dimensions in millimeters

d	l_1	l	a	l_4	d	l_1	l	a	l_4
(1,5)	20	41	1,12		22	107	215	18,00	22
1,6	21	44	1,25		(23)				
1,8	23	47	1,40	4	(24)	115	231	20,00	24
2,0	25	50	1,60		25				
2,2	27	54	1,80		(26)				
2,5	29	58	2,00		(27)	124	247	22,40	26
2,8	31	62	2,24		28				
3,0				5	(30)				
3,5	35	71	2,80		32	133	265	25,00	28
4,0	38	76	3,15		(34)	142	284	28,00	31
4,5	41	81	3,55	6	(35)				
5,0	44	87	4,00		36				
5,5	47	93	4,50	7	(38)	152	305	31,50	34
6,0					40				
7,0	54	107	5,60	8	(42)	163	326	35,50	38
8,0	58	115	6,30	9	(44)				
9,0	62	124	7,10	10	45				
10,0	66	133	8,00	11	(46)				
11,0	71	142	9,00	12	(48)				
12,0	76	152	10,00	13	50	174	347	40,00	42
(13,0)					(52)				
14,0	81	163	11,20	14	(55)	184	367	45,00	46
(15,0)					56				
16,0	87	175	12,50	16	(58)				
(17,0)					(60)				
18,0	93	188	14,00	18	(62)	194	387	50,00	51
(19,0)					63				
20,0	100	201	16,00	20	67				
(21,0)					71	203	406	56,00	56

NOTES:
1. Diameters shown in brackets ( ) are second choices.
2. Dimensions and tolerances for squares are shown in Table 17-13.
3. Tolerance for shank diameter is h8 (see Table 17-12) and cutting diameter is m6 (see Table 17-12). Length tolerances are defined under "Reamers."
4. The decimal marker used in the ISO standard and in this table is a comma ",".

METAL CUTTING TOOLS

Cylindrical Cutters with Plain Bore and Key Drive

The ISO 2584 Standard describes four solid and four interlocking cutters in sizes ranging from 50 to 160 mm outside diameters. Nominal dimensions for cutters have been based on ISO 523 and ISO 240 standards, and interchangeability with other types of similar ISO cutters is secured worldwide. Nominal dimensions and other national standards for the above cutters are shown in Table 17-32.

Slotting Cutters with Plain Bore and Key Drive

Slotting cutters with outside diameters ranging from 50 to 200 mm, and standard widths from 4 to 40 mm, are specified in the ISO 2585 Standard. Nominal dimensions for slotting cutters are shown in Table 17-33.

Side and Face Milling Cutters with Plain Bore and Key Drive

The ISO 2587 Standard describes side and face milling cutters with plain bore and key drive in sizes ranging from 50 to 200 mm outside diameters and widths from 4 to 40 mm. Nominal dimensions and national standards for cutters are shown in Table 17-34.

SPECIAL METRIC COMPONENT MILLING CUTTERS

Metric Module Gear Milling Cutters

World standards for metric module gears are shown in Table 13-9. National standards for milling cutters to produce metric module gears as well as some basic cutter dimensions are shown in Table 17-35.

Table 17-15. Parallel Machine Reamers with Morse Taper Shanks (ISO 236/II)

	ISO	236/II
U.S.A.	ANSI	
JAPAN	JIS	B4403
GERMANY	DIN	
FRANCE	NF	E66-015
U.K.	BS	122/II
ITALY	UNI	6839
AUSTRALIA	AS	

Dimensions in millimeters

d	l_1	l	M.T.	d	l_1	l	M.T.
7	54	134	1	32	133	293	4
8	58	138	1	(34)	142	302	4
9	62	142	1	(35)	142	302	4
10	66	146	1	36			4
11	71	151	1	(38)			4
12	76	156	1	40	152	312	4
(13)	76	156	1	(42)			4
14	81	161		(44)			4
(15)	81	181		45	163	323	4
16	87	187		(46)			4
(17)	87	187		(48)		334	4
18	93	193	2	50	174	334	
(19)	93	193	2	(52)		371	
20	100	200		(55)			
(21)	100	200		56	184	381	
22	107	207		(58)	184	381	5
(23)	107	207		(60)			5
(24)				(62)			5
25	115	242		63	194	391	5
(26)				67			5
(27)			3	71	203	400	5
28	124	251					
(30)							

NOTES:
1. Diameters shown in brackets ( ) are second choices.
2. MT = Morse Taper Number. See Table 17-18 for taper details and std. ref.
3. Tolerance for cutting diameters is m6 (see Table 17-12). Length tolerances are defined in the text under "Reamers."

Table 17-16. Machine (Chucking) Reamers with Taper Shanks (ISO 521)

U.S.A.	ISO	521
JAPAN	ANSI	
GERMANY	JIS	
GERMANY	DIN	208 (Reference only)
FRANCE	NF	E66-015
U.K.	BS	122/II
ITALY	UNI	6854
AUSTRALIA	AS	

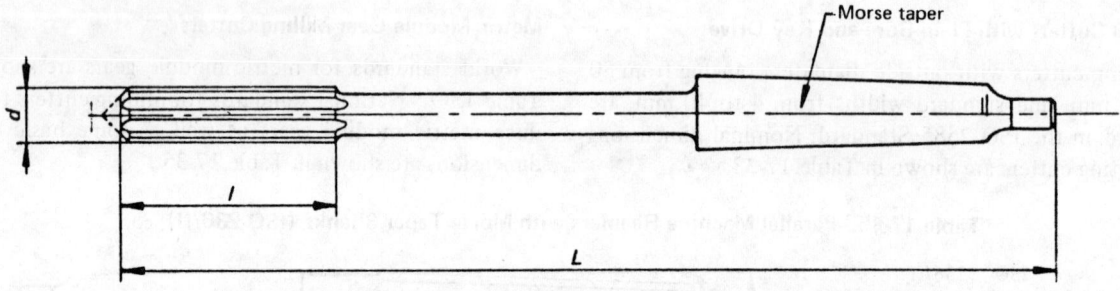

*Dimensions in millimeters.

d	L	l	Morse taper No.	d	L	l	Morse taper No.
5,5	138	26	1	(24)	268	68	3
6	138	26	1	25	268	68	3
7	150	31	1	(26)	273	70	3
8	156	33	1	28	277	71	3
9	162	36	1	(30)	281	73	3
10	168	38	1	32	317	77	4
11	175	41	1	(34)	321	78	4
12	182	44	1	(35)	321	78	4
(13)	182	44	1	36	325	79	4
14	189	47	2	(38)	329	81	4
15	204	50	2	40	329	81	4
16	210	52	2	(42)	333	82	4
(17)	214	54	2	(44)	336	83	4
18	219	56	2	45	336	83	4
(19)	223	58	2	(46)	340	84	4
20	228	60	2	(48)	344	86	4
22	237	64	2	50	344	86	4

NOTES:
1. Diameters shown in brackets ( ) are second choices.
2. Morse taper shanks are according to ISO 296. See Table 17-18 for taper details and national standard references.
3. Tolerance for cutting diameters is m6 (see Table 17-12). Length tolerances are defined in the text under "Reamers."

METAL CUTTING TOOLS

17-23

Table 17-17. Machine (Chucking) Reamers with Parallel Shanks (ISO 521)

	ISO	521
U.S.A.	ANSI	
JAPAN	JIS	
GERMANY	DIN	212 (Reference only)
FRANCE	NF	E66-014
U.K.	BS	122/II
ITALY	UNI	6853
AUSTRALIA	AS	

For d up to 3,75 mm

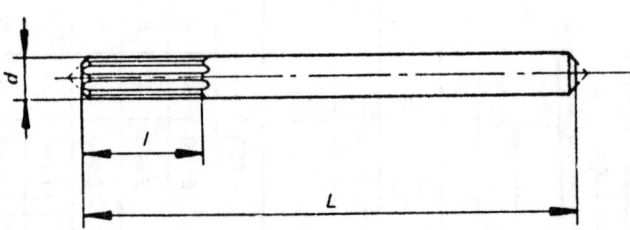

For d over 3,75 mm

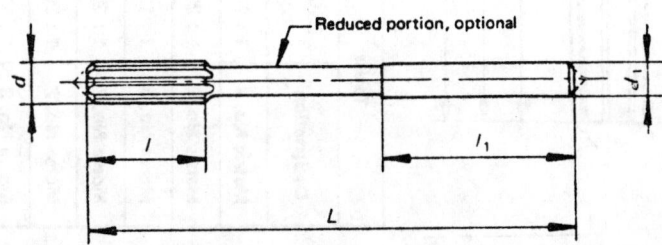

Dimensions in millimeters

d	d_1	L	l	l_1
1,4	1,4	40	8	
(1,5)	1,5	40	8	
1,6	1,6	43	9	
1,8	1,8	46	10	
2,0	2,0	49	11	
2,2	2,2	53	12	
2,5	2,5	57	14	
2,8	2,8	61	15	
3,0	3,0	61	15	
3,2	3,2	65	16	
3,5	3,5	70	18	
4,0	4,0	75	19	32
4,5	4,5	80	21	33
5,0	5,0	86	23	34
5,5	5,6	93	26	36

d	d_1	L	l	l_1
6	5,6	93	26	36
7	7,1	109	31	40
8	8,0	117	33	42
9	9,0	125	36	44
10	10,0	133	38	46
11	10,0	142	41	46
12	10,0			46
(13)	12,5	151	44	50
14	12,5	160	47	50
(15)	12,5	162	50	50
16	12,5	170	52	50
(17)	14,0	175	54	52
18	14,0	182	56	52
(19)	16,0	189	58	58
20	16,0	195	60	58

NOTES:
1. Diameters shown in brackets ( ) are second choices.
2. Tolerance on the cutting diameter is m6 (see Table 17-12).and on the shank diameter h9 (see Table 17-12). Length tolerances are defined in the text under "Reamers."
3. The decimal marker used in the ISO standard and in this table is a comma ",".

METAL CUTTING TOOLS

Table 17-18. Finishing Reamers for Morse and Metric Tapers (ISO 2250)

	Reamer	Tool Taper
U.S.A. ISO	2250	296
ANSI	B94.2	B5.10
JAPAN JIS	B4401	B4003
GERMANY* DIN		228
FRANCE NF	E66-017	E66-017
U.K. BS		1660
ITALY UNI		533
AUSTRALIA AS		

Parallel shank reamers

Dimensions in millimeters

Taper				mm			
Designation	Rate of taper	d	L	l	l₁	d₁ h9
Metric No. 4	1 : 20.000	4.000	48	30	22	4.0
Metric No. 6	1 : 20.000	6.000	63	40	30	5.0
Morse No. 0	1 : 19.212	9.045	93	61	48	8.0
Morse No. 1	1 : 20.047	12.065	102	66	50	10.0
Morse No. 2	1 : 20.020	17.780	121	79	61	14.0
Morse No. 3	1 : 19.922	23.825	146	96	76	20.0
Morse No. 4	1 : 19.254	31.267	179	119	97	25.0
Morse No. 5	1 : 19.002	44.399	222	150	124	31.5
Morse No. 6	1 : 19.180	63.348	300	208	176	45.0

Morse taper shank reamers

Taper				mm			Morse taper shank No.
Designation	Rate of taper	d	L	l	l₁	
Metric No. 4	1 : 20.000	4.000	106	30	22	1
Metric No. 6	1 : 20.000	6.000	116	40	30	1
Morse No. 0	1 : 19.212	9.045	137	61	48	1
Morse No. 1	1 : 20.047	12.065	142	66	50	1
Morse No. 2	1 : 20.020	17.780	173	79	61	2
Morse No. 3	1 : 19.922	23.825	212	96	76	3
Morse No. 4	1 : 19.254	31.267	263	119	97	4
Morse No. 5	1 : 19.002	44.399	331	150	124	5
Morse No. 6	1 : 19.180	63.348	389	208	176	5

NOTES:
1. Dimensions and tolerances for squares are shown in Table 17-13.
2. Morse taper shanks are according to ISO 296. See above for taper details and national standard references.
3. Limits for the ISO tolerance h9 are shown in Table 17-12.

METAL CUTTING TOOLS

Woodruff Key-Seat Cutters

T-slot cutters to produce Woodruff key seats and their corresponding nominal Woodruff key sizes and standards are shown in Table 17-36.

T-Slot Cutters with Parallel or Morse Taper Shanks

The ISO international standard 3337 describes T-slot cutters with parallel shafts in sizes from 11 to 60 mm, nominal diameters, and cutter widths from 3.5 to 28 mm;

Table 17-19. Taper Pin Reamers* (ISO 3465)

	Type	A	B	C
	ISO	3465	3466	3467
U.S.A.	ANSI			
JAPAN	JIS	B4411	B4410	
GERMANY	DIN	9	1898A	1898B
FRANCE	NF	E66-011		
U.K.	BS			
ITALY	UNI	6856		
AUSTRALIA	AS			

Type A = Hand taper pin reamer (shown below)
Type B = Machine taper pin reamers with parallel shanks
Type C = Machine taper pin reamers with Morse taper shanks

Taper 1 : 50

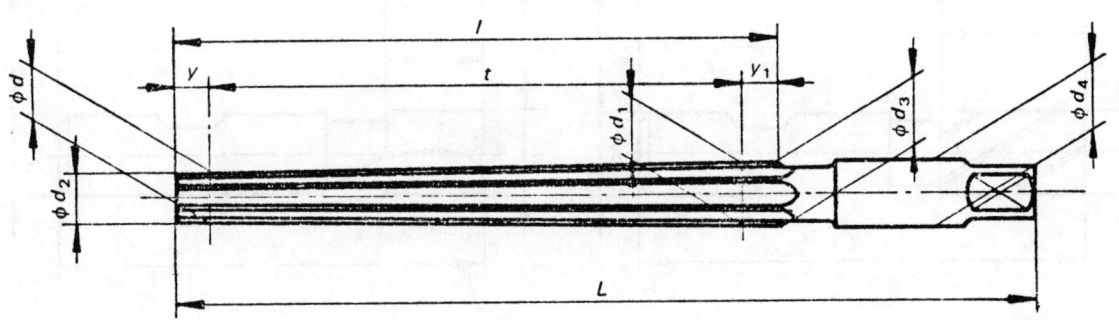

*Dimensions in millimeters.

d nominal	d_1	t	y	y_1	d_2	d_3	l	d_4 n 11	L
0,6	0,76	8	5	7	0,5	0,90	20	3,15	38
0,8	1,04	12	5	7	0,7	1,18	24	3,15	42
1,0	1,32	16	5	7	0,9	1,46	28	3,15	46
1,2	1,60	20	5	7	1,1	1,74	32	3,15	50
1,5	2,00	25	5	7	1,4	2,14	37	3,15	57
2,0	2,70	35	5	8	1,9	2,86	48	3,15	68
2,5	3,20	35	5	8	2,4	3,36	48	3,15	68
3,0	3,90	45	5	8	2,9	4,06	58	4,0	80
4,0	5,10	55	5	8	3,9	5,26	68	5,0	93
5,0	6,20	60	5	8	4,9	6,36	73	6,3	100
6,0	7,80	90	5	10	5,9	8,00	105	8,0	135
8,0	10,60	130	5	10	7,9	10,80	145	10,0	180
10,0	13,20	160	5	10	9,9	13,40	175	12,5	215
12,0	15,60	180	10	20	11,8	16,00	210	14,0	255
16,0	20,00	200	10	20	15,8	20,40	230	18,0	280
20,0	24,40	220	10	20	19,8	24,80	250	22,4	310
25,0	29,80	240	15	45	24,7	30,70	300	28,0	370
30,0	35,20	260	15	45	29,7	36,10	320	31,5	400
40,0	45,60	280	15	45	39,7	46,50	340	40,0	430
50,0	56,00	300	15	45	49,7	56,90	360	50,0	460

NOTE:
1. For limit dimensions to the ISO tolerance h11 see Table 6-26.
2. Detail dimensions to square drive are shown in Table 17-13.
3. The decimal marker used in the ISO standard and in this table is a comma ",".

Table 17-20. End Mills with Parallel Shanks (ISO 1641/I)

	ISO	1641/I
U.S.A.	ANSI	
JAPAN	JIS	B4211, B4208
GERMANY	DIN	844
FRANCE	NF	E66-211
U.K.	BS	
ITALY	UNI	3912
AUSTRALIA	AS	

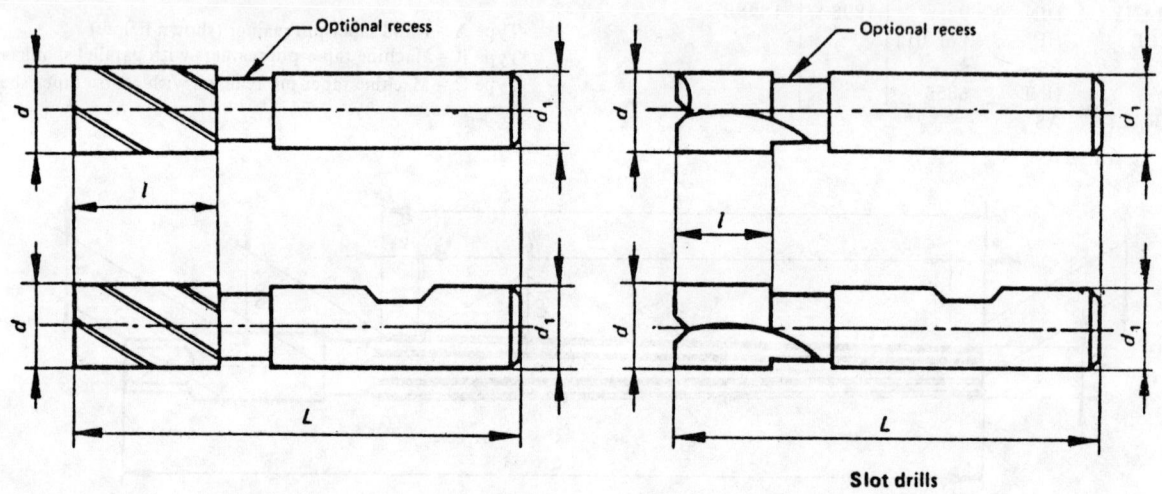

Slot drills

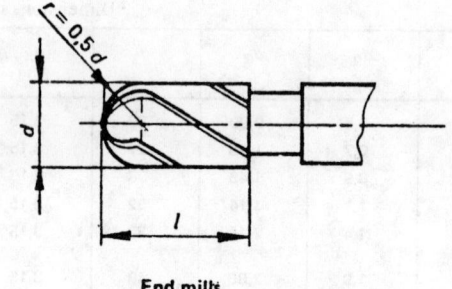

End mills

Designation: Milling cutters are designated by their type and their cutting diameter d together with their shank diameter in the case where the diameter is different from the cutting diameter.

METAL CUTTING TOOLS

Table 17-20 (Continued). End Mills with Parallel Shanks (ISO 1641/I)

Ranges of diameters d		Recommended diameters d		Length l			Shank $d_1 \times l_1$		$L - l$	
over	up to (including)			Short series	Standard series	Long series	Alternative I	II	Alternative I	II
1,9	2,36	2	–	4	7	10	$4 \times 28^{1)}$	6×36	32	44
2,36	3	2,5	–	5	8	12				
		3								
3	3,75	–	3 5	6	10	15				
3,75	4	4	–	7	11	19				
4	4,75	–					$5 \times 28^{1)}$	6×36	34	44
4,75	5	5	–	8	13	24				
5	6	6	–				6×36		44	
6	7,5	–	7	10	16	30	8×36	10×40	44	50
7,5	8	8	–	11	19	38				
8	9,5	–	9				10×40		50	
9,5	10	10	–	13	22	45				
10	11,8	–	11				12×45		57	
11,8	15	12	14	16	26	53				
15	19	16	18	19	32	63	16×48		60	
19	23,6	20	22	22	38	75	20×50		66	
23,6	30	25	28	26	45	90	25×56		76	
30	37,5	32	36	32	53	106	32×60		80	
37,5	47,5	40	45	38	63	125	40×70		92	
47,5	60	50	56	45	75	150	50×80		102	
60	67	63	–	53	90	180				
67	75	–	75				63×90		112	

NOTES:
1. Only for plain parallel shank.
2. Tolerances on cutting diameters d:
   End mills: $j_s 14$ (see Table 6-27)
   Slot drills: e8 (see Table 6-24)
3. In the case of a double-ended milling cutter having a cutting diameter equal to the shank diameter, the tolerance level on wedge shall apply to a nominal cutting diameter slightly smaller than the shank diameter.
4. The decimal marker used in the ISO standard and in this table is a comma ",".

Table 17-21. End Mills with Morse Taper Shanks (ISO 1641/II)

	ISO	1641/II
U.S.A.	ANSI	
JAPAN	JIS	B4212, B4209
GERMANY	DIN	845B
FRANCE	NF	E66-212
U.K.	BS	
ITALY	UNI	3913B, 3915B
AUSTRALIA	AS	

Designation: Milling cutters are designated by their type and their cutting diameter d.

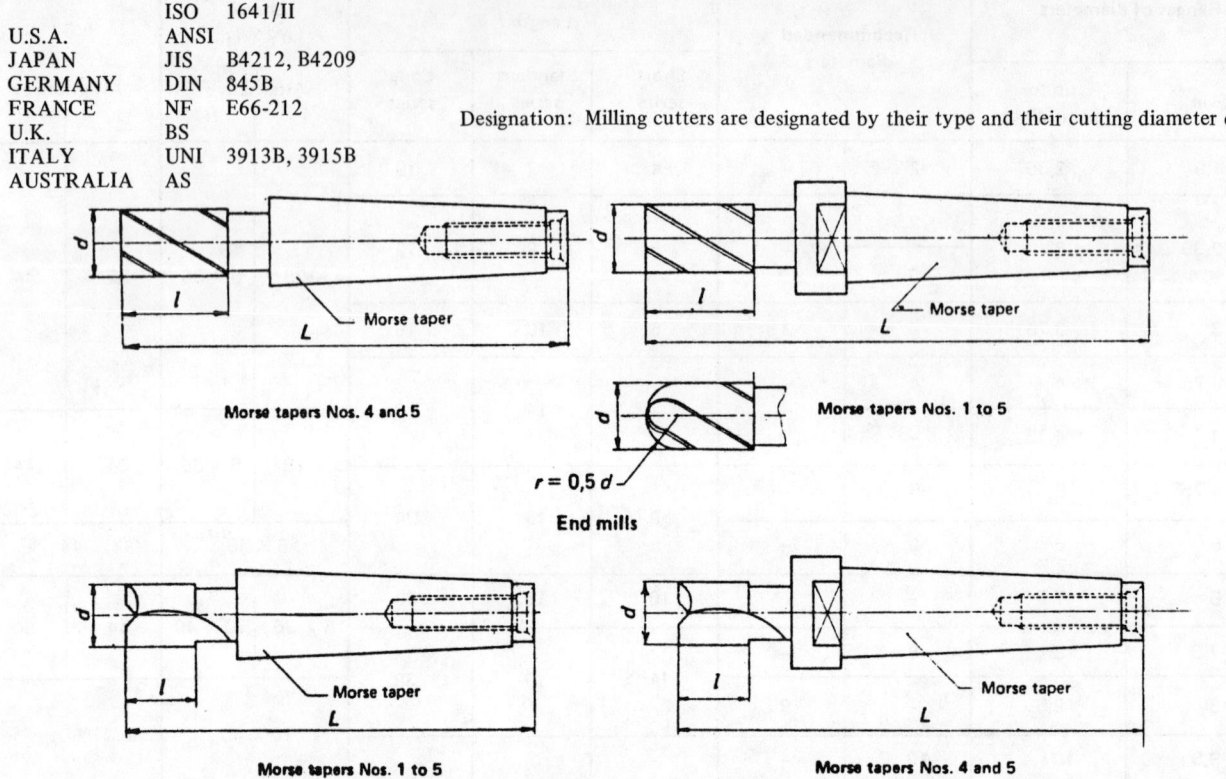

End mills

Slot drills

Ranges of diameters d		Recommended diameters d		Length l			L − l						
							Morse taper No						
over (excluded)	up to (included)			Short series	Standard series	Long series	1	2	3	4		5	
										without positive drive	with positive drive	without positive drive	with positive drive
5	6	6	−	8	13	24	70						
6	7.5	−	7	10	16	30	70						
7.5	9.5	8	9	11	19	38	70						
9.5	11.8	10	11	13	22	45	70						
11.8	15	12	14	16	26	53	70	85					
15	19	16	18	19	32	63		85					
19	23.6	20	22	22	38	75		85	102				
23.6	30	25	28	26	45	90			102				
30	37.5	32	36	32	53	106			102	125	148		
37.5	47.5	40	45	38	63	125				125	148	158	186
47.5	60	50	56	45	75	150				125	148	158	186
60	75	63	−	53	90	180						158	186

NOTES:
1. Tolerances on cutting diameters d:
   End mills: $j_s 14$ (see Table 6-27)
   Slot drills: e8 (see Table 6-24).

METAL CUTTING TOOLS

Table 17-22. End Mills with 7/24 Taper Shanks (ISO 1641/III)

	ISO 1641/III
U.S.A.	ANSI
JAPAN	JIS
GERMANY	DIN
FRANCE	NF
U.K.	BS
ITALY	UNI
AUSTRALIA	AS

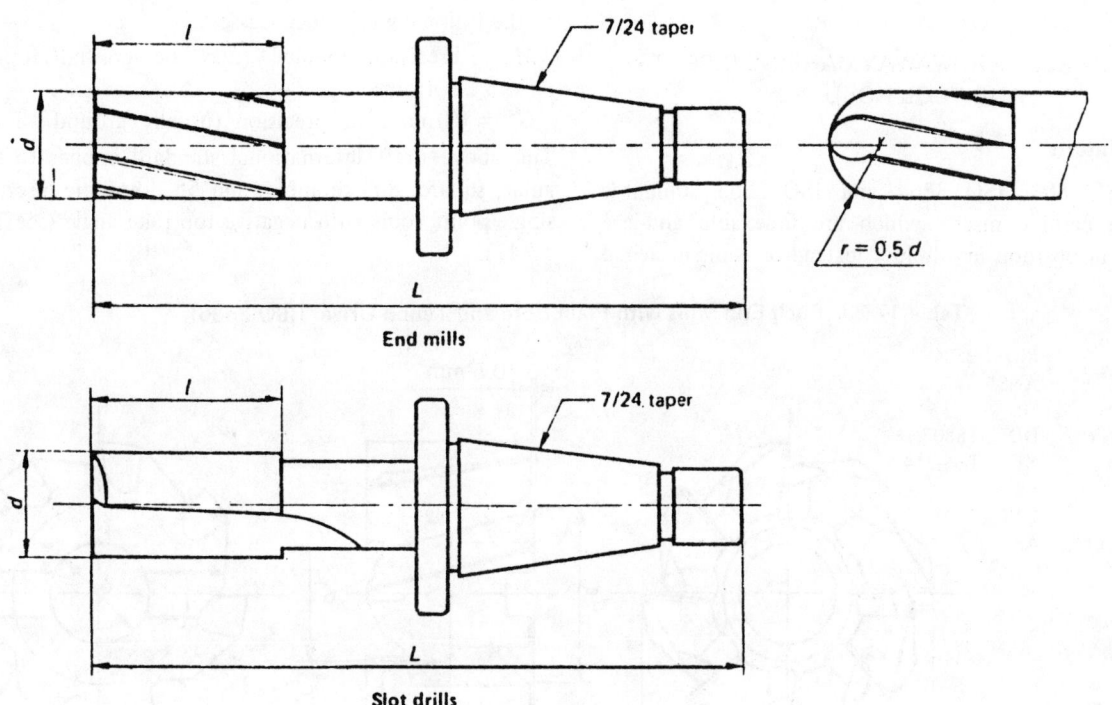

End mills

Slot drills

Designation: Milling cutters are designated by their type and their cutting diameter d.

Dimensions in millimeters

Ranges of diameters d		Recommended diameters d		Length l			L – l			
over	up to (including)			Short series	Standard series	Long series	7/24 taper No.			
							30	40	45	50
23,6	30	25	28	26	45	90	105			
30	37,5	32	36	32	53	106	105	135	155	
37,5	47,5	40	45	38	63	125		135	155	177
47,5	60	50	56	45	75	150		135	155	177
60	75	63	71	53	90	180			155	177
75	95	80	–	63	106	212				177

NOTES:
1. Tolerances on cutting diameters d:
2. End mills: j_s14 (see Table 6-27).
3. Slot drills: e8 (see Table 6-24).

17-30   METAL CUTTING TOOLS

basic dimensions are shown in Table 17-37. T-slot cutters with Morse taper shanks in sizes from 18 to 95 mm nominal diameters, and cutter widths from 8 to 44 mm, as well as standard sizes with the corresponding basic dimensions are shown in Table 17-38.

Thread-Cutting Tools

A worldwide comparison of ISO and national screw thread standards is shown in Table 8-1, while some important ISO metric cutting-tool standards are quoted in Table 17-39

INDEXABLE THROWAWAY CARBIDE INSERTS AND TOOL HOLDERS

Carbide Inserts

The ISO 883, ISO 3364 and ISO 3365 standards relate to carbide inserts which are indexable and are clamped in position in a holder instead of being attached to the shank by brazing. The ISO 883 Standard describes inserts to two tolerance classes as follows:

U = Utility or least precision (usually ground tip and bottom faces)

G = Ground or precision (usually ground all over)

Dimensions are specified for utility and precision triangular and square inserts for single-point tools with negative and positive top-rake angles (see Table 17-40).

The ISO 3364 and ISO 3365 Standards specifies inserts of the following tolerance classes:

M = Precision molded (may be ground top and bottom)

G = Ground or precision (usually ground all over)

The above draft international standard applies to triangular, square, 80° rhombic, and 55° rhombic inserts for single-point tools with negative top rake angle (see Table 17-41).

Table 17-23. Shell End Mills with Plain Bore and Tenon Drive (ISO 2586)

	ISO	2586
U.S.A.	ANSI	
JAPAN	JIS	
GERMANY	DIN	1880
FRANCE	NF	E66-214
U.K.	BS	
ITALY	UNI	3903
AUSTRALIA	AS	

Dimensions in millimeters

D j_s16	d H7	L min.	l min.	d_1 min.	d_5* min.
40	16	32	18	22	33
50	22	36	20	30	41
63	27	40	22	38	49
80	27	45	22	38	49
100	32	50	25	45	59
125	40	56	28	56	71
160	50	63	31	67	91

* The disengagement of 0,5 mm on the rear face is optional.

First angle projection

ISO tolerances js16 and H7 are defined in Section 6. The tenon seatings shall be in accordance with the metric series of ISO 240. (See Table 17-28B)

These cutters are with helicoidal teeth angled to the right or left.

METAL CUTTING TOOLS

Table 17-24. Ends of Arbor and Adapter with Taper 7/24 (Quick release) (ISO 297)

Dimensions in millimeters

Designation No.	Taper D_1 1)	z	L $0\atop-0.25$	l_1	Cylindrical tenor d_1 a10	p	d_3	Collar y	b H12	t max.	w	Thread g 21	l_2	l_3 min.	l_4 $0\atop-0.25$
30	31,750	0,4	63,4	48,4	17,4	3	16,5	1,6	16,1	16,2	0,12	M12	24	50	63,5
40	44,450	0,4	93,4	65,4	25,3	5	24	1,6	16,1	22,5	0,12	M16	30	70	88,9
45	57,150	0,4	106,8	82,8	32,4	6	30	3,2	19,3	29	0,12	M20	33	70	103,1
50	69,850	0,4	126,8	101,8	39,6	8	38	3,2	25,7	35,3	0,2	M24	45	90	120,6
55	88,900	0,4	164,8	126,8	50,4	9	48	3,2	25,7	45	0,2	M24	45	90	158,7
60	107,950	0,4	206,8	161,8	60,2	10	58	3,2	25,7	60	0,2	M30	56	110	198,4
65	133,350	0,4	246	202	75	12	72	4	32,4	72	0,3	M36	70	160	230
70	165,100	0,4	296	252	92	14	90	4	32,4	85	0.3	M36	90	160	280
75	203,200	0,4	370	307	114	16	110	5	40,5	104	0,3	M48	90	180	360
80	254,000	0,4	469	394	140	18	136	6	40,5	132	0,3	M48	90	180	449

NOTES: 1. D_1: Basic diameter enclosed in the guage plane.
2. Thread diameter g: This is either a metric thread M with coarse pitch or, if expressly stated, a UN thread. In every case, the appropriate symbol M or UN should be marked on the component.
3. For tolerances a10 see Table 6-22 and H12 see Table 6-14.
4. The decimal marker used in the ISO standard and in this table is a comma ",".

Carbide Insert Designation System

A summary of the ISO 1832 recommendation on carbide insert designation systems is shown in Table 17-42, and the letter symbols, with minor exceptions, conform to these recommendations worldwide. The designation code number in metrics is a six-digit number describing the size, thickness, and cutting-point radius for inserts rounded off, as shown in Table 17-42. This metric code is used in those major industrial countries of the world already on the metric system. See ANSI B94.4.

Carbide Insert Tool Holders

Holders for carbide inserts are not standardized worldwide as yet, and reference to the American ANSI B94.45 Precision Indexable Insert Holders, where tool holders for triangular, square, and rhombic (diamond) carbide inserts are described, should be useful.

Table 17-25. Milling Machine Arbors with 7/24 Tapers—Metric Series (ISO 839/I)

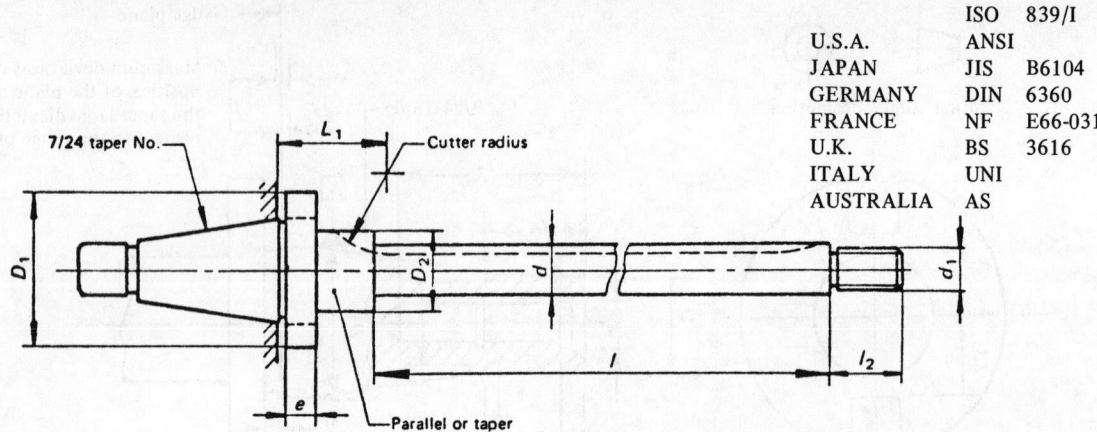

ISO		839/I
U.S.A.	ANSI	
JAPAN	JIS	B6104
GERMANY	DIN	6360
FRANCE	NF	E66-031
U.K.	BS	3616
ITALY	UNI	
AUSTRALIA	AS	

Dimensions in millimeters

7/24 taper No.	D_1	e	L_1 max.	D_2 min.	d h6	63	100	160	200	250	315	400	500 (450)	630 (560)	800 (710)	1000 (900)	d_1	l_2
30	50	8	32	27	16	–	–	–	–	–							M 16 × 1,5	
				34	22	–	–	–	–	–	–						M 20 × 2	
				41	27	–	–	–	–	–	–						M 24 × 2	
40	63	10	36	27	16	–	–	–	–	–							M 16 × 1,5	
				34	22	–	–	–	–	–	–	–	–				M 20 × 2	
				41	27	–	–	–	–	–	–	–	–	–			M 24 × 2	1 to 1,25 d_1 + 2 mm
				47	32	–	–	–	–	–	–	–	–				M 27 × 2	
				55	40	–	–	–	–	–	–						M 33 × 2	
50	100	12	45	34	22	–	–	–	–	–	–	–	–	–			M 20 × 2	
				41	27	–	–	–	–	–	–	–	–	–	–		M 24 × 2	
				47	32	–	–	–	–	–	–	–	–				M 27 × 2	
				55	40	–		–	–	–	–	–					M 33 × 2	
				69	50								–	–	–	–	M 39 × 3	
				84	60								–	–	–	–	M 45 × 3	
60	160	16	56	69	50									–	–	–	M 39 × 3	
				84	60									–	–	–	M 45 × 3	
				109	80										–	–	M 56 × 4	
				134	100										–	–	M 68 × 4	

The lengths l in brackets should be avoided as far as possible.
7/24 tapers and drive seating conform to ISO 297 (see Table 17-24).
Body of arbor and keys and keyways conform to ISO 240 (see Table 17-28A and 17-28B).

METAL CUTTING TOOLS

Table 17-26. Pilot at End of Arbor (ISO 839/I)

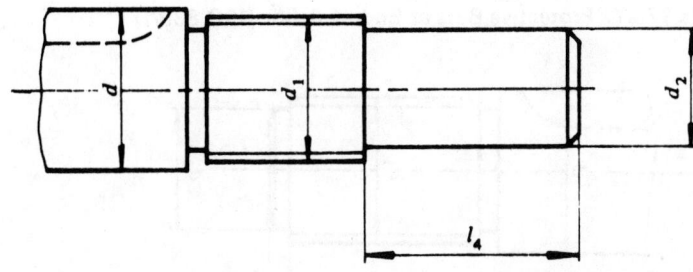

Dimensions in millimeters

d	d_1	d_2 g6	l_4
16	M 16 × 1.5	13	20
22	M 20 × 2	16	25
27	M 24 × 2	20	32
32	M 27 × 2	23	32
40	M 33 × 2	29	
50	M 39 × 3	34	56
60	M 45 × 3	40	

Tolerance on d_2: g6 (see Table 6-25)

Table 17-27. Protective Boss at End of Arbor (ISO 839/I)

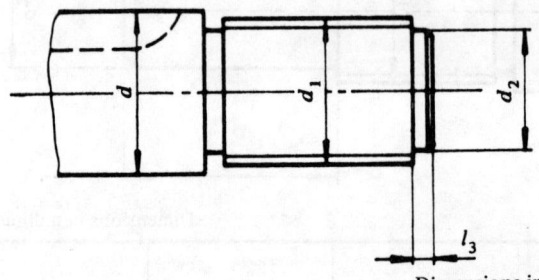

Dimensions in millimeters

d	d_1	d_2	l_3 min.
16	M 16 × 1.5	13	2
22	M 20 × 2	16	
27	M 24 × 2	20	
32	M 27 × 2	23	
40	M 33 × 2	29	
50	M 39 × 3	34	3
60	M 45 × 3	40	
80	M 56 × 4	49	5
100	M 68 × 4	61	

METAL CUTTING TOOLS

Table 17-28A. Standard Arbor Drives—Keyways (ISO 240)

	ISO	240
U.S.A.	ANSI	
JAPAN	JIS	
GERMANY	DIN	138
FRANCE	NF	E66-202
U.K.	BS	3616
ITALY	UNI	
AUSTRALIA	AS	

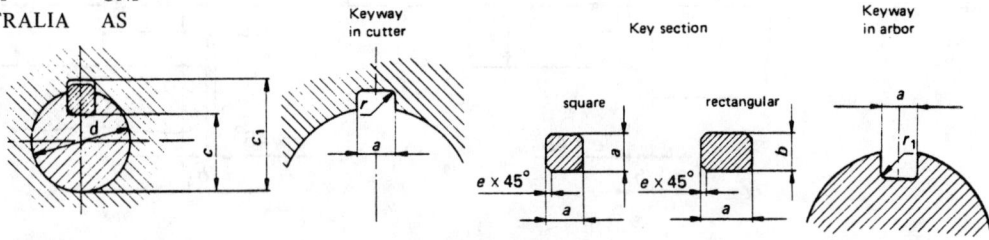

Dimensions in millimeters

d	a	b	c	Tol.	c_1	Tol.	e	Tol.	r	Tol.	r_1	Tol.
8	2		6,7		8,9				0,4	0 / −0,1		
10	3		8,2		11,5		0,16	+0,09 / 0			0,16	0 / −0,08
13	3		11,2	0 / −0,1	14,6	+0,1 / 0						
16	4		13,2		17,7				0,6	0 / −0,2		
19	5		15,6		21,1				1,0			
22	6		17,6		24,1		0,25	+0,15 / 0		0 / −0,3	0,25	0 / −0,09
27	7		22,0		29,8							
32	8	7	27,0		34,8				1,2			
40	10	8	34,5		43,5							
50	12	8	44,5	0 / −0,2	53,5	+0,2 / 0			1,6			
60	14	9	54,0		64,2		0,40	+0,20 / 0			0,40	0 / −0,15
70	16	10	63,5		75,0				2,0	0 / −0,5		
80	18	11	73,0		85,5							
100	25*	14	91,0		107,0		0,60		2,5		0,60	0 / −0,20

*The 24 × 14 key for a diameter of 100 has been replaced by the 25 × 14 key specified in ISO/R 733, *Rectangular or square parallel keys and their corresponding keyways (Dimensions in millimeters)*

Tolerances
—on d (except for gear hobs):
  on the arbor: h6 (see Table 6-26)
  on the cutter: H7 (see Table 6-14)

—on b: h11

—on a:
  for keyway in arbor:
    free keying: H9 (see Table 6-14)
    close keying: N9 (see Table 6-17)
  for keyway in cutter: C11 (see Table 6-11)
  key: h9 (see Table 6-26)

The decimal marker used in the ISO standard and in this table is a comma ",".

17-36   METAL CUTTING TOOLS

Table 17-28B. Standard Arbor Drives—Tenon (ISO 240)

ISO	240
U.S.A. ANSI	
JAPAN JIS	
GERMANY DIN	138
FRANCE NF	E66-203
U.K. BS	3616
ITALY UNI	
AUSTRALIA AS	

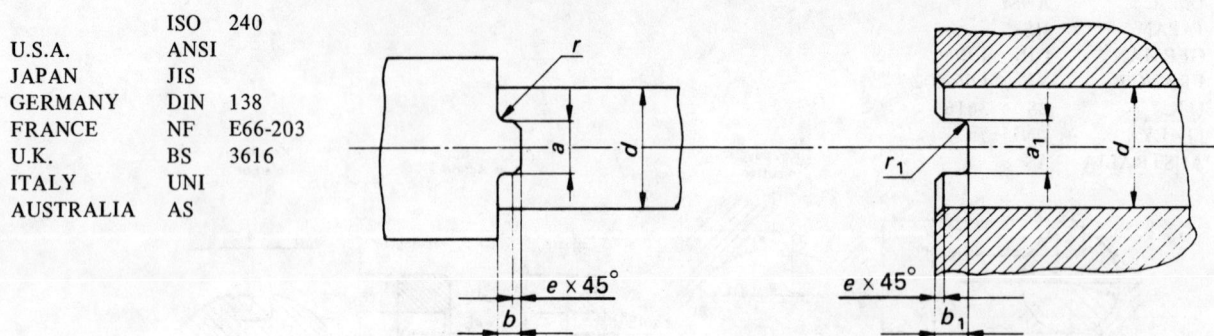

Dimensions in millimeters

d	Arbor a	Arbor b	Arbor r max.	Cutter a_1	Cutter b_1	Cutter r_1 max.	e	e Tol.	$z^{1)}$
5	3	2,0	0,3	3,3	2,5	0,6	0,3	+0,1 / 0	0,075
8	5	3,5	0,4	5,4	4,0	0,6	0,4	+0,1 / 0	0,075
10	6	4,0	0,5	6,4	4,5	0,8	0,5	+0,1 / 0	0,075
13	8	4,5	0,5	8,4	5,0	1,0	0,5	+0,1 / 0	0,075
16	8	5,0	0,5	8,4	5,6	1,0	0,5	+0,1 / 0	0,075
19	10	5,6	0,6	10,4	6,3	1,2	0,6	+0,2 / 0	0,100
22	10	5,6	0,6	10,4	6,3	1,2	0,6	+0,2 / 0	0,100
27	12	6,3	0,8	12,4	7,0	1,6	0,8	+0,2 / 0	0,100
32	14	7,0	0,8	14,4	8,0	1,6	0,8	+0,2 / 0	0,100
40	16	8,0	1,0	16,4	9,0	2,0	1,0	+0,3 / 0	0,100
50	18	9,0	1,0	18,4	10,0	2,0	1,0	+0,3 / 0	0,100
60	20	10,0	1,0	20,5	11,2	2,0	1,0	+0,3 / 0	0,125

1. $+z$ = maximum permissible deviation between the axial plane of the tenon and the axis of arbor of diameter d.

Tolerances
—on d (except for gear hobs):   —on a and b: h11 (see Table 6-26)
   on the arbor: h6 (see Table 6-26)   —on a_1: H11 (see Table 6-14)
   on the cutter: H7 (see Table 6-14)   —on b_1: H13 (see Table 6-14)

2. The decimal marker used in the ISO standard and in this table is a comma ",".

METAL CUTTING TOOLS

Table 17-29. Spacing for Milling Machine Arbors (ISO 839/II)

U.S.A.	ISO 839/II
U.S.A.	ANSI
JAPAN	JIS B 6105
GERMANY	DIN 2084
FRANCE	NF
U.K.	BS 3616
ITALY	UNI
AUSTRALIA	AS

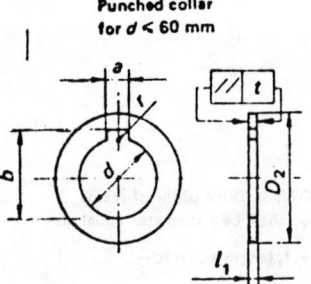
Punched collar for $d \leq 60$ mm

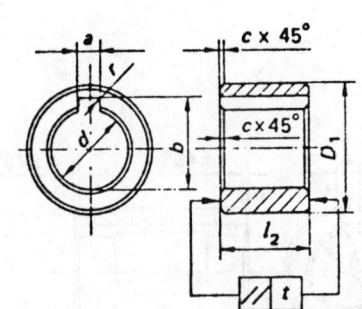
Machined collar

NOTE — For long collars, a recess may be provided, of diameter equal to $d + 1$ and of length equal to $\frac{l_2}{2}$.

Dimensions in millimeters

d	C11	16	22	27	32	40	50	60	80	100
D_1	h11	27	34	41	47	55	69	84	109	134
D_2	h11	26	33	40	46	54	68	83	–	–
a	C11	4	6	7	8	10	12	14	18	25
b		17,7	24,1	29,8	34,8	43,5	53,5	64,2	85,5	107,0
b		+0,1 / 0					+0,2 / 0			
r	max.	0,6	1	1,2	1,2	1,2	1,6	1,6	2	2,5
c		0,4	0,4	0,4	0,6	0,6	0,6	0,6	1	1
l_1		(0,03) – (0,04) – 0,05 – 0,1 – 0,2 – 0,3 – 0,6 – 1								
l_2	±0,1	2 – 3 – 6								
		10								
		–							(12) (13) (16)	
		20								
		30								
		–			60					
		–			100					
Parallelism of faces, t		0,004					0,005			0,006

The dimensions in parentheses should be avoided as far as possible.
In the case of punched collars, the periphery, the bore and the key seating should be carefully deburred.
For tolerances C11 see Table 6-11 and h11 see Table 6-26.
The decimal marker used in the ISO standard and in this table is a comma ",".

Table 17-30. Bearing Collars for Milling Machine Arbors (ISO 839/II)

	ISO	839/II
U.S.A.	ANSI	
JAPAN	JIS	B6106
GERMANY	DIN	2083
FRANCE	NF	
U.K.	BS	3616
ITALY	UNI	
AUSTRALIA	AS	

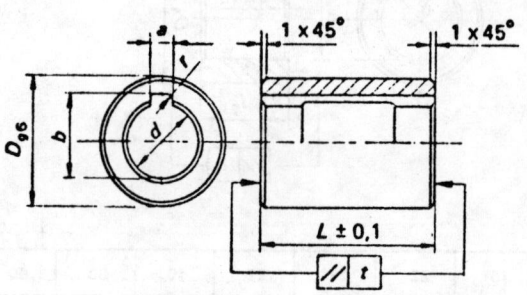

NOTE — Dimensions of recess, if any, shall be : diameter equal to $d + 1$, length equal to $\frac{L}{2}$.

Values in millimeters

d H7	a C11	b	r max.	D 42	L 60	D 48	L 70	D 56	L 80	D 70	L 100	D 85	L 120	D 110	L 140	D 140	L 160
16	4	17,7	+0,1 / 0	0,6	x		x		x									
22	6	24,1		1,0	x		x		x		x							
27	7	29,8					x		x		x		x		x				
32	8	34,8		1,2					x		x		x		x				
40	10	43,5									x		x		x				
50	12	53,5	+0,2 / 0	1,6									x		x		x		
60	14	64,2															x		
80	18	85,5		2,0													x		
100	25	107,0		2,5															x
Parallelism of faces, t					0,004						0,005				0,006			

For tolerances C11 see Table 6-11 and H7 see Table 6-14.

The decimal marker used in the ISO standard and in this table is a comma ",".

METAL CUTTING TOOLS

Table 17-31. Nuts for Milling Machine Arbors (ISO 839/II)

	ISO	839/II
U.S.A.	ANSI	
JAPAN	JIS	B6108
GERMANY	DIN	2082
FRANCE	NF	
U.K.	BS	3616
ITALY	UNI	
AUSTRALIA	AS	

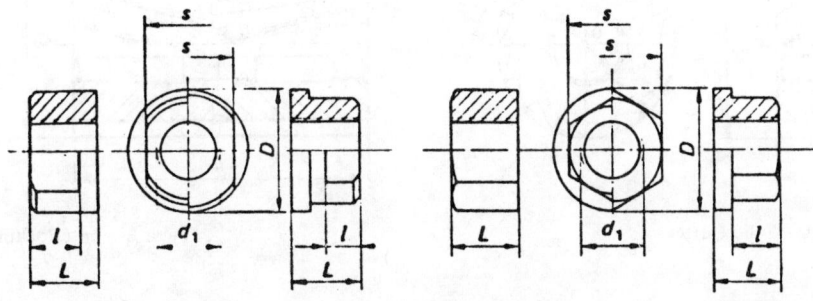

First angle projection

Dimensions in millimeters

For arbor d	16	22	27	32	40	50	60	80	100
d_1	M 16 x 1,5	M 20 x 2	M 24 x 2	M 27 x 2	M 33 x 2	M 39 x 3	M 45 x 3	M 56 x 4	M 68 x 4
D	27	34	41	47	55	69	84	109	134
L	1 to 1,25 d_1									
s	It is essential that values of standard widths across flats be used (see ISO/R 272)								
l	> thickness of appropriate spanner (see ISO/R 272)								

Table 17-32. Cylindrical cutters with Plain Bore and Key Drive (ISO 2584)

U.S.A.	ISO 2584
U.S.A.	ANSI
JAPAN	JIS
GERMANY	DIN
FRANCE	NF E66-226
U.K.	BS
ITALY	UNI 3902
AUSTRALIA	AS

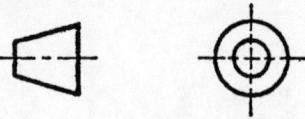

First angle projection

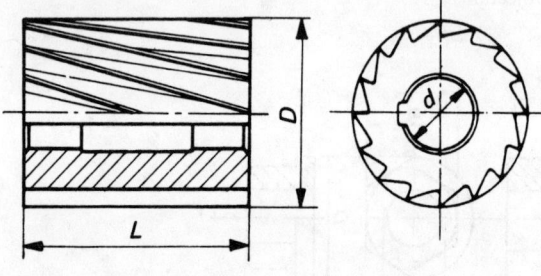

Type A – Solid Cutter

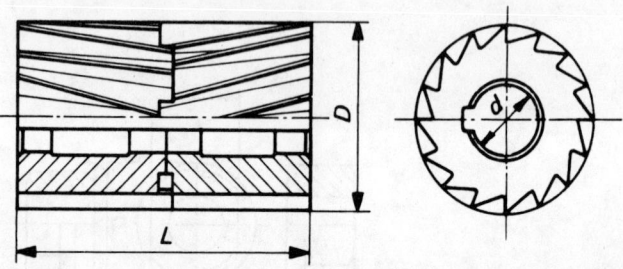

Type B – Interlocking cutter

Dimensions in millimeters

D	d*	L						
j_s16	H7	40	50	63	70	80	100	125
50	22	X		X		X		
63	27		X		X			
80	32			X			X	
100	40					X		X

Dimensions in millimeters

D	d*	L					
j_s16	H7	80	100	125	160	200	250
80	32	X		X			
100	40		X		X		
125	50			X		X	
160	60				X		X

*The bore and keyway dimensions shall be in accordance with the metric series of ISO 240.
For tolerances H7 see Table 6-14 and j_s16 see Table 6-27.

METAL CUTTING TOOLS

Table 17-33. Slotting Cutters with Plain Bore and Key Drive (ISO 2585)

	ISO	2585
U.S.A.	ANSI	
JAPAN	JIS	
GERMANY	DIN	
FRANCE	NF	E66-225
U.K.	BS	
ITALY	UNI	3903
AUSTRALIA	AS	

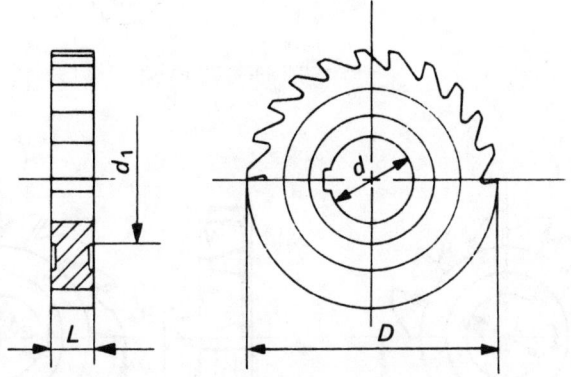

First angle projection

Dimensions in millimeters

D	d*	d₁	L**													
js16	H7	min.	4	5	6	8	10	12	14	16	18	20	22	25	28	32	36	40
50	16	27	X	X	X	X	X											
63	22	34	X	X	X	X	X	X	X									
80	27	41		X	X	X	X	X	X	X	X							
100	32	47				X	X	X	X	X	X	X	X	X				
125							X	X	X	X	X	X	X	X				
160	40	55						X	X	X	X	X	X	X	X	X		
200								X	X	X	X	X	X	X	X	X	X	X

*The bore and keyway dimensions shall be in accordance with the metric series of ISO 240. (See Table 17-28A).
**The tolerance on thickness L of the cutter is to be determined by agreement between the interested parties as a function of the tolerance of the part to be produced.
For tolerances H7 see Table 6-14 and j_s16 see Table 6-27.

METAL CUTTING TOOLS

Table 17-34. Side and Face Milling Cutters with Plain Bore and Key Drive (ISO 2587)

	ISO	2587
U.S.A.	ANSI	
JAPAN	JIS	
GERMANY	DIN	885
FRANCE	NF	E66-246
U.K.	BS	
ITALY	UNI	3905
AUSTRALIA	AS	

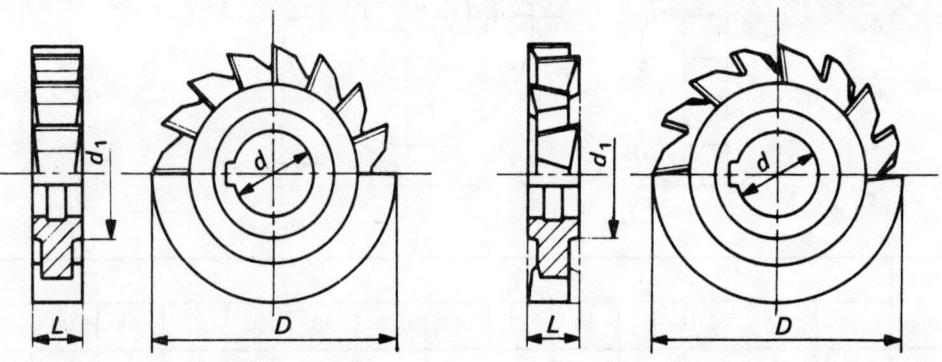

First angle projection

Type A—Cutter with straight teeth   Type B—Cutter with double alternate helix

Dimensions in millimeters

D j_s16	d H7	d_1 min.	4	5	6	8	10	12	14	16	18	20	22	25	28	32	36	40
50	16	27	X	X	X	X	X											
63	22	34	X	X	X	X	X	X	X	X								
80	27	41		X	X	X	X	X	X	X	X	X						
100	32	47			X	X	X	X	X	X	X	X	X	X				
125	32	47				X	X	X	X	X	X	X	X	X	X			
160	40	55					X	X	X	X	X	X	X	X	X	X		
200	40	55						X	X	X	X	X	X	X	X	X	X	X

L column header: k11

The bore and keyway dimensions shall be in accordance with the metric series of ISO 240. (See Table 17-28A) (ISO tolerances j_s16, H7 and k11 are defined in section 6).

METAL CUTTING TOOLS

17-43

Table 17-35. Metric Module Gear Milling Cutters (ISO 2490)

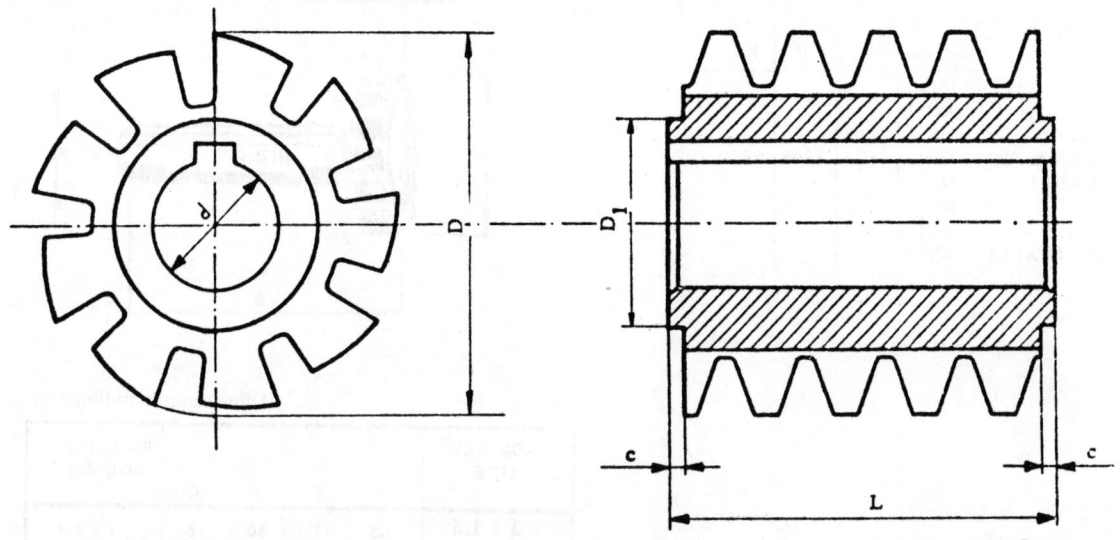

Dimensions in millimeters

Module 1st choice	Module 2nd choice	Module 3rd choice	Outside Diameter D	Bore diameter d	Overall length L	Hub depth c min.
1			50	22	32	3
	1.125		50	22	32	3
1.25			50	22	40	3
	1.375		50	22	40	3
1.5			63	27	45	3
	1.75		63	27	50	3
2			63	27	50	3
	2.25		71	27	56	3
2.5			71	27	63	3
	2.75		71	27	63	3
3			80	32	71	3
		3.25	80	32	71	3
	3.5		80	32	71	4
		3.75	90	32	80	4
4			90	32	80	4
	4.5		90	32	90	4
5			100	32	100	4
	5.5		112	40	112	4
6			112	40	112	4
		6.5	112	40	118	4
	7		118	40	125	4
8			125	40	140	4
	9		140	40	150	4
10			150	40	170	4
	11		160	50	180	5
12			170	50	200	5
	14		190	50	224	5
16			212	60	250	5
	18		236	60	280	5
20			250	60	300	5

NOTES:
1. Axial gashing is permitted up to 6° lead angle.
2. Hubs may be either parallel or conical.
3. Hub diameter D_1 is determined at the manufacturer's discretion. The diameter shall be as large as possible and in all cases greater than the spacing ring diameter as given in ISO 839, Milling machine arbors with 7/24 tapers and milling machine accessories (see Table 17-29).

METAL CUTTING TOOLS

Table 17-36. Woodruff Key Seat Cutters*

	Type	A	B	Key
	ISO			
U.S.A.	ANSI			
Japan	JIS			
GERMANY	DIN	850A	850B	6880
FRANCE	NF			
U.K.	BS			
ITALY	UNI			3916
AUSTRALIA	AS			

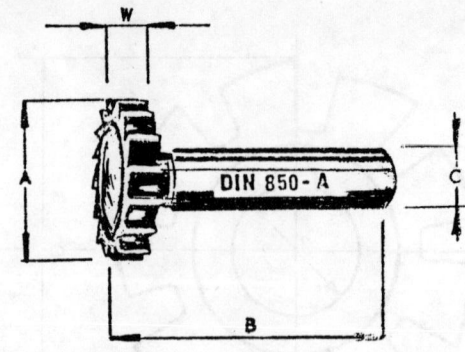

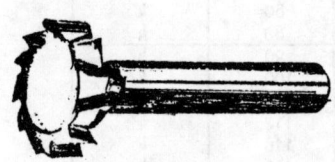

Type A—Cutter with straight teeth

Type B—Cutter with double alternate helix

Dimensions in millimeters

NOMINAL SIZE	A	W	B	C	Woodruff Key NOM. SIZE
1 x 1.4	4.5	1	50	6	1 x 1.4
1.5 x 2.6	7.5	1.5	50	6	1.5 x 2.6
2 x 2.6	7.5	2	50	6	2 x 2.6
2 x 3.7	10.5	2	50	6	2 x 3.7
(2 x 5)	13.5	2	50	10	2 x 5
2.5 x 3.7	10.5	2.5	50	6	2.5 x 3.7
3 x 3.7	10.5	3	50	6	3 x 3.7
3 x 5	13.5	3	55	10	3 x 5
3 x 6.5	16.5	3	55	10	3 x 6.5
(3 x 7.5)	19.5	3	55	10	3 x 7.5
4 x 5	13.5	4	55	10	4 x 5
4 x 6.5	16.5	4	55	10	4 x 6.5
4 x 7.5	19.5	4	55	10	4 x 7.5
(4 x 9)	22.5	4	60	10	4 x 9
5 x 6.5	16.5	5	55	10	5 x 6.5
5 x 7.5	19.5	5	55	10	5 x 7.5
5 x 9	22.5	5	60	10	5 x 9
5 x 10	25.5	5	60	10	5 x 10
6 x 7.5	19.5	6	60	10	6 x 7.5
6 x 9	22.5	6	60	10	6 x 9
6 x 10	25.5	6	60	10	6 x 10
6 x 11	28.5	6	60	10	6 x 11
(6 x 13)	32.5	6	65	10	6 x 13
8 x 9	22.5	8	60	10	8 x 9
8 x 11	28.5	8	60	10	8 x 11
8 x 13	32.5	8	60	10	8 x 13
(8 x 15)	38.5	8	65	12	8 x 15
(8 x 16)	45.5	8	65	12	8 x 16
10 x 11	28.5	10	65	12	10 x 11
10 x 13	32.5	10	65	12	10 x 13
10 x 16	45.5	10	65	12	10 x 16

*Courtesy: METRIC & MULTISTANDARD COMPONENTS CORPORATION
120 Old Saw Mill River Road, HAWTHORNE, N.Y. 10532

NOTE: Sizes shown in brackets ( ) are second choice sizes.

METAL CUTTING TOOLS

Table 17-37. T-Slot Cutters with Parallel Shanks (ISO 3337)

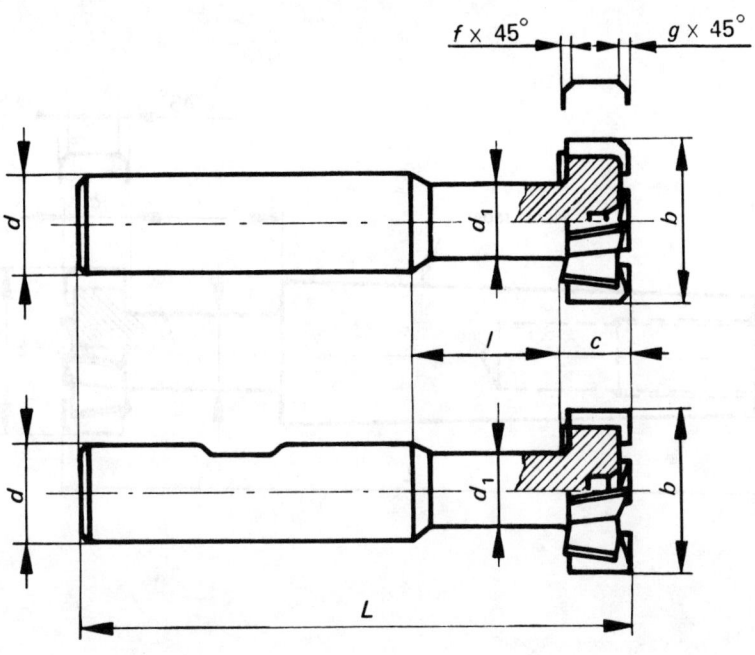

Dimensions in millimeters

b h12	c h12	d_1 max.	l +1 0	$d^{1)}$	L	f max.	g max.	For slot of
11	3,5	4	10	10,0	53,5	0,6	1,0	5
12,5	6	5	11		57			6
16	8	7	14		62			8
18		8	17	12,5	70			10
21	9	10	20		74			12
25	11	12	23	16,0	82		1,6	14
32	14	15	28		90			18
40	18	19	34	25,0	108	1,0	2,5	22
50	22	25	42	31,5	124			28
60	28	30	51		139			36

NOTES:
1. Tolerance on d: h8 for plain parallel shanks, h6 for flatted parallel shanks.
The ISO tolerances h6, h8 and h12 are shown in Table 6-26.
Chamfers f and g may be replaced by radii of the same value. These are optional configurations.
Parallel shanks and flatted parallel shanks are in accordance with ISO 3338
Designation of cutters: The cutters are designated by the values given in the column "For slot of."
2. The decimal marker used in the ISO standard and in this table is a comma ",".

METAL CUTTING TOOLS

Table 17-38. T-Slot Cutters with Morse Taper Shanks (ISO 3337)

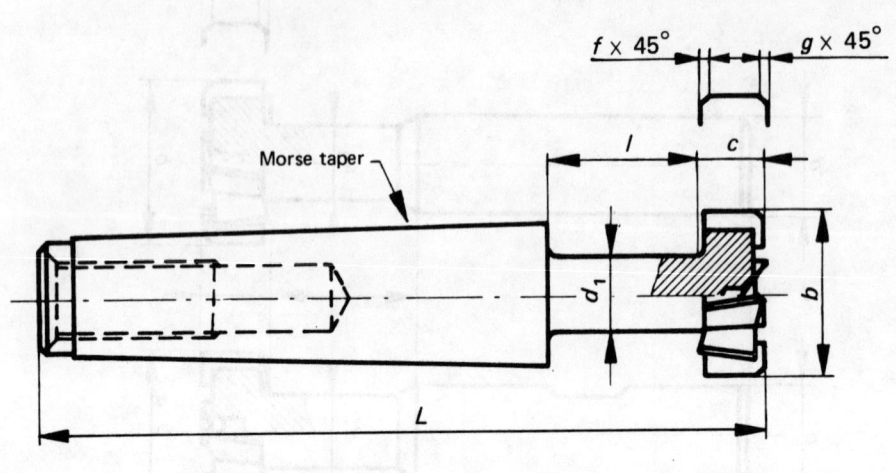

Dimensions in millimeters

b h12	c h12	d_1 max.	l +1 0	L	f max.	g max.	Morse taper No.	For slot of
18	8	8	17	82	0,6	1,0	1	10
21	9	10	20	98	0,6	1,0	2	12
25	11	12	23	103	0,6	1,6	2	14
32	14	15	28	111	0,6	1,6	2	18
40	18	19	34	138	1,0	2,5	3	22
50	22	25	42	173	1,0	2,5	4	28
60	28	30	51	188	1,0	2,5	4	36
72	35	36	58	229	1,6	4,0	5	42
85	40	42	64	240	2,0	6,0	5	48
95	44	44	71	251	2,0	6,0	5	54

The ISO tolerance h12 is specified in Table 6-26.
Chamfers f and g may be replaced by radii of the same value. These are optional configurations.
Morse taper shanks: Tapers having tapped in accordance with ISO 296.
Designation of cutters: The cutters are designated by the values given in the column "For slot of."
The decimal marker used in the ISO standard and in this table is a comma ",".

METAL CUTTING TOOLS

17-47

Table 17-39. World Metric Screw Thread Cutting Tool Standards

Standard	Figure*	Name
U.S.A. ANSI JAPAN JIS GERMANY DIN 371,376 FRANCE NF E66-105 U.K. BS ITALY UNI AUSTRALIA AS ISO 2283		(E) LONG SHANK MACHINE TAPS WITH NOMINAL DIAMETERS FROM 3 TO 24 mm (COARSE THREAD) (G) MASCHINENGEWINDEBOHRER MIT VERSTARKTEM SCHAFT FUR DURCHMESSER VOS 3 BIS 24 mm (REGELGEWINDE) (F) TARAUDS A MACHINE, A QUEUE LONGUE, DE DIAMETRE NOMINAL 3 A 24 mm (PAS GROS) (I)
U.S.A. ANSI JAPAN JIS GERMANY DIN 374 FRANCE NF E66-105 U.K. BS ITALY UNI AUSTRALIA AS ISO 2283		(E) LONG SHANK MACHINE TAPS FOR ISO METRIC FINE THREAD (G) MASCHINENGEWINDEBOHRER FUR METRISCHES ISO FEINGEWINDE (F) TARAUDS A MACHINE, A QUEUE LONGUE (PAS FINS) (I)
U.S.A. ANSI JAPAN JIS GERMANY DIN 352 FRANCE NF E66-103 U.K. BS 949 ITALY UNI 3801 AUSTRALIA AS ISO 529		(E) SHORT MACHINE TAPS AND HAND TAPS (COARSE THREAD) (G) SATZGEWINDEBOHRER FUR METRISCHES ISO REGELGEWINDE (F) TARAUDS COURTS, A MACHINE ET A MAIN (PAS GROS) (I)
U.S.A. ANSI JAPAN JIS GERMANY DIN 2181 FRANCE NF E66-103 U.K. BS 949 ITALY UNI 3802,3803 AUSTRALIA AS ISO 529		(E) SHORT MACHINE TAPS AND HAND TAPS (FINE THREAD) (G) SATZGEWINDEBOHRER FUR METRISCHES ISO FEINGEWINDE (F) TARAUDS COURTS, A MACHINE ET A MAIN (PAS FINS) (I)
U.S.A. ANSI JAPAN JIS GERMANY DIN 223 FRANCE NF E66-150 U.K. BS 1127 ITALY UNI 539-542 AUSTRALIA AS ISO 2568		(E) HAND AND MACHINE OPERATED CIRCULAR SCREWING DIES (G) RUNDE SCHNEIDEISEN (F) FILIERES RONDES DE FILETAGE, A MAIN ET A MACHINE (I)
U.S.A. ANSI JAPAN JIS GERMANY DIN 225 FRANCE NF E66-152 U.K. BS ITALY UNI 549 AUSTRALIA AS ISO 2568		(E) CIRCULAR SCREWING DIE HOLDERS (G) SCHNEIDEISENHALTER FUR RUNDE SCHNEIDEISEN (F) PORTS-FILIERE AMAIN (I)

*Courtesy: METRIC & MULTISTANDARD COMPONENTS CORPORATION
120 Old Saw Mill River Road, HAWTHORNE, N.Y. 10532

NOTES:
1. See Table 8-1 for World Screw Thread Standard diameter pitch combinations.
2. See Table 9-26 for recommended metric tap drill sizes.

17-48
METAL CUTTING TOOLS

Table 17-40. Throwaway Carbide Indexable Inserts Without Holes

STANDARD	FIGURE	NAME	DIMENSION (1)			In Millimeters
ISO 883 U.S.A. ANSI B94.25 JAPAN JIS B4104 GERMANY DIN 4968 FRANCE NF E66-307 U.K. BS 4193/I ITALY UNI AUSTRAL AS	(triangular figure)	(E) TRIANGULAR INSERTS FOR TOOLS WITH NEGATIVE TOP RAKE ANGLE (G) DREIECKIGE.......... (F) TRIANGULAIRES......... (I)	DESIG. (3) TNUN TNUN TNUN TNUN TNUN TNUN TNGN TNGN TNGN TNGN TNGN TNGN	A 6.35 9.52 12.70 (15.88) (19.05) (25.40) 6.35 9.52 12.70 (15.88) (19.05) (25.40)	T 3.18 (3.18) 4.78 4.78 (6.35) (6.35 7.92) (7.92 9.52) (11.13) 3.18 (3.18) 4.78 4.78 (6.35) (6.35 7.92) (6.35 7.92) (9.52)	R 0.4 0.8 (0.4) 0.8 1.2 (1.6) (0.4 0.8) 1.2 1.6 (3.2) (1.2 1.6 2.4) (1.6 2.4 3.2) (3.2) 0.4 (0.8) (0.4) 0.8 1.2 (1.6 2.4) (0.4 0.8) 1.2 (1.6 3.2) (0.8 1.2 1.6 2.4) (1.6) (2.4)
ISO 883 U.S.A. ANSI B94.25 JAPAN JIS B4104 GERMANY DIN 4968 FRANCE NF E66-307 U.K. BS 4193/I ITALY UNI AUSTRAL AS	(triangular figure with 11°)	(E) TRIANGULAR INSERTS FOR TOOLS WITH POSITIVE TOP RAKE ANGLE (G) DREIECKIGE WENDESCHNEIDPLATTEN AUS HARTMETALL (F) TRIANGULAIRES PLAQUETTES AMOVIBLES EN CARBURES METALLIQUES (I) SANS TROU DE FIXATION	DESIG. TPUN TPUN TPUN TPUN TPUN TPGN TPGN TPGN TPGN	A (6.35) 9.52 12.70 (15.88) (19.05) (6.35) 9.52 12.70 (15.88)	T (3.18) 3.18 4.78 (6.35) (7.92) 3.18 3.18 (4.78) (3.18) 4.78 (6.35)	R (0.8) 0.8 1.2 (0.8) 1.2 1.6 (1.2 1.6 2.4) (3.2) 0.4 0.8 (0.4) 0.8 1.2 (1.6) (0.4 0.8) 1.2 (1.6 3.2) (1.2 1.6 2.4)
ISO 883 U.S.A. ANSI B94.25 JAPAN JIS B4104 GERMANY DIN 4968 FRANCE NF E66-307 U.K. BS 4193/I ITALY UNI AUSTRAL AS	(square figure)	(E) SQUARE INSERTS FOR TOOLS WITH NEGATIVE TOP RAKE ANGLE (G) VIERECKIGE......... (F) CARREES........... (I)	DESIG. SNUN SNUN SNUN SNUN SNUN SNGN SNGN SNGN SNGN SNGN	A 9.52 12.70 15.88 19.05 (25.40) 9.52 12.70 (15.88) (19.05) (25.40)	T 3.18 (3.18) 4.78 4.78 (6.35) 4.78 (6.35) (7.92 9.52) 3.18 (3.18) 4.78 (4.78) (4.78 6.35) (7.92 9.52)	R 0.4 0.8 (0.4) 0.8 1.2 (1.6) 1.2 1.6 (0.8) 1.2 1.6 1.6 2.4 0.8 (1.2 1.6 2.4) (0.4) 0.8 1.2 (1.6 2.4) (0.8 1.2 1.6) (0.4 0.8 1.2 1.6 3.2) (1.6)
ISO 883 U.S.A. ANSI B94.25 JAPAN JIS B4104 GERMANY DIN 4968 FRANCE NF E66-307 U.K. BS 4193/I ITALY UNI AUSTRAL AS	(square figure with 11°)	(E) SQUARE INSERTS FOR TOOLS WITH POSITIVE TOP RAKE ANGLE (G) VIERECKIGE .... (F) CARREES............... (I)	DESIG. SPUN SPUN SPUN SPGN SPGN SPGN SPGN	A (9.52) 12.70 19.05 (9.52) 12.70 (15.88) (19.05)	T (3.18) 3.18 (4.78) 4.78 (3.18) 3.18 (4.78) (4.78) (4.78)	R (0.8) 0.8 1.2 (1.2) 1.6 (0.8 1.2) (0.4) 0.8 1.2 (1.6) (0.8 1.2 1.6) (0.4 0.8 1.2 1.6 3.2)
ISO U.S.A. ANSI B94.25 JAPAN JIS GERMANY DIN FRANCE NF U.K. BS 4193/I ITALY UNI AUSTRAL AS	(rhombic figure TYPE CN, TYPE CPG, 80°, 11°)	(E) RHOMBIC INSERTS FOR TOOLS WITH NEGATIVE AND POSITIVE (CPG) TOP RAKE ANGLE (G) (F) (I)	DESIG. (2) CNUN CNUN CNUN CNGN CNGN CNGN CPGN CPGN	A 12.70 15.88 19.05 12.70 15.88 19.05 12.70 15.05	T 3.18 4.78 6.35 4.78 6.35 3.18 4.78 6.35 4.78 6.35 3.18 4.78 4.78 6.35	R 0.4 0.8 1.2 0.8 1.2 1.6 0.8 1.2 1.6 0.4 0.8 1.2 0.8 1.2 1.6 0.8 1.2 1.6 0.4 0.8 1.2 0.8 1.2 1.6

NOTES:
1. Dimensions shown in brackets ( ) are not ISO standard sizes.
2. These types are not covered in the standard.
3. Tolerance Class U = least precision (utility); G = Precision (ground) see Table 17-42.

METAL CUTTING TOOLS

Table 17-41. Throwaway Carbide Indexable Inserts With Holes

STANDARD	FIGURE	NAME	DIMENSION (1)			In Millimeters	
ISO 3364 U.S.A. ANSI B94.25 JAPAN JIS GERMANY DIN FRANCE NF U.K. BS ITALY UNI AUSTRAL AS		(E) TRIANGULAR INSERTS WITH HOLE FOR TOOLS WITH NEGATIVE TOP RAKE ANGLE (G) (F) (I)	DESIG. (3) TNMA TNMA TNMA TNMA TNMA TNMG TNMG TNMG TNMG TNMG	A (6.35) 9.52 12.70 (15.88) (19.05) (6.35) 9.52 12.70 (15.88) (19.05)	T (3.18) (3.18) 4.78 4.78 (6.35) (9.52) (3.18) (3.18) 4.78 4.78 (6.35) (9.52)	H (2.26) 3.81 5.16 (6.35) (7.92) (2.26) 3.81 5.16 (6.35) (7.92)	R (0.8) 0.8 1.2 0.8 1.2 1.6 (0.8 1.2 1.6) (2.4) (0.8) 0.8 1.2 (1.6) 0.8 1.2 1.6 (3.2) (0.8 1.2 1.6) (2.4)
ISO 3364 U.S.A. ANSI B94.25 JAPAN JIS GERMANY DIN FRANCE NF U.K. BS ITALY UNI AUSTRAL AS		(E) SQUARE INSERTS WITH HOLE FOR TOOLS WITH NEGATIVE TOP RAKE ANGLE (G) (F) (I)	DESIG. SNMA SNMA SNMA SNMA SNMA SNMG SNMG SNMG SNMG SNMG	A (9.52) 12.70 (15.88) 19.05 (25.40) 9.52 12.70 (15.88) 19.05 (25.40)	T (3.18) 4.78 (6.35) 6.35 (9.52) 3.18 4.78 (6.35) 6.35 (9.52)	H (3.81) 5.16 (6.35) 7.92 (9.12) 3.81 5.16 (6.35) 7.92 (9.12)	R (0.8) 0.8 1.2 (1.6) (1.2) 1.2 1.6 (2.4) 0.4 0.8 (1.2) 0.8 1.2 (1.6) (1.2) 1.2 1.6 (2.4)
ISO 3364 U.S.A. ANSI B94.25 JAPAN JIS GERMANY DIN FRANCE NF U.K. BS ITALY UNI AUSTRAL AS		(E) 80° RHOMBIC INSERTS WITH HOLE FOR TOOLS WITH NEGATIVE TOP RAKE ANGLE (G) (F) (I)	DESIG. CNMA CNMA CNMA CNGA (2) CNGA CNGA CNMG CNMG CNMG CNGG (2) CNGG CNGG	A 12.70 (15.88) 19.05 12.70 15.88 19.05 12.70 (15.88) 19.05 12.70 15.88 19.05	T (3.18) 4.78 (6.35) (4.78) 6.35 3.18 4.78 6.35 4.78 6.35 (3.18) 4.78 (6.35) (4.78) 6.35 3.18 4.78 6.35 4.78 6.35	H 5.16 (6.35) 7.92 5.16 6.35 7.92 5.16 (6.35) 7.92 5.16 6.35 7.92	R (0.4) 0.8 1.2 (0.8 1.2 1.6) (0.8) 1.2 1.6 0.4 0.8 1.2 0.8 1.2 1.6 0.8 1.2 1.6 (0.4) 0.8 1.2 (0.8 1.2 1.6) (0.8) 1.2 1.6 0.4 0.8 1.2 0.8 1.2 1.6 0.8 1.2 1.6
ISO 3364 U.S.A. ANSI B94.25 JAPAN JIS GERMANY DIN FRANCE NF U.K. BS ITALY UNI AUSTRAL AS		(E) 55° RHOMBIC INSERTS WITH HOLE FOR TOOLS WITH NEGATIVE TOP RAKE ANGLE (G) (F) (I)	DESIG. DNMA DNMA DNGA (2) DNGA DNMG DNMG DNGG (2) DNGG	A 12.70 15.88 12.70 15.88 12.70 15.88 12.70 15.88	T (3.18) 4.78 (4.78) 6.35 3.18 4.78 4.78 6.35 (3.18) 4.78 (4.78) 6.35 3.18 4.78 4.78 6.35	H 5.16 6.35 5.16 6.35 5.16 6.35 5.16 6.35	R (0.4) 0.8 1.2 0.8 1.2 1.6 0.4 0.8 1.2 0.8 1.2 1.6 (0.4) 0.8 1.2 0.8 1.2 1.6 0.4 0.8 1.2 0.8 1.2 1.6
ISO U.S.A. ANSI B94.25 JAPAN JIS GERMANY DIN FRANCE NF U.K. BS ITALY UNI AUSTRAL AS		(E) 35° RHOMBIC INSERTS WITH HOLE FOR TOOLS WITH NEGATIVE TOP RAKE ANGLE (G) (F) (I)	DESIG. (2) VNMA VNMA VNGA VNGA VNMG VNMG VNGG VNGG	A 9.52 12.70 9.52 12.70 9.52 12.70 9.52 12.70	T 4.78 4.78 4.78 4.78 4.78 4.78 4.78 4.78	H 3.81 5.16 3.81 5.16 3.81 5.16 3.81 5.16	R 0.4 0.8 0.4 0.8 1.2 0.4 0.8 0.4 0.8 1.2 0.4 0.8 0.4 0.8 1.2 0.4 0.8 0.4 0.8 1.2

NOTES:
1. Dimensions shown in brackets ( ) are not ISO standard sizes.
2. These types are not covered in the ISO standard 3364.
3. Tolerance class M = Precision molded; G = Ground precision and special condition A = with hole;
   G = with hole and chip grooves (both sides) see Table 17-42.

17-50   METAL CUTTING TOOLS

Table 17-42. ISO Indexable Throwaway Insert Designation System (ISO 1832)

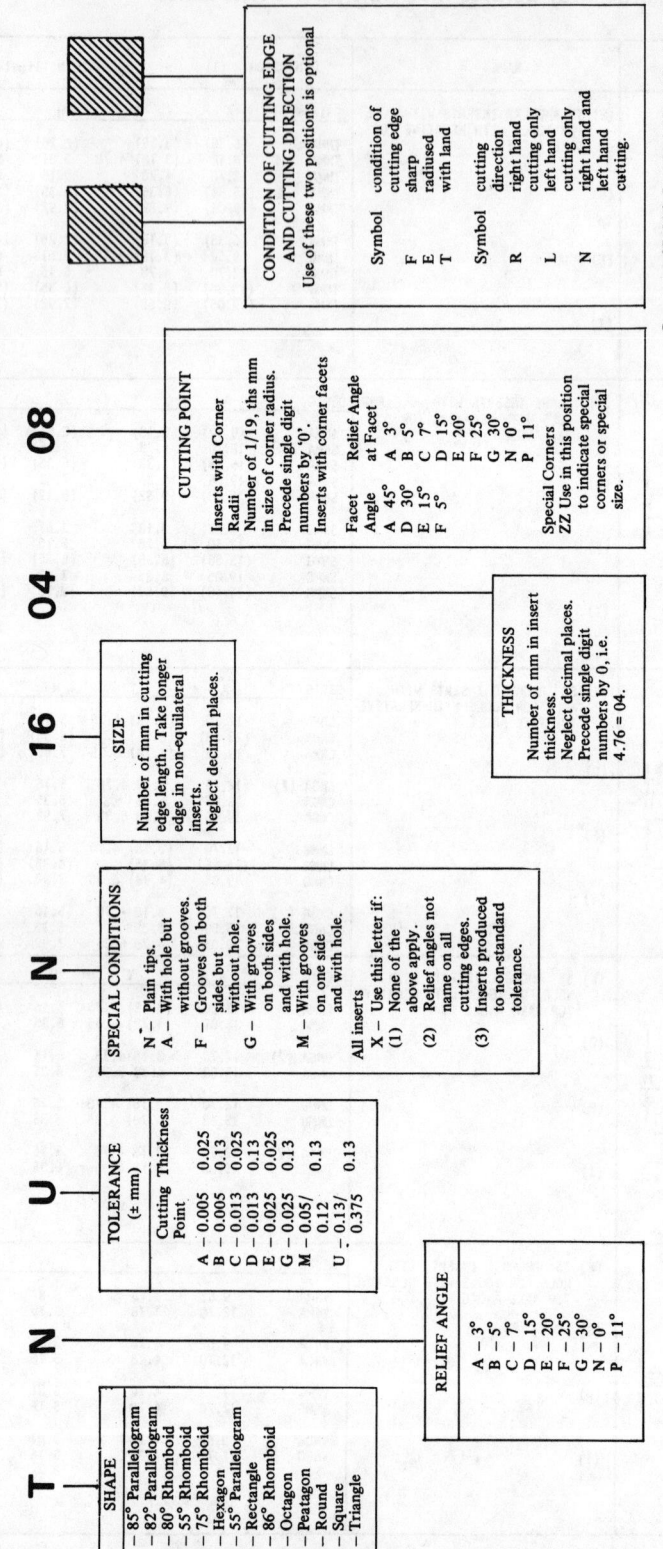

Courtesy of British Standard BS 4193: Part 2.

Designation Examples

1. Triangular insert for tools with negative top rake angle, tolerance, class "U", plain tip "N" and dimensions A = 12.7 mm; T = 4.78 mm; R = 1.6 (see Table 17-40)
   TNUN-120416

2. Square insert for holders with positive top rake angle, tolerance Class "G", plain tip "N", and dimensions A = 12.7 mm; T = 3.18 mm; R = 1.2 mm (see Table 17-40)
   SPGN-120312

3. 55° rhombic insert for tools with negative top rake angle, tolerance Class "M", with hole but without chip curler grooves, and dimensions A = 15.88 mm; T = 4.78 mm; R = 1.2 mm (see Table 17-41)
   DNMA-150412

METAL CUTTING TOOLS

RELATED ISO STANDARDS (TC 29, TC 39, TC 60)

Small Tools (TC 29)

ISO 234-1975	Files and rasps—lengths and cross sections
ISO 235/I-1975	Parallel shank twist drills jobber and stub series and Morse taper shank twist drills and core drills
ISO 235/II-1972	Core drills with parallel shanks and Morse taper shanks—Recommended stocked sizes
ISO 236/I-1976	Hand reamers.
ISO 236/II-1976	Long fluted machine reamers, Morse taper shank
ISO 237-1975	Diameters of shanks and sizes of driving squares for rotating tools with parallel shanks
ISO 238-1974	Reduction sleeves and extension sockets for tools with Morse taper shanks
ISO 239-1974	Drill chuck tapers
ISO 240-1975	Interchangeability dimensions for milling cutters and cutter arbors or cutter mandrels—Metric series and inch series—2nd Edition
ISO 241-1975	Shanks for turning and planing tools—Sections and tolerances
ISO 242-1975	Carbide tips for turning tools—Metric series
ISO 243-1975	Turning tools with carbide tips—Metric series
ISO 494-1975	Parallel shank twist drills—Long series
ISO 504-1975	Turning tools with carbide tips—Designation and marking
ISO 513-1975	Application of carbides for machining by chip removal—Designation of main groups of chip removal and groups of applications
ISO 514-1975	Turning tools with carbide tips—Internal tools (Metric Series)
ISO 521-1975	Machine chucking reamers with parallel shanks or Morse taper shanks
ISO 522-1975	Special tolerances for reamers
ISO 523-1974	Recommended range of outside diameters for milling cutters
ISO/R 524-1966	Hard metal wire drawing dies—Interchangeability—Dimension of pellets and cases
ISO 525-1975	Bonded abrasive products—General features—Designation—Ranges of dimensions—Profiles
ISO 529-1975	Short machine taps and hand taps
ISO/R 603-1967	Bonded abrasive products—Grinding wheel dimensions: Part I
ISO 691-1975	Tolerances on spanner gaps and sockets (Metric values for general use)
ISO 839/I-1976	Milling machine arbors with 7/24 tapers and milling machine accessories—Part 1 Dimensions
ISO 866-1975	Center drills for center holes without protecting chamfers—Type A
ISO 883-1976	Throwaway carbide indexable inserts—Dimensions
ISO 1117-1975	Bonded abrasive products—Grinding wheel dimensions—Part 2
ISO/R 1180-1970	Shanks for pneumatic tools and fitting dimensions of chuck-bushings—Part I
ISO/R 1571-1970	Shanks for pneumatic tools and fitting dimensions of chuck bushings—Part II
ISO/R 1641-1970	End mills with parallel shanks and with Morse taper shanks—Standard series and long series
ISO 1651-1974	Tube drawing mandrels
ISO 1684-1975	Wire, bar and tube drawing dies—Designation—Marking—Dimensions
ISO/R 1832-1971	Indexable (Throwaway) inserts—Designation—Code of symbolization
ISO 1929-1974	Abrasive belts—Designations—Dimensions—Tolerances
ISO 2220-1972	Hand finishing sticks and oil stones—Dimensions
ISO 2235-1972	Abrasive sheets and abrasive discs—Dimensions
ISO 2238-1972	Machine bridge reamers
ISO 2250-1972	Finishing reamers for Morse and metric tapers, with parallel shanks and Morse taper shanks
ISO 2283-1972	Long shank machine taps with nominal diameters from 3 to 24 mm and 1/8 to 1 inch
ISO 2284-1976	Hand taps for pipe threads for parallel and taper threads—General dimensions and marking
ISO 2296-1972	Metal slitting saws with fine and coarse teeth—Metric series
ISO 2306-1972	Drills for use prior to tapping screw threads

ISO 2324-1972	End mills with 7/24 taper shanks—Standard series and long series
ISO 2336-1972	Hand and machine hacksaw blades—Dimensions for lengths up to 450 mm and pitches up to 6.3 mm
ISO 2402-1972	Shell reamers with taper bore (taper bore 1:30 [included]) with slot drive and arbors for shell reamers
ISO 2421-1972	Cylindrical abrasive sleeves—Designation—Dimensions—Tolerances
ISO 2422-1972	Truncated cone abrasive sleeves—Designation—Dimensions—Tolerances
ISO 2540-1973	Center drills for center holes with protecting chamfer—Type B
ISO 2541-1973	Center drills for center holes with radius form—Type R
ISO 2568-1973	Hand and machine-operated circular screwing dies and hand-operated die stocks
ISO 2583-1972	Tool shanks and equipment with 7/24 tapers—Collar dimensions
ISO 2584-1972	Cylindrical cutters with plain bore and key drive—Metric series
ISO 2585-1972	Slotting cutters with plain bore and key drive—Metric series
ISO 2586-1973	Shell end mills with plain bore and tenon drive—Metric series
ISO 2587-1972	Side and face milling cutters with plain bore and key drive—Metric series
ISO 2780-1973	Milling cutters with tenon drive—Interchangeability dimensions with cutters arbors—Metric series
ISO 2804-1973	Wire, bar or tube drawing dies - As sintered pellets of hard metal (carbide)—Dimensions
ISO 2857-1973	Ground thread taps for ISO metric threads of tolerances 4H to 8H and 4G to 6G coarse and fine pitches—Manufacturing tolerances on the threaded portion
ISO 2924-1973	Solid and segmental circular saws for cold cutting of metals—Interchangeability dimensions of the drive—Saw diameter range 224 to 2240 mm.
ISO 2933-1974	Bonded abrasive products—Grinding wheel dimensions (Part 3)
ISO 2940/I-1974	Milling cutters mounted on centering arbors having a 7/24 taper—Fitting dimensions-centering arbors
ISO 2940/II-1974	Milling cutters mounted on centering arbors having a 7/24 taper—Inserted tooth cutters
ISO 2976-1973	Abrasive belts—Selection of width/length combinations
DIS 3002/1	Cutting tools – Terms – Reference systems—Working Angles
ISO DIS 3002/2	General conversion formulae to relate tool and working angles
ISO 3017-1973	Abrasive discs—Selection of disc outside diameter/center hole diameter combinations
ISO 3286-1976	Corner radius of single-point cutting tools
ISO 3291-1975	Morse taper shank twist drills extra-long series
ISO 3292-1975	Parallel shank twist drills-extra long series
ISO 3293-1975	Morse taper shank countersinks
ISO 3294-1975	Parallel shank countersinks
ISO 3314-1975	Shell drills with taper bore with slot drive
ISO 3337-1975	T-slot cutters—Metric series
3338/1-1977	Parallel Shanks for Milling Cutters—Part I: Dimensional Characteristics of Plain Parallel Shanks
ISO 3364-1977	Indexable throwaway carbide inserts—Dimensions
ISO 3365/I-1977	Indexable/throwaway/carbide inserts for milling cutters—Part I: Square inserts
ISO 3366-1975	Coated abrasives—Designation—Dimensions
ISO 3367-1975	Coated abrasives—Rolls for widths of 50 mm—Designation-Dimensions
ISO 3368-1975	Coated abrasives-cloth rolls up to 40 mm width—Designation—Dimensions
ISO 3438-1975	Subland twist drills with Morse taper shanks
ISO 3439-1975	Subland twist drills with parallel shanks
DIS 3440	Carbide-tipped twist drills with parallel shanks for use on metal and plaster
DIS 3441	Carbide tipped twist drills with Morse taper shanks
ISO 3465-1975	Hand taper pin reamers
ISO 3466-1975	Machine taper pin reamers with parallel shanks

METAL CUTTING TOOLS

ISO 3467-1975	Machine taper pin reamers with Morse taper shanks
ISO 3685-1977	Tool-life testing with single-point turning tools
DIS 3855	Milling cutter terminology
DIS 3859	Inverse dovetail cutters and dovetail cutters with parallel shanks
ISO 3860-1976	Form milling cutters with a constant profile
DIS 3919	Coated abrasive-flap wheels with shafts
ISO 3920-1976	Honing stones with square sections
ISO 3921-1976	Honing stones with rectangular sections
ISO 3936-1976	Reduction sleeves with tenon drive with 7/24 taper
ISO 3937-1976	Cutter arbors with tenon drive
ISO 3940-1977	Tapered die sinking cutters with parallel shanks
ISO 4202-1976	Reduction sleeves with external 7/24 taper for tools with Morse taper shanks
ISO DIS 4203	Dimensions of driving tenons and sockets for parallel shank tools
ISO DIS 4204	Countersinks, 90 degrees, with Morse taper shanks and detachable pilots
ISO DIS 4205	Countersinks, 90 degrees, with parallel shanks and solid pilot
ISO DIS 4206	Counterbores with parallel shanks and solid pilots
ISO DIS 4207	Counterbores with Morse taper shanks and detachable pilots
ISO DIS 4208	Detachable pilots for use with counterbores and 90 degrees countersinks
DIS 4230	Circular screwing dies for taper pipe threads - "R" series
DIS 4231	Circular screwing dies for parallel pipe threads - "G" series
ISO DIS 4247	Dimensionss of jig bushes and accessories for drilling purposes
ISO DIS 4248	Jig bush definitions and nomenclature
ISO DPR 4857/1	Metal cutting band saws—Part I—Definitions and terminology
ISO DPR 4857/2	Metal cutting band saws—Part II—Basic dimensions and tolerances
ISO DPR 4857/3	Metal cutting band saws—Part III—Characteristics relating to each type of blade
ISO DIS 4875/1	Metal cutting band saw blades—Part 1
ISO DIS 4875/2	Metal cutting band saw blades—Part 2
ISO DIS 4875/3	Metal cutting band saw blades—Part 3: Characteristics relating to
ISO DIS 5396	Hard metal heading dies—Terminology
ISO DPR 5407	Hard metal as sintered pellets for cold forming tools—Tolerances
ISO 5413-1976	Machine tools—Positive drive of Morse tapers
ISO DPR 5414	Chucks (end mill holders) with clamp screws for flatted parallel shank tools
ISO DPR 5419	Definitions and nomenclature of drills
ISO DPR 5420	Definitions and nomenclature of reamers
ISO DIS 5415	Reduction sleeves with 7/24 external and Morse internal taper and incorporated screw
ISO DIS 5421	High speed steel tool bits
ISO DIS 5429	Coated abrasives—Flap wheels with separate flanges or incorporated flanges—Designation—Dimensions
ISO DPR 5465	Cams for control of limit switches and corresponding rails
ISO DIS 5468	Carbide tipped masonry drills—Diameters and lengths
ISO DIS 5608	Turning and copying tool holders and cartridges for indexable (throwaway) inserts—Designation—Code of symbolization
ISO DIS 5609	Boring bars for indexable (throwaway) inserts—Dimensions
ISO DIS 5610	Single point tool holders for turning and copying, for indexable (throwaway) inserts—Dimensions
ISO DIS 5611	Cartridges type A for indexable (throwaway) inserts—Dimensions
ISO DPR 5967	Terminology of taps
ISO DPR 5968	Terminology of circular screwing dies
ISO DPR 5969	Tolerances on the threaded portion of taps for pipe threads
ISO DPR 6103	Balancing of grinding wheels
ISO DIS 6104	Abrasive products—Diamond or cubic boron nitride grinding—Wheels and saws—General—Designation—Multilingual nomenclature

ISO DIS 6105	Abrasive products—Diamond or cubic boron nitride saws—Dimensions
ISO DPR 6106	Diamond grainsizes
ISO DIS 6108	Double equal angle cutters with plain bore and key drive
ISO DIS 6168	Abrasive products—Diamond or cubic boron nitride grinding wheels—Dimensions

Machine Tools (TC 39)

ISO/R 213-1961	Lathe tools posts (overall internal height)
ISO 229-1973	Machine tools—Speeds and feeds
ISO/R 230-1961	Machine tool test code
ISO 296-1974	Machine tools—Self-holding tapers for tool shanks
ISO/R 297-1963	7/24 tapers for tool shanks
ISO 298-1973	Machine tools—Lathe centres—Sizes for interchangeability
ISO 299-1973	Machine tool tables—T slots and corresponding bolts
ISO/R 369-1964	Symbols for indications appearing on machine tools
ISO 447-1973	Machine tools—Direction of operation of controls
ISO 666-1975	Machine tools—Mounting of plain grinding wheels by means of hub flanges
ISO 702/I-1975	Machine tools—Spindle noses and face plates—Sizes for interchangeability—Part I: Type A
ISO 702/II-1975	Machine tools—Spindle noses and face plates—Sizes for interchangeability—Part II: Camlock type
ISO 702/III-1975	Machine tools—Spindle noses and face plates—Sizes for interchangeability—Part III: Bayonet type
ISO/R 867-1968	Spindle noses and face plates—Bayonet type—Sizes for interchangeability—Metric series
ISO 1080-1975	Machine tools—Morse taper shanks—Cotter slots with taper keys
ISO 1701-1974	Test conditions for milling machines with table of variable height, with horizontal or vertical spindle—Testing of the accuracy
ISO 1708-1975	Test conditions for general purpose parallel lathes—Testing of the accuracy
ISO 1984-1974	Test conditions for milling machines with table of fixed height with horizontal or vertical spindle—Testing of accuracy
ISO 1985-1974	Test conditions for surface grinding machines with vertical grinding wheel spindle and reciprocating table—Testing of accuracy
ISO 1986-1974	Test conditions for surface grinding machines with horizontal grinding wheel spindle and reciprocating table—Testing of accuracy
ISO 2407-1973	Test conditions for internal cylindrical grinding machines with horizontal spindle—Testing of accuracy
ISO 2423-1974	Test conditions for radial drilling machines with the arm adjustable in height—Testing of the accuracy
ISO 2433-1973	Test conditions for external cylindrical grinding machines with a movable table—Testing of accuracy
ISO 2562-1973	Modular units for machine tool construction—Slide units
ISO 2727-1973	Modular units for machine tool construction—Headstocks
ISO 2769-1973	Modular units for machine tool construction—Wing bases for slide units
ISO 2772/I-1973	Test conditions for box type vertical drilling machines—Testing of the accuracy—Part I: Geometrical tests
ISO 2772/II-1974	Test conditions for box type vertical drilling machines—Testing of the accuracy—Part II: Practical test
ISO 2773/I-1973	Test conditions for pillar type vertical drilling machines—Testing of the accuracy—Part I: Geometrical tests
ISO 2773/II-1973	Test conditions for pillar type vertical drilling machines—Testing of the accuracy—Part II: Practical test
ISO 2891-1973	Modular units for machine tool construction—Centre bases and columns
ISO 2905-1974	Modular units for machine tool construction—Spindle noses and adjustable adaptors for multi-spindle heads
ISO 2912-1973	Modular units for machine tool construction—Multi-spindle heads—Casing and input drive shaft dimensions
ISO 2934-1973	Modular units for machine tool construction—Wing bases for columns

METAL CUTTING TOOLS

ISO 3070/0-1975	Test conditions for boring and milling machines with horizontal spindle—Testing of the accuracy—Part 0: General introduction
ISO 3070/1-1975	Test conditions for boring and milling machines with horizontal spindle—Testing of the accuracy—Part I: Table type machines
ISO DIS 3070/2	Test conditions for boring and milling machines with horizontal spindle—Testing of the accuracy—Part II: Floor type machines
ISO 3089-1974	Self-centering, manually operated chucks for machine tools—Normal accuracy—Acceptance test specifications (geometrical tests)
ISO 3190-1975	Test conditions for turret and single spindle co-ordinate drilling machines with vertical spindle—Testing of the accuracy
ISO 3371-1975	Modular units for machine tools construction—Rotary tables and multi-sited centre bases for rotary tables
ISO 3408-1975	Ball screws—Nominal diameters and basic leads—Metric and inch series
ISO 3442-1975	Self-centering chucks for machine tools with two-piece jaws (tongue and groove type)—Sizes for interchangeability and acceptance test specifications
ISO 3476-1975	Modular units for machine tool construction—Tenon drive and flanges for mounting multi-spindle heads
ISO/TR 3498-1974	Lubricants for machine tools
ISO 3589-1975	Modular units for machine tool construction—Integral way columns
ISO 3590-1976	Modular units for machine tool construction—Spindle units
ISO 3610-1976	Modular units for machine tool construction—Support brackets
ISO 3655/0-1976	Test conditions for vertical turning and boring lathes with one or two columns—Testing of the accuracy—Part 0: General introduction
ISO 3655/I-1976	Test conditions for vertical turning and boring lathes with one or two columns—Testing of the accuracy—Part I: Lathes with a single fixed or movable table
ISO 3686-1976	Test conditions for turret and spindle co-ordinate drilling and boring machines with tables or fixed height with vertical spindle—High accuracy machines—Testing of the accuracy
ISO DIS 3875	Test conditions for external cylindrical centreless grinding machines—Testing of the accuracy
ISO DIS 3970	Modular units for machine tool construction—Integral way columns—Floor-mounted type
ISO DIS 4703	Test conditions for surface grinding machines with two columns—Machines for grinding sideways—Testing of the accuracy
ISO DIS 5169	Machine tools—Presentation of lubrication instructions
ISO DIS 5170	Machine tools—Lubrication systems
ISO DIS 5734	Test conditions of mechanical dividing heads for machine tools testing of the accuracy
ISO DIS 6102	Cylindrical holders for tools
ISO DIS 6155	Test conditions for horizontal spindle capstan, turret and single spindle automatic lathes—Testing of the accuracy

Gears (TC 60)

ISO DIS 2490	Nominal dimensions of single start gear hobs with axial keyway

18 Measuring Tools and Instruments

General

The ISO Technical Committee TC3 is working diligently to publish several new world standards[1] on measuring tools and instruments. A listing of the ISO standards and drafts published to date is shown at the end of this section. Many standards on this subject have been published in Great Britain, and are based on metric practices in Continental Europe and on available ISO standards. The material presented in this section may not fit in with our American requirements and practices in all cases, but might be used as a rough reference guide.

SCALES AND MEASURING RULES

Scales for Draftsmen

British Standard BS 1347: Part 3 specifies a series of metric scales recommended for use by engineers, architects, and surveyors. The scales are not restricted to any particular industry but because some users prefer open divided scales for the purpose of taking measurements from a drawing, this type of scale and the more widely used, fully divided scale, are both recognized.

The shape and dimensions for metric scales in the British standard are shown in Fig. 18-1 and Table 18-1. The parallelogram section scales are widely used in America, and it is shown together with the three standard British sections for reference only.

The dimensions and sections shown are applicable to hand scales. Machine scales for attachment to drafting machines are dependent upon the design of the machine head, and scales of this type are not covered in the BS 1347: Part 3 Standard.

Typical examples of open and fully divided scales are shown in Fig. 18-2. The open divided scale should be divided by lines at intervals, indicating the main unit of the scale, and the first main unit is subdivided. The first graduation line has no figure. The zero of the scale is at the *second* major graduation line and marked with the symbol "0." The fully divided scale is subdivided throughout, and the zero of the scale is at the *first* major graduation line. See Tables 18-2 and 18-3.

The ISO standard recommended scales are shown in the first part of SECTION 3, and are normally shown in ratios as follows: 2:1, 1:1, 1:2, 1:5, etc.

Examples of Scale Rules

Example of a metric scale rule, from 0 to 600 mm, is shown in Fig. 18-3.

Metric Steel Measuring Rules

The British standard BS 4372 specifies requirements for steel rules up to 1 meter nominal length, and also includes folding steel rules. Some of the data presented in the above standard are the recommended lengths for steel rules as shown below, with preference given to the first-choice series, whenever possible:

(BS 4372)

First Choice		150 mm		300 mm	500 mm		1000 mm
Second Choice	100 mm		200 mm			600 mm	

[1] For information about the term "standard" as used in this book, please see page 1-2.

18-2   MEASURING TOOLS AND INSTRUMENTS

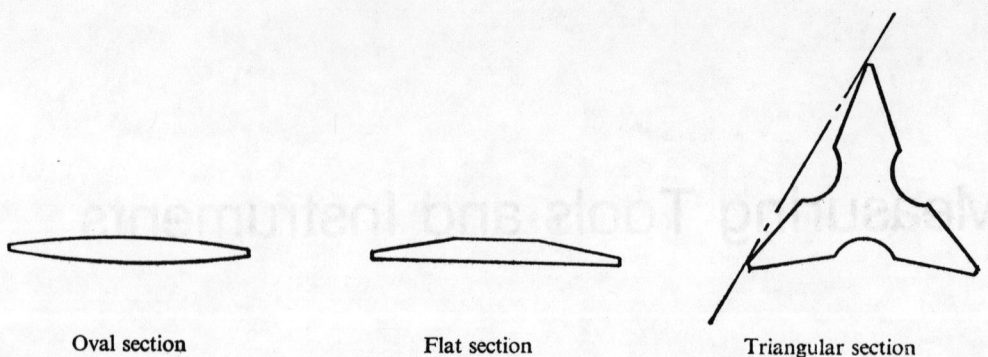

Fig. 1. Hand measuring scale sections (BS 1347/3).

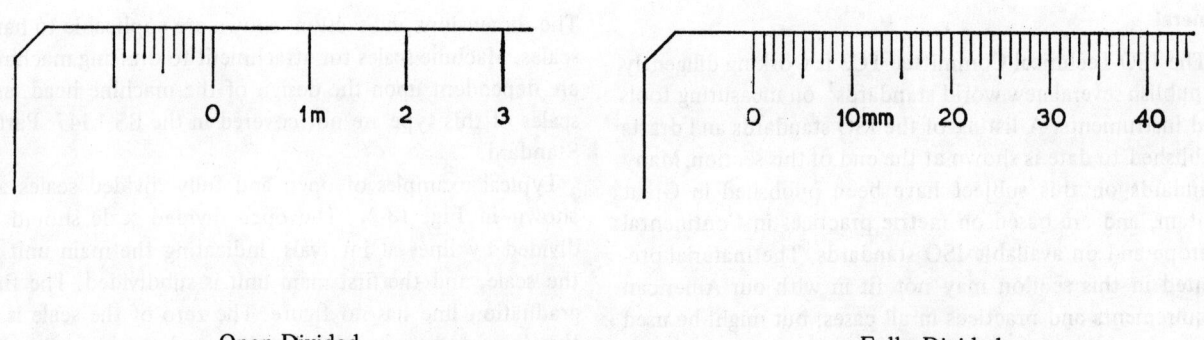

Fig. 2. Open and fully divided scales (BS 1347/3).

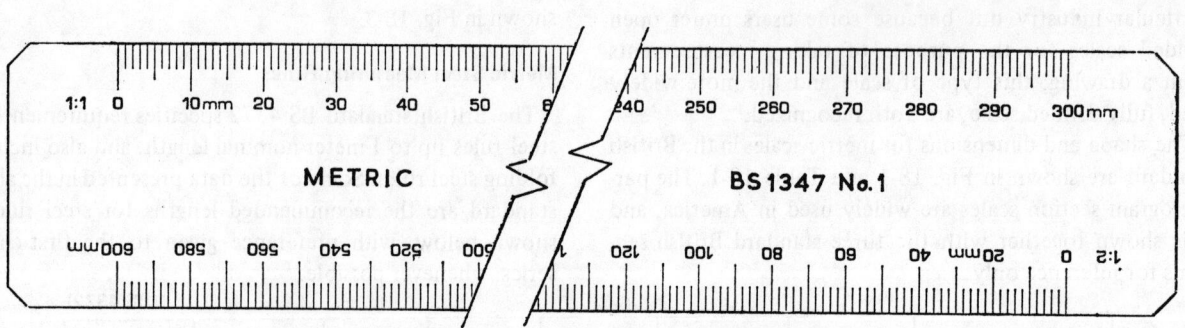

Fig. 18-3. Oval section scale rule BS 1347 No. 1.

MEASURING TOOLS AND INSTRUMENTS

Table 18-1. Principal Dimensions of Scale Rules—Flat and Oval Section* (BS 1347/3)

Nominal length	Minimum overall length	Width min.	Width max.	Thickness of mid-section min.	Thickness of mid-section max.	Minimum width of bevelled edge
150	170	32	36	3.0	3.4	10
300**	320	32	36	3.0	3.4	10
500**	525	36	40	4.0	4.5	12

*All dimensions are in millimeters.
**Preferred lengths of scale rules for drafting machines. Drafting machine scale rules are made to dimensions to suit the requirements of the individual machines.

Table 18-2. Recommended Scale Rules for General Use—Oval Section (BS 1347/3)

Scale rule reference No.	Designation	Sub-division	Figuring	Dividing
BS 1347 No. 1	1 : 1	1 mm	0, 10 mm, 20, 30 . . .	All scales fully divided
	1 : 2	2 mm	0, 20 mm, 40, 60 . . .	
	1 : 5	10 mm	0, 100 mm, 200, 300 . . .	
	1 : 10	10 mm	0, 200 mm, 400, 600 . . .	
BS 1347 No. 2	1 : 5	10 mm	0, 100 mm, 200, 300 . . .	All scales fully divided
	1 : 50	100 mm	0, 1 m, 2, 3 . . .	
	1 : 10	10 mm	0, 100 mm, 200, 300 . . .	
	1 : 100	100 mm	0, 1 m, 2, 3 . . .	
	1 : 20	20 mm	0, 200 mm, 400, 600 . . .	
	1 : 200	200 mm	0, 2 m, 4, 6 . . .	
	1 : 500	1/2 m	0, 5 m, 10, 15 . . .	
	1 : 1000	1 m	0, 10 m, 20, 30 . . .	
BS 1347 No. 3	1 : 1	1 mm	0, 10 mm, 20, 30 . . .	All scales fully divided
	1 : 100	100 mm	0, 1 m, 2, 3 . . .	
	1 : 20	20 mm	0, 200 mm, 400, 600 . . .	
	1 : 200	200 mm	0, 2 m, 4, 6 . . .	
	1 : 5	10 mm	0, 100 mm, 200, 300 . . .	
	1 : 50	100 mm	0, 1 m, 2, 3 . . .	
	1 : 1250	1 m	0, 10 m, 20, 30 . . .	
	1 : 2500	2 m	0, 20 m, 40, 60 . . .	
BS 1347 No. 4	1 : 1250	1 m	0, 10 m, 20, 30 . . .	All scales fully divided
	1 : 2500	2 m	0, 20 m, 40, 60 . . .	
	1 : 10 000	10 m	0, 200 m, 400, 600 . . .	
	1 : 10 560	10 m	0, 200 m, 400, 600 . . .	

NOTE:
Where two scales are shown bracketed together as appearing on one edge, the scale appearing first is to be divided on the outside edge.

Table 18-3. Recommended Scale Rules for Detail Work—Flat Section (BS 1347/3)

Scale rule reference No.	Designation	Sub-division	Figuring	Dividing
BS 1347 No. 10 and	1 : 1	½ mm	0, 10 mm, 20, 30 . . .	Fully
BS 1347 No. 10D	1 : 2	2 mm	0, 20 mm, 40, 60 . . .	Fully
BS 1347 No. 11 and	1 : 1	1 mm	0, 10 mm, 20, 30 . . .	Fully
BS 1347 No. 11D	1 : 2	2 mm	0, 20 mm, 40, 60 . . .	Fully
BS 1347 No. 12	1 : 1 1 : 5	1 mm 10 mm	0, 10 mm, 20, 30 . . . 0, 100 mm, 200, 300 . . .	Fully Fully
BS 1347 No. 13 and	1 : 5	10 mm	0, 100 mm, 200, 300 . . .	Fully
BS 1347 No. 13D	1 : 10	10 mm	0, 200 mm, 400, 600 . . .	Fully
BS 1347 No. 14 and	1 : 20	20 mm	0, 200 mm, 400, 600 . . .	Fully
BS 1347 No. 14D	1 : 50	100 mm	0, 1 m, 2, 3 . . .	Fully
BS 1347 No. 15 and	1 : 100	100 mm	0, 1 m, 2, 3 . . .	Fully
BS 1347 No. 15D	1 : 200	200 mm	0, 2 m, 4, 6 . . .	Fully
BS 1347 No. 16	1 : 500 1 : 1000	1 m 1 m	0, 10 m, 20, 30 . . . 0, 10 m, 20, 30 . . .	Fully Fully
BS 1347 No. 17	1 : 1250 1 : 2500	1 m 2 m	0, 10 m, 20, 30 . . . 0, 20 m, 40, 60 . . .	Fully Fully
BS 1347 No. 18	1 : 10 000 1 : 10 560	10 m 10 m	0, 200 m, 400, 600 . . . 0, 200 m, 400, 600 . . .	Fully Fully
BS 1347 No. 30	1 : 5 1 : 50	10 mm 100 mm	0, 100 mm, 200, 300 . . . 0, 1 m, 2 . . .	Fully Open
BS 1347 No. 31	1 : 10 1 : 100	10 mm 100 mm	0.9, 0.8, . . . 0, 1 m, 2 . . . 0, 1 m, 2 . . .	Open Open
BS 1347 No. 32	1 : 20 1 : 200	10 mm 100 mm	0.8, 0.6, . . . 0, 1 m, 2 . . . 0, 5 m, 10 . . .	Open Open

NOTE:
Scale rules shown with a reference number suffixed D are recommended for drafting machines.

MEASURING TOOLS AND INSTRUMENTS

Table 18-4. Suggested Sections for Steel Rules (BS 4372)

Nominal rule length	Material section	
	narrow	standard
	mm	mm
100 mm 150 mm 200 mm	13 × 0.4	20 × 0.75
300 mm	13 × 0.4	28 × 1.0
500 mm 600 mm 1 m	20 × 0.5*	30 × 1.5

*This section is not normally available in the 1 m length.

The British standard specifies the various recommended width and thickness combinations as shown in Table 18-4.

Graduation Lines and Figuring

Graduation Lines are of uniform width, square to the edges. They are extended from the rule edge for a distance of at least 1.5 mm and must differ in length so as to clearly indicate the subdivisions of the basic measurement unit. The width of these lines shall not be less than 0.1 mm nor more than 0.2 mm, and shall not exceed half the width of the scale interval.

Figuring, which may refer to either one or both scales on a rule face, shall indicate:
(1) The distance, in millimeters of each 10 mm graduation line, or
(2) The distance in centimeters of each 1 cm graduation line from a square end of the rule.

When rules are *graduated* in cm, then the first figured graduation line should be marked "cm."

NOTE: Recommended SI units* are the meter and the millimeter and it is officially recommended that preference be given to figuring in millimeters wherever possible.

The degree of scale interval shall be marked within 25 mm of the commencement of, and be aligned with, such scale intervals.

Marking of Graduation Lines and Figuring. Graduation lines and figuring of rules shall be effected by metal removal; this shall have a minimum depth of 0.02 mm for carbon steel and 0.01 mm for rust-resistant steel rules. The graduation lines and figuring of rules shall provide a durable contrast on the surface.

---

*Me*tre*, millime*tre*, and centime*tre*, etc., spelling is accepted internationally.

Accuracy

When referred to the standard reference temperature of 20°C, the tolerances listed below shall apply:
(1) Rule edges shall be straight to within 0.1 mm on any length up to 300 mm.
(2) Rule edges shall be parallel to within 0.1 mm on any length up to 300 mm.
(3) Flat ends shall be square to the rule edges to within 0.05 mm over the width of the rule.
(4) Graduation lines shall be accurate to within the tolerances specified in Table 18-5. These tolerances apply to the distances between the center lines of the graduation lines.

GAGE BLOCKS (ISO DIS 3650)

The ISO Draft International Standard 3650, on which this material has been based, specifies requirements for gage blocks of rectangular form in metric sizes up to 1000 mm.

Provision is made for four grades of accuracy. Those intended for normal purposes in inspection, tool-making, and manufacturing departments are designated as Grades 0, 1, and 2.

The calibration grades and Grade 00 are intended as standard gage lengths when measuring other grades of gage blocks by comparison. Nomenclature and definition of gage length is shown in Fig. 18-4 and 18-5. The standard reference temperature is 20°C, as specified in ISO 1 Standard.

General Dimensions of Gage Blocks

The section of each gage block shall be rectangular. The face width shall be $9^{-0.05}_{-0.2}$ mm and the face length shall be $30^{+0}_{-0.3}$ mm for sizes up to and including 10 mm and $35^{+0}_{-0.3}$ mm for larger sizes.

The dimensions of the cross-section and their tolerances are given in Table 18-6.

Accuracy Requirements

The gage blocks shall conform to the requirements for accuracy appropriate to the grade specified, as given below. It should be noted, however, that the requirements for accuracy apply to an area of the measuring face omitting a border zone of 0.8 mm, as measured from the side faces. In this excluded border zone the surface shall not lie above the plane of the measuring face.

Flatness and Parallelism of Measuring Faces

The two measuring faces of any gage block shall be flat and parallel to each other, to within the tolerances specified in Table 18-7 and Table 18-8.

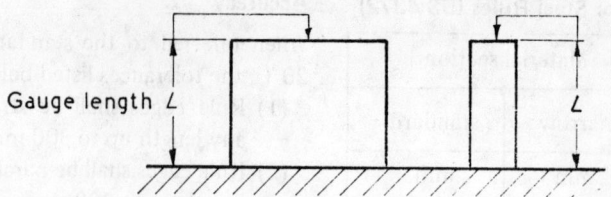

Fig. 18-4. Gage length L, for gage blocks (BS 4311).

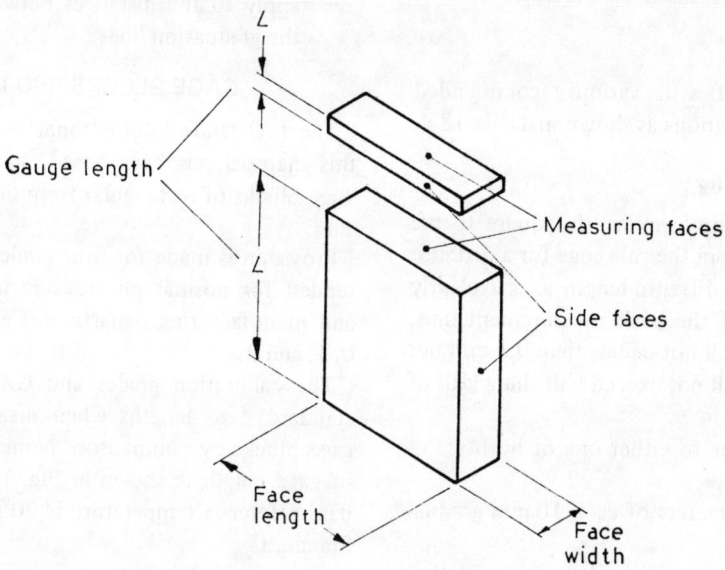

Fig. 5. Nomenclature for gage blocks (BS 4311).

Flatness Tolerance

Gage blocks longer than 2.5 mm—The deviation from flatness of each measuring face of a gage block with length greater than 2.5 mm shall not exceed the value given in Table 18-7 for the appropriate grade. This requirement shall apply whether the gage block is wrung to a rigid plane surface or is in the unwrung state.

Gage blocks 2.5 mm in length or smaller—Gage blocks of 2.5 mm length or smaller shall be tested when wrung to the rigid plane surface of an auxiliary body with a thickness of at least 11 mm. With the gage block in this condition, the deviation from flatness of each measuring face shall not exceed the value given in Table 18-7 for the appropriate grade.

With the gage block in the unwrung state, each measuring face shall be flat to within 0.004 mm.

Thin gage blocks of all grades shall wring readily.

Squareness

The gage block side faces shall be square to the measuring faces within the tolerances stated in Table 18-9.

Gage Length

Gage length of calibration grades and Grade 00 gage blocks. The gage length, L, of a calibration grade or Grade 00 gage block, referred to the standard temperature of 20° C, normally shall be measured by interference of light methods and include the thickness of one wringing film and shall not depart from the length on it by more than the amount specified in Table 18-8.

Gage length of Grades 0, 1, and 2 gage blocks. The gage length of a Grade 0, 1, or 2 gage block, referred to the standard temperature of 20°C, shall be measured by comparison with a standard of similar material and of known

MEASURING TOOLS AND INSTRUMENTS

Table 18-5. Accuracy of Graduation (BS 4372)

Rule length	Departure from nominal		
	up to and inc. 300 mm	over 300 mm up to and inc. 500 mm	over 500 mm up to and inc. 1 m
	mm	mm	mm
Distance between any two graduation lines on a single scale	0.1	0.2	0.25
Distance between any two adjacent graduation lines	0.05	0.05	0.05
Position of the 10 mm graduation line from its flat end datum	0.08	0.08	0.08

length and shall not depart from the length marked on it by more than the amount specified for the appropriate grade in Table 18-8.

Recommended Sets of Gage Blocks

The recommendations for sets of metric gage blocks, given below, are offered only for the guidance of prospective users. They are based on information concerning the sets most in demand at present, but it should be noted that other sets are also available from the manufacturers.

The economic advantages offered by buying the small sets, e.g., M 32, should be carefully compared with the technical benefits of owning the more comprehensive but more expensive larger sets, e.g., M 88 and M 112. This is particularly important as the purchase of additional sets may then be unnecessary when one or more large sets are already available.

The 2-mm-based size series is recommended as it is less likely to suffer flatness deterioration than similar size series based on 1-mm-thin slips.

Protector Gage Blocks

Grades 0, 1, and 2 sets of gages may include a pair of extra gages, usually of 2-mm length, known as "protector gages." These gages are of the same general dimensions and accuracy as the other gages in the set and are customarily marked with the letter 'P' on one measuring face. Protector gages of steel, or a wear-resisting material such as tungsten carbide, may be obtained as separate accessories.

Typical Sets, 1-mm Base (BS 4311)

Set Number	Size or series (mm)	Increment (mm)	Number of pieces
Set M 32/1	1.005	...	1
	1.01 to 1.09	0.01	9
	1.1 to 1.9	0.1	9
	1 to 9	1.0	19
	10, 20, 30	10.0	3
	60	...	1
			Total 32 pieces
Set M 41/1	1.001 to 1.009	0.001	9
	1.01 to 1.09	0.01	9
	1.1 to 1.9	0.1	9
	1 to 9	1.0	9
	10, 20, 30	10.0	3
	60	...	1
	100	...	1
			Total 41 pieces
Set M 47/1	1.005	0.005	1
	1.01 to 1.09	0.01	9
	1.1 to 1.9	0.1	9
	1 to 24	1.0	24
	25 to 100	25	4
			Total 47 pieces

```
Set M 88/1        1.0005   ...     1
            1.001 to 1.009   0.001   9
            1.01  to 1.49    0.01   49
            0.5   to 9.5     0.5    19
            10    to 100     10     10
                                   ────
                              Total 88 pieces

Set M 112/1       1.0005   ...     1
            1.001 to 1.009   0.001   9
            1.01  to 1.49    0.01   49
            0.5   to 24.5    0.5    49
            25    to 100     25      4
                                   ────
                              Total 112 pieces
```

NOTE: *Set* M 46/1 As set M 41/1 plus 5 slips: 40, 50, 70, 80, and 90 mm. The economic value of this set is dubious, particularly as the additional slips are expensive.

Table 18-6. Tolerances for General Gage Block Dimensions (ISO 3650)

Values in millimetres

Cross-section	Nominal length over	Nominal length up to and including	a	b
b / a	—	10,1	30 $^{\ 0}_{-0,3}$	9 $^{-0,05}_{-0,2}$
	10,1	1 000	35 $^{\ 0}_{-0,3}$	

Table 18-7. Flatness Tolerances (ISO 3650)

Deviations in micrometres

Nominal length mm over	up to and including	Maximum permitted deviation from flatness Grade 00	0	1	2
—	150	0,05	0,10	0,15	0,25
150	500	0,10	0,15	0,18	0,25
500	1 000	0,15	0,18	0,20	0,25

Typical Sets, 2-mm Base (BS 4311)

```
Set        Size or series   Increment   Number of
Numbers       (mm)            (mm)      pieces

Set M 33/2      2.005        ...           1
         2.01  to 2.09       0.01          9
         2.1   to 2.9        0.1           9
         1     to 9          1.0           9
         10, 20, 30          10.0          3
         60                  ...           1
         100                 ...           1
                                          ───
                              Total 33 pieces

Set M 46/2  2.001 to 2.009   0.001         9
         2.01  to 2.09       0.01          9
         2.1   to 2.9        0.1           9
         1     to 9          1.0           9
         10    to 100        10.0         10
                                          ───
                              Total 46 pieces

Set M 88/1       1.0005      ...           1
         1.001 to 1.009      0.001         9
         1.01  to 1.49       0.01         49
         0.5   to 9.5        0.5          19
         10    to 100        10           10
                                          ───
                              Total 88 pieces

Set M 112/1      1.0005      ...           1
         1.001 to 1.009      0.001         9
         1.01  to 1.49       0.01         49
         0.5   to 24.5       0.5          49
         25    to 100        25            4
                                          ───
                              Total 112 pieces
```

STANDARD GAGES FOR ISO LIMITS AND FITS

A series of standard gages to be used in connection with the ISO system of Limits and Fits is available throughout Europe. (See Section 6 for a description of the ISO system of Limits and Fits.)

Fixed Limit Gages for Holes

Standard limit gages for holes are made to control bores held to an ISO tolerance of H7 (+0 —a constant, see Table 6-1), and coarser, tolerances such as H8, H9, H10, etc. It could, therefore, be to advantage to specify the limits for drilled holes with ISO fundamental deviation of H, since tooling and checking gages are based on these limits in countries already on the metric system. Plain cylindrical plug gages with go and no-go ends made from high quality

MEASURING TOOLS AND INSTRUMENTS

Table 18-8. Tolerances on Flatness, Parallelism, and Length of Gage Blocks (ISO 3650)

Deviations in micrometres

1	2	3	4	5	6	7	8	9	10
Range of nominal sizes mm		Grade 00		Grade 0		Grade 1		Grade 2	
Over	Up to and including	Deviation from nominal size at any point	Variation in length	Deviation from nominal size at any point	Variation in length	Deviation from nominal size at any point	Variation in length	Deviation from nominal size at any point	Variation in length
–	10	± 0,06	0,05	± 0,12	0,10	± 0,20	0,16	± 0,45	0,30
10	25	± 0,07	0,05	± 0,14	0,10	± 0,30	0,16	± 0,60	0,30
25	50	± 0,10	0,06	± 0,20	0,10	± 0,40	0,18	± 0,80	0,30
50	75	± 0,12	0,06	± 0,25	0,12	± 0,50	0,18	± 1,00	0,35
75	100	± 0,14	0,07	± 0,30	0,12	± 0,60	0,20	± 1,20	0,35
100	150	± 0,20	0,08	± 0,40	0,14	± 0,80	0,20	± 1,60	0,40
150	200	± 0,25	0,09	± 0,50	0,16	± 1,00	0,25	± 2,00	0,40
200	250	± 0,30	0,10	± 0,60	0,16	± 1,20	0,25	± 2,40	0,45
250	300	± 0,35	0,10	± 0,70	0,18	± 1,40	0,25	± 2,80	0,50
300	400	± 0,45	0,12	± 0,90	0,20	± 1,80	0,30	± 3,60	0,50
400	500	± 0,50	0,14	± 1,10	0,25	± 2,20	0,35	± 4,40	0,60
500	600	± 0,60	0,16	± 1,30	0,25	± 2,60	0,40	± 5,00	0,70
600	700	± 0,70	0,18	± 1,50	0,30	± 3,00	0,45	± 6,00	0,70
700	800	± 0,80	0,20	± 1,70	0,30	± 3,40	0,50	± 6,50	0,80
800	900	± 0,90	0,20	± 1,90	0,35	± 3,80	0,50	± 7,50	0,90
900	1 000	± 1,00	0,25	± 2,00	0,40	± 4,20	0,60	± 8,00	1,00

steels, and also with sintered metal ends, are available in nominal diameters shown in Table 18-10.

Slot Limit Gages

Flat end Go No-Go gages to check the widths of slots made to the ISO bilateral tolerance J9 are available for tolerances with IT Grade 6, and coarser, qualities. Gages for checking the nominal widths in Table 18-10 are available in Europe; also available are gages to control the unilateral tolerance H7, in slots.

Fixed Snap Gages for Shafts

Snap gages with fixed ends, of the Go-No-Go type, are marketed with one or two gaps; they are generally produced to check shafts to the ISO tolerance H7 and coarser, IT qualities. The standard nominal diamters these gages are made to control are also shown in Table 18-10.

The gages mentioned above, are only a few of the many types produced in Europe, to control sizes made to ISO Standard limits.

Table 18-9. Squareness Tolerances (ISO 3650)

Nominal length mm		Deviation from squareness	
over	up to and including	µm	mm
10	25	50	0,05
25	60	70	0,07
60	150	100	0,10
150	400	140	0,14
400	1 000	180	0,18

The decimal marker used in the ISO standard and in this table is a comma ",".

MEASURING TOOLS AND INSTRUMENTS

Table 18-10. Standard Gages for Checking of ISO Limits

STANDARD	FIGURE	NAME	DIMENSIONS (IN MILLIMETERS)
ISO U.S.A. ANSI JAPAN JIS GERMANY DIN 2245/306 FRANCE NF U.K. BS ITALY UNI AUSTRAL AS	ISO Tolerance H7	(E) Hole Limit Gage (G) GRENZLEHRDORNE (F) (I)	NOMINAL DIAMETERS 1 1.2 1.5 1.8 2 2.2 2.5 2.8 3 3.5 4 4.5 5 6 7 8 9 10 11 12 13 14 15 16 17 18 19 20 21 22 23 24 25 26 27 28 30 32 33 34 35 36 37 38 40 42 44 45 46 47 48 50 52 55 58 60 62 65 68 70 72 75 78 80 82 85 88 90 92 95 98 100
ISO U.S.A. ANSI JAPAN JIS GERMANY DIN FARNCE NF U.K. BS ITALY UNI AUSTRAL AS	ISO Tolerances H7, H8 and coarser	(E) Hole Limit Gage- Sintered Metal Ends (G) HARTMETALL - GRENZLEHRDORNE (F) (I)	NOMINAL DIAMETERS 1 1.5 2 3 5 6 8 10 12 14 16 18 20 22 24 26 28 30 40 50 60 65 70 80 90 100
ISO U.S.A. ANSI JAPAN JIS GERMANY DIN FRANCE NF U.K. BS ITALY UNI AUSTRAL AS	ISO Tolerance J9 Available with IT Grade 6 and coarser	(E) Slot Limit Gage (G) NUTEN-GRENZ- FLACHLEHREN (F) (I)	NOMINAL THICKNESSES 2 3 4 5 6 8 10 12 14 16 18 20 22 25 28 32 35 40 45 50 56 63 70 80 90 100
ISO U.S.A. ANSI JAPAN JIS GERMANY DIN FRANCE NF U.K. BS ITALY UNI AUSTRAL AS	ISO Tolerance h6 IT 7 and coarser Standard	(E) GO, NO-GO Snap Gages (G) GRENZRACHENLEHREN (F) (I)	NOMINAL WIDTHS 3 3.5 4 4.5 5 6 7 8 9 10 11 12 13 14 15 16 17 18 19 20 21 22 23 24 25 26 27 28 30 32 33 34 35 36 37 38 40 42 44 45 46 47 48 50 52 55 58 60 62 65 68 70 72 75 78 80 82 85 88 90 92 95 98 100 110 125 140 155 170 185 200 220 240 265 290 315 365 390 415 445 475 500
ISO U.S.A. ANSI JAPAN JIS GERMANY DIN 2235 FRANCE NF U.K. BS ITALY UNI AUSTRAL AS	ISO Tolerance h6 IT 7 and coarser Standard	(E) GO, NO-GO Snap Gages (G) GRENZRACHENLEHREN AUS FLACHSTAHL (F) (I)	NOMINAL WIDTHS 2 2.2 2.5 2.8 3 3.5 4 4.5 5 6 7 8 9 10 11 12 13 14 15 16 17 18 19 20 21 22 23 24 25 26 27 28 30 32 33 34 35 36 37 38 40 42 44 45 46 47 48 50 52 55 58 60 62 65 68 70 72 75 78 80 82 85 88 90 92 95 98 100 105 110 115 120 125 130 135 140 145 150 155 160

NOTE: Figures and nominal sizes reproduced with permission from Hahn & Kolb, Stuttgard, W. Germany.

MEASURING TOOLS AND INSTRUMENTS

RELATED INTERNATIONAL (TC3) AND NATIONAL STANDARDS

ISO 1-1975	Standard reference temperature for industrial length measurements
ISO/R 286-1962	ISO system for limits and fits—Part 1: General, tolerances and deviations
ISO 370-1975	Conversion of toleranced dimensions from inches into millimeters, and vice versa
ISO/R 463-1965	Dial Gages Reading in 0.01 mm, 0.001 in. and 0.0001 in.
ISO/R 1047-1973	Architectural and building drawings, presentation of drawings, scales
ISO 1119-1975	Series of conical tapers and taper angles
ISO 1829-1975	Selection of tolerance zones for general purposes
ISO/R 1938-1971	ISO system of limits and fits—Part 2: Inspection of plain workpieces
ISO 1947-1973	System of cone tolerances for conical workpieces from $C = 1 : 3$ to $1 : 500$ and lengths from 6 to 630 mm
ISO 2538-1974	Limits and fits—Series of angles and slopes on wedges and prisms
ISO 2768-1973	Permissible machining variations in dimensions without tolerance indication
ISO 3599-1976	Vernier calipers reading to 0.1 and 0.05 mm
ISO/DIS 3611	Micrometer calipers for external measurement
ISO/DIS 3650	Gage blocks
ISO/DIS 3670	Blanks for plug and ring gages taper lock and trilock
ISO/DPR 4910	Vocabulary for terms used for cones
ISO/DPR 5166	System of conical fits

Japan

JIS B 7502-1963	External micrometers
JIS B 7504-1963	Micrometer heads
JIS B 7505-1968	Bourdon tube pressure gages
JIS B 7506-1961	Block gages
JIS B 7507-1963	Vernier calipers
JIS B 7508-1963	Tubular internal micrometers
JIS B 7516-1959	Steel ruler
JIS B 7519-1961	Microindicators
JIS B 7523-1961	Sine bars
JIS B 7524-1962	Feeler gages
JIS B 7526-1962	Squares
JIS B 7533-1966	Lever type dial test indicators
JIS B 7535-1967	Flow type air gages

Germany

DIN 13	Sheet 16: ISO metric threads; gages for bolt and nut threads, gaging system and denominations (5)
	Sheet 17: ISO metric threads; gages for bolt and nut threads, gage dimensions and design features (9)
DIN 259	Sheet 4: Whitworth pipe threads; parallel internal and external threads, gaging of external threads, gage dimensions
	Sheet 5: -; Parallel internal and external threads, gaging of internal threads, gage dimensions
DIN 861	Sheet 1: Gage blocks measuring jaws, slip gage holders; slip gages, definitions, finish, permissible variations
	Sheet 2: -; Measuring jaws, gage block holders, definitions, finish, permissible variations
DIN 862	Vernier calipers and depth gages; definitions, requirements, permissible variations, testing
DIN 863	External micrometers; definitions, requirements, permissible variations, testing
DIN 878	Dial gages (5)
DIN 879	Dial indicator for linear measurement (5)
DIN 2231	Limit gap gages and forged gage bodies for nominal dimension range above 3, up to 100 mm
DIN 2232	"Go" gap gages with forged gage body for nominal dimension range above 3, up to 100 mm
DIN 2233	"Go" gap gages with forged gage body for nominal dimension range above 3, up to 100 mm
DIN 2243	Measuring instruments; handles for radial-end measuring rods
DIN 2245	Sheet 1: Limit plug gages for bores from 1 to 30 mm, nominal diameter
	Sheet 2: Limit plug gages for bores over 30 mm and up to 50 mm, nominal diameter
DIN 2250	"Go" ring gages and setting ring gages
DIN 2253	Sheet 1: Measuring instruments; check gages for gap gages from 1 to 10 mm, nominal dimension

	Sheet 2: -; Check gages for gap gages above 10, and up to 315 mm, nominal dimension
DIN 2280	Measuring instruments; thread limit plug gages for threads with metric profile from 1 to 30 mm, nominal diameter of thread
DIN 2281	Sheet 1: -; Thread "Go" plug gages and thread mating plug gages for threads with metric profile from 1 to 30 mm, nominal diameter of thread
DIN 3970	Sheet 1: Master gears for checking spur gears; gear blank and tooth system
DIN 7150	Sheet 1: ISO system of limits and fits for sizes from 1 to 500 mm, introduction
DIN 7151	ISO standard tolerances for sizes from 1 to 500 mm
DIN 7152	Formation of tolerance zones from ISO fundamental deviations for nominal sizes from 1 to 500 mm
DIN 7157	Recommended selection of fits; tolerance zones, allowances, fit tolerances
DIN 7160	ISO allowances for external dimensions (shafts) for nominal dimensions from 1 to 500 mm
DIN 7161	ISO allowances for internal dimensions (holes) for nominal dimensions from 1 to 500 mm
DIN 7162	Plain workshop and inspection gages; manufacturing tolerances and permissible wear
DIN 7163	Workshop gap gages and check gages for ISO fit sizes from 1 to 500 mm, nominal dimensions; gage dimensions, and manufacturing tolerances
DIN 7164	Workshop plug gages and spherical end measuring rods for ISO fit dimensions from 1 to 500 mm, nominal dimension; gage dimensions, and manufacturing tolerances
DIN 7168	Sheet 1: Permissible variations for dimensions without tolerance indication; variations on length measurements, radii of curvature and chamfers, angular dimensions
DIN 7172	Sheet 1: ISO tolerances and ISO variations for linear dimensions above 500, up to 3150 mm; standard tolerances
DIN 7182	Sheet 2: Classes of fit; clearance fits, terminology and notation
DIN 7186	Sheet 1: Statistical tolerancing; distribution of actual sizes in the tolerance zone
DIN 7962	Sheet 2: Recesses; recess penetration gages (Kreuzschlitze;Tiefenlehren)
DIN 40401	Sheet 1: Electro screw thread; "go" and "not-go" thread ring gages Sheet 2: -; thread plug gages
DIN 40437	Sheet 1: Steel conduit thread; thread gages, "go" and "not-go" thread ring gages

United Kingdom

BS 817	Surface plates and tables
BS 818	Cast iron straightedges
BS 887	Vernier calipers
BS 870	External micrometers
BS 906: Part 1	Engineers' parallels
BS 957: Part 2	Feeler gages
BS 958	Spirit levels for use in precision engineering
BS 959	Internal micrometers
BS 1347: Part 3	Architects', engineers', and surveyors' scales
BS 1643	Vernier height gages
BS 1734	Micrometer heads
BS 1790	Length bars and their accessories
BS 2795: Part 1	Dial test indicators (level type) for linear measurement
BS 3064	Sine bars and sine tables
BS 4311	Gage blocks
BS 4372	Engineers' steel measuring rules
BS Draft	Inspection of plain products (similar to ISO 1938)

19 Conversion Factors and Tables

Introduction

The customary (inch), SI (metric)* and other conversion factors and tables are intended to help engineers relate physical quantities expressed in other units to the SI system and vice versa. A description of the proper use of the SI system can be found in Section 2. Until all our standards, technical specifications, and books are expressed in SI units, this section will provide useful information to all users of the above-mentioned reference materials.

The conversion factors and tables will serve the three following purposes:

1. Provide a complete set of conversion factors between the customary (inch) and the SI (metric) measuring systems or vice versa.
2. Relate the "old" metric to the corresponding SI (metric) units of measure.
3. Identify "non-metric" units of measure used in countries outside the United States and provide their ratio relative to the proper SI (metric) units.

The conversion factors and two examples here are copied from the ASTM E380-76 (ANSI Z210.1) Standard for Metric Practice with permission from the American Society for Testing and Materials, 1916 Race Street, Philadelphia, Pennsylvania 19103. The above referenced standard is available from ASTM or ANSI.

A more complete set of conversion tables than shown here is published in the book *Conversion Tables for SI Metrication* available from Industrial Press, Inc., 200 Madi- Avenue, New York, N.Y. 10016

---

*NOTE: The British spelling of METRE and LITRE is used in parts of Section 19.

USE OF CONVERSION FACTORS

The conversion factors are listed in alphabetical order and given with six places accuracy or less. The conversion factors marked with an asterisk (*) are exact.

The computer E format is used to indicate the power of 10 by which the number must be multiplied to obtain the correct value. For an exact definition of the E format see Section 2, Table 2-2.

Example—To find proper decimal place for conversion factor:

to convert from   to   multiple by
inch   meter   2.540 000* E-02

$2.540\ 000\ \text{E-02} = 2.540\ 000 \times 10^{-2} = 0.0254$
thus 1 inch = 0.0254 m

Example—To find conversion factor lb·ft/s to kg·m/s:

first convert   1 lb to 0.453 592 4 kg
and   1 ft to 0.3048 m
then substitute:
     (0.453 592 4 kg)·(0.3048 m)/s = 0.138 255 kg·m/s
thus the factor is 1.382 55 E-01

Example—To find conversion factor of oz·in² to kg·m²:

first convert   1 oz to 0.028 349 42 kg
and   1 in² to 0.000 645 16 m²
then substitute:
     (0.028 349 52 kg)·(0.000 645 16 m²)
                    = 0.000 018 289 98 kg·m²
thus the factor is 1.828 998 E-05

Conversion Factors (*Courtesy of the American Society for Testing and Materials*)

(Symbols of SI units given in parentheses)

To convert from	to	Multiply by
abampere	ampere (A)	1.000 000*E+01
abcoulomb	coulomb (C)	1.000 000*E+01
abfarad	farad (F)	1.000 000*E+09
abhenry	henry (H)	1.000 000*E−09
abmho	siemens (S)	1.000 000*E+09
abohm	ohm (Ω)	1.000 000*E−09
abvolt	volt (V)	1.000 000*E−08
acre foot (U.S. survey)[1]	metre³ (m³)	1.233 489 E+03
acre (U.S. survey)[1]	metre² (m²)	4.046 873 E+03
ampere hour	coulomb (C)	3.600 000*E+03
are	metre² (m²)	1.000 000*E+02
angstrom	metre (m)	1.000 000*E−10
astronomical unit[2]	metre (m)	1.495 979 E+11
atmosphere (standard)	pascal (Pa)	1.013 250*E+05
atmosphere (technical = 1 kgf/cm²)	pascal (Pa)	9.806 650*E+04
bar	pascal (Pa)	1.000 000*E+05
barn	metre² (m²)	1.000 000*E−28
barrel (for petroleum, 42 gal)	metre³ (m³)	1.589 873 E−01
board foot	metre³ (m³)	2.359 737 E−03
British thermal unit (International Table)[3]	joule (J)	1.055 056 E+03
British thermal unit (mean)	joule (J)	1.055 87 E+03
British thermal unit (thermochemical)	joule (J)	1.054 350 E+03
British thermal unit (39°F)	joule (J)	1.059 67 E+03
British thermal unit (59°F)	joule (J)	1.054 80 E+03
British thermal unit (60°F)	joule (J)	1.054 68 E+03
Btu (International Table)·ft/h·ft²·°F (*k*, thermal conductivity)	watt per metre kelvin (W/m·K)	1.730 735 E+00
Btu (thermochemical)·ft/h·ft²·°F (*k*, thermal conductivity)	watt per metre kelvin (W/m·K)	1.729 577 E+00
Btu (International Table)·in/h·ft²·°F (*k*, thermal conductivity)	watt per metre kelvin (W/m·K)	1.442 279 E−01
Btu (thermochemical)·in/h·ft²·°F (*k*, thermal conductivity)	watt per metre kelvin (W/m·K)	1.441 314 E−01
Btu (International Table)·in/s·ft²·°F (*k*, thermal conductivity)	watt per metre kelvin (W/m·K)	5.192 204 E+02
Btu (thermochemical)·in/s·ft²·°F (*k*, thermal conductivity)	watt per metre kelvin (W/m·K)	5.188 732 E+02
Btu (International Table)/h	watt (W)	2.930 711 E−01
Btu (thermochemical)/h	watt (W)	2.928 751 E−01
Btu (thermochemical)/min	watt (W)	1.757 250 E+01
Btu (thermochemical)/s	watt (W)	1.054 350 E+03
Btu (International Table)/ft²	joule per metre² (J/m²)	1.135 653 E+04
Btu (thermochemical)/ft²	joule per metre² (J/m²)	1.134 893 E+04
Btu (thermochemical)/ft²·h	watt per metre² (W/m²)	3.152 481 E+00 †
Btu (thermochemical)/ft²·min	watt per metre² (W/m²)	1.891 489 E+02
Btu (thermochemical)/ft²·s	watt per metre² (W/m²)	1.134 893 E+04
Btu (thermochemical)/in²·s	watt per metre² (W/m²)	1.634 246 E+06
Btu (International Table)/h·ft²·°F (*C*, thermal conductance)	watt per metre² kelvin (W/m²·K)	5.678 263 E+00

---

[1] Since 1893 the U.S. basis of length measurement has been derived from metric standards. In 1959 a small refinement was made in the definition of the yard to resolve discrepancies both in this country and abroad, which changed its length from 3600/3937 m to 0.9144 m exactly. This resulted in the new value being shorter by two parts in a million.

At the same time it was decided that any data in feet derived from and published as a result of geodetic surveys within the U.S. would remain with the old standard (1 ft = 1200/3937 m) until further decision. This foot is named the U.S. survey foot.

As a result all U.S. land measurements in U.S. customary units will relate to the metre by the old standard. All the conversion factors in these tables for units referenced to this footnote are based on the U.S. survey foot, rather than the international foot.

Conversion factors for the land measures given below may be determined from the following relationships:

$$1 \text{ league} = 3 \text{ miles (exactly)}$$
$$1 \text{ rod} = 16\tfrac{1}{2} \text{ feet (exactly)}$$
$$1 \text{ section} = 1 \text{ square mile (exactly)}$$
$$1 \text{ township} = 36 \text{ square miles (exactly)}$$

[2] This value conflicts with the value printed in NBS 330 (**17**). The value requires updating in NBS 330.

[3] This value was adopted in 1956. Some of the older International Tables use the value 1.055 04 E+03. The exact conversion factor is 1.055 055 852 62*E+03.

†Editorially corrected.

METRIC CONVERSION FACTORS AND TABLES

To convert from	to	Multiply by
Btu (thermochemical)/h·ft²·°F (C, thermal conductance)	watt per metre² kelvin (W/m²·K)	5.674 466 E+00
Btu (International Table)/s·ft²·°F	watt per metre² kelvin (W/m²·K)	2.044 175 E+04
Btu (thermochemical)/s·ft²·°F	watt per metre² kelvin (W/m²·K)	2.042 808 E+04
Btu (International Table)/lb	joule per kilogram (J/kg)	2.326 000*E+03
Btu (thermochemical)/lb	joule per kilogram (J/kg)	2.324 444 E+03
Btu (International Table)/lb·°F (c, heat capacity)	joule per kilogram kelvin (J/kg·K)	4.186 800*E+03
Btu (thermochemical)/lb·°F (c, heat capacity)	joule per kilogram kelvin (J/kg·K)	4.184 000 E+03
bushel (U.S.)	metre³ (m³)	3.523 907 E−02
caliber (inch)	metre (m)	2.540 000*E−02
calorie (International Table)	joule (J)	4.186 800*E+00
calorie (mean)	joule (J)	4.190 02 E+00
calorie (thermochemical)	joule (J)	4.184 000*E+00
calorie (15°C)	joule (J)	4.185 80 E+00
calorie (20°C)	joule (J)	4.181 90 E+00
calorie (kilogram, International Table)	joule (J)	4.186 800*E+03
calorie (kilogram, mean)	joule (J)	4.190 02 E+03
calorie (kilogram, thermochemical)	joule (J)	4.184 000*E+03
cal (thermochemical)/cm²	joule per metre² (J/m²)	4.184 000*E+04
cal (International Table)/g	joule per kilogram (J/kg)	4.186 800*E+03
cal (thermochemical)/g	joule per kilogram (J/kg)	4.184 000*E+03
cal (International Table)/g·°C	joule per kilogram kelvin (J/kg·K)	4.186 800*E+03
cal (thermochemical)/g·°C	joule per kilogram kelvin (J/kg·K)	4.184 000*E+03
cal (thermochemical)/min	watt (W)	6.973 333 E−02
cal (thermochemical)/s	watt (W)	4.184 000*E+00
cal (thermochemical)/cm²·min	watt per metre² (W/m²)	6.973 333 E+02
cal (thermochemical)/cm²·s	watt per metre² (W/m²)	4.184 000*E+04
cal (thermochemical)/cm·s·°C	watt per metre kelvin (W/m·K)	4.184 000*E+02
carat (metric)	kilogram (kg)	2.000 000*E−04
centimetre of mercury (0°C)	pascal (Pa)	1.333 22 E+03
centimetre of water (4°C)	pascal (Pa)	9.806 38 E+01
centipoise	pascal second (Pa·s)	1.000 000*E−03
centistokes	metre² per second (m²/s)	1.000 000*E−06
circular mil	metre² (m²)	5.067 075 E−10
clo	kelvin metre² per watt (K·m²/W)	2.003 712 E−01
cup	metre³ (m³)	2.365 882 E−04
curie	becquerel (Bq)	3.700 000*E+10
day (mean solar)	second (s)	8.640 000 E+04
day (sidereal)	second (s)	8.616 409 E+04
degree (angle)	radian (rad)	1.745 329 E−02
degree Celsius	kelvin (K)	$t_K = t_{°C} + 273.15$
degree centigrade	[see 3.4.2]	
degree Fahrenheit	degree Celsius	$t_{°C} = (t_{°F} - 32)/1.8$
degree Fahrenheit	kelvin (K)	$t_K = (t_{°F} + 459.67)/1.8$
degree Rankine	kelvin (K)	$t_K = t_{°R}/1.8$
°F·h·ft²/Btu (International Table) (R, thermal resistance)	kelvin metre² per watt (K·m²/W)	1.761 102 E−01
°F·h·ft²/Btu (thermochemical) (R, thermal resistance)	kelvin metre² per watt (K·m²/W)	1.762 280 E−01
denier	kilogram per metre (kg/m)	1.111 111 E−07
dyne	newton (N)	1.000 000*E−05
dyne·cm	newton metre (N·m)	1.000 000*E−07
dyne/cm²	pascal (Pa)	1.000 000*E−01
electronvolt	joule (J)	1.602 19 E−19
EMU of capacitance	farad (F)	1.000 000*E+09
EMU of current	ampere (A)	1.000 000*E+01
EMU of electric potential	volt (V)	1.000 000*E−08
EMU of inductance	henry (H)	1.000 000*E−09
EMU of resistance	ohm (Ω)	1.000 000*E−09
ESU of capacitance	farad (F)	1.112 650 E−12
ESU of current	ampere (A)	3.335 6 E−10
ESU of electric potential	volt (V)	2.997 9 E+02
ESU of inductance	henry (H)	8.987 554 E+11

To convert from	to	Multiply by
ESU of resistance	ohm (Ω)	8.987 554 E+11
erg	joule (J)	1.000 000*E−07
erg/cm²·s	watt per metre² (W/m²)	1.000 000*E−03
erg/s	watt (W)	1.000 000*E−07
faraday (based on carbon-12)	coulomb (C)	9.648 70 E+04
faraday (chemical)	coulomb (C)	9.649 57 E+04
faraday (physical)	coulomb (C)	9.652 19 E+04
fathom	metre (m)	1.828 8 E+00
fermi (femtometer)	metre (m)	1.000 000*E−15
fluid ounce (U.S.)	metre³ (m³)	2.957 353 E−05
foot	metre (m)	3.048 000*E−01
foot (U.S. survey)¹	metre (m)	3.048 006 E−01
foot of water (39.2°F)	pascal (Pa)	2.988 98 E+03
ft²	metre² (m²)	9.290 304*E−02
ft²/h (thermal diffusivity)	metre² per second (m²/s)	2.580 640*E−05
ft²/s	metre² per second (m²/s)	9.290 304*E−02
ft³ (volume; section modulus)	metre³ (m³)	2.831 685 E−02
ft³/min	metre³ per second (m³/s)	4.719 474 E−04
ft³/s	metre³ per second (m³/s)	2.831 685 E−02
ft⁴ (moment of section)⁴	metre⁴ (m⁴)	8.630 975 E−03
ft/h	metre per second (m/s)	8.466 667 E−05
ft/min	metre per second (m/s)	5.080 000*E−03
ft/s	metre per second (m/s)	3.048 000*E−01
ft/s²	metre per second² (m/s²)	3.048 000*E−01
footcandle	lux (lx)	1.076 391 E+01
footlambert	candela per metre² (cd/m²)	3.426 259 E+00
ft·lbf	joule (J)	1.355 818 E+00
ft·lbf/h	watt (W)	3.766 161 E−04
ft·lbf/min	watt (W)	2.259 697 E−02
ft·lbf/s	watt (W)	1.355 818 E+00
ft·poundal	joule (J)	4.214 011 E−02
free fall, standard (g)	metre per second² (m/s²)	9.806 650*E+00
gal	metre per second² (m/s²)	1.000 000*E−02
gallon (Canadian liquid)	metre³ (m³)	4.546 090 E−03
gallon (U.K. liquid)	metre³ (m³)	4.546 092 E−03
gallon (U.S. dry)	metre³ (m³)	4.404 884 E−03
gallon (U.S. liquid)	metre³ (m³)	3.785 412 E−03
gal (U.S. liquid)/day	metre³ per second (m³/s)	4.381 264 E−08
gal (U.S. liquid)/min	metre³ per second (m³/s)	6.309 020 E−05
gal (U.S. liquid)/hp·h (SFC, specific fuel consumption)	metre³ per joule (m³/J)	1.410 089 E−09
gamma	tesla (T)	1.000 000*E−09
gauss	tesla (T)	1.000 000*E−04
gilbert	ampere	7.957 747 E−01
gill (U.K.)	metre³ (m³)	1.420 654 E−04
gill (U.S.)	metre³ (m³)	1.182 941 E−04
grad	degree (angular)	9.000 000*E−01
grad	radian (rad)	1.570 796 E−02
grain (1/7000 lb avoirdupois)	kilogram (kg)	6.479 891*E−05
grain (lb avoirdupois/7000)/gal (U.S. liquid)	kilogram per metre³ (kg/m³)	1.711 806 E−02
gram	kilogram (kg)	1.000 000*E−03
g/cm³	kilogram per metre³ (kg/m³)	1.000 000*E+03
gram-force/cm²	pascal (Pa)	9.806 650*E+01
hectare	metre² (m²)	1.000 000*E+04
horsepower (550 ft·lbf/s)	watt (W)	7.456 999 E+02
horsepower (boiler)	watt (W)	9.809 50 E+03
horsepower (electric)	watt (W)	7.460 000*E+02
horsepower (metric)	watt (W)	7.354 99 E+02
horsepower (water)	watt (W)	7.460 43 E+02
horsepower (U.K.)	watt (W)	7.457 0 E+02
hour (mean solar)	second (s)	3.600 000 E+03
hour (sidereal)	second (s)	3.590 170 E+03
hundredweight (long)	kilogram (kg)	5.080 235 E+01
hundredweight (short)	kilogram (kg)	4.535 924 E+01
inch	metre (m)	2.540 000*E−02
inch of mercury (32°F)	pascal (Pa)	3.386 38 E+03
inch of mercury (60°F)	pascal (Pa)	3.376 85 E+03

⁴ This is sometimes called the moment of inertia of a plane section about a specified axis.

To convert from	to	Multiply by
inch of water (39.2°F)	pascal (Pa)	2.490 82 E+02
inch of water (60°F)	pascal (Pa)	2.488 4 E+02
in²	metre² (m²)	6.451 600*E−04
in³ (volume; section modulus)[5]	metre³ (m³)	1.638 706 E−05
in³/min	metre³ per second (m³/s)	2.731 177 E−07
in⁴ (moment of section)[4]	metre⁴ (m⁴)	4.162 314 E−07
in/s	metre per second (m/s)	2.540 000*E−02
in/s²	metre per second² (m/s²)	2.540 000*E−02
kayser	1 per metre (1/m)	1.000 000*E+02
kelvin	degree Celsius	$t_{°C} = t_K - 273.15$
kilocalorie (International Table)	joule (J)	4.186 800*E+03
kilocalorie (mean)	joule (J)	4.190 02 E+03
kilocalorie (thermochemical)	joule (J)	4.184 000*E+03
kilocalorie (thermochemical)/min	watt (W)	6.973 333 E+01
kilocalorie (thermochemical)/s	watt (W)	4.184 000*E+03
kilogram-force (kgf)	newton (N)	9.806 650*E+00
kgf·m	newton metre (N·m)	9.806 650*E+00
kgf·s²/m (mass)	kilogram (kg)	9.806 650*E+00
kgf/cm²	pascal (Pa)	9.806 650*E+04
kgf/m²	pascal (Pa)	9.806 650*E+00
kgf/mm²	pascal (Pa)	9.806 650*E+06
km/h	metre per second (m/s)	2.777 778 E−01
kilopond	newton (N)	9.806 650*E+00
kW·h	joule (J)	3.600 000*E+06
kip (1000 lbf)	newton (N)	4.448 222 E+03
kip/in² (ksi)	pascal (Pa)	6.894 757 E+06
knot (international)	metre per second (m/s)	5.144 444 E−01
lambert	candela per metre² (cd/m²)	$1/\pi$ *E+04
lambert	candela per metre² (cd/m²)	3.183 099 E+03
langley	joule per metre² (J/m²)	4.184 000*E+04
league	metre (m)	[see footnote 1]
light year	metre (m)	9.460 55 E+15
litre [6]	metre³ (m³)	1.000 000*E−03
maxwell	weber (Wb)	1.000 000*E−08
mho	siemens (S)	1.000 000*E+00
microinch	metre (m)	2.540 000*E−08
micron	metre (m)	1.000 000*E−06
mil	metre (m)	2.540 000*E−05
mile (international)	metre (m)	1.609 344*E+03
mile (statute)	metre (m)	1.609 3 E+03
mile (U.S. survey)[1]	metre (m)	1.609 347 E+03
mile (international nautical)	metre (m)	1.852 000*E+03
mile (U.K. nautical)	metre (m)	1.853 184*E+03
mile (U.S. nautical)	metre (m)	1.852 000*E+03
mi² (international)	metre² (m²)	2.589 988 E+06
mi² (U.S. survey)[1]	metre² (m²)	2.589 998 E+06
mi/h (international)	metre per second (m/s)	4.470 400*E−01
mi/h (international)	kilometre per hour (km/h)	1.609 344*E+00
mi/min (international)	metre per second (m/s)	2.682 240 E+01
mi/s (international)	metre per second (m/s)	1.609 344*E+03
millibar	pascal (Pa)	1.000 000*E+02
millimetre of mercury (0°C)	pascal (Pa)	1.333 22 E+02
minute (angle)	radian (rad)	2.908 882 E−04
minute (mean solar)	second (s)	6.000 000 E+01
minute (sidereal)	second (s)	5.983 617 E+01
month (mean calendar)	second (s)	2.628 000 E+06
oersted	ampere per metre (A/m)	7.957 747 E+01
ohm centimetre	ohm metre (Ω·m)	1.000 000*E−02
ohm circular-mil per foot	ohm millimetre² per metre (Ω·mm²/m)	1.662 426 E−03
ounce (avoirdupois)	kilogram (kg)	2.834 952 E−02
ounce (troy or apothecary)	kilogram (kg)	3.110 348 E−02
ounce (U.K. fluid)	metre³ (m³)	2.841 307 E−05
ounce (U.S. fluid)	metre³ (m³)	2.957 353 E−05
ounce-force	newton (N)	2.780 139 E−01
ozf·in	newton metre (N·m)	7.061 552 E−03

---

[5] The exact conversion factor is 1.638 706 4*E−05.
[6] In 1964 the General Conference on Weights and Measures adopted the name litre as a special name for the cubic decimetre. Prior to this decision the litre differed slightly (previous value, 1.000028 dm³) and in expression of precision volume measurement this fact must be kept in mind.

To convert from	to	Multiply by
oz (avoirdupois)/gal (U.K. liquid)	kilogram per metre³ (kg/m³)	6.236 021 E+00
oz (avoirdupois)/gal (U.S. liquid)	kilogram per metre³ (kg/m³)	7.489 152 E+00
oz (avoirdupois)/in³	kilogram per metre³ (kg/m³)	1.729 994 E+03
oz (avoirdupois)/ft²	kilogram per metre² (kg/m²)	3.051 517 E−01
oz (avoirdupois)/yd²	kilogram per metre² (kg/m²)	3.390 575 E−02
parsec	metre (m)	3.085 678 E+16
peck (U.S.)	metre³ (m³)	8.809 768 E−03
pennyweight	kilogram (kg)	1.555 174 E−03
perm (0°C)	kilogram per pascal second metre² (kg/Pa·s·m²)	5.721 35 E−11
perm (23°C)	kilogram per pascal second metre² (kg/Pa·s·m²)	5.745 25 E−11
perm·in (0°C)	kilogram per pascal second metre (kg/Pa·s·m)	1.453 22 E−12
perm·in (23°C)	kilogram per pascal second metre (kg/Pa·s·m)	1.459 29 E−12
phot	lumen per metre² (lm/m²)	1.000 000*E+04
pica (printer's)	metre (m)	4.217 518 E−03
pint (U.S. dry)	metre³ (m³)	5.506 105 E−04
pint (U.S. liquid)	metre³ (m³)	4.731 765 E−04
point (printer's)	metre (m)	3.514 598*E−04
poise (absolute viscosity)	pascal second (Pa·s)	1.000 000*E−01
pound (lb avoirdupois)[7]	kilogram (kg)	4.535 924 E−01
pound (troy or apothecary)	kilogram (kg)	3.732 417 E−01
lb·ft² (moment of inertia)	kilogram metre² (kg·m²)	4.214 011 E−02
lb·in² (moment of inertia)	kilogram metre² (kg·m²)	2.926 397 E−04
lb/ft·h	pascal second (Pa·s)	4.133 789 E−04
lb/ft·s	pascal second (Pa·s)	1.488 164 E+00
lb/ft²	kilogram per metre² (kg/m²)	4.882 428 E+00
lb/ft³	kilogram per metre³ (kg/m³)	1.601 846 E+01
lb/gal (U.K. liquid)	kilogram per metre³ (kg/m³)	9.977 633 E+01
lb/gal (U.S. liquid)	kilogram per metre³ (kg/m³)	1.198 264 E+02
lb/h	kilogram per second (kg/s)	1.259 979 E−04
lb/hp·h (SFC, specific fuel consumption)	kilogram per joule (kg/J)	1.689 659 E−07
lb/in³	kilogram per metre³ (kg/m³)	2.767 990 E+04
lb/min	kilogram per second (kg/s)	7.559 873 E−03
lb/s	kilogram per second (kg/s)	4.535 924 E−01
lb/yd³	kilogram per metre³ (kg/m³)	5.932 764 E−01
poundal	newton (N)	1.382 550 E−01
poundal/ft²	pascal (Pa)	1.488 164 E+00
poundal·s/ft²	pascal second (Pa·s)	1.488 164 E+00
pound-force (lbf)[8]	newton (N)	4.448 222 E+00
lbf·ft	newton metre (N·m)	1.355 818 E+00
lbf·ft/in	newton metre per metre (N·m/m)	5.337 866 E+01
lbf·in	newton metre (N·m)	1.129 848 E−01
lbf·in/in	newton metre per metre (N·m/m)	4.448 222 E+00
lbf·s/ft²	pascal second (Pa·s)	4.788 026 E+01
lbf/ft	newton per metre (N/m)	1.459 390 E+01
lbf/ft²	pascal (Pa)	4.788 026 E+01
lbf/in	newton per metre (N/m)	1.751 268 E+02
lbf/in² (psi)	pascal (Pa)	6.894 757 E+03
lbf/lb (thrust/weight [mass] ratio)	newton per kilogram (N/kg)	9.806 650 E+00
quart (U.S. dry)	metre³ (m³)	1.101 221 E−03
quart (U.S. liquid)	metre³ (m³)	9.463 529 E−04
rad (radiation dose absorbed)	gray (Gy)	1.000 000*E−02
rhe	1 per pascal second (1/Pa·s)	1.000 000*E+01
rod	metre (m)	[see footnote 1]
roentgen	coulomb per kilogram (C/kg)	2.58 E−04
second (angle)	radian (rad)	4.848 137 E−06
second (sidereal)	second (s)	9.972 696 E−01
section	metre² (m²)	[see footnote 1]
shake	second (s)	1.000 000*E−08
slug	kilogram (kg)	1.459 390 E+01
slug/ft·s	pascal second (Pa·s)	4.788 026 E+01
slug/ft³	kilogram per metre³ (kg/m³)	5.153 788 E+02
statampere	ampere (A)	3.335 640 E−10
statcoulomb	coulomb (C)	3.335 640 E−10

[7] The exact conversion factor is 4.535 923 7*E−01.
[8] The exact conversion factor is 4.448 221 615 260 5*E+00.

METRIC CONVERSION FACTORS AND TABLES

To convert from	to	Multiply by
statfarad	farad (F)	1.112 650 E−12
stathenry	henry (H)	8.987 554 E+11
statmho	siemens (S)	1.112 650 E−12
statohm	ohm (Ω)	8.987 554 E+11
statvolt	volt (V)	2.997 925 E+02
stere	metre^3 (m^3)	1.000 000*E+00
stilb	candela per metre^2 (cd/m^2)	1.000 000*E+04
stokes (kinematic viscosity)	metre^2 per second (m^2/s)	1.000 000*E−04
tablespoon	metre^3 (m^3)	1.478 676 E−05
teaspoon	metre^3 (m^3)	4.928 922 E−06
tex	kilogram per metre (kg/m)	1.000 000*E−06
therm	joule (J)	1.055 056 E+08
ton (assay)	kilogram (kg)	2.916 667 E−02
ton (long, 2240 lb)	kilogram (kg)	1.016 047 E+03
ton (metric)	kilogram (kg)	1.000 000*E+03
ton (nuclear equivalent of TNT)	joule (J)	4.184 E+09[9]
ton (refrigeration)	watt (W)	3.516 800 E+03
ton (register)	metre^3 (m^3)	2.831 685 E+00
ton (short, 2000 lb)	kilogram (kg)	9.071 847 E+02
ton (long)/yd^3	kilogram per metre^3 (kg/m^3)	1.328 939 E+03
ton (short)/yd^3	kilogram per metre^3 (kg/m^3)	1.186 553 E+03
ton (short)/h	kilogram per second (kg/s)	2.519 958 E−01
ton-force (2000 lbf)	newton (N)	8.896 444 E+03
tonne	kilogram (kg)	1.000 000*E+03
torr (mm Hg, 0°C)	pascal (Pa)	1.333 22 E+02
township	metre^2 (m^2)	[see footnote 1]
unit pole	weber (Wb)	1.256 637 E−07
W·h	joule (J)	3.600 000*E+03
W·s	joule (J)	1.000 000*E+00
W/cm^2	watt per metre^2 (W/m^2)	1.000 000*E+04
W/in^2	watt per metre^2 (W/m^2)	1.550 003 E+03
yard	metre (m)	9.144 000*E−01
yd^2	metre^2 (m^2)	8.361 274 E−01
yd^3	metre^3 (m^3)	7.645 549 E−01
yd^3/min	metre^3 per second (m^3/s)	1.274 258 E−02
year (calendar)	second (s)	3.153 600 E+07
year (sidereal)	second (s)	3.155 815 E+07
year (tropical)	second (s)	3.155 693 E+07

[9] Defined (not measured) value.

USE OF CONVERSION TABLES

Table 19-1. Millimeters (from 1 through 1009 mm) to decimal inches conversions.

Example—To convert 53.2 millimeters to inches:

    Page 19-7: 532 mm = 20.9449 inches
    Divide by 10: 53.2 mm = 2.09449 inches

Table 19-2. Decimal inches (from .001 through 1.009 inches) to millimeters conversions.

Example—To convert 3.526 inches to millimeters:

    Page 19-11:   .030 inch = 0.7620 mm
    Multiply by 100: 3.00 inches = 76.20 mm
    Table 19-2:   .526 inch = 13.3604 mm
    Values added: 3.526 inches = 89.5604 mm

Table 19-3. Fractional inches (from 1/64 through 19-63/64 inches) to millimeter conversions.

Example—To convert 5-17/32 inches to millimeters

    Page 19-13: 5-17/32 inches = 140.4937 mm

Table 19-4. Degrees Celsius or Fahrenheit (from -459.67°F through 4850°C or °F) to Kelvins, degrees Fahrenheit or Celsius.

Example—To convert 50°F to degrees Celsius:

    Page 19-16: Start at 50 in column marked °F and °C and follow arrow to the left:
    50°F = 10°C or 283.1 K, or

To convert 50°C to degrees Fahrenheit:

    Page 19-16: Start at 50 in column marked °F and °C and follow arrow to the right:
    50°C = 122°F

Table 19-5. Kilopascals (from 1 through 1009 kPa) to psi conversions.

Example—To convert 492 kPa to psi:

    Page 19-19: 492 kPa = 71.359 psi, or
    To convert 621 MPa to ksi (1000 psi):

    Page 19-20: 621 MPa = 90.068 ksi
                          = 90068 psi

Table 19-6. PSI (from 1 through 1009 psi) to kilopascals conversions.

Table 19-7. Kilograms (from 1 through 1009 kg) to pounds conversions.

Table 19-8. Pounds (from 1 through 1009 lbs) to kilograms conversions.

Table 19-9. Newtons (from 1 through 1009 N) to pounds force conversions.

Table 19-10. Pounds force (from 1 through 1009 lbs) to newtons conversions.

METRIC CONVERSION FACTORS AND TABLES

Table 19-1. Length: Millimeters to Inches (from 1 through 499 Millimeters)

1 INCH = 25.4 MILLIMETERS (EXACTLY)

MILLIMETERS	0	1	2	3	4	5	6	7	8	9
					INCHES					
0	0.0000	0.0394	0.0787	0.1181	0.1575	0.1969	0.2362	0.2756	0.3150	0.3543
10	0.3937	0.4331	0.4724	0.5118	0.5512	0.5906	0.6299	0.6693	0.7087	0.7480
20	0.7874	0.8268	0.8661	0.9055	0.9449	0.9843	1.0236	1.0630	1.1024	1.1417
30	1.1811	1.2205	1.2598	1.2992	1.3386	1.3780	1.4173	1.4567	1.4961	1.5354
40	1.5748	1.6142	1.6535	1.6929	1.7323	1.7717	1.8110	1.8504	1.8898	1.9291
50	1.9685	2.0079	2.0472	2.0866	2.1260	2.1654	2.2047	2.2441	2.2835	2.3228
60	2.3622	2.4016	2.4409	2.4803	2.5197	2.5591	2.5984	2.6378	2.6772	2.7165
70	2.7559	2.7953	2.8346	2.8740	2.9134	2.9528	2.9921	3.0315	3.0709	3.1102
80	3.1496	3.1890	3.2283	3.2677	3.3071	3.3465	3.3858	3.4252	3.4646	3.5039
90	3.5433	3.5827	3.6220	3.6614	3.7008	3.7402	3.7795	3.8189	3.8583	3.8976
100	3.9370	3.9764	4.0157	4.0551	4.0945	4.1339	4.1732	4.2126	4.2520	4.2913
110	4.3307	4.3701	4.4094	4.4488	4.4882	4.5276	4.5669	4.6063	4.6457	4.6850
120	4.7244	4.7638	4.8031	4.8425	4.8819	4.9213	4.9606	5.0000	5.0394	5.0787
130	5.1181	5.1575	5.1969	5.2362	5.2756	5.3150	5.3543	5.3937	5.4331	5.4724
140	5.5118	5.5512	5.5906	5.6299	5.6693	5.7087	5.7480	5.7874	5.8268	5.8661
150	5.9055	5.9449	5.9843	6.0236	6.0630	6.1024	6.1417	6.1811	6.2205	6.2598
160	6.2992	6.3386	6.3780	6.4173	6.4567	6.4961	6.5354	6.5748	6.6142	6.6535
170	6.6929	6.7323	6.7717	6.8110	6.8504	6.8898	6.9291	6.9685	7.0079	7.0472
180	7.0866	7.1260	7.1654	7.2047	7.2441	7.2835	7.3228	7.3622	7.4016	7.4409
190	7.4803	7.5197	7.5591	7.5984	7.6378	7.6772	7.7165	7.7559	7.7953	7.8346
200	7.8740	7.9134	7.9528	7.9921	8.0315	8.0709	8.1102	8.1496	8.1890	8.2283
210	8.2677	8.3071	8.3465	8.3858	8.4252	8.4646	8.5039	8.5433	8.5827	8.6220
220	8.6614	8.7008	8.7402	8.7795	8.8189	8.8583	8.8976	8.9370	8.9764	9.0157
230	9.0551	9.0945	9.1339	9.1732	9.2126	9.2520	9.2913	9.3307	9.3701	9.4094
240	9.4488	9.4882	9.5276	9.5669	9.6063	9.6457	9.6850	9.7244	9.7638	9.8031
250	9.8425	9.8819	9.9213	9.9606	10.0000	10.0394	10.0787	10.1181	10.1575	10.1969
260	10.2362	10.2756	10.3150	10.3543	10.3937	10.4331	10.4724	10.5118	10.5512	10.5906
270	10.6299	10.6693	10.7087	10.7480	10.7874	10.8268	10.8661	10.9055	10.9449	10.9843
280	11.0236	11.0630	11.1024	11.1417	11.1811	11.2205	11.2598	11.2992	11.3386	11.3780
290	11.4173	11.4567	11.4961	11.5354	11.5748	11.6142	11.6535	11.6929	11.7323	11.7717
300	11.8110	11.8504	11.8898	11.9291	11.9685	12.0079	12.0472	12.0866	12.1260	12.1654
310	12.2047	12.2441	12.2835	12.3228	12.3622	12.4016	12.4409	12.4803	12.5197	12.5591
320	12.5984	12.6378	12.6772	12.7165	12.7559	12.7953	12.8346	12.8740	12.9134	12.9528
330	12.9921	13.0315	13.0709	13.1102	13.1496	13.1890	13.2283	13.2677	13.3071	13.3465
340	13.3858	13.4252	13.4646	13.5039	13.5433	13.5827	13.6220	13.6614	13.7008	13.7402
350	13.7795	13.8189	13.8583	13.8976	13.9370	13.9764	14.0157	14.0551	14.0945	14.1339
360	14.1732	14.2126	14.2520	14.2913	14.3307	14.3701	14.4094	14.4488	14.4882	14.5276
370	14.5669	14.6063	14.6457	14.6850	14.7244	14.7638	14.8031	14.8425	14.8819	14.9213
380	14.9606	15.0000	15.0394	15.0787	15.1181	15.1575	15.1969	15.2362	15.2756	15.3150
390	15.3543	15.3937	15.4331	15.4724	15.5118	15.5512	15.5906	15.6299	15.6693	15.7087
400	15.7480	15.7874	15.8268	15.8661	15.9055	15.9449	15.9843	16.0236	16.0630	16.1024
410	16.1417	16.1811	16.2205	16.2598	16.2992	16.3386	16.3780	16.4173	16.4567	16.4961
420	16.5354	16.5748	16.6142	16.6535	16.6929	16.7323	16.7717	16.8110	16.8504	16.8898
430	16.9291	16.9685	17.0079	17.0472	17.0866	17.1260	17.1654	17.2047	17.2441	17.2835
440	17.3228	17.3622	17.4016	17.4409	17.4803	17.5197	17.5591	17.5984	17.6378	17.6772
450	17.7165	17.7559	17.7953	17.8346	17.8740	17.9134	17.9528	17.9921	18.0315	18.0709
460	18.1102	18.1496	18.1890	18.2283	18.2677	18.3071	18.3465	18.3858	18.4252	18.4646
470	18.5039	18.5433	18.5827	18.6220	18.6614	18.7008	18.7402	18.7795	18.8189	18.8583
480	18.8976	18.9370	18.9764	19.0157	19.0551	19.0945	19.1339	19.1732	19.2126	19.2520
490	19.2913	19.3307	19.3701	19.4094	19.4488	19.4882	19.5276	19.5669	19.6063	19.6457

Table 19-1 *(Continued)*. Length: Millimeters to Inches (from 500 through 1009 Millimeters)

MILLIMETERS	0	1	2	3	4	5	6	7	8	9
					INCHES					
500	19.6850	19.7244	19.7638	19.8031	19.8425	19.8819	19.9213	19.9606	20.0000	20.0394
510	20.0787	20.1181	20.1575	20.1969	20.2362	20.2756	20.3150	20.3543	20.3937	20.4331
520	20.4724	20.5118	20.5512	20.5906	20.6299	20.6693	20.7087	20.7480	20.7874	20.8268
530	20.8661	20.9055	20.9449	20.9843	21.0236	21.0630	21.1024	21.1417	21.1811	21.2205
540	21.2598	21.2992	21.3386	21.3780	21.4173	21.4567	21.4961	21.5354	21.5748	21.6142
550	21.6535	21.6929	21.7323	21.7717	21.8110	21.8504	21.8898	21.9291	21.9685	22.0079
560	22.0472	22.0866	22.1260	22.1654	22.2047	22.2441	22.2835	22.3228	22.3622	22.4016
570	22.4409	22.4803	22.5197	22.5591	22.5984	22.6378	22.6772	22.7165	22.7559	22.7953
580	22.8346	22.8740	22.9134	22.9528	22.9921	23.0315	23.0709	23.1102	23.1496	23.1890
590	23.2283	23.2677	23.3071	23.3465	23.3858	23.4252	23.4646	23.5039	23.5433	23.5827
600	23.6220	23.6614	23.7008	23.7402	23.7795	23.8189	23.8583	23.8976	23.9370	23.9764
610	24.0157	24.0551	24.0945	24.1339	24.1732	24.2126	24.2520	24.2913	24.3307	24.3701
620	24.4094	24.4488	24.4882	24.5276	24.5669	24.6063	24.6457	24.6850	24.7244	24.7638
630	24.8031	24.8425	24.8819	24.9213	24.9606	25.0000	25.0394	25.0787	25.1181	25.1575
640	25.1969	25.2362	25.2756	25.3150	25.3543	25.3937	25.4331	25.4724	25.5118	25.5512
650	25.5906	25.6299	25.6693	25.7087	25.7480	25.7874	25.8268	25.8661	25.9055	25.9449
660	25.9843	26.0236	26.0630	26.1024	26.1417	26.1811	26.2205	26.2598	26.2992	26.3386
670	26.3780	26.4173	26.4567	26.4961	26.5354	26.5748	26.6142	26.6535	26.6929	26.7323
680	26.7717	26.8110	26.8504	26.8898	26.9291	26.9685	27.0079	27.0472	27.0866	27.1260
690	27.1654	27.2047	27.2441	27.2835	27.3228	27.3622	27.4016	27.4409	27.4803	27.5197
700	27.5591	27.5984	27.6378	27.6772	27.7165	27.7559	27.7953	27.8346	27.8740	27.9134
710	27.9528	27.9921	28.0315	28.0709	28.1102	28.1496	28.1890	28.2283	28.2677	28.3071
720	28.3465	28.3858	28.4252	28.4646	28.5039	28.5433	28.5827	28.6220	28.6614	28.7008
730	28.7402	28.7795	28.8189	28.8583	28.8976	28.9370	28.9764	29.0157	29.0551	29.0945
740	29.1339	29.1732	29.2126	29.2520	29.2913	29.3307	29.3701	29.4094	29.4488	29.4882
750	29.5276	29.5669	29.6063	29.6457	29.6850	29.7244	29.7638	29.8031	29.8425	29.8819
760	29.9213	29.9606	30.0000	30.0394	30.0787	30.1181	30.1575	30.1969	30.2362	30.2756
770	30.3150	30.3543	30.3937	30.4331	30.4724	30.5118	30.5512	30.5906	30.6299	30.6693
780	30.7087	30.7480	30.7874	30.8268	30.8661	30.9055	30.9449	30.9843	31.0236	31.0630
790	31.1024	31.1417	31.1811	31.2205	31.2598	31.2992	31.3386	31.3780	31.4173	31.4567
800	31.4961	31.5354	31.5748	31.6142	31.6535	31.6929	31.7323	31.7717	31.8110	31.8504
810	31.8898	31.9291	31.9685	32.0079	32.0472	32.0866	32.1260	32.1654	32.2047	32.2441
820	32.2835	32.3228	32.3622	32.4016	32.4409	32.4803	32.5197	32.5591	32.5984	32.6378
830	32.6772	32.7165	32.7559	32.7953	32.8346	32.8740	32.9134	32.9528	32.9921	33.0315
840	33.0709	33.1102	33.1496	33.1890	33.2283	33.2677	33.3071	33.3465	33.3858	33.4252
850	33.4646	33.5039	33.5433	33.5827	33.6220	33.6614	33.7008	33.7402	33.7795	33.8189
860	33.8583	33.8976	33.9370	33.9764	34.0157	34.0551	34.0945	34.1339	34.1732	34.2126
870	34.2520	34.2913	34.3307	34.3701	34.4094	34.4488	34.4882	34.5276	34.5669	34.6063
880	34.6457	34.6850	34.7244	34.7638	34.8031	34.8425	34.8819	34.9213	34.9606	35.0000
890	35.0394	35.0787	35.1181	35.1575	35.1969	35.2362	35.2756	35.3150	35.3543	35.3937
900	35.4331	35.4724	35.5118	35.5512	35.5906	35.6299	35.6693	35.7087	35.7480	35.7874
910	35.8268	35.8661	35.9055	35.9449	35.9843	36.0236	36.0630	36.1024	36.1417	36.1811
920	36.2205	36.2598	36.2992	36.3386	36.3780	36.4173	36.4567	36.4961	36.5354	36.5748
930	36.6142	36.6535	36.6929	36.7323	36.7717	36.8110	36.8504	36.8898	36.9291	36.9685
940	37.0079	37.0472	37.0866	37.1260	37.1654	37.2047	37.2441	37.2835	37.3228	37.3622
950	37.4016	37.4409	37.4803	37.5197	37.5591	37.5984	37.6378	37.6772	37.7165	37.7559
960	37.7953	37.8346	37.8740	37.9134	37.9528	37.9921	38.0315	38.0709	38.1102	38.1496
970	38.1890	38.2283	38.2677	38.3071	38.3465	38.3858	38.4252	38.4646	38.5039	38.5433
980	38.5827	38.6220	38.6614	38.7008	38.7402	38.7795	38.8189	38.8583	38.8976	38.9370
990	38.9764	39.0157	39.0551	39.0945	39.1339	39.1732	39.2126	39.2520	39.2913	39.3307
1000	39.3701	39.4094	39.4488	39.4882	39.5276	39.5669	39.6063	39.6457	39.6850	39.7244

METRIC CONVERSION FACTORS AND TABLES

Table 19-2. Length: Decimal Inches to Millimeters (from .001 through .499 Inches)

1 INCH = 25.4 MILLIMETERS (EXACTLY)

INCHES	.000	.001	.002	.003	.004	.005	.006	.007	.008	.009
					MILLIMETERS					
0.000	0.0000	0.0254	0.0508	0.0762	0.1016	0.1270	0.1524	0.1778	0.2032	0.2286
0.010	0.2540	0.2794	0.3048	0.3302	0.3556	0.3810	0.4064	0.4318	0.4572	0.4826
0.020	0.5080	0.5334	0.5588	0.5842	0.6096	0.6350	0.6604	0.6858	0.7112	0.7366
0.030	0.7620	0.7874	0.8128	0.8382	0.8636	0.8890	0.9144	0.9398	0.9652	0.9906
0.040	1.0160	1.0414	1.0668	1.0922	1.1176	1.1430	1.1684	1.1938	1.2192	1.2446
0.050	1.2700	1.2954	1.3208	1.3462	1.3716	1.3970	1.4224	1.4478	1.4732	1.4986
0.060	1.5240	1.5494	1.5748	1.6002	1.6256	1.6510	1.6764	1.7018	1.7272	1.7526
0.070	1.7780	1.8034	1.8288	1.8542	1.8796	1.9050	1.9304	1.9558	1.9812	2.0066
0.080	2.0320	2.0574	2.0828	2.1082	2.1336	2.1590	2.1844	2.2098	2.2352	2.2606
0.090	2.2860	2.3114	2.3368	2.3622	2.3876	2.4130	2.4384	2.4638	2.4892	2.5146
0.100	2.5400	2.5654	2.5908	2.6162	2.6416	2.6670	2.6924	2.7178	2.7432	2.7686
0.110	2.7940	2.8194	2.8448	2.8702	2.8956	2.9210	2.9464	2.9718	2.9972	3.0226
0.120	3.0480	3.0734	3.0988	3.1242	3.1496	3.1750	3.2004	3.2258	3.2512	3.2766
0.130	3.3020	3.3274	3.3528	3.3782	3.4036	3.4290	3.4544	3.4798	3.5052	3.5306
0.140	3.5560	3.5814	3.6068	3.6322	3.6576	3.6830	3.7084	3.7338	3.7592	3.7846
0.150	3.8100	3.8354	3.8608	3.8862	3.9116	3.9370	3.9624	3.9878	4.0132	4.0386
0.160	4.0640	4.0894	4.1148	4.1402	4.1656	4.1910	4.2164	4.2418	4.2672	4.2926
0.170	4.3180	4.3434	4.3688	4.3942	4.4196	4.4450	4.4704	4.4958	4.5212	4.5466
0.180	4.5720	4.5974	4.6228	4.6482	4.6736	4.6990	4.7244	4.7498	4.7752	4.8006
0.190	4.8260	4.8514	4.8768	4.9022	4.9276	4.9530	4.9784	5.0038	5.0292	5.0546
0.200	5.0800	5.1054	5.1308	5.1562	5.1816	5.2070	5.2324	5.2578	5.2832	5.3086
0.210	5.3340	5.3594	5.3848	5.4102	5.4356	5.4610	5.4864	5.5118	5.5372	5.5626
0.220	5.5880	5.6134	5.6388	5.6642	5.6896	5.7150	5.7404	5.7658	5.7912	5.8166
0.230	5.8420	5.8674	5.8928	5.9182	5.9436	5.9690	5.9944	6.0198	6.0452	6.0706
0.240	6.0960	6.1214	6.1468	6.1722	6.1976	6.2230	6.2484	6.2738	6.2992	6.3246
0.250	6.3500	6.3754	6.4008	6.4262	6.4516	6.4770	6.5024	6.5278	6.5532	6.5786
0.260	6.6040	6.6294	6.6548	6.6802	6.7056	6.7310	6.7564	6.7818	6.8072	6.8326
0.270	6.8580	6.8834	6.9088	6.9342	6.9596	6.9850	7.0104	7.0358	7.0612	7.0866
0.280	7.1120	7.1374	7.1628	7.1882	7.2136	7.2390	7.2644	7.2898	7.3152	7.3406
0.290	7.3660	7.3914	7.4168	7.4422	7.4676	7.4930	7.5184	7.5438	7.5692	7.5946
0.300	7.6200	7.6454	7.6708	7.6962	7.7216	7.7470	7.7724	7.7978	7.8232	7.8486
0.310	7.8740	7.8994	7.9248	7.9502	7.9756	8.0010	8.0264	8.0518	8.0772	8.1026
0.320	8.1280	8.1534	8.1788	8.2042	8.2296	8.2550	8.2804	8.3058	8.3312	8.3566
0.330	8.3820	8.4074	8.4328	8.4582	8.4836	8.5090	8.5344	8.5598	8.5852	8.6106
0.340	8.6360	8.6614	8.6868	8.7122	8.7376	8.7630	8.7884	8.8138	8.8392	8.8646
0.350	8.8900	8.9154	8.9408	8.9662	8.9916	9.0170	9.0424	9.0678	9.0932	9.1186
0.360	9.1440	9.1694	9.1948	9.2202	9.2456	9.2710	9.2964	9.3218	9.3472	9.3726
0.370	9.3980	9.4234	9.4488	9.4742	9.4996	9.5250	9.5504	9.5758	9.6012	9.6266
0.380	9.6520	9.6774	9.7028	9.7282	9.7536	9.7790	9.8044	9.8298	9.8552	9.8806
0.390	9.9060	9.9314	9.9568	9.9822	10.0076	10.0330	10.0584	10.0838	10.1092	10.1346
0.400	10.1600	10.1854	10.2108	10.2362	10.2616	10.2870	10.3124	10.3378	10.3632	10.3886
0.410	10.4140	10.4394	10.4648	10.4902	10.5156	10.5410	10.5664	10.5918	10.6172	10.6426
0.420	10.6680	10.6934	10.7188	10.7442	10.7696	10.7950	10.8204	10.8458	10.8712	10.8966
0.430	10.9220	10.9474	10.9728	10.9982	11.0236	11.0490	11.0744	11.0998	11.1252	11.1506
0.440	11.1760	11.2014	11.2268	11.2522	11.2776	11.3030	11.3284	11.3538	11.3792	11.4046
0.450	11.4300	11.4554	11.4808	11.5062	11.5316	11.5570	11.5824	11.6078	11.6332	11.6586
0.460	11.6840	11.7094	11.7348	11.7602	11.7856	11.8110	11.8364	11.8618	11.8872	11.9126
0.470	11.9380	11.9634	11.9888	12.0142	12.0396	12.0650	12.0904	12.1158	12.1412	12.1666
0.480	12.1920	12.2174	12.2428	12.2682	12.2936	12.3190	12.3444	12.3698	12.3952	12.4206
0.490	12.4460	12.4714	12.4968	12.5222	12.5476	12.5730	12.5984	12.6238	12.6492	12.6746

Table 9-2 (*Continued*). Length: Decimal Inches to Millimeters (from .500 through 1.009 Inches)

INCHES	.000	.001	.002	.003	.004	.005	.006	.007	.008	.009
					MILLIMETERS					
0.500	12.7000	12.7254	12.7508	12.7762	12.8016	12.8270	12.8524	12.8778	12.9032	12.9286
0.510	12.9540	12.9794	13.0048	13.0302	13.0556	13.0810	13.1064	13.1318	13.1572	13.1826
0.520	13.2080	13.2334	13.2588	13.2842	13.3096	13.3350	13.3604	13.3858	13.4112	13.4366
0.530	13.4620	13.4874	13.5128	13.5382	13.5636	13.5890	13.6144	13.6398	13.6652	13.6906
0.540	13.7160	13.7414	13.7668	13.7922	13.8176	13.8430	13.8684	13.8938	13.9192	13.9446
0.550	13.9700	13.9954	14.0208	14.0462	14.0716	14.0970	14.1224	14.1478	14.1732	14.1986
0.560	14.2240	14.2494	14.2748	14.3002	14.3256	14.3510	14.3764	14.4018	14.4272	14.4526
0.570	14.4780	14.5034	14.5288	14.5542	14.5796	14.6050	14.6304	14.6558	14.6812	14.7066
0.580	14.7320	14.7574	14.7828	14.8082	14.8336	14.8590	14.8844	14.9098	14.9352	14.9606
0.590	14.9860	15.0114	15.0368	15.0622	15.0876	15.1130	15.1384	15.1638	15.1892	15.2146
0.600	15.2400	15.2654	15.2908	15.3162	15.3416	15.3670	15.3924	15.4178	15.4432	15.4686
0.610	15.4940	15.5194	15.5448	15.5702	15.5956	15.6210	15.6464	15.6718	15.6972	15.7226
0.620	15.7480	15.7734	15.7988	15.8242	15.8496	15.8750	15.9004	15.9258	15.9512	15.9766
0.630	16.0020	16.0274	16.0528	16.0782	16.1036	16.1290	16.1544	16.1798	16.2052	16.2306
0.640	16.2560	16.2814	16.3068	16.3322	16.3576	16.3830	16.4084	16.4338	16.4592	16.4846
0.650	16.5100	16.5354	16.5608	16.5862	16.6116	16.6370	16.6624	16.6878	16.7132	16.7386
0.660	16.7640	16.7894	16.8148	16.8402	16.8656	16.8910	16.9164	16.9418	16.9672	16.9926
0.670	17.0180	17.0434	17.0688	17.0942	17.1196	17.1450	17.1704	17.1958	17.2212	17.2466
0.680	17.2720	17.2974	17.3228	17.3482	17.3736	17.3990	17.4244	17.4498	17.4752	17.5006
0.690	17.5260	17.5514	17.5768	17.6022	17.6276	17.6530	17.6784	17.7038	17.7292	17.7546
0.700	17.7800	17.8054	17.8308	17.8562	17.8816	17.9070	17.9324	17.9578	17.9832	18.0086
0.710	18.0340	18.0594	18.0848	18.1102	18.1356	18.1610	18.1864	18.2118	18.2372	18.2626
0.720	18.2880	18.3134	18.3388	18.3642	18.3896	18.4150	18.4404	18.4658	18.4912	18.5166
0.730	18.5420	18.5674	18.5928	18.6182	18.6436	18.6690	18.6944	18.7198	18.7452	18.7706
0.740	18.7960	18.8214	18.8468	18.8722	18.8976	18.9230	18.9484	18.9738	18.9992	19.0246
0.750	19.0500	19.0754	19.1008	19.1262	19.1516	19.1770	19.2024	19.2278	19.2532	19.2786
0.760	19.3040	19.3294	19.3548	19.3802	19.4056	19.4310	19.4564	19.4818	19.5072	19.5326
0.770	19.5580	19.5834	19.6088	19.6342	19.6596	19.6850	19.7104	19.7358	19.7612	19.7866
0.780	19.8120	19.8374	19.8628	19.8882	19.9136	19.9390	19.9644	19.9898	20.0152	20.0406
0.790	20.0660	20.0914	20.1168	20.1422	20.1676	20.1930	20.2184	20.2438	20.2692	20.2946
0.800	20.3200	20.3454	20.3708	20.3962	20.4216	20.4470	20.4724	20.4978	20.5232	20.5486
0.810	20.5740	20.5994	20.6248	20.6502	20.6756	20.7010	20.7264	20.7518	20.7772	20.8026
0.820	20.8280	20.8534	20.8788	20.9042	20.9296	20.9550	20.9804	21.0058	21.0312	21.0566
0.830	21.0820	21.1074	21.1328	21.1582	21.1836	21.2090	21.2344	21.2598	21.2852	21.3106
0.840	21.3360	21.3614	21.3868	21.4122	21.4376	21.4630	21.4884	21.5138	21.5392	21.5646
0.850	21.5900	21.6154	21.6408	21.6662	21.6916	21.7170	21.7424	21.7678	21.7932	21.8186
0.860	21.8440	21.8694	21.8948	21.9202	21.9456	21.9710	21.9964	22.0218	22.0472	22.0726
0.870	22.0980	22.1234	22.1488	22.1742	22.1996	22.2250	22.2504	22.2758	22.3012	22.3266
0.880	22.3520	22.3774	22.4028	22.4282	22.4536	22.4790	22.5044	22.5298	22.5552	22.5806
0.890	22.6060	22.6314	22.6568	22.6822	22.7076	22.7330	22.7584	22.7838	22.8092	22.8346
0.900	22.8600	22.8854	22.9108	22.9362	22.9616	22.9870	23.0124	23.0378	23.0632	23.0886
0.910	23.1140	23.1394	23.1648	23.1902	23.2156	23.2410	23.2664	23.2918	23.3172	23.3426
0.920	23.3680	23.3934	23.4188	23.4442	23.4696	23.4950	23.5204	23.5458	23.5712	23.5966
0.930	23.6220	23.6474	23.6728	23.6982	23.7236	23.7490	23.7744	23.7998	23.8252	23.8506
0.940	23.8760	23.9014	23.9268	23.9522	23.9776	24.0030	24.0284	24.0538	24.0792	24.1046
0.950	24.1300	24.1554	24.1808	24.2062	24.2316	24.2570	24.2824	24.3078	24.3332	24.3586
0.960	24.3840	24.4094	24.4348	24.4602	24.4856	24.5110	24.5364	24.5618	24.5872	24.6126
0.970	24.6380	24.6634	24.6888	24.7142	24.7396	24.7650	24.7904	24.8158	24.8412	24.8666
0.980	24.8920	24.9174	24.9428	24.9682	24.9936	25.0190	25.0444	25.0698	25.0952	25.1206
0.990	25.1460	25.1714	25.1968	25.2222	25.2476	25.2730	25.2984	25.3238	25.3492	25.3746
1.000	25.4000	25.4254	25.4508	25.4762	25.5016	25.5270	25.5524	25.5778	25.6032	25.6286

METRIC CONVERSION FACTORS AND TABLES

Table 19-3. Length: Fractional Inches to Millimeters (from 1/64 to 9 63/64 Inches)

1 INCH = 25.4 MILLIMETERS (EXACTLY)

INCHES		0	1	2	3	4	5	6	7	8	9
					MILLIMETERS						
0.000000	0/64	0.0000	25.4000	50.8000	76.2000	101.6000	127.0000	152.4000	177.8000	203.2000	228.5999
0.015625	1/64	0.3969	25.7969	51.1969	76.5969	101.9969	127.3969	152.7968	178.1968	203.5968	228.9968
0.031250	1/32	0.7937	26.1937	51.5937	76.9937	102.3937	127.7937	153.1937	178.5937	203.9937	229.3937
0.046875	3/64	1.1906	26.5906	51.9906	77.3906	102.7906	128.1906	153.5906	178.9906	204.3906	229.7906
0.062500	1/16	1.5875	26.9875	52.3875	77.7875	103.1875	128.5875	153.9875	179.3875	204.7874	230.1874
0.078125	5/64	1.9844	27.3844	52.7844	78.1844	103.5844	128.9843	154.3843	179.7843	205.1843	230.5843
0.093750	3/32	2.3812	27.7812	53.1812	78.5812	103.9812	129.3812	154.7812	180.1812	205.5812	230.9812
0.109375	7/64	2.7781	28.1781	53.5781	78.9781	104.3781	129.7781	155.1781	180.5781	205.9781	231.3781
0.125000	1/8	3.1750	28.5750	53.9750	79.3750	104.7750	130.1750	155.5750	180.9750	206.3750	231.7749
0.140625	9/64	3.5719	28.9719	54.3719	79.7719	105.1718	130.5718	155.9718	181.3718	206.7718	232.1718
0.156250	5/32	3.9687	29.3687	54.7687	80.1687	105.5687	130.9687	156.3687	181.7687	207.1687	232.5687
0.171875	11/64	4.3656	29.7656	55.1656	80.5656	105.9656	131.3656	156.7656	182.1656	207.5656	232.9656
0.187500	3/16	4.7625	30.1625	55.5625	80.9625	106.3625	131.7625	157.1625	182.5625	207.9624	233.3624
0.203125	13/64	5.1594	30.5594	55.9594	81.3594	106.7594	132.1593	157.5593	182.9593	208.3593	233.7593
0.218750	7/32	5.5562	30.9562	56.3562	81.7562	107.1562	132.5562	157.9562	183.3562	208.7562	234.1562
0.234375	15/64	5.9531	31.3531	56.7531	82.1531	107.5531	132.9531	158.3531	183.7531	209.1531	234.5531
0.250000	1/4	6.3500	31.7500	57.1500	82.5500	107.9500	133.3500	158.7500	184.1500	209.5500	234.9500
0.265625	17/64	6.7469	32.1469	57.5469	82.9469	108.3468	133.7468	159.1468	184.5468	209.9468	235.3468
0.281250	9/32	7.1437	32.5437	57.9437	83.3437	108.7437	134.1437	159.5437	184.9437	210.3437	235.7437
0.296875	19/64	7.5406	32.9406	58.3406	83.7406	109.1406	134.5406	159.9406	185.3406	210.7406	236.1406
0.312500	5/16	7.9375	33.3375	58.7375	84.1375	109.5375	134.9375	160.3375	185.7375	211.1375	236.5374
0.328125	21/64	8.3344	33.7344	59.1344	84.5343	109.9343	135.3343	160.7343	186.1343	211.5343	236.9343
0.343750	11/32	8.7312	34.1312	59.5312	84.9312	110.3312	135.7312	161.1312	186.5312	211.9312	237.3312
0.359375	23/64	9.1281	34.5281	59.9281	85.3281	110.7281	136.1281	161.5281	186.9281	212.3281	237.7281
0.375000	3/8	9.5250	34.9250	60.3250	85.7250	111.1250	136.5250	161.9250	187.3250	212.7249	238.1249
0.390625	25/64	9.9219	35.3219	60.7219	86.1219	111.5219	136.9218	162.3218	187.7218	213.1218	238.5218
0.406250	13/32	10.3187	35.7187	61.1187	86.5187	111.9187	137.3187	162.7187	188.1187	213.5187	238.9187
0.421875	27/64	10.7156	36.1156	61.5156	86.9156	112.3156	137.7156	163.1156	188.5156	213.9156	239.3156
0.437500	7/16	11.1125	36.5125	61.9125	87.3125	112.7125	138.1125	163.5125	188.9125	214.3125	239.7124
0.453125	29/64	11.5094	36.9094	62.3094	87.7094	113.1093	138.5093	163.9093	189.3093	214.7093	240.1093
0.468750	15/32	11.9062	37.3062	62.7062	88.1062	113.5062	138.9062	164.3062	189.7062	215.1062	240.5062
0.484375	31/64	12.3031	37.7031	63.1031	88.5031	113.9031	139.3031	164.7031	190.1031	215.5031	240.9031
0.500000	1/2	12.7000	38.1000	63.5000	88.9000	114.3000	139.7000	165.1000	190.5000	215.8999	241.2999
0.515625	33/64	13.0969	38.4969	63.8969	89.2969	114.6969	140.0968	165.4968	190.8968	216.2968	241.6968
0.531250	17/32	13.4937	38.8937	64.2937	89.6937	115.0937	140.4937	165.8937	191.2937	216.6937	242.0937
0.546875	35/64	13.8906	39.2906	64.6906	90.0906	115.4906	140.8906	166.2906	191.6906	217.0906	242.4906
0.562500	9/16	14.2875	39.6875	65.0875	90.4875	115.8875	141.2875	166.6875	192.0874	217.4874	242.8874
0.578125	37/64	14.6844	40.0844	65.4844	90.8844	116.2843	141.6843	167.0843	192.4843	217.8843	243.2843
0.593750	19/32	15.0812	40.4812	65.8812	91.2812	116.6812	142.0812	167.4812	192.8812	218.2812	243.6812
0.609375	39/64	15.4781	40.8781	66.2781	91.6781	117.0781	142.4781	167.8781	193.2781	218.6781	244.0781
0.625000	5/8	15.8750	41.2750	66.6750	92.0750	117.4750	142.8750	168.2750	193.6750	219.0750	244.4749
0.640625	41/64	16.2719	41.6719	67.0719	92.4718	117.8718	143.2718	168.6718	194.0718	219.4718	244.8718
0.656250	21/32	16.6687	42.0687	67.4687	92.8687	118.2687	143.6687	169.0687	194.4687	219.8687	245.2687
0.671875	43/64	17.0656	42.4656	67.8656	93.2656	118.6656	144.0656	169.4656	194.8656	220.2656	245.6656
0.687500	11/16	17.4625	42.8625	68.2625	93.6625	119.0625	144.4625	169.8625	195.2625	220.6624	246.0624
0.703125	45/64	17.8594	43.2594	68.6594	94.0594	119.4594	144.8593	170.2593	195.6593	221.0593	246.4593
0.718750	23/32	18.2562	43.6562	69.0562	94.4562	119.8562	145.2562	170.6562	196.0562	221.4562	246.8562
0.734375	47/64	18.6531	44.0531	69.4531	94.8531	120.2531	145.6531	171.0531	196.4531	221.8531	247.2531
0.750000	3/4	19.0500	44.4500	69.8500	95.2500	120.6500	146.0500	171.4500	196.8500	222.2500	247.6499
0.765625	49/64	19.4469	44.8469	70.2469	95.6469	121.0468	146.4468	171.8468	197.2468	222.6468	248.0468
0.781250	25/32	19.8438	45.2437	70.6437	96.0437	121.4437	146.8437	172.2437	197.6437	223.0437	248.4437
0.796875	51/64	20.2406	45.6406	71.0406	96.4406	121.8406	147.2406	172.6406	198.0406	223.4406	248.8406
0.812500	13/16	20.6375	46.0375	71.4375	96.8375	122.2375	147.6375	173.0375	198.4375	223.8374	249.2374
0.828125	53/64	21.0344	46.4344	71.8344	97.2343	122.6343	148.0343	173.4343	198.8343	224.2343	249.6343
0.843750	27/32	21.4312	46.8312	72.2312	97.6312	123.0312	148.4312	173.8312	199.2312	224.6312	250.0312
0.859375	55/64	21.8281	47.2281	72.6281	98.0281	123.4281	148.8281	174.2281	199.6281	225.0281	250.4281
0.875000	7/8	22.2250	47.6250	73.0250	98.4250	123.8250	149.2250	174.6250	200.0249	225.4249	250.8249
0.890625	57/64	22.6219	48.0219	73.4219	98.8219	124.2218	149.6218	175.0218	200.4218	225.8218	251.2218
0.906250	29/32	23.0187	48.4187	73.8187	99.2187	124.6187	150.0187	175.4187	200.8187	226.2187	251.6187
0.921875	59/64	23.4156	48.8156	74.2156	99.6156	125.0156	150.4156	175.8156	201.2156	226.6156	252.0156
0.937500	15/16	23.8125	49.2125	74.6125	100.0125	125.4125	150.8125	176.2125	201.6125	227.0125	252.4124
0.953125	61/64	24.2094	49.6094	75.0094	100.4093	125.8093	151.2093	176.6093	202.0093	227.4093	252.8093
0.968750	31/32	24.6062	50.0062	75.4062	100.8062	126.2062	151.6062	177.0062	202.4062	227.8062	253.2062
0.984375	63/64	25.0031	50.4031	75.8031	101.2031	126.6031	152.0031	177.4031	202.8031	228.2031	253.6031

Table 19-3 (*Continued*). Length: Fractional Inches to Millimeters (from 10 1/64 to 19 63/64 Inches)

INCHES		10	11	12	13	14	15	16	17	18	19
					MILLIMETERS						
0.000000	0/64	253.9999	279.3999	304.8000	330.2000	355.5999	381.0000	406.3999	431.7998	457.2000	482.5999
0.015625	1/64	254.3968	279.7969	305.1968	330.5967	355.9968	381.3967	406.7969	432.1968	457.5967	482.9968
0.031250	1/32	254.7937	280.1936	305.5937	330.9937	356.3936	381.7937	407.1936	432.5937	457.9937	483.3936
0.046875	3/64	255.1906	280.5906	305.9905	331.3906	356.7905	382.1904	407.5906	432.9905	458.3906	483.7905
0.062500	1/16	255.5874	280.9875	306.3875	331.7874	357.1875	382.5874	407.9873	433.3875	458.7874	484.1875
0.078125	5/64	255.9843	281.3843	306.7842	332.1843	357.5842	382.9844	408.3843	433.7842	459.1843	484.5842
0.093750	3/32	256.3811	281.7813	307.1812	332.5811	357.9812	383.3811	408.7813	434.1812	459.5811	484.9812
0.109375	7/64	256.7781	282.1780	307.5781	332.9780	358.3779	383.7781	409.1780	434.5781	459.9780	485.3779
0.125000	1/8	257.1750	282.5750	307.9749	333.3750	358.7749	384.1748	409.5750	434.9749	460.3750	485.7749
0.140625	9/64	257.5718	282.9719	308.3718	333.7717	359.1719	384.5718	409.9717	435.3718	460.7717	486.1719
0.156250	5/32	257.9687	283.3687	308.7686	334.1687	359.5686	384.9687	410.3687	435.7686	461.1687	486.5686
0.171875	11/64	258.3655	283.7656	309.1655	334.5654	359.9656	385.3655	410.7656	436.1655	461.5654	486.9656
0.187500	3/16	258.7625	284.1624	309.5625	334.9624	360.3623	385.7625	411.1624	436.5625	461.9624	487.3623
0.203125	13/64	259.1594	284.5593	309.9592	335.3594	360.7593	386.1592	411.5593	436.9592	462.3594	487.7593
0.218750	7/32	259.5562	284.9563	310.3562	335.7561	361.1563	386.5562	411.9561	437.3562	462.7561	488.1563
0.234375	15/64	259.9531	285.3530	310.7529	336.1531	361.5530	386.9531	412.3530	437.7529	463.1531	488.5530
0.250000	1/4	260.3499	285.7500	311.1499	336.5498	361.9500	387.3499	412.7500	438.1499	463.5498	488.9500
0.265625	17/64	260.7468	286.1467	311.5469	336.9468	362.3467	387.7468	413.1467	438.5469	463.9468	489.3467
0.281250	9/32	261.1438	286.5437	311.9436	337.3438	362.7437	388.1436	413.5437	438.9436	464.3437	489.7437
0.296875	19/64	261.5405	286.9407	312.3406	337.7405	363.1406	388.5405	413.9404	439.3406	464.7405	490.1406
0.312500	5/16	261.9375	287.3374	312.7373	338.1375	363.5374	388.9375	414.3374	439.7373	465.1375	490.5374
0.328125	21/64	262.3342	287.7344	313.1343	338.5342	363.9343	389.3342	414.7344	440.1343	465.5342	490.9343
0.343750	11/32	262.7312	288.1311	313.5312	338.9312	364.3311	389.7312	415.1311	440.5313	465.9312	491.3311
0.359375	23/64	263.1282	288.5281	313.9280	339.3281	364.7280	390.1279	415.5281	440.9280	466.3281	491.7280
0.375000	3/8	263.5249	288.9250	314.3250	339.7249	365.1250	390.5249	415.9248	441.3250	466.7249	492.1250
0.390625	25/64	263.9219	289.3218	314.7217	340.1218	365.5217	390.9219	416.3218	441.7217	467.1218	492.5217
0.406250	13/32	264.3186	289.7187	315.1187	340.5186	365.9187	391.3186	416.7188	442.1187	467.5186	492.9187
0.421875	27/64	264.7156	290.1155	315.5156	340.9155	366.3154	391.7156	417.1155	442.5156	467.9155	493.3154
0.437500	7/16	265.1125	290.5125	315.9124	341.3125	366.7124	392.1123	417.5125	442.9124	468.3125	493.7124
0.453125	29/64	265.5093	290.9094	316.3093	341.7092	367.1094	392.5093	417.9092	443.3093	468.7092	494.1094
0.468750	15/32	265.9062	291.3062	316.7061	342.1062	367.5061	392.9062	418.3062	443.7061	469.1062	494.5061
0.484375	31/64	266.3030	291.7031	317.1030	342.5029	367.9031	393.3030	418.7031	444.1030	469.5029	494.9031
0.500000	1/2	266.7000	292.0999	317.5000	342.8999	368.2998	393.7000	419.0999	444.5000	469.8999	495.2998
0.515625	33/64	267.0969	292.4968	317.8967	343.2969	368.6968	394.0967	419.4968	444.8967	470.2969	495.6968
0.531250	17/32	267.4937	292.8938	318.2937	343.6936	369.0937	394.4937	419.8936	445.2937	470.6936	496.0938
0.546875	35/64	267.8906	293.2905	318.6904	344.0906	369.4905	394.8906	420.2905	445.6904	471.0906	496.4905
0.562500	9/16	268.2874	293.6875	319.0874	344.4873	369.8875	395.2874	420.6875	446.0874	471.4873	496.8875
0.578125	37/64	268.6843	294.0842	319.4844	344.8843	370.2842	395.6843	421.0842	446.4844	471.8843	497.2842
0.593750	19/32	269.0813	294.4812	319.8811	345.2812	370.6812	396.0811	421.4812	446.8811	472.2812	497.6812
0.609375	39/64	269.4780	294.8782	320.2781	345.6780	371.0781	396.4780	421.8779	447.2781	472.6780	498.0781
0.625000	5/8	269.8750	295.2749	320.6748	346.0750	371.4749	396.8750	422.2749	447.6748	473.0750	498.4749
0.640625	41/64	270.2717	295.6719	321.0718	346.4717	371.8718	397.2717	422.6719	448.0718	473.4717	498.8718
0.656250	21/32	270.6687	296.0686	321.4688	346.8687	372.2686	397.6687	423.0686	448.4688	473.8687	499.2686
0.671875	43/64	271.0657	296.4656	321.8655	347.2656	372.6655	398.0654	423.4656	448.8655	474.2656	499.6655
0.687500	11/16	271.4624	296.8625	322.2625	347.6624	373.0625	398.4624	423.8623	449.2625	474.6624	500.0625
0.703125	45/64	271.8594	297.2593	322.6592	348.0593	373.4592	398.8594	424.2593	449.6592	475.0593	500.4592
0.718750	23/32	272.2561	297.6562	323.0562	348.4561	373.8562	399.2561	424.6563	450.0562	475.4561	500.8562
0.734375	47/64	272.6531	298.0530	323.4531	348.8530	374.2529	399.6531	425.0530	450.4531	475.8530	501.2529
0.750000	3/4	273.0500	298.4500	323.8499	349.2500	374.6499	400.0498	425.4500	450.8499	476.2500	501.6499
0.765625	49/64	273.4468	298.8469	324.2468	349.6467	375.0469	400.4468	425.8467	451.2468	476.6467	502.0469
0.781250	25/32	273.8437	299.2437	324.6436	350.0437	375.4436	400.8438	426.2437	451.6436	477.0437	502.4436
0.796875	51/64	274.2405	299.6406	325.0405	350.4404	375.8406	401.2405	426.6406	452.0405	477.4404	502.8406
0.812500	13/16	274.6375	300.0374	325.4375	350.8374	376.2373	401.6375	427.0374	452.4375	477.8374	503.2373
0.828125	53/64	275.0344	300.4343	325.8342	351.2344	376.6343	402.0342	427.4343	452.8342	478.2344	503.6343
0.843750	27/32	275.4312	300.8313	326.2312	351.6311	377.0312	402.4312	427.8311	453.2312	478.6311	504.0313
0.859375	55/64	275.8281	301.2280	326.6279	352.0281	377.4280	402.8281	428.2280	453.6279	479.0281	504.4280
0.875000	7/8	276.2249	301.6250	327.0249	352.4248	377.8250	403.2249	428.6250	454.0249	479.4248	504.8250
0.890625	57/64	276.6218	302.0217	327.4219	352.8218	378.2217	403.6218	429.0217	454.4219	479.8218	505.2217
0.906250	29/32	277.0188	302.4187	327.8186	353.2187	378.6187	404.0186	429.4187	454.8186	480.2188	505.6187
0.921875	59/64	277.4155	302.8157	328.2156	353.6155	379.0156	404.4155	429.8154	455.2156	480.6155	506.0156
0.937500	15/16	277.8125	303.2124	328.6123	354.0125	379.4124	404.8125	430.2124	455.6123	481.0125	506.4124
0.953125	61/64	278.2092	303.6094	329.0093	354.4092	379.8093	405.2092	430.6094	456.0093	481.4092	506.8093
0.968750	31/32	278.6062	304.0061	329.4063	354.8062	380.2061	405.6062	431.0061	456.4062	481.8062	507.2061
0.984375	63/64	279.0032	304.4031	329.8030	355.2031	380.6030	406.0029	431.4031	456.8030	482.2031	507.6030

METRIC CONVERSION FACTORS AND TABLES

19-15

Table 19-4. Temperature: Kelvins, Degrees Celsius and Degrees Fahrenheit Conversions

1 DEG CELSIUS = (DEG F - 32)/1.8 (EXACTLY)

K	°C	°F → °C	°C → °F
0.0	-273.15	-459.67	
2.0	-271.1	-456.0	
3.1	-270.0	-454.0	
4.3	-268.9	-452.0	
5.4	-267.8	-450.0	
6.5	-266.7	-448.0	
7.6	-265.6	-446.0	
8.7	-264.4	-444.0	
9.8	-263.3	-442.0	
10.9	-262.2	-440.0	
12.0	-261.1	-438.0	
13.1	-260.0	-436.0	
14.3	-258.9	-434.0	
15.4	-257.8	-432.0	
16.5	-256.7	-430.0	
17.6	-255.6	-428.0	
18.7	-254.4	-426.0	
19.8	-253.3	-424.0	
20.9	-252.2	-422.0	
22.0	-251.1	-420.0	
23.1	-250.0	-418.0	
24.3	-248.9	-416.0	
25.4	-247.8	-414.0	
26.5	-246.7	-412.0	
27.6	-245.6	-410.0	
28.7	-244.4	-408.0	
29.8	-243.3	-406.0	
30.9	-242.2	-404.0	
32.0	-241.1	-402.0	
33.1	-240.0	-400.0	
34.3	-238.9	-398.0	
35.4	-237.8	-396.0	
36.5	-236.7	-394.0	
37.6	-235.6	-392.0	
38.7	-234.4	-390.0	
39.8	-233.3	-388.0	
40.9	-232.2	-386.0	
42.0	-231.1	-384.0	
43.1	-230.0	-382.0	
44.3	-228.9	-380.0	
45.4	-227.8	-378.0	
46.5	-226.7	-376.0	
47.6	-225.6	-374.0	
48.7	-224.4	-372.0	
49.8	-223.3	-370.0	
50.9	-222.2	-368.0	
52.0	-221.1	-366.0	
53.1	-220.0	-364.0	
54.3	-218.9	-362.0	
55.4	-217.8	-360.0	
56.5	-216.7	-358.0	
57.6	-215.6	-356.0	
58.7	-214.4	-354.0	
59.8	-213.3	-352.0	
60.9	-212.2	-350.0	
62.0	-211.1	-348.0	
63.1	-210.0	-346.0	
64.3	-208.9	-344.0	
65.4	-207.8	-342.0	
66.5	-206.7	-340.0	
67.6	-205.6	-338.0	
68.7	-204.4	-336.0	

K	°C	°F → °C	°C → °F
69.8	-203.3	-334.0	
70.9	-202.2	-332.0	
72.0	-201.1	-330.0	
73.1	-200.0	-328.0	
74.3	-198.9	-326.0	
75.4	-197.8	-324.0	
76.5	-196.7	-322.0	
77.6	-195.6	-320.0	
78.7	-194.4	-318.0	
79.8	-193.3	-316.0	
80.9	-192.2	-314.0	
82.0	-191.1	-312.0	
83.1	-190.0	-310.0	
84.3	-188.9	-308.0	
85.4	-187.8	-306.0	
86.5	-186.7	-304.0	
87.6	-185.6	-302.0	
88.7	-184.4	-300.0	
89.8	-183.3	-298.0	
90.9	-182.2	-296.0	
92.0	-181.1	-294.0	
93.1	-180.0	-292.0	
94.3	-178.9	-290.0	
95.4	-177.8	-288.0	
96.5	-176.7	-286.0	
97.6	-175.6	-284.0	
98.7	-174.4	-282.0	
99.8	-173.3	-280.0	
100.9	-172.2	-278.0	
102.0	-171.1	-276.0	
103.1	-170.0	-274.0	
104.3	-168.9	-272.0	-457.6
105.4	-167.8	-270.0	-454.0
106.5	-166.7	-268.0	-450.4
107.6	-165.6	-266.0	-446.8
108.7	-164.4	-264.0	-443.2
109.8	-163.3	-262.0	-439.6
110.9	-162.2	-260.0	-436.0
112.0	-161.1	-258.0	-432.4
113.1	-160.0	-256.0	-428.8
114.3	-158.9	-254.0	-425.2
115.4	-157.8	-252.0	-421.6
116.5	-156.7	-250.0	-418.0
117.6	-155.6	-248.0	-414.4
118.7	-154.4	-246.0	-410.8
119.8	-153.3	-244.0	-407.2
120.9	-152.2	-242.0	-403.6
122.0	-151.1	-240.0	-400.0
123.1	-150.0	-238.0	-396.4
124.3	-148.9	-236.0	-392.8
125.4	-147.8	-234.0	-389.2
126.5	-146.7	-232.0	-385.6
127.6	-145.6	-230.0	-382.0
128.7	-144.4	-228.0	-378.4
129.8	-143.3	-226.0	-374.8
130.9	-142.2	-224.0	-371.2
132.0	-141.1	-222.0	-367.6
133.1	-140.0	-220.0	-364.0
134.3	-138.9	-218.0	-360.4
135.4	-137.8	-216.0	-356.8
136.5	-136.7	-214.0	-353.2
137.6	-135.6	-212.0	-349.6

K	°C	°F → °C	°C → °F
138.7	-134.4	-210.0	-346.0
139.8	-133.3	-208.0	-342.4
140.9	-132.2	-206.0	-338.8
142.0	-131.1	-204.0	-335.2
143.1	-130.0	-202.0	-331.6
144.3	-128.9	-200.0	-328.0
145.4	-127.8	-198.0	-324.4
146.5	-126.7	-196.0	-320.8
147.6	-125.6	-194.0	-317.2
148.7	-124.4	-192.0	-313.6
149.8	-123.3	-190.0	-310.0
150.9	-122.2	-188.0	-306.4
152.0	-121.1	-186.0	-302.8
153.1	-120.0	-184.0	-299.2
154.3	-118.9	-182.0	-295.6
155.4	-117.8	-180.0	-292.0
156.5	-116.7	-178.0	-288.4
157.6	-115.6	-176.0	-284.8
158.7	-114.4	-174.0	-281.2
159.8	-113.3	-172.0	-277.6
160.9	-112.2	-170.0	-274.0
162.0	-111.1	-168.0	-270.4
163.1	-110.0	-166.0	-266.8
164.3	-108.9	-164.0	-263.2
165.4	-107.8	-162.0	-259.6
166.5	-106.7	-160.0	-256.0
167.6	-105.6	-158.0	-252.4
168.7	-104.4	-156.0	-248.8
169.8	-103.3	-154.0	-245.2
170.9	-102.2	-152.0	-241.6
172.0	-101.1	-150.0	-238.0
173.1	-100.0	-148.0	-234.4
174.3	-98.9	-146.0	-230.8
175.4	-97.8	-144.0	-227.2
176.5	-96.7	-142.0	-223.6
177.6	-95.6	-140.0	-220.0
178.7	-94.4	-138.0	-216.4
179.8	-93.3	-136.0	-212.8
180.9	-92.2	-134.0	-209.2
182.0	-91.1	-132.0	-205.6
183.1	-90.0	-130.0	-202.0
184.3	-88.9	-128.0	-198.4
185.4	-87.8	-126.0	-194.8
186.5	-86.7	-124.0	-191.2
187.6	-85.6	-122.0	-187.6
188.7	-84.4	-120.0	-184.0
189.8	-83.3	-118.0	-180.4
190.9	-82.2	-116.0	-176.8
192.0	-81.1	-114.0	-173.2
193.1	-80.0	-112.0	-169.6
194.3	-78.9	-110.0	-166.0
195.4	-77.8	-108.0	-162.4
196.5	-76.7	-106.0	-158.8
197.6	-75.6	-104.0	-155.2
198.7	-74.4	-102.0	-151.6
199.8	-73.3	-100.0	-148.0
200.9	-72.2	-98.0	-144.4
202.0	-71.1	-96.0	-140.8
203.1	-70.0	-94.0	-137.2
204.3	-68.9	-92.0	-133.6
205.4	-67.8	-90.0	-130.0
206.5	-66.7	-88.0	-126.4

Table 19-4 (Continued). Temperature: Kelvins, Degrees Celsius and Degrees Fahrenheit Conversions

1 DEG CELSIUS = (DEG F - 32)/1.8 (EXACTLY)

K	°C	°C → °F	°F
207.6	-65.6	-86.0	-122.8
208.7	-64.4	-84.0	-119.2
209.8	-63.3	-82.0	-115.6
210.9	-62.2	-80.0	-112.0
212.0	-61.1	-78.0	-108.4
213.1	-60.0	-76.0	-104.8
214.3	-58.9	-74.0	-101.2
215.4	-57.8	-72.0	-97.6
216.5	-56.7	-70.0	-94.0
217.6	-55.6	-68.0	-90.4
218.7	-54.4	-66.0	-86.8
219.8	-53.3	-64.0	-83.2
220.9	-52.2	-62.0	-79.6
222.0	-51.1	-60.0	-76.0
223.1	-50.0	-58.0	-72.4
224.3	-48.9	-56.0	-68.8
225.4	-47.8	-54.0	-65.2
226.5	-46.7	-52.0	-61.6
227.6	-45.6	-50.0	-58.0
228.7	-44.4	-48.0	-54.4
229.8	-43.3	-46.0	-50.8
230.9	-42.2	-44.0	-47.2
232.0	-41.1	-42.0	-43.6
233.1	-40.0	-40.0	-40.0
234.3	-38.9	-38.0	-36.4
235.4	-37.8	-36.0	-32.8
236.5	-36.7	-34.0	-29.2
237.6	-35.6	-32.0	-25.6
238.7	-34.4	-30.0	-22.0
239.8	-33.3	-28.0	-18.4
240.9	-32.2	-26.0	-14.8
242.0	-31.1	-24.0	-11.2
243.1	-30.0	-22.0	-7.6
244.3	-28.9	-20.0	-4.0
244.8	-28.3	-19.0	-2.2
245.4	-27.8	-18.0	-0.4
245.9	-27.2	-17.0	1.4
246.5	-26.7	-16.0	3.2
247.0	-26.1	-15.0	5.0
247.6	-25.6	-14.0	6.8
248.1	-25.0	-13.0	8.6
248.7	-24.4	-12.0	10.4
249.3	-23.9	-11.0	12.2
249.8	-23.3	-10.0	14.0
250.4	-22.8	-9.0	15.8
250.9	-22.2	-8.0	17.6
251.5	-21.7	-7.0	19.4
252.0	-21.1	-6.0	21.2
252.6	-20.6	-5.0	23.0
253.1	-20.0	-4.0	24.8
253.7	-19.4	-3.0	26.6
254.3	-18.9	-2.0	28.4
254.8	-18.3	-1.0	30.2
255.4	-17.8	0.0	32.0
255.9	-17.2	1.0	33.8
256.5	-16.7	2.0	35.6
257.0	-16.1	3.0	37.4
257.6	-15.6	4.0	39.2
258.1	-15.0	5.0	41.0
258.7	-14.4	6.0	42.8
259.3	-13.9	7.0	44.6
259.8	-13.3	8.0	46.4

K	°C	°C → °F	°F
260.4	-12.8	9.0	48.2
260.9	-12.2	10.0	50.0
261.5	-11.7	11.0	51.8
262.0	-11.1	12.0	53.6
262.6	-10.6	13.0	55.4
263.1	-10.0	14.0	57.2
263.7	-9.4	15.0	59.0
264.3	-8.9	16.0	60.8
264.8	-8.3	17.0	62.6
265.4	-7.8	18.0	64.4
265.9	-7.2	19.0	66.2
266.5	-6.7	20.0	68.0
267.0	-6.1	21.0	69.8
267.6	-5.6	22.0	71.6
268.1	-5.0	23.0	73.4
268.7	-4.4	24.0	75.2
269.3	-3.9	25.0	77.0
269.8	-3.3	26.0	78.8
270.4	-2.8	27.0	80.6
270.9	-2.2	28.0	82.4
271.5	-1.7	29.0	84.2
272.0	-1.1	30.0	86.0
272.6	-0.6	31.0	87.8
273.1	0.0	32.0	89.6
273.7	0.6	33.0	91.4
274.3	1.1	34.0	93.2
274.8	1.7	35.0	95.0
275.4	2.2	36.0	96.8
275.9	2.8	37.0	98.6
276.5	3.3	38.0	100.4
277.0	3.9	39.0	102.2
277.6	4.4	40.0	104.0
278.1	5.0	41.0	105.8
278.7	5.6	42.0	107.6
279.3	6.1	43.0	109.4
279.8	6.7	44.0	111.2
280.4	7.2	45.0	113.0
280.9	7.8	46.0	114.8
281.5	8.3	47.0	116.6
282.0	8.9	48.0	118.4
282.6	9.4	49.0	120.2
283.1	10.0	50.0	122.0
283.7	10.6	51.0	123.8
284.3	11.1	52.0	125.6
284.8	11.7	53.0	127.4
285.4	12.2	54.0	129.2
285.9	12.8	55.0	131.0
286.5	13.3	56.0	132.8
287.0	13.9	57.0	134.6
287.6	14.4	58.0	136.4
288.1	15.0	59.0	138.2
288.7	15.6	60.0	140.0
289.3	16.1	61.0	141.8
289.8	16.7	62.0	143.6
290.4	17.2	63.0	145.4
290.9	17.8	64.0	147.2
291.5	18.3	65.0	149.0
292.0	18.9	66.0	150.8
292.6	19.4	67.0	152.6
293.1	20.0	68.0	154.4
293.7	20.6	69.0	156.2
294.3	21.1	70.0	158.0

K	°C	°C → °F	°F
294.8	21.7	71.0	159.8
295.4	22.2	72.0	161.6
295.9	22.8	73.0	163.4
296.5	23.3	74.0	165.2
297.0	23.9	75.0	167.0
297.6	24.4	76.0	168.8
298.1	25.0	77.0	170.6
298.7	25.6	78.0	172.4
299.3	26.1	79.0	174.2
299.8	26.7	80.0	176.0
300.4	27.2	81.0	177.8
300.9	27.8	82.0	179.6
301.5	28.3	83.0	181.4
302.0	28.9	84.0	183.2
302.6	29.4	85.0	185.0
303.1	30.0	86.0	186.8
303.7	30.6	87.0	188.6
304.3	31.1	88.0	190.4
304.8	31.7	89.0	192.2
305.4	32.2	90.0	194.0
305.9	32.8	91.0	195.8
306.5	33.3	92.0	197.6
307.0	33.9	93.0	199.4
307.6	34.4	94.0	201.2
308.1	35.0	95.0	203.0
308.7	35.6	96.0	204.8
309.3	36.1	97.0	206.6
309.8	36.7	98.0	208.4
310.4	37.2	99.0	210.2
310.9	37.8	100.0	212.0
312.0	38.9	102.0	215.6
313.1	40.0	104.0	219.2
314.3	41.1	106.0	222.8
315.4	42.2	108.0	226.4
316.5	43.3	110.0	230.0
317.6	44.4	112.0	233.6
318.7	45.6	114.0	237.2
319.8	46.7	116.0	240.8
320.9	47.8	118.0	244.4
322.0	48.9	120.0	248.0
323.1	50.0	122.0	251.6
324.3	51.1	124.0	255.2
325.4	52.2	126.0	258.8
326.5	53.3	128.0	262.4
327.6	54.4	130.0	266.0
328.7	55.6	132.0	269.6
329.8	56.7	134.0	273.2
330.9	57.8	136.0	276.8
332.0	58.9	138.0	280.4
333.1	60.0	140.0	284.0
334.3	61.1	142.0	287.6
335.4	62.2	144.0	291.2
336.5	63.3	146.0	294.8
337.6	64.4	148.0	298.4
338.7	65.6	150.0	302.0
339.8	66.7	152.0	305.6
340.9	67.8	154.0	309.2
342.0	68.9	156.0	312.8
343.1	70.0	158.0	316.4
344.3	71.1	160.0	320.0
345.4	72.2	162.0	323.6
346.5	73.3	164.0	327.2

METRIC CONVERSION FACTORS AND TABLES

Table 19-4 (*Continued*). Temperature: Kelvins, Degrees Celsius and Degrees Fahrenheit Conversions

1 DEG CELSIUS = (DEG F - 32)/1.8 (EXACTLY)

K	°C	°C	°F	K	°C	°C	°F	K	°C	°C	°F
347.6	74.4	166.0	330.8	474.8	201.7	395.0	743.0	755.4	482.2	900.0	1652.0
348.7	75.6	168.0	334.4	477.6	204.4	400.0	752.0	760.9	487.8	910.0	1670.0
349.8	76.7	170.0	338.0	480.4	207.2	405.0	761.0	766.5	493.3	920.0	1688.0
350.9	77.8	172.0	341.6	483.1	210.0	410.0	770.0	772.0	498.9	930.0	1706.0
352.0	78.9	174.0	345.2	485.9	212.8	415.0	779.0	777.6	504.4	940.0	1724.0
353.1	80.0	176.0	348.8	488.7	215.6	420.0	788.0	783.2	510.0	950.0	1742.0
354.3	81.1	178.0	352.4	491.5	218.3	425.0	797.0	788.7	515.6	960.0	1760.0
355.4	82.2	180.0	356.0	494.3	221.1	430.0	806.0	794.3	521.1	970.0	1778.0
356.5	83.3	182.0	359.6	497.0	223.9	435.0	815.0	799.8	526.7	980.0	1796.0
357.6	84.4	184.0	363.2	499.8	226.7	440.0	824.0	805.4	532.2	990.0	1814.0
358.7	85.6	186.0	366.8	502.6	229.4	445.0	833.0	810.9	537.8	1000.0	1832.0
359.8	86.7	188.0	370.4	505.4	232.2	450.0	842.0	816.5	543.3	1010.0	1850.0
360.9	87.8	190.0	374.0	508.1	235.0	455.0	851.0	822.0	548.9	1020.0	1868.0
362.0	88.9	192.0	377.6	510.9	237.8	460.0	860.0	827.6	554.4	1030.0	1886.0
363.1	90.0	194.0	381.2	513.7	240.6	465.0	869.0	833.2	560.0	1040.0	1904.0
364.3	91.1	196.0	384.8	516.5	243.3	470.0	878.0	838.7	565.6	1050.0	1922.0
365.4	92.2	198.0	388.4	519.3	246.1	475.0	887.0	844.3	571.1	1060.0	1940.0
366.5	93.3	200.0	392.0	522.0	248.9	480.0	896.0	849.8	576.7	1070.0	1958.0
367.6	94.4	202.0	395.6	524.8	251.7	485.0	905.0	855.4	582.2	1080.0	1976.0
368.7	95.6	204.0	399.2	527.6	254.4	490.0	914.0	860.9	587.8	1090.0	1994.0
369.8	96.7	206.0	402.8	530.4	257.2	495.0	923.0	866.5	593.3	1100.0	2012.0
370.9	97.8	208.0	406.4	533.1	260.0	500.0	932.0	872.0	598.9	1110.0	2030.0
372.0	98.9	210.0	410.0	535.9	262.8	505.0	941.0	877.6	604.4	1120.0	2048.0
373.1	100.0	212.0	413.6	538.7	265.6	510.0	950.0	883.2	610.0	1130.0	2066.0
374.3	101.1	214.0	417.2	544.3	271.1	520.0	968.0	888.7	615.6	1140.0	2084.0
375.4	102.2	216.0	420.8	549.8	276.7	530.0	986.0	894.3	621.1	1150.0	2102.0
376.5	103.3	218.0	424.4	555.4	282.2	540.0	1004.0	899.8	626.7	1160.0	2120.0
377.6	104.4	220.0	428.0	560.9	287.8	550.0	1022.0	905.4	632.2	1170.0	2138.0
380.4	107.2	225.0	437.0	566.5	293.3	560.0	1040.0	910.9	637.8	1180.0	2156.0
383.1	110.0	230.0	446.0	572.0	298.9	570.0	1058.0	916.5	643.3	1190.0	2174.0
385.9	112.8	235.0	455.0	577.6	304.4	580.0	1076.0	922.0	648.9	1200.0	2192.0
388.7	115.6	240.0	464.0	583.2	310.0	590.0	1094.0	927.6	654.4	1210.0	2210.0
391.5	118.3	245.0	473.0	588.7	315.6	600.0	1112.0	933.2	660.0	1220.0	2228.0
394.3	121.1	250.0	482.0	594.3	321.1	610.0	1130.0	938.7	665.6	1230.0	2246.0
397.0	123.9	255.0	491.0	599.8	326.7	620.0	1148.0	944.3	671.1	1240.0	2264.0
399.8	126.7	260.0	500.0	605.4	332.2	630.0	1166.0	949.8	676.7	1250.0	2282.0
402.6	129.4	265.0	509.0	610.9	337.8	640.0	1184.0	955.4	682.2	1260.0	2300.0
405.4	132.2	270.0	518.0	616.5	343.3	650.0	1202.0	960.9	687.8	1270.0	2318.0
408.1	135.0	275.0	527.0	622.0	348.9	660.0	1220.0	966.5	693.3	1280.0	2336.0
410.9	137.8	280.0	536.0	627.6	354.4	670.0	1238.0	972.0	698.9	1290.0	2354.0
413.7	140.6	285.0	545.0	633.2	360.0	680.0	1256.0	977.6	704.4	1300.0	2372.0
416.5	143.3	290.0	554.0	638.7	365.6	690.0	1274.0	983.2	710.0	1310.0	2390.0
419.3	146.1	295.0	563.0	644.3	371.1	700.0	1292.0	988.7	715.6	1320.0	2408.0
422.0	148.9	300.0	572.0	649.8	376.7	710.0	1310.0	994.3	721.1	1330.0	2426.0
424.8	151.7	305.0	581.0	655.4	382.2	720.0	1328.0	999.8	726.7	1340.0	2444.0
427.6	154.4	310.0	590.0	660.9	387.8	730.0	1346.0	1005.4	732.2	1350.0	2462.0
430.4	157.2	315.0	599.0	666.5	393.3	740.0	1364.0	1010.9	737.8	1360.0	2480.0
433.1	160.0	320.0	608.0	672.0	398.9	750.0	1382.0	1016.5	743.3	1370.0	2498.0
435.9	162.8	325.0	617.0	677.6	404.4	760.0	1400.0	1022.0	748.9	1380.0	2516.0
438.7	165.6	330.0	626.0	683.2	410.0	770.0	1418.0	1027.6	754.4	1390.0	2534.0
441.5	168.3	335.0	635.0	688.7	415.6	780.0	1436.0	1033.2	760.0	1400.0	2552.0
444.3	171.1	340.0	644.0	694.3	421.1	790.0	1454.0	1038.7	765.6	1410.0	2570.0
447.0	173.9	345.0	653.0	699.8	426.7	800.0	1472.0	1044.3	771.1	1420.0	2588.0
449.8	176.7	350.0	662.0	705.4	432.2	810.0	1490.0	1049.8	776.7	1430.0	2606.0
452.6	179.4	355.0	671.0	710.9	437.8	820.0	1508.0	1055.4	782.2	1440.0	2624.0
455.4	182.2	360.0	680.0	716.5	443.3	830.0	1526.0	1060.9	787.8	1450.0	2642.0
458.1	185.0	365.0	689.0	722.0	448.9	840.0	1544.0	1066.5	793.3	1460.0	2660.0
460.9	187.8	370.0	698.0	727.6	454.4	850.0	1562.0	1072.0	798.9	1470.0	2678.0
463.7	190.6	375.0	707.0	733.2	460.0	860.0	1580.0	1077.6	804.4	1480.0	2696.0
466.5	193.3	380.0	716.0	738.7	465.6	870.0	1598.0	1083.2	810.0	1490.0	2714.0
469.3	196.1	385.0	725.0	744.3	471.1	880.0	1616.0	1088.7	815.6	1500.0	2732.0
472.0	198.9	390.0	734.0	749.8	476.7	890.0	1634.0	1094.3	821.1	1510.0	2750.0

Table 19-4 (*Continued*). Temperature: Kelvins, Degrees Celsius and Degrees Fahrenheit Conversions

1 DEG CELSIUS = (DEG F - 32)/1.8 (EXACTLY)

K	°C	°C	°F	K	°C	°C	°F	K	°C	°C	°F
1099.8	826.7	1520.0	2768.0	1444.3	1171.1	2140.0	3884.0	1788.7	1515.6	2760.0	5000.0
1105.4	832.2	1530.0	2786.0	1449.8	1176.7	2150.0	3902.0	1794.3	1521.1	2770.0	5018.0
1110.9	837.8	1540.0	2804.0	1455.4	1182.2	2160.0	3920.0	1799.8	1526.7	2780.0	5036.0
1116.5	843.3	1550.0	2822.0	1460.9	1187.8	2170.0	3938.0	1805.4	1532.2	2790.0	5054.0
1122.0	848.9	1560.0	2840.0	1466.5	1193.3	2180.0	3956.0	1810.9	1537.8	2800.0	5072.0
1127.6	854.4	1570.0	2858.0	1472.0	1198.9	2190.0	3974.0	1816.5	1543.3	2810.0	5090.0
1133.2	860.0	1580.0	2876.0	1477.6	1204.4	2200.0	3992.0	1822.0	1548.9	2820.0	5108.0
1138.7	865.6	1590.0	2894.0	1483.2	1210.0	2210.0	4010.0	1827.6	1554.4	2830.0	5126.0
1144.3	871.1	1600.0	2912.0	1488.7	1215.6	2220.0	4028.0	1833.2	1560.0	2840.0	5144.0
1149.8	876.7	1610.0	2930.0	1494.3	1221.1	2230.0	4046.0	1838.7	1565.6	2850.0	5162.0
1155.4	882.2	1620.0	2948.0	1499.8	1226.7	2240.0	4064.0	1844.3	1571.1	2860.0	5180.0
1160.9	887.8	1630.0	2966.0	1505.4	1232.2	2250.0	4082.0	1849.8	1576.7	2870.0	5198.0
1166.5	893.3	1640.0	2984.0	1510.9	1237.8	2260.0	4100.0	1855.4	1582.2	2880.0	5216.0
1172.0	898.9	1650.0	3002.0	1516.5	1243.3	2270.0	4118.0	1860.9	1587.8	2890.0	5234.0
1177.6	904.4	1660.0	3020.0	1522.0	1248.9	2280.0	4136.0	1866.5	1593.3	2900.0	5252.0
1183.2	910.0	1670.0	3038.0	1527.6	1254.4	2290.0	4154.0	1872.0	1598.9	2910.0	5270.0
1188.7	915.6	1680.0	3056.0	1533.2	1260.0	2300.0	4172.0	1877.6	1604.4	2920.0	5288.0
1194.3	921.1	1690.0	3074.0	1538.7	1265.6	2310.0	4190.0	1883.2	1610.0	2930.0	5306.0
1199.8	926.7	1700.0	3092.0	1544.3	1271.1	2320.0	4208.0	1888.7	1615.6	2940.0	5324.0
1205.4	932.2	1710.0	3110.0	1549.8	1276.7	2330.0	4226.0	1894.3	1621.1	2950.0	5342.0
1210.9	937.8	1720.0	3128.0	1555.4	1282.2	2340.0	4244.0	1899.8	1626.7	2960.0	5360.0
1216.5	943.3	1730.0	3146.0	1560.9	1287.8	2350.0	4262.0	1905.4	1632.2	2970.0	5378.0
1222.0	948.9	1740.0	3164.0	1566.5	1293.3	2360.0	4280.0	1910.9	1637.8	2980.0	5396.0
1227.6	954.4	1750.0	3182.0	1572.0	1298.9	2370.0	4298.0	1916.5	1643.3	2990.0	5414.0
1233.2	960.0	1760.0	3200.0	1577.6	1304.4	2380.0	4316.0	1922.0	1648.9	3000.0	5432.0
1238.7	965.6	1770.0	3218.0	1583.2	1310.0	2390.0	4334.0	1949.8	1676.7	3050.0	5522.0
1244.3	971.1	1780.0	3236.0	1588.7	1315.6	2400.0	4352.0	1977.6	1704.4	3100.0	5612.0
1249.8	976.7	1790.0	3254.0	1594.3	1321.1	2410.0	4370.0	2005.4	1732.2	3150.0	5702.0
1255.4	982.2	1800.0	3272.0	1599.8	1326.7	2420.0	4388.0	2033.2	1760.0	3200.0	5792.0
1260.9	987.8	1810.0	3290.0	1605.4	1332.2	2430.0	4406.0	2060.9	1787.8	3250.0	5882.0
1266.5	993.3	1820.0	3308.0	1610.9	1337.8	2440.0	4424.0	2088.7	1815.6	3300.0	5972.0
1272.0	998.9	1830.0	3326.0	1616.5	1343.3	2450.0	4442.0	2116.5	1843.3	3350.0	6062.0
1277.6	1004.4	1840.0	3344.0	1622.0	1348.9	2460.0	4460.0	2144.3	1871.1	3400.0	6152.0
1283.2	1010.0	1850.0	3362.0	1627.6	1354.4	2470.0	4478.0	2172.0	1898.9	3450.0	6242.0
1288.7	1015.6	1860.0	3380.0	1633.2	1360.0	2480.0	4496.0	2199.8	1926.7	3500.0	6332.0
1294.3	1021.1	1870.0	3398.0	1638.7	1365.6	2490.0	4514.0	2227.6	1954.4	3550.0	6422.0
1299.8	1026.7	1880.0	3416.0	1644.3	1371.1	2500.0	4532.0	2255.4	1982.2	3600.0	6512.0
1305.4	1032.2	1890.0	3434.0	1649.8	1376.7	2510.0	4550.0	2283.2	2010.0	3650.0	6602.0
1310.9	1037.8	1900.0	3452.0	1655.4	1382.2	2520.0	4568.0	2310.9	2037.8	3700.0	6692.0
1316.5	1043.3	1910.0	3470.0	1660.9	1387.8	2530.0	4586.0	2338.7	2065.6	3750.0	6782.0
1322.0	1048.9	1920.0	3488.0	1666.5	1393.3	2540.0	4604.0	2366.5	2093.3	3800.0	6872.0
1327.6	1054.4	1930.0	3506.0	1672.0	1398.9	2550.0	4622.0	2394.3	2121.1	3850.0	6962.0
1333.2	1060.0	1940.0	3524.0	1677.6	1404.4	2560.0	4640.0	2422.0	2148.9	3900.0	7052.0
1338.7	1065.6	1950.0	3542.0	1683.2	1410.0	2570.0	4658.0	2449.8	2176.7	3950.0	7142.0
1344.3	1071.1	1960.0	3560.0	1688.7	1415.6	2580.0	4676.0	2477.6	2204.4	4000.0	7232.0
1349.8	1076.7	1970.0	3578.0	1694.3	1421.1	2590.0	4694.0	2505.4	2232.2	4050.0	7322.0
1355.4	1082.2	1980.0	3596.0	1699.8	1426.7	2600.0	4712.0	2533.2	2260.0	4100.0	7412.0
1360.9	1087.8	1990.0	3614.0	1705.4	1432.2	2610.0	4730.0	2560.9	2287.8	4150.0	7502.0
1366.5	1093.3	2000.0	3632.0	1710.9	1437.8	2620.0	4748.0	2588.7	2315.6	4200.0	7592.0
1372.0	1098.9	2010.0	3650.0	1716.5	1443.3	2630.0	4766.0	2616.5	2343.3	4250.0	7682.0
1377.6	1104.4	2020.0	3668.0	1722.0	1448.9	2640.0	4784.0	2644.3	2371.1	4300.0	7772.0
1383.2	1110.0	2030.0	3686.0	1727.6	1454.4	2650.0	4802.0	2672.0	2398.9	4350.0	7862.0
1388.7	1115.6	2040.0	3704.0	1733.2	1460.0	2660.0	4820.0	2699.8	2426.7	4400.0	7952.0
1394.3	1121.1	2050.0	3722.0	1738.7	1465.6	2670.0	4838.0	2727.6	2454.4	4450.0	8042.0
1399.8	1126.7	2060.0	3740.0	1744.3	1471.1	2680.0	4856.0	2755.4	2482.2	4500.0	8132.0
1405.4	1132.2	2070.0	3758.0	1749.8	1476.7	2690.0	4874.0	2783.2	2510.0	4550.0	8222.0
1410.9	1137.8	2080.0	3776.0	1755.4	1482.2	2700.0	4892.0	2810.9	2537.8	4600.0	8312.0
1416.5	1143.3	2090.0	3794.0	1760.9	1487.8	2710.0	4910.0	2838.7	2565.6	4650.0	8402.0
1422.0	1148.9	2100.0	3812.0	1766.5	1493.3	2720.0	4928.0	2866.5	2593.3	4700.0	8492.0
1427.6	1154.4	2110.0	3830.0	1772.0	1498.9	2730.0	4946.0	2894.3	2621.1	4750.0	8582.0
1433.2	1160.0	2120.0	3848.0	1777.6	1504.4	2740.0	4964.0	2922.0	2648.9	4800.0	8672.0
1438.7	1165.6	2130.0	3866.0	1783.2	1510.0	2750.0	4982.0	2949.8	2676.7	4850.0	8762.0

METRIC CONVERSION FACTORS AND TABLES

Table 19-5. Pressure, Stress: Kilopascals to PSI (from 1 through 499 Kilopascals)

1 PSI = 6.894757 KILOPASCALS

KILOPASCALS	0	1	2	3	4	5	6	7	8	9
					PSI					
0	0.000	0.145	0.290	0.435	0.580	0.725	0.870	1.015	1.160	1.305
10	1.450	1.595	1.740	1.885	2.031	2.176	2.321	2.466	2.611	2.756
20	2.901	3.046	3.191	3.336	3.481	3.626	3.771	3.916	4.061	4.206
30	4.351	4.496	4.641	4.786	4.931	5.076	5.221	5.366	5.511	5.656
40	5.802	5.947	6.092	6.237	6.382	6.527	6.672	6.817	6.962	7.107
50	7.252	7.397	7.542	7.687	7.832	7.977	8.122	8.267	8.412	8.557
60	8.702	8.847	8.992	9.137	9.282	9.427	9.572	9.718	9.863	10.008
70	10.153	10.298	10.443	10.588	10.733	10.878	11.023	11.168	11.313	11.458
80	11.603	11.748	11.893	12.038	12.183	12.328	12.473	12.618	12.763	12.908
90	13.053	13.198	13.343	13.489	13.634	13.779	13.924	14.069	14.214	14.359
100	14.504	14.649	14.794	14.939	15.084	15.229	15.374	15.519	15.664	15.809
110	15.954	16.099	16.244	16.389	16.534	16.679	16.824	16.969	17.114	17.259
120	17.405	17.550	17.695	17.840	17.985	18.130	18.275	18.420	18.565	18.710
130	18.855	19.000	19.145	19.290	19.435	19.580	19.725	19.870	20.015	20.160
140	20.305	20.450	20.595	20.740	20.885	21.030	21.176	21.321	21.466	21.611
150	21.756	21.901	22.046	22.191	22.336	22.481	22.626	22.771	22.916	23.061
160	23.206	23.351	23.496	23.641	23.786	23.931	24.076	24.221	24.366	24.511
170	24.656	24.801	24.946	25.092	25.237	25.382	25.527	25.672	25.817	25.962
180	26.107	26.252	26.397	26.542	26.687	26.832	26.977	27.122	27.267	27.412
190	27.557	27.702	27.847	27.992	28.137	28.282	28.427	28.572	28.717	28.863
200	29.008	29.153	29.298	29.443	29.588	29.733	29.878	30.023	30.168	30.313
210	30.458	30.603	30.748	30.893	31.038	31.183	31.328	31.473	31.618	31.763
220	31.908	32.053	32.198	32.343	32.488	32.633	32.779	32.924	33.069	33.214
230	33.359	33.504	33.649	33.794	33.939	34.084	34.229	34.374	34.519	34.664
240	34.809	34.954	35.099	35.244	35.389	35.534	35.679	35.824	35.969	36.114
250	36.259	36.404	36.550	36.695	36.840	36.985	37.130	37.275	37.420	37.565
260	37.710	37.855	38.000	38.145	38.290	38.435	38.580	38.725	38.870	39.015
270	39.160	39.305	39.450	39.595	39.740	39.885	40.030	40.175	40.320	40.466
280	40.611	40.756	40.901	41.046	41.191	41.336	41.481	41.626	41.771	41.916
290	42.061	42.206	42.351	42.496	42.641	42.786	42.931	43.076	43.221	43.366
300	43.511	43.656	43.801	43.946	44.091	44.237	44.382	44.527	44.672	44.817
310	44.962	45.107	45.252	45.397	45.542	45.687	45.832	45.977	46.122	46.267
320	46.412	46.557	46.702	46.847	46.992	47.137	47.282	47.427	47.572	47.717
330	47.862	48.007	48.153	48.298	48.443	48.588	48.733	48.878	49.023	49.168
340	49.313	49.458	49.603	49.748	49.893	50.038	50.183	50.328	50.473	50.618
350	50.763	50.908	51.053	51.198	51.343	51.488	51.633	51.778	51.924	52.069
360	52.214	52.359	52.504	52.649	52.794	52.939	53.084	53.229	53.374	53.519
370	53.664	53.809	53.954	54.099	54.244	54.389	54.534	54.679	54.824	54.969
380	55.114	55.259	55.404	55.549	55.694	55.840	55.985	56.130	56.275	56.420
390	56.565	56.710	56.855	57.000	57.145	57.290	57.435	57.580	57.725	57.870
400	58.015	58.160	58.305	58.450	58.595	58.740	58.885	59.030	59.175	59.320
410	59.465	59.611	59.756	59.901	60.046	60.191	60.336	60.481	60.626	60.771
420	60.916	61.061	61.206	61.351	61.496	61.641	61.786	61.931	62.076	62.221
430	62.366	62.511	62.656	62.801	62.946	63.091	63.236	63.381	63.527	63.672
440	63.817	63.962	64.107	64.252	64.397	64.542	64.687	64.832	64.977	65.122
450	65.267	65.412	65.557	65.702	65.847	65.992	66.137	66.282	66.427	66.572
460	66.717	66.862	67.007	67.152	67.298	67.443	67.588	67.733	67.878	68.023
470	68.168	68.313	68.458	68.603	68.748	68.893	69.038	69.183	69.328	69.473
480	69.618	69.763	69.908	70.053	70.198	70.343	70.488	70.633	70.778	70.923
490	71.068	71.214	71.359	71.504	71.649	71.794	71.939	72.084	72.229	72.374

METRIC CONVERSION FACTORS AND TABLES

Table 19-5 (*Continued*). Pressure. Stress: Kilopascals to PSI (from 500 through 1009 Kilopascals)

KILOPASCALS	0	1	2	3	4	5	6	7	8	9
					PSI					
500	72.519	72.664	72.809	72.954	73.099	73.244	73.389	73.534	73.679	73.824
510	73.969	74.114	74.259	74.404	74.549	74.694	74.839	74.985	75.130	75.275
520	75.420	75.565	75.710	75.855	76.000	76.145	76.290	76.435	76.580	76.725
530	76.870	77.015	77.160	77.305	77.450	77.595	77.740	77.885	78.030	78.175
540	78.320	78.465	78.610	78.755	78.901	79.046	79.191	79.336	79.481	79.626
550	79.771	79.916	80.061	80.206	80.351	80.496	80.641	80.786	80.931	81.076
560	81.221	81.366	81.511	81.656	81.801	81.946	82.091	82.236	82.381	82.526
570	82.672	82.817	82.962	83.107	83.252	83.397	83.542	83.687	83.832	83.977
580	84.122	84.267	84.412	84.557	84.702	84.847	84.992	85.137	85.282	85.427
590	85.572	85.717	85.862	86.007	86.152	86.297	86.442	86.588	86.733	86.878
600	87.023	87.168	87.313	87.458	87.603	87.748	87.893	88.038	88.183	88.328
610	88.473	88.618	88.763	88.908	89.053	89.198	89.343	89.488	89.633	89.778
620	89.923	90.068	90.213	90.359	90.504	90.649	90.794	90.939	91.084	91.229
630	91.374	91.519	91.664	91.809	91.954	92.099	92.244	92.389	92.534	92.679
640	92.824	92.969	93.114	93.259	93.404	93.549	93.694	93.839	93.984	94.129
650	94.275	94.420	94.565	94.710	94.855	95.000	95.145	95.290	95.435	95.580
660	95.725	95.870	96.015	96.160	96.305	96.450	96.595	96.740	96.885	97.030
670	97.175	97.320	97.465	97.610	97.755	97.900	98.046	98.191	98.336	98.481
680	98.626	98.771	98.916	99.061	99.206	99.351	99.496	99.641	99.786	99.931
690	100.076	100.221	100.366	100.511	100.656	100.801	100.946	101.091	101.236	101.381
700	101.526	101.671	101.816	101.962	102.107	102.252	102.397	102.542	102.687	102.832
710	102.977	103.122	103.267	103.412	103.557	103.702	103.847	103.992	104.137	104.282
720	104.427	104.572	104.717	104.862	105.007	105.152	105.297	105.442	105.587	105.733
730	105.878	106.023	106.168	106.313	106.458	106.603	106.748	106.893	107.038	107.183
740	107.328	107.473	107.618	107.763	107.908	108.053	108.198	108.343	108.488	108.633
750	108.778	108.923	109.068	109.213	109.358	109.503	109.649	109.794	109.939	110.084
760	110.229	110.374	110.519	110.664	110.809	110.954	111.099	111.244	111.389	111.534
770	111.679	111.824	111.969	112.114	112.259	112.404	112.549	112.694	112.839	112.984
780	113.129	113.274	113.420	113.565	113.710	113.855	114.000	114.145	114.290	114.435
790	114.580	114.725	114.870	115.015	115.160	115.305	115.450	115.595	115.740	115.885
800	116.030	116.175	116.320	116.465	116.610	116.755	116.900	117.045	117.190	117.336
810	117.481	117.626	117.771	117.916	118.061	118.206	118.351	118.496	118.641	118.786
820	118.931	119.076	119.221	119.366	119.511	119.656	119.801	119.946	120.091	120.236
830	120.381	120.526	120.671	120.816	120.961	121.107	121.252	121.397	121.542	121.687
840	121.832	121.977	122.122	122.267	122.412	122.557	122.702	122.847	122.992	123.137
850	123.282	123.427	123.572	123.717	123.862	124.007	124.152	124.297	124.442	124.587
860	124.732	124.877	125.023	125.168	125.313	125.458	125.603	125.748	125.893	126.038
870	126.183	126.328	126.473	126.618	126.763	126.908	127.053	127.198	127.343	127.488
880	127.633	127.778	127.923	128.068	128.213	128.358	128.503	128.648	128.794	128.939
890	129.084	129.229	129.374	129.519	129.664	129.809	129.954	130.099	130.244	130.389
900	130.534	130.679	130.824	130.969	131.114	131.259	131.404	131.549	131.694	131.839
910	131.984	132.129	132.274	132.419	132.564	132.710	132.855	133.000	133.145	133.290
920	133.435	133.580	133.725	133.870	134.015	134.160	134.305	134.450	134.595	134.740
930	134.885	135.030	135.175	135.320	135.465	135.610	135.755	135.900	136.045	136.190
940	136.335	136.481	136.626	136.771	136.916	137.061	137.206	137.351	137.496	137.641
950	137.786	137.931	138.076	138.221	138.366	138.511	138.656	138.801	138.946	139.091
960	139.236	139.381	139.526	139.671	139.816	139.961	140.106	140.251	140.397	140.542
970	140.687	140.832	140.977	141.122	141.267	141.412	141.557	141.702	141.847	141.992
980	142.137	142.282	142.427	142.572	142.717	142.862	143.007	143.152	143.297	143.442
990	143.587	143.732	143.877	144.022	144.168	144.313	144.458	144.603	144.748	144.893
1000	145.038	145.183	145.328	145.473	145.618	145.763	145.908	146.053	146.198	146.343

METRIC CONVERSION FACTORS AND TABLES

Table 19-6. Pressure, Stress: PSI to Kilopascals (from 1 through 499 PSI)

1 PSI = 6.894757 KILOPASCALS

PSI	0	1	2	3	4	5	6	7	8	9
					KILOPASCALS					
0	0.000	6.895	13.790	20.684	27.579	34.474	41.369	48.263	55.158	62.053
10	68.948	75.842	82.737	89.632	96.527	103.421	110.316	117.211	124.106	131.000
20	137.895	144.790	151.685	158.579	165.474	172.369	179.264	186.158	193.053	199.948
30	206.843	213.737	220.632	227.527	234.422	241.316	248.211	255.106	262.001	268.896
40	275.790	282.685	289.580	296.475	303.369	310.264	317.159	324.054	330.948	337.843
50	344.738	351.633	358.527	365.422	372.317	379.212	386.106	393.001	399.896	406.791
60	413.685	420.580	427.475	434.370	441.264	448.159	455.054	461.949	468.843	475.738
70	482.633	489.528	496.423	503.317	510.212	517.107	524.002	530.896	537.791	544.686
80	551.581	558.475	565.370	572.265	579.160	586.054	592.949	599.844	606.739	613.633
90	620.528	627.423	634.318	641.212	648.107	655.002	661.897	668.791	675.686	682.581
100	689.476	696.370	703.265	710.160	717.055	723.949	730.844	737.739	744.634	751.529
110	758.423	765.318	772.213	779.108	786.002	792.897	799.792	806.687	813.581	820.476
120	827.371	834.266	841.160	848.055	854.950	861.845	868.739	875.634	882.529	889.424
130	896.318	903.213	910.108	917.003	923.897	930.792	937.687	944.582	951.476	958.371
140	965.266	972.161	979.055	985.950	992.845	999.740	1006.635	1013.529	1020.424	1027.319
150	1034.214	1041.108	1048.003	1054.898	1061.793	1068.687	1075.582	1082.477	1089.372	1096.266
160	1103.161	1110.056	1116.951	1123.845	1130.740	1137.635	1144.530	1151.424	1158.319	1165.214
170	1172.109	1179.003	1185.898	1192.793	1199.688	1206.582	1213.477	1220.372	1227.267	1234.162
180	1241.056	1247.951	1254.846	1261.741	1268.635	1275.530	1282.425	1289.320	1296.214	1303.109
190	1310.004	1316.899	1323.793	1330.688	1337.583	1344.478	1351.372	1358.267	1365.162	1372.057
200	1378.951	1385.846	1392.741	1399.636	1406.530	1413.425	1420.320	1427.215	1434.109	1441.004
210	1447.899	1454.794	1461.688	1468.583	1475.478	1482.373	1489.268	1496.162	1503.057	1509.952
220	1516.847	1523.741	1530.636	1537.531	1544.426	1551.320	1558.215	1565.110	1572.005	1578.899
230	1585.794	1592.689	1599.584	1606.478	1613.373	1620.268	1627.163	1634.057	1640.952	1647.847
240	1654.742	1661.636	1668.531	1675.426	1682.321	1689.215	1696.110	1703.005	1709.900	1716.794
250	1723.689	1730.584	1737.479	1744.374	1751.268	1758.163	1765.058	1771.953	1778.847	1785.742
260	1792.637	1799.532	1806.426	1813.321	1820.216	1827.111	1834.005	1840.900	1847.795	1854.690
270	1861.584	1868.479	1875.374	1882.269	1889.163	1896.058	1902.953	1909.848	1916.742	1923.637
280	1930.532	1937.427	1944.321	1951.216	1958.111	1965.006	1971.901	1978.795	1985.690	1992.585
290	1999.480	2006.374	2013.269	2020.164	2027.059	2033.953	2040.848	2047.743	2054.638	2061.532
300	2068.427	2075.322	2082.217	2089.111	2096.006	2102.901	2109.796	2116.690	2123.585	2130.480
310	2137.375	2144.269	2151.164	2158.059	2164.954	2171.848	2178.743	2185.638	2192.533	2199.427
320	2206.322	2213.217	2220.112	2227.007	2233.901	2240.796	2247.691	2254.586	2261.480	2268.375
330	2275.270	2282.165	2289.059	2295.954	2302.849	2309.744	2316.638	2323.533	2330.428	2337.323
340	2344.217	2351.112	2358.007	2364.902	2371.796	2378.691	2385.586	2392.481	2399.375	2406.270
350	2413.165	2420.060	2426.954	2433.849	2440.744	2447.639	2454.533	2461.428	2468.323	2475.218
360	2482.113	2489.007	2495.902	2502.797	2509.692	2516.586	2523.481	2530.376	2537.271	2544.165
370	2551.060	2557.955	2564.850	2571.744	2578.639	2585.534	2592.429	2599.323	2606.218	2613.113
380	2620.008	2626.902	2633.797	2640.692	2647.587	2654.481	2661.376	2668.271	2675.166	2682.060
390	2688.955	2695.850	2702.745	2709.640	2716.534	2723.429	2730.324	2737.219	2744.113	2751.008
400	2757.903	2764.798	2771.692	2778.587	2785.482	2792.377	2799.271	2806.166	2813.061	2819.956
410	2826.850	2833.745	2840.640	2847.535	2854.429	2861.324	2868.219	2875.114	2882.008	2888.903
420	2895.798	2902.693	2909.587	2916.482	2923.377	2930.272	2937.166	2944.061	2950.956	2957.851
430	2964.746	2971.640	2978.535	2985.430	2992.325	2999.219	3006.114	3013.009	3019.904	3026.798
440	3033.693	3040.588	3047.483	3054.377	3061.272	3068.167	3075.062	3081.956	3088.851	3095.746
450	3102.641	3109.535	3116.430	3123.325	3130.220	3137.114	3144.009	3150.904	3157.799	3164.693
460	3171.588	3178.483	3185.378	3192.272	3199.167	3206.062	3212.957	3219.852	3226.746	3233.641
470	3240.536	3247.431	3254.325	3261.220	3268.115	3275.010	3281.904	3288.799	3295.694	3302.589
480	3309.483	3316.378	3323.273	3330.168	3337.062	3343.957	3350.852	3357.747	3364.641	3371.536
490	3378.431	3385.326	3392.220	3399.115	3406.010	3412.905	3419.799	3426.694	3433.589	3440.484

Table 19-6 (*Continued*). Pressure, Stress: PSI to Kilopascals (from 500 through 1009 PSI)

PSI	0	1	2	3	4	5	6	7	8	9
					KILOPASCALS					
500	3447.379	3454.273	3461.168	3468.063	3474.958	3481.852	3488.747	3495.642	3502.537	3509.431
510	3516.326	3523.221	3530.116	3537.010	3543.905	3550.800	3557.695	3564.589	3571.484	3578.379
520	3585.274	3592.168	3599.063	3605.958	3612.853	3619.747	3626.642	3633.537	3640.432	3647.326
530	3654.221	3661.116	3668.011	3674.905	3681.800	3688.695	3695.590	3702.485	3709.379	3716.274
540	3723.169	3730.064	3736.958	3743.853	3750.748	3757.643	3764.537	3771.432	3778.327	3785.222
550	3792.116	3799.011	3805.906	3812.801	3819.695	3826.590	3833.485	3840.380	3847.274	3854.169
560	3861.064	3867.959	3874.853	3881.748	3888.643	3895.538	3902.432	3909.327	3916.222	3923.117
570	3930.011	3936.906	3943.801	3950.696	3957.591	3964.485	3971.380	3978.275	3985.170	3992.064
580	3998.959	4005.854	4012.749	4019.643	4026.538	4033.433	4040.328	4047.222	4054.117	4061.012
590	4067.907	4074.801	4081.696	4088.591	4095.486	4102.380	4109.275	4116.170	4123.065	4129.959
600	4136.854	4143.749	4150.644	4157.538	4164.433	4171.328	4178.223	4185.117	4192.012	4198.907
610	4205.802	4212.697	4219.591	4226.486	4233.381	4240.276	4247.170	4254.065	4260.960	4267.855
620	4274.749	4281.644	4288.539	4295.434	4302.328	4309.223	4316.118	4323.013	4329.907	4336.802
630	4343.697	4350.592	4357.486	4364.381	4371.276	4378.171	4385.065	4391.960	4398.855	4405.750
640	4412.644	4419.539	4426.434	4433.329	4440.224	4447.118	4454.013	4460.908	4467.803	4474.697
650	4481.592	4488.487	4495.382	4502.276	4509.171	4516.066	4522.961	4529.855	4536.750	4543.645
660	4550.540	4557.434	4564.329	4571.224	4578.119	4585.013	4591.908	4598.803	4605.698	4612.592
670	4619.487	4626.382	4633.277	4640.171	4647.066	4653.961	4660.856	4667.750	4674.645	4681.540
680	4688.435	4695.330	4702.224	4709.119	4716.014	4722.909	4729.803	4736.698	4743.593	4750.488
690	4757.382	4764.277	4771.172	4778.067	4784.961	4791.856	4798.751	4805.646	4812.540	4819.435
700	4826.330	4833.225	4840.119	4847.014	4853.909	4860.804	4867.698	4874.593	4881.488	4888.383
710	4895.277	4902.172	4909.067	4915.962	4922.856	4929.751	4936.646	4943.541	4950.436	4957.330
720	4964.225	4971.120	4978.015	4984.909	4991.804	4998.699	5005.594	5012.488	5019.383	5026.278
730	5033.173	5040.067	5046.962	5053.857	5060.752	5067.646	5074.541	5081.436	5088.331	5095.225
740	5102.120	5109.015	5115.910	5122.804	5129.699	5136.594	5143.489	5150.383	5157.278	5164.173
750	5171.068	5177.963	5184.857	5191.752	5198.647	5205.542	5212.436	5219.331	5226.226	5233.121
760	5240.015	5246.910	5253.805	5260.700	5267.594	5274.489	5281.384	5288.279	5295.173	5302.068
770	5308.963	5315.858	5322.752	5329.647	5336.542	5343.437	5350.331	5357.226	5364.121	5371.016
780	5377.910	5384.805	5391.700	5398.595	5405.489	5412.384	5419.279	5426.174	5433.069	5439.963
790	5446.858	5453.753	5460.648	5467.542	5474.437	5481.332	5488.227	5495.121	5502.016	5508.911
800	5515.806	5522.700	5529.595	5536.490	5543.385	5550.279	5557.174	5564.069	5570.964	5577.858
810	5584.753	5591.648	5598.543	5605.437	5612.332	5619.227	5626.122	5633.016	5639.911	5646.806
820	5653.701	5660.595	5667.490	5674.385	5681.280	5688.175	5695.069	5701.964	5708.859	5715.754
830	5722.648	5729.543	5736.438	5743.333	5750.227	5757.122	5764.017	5770.912	5777.806	5784.701
840	5791.596	5798.491	5805.385	5812.280	5819.175	5826.070	5832.964	5839.859	5846.754	5853.649
850	5860.543	5867.438	5874.333	5881.228	5888.122	5895.017	5901.912	5908.807	5915.702	5922.596
860	5929.491	5936.386	5943.281	5950.175	5957.070	5963.965	5970.860	5977.754	5984.649	5991.544
870	5998.439	6005.333	6012.228	6019.123	6026.018	6032.912	6039.807	6046.702	6053.597	6060.491
880	6067.386	6074.281	6081.176	6088.070	6094.965	6101.860	6108.755	6115.649	6122.544	6129.439
890	6136.334	6143.228	6150.123	6157.018	6163.913	6170.808	6177.702	6184.597	6191.492	6198.387
900	6205.281	6212.176	6219.071	6225.966	6232.860	6239.755	6246.650	6253.545	6260.439	6267.334
910	6274.229	6281.124	6288.018	6294.913	6301.808	6308.703	6315.597	6322.492	6329.387	6336.282
920	6343.176	6350.071	6356.966	6363.861	6370.755	6377.650	6384.545	6391.440	6398.334	6405.229
930	6412.124	6419.019	6425.914	6432.808	6439.703	6446.598	6453.493	6460.387	6467.282	6474.177
940	6481.072	6487.966	6494.861	6501.756	6508.651	6515.545	6522.440	6529.335	6536.230	6543.124
950	6550.019	6556.914	6563.809	6570.703	6577.598	6584.493	6591.388	6598.282	6605.177	6612.072
960	6618.967	6625.861	6632.756	6639.651	6646.546	6653.441	6660.335	6667.230	6674.125	6681.020
970	6687.914	6694.809	6701.704	6708.599	6715.493	6722.388	6729.283	6736.178	6743.072	6749.967
980	6756.862	6763.757	6770.651	6777.546	6784.441	6791.336	6798.230	6805.125	6812.020	6818.915
990	6825.809	6832.704	6839.599	6846.494	6853.388	6860.283	6867.178	6874.073	6880.967	6887.862
1000	6894.757	6901.652	6908.547	6915.441	6922.336	6929.231	6936.126	6943.020	6949.915	6956.810

METRIC CONVERSION FACTORS AND TABLES

Table 19-7. Mass: Kilograms to Pounds (from 1 through 499 Kilograms)

1 POUND = 0.4535924 KILOGRAMS

KILOGRAMS	0	1	2	3	4	5	6	7	8	9
					POUNDS					
0	0.000	2.205	4.409	6.614	8.818	11.023	13.228	15.432	17.637	19.842
10	22.046	24.251	26.455	28.660	30.865	33.069	35.274	37.479	39.683	41.888
20	44.092	46.297	48.502	50.706	52.911	55.116	57.320	59.525	61.729	63.934
30	66.139	68.343	70.548	72.753	74.957	77.162	79.366	81.571	83.776	85.980
40	88.185	90.390	92.594	94.799	97.003	99.208	101.413	103.617	105.822	108.027
50	110.231	112.436	114.640	116.845	119.050	121.254	123.459	125.663	127.868	130.073
60	132.277	134.482	136.687	138.891	141.096	143.300	145.505	147.710	149.914	152.119
70	154.324	156.528	158.733	160.937	163.142	165.347	167.551	169.756	171.961	174.165
80	176.370	178.574	180.779	182.984	185.188	187.393	189.598	191.802	194.007	196.211
90	198.416	200.621	202.825	205.030	207.235	209.439	211.644	213.848	216.053	218.258
100	220.462	222.667	224.872	227.076	229.281	231.485	233.690	235.895	238.099	240.304
110	242.508	244.713	246.918	249.122	251.327	253.532	255.736	257.941	260.145	262.350
120	264.555	266.759	268.964	271.169	273.373	275.578	277.782	279.987	282.192	284.396
130	286.601	288.806	291.010	293.215	295.419	297.624	299.829	302.033	304.238	306.443
140	308.647	310.852	313.056	315.261	317.466	319.670	321.875	324.080	326.284	328.489
150	330.693	332.898	335.103	337.307	339.512	341.717	343.921	346.126	348.330	350.535
160	352.740	354.944	357.149	359.353	361.558	363.763	365.967	368.172	370.377	372.581
170	374.786	376.990	379.195	381.400	383.604	385.809	388.014	390.218	392.423	394.627
180	396.832	399.037	401.241	403.446	405.651	407.855	410.060	412.264	414.469	416.674
190	418.878	421.083	423.288	425.492	427.697	429.901	432.106	434.311	436.515	438.720
200	440.925	443.129	445.334	447.538	449.743	451.948	454.152	456.357	458.562	460.766
210	462.971	465.175	467.380	469.585	471.789	473.994	476.198	478.403	480.608	482.812
220	485.017	487.222	489.426	491.631	493.835	496.040	498.245	500.449	502.654	504.859
230	507.063	509.268	511.472	513.677	515.882	518.086	520.291	522.496	524.700	526.905
240	529.109	531.314	533.519	535.723	537.928	540.133	542.337	544.542	546.746	548.951
250	551.156	553.360	555.565	557.770	559.974	562.179	564.383	566.588	568.793	570.997
260	573.202	575.407	577.611	579.816	582.020	584.225	586.430	588.634	590.839	593.043
270	595.248	597.453	599.657	601.862	604.067	606.271	608.476	610.680	612.885	615.090
280	617.294	619.499	621.704	623.908	626.113	628.317	630.522	632.727	634.931	637.136
290	639.341	641.545	643.750	645.954	648.159	650.364	652.568	654.773	656.978	659.182
300	661.387	663.591	665.796	668.001	670.205	672.410	674.615	676.819	679.024	681.228
310	683.433	685.638	687.842	690.047	692.252	694.456	696.661	698.865	701.070	703.275
320	705.479	707.684	709.888	712.093	714.298	716.502	718.707	720.912	723.116	725.321
330	727.525	729.730	731.935	734.139	736.344	738.549	740.753	742.958	745.162	747.367
340	749.572	751.776	753.981	756.186	758.390	760.595	762.799	765.004	767.209	769.413
350	771.618	773.823	776.027	778.232	780.436	782.641	784.846	787.050	789.255	791.460
360	793.664	795.869	798.073	800.278	802.483	804.687	806.892	809.097	811.301	813.506
370	815.710	817.915	820.120	822.324	824.529	826.733	828.938	831.143	833.347	835.552
380	837.757	839.961	842.166	844.370	846.575	848.780	850.984	853.189	855.394	857.598
390	859.803	862.007	864.212	866.417	868.621	870.826	873.031	875.235	877.440	879.644
400	881.849	884.054	886.258	888.463	890.668	892.872	895.077	897.281	899.486	901.691
410	903.895	906.100	908.305	910.509	912.714	914.918	917.123	919.328	921.532	923.737
420	925.942	928.146	930.351	932.555	934.760	936.965	939.169	941.374	943.578	945.783
430	947.988	950.192	952.397	954.602	956.806	959.011	961.215	963.420	965.625	967.829
440	970.034	972.239	974.443	976.648	978.852	981.057	983.262	985.466	987.671	989.876
450	992.080	994.285	996.489	998.694	1000.899	1003.103	1005.308	1007.513	1009.717	1011.922
460	1014.126	1016.331	1018.536	1020.740	1022.945	1025.150	1027.354	1029.559	1031.763	1033.968
470	1036.173	1038.377	1040.582	1042.787	1044.991	1047.196	1049.400	1051.605	1053.810	1056.014
480	1058.219	1060.423	1062.628	1064.833	1067.037	1069.242	1071.447	1073.651	1075.856	1078.060
490	1080.265	1082.470	1084.674	1086.879	1089.084	1091.288	1093.493	1095.697	1097.902	1100.107

Table 19-7 (*Continued*). Mass: Kilograms to Pounds (from 500 through 1009 Kilograms)

KILOGRAMS	0	1	2	3	4	5	6	7	8	9
					POUNDS					
500	1102.311	1104.516	1106.721	1108.925	1111.130	1113.334	1115.539	1117.744	1119.948	1122.153
510	1124.358	1126.562	1128.767	1130.971	1133.176	1135.381	1137.585	1139.790	1141.995	1144.199
520	1146.404	1148.608	1150.813	1153.018	1155.222	1157.427	1159.631	1161.836	1164.041	1166.245
530	1168.450	1170.655	1172.859	1175.064	1177.268	1179.473	1181.678	1183.882	1186.087	1188.292
540	1190.496	1192.701	1194.905	1197.110	1199.315	1201.519	1203.724	1205.929	1208.133	1210.338
550	1212.542	1214.747	1216.952	1219.156	1221.361	1223.566	1225.770	1227.975	1230.179	1232.384
560	1234.589	1236.793	1238.998	1241.203	1243.407	1245.612	1247.816	1250.021	1252.226	1254.430
570	1256.635	1258.840	1261.044	1263.249	1265.453	1267.658	1269.863	1272.067	1274.272	1276.476
580	1278.681	1280.886	1283.090	1285.295	1287.500	1289.704	1291.909	1294.113	1296.318	1298.523
590	1300.727	1302.932	1305.137	1307.341	1309.546	1311.750	1313.955	1316.160	1318.364	1320.569
600	1322.774	1324.978	1327.183	1329.387	1331.592	1333.797	1336.001	1338.206	1340.411	1342.615
610	1344.820	1347.024	1349.229	1351.434	1353.638	1355.843	1358.048	1360.252	1362.457	1364.661
620	1366.866	1369.071	1371.275	1373.480	1375.685	1377.889	1380.094	1382.298	1384.503	1386.708
630	1388.912	1391.117	1393.321	1395.526	1397.731	1399.935	1402.140	1404.345	1406.549	1408.754
640	1410.958	1413.163	1415.368	1417.572	1419.777	1421.982	1424.186	1426.391	1428.595	1430.800
650	1433.005	1435.209	1437.414	1439.619	1441.823	1444.028	1446.232	1448.437	1450.642	1452.846
660	1455.051	1457.256	1459.460	1461.665	1463.869	1466.074	1468.279	1470.483	1472.688	1474.893
670	1477.097	1479.302	1481.506	1483.711	1485.916	1488.120	1490.325	1492.530	1494.734	1496.939
680	1499.143	1501.348	1503.553	1505.757	1507.962	1510.166	1512.371	1514.576	1516.780	1518.985
690	1521.190	1523.394	1525.599	1527.803	1530.008	1532.213	1534.417	1536.622	1538.827	1541.031
700	1543.236	1545.440	1547.645	1549.850	1552.054	1554.259	1556.464	1558.668	1560.873	1563.077
710	1565.282	1567.487	1569.691	1571.896	1574.101	1576.305	1578.510	1580.714	1582.919	1585.124
720	1587.328	1589.533	1591.738	1593.942	1596.147	1598.351	1600.556	1602.761	1604.965	1607.170
730	1609.375	1611.579	1613.784	1615.988	1618.193	1620.398	1622.602	1624.807	1627.011	1629.216
740	1631.421	1633.625	1635.830	1638.035	1640.239	1642.444	1644.648	1646.853	1649.058	1651.262
750	1653.467	1655.672	1657.876	1660.081	1662.285	1664.490	1666.695	1668.699	1671.104	1673.309
760	1675.513	1677.718	1679.922	1682.127	1684.332	1686.536	1688.741	1690.946	1693.150	1695.355
770	1697.559	1699.764	1701.969	1704.173	1706.378	1708.583	1710.787	1712.992	1715.196	1717.401
780	1719.606	1721.810	1724.015	1726.220	1728.424	1730.629	1732.833	1735.038	1737.243	1739.447
790	1741.652	1743.856	1746.061	1748.266	1750.470	1752.675	1754.880	1757.084	1759.289	1761.493
800	1763.698	1765.903	1768.107	1770.312	1772.517	1774.721	1776.926	1779.130	1781.335	1783.540
810	1785.744	1787.949	1790.154	1792.358	1794.563	1796.767	1798.972	1801.177	1803.381	1805.586
820	1807.791	1809.995	1812.200	1814.404	1816.609	1818.814	1821.018	1823.223	1825.428	1827.632
830	1829.837	1832.041	1834.246	1836.451	1838.655	1840.860	1843.065	1845.269	1847.474	1849.678
840	1851.883	1854.088	1856.292	1858.497	1860.701	1862.906	1865.111	1867.315	1869.520	1871.725
850	1873.929	1876.134	1878.338	1880.543	1882.748	1884.952	1887.157	1889.362	1891.566	1893.771
860	1895.975	1898.180	1900.385	1902.589	1904.794	1906.999	1909.203	1911.408	1913.612	1915.817
870	1918.022	1920.226	1922.431	1924.636	1926.840	1929.045	1931.249	1933.454	1935.659	1937.863
880	1940.068	1942.273	1944.477	1946.682	1948.886	1951.091	1953.296	1955.500	1957.705	1959.910
890	1962.114	1964.319	1966.523	1968.728	1970.933	1973.137	1975.342	1977.546	1979.751	1981.956
900	1984.160	1986.365	1988.570	1990.774	1992.979	1995.183	1997.388	1999.593	2001.797	2004.002
910	2006.207	2008.411	2010.616	2012.820	2015.025	2017.230	2019.434	2021.639	2023.844	2026.048
920	2028.253	2030.457	2032.662	2034.867	2037.071	2039.276	2041.481	2043.685	2045.890	2048.094
930	2050.299	2052.504	2054.708	2056.913	2059.118	2061.322	2063.527	2065.731	2067.936	2070.141
940	2072.345	2074.550	2076.755	2078.959	2081.164	2083.368	2085.573	2087.778	2089.982	2092.187
950	2094.391	2096.596	2098.801	2101.005	2103.210	2105.415	2107.619	2109.824	2112.028	2114.233
960	2116.438	2118.642	2120.847	2123.052	2125.256	2127.461	2129.665	2131.870	2134.075	2136.279
970	2138.484	2140.689	2142.893	2145.098	2147.302	2149.507	2151.712	2153.916	2156.121	2158.326
980	2160.530	2162.735	2164.939	2167.144	2169.349	2171.553	2173.758	2175.963	2178.167	2180.372
990	2182.576	2184.781	2186.986	2189.190	2191.395	2193.600	2195.804	2198.009	2200.213	2202.418
1000	2204.623	2206.827	2209.032	2211.236	2213.441	2215.646	2217.850	2220.055	2222.260	2224.464

METRIC CONVERSION FACTORS AND TABLES

Table 19-8. Mass: Pounds to Kilograms (from 1 through 499 Pounds)

1 POUND = 0.4535924 KILOGRAMS

POUNDS	0	1	2	3	4	5	6	7	8	9
					KILOGRAMS					
0	0.000	0.454	0.907	1.361	1.814	2.268	2.722	3.175	3.629	4.082
10	4.536	4.990	5.443	5.897	6.350	6.804	7.257	7.711	8.165	8.618
20	9.072	9.525	9.979	10.433	10.886	11.340	11.793	12.247	12.701	13.154
30	13.608	14.061	14.515	14.969	15.422	15.876	16.329	16.783	17.237	17.690
40	18.144	18.597	19.051	19.504	19.958	20.412	20.865	21.319	21.772	22.226
50	22.680	23.133	23.587	24.040	24.494	24.948	25.401	25.855	26.308	26.762
60	27.216	27.669	28.123	28.576	29.030	29.484	29.937	30.391	30.844	31.298
70	31.751	32.205	32.659	33.112	33.566	34.019	34.473	34.927	35.380	35.834
80	36.287	36.741	37.195	37.648	38.102	38.555	39.009	39.463	39.916	40.370
90	40.823	41.277	41.731	42.184	42.638	43.091	43.545	43.998	44.452	44.906
100	45.359	45.813	46.266	46.720	47.174	47.627	48.081	48.534	48.988	49.442
110	49.895	50.349	50.802	51.256	51.710	52.163	52.617	53.070	53.524	53.977
120	54.431	54.885	55.338	55.792	56.245	56.699	57.153	57.606	58.060	58.513
130	58.967	59.421	59.874	60.328	60.781	61.235	61.689	62.142	62.596	63.049
140	63.503	63.957	64.410	64.864	65.317	65.771	66.224	66.678	67.132	67.585
150	68.039	68.492	68.946	69.400	69.853	70.307	70.760	71.214	71.668	72.121
160	72.575	73.028	73.482	73.936	74.389	74.843	75.296	75.750	76.204	76.657
170	77.111	77.564	78.018	78.471	78.925	79.379	79.832	80.286	80.739	81.193
180	81.647	82.100	82.554	83.007	83.461	83.915	84.368	84.822	85.275	85.729
190	86.183	86.636	87.090	87.543	87.997	88.451	88.904	89.358	89.811	90.265
200	90.718	91.172	91.626	92.079	92.533	92.986	93.440	93.894	94.347	94.801
210	95.254	95.708	96.162	96.615	97.069	97.522	97.976	98.430	98.883	99.337
220	99.790	100.244	100.698	101.151	101.605	102.058	102.512	102.965	103.419	103.873
230	104.326	104.780	105.233	105.687	106.141	106.594	107.048	107.501	107.955	108.409
240	108.862	109.316	109.769	110.223	110.677	111.130	111.584	112.037	112.491	112.945
250	113.398	113.852	114.305	114.759	115.212	115.666	116.120	116.573	117.027	117.480
260	117.934	118.388	118.841	119.295	119.748	120.202	120.656	121.109	121.563	122.016
270	122.470	122.924	123.377	123.831	124.284	124.738	125.192	125.645	126.099	126.552
280	127.006	127.459	127.913	128.367	128.820	129.274	129.727	130.181	130.635	131.088
290	131.542	131.995	132.449	132.903	133.356	133.810	134.263	134.717	135.171	135.624
300	136.078	136.531	136.985	137.438	137.892	138.346	138.799	139.253	139.706	140.160
310	140.614	141.067	141.521	141.974	142.428	142.882	143.335	143.789	144.242	144.696
320	145.150	145.603	146.057	146.510	146.964	147.418	147.871	148.325	148.778	149.232
330	149.685	150.139	150.593	151.046	151.500	151.953	152.407	152.861	153.314	153.768
340	154.221	154.675	155.129	155.582	156.036	156.489	156.943	157.397	157.850	158.304
350	158.757	159.211	159.665	160.118	160.572	161.025	161.479	161.932	162.386	162.840
360	163.293	163.747	164.200	164.654	165.108	165.561	166.015	166.468	166.922	167.376
370	167.829	168.283	168.736	169.190	169.644	170.097	170.551	171.004	171.458	171.912
380	172.365	172.819	173.272	173.726	174.179	174.633	175.087	175.540	175.994	176.447
390	176.901	177.355	177.808	178.262	178.715	179.169	179.623	180.076	180.530	180.983
400	181.437	181.891	182.344	182.798	183.251	183.705	184.159	184.612	185.066	185.519
410	185.973	186.426	186.880	187.334	187.787	188.241	188.694	189.148	189.602	190.055
420	190.509	190.962	191.416	191.870	192.323	192.777	193.230	193.684	194.138	194.591
430	195.045	195.498	195.952	196.406	196.859	197.313	197.766	198.220	198.673	199.127
440	199.581	200.034	200.488	200.941	201.395	201.849	202.302	202.756	203.209	203.663
450	204.117	204.570	205.024	205.477	205.931	206.385	206.838	207.292	207.745	208.199
460	208.653	209.106	209.560	210.013	210.467	210.920	211.374	211.828	212.281	212.735
470	213.188	213.642	214.096	214.549	215.003	215.456	215.910	216.364	216.817	217.271
480	217.724	218.178	218.632	219.085	219.539	219.992	220.446	220.899	221.353	221.807
490	222.260	222.714	223.167	223.621	224.075	224.528	224.982	225.435	225.889	226.343

Table 19-8 (*Continued*). Mass: Pounds to Kilograms (from 500 through 1009 Pounds)

POUNDS	0	1	2	3	4	5	6	7	8	9
					KILOGRAMS					
500	226.796	227.250	227.703	228.157	228.611	229.064	229.518	229.971	230.425	230.879
510	231.332	231.786	232.239	232.693	233.146	233.600	234.054	234.507	234.961	235.414
520	235.868	236.322	236.775	237.229	237.682	238.136	238.590	239.043	239.497	239.950
530	240.404	240.858	241.311	241.765	242.218	242.672	243.126	243.579	244.033	244.486
540	244.940	245.393	245.847	246.301	246.754	247.208	247.661	248.115	248.569	249.022
550	249.476	249.929	250.383	250.837	251.290	251.744	252.197	252.651	253.105	253.558
560	254.012	254.465	254.919	255.373	255.826	256.280	256.733	257.187	257.640	258.094
570	258.548	259.001	259.455	259.908	260.362	260.816	261.269	261.723	262.176	262.630
580	263.084	263.537	263.991	264.444	264.898	265.352	265.805	266.259	266.712	267.166
590	267.620	268.073	268.527	268.980	269.434	269.887	270.341	270.795	271.248	271.702
600	272.155	272.609	273.063	273.516	273.970	274.423	274.877	275.331	275.784	276.238
610	276.691	277.145	277.599	278.052	278.506	278.959	279.413	279.867	280.320	280.774
620	281.227	281.681	282.134	282.588	283.042	283.495	283.949	284.402	284.856	285.310
630	285.763	286.217	286.670	287.124	287.578	288.031	288.485	288.938	289.392	289.846
640	290.299	290.753	291.206	291.660	292.114	292.567	293.021	293.474	293.928	294.381
650	294.835	295.289	295.742	296.196	296.649	297.103	297.557	298.010	298.464	298.917
660	299.371	299.825	300.278	300.732	301.185	301.639	302.093	302.546	303.000	303.453
670	303.907	304.361	304.814	305.268	305.721	306.175	306.628	307.082	307.536	307.989
680	308.443	308.896	309.350	309.804	310.257	310.711	311.164	311.618	312.072	312.525
690	312.979	313.432	313.886	314.340	314.793	315.247	315.700	316.154	316.607	317.061
700	317.515	317.968	318.422	318.875	319.329	319.783	320.236	320.690	321.143	321.597
710	322.051	322.504	322.958	323.411	323.865	324.319	324.772	325.226	325.679	326.133
720	326.587	327.040	327.494	327.947	328.401	328.854	329.308	329.762	330.215	330.669
730	331.122	331.576	332.030	332.483	332.937	333.390	333.844	334.298	334.751	335.205
740	335.658	336.112	336.566	337.019	337.473	337.926	338.380	338.834	339.287	339.741
750	340.194	340.648	341.101	341.555	342.009	342.462	342.916	343.369	343.823	344.277
760	344.730	345.184	345.637	346.091	346.545	346.998	347.452	347.905	348.359	348.813
770	349.266	349.720	350.173	350.627	351.081	351.534	351.988	352.441	352.895	353.348
780	353.802	354.256	354.709	355.163	355.616	356.070	356.524	356.977	357.431	357.884
790	358.338	358.792	359.245	359.699	360.152	360.606	361.060	361.513	361.967	362.420
800	362.874	363.328	363.781	364.235	364.688	365.142	365.595	366.049	366.503	366.956
810	367.410	367.863	368.317	368.771	369.224	369.678	370.131	370.585	371.039	371.492
820	371.946	372.399	372.853	373.307	373.760	374.214	374.667	375.121	375.575	376.028
830	376.482	376.935	377.389	377.842	378.296	378.750	379.203	379.657	380.110	380.564
840	381.018	381.471	381.925	382.378	382.832	383.286	383.739	384.193	384.646	385.100
850	385.554	386.007	386.461	386.914	387.368	387.822	388.275	388.729	389.182	389.636
860	390.089	390.543	390.997	391.450	391.904	392.357	392.811	393.265	393.718	394.172
870	394.625	395.079	395.533	395.986	396.440	396.893	397.347	397.801	398.254	398.708
880	399.161	399.615	400.068	400.522	400.976	401.429	401.883	402.336	402.790	403.244
890	403.697	404.151	404.604	405.058	405.512	405.965	406.419	406.872	407.326	407.780
900	408.233	408.687	409.140	409.594	410.048	410.501	410.955	411.408	411.862	412.315
910	412.769	413.223	413.676	414.130	414.583	415.037	415.491	415.944	416.398	416.851
920	417.305	417.759	418.212	418.666	419.119	419.573	420.027	420.480	420.934	421.387
930	421.841	422.295	422.748	423.202	423.655	424.109	424.562	425.016	425.470	425.923
940	426.377	426.830	427.284	427.738	428.191	428.645	429.098	429.552	430.006	430.459
950	430.913	431.366	431.820	432.274	432.727	433.181	433.634	434.088	434.542	434.995
960	435.449	435.902	436.356	436.809	437.263	437.717	438.170	438.624	439.077	439.531
970	439.985	440.438	440.892	441.345	441.799	442.253	442.706	443.160	443.613	444.067
980	444.521	444.974	445.428	445.881	446.335	446.789	447.242	447.696	448.149	448.603
990	449.056	449.510	449.964	450.417	450.871	451.324	451.778	452.232	452.685	453.139
1000	453.592	454.046	454.500	454.953	455.407	455.860	456.314	456.768	457.221	457.675

METRIC CONVERSION FACTORS AND TABLES

Table 19-9. Force: Newtons to Pounds Force (from 1 through 499 Newtons)

1 POUND FORCE = 4.448222 NEWTONS

NEWTONS	0	1	2	3	4	5	6	7	8	9
				POUNDS FORCE						
0	0.000	0.225	0.450	0.674	0.899	1.124	1.349	1.574	1.798	2.023
10	2.248	2.473	2.698	2.923	3.147	3.372	3.597	3.822	4.047	4.271
20	4.496	4.721	4.946	5.171	5.395	5.620	5.845	6.070	6.295	6.519
30	6.744	6.969	7.194	7.419	7.644	7.868	8.093	8.318	8.543	8.768
40	8.992	9.217	9.442	9.667	9.892	10.116	10.341	10.566	10.791	11.016
50	11.240	11.465	11.690	11.915	12.140	12.364	12.589	12.814	13.039	13.264
60	13.489	13.713	13.938	14.163	14.388	14.613	14.837	15.062	15.287	15.512
70	15.737	15.961	16.186	16.411	16.636	16.861	17.085	17.310	17.535	17.760
80	17.985	18.210	18.434	18.659	18.884	19.109	19.334	19.558	19.783	20.008
90	20.233	20.458	20.682	20.907	21.132	21.357	21.582	21.806	22.031	22.256
100	22.481	22.706	22.931	23.155	23.380	23.605	23.830	24.055	24.279	24.504
110	24.729	24.954	25.179	25.403	25.628	25.853	26.078	26.303	26.527	26.752
120	26.977	27.202	27.427	27.651	27.876	28.101	28.326	28.551	28.776	29.000
130	29.225	29.450	29.675	29.900	30.124	30.349	30.574	30.799	31.024	31.248
140	31.473	31.698	31.923	32.148	32.372	32.597	32.822	33.047	33.272	33.497
150	33.721	33.946	34.171	34.396	34.621	34.845	35.070	35.295	35.520	35.745
160	35.969	36.194	36.419	36.644	36.869	37.093	37.318	37.543	37.768	37.993
170	38.218	38.442	38.667	38.892	39.117	39.342	39.566	39.791	40.016	40.241
180	40.466	40.690	40.915	41.140	41.365	41.590	41.814	42.039	42.264	42.489
190	42.714	42.939	43.163	43.388	43.613	43.838	44.063	44.287	44.512	44.737
200	44.962	45.187	45.411	45.636	45.861	46.086	46.311	46.535	46.760	46.985
210	47.210	47.435	47.659	47.884	48.109	48.334	48.559	48.784	49.008	49.233
220	49.458	49.683	49.908	50.132	50.357	50.582	50.807	51.032	51.256	51.481
230	51.706	51.931	52.156	52.380	52.605	52.830	53.055	53.280	53.505	53.729
240	53.954	54.179	54.404	54.629	54.853	55.078	55.303	55.528	55.753	55.977
250	56.202	56.427	56.652	56.877	57.101	57.326	57.551	57.776	58.001	58.226
260	58.450	58.675	58.900	59.125	59.350	59.574	59.799	60.024	60.249	60.474
270	60.698	60.923	61.148	61.373	61.598	61.622	62.047	62.272	62.497	62.722
280	62.946	63.171	63.396	63.621	63.846	64.071	64.295	64.520	64.745	64.970
290	65.195	65.419	65.644	65.869	66.094	66.319	66.543	66.768	66.993	67.218
300	67.443	67.667	67.892	68.117	68.342	68.567	68.792	69.016	69.241	69.466
310	69.691	69.916	70.140	70.365	70.590	70.815	71.040	71.264	71.489	71.714
320	71.939	72.164	72.388	72.613	72.838	73.063	73.288	73.513	73.737	73.962
330	74.187	74.412	74.637	74.861	75.086	75.311	75.536	75.761	75.985	76.210
340	76.435	76.660	76.885	77.109	77.334	77.559	77.784	78.009	78.234	78.458
350	78.683	78.908	79.133	79.358	79.582	79.807	80.032	80.257	80.482	80.706
360	80.931	81.156	81.381	81.606	81.830	82.055	82.280	82.505	82.730	82.954
370	83.179	83.404	83.629	83.854	84.079	84.303	84.528	84.753	84.978	85.203
380	85.427	85.652	85.877	86.102	86.327	86.551	86.776	87.001	87.226	87.451
390	87.675	87.900	88.125	88.350	88.575	88.800	89.024	89.249	89.474	89.699
400	89.924	90.148	90.373	90.598	90.823	91.048	91.272	91.497	91.722	91.947
410	92.172	92.396	92.621	92.846	93.071	93.296	93.521	93.745	93.970	94.195
420	94.420	94.645	94.869	95.094	95.319	95.544	95.769	95.993	96.218	96.443
430	96.668	96.893	97.117	97.342	97.567	97.792	98.017	98.241	98.466	98.691
440	98.916	99.141	99.366	99.590	99.815	100.040	100.265	100.490	100.714	100.939
450	101.164	101.389	101.614	101.838	102.063	102.288	102.513	102.738	102.962	103.187
460	103.412	103.637	103.862	104.087	104.311	104.536	104.761	104.986	105.211	105.435
470	105.660	105.885	106.110	106.335	106.559	106.784	107.009	107.234	107.459	107.683
480	107.908	108.133	108.358	108.583	108.808	109.032	109.257	109.482	109.707	109.932
490	110.156	110.381	110.606	110.831	111.056	111.280	111.505	111.730	111.955	112.180

Table 19-9 (*Continued*). Force: Newtons to Pounds Force (from 500 through 1009 Newtons)

NEWTONS	0	1	2	3	4	5	6	7	8	9
					POUNDS FORCE					
500	112.404	112.629	112.854	113.079	113.304	113.529	113.753	113.978	114.203	114.428
510	114.653	114.877	115.102	115.327	115.552	115.777	116.001	116.226	116.451	116.676
520	116.901	117.125	117.350	117.575	117.800	118.025	118.249	118.474	118.699	118.924
530	119.149	119.374	119.598	119.823	120.048	120.273	120.498	120.722	120.947	121.172
540	121.397	121.622	121.846	122.071	122.296	122.521	122.746	122.970	123.195	123.420
550	123.645	123.870	124.095	124.319	124.544	124.769	124.994	125.219	125.443	125.668
560	125.893	126.118	126.343	126.567	126.792	127.017	127.242	127.467	127.691	127.916
570	128.141	128.366	128.591	128.816	129.040	129.265	129.490	129.715	129.940	130.164
580	130.389	130.614	130.839	131.064	131.288	131.513	131.738	131.963	132.188	132.412
590	132.637	132.862	133.087	133.312	133.537	133.761	133.986	134.211	134.436	134.661
600	134.885	135.110	135.335	135.560	135.785	136.009	136.234	136.459	136.684	136.909
610	137.133	137.358	137.583	137.808	138.033	138.257	138.482	138.707	138.932	139.157
620	139.382	139.606	139.831	140.056	140.281	140.506	140.730	140.955	141.180	141.405
630	141.630	141.854	142.079	142.304	142.529	142.754	142.978	143.203	143.428	143.653
640	143.878	144.103	144.327	144.552	144.777	145.002	145.227	145.451	145.676	145.901
650	146.126	146.351	146.575	146.800	147.025	147.250	147.475	147.699	147.924	148.149
660	148.374	148.599	148.824	149.048	149.273	149.498	149.723	149.948	150.172	150.397
670	150.622	150.847	151.072	151.296	151.521	151.746	151.971	152.196	152.420	152.645
680	152.870	153.095	153.320	153.544	153.769	153.994	154.219	154.444	154.669	154.893
690	155.118	155.343	155.568	155.793	156.017	156.242	156.467	156.692	156.917	157.141
700	157.366	157.591	157.816	158.041	158.265	158.490	158.715	158.940	159.165	159.390
710	159.614	159.839	160.064	160.289	160.514	160.738	160.963	161.188	161.413	161.638
720	161.862	162.087	162.312	162.537	162.762	162.986	163.211	163.436	163.661	163.886
730	164.111	164.335	164.560	164.785	165.010	165.235	165.459	165.684	165.909	166.134
740	166.359	166.583	166.808	167.033	167.258	167.483	167.707	167.932	168.157	168.382
750	168.607	168.832	169.056	169.281	169.506	169.731	169.956	170.180	170.405	170.630
760	170.855	171.080	171.304	171.529	171.754	171.979	172.204	172.428	172.653	172.878
770	173.103	173.328	173.552	173.777	174.002	174.227	174.452	174.677	174.901	175.126
780	175.351	175.576	175.801	176.025	176.250	176.475	176.700	176.925	177.149	177.374
790	177.599	177.824	178.049	178.273	178.498	178.723	178.948	179.173	179.398	179.622
800	179.847	180.072	180.297	180.522	180.746	180.971	181.196	181.421	181.646	181.870
810	182.095	182.320	182.545	182.770	182.994	183.219	183.444	183.669	183.894	184.119
820	184.343	184.568	184.793	185.018	185.243	185.467	185.692	185.917	186.142	186.367
830	186.591	186.816	187.041	187.266	187.491	187.715	187.940	188.165	188.390	188.615
840	188.839	189.064	189.289	189.514	189.739	189.964	190.188	190.413	190.638	190.863
850	191.088	191.312	191.537	191.762	191.987	192.212	192.436	192.661	192.886	193.111
860	193.336	193.560	193.785	194.010	194.235	194.460	194.685	194.909	195.134	195.359
870	195.584	195.809	196.033	196.258	196.483	196.708	196.933	197.157	197.382	197.607
880	197.832	198.057	198.281	198.506	198.731	198.956	199.181	199.406	199.630	199.855
890	200.080	200.305	200.530	200.754	200.979	201.204	201.429	201.654	201.878	202.103
900	202.328	202.553	202.778	203.002	203.227	203.452	203.677	203.902	204.127	204.351
910	204.576	204.801	205.026	205.251	205.475	205.700	205.925	206.150	206.375	206.599
920	206.824	207.049	207.274	207.499	207.723	207.948	208.173	208.398	208.623	208.847
930	209.072	209.297	209.522	209.747	209.972	210.196	210.421	210.646	210.871	211.096
940	211.320	211.545	211.770	211.995	212.220	212.444	212.669	212.894	213.119	213.344
950	213.568	213.793	214.018	214.243	214.468	214.693	214.917	215.142	215.367	215.592
960	215.817	216.041	216.266	216.491	216.716	216.941	217.165	217.390	217.615	217.840
970	218.065	218.289	218.514	218.739	218.964	219.189	219.414	219.638	219.863	220.088
980	220.313	220.538	220.762	220.987	221.212	221.437	221.662	221.886	222.111	222.336
990	222.561	222.786	223.010	223.235	223.460	223.685	223.910	224.134	224.359	224.584
1000	224.809	225.034	225.259	225.483	225.708	225.933	226.158	226.383	226.607	226.832

METRIC CONVERSION FACTORS AND TABLES

Table 19-10. Force: Pounds Force to Newtons (from 1 through 499 Pounds Force)

1 POUND FORCE = 4.448222 NEWTONS

POUNDS FORCE	0	1	2	3	4	5	6	7	8	9
					NEWTONS					
0	0.000	4.448	8.896	13.345	17.793	22.241	26.689	31.138	35.586	40.034
10	44.482	48.930	53.379	57.827	62.275	66.723	71.172	75.620	80.068	84.516
20	88.964	93.413	97.861	102.309	106.757	111.206	115.654	120.102	124.550	128.998
30	133.447	137.895	142.343	146.791	151.240	155.688	160.136	164.584	169.032	173.481
40	177.929	182.377	186.825	191.274	195.722	200.170	204.618	209.066	213.515	217.963
50	222.411	226.859	231.308	235.756	240.204	244.652	249.100	253.549	257.997	262.445
60	266.893	271.342	275.790	280.238	284.686	289.134	293.583	298.031	302.479	306.927
70	311.376	315.824	320.272	324.720	329.168	333.617	338.065	342.513	346.961	351.410
80	355.858	360.306	364.754	369.202	373.651	378.099	382.547	386.995	391.444	395.892
90	400.340	404.788	409.236	413.685	418.133	422.581	427.029	431.478	435.926	440.374
100	444.822	449.270	453.719	458.167	462.615	467.063	471.512	475.960	480.408	484.856
110	489.304	493.753	498.201	502.649	507.097	511.546	515.994	520.442	524.890	529.338
120	533.787	538.235	542.683	547.131	551.580	556.028	560.476	564.924	569.372	573.821
130	578.269	582.717	587.165	591.614	596.062	600.510	604.958	609.406	613.855	618.303
140	622.751	627.199	631.648	636.096	640.544	644.992	649.440	653.889	658.337	662.785
150	667.233	671.682	676.130	680.578	685.026	689.474	693.923	698.371	702.819	707.267
160	711.716	716.164	720.612	725.060	729.508	733.957	738.405	742.853	747.301	751.750
170	756.198	760.646	765.094	769.542	773.991	778.439	782.887	787.335	791.784	796.232
180	800.680	805.128	809.576	814.025	818.473	822.921	827.369	831.818	836.266	840.714
190	845.162	849.610	854.059	858.507	862.955	867.403	871.852	876.300	880.748	885.196
200	889.644	894.093	898.541	902.989	907.437	911.886	916.334	920.782	925.230	929.678
210	934.127	938.575	943.023	947.471	951.920	956.368	960.816	965.264	969.712	974.161
220	978.609	983.057	987.505	991.954	996.402	1000.850	1005.298	1009.746	1014.195	1018.643
230	1023.091	1027.539	1031.988	1036.436	1040.884	1045.332	1049.780	1054.229	1058.677	1063.125
240	1067.573	1072.022	1076.470	1080.918	1085.366	1089.814	1094.263	1098.711	1103.159	1107.607
250	1112.055	1116.504	1120.952	1125.400	1129.848	1134.297	1138.745	1143.193	1147.641	1152.089
260	1156.538	1160.986	1165.434	1169.882	1174.331	1178.779	1183.227	1187.675	1192.123	1196.572
270	1201.020	1205.468	1209.916	1214.365	1218.813	1223.261	1227.709	1232.157	1236.606	1241.054
280	1245.502	1249.950	1254.399	1258.847	1263.295	1267.743	1272.191	1276.640	1281.088	1285.536
290	1289.984	1294.433	1298.881	1303.329	1307.777	1312.225	1316.674	1321.122	1325.570	1330.018
300	1334.467	1338.915	1343.363	1347.811	1352.259	1356.708	1361.156	1365.604	1370.052	1374.501
310	1378.949	1383.397	1387.845	1392.293	1396.742	1401.190	1405.638	1410.086	1414.535	1418.983
320	1423.431	1427.879	1432.327	1436.776	1441.224	1445.672	1450.120	1454.569	1459.017	1463.465
330	1467.913	1472.361	1476.810	1481.258	1485.706	1490.154	1494.603	1499.051	1503.499	1507.947
340	1512.395	1516.844	1521.292	1525.740	1530.188	1534.637	1539.085	1543.533	1547.981	1552.429
350	1556.878	1561.326	1565.774	1570.222	1574.671	1579.119	1583.567	1588.015	1592.463	1596.912
360	1601.360	1605.808	1610.256	1614.705	1619.153	1623.601	1628.049	1632.497	1636.946	1641.394
370	1645.842	1650.290	1654.739	1659.187	1663.635	1668.083	1672.531	1676.980	1681.428	1685.876
380	1690.324	1694.773	1699.221	1703.669	1708.117	1712.565	1717.014	1721.462	1725.910	1730.358
390	1734.807	1739.255	1743.703	1748.151	1752.599	1757.048	1761.496	1765.944	1770.392	1774.841
400	1779.289	1783.737	1788.185	1792.633	1797.082	1801.530	1805.978	1810.426	1814.875	1819.323
410	1823.771	1828.219	1832.667	1837.116	1841.564	1846.012	1850.460	1854.909	1859.357	1863.805
420	1868.253	1872.701	1877.150	1881.598	1886.046	1890.494	1894.943	1899.391	1903.839	1908.287
430	1912.735	1917.184	1921.632	1926.080	1930.528	1934.977	1939.425	1943.873	1948.321	1952.769
440	1957.218	1961.666	1966.114	1970.562	1975.011	1979.459	1983.907	1988.355	1992.803	1997.252
450	2001.700	2006.148	2010.596	2015.045	2019.493	2023.941	2028.389	2032.837	2037.286	2041.734
460	2046.182	2050.630	2055.079	2059.527	2063.975	2068.423	2072.871	2077.320	2081.768	2086.216
470	2090.664	2095.113	2099.561	2104.009	2108.457	2112.905	2117.354	2121.802	2126.250	2130.698
480	2135.147	2139.595	2144.043	2148.491	2152.939	2157.388	2161.836	2166.284	2170.732	2175.181
490	2179.629	2184.077	2188.525	2192.973	2197.422	2201.870	2206.318	2210.766	2215.215	2219.663

Table 19-10 (*Continued*). Force: Pounds Force to Newtons (from 500 through 1009 Pounds Force)

POUNDS FORCE	0	1	2	3	4	5	6	7	8	9
					NEWTONS					
500	2224.111	2228.559	2233.007	2237.456	2241.904	2246.352	2250.800	2255.249	2259.697	2264.145
510	2268.593	2273.041	2277.490	2281.938	2286.386	2290.834	2295.283	2299.731	2304.179	2308.627
520	2313.075	2317.524	2321.972	2326.420	2330.868	2335.317	2339.765	2344.213	2348.661	2353.109
530	2357.558	2362.006	2366.454	2370.902	2375.351	2379.799	2384.247	2388.695	2393.143	2397.592
540	2402.040	2406.488	2410.936	2415.385	2419.833	2424.281	2428.729	2433.177	2437.626	2442.074
550	2446.522	2450.970	2455.419	2459.867	2464.315	2468.763	2473.211	2477.660	2482.108	2486.556
560	2491.004	2495.453	2499.901	2504.349	2508.797	2513.245	2517.694	2522.142	2526.590	2531.038
570	2535.487	2539.935	2544.383	2548.831	2553.279	2557.728	2562.176	2566.624	2571.072	2575.521
580	2579.969	2584.417	2588.865	2593.313	2597.762	2602.210	2606.658	2611.106	2615.555	2620.003
590	2624.451	2628.899	2633.347	2637.796	2642.244	2646.692	2651.140	2655.589	2660.037	2664.485
600	2668.933	2673.381	2677.830	2682.278	2686.726	2691.174	2695.623	2700.071	2704.519	2708.967
610	2713.415	2717.864	2722.312	2726.760	2731.208	2735.657	2740.105	2744.553	2749.001	2753.449
620	2757.898	2762.346	2766.794	2771.242	2775.691	2780.139	2784.587	2789.035	2793.483	2797.932
630	2802.380	2806.828	2811.276	2815.725	2820.173	2824.621	2829.069	2833.517	2837.966	2842.414
640	2846.862	2851.310	2855.759	2860.207	2864.655	2869.103	2873.551	2878.000	2882.448	2886.896
650	2891.344	2895.793	2900.241	2904.689	2909.137	2913.585	2918.034	2922.482	2926.930	2931.378
660	2935.827	2940.275	2944.723	2949.171	2953.619	2958.068	2962.516	2966.964	2971.412	2975.861
670	2980.309	2984.757	2989.205	2993.653	2998.102	3002.550	3006.998	3011.446	3015.895	3020.343
680	3024.791	3029.239	3033.687	3038.136	3042.584	3047.032	3051.480	3055.929	3060.377	3064.825
690	3069.273	3073.721	3078.170	3082.618	3087.066	3091.514	3095.963	3100.411	3104.859	3109.307
700	3113.755	3118.204	3122.652	3127.100	3131.548	3135.997	3140.445	3144.893	3149.341	3153.789
710	3158.238	3162.686	3167.134	3171.582	3176.031	3180.479	3184.927	3189.375	3193.823	3198.272
720	3202.720	3207.168	3211.616	3216.065	3220.513	3224.961	3229.409	3233.857	3238.306	3242.754
730	3247.202	3251.650	3256.099	3260.547	3264.995	3269.443	3273.891	3278.340	3282.788	3287.236
740	3291.684	3296.133	3300.581	3305.029	3309.477	3313.925	3318.374	3322.822	3327.270	3331.718
750	3336.166	3340.615	3345.063	3349.511	3353.959	3358.408	3362.856	3367.304	3371.752	3376.200
760	3380.649	3385.097	3389.545	3393.993	3398.442	3402.890	3407.338	3411.786	3416.234	3420.683
770	3425.131	3429.579	3434.027	3438.476	3442.924	3447.372	3451.820	3456.268	3460.717	3465.165
780	3469.613	3474.061	3478.510	3482.958	3487.406	3491.854	3496.302	3500.751	3505.199	3509.647
790	3514.095	3518.544	3522.992	3527.440	3531.888	3536.336	3540.785	3545.233	3549.681	3554.129
800	3558.578	3563.026	3567.474	3571.922	3576.370	3580.819	3585.267	3589.715	3594.163	3598.612
810	3603.060	3607.508	3611.956	3616.404	3620.853	3625.301	3629.749	3634.197	3638.646	3643.094
820	3647.542	3651.990	3656.438	3660.887	3665.335	3669.783	3674.231	3678.680	3683.128	3687.576
830	3692.024	3696.472	3700.921	3705.369	3709.817	3714.265	3718.714	3723.162	3727.610	3732.058
840	3736.506	3740.955	3745.403	3749.851	3754.299	3758.748	3763.196	3767.644	3772.092	3776.540
850	3780.989	3785.437	3789.885	3794.333	3798.782	3803.230	3807.678	3812.126	3816.574	3821.023
860	3825.471	3829.919	3834.367	3838.816	3843.264	3847.712	3852.160	3856.608	3861.057	3865.505
870	3869.953	3874.401	3878.850	3883.298	3887.746	3892.194	3896.642	3901.091	3905.539	3909.987
880	3914.435	3918.884	3923.332	3927.780	3932.228	3936.676	3941.125	3945.573	3950.021	3954.469
890	3958.918	3963.366	3967.814	3972.262	3976.710	3981.159	3985.607	3990.055	3994.503	3998.952
900	4003.400	4007.848	4012.296	4016.744	4021.193	4025.641	4030.089	4034.537	4038.986	4043.434
910	4047.882	4052.330	4056.778	4061.227	4065.675	4070.123	4074.571	4079.020	4083.468	4087.916
920	4092.364	4096.812	4101.261	4105.709	4110.157	4114.605	4119.054	4123.502	4127.950	4132.398
930	4136.846	4141.295	4145.743	4150.191	4154.639	4159.088	4163.536	4167.984	4172.432	4176.880
940	4181.329	4185.777	4190.225	4194.673	4199.122	4203.570	4208.018	4212.466	4216.914	4221.363
950	4225.811	4230.259	4234.707	4239.156	4243.604	4248.052	4252.500	4256.948	4261.397	4265.845
960	4270.293	4274.741	4279.190	4283.638	4288.086	4292.534	4296.982	4301.431	4305.879	4310.327
970	4314.775	4319.224	4323.672	4328.120	4332.568	4337.016	4341.465	4345.913	4350.361	4354.809
980	4359.258	4363.706	4368.154	4372.602	4377.050	4381.499	4385.947	4390.395	4394.843	4399.292
990	4403.740	4408.188	4412.636	4417.084	4421.533	4425.981	4430.429	4434.877	4439.326	4443.774
1000	4448.222	4452.670	4457.118	4461.567	4466.015	4470.463	4474.911	4479.360	4483.808	4488.256

Index

ABCA, address of, 1-7
AFBMA, address of, 1-7
AFNOR, address of, 1-7
AGMA, address of, 1-7
AGMA, gear standards, 13-41
AGMA rating formulas, 13-23 to 13-27
AISI, address of, 1-7
American gear standards, 13-16
American National Standards Institute (ANSI), 1-6
ANMC, address of, 1-7
ANSI, 1-6
  address of, 1-7
  catalog, 1-12
  fastener standards, 9-125, 9-126
  gear standards, 13-41
  publications available from, 1-12
ANSI-OMFS fastener screw threads, development of, 8-1
API, address of. 1-8
AS, see SAA
ASAC, 1-6
  address of, 1-8
ASAE, address of, 1-8
Asian Standards Advisory Committee (ASAC), 1-6
ASM, address of, 1-8
ASME, address of, 1-8
ASTM, address of, 1-8
Australian gear standards, 13-16, 13-43

Ball and cylindrical roller bearings, 12-1 to 12-4
Bar standards, world non-ferrous, 11-2
Bearings, 12-1 to 12-21
  ball and cylindrical roller, 12-1 to 12-4
  ISO roller, 12-19, 12-20
  needle roller, 12-8, 12-9
  plain, 12-13
  tapered roller, 12-9 to 12-13
Bearing standards, ISO plain, 12-20, 12-21
  ISO roller, 12-19, 12-20
Belt drives, agricultural V-, 13-12
  automotive V-, 13-12
  endless-, 13-11 to 13-13
  industrial V-, 13-12
  synchronous, 13-11 to 13-13
  V-, 13-11
  V-ribbed, 13-13
Blind rivets, 9-98, 9-99
Blocks, gage, 18-5 to 18-8
Bolt loads, basis for calculating proof and ultimate, 9-20
Bolts, clearance holes for metric, 9-29
  flat countersunk square neck, 9-50
  hexagon head cap screws and, 9-33 to 9-45
  plow, 9-50
  round head square neck, 9-45 to 9-50
Break mandrel blind rivets, 9-98
Break mandrel closed end blind rivets, 9-98, 9-99
British gear standards, 13-16, 13-42, 13-43
British measuring tools and instruments standards, 18-12
BS, see BSI
BSI, address of, 1-8
Buckingham equations for gearing, 13-23
Bushings, self-aligning, 12-13

Cap screws, hexagon head, 9-33 to 9-38
  socket head, 9-54
Carbide inserts and tool holders, indexable throwaway, 17-30, 17-31
CCPA, address of, 1-8
CDA, address of, 1-9
CEE, address of, 1-9
CEN, 1-1, 1-6
  address of, 1-9
CENELEC, 1-1, 1-6
  address of, 1-9
CGPM, address of, 1-9
CIPM, address of, 1-9
Classes of thread fit (ISO), 8-13
Clearance holes for metric bolts and screws, 9-29
Clevis pins, 9-103
Coated threads, designations for, 8-25
Coatings standards, ISO, 10-179, 10-180
Coiled spring pins, 9-99, 9-106 to 9-109
Combined drills and countersinks, 17-4
Comparison of R_{max} to roughness number, 5-4
Conductors, fluid power, 14-11
Conversion of prime dimensions on an engineering drawing, 3-8
COPANT, 1-6
  address of, 1-9
Copper and copper alloy standards, ISO, 11-33, 11-34
Cotter pins, split, 9-99
Countersinks, combined drills and, 17-4
Cranes, standard lifting capacity of, 4-9
Cross references, international material, 10-127, 10-137
  world steel designation, 10-127
CSA, address of, 1-9
Cutters, metric milling, 17-18, 17-21
  milling, 17-18, 17-21
  T-slot, 17-25
  Woodruff key-seat, 17-25
Cutting tools, metal, 17-1 tp 17-55
Cylinders, fluid power, 14-8 to 14-11
Cylindrical roller bearings, ball and, 12-1 to 12-4

Designations for steel, world, 10-127
DIN, see DNA
Direction of lay, symbols for, 5-3
DNA, address of, 1-9
Double end studs, 9-66
Drawing practice, engineering, 3-1 to 3-19
Drills and countersinks, combined, 17-4
Drills, twist, 17-1 to 17-4
Dryseal pipe threads, 8-48
Dual systems of measure on a drawing, 3-8

ECAFE, 1-6
Economic Commission for Asia and the Far East (ECAFE), 1-6
ECSC, 1-6
  address of, 1-9
Electrical components, 15-1 to 15-3
Electrical component standards, German DIN, 15-3
  IEC, 15-2
  ISO, 15-1, 15-2

Electrical component standards, (*Cont.*)
   Italian, 15-3
   Japanese JIS, 15-2, 15-3
   SAE, 15-2
Electrical currents, standard, 4-8
Electric motors, 13-1 to 13-3
   flange-mounted, 13-3
   foot-mounted, 13-1 to 13-3
   IEC standards for, 13-39
   ISO standards for, 13-39
Endless-belt drives, 13-11 to 13-13
End mills, 17-18
Engineering drawing, conversion of prime dimensions on, 3-8
   dual systems of measure on an, 3-8
   German method of indicating surface texture on an, 5-3
   prime measuring unit on, 3-8
Engineering drawing practice, 3-1 to 3-19
   ISO standards for, 3-18, 3-19
   systems of measure for, 3-5
   tolerancing in, 3-11
E-rings for shafts, 9-109
EURONORM, see ECSC
European Coal and Steel Community (ECSC), 1-6
European Committee for Standardization (CEN), 1-1, 1-6
European Electrical Standards Coordinating Committee (CENELEC), 1-1, 1-6
European steel section standards, 10-127
External screw thread criteria, 8-26

Fasteners, 9-1 to 9-126
   general specification for, 9-1, 9-17 to 9-20
   installation of threaded, 9-29 to 9-33
   length specifications for, 9-18
   marking, 9-20
   nuts, 9-23 to 9-29
   strength properties for threaded, 9-18 to 9-20
Fastener standard handbooks metric, 9-126
Fastener standards, 9-124, 9-126
   ANSI, 9-125, 9-126
Fastener standards index, world metric, 9-2 to 9-16
Ferrous material data, 10-1 to 10-182
Ferrous materials index, world, 10-150 to 10-173
FIEI, address of, 1-10
First angle projection, 3-5
Fittings, ISO metal pipes and, 10-178, 10-179
Flange head screws, 12-point spline, 9-50 to 9-54
Flange-mounted electric motors, 13-3
Flat countersunk head machine screws, 9-54
Flat countersunk square neck bolts, 9-50
Flat washers, 9-93
Fluid power, conductors, 14-11
   cylinders, 14-8 to 14-11
   pressures, 14-1
   pumps and motors, 14-1 to 14-8
   sealing devices, 14-11 to 14-18
   SI units for, 14-1
   ISO standards for, 14-18 to 14-20
   systems and components, 14-1 to 14-20
Foot-mounted electric motors, 13-1 to 13-3
Foreign steel suppliers, 10-174
French gear standards, 13-16, 13-42

Gage blocks, 18-5 to 18-8
Gage design, screw thread, 8-27
Gage marking, examples of, 8-19, 8-20
Gages, for ISO limits and fits, standard, 18-8, 18-9
   for ISO metric screw threads, 8-15
   for verification of product external thread, threaded and plain ring, snap and indicating thread, 8-18, 8-19
   for verification of product internal threads, threaded and plain, 8-17, 8-18
   marking of, 8-19
   thread setting plug, 8-18
Gage thread profile dimensions, 8-27, 8-28
Gaging and verification of product threads, 8-17 to 8-20
Gaging, boundary profiles for screw thread, 8-25
Gaging product threads, limitations in, 8-19

Gearing, 13-13 to 13-27
   AGMA rating formulas, 13-23 to 13-27
   Buckingham equations, 13-23
   conversion of classical strength formulas for metric usage, 13-21 to 13-27
   converting practices between metric and inch systems, 13-17
   formulas for kinematic, 13-17 to 13-21
   Lewis formula, 13-21 to 13-23
   metric, 13-14 to 13-27
   preferred modules and national practices, 13-17
Gear standards, AGMA, 13-41
   American, 13-16
   ANSI, 13-41
   Australian, 13-16, 13-43
   British, 13-16, 13-42, 13-43
   French, 13-16, 13-42
   German DIN, 13-16, 13-41, 13-42
   ISO, 13-40, 17-55
   Italian, 13-16, 13-43
   Japanese JIS, 13-16, 13-41
Gear strength and durability rating formulas, 13-21
Gear-tooth proportions, metric, 13-14, 13-15
Geometric tolerancing training charts, 3-11 to 3-16
German DIN, electrical component standards, 15-3
   gear standards, 13-16, 13-41, 13-42
   measuring tool and instrument standards, 18-11, 18-12

Helical spring lock washers, 9-93
Hexagon head cap screws and bolts, 9-33 to 9-45
Hexagon head machine screws, 9-59
Hexagon nuts, 9-80
Hexagon nuts, slotted, 9-89
Hexagon socket set screws, 9-66
Hexagon steel locknuts, 9-80 to 9-89
Holes for tapping screws, 9-30

IEC, 1-6
   address of, 1-10
   electrical component standards, 15-1 to 15-3
   standards for electric motors, 13-39
IFI, address of, 1-10
Index, world ferrous materials, 10-150 to 10-173
   world metric fastener standards, 9-2 to 9-16
Inflation pressures, tire, 16-5
Inserts, carbide, 17-30
Inspection of plain workpieces, ISO system of limits anf fits for, 7-1 to 7-24
Installation of threaded fasteners, 9-29 to 9-33
Instruments, measuring tools and, 18-1 to 18-12
Interchangeability of metric module and inch diametral gears, 13-16, 13-17
Internal screw thread criteria, 8-27
International Electrotechnical Commission (IEC), 1-6
International material cross references, 10-127, 10-137
International Organization for Standardization (ISO), 1-4 to 1-6
International pipe threads, 8-48
International standard, evolution of an, 1-6
International system of measuring units (SI), 2-1 to 2-7
Involute, splines, 13-27
ISO, 1-4 to 1-6
   address of, 1-10
   bulletin of, 1-12
   catalog of, 1-12
   directives of, 1-12
   member structure of, 1-4
   memento of, 1-12
   objectives of, 1-4
   organization of, 1-4
   participation of, 1-12
   status report of, 1-12
ISO basic thread profile, 8-12
ISO classes of thread fit, 8-13
ISO copper and copper alloys standards, 11-33, 11-34
ISO electrical component standards, 15-1, 15-2
ISO fastener screw threads, modified, 8-20 to 8-31
ISO fastener standards, 9-124, 9-125

INDEX

ISO fluid power standards, 14-18 to 14-20
ISO gear standards, 13-40, 17-55
ISO general metric screw threads, 8-11
ISO general metric screw threads, development of, 8-1
ISO light metals and their alloys standards, 11-34 to 11-36
ISO limits and fits, see ISO system of limits and fits
ISO machine tools standards, 17-54, 17-55
ISO measuring tools and instruments standards, 18-11
ISO metallic coatings standards, 10-179, 10-180
ISO metal pipes and fittings standards, 10-178, 10-179
ISO metric module tooth proportions, 13-15, 13-16
ISO metric screw threads, gages for, 8-15 to 8-30
ISO paper sizes, 3-1
ISO plain bearings standards, 12-20, 12-21
ISO roller bearings standard, 12-19, 12-20
ISO roller chain standards, 13-3, 13-11, 13-39, 13-40
ISO screw threads, existing versus modified, 8-1, 8-11
ISO small tool standards, 9-125, 17-51 to 17-54
ISO standards, accelerating pace in publication of, 1-2
  for electric motors, 13-39
  for engineering drawing practice, 3-18, 3-19
  for preferred numbers, 4-9
  for steel, 10-174 to 10-178
ISO system of limits and fits, bilateral tolerances in, 6-35
  calculation of limits in, 6-32
  conversion of fits in, 6-38 to 6-40
  definitions in, 6-1, 6-2, 6-12
  description of the, 6-12
  designation in, 6-32
  fundamental deviations in, 6-12
  hole-basis or shaft-basis fits in, 6-35
  non-toleranced dimensions in, 6-40
  practical application of the, 6-32
  selected fits in, 6-35 to 6-37
  standards of, 18-11
  tolerances and deviations, 6-1 to 6-65
  tolerances on angles in, 6-40
  unilateral tolerances in, 6-35
ISO system of limits and fits for inspection of plain workpieces, 7-1 to 7-24
  general rules for, 7-1 to 7-3
  indicating measuring instruments in, 7-14 to 7-16
  limit gages in, 7-4 to 7-14
  standards of, 18-11
ISO thread designations, 8-15
ISO tire and rim standards, 16-8
ISO V-belt standards, 13-40
ISO zinc and zinc alloys standards, 11-33
Italian electrical component standards, 15-3
Italian gear standards, 13-16, 13-43

Japanese JIS electrical component standards, 15-2, 15-3
Japanese JIS gear standards, 13-16, 13-41
Japanese JIS measuring tool and instrument standards, 18-11
JIS, see JISC
JISC, address of, 1-10
JSA, address of, 1-10

Keys and keyways, 13-29
  standards for, 13-39
Kinematic gearing formulas, 13-17 to 13-21

Length specifications, fastener, 9-18
Lettering conventions in engineering drawing practice, 3-3, 3-4
Lewis formula, 13-21
Light metal and light metal alloy standards, ISO, 11-34 to 11-36
Lifting capacity of cranes, standard, 4-9
Limits and fits for inspection of plain workpieces, ISO system of, 7-1 to 7-24
Limits and fits, history of the ISO system of, 6-1
  ISO system for tolerances and deviations, 6-1 to 6-65
  standard gages for, ISO, 18-8, 18-9
Line conventions in engineering drawing practice, 3-2, 3-3
Locknuts, hexagon steel, 9-80 to 9-89
Lock washers, helical spring, 9-93
Low nuts, 9-27, 9-28

Machine screws, 9-54 to 9-66
  flat countersunk head, 9-54
  general notes on, 9-59 to 9-66
  hexagon head, 9-59
  oval countersunk head, 9-54
  pan head, 9-54
Machine tool standards, ISO, 17-54, 17-55
Marking fasteners, 9-20
Marking of gages, 8-19
Materials index, world ferrous, 10-150 to 10-173
MCTI, address of, 1-10
Measuring force for wire measurements of 60° threads, 8-17
Measuring rules, scales and, 18-1 to 18-5
Measuring tools and instruments, 18-1 to 18-12
  British standards for, 18-12
  German DIN standards for, 18-11, 18-12
  ISO standards for, 18-11
  Japanese JIS standards for, 18-11
Mechanical power transmission systems, 13-1 to 13-43
Metal cutting tools, 17-1 to 17-55
Metal pipes and fittings standards, ISO, 10-178, 10-179
Metric fastener standards index, world, 9-2 to 9-16
Metric fastener standard handbooks, 9-126
Metric gearing, 13-14 to 13-27
Metric gear-tooth proportions, 13-14, 13-15
Metric hardware, ordering, 9-1, 9-17, 9-18
Metric module, 13-14
Metric module tooth proportions, ISO, 13-15
Metric system (SI), 2-1 to 2-7 see also SI
Milling cutters, 17-18, 17-21
Miniature screw threads, 8-42, 8-46
Miniature screw threads standard, world metric, 8-46
Modified ISO fastener screw threads, 8-20 to 8-31
  basic thread profile of, 8-20
  external, 8-21
  internal, 8-21
Motors, electric, 13-1 to 13-3
  flange-mounted electric, 13-3
  fluid power, 14-1 to 14-8
  foot-mounted electric, 13-1 to 13-3
  IEC standards for electric, 13-39
  ISO standards for electric, 13-39

National standards organizations, 1-6 to 1-11
NBS, address of, 1-10
Needle roller bearings, 12-8, 12-9
NEMA, address of, 1-11
NF, see AFNOR
NFPA, address of, 1-11
Non-ferrous bar standards, world, 11-2
Non-ferrous material, 11-1 to 11-37
Non-ferrous material standards, other sources for national, 11-37
Non-ferrous plate and sheet standards, world, 11-1
Non-ferrous tube standards, world, 11-2
Non-ferrous wire standards, world, 11-1, 11-2
Non-sealing pipe threads, 8-48
NSC, address of, 1-11
Nuts, 9-23 to 9-29
  designation of, 9-26
  hardness of, 9-26
  hexagon, 9-80
  low, 9-27, 9-28
  marking, 9-27
  materials and processes for, 9-26
  projection weld, 9-89
  proof load of, 9-26
  requirements for, 9-26
  slotted hexagon, 9-89
  steel locknuts, 9-80 to 9-89
  without specified load requirements, 9-27

OMFS classes of thread fit, 8-20
Optimum metric fastener system (OMFS) classes of thread fit, 8-20
Ordering metric hardware, 9-1, 9-17, 9-18
Organizations concerned with tires, rims, and valves, 16-8
O-ring sizes and tolerances, 14-11, 14-12
Oval countersunk head machine screws, 9-54

Pacific Area Standards Congress (PASC), 1-6
Pan American Standards Commission (COPANT), 1-6
Pan head machine screws, 9-54
Paper sizes, ISO, 3-1
Parallel pins, 9-103
PASC, 1-6
  address of, 1-11
Pins, 9-99 to 9-109
  clevis, 9-103
  coiled spring, 9-99, 9-106 to 9-109
  parallel, 9-103
  roll, 9-99, 9-104 to 106
  slotted spring, see roll pins
  split cotter, 9-99
  taper, 9-103
Pipe threads, 8-48
Plain bearings, 12-13
Plain bearings standards, ISO, 12-20, 12-21
Plate and sheet standards, world, non-ferrous, 11-1
Plow bolts, 9-50
Plug gages, thread setting, 8-18
Power transmission systems, mechanical, 13-1 to 13-43
Preferred metric sizes, 4-7
Preferred numbers, 4-1 to 4-9
  derivation of, 4-1
  designation of, 4-3
  ISO standards for, 4-9
  more rounded values of, 4-5
  multiplication or division of series of, 4-4
  nomenclature and definitions, 4-1, 4-3
  practical examples of internationally used, 4-8, 4-9
  preferred metric sizes of, 4-7
  producing a logarithmic scale from, 4-5
  series of, 4-3
Pressure bases, miscellaneous, 2-7
Pressure containers, standard, 4-9
Pressures, for fluid power, 14-1
  tire inflation, 16-5
Prime measuring unit on drawing, 3-8
Production costs, surface texture versus, 5-6
Product threads acceptability, 8-26
  conformance gaging in, 8-16
  gaging and verification of, 8-17 to 8-20
  limitations in gaging, 8-19
  screw thread conformance in, 8-16
  types of gages for verification of, 8-17
  verification of external, 8-18, 8-19
  verification of internal, 8-17, 8-18
Projection in engineering drawing practice, 3-5
Projection weld nuts, 9-89
Proof and ultimate bolt loads, basis for calculating, 9-20
Publications available from ANSI, 1-12
Pumps and motors, fluid power, 14-1 to 14-8

Reamers, 17-4 to 17-18
  designation, 17-11
  hand and long-fluted machine, 17-11
  hand and machine pin, 17-17
  machine chucking, 17-16
  Morse taper, 17-16
  tolerance on cutting diameter, 17-4, 17-6
  tolerance on length, 17-6
  tolerance on parallel shank diameter, 17-6
Retaining rings, 9-109
Rim standards, ISO tire and, 16-8
Rivets, 9-98, 9-99
RMA, address of, 1-11
Roller bearings, ball and cylindrical, 12-1 to 12-4
  ISO standard, 12-19, 12-20
  needle, 12-8, 12-9
  tapered, 12-9 to 12-13
Roller chain standards, ISO, 13-3, 13-11, 13-39, 13-40
Roll pins, 9-99, 9-104 to 9-106
Roughness number, comparison of R_{max} to, 5-4
Round head square neck bolts, 9-45 to 9-50
Rules, scales and measuring, 18-1 to 18-5

SAA, address of, 1-11
SAE, address of, 1-11
SAE electrical component standards, 15-2
Safety standards for tires, 16-7, 16-8
Sampling lengths, 5-7
Scales and measuring rules, 18-1 to 18-5
Screws, clearance holes for metric bolts and, 9-29
  machine, 9-54 to 9-66
  socket head cap, 9-54
  tapping, 9-74 to 9-79
  12-point spline flange head, 9-50 to 9-54
Screw thread acceptability, product, 8-26
Screw thread criteria, external, 8-26
  internal, 8-27
Screw thread gage design, 8-27
Screw thread gaging, boundary profiles for, 8-25
  standard temperature in, 8-17
Screw threads, 8-1 to 8-50
  development of ANSI-OMFS fastener, 8-1
  development of ISO general metric, 8-1
  dryseal pipe, 8-48
  existing versus modified ISO, 8-1, 8-11
  gages for ISO metric, 8-15
  international pipe, 8-48
  ISO general metric, 8-11
  miniature, 8-42 to 8-46
  modified ISO fastener, 8-20 to 8-31
  non-sealing pipe, 8-48
  standards for, 8-49, 8-50
  trapezoidal, 8-31 to 8-42
Screw thread standards, world metric, 8-12
  world metric miniature, 8-46
Screw thread strength data, 8-30
Screw thread tolerances, trapezoidal, 8-39 to 8-41
Sealing devices, fluid power, 14-11 to 14-18
Self-aligning bushings, 12-13
SES, address of, 1-11
Set screws, 9-66
Sheet metal standards, world, 10-2
Sheet standards, world non-ferrous plate and, 11-1
SI metric units, 2-1 to 2-7
SI base units, definition of, 2-1
SI supplementary units, definition of, 2-1
SI metric units, 2-1 to 2-7
  and symbols, 2-1
  for fluid power, 14-1
Slotted hexagon nuts, 9-89
Slotted spring pins, see roll pins
SME, address of, 1-11
Small tool standards, ISO, 9-125, 17-51 to 17-54
Socket head cap screws, 9-54
Speeds of machine-tool spindles, standard rotating, 4-8
Splines, 13-27 to 13-29
  involute, 13-27
  standards for 13-43
  straight, 13-27
Split cotter pins, 9-99
Square neck bolts, flat countersunk, 9-50
  round head, 9-45 to 9-50
Standard conditions and physical constants, 2-6
Standard copper wire, diameters of 4-8
Standard electrical currents, 4-8
Standardization, analytical, 1-3
  conservative, 1-3
  ISO definition of, 1-2, 1-3
  levels of, 1-4
  objectives of, 1-3
  role of, 1-1
Standard lifting capacity of cranes, 4-9
Standard pressure containers, 4-9
Standard rotating speeds of machine-tool spindles, 4-8
Standards, accelerating pace in publication of ISO, 1-2
  development of, 1-3
  evolution of international, 1-6
  increase in demand for mandatory world, 1-1, 1-2
  ISO definition of, 1-2, 1-3
  metric and inch, 1-1

INDEX

Standards (*Cont.*)
  user acceptance of, 1-4
Standards index, world metric fastener, 9-2 to 9-16
Standards organizations, acronyms and addresses to important,
    1-7 to 1-11
  national, 1-6 to 1-12
  regional, 1-6
  world, 1-1 to 1-12
Standard temperature in screw thread gaging, 8-17
Steel bar materials, 10-6
Steel bar standards, world, 10-3, 10-4
Steel bar tolerances, 10-4
Steel designation cross references, world, 10-127
Steel plate standards, world, 10-2
Steel section standards, European, 10-127
Steel standards, ISO, 10-174 to 10-178
Steel suppliers, foreign, 10-174
Steel tube materials, 10-127
Steel tube standards, world, 10-6
Steel tube tolerances, 10-127
Steel wire standards, world, 10-3
Steel, world designation systems for, 10-127
Straight splines, 13-27
Strength properties for threaded fasteners, 9-18 to 9-20
Studs, set screws and double end, 9-66 to 9-74
Surface finish and tolerances, choice of, 5-4 to 5-6
Surface roughness, commonly produced, 5-6
Surface texture, 5-1 to 5-10
  definitions of, 5-1, 5-2
  German method of indicating, 5-3
  sampling lengths in, 5-7
  standards for, 5-10
  symbols for, 5-2 to 5-4
  versus production costs, 5-6
Symbols for the direction of lay, 5-3
Symbols for surface texture, 5-2 to 5-4
Synchronous belt drives, 13-11 to 13-13
Systems of measure for engineering drawings, 3-5 to 3-8

Tap drill sizes, recommended metric, 9-30
Tapered roller bearings, 12-9 to 12-13
Taper pins, 9-103
Tapping screws, 9-74 to 9-79
  head types for, 9-74, 9-75
  screw threads for, 9-76 to 9-78
  strength grades for, 9-78
Taylor principle, 7-1, 7-2, 7-3
Third angle projection, 3-5
Thread-cutting tools, 17-30
Thread designations (ISO), 8-15
Threaded fasteners, installation of, 9-29 to 9-33
  strength properties for, 9-18 to 9-20
Thread profile dimensions, gage, 8-27, 8-28
Threads, *see* Screw threads
Thread setting plug gages, 8-18
Threads, screw, 8-1 to 8-50, *see* Screw threads
Tire and rim standards, ISO, 16-8
Tire designations, 16-3 to 16-5
Tire inflation pressures, 16-5
Tire performance and safety standards, 16-7, 16-8

Tires, dimensional standards for, 16-5 to 16-7
  rims, and valves,, 16-1 to 16-8
  rims, and valves organizations, 16-8
  units of measure for, 16-3
Tire standards, from soft to hard conversion in, 16-5
Title block information in engineering drawing practice, 3-4, 3-5
Tolerances and deviations, ISO system of limits and fits,
    6-1 to 6-65
Tolerances, choice of surface finish and, 5-4 to 5-6
Tolerances for steel bars, 10-4, 10-6
Tolerances for steel tubes, 10-127
Tolerancing in engineering drawing practice, 3-11
Tolerancing training charts, geometric, 3-11 to 3-16
Tool holders, indexable throwaway carbide inserts and, 17-30,
    17-31
Tools and instruments, measuring, 18-1 to 18-12
Tools, metal cutting, 17-1 to 17-55
  thread-cutting, 17-30
Tool standard, ISO machine, 17-54, 17-55
  ISO small, 17-51 to 17-54
Torque values for metric fasteners, recommended, 9-33
TRA, address of, 1-11
Trapezoidal screw threads, 8-31 to 8-42
  tolerances for, 8-39 to 8-41
T-slot cutters, 17-25, 17-30
Tube standards, world non-ferrous, 11-2
12-point spline flange head screws, 9-50 to 9-54
Twist drills, 17-1 to 17-4

ULI, address of, 1-11
UNI, address of, 1-12

V-belt drives, 13-11
  agricultural, 13-12
  automotive, 13-12
  industrial, 13-12
V-belt standards, ISO, 13-40
Verification of product external thread, threaded and plain ring,
    snap and indicating thread gages for, 8-18, 8-19
Verification of product internal threads, threaded and plain gages
    for, 8-17, 8-18
V-ribbed belt drives, 13-13

Washers, 9-93
Wire standards, world non-ferrous, 11-1
Woodruff key-seat cutters, 17-25, 17-30
World designation systems for steel, 10-127
World ferrous materials index, 10-150 to 10-173
World metric fastener standards index, 9-2 to 9-16
World metric miniature screw threads standard, 8-46
World metric screw thread standards, 8-12
World sheet metal standards, 10-2
World standards, increase in demand for mandatory, 1-1, 1-2
World standards organizations, 1-1 to 1-12
World steel bar standards, 10-3, 10-4
World steel designation cross references, 10-127
World steel plate standards, 10-2
World steel tube standards, 10-6
World steel wire standards, 10-3

Zinc and zinc alloys standards, ISO, 11-33